# Student Study Guide/
## Solutions Manual

to accompany

# Organic Chemistry

## Second Edition

### Janice Gorzynski Smith
*University of Hawai·i at Mānoa*

and

### Erin R. Smith

 **Higher Education**

Boston   Burr Ridge, IL   Dubuque, IA   New York   San Francisco   St. Louis
Bangkok   Bogotá   Caracas   Kuala Lumpur   Lisbon   London   Madrid   Mexico City
Milan   Montreal   New Delhi   Santiago   Seoul   Singapore   Sydney   Taipei   Toronto

The **McGraw·Hill** Companies

Student Study Guide/Solutions Manual to accompany
ORGANIC CHEMISTRY, SECOND EDITION
JANICE GORZYNKI SMITH, ERIN R. SMITH

Published by McGraw-Hill Higher Education, an imprint of The McGraw-Hill Companies, Inc., 1221 Avenue of the Americas,
New York, NY 10020. Copyright © 2008 by The McGraw-Hill Companies, Inc. All rights reserved.

4 5 6 7 8 9 0 QPD/QPD 0 9 8

ISBN: 978-0-07-304987-8
MHID: 0-07-304987-5

www.mhhe.com

# Contents

# Chapter 1: Structure and Bonding

♦ Important facts

- **The general rule of bonding.** Atoms strive to attain a complete outer shell of valence electrons (Section 1.2). H "wants" 2 electrons. Second row elements "want" 8 electrons.

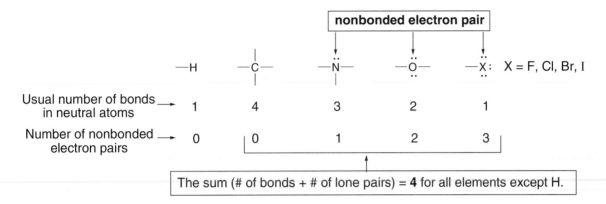

| nonbonded electron pair | | | |

| —H | —C— | —N— | —O— | —X :  X = F, Cl, Br, I |
|---|---|---|---|---|
| Usual number of bonds in neutral atoms → 1 | 4 | 3 | 2 | 1 |
| Number of nonbonded electron pairs → 0 | 0 | 1 | 2 | 3 |

The sum (# of bonds + # of lone pairs) = **4** for all elements except H.

- **Formal charge** (FC) is the difference between the number of valence electrons of an atom and the number of electrons it "owns" (Section 1.3C). See Sample Problem 1.4 for a stepwise example.

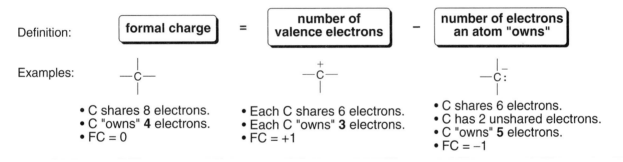

Definition:   **formal charge**  =  **number of valence electrons**  −  **number of electrons an atom "owns"**

Examples:

—C—
- C shares 8 electrons.
- C "owns" **4** electrons.
- FC = 0

—C— (+)
- Each C shares 6 electrons.
- Each C "owns" **3** electrons.
- FC = +1

—C: (−)
- C shares 6 electrons.
- C has 2 unshared electrons.
- C "owns" **5** electrons.
- FC = −1

- **Curved arrow notation** shows the movement of an electron pair. The tail of the arrow always begins at an electron pair, either in a bond or a lone pair. The head points to where the electron pair "moves" (Section 1.5).

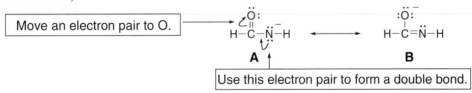

Move an electron pair to O.

H–C–N–H    ⟷    H–C=N–H

A                         B

Use this electron pair to form a double bond.

- **Electrostatic potential plots** are color-coded maps of electron density, indicating electron rich and electron deficient regions (Section 1.11).

♦ **The importance of Lewis structures (Section 1.3, 1.4)**

A properly drawn Lewis structure shows the number of bonds and lone pairs present around each atom in a molecule. In a valid Lewis structure, each H has two electrons, and each second-row element has no more than eight. This is the first step needed to determine many properties of a molecule.

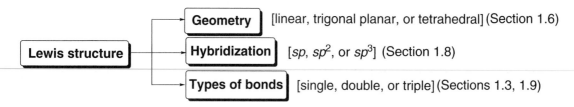

♦ **Resonance (Section 1.5)**

The basic principles:

- Resonance occurs when a compound cannot be represented by a single Lewis structure.
- Two resonance structures differ *only* in the position of nonbonded electrons and π bonds.
- The resonance hybrid is the only accurate representation for a resonance-stabilized compound. A hybrid is more stable than any single resonance structure because electron density is delocalized.

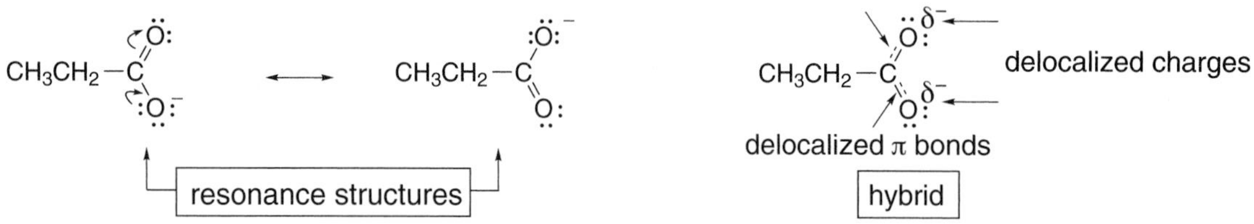

The difference between resonance structures and isomers:

- Two **isomers** differ in the arrangement of *both* atoms and electrons.
- **Resonance structures** differ *only* in the *arrangement of electrons*.

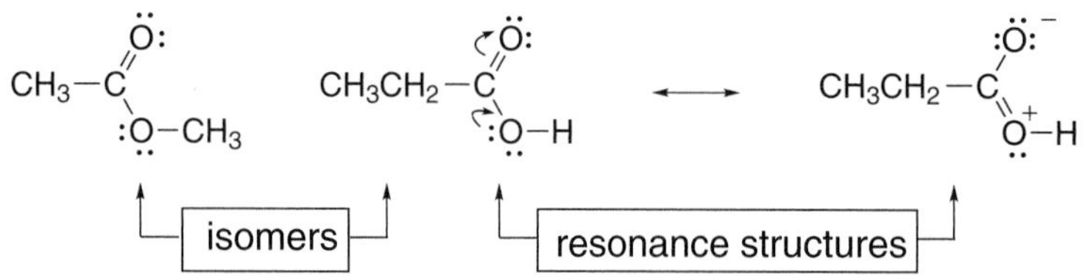

♦ **Geometry and hybridization**

The number of groups around an atom determines both its geometry (Section 1.6) and hybridization (Section 1.8).

| Number of groups | Geometry | Bond angle ($^0$) | Hybridization | Examples |
|---|---|---|---|---|
| 2 | linear | 180 | $sp$ | $BeH_2$, $HC{\equiv}CH$ |
| 3 | trigonal planar | 120 | $sp^2$ | $BF_3$, $CH_2{=}CH_2$ |
| 4 | tetrahedral | 109.5 | $sp^3$ | $CH_4$, $NH_3$, $H_2O$ |

**[2] Count up the number of valence electrons.**

$$
\begin{array}{llll}
\text{3 H's} & \times & \text{1 electron per H} & = & \text{3 electrons} \\
\text{2 C's} & \times & \text{4 electrons per C} & = & \text{8 electrons} \\
\text{1 N} & \times & \text{5 electrons per N} & = & \text{5 electrons} \\
\text{1 O} & \times & \text{6 electrons per O} & = & +\ \text{6 electrons} \\
\end{array}
$$

**22 electrons total**

**[3] Use the shortcut to figure out how many bonds are needed.**

- Number of electrons needed if there were no bonds:

$$
\begin{array}{llll}
\text{3 H's} & \times & \text{2 electrons per H} & = & \text{6 electrons} \\
\text{4 second-row elements} & \times & \text{8 electrons per element} & = & +\ \text{32 electrons} \\
\end{array}
$$

**38 electrons needed if
there were no bonds**

- Number of electrons that must be shared:

$$
\begin{array}{l}
\text{38 electrons} \\
-\ \text{22 electrons} \\
\hline
\textbf{16 electrons must be shared}
\end{array}
$$

- Since every bond takes two electrons, 16/2 = **8 bonds are needed.**

**[4] Draw all possible Lewis structures.**

- Draw the bonds to the H's first (three bonds). Then add five more bonds. Arrange them between the C's, N, and O, making sure that no atom gets more than eight electrons. There are three possible arrangements of bonds; i.e., there are three resonance structures.
- Add additional electron pairs to give each atom an octet and check that all 22 electrons are used.

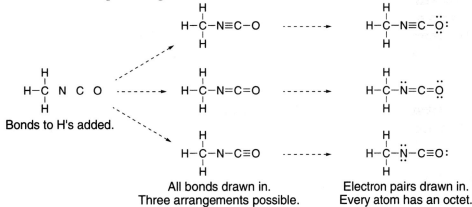

Bonds to H's added.

All bonds drawn in.
Three arrangements possible.

Electron pairs drawn in.
Every atom has an octet.

- Calculate the formal charge on each atom.

$$
\overset{\displaystyle H}{\underset{\displaystyle H}{H-\overset{|}{\underset{|}{C}}-N\equiv C-\overset{..}{\underset{..}{O}}:}} \quad\longleftrightarrow\quad \overset{\displaystyle H}{\underset{\displaystyle H}{H-\overset{|}{\underset{|}{C}}-\overset{..}{N}=C=\overset{..}{\underset{..}{O}}}} \quad\longleftrightarrow\quad \overset{\displaystyle H}{\underset{\displaystyle H}{H-\overset{|}{\underset{|}{C}}-\overset{..}{N}-C\equiv O:}}
$$

+1  −1                                      −1  +1

- You can evaluate the Lewis structures you have drawn. The middle structure is the best resonance structure, since it has no charged atoms.

**Note:** This method works for compounds that contain second-row elements in which every element gets an octet of electrons. It does NOT necessarily work for compounds with an atom that does not have an octet (such as $BF_3$), or compounds that have elements located in the third row and later in the periodic table.

## Chapter 1: Answers to Problems

**1.1** The **mass number** is the number of protons and neutrons. The **atomic number** is the number of protons and is the same for all isotopes.

| | Mass Number | | |
|---|---|---|---|
| | 16 | 17 | 18 |
| a. number of protons = atomic number for O = 8 | 8 | 8 | 8 |
| b. number of neutrons = mass number – atomic number | 8 | 9 | 10 |
| c. number for electrons = number of protons | 8 | 8 | 8 |
| d. The group number is the same for all isotopes. | 6A | 6A | 6A |

**1.2** The **atomic number** is the number of protons. The **total number of electrons** in the neutral atom is equal to the number of protons. The number of **valence electrons** is equal to the group number for second-row elements. The **group number** is located above each column in the periodic table.

| | a. atomic number | b. total number of $e^-$ | c. valence $e^-$ | d. group number |
|---|---|---|---|---|
| [1] $^{14}_{7}N$ | 7 | 7 | 5 | 5A |
| [2] $^{19}_{9}F$ | 9 | 9 | 7 | 7A |
| [3] $^{2}_{1}H$ | 1 | 1 | 1 | 1A |

**1.3** **Ionic bonds** form when an element on the far left side of the periodic table transfers an electron to an element on the far right side of the periodic table. **Covalent bonds** result when two atoms *share* electrons.

a.  F–F
covalent

b.  $Li^+$ $Br^-$
ionic

c.  H–C–C–H (with H H top and H H bottom)
All C–H and C–C bonds are covalent.

d.  $Na^+$ :N–H
ionic   Both N–H bonds are covalent.

**1.4** Atoms with one, two, or three valence electrons form one, two, or three bonds respectively. Atoms with four or more valence electrons form [8 – (number of valence electrons)] bonds.

a.  O  8 – 6 valence $e^-$ = 2 bonds     b.  Al  3 valence $e^-$ = 3 bonds     c.  Br  8 –7 valence $e^-$ = 1 bond

**1.5** [1] Arrange the atoms with the H's on the periphery.
[2] Count the valence electrons.
[3] Arrange the electrons around the atoms. Give the H's 2 electrons first, and then fill the octets of the other atoms.
[4] Assign formal charges (Section 1.3C).

a.
[1]  H H
    H C C H
      H H

[2]  Count valence $e^-$.
    2C x 4 $e^-$ = 8
    6H x 1 $e^-$ = 6
    total $e^-$   = 14

[3]  H H
    H–C–C–H
      H H

All 14 $e^-$ used.
All second-row elements have an octet.

b.
[1]
```
    H
H   C   N   H
    H   H
```

[2] Count valence e⁻.
1C x 4 e⁻ = 4
5H x 1 e⁻ = 5
1N x 5 e⁻ = 5
_____
total e⁻   = 14

[3]
```
      H
      |
H–C–N–H
  |   |
  H   H
```
⟶
```
      H
      |
H–C–N̈–H
  |   |
  H   H
```

12 e⁻ used.
N needs 2 more
electrons for an octet.

c.
[1]
```
    H
H   C
    H
```

[2] Count valence e⁻.
1C x 4 e⁻     = 4
3H x 1 e⁻     = 3
negative charge = 1
_____
total e⁻      = 8

[3]
```
    H
    |
H–C
    |
    H
```
⟶
```
      H
      |
H–C:⁻
      |
      H
```

6 e⁻ used.
C needs 2 more
electrons for an octet.

[The –1 charge on C is
explained in Section 1.3C.]

**1.6**   Follow the directions from Answer 1.5.

a. HCN     H  C  N

Count valence e⁻.
1C x 4 e⁻ = 4
1H x 1 e⁻ = 1
1N x 5 e⁻ = 5
_____
total e⁻   = 10

H–C–N

4 e⁻ used.

H–C≡N:

Complete N and C octets.

b. $H_2CO$
```
H  C  O
   H
```

Count valence e⁻.
1C x 4 e⁻ = 4
2H x 1 e⁻ = 2
1O x 6 e⁻ = 6
_____
total e⁻   = 12

```
H–C–O
   |
   H
```

6 e⁻ used.

```
H–C=Ö:
   |
   H
```

Complete O and C octets.

c. $(CH_3CO)^+$
```
   H
H  C  C  O
   H
```

Count valence e⁻.
2C x 4 e⁻ = 8
3H x 1 e⁻ = 3
1O x 6 e⁻ = 6
_____
total e⁻   = 17
Subtract 1 for (+) charge = 16

```
  H
  |
H–C–C–O
  |
  H
```

10 e⁻ used.

```
  H
  |
H–C–C=Ö:   or
  |   +
  H
```
```
  H
  |
H–C–C≡Ö+
  |   ··
  H
```

Complete octets.

[The (+) charge in these structures
is explained in Section 1.3C.]

**1.7**   **Formal charge** (FC) = number of valence electrons – [number of unshared electrons + 1/2 (number of shared electrons)]

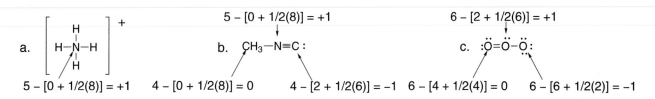

a.
$$\begin{bmatrix} \text{H} \\ \text{H–N–H} \\ \text{H} \end{bmatrix}^+$$

5 – [0 + 1/2(8)] = +1

5 – [0 + 1/2(8)] = +1

b. $CH_3$–N≡C:

4 – [0 + 1/2(8)] = 0     4 – [2 + 1/2(6)] = –1

c. :Ö=Ö–Ö:

6 – [2 + 1/2(6)] = +1

6 – [4 + 1/2(4)] = 0     6 – [6 + 1/2(2)] = –1

**1.8**

a. $CH_3O^-$
[1]
```
   H
H  C  O
   H
```

[2] Count valence e⁻.
1C x 4 e⁻ = 4
3H x 1 e⁻ = 3
1O x 6 e⁻ = 6
_____
total e⁻   = 13
Add 1 for (–) charge = 14

[3]
```
  H
  |
H–C–O
  |
  H
```
⟶
```
  H
  |
H–C–Ö:
  |
  H
```

8 e⁻ used.

[4]
```
  H
  |
H–C–Ö:⁻
  |  ··
  H
```

Assign charge.

b. $HC_2^-$    [1] H C C    [2] Count valence e⁻.    [3] H–C–C $\longrightarrow$ H–C≡C:    [4] H–C≡C:⁻

[2] Count valence e⁻.
$2C \times 4\,e^- = 8$
$1H \times 1\,e^- = 1$
total e⁻ $= 9$
Add 1 for (–) charge $= 10$

4 e⁻ used.

Assign charge.

c. $(CH_3NH_3)^+$    [1] H H H C N H H H

[2] Count valence e⁻.
$1C \times 4\,e^- = 4$
$6H \times 1\,e^- = 6$
$1N \times 5\,e^- = 5$
total e⁻ $= 15$
Subtract 1 for (+) charge $= 14$

[3]

H H
H–C–N–H
H H

14 e⁻ used.

[4]

H H
H–C–N⁺–H
H H

Assign charge.

## 1.9

a. $C_2H_4Cl_2$ (two isomers)

Count valence e⁻.
$2C \times 4\,e^- = 8$
$4H \times 1\,e^- = 4$
$2Cl \times 7\,e^- = 14$
total e⁻ $= 26$

H H
H–C–C–Cl:
H :Cl:

H H
H–C–C–Cl:
:Cl: H

b. $C_3H_8O$ (three isomers)

Count valence e⁻.
$3C \times 4\,e^- = 12$
$8H \times 1\,e^- = 8$
$1O \times 6\,e^- = 6$
total e⁻ $= 26$

H H H
H–C–C–C–O–H
H H H

H :O: H
H–C–C–C–H
H H H

H H H
H–C–C–O–C–H
H H H

c. $C_3H_6$ (two isomers)

Count valence e⁻.
$3C \times 4\,e^- = 12$
$6H \times 1\,e^- = 6$
total e⁻ $= 18$

H H
H–C–C=C
H H H

H H
C
H–C–C–H
H H

## 1.10 Two different definitions:

- **Isomers** have the same molecular formula and a *different* arrangement of atoms.
- **Resonance structures** have the same molecular formula and the *same* arrangement of atoms.

N at the end

N in the middle

a.   ⁻:N=C=O:   and   :C≡N–O:⁻

different arrangement of atoms = **isomers**

2 lone pairs    3 lone pairs

:O:
b.   HO–C–O:⁻   and   HO–C=O:

same arrangement of atoms =
**resonance structures**

## 1.11 Curved arrow notation shows the movement of an electron pair. The tail begins at an electron pair (a bond or a lone pair) and the head points to where the electron pair moves.

a.   H–C=O: $\longleftrightarrow$ H–C⁺–O:⁻
     H          H

The net charge is the same
in both resonance structures.

b.   $CH_3=C–C=CH_2$ $\longleftrightarrow$ $CH_3–C=C–CH_2^-$
     H H           H H

The net charge is the same
in both resonance structures.

**1.12** Compare the resonance structures to see what electrons have "moved." **Use one curved arrow to show the movement of each electron pair.**

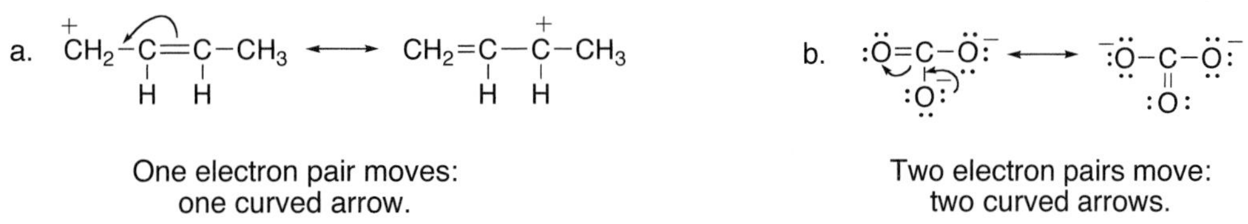
a.
One electron pair moves:
one curved arrow.
b.
Two electron pairs move:
two curved arrows.

**1.13** To draw another resonance structure, **only move electrons in multiple bonds and lone pairs** and keep the number of unpaired electrons constant.

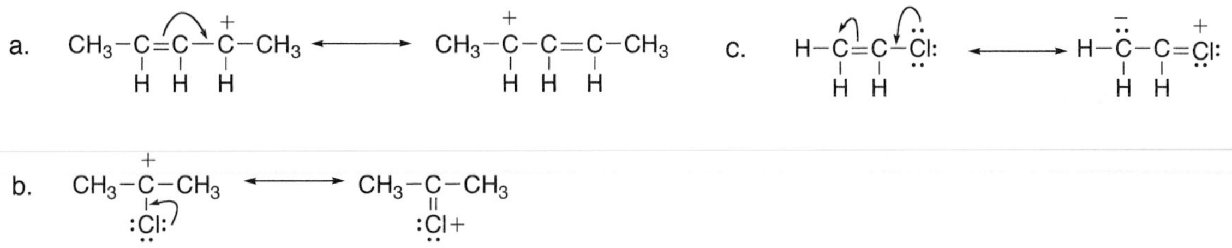
a.
c.
b.

**1.14** To draw the resonance hybrid:

- Use solid lines for bonds that appear in both resonance structures and dashed lines for bonds that are only in one of the resonance structures.
- Use partial charges when the charge is on different atoms in the resonance structures.

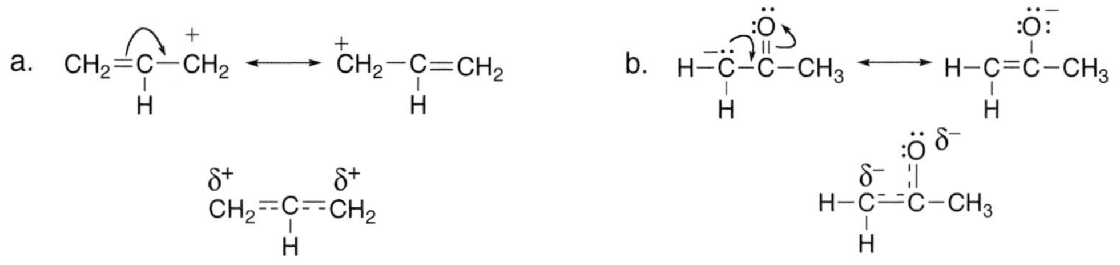
a.
b.

**1.15 A "better" resonance structure is one that has more bonds and fewer charges.** The better structure is the major contributor and all others are minor contributors. To draw the resonance hybrid, use the rules in Answer 1.14.

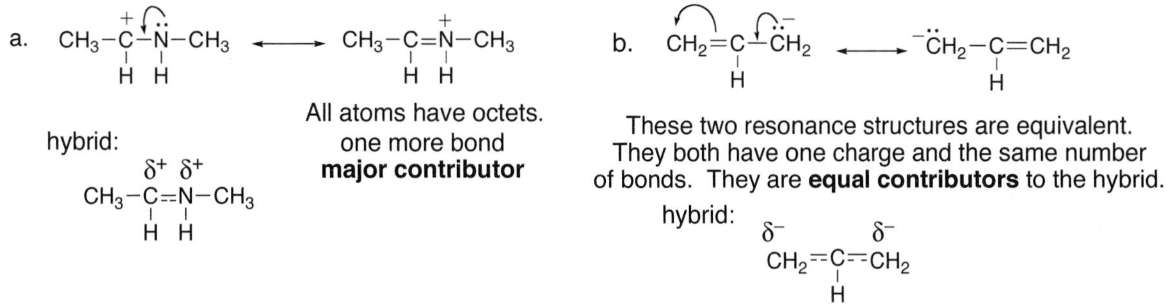
a.
hybrid:
All atoms have octets.
one more bond
major contributor
b.
These two resonance structures are equivalent.
They both have one charge and the same number
of bonds. They are equal contributors to the hybrid.
hybrid:

**1.16** All representations have a carbon with two bonds in the plane of the page, one in front of the page (solid wedge) and one behind the page (dashed line). Four possibilities:

**1.17** To predict the geometry around an atom, **count the number of groups (atoms + lone pairs),** making sure to draw in any needed lone pairs or hydrogens: 2 groups = linear, 3 groups = trigonal planar, 4 groups = tetrahedral.

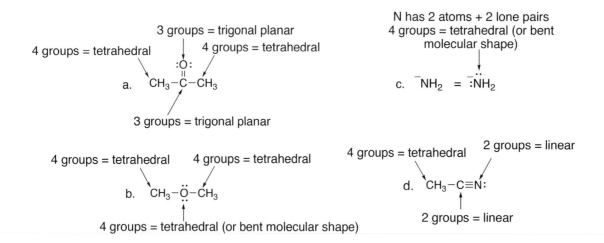

**1.18** To predict the bond angle around an atom, **count the number of groups (atoms + lone pairs),** making sure to draw in any needed lone pairs or hydrogens: 2 groups = 180°, 3 groups = 120°, 4 groups = 109.5°.

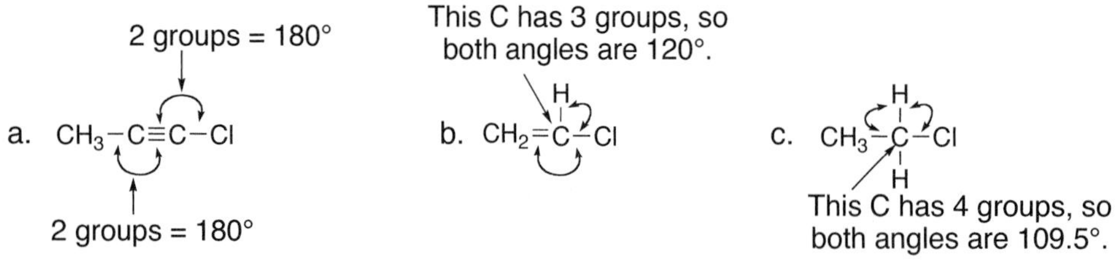

a. CH₃–C≡C–Cl
2 groups = 180°
2 groups = 180°

b. CH₂=C–Cl
This C has 3 groups, so both angles are 120°.

c. CH₃–C–Cl
This C has 4 groups, so both angles are 109.5°.

**1.19** To predict the geometry around an atom, use the rules in Answer 1.17.

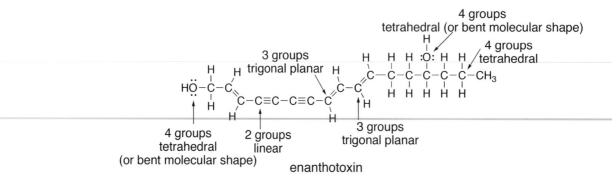

enanthotoxin

**1.20** Reading from left to right, draw the molecule as a Lewis structure. Always check that carbon has four bonds and all heteroatoms have an octet by adding any needed lone pairs.

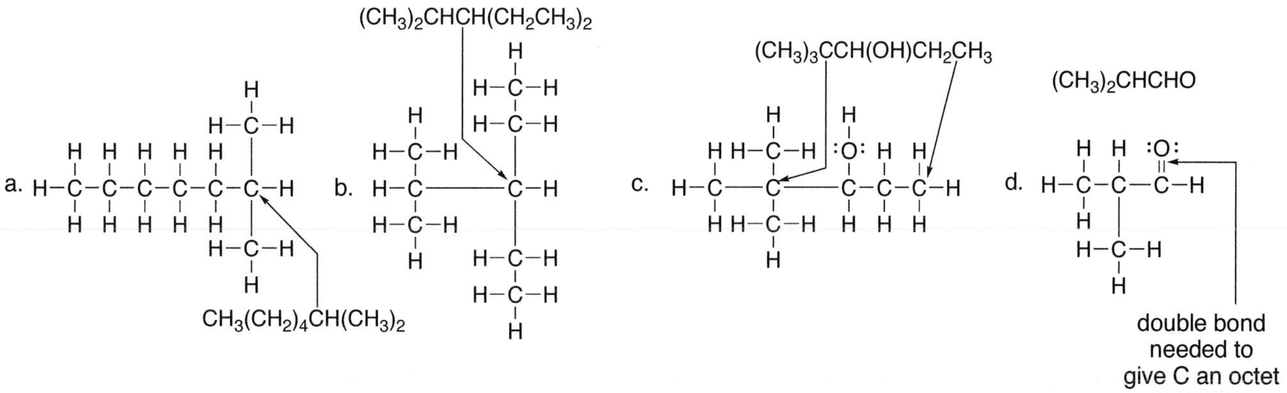

(CH₃)₂CHCH(CH₂CH₃)₂ — represented as $(CH_3)_2CHCH(CH_2CH_3)_2$

(CH₃)₃CCH(OH)CH₂CH₃ — $(CH_3)_3CCH(OH)CH_2CH_3$

(CH₃)₂CHCHO — $(CH_3)_2CHCHO$

a.  CH₃(CH₂)₄CH(CH₃)₂

b.

c.

d.
double bond
needed to
give C an octet

**1.21** In shorthand or skeletal drawings, **all line junctions or ends of lines represent carbon atoms**. The carbons are all tetravalent.

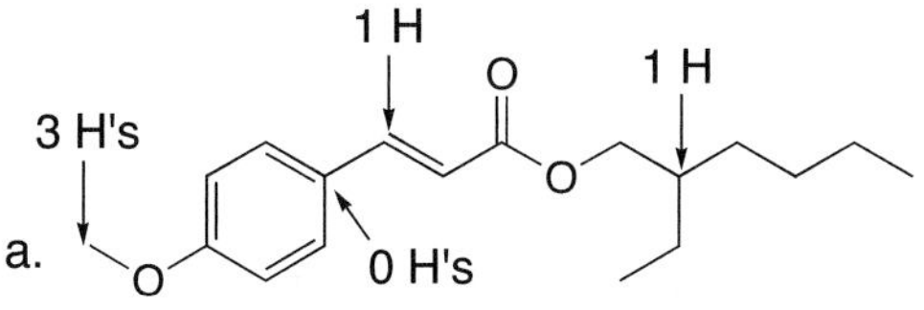

a.
3 H's    1 H    1 H    0 H's
octinoxate
(2-ethylhexyl 4-methoxycinnamate)

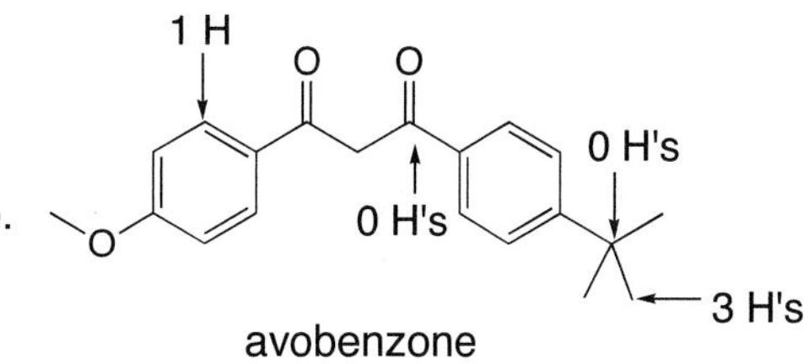

b.
1 H    0 H's    0 H's    3 H's
avobenzone

**1.22** In shorthand or skeletal drawings, **all line junctions or ends of lines represent carbon atoms**. Convert by writing in all carbons, and then adding hydrogen atoms to make the carbons tetravalent.

a.

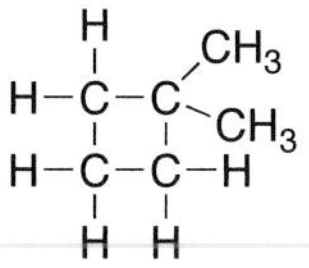

b.

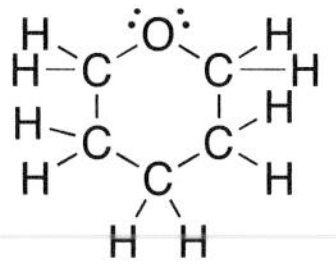

c.

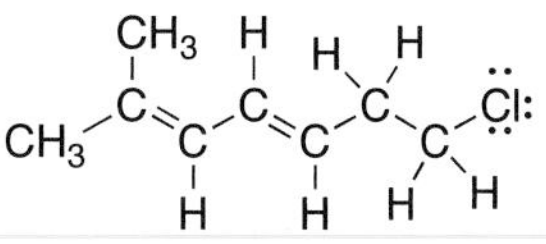

d.

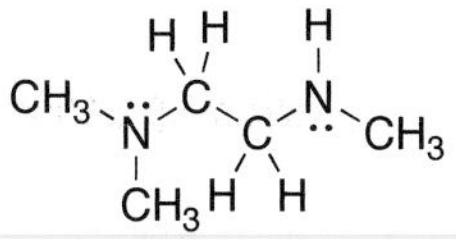

**1.23** A charge on a carbon atom takes the place of one hydrogen atom. **A negatively charged C has one lone pair, and a positively charged C has none.**

a. 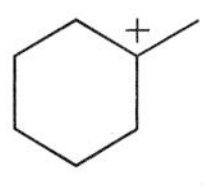
positive charge
no lone pairs
no H's needed

b. 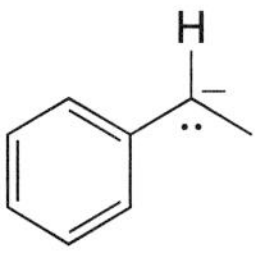
negative charge
one lone pair
one H needed

c. 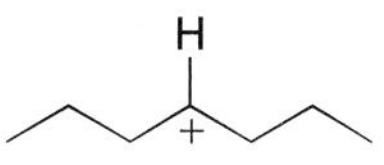
positive charge
no lone pairs
one H needed

d. 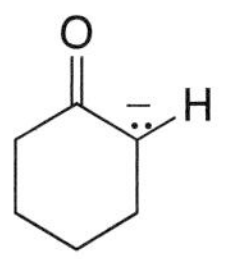
negative charge
one lone pair
one H needed

**1.24**

a. 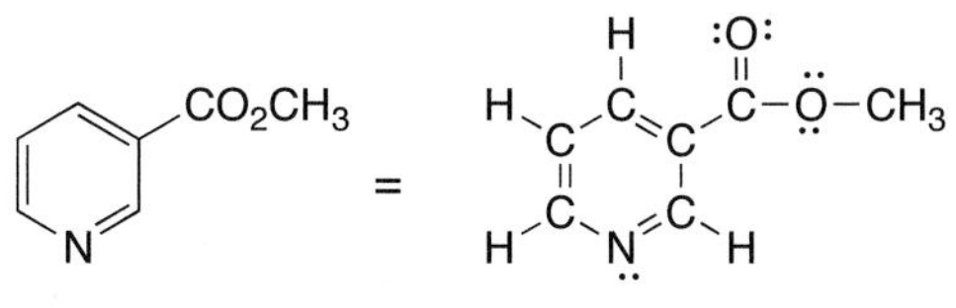  $=$

b.  = 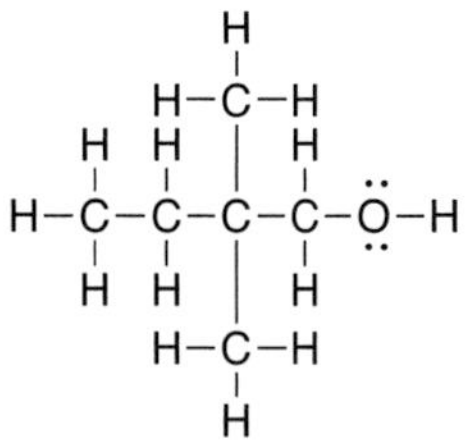 = $CH_3CH_2C(CH_3)_2CH_2OH$

c. 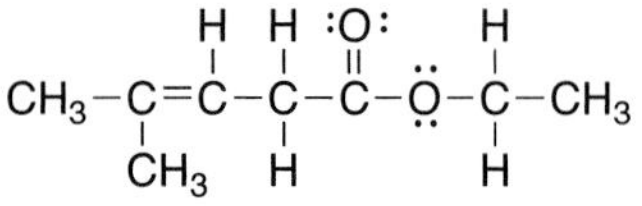 = $(CH_3)_2C=CHCH_2COOCH_2CH_3$
or
$(CH_3)_2C=CHCH_2CO_2CH_2CH_3$ =

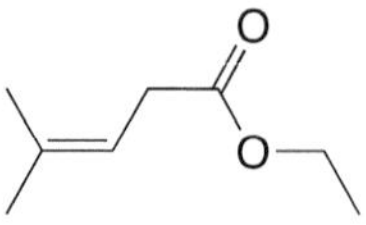

**1.25**

a. 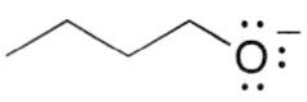  b. 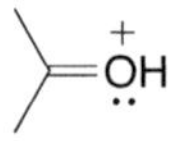  c. 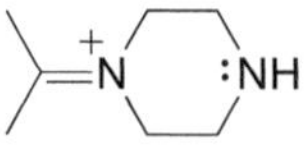  d.

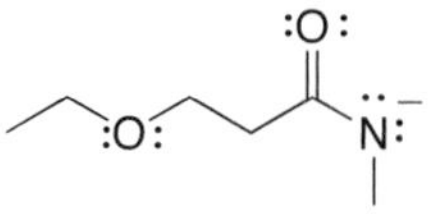

**1.26** To determine the orbitals used in bonding, **count the number of groups** (atoms + lone pairs): 4 groups = $sp^3$, 3 groups = $sp^2$, 2 groups = $sp$, H atom = $1s$ (no hybridization).

All covalent single bonds are σ, and all double bonds contain one σ and one π bond.

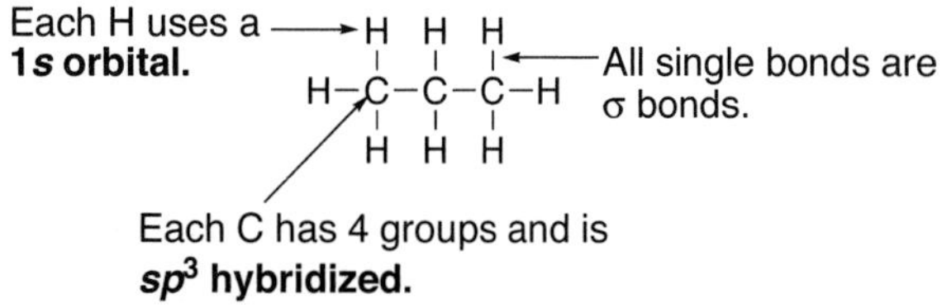

**1.27** [1] Draw a valid Lewis structure for each molecule.

[2] **Count the number of groups** around each atom: 4 groups = $sp^3$, 3 groups = $sp^2$, 2 groups = $sp$, H atom = $1s$ (no hybridization).

*Note:* **Be and B** (Groups 2A and 3A) do not have enough valence e⁻ to form an octet, **and do not form an octet in neutral molecules**.

a. [1]
Be has 2 bonds.

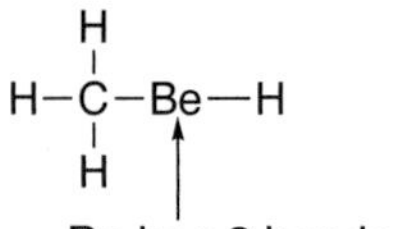

[2] Count groups around each atom:
4 groups  2 groups
$sp^3$   $sp$

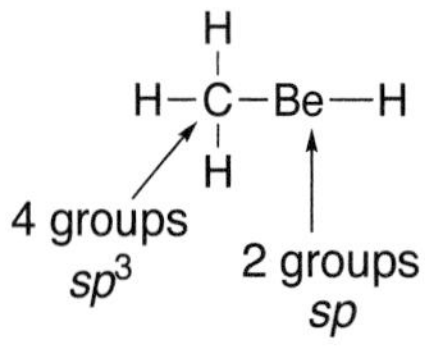

[3] All C–H bonds: $Csp^3$–H$1s$
C–Be bond: $Csp^3$–Be$sp$
Be–H bond: Be$sp$–H$1s$

b. [1]
B forms 3 bonds.

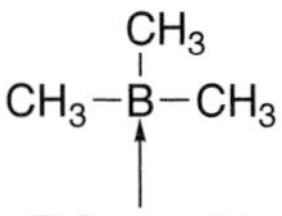

[2] Count groups around each atom:
4 groups  3 groups
$sp^3$   $sp^2$

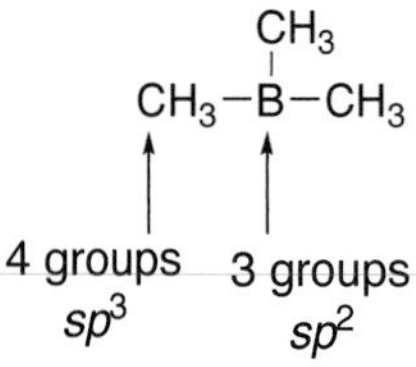

[3] All C–H bonds: $Csp^3$–H$1s$
C–B bonds: $Csp^3$–B$sp^2$

c. [1] $H-\overset{\overset{\textstyle H}{|}}{\underset{\underset{\textstyle H}{|}}{C}}-\overset{\cdot\cdot}{\underset{\cdot\cdot}{O}}-\overset{\overset{\textstyle H}{|}}{\underset{\underset{\textstyle H}{|}}{C}}-H$     [2] Count groups around each atom:

$$H-\underset{\underset{H}{|}}{\overset{\overset{H}{|}}{C}}-\overset{\cdot\cdot}{\underset{\cdot\cdot}{O}}-\underset{\underset{H}{|}}{\overset{\overset{H}{|}}{C}}-H$$

4 groups $sp^3$    4 groups $sp^3$

[3] All C–H bonds: $C_{sp^3}$–$H_{1s}$
C–O bonds: $C_{sp^3}$–$O_{sp^3}$

**1.28** To determine the hybridization, **count the number of groups** around each atom: 4 groups = $sp^3$, 3 groups = $sp^2$, 2 groups = $sp$, H atom = $1s$ (no hybridization).

a.   $CH_3-C\equiv CH$

4 groups $sp^3$    2 groups $sp$

b.   (cyclohexane ring)$=\overset{\cdot\cdot}{N}-CH_3$

3 groups $sp^2$    3 groups $sp^2$

c.   $CH_2=C=CH_2$

3 groups $sp^2$    2 groups $sp$

**1.29** All single bonds are σ. Multiple bonds contain one σ bond, and all others are π bonds.

| All C–H bonds are σ bonds. |
|---|

a.   $CH_3-\overset{\overset{O}{\|}}{C}-H$  — one σ bond, one π bond

σ bond   σ bond

b.   $CH_3-C\equiv N$

σ bond    one σ + two π bonds

c.   $H-\overset{\overset{O}{\|}}{C}\overset{}{\diagdown}O-CH_3$

σ bond ← one σ bond, one π bond

σ bond    σ bond

**1.30** Bond length and bond strength are inversely related: **longer bonds are weaker bonds**. Single bonds are weaker and longer than double bonds, which are weaker and longer than triple bonds.

a.   (cyclohexane)$-C\equiv C-$(cyclohexene)

bond 1: single bond
bond 2: triple bond
bond 3: double bond

increasing bond strength: $1 < 3 < 2$
increasing bond length: $2 < 3 < 1$

b.   $CH_3-\overset{\overset{H}{|}}{N}-$ (ring with N) $-C\equiv N$

bond 1: single bond
bond 2: double bond
bond 3: triple bond

increasing bond strength: $1 < 2 < 3$
increasing bond length: $3 < 2 < 1$

**1.31** Bond length and bond strength are inversely related: **longer bonds are weaker bonds.** Single bonds are weaker and longer than double bonds, which are weaker and longer than triple bonds. Increasing percent *s*-character increases bond strength and decreases bond length.

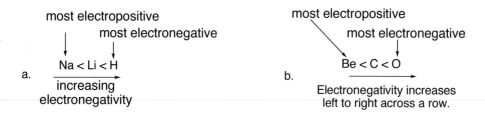

a. $CH_3-C\equiv C-H$  and

$C_{sp}-H_{1s}$
50% *s*-character
shorter bond

$\overset{CH_3}{\underset{H}{C}}=CH_2$

$C_{sp^2}-H_{1s}$
33% *s*-character

c. $CH_2=\overset{..}{N}-H$  and  $CH_3-\overset{H}{\underset{..}{N}}-H$

$N_{sp^2}-H_{1s}$
33% *s*-character
shorter bond

$N_{sp^3}-H_{1s}$
25% *s*-character

b. $\overset{H}{\underset{H}{C}}=\overset{..}{\overset{..}{O}}$  and  $H-\overset{H}{\underset{H}{C}}-\overset{..}{\overset{..}{O}}H$

$C_{sp^2}-H_{1s}$
33% *s*-character
shorter bond

$C_{sp^3}-H_{1s}$
25% *s*-character

**1.32** **Electronegativity increases across a row of the periodic table, and decreases down a column.** Look at the relative position of the atoms to determine their relative electronegativity.

a.
most electropositive

most electronegative

$Na < Li < H$

increasing
electronegativity

b.
most electropositive

most electronegative

$Be < C < O$

Electronegativity increases
left to right across a row.

**1.33** Dipoles result from unequal sharing of electrons in covalent bonds. More electronegative atoms "pull" electron density towards them, making a dipole. **Dipole arrows point towards the atom of higher electron density.**

a. $\overset{\delta^+ \ \ \delta^-}{H-F}$
b. $\overset{\delta^+}{-B}-\overset{\delta^-}{C}-$
c. $\overset{\delta^-}{-C}-\overset{\delta^+}{Li}$
d. $\overset{\delta^+}{-C}-\overset{\delta^-}{Cl}$

**1.34** Polar molecules result from a net dipole. To determine polarity, draw the molecule in three dimensions around any polar bonds, draw in the dipoles, and look to see whether the dipoles cancel or reinforce.

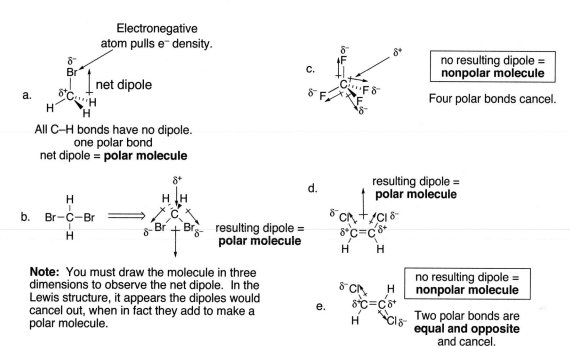

a. Electronegative atom pulls e⁻ density.
All C–H bonds have no dipole.
one polar bond
net dipole = **polar molecule**

c. no resulting dipole = **nonpolar molecule**
Four polar bonds cancel.

b. resulting dipole = **polar molecule**

**Note:** You must draw the molecule in three dimensions to observe the net dipole. In the Lewis structure, it appears the dipoles would cancel out, when in fact they add to make a polar molecule.

d. resulting dipole = **polar molecule**

e. no resulting dipole = **nonpolar molecule**
Two polar bonds are **equal and opposite** and cancel.

**1.35**

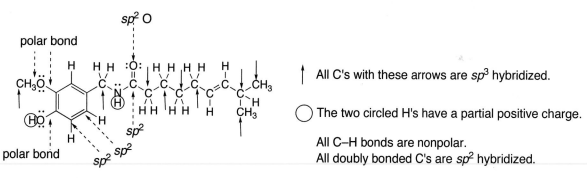

polar bond

polar bond

↑ All C's with these arrows are $sp^3$ hybridized.

◯ The two circled H's have a partial positive charge.

All C–H bonds are nonpolar.
All doubly bonded C's are $sp^2$ hybridized.

In addition to the labeled polar bonds, all C–N, N–H, and O–H bonds are also polar.

**1.36** Use bonding rules in Answer 1.3.

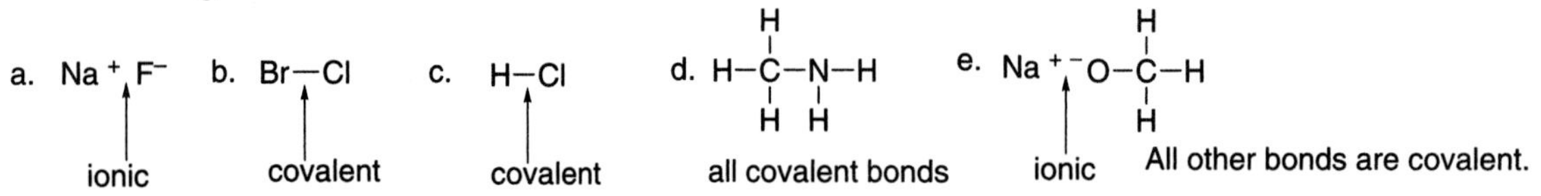

a. $Na^+ \ F^-$    b. $Br-Cl$    c. $H-Cl$    d. $H-\overset{H}{\underset{H}{\overset{|}{C}}}-\overset{}{\underset{H}{\overset{|}{N}}}-H$    e. $Na^{+\,-}O-\overset{H}{\underset{H}{\overset{|}{C}}}-H$

   ionic      covalent      covalent      all covalent bonds    ionic    All other bonds are covalent.

**1.37 Formal charge** (FC) = number of valence electrons – [number of unshared electrons + 1/2 (number of shared electrons)]. C is in group 4A.

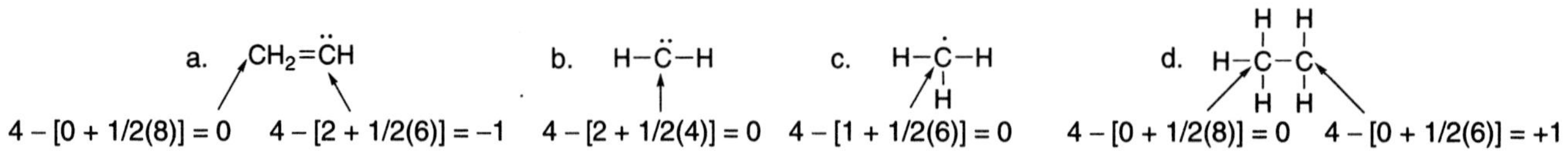

a. $CH_2=\ddot{C}H$      b. $H-\ddot{C}-H$      c. $H-\dot{C}-H$      d. $H-\overset{H\ \ H}{\underset{H\ \ H}{\overset{|\ \ \ |}{C-C}}}$

$4 - [0 + 1/2(8)] = 0$    $4 - [2 + 1/2(6)] = -1$    $4 - [2 + 1/2(4)] = 0$    $4 - [1 + 1/2(6)] = 0$    $4 - [0 + 1/2(8)] = 0$    $4 - [0 + 1/2(6)] = +1$

**1.38 Formal charge** (FC) = number of valence electrons – [number of unshared electrons + 1/2 (number of shared electrons)]. N is in group 5A and O is in group 6A.

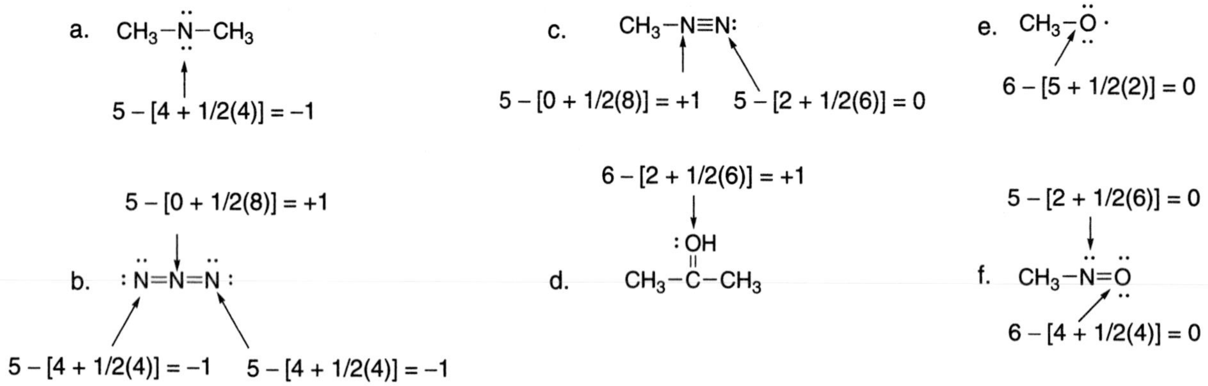

a. $CH_3-\ddot{N}-CH_3$

$5 - [4 + 1/2(4)] = -1$

$5 - [0 + 1/2(8)] = +1$

b. $:N=N=N:$

$5 - [4 + 1/2(4)] = -1$    $5 - [4 + 1/2(4)] = -1$

c. $CH_3-N\equiv N:$

$5 - [0 + 1/2(8)] = +1$    $5 - [2 + 1/2(6)] = 0$

$6 - [2 + 1/2(6)] = +1$

$:OH$

d. $CH_3-\overset{||}{C}-CH_3$

e. $CH_3-\ddot{O}\cdot$

$6 - [5 + 1/2(2)] = 0$

$5 - [2 + 1/2(6)] = 0$

f. $CH_3-\ddot{N}=\ddot{O}$

$6 - [4 + 1/2(4)] = 0$

**1.39** Follow the steps in Answer 1.5 to draw Lewis structures.

a. $CH_2N_2$

$$H-\overset{|}{\underset{H}{C}}=\overset{+}{N}=\overset{..}{\underset{..}{N}}{}^{-}$$

or

$$H-\overset{..}{\underset{H}{\underset{|}{C}}}-\overset{+}{N}\equiv N:$$

| valence e⁻ | |
| --- | --- |
| 1C x 4 e⁻ | = 4 |
| 2H x 1 e⁻ | = 2 |
| 2N x 5 e⁻ | = 10 |
| total e⁻ | = 16 |

b. $CH_3NO_2$

$$H-\overset{H}{\underset{H}{\overset{|}{\underset{|}{C}}}}-\overset{+}{N}-\overset{..}{\underset{..}{O}}{}^{-} \quad or \quad H-\overset{H}{\underset{H}{\overset{|}{\underset{|}{C}}}}-\overset{+}{N}=\overset{..}{\underset{..}{O}}:$$
$$\overset{}{:}\overset{..}{O}:\qquad\qquad :\overset{..}{O}{}^{-}$$

| valence e⁻ | |
| --- | --- |
| 1C x 4 e⁻ | = 4 |
| 3H x 1 e⁻ | = 3 |
| 1N x 5 e⁻ | = 5 |
| 2O x 6 e⁻ | = 12 |
| total e⁻ | = 24 |

c. $CH_3CNO$

$$H-\overset{H}{\underset{H}{\overset{|}{\underset{|}{C}}}}-C\equiv\overset{+}{N}-\overset{..}{\underset{..}{O}}:^{-} \quad or \quad H-\overset{H}{\underset{H}{\overset{|}{\underset{|}{C}}}}-\overset{-}{C}=\overset{+}{N}=\overset{..}{O}:$$

$$or \quad H-\overset{H}{\underset{H}{\overset{|}{\underset{|}{C}}}}-\overset{..}{\underset{..}{C}}{}^{2-}\overset{+}{-}\overset{+}{N}\equiv O:$$

| valence e⁻ | |
| --- | --- |
| 2C x 4 e⁻ | = 8 |
| 3H x 1 e⁻ | = 3 |
| 1N x 5 e⁻ | = 5 |
| 1O x 6 e⁻ | = 6 |
| total e⁻ | = 22 |

d. $HCO_2^-$

$$\overset{:\overset{..}{O}:^{-}}{H-C=\overset{..}{\underset{..}{O}}:} \quad or \quad \overset{:O:}{H-\overset{||}{C}-\overset{..}{\underset{..}{O}}:^{-}}$$

| valence e⁻ | |
| --- | --- |
| 1C x 4 e⁻ | = 4 |
| 1H x 1 e⁻ | = 1 |
| 2O x 6 e⁻ | = 12 |
| 1 for (–) charge | = 1 |
| total e⁻ | = 18 |

e. $HCO_3^-$

$$\overset{:\overset{..}{O}:^{-}}{H-\overset{..}{\underset{..}{O}}-C=\overset{..}{\underset{..}{O}}:} \quad or \quad \overset{:O:}{H-\overset{..}{\underset{..}{O}}-\overset{||}{C}-\overset{..}{\underset{..}{O}}:^{-}}$$

$$or \quad \overset{:\overset{..}{O}:^{-}}{H-\overset{..}{\underset{..}{O}}\overset{+}{=}C-\overset{..}{\underset{..}{O}}:^{-}}$$

| valence e⁻ | |
| --- | --- |
| 1C x 4 e⁻ | = 4 |
| 1H x 1 e⁻ | = 1 |
| 3O x 6 e⁻ | = 18 |
| 1 for (–) charge | = 1 |
| total e⁻ | = 24 |

f. $^-CH_2CN$

$$H-C=C=\overset{..}{N}:^{-} \quad or \quad H-\overset{..}{\underset{H}{\underset{|}{C}}}{}^{-}C\equiv N:$$
$$\underset{H}{|}$$

| valence e⁻ | |
| --- | --- |
| 2C x 4 e⁻ | = 8 |
| 2H x 1 e⁻ | = 2 |
| 1N x 5 e⁻ | = 5 |
| 1 for (–) charge | = 1 |
| total e⁻ | = 16 |

**1.40** Follow the steps in Answer 1.5 to draw Lewis structures.

a. $N_2$

[1] N N

[2] Count valence e⁻.

| 2N x 5 e⁻ | = 10 |
| --- | --- |
| total e⁻ | = 10 |

[3] N—N $\longrightarrow$ :N≡N:

2 e⁻ used.

Complete N octet.

b. $(CH_3OH_2)^+$

[1]
$$\underset{\underset{H\ \ H}{}}{H\ \ C\ \ O\ \ H}^{H}$$

[2] Count valence e⁻.

| 1C x 4 e⁻ | = 4 |
| --- | --- |
| 5H x 1 e⁻ | = 5 |
| 1O x 6 e⁻ | = 6 |
| total e⁻ | = 15 |
| Subtract 1 for (+) charge | = 14 |

[3]
$$H-\overset{H}{\underset{H}{\overset{|}{\underset{|}{C}}}}-O-H$$
$$\underset{H\ \ H}{}$$

12 e⁻ used.

[4]
$$H-\overset{H}{\underset{H}{\overset{|}{\underset{|}{C}}}}-\overset{+}{\overset{..}{O}}-H$$
$$\underset{H\ \ H}{}$$

Add charge and lone pair.

c. $(CH_3CH_2)^-$

[1]
$$\underset{\underset{H\ \ H}{}}{H\ \ C\ \ C\ \ H}^{H}$$

[2] Count valence e⁻.

| 2C x 4 e⁻ | = 8 |
| --- | --- |
| 5H x 1 e⁻ | = 5 |
| total e⁻ | = 13 |
| Add 1 for (–) charge | = 14 |

[3]
$$H-\overset{H}{\underset{H}{\overset{|}{\underset{|}{C}}}}-C-H$$
$$\underset{H\ \ H}{}$$

12 e⁻ used.

[4]
$$H-\overset{H}{\underset{H}{\overset{|}{\underset{|}{C}}}}-\overset{..}{C}-H^{-}$$
$$\underset{H\ \ H}{}$$

Add charge and lone pair.

d. HNNH

[1] H N N H

[2] Count valence e⁻.

| 2H x 1 e⁻ | = 2 |
| --- | --- |
| 2N x 5 e⁻ | = 10 |
| total e⁻ | = 12 |

[3] H—N—N—H $\longrightarrow$ $H-\overset{..}{N}=\overset{..}{N}-H$

6 e⁻ used.

Complete N octet.

e. $H_6BN$

[1]
```
      H   H
      |   |
  H   B   N   H
      |   |
      H   H
```

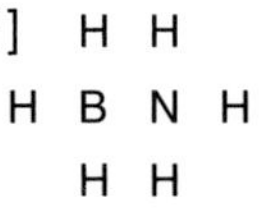

[2] Count valence e⁻.

1B x 3 e⁻ = 3
6H x 1 e⁻ = 6
1N x 5 e⁻ = 5
─────────────
total e⁻ = 14

[3] 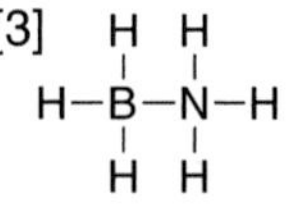
```
      H   H
      |   |
  H—B—N—H
      |   |
      H   H
```
14 e⁻ used.

[4] 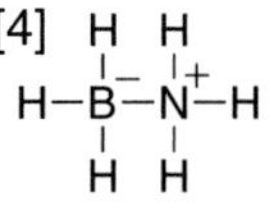
```
      H   H
      |⁻  |⁺
  H—B—N—H
      |   |
      H   H
```
Add charges.

## 1.41 Isomers must have a different arrangement of atoms.

a. Two isomers of molecular formula: $C_3H_7Cl$

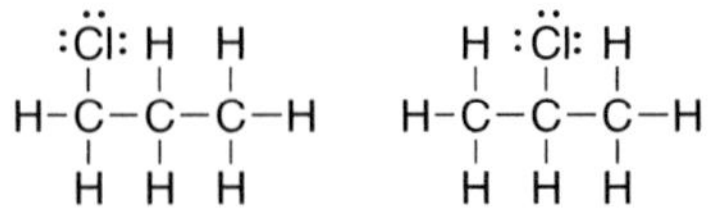

c. Four isomers of molecular formula: $C_3H_9N$

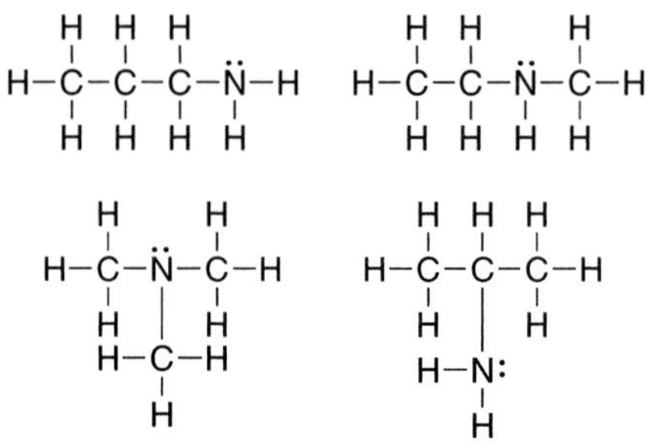

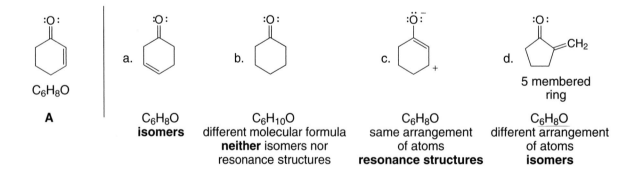

b. Three isomers of molecular formula: $C_2H_4O$

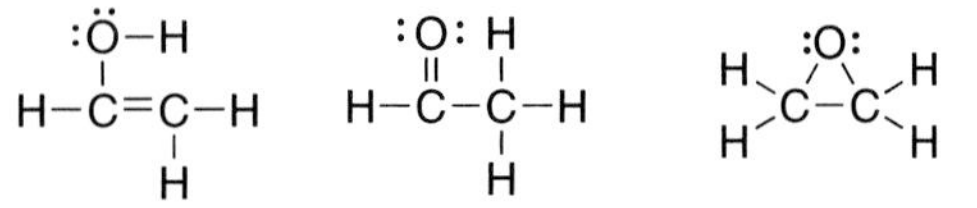

## 1.42

### Nine isomers of $C_3H_6O$:

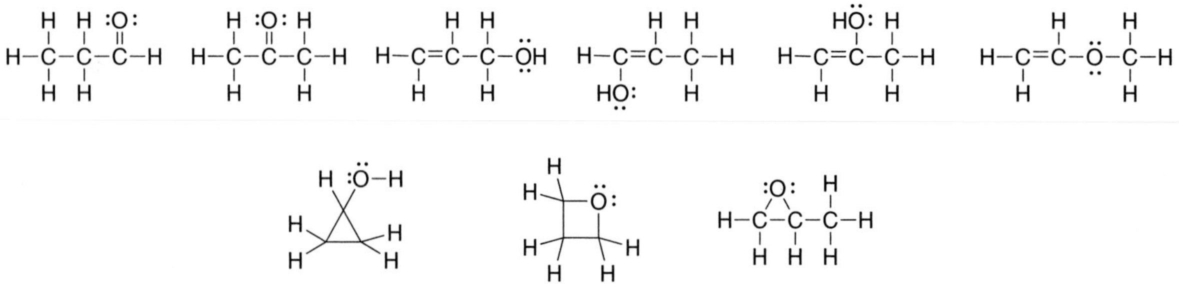

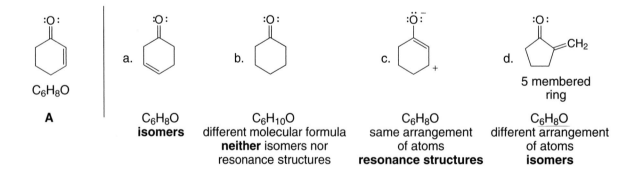

## 1.43 Use the definition of isomers and resonance structures in Answer 1.10.

:O:

$C_6H_8O$

**A**

a. 
$C_6H_8O$
**isomers**

b. 
$C_6H_{10}O$
different molecular formula
**neither** isomers nor
resonance structures

c. 
:Ö:⁻
$C_6H_8O$
same arrangement
of atoms
**resonance structures**

d. 
:O:
=CH₂
5 membered
ring
$C_6H_8O$
different arrangement
of atoms
**isomers**

**1.44** Use the definitions of isomers and resonance structures in Answer 1.10.

a.  $CH_3-\ddot{O}-CH_2CH_3$ and $CH_3-\overset{\overset{\displaystyle CH_3}{|}}{\underset{\underset{\displaystyle H}{|}}{C}}-\ddot{O}H$

two C–O bonds     one O–H bond

Different arrangement of atoms.
Both have molecular formula $C_3H_8O$ = **isomers**

b.  ▢    and    $CH_3-\overset{\underset{\displaystyle H}{|}}{C}=\overset{\underset{\displaystyle H}{|}}{C}-CH_3$

ring          double bond

Different arrangement of atoms.
Both have molecular formula $C_4H_8$ = **isomers**

c.  $CH_2=CH-\overset{-}{\ddot{C}H}-CH_3$ and $^-\ddot{C}H_2-CH=CH-CH_3$

Same arrangement of atoms.
Both have molecular formula $(C_4H_7)^-$.
Different arrangement of electrons = **resonance structures**

d.  $CH_3CH_2CH_3$ and $CH_3CH_2\overset{-}{\ddot{C}H_2}$

molecular formula $C_3H_8$    molecular formula $(C_3H_7)^-$

different molecular formulas = **neither**

**1.45** Compare the resonance structures to see what electrons have "moved." **Use one curved arrow to show the movement of each electron pair.**

a.  $CH_3-\overset{\overset{\displaystyle H}{|}}{\underset{\underset{\displaystyle H}{|}}{\overset{+}{C}}}-\overset{}{\ddot{N}}-CH_3 \longleftrightarrow CH_3-\overset{\overset{\displaystyle H}{|}}{C}=\overset{+}{\underset{\underset{\displaystyle H}{|}}{N}}-CH_3$

One electron pair moves = one arrow

b.  $H-\overset{\overset{\displaystyle \ddot{O}:}{||}}{C}-\overset{..}{N}H_2 \longleftrightarrow H-\overset{\overset{\displaystyle :\ddot{O}:^-}{|}}{C}=\overset{+}{N}H_2$

Two electron pairs move = two arrows

c.  (ring) $-\overset{-}{\ddot{C}H_2} \longleftrightarrow$ (ring) $=CH_2$

Two electron pairs move = two arrows

**1.46** Curved arrow notation shows the movement of an electron pair. The tail begins at an electron pair (a bond or a lone pair) and the head points to where the electron pair moves.

a.  $CH_3-\overset{+}{N}\equiv N: \longleftrightarrow CH_3-\overset{..}{N}=\overset{+}{N}:$

b.  $CH_3-\overset{\overset{\displaystyle :\ddot{O}:^-}{|}}{C}=CH-\overset{+}{C}H_2 \longleftrightarrow CH_3-\overset{\overset{\displaystyle :O:}{||}}{C}-CH=CH_2$

c.  (ring with $CH_3$, +) $\longleftrightarrow$ (ring with $CH_3$)

d.  (ring with $\overset{+}{N}H_2$) $\longleftrightarrow$ (ring with $\overset{..}{N}H_2$)

**1.47** Use the rules in Answer 1.13.

a.  $CH_3-\overset{\overset{\displaystyle :O:}{||}}{C}-\overset{..}{\ddot{O}:}^- \longleftrightarrow CH_3-\overset{\overset{\displaystyle :\ddot{O}:^-}{|}}{C}=\overset{..}{O}:$

Two electron pairs move = two arrows

b.  $CH_2=\overset{+}{N}H_2 \longleftrightarrow \overset{+}{C}H_2-\ddot{N}H_2$

One electron pair moves = one arrow

c.  (ring with $:O:$ double bond) $\longleftrightarrow$ (ring with $:\ddot{O}:^-$)

Two electron pairs move = two arrows

d.  $H-\overset{\overset{\displaystyle :\overset{+}{O}H}{||}}{C}-H \longleftrightarrow H-\overset{\overset{\displaystyle :\ddot{O}H}{|}}{\underset{+}{C}}-H$

One electron pair moves = one arrow

**1.48** To draw the **resonance hybrid**, use the rules in Answer 1.14.

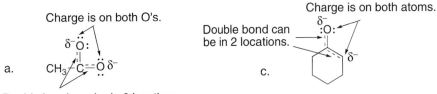

Charge is on both O's.

a. $CH_3$—C=O $\delta^-$   $\delta^-\ddot{O}$:

Double bond can be in 2 locations.

b. partial double bond character

$\delta^+ \downarrow \delta^+$
$CH_2$=$NH_2$
$\uparrow \quad \uparrow$

Charge is on both atoms.

c. Charge is on both atoms.

Double bond can be in 2 locations.

d. Charge is on both atoms.

:OH
H—C—H
$\delta^+$

C—O bond has
partial double bond character.

**1.49** For the compounds where the arrangement of atoms is not given, first draw a Lewis structure. Then use the rules in Answer 1.13.

a. $O_3$

Count valence e⁻.
3 O x 6 e⁻ = 18
total e⁻  = 18

$:\ddot{O}-\overset{+}{O}=\ddot{O}$ ⟷ $\ddot{O}=\overset{+}{O}-\ddot{O}:$

b. $NO_3^-$ (a central N atom)

Count valence e⁻.
1 N x 5 e⁻  = 5
3 O x 6 e⁻  = 18
(−) charge  = 1
total e⁻    = 24

$\ddot{O}=\overset{+}{N}+$ ⟷ $:\ddot{O}-\overset{+}{N}+$ ⟷ $:\ddot{O}-\overset{+}{N}+$

c. $N_3^-$

Count valence e⁻.
3 N x 5 e⁻  = 15
(−) charge  = 1
total e⁻    = 16

$N=\overset{+}{N}=N$ ⟷ $N≡\overset{+}{N}-N:$ ⟷ $:N-\overset{+}{N}≡N$

d.

$\begin{matrix} H & :O: & H & :O: & H \\ | & \| & | & \| & | \\ H-C-C-C-C-C-H \\ | & & & & | \\ H & & & & H \end{matrix}$ ⟷ $\begin{matrix} H & :O: & H & :\ddot{O}: & H \\ | & \| & | & | & | \\ H-C-C-C=C-C-H \\ | & & & & | \\ H & & & & H \end{matrix}$ ⟷ $\begin{matrix} H & :\ddot{O}: & H & :O: & H \\ | & | & | & \| & | \\ H-C-C=C-C-C-H \\ | & & & & | \\ H & & & & H \end{matrix}$

e.

f. $CH_2=CH-CH-CH=CH_2$ ⟷ $CH_2-CH=CH-CH=CH_2$ ⟷ $CH_2=CH-CH=CH-CH_2$

g.

**1.50** To draw the **resonance hybrid**, use the rules in Answer 1.14.

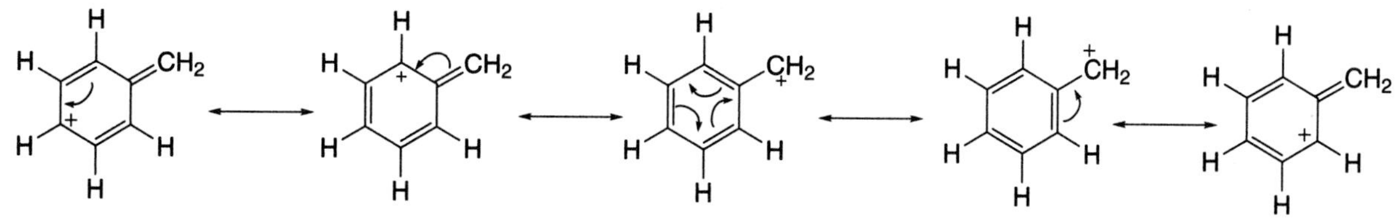

resonance hybrid

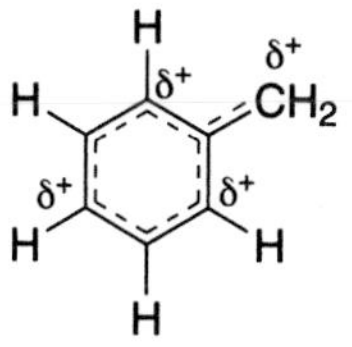

**1.51 A "better" resonance structure is one that has more bonds and fewer charges.** The better structure is the major contributor and all others are minor contributors.

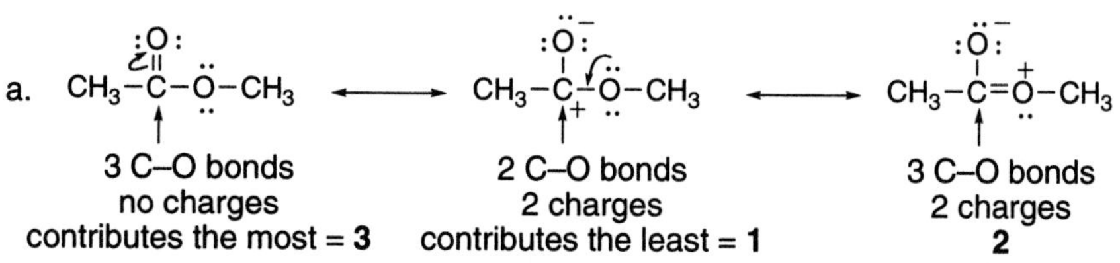

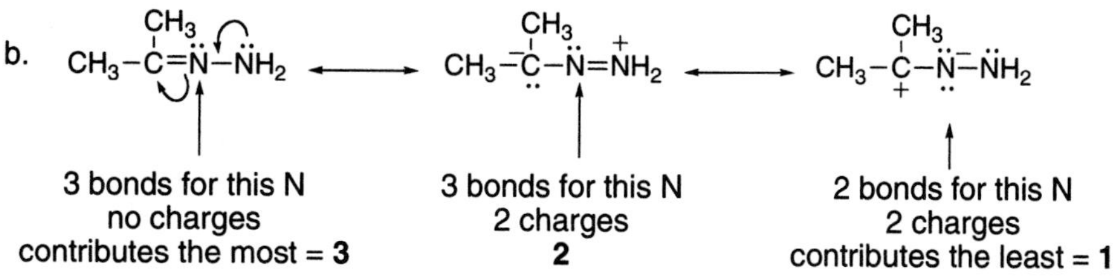

a.

| 3 C–O bonds | 2 C–O bonds | 3 C–O bonds |
| no charges | 2 charges | 2 charges |
| contributes the most = **3** | contributes the least = **1** | **2** |

b.

| 3 bonds for this N | 3 bonds for this N | 2 bonds for this N |
| no charges | 2 charges | 2 charges |
| contributes the most = **3** | **2** | contributes the least = **1** |

**1.52**

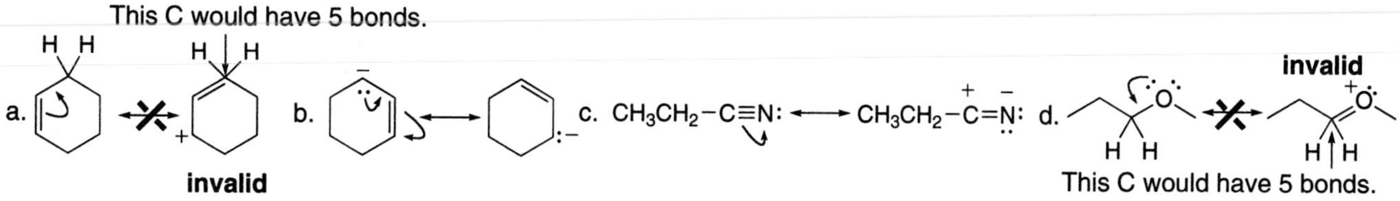

This C would have 5 bonds.

a. **invalid**

b.

c. $CH_3CH_2-C\equiv N:$ ⟷ $CH_3CH_2-\overset{+}{C}=\overset{-}{N}:$

d. **invalid**

This C would have 5 bonds.

[Note: The pentavalent C's in (a) and (d) bear a (–1) formal charge.]

**1.53** Follow the steps in Answer 1.5 to draw Lewis structures.

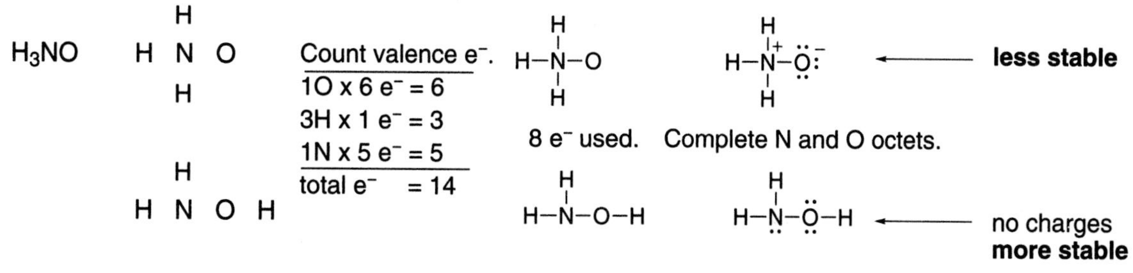

$H_3NO$

Count valence e⁻.

1 O x 6 e⁻ = 6
3 H x 1 e⁻ = 3
1 N x 5 e⁻ = 5
total e⁻ = 14

8 e⁻ used.   Complete N and O octets.

H–N–O ⟵ H–Ṅ–Ö: ⟵ **less stable**

H–N–O–H ⟵ H–N–Ö–H ⟵ no charges **more stable**

**1.54** Use the rules in Answer 1.18.

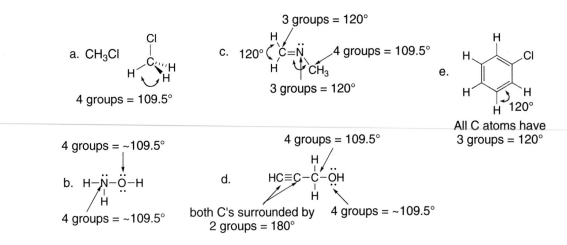

a. CH₃Cl — 4 groups = 109.5°

c. 3 groups = 120°, 4 groups = 109.5°, 3 groups = 120°

e. All C atoms have 3 groups = 120°

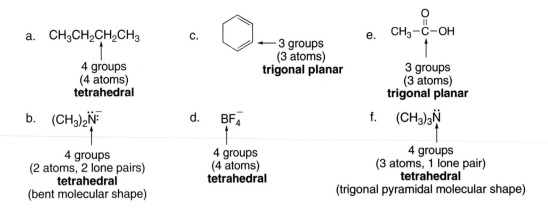

b. H–N̈–Ö–H — 4 groups = ~109.5°, 4 groups = ~109.5°

d. HC≡C–C–ÖH — 4 groups = 109.5°, both C's surrounded by 2 groups = 180°, 4 groups = ~109.5°

**1.55** To predict the geometry around an atom, use the rules in Answer 1.17.

a. CH₃CH₂CH₂CH₃
4 groups
(4 atoms)
**tetrahedral**

c.
← 3 groups
(3 atoms)
**trigonal planar**

e. CH₃–C(=O)–OH
3 groups
(3 atoms)
**trigonal planar**

b. (CH₃)₂N̈:⁻
4 groups
(2 atoms, 2 lone pairs)
**tetrahedral**
(bent molecular shape)

d. BF₄⁻
4 groups
(4 atoms)
**tetrahedral**

f. (CH₃)₃N̈
4 groups
(3 atoms, 1 lone pair)
**tetrahedral**
(trigonal pyramidal molecular shape)

**1.56** Each C has two bonds in the plane of the page, one in front of the page (solid wedge) and one behind the page (dashed line).

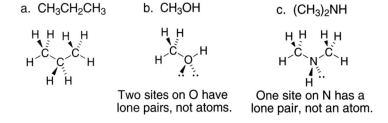

a. CH₃CH₂CH₃

b. CH₃OH
Two sites on O have lone pairs, not atoms.

c. (CH₃)₂NH
One site on N has a lone pair, not an atom.

**1.57** In shorthand or skeletal drawings, **all line junctions or ends of lines represent carbon atoms**. The C's are all tetravalent.

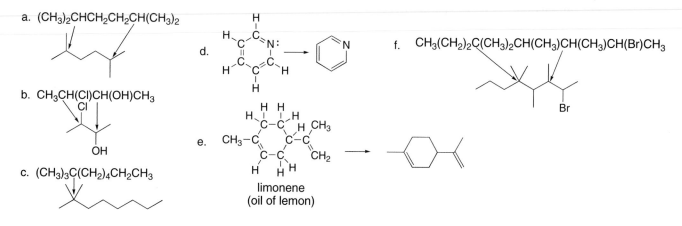

a.

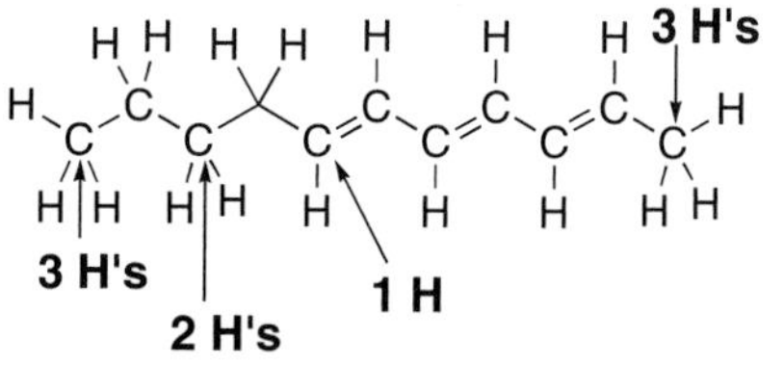

**2,4,6-undecatriene**
(isolated from limu lipoa,
a common brown Hawaiian seaweed)

b.

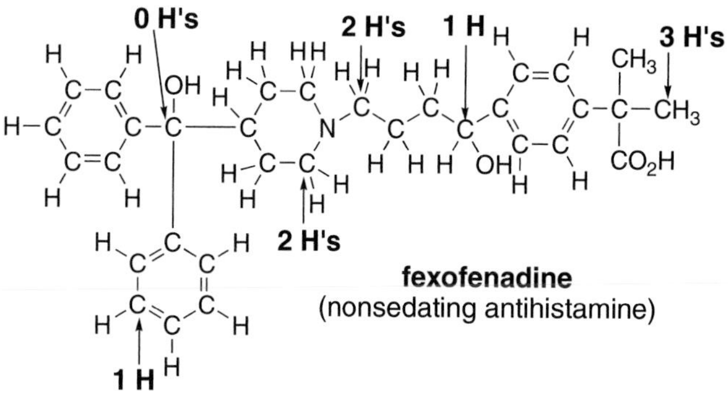

**fexofenadine**
(nonsedating antihistamine)

**1.58** In shorthand or skeletal drawings, **all line junctions or ends of lines represent carbon atoms**. Convert by writing in all C's, and then adding H's to make the C's tetravalent.

a.

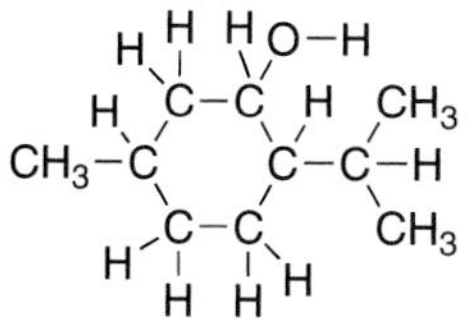

**menthol**
(isolated from peppermint oil)

b.

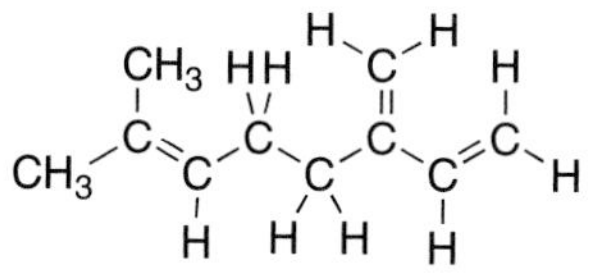

**myrcene**
(isolated from bayberry)

c.

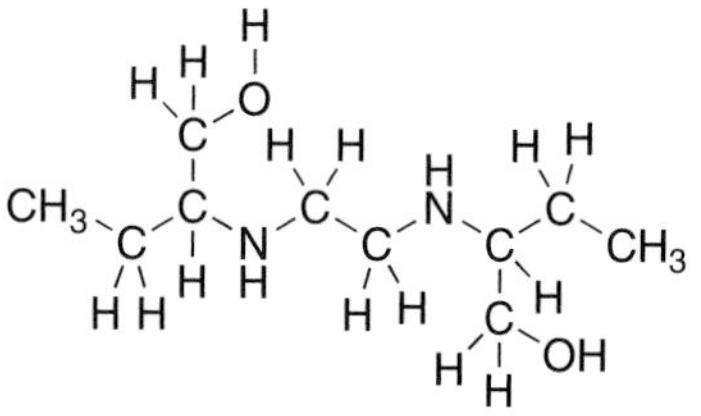

**ethambutol**
(drug used to treat tuberculosis)

d.

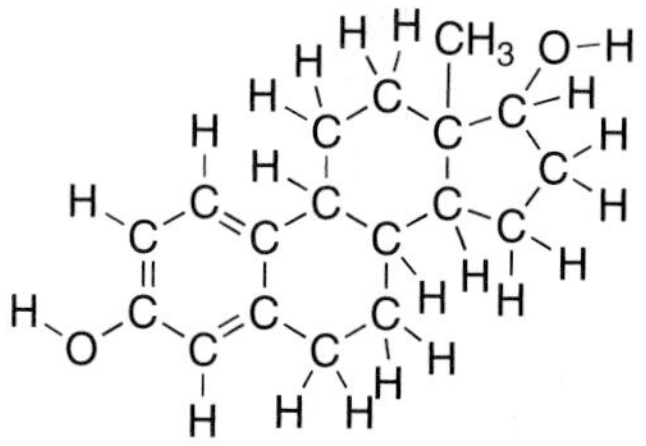

**estradiol**
(a female sex hormone)

**1.59** In skeletal formulas, leave out all C's and H's, except H's bonded to heteroatoms.

a. $(CH_3)_2CHCH_2CH_2CH(CH_3)_2$

b. $CH_3CH(Cl)CH(OH)CH_3$

c. $(CH_3)_3C(CH_2)_4CH_2CH_3$

d.

e.

**limonene**
(oil of lemon)

f. $CH_3(CH_2)_2C(CH_3)_2CH(CH_3)CH(CH_3)CH(Br)CH_3$

**1.60** For Lewis structures, all atoms including H's and all lone pairs must be drawn in.

a. $CH_3CH_2COOH$    b. $CH_3CONHCH_3$    c. $CH_3COCH_2Br$    d. $(CH_3)_3COH$    e. $(CH_3)_3CCHO$    f. $CH_3COCl$

**1.61** A charge on a C atom takes the place of one H atom. A negatively charged C has one lone pair, and a positively charged C has none.

a.    b.    c.

**1.62**

a.    b. $CH_3-C\equiv\overset{+}{N}H$    c.    d.

**1.63** To determine the hybridization around the labeled atoms, use the procedure in Answer 1.28.

a. $CH_3\overset{..}{C}H_2$

↑
4 groups
(3 atoms, 1 lone pair)
**$sp^3$, tetrahedral**

c. $(CH_3)_3\overset{..}{O}{}^+$

↑
4 groups
(3 atoms, 1 lone pair)
**$sp^3$, tetrahedral**

e. $CH_3-C\equiv C-H$

↑
2 groups
(2 atoms)
**sp, linear**

g. $CH_3CH=C=CH_2$

3 groups          2 groups
(3 atoms)         (2 atoms)
**$sp^2$**          **sp, linear**
**trigonal planar**

b.

4 groups
(4 atoms)
**$sp^3$, tetrahedral**
(Each C has 2 H's.)

d.

$-CH_2Cl$

4 groups
(4 atoms)
**$sp^3$, tetrahedral**

f. $CH_2=\overset{..}{N}OCH_3$

↑
3 groups
(2 atoms,1 lone pair)
**$sp^2$, trigonal planar**

**1.64** To determine what orbitals are involved in bonding, use the procedure in Answer 1.26.

a.

$C_{sp^3}-H_{1s}$
$C_{sp^3}-C_{sp^3}$

c.

$C_{sp^2}-C_{sp^3}$    :O:    σ: $C_{sp^2}-O_{sp^2}$
π: $C_p-O_p$

b.

$C_{sp^2}-H_{1s} \rightarrow$    σ: $C_{sp^2}-C_{sp^2}$
π: $C_p-C_p$
$C_{sp^2}-C_{sp^3}$

d.

$H-C\equiv C-\overset{C_{sp}-C_{sp^2}}{C}=\overset{..}{N}-CH_3$

$C_{sp}-H_{1s}$    σ: $C_{sp}-C_{sp}$
π: $C_p-C_p$
π: $C_p-C_p$
σ: $C_{sp^3}-N_{sp^2}$

**1.65** To determine what orbitals are involved in bonding, use the procedure in Answer 1.26.

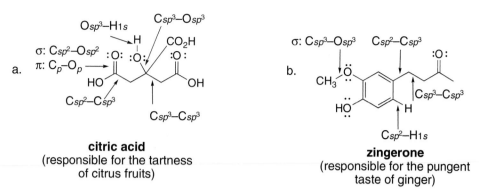

a.

**citric acid**
(responsible for the tartness
of citrus fruits)

b.

**zingerone**
(responsible for the pungent
taste of ginger)

**1.66**

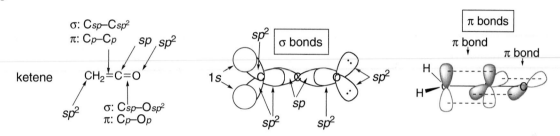

ketene

[For clarity, only the large bonding lobe of the hybrid
orbitals is drawn.]

**1.67**

$CH_2=\overset{+}{CH}$     σ: $C_{sp^2}-C_{sp}$
                      π: $C_p-C_p$

$sp^2$  $sp$

$CH_2=\overset{..}{\overset{-}{CH}}$     σ: $C_{sp^2}-C_{sp^2}$
                      π: $C_p-C_p$

$sp^2$  $sp^2$

**1.68** To determine relative bond length, use the rules in Answers 1.31.

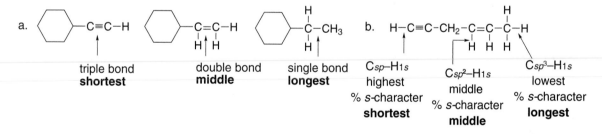

a.

triple bond
**shortest**

double bond
**middle**

single bond
**longest**

b.

$C_{sp}-H_{1s}$
highest
% s-character
**shortest**

$C_{sp^2}-H_{1s}$
middle
% s-character
**middle**

$C_{sp^3}-H_{1s}$
lowest
% s-character
**longest**

**1.69** To determine relative bond length, use the rules in Answers 1.31.

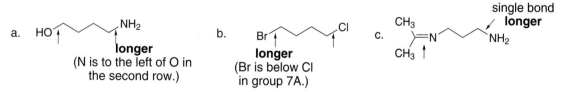

a.

**longer**
(N is to the left of O in
the second row.)

b.

**longer**
(Br is below Cl
in group 7A.)

c.

single bond
**longer**

**1.70**

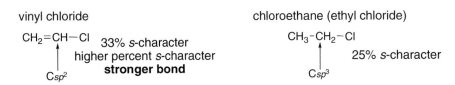

c. shortest, **strongest** C–C bond

b. and d. bond (1)
**longest, weakest** C–C single bond

bond (2)

a. **shortest** C–C single bond

e. strongest C–H bond

f. Bond (1) is a $C_{sp^3}$–$C_{sp^3}$ bond, and bond (2) is a $C_{sp^3}$–$C_{sp^2}$ bond. Bond (2) is shorter due to the increased percent *s*-character in the $sp^2$ hybridized carbon.

**1.71** Remember shorter bonds are stronger bonds. A $\sigma$ bond formed from two $sp^2$ hybridized C's is stronger than a $\sigma$ bond formed from two $sp^3$ hybridized C's because the $sp^2$ hybridized C orbitals have a higher percent *s*-character.

**1.72 Percent *s*-character** determines the strength of a bond. **The higher percent *s*-character of an orbital used to form a bond, the stronger the bond.**

vinyl chloride

$CH_2=CH-Cl$  33% *s*-character
higher percent *s*-character
**stronger bond**

$C_{sp^2}$

chloroethane (ethyl chloride)

$CH_3-CH_2-Cl$  25% *s*-character

$C_{sp^3}$

**1.73** Dipoles result from unequal sharing of electrons in covalent bonds. More electronegative atoms "pull" electron density towards them, making a dipole.

a.  $\delta^+$ Br–Cl $\delta^-$

b.  $\delta^+$ $\delta^-$
    $NH_2$–OH

c.  $\delta^+$ $\delta^-$
    $CH_3$–$NH_2$

d.  $\delta^-$ —Li $\delta^+$

**1.74** Use the directions from Answer 1.34.

a. $CHBr_3$

net dipole

b. $CH_3CH_2OCH_2CH_3$

net dipole

c. $CBr_4$

no net dipole

d. $CH_3$
   C=O
   $CH_3$

net dipole

e.
Br
Br

net dipole

f.
Cl
Cl

no net dipole

**1.75**

The dipole of any one Cl atom opposes the resultant dipole of the other 2 Cl's, thus decreasing the net dipole on the whole molecule.

Cl
Cl   C   Cl
     H

H
Cl   C   Cl
     H

net dipole

This net dipole is the result of 2 Cl atoms pulling down. They reinforce each other more to result in a **stronger net dipole.**

**1.76**

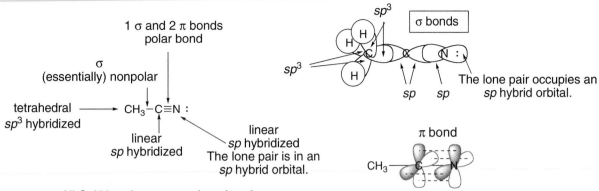

1 σ and 2 π bonds
polar bond

σ
(essentially) nonpolar

tetrahedral ⟶ CH₃–C≡N :
sp³ hybridized

linear
sp hybridized

linear
sp hybridized
The lone pair is in an
sp hybrid orbital.

All C–H bonds are nonpolar σ bonds.
All H's use a 1s orbital in bonding.

σ bonds

The lone pair occupies an
sp hybrid orbital.

π bond

π bond
[For clarity, only the large bonding lobe
of the hybrid orbitals is drawn.]

**1.77**

a. $sp^2$

b. Each C is trigonal planar; the ring is flat, drawn as a hexagon.

c.

sp² hybridized C

p orbitals on C's overlap

π bonds

sp² hybrid orbitals on C

σ bonds

[Only the larger bonding lobe
of each orbital is drawn.]

d.

e. Benzene is stable because of its two resonance structures that contribute
equally to the hybrid. [This is only part of the story. We'll learn more about
benzene's unusual stability in Chapter 17.]

**1.78**

a.

3 groups
$sp^2$
trigonal planar

3 groups
$sp^2$
trigonal planar

4 groups
$sp^3$
tetrahedral

4 groups
$sp^3$
tetrahedral

3 groups
$sp^2$
trigonal planar

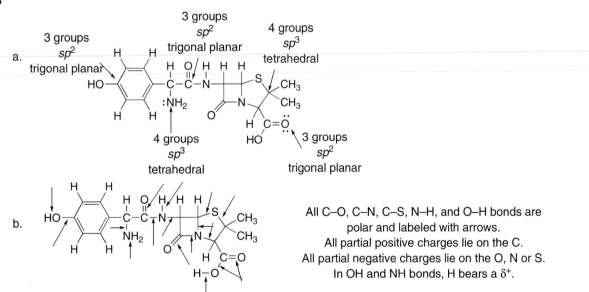

b.

All C–O, C–N, C–S, N–H, and O–H bonds are
polar and labeled with arrows.
All partial positive charges lie on the C.
All partial negative charges lie on the O, N or S.
In OH and NH bonds, H bears a δ⁺.

**skeletal structure:**

c.

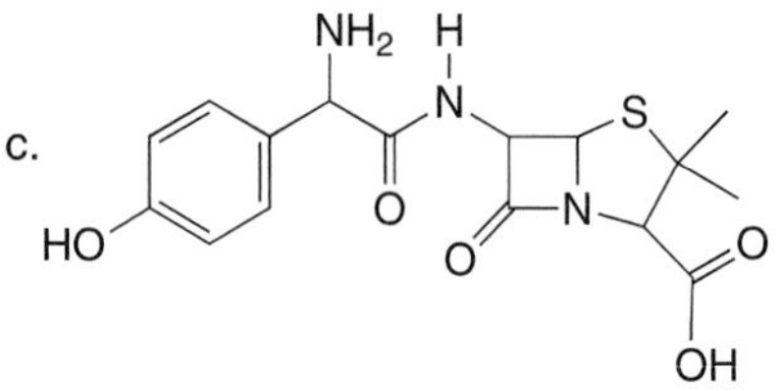

d.

6 π bonds

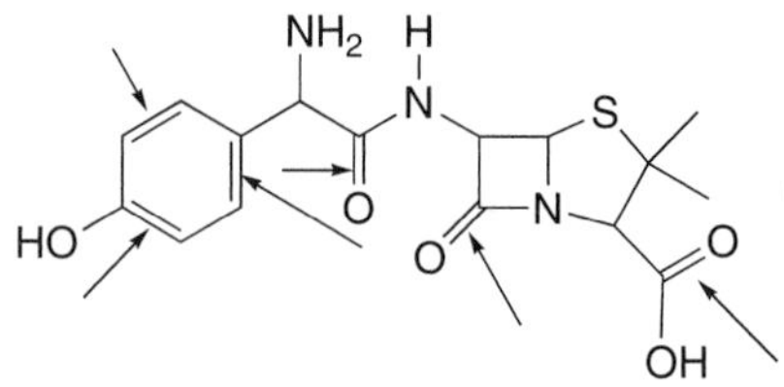

e. **33% *s*-character = *sp*² hybridized**

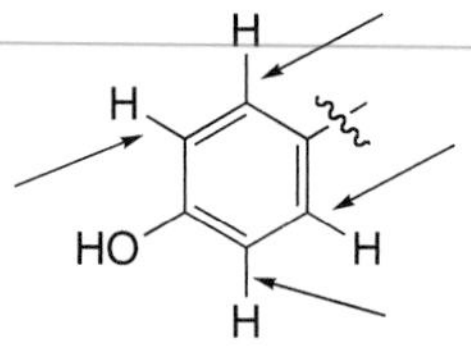

These C–H bonds have 33% *s*-character.

**1.79**

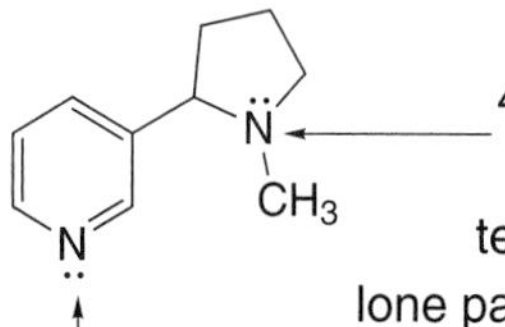

**nicotine**

4 groups
*sp*³
tetrahedral
lone pair in *sp*³ orbital

3 groups
*sp*²
trigonal planar
lone pair in *sp*² orbital

**1.80**

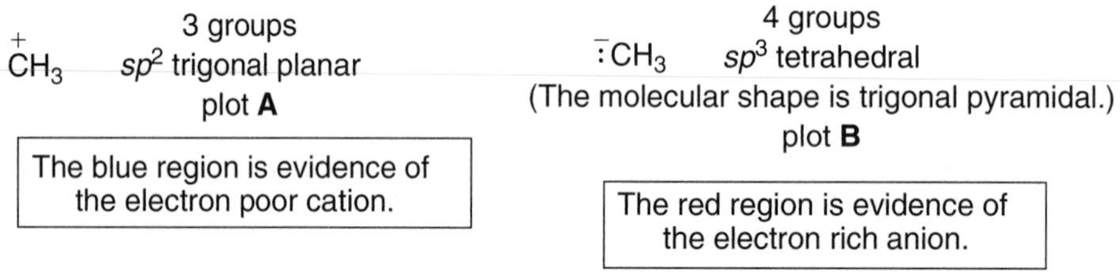

$\overset{+}{C}H_3$  3 groups
*sp*² trigonal planar
plot **A**

The blue region is evidence of
the electron poor cation.

$^-CH_3$  4 groups
*sp*³ tetrahedral
(The molecular shape is trigonal pyramidal.)
plot **B**

The red region is evidence of
the electron rich anion.

**1.81 Polar bonds result from unequal sharing of electrons in covalent bonds**. Normally we think of more electronegative atoms "pulling" more of the electron density towards them, making a dipole. In looking at a $Csp^2$–$Csp^3$ bond, the atom with a higher percent *s*-character will "pull" more of the electron density towards it, creating a small dipole.

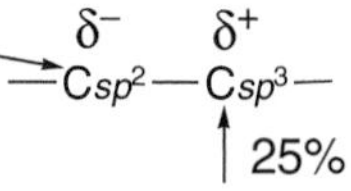

33% *s*-character
higher percent *s*-character
pulls more electron density
**more electronegative**

$\overset{\delta^-}{C}sp^2$—$\overset{\delta^+}{C}sp^3$—

↑ 25% *s*-character

**1.82**

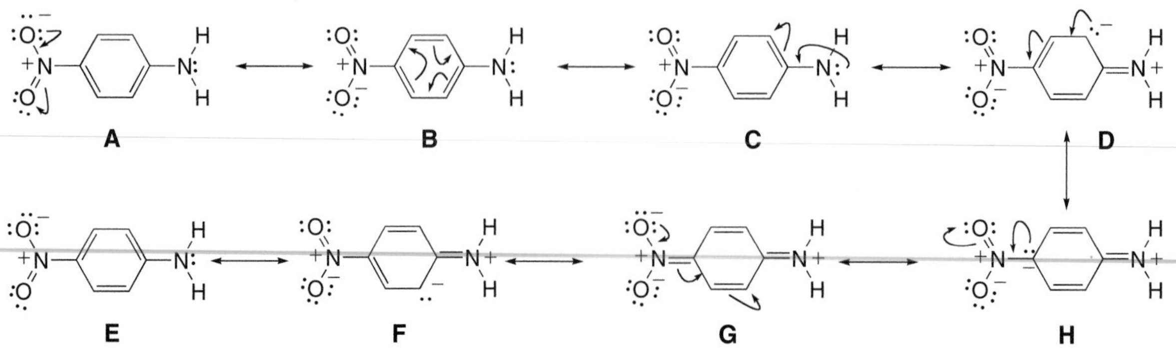

A  ⟷  B  ⟷  C  ⟷  D

↕

E  ⟷  F  ⟷  G  ⟷  H

Structures **D**, **F**, and **H** have an additional resonance structure.

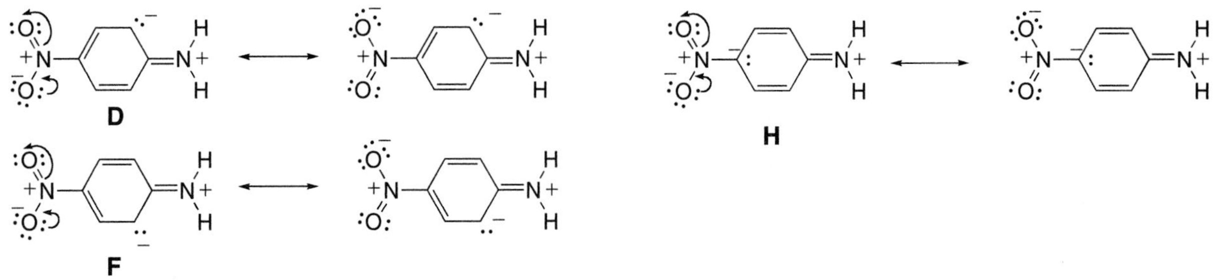

**1.83**

Isomers of $C_4H_8$:

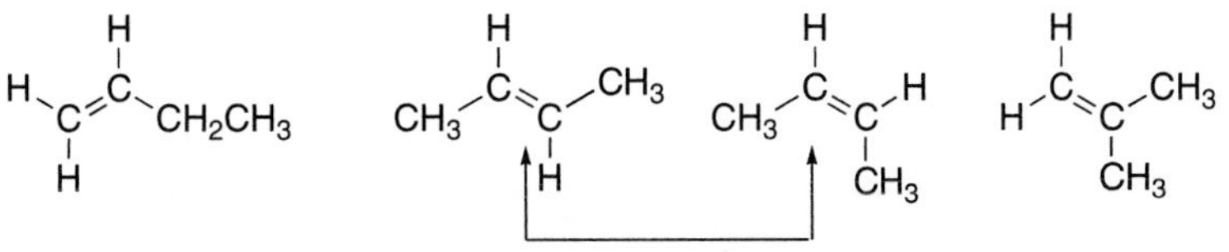

These two compounds are different
because of restricted rotation around
the C=C (Section 8.2B).

**1.84**

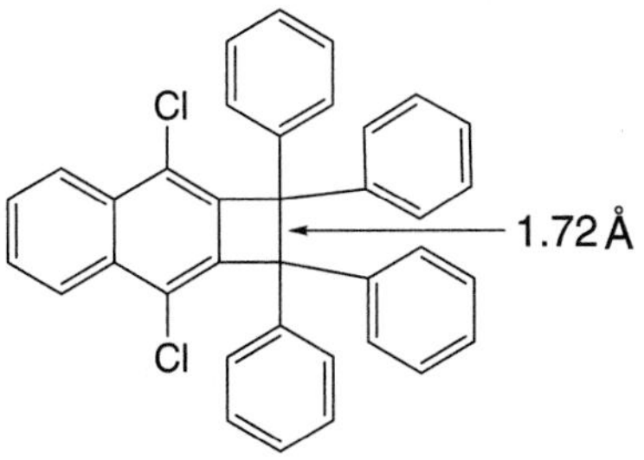

1.72 Å

The four rings bonded to the C–C bond are very crowded.
With a longer C–C bond, the rings have somewhat more
space, and this adds stability to the molecule.

Chapter 2: Acids and Bases

## ◆ A comparison of Brønsted–Lowry and Lewis acids and bases

| Type | Definition | Structural feature | Examples |
|---|---|---|---|
| Brønsted–Lowry acid (2.1) | proton donor | a proton | $HCl$, $H_2SO_4$, $H_2O$, $CH_3COOH$, TsOH |
| Brønsted–Lowry base (2.1) | proton acceptor | a lone pair *or* a π bond | $^-OH$, $^-OCH_3$, $H^-$, $^-NH_2$, $CH_2=CH_2$ |
| Lewis acid (2.8) | electron pair acceptor | a proton, *or* an unfilled valence shell, *or* a partial (+) charge | $BF_3$, $AlCl_3$, $HCl$, $CH_3COOH$, $H_2O$ |
| Lewis base (2.8) | electron pair donor | a lone pair *or* a π bond | $^-OH$, $^-OCH_3$, $H^-$, $^-NH_2$, $CH_2=CH_2$ |

## ◆ Acid–base reactions

[1] A Brønsted–Lowry acid donates a proton to a Brønsted–Lowry base (2.2).

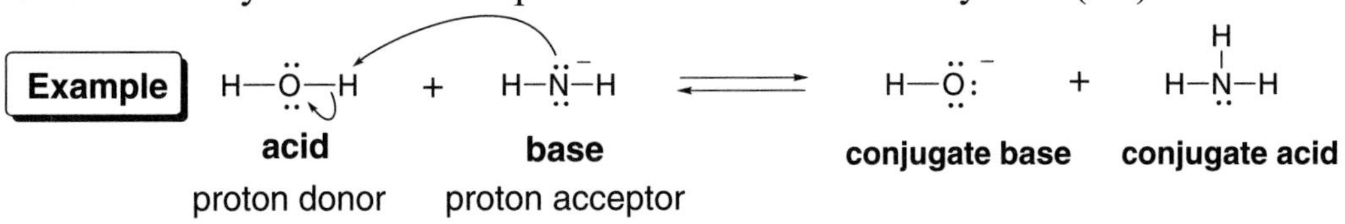

**acid**      **base**     **conjugate base**    **conjugate acid**

proton donor    proton acceptor

[2] A Lewis base donates an electron pair to a Lewis acid (2.8).

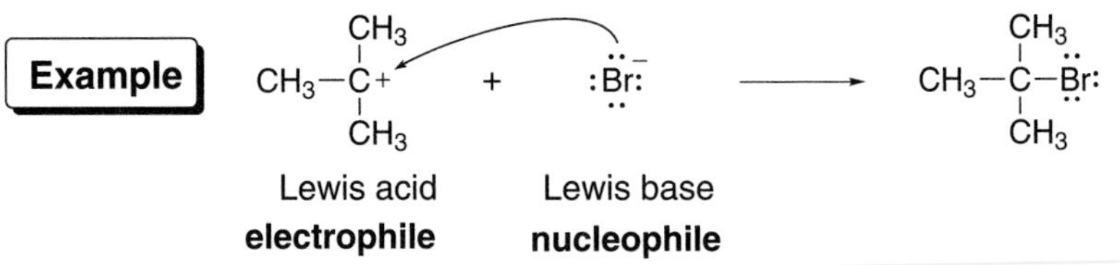

Lewis acid     Lewis base

**electrophile**     **nucleophile**

- Electron rich species react with electron poor ones.
- Nucleophiles react with electrophiles.

## ◆ Important facts

- Definition: $pK_a = -\log K_a$. The **lower the $pK_a$**, the **stronger** the acid (2.3).

$NH_3$        versus        $H_2O$

$pK_a = 38$                  $pK_a = 15.7$

lower $pK_a$ = stronger acid

- The stronger the acid, the weaker the conjugate base (2.3).

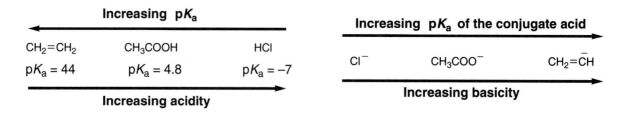

- In proton transfer reactions, equilibrium favors the weaker acid and weaker base (2.4).

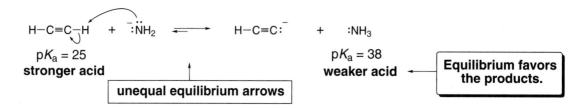

- An acid can be deprotonated by the conjugate base of any acid having a **higher pK$_a$** (2.4).

| Acid | pK$_a$ | Conjugate base | |
|---|---|---|---|
| CH$_3$COO–H | 4.8 | CH$_3$COO$^-$ | |
| CH$_3$CH$_2$O–H | 16 | CH$_3$CH$_2$O$^-$ | These bases |
| HC≡CH | 25 | HC≡C$^-$ | can deprotonate |
| H–H | 35 | H$^-$ | CH$_3$COO–H. |

higher pK$_a$ than CH$_3$COO–H

♦ Factors that determine acidity (2.5)

[1] **Element effects** (2.5A)    The acidity of H–A increases both across a row and down a column of the periodic table.

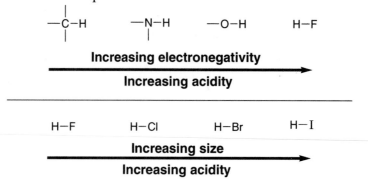

[2] **Inductive effects** (2.5B)    The acidity of H–A increases with the presence of electron-withdrawing groups in A.

$CH_3CH_2OH$ → $CH_3CH_2O^-$
weaker acid

**No additional electronegative atoms stabilize the conjugate base.**

$CF_3CH_2OH$ →
stronger acid

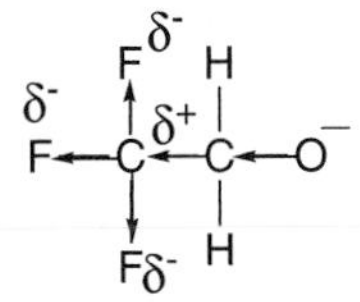

**CF$_3$ withdraws electron density, stabilizing the conjugate base.**

[3] **Resonance effects** (2.5C)    The acidity of H–A increases when the conjugate base A:$^-$ is resonance stabilized.

$CH_3CH_2\ddot{O}-H$ → $CH_3CH_2\ddot{O}:^-$
ethanol          ethoxide
               conjugate base

only **one** Lewis structure

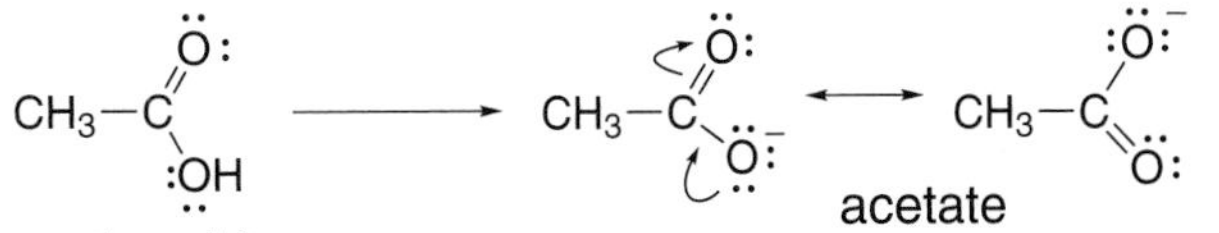

acetic acid          acetate
more acidic          conjugate base

**two** resonance structures

[4] **Hybridization effects** (2.5D)    The acidity of H–A increases as the percent *s*-character of the A:$^-$ increases.

| $CH_3CH_3$ | $CH_2=CH_2$ | $H-C\equiv C-H$ |
|---|---|---|
| **ethane** | **ethylene** | **acetylene** |
| $pK_a = 50$ | $pK_a = 44$ | $pK_a = 25$ |

**Increasing acidity** →

## Chapter 2: Answers to Problems

**2.1** **Brønsted–Lowry acids** are **proton donors** and must contain a hydrogen atom.
**Brønsted–Lowry bases** are **proton acceptors** and must have an available electron pair (either a lone pair or a $\pi$ bond).

a.
HB̈r:
**acid**

N̈H₃
**acid**

CCl₄
not an acid—no H

b.
CH₃CH₃
no lone pairs
or $\pi$ bonds
not a base

(CH₃)₃CÖ:⁻
lone pairs
on O
**base**

H–C≡C–H
**base**—$\pi$ bonds

c.
CH₃CH₂CH₂ÖH
**base**—lone pairs on O
**acid**—contains H atoms

CH₃CH₂CH₂CH₃
not a base—no lone pairs
or $\pi$ bonds
**acid**—contains H atoms

$$CH_3-C\overset{\ddot{O}:}{\underset{:\ddot{O}CH_3}{}}$$
**base**—lone pairs on O's, $\pi$ bond
**acid**—contains H atoms

**2.2** The arrows indicate basic sites.

morphine

The lone pairs and $\pi$ bonds are basic sites.

**2.3** A Brønsted–Lowry base accepts a proton to form the conjugate acid. A Brønsted–Lowry acid loses a proton to form the conjugate base.

a.   NH₃ ⟶ NH₄⁺

Cl⁻ ⟶ HCl

(CH₃)₂C=O ⟶ (CH₃)₂C=ÖH⁺

b.   HBr ⟶ Br⁻

HSO₄⁻ ⟶ SO₄²⁻

CH₃OH ⟶ CH₃O⁻

**2.4** Use the definitions from Answer 2.3.

CH₂=CH₂ ⟶ CH₂–CH₃⁺
ethylene    accepts a proton
conjugate acid

CH₂=CH₂ ⟶ CH₂=C̈H⁻
loses a proton
conjugate base

**2.5** The Brønsted–Lowry base accepts a proton to form the conjugate acid. The Brønsted–Lowry acid loses a proton to form the conjugate base. Use curved arrows to show the movement of electrons (**NOT protons**). Re-draw the starting materials if necessary to clarify the electron movement.

a.   H–C̈l: + H₂Ö ⇌ :C̈l:⁻ + H₃Ö⁺

acid    base    conjugate base   conjugate acid

b.

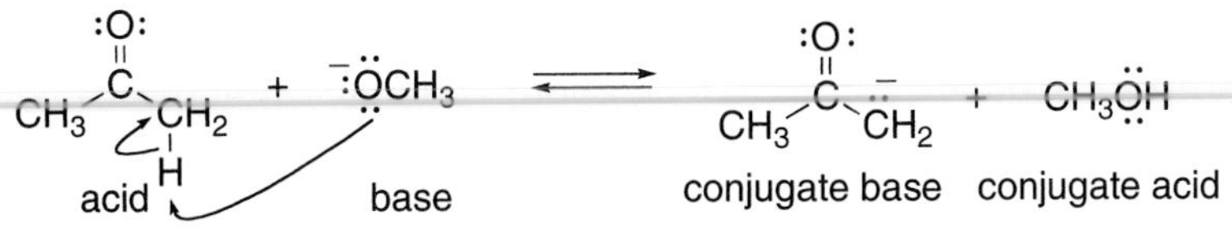

acid + base ⇌ conjugate base + conjugate acid

## 2.6 To draw the products:
[1] Find the acid and base.
[2] Transfer a proton from the acid to the base.
[3] Check that the charges on each side of the arrows are balanced.

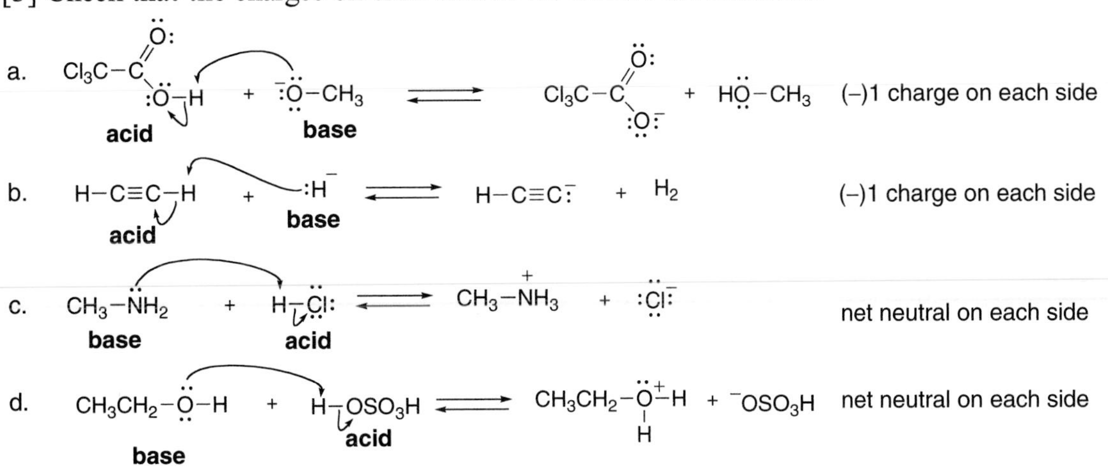

a.   (–)1 charge on each side

b.   (–)1 charge on each side

c.   net neutral on each side

d.   net neutral on each side

## 2.7 The smaller the $pK_a$, the stronger the acid. The larger $K_a$, the stronger the acid.

a. $CH_3CH_2CH_3$   and   $CH_3CH_2OH$

$pK_a = 50$     $pK_a = 16$

smaller $pK_a$
**stronger acid**

b.   and

$K_a = 10^{-10}$     $K_a = 10^{-41}$

larger $K_a$
**stronger acid**

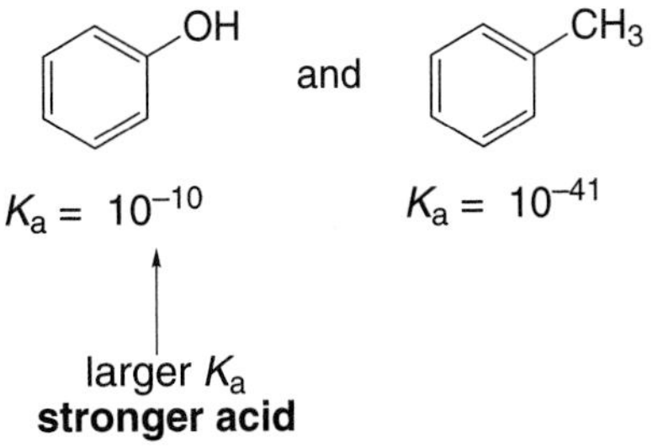

## 2.8 To convert from $K_a$ to $pK_a$, take (–) the log of the $K_a$. $\mathbf{pK_a = -log K_a}$
To convert $pK_a$ to $K_a$, take the antilog of (–) the $pK_a$.

a. $K_a = 10^{-10}$     $K_a = 10^{-21}$     $K_a = 5.2 \times 10^{-5}$     b. $pK_a = 7$   $pK_a = 11$   $pK_a = 3.2$

$pK_a = 10$     $pK_a = 21$     $pK_a = 4.3$     $K_a = 10^{-7}$   $K_a = 10^{-11}$   $K_a = 6.3 \times 10^{-4}$

**2.9** Since **strong acids form weak conjugate bases**, the basicity of conjugate bases increases with increasing p$K_a$ of their acids. Find the p$K_a$ of each acid from Table 2.1 and then rank the acids in order of increasing p$K_a$. This will also be the order of increasing basicity of their conjugate bases.

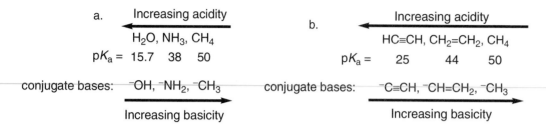

a.

Increasing acidity ⟵

$H_2O$, $NH_3$, $CH_4$

p$K_a$ = 15.7　38　50

b.

Increasing acidity ⟵

$HC\equiv CH$, $CH_2=CH_2$, $CH_4$

p$K_a$ =　25　　44　　50

conjugate bases:　$^-OH$, $^-NH_2$, $^-CH_3$

Increasing basicity ⟶

conjugate bases:　$^-C\equiv CH$, $^-CH=CH_2$, $^-CH_3$

Increasing basicity ⟶

**2.10** To estimate the p$K_a$ of the indicated bond, find a similar bond in the p$K_a$ table (H bonded to the same atom with the same hybridization).

a.

For $NH_3$, p$K_a$ is 38.
**estimated p$K_a$ = 38**

b.

For $CH_3CH_2OH$,
p$K_a$ is 16.
**estimated p$K_a$ = 16**

c. $BrCH_2COO{-}H$

For $CH_3COOH$, p$K_a$ is 4.8.
**estimated p$K_a$ = 5**

**2.11** Label the acid and the base and then transfer a proton from the acid to the base. To determine if the reaction will proceed as written, compare the p$K_a$ of the acid on the left with the conjugate acid on the right. **The equilibrium always favors the formation of the weaker acid and the weaker base.**

a.　$H-C\equiv C-H$　+　$H:^-$　⇌　$H-C\equiv C:^-$　+　$H_2$

　　acid　　　　　base　　　conjugate base　conjugate acid
　　p$K_a$ = 25　　　　　　　　　　　　　　p$K_a$ = 35
　　　　　　　　　　　　　　　　　　　　**weaker acid**

Equilibrium favors
the **products.**

b.　$CH_4$　+　$:\ddot{O}H^-$　⇌　$:CH_3$　+　$H_2\ddot{O}:$

　　acid　　base　　conjugate base　conjugate acid
　　p$K_a$ = 50　　　　　　　　　　　　p$K_a$ = 15.7
　**weaker acid**

Equilibrium favors
the **starting materials.**

c.　$CH_3COOH$　+　$CH_3CH_2\ddot{O}:^-$　⇌　$CH_3COO^-$　+　$CH_3CH_2\ddot{O}H$

　　acid　　　　　base　　　　　　conjugate base　conjugate acid
　　p$K_a$ = 4.8　　　　　　　　　　　　　　　　p$K_a$ = 16
　　　　　　　　　　　　　　　　　　　　　　**weaker acid**

Equilibrium favors
the **products.**

d.　$:\ddot{Cl}:^-$　+　$CH_3CH_2\ddot{O}H$　⇌　$H\ddot{Cl}:$　+　$CH_3CH_2\ddot{O}:^-$

　　base　　　　acid　　　　conjugate acid　conjugate base
　　　　　　　p$K_a$ = 16　　　p$K_a$ = –7
　　　　　　**weaker acid**

Equilibrium favors
the **starting materials.**

**2.12**   An acid can be deprotonated by the conjugate base of any acid with a higher $pK_a$.

$CH_3COOH$
**$pK_a$ = 4.8**
Any base having a conjugate
acid with a $pK_a$ higher than
4.8 can deprotonate this acid.

| Acid | $pK_a$ | Conjugate base | |
|------|--------|----------------|---|
| HCl | –7 | Cl⁻ | not strong enough |
| HC≡CH | 25 | HC≡C⁻ | strong enough |
| $H_2$ | 35 | H⁻ | |

HC≡CH
**$pK_a$ = 25**
All of these acids have a higher $pK_a$
than HC≡CH, and a conjugate
base that can deprotonate HC≡CH.

| Acid | $pK_a$ | Conjugate base |
|------|--------|----------------|
| $H_2$ | 35 | H⁻ |
| $NH_3$ | 38 | ⁻$NH_2$ |
| $CH_2=CH_2$ | 44 | $CH_2=CH⁻$ |
| $CH_4$ | 50 | $CH_3⁻$ |

**2.13**   Acidity of H–Z **increases across a row and down a column** of the periodic table.

a. $NH_3$, $H_2O$

O is further to the right in
the periodic table.
**stronger acid**

b. HBr, HCl

Br is further down
the periodic table.
**stronger acid**

c. $H_2S$, HBr

Br is further across and down
the periodic table.
**stronger acid**

**2.14**   Look at the element bonded to the acidic H and decide its acidity based on the periodic trends.
**Further right and down the periodic table is more acidic.**

a.   most acidic
$CH_3CH_2CH_2CH_2O\overset{\downarrow}{H}$

Molecule contains C–H
and O–H bonds.
O is further right; therefore,
O–H hydrogen is the most acidic.

b.   most acidic
$H\overset{\downarrow}{O}CH_2CH_2CH_2NH_2$

Molecule contains C–H,
N–H and O–H bonds.
O is furthest right; therefore,
O–H hydrogen is the most acidic.

c.   most acidic
$(CH_3)_2NCH_2CH_2CH_2N\overset{\downarrow}{H}_2$

Molecule contains C–H and N–H
bonds.
N is further right; therefore,
N–H hydrogen is the most acidic.

**2.15**   **More electronegative groups stabilize the conjugate base, making the acid stronger.**
Compare the electron-withdrawing groups on the acids below to decide which is a stronger acid
(**more electronegative groups = more acidic**).

a.   $ClCH_2COOH$   and   $FCH_2COOH$
                                        more acidic

F is more electronegative than Cl making the
O–H bond in the acid on the right **more acidic.**

c.   $CH_3COOH$   and   $O_2NCH_2COOH$
                                        more acidic

$NO_2$ is electron withdrawing making the
O–H bond in the acid on the
right **more acidic.**

b.   $Cl_2CHCH_2OH$   and   $Cl_2CHCH_2CH_2OH$

Cl is closer
to the acidic O–H bond.
**more acidic**

Cl is further from the
O–H bond.

**2.16** The acidity of an acid increases when the conjugate base is resonance stabilized. Compare the conjugate bases of acetone and propane to explain why acetone is more acidic.

acetone
$pK_a = 19.2$

2 resonance structures
more stable conjugate base
**Acetone is more acidic.**

One resonance structure places the (–) charge on the more electronegative O atom. This is especially good.

CH$_3$CH$_2$CH$_3$ → CH$_3$CH$_2$ĊH$_2$    only one Lewis structure
less stable conjugate base

propane
$pK_a = 50$

(Any C–H bond in the starting material can be removed.)

**2.17** The acidity of an acid increases when the conjugate base is resonance stabilized. Acetonitrile has a resonance-stabilized conjugate base, which accounts for its acidity.

acetonitrile
(one Lewis structure)

The negative charge is stabilized by delocalization on the C and N atoms.

Having the (–) charge on the electronegative N atom adds stability.

**2.18** **Increasing percent s-character makes an acid more acidic.** Compare the percent s-character of the carbon atoms in each of the C–H bonds in question. A stronger acid has a weaker conjugate base.

a.   CH$_3$CH$_2$—C≡C—H   and   CH$_3$CH$_2$CH$_2$CH$_2$—H     b.

*sp* hybridized C
50% *s*-character
**more acidic**

*sp*$^3$ hybridized C
25% *s*-character

*sp*$^2$ hybridized C
33% *s*-character
**more acidic**

*sp*$^3$ hybridized C
25% *s*-character

CH$_3$CH$_2$—C≡C:⁻   and   CH$_3$CH$_2$CH$_2$ĊH$_2$

**stronger conjugate base**

**stronger conjugate base**

**2.19** To compare the acids, first **look for element effects**. Then identify electron-withdrawing groups, resonance, or hybridization differences.

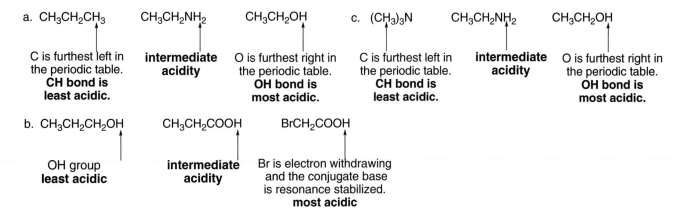

a. CH₃CH₂CH₃          CH₃CH₂NH₂          CH₃CH₂OH          c. (CH₃)₃N          CH₃CH₂NH₂          CH₃CH₂OH

C is furthest left in the periodic table. **CH bond is least acidic.**   |   **intermediate acidity**   |   O is furthest right in the periodic table. **OH bond is most acidic.**   |   C is furthest left in the periodic table. **CH bond is least acidic.**   |   **intermediate acidity**   |   O is furthest right in the periodic table. **OH bond is most acidic.**

b. CH₃CH₂CH₂OH          CH₃CH₂COOH          BrCH₂COOH

**OH group least acidic**   |   **intermediate acidity**   |   Br is electron withdrawing and the conjugate base is resonance stabilized. **most acidic**

**2.20** Look at the element bonded to the acidic H and decide its acidity based on the periodic trends. **Further right and down the periodic table is more acidic.**

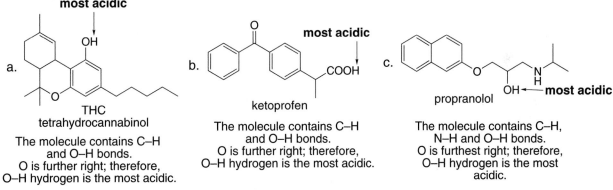

a. **most acidic** — OH

THC
tetrahydrocannabinol

The molecule contains C–H and O–H bonds. O is further right; therefore, O–H hydrogen is the most acidic.

b. **most acidic** — COOH

ketoprofen

The molecule contains C–H and O–H bonds. O is further right; therefore, O–H hydrogen is the most acidic.

c. propranolol          OH ← **most acidic**

The molecule contains C–H, N–H and O–H bonds. O is furthest right; therefore, O–H hydrogen is the most acidic.

**2.21** Label the acid and the base in the starting materials and then draw the products of proton transfer from the acid to the base.

a. CH₃CH₂Ö–H  +  Na⁺ H:⁻   ⇌   CH₃CH₂Ö:⁻ Na⁺   +   H₂
   acid              base              conjugate base      conjugate acid

b. CH₃COO–H  +  Na⁺ :ÖCH₂CH₃   ⇌   CH₃COO⁻Na⁺   +   HÖCH₂CH₃
   acid              base                    conjugate base      conjugate acid

c. CH₃CH₂CH₂C̈H₂ Li⁺  +  H–ÖH   ⇌   CH₃CH₂CH₂CH₃   +   Li⁺ :ÖH
   base                    acid              conjugate acid        conjugate base

d. CH₃––S–O–H  +  N̈(CH₂CH₃)₃   ⇌   CH₃––SO₃⁻  +  ⁺HN(CH₂CH₃)₃
   acid                 base                 conjugate base      conjugate acid

**2.22** To cross a cell membrane, amphetamine must be in its neutral (not ionic) form.

amphetamine

**absorption here** ↑
in the neutral form

**2.23** **Lewis bases are electron pair donors**: they contain a lone pair or a π bond.

a. $\overset{..}{N}H_3$

b. $CH_3CH_2CH_3$

c. $H\overset{..}{:}^-$

d. $H-C\equiv C-H$

**yes** - has
lone pair

**no** - no lone pair
or π bond

**yes** - has
lone pair

**yes** - has
2 π bonds

**2.24** **Lewis acids are electron pair acceptors.** Most Lewis acids contain a proton or an unfilled valence shell of electrons.

a. $BBr_3$

b. $CH_3CH_2OH$

c. $(CH_3)_3C^+$

d. $Br^-$

**yes**
unfilled valence shell
on B

**yes**
contains a proton

**yes**
unfilled valence shell
on C

**no**
no proton
no unfilled valence shell

**2.25** Label the Lewis acid and Lewis base and then draw the curved arrows.

new bond

a. $BF_3$ + $CH_3-\overset{..}{O}-CH_3$ ⟶

b. $(CH_3)_2CH^+$ + $\overset{..}{:}\overset{-}{O}H$ ⟶ $(CH_3)_2CH\overset{..}{O}H$

**Lewis acid**
unfilled valence shell
on B

**Lewis base**
lone pairs
on O

**Lewis acid**
unfilled valence
shell on C

**Lewis base**
lone pairs
on O

**2.26** A Lewis acid is also called an **electrophile**. When a Lewis base reacts with an electrophile other than a proton, it is called a **nucleophile**. Label the electrophile and nucleophile in the starting materials and then draw the products.

a. $CH_3CH_2-\overset{..}{O}-CH_2CH_3$ + $BBr_3$ ⟶ $CH_3CH_2-\overset{..}{\overset{+}{O}}-CH_2CH_3$

**Lewis base**
**nucleophile**
lone pairs
on O

**Lewis acid**
**electrophile**
unfilled valence shell
on B

b. ⟶

**Lewis base**
**nucleophile**
lone pairs
on O

**Lewis acid**
**electrophile**
unfilled valence shell
on Al

**2.27** Curved arrows begin at the Lewis base and point toward the Lewis acid.

new C–H bond

$CH_2=C$ + $H_2\overset{..}{O}^+$ ⟶ + $H_2\overset{..}{O}$

**Lewis base**
contains a π bond

**Lewis acid**
contains a proton

**2.28** To draw the conjugate acid of a Brønsted–Lowry base, **add a proton to the base**.

a. $H_2\ddot{O}:$ $\xrightarrow{\;\overset{+}{H}\;}$ $H_3\ddot{O}^+$

b. $\overset{-}{:}\!NH_2$ $\xrightarrow{\;\overset{+}{H}\;}$ $\ddot{N}H_3$

c. $HCO_3^-$ $\xrightarrow{\;\overset{+}{H}\;}$ $H_2CO_3$

d. $CH_3CH_2\overset{..}{N}HCH_3$ $\xrightarrow{\;\overset{+}{H}\;}$ $CH_3CH_2\overset{+}{N}H_2CH_3$

e. $CH_3\ddot{O}CH_3$ $\xrightarrow{\;\overset{+}{H}\;}$ $CH_3-\overset{\overset{H}{|}}{\underset{}{\overset{+}{O}}}-CH_3$

f. $CH_3COO^-$ $\xrightarrow{\;\overset{+}{H}\;}$ $CH_3COOH$

**2.29** To draw the conjugate base of a Brønsted–Lowry acid, **remove a proton from the acid**.

a. $HCN$ $\xrightarrow{\;-\overset{+}{H}\;}$ $^-CN$

b. $HCO_3^-$ $\xrightarrow{\;-\overset{+}{H}\;}$ $CO_3^{2-}$

c. $(CH_3)_2\overset{+}{N}H_2$ $\xrightarrow{\;-\overset{+}{H}\;}$ $(CH_3)_2NH$

d. $HC\equiv CH$ $\xrightarrow{\;-\overset{+}{H}\;}$ $HC\equiv C^-$

e. $CH_3CH_2COOH$ $\xrightarrow{\;-\overset{+}{H}\;}$ $CH_3CH_2COO^-$

f. $CH_3SO_3H$ $\xrightarrow{\;-\overset{+}{H}\;}$ $CH_3SO_3^-$

**2.30** Label the Brønsted–Lowry acid and Brønsted–Lowry base in the starting materials and **transfer a proton from the acid to the base** for the products.

a. $CH_3\ddot{O}-H$ + $\overset{-}{:}\!\ddot{N}H_2$ &rlarr; $CH_3\ddot{O}:^-$ + $\ddot{N}H_3$
    **acid**    **base**        **conjugate base**  **conjugate acid**

b. $CH_3CH_2-C\!\!\begin{smallmatrix}\ddot{O}:\\ \ddot{O}-H\end{smallmatrix}$ + $:\ddot{O}-CH_3$ &rlarr; $CH_3CH_2-C\!\!\begin{smallmatrix}\ddot{O}:\\ :\ddot{O}:^-\end{smallmatrix}$ + $H\ddot{O}-CH_3$
    **acid**          **base**        **conjugate base**      **conjugate acid**

c. $CH_3CH_2-C\equiv C-H$ + $:H^-$ &rlarr; $CH_3CH_2-C\equiv C:^-$ + $H_2$
        **acid**        **base**      **conjugate base**    **conjugate acid**

d. $(CH_3CH_2)_3\ddot{N}$ + $H-\ddot{C}l:$ &rlarr; $(CH_3CH_2)_3\overset{+}{N}H$ + $:\ddot{C}l:^-$
    **base**         **acid**    **conjugate acid**    **conjugate base**

e. $CH_3CH_2-\ddot{O}-H$ + $H-\ddot{B}r:$ &rlarr; $CH_3CH_2-\overset{+}{\underset{H}{\ddot{O}}}-H$ + $:\ddot{B}r:^-$
    **base**        **acid**         **conjugate acid**    **conjugate base**

f. $CH_3C\equiv C:^-$ + $H-\ddot{O}H$ &rlarr; $CH_3C\equiv CH$ + $H\ddot{O}:^-$
    **base**      **acid**     **conjugate acid**  **conjugate base**

**2.31** Label the acid and base in the starting materials and then draw the products of proton transfer from acid to base.

a. (benzoic acid structure) $C\!\!\begin{smallmatrix}:O:\\ \ddot{O}-H\end{smallmatrix}$ + $\ddot{N}H_3$ &rlarr; (benzoate structure) $C\!\!\begin{smallmatrix}:O:\\ \ddot{O}:^-\end{smallmatrix}$ + $\overset{+}{N}H_4$
    **acid**         **base**

b.

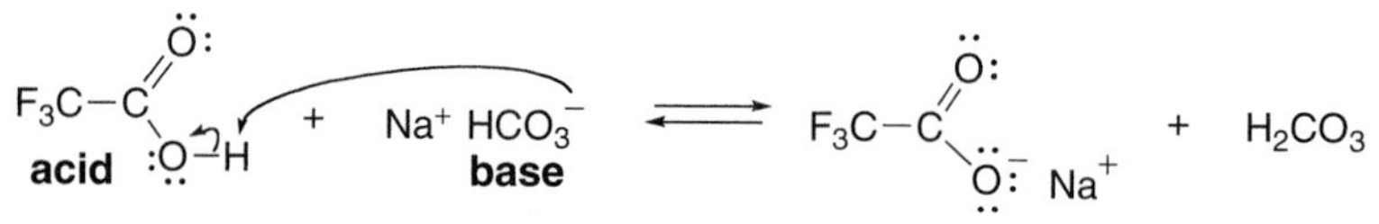

c.

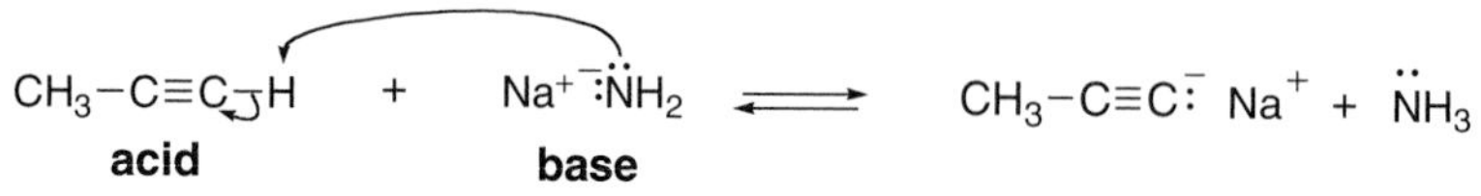

d.

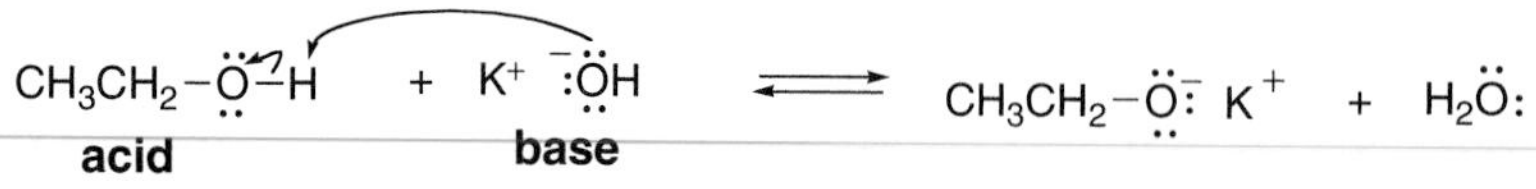

e.

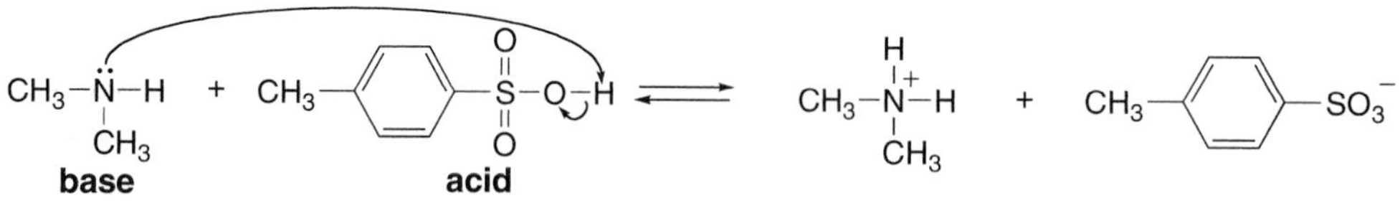

f.

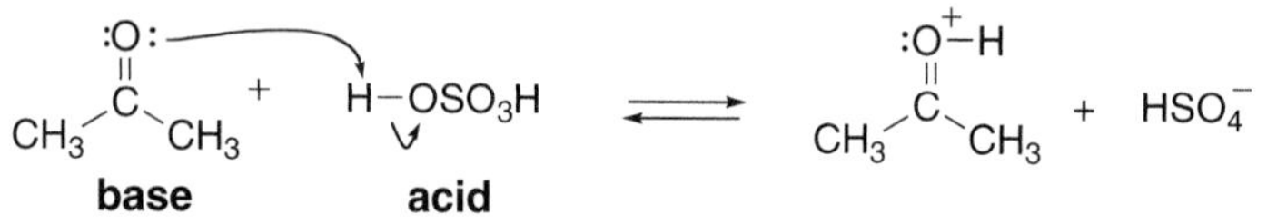

**2.32**   Draw the products of proton transfer from acid to base.

a.

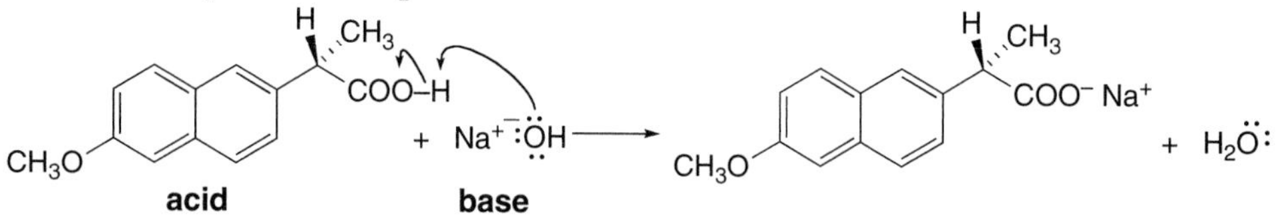

b.

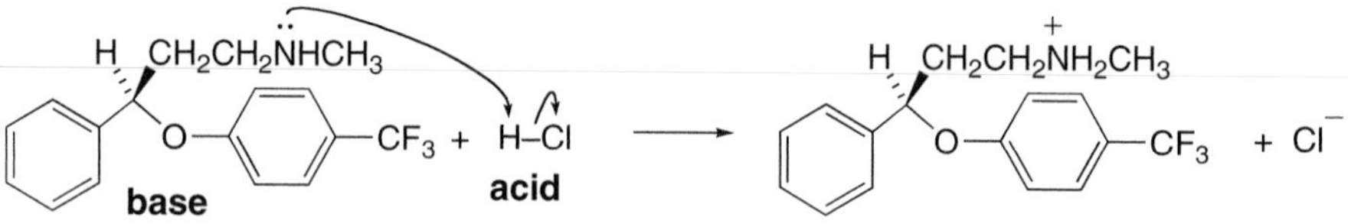

**2.33**   To convert p$K_a$ to $K_a$, take the antilog of (–) the p$K_a$.

a.  $H_2S$
   p$K_a = 7.0$
   $K_a = 10^{-7}$

b.  $ClCH_2COOH$
   p$K_a = 2.8$
   $K_a = 1.6 \times 10^{-3}$

c.  HCN
   p$K_a = 9.1$
   $K_a = 7.9 \times 10^{-10}$

**2.34**   To convert from $K_a$ to p$K_a$, take (–) the log of the $K_a$.  **p$K_a$ = –log$K_a$**

a.

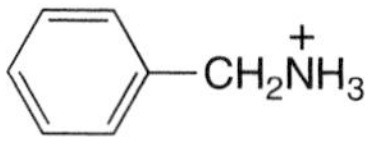

   $K_a = 4.7 \times 10^{-10}$
   p$K_a = 9.3$

b.

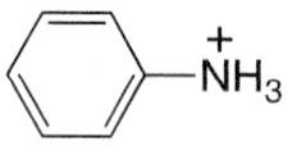

   $K_a = 2.3 \times 10^{-5}$
   p$K_a = 4.6$

c.  $CF_3COOH$

   $K_a = 5.9 \times 10^{-1}$
   p$K_a = 0.23$

**2.35**  An acid can be deprotonated by the conjugate base of any acid with a higher $pK_a$.

**a. $H_2O$**
**$pK_a$ = 15.7**
Any base with a conjugate acid having a $pK_a$ higher than 15.7 can deprotonate it.

| Acid | $pK_a$ | Conjugate base | |
|---|---|---|---|
| $CH_3CH_2OH$ | 16 | $CH_3CH_2O^-$ | |
| $HC\equiv CH$ | 25 | $HC\equiv C^-$ | Strong enough to deprotonate $H_2O$. |
| $H_2$ | 35 | $H^-$ | |
| $NH_3$ | 38 | $^-NH_2$ | |
| $CH_2=CH_2$ | 44 | $CH_2=CH^-$ | |
| $CH_4$ | 50 | $CH_3^-$ | |

**b. $NH_3$**
**$pK_a$ = 38**
Any base with a conjugate acid having a $pK_a$ higher than 38 can deprotonate it.

| Acid | $pK_a$ | Conjugate base | |
|---|---|---|---|
| $CH_2=CH_2$ | 44 | $CH_2=CH^-$ | Strong enough to deprotonate $NH_3$. |
| $CH_4$ | 50 | $CH_3^-$ | |

**c. $CH_4$**
**$pK_a$ = 50**
There is no base with a conjugate acid having a $pK_a$ higher than 50 in the table.

**2.36**  $^-OH$ can deprotonate any acid with a $pK_a < 15.7$.

a.  HCOOH

$pK_a$ = 3.8
stronger acid
deprotonated

b.  $H_2S$

$pK_a$ = 7.0
stronger acid
deprotonated

c.  ⬡—$CH_3$

$pK_a$ = 41
weaker acid
↑

d.  $CH_3NH_2$

$pK_a$ = 40
weaker acid
↑

These acids are too weak to be deprotonated by $^-OH$.

**2.37**  Draw the products and then compare the $pK_a$ of the acid on the left, and the conjugate acid on the right.  **The equilibrium lies towards the side having the acid with a higher $pK_a$ (weaker acid).**

a.  $CF_3-C(=\ddot{O})-\ddot{O}-H$  +  $^-\ddot{O}CH_2CH_3$  ⇌  $CF_3-C(=\ddot{O})-\ddot{O}^-$  +  $H\ddot{O}CH_2CH_3$   **products favored**
$pK_a$ = 0.2 ············ $pK_a$ = 16

b.  $CH_3CH_2-C(=\ddot{O})-\ddot{O}-H$  +  $Na^+$ $:\ddot{Cl}:^-$  ⇌  $CH_3CH_2-C(=\ddot{O})-\ddot{O}:^-$ $Na^+$  +  $H\ddot{Cl}:$   **starting material favored**
$pK_a$ = ~5 ············ $pK_a$ = −7

c.  $(CH_3)_3C\ddot{O}H$  +  $H-OSO_3H$  ⇌  $(CH_3)_3C\overset{+}{O}H_2$  +  $HSO_4^-$   **products favored**
$pK_a$ = −9 ············ $pK_a$ = ~ −3

d.  ⬡—$\ddot{O}-H$  +  $Na^+$ $HCO_3^-$  ⇌  ⬡—$\ddot{O}:^-$ $Na^+$  +  $H_2CO_3$   **starting material favored**
$pK_a$ = 10 ············ $pK_a$ = 6.4

e.  $H-C\equiv C-H$  +  $Li^+$ $^-\ddot{C}H_2CH_3$  ⇌  $H-C\equiv C:^-$ $Li^+$  +  $CH_3CH_3$   **products favored**
$pK_a$ = 25 ············ $pK_a$ = 50

f.  $CH_3\ddot{N}H_2$  +  $H-OSO_3H$  ⇌  $CH_3\overset{+}{N}H_3$  +  $HSO_4^-$   **products favored**
$pK_a$ = −9 ············ $pK_a$ = 10.7

**2.38** Compare element effects first and then, resonance, hybridization and electron-withdrawing groups to determine the relative strengths of the acids.

a. Acidity increases across a row:
$$NH_3 < H_2O < HF$$

b. Acidity increases down a column:
$$HF < HCl < HBr$$

c. increasing acidity: $^-OH < H_2O < H_3O^+$

d. increasing acidity: $NH_3 < H_2O < H_2S$

Compare NH and OH bonds first:
acidity increases across a row.
OH is more acidic.

Then compare OH and SH bonds:
acidity increases down a column.
SH is more acidic.

e. Acidity increases across a row:
$$CH_3CH_3 < CH_3NH_2 < CH_3OH$$

f. increasing acidity: $H_2O < H_2S < HCl$

Compare HCl and SH bonds first:
acidity increases across a row.
H–Cl is more acidic.

Compare OH and SH bonds:
acidity increases down a column.
SH is more acidic.

g. $CH_3CH_2CH_3$,  $ClCH_2CH_2OH$,  $CH_3CH_2OH$

only C–H bonds       O–H bond and       O–H bond
**weakest acid**   electron-withdrawing Cl
                     **strongest acid**

increasing acidity: $CH_3CH_2CH_3 < CH_3CH_2OH < ClCH_2CH_2OH$

h. $HC{\equiv}CCH_2CH_3$    $CH_3CH_2CH_2CH_3$    $CH_3C{=}CCH_3$
                                                                       |  |
                                                                       H  H

*sp* C–H            all *sp*$^3$ C–H        *sp*$^2$ C–H
**strongest acid**   **weakest acid**

increasing acidity: $CH_3CH_2CH_2CH_3 < CH_3CH{=}CHCH_3 < HC{\equiv}CCH_2CH_3$

**2.39** The strongest acid has the weakest conjugate base.

a. Draw the conjugate acid.
Increasing acidity of conjugate acids:
$$CH_3CH_3 < CH_3NH_2 < CH_3OH$$

**increasing basicity:** $CH_3O^- < CH_3NH^- < CH_3CH_2^-$

b. Draw the conjugate acid.
Increasing acidity of conjugate acids:
$$CH_4 < H_2O < HBr$$

**increasing basicity:** $Br^- < HO^- < {}^-CH_3$

c. Draw the conjugate acid.
Increasing acidity of conjugate acids:
$$CH_3CH_2OH < CH_3COOH < ClCH_2COOH$$

**increasing basicity:** $ClCH_2COO^- < CH_3COO^- < CH_3CH_2O^-$

d. Draw the conjugate acid.
Increasing acidity of conjugate acids:

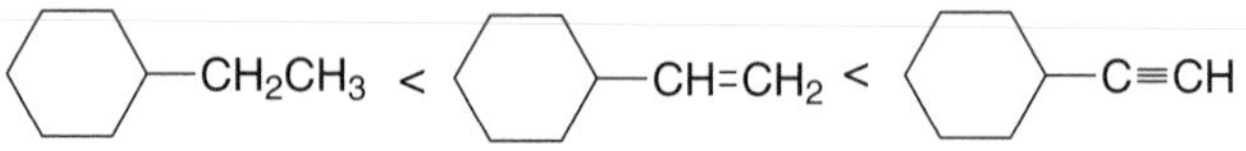

**increasing basicity:**

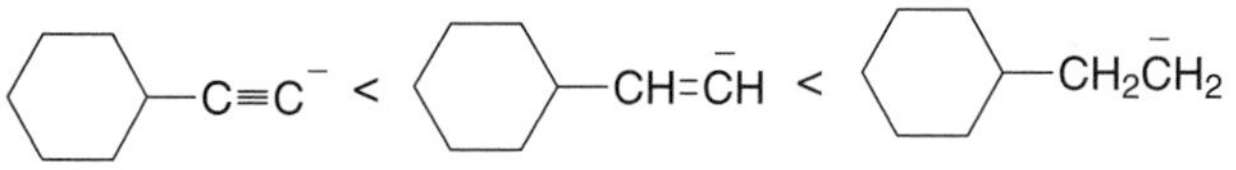

**2.40**

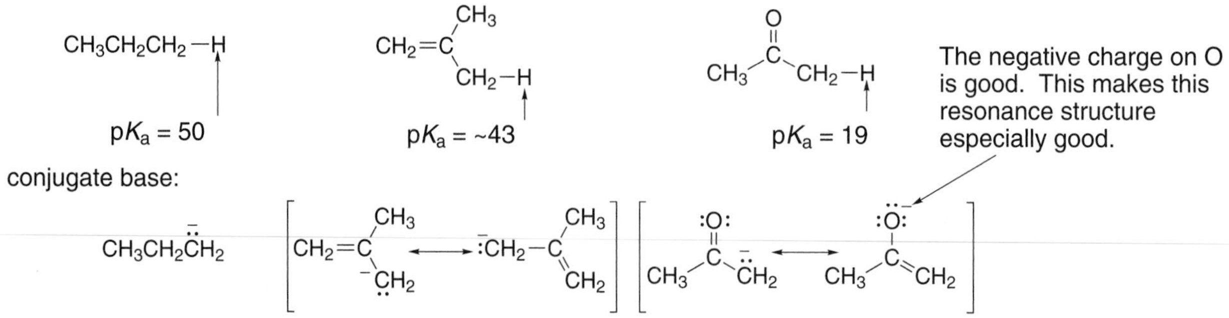

$pK_a = 50$        $pK_a = {\sim}43$        $pK_a = 19$

The negative charge on O is good. This makes this resonance structure especially good.

conjugate base:

one resonance structure       two resonance structures       two resonance structures
**weakest acid**       negative charge delocalized       negative charge delocalized
                            on two carbons               on one O and one C
                                                          **strongest acid**

**2.41** To draw the conjugate acid, look for the most basic site and protonate it. To draw the conjugate base, look for the most acidic site and remove a proton.

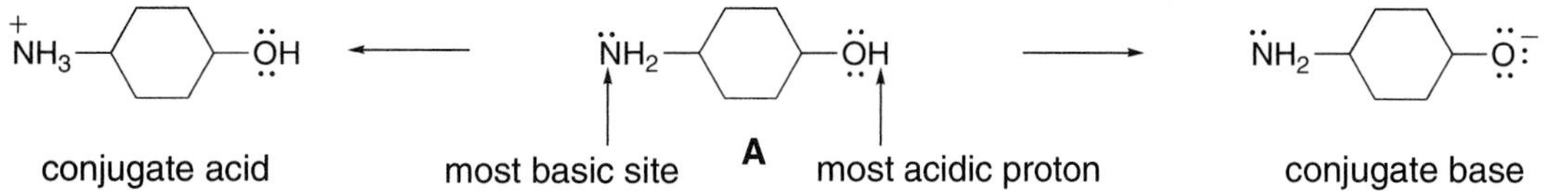

conjugate acid     most basic site    **A**    most acidic proton     conjugate base

**2.42** Remove the most acidic proton to form the conjugate base. Protonate the most basic electron pair to form the conjugate acid.

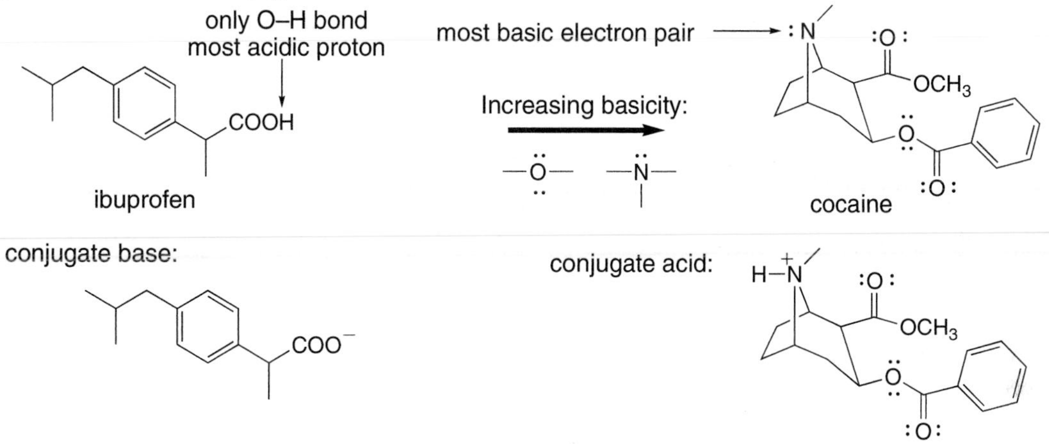

only O–H bond
most acidic proton

ibuprofen

most basic electron pair

Increasing basicity:

cocaine

conjugate base:

conjugate acid:

**2.43** **A lower p$K_a$ means a stronger acid.** The p$K_a$ is low for the C–H bond in $CH_3NO_2$ due to resonance stabilization of the conjugate base.

The negative charge is delocalized on the electronegative O atom. This stabilizes the conjugate base.

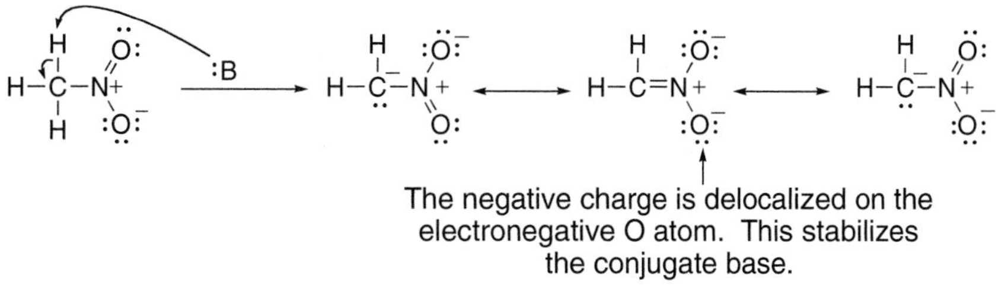

**2.44** Draw the conjugate bases and compare.

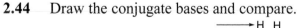

more acidic acid     1,4-pentadiene     pentane     less acidic acid

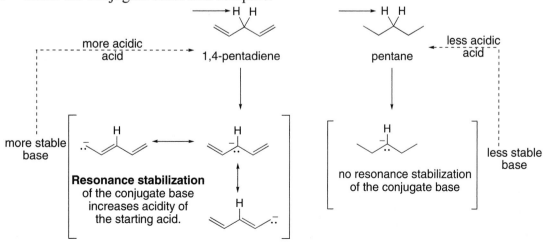

more stable base

**Resonance stabilization** of the conjugate base increases acidity of the starting acid.

no resonance stabilization of the conjugate base

less stable base

**2.45**    Compare the isomers.

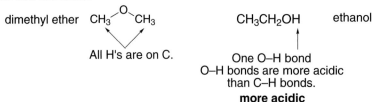

dimethyl ether    CH$_3$—O—CH$_3$                    CH$_3$CH$_2$OH        ethanol

All H's are on C.

One O–H bond
O–H bonds are more acidic
than C–H bonds.
**more acidic**

**2.46**    Recall that in going from *sp* to *sp*$^2$ to *sp*$^3$ hybridization for an atom A, H–A becomes less acidic and A:¯ becomes more basic.

CH$_3$C≡N:                CH$_2$=N̈CH$_3$                CH$_3$—N(H)—CH$_3$

*sp* hybridized      *sp*$^2$ hybridized      *sp*$^3$ hybridized
**least basic**                                          **most basic**

⟶ Increasing basicity

**2.47**    Use the rules from Answer 2.19 to determine which protons are the most acidic.

a.
O–H hydrogen is
**more acidic** than C–H.

c.
O–H hydrogen is
**more acidic** than C–H.

Removal of this H gives a
resonance-stabilized anion,
making it **more acidic**.

b.

No resonance stabilization makes this the less stable
conjugate base.

**2.48** Use the rules from Answer 2.19 to determine which protons are the most acidic.

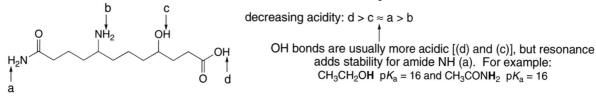

decreasing acidity: d > c ≈ a > b

OH bonds are usually more acidic [(d) and (c)], but resonance adds stability for amide NH (a). For example:
CH₃CH₂OH p$K_a$ = 16 and CH₃CONH₂ p$K_a$ = 16

Conjugate bases:

loss of (a)

Resonance stabilization of the conjugate base with the negative charge on N or O makes (a) more acidic than (b).

loss of (b)

no resonance stabilization

loss of (c)

No resonance stabilization, but the negative charge is on O.

loss of (d)

Resonance stabilization of the conjugate base with the negative charge on O in both resonance structures makes (d) the most acidic.

**2.49** *Lewis bases* **are electron pair donors**: they contain a lone pair or a π bond. *Brønsted–Lowry bases* **are proton acceptors**: to accept a proton they need a lone pair or a π bond. This means all Lewis bases are Brønsted–Lowry bases.

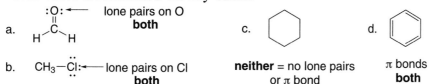

a. :O: ←— lone pairs on O
**both**

b. CH₃—Cl: ←— lone pairs on Cl
**both**

c. **neither** = no lone pairs or π bond

d. π bonds **both**

**2.50** **A *Lewis acid* is an electron pair acceptor** and usually contains a proton or an unfilled valence shell of electrons. A *Brønsted–Lowry acid* **is a proton donor** and must contain a hydrogen atom. All Brønsted–Lowry acids are Lewis acids, though the reverse may not be true.

a. H₃O⁺

**both** -
contains a H

b. Cl₃C⁺

**Lewis acid** -
unfilled valence
shell on C

c. BCl₃

**Lewis acid** -
unfilled valence
shell on B

d. BF₄⁻

**neither** -
no H or unfilled
valence shell

**2.51** Label the Lewis acid and Lewis base and then draw the products.

a. :Cl:⁻  +  BCl₃  ⟶  Cl–B–Cl

**Lewis base**   **Lewis acid**

new bond

c.

$$CH_3-\overset{O}{\overset{||}{C}}-\overset{..}{\underset{..}{Cl}}:  +  :OH  \longrightarrow  CH_3-\overset{O}{\overset{||}{C}}-\overset{..}{\underset{..}{Cl}}:$$

**Lewis acid**   **Lewis base**

:OH   new bond

b.

CH₃   CH₃
    C=C        +   H–OSO₃H   ⟶
CH₃   CH₃                **Lewis acid**

**Lewis base**

CH₃   CH₃
    C–C–H   +   HSO₄⁻
CH₃   CH₃

new bond

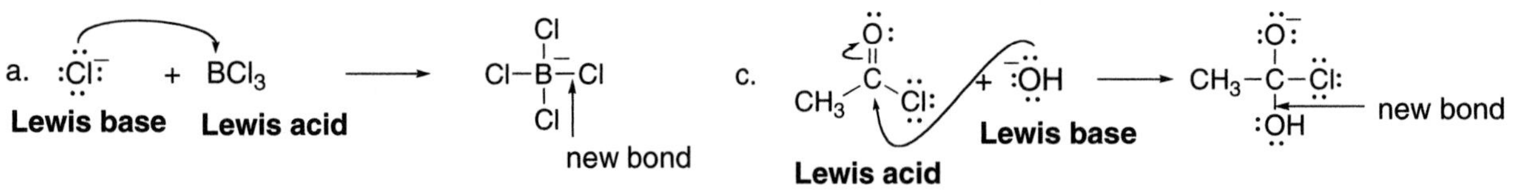

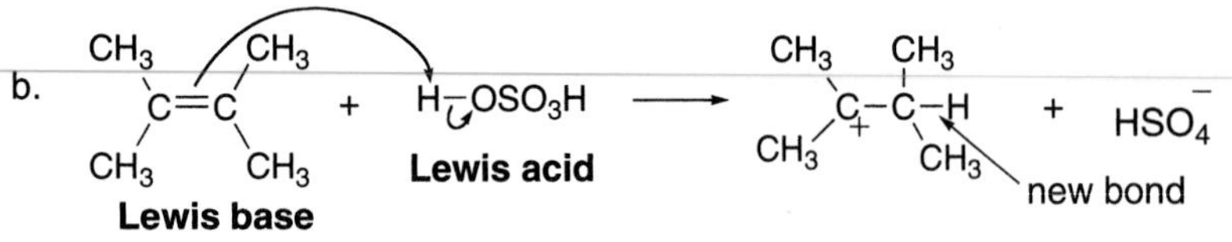

**2.52** A Lewis acid is also called an **electrophile**. When a Lewis base reacts with an electrophile other than a proton, it is called a **nucleophile**. Label the electrophile and nucleophile in the starting materials and then draw the products.

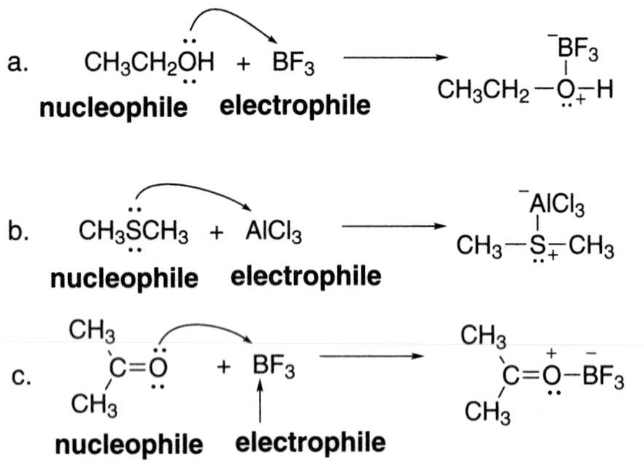

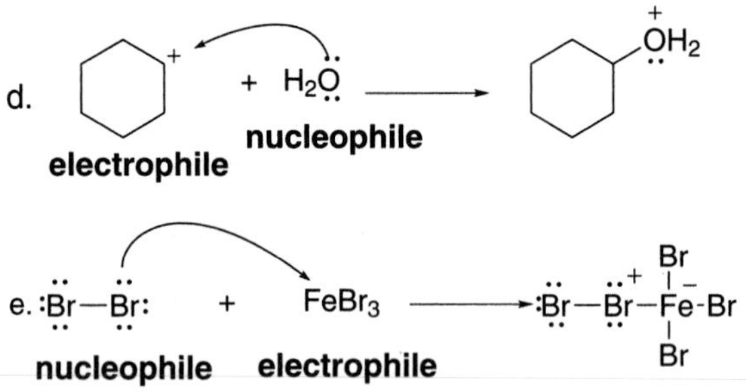

a.   CH₃CH₂OH  +  BF₃   ⟶   ⁻BF₃

**nucleophile**   **electrophile**   CH₃CH₂–O–H
                                              :⁺

b.   CH₃SCH₃  +  AlCl₃   ⟶   ⁻AlCl₃

**nucleophile**   **electrophile**   CH₃–S–CH₃
                                            ⁺

c.

CH₃
    C=O   +  BF₃   ⟶
CH₃

**nucleophile**   **electrophile**

CH₃
    C=O–BF₃
CH₃
    ⁺   ⁻

d.

⬡⁺  +  H₂O   ⟶   ⬡ OH₂⁺

**electrophile**   **nucleophile**

e. :Br–Br:  +  FeBr₃   ⟶   :Br–Br–Fe–Br
                                              Br
                                              Br

**nucleophile**   **electrophile**

**2.53**

a.

:OH   +   H–Br   **proton transfer**⟶   ⁺OH₂   +  Br⁻
                                                                      **nucleophile**

⟶   Br   +  H₂O:

**electrophile**

b.

⬡   +   H–Br   **proton transfer**⟶   ⬡⁺  +  Br⁻

**electrophile**   **nucleophile**

⟶   ⬡ Br

c.

⬡ H   +  Br⁺   ⟶   ⬡ H Br   Br⁻   **proton transfer**⟶   ⬡ Br   +  HBr
   H                              H ⁺                              H

**nucleophile**   **electrophile**

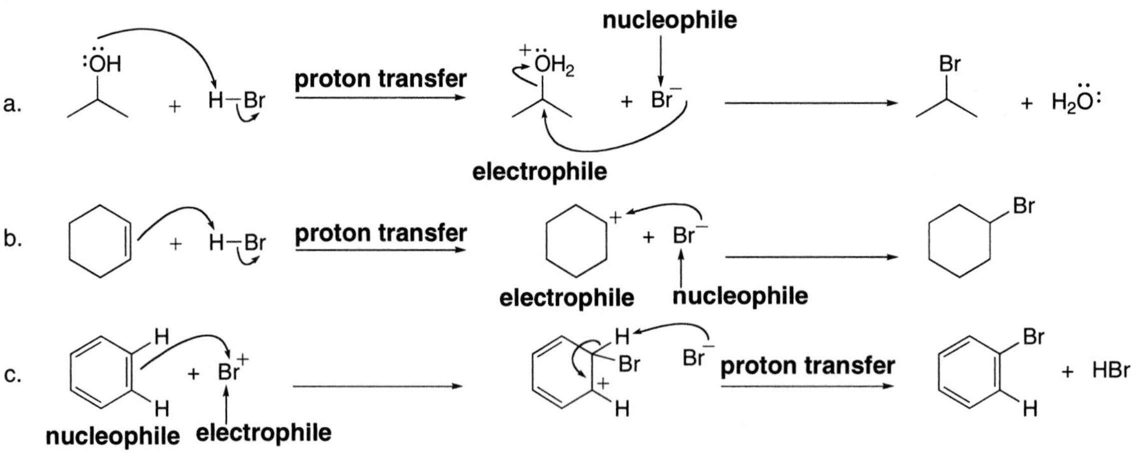

**2.54** Draw the products of each reaction.

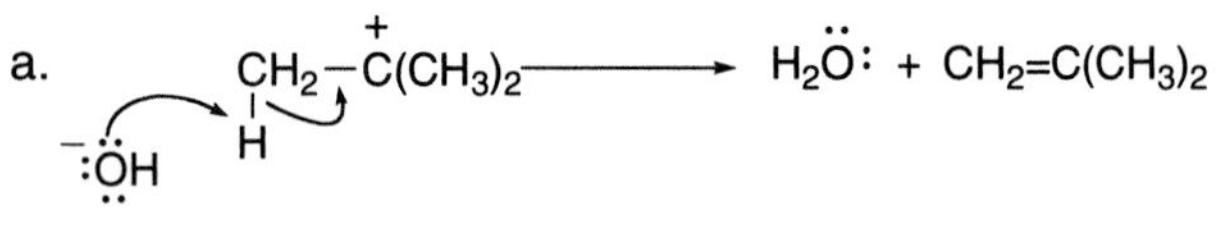

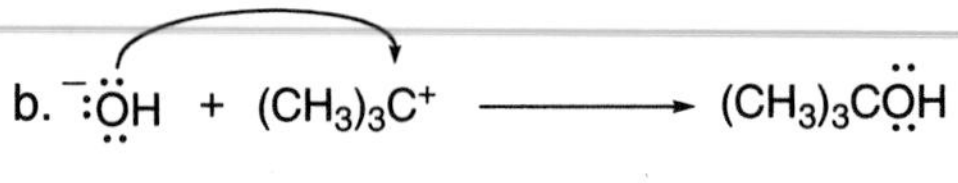

a.

CH₂–C(CH₃)₂   ⟶   H₂O:  +  CH₂=C(CH₃)₂
   H
:OH⁻

b.   ⁻:OH  +  (CH₃)₃C⁺   ⟶   (CH₃)₃COH

**2.55**   Draw the product of protonation of either O or N and compare the conjugate acids.

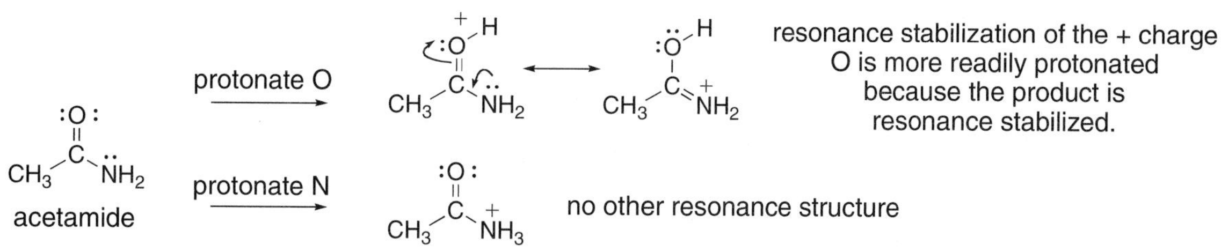

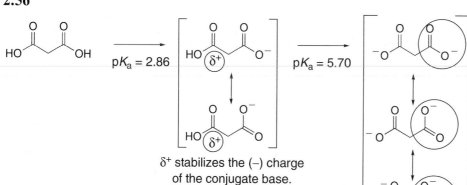

protonate O

resonance stabilization of the + charge
O is more readily protonated
because the product is
resonance stabilized.

acetamide   protonate N

no other resonance structure

When acetamide reacts with an acid, the O atom is protonated because it
results in a resonance-stabilized conjugate acid.

**2.56**

$pK_a = 2.86$   $pK_a = 5.70$

This group destabilizes the
second negative charge.

$\delta^+$ stabilizes the (−) charge
of the conjugate base.

The nearby COOH group serves
as an electron-**withdrawing** group to
stabilize the negative charge.  This
makes the first proton **more** acidic
than $CH_3COOH$.

$COO^-$ now acts as an electron-**donor** group
which destabilizes the conjugate base,
making removal of the second proton more
difficult and thus it is **less** acidic than $CH_3COOH$.

**2.57**

The COOH group of glycine gives up a proton to the basic $NH_2$ group to form the zwitterion.

a. 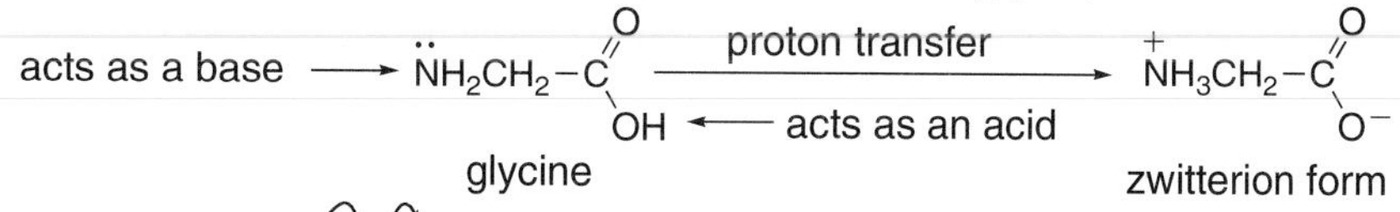   acts as a base $\longrightarrow$

proton transfer

acts as an acid $\longleftarrow$

glycine

zwitterion form

b.

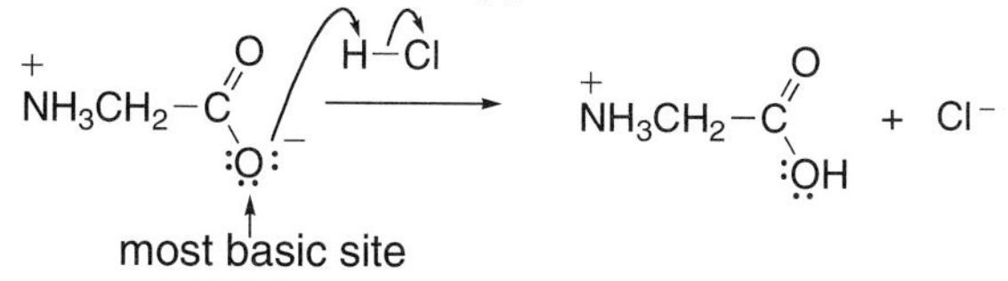

most basic site

c.

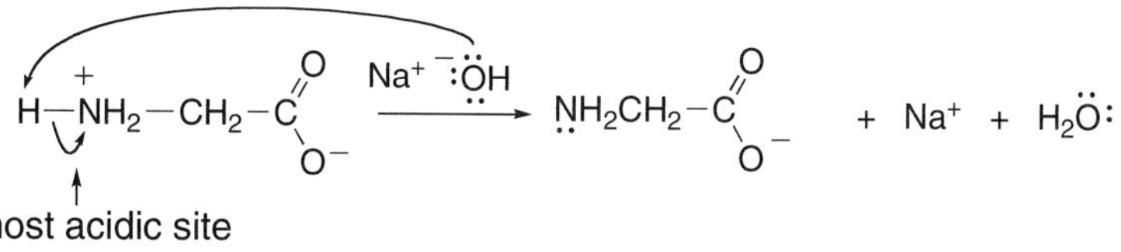

most acidic site

**2.58**    Use curved arrows to show how the reaction occurs.

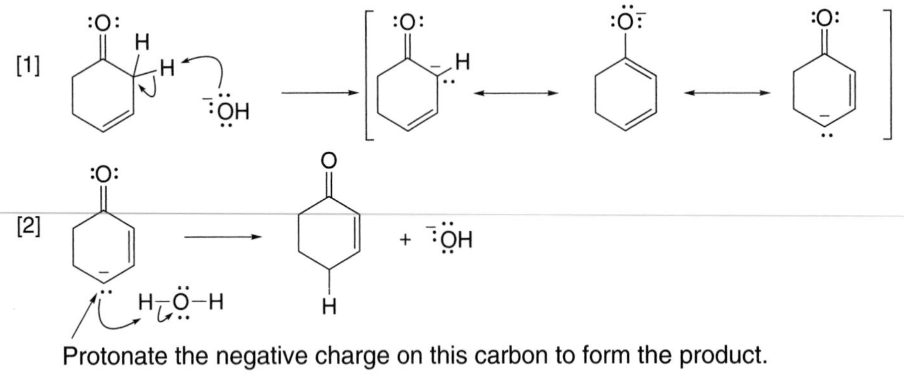

[1]

[2]

Protonate the negative charge on this carbon to form the product.

**2.59**    Compare the OH bonds in Vitamin C and decide which one is the most acidic.

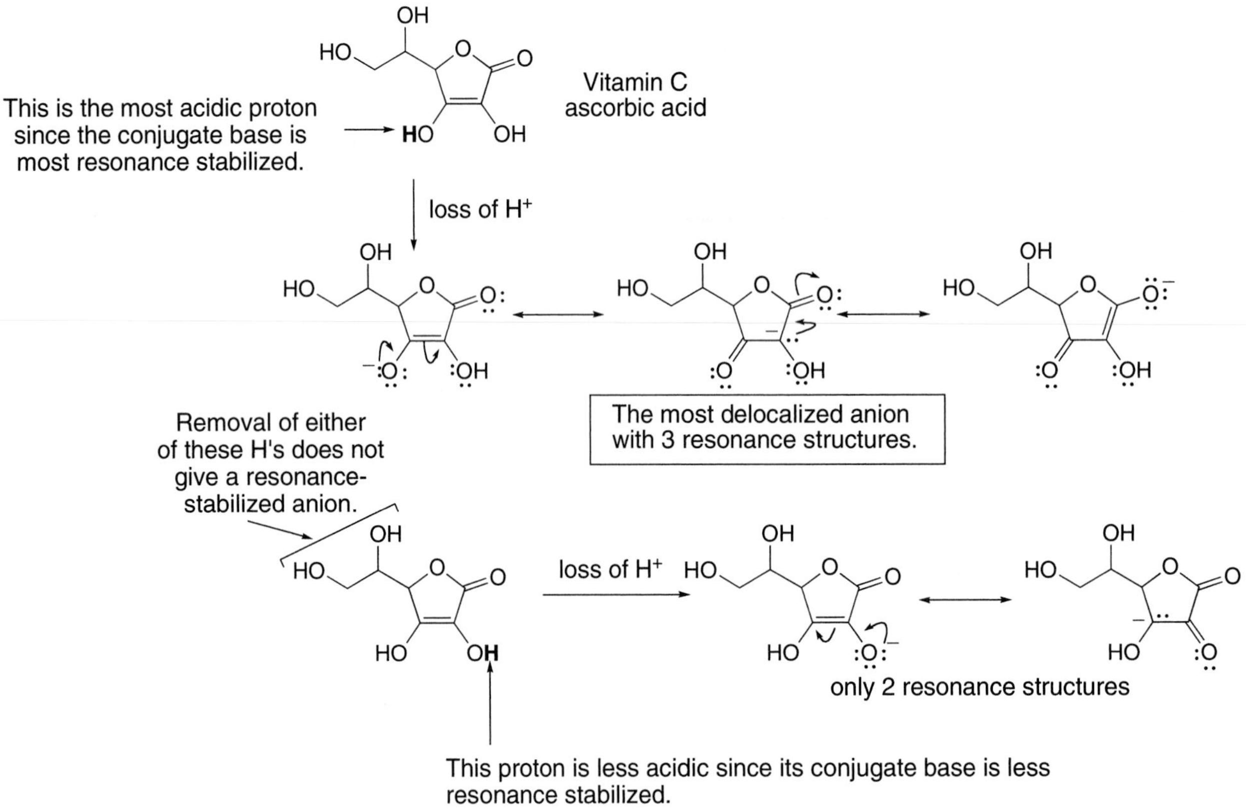

Vitamin C
ascorbic acid

This is the most acidic proton
since the conjugate base is
most resonance stabilized.

loss of H⁺

Removal of either
of these H's does not
give a resonance-
stabilized anion.

The most delocalized anion
with 3 resonance structures.

loss of H⁺

only 2 resonance structures

This proton is less acidic since its conjugate base is less
resonance stabilized.

Chapter 3: Introduction to Organic Molecules and Functional Groups

♦ Types of intermolecular forces (3.3A)

| | Type of force | Cause | Examples |
|---|---|---|---|
| **Increasing strength** ↓ | van der Waals (VDW) | Due to the interaction of temporary dipoles <br> • Larger surface area, stronger forces <br> • Larger, more polarizable atoms, stronger forces | All organic compounds |
| | dipole–dipole (DD) | Due to the interaction of permanent dipoles | $(CH_3)_2C=O$, $H_2O$ |
| | hydrogen bonding (HB or H-bonding) | Due to the electrostatic interaction of a H atom in an O–H, N–H, or H–F bond with another N, O, or F atom. | $H_2O$ |
| | ion–ion | Due to the interaction of two ions | NaCl, LiF |

♦ Physical properties

| Property | Observation |
|---|---|
| **Boiling point (3.4A)** | • For compounds of comparable molecular weight, the stronger the forces the higher the bp. <br><br> $CH_3CH_2CH_2CH_2CH_3$  $CH_3CH_2CH_2CHO$  $CH_3CH_2CH_2CH_2OH$ <br> VDW  VDW, DD  VDW, DD, HB <br> MW = 72  MW = 72  MW = 74 <br> bp = 36 °C  bp = 76 °C  bp = 118 °C <br><br> → Increasing strength of intermolecular forces <br> Increasing boiling point <br><br> • For compounds with similar functional groups, the larger the surface area, the higher the bp. <br><br> $CH_3CH_2CH_2CH_3$  $CH_3CH_2CH_2CH_2CH_3$ <br> bp = 0 °C  bp = 36 °C <br><br> → Increasing surface area <br> Increasing boiling point <br><br> • For compounds with similar functional groups, the more polarizable the atoms, the higher the bp. <br><br> $CH_3F$  $CH_3I$ <br> bp = −78 °C  bp = 42 °C <br><br> → Increasing polarizability <br> Increasing boiling point |

| **Melting point (3.4B)** | • For compounds of comparable molecular weight, the stronger the forces the higher the mp. |
|---|---|

$$CH_3CH_2CH_2CH_2CH_3 \qquad CH_3CH_2CH_2CHO \qquad CH_3CH_2CH_2CH_2OH$$

VDW $\qquad\qquad$ VDW, DD $\qquad\qquad$ VDW, DD, HB

MW = 72 $\qquad\qquad$ MW = 72 $\qquad\qquad$ MW = 74

mp = –130 °C $\qquad\qquad$ mp = –96 °C $\qquad\qquad$ mp = –90 °C

→

**Increasing strength of intermolecular forces**
**Increasing melting point**

• For compounds with similar functional groups, the more symmetrical the compound, the higher the mp.

$$CH_3CH_2CH(CH_3)_2 \qquad (CH_3)_4C$$

mp = –160 °C $\qquad\qquad$ mp = –17 °C

→

**Increasing symmetry**
**Increasing melting point**

| **Solubility (3.4C)** | Types of $H_2O$-soluble compounds: |
|---|---|

• Ionic compounds
• Organic compounds having ≤ 5 C's, and an O or N atom for hydrogen bonding (for a compound with one functional group).

Types of compounds soluble in organic solvents:
• Organic compounds regardless of size or functional group.
• Examples:

Key: VDW = van der Waals, DD = dipole–dipole, HB = hydrogen bonding
MW = molecular weight

♦ Reactivity (3.8)

- **Nucleophiles react with electrophiles.**
- Electronegative heteroatoms create electrophilic carbon atoms, which tend to react with nucleophiles.
- Lone pairs and π bonds are nucleophilic sites that tend to react with electrophiles.

**3.1**

$$CH_3CH_2-\ddot{O}H \xrightarrow{H-OSO_3H} CH_3CH_2-\overset{+}{\underset{..}{O}}H_2 \quad + \quad HSO_4^- \qquad CH_3CH_3 \xrightarrow{H_2SO_4} \text{no reaction}$$

$$CH_3CH_2-\ddot{\underset{..}{O}}H \xrightarrow{Na^+H^-} CH_3CH_2-\ddot{\underset{..}{O}}\colon Na^+ \quad + \quad H_2 \qquad CH_3CH_3 \xrightarrow{NaH} \text{no reaction}$$

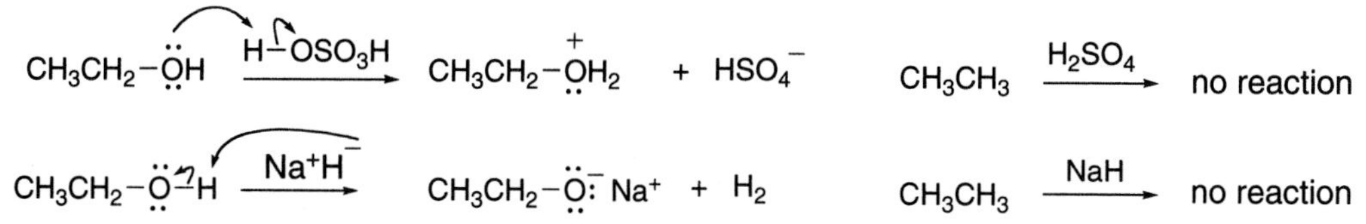

**3.2** Identify the functional groups based on Tables 3.1, 3.2, and 3.3.

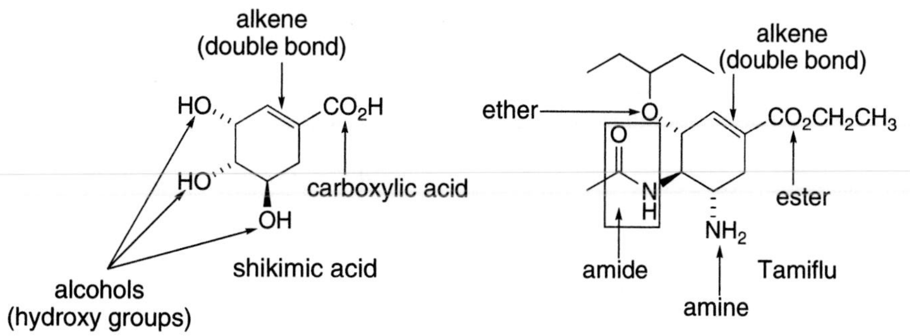

**3.3** One possible structure for each functional group:

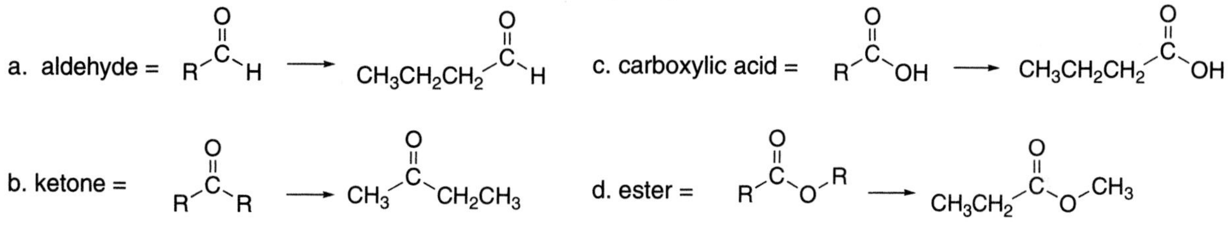

a. aldehyde =

c. carboxylic acid =

b. ketone =

d. ester =

**3.4** Summary of forces:

- **All compounds exhibit van der Waals forces (VDW).**
- **Polar molecules have dipole–dipole forces (DD).**
- **Hydrogen bonding (H-bonding)** can only occur when a **H is bonded to an O, N, or F.**

a.

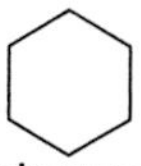

only nonpolar C–C
and C–H bonds
**VDW only**

c.   $(CH_3CH_2)_3N$

- **VDW forces**
- polar C–N bonds - **DD**
- no H on N so
  no H-bonding

e.   $CH_3CH_2CH_2COOH$

- **VDW forces**
- polar C–O bonds
  and a net dipole - **DD**
- H bonded to O -
  **H-bonding**

b.

- **VDW forces**
- 2 polar C–O bonds
  and a net dipole - **DD**
- no H on O so
  no H-bonding

d.   $CH_2{=}CHCl$

- **VDW forces**
- polar C–Cl bond - **DD**

f.   $CH_3-C{\equiv}C-CH_3$

only nonpolar C–H and
C–C bonds
**VDW only**

**3.5** One principle governs boiling point:

- **Stronger intermolecular forces = higher bp.**

  Increasing intermolecular forces: van der Waals < dipole–dipole < hydrogen bonding

  Two factors affect the strength of van der Waals forces, and thus affect bp:

- **Increasing surface area = increasing bp.**

  Longer molecules have a higher surface area. Any branching decreases the surface area of a molecule.

- **Increasing polarizability = increasing bp.**

a. $(CH_3)_2C=CH_2$ and $(CH_3)_2C=O$

    ↑ only VDW      ↑ VDW and DD
            polar, stronger intermolecular forces
            **higher boiling point**

c. $CH_3(CH_2)_4CH_3$ and $CH_3(CH_2)_5CH_3$

            ↑ longer molecule, more surface area
            **higher boiling point**

b. $CH_3CH_2COOH$ and $CH_3COOCH_3$

    ↑           ↑ no H-bonding
  VDW, DD, and H-bonding
  stronger intermolecular forces
     **higher boiling point**

d. $CH_2=CHCl$ and $CH_2=CHI$

            ↑ I is more polarizable.
            **higher boiling point**

**3.6**

a.    and NH₂

           ↑ more polar
      stronger intermolecular forces
         (H-bonding)
        **higher mp**

b. and

    ↑ more spherical
    packs better
    **higher mp**

**3.7** In the more ordered solid phase, molecules are much closer together than in the less ordered liquid phase. The shape of a molecule determines how closely it can pack in the solid phase so symmetry is important. In the liquid phase, molecules are already further apart, so symmetry is less important and thus it doesn't affect boiling point.

**3.8** A compound is water soluble if it is ionic or if it has an O or N atom and ≤ 5 C's.

a. $CH_3CH_2OCH_2CH_3$

      ↑ an O atom that
   can H-bond with water
        ≤ 5 C's
     **water soluble**

b. $CH_3CH_2CH_2CH_2CH_3$

      ↑ nonpolar
    **not water soluble**

c. $(CH_3CH_2CH_2CH_2)_3N$

      ↑ an N atom that can
   H-bond to $H_2O$, but
       > 5 C's
    **not water soluble**

**3.9** **Hydrophobic** portions will primarily be hydrocarbon chains. **Hydrophilic** portions will be polar.

Circled regions are **hydrophilic** because they are polar.
All other regions are **hydrophobic** since they have only C and H.

a.

norethindrone

b.

arachidonic acid

c.

benzo[*a*]pyrene derivative

**3.10 Like dissolves like.**

- To be **soluble in water**, a molecule must be ionic, or have a polar functional group capable of H-bonding for every 5 C's.
- Organic compounds are generally **soluble in organic solvents** regardless of size or functional group.

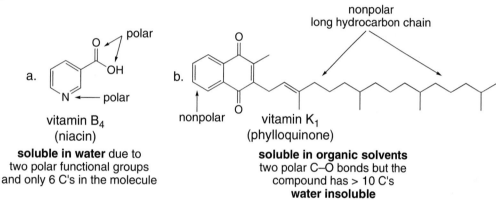

a.

vitamin B$_4$
(niacin)

**soluble in water** due to
two polar functional groups
and only 6 C's in the molecule

b.

vitamin K$_1$
(phylloquinone)

**soluble in organic solvents**
two polar C–O bonds but the
compound has > 10 C's
**water insoluble**

**3.11** Detergents have a polar head consisting of oppositely charged ions, and a nonpolar tail consisting of C–C and C–H bonds, just like soaps do. Detergents clean by having the **hydrophobic ends of molecules surround grease**, while the **hydrophilic portion of the molecule interacts with the polar solvent** (usually water).

a detergent

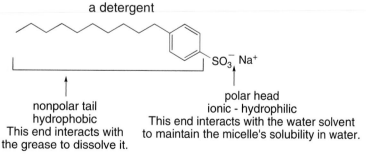

nonpolar tail
hydrophobic
This end interacts with
the grease to dissolve it.

polar head
ionic - hydrophilic
This end interacts with the water solvent
to maintain the micelle's solubility in water.

**3.12**

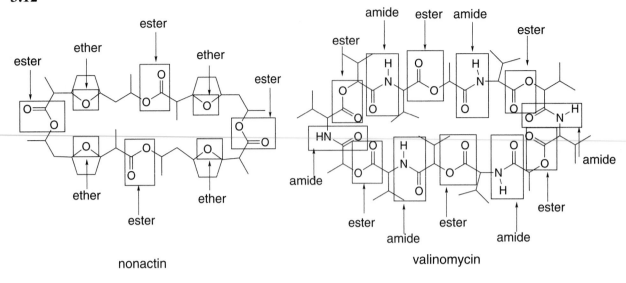

nonactin                              valinomycin

**3.13** Electronegative heteroatoms like N, O, or X make a carbon atom an *electrophile*.
A lone pair on a heteroatom makes it basic and nucleophilic.
π Bonds create *nucleophilic* sites and are more easily broken than σ bonds.

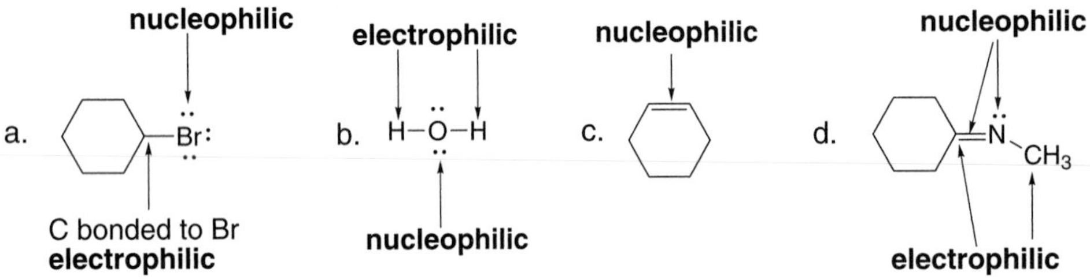

a.   **nucleophilic** →
C bonded to Br
**electrophilic**

b.   **electrophilic** ↓   H–O–H   ↓
**nucleophilic**

c.   **nucleophilic** ↓

d.   **nucleophilic** →
**electrophilic**

**3.14** Electrophiles and nucleophiles react with each other.

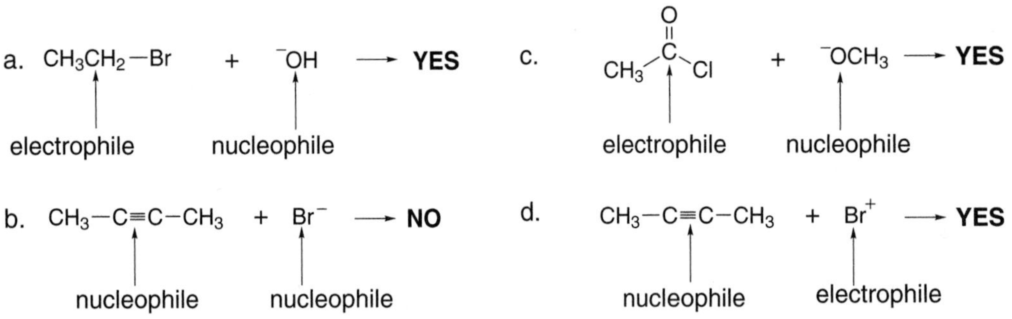

a.   CH₃CH₂—Br   +   ⁻OH   → **YES**

electrophile        nucleophile

b.   CH₃—C≡C—CH₃   +   Br⁻   → **NO**

nucleophile        nucleophile

c.   CH₃—C(=O)—Cl   +   ⁻OCH₃   → **YES**

electrophile        nucleophile

d.   CH₃—C≡C—CH₃   +   Br⁺   → **YES**

nucleophile        electrophile

**3.15**  Identify the functional groups based on Tables 3.1, 3.2, and 3.3.

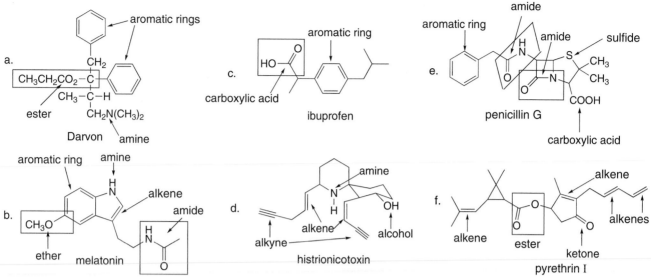

**3.16**

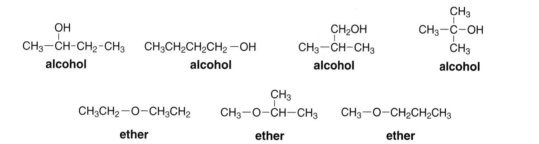

**3.17**  Use the rules from Answer 3.4.

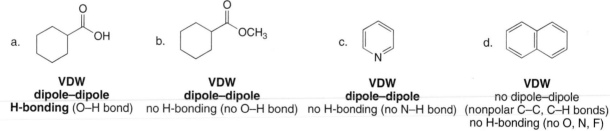

a.  **VDW**
**dipole–dipole**
**H-bonding** (O–H bond)

b.  **VDW**
**dipole–dipole**
no H-bonding (no O–H bond)

c.  **VDW**
**dipole–dipole**
no H-bonding (no N–H bond)

d.  **VDW**
no dipole–dipole
(nonpolar C–C, C–H bonds)
no H-bonding (no O, N, F)

**3.18**  **Increasing intermolecular forces**: van der Waals < dipole–dipole < H-bonding

a. **increasing intermolecular forces:**

$CH_3CH_3 < CH_3Cl < CH_3NH_2$

VDW     VDW     VDW
      dipole–dipole   dipole–dipole
                      H-bonding

b. **increasing intermolecular forces:**

$CH_3Cl < CH_3Br < CH_3I$

Increasing polarizability
stronger intermolecular forces

c. **increasing intermolecular forces:**

$(CH_3)_2C=C(CH_3)_2 < (CH_3)_2CHCOCH_3 < (CH_3)_2CHCOOH$

VDW                VDW                VDW
                   dipole–dipole      dipole–dipole
                                      H-bonding

d. **increasing intermolecular forces:**

$CH_3Cl < CH_3OH < NaCl$

VDW          VDW          ionic
dipole–dipole   dipole–dipole
             H-bonding

**3.19**

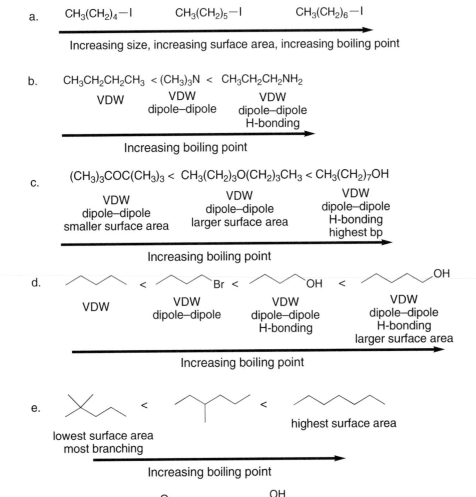

$CH_3-C$ hydrogen bonding between two acetic acid molecules

**3.20**   **A** = VDW forces; **B** = H-bonding; **C** = ion–ion interactions; **D** = H-bonding; **E** = H-bonding; **F** = VDW forces.

**3.21**   Use the principles from Answer 3.5.

a.   $CH_3(CH_2)_4-I$    $CH_3(CH_2)_5-I$    $CH_3(CH_2)_6-I$

Increasing size, increasing surface area, increasing boiling point

b.   $CH_3CH_2CH_2CH_3$ < $(CH_3)_3N$ < $CH_3CH_2CH_2NH_2$
        VDW          VDW              VDW
                dipole–dipole      dipole–dipole
                                   H-bonding

Increasing boiling point

c.   $(CH_3)_3COC(CH_3)_3$ < $CH_3(CH_2)_3O(CH_2)_3CH_3$ < $CH_3(CH_2)_7OH$
        VDW               VDW                      VDW
        dipole–dipole     dipole–dipole            dipole–dipole
        smaller surface area   larger surface area  H-bonding
                                                    highest bp

Increasing boiling point

d.   VDW   <   VDW   Br   <   VDW   OH   <   VDW   OH
            dipole–dipole    dipole–dipole      dipole–dipole
                             H-bonding          H-bonding
                                                larger surface area

Increasing boiling point

e.   lowest surface area   <        <   highest surface area
     most branching

Increasing boiling point

f.   VDW   <   VDW   <   OH
            dipole–dipole   VDW
                            dipole–dipole
                            H-bonding

Increasing boiling point

**3.22**   In $CH_3CH_2NHCH_3$, there is a N–H bond so the molecules exhibit intermolecular hydrogen bonding, whereas in $(CH_3)_3N$ the N is bonded only to C, so there is no hydrogen bonding.  The hydrogen bonding in $CH_3CH_2NHCH_3$ makes it have much **stronger intermolecular forces** than $(CH_3)_3N$.  As intermolecular forces increase, the boiling point of a molecule of the same molecular weight is higher.

**3.23**   Stronger forces, higher mp.  More symmetrical compounds, higher mp.

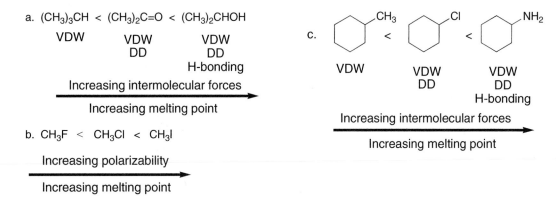

a. $(CH_3)_3CH$  <  $(CH_3)_2C{=}O$  <  $(CH_3)_2CHOH$

　　　VDW　　　　　VDW　　　　　　VDW
　　　　　　　　　　　DD　　　　　　　DD
　　　　　　　　　　　　　　　　　　H-bonding

Increasing intermolecular forces →

Increasing melting point

c.

　　　VDW　　　　　VDW　　　　　VDW
　　　　　　　　　　　DD　　　　　　DD
　　　　　　　　　　　　　　　　　H-bonding

Increasing intermolecular forces →

Increasing melting point

b. $CH_3F$  <  $CH_3Cl$  <  $CH_3I$

Increasing polarizability →

Increasing melting point

**3.24**

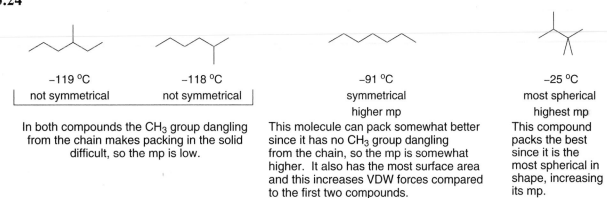

| –119 °C | –118 °C | –91 °C | –25 °C |
|---|---|---|---|
| not symmetrical | not symmetrical | symmetrical | most spherical |
| | | higher mp | highest mp |

In both compounds the CH₃ group dangling from the chain makes packing in the solid difficult, so the mp is low.

This molecule can pack somewhat better since it has no CH₃ group dangling from the chain, so the mp is somewhat higher.  It also has the most surface area and this increases VDW forces compared to the first two compounds.

This compound packs the best since it is the most spherical in shape, increasing its mp.

**3.25**   **Boiling point is determined solely by the strength of the intermolecular forces.**  Since benzene has a smaller size, it has less surface area and weaker VDW interactions and therefore a lower boiling point than toluene.  The increased melting point for benzene can be explained by symmetry: benzene is much more symmetrical than toluene.  More symmetrical molecules can pack more tightly together, increasing their melting point. Symmetry has no effect on boiling point.

benzene
bp = 80 °C
mp = 5 °C

very symmetrical
closer packing in solid form
**higher mp**

and

toluene
bp = 111 °C
mp = –93 °C

less symmetrical
**lower mp**

**3.26** Increasing polarity = increasing water solubility.

Neither compound is very H₂O soluble.

a. $\overline{CH_3CH_2CH_2CH_3 \quad < \quad (CH_3)_3CH}$ < $CH_3OCH_2CH_3$ < $CH_3CH_2CH_2OH$

|   VDW   |       VDW        |    VDW    |    VDW     |
|---------|------------------|-----------|------------|
|         |  more spherical  |    DD     |     DD     |

(This nonpolar, hydrophobic molecule is more compact, making it more water soluble than its straight-chain isomer, drawn to the left.)

b.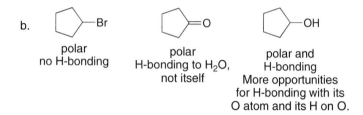

| polar | polar | polar and |
|-------|-------|-----------|
| no H-bonding | H-bonding to H₂O, not itself | H-bonding |

More opportunities for H-bonding with its O atom and its H on O.

**3.27** Look for two things:
- To H-bond to another molecule like itself, the molecule must contain a **H bonded to O, N, or F.**
- To H-bond with water, a molecule need **only contain an O, N, or F.**

Each of these molecules can H-bond to another molecule like itself. Both compounds have N–H bonds.
    b. CH₃NH₂, e. CH₃CH₂CH₂CONH₂

These molecules can H-bond with water. All of these molecules have an O or N atom.
    b. CH₃NH₂, c. CH₃OCH₃, d. (CH₃CH₂)₃N,
    e. CH₃CH₂CH₂CONH₂, g. CH₃SOCH₃,
    h. CH₃CH₂COOCH₃

**3.28** Draw the molecules in question and look at the intermolecular forces involved.

no H bonded to O

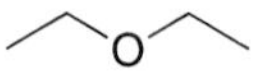

diethyl ether

1-butanol   ←——   H bonded to O: hydrogen bonding

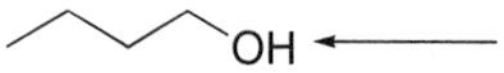

VDW forces
dipole–dipole forces

VDW forces
dipole–dipole forces
**H-bonding**

- Both have ≤ 5 C's and an electronegative O atom, so they can H-bond to water, making them soluble in water.
- Only 1-butanol can H-bond to another molecule life itself, and this increases its boiling point.

**3.29**

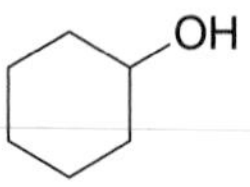

cyclohexanol

1-hexanol

The nonpolar hydrocarbon part is more compact, so it is easier for the OH group to solubilize it in water.

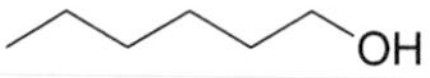

**3.30**   Use the solubility rule from Answer 3.8.

a.

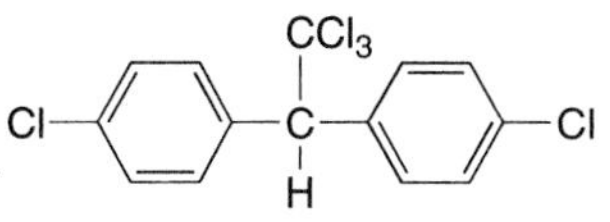

DDT
no N or O
**not water soluble**

b.

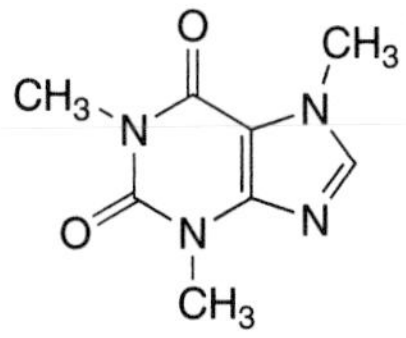

caffeine
many polar bonds with N and O
atoms
many opportunities for H-bonding
**water soluble**

c.

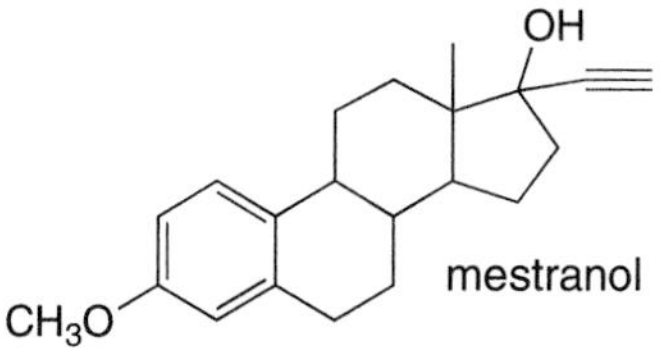

mestranol
2 polar functional groups
but > 10 C's
**not water soluble**

d.

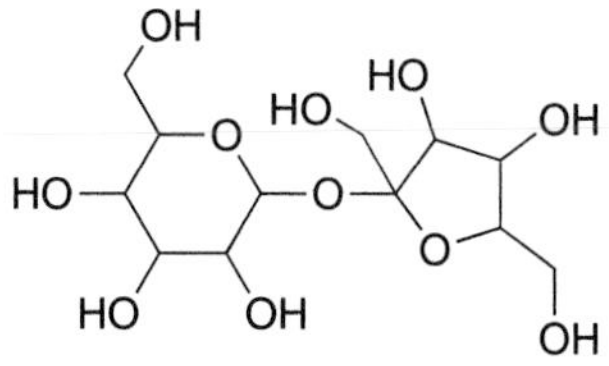

sucrose
many polar bonds with O
11 O's and 12 C's
many opportunities for H-bonding with H₂O
**water soluble**

e.

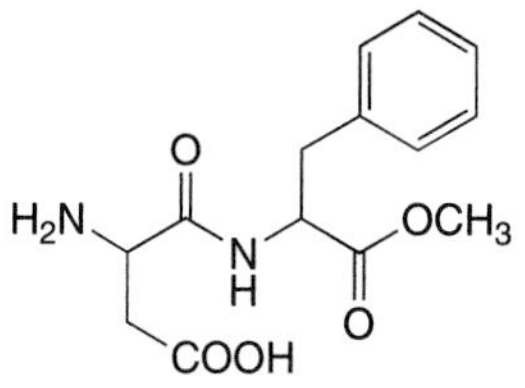

aspartame
many polar bonds with N and O atoms
many opportunities for H-bonding
**water soluble**

f.

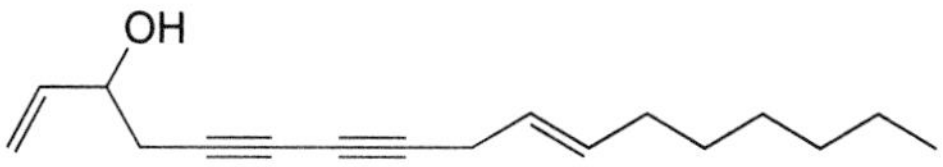

carotatoxin
1 polar functional group
but > 10 C's
**not water soluble**

**3.31**

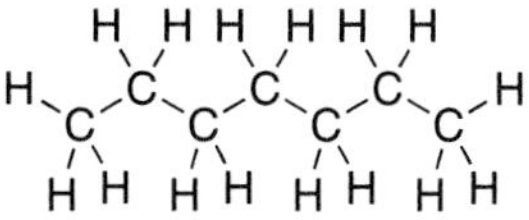

heptane
bp = 98 °C

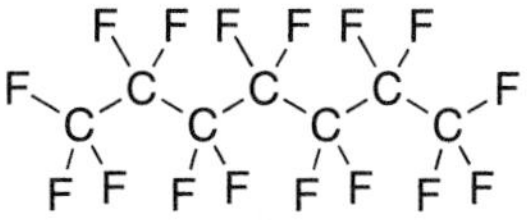

perfluoroheptane
bp = 82–84 °C

molecular weight = 100 g/mol    molecular weight = 388 g/mol

F atoms are very electronegative and small, and their electron clouds
are held tightly making them very poorly polarizable.  This means there
is little force of attraction between polyflourinated molecules, giving
them much lower bp's than you would expect based on their molecular
weights.

**3.32**

| $(CH_3)_2CHCH(CH_3)_2$ | $CH_3(CH_2)_4CH_3$ | $CH_3(CH_2)_5CH_3$ | $CH_3(CH_2)_6CH_3$ |
|---|---|---|---|
| **B** | **C** | **D** | **A** |
| 6 C's | 6 C's | 7 C's | 8 C's |
| Branching makes less surface area, weaker VDW. **lowest bp** | no branching | | **highest bp** |

**C**, **D**, and **A** are all long chain hydrocarbons,
but the size increases from **C** to **D** to **A**, increasing
the VDW forces and increasing bp.

**3.33**   Water solubility is determined by polarity.  Polar molecules are soluble in water, while nonpolar molecules are soluble in organic solvents.

a.

| Arrows indicate polar functional groups. |

b.

**vitamin E**

only 2 polar functional groups
many nonpolar C–C and C–H bonds (29 C's)
**soluble in organic solvents**
**insoluble in H$_2$O**

pyridoxine
**vitamin B$_6$**

many polar bonds and few nonpolar bonds
**soluble in H$_2$O**
It is also soluble in organic solvents since it
is organic, but is probably more soluble in H$_2$O.

**3.34**   Molecules that dissolve in water are readily excreted from the body in urine whereas less polar molecules that dissolve in organic solvents are soluble in fatty tissue and are retained for longer periods.  Compare the solubility properties of THC and ethanol to determine why drug screenings can detect THC and not ethanol weeks after introduction to the body.

ethanol

tetrahydrocannabinol
THC

THC has relatively few polar
bonds compared to the number
of nonpolar bonds making it
**soluble in organic solvents**
and therefore **soluble in fatty tissue.**

Ethanol has 1 O atom and
only 2 C's making it
**soluble in water.**

Due to their solubilities, **THC is retained much longer in the fatty tissue of the body,** being
slowly excreted over many weeks, while ethanol is excreted rapidly in urine after ingestion.

**3.35**   Compare the intermolecular forces of crack and cocaine hydrochloride.  Higher intermolecular forces increase both the boiling point and the water solubility.

ionic bond

cocaine (crack)
neutral organic molecule

cocaine hydrochloride
a salt

The molecules are identical except for the ionic bond in cocaine hydrochloride. Ionic forces are extremely strong forces, and therefore the cocaine hydrochloride salt has a much **higher boiling point and is more water soluble.** Since the salt is highly water soluble, it can be injected directly into the blood stream where it dissolves. Crack is smoked because it can dissolve in the organic tissues of the nasal passage and lungs.

**3.36** A laundry detergent must have both a highly polar end of the molecule, and a nonpolar end of the molecule. The polar end will interact with water, while the nonpolar end surrounds the grease/organic material.

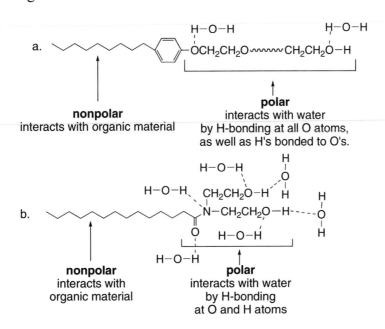

**3.37** An emulsifying agent is one that dissolves a compound in a solvent in which it is not normally soluble. In this case the phospholipids can dissolve the oil in its nonpolar tails and bring it into solution in the aqueous vinegar solution. Or, the nonpolar tails dissolve in the oil, and the polar head brings the water-soluble compounds into solution. In any case, the phospholipids make a uniform medium, mayonnaise, from two insoluble layers.

|  |  |
|---|---|
| vinegar | oil |
| aqueous | organic |
| hydrophilic | hydrophobic |

These two ingredients will not mix. The emulsifying agent (egg yolk) has phospholipids that have both hydrophobic and hydrophilic portions, making the mayonnaise uniform.

**3.38** Use the rules from Answer 3.13.

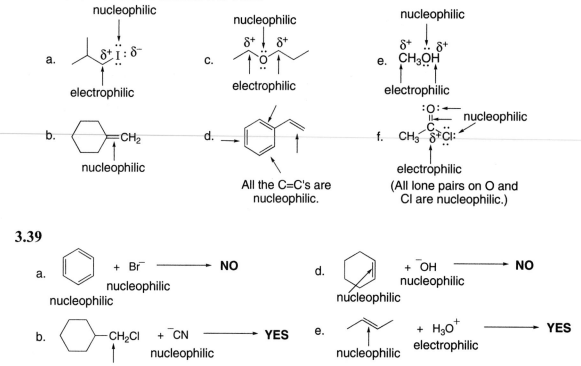

a. nucleophilic
$\delta^+$ I : $\delta^-$
electrophilic

b. =CH₂
nucleophilic

c. nucleophilic
$\delta^+$ O $\delta^+$
electrophilic

d. All the C=C's are nucleophilic.

e. nucleophilic
$\delta^+$ .. $\delta^+$
CH₃OH
electrophilic

f. :O: ← nucleophilic
CH₃ $\delta^+$Cl:
electrophilic
(All lone pairs on O and Cl are nucleophilic.)

**3.39**

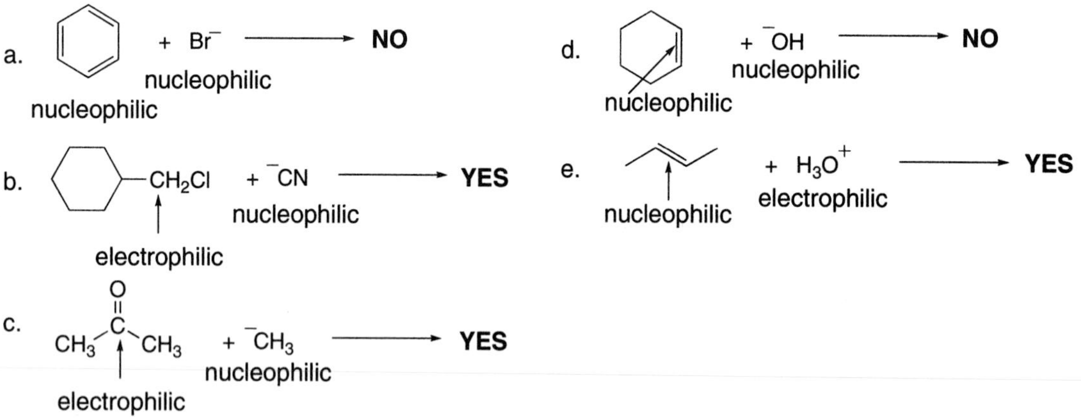

a. + Br⁻ ⟶ **NO**
nucleophilic
nucleophilic

b. —CH₂Cl + ⁻CN ⟶ **YES**
nucleophilic
electrophilic

c. 
O
CH₃—C—CH₃ + ⁻CH₃ ⟶ **YES**
nucleophilic
electrophilic

d. + ⁻OH ⟶ **NO**
nucleophilic
nucleophilic

e. + H₃O⁺ ⟶ **YES**
nucleophilic electrophilic

**3.40** More rigid cell membranes have phospholipids with *fewer* C=C's. Each C=C introduces a bend in the molecule, making the phospholipids pack less tightly. Phospholipids without C=C's can pack very tightly, making the membrane less fluid, and more rigid.

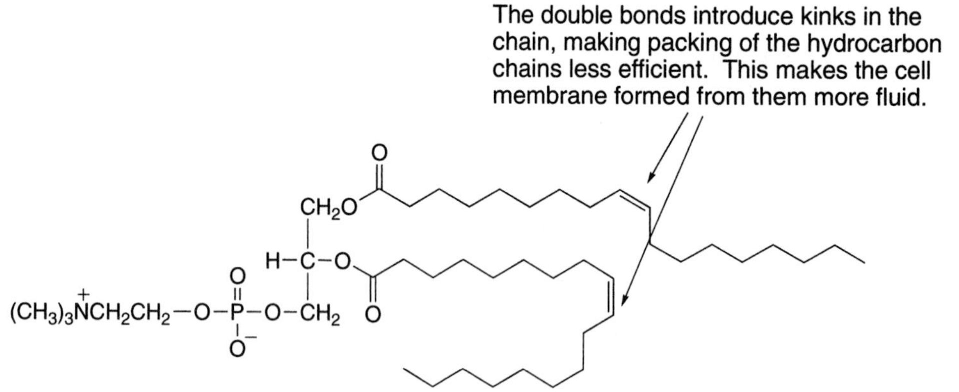

The double bonds introduce kinks in the chain, making packing of the hydrocarbon chains less efficient. This makes the cell membrane formed from them more fluid.

(CH₃)₃N⁺CH₂CH₂—O—P—O—CH₂
O
‖
O
|
O⁻

**3.41**

amine
can H-bond

hydroxy group
can H-bond

a. 7 amide groups [regular (unbolded) arrows]
b. OH groups bonded to $sp^3$ C's are circled.
   OH groups bonded to $sp^2$ C's have a square.
c. Despite its size, vancomycin is water soluble
   because it contains many polar groups and many
   N and O atoms that can H-bond to $H_2O$.
d. The most acidic proton is labeled (COOH group).
e. Four functional groups capable of H-bonding are
   ROH, RCOOH, amides, and amines.

most
acidic
proton

amide
can H-bond

vancomycin

**3.42**

A

B

The OH and CHO groups are close enough that they can intramolecularly H-bond to each other.  Since the two polar functional groups are involved in intramolecular H-bonding, they are less available for H-bonding to $H_2O$. This makes **A** less $H_2O$ soluble than **B**, whose two functional groups are both available for H-bonding to the $H_2O$ solvent.

The OH and the CHO are too far apart to intramolecularly H-bond to each other, leaving more opportunity to H-bond with solvent.

**3.43**

a. melting point

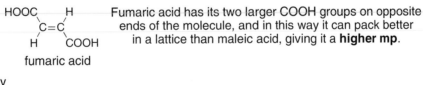

fumaric acid

Fumaric acid has its two larger COOH groups on opposite ends of the molecule, and in this way it can pack better in a lattice than maleic acid, giving it a **higher mp**.

b. solubility

net dipole

maleic acid

Maleic acid is more polar, giving it
greater **$H_2O$ solubility.**

c. removal of the first proton (p$K_{a1}$)

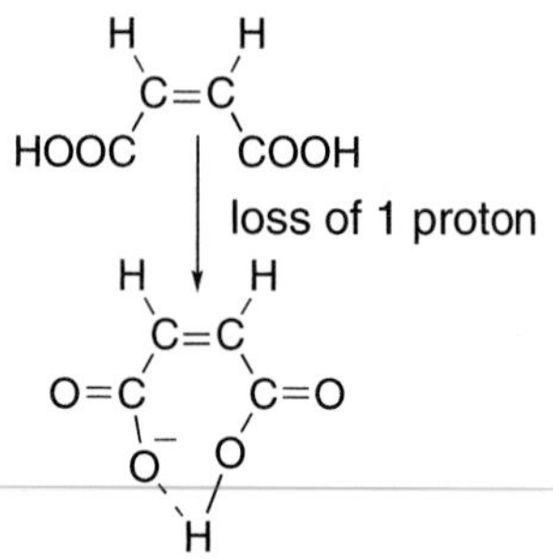

In maleic acid, intramolecular H-bonding stabilizes the conjugate base after one H is removed, making maleic acid more acidic than fumaric acid.

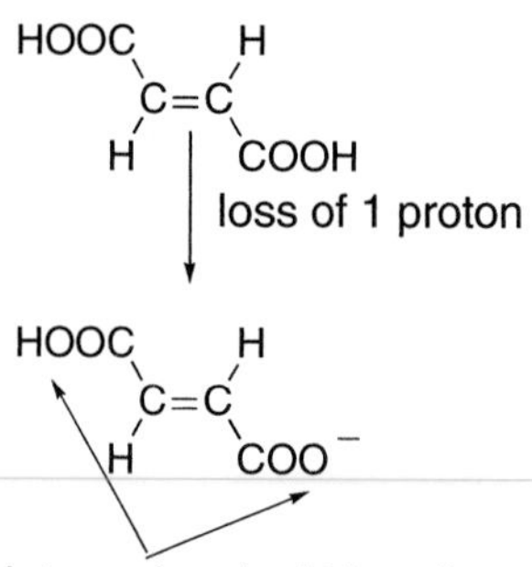

Intramolecular H-bonding is not possible here.

d. removal of the second proton (p$K_{a2}$)

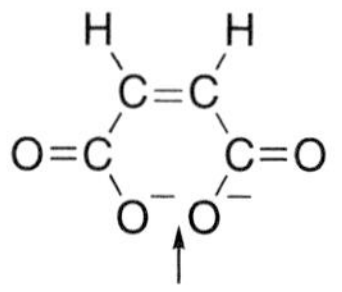

Now the dianion is held in close proximity in maleic acid, and this destabilizes the conjugate base. So removing the second H in maleic acid is harder, making it a weaker acid than fumaric acid for removal of the second proton.

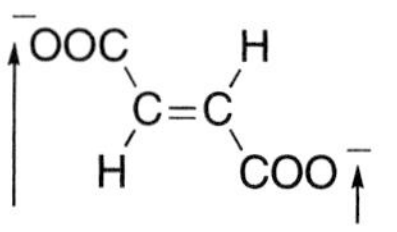

The two negative charges are much further apart. This makes the dianion from fumaric acid more stable and thus p$K_{a2}$ is lower for fumaric acid than maleic acid.

## Chapter 4: Alkanes

### ◆ General facts about alkanes (4.1–4.3)

- Alkanes are composed of **tetrahedral,** $sp^3$ hybridized C's.
- There are two types of alkanes: acyclic alkanes having molecular formula $C_nH_{2n+2}$, and cycloalkanes having molecular formula $C_nH_{2n}$.
- Alkanes have only **nonpolar C−C and C−H bonds** and no functional group so they undergo few reactions.
- Alkanes are named with the suffix **-ane.**

### ◆ Classifying C's and H's (4.1A)

- Carbon atoms are classified by the number of C's bonded to them; **a $1^\circ$ C is bonded to one other C,** and so forth.

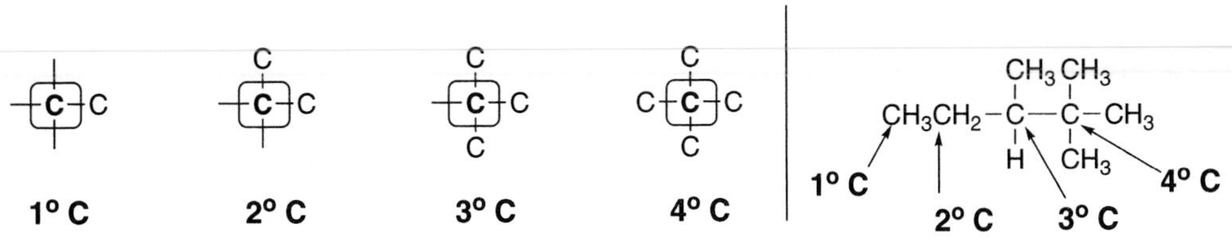

- Hydrogen atoms are classified by the type of carbon atom to which they are bonded; **a $1^\circ$ H is bonded to a $1^\circ$ C,** and so forth.

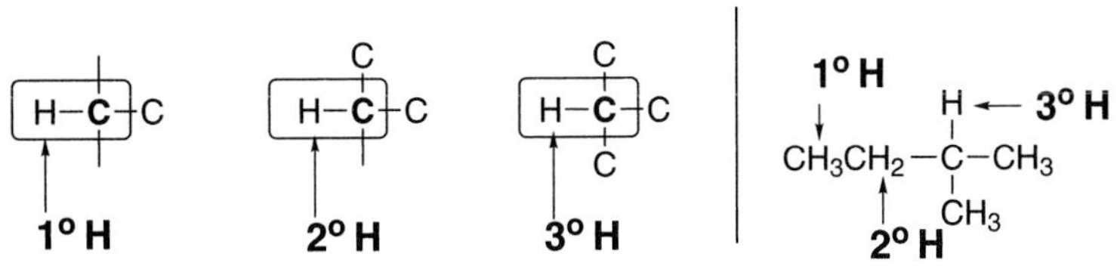

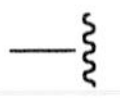

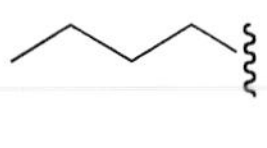

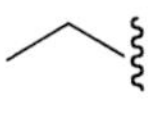

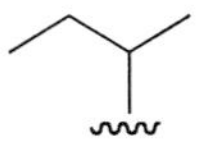

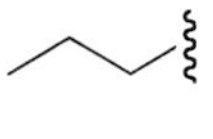

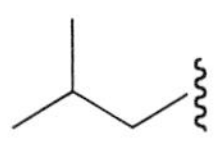

### ◆ Names of Alkyl Groups (4.4A)

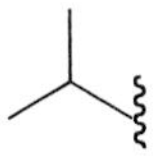

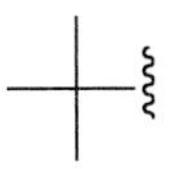

♦ Conformations in acyclic alkanes (4.9, 4.10)

- Alkane conformations can be classified as **staggered**, **eclipsed**, **anti,** or **gauche** depending on the relative orientation of the groups on adjacent carbons.

| eclipsed | staggered | anti | gauche |
|---|---|---|---|
| 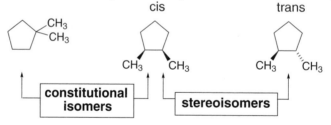 | | | |
| • Dihedral angle = 0° | • Dihedral angle = 60° | • Dihedral angle of 2 CH$_3$'s = 180° | • Dihedral angle of 2 CH$_3$'s = 60° |

- A staggered conformation is **lower in energy** than an eclipsed conformation.
- An anti conformation is **lower in energy** than a gauche conformation.

♦ Types of strain

- **Torsional strain**—an increase in energy due to eclipsing interactions (4.9).
- **Steric strain**—an increase in energy when atoms are forced too close to each other (4.10).
- **Angle strain**—an increase in energy when bond angles deviate from 109.5° (4.11).

♦ Two types of isomers

[1] **Constitutional isomers** – isomers that differ in the way the atoms are connected to each other (4.1A).

[2] **Stereoisomers** – isomers that differ only in the way atoms are oriented in space (4.13B).

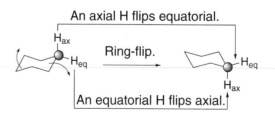

♦ Conformations in cyclohexane (4.12, 4.13)

- Cyclohexane exists as **two chair conformations** in rapid equilibrium at room temperature.
- Each carbon atom on a cyclohexane ring has **one axial** and **one equatorial hydrogen**. Ring-flipping converts axial to equatorial H's, and vice versa.

An axial H flips equatorial.

Ring-flip.

An equatorial H flips axial.

- In substituted cyclohexanes, groups larger than hydrogen are more stable in the **more roomy equatorial position.**

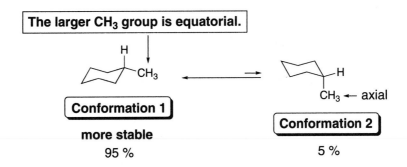

- Disubstituted cyclohexanes with substituents on different atoms, exist as two possible stereoisomers.
  - The **cis** isomer has two groups on the **same side** of the ring, either both up or both down.
  - The **trans** isomer has two groups on **opposite sides** of the ring, one up and one down.

**trans** isomer          **cis** isomer

♦ Oxidation–reduction reactions (4.14)

- **Oxidation** results in an **increase in the number of C–Z bonds** or a **decrease in the number of C–H bonds.**

$CH_3CH_2-OH$  ⟶  $CH_3\overset{\displaystyle O}{\underset{}{C}}{-}OH$

ethanol          acetic acid

Increase in C–O bonds = **oxidation**

- **Reduction** results in a **decrease in the number of C–Z bonds** or an **increase in the number of C–H bonds.**

ethylene          ethane

Increase in C–H bonds = **reduction**

## Chapter 4: Answers to Problems

**4.1** The general molecular formula for an acyclic alkane is $C_nH_{2n+2}$.

| a. $C_{12}H_{26}$ | b. $C_8H_{16}$ | c. $C_{30}H_{64}$ |
|---|---|---|
| $2n + 2 = $ # H's | $2n + 2 = $ # H's | $2n + 2 = $ # H's |
| $2(12) + 2 = 26$ | $2(8) + 2 = 18$ | $2(30) + 2 = 62$ |
| **yes** | **no** | **no** |

**4.2** Isopentane has 4 C's in a row with a 1 C branch.

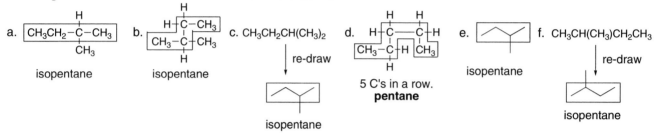

**4.3** To classify a carbon atom as 1°, 2°, 3°, or 4° **determine how many carbon atoms it is bonded to** (1° C = bonded to **one** other C, 2° C = bonded to **two** other C's, 3° C = bonded to **three** other C's, 4° C = bonded to **four** other C's). Re-draw if necessary to see each carbon clearly.

1° C          1° C

a. $CH_3CH_2CH_2CH_3$          b. $(CH_3)_3CH$          c.          d.

2° C's

1° C's ▸ $CH_3$–$\overset{\underset{CH_3}{|}}{\underset{3°\ C}{C}}$–H

4° C's
All other C's are 1° C's.

1° C's
4° C
1° C
All other C's are 2° C's.
3°C

**4.4** To classify a hydrogen atom as 1°, 2°, or 3°, **determine if it is bonded to a 1°, 2°, or 3° C (A 1° H is bonded to a 1° C; a 2° H is bonded to a 2° C; a 3° H is bonded to a 3° C).** Re-draw if necessary.

1° H          1° H

a. $CH_3CH_2CH_3$          b. $CH_3CH_2CH(CH_3)C(CH_3)_3$          c.          re-draw

2° H's

re-draw

1° H's          3° H

$\overset{H}{|}$   $\overset{CH_3}{}$

$CH_3$–$CH_2$–$\overset{\underset{CH_3}{|}}{C}$—$\overset{\underset{CH_3}{|}}{C}$–$CH_3$

2° H's

1° H's

2° H's

3° H
$CH_3$—$CH$
$CH_2$

$CH_2$

$CH_3$   $CH_3$
$\overset{|}{C}$
$CH_2$   $CH$   $CH_3$

$CH$   3° H's
$CH_3$

All other H's are 1° H's.

**4.5 Constitutional isomers differ in the way the atoms are connected to each other.** To draw all the constitutional isomers:

[1] Draw all of the C's in a long chain.

[2] Take off one C and use it as a substituent. (Don't add it to the end carbon: this re-makes the long chain.)

[3] Take off two C's and use these as substituents, etc.

Five **constitutional isomers** of molecular formula $C_6H_{14}$:

[1] long chain

$CH_3CH_2CH_2CH_2CH_2CH_3$

[2] with one C as a substituent

$CH_3CH_2CH_2\overset{\overset{\displaystyle CH_3}{|}}{\underset{\underset{\displaystyle H}{|}}{C}}CH_3$    $CH_3CH_2\overset{\overset{\displaystyle CH_3}{|}}{\underset{\underset{\displaystyle H}{|}}{C}}CH_2CH_3$

[3] using 2 C's as substituents

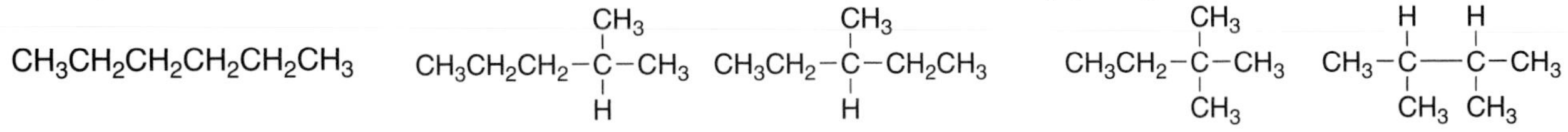

$CH_3CH_2\overset{\overset{\displaystyle CH_3}{|}}{\underset{\underset{\displaystyle CH_3}{|}}{C}}CH_3$    $CH_3\overset{\overset{\displaystyle H}{|}}{\underset{\underset{\displaystyle CH_3}{|}}{C}}\overset{\overset{\displaystyle H}{|}}{\underset{\underset{\displaystyle CH_3}{|}}{C}}CH_3$

**4.6**

Molecular formula $C_8H_{18}$ with one $CH_3$ substituent:

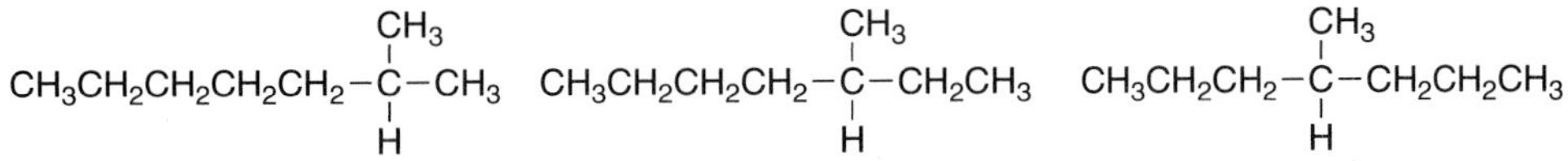

$CH_3CH_2CH_2CH_2CH_2\overset{\overset{\displaystyle CH_3}{|}}{\underset{\underset{\displaystyle H}{|}}{C}}CH_3$    $CH_3CH_2CH_2CH_2\overset{\overset{\displaystyle CH_3}{|}}{\underset{\underset{\displaystyle H}{|}}{C}}CH_2CH_3$    $CH_3CH_2CH_2\overset{\overset{\displaystyle CH_3}{|}}{\underset{\underset{\displaystyle H}{|}}{C}}CH_2CH_2CH_3$

**4.7** Draw each alkane to satisfy the requirements.

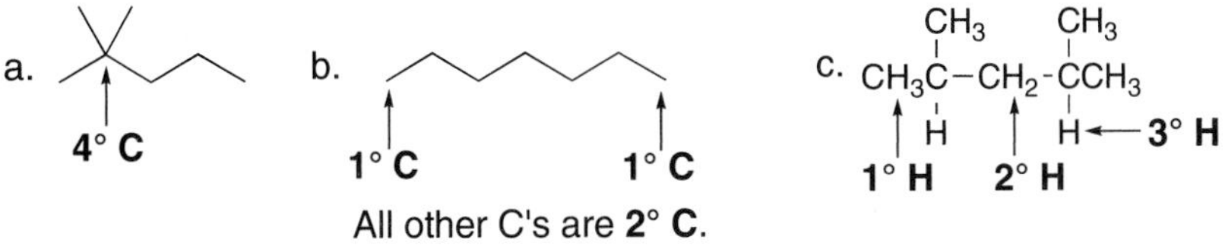

a. 

**4° C**

b. 

**1° C**    **1° C**

All other C's are **2° C**.

c. $CH_3\overset{\overset{\displaystyle CH_3}{|}}{C}-CH_2-\overset{\overset{\displaystyle CH_3}{|}}{C}CH_3$

**1° H    2° H**    **H ← 3° H**

**4.8** Compare the structures. Recall the definition of constitutional isomers from Answer 4.5.

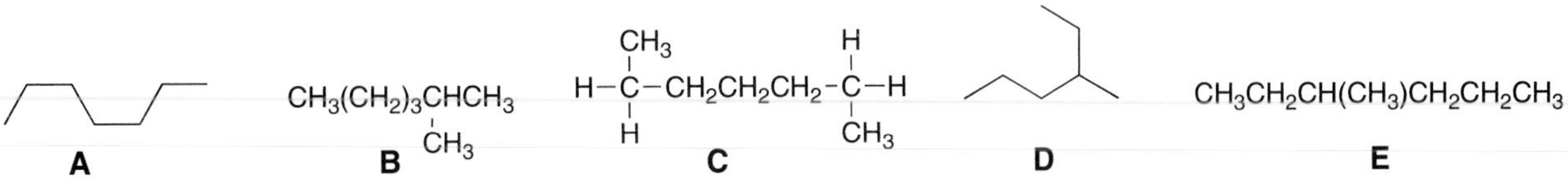

**A**

$CH_3(CH_2)_3CHCH_3$
      $\underset{\displaystyle CH_3}{|}$
**B**

$H-\overset{\overset{\displaystyle CH_3}{|}}{\underset{\underset{\displaystyle H}{|}}{C}}-CH_2CH_2CH_2-\overset{\overset{\displaystyle H}{|}}{\underset{\underset{\displaystyle CH_3}{|}}{C}}-H$
**C**

**D**

$CH_3CH_2CH(CH_3)CH_2CH_2CH_3$
**E**

a. **A** and **B** are constitutional isomers.
b. **A** and **C** are identical molecules.
c. **B** and **D** are constitutional isomers.

d. **D** and **E** are identical molecules.
e. **B** and **E** are constitutional isomers.

**4.9** Use the steps from Answer 4.5 to draw the constitutional isomers.

Five **constitutional isomers** of molecular formula $C_5H_{10}$ having one ring:

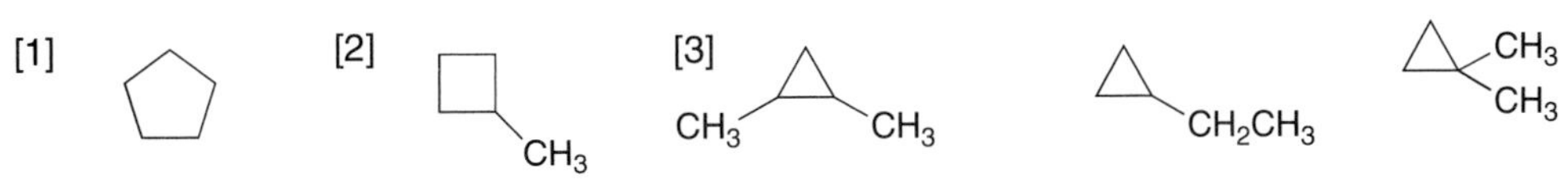

[1]    [2]    [3]

$CH_3$    $CH_3$    $CH_3$    $CH_2CH_3$    $\overset{CH_3}{CH_3}$

**4.10** Cycloalkanes have molecular formula $C_nH_{2n}$. For a cycloalkane with 288 C's, there would be $2(288) = 576$ H's. **Molecular formula = $C_{288}H_{576}$.**

**4.11** Follow these steps to name an alkane:
  [1] **Name the parent chain** by finding the longest C chain.
  [2] **Number the chain** so that the first substituent gets the lower number.  Then **name and number all substituents**, giving like substituents a prefix (di, tri, etc.).
  [3] **Combine all parts**, alphabetizing the substituents, ignoring all prefixes except *iso*.

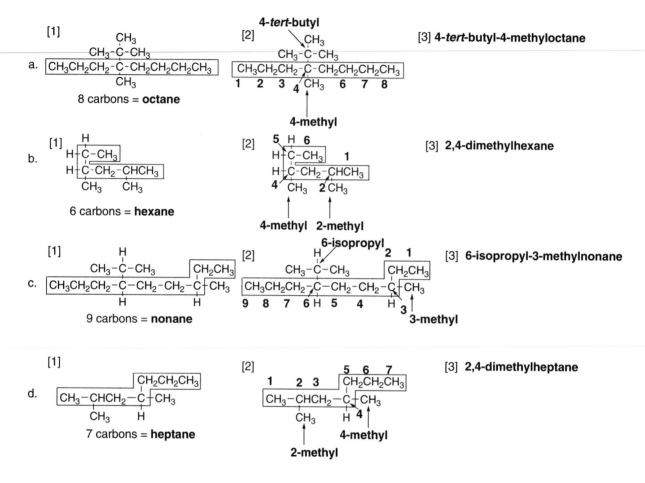

**4.12** Use the steps in Answer 4.11 to name each alkane.

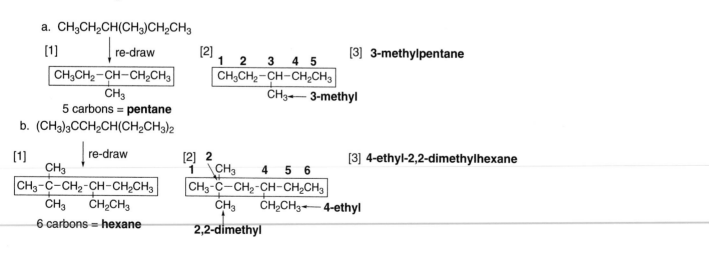

c. CH$_3$(CH$_2$)$_3$CH(CH$_2$CH$_2$CH$_3$)CH(CH$_3$)$_2$

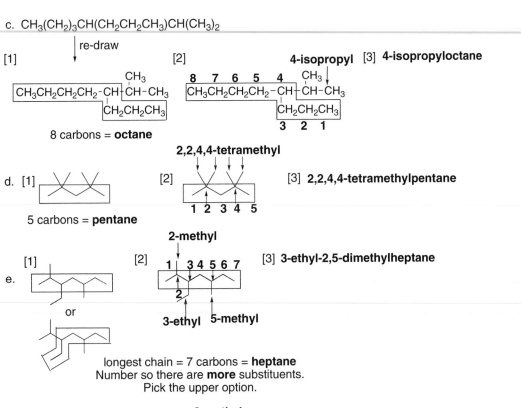

[1]

re-draw

CH$_3$
CH$_3$CH$_2$CH$_2$CH$_2$-CH$\vert$CH-CH$_3$
CH$_2$CH$_2$CH$_3$

8 carbons = **octane**

[2]

8  7  6  5  4   CH$_3$
CH$_3$CH$_2$CH$_2$CH$_2$-CH$\vert$CH-CH$_3$
CH$_2$CH$_2$CH$_3$
3  2  1

**4-isopropyl**   [3] **4-isopropyloctane**

**2,2,4,4-tetramethyl**

d. [1]

5 carbons = **pentane**

[2]

1 2 3 4 5

[3] **2,2,4,4-tetramethylpentane**

e. [1]

or

**2-methyl**

[2]

1 $\vert$ 3 4 5 6 7

2

**3-ethyl** **5-methyl**

[3] **3-ethyl-2,5-dimethylheptane**

longest chain = 7 carbons = **heptane**
Number so there are **more** substituents.
Pick the upper option.

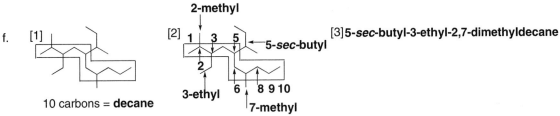

f. [1]

10 carbons = **decane**

**2-methyl**

[2]

1  3  5
2

**5-sec-butyl**

**3-ethyl**  6  8 9 10
**7-methyl**

[3] **5-*sec*-butyl-3-ethyl-2,7-dimethyldecane**

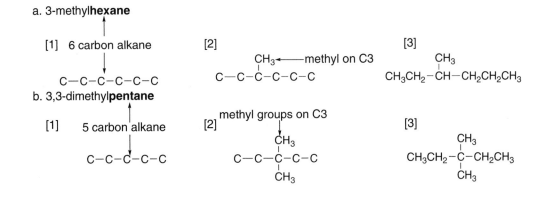

**4.13  To work backwards from a name to a structure:**

[1] Find the parent name and draw that number of C's.  Use the suffix to identify the functional group.  (**-ane = alkane**)

[2] Arbitrarily number the C's in the chain.  Add the substituents to the appropriate C's.

[3] Re-draw with H's to make C's have four bonds.

a. 3-methyl**hexane**

[1]  6 carbon alkane

C—C—C—C—C—C

[2]

CH$_3$←——methyl on C3
C—C—C—C—C—C

[3]

CH$_3$
CH$_3$CH$_2$—CH—CH$_2$CH$_2$CH$_3$

b. 3,3-dimethyl**pentane**

[1]  5 carbon alkane

C—C—C—C—C

[2]

methyl groups on C3
CH$_3$
C—C—C—C—C
CH$_3$

[3]

CH$_3$
CH$_3$CH$_2$—C—CH$_2$CH$_3$
CH$_3$

c. 3,5,5-trimethyl**octane**

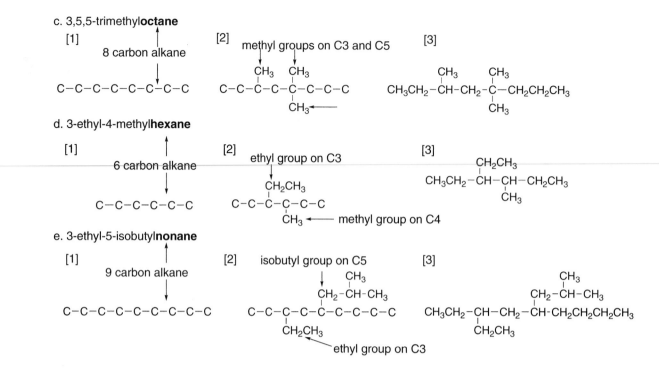

d. 3-ethyl-4-methyl**hexane**

e. 3-ethyl-5-isobutyl**nonane**

**4.14** Use the steps in Answer 4.11 to name each alkane.

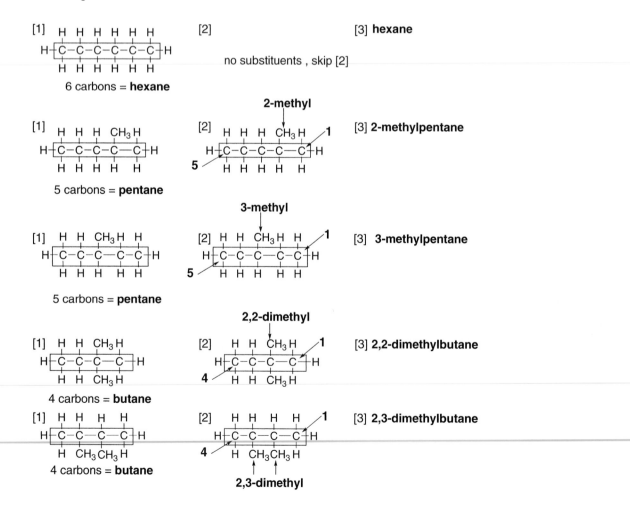

**4.15** Follow these steps to name a cycloalkane:

[1] **Name the parent cycloalkane** by counting the C's in the ring and adding cyclo-.

[2] **Numbering:**

[2a] **Number around the ring** beginning at a substituent and giving the second substituent the lower number.

[2b] **Number to assign the lower number to the substituents alphabetically**.

[2c] **Name and number all substituents**, giving like substituents a prefix (di, tri, etc.).

[3] **Combine all parts**, alphabetizing the substituents, ignoring all prefixes except *iso*.
(Remember: If a carbon chain has more C's than the ring, the chain is the parent, and the ring is a substituent.)

a.
[1] 6 carbons in ring = **cyclohexane**
[2] Number so the substituents are at C1. **1,1-dimethyl**
[3] **1,1-dimethylcyclohexane**

b.
[1] 5 carbons in ring = **cyclopentane**
[2] Number so the first substituent is at C1, second at C2. **1,2,3-trimethyl**
[3] **1,2,3-trimethylcyclopentane**

c.
[1] 6 carbons in ring = **cyclohexane**
[2] Number so the earliest alphabetical substituent is at C1, **b**utyl before **m**ethyl. **4-methyl 1-butyl**
[3] **1-butyl-4-methylcyclohexane**

d.
[1] 6 carbons in ring = **cyclohexane**
[2] Number so the earliest alphabetical substituent is at C1, **b**utyl before **i**sopropyl. **1-sec-butyl 2-isopropyl**
[3] **1-sec-butyl-2-isopropylcyclohexane**

e.
[1] longest chain = 5 carbons = **pentane**
[2] Number so the cyclopropyl is at C1. **1-cyclopropyl**
[3] **1-cyclopropylpentane**

**4.16** To draw the structures, use the steps in Answer 4.13.

a. 1,2-dimethyl**cyclobutane**

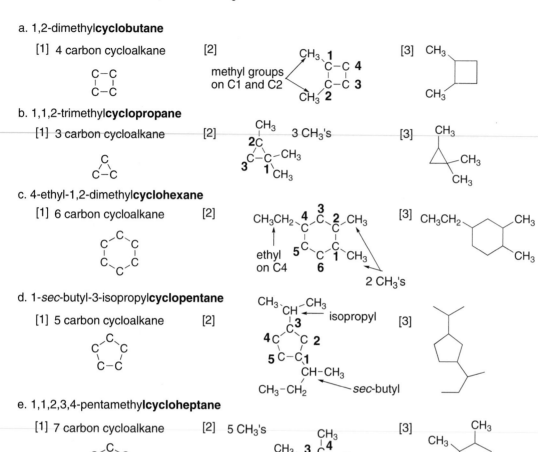

b. 1,1,2-trimethyl**cyclopropane**

c. 4-ethyl-1,2-dimethyl**cyclohexane**

d. 1-*sec*-butyl-3-isopropyl**cyclopentane**

e. 1,1,2,3,4-pentamethyl**cycloheptane**

**4.17** To name the cycloalkanes, use the steps from Answer 4.15.

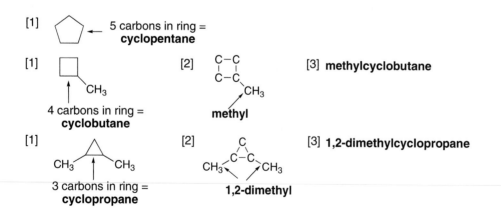

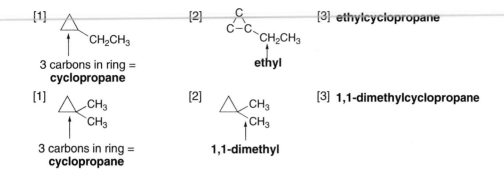

**4.18** Compare the molecular weights to determine relative boiling points.

**gasoline:** $C_5H_{12} - C_{12}H_{26}$     **kerosene:** $C_{12}H_{26} - C_{16}H_{34}$     **diesel fuel:** $C_{15}H_{32} - C_{18}H_{38}$

lowest molecular weight:          middle molecular weight:          highest molecular weight:
**lowest boiling point**          **intermediate boiling point**          **highest boiling point**

**4.19** **Compare the number of C's and surface area to determine relative boiling points.** Rules:
  [1] Increasing number of C's = increasing boiling point.
  [2] Increasing surface area = increasing boiling point (branching decreases surface area).

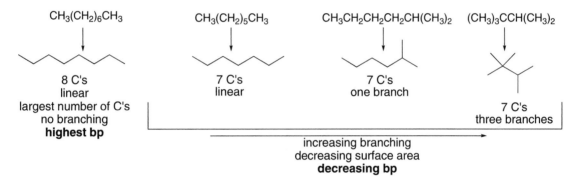

$CH_3(CH_2)_6CH_3$          $CH_3(CH_2)_5CH_3$          $CH_3CH_2CH_2CH_2CH(CH_3)_2$     $(CH_3)_3CCH(CH_3)_2$

8 C's          7 C's          7 C's          7 C's
linear          linear          one branch          three branches
largest number of C's
no branching
**highest bp**

increasing branching
decreasing surface area
**decreasing bp**

**Increasing boiling point:**  $(CH_3)_3CCH(CH_3)_2 < CH_3CH_2CH_2CH_2CH(CH_3)_2 < CH_3(CH_2)_5CH_3 < CH_3(CH_2)_6CH_3$

**4.20** To draw a Newman projection, visualize the carbons as one in front and one in back of each other. The C–C bond is not drawn. There is only one staggered and one eclipsed conformation.

**4.21** Staggered conformations are more stable than eclipsed conformations.

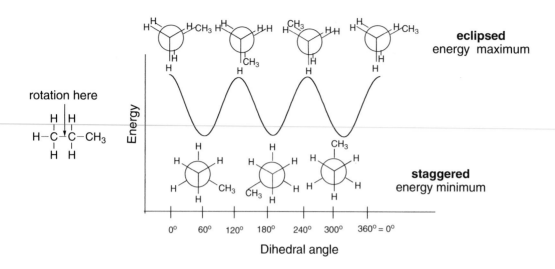

**4.22**

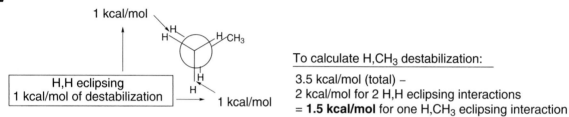

To calculate H,CH₃ destabilization:

3.5 kcal/mol (total) –
2 kcal/mol for 2 H,H eclipsing interactions
= **1.5 kcal/mol** for one H,CH₃ eclipsing interaction

**4.23** To determine the energy of conformations keep two things in mind:
   [1] Staggered conformations are more stable than eclipsed conformations.
   [2] Minimize steric interactions: keep large groups away from each other.
   **The highest energy conformation is the eclipsed conformation in which the two largest groups are eclipsed. The lowest energy conformation is the staggered conformation in which the two largest groups are anti.**

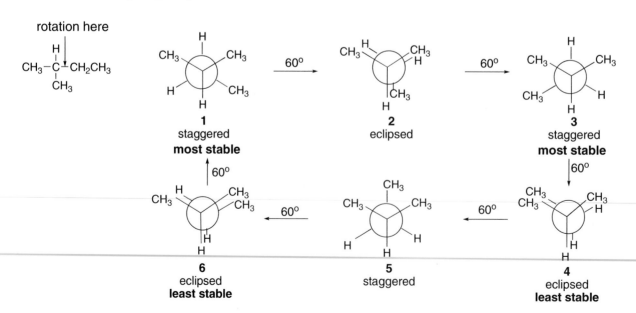

**4.24** To determine the most and least stable conformations, use the rules from Answer 4.23.

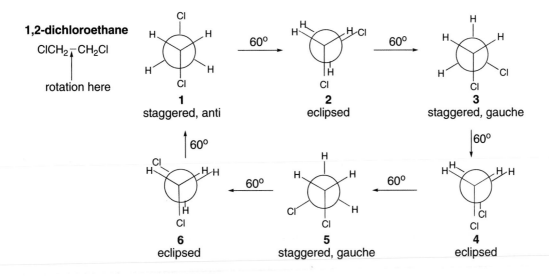

**1,2-dichloroethane**

$ClCH_2 \text{—} CH_2Cl$

rotation here

| | | |
|---|---|---|
| **1** | **2** | **3** |
| staggered, anti | eclipsed | staggered, gauche |

| | | |
|---|---|---|
| **6** | **5** | **4** |
| eclipsed | staggered, gauche | eclipsed |

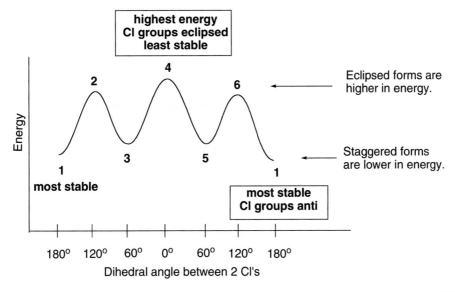

highest energy
Cl groups eclipsed
least stable

**4**

**2**

**6**

Eclipsed forms are
higher in energy.

Energy

**1**
most stable

**3**

**5**

**1**

Staggered forms
are lower in energy.

most stable
Cl groups anti

180°  120°  60°  0°  60°  120°  180°

Dihedral angle between 2 Cl's

**4.25** Add the energy increase for each eclipsing interaction to determine the destabilization.

a.

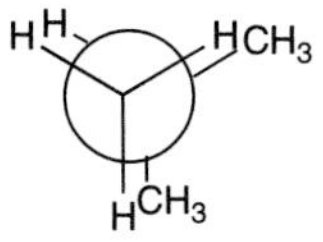

b.

1 H,H interaction =     1 kcal/mol
2 H,CH$_3$ interactions
  (2 x 1.5 kcal/mol) =     3 kcal/mol
──────────────────────────────
**Total destabilization =     4 kcal/mol**

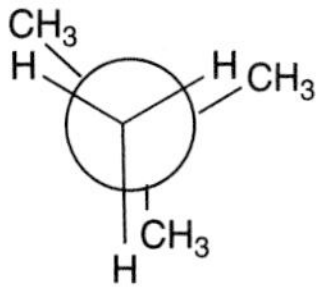

3 H,CH$_3$ interactions
  (3 x 1.5 kcal/mol) =     **4.5 kcal/mol**

**Total destabilization**

**4.26** Two points:
- Axial bonds point up or down, while equatorial bonds point out.
- An *up* carbon has an axial *up* bond, and a *down* carbon has an axial *down* bond.

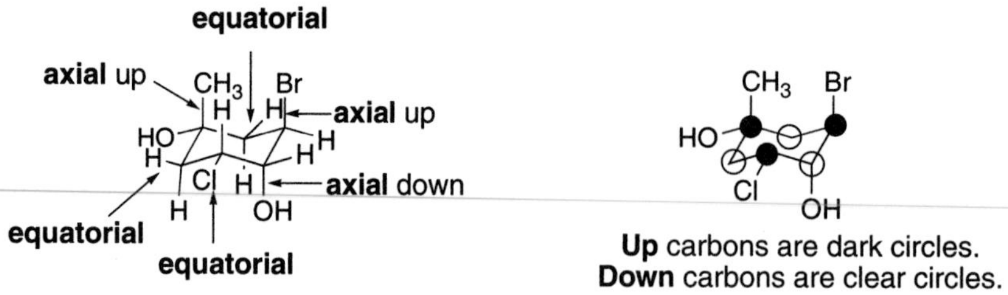

**Up** carbons are dark circles.
**Down** carbons are clear circles.

**4.27** Draw the second chair conformation by flipping the ring.

- **The *up* carbons become *down* carbons, and the axial bonds become equatorial bonds.**
- **Axial bonds become equatorial, but *up* bonds stay *up*;** i.e., an axial *up* bond becomes an equatorial *up* bond.
- **The conformation with larger groups equatorial is the more stable** conformation and is present in higher concentration at equilibrium.

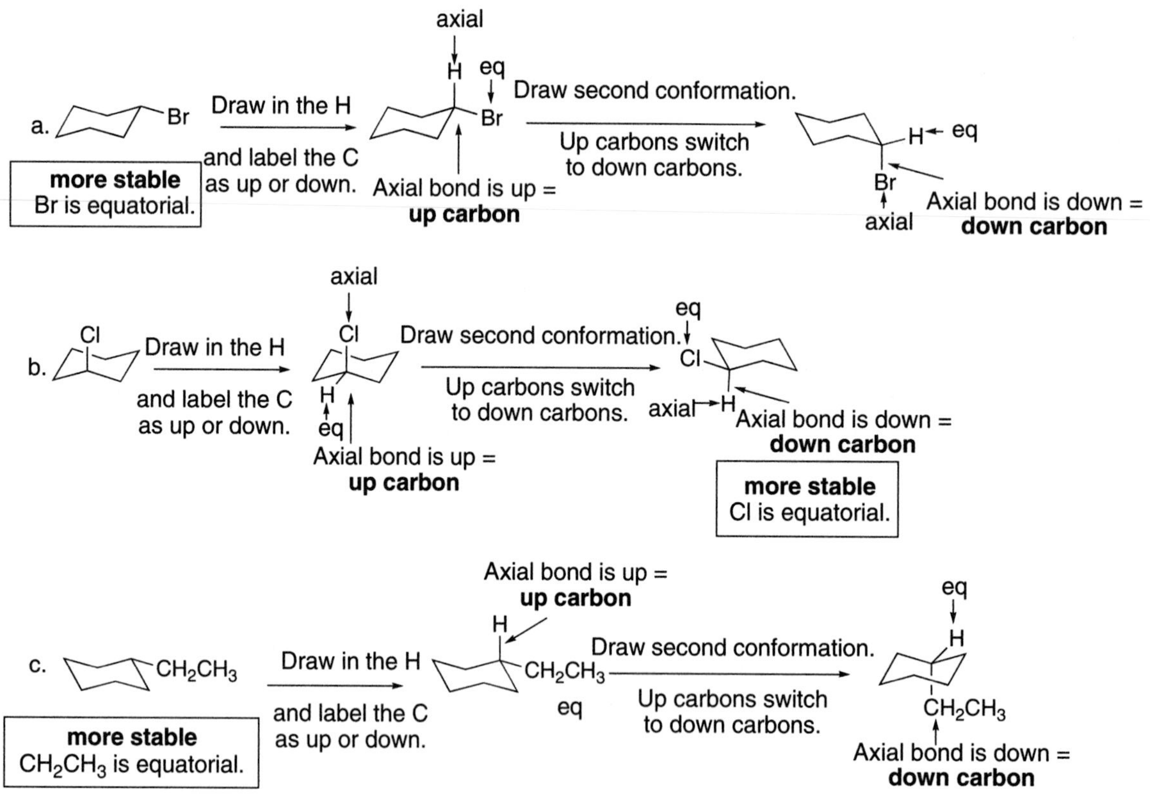

**4.28** Larger axial substituents create unfavorable diaxial interactions, whereas equatorial groups have more room and are favored.

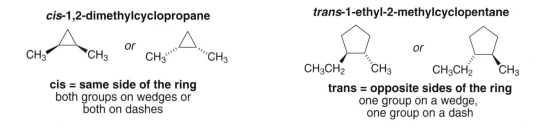

**larger substituent**
**more important to be equatorial**

equatorial $CH_2CH_3$

The H's and $CH_3$ of the $sp^3$ hybridized C have severe 1,3-diaxial interactions with the two other axial H's.

—C≡C—H

**more compact substituent**
**less important to be equatorial**

equatorial C≡CH

The $sp$ hybridized C's are linear and point down. The 1,3-diaxial interactions with the two other axial H's are less severe.

The axial conformation containing the C≡CH group is not as unstable as the axial conformation containing the $CH_2CH_3$, so it is present in higher concentration at equilibrium.

**4.29** Wedges represent "up" groups in front of the page, and dashes are "down" groups in back of the page. Cis groups are on the same side of the ring, and trans groups are on opposite sides of the ring.

**cis-1,2-dimethylcyclopropane**

$CH_3$      $CH_3$    or    $CH_3$      $CH_3$

**cis = same side of the ring**
both groups on wedges or
both on dashes

**trans-1-ethyl-2-methylcyclopentane**

$CH_3CH_2$      $CH_3$    or    $CH_3CH_2$      $CH_3$

**trans = opposite sides of the ring**
one group on a wedge,
one group on a dash

**4.30** To classify a compound as a cis or trans isomer, **classify each non-hydrogen group as up or down. Groups on the same side = cis isomer, groups on opposite sides = trans isomer.**

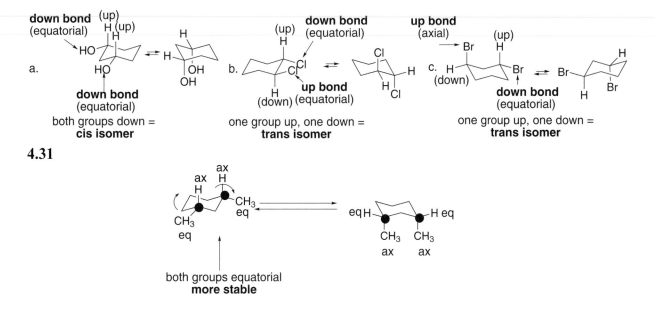

**down bond** (up)
(equatorial) H (up)
HO
a.
HO

**down bond**
(equatorial)
both groups down =
**cis isomer**

**down bond** (up)
(equatorial) H
b.
Cl
H
(down) (equatorial)

**up bond**
(equatorial)

one group up, one down =
**trans isomer**

**up bond**
(axial)
Br H (up)
c. H
(down)
Br

**down bond**
(equatorial)
one group up, one down =
**trans isomer**

**4.31**

both groups equatorial
**more stable**

**4.32**

a.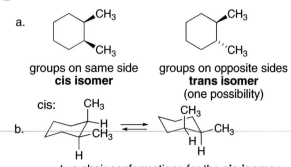

groups on same side
**cis isomer**

groups on opposite sides
**trans isomer**
(one possibility)

trans: CH3

c.

both groups equatorial
**more stable**
two chair conformations for the **trans isomer**

cis: CH3

b.

two chair conformations for the **cis isomer**

**Same stability** since they are identical groups with one equatorial, one axial.

d. The **trans isomer is more stable** because it can have both methyl groups in the more roomy **equatorial** position.

**4.33** *Oxidation* results in an *increase* in the number of C–Z bonds, or a *decrease* in the number of C–H bonds.

*Reduction* results in a *decrease* in the number of C–Z bonds, or an *increase* in the number of C–H bonds.

a.

Decrease in the number of C–H bonds.
Increase in the number of C–O bonds.
**Oxidation**

c.

No change in the number of C–O or C–H bonds. **Neither**

b.

$CH_3CH_2CH_3$

Decrease in the number of C–O bonds.
Increase in the number of C–H bonds.
**Reduction**

d.

Decrease in the number of C–O bonds.
Increase in the number of C–H bonds.
**Reduction**

**4.34** The products of a combustion reaction of a hydrocarbon are always the same: **$CO_2$ and $H_2O$.**

a. $CH_3CH_2CH_3$ + 5 $O_2$  $\xrightarrow{\text{flame}}$  3 $CO_2$ + 4 $H_2O$ + heat

b. ⬡ + 9 $O_2$  $\xrightarrow{\text{flame}}$  6 $CO_2$ + 6 $H_2O$ + heat

**4.35** Lipids contain many nonpolar C–C and C–H bonds and few polar functional groups.

a. $CH_3(CH_2)_7CH=CH(CH_2)_7COOH$

oleic acid

only one polar functional group
18 carbons
**a lipid**

b.

aspartame

many polar functional groups
only 14 carbons
**not a lipid**

c.

$CH_2-O-\overset{O}{\overset{\|}{C}}-(CH_2)_{16}CH_3$
$CH-O-\overset{O}{\overset{\|}{C}}-(CH_2)_{16}CH_3$
$CH_2-O-\overset{O}{\overset{\|}{C}}-(CH_2)_{16}CH_3$

tristearin

three polar functional groups
57 carbons
**a lipid**

**4.36** "Like dissolves like." Beeswax is a lipid, and therefore, it will be more soluble in nonpolar solvents. $H_2O$ is very polar, ethanol is slightly less polar, and chloroform is least polar. Beeswax is most soluble in the least polar solvent.

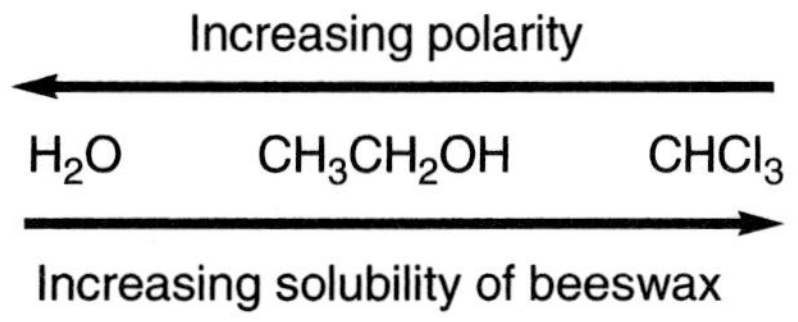

Increasing polarity

$H_2O$    $CH_3CH_2OH$    $CHCl_3$

Increasing solubility of beeswax

**4.37** Use the rules from Answers 4.3 and 4.4.

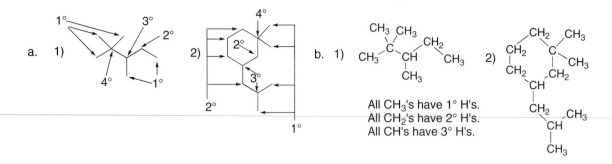

a. 1)    2)

b. 1)

All CH₃'s have 1° H's.
All CH₂'s have 2° H's.
All CH's have 3° H's.

2)

**4.38**

One possibility:

a. 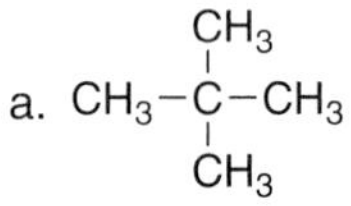 $CH_3-\underset{\underset{CH_3}{|}}{\overset{\overset{CH_3}{|}}{C}}-CH_3$    b. 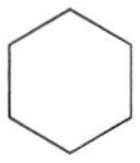    c. $CH_3CH_2CH_3$    d. $(CH_3)_3CH$

**4.39**

a. Five constitutional isomers of molecular formula $C_4H_8$:

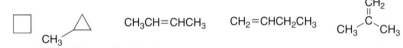

$CH_3CH=CHCH_3$    $CH_2=CHCH_2CH_3$    $CH_3-\overset{\overset{CH_2}{\|}}{C}-CH_3$

b. Nine constitutional isomers of molecular formula $C_7H_{16}$:

$CH_3CH_2CH_2CH_2CH_2CH_2CH_3$    $CH_3-\underset{\underset{CH_3}{|}}{\overset{\overset{H}{|}}{C}}-CH_2CH_2CH_2CH_3$    $CH_3CH_2-\underset{\underset{CH_3}{|}}{\overset{\overset{H}{|}}{C}}-CH_2CH_2CH_3$

$CH_3-\underset{\underset{CH_3}{|}}{\overset{\overset{CH_3}{|}}{C}}-CH_2CH_2CH_3$    $CH_3CH_2-\underset{\underset{CH_3}{|}}{\overset{\overset{CH_3}{|}}{C}}-CH_2CH_3$    $CH_3-\underset{\underset{CH_3}{|}}{\overset{\overset{H}{|}}{C}}-\underset{\underset{CH_3}{|}}{\overset{\overset{H}{|}}{C}}-CH_2CH_3$    $CH_3-\underset{\underset{CH_3}{|}}{\overset{\overset{H}{|}}{C}}-CH_2-\underset{\underset{CH_3}{|}}{\overset{\overset{H}{|}}{C}}-CH_3$

$CH_3CH_2-\underset{\underset{CH_2CH_3}{|}}{\overset{\overset{H}{|}}{C}}-CH_2CH_3$    $CH_3-\underset{\underset{CH_3}{|}}{\overset{\overset{H}{|}}{C}}-\underset{\underset{CH_3}{|}}{\overset{\overset{CH_3}{|}}{C}}-CH_3$

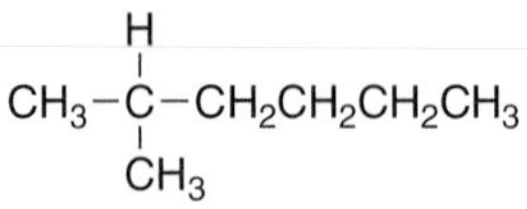

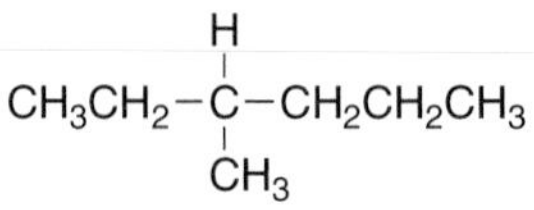

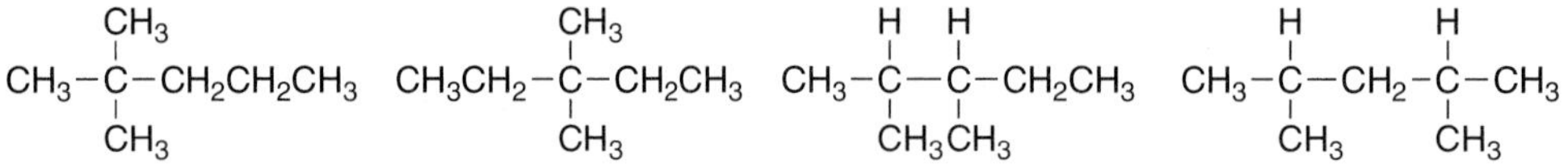

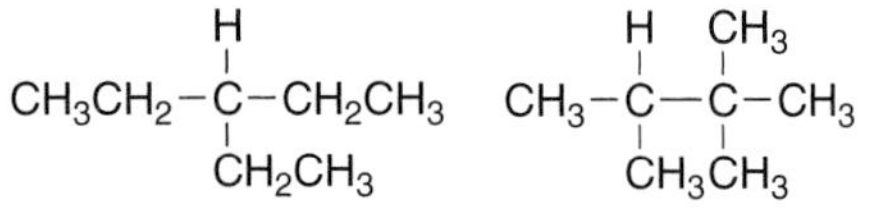

c. Twelve constitutional isomers of molecular formula $C_6H_{12}$ containing one ring:

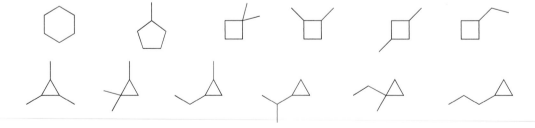

**4.40** Use the steps in Answers 4.11 and 4.15 to name the alkanes.

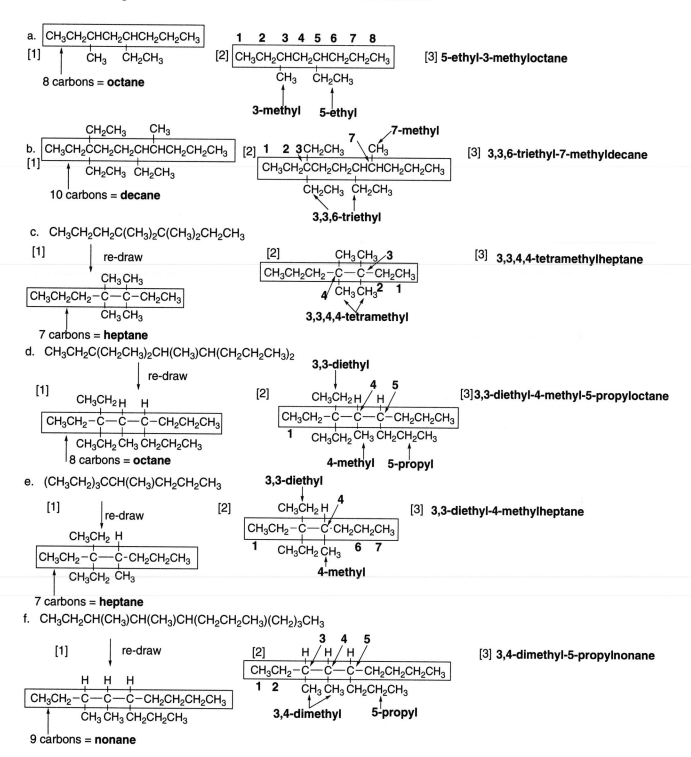

a. $CH_3CH_2CHCH_2CHCH_2CH_2CH_3$

[1]   $CH_3$   $CH_2CH_3$

8 carbons = **octane**

**1 2 3 4 5 6 7 8**

[2] $CH_3CH_2CHCH_2CHCH_2CH_2CH_3$
     $CH_3$    $CH_2CH_3$

**3-methyl**   **5-ethyl**

[3] **5-ethyl-3-methyloctane**

b.   $CH_2CH_3$    $CH_3$

[1] $CH_3CH_2CCH_2CH_2CHCHCH_2CH_2CH_3$
     $CH_2CH_3$   $CH_2CH_3$

10 carbons = **decane**

[2] **1 2 3** $CH_2CH_3$    **7**   **7-methyl**
     $CH_3$

$CH_3CH_2CCH_2CH_2CHCHCH_2CH_2CH_3$
     $CH_2CH_3$   $CH_2CH_3$

**3,3,6-triethyl**

[3] **3,3,6-triethyl-7-methyldecane**

c.   $CH_3CH_2CH_2C(CH_3)_2C(CH_3)_2CH_2CH_3$

[1]    re-draw

     $CH_3$ $CH_3$

$CH_3CH_2CH_2-C-C-CH_2CH_3$
     $CH_3$ $CH_3$

7 carbons = **heptane**

[2]    $CH_3$ $CH_3$ **3**

$CH_3CH_2CH_2-C-C-CH_2CH_3$
  **4** $CH_3$ $CH_3$ **2**   **1**

**3,3,4,4-tetramethyl**

[3] **3,3,4,4-tetramethylheptane**

d.   $CH_3CH_2C(CH_2CH_3)_2CH(CH_3)CH(CH_2CH_2CH_3)_2$

[1]    re-draw

     $CH_3CH_2$ H   H

$CH_3CH_2-C-C-C-CH_2CH_2CH_3$
     $CH_3CH_2$ $CH_3$ $CH_2CH_2CH_3$

8 carbons = **octane**

[2] **3,3-diethyl**

     $CH_3CH_2$ H **4** H **5**

$CH_3CH_2-C-C-C-CH_2CH_2CH_3$
  **1** $CH_3CH_2$ $CH_3$ $CH_2CH_2CH_3$

**4-methyl**   **5-propyl**

[3] **3,3-diethyl-4-methyl-5-propyloctane**

e.   $(CH_3CH_2)_3CCH(CH_3)CH_2CH_2CH_3$

[1]    re-draw

     $CH_3CH_2$ H

$CH_3CH_2-C-C-CH_2CH_2CH_3$
     $CH_3CH_2$ $CH_3$

7 carbons = **heptane**

[2] **3,3-diethyl**

     $CH_3CH_2$ H **4**

$CH_3CH_2-C-C-CH_2CH_2CH_3$
  **1** $CH_3CH_2$ $CH_3$ **6 7**

**4-methyl**

[3] **3,3-diethyl-4-methylheptane**

f.   $CH_3CH_2CH(CH_3)CH(CH_3)CH(CH_2CH_2CH_3)(CH_2)_3CH_3$

[1]    re-draw

     H   H   H

$CH_3CH_2-C-C-C-CH_2CH_2CH_2CH_3$
     $CH_3$ $CH_3$ $CH_2CH_2CH_3$

9 carbons = **nonane**

[2]    **3 4 5**
   H $/$ H $/$ H

$CH_3CH_2-C-C-C-CH_2CH_2CH_2CH_3$
  **1 2** $CH_3$ $CH_3$ $CH_2CH_2CH_3$

**3,4-dimethyl**   **5-propyl**

[3] **3,4-dimethyl-5-propylnonane**

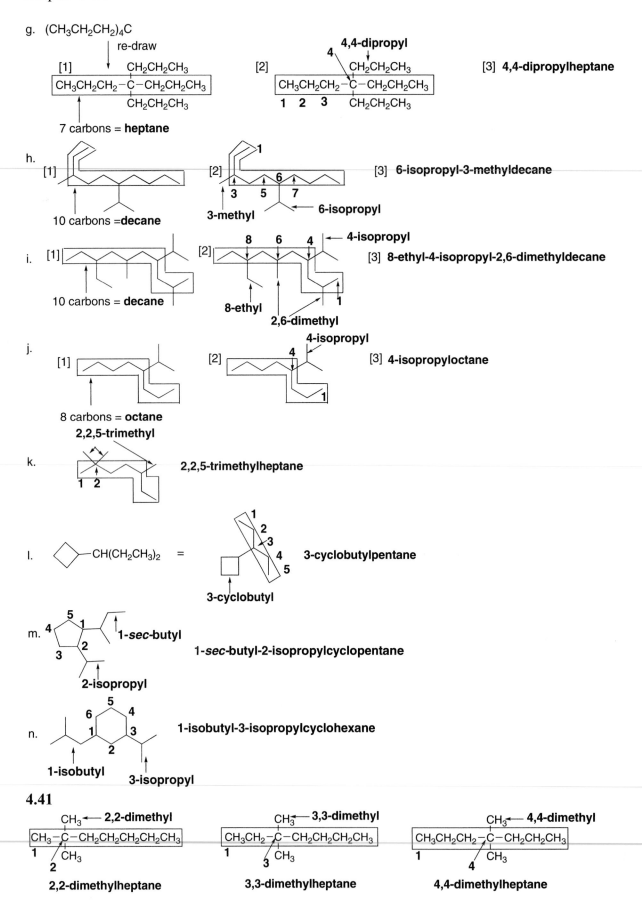

g. (CH₃CH₂CH₂)₄C

[1] 7 carbons = heptane

[2] 4,4-dipropyl

[3] **4,4-dipropylheptane**

h. [1] 10 carbons = decane

[2] 3-methyl, 6-isopropyl

[3] **6-isopropyl-3-methyldecane**

i. [1] 10 carbons = decane

[2] 8-ethyl, 2,6-dimethyl, 4-isopropyl

[3] **8-ethyl-4-isopropyl-2,6-dimethyldecane**

j. [1] 8 carbons = octane

[2] 4-isopropyl

[3] **4-isopropyloctane**

k. 2,2,5-trimethyl

**2,2,5-trimethylheptane**

l. 3-cyclobutyl

**3-cyclobutylpentane**

m. 1-*sec*-butyl, 2-isopropyl

**1-*sec*-butyl-2-isopropylcyclopentane**

n. 1-isobutyl, 3-isopropyl

**1-isobutyl-3-isopropylcyclohexane**

## 4.41

2,2-dimethyl — **2,2-dimethylheptane**

3,3-dimethyl — **3,3-dimethylheptane**

4,4-dimethyl — **4,4-dimethylheptane**

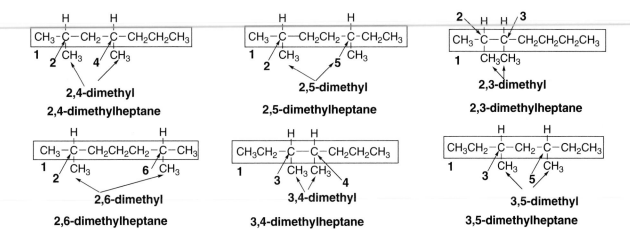

2,4-dimethyl
2,4-dimethylheptane

2,5-dimethyl
2,5-dimethylheptane

2,3-dimethyl
2,3-dimethylheptane

2,6-dimethyl
2,6-dimethylheptane

3,4-dimethyl
3,4-dimethylheptane

3,5-dimethyl
3,5-dimethylheptane

**4.42** Use the steps in Answer 4.13 to draw the structures.

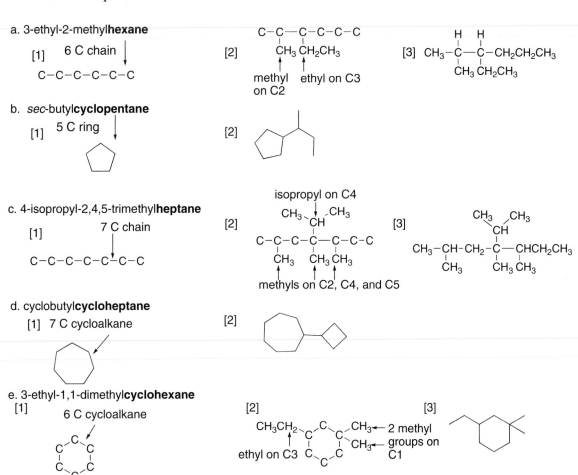

a. 3-ethyl-2-methyl**hexane**

[1] 6 C chain

C—C—C—C—C—C

[2] C—C—C—C—C—C
methyl on C2   ethyl on C3

[3] structure with CH₃ and CH₂CH₃ groups

b. *sec*-butyl**cyclopentane**

[1] 5 C ring

[2] structure

c. 4-isopropyl-2,4,5-trimethyl**heptane**

[1] 7 C chain

C—C—C—C—C—C—C

[2] isopropyl on C4
methyls on C2, C4, and C5

[3] structure

d. cyclobutyl**cycloheptane**

[1] 7 C cycloalkane

[2] structure

e. 3-ethyl-1,1-dimethyl**cyclohexane**

[1] 6 C cycloalkane

[2] ethyl on C3; 2 methyl groups on C1

[3] structure

f. 4-butyl-1,1-diethyl**cyclooctane**

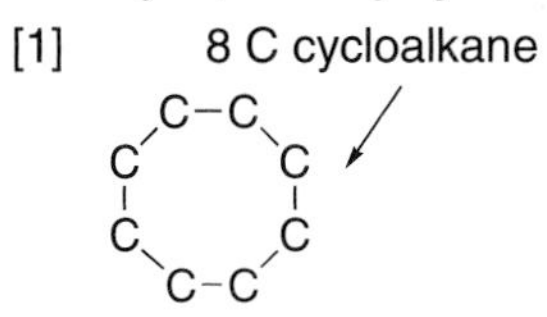

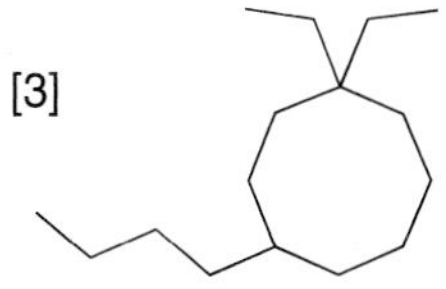

[1] 8 C cycloalkane

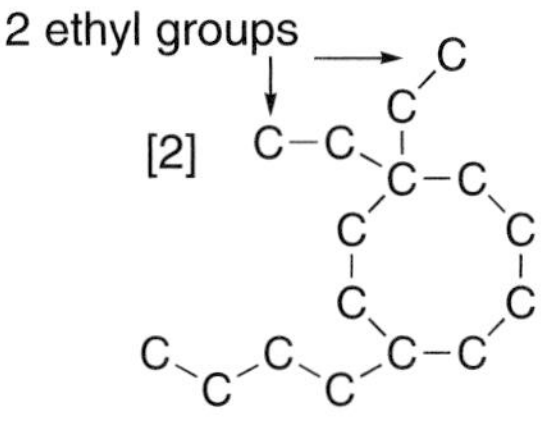

[2] 2 ethyl groups

[3] structure

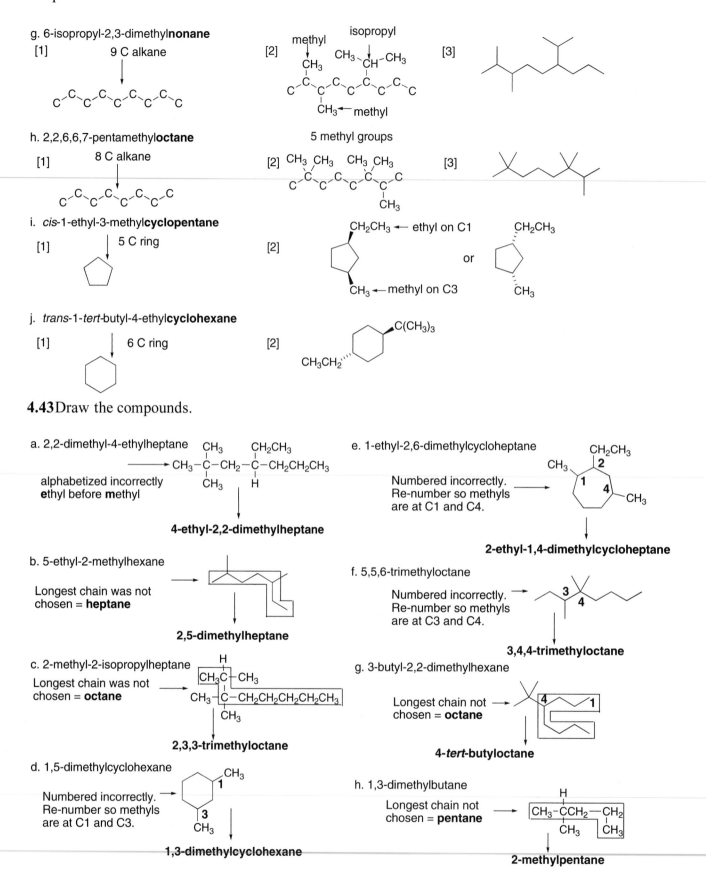

g. 6-isopropyl-2,3-dimethyl**nonane**

[1] 9 C alkane

[2] methyl   isopropyl

5 methyl groups

[3]

h. 2,2,6,6,7-pentamethyl**octane**

[1] 8 C alkane

[2]

[3]

i. *cis*-1-ethyl-3-methyl**cyclopentane**

[1] 5 C ring

[2] CH₂CH₃ ← ethyl on C1

CH₃ ← methyl on C3

or

j. *trans*-1-*tert*-butyl-4-ethyl**cyclohexane**

[1] 6 C ring

[2]

**4.43** Draw the compounds.

a. 2,2-dimethyl-4-ethylheptane

alphabetized incorrectly
**e**thyl before **m**ethyl

**4-ethyl-2,2-dimethylheptane**

b. 5-ethyl-2-methylhexane

Longest chain was not
chosen = **heptane**

**2,5-dimethylheptane**

c. 2-methyl-2-isopropylheptane

Longest chain was not
chosen = **octane**

**2,3,3-trimethyloctane**

d. 1,5-dimethylcyclohexane

Numbered incorrectly.
Re-number so methyls
are at C1 and C3.

**1,3-dimethylcyclohexane**

e. 1-ethyl-2,6-dimethylcycloheptane

Numbered incorrectly.
Re-number so methyls
are at C1 and C4.

**2-ethyl-1,4-dimethylcycloheptane**

f. 5,5,6-trimethyloctane

Numbered incorrectly.
Re-number so methyls
are at C3 and C4.

**3,4,4-trimethyloctane**

g. 3-butyl-2,2-dimethylhexane

Longest chain not
chosen = **octane**

**4-*tert*-butyloctane**

h. 1,3-dimethylbutane

Longest chain not
chosen = **pentane**

**2-methylpentane**

**4.44**

a.

$CH_3$

H

$CH_2CH_2CH_3$

$CH_3$

H

$CH_2CH_2CH_3$

↓ re-draw

**4**

**4-isopropylheptane**

b.

$CH_3$

$CH_3$

$CH_2CH_3$

H

H

$CH_2CH_3$

↓ re-draw

**3**

**3-ethyl-3-methylpentane**

c.

$CH_3$

$CH_3CH_2$

$CH_2CH_2CH_3$

$CH_3CH_2CH_2$

H

$CH_2CH_3$

↓ re-draw

**4** **5**

**4,4-diethyl-5-methyloctane**

**4.45** Use the rules from Answer 4.19.

a.  $CH_3CH_2CH_3$     $CH_3CH_2CH_2CH_3$     $CH_3CH_2CH_2CH_2CH_3$
     3C's              4 C's                  5 C's
  **lowest boiling point**                 **highest boiling point**

b.  $(CH_3)_2CHCH(CH_3)_2$     $CH_3CH_2CH_2CH(CH_3)_2$     $CH_3(CH_2)_4CH_3$
     most branching                            least branching
  **lowest boiling point**                     **highest boiling point**

**4.46**

$CH_3(CH_2)_6CH_3$
no branching = higher surface area
**higher boiling point**

$(CH_3)_3CC(CH_3)_3$
branching = lower surface area
**lower boiling point**
more spherical, better packing =
**higher melting point**

**4.47**

a.

and

1 gauche $CH_3,CH_3$
= 0.9 kcal/mol
of destabilization

**higher energy**
2 gauche $CH_3,CH_3$
0.9 kcal/mol x 2 = 1.8 kcal/mol
of destabilization

**Energy difference =**

1.8 kcal/mol − 0.9 kcal/mol = | **0.9 kcal/mol** |

b.

and

2 gauche $CH_3,CH_3$
0.9 kcal/mol x 2 =
1.8 kcal/mol
of destabilization

**higher energy**
3 eclipsed H,$CH_3$
1.5 kcal/mol x 3 = 4.5 kcal/mol
of destabilization

**Energy difference =**

4.5 kcal/mol − 1.8 kcal/mol = | **2.7 kcal/mol** |

**4.48** Use the rules from Answer 4.23 to determine the most and least stable conformations.

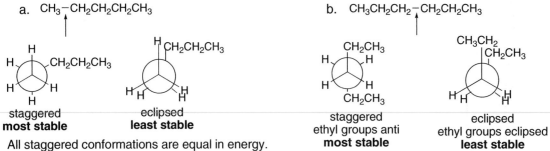

a. $CH_3-CH_2CH_2CH_2CH_3$

staggered
**most stable**

eclipsed
**least stable**

All staggered conformations are equal in energy.
All eclipsed conformations are equal in energy.

b. $CH_3CH_2CH_2-CH_2CH_2CH_3$

staggered
ethyl groups anti
**most stable**

eclipsed
ethyl groups eclipsed
**least stable**

**4.49**

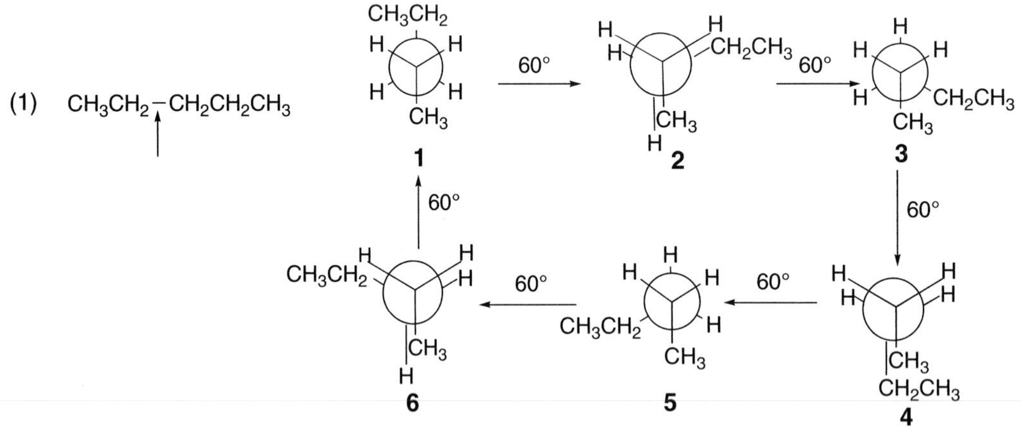

(1)   $CH_3CH_2-CH_2CH_2CH_3$

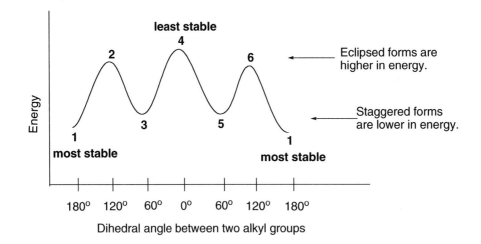

least stable
4

Eclipsed forms are
higher in energy.

2                    6

3        5

1                    1
**most stable**         **most stable**

Energy

180°   120°   60°   0°   60°   120°   180°

Dihedral angle between two alkyl groups

Staggered forms
are lower in energy.

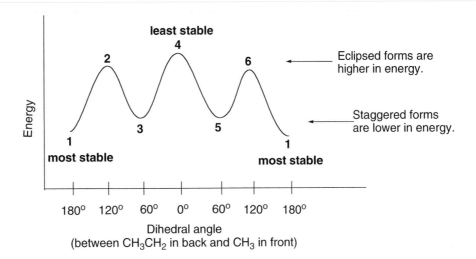

(2)  CH₃CH₂–CHCH₂CH₃
              |
              CH₃

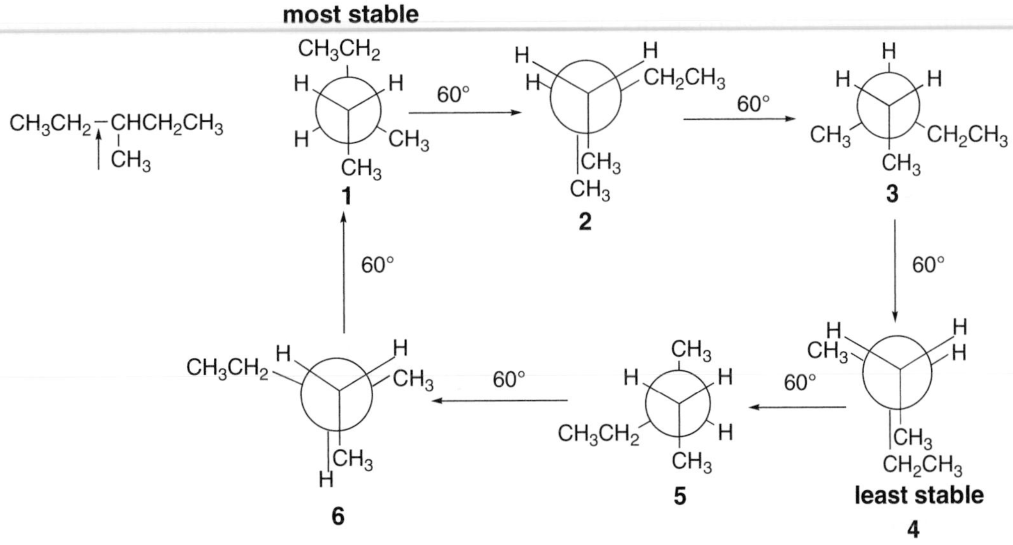

**4.50** Two types of strain:
- *Torsional strain* is due to eclipsed groups on adjacent carbon atoms.
- *Steric strain* is due to overlapping electron clouds of large groups (ex: gauche interactions).

a.

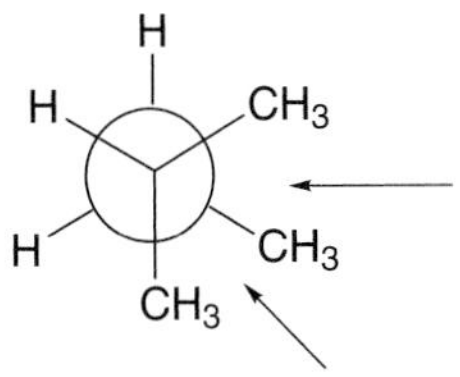

two sites
three bulky methyl groups close =
**steric strain**

b.

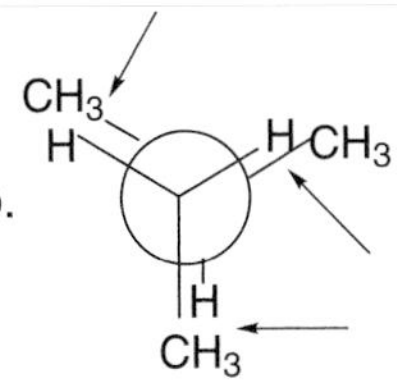

eclipsed conformation =
**torsional strain**

c.

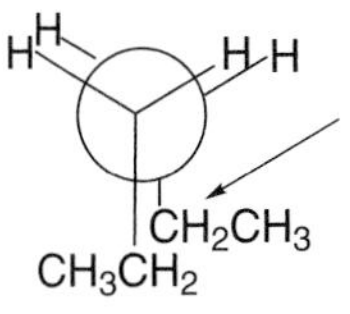

two bulky ethyl groups close =
**steric strain**
eclipsed conformation =
**torsional strain**

**4.51** The barrier to rotation is equal to the difference in energy between the highest energy eclipsed and lowest energy staggered conformations of the molecule.

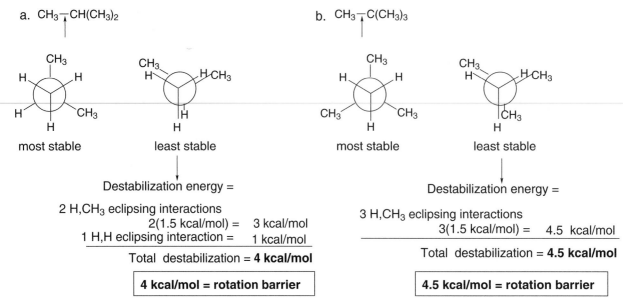

a. $CH_3{-}CH(CH_3)_2$

most stable          least stable

Destabilization energy =

2 H,CH₃ eclipsing interactions
$2(1.5 \text{ kcal/mol}) =$   3 kcal/mol
1 H,H eclipsing interaction =   1 kcal/mol

Total destabilization = **4 kcal/mol**

| **4 kcal/mol = rotation barrier** |

b. $CH_3{-}C(CH_3)_3$

most stable          least stable

Destabilization energy =

3 H,CH₃ eclipsing interactions
$3(1.5 \text{ kcal/mol}) =$   4.5 kcal/mol

Total destabilization = **4.5 kcal/mol**

| **4.5 kcal/mol = rotation barrier** |

**4.52**

most stable          least stable

2 H,H eclipsing interactions = 2(1 kcal/mol) = 2 kcal/mol

Since the barrier to rotation is 3.7 kcal/mol, the difference between this value and the destabilization due to H,H eclipsing is the destabilization due to H,Cl eclipsing.

3.7 kcal/mol – 2 kcal/mol = **1.7 kcal/mol**
**destabilization due to H,Cl eclipsing**

**4.53** The gauche conformation can intramolecularly hydrogen bond, making it the more stable conformation.

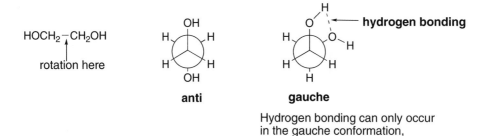

$HOCH_2{-}CH_2OH$

rotation here

**anti**

**gauche**

hydrogen bonding

Hydrogen bonding can only occur in the gauche conformation, making it **more stable.**

**4.54**

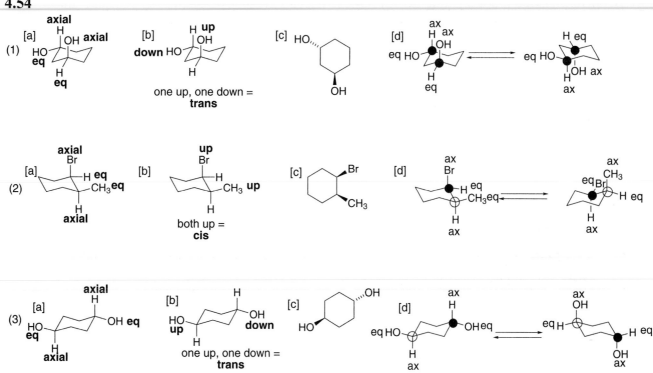

**4.55** A **cis isomer** has two groups on the **same side** of the ring. The two groups can be drawn both up or both down. Only one possibility is drawn. A **trans isomer** has one group on one side of the ring and one group on the other side. Either group can be drawn on either side. Only one possibility is drawn.

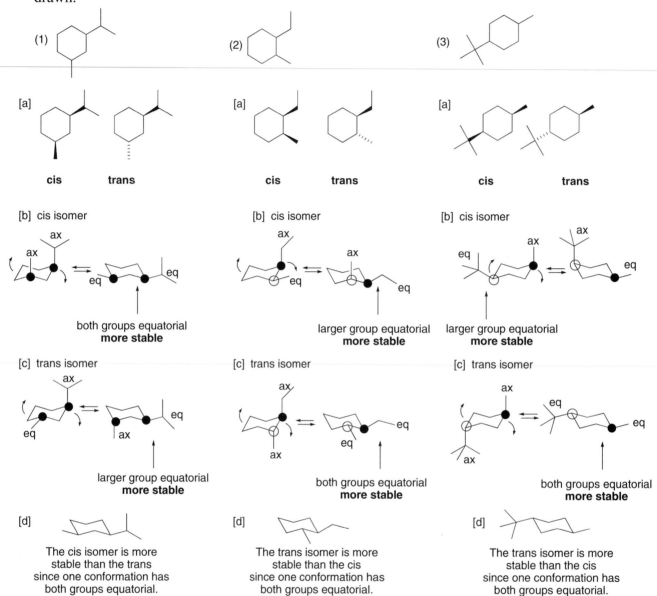

**4.56** Compare the isomers by drawing them in chair conformations. Equatorial substituents are more stable. See the definitions in Problem 4.55.

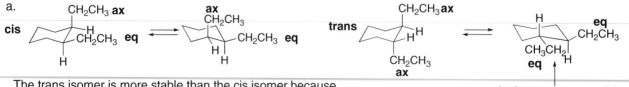

The trans isomer is more stable than the cis isomer because its more stable conformation has two groups equatorial.

both groups equatorial
**most stable of all conformations**
**trans isomer**

**b. 1-ethyl-3-isopropylcyclohexane**

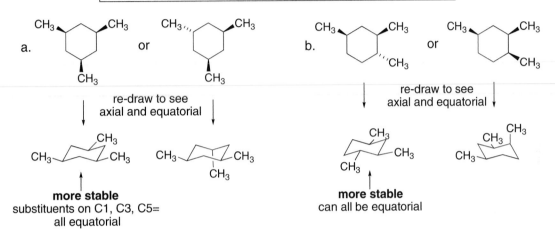

**cis**

$(CH_3)_2CH$ — — $CH_2CH_3$
**eq**          **eq**
H    H

both groups equatorial
**most stable of all conformations**
**cis isomer**

The cis isomer is more stable than the trans isomer because
its more stable conformation has two groups equatorial.

**4.57**

| Only the more stable conformation of each compound is drawn. |

a.   $CH_3$   $CH_3$   or   $CH_3$   $CH_3$     b.   $CH_3$   $CH_3$   or   $CH_3$   $CH_3$
              $CH_3$              $CH_3$                   $CH_3$                   $CH_3$

re-draw to see
axial and equatorial

re-draw to see
axial and equatorial

$CH_3$
$CH_3$ — $CH_3$     $CH_3$ — $CH_3$
                      $CH_3$

$CH_3$ — $CH_3$     $CH_3$ $CH_3$
$CH_3$                $CH_3$

**more stable**
substituents on C1, C3, C5=
all equatorial

**more stable**
can all be equatorial

**4.58**

a.  HO — OH   or   HO   OH
    HO   O          HO
    HO   OH         HO   O
                         OH
**most stable**
All groups are equatorial.

b.  HO   or   HO
    HO   O          HO   O
    HO   OH    HO   OH
    HO   OH         HO   OH

**4.59**

a.  [propene] and [cyclobutane]

same molecular formula $C_4H_8$
different connectivity
**constitutional isomers**

b.  [structures] and [structures]

different arrangement in three dimensions
**stereoisomers**

c.  $CH_3$  H  and  H — $CH_3$  $CH_3$  H
    H         $CH_3$ H        H
    $CH_3$              $CH_3$   $CH_3$

1 down, 1 up =        1 down, 1 up =
**trans**                **trans**

same arrangement in three dimensions
**identical**

d.  $CH_2CH_3$ and  $CH_2CH_3$
    $CH_2CH_3$        $CH_2CH_3$

same molecular formula $C_{10}H_{20}$
different connectivity
**constitutional isomers**

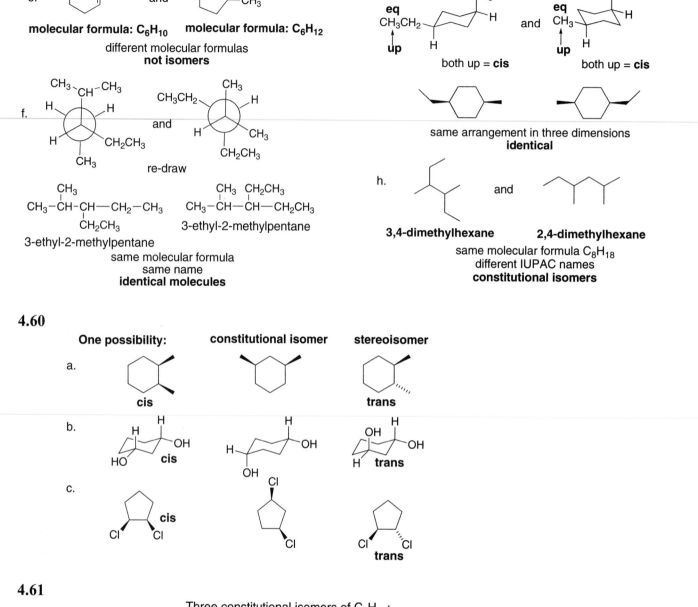

e. molecular formula: $C_6H_{10}$   molecular formula: $C_6H_{12}$
different molecular formulas
**not isomers**

f. re-draw

3-ethyl-2-methylpentane   3-ethyl-2-methylpentane
same molecular formula
same name
**identical molecules**

g. both up = **cis**   both up = **cis**
same arrangement in three dimensions
**identical**

h. **3,4-dimethylhexane**   **2,4-dimethylhexane**
same molecular formula $C_8H_{18}$
different IUPAC names
**constitutional isomers**

**4.60**

One possibility:   constitutional isomer   stereoisomer

a. **cis**   **trans**

b. **cis**   **trans**

c. **cis**   **trans**

**4.61**

Three constitutional isomers of $C_7H_{14}$:

1,1-dimethylcyclopentane   1,2-dimethylcyclopentane   1,3-dimethylcyclopentane

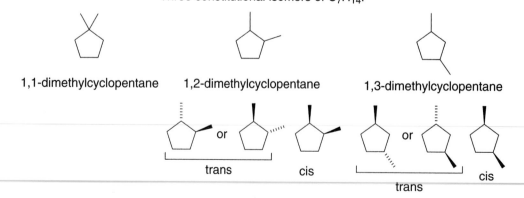

trans   cis   trans   cis

**4.62** Use the definitions from Answer 4.33 to classify the reactions.

a. $CH_3CHO$ = $\longrightarrow$ $CH_3CH_2OH$

Decrease in the number of C–O
bonds. **Reduction**

d. $CH_2=CH_2$ $\longrightarrow$ $H-C\equiv C-H$

Decrease in the number of C–H
bonds. **Oxidation**

b. $\longrightarrow$

Increase in the number of C–O
bonds. **Oxidation**

e. $\longrightarrow$

Increase in the number of C–Z
bonds. **Oxidation**

c. $CH_2=CH_2$ $\longrightarrow$ $HOCH_2CH_2OH$

Two new C–O
bonds. **Oxidation**

f. $CH_3CH_2OH$ $\longrightarrow$ $CH_2=CH_2$

Loss of one C–O
bond *and* one C–H
bond. **Neither**

**4.63** Use the rule from Answer 4.34.

a. $CH_3CH_2CH_2CH_2CH(CH_3)_2$ $\xrightarrow[\text{11 } O_2]{\text{flame}}$ $7\ CO_2 + 8\ H_2O +$ heat

b. $\xrightarrow[\text{(13/2) } O_2]{\text{flame}}$ $4\ CO_2 + 5\ H_2O +$ heat

**4.64**

a.

benzene   an arene oxide   phenol

[1] increase in C–O bonds
**oxidation reaction**

[2] loss of 1 C–O bond,
loss of 1 C–H bond
**neither**

b. Phenol is more water soluble than benzene because it is **polar (contains an O–H group) and can hydrogen bond with water,** whereas benzene is nonpolar and cannot hydrogen bond.

**4.65** Lipids contain many nonpolar C–C and C–H bonds and few polar functional groups.

a.
mevalonic acid

many polar functional groups
**not a lipid**

c.
estradiol

few polar functional groups
**a lipid**

d.
sucrose

many polar functional groups
**not a lipid**

b.
squalene

no polar functional groups
**a lipid**

**4.66**

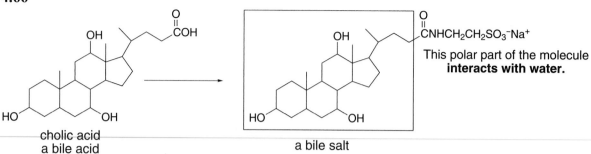

cholic acid
a bile acid

a bile salt

This polar part of the molecule
**interacts with water.**

This nonpolar part of the molecule
can **interact with lipids** to create
micelles that allow for transport
of lipids through aqueous environments.

**4.67** The amide in the four-membered ring has 90° bond angles giving it angle strain, and therefore making it more reactive.

amide

penicillin G

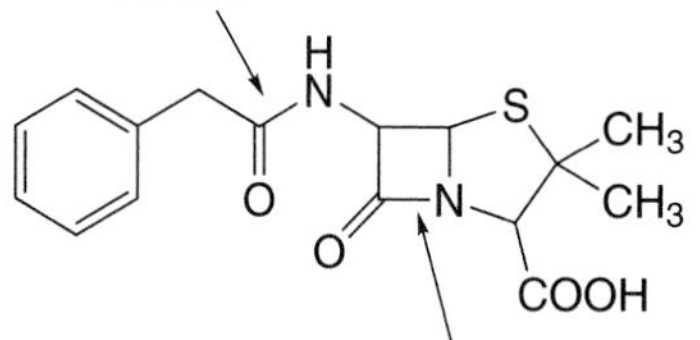

CH₃
CH₃
COOH

strained amide
**more reactive**

**4.68**

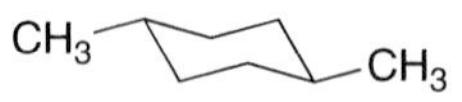

CH₃ ⟍ CH₃

CH₃ ⟍ CH₃

*trans*-1,4-dimethylcyclohexane
more symmetrical
better packing
**higher melting point**

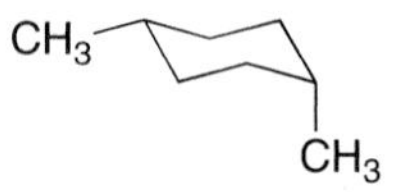

*cis*-1,4-dimethylcyclohexane

**4.69**

Example:

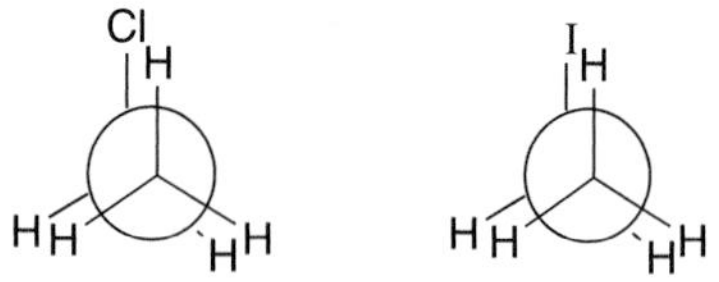

Although I is a much bigger atom than Cl, the C–I bond is also much longer than the C–Cl bond. As a result the eclipsing interaction of the H and I atoms is not very much different from the H,Cl eclipsing interaction in magnitude.

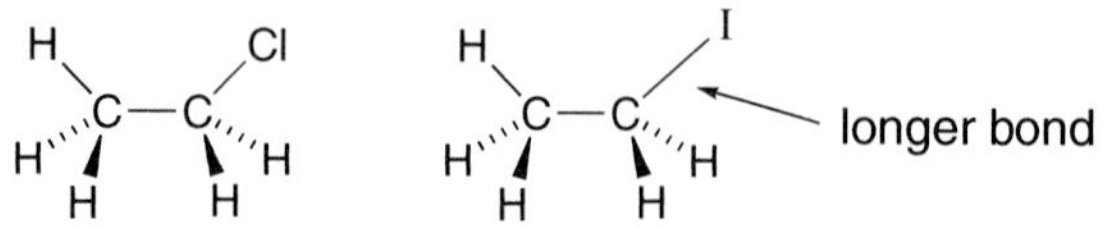

longer bond

**4.70**

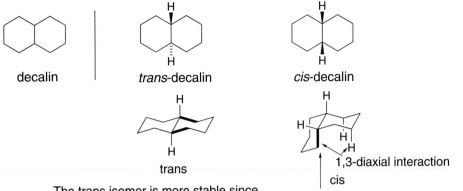

decalin    |    *trans*-decalin    *cis*-decalin

trans

cis    1,3-diaxial interaction

The trans isomer is more stable since the carbon groups at the ring junction are both in the favorable equatorial position.

This bond is axial, creating unfavorable 1,3-diaxial interactions.

## Chapter 5: Stereochemistry

♦ **Isomers are different compounds with the same molecular formula. (5.2, 5.11)**

**[1] Constitutional isomers**—isomers that differ in the way the atoms are connected to each other. They have:
- different IUPAC names;
- the same or different functional groups;
- different physical and chemical properties.

**[2] Stereoisomers**—isomers that differ only in the way atoms are oriented in space. They have the same functional group and the same IUPAC name except for prefixes such as cis, trans, *R*, and *S*.
- **Enantiomers**—stereoisomers that are mirror images of each other (5.4).
- **Diastereomers**—stereoisomers that are not mirror images of each other (5.7).

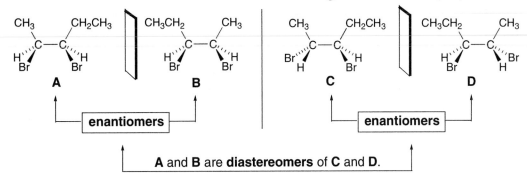

♦ **Assigning priority**

- Assign priorities (1, 2, 3, or 4) to the atoms directly bonded to the stereogenic center in order of decreasing atomic number. The atom of *highest* atomic number gets the *highest* priority (1).
- If two atoms on a stereogenic center are the *same*, assign priority based on the atomic number of the atoms bonded to these atoms. *One* atom of higher atomic number determines a higher priority.
- If two isotopes are bonded to the stereogenic center, assign priorities in order of decreasing *mass* number.
- To assign a priority to an atom that is part of a multiple bond, consider a multiply bonded atom as an equivalent number of singly bonded atoms.

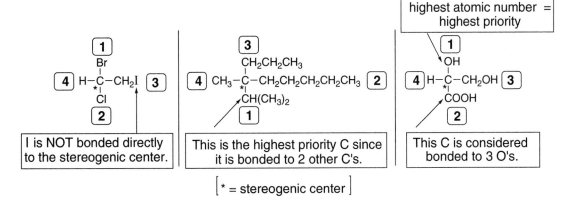

### ♦ Some basic principles

- When a compound and its mirror image are **superimposable**, they are **identical achiral compounds**. A plane of symmetry in one conformation makes a compound achiral (5.3).
- When a compound and its mirror image are **not superimposable**, they are **different chiral compounds** called **enantiomers**. A chiral compound has no plane of symmetry in any conformation (5.3).
- A **tetrahedral stereogenic center** is a carbon atom bonded to four different groups (5.4, 5.5).
- For *n* **stereogenic centers**, the maximum number of stereoisomers is $2^n$ (5.7).

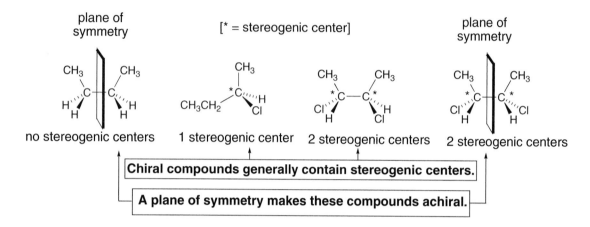

### ♦ Optical activity is the ability of a compound to rotate plane-polarized light (5.12).

- An optically active solution contains a chiral compound.
- An optically inactive solution contains one of the following:
  - An achiral compound with no stereogenic centers.
  - A meso compound—an achiral compound with two or more stereogenic centers.
  - A racemic mixture—an equal amount of two enantiomers.

### ♦ The prefixes *R* and *S* compared with *d* and *l*

The prefixes *R* and *S* are labels used in nomenclature. Rules on assigning *R,S* are found in Section 5.6.
- An enantiomer has every stereogenic center opposite in configuration. If a compound with two stereogenic centers has the *R,R* configuration, its enantiomer has the *S,S* configuration.
- A diastereomer of this same compound has either the *R,S* or *S,R* configuration; one stereogenic center has the same configuration and one is opposite.

The prefixes *d* (or +) and *l* (or –) tell the direction a compound rotates plane-polarized light (5.12).
- *d* (or +) stands for dextrorotatory, rotating polarized light clockwise.
- *l* (or –) stands for levorotatory, rotating polarized light counterclockwise.

### ♦ The physical properties of isomers compared (5.12)

| Type of isomer | Physical properties |
| --- | --- |
| Constitutional isomers | Different |
| Enantiomers | Identical except the direction of rotation of polarized light |
| Diastereomers | Different |
| Racemic mixture | Possibly different from either enantiomer |

♦ Equations

- Specific rotation (5.12C):

$$\text{specific rotation} = [\alpha] = \frac{\alpha}{l \times c}$$

$\alpha$ = observed rotation ($^\circ$)
$l$ = length of sample tube (dm)
$c$ = concentration (g/mL)

$\left[ \begin{array}{l} \text{dm = decimeter} \\ \text{1 dm = 10 cm} \end{array} \right]$

- Enantiomeric excess (5.12D):

$$ee = \% \text{ of one enantiomer} - \% \text{ other enantiomer}$$

$$= \frac{[\alpha] \text{ mixture}}{[\alpha] \text{ pure enantiomer}} \times 100\%$$

## Chapter 5: Answers to Problems

**5.1** Cellulose consists of long chains held together by intermolecular hydrogen bonds forming sheets that stack in extensive three-dimensional arrays. Most of the OH groups in cellulose are in the interior of this three-dimensional network, unavailable for hydrogen bonding to water. Thus, even though cellulose has many OH groups, its three-dimensional structure prevents many of the OH groups from hydrogen bonding with the solvent and this makes it water insoluble.

**5.2 Constitutional isomers** have atoms bonded to different atoms.
**Stereoisomers** differ only in the three-dimensional arrangement of atoms.

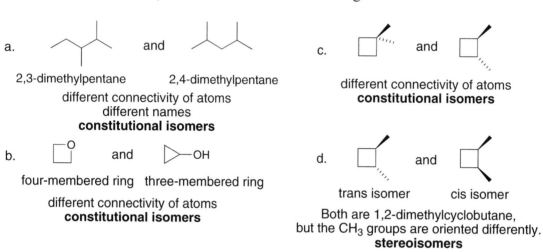

a.

2,3-dimethylpentane     2,4-dimethylpentane
different connectivity of atoms
different names
**constitutional isomers**

b.

four-membered ring    three-membered ring

different connectivity of atoms
**constitutional isomers**

c.

different connectivity of atoms
**constitutional isomers**

d.

trans isomer     cis isomer

Both are 1,2-dimethylcyclobutane,
but the $CH_3$ groups are oriented differently.
**stereoisomers**

**5.3** Draw the mirror image of each molecule by drawing a mirror plane and then drawing the molecule's reflection. **A chiral molecule is one that is not superimposable on its mirror image.** A molecule with one stereogenic center is always chiral. A molecule with zero stereogenic centers is not chiral (in general).

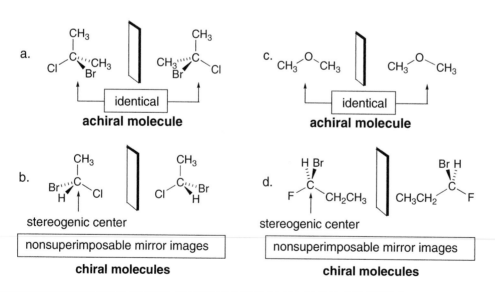

a.

identical
**achiral molecule**

c.

identical
**achiral molecule**

b.

stereogenic center
nonsuperimposable mirror images
**chiral molecules**

d.

stereogenic center
nonsuperimposable mirror images
**chiral molecules**

**5.4** A plane of symmetry cuts the molecule into **two identical halves**.

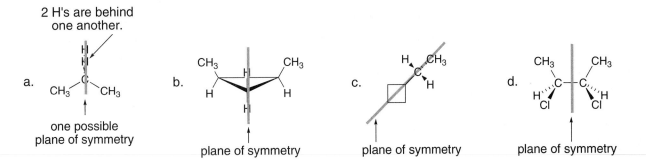

a.

2 H's are behind
one another.

one possible
plane of symmetry

b.

plane of symmetry

c.

plane of symmetry

d.

plane of symmetry

**5.5** Rotate around the middle C–C bond so that the Br atoms are eclipsed.

rotate
CH₃ here

C2  C3

plane of symmetry

**5.6** To locate a stereogenic center, omit:

All C's with 2 or more H's, all *sp* and *sp²* hybridized atoms, and all heteroatoms. (In Chapter 25, we will learn that the N atoms of tetraalkylammonium salts [R₄N⁺X⁻] can sometimes be stereogenic centers.)

Then evaluate any remaining atoms: a tetrahedral stereogenic center has a carbon bonded to **four different groups**.

a.   $CH_3CH_2-\overset{\displaystyle H}{\underset{\displaystyle Cl}{C}}-CH_2CH_3$

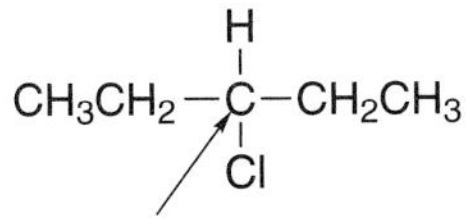

bonded to 2 identical
ethyl groups
**0 stereogenic centers**

b.  (CH₃)₃CH

**0 stereogenic centers**

c.   $CH_3-\overset{\displaystyle H}{\underset{\displaystyle OH}{C}}-CH=CH_2$

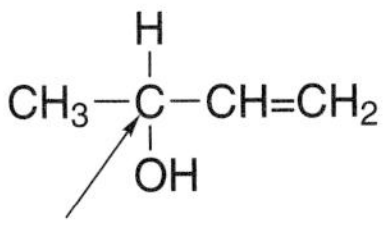

This C is bonded to
4 different groups.
**1 stereogenic center**

d.  CH₃CH₂CH₂OH

**0 stereogenic centers**

e.  $(CH_3)_2CHCH_2CH_2-\overset{\displaystyle CH_3}{\underset{\displaystyle H}{C}}-CH_2CH_3$

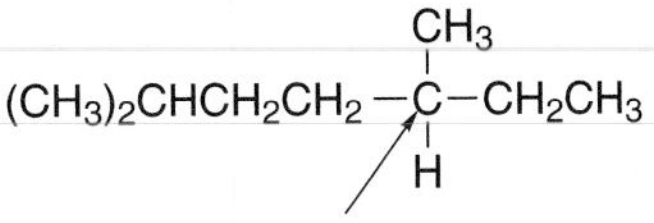

This C is bonded to
4 different groups.
**1 stereogenic center**

f.   $CH_3CH_2-\overset{\displaystyle H}{\underset{\displaystyle CH_3}{C}}-CH_2CH_2CH_3$

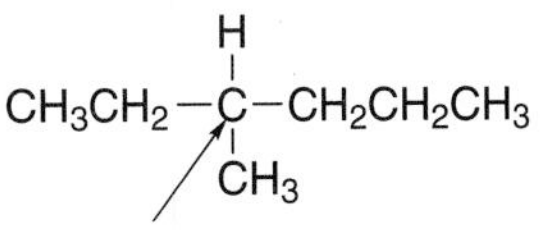

This C is bonded to
4 different groups.
**1 stereogenic center**

**5.7** Use the directions from Answer 5.6 to locate the stereogenic centers.

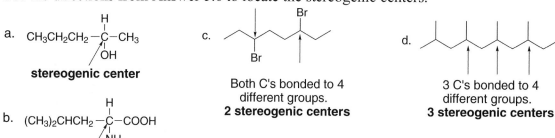

a. $CH_3CH_2CH_2 - \overset{\overset{\displaystyle H}{|}}{\underset{\diagup OH}{C}} - CH_3$

**stereogenic center**

c.

Br

Br

Both C's bonded to 4
different groups.
**2 stereogenic centers**

d.

3 C's bonded to 4
different groups.
**3 stereogenic centers**

b. $(CH_3)_2CHCH_2 - \overset{\overset{\displaystyle H}{|}}{\underset{\diagup NH_2}{C}} - COOH$

**stereogenic center**

**5.8** Use the directions from Answer 5.6 to locate the stereogenic centers.

a.

CHO
HO—C—H
HO—C—H
H—C—OH
H—C—OH
CH$_2$OH

mannose

4 C's bonded to
4 different groups:
**4 stereogenic centers.**

b. H$_2$N

SH

COOH

O

H
N

H
N

O

O

OH

glutathione

2 C's bonded to 4 different groups:
**2 stereogenic centers.**

**5.9** Find the C bonded to four different groups in each molecule. At the stereogenic center, draw two
bonds in the plane of the page, one in front (on a wedge) and one behind (on a dash). Then draw
the mirror image (enantiomer).

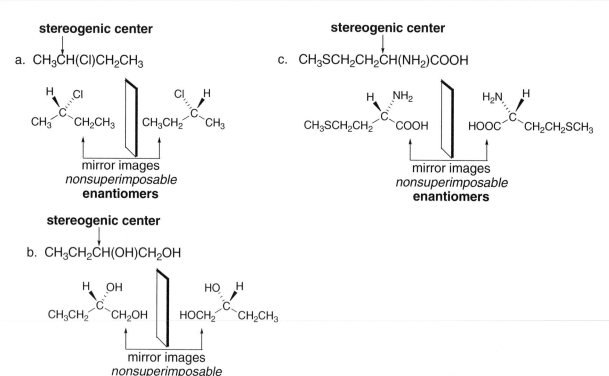

**stereogenic center**

a. CH$_3$CH(Cl)CH$_2$CH$_3$

H      Cl

CH$_3$—C—CH$_2$CH$_3$

Cl      H

CH$_3$CH$_2$—C—CH$_3$

mirror images
*nonsuperimposable*
**enantiomers**

**stereogenic center**

c. CH$_3$SCH$_2$CH$_2$CH(NH$_2$)COOH

H      NH$_2$

CH$_3$SCH$_2$CH$_2$—C—COOH

H$_2$N      H

HOOC—C—CH$_2$CH$_2$SCH$_3$

mirror images
*nonsuperimposable*
**enantiomers**

**stereogenic center**

b. CH$_3$CH$_2$CH(OH)CH$_2$OH

H      OH

CH$_3$CH$_2$—C—CH$_2$OH

HO      H

HOCH$_2$—C—CH$_2$CH$_3$

mirror images
*nonsuperimposable*
**enantiomers**

**5.10** Use the directions from Answer 5.6 to locate the stereogenic centers.

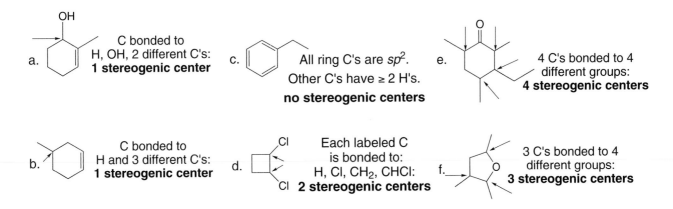

a. **C bonded to H, OH, 2 different C's: 1 stereogenic center**

c. **All ring C's are *sp²*. Other C's have ≥ 2 H's. no stereogenic centers**

e. **4 C's bonded to 4 different groups: 4 stereogenic centers**

b. **C bonded to H and 3 different C's: 1 stereogenic center**

d. **Each labeled C is bonded to: H, Cl, CH₂, CHCl: 2 stereogenic centers**

f. **3 C's bonded to 4 different groups: 3 stereogenic centers**

**5.11**

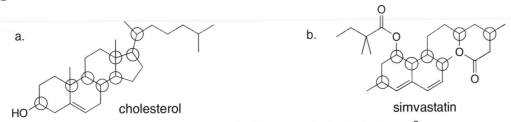

a.

cholesterol

b.

simvastatin

All stereogenic C's are circled. Each C is *sp³* hybridized and bonded to 4 different groups.

**5.12** Assign priority based on atomic number: atoms with a higher atomic number get a higher priority. If two atoms are the same, look at what they are bonded to and assign priority based on the atomic number of these atoms.

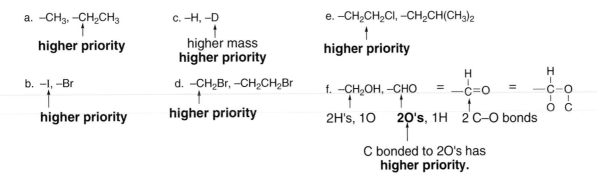

a. $-CH_3$, $-CH_2CH_3$

**higher priority**

c. $-H$, $-D$

**higher mass higher priority**

e. $-CH_2CH_2Cl$, $-CH_2CH(CH_3)_2$

**higher priority**

b. $-I$, $-Br$

**higher priority**

d. $-CH_2Br$, $-CH_2CH_2Br$

**higher priority**

f. $-CH_2OH$, $-CHO$

2H's, 1O     2O's, 1H    2 C–O bonds

$= -\overset{H}{\underset{}{C}}=O = -\overset{H}{\underset{O}{\overset{|}{C}}}-\overset{}{\underset{C}{\overset{O}{|}}}$

C bonded to 2O's has **higher priority.**

**5.13** Rank by decreasing priority. Lower atomic number = lower priority.

**Highest priority = 1, Lowest priority = 4**

a. –COOH  C = second lowest atomic number — priority **3**

–H  H = lowest atomic number  **4**

–NH$_2$  N = second highest atomic number  **2**

–OH  O = highest atomic number  **1**

**decreasing priority: –OH, –NH$_2$, –COOH, –H**

c. –CH$_2$CH$_3$  C bonded to 2H's + **1C**  priority **2**

–CH$_3$  C bonded to 3H's  **3**

–H  H = lowest atomic number  **4**

–CH(CH$_3$)$_2$  C bonded to 1H + **2C's**  **1**

**decreasing priority: –CH(CH$_3$)$_2$, –CH$_2$CH$_3$, –CH$_3$, –H**

b. –H  H = lowest atomic number  priority **4**

–CH$_3$  C bonded to 3H's  **3**

–Cl  Cl = highest atomic number  **1**

–CH$_2$Cl  C bonded to 2H's + **1 Cl**  **2**

**decreasing priority: –Cl, –CH$_2$Cl, –CH$_3$, –H**

d. –CH=CH$_2$  C bonded to 1H + **2C's**  priority **2**

–CH$_3$  C bonded to 3H's  **3**

–C≡CH  C bonded to **3C's**  **1**

–H  H = lowest atomic number  **4**

**decreasing priority: –C≡CH, –CH=CH$_2$, –CH$_3$, –H**

**5.14** To assign $R$ or $S$ to the molecule, first rank the groups. The lowest priority group must be oriented behind the page. If tracing a circle from (1) → (2) → (3) proceeds in the clockwise direction, the stereogenic center is labeled $R$; if the circle is counterclockwise, it is labeled $S$.

a. counterclockwise **S isomer**

b. counterclockwise **S isomer**

c. rotate — lowest priority group now back — clockwise **R isomer**

d. counterclockwise **S isomer**

**5.15**

fenfluramine

counterclockwise **S isomer** **dexfenfluramine**

counterclockwise **R isomer**

**5.16** The maximum number of stereoisomers = $2^n$ where $n$ = the number of stereogenic centers.

a. 3 stereogenic centers
$2^3$ = 8 stereoisomers

b. 8 stereogenic centers
$2^8$ = 256 stereoisomers

**5.17**

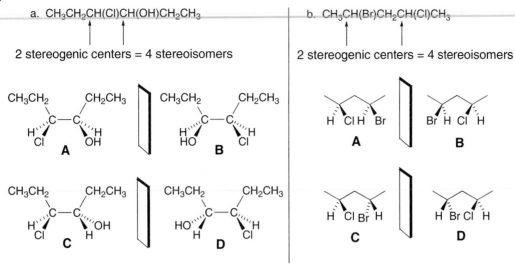

a. CH₃CH₂CH(Cl)CH(OH)CH₂CH₃

2 stereogenic centers = 4 stereoisomers

b. CH₃CH(Br)CH₂CH(Cl)CH₃

2 stereogenic centers = 4 stereoisomers

**5.18**

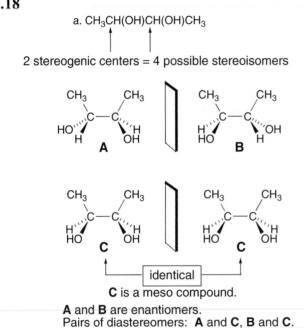

a. CH₃CH(OH)CH(OH)CH₃

2 stereogenic centers = 4 possible stereoisomers

identical

**C** is a meso compound.
**A** and **B** are enantiomers.
Pairs of diastereomers: **A** and **C**, **B** and **C**.

b. CH₃CH(OH)CH(Cl)CH₃

2 stereogenic centers = 4 possible stereoisomers

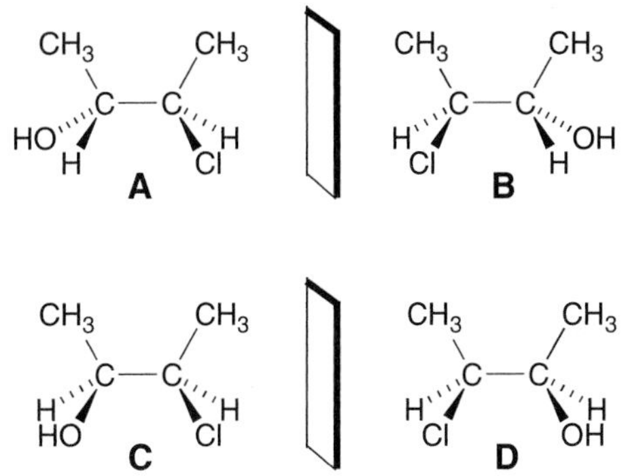

Pairs of enantiomers: **A** and **B**, **C** and **D**.
Pairs of diastereomers: **A** and **C**, **A** and **D**,
**B** and **C**, **B** and **D**.

**5.19** An **enantiomer** is a stereoisomer that is a nonsuperimposable mirror image. A **diastereomer** is a stereoisomer that is not a mirror image.

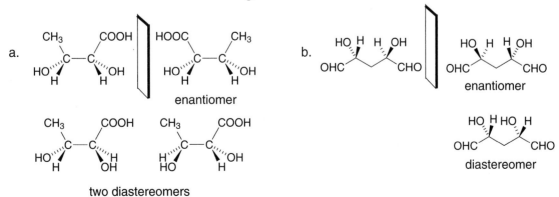

a.

enantiomer

two diastereomers

b.

enantiomer

diastereomer

**5.20**   **A meso compound must have at least two stereogenic centers.  Usually a meso compound has a plane of symmetry.**  You may have to rotate around a C–C bond to see the plane of symmetry clearly.

a.

| 2 stereogenic centers |
| plane of symmetry |
| **meso compound** |

b.

| 2 stereogenic centers |
| no plane of symmetry |
| **not a meso compound** |

c.

| 2 stereogenic centers |
| no plane of symmetry |
| **not a meso compound** |

d.

rotate

| 2 stereogenic centers |
| plane of symmetry |
| **meso compound** |

**5.21**   The enantiomer must have the exact opposite $R,S$ designations.  Diastereomers with two stereogenic centers have one center the same and one different.

If a compound is  **R,S:**

Its enantiomer is: **S,R** ◄─────────── Exact opposite: $R$ and $S$ interchanged.

Its diastereomers are: **R,R and S,S** ◄───── One designation remains the same, the other changes.

**5.22**   The enantiomer must have the exact opposite $R,S$ designations.  For diastereomers, at least one of the $R,S$ designations is the same, but not all of them.

    a. (2R,3S)-2,3-hexanediol and (2R,3R)-2,3-hexanediol
        One changes; one remains the same:
            **diastereomers**

    b. (2R,3R)-2,3-hexanediol and (2S,3S)-2,3-hexanediol
        Both $R$'s change to $S$'s:
            **enantiomers**

    c. (2R,3S,4R)-2,3,4-hexanetriol and (2S,3R,4R)-2,3,4-hexanetriol
        Two change; one remains the same:
            **diastereomers**

**5.23**

a. The five tetrahedral stereogenic centers are labeled in each compound above.
b. PGF$_{2\alpha}$ and **A** are **enantiomers.**
c. PGF$_{2\alpha}$ and **B** are **diastereomers.**

**5.24** **Meso compounds generally have a plane of symmetry.** They cannot have just one stereogenic center.

a.

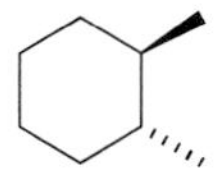

no plane of symmetry
**not a meso compound**

b.

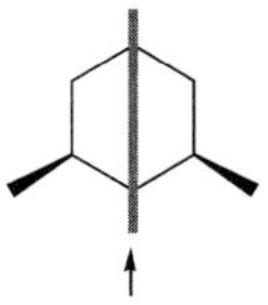

plane of symmetry
**meso compound**

c.

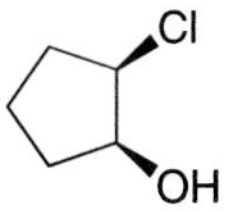

no plane of symmetry
**not a meso compound**

**5.25**

a. 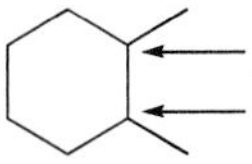 ← 2 stereogenic centers = 4 stereoisomers maximum

Draw the cis and trans isomers:

**cis**

CH₃  CH₃
CH₃  CH₃
**A**
identical

**trans**

CH₃  CH₃
CH₃  CH₃
**B**    **C**

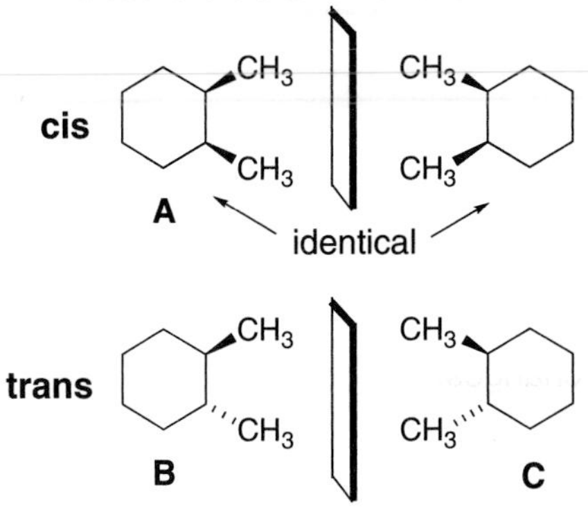

Pair of enantiomers: **B** and **C**.
Pairs of diastereomers: **A** and **B**, **A** and **C**.

**Only 3 stereoisomers exist.**

c.

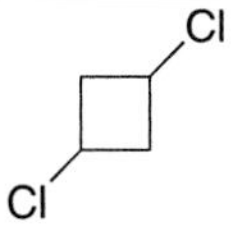

Cl
Cl

Draw the cis and trans isomers:

Cl  Cl
Cl  Cl
**A**
identical

Cl  Cl
Cl  Cl
**B** 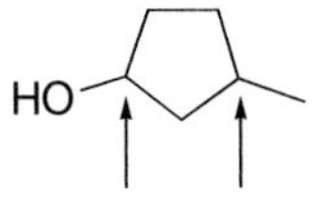
identical

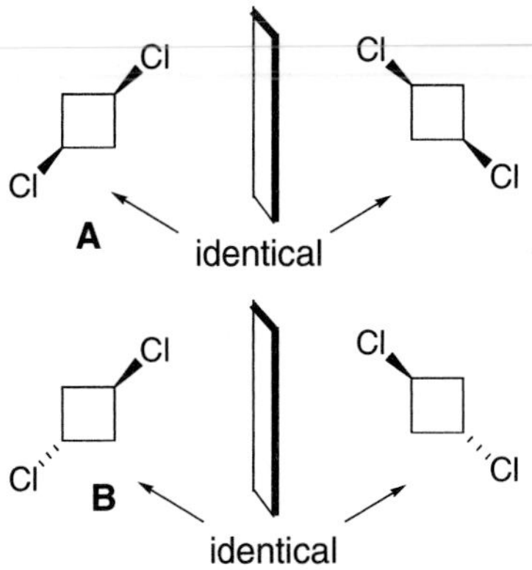

Pair of diastereomers: **A** and **B**.

**Only 2 stereoisomers exist.**

b. HO ← → 2 stereogenic centers = 4 stereoisomers maximum

Draw the cis and trans isomers:

**cis**

CH₃ OH    HO CH₃
**A**        **B**

**trans**

CH₃ OH    HO CH₃
**C**        **D**

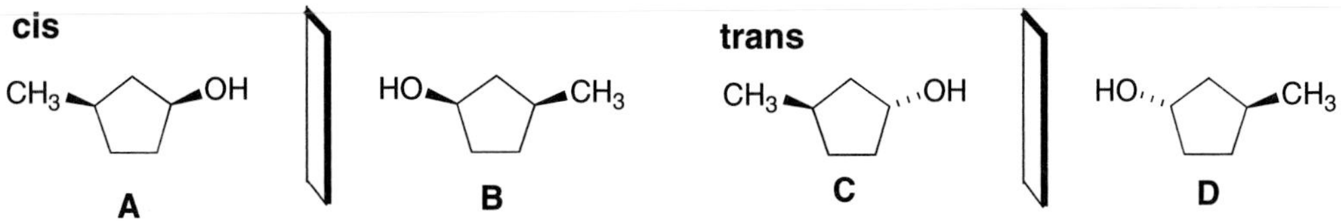

Pairs of enantiomers: **A** and **B**, **C** and **D**.
Pairs of diastereomers: **A** and **C**, **A** and **D**,
**B** and **C**, **B** and **D**.

**All 4 stereoisomers exist.**

**5.26** Four facts:
- **Enantiomers** are mirror image isomers.
- **Diastereomers** are stereoisomers that are not mirror images.
- **Constitutional isomers** have the same molecular formula but the atoms are bonded to different atoms.
- **Cis and trans isomers** are always diastereomers.

a.

and

same molecular formula
same *R,S* designation:
**identical**

c. and

1,4- isomer          1,3-isomer
**constitutional isomers**

b. and

same molecular formula,
opposite configuration at one
stereogenic center
**enantiomers**

d. and

**trans**                **cis**

Both 1,3 isomers,
cis and trans:
**diastereomers**

## 5.27

(*S*)-alanine

$[\alpha] = +8.5$

mp = 297 °C

a. Mp = same as the *S* isomer
b. The mp of a racemic mixture is often different from the melting
   point of the enantiomers.
c. –8.5, same as *S* but opposite sign
d. Zero. A racemic mixture is optically inactive.
e. Solution of pure (*S*)-alanine: **optically active**
   Equal mixture of (*R*) and (*S*)-alanine: **optically inactive**
   75% (*S*) and 25% (*R*)-alanine: **optically active**

## 5.28

$$[\alpha] = \frac{\alpha}{l \times c}$$

$\alpha$ = observed rotation
$l$ = length of tube (dm)
$c$ = concentration (g/mL)

$$[\alpha] = \frac{10°}{1 \text{ dm} \times (1 \text{ g}/10 \text{ mL})} = +100 = \textbf{specific rotation}$$

**5.29** **Enantiomeric excess** = *ee* = % of one enantiomer − % of other enantiomer.

a. 95 – 5 = **90% *ee***          b. 85 – 15 = **70% *ee***

## 5.30

a. 90% *ee* means 90% excess of **A**, and 10% racemic mixture of **A** and **B** (5% each); therefore,
   **95% A and 5% B.**
b. 99% *ee* means 99% excess of **A**, and 1% racemic mixture of **A** and **B** (0.5% each); therefore,
   **99.5% A and 0.5% B.**
c. 60% *ee* means 60% excess of **A**, and 40% racemic mixture of **A** and **B** (20% each); therefore,
   **80% A and 20% B.**

**5.31**

$$ee = \frac{[\alpha]\ \text{mixture}}{[\alpha]\ \text{pure enantiomer}} \times 100\%$$

a. $\dfrac{+10}{+24} \times 100\% = 42\%\ ee$

b. $\dfrac{[\alpha]\ \text{solution}}{+24} \times 100\% = 80\%\ ee$

$[\alpha]\ \text{solution} = +19.2$

**5.32 • Enantiomers have the same physical properties** (mp, bp, solubility), and rotate the plane of polarized light to an equal but opposite extent.
  • **Diastereomers have different physical properties.**
  • **A racemic mixture is optically inactive.**

**trans isomers**          **cis isomer**

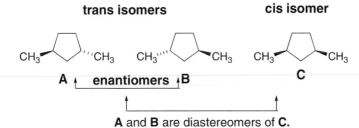

A ⥮ **enantiomers** ⥮ B          **C**

**A** and **B** are diastereomers of **C.**

a. The bp's of **A** and **B** are the same. The bp's of **A** and **C** are different.
b. Pure **A**: optically active
   Pure **B**: optically active
   Pure **C**: optically inactive
   Equal mixture of **A** and **B**: optically inactive
   Equal mixture of **A** and **C**: optically active
c. There would be two fractions: one containing **A** and **B** (optically inactive), and one containing **C** (optically inactive).

**5.33** Use the definitions from Answer 5.2.

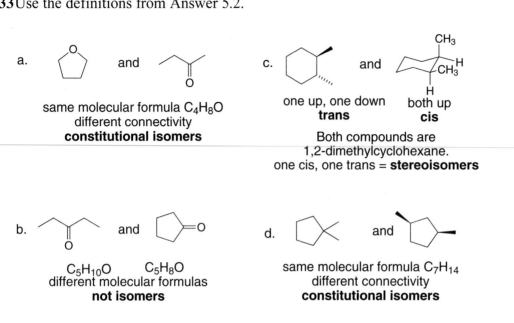

**a.** same molecular formula C$_4$H$_8$O
different connectivity
**constitutional isomers**

**c.** one up, one down  both up
**trans**  **cis**

Both compounds are
1,2-dimethylcyclohexane.
one cis, one trans = **stereoisomers**

**b.** C$_5$H$_{10}$O  C$_5$H$_8$O
different molecular formulas
**not isomers**

**d.** same molecular formula C$_7$H$_{14}$
different connectivity
**constitutional isomers**

**5.34** Use the definitions from Answer 5.3.

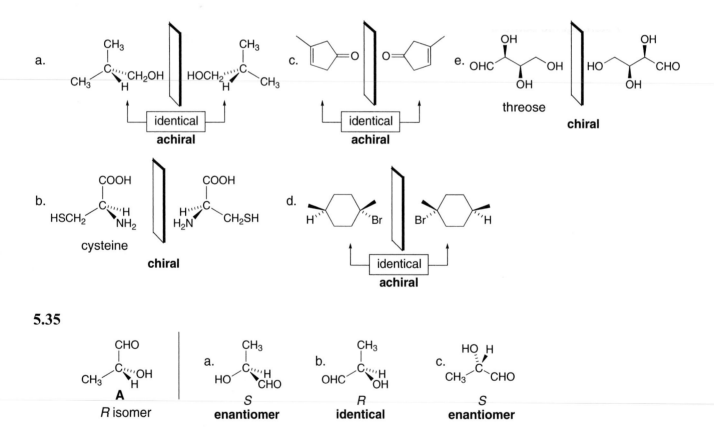

**a.** identical
**achiral**

**c.** identical
**achiral**

**e.** threose
**chiral**

**b.** cysteine
**chiral**

**d.** identical
**achiral**

**5.35**

A
*R* isomer

**a.** *S*
**enantiomer**

**b.** *R*
**identical**

**c.** *S*
**enantiomer**

**5.36** A plane of symmetry cuts the molecule into **two identical halves**.

a.

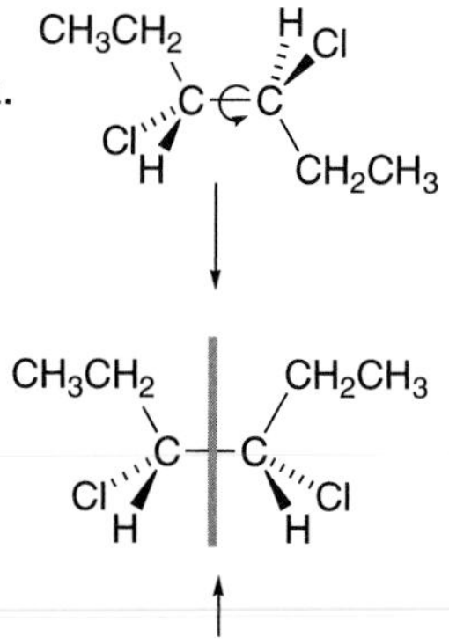

plane of symmetry

b.

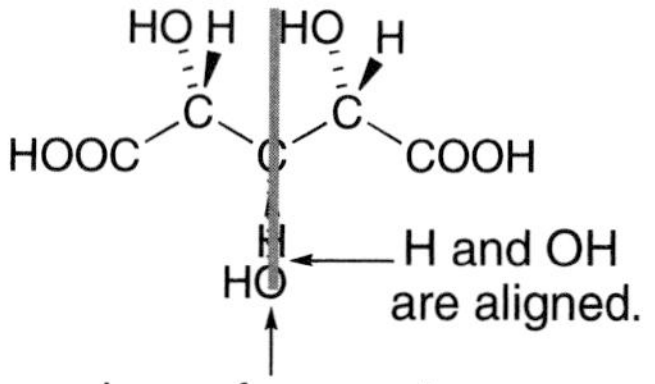

H and OH are aligned.

plane of symmetry

c.

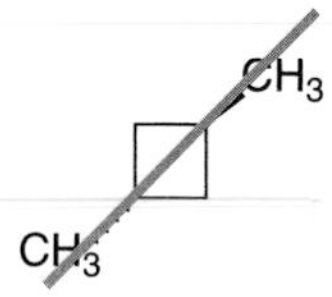

The plane of symmetry bisects the molecule.

d.

no plane of symmetry

e.

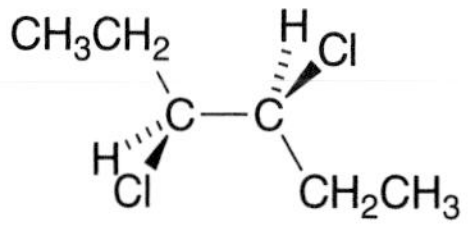

no plane of symmetry

**5.37** Use the directions from Answer 5.6 to locate the stereogenic centers.

a. $CH_3CH_2CH_2CH_2CH_2CH_3$
All C's have 2 or more H's.
**0 stereogenic centers**

b. 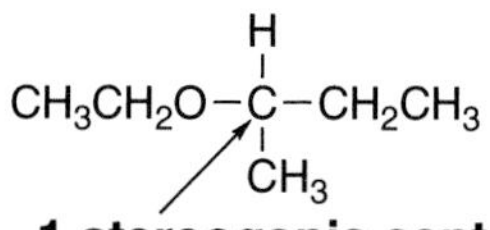
$$CH_3CH_2O-\overset{\overset{H}{|}}{\underset{\underset{CH_3}{|}}{C}}-CH_2CH_3$$
**1 stereogenic center**

c. $(CH_3)_2CHCH(OH)CH(CH_3)_2$
**0 stereogenic centers**

d. $(CH_3)_2CHCH_2-\overset{\overset{H}{|}}{\underset{\underset{CH_3}{|}}{C}}-CH_2-\overset{\overset{H}{|}}{\underset{\underset{CH_3}{|}}{C}}-\overset{\overset{H}{|}}{\underset{\underset{CH_3}{|}}{C}}-CH_2CH_3$
**3 stereogenic centers**

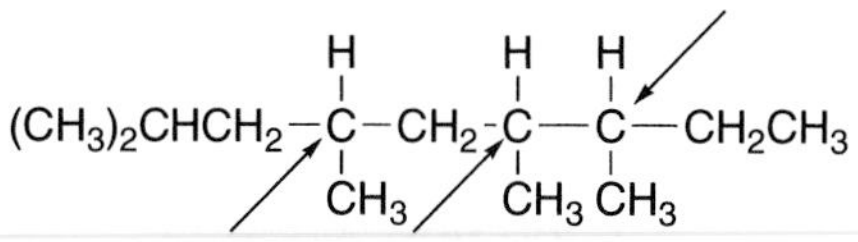

e. $CH_3-\overset{\overset{H}{|}}{\underset{\underset{D}{|}}{C}}-CH_2CH_3$
bonded to 4 different groups
**1 stereogenic center**

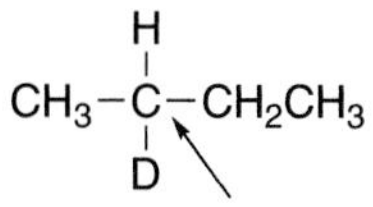

f. 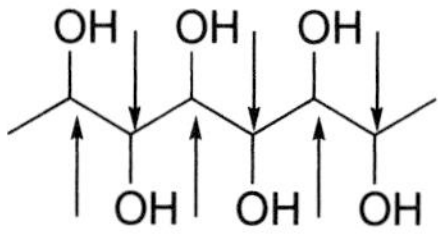
Each labeled C bonded to 4 different groups =
**6 stereogenic centers**

g. 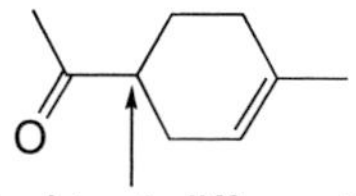
bonded to 4 different groups
**1 stereogenic center**

h. 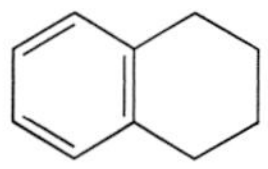
All C's have 2 or more H's, or are $sp^2$ hybridized.
**0 stereogenic centers**

i. 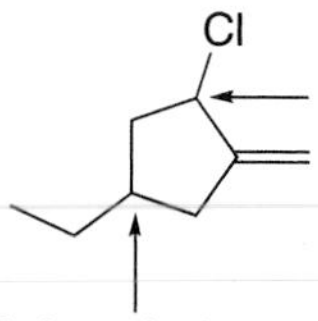
Each bonded to 4 different groups =
**2 stereogenic centers**

j. 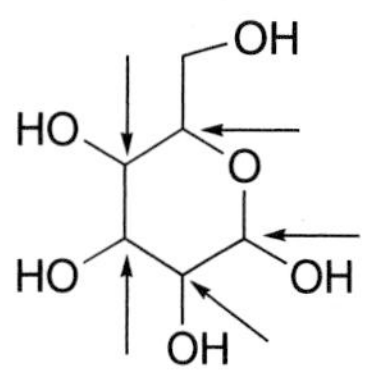
Each labeled C bonded to 4 different groups =
**5 stereogenic centers**

**5.38** Stereogenic centers are circled.

Eight constitutional isomers:

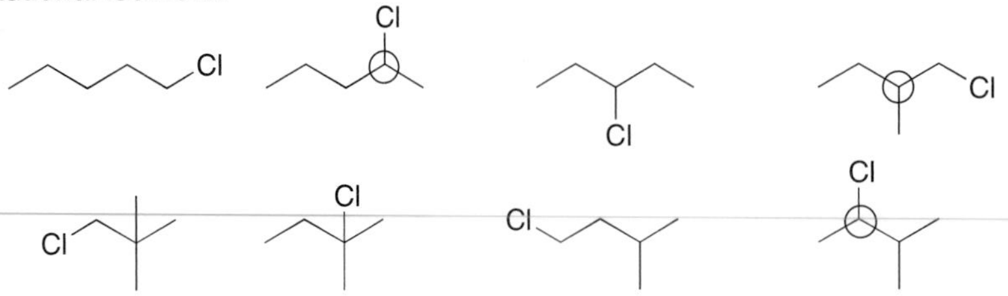

**5.39**

a.

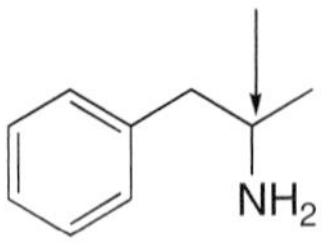

amphetamine

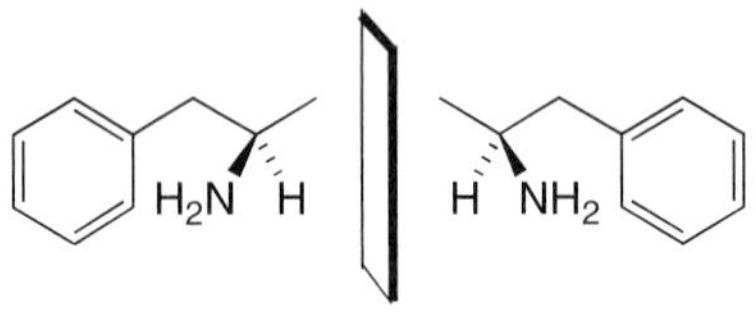

b.

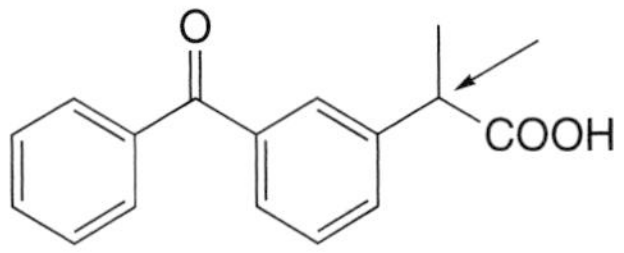

ketoprofen

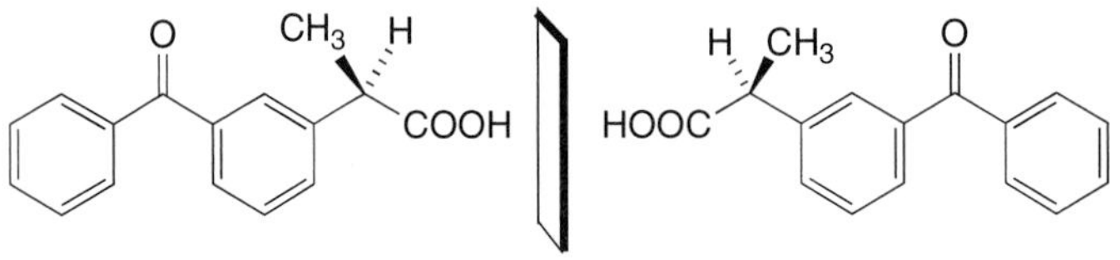

**5.40**

$$CH_3-\underset{\underset{H}{|}}{\overset{\overset{CH_2CH_3}{|}}{C}}-CH_2CH_2CH_3 \quad \text{or} \quad CH_3-\underset{\underset{H}{|}}{\overset{\overset{CH_2CH_3}{|}}{C}}-CH(CH_3)_2$$

**5.41** Assign priority based on the rules in Answer 5.12.

a. –OH, –NH₂
   higher atomic number
   **higher priority**

b. –CD₃, –CH₃
   D higher mass than H
   **higher priority**

c. –CH(CH₃)₂, –CH₂OH
   C bonded to O
   **higher priority**

d. –CH₂Cl, –CH₂CH₂CH₂Br
   C bonded to Cl
   **higher priority**

e. –CHO, –COOH
   C has 3 bonds to O
   **higher priority**

f. –CH₂NH₂, –NHCH₃
   higher atomic number
   **higher priority**

**5.42** Assign priority based on the rules in Answer 5.12.

a. –F > –OH > –NH₂ > –CH₃

b. –(CH₂)₃CH₃ > –CH₂CH₂CH₃ > –CH₂CH₃ > –CH₃

c. –NH₂ > –CH₂NHCH₃ > –CH₂NH₂ > –CH₃

d. –COOH > –CHO > –CH₂OH > –H

e. –Cl > –SH > –OH > –CH₃

f.  –C≡CH >  –CH=CH₂ >  –CH(CH₃)₂ > –CH₂CH₃

**5.43** Use the rules in Answer 5.14 to assign *R* or *S* to each stereogenic center.

a.

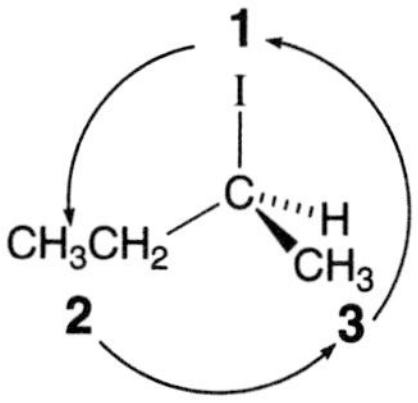

counterclockwise
**S isomer**

c.

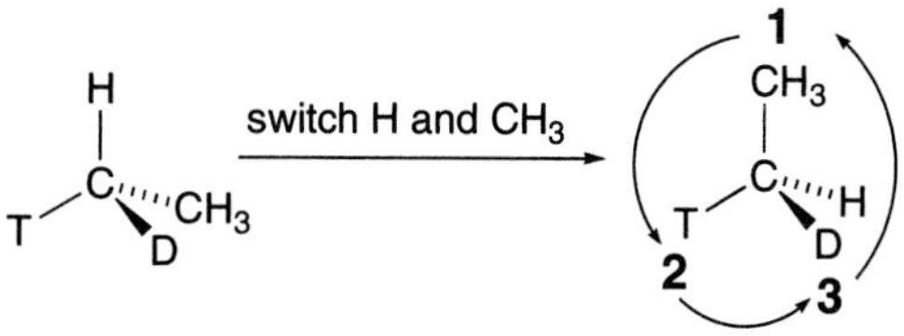

switch H and CH₃

counterclockwise
It looks like an *S* isomer, but we
must reverse the answer, *S* to *R*.
**R isomer**

b.

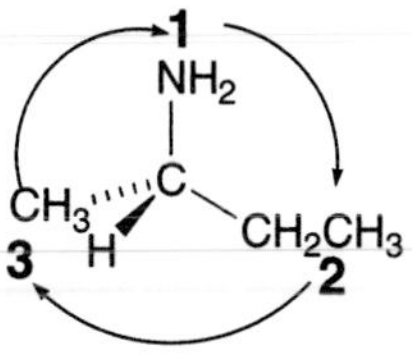

clockwise, but H in front
**S isomer**

d.

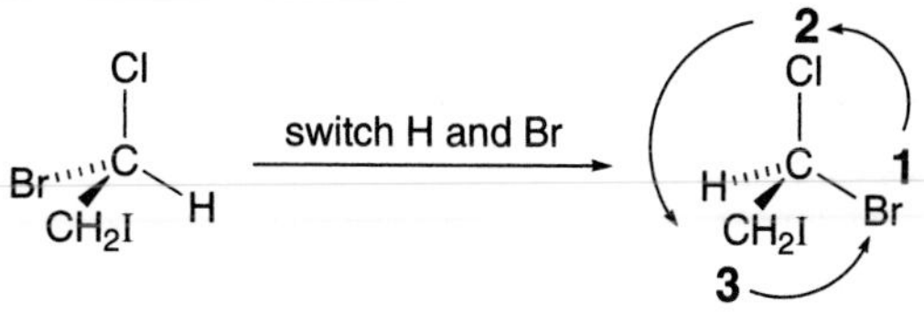

switch H and Br

counterclockwise
It looks like an *S* isomer, but we
must reverse the answer, *S* to *R*.
**R isomer**

e.

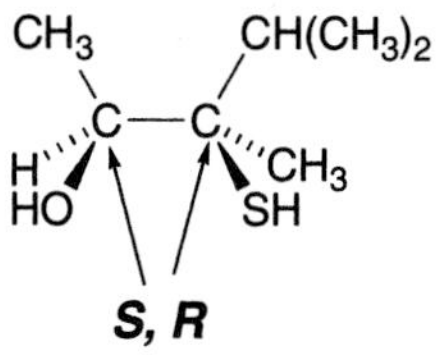

**S, R**

f.

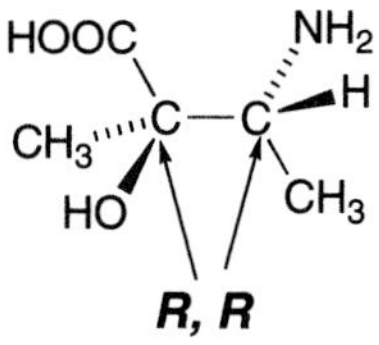

**R, R**

g.

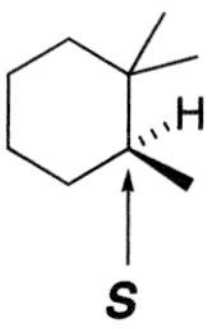

**S**

h.

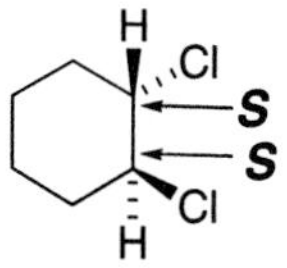

**S**
**S**

**5.44**

a. (3*R*)-3-methylhexane

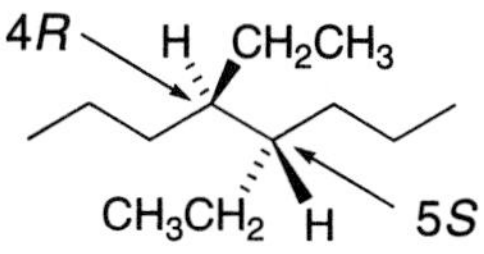

c. (3*R*,5*S*,6*R*)-5-ethyl-3,6-dimethylnonane

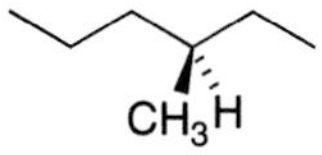

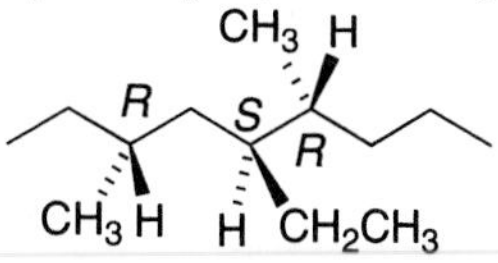

b. (4*R*,5*S*)-4,5-diethyloctane

d. (3*S*,6*S*)-6-isopropyl-3-methyldecane

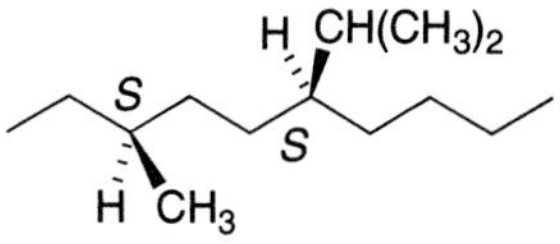

**5.45** Two enantiomers of the amino acid leucine.

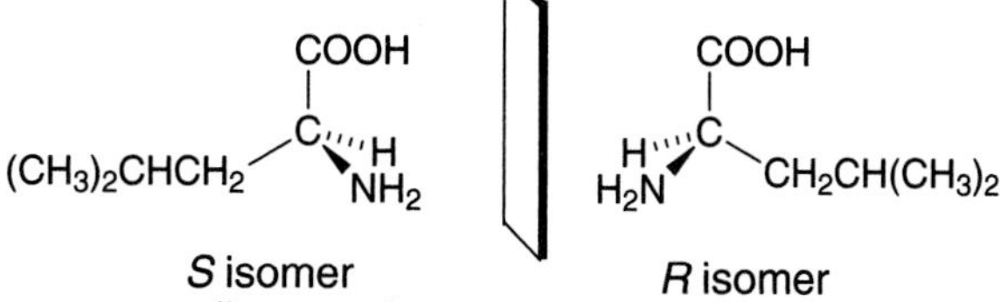

*S* isomer
naturally occurring

*R* isomer

**5.46**

a. 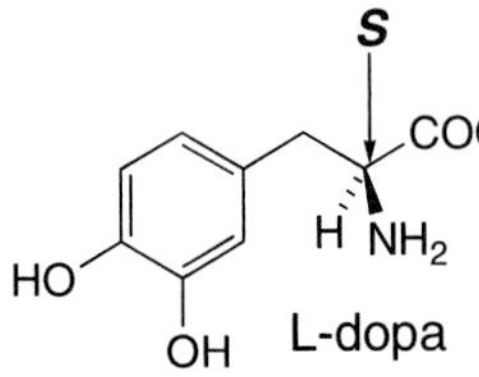 L-dopa

b. 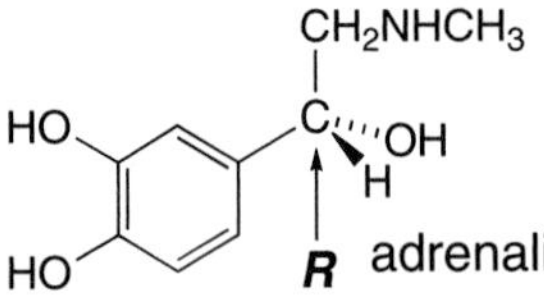 adrenaline

c. 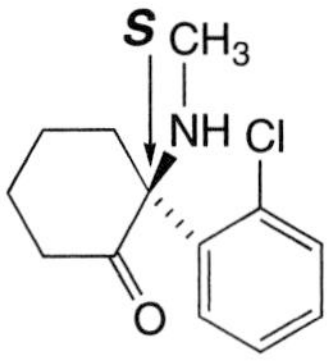 ketamine

**5.47**

methylphenidate  *R, R* 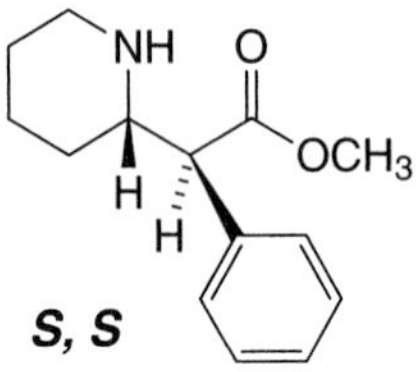  *S, S*

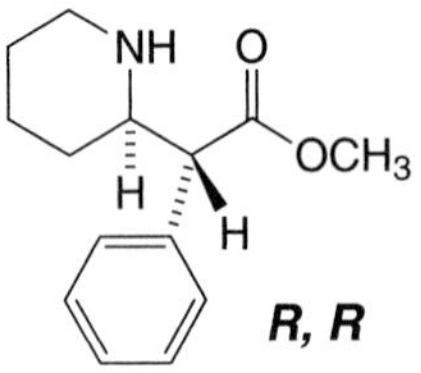

**5.48**

a. 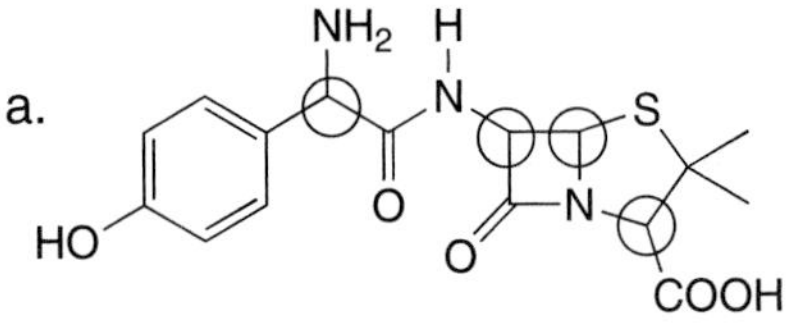 amoxicillin

b. 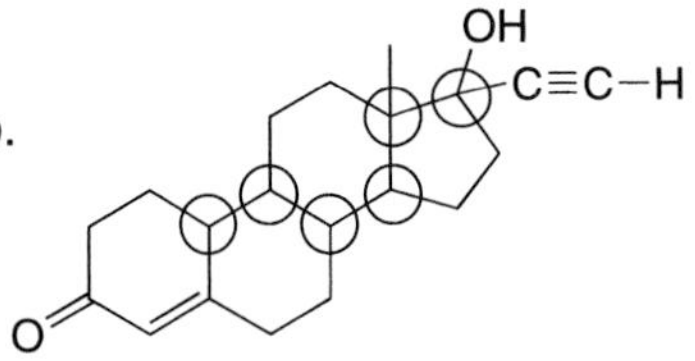 norethindrone

c.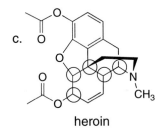
heroin

**5.49**

a. CH₃CH(OH)CH(OH)CH₂CH₃

2 stereogenic centers
$2^2$ = 4 possible stereoisomers

b. CH₃CH₂CH₂CH(CH₃)₂

0 stereogenic centers

c. 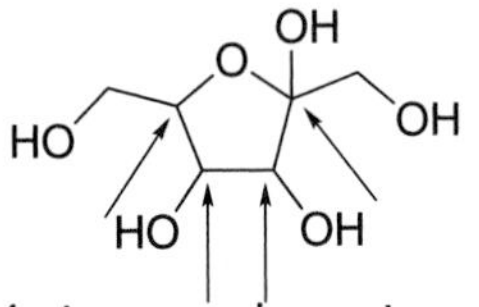 HO

4 stereogenic centers
$2^4$ = 16 possible stereoisomers

**5.50**

a. CH₃CH(OH)CH(OH)CH₂CH₃

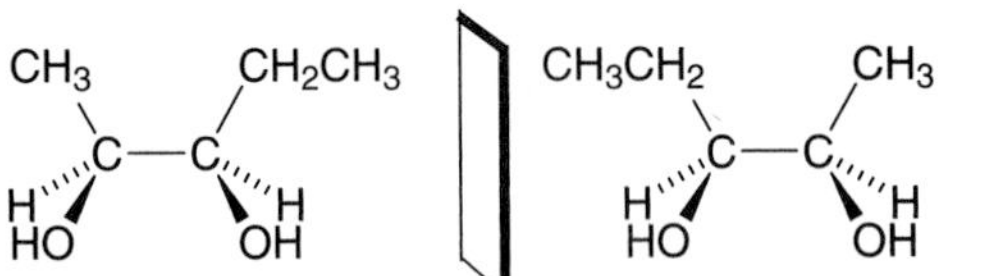

**A**   **B** 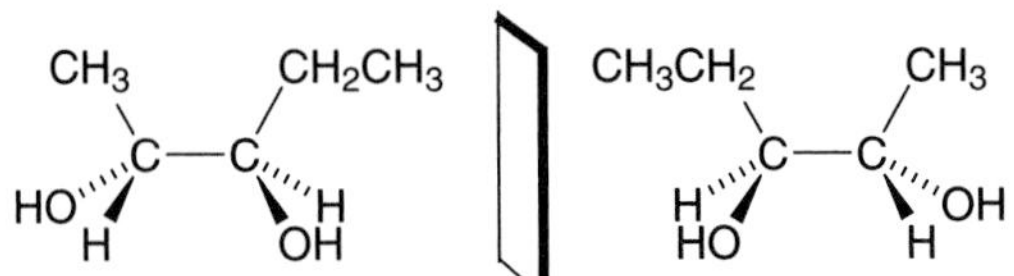   **C**   **D**

Pairs of enantiomers: **A** and **B**, **C** and **D**.
Pairs of diastereomers: **A** and **C**, **A** and **D**, **B** and **C**, **B** and **D**.

b.  $CH_3CH(OH)CH_2CH_2CH(OH)CH_3$

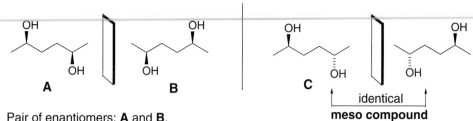

Pair of enantiomers: **A** and **B**.
Pairs of diastereomers: **A** and **C**, **B** and **C**.

c.  $CH_3CH(Cl)CH_2CH(Br)CH_3$

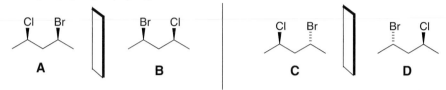

Pairs of enantiomers: **A** and **B**, **C** and **D**.
Pairs of diastereomers: **A** and **C**, **A** and **D**, **B** and **C**, **B** and **D**.

d.  $CH_3CH(Br)CH(Br)CH(Br)CH_3$

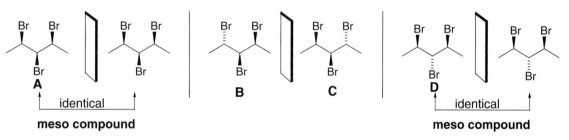

Pair of enantiomers: **B** and **C**.
Pairs of diastereomers: **A** and **B**, **A** and **C**, **A** and **D**, **B** and **D**,  **C** and **D**

**5.51**

a.

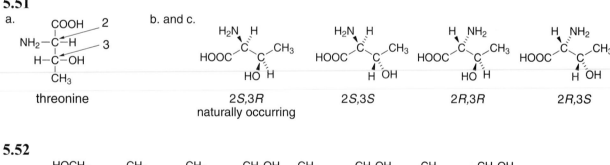

threonine

2S,3R
naturally occurring

2S,3S

2R,3R

2R,3S

**5.52**

a.

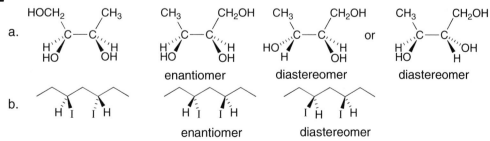

enantiomer

diastereomer

or

diastereomer

b.

enantiomer

diastereomer

c.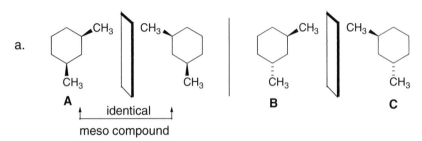

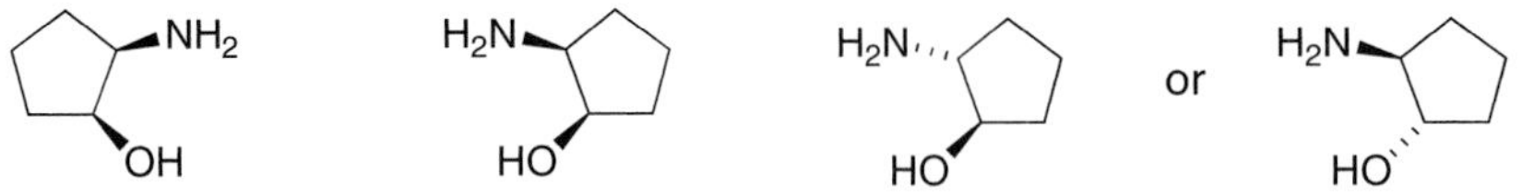

enantiomer     diastereomer      or      diastereomer

d.

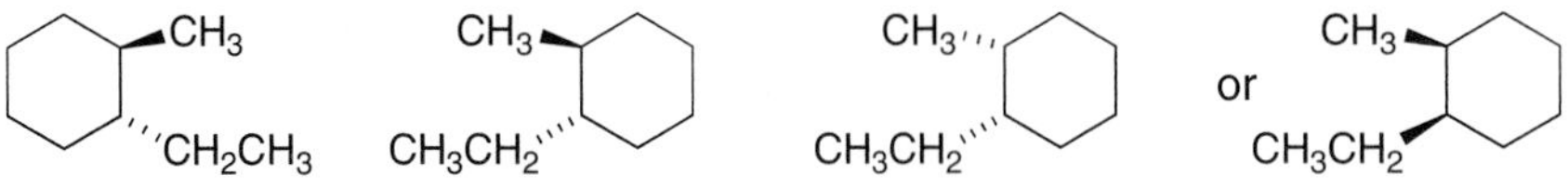

enantiomer     diastereomer      or      diastereomer

**5.53**

a.

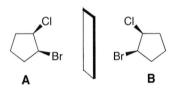

**A**      identical

meso compound

Pair of enantiomers: **B** and **C**.
Pairs of diastereomers: **A** and **B**, **A** and **C**.

b.

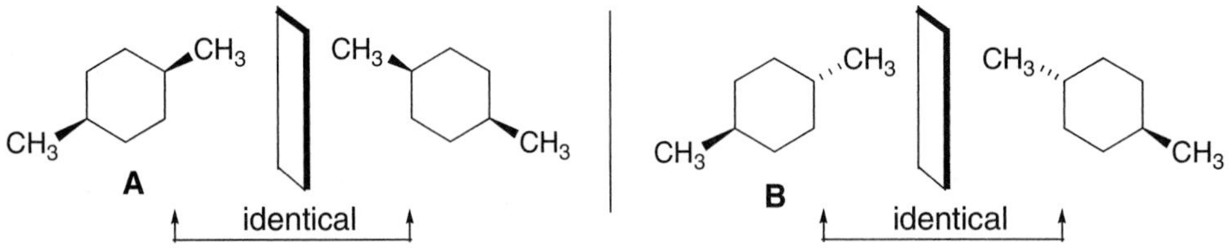

**A**     identical        **B**     identical

Pair of diastereomers: **A** and **B**.
Meso compounds: **A** and **B**.

c.

**A**      **B**      **C**      **D**

Pairs of enantiomers: **A** and **B**, **C** and **D**.
Pairs of diastereomers: **A** and **C**, **A** and **D**, **B** and **C**, **B** and **D**.

**5.54**

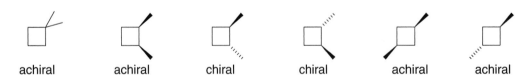

achiral      achiral      chiral      chiral      achiral      achiral

**5.55** **A** has two stereogenic centers and a plane of symmetry, making it an achiral meso compound. Since it superimposable on its mirror image it has no enantiomer. **B** has only one stereogenic center. Its two possible stereoisomers consist of a pair of enantiomers, but no diastereomer.

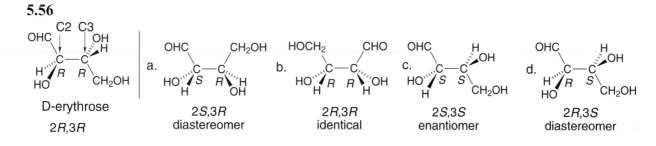

**A**

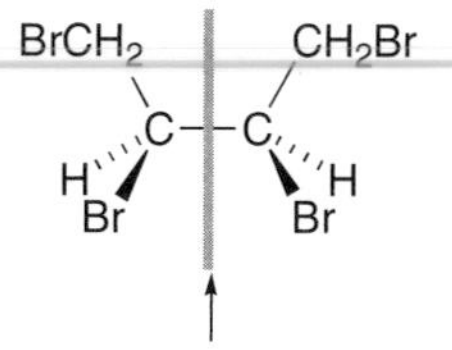

BrCH₂     CH₂Br

plane of symmetry
an achiral compound

**B**

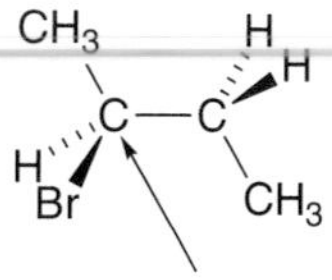

only one stereogenic center
**no diastereomer**

**5.56**

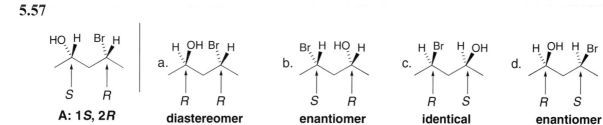

D-erythrose
**2R,3R**

a.
2S,3R
diastereomer

b.
2R,3R
identical

c.
2S,3S
enantiomer

d.
2R,3S
diastereomer

**5.57**

a.
R  R
**diastereomer**

b.
S  R
**enantiomer**

c.
R  S
**identical**

d.
R  S
**enantiomer**

A: **1S, 2R**

**5.58**

a.  **enantiomers**

b. 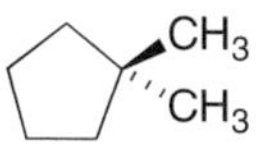 **same molecular formula
different connectivity
constitutional isomers**

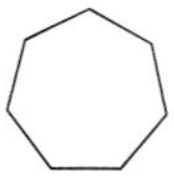

c.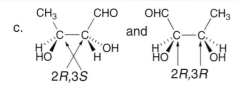

and

2R,3S     2R,3R

one different configuration
**diastereomers**

d. 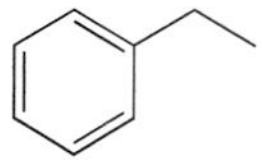 and 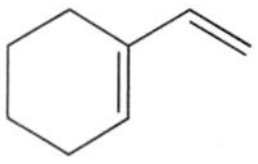
different molecular formulas
**not isomers**

e.  and 
mirror images
not superimposable
**enantiomers**

f. 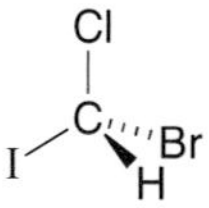 and 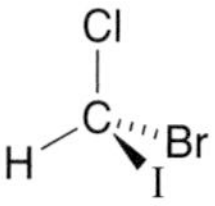
**enantiomers**

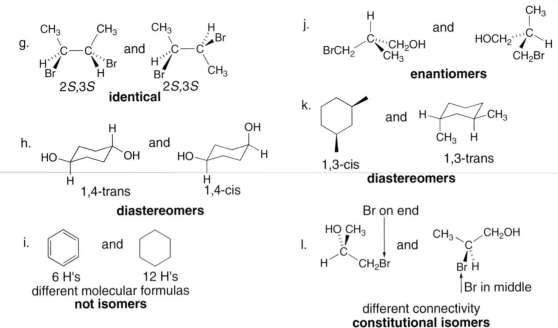

g. ... 2S,3S **identical** 2S,3S

j. **enantiomers**

h. 1,4-trans **diasteromers** 1,4-cis

k. 1,3-cis **diasteromers** 1,3-trans

i. 6 H's / 12 H's different molecular formulas **not isomers**

l. Br on end ... different connectivity **constitutional isomers** ... Br in middle

**5.59**

a. **A** and **B** are constitutional isomers.
A and **C** are constitutional isomers.
**B** and **C** are diastereomers (cis and trans).
**C** and **D** are enantiomers.

b.

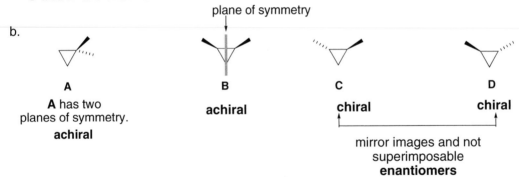

plane of symmetry

A — **A** has two planes of symmetry. **achiral**

B — **achiral**

C — **chiral**

D — **chiral**

mirror images and not superimposable **enantiomers**

c. Alone, **C** and **D** would be optically active.
d. **A** and **B** have a plane of symmetry.
e. **A** and **B** have different boiling points.
**B** and **C** have different boiling points.
**C** and **D** have the same boiling point.
f. **B** is a meso compound.
g. An equal mixture of **C** and **D** is optically inactive because it is a racemic mixture.
An equal mixture of **B** and **C** would be optically active.

**5.60**

quinine

$$ee = \frac{[\alpha] \text{ mixture}}{[\alpha] \text{ pure enantiomer}} \times 100\%$$

quinine = **A**
quinine's enantiomer = **B**

a.

$$\frac{-50}{-165} \times 100\% = 30\% \; ee$$

b. 30% *ee* = 30% excess one compound (**A**)
remaining 70% = mixture of 2 compounds (35% each **A** and **B**)
Amount of **A** = 30 + 35 = **65%**
Amount of **B** = **35%**

$$\frac{-83}{-165} \times 100\% = 50\% \; ee$$

50% *ee* = 50% excess one compound (**A**)
remaining 50% = mixture of 2 compounds (25% each **A** and **B**)
Amount of **A** = 50 + 25 = **75%**
Amount of **B** = 25%

$$\frac{-120}{-165} \times 100\% = 73\% \; ee$$

73% *ee* = 73% excess of one compound (**A**)
remaining 27% = mixture of 2 compounds (13.5% each **A** and **B**)
Amount of **A** = 73 + 13.5 = **86.5%**
Amount of **B** = **13.5%**

c. $[\alpha] = +165$
d. $80\% - 20\% = 60\% \; ee$

e. $60\% = \dfrac{[\alpha] \text{ mixture}}{-165°} \times 100\%$

$[\alpha]$ mixture = −99

**5.61**

amygdalin
(laetrile)

HCl, H₂O

mandelic acid

only one of the
products formed

a. The 11 stereogenic centers are circled. Maximum number of stereoisomers = $2^{11}$ = 2048
b. Enantiomers of mandelic acid:

*R*          *S*

c. $60\% - 40\% = 20\% \; ee$
$20\% = [\alpha] \text{ mixture}/-154 \times 100\%$
$[\alpha]$ mixture = −31

d. $ee = \dfrac{+50}{+154} \times 100\% = 32\% \; ee$

$[\alpha]$ for (*S*)-mandelic acid = +154

32% excess of the *S* enantiomer
68% of racemic *R* and *S* = 34% *S* and 34% *R*

*S* enantiomer: 32% + 34% = 66%
*R* enantiomer = 34%

**5.62** Allenes contain an *sp* hybridized carbon atom doubly bonded to two other carbons. This makes the double bonds of an allene perpendicular to each other. When each end of the allene has two like substituents, the allene contains two planes of symmetry and it is achiral. When each end of the allene has two different groups, the allene has no plane of symmetry and it becomes chiral.

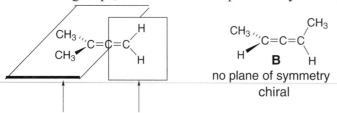

no plane of symmetry
chiral

These two substituents are at 90° to these two substituents.
Allene **A** contains two planes of symmetry,
making it **achiral.**

**5.63**

discodermolide

a. The 13 tetrahedral stereogenic centers are circled.

b. Because there is restricted rotation around a C–C double bond, groups on the end of the double bond cannot interconvert. Whenever the substituents on each end of the double bond are different from each other, the double bond is a stereogenic site. Thus, the following two double bonds are isomers:

These compounds are isomers.

There are three stereogenic double bonds in discodermolide, labeled with arrows.

c. The maximum number of stereoisomers for discodermolide must include the 13 tetrahedral stereogenic centers and the three double bonds. Maximum number of stereoisomers = $2^{16}$ = 65,536.

**5.64**

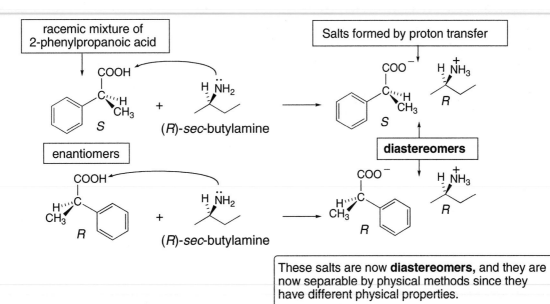

racemic mixture of
2-phenylpropanoic acid
COOH
C
H
CH3
S
+
H NH2
(R)-sec-butylamine
enantiomers
COOH
H
C
CH3
R
+
H NH2
(R)-sec-butylamine
Salts formed by proton transfer
COO−
C
H
CH3
S
H NH3+
R
diastereomers
COO−
H
C
CH3
R
H NH3+
R
These salts are now diastereomers, and they are now separable by physical methods since they have different physical properties.

# Chapter 6: Understanding Organic Reactions

## ◆ Writing organic reactions (6.1)

- Use curved arrows to show the movement of electrons. Full-headed arrows are used for electron pairs and half-headed arrows are used for single electrons.

$$-\overset{|}{\underset{|}{C}}\!\!-\!\!Z \longrightarrow -\overset{|}{\underset{|}{C}}\cdot \ + \ \cdot Z \qquad\qquad -\overset{|}{\underset{|}{C}}\!\!-\!\!Z \longrightarrow -\overset{|}{\underset{|}{C}}+ \ + \ :Z^{-}$$

half-headed arrows                          full-headed arrow

- Reagents can be drawn either on the left side of an equation or over an arrow. Catalysts are drawn over or under an arrow.

## ◆ Types of reactions (6.2)

| [1] Substitution | $-\overset{|}{\underset{|}{C}}\!\!-\!\!Z \ + \ Y \longrightarrow -\overset{|}{\underset{|}{C}}\!\!-\!\!Y \ + \ Z$  $\left[\, Z = H \text{ or a heteroatom} \,\right]$ <br><br> Y replaces Z |
|---|---|
| [2] Elimination | $-\overset{|}{\underset{\underset{X}{|}}{C}}\!\!-\!\!\overset{|}{\underset{\underset{Y}{|}}{C}}\!\!- \ + \ \text{reagent} \longrightarrow \overset{\diagdown}{\diagup}C\!\!=\!\!C\overset{\diagup}{\diagdown} \ + \ X\!\!-\!\!Y$ <br><br> Two σ bonds are broken.       π bond |
| [3] Addition | $\overset{\diagdown}{\diagup}C\!\!=\!\!C\overset{\diagup}{\diagdown} \ + \ X\!\!-\!\!Y \longrightarrow -\overset{|}{\underset{\underset{X}{|}}{C}}\!\!-\!\!\overset{|}{\underset{\underset{Y}{|}}{C}}\!\!-$ <br><br> This π bond is broken.       Two σ bonds are formed. |

## ◆ Important trends

| Values compared | Trend |
|---|---|
| **Bond dissociation energy** and **bond strength** | The *higher* the bond dissociation energy, the *stronger* the bond (6.4). <br><br> **Increasing size of the halogen** ⟶ <br><br> $CH_3-F$    $CH_3-Cl$    $CH_3-Br$    $CH_3-I$ <br><br> $\Delta H^{\circ} = 109$ kcal/mol    84 kcal/mol    70 kcal/mol    56 kcal/mol <br><br> ⟵ **Increasing bond strength** |

| $E_a$ and **reaction rate** | The *larger* the energy of activation, the *slower* the reaction (6.9A). |
|---|---|
| | 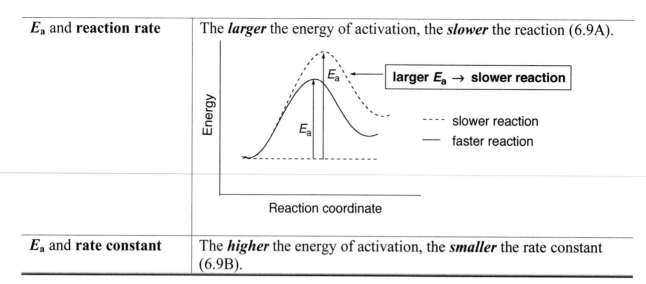 |
| $E_a$ and **rate constant** | The *higher* the energy of activation, the *smaller* the rate constant (6.9B). |

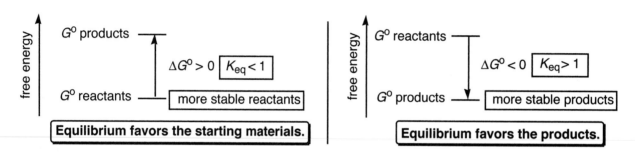

♦ Reactive intermediates (6.3)

- Breaking bonds generates reactive intermediates.
- Homolysis generates radicals with unpaired electrons.
- Heterolysis generates ions.

| Reactive intermediate | General structure | Reactive feature | Reactivity |
|---|---|---|---|
| radical | —Ċ· | unpaired electron | electrophilic |
| carbocation | —C+ | positive charge; only six electrons around C | electrophilic |
| carbanion | —C:⁻ | net negative charge; lone electron pair on C | nucleophilic |

◆ Energy diagrams (6.7, 6.8)

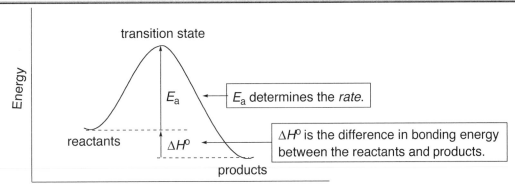

◆ Conditions favoring product formation (6.5, 6.6)

| Variable | Value | Meaning |
|----------|-------|---------|
| $K_{eq}$ | $K_{eq} > 1$ | More product than starting material is present at equilibrium. |
| $\Delta G^\circ$ | $\Delta G^\circ < 0$ | The energy of the products is **lower** than the energy of the reactants. |
| $\Delta H^\circ$ | $\Delta H^\circ < 0$ | Bonds in the products are **stronger** than bonds in the reactants. |
| $\Delta S^\circ$ | $\Delta S^\circ > 0$ | The product is **more disordered** than the reactant. |

◆ Equations (6.5, 6.6)

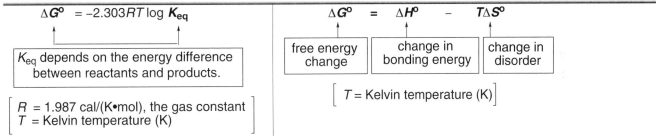

$$\Delta G^\circ = -2.303RT \log K_{eq}$$

$K_{eq}$ depends on the energy difference between reactants and products.

$\left[ \begin{array}{l} R = 1.987 \text{ cal/(K•mol), the gas constant} \\ T = \text{Kelvin temperature (K)} \end{array} \right]$

$$\Delta G^\circ = \Delta H^\circ - T\Delta S^\circ$$

free energy change    change in bonding energy    change in disorder

$\left[ T = \text{Kelvin temperature (K)} \right]$

◆ Factors affecting reaction rate (6.9)

| Factor | Effect |
|--------|--------|
| energy of activation | higher $E_a \rightarrow$ slower reaction |
| concentration | higher concentration $\rightarrow$ faster reaction |
| temperature | higher temperature $\rightarrow$ faster reaction |

## Chapter 6: Answers to Problems

**6.1** [1] In a **substitution reaction**, one group replaces another.
[2] In an **elimination reaction**, elements of the starting material are lost and a π bond is formed.
[3] In an **addition reaction**, elements are added to the starting material.

a.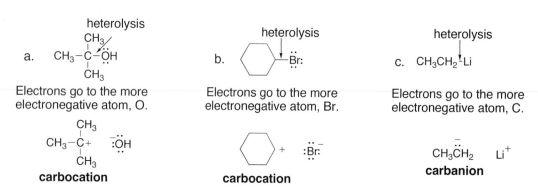

Br replaces OH =
**substitution reaction**

c.

Cl replaces H =
**substitution reaction**

b.

addition of 2 H's
**addition reaction**

d.   $CH_3CH_2CH(OH)CH_3$ ⟶ $CH_3CH=CHCH_3$

elements lost
(H + OH)

π bond formed

**elimination reaction**

**6.2** **Heterolysis** means one atom gets both of the electrons when a bond is broken. A carbocation is a C with a positive charge, and a carbanion is a C with a negative charge.

heterolysis

a.   $CH_3-\overset{\underset{CH_3}{|}}{\overset{CH_3}{|}}C-\ddot{O}H$

Electrons go to the more electronegative atom, O.

$CH_3-\overset{\underset{CH_3}{|}}{\overset{CH_3}{|}}C+$   $:\ddot{O}H$

**carbocation**

heterolysis

b.   —Br:

Electrons go to the more electronegative atom, Br.

+   $:\ddot{Br}:^-$

**carbocation**

heterolysis

c.   $CH_3CH_2-Li$

Electrons go to the more electronegative atom, C.

$CH_3\bar{\ddot{C}}H_2$   $Li^+$

**carbanion**

**6.3** Use **full-headed arrows** to show the movement of electron pairs, and **half-headed arrows** to show the movement of single electrons.

a. $(CH_3)_3C-\overset{+}{N}\equiv N:$ ⟶ $(CH_3)_3C^+$ + $:N\equiv N:$

c. $CH_3-\overset{\underset{CH_3}{|}}{\overset{CH_3}{|}}\overset{+}{C}$ + $:\ddot{Br}:^-$ ⟶ $CH_3-\overset{\underset{CH_3}{|}}{\overset{CH_3}{|}}C-\ddot{Br}:$

b. $\cdot CH_3$ + $\cdot CH_3$ ⟶ $CH_3-CH_3$

d. $H\ddot{O}-\ddot{O}H$ ⟶ 2 $H\ddot{O}\cdot$

**6.4** Increasing number of electrons between atoms = increasing bond strength = increasing bond dissociation energy = decreasing bond length.
Increasing size of an atom = increasing bond length = decreasing bond strength.

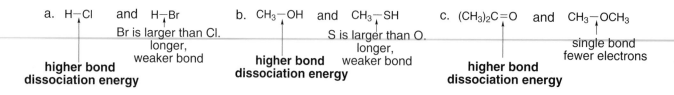

a.  H—Cl   and   H—Br

Br is larger than Cl.
longer,
weaker bond

**higher bond dissociation energy**

b.  $CH_3$—OH   and   $CH_3$—SH

S is larger than O.
longer,
weaker bond

**higher bond dissociation energy**

c.  $(CH_3)_2C=O$   and   $CH_3$—$OCH_3$

single bond
fewer electrons

**higher bond dissociation energy**

**6.5  To determine $\Delta H^0$ for a reaction**:

[1] Add the bond dissociation energies for all bonds *broken* in the equation (+ values).
[2] Add the bond dissociation energies for all of the bonds *formed* in the equation (– values).
[3] *Add the energies together* to get the $\Delta H^0$ for the reaction.
**A positive $\Delta H^0$ means the reaction is *endothermic*.  A negative $\Delta H^0$ means the reaction is *exothermic*.**

a.  $CH_3CH_2-Br + H_2O \longrightarrow CH_3CH_2-OH + HBr$

**[1] Bonds broken**

| | $\Delta H^0$ (kcal/mol) |
|---|---|
| $CH_3CH_2-Br$ | + 68 |
| H–OH | + 119 |
| Total | + 187  kcal/mol |

**[2] Bonds formed**

| | $\Delta H^0$ (kcal/mol) |
|---|---|
| $CH_3CH_2-OH$ | – 94 |
| H–Br | – 88 |
| Total | – 182  kcal/mol |

**[3] Overall $\Delta H^0$ =**

| sum in Step [1] |
|---|
| + |
| sum in Step [2] |

+ 187  kcal/mol
– 182  kcal/mol

ANSWER: + 5  kcal/mol
**endothermic**

b.  $CH_4 + Cl_2 \longrightarrow CH_3Cl + HCl$

**[1] Bonds broken**

| | $\Delta H^0$ (kcal/mol) |
|---|---|
| $CH_3-H$ | + 104 |
| Cl–Cl | + 58 |
| Total | + 162  kcal/mol |

**[2] Bonds formed**

| | $\Delta H^0$ (kcal/mol) |
|---|---|
| $CH_3-Cl$ | – 84 |
| H–Cl | – 103 |
| Total | – 187  kcal/mol |

**[3] Overall $\Delta H^0$ =**

| sum in Step [1] |
|---|
| + |
| sum in Step [2] |

+ 162  kcal/mol
– 187  kcal/mol

ANSWER: – 25  kcal/mol
**exothermic**

**6.6**  Use the directions from Answer 6.5.  In determining the number of bonds broken or formed, you must take into account the coefficients needed to  balance an equation.

a.  $CH_4 + 2 O_2 \longrightarrow CO_2 + 2 H_2O$

**[1] Bonds broken**

| | $\Delta H^0$ (kcal/mol) |
|---|---|
| $CH_3-H$ | + 104 x 4 = + 416 |
| O–O | + 119 x 2 = + 238 |
| Total | + 654 kcal/mol |

**[2] Bonds formed**

| | $\Delta H^0$ (kcal/mol) |
|---|---|
| OC–O | – 128 x 2 = – 256 |
| HO–H | – 119 x 4 = – 476 |
| Total | – 732 kcal/mol |

**[3] Overall $\Delta H^0$ =**

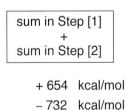

| sum in Step [1] |
|---|
| + |
| sum in Step [2] |

+ 654  kcal/mol
– 732  kcal/mol

ANSWER: – 78  kcal/mol

b. $CH_3CH_3 + (7/2) O_2 \longrightarrow 2 CO_2 + 3 H_2O$

| **[1] Bonds broken** | | **[2] Bonds formed** | | **[3] Overall $\Delta H^\circ =$** |
|---|---|---|---|---|
| | $\Delta H^\circ$ (kcal/mol) | | $\Delta H^\circ$ (kcal/mol) | sum in Step [1] |
| $CH_3CH_2-H$ | + 98 x 6 = + 588 | OC−O | − 128 x 4 = − 512 | + |
| O−O | + 119 x 3.5 = + 417 | HO−H | − 119 x 6 = − 714 | sum in Step [2] |
| C−C | + 88 | | | |
| | | Total | − 1226 kcal/mol | + 1093 kcal/mol |
| Total | + 1093 kcal/mol | | | − 1226 kcal/mol |

ANSWER: − 133 kcal/mol

**6.7** Use the following relationships to answer the questions:
$K_{eq} = 1$ then $\Delta G^\circ = 0$; $K_{eq} > 1$ then $\Delta G^\circ < 0$; $K_{eq} < 1$ then $\Delta G^\circ > 0$

   a. A negative value of $\Delta G^\circ$ means the equilibrium favors the product and $K_{eq}$ is $> 1$. Therefore $K_{eq} = 1000$ is the answer.
   b. A lower value of $\Delta G^\circ$ means a larger value of $K_{eq}$, and the products are more favored. $K_{eq} = 10^{-2}$ is larger than $K_{eq} = 10^{-5}$, so $\Delta G^\circ$ is lower.

**6.8** Use the relationships from Answer 6.7.

   a. $K_{eq} = 5.5$. $K_{eq} > 1$ means that the equilibrium favors the **product**.
   b. $\Delta G^\circ = 10.3$ kcal. A positive $\Delta G^\circ$ means the equilibrium favors the **starting material**.

**6.9** When the product is lower in energy than the starting material, the equilibrium favors the product. When the starting material is lower in energy than the product, the equilibrium favors the starting material.

   a. $\Delta G^\circ$ **is positive** so the equilibrium favors the starting material. Therefore the *starting material is lower in energy than the product.*
   b. $K_{eq}$ **is $> 1$** so the equilibrium favors the product. Therefore the *product is lower in energy than the starting material.*
   c. $\Delta G^\circ$ **is negative** so the equilibrium favors the products. Therefore the *products are lower in energy than the starting material.*
   d. $K_{eq}$ **is $< 1$** so the equilibrium favors the starting material. Therefore *the starting material is lower in energy than the product.*

**6.10**

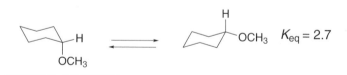

   a. The $K_{eq}$ is $> 1$ and therefore the **product** (the conformation on the right) is favored at equilibrium.
   b. The $\Delta G^\circ$ for this process must be **negative** since the product is favored.
   c. $\Delta G^\circ$ is somewhere between 0 and −1.4 kcal/mol.

**6.11** A positive $\Delta H°$ favors the starting material. A negative $\Delta H°$ favors the product.
    a. $\Delta H°$ is positive (20 kcal). The starting material is favored.
    b. $\Delta H°$ is negative (–10 kcal). The product is favored.

**6.12**
    a. **False.** The reaction is endothermic.
    b. **True.** This assumes that $\Delta G°$ is approximately equal to $\Delta H°$.
    c. **False.** $K_{eq} < 1$
    d. **True**
    e. **False.** The starting material is favored at equilibrium.

**6.13**

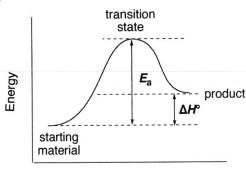

**6.14** A transition state is drawn with dashed lines to indicate the partially broken and partially formed bonds. Any atom that gains or loses a charge contains a partial charge in the transition state.

    a.

$$CH_3-\underset{\underset{CH_3}{|}}{\overset{\overset{CH_3}{|}}{C}}-\overset{+}{O}H_2 \longrightarrow CH_3-\underset{\underset{CH_3}{|}}{\overset{\overset{CH_3}{|}}{C}}+ \quad + \quad H_2O$$

transition state: $\left[ CH_3 \overset{\overset{CH_3}{|}}{\underset{\underset{CH_3}{|}}{\overset{\delta+}{C}}}---\overset{\delta+}{O}H_2 \right]^{\ddagger}$

    b. $CH_3O-H \ + \ {}^-OH \longrightarrow CH_3O^- \ + \ H_2O$

transition state: $\left[ CH_3\overset{\delta-}{\ddot{O}}---H---\overset{\delta-}{\ddot{O}}H \right]^{\ddagger}$

**6.15**

$$\underset{CH_3}{\overset{O}{\overset{\|}{C}}}{-}OH \ + \ CH_3-\underset{\underset{CH_3}{|}}{\overset{\overset{CH_3}{|}}{C}}-O^- \longrightarrow \underset{CH_3}{\overset{O}{\overset{\|}{C}}}{-}O^- \ + \ CH_3-\underset{\underset{CH_3}{|}}{\overset{\overset{CH_3}{|}}{C}}-OH$$

Since $pK_a$ ($CH_3CO_2H$) = 4.8 and $pK_a$ [$(CH_3)_3COH$] = 18, the weaker acid is formed as product, and equilibrium favors the products. Thus, $\Delta H°$ is negative, and the products are lower in energy than the starting materials.

transition state:

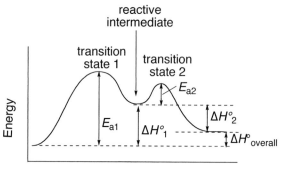

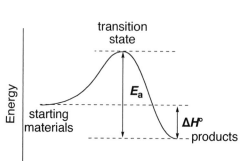

**6.16**

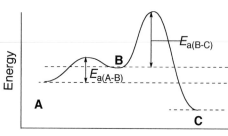

a. Two steps since there are two energy barriers.
b. See labels.
c. See labels.
d. One reactive intermediate is formed (see label).
e. The first step is rate-determining since its transition state is at higher energy.
f. The overall reaction is endothermic since the energy of the products is higher than the energy of the reactants.

**6.17**

relative energies: C < A < B
B ⟶ C is rate-determining.

**6.18** $E_a$, **concentration, and temperature affect reaction rate.** $\Delta H°$, $\Delta G°$, and $K_{eq}$ do not affect reaction rate.

a. $E_a = 1$ **kcal** corresponds to a faster reaction rate.
b. A temperature of **25 °C** will have a faster reaction rate since a higher temperature corresponds to a faster reaction.
c. **No change**: $K_{eq}$ does not affect reaction rate.
d. **No change**: $\Delta H°$ does not affect reaction rate.

**6.19** The $E_a$ of an endothermic reaction is at least as large as its $\Delta H°$ because the $E_a$ essentially "includes" the $\Delta H°$ in its total. The $E_a$ measures the difference between the energy of the starting material and the energy of the transition state, and in an endothermic reaction, the energy of the products is somewhere in between these two values.

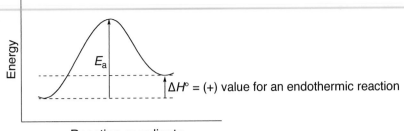

$E_a$

$\uparrow \Delta H° = (+)$ value for an endothermic reaction

Reaction coordinate

**6.20** All reactants in the rate equation determine the rate of the reaction.

(1) rate = $k[CH_3CH_2Br][^-OH]$

a. Tripling the concentration of $CH_3CH_2Br$ only → **The rate is tripled.**

b. Tripling the concentration of $^-OH$ only → **The rate is tripled.**

c. Tripling the concentration of both $CH_3CH_2CH_2Br$ and $^-OH$ → **The rate increases by a factor of 9 (3 x 3 = 9).**

(2) rate = $k[(CH_3)_3COH]$

a. Doubling the concentration of $(CH_3)_3COH$ → **The rate is doubled.**

b. Increasing the concentration of $(CH_3)_3COH$ by a factor of 10 → **The rate increases by a factor of 10.**

**6.21** The rate equation is determined by the rate-determining step.

a. $CH_3CH_2-Br + {}^-OH \longrightarrow CH_2=CH_2 + H_2O + Br^-$     one step     **rate = $k[CH_3CH_2Br][^-OH]$**

b. $(CH_3)_3C-Br \xrightarrow{slow} (CH_3)_3C+ {}_{+ Br^-} \xrightarrow[fast]{^-OH} (CH_3)_2C=CH_2 + H_2O$   two steps  The slow step determines the rate equation.
**rate = $k[(CH_3)_3CBr]$**

**6.22** A catalyst is not used up or changed in the reaction. It only speeds up the reaction rate.

a. OH and H are added to the starting material.

$CH_2=CH_2 \xrightarrow[H_2SO_4]{H_2O} CH_3CH_2OH$

$H_2SO_4$ is not used up = **catalyst.**

b. $I^-$ not used up = **catalyst.**

$CH_3Cl \xrightarrow[^-OH]{I^-} CH_3OH$

$^-OH$ substitutes for $Cl^-$.

c. $H_2$ adds to the starting material.

Pt not used up = **catalyst.**

**6.23** Use the directions from Answer 6.1.

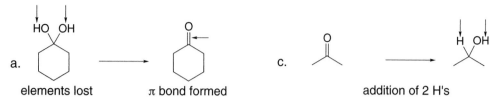

a. 

elements lost
(H + OH)
**elimination reaction**

π bond formed

c. 

addition of 2 H's
**addition reaction**

b. 

Cl replaces H =
**substitution reaction**

d. 

H replaces Cl =
**substitution reaction**

**6.24**

a. homolysis of $CH_3-\overset{\overset{\displaystyle H}{|}}{\underset{\underset{\displaystyle H}{|}}{C}}-H$

$CH_3-\overset{\overset{\displaystyle H}{|}}{\underset{\underset{\displaystyle H}{|}}{C}}\cdot \ + \ \cdot H$

radical

b. heterolysis of $CH_3-O\!-\!H$

$CH_3-\overset{..}{\underset{..}{O}}\!:^- \ + \ H^+$

c. heterolysis of $CH_3\!-\!MgBr$

$^-\overset{..}{C}H_3 \ + \ ^+MgBr$

carbanion

**6.25** Use the rules in Answer 6.3 to draw the arrows.

a. 

$+ \ :\overset{..}{\underset{..}{Br}}:^-$

b. $CH_3-\overset{\overset{\displaystyle :\overset{..}{O}:}{|}}{\underset{\underset{\displaystyle :\overset{..}{Cl}:}{|}}{C}}-CH_3 \longrightarrow \overset{\overset{\displaystyle :O:}{||}}{\underset{}{C}}\overset{}{\underset{CH_3 \quad CH_3}{}} \ + \ :\overset{..}{\underset{..}{Cl}}:^-$

c. $\cdot CH_3 \ + \ \cdot \overset{..}{\underset{..}{Cl}}: \longrightarrow CH_3-\overset{..}{\underset{..}{Cl}}:$

d. 

$+ \ :\overset{..}{\underset{..}{Br}}-\overset{..}{\underset{..}{Br}}: \longrightarrow$

$+ \ :\overset{..}{\underset{..}{Br}}\cdot$

e. $CH_3CH_2\!-\!\overset{..}{\underset{..}{Br}}: \ + \ ^-:\overset{..}{O}H \longrightarrow CH_3CH_2OH \ + :\overset{..}{\underset{..}{Br}}:^-$

f. $CH_3-\overset{\overset{\displaystyle CH_3 \ H}{|}}{\underset{\underset{\displaystyle \overset{+}{} \quad H}{|}}{C}}-\overset{}{\underset{}{C}}-H \ + \ ^-:\overset{..}{O}H \longrightarrow \overset{\overset{\displaystyle CH_3}{}}{\underset{\underset{\displaystyle CH_3}{}}{}}C=C\overset{\overset{\displaystyle H}{}}{\underset{\underset{\displaystyle H}{}}{}} \ + \ H_2\overset{..}{O}:$

**6.26**

a. 

$\!-\!\overset{..}{\underset{..}{I}}: \ + \ ^-:\overset{..}{O}H \longrightarrow$ 

$-\overset{..}{O}H \ + :\overset{..}{\underset{..}{I}}:^-$

b. $CH_3-\overset{\overset{\displaystyle :\overset{..}{O}:}{|}}{\underset{\underset{\displaystyle :\overset{..}{O}CH_2CH_3}{|}}{C}}-CH_2CH_2CH_3 \longrightarrow \overset{\overset{\displaystyle :O:}{||}}{\underset{}{C}}\overset{}{\underset{CH_3 \quad CH_2CH_2CH_3}{}}$

$+ \ ^-:\overset{..}{O}CH_2CH_3$

c. $^-:\overset{..}{O}H \quad H-\overset{\overset{\displaystyle H}{|}}{\underset{\underset{\displaystyle H}{|}}{C}}-\overset{\overset{\displaystyle H}{|}}{\underset{\underset{\displaystyle :\overset{..}{Br}:}{|}}{C}}-H \longrightarrow \overset{\overset{\displaystyle H}{}}{\underset{\underset{\displaystyle H}{}}{}}C=C\overset{\overset{\displaystyle H}{}}{\underset{\underset{\displaystyle H}{}}{}} \ + \ H_2\overset{..}{O}: \ + :\overset{..}{\underset{..}{Br}}:^-$

d. 

$\overset{\overset{\displaystyle H}{|}}{\underset{\underset{\displaystyle H}{|}}{C}}-H \ + \ \cdot\overset{..}{\underset{..}{Cl}}: \longrightarrow$ 

$\overset{\overset{\displaystyle H}{|}}{\underset{\underset{\displaystyle H}{|}}{C}}\cdot \ + \ H\overset{..}{\underset{..}{Cl}}:$

**6.27** Use the rules from Answer 6.4.

a. $I\!-\!CCl_3$     $Br\!-\!CCl_3$     $Cl\!-\!CCl_3$

largest halogen    **intermediate**     smallest halogen
**weakest bond**    **bond strength**     **strongest bond**

b. $H_2N\!-\!NH_2$     $HN\!=\!NH$     $N\!\equiv\!N$

single bond     double bond     triple bond
**weakest bond**    **intermediate**     **strongest bond**
             **bond strength**

**6.28** Use the directions from Answer 6.5.

a. $CH_3CH_2-H + Br_2 \longrightarrow CH_3CH_2-Br + HBr$

| **[1] Bonds broken** | | **[2] Bonds formed** | | **[3] Overall $\Delta H^o$ =** |
|---|---|---|---|---|
| | $\Delta H^o$ (kcal/mol) | | $\Delta H^o$ (kcal/mol) | |
| $CH_3CH_2-H$ | + 98 | $CH_3CH_2-Br$ | − 68 | + 144  kcal/mol |
| $Br-Br$ | + 46 | $H-Br$ | − 88 | − 156  kcal/mol |
| Total | + 144  kcal/mol | Total | − 156 kcal/mol | ANSWER: − 12 kcal/mol |

b. $\cdot OH + CH_4 \longrightarrow \cdot CH_3 + H_2O$

| **[1] Bonds broken** | | **[2] Bonds formed** | | **[3] Overall $\Delta H^o$ =** |
|---|---|---|---|---|
| | | | | + 104 kcal/mol |
| | $\Delta H^o$ (kcal/mol) | | $\Delta H^o$ (kcal/mol) | − 119 kcal/mol |
| $CH_3-H$ | + 104  kcal/mol | $H-OH$ | − 119 kcal/mol | ANSWER: − 15 kcal/mol |

c. $CH_3-OH + HBr \longrightarrow CH_3-Br + H_2O$

| **[1] Bonds broken** | | **[2] Bonds formed** | | **[3] Overall $\Delta H^o$ =** |
|---|---|---|---|---|
| | $\Delta H^o$ (kcal/mol) | | $\Delta H^o$ (kcal/mol) | |
| $CH_3-OH$ | + 93 | $CH_3-Br$ | − 70 | + 181  kcal/mol |
| $H-Br$ | + 88 | $H-Oh$ | − 119 | − 189  kcal/mol |
| Total | + 181  kcal/mol | Total | − 189 kcal/mol | ANSWER: − 8  kcal/mol |

d. $\cdot Br + CH_4 \longrightarrow \cdot H + CH_3Br$

| **[1] Bonds broken** | | **[2] Bonds formed** | | **[3] Overall $\Delta H^o$ =** |
|---|---|---|---|---|
| | | | | + 104 kcal/mol |
| | $\Delta H^o$ (kcal/mol) | | $\Delta H^o$ (kcal/mol) | − 70  kcal/mol |
| $CH_3-H$ | + 104  kcal/mol | $CH_3-Br$ | − 70  kcal/mol | ANSWER: + 34 kcal/mol |

**6.29**

propane

$CH_3-CH_2CH_3 \longrightarrow \cdot CH_3 + \cdot CH_2CH_3$

$\Delta H^o$ = 85 kcal/mol
This bond is formed from two $sp^3$ hybridized C's.

propene

$CH_3-CH=CH_2 \longrightarrow \cdot CH_3 + \cdot CH=CH_2$

$\Delta H^o$ = 92 kcal/mol
This bond is formed from one $sp^2$ and one $sp^3$ hybridized C.  The higher percent s-character in one C makes a stronger bond; thus the bond dissociation energy is higher.

**6.30**

$$CH_2=CH-\overset{\overset{\textstyle H}{|}}{\underset{\underset{\textstyle H}{|}}{C}}-H \longrightarrow CH_2=CH-\overset{\overset{\textstyle H}{|}}{C}-H \longleftrightarrow \cdot CH_2-CH=\overset{\overset{\textstyle H}{|}}{C}-H$$

hybrid:

$$CH_2=CH=\overset{\overset{\textstyle H}{|}}{\underset{\underset{\textstyle \delta\cdot}{}}{C}}-H \quad \overset{\delta\cdot}{}$$

**6.31** The more stable radical is formed by a reaction with a smaller $\Delta H°$.

$$CH_3-CH_2-\overset{\overset{\textstyle H}{|}}{\underset{\underset{\textstyle H}{|}}{C}}-H \longrightarrow CH_3-CH_2-\overset{\overset{\textstyle \cdot}{}}{\underset{\underset{\textstyle H}{|}}{C}}-H \quad \Delta H° = 98 \text{ kcal/mol} = \text{less stable radical}$$

This C–H bond is stronger.  **A**

$$CH_3-\overset{\overset{\textstyle H}{|}}{\underset{\underset{\textstyle H}{|}}{C}}-CH_3 \longrightarrow CH_3-\overset{\overset{\textstyle \cdot}{}}{\underset{\underset{\textstyle H}{|}}{C}}-CH_3 \quad \Delta H° = 95 \text{ kcal/mol} = \text{more stable radical}$$

This C–H bond is weaker.  **B**

Since the bond dissociation for cleavage of the C–H bond to form radical **A** is higher, more energy must be added to form it. This makes **A** higher in energy and therefore less stable than **B**.

**6.32** Use the bond dissociation energy for the C–C σ bond in ethane as an estimate of the σ bond strength in ethylene. Then you can estimate the π bond strength as well.

$$CH_3-CH_3 \qquad CH_2=CH_2$$

$$\Delta H° = 88 \text{ kcal/mol} \qquad \Delta H° = 152 \text{ kcal/mol}$$

152 – 88 = 64 kcal/mol = π bond

**6.33** Use the rules from Answer 6.9.
   a. $K_{eq} = 0.5$. $K_{eq}$ is less than one so the **starting material** is favored.
   b. $\Delta G° = -25$ kcal/mol. $\Delta G°$ is less than 0 so the **product** is favored.
   c. $\Delta H° = 2.0$ kcal/mol. $\Delta H°$ is positive, so the **starting material** is favored.
   d. $K_{eq} = 16$. $K_{eq}$ is greater than one so the **product** is favored.
   e. $\Delta G° = 0.5$ kcal/mol. $\Delta G°$ is greater than zero so the **starting material** is favored.
   f. $\Delta H° = 100$ kcal/mol. $\Delta H°$ is positive so the **starting material** is favored.
   g. $\Delta S° = 2$ cal/(K•mol). $\Delta S°$ is greater than zero so the **product** is more disordered and favored.
   h. $\Delta S° = -2$ cal/(K•mol). $\Delta S°$ is less than zero so the **starting material** is more disordered and favored.

**6.34**
   a. A negative $\Delta G°$ must have $K_{eq} > 1$. $K_{eq} = 10^2$.
   b. $K_{eq} = [\text{products}]/[\text{reactants}] = [1]/[5] = 0.2 = K_{eq}$. $\Delta G°$ is positive.
   c. A negative $\Delta G°$ has $K_{eq} > 1$, and a positive $\Delta G°$ has $K_{eq} < 1$. $\Delta G° = -2$ kcal/mol will have a larger $K_{eq}$.

**6.35**

R axial    equatorial

| R | $K_{eq}$ |
|---|---|
| $-CH_3$ | 18 |
| $-CH_2CH_3$ | 23 |
| $-CH(CH_3)_2$ | 38 |
| $-C(CH_3)_3$ | 4000 |

a. The equatorial conformation is always present in the larger amount at equilibrium since the $K_{eq}$ for all R groups is greater than 1.

b. The cyclohexane with the $-C(CH_3)_3$ group will have the greatest amount of equatorial conformation at equilibrium since this group has the highest $K_{eq}$.

c. The cyclohexane with the $-CH_3$ group will have the greatest amount of axial conformation at equilibrium since this group has the lowest $K_{eq}$.

d. The cyclohexane with the $-C(CH_3)_3$ group will have the most negative $\Delta G°$ since it has the largest $K_{eq}$.

e. The larger the R group, the more favored the equatorial conformation.

f. The $K_{eq}$ for *tert*-butylcyclohexane is much higher because the *tert*-butyl group is bulkier than the other groups. With a *tert*-butyl group, a $CH_3$ group is always oriented over the ring when the group is axial, creating severe 1,3-diaxial interactions. With all other substituents, the larger $CH_3$ groups can be oriented away from the ring, placing a H over the ring, making the 1,3-diaxial interactions less severe. Compare:

*tert*-butylcyclohexane

isopropylcyclohexane

severe 1,3-diaxial interactions with
the $CH_3$ groups and the axial H

less severe 1,3-diaxial interactions

**6.36** Calculate $K_{eq}$, and then find the percentage of axial and equatorial conformations present at equilibrium.

fluorocyclohexane
1 part

1.5 parts

a. $\Delta G° = -1.4\log K_{eq}$
$\Delta G° = -0.25$ kcal/mol
$-0.25$ kcal/mol $= -1.4\log K_{eq}$
$K_{eq} = 1.5$

b. $K_{eq} = $ [products]/[reactants]
$1.5 = $ [products]/[reactants]
$1.5$[reactants] $= $ [products]
[reactants] $= 0.4 = 40\%$ axial
[products] $= 0.6 = 60\%$ equatorial

**6.37** Reactions resulting in an increase in entropy are favored. When a single molecule forms two molecules, there is an increase in entropy.

a.
increased number of molecules
$\Delta S°$ is positive.
products favored

b. $CH_3\cdot$ + $CH_3\cdot$ ⟶ $CH_3CH_3$
decreased number of molecules
$\Delta S°$ is negative.
starting material favored

c. $(CH_3)_2C(OH)_2$ ⟶ $(CH_3)_2C=O$ + $H_2O$
increased number of molecules
$\Delta S°$ is positive.
products favored

d. $CH_3COOCH_3$ + $H_2O$ ⟶ $CH_3COOH$ + $CH_3OH$
no change in the number of molecules
neither favored

**6.38** Use the directions in Answer 6.14 to draw the transition state. Nonbonded electron pairs are drawn in at reacting sites.

a.

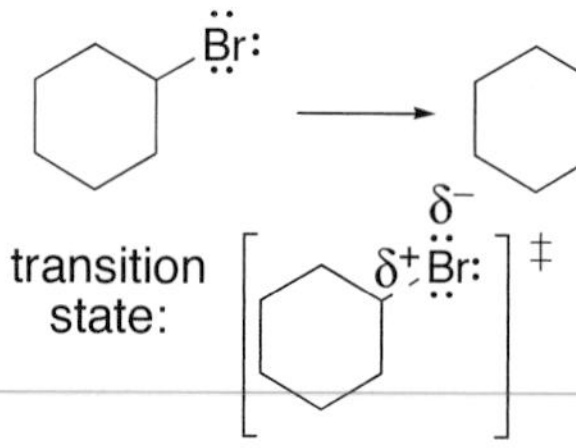

transition
state:

c.

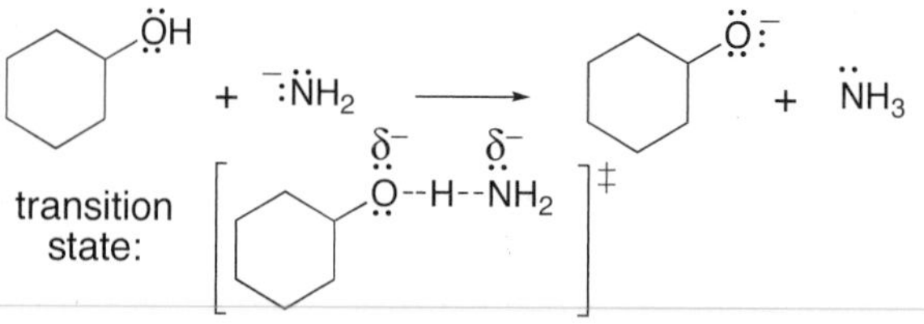

transition
state:

b.  BF$_3$  +  :Cl:$^-$  ⟶  F—B—Cl:

transition
state:

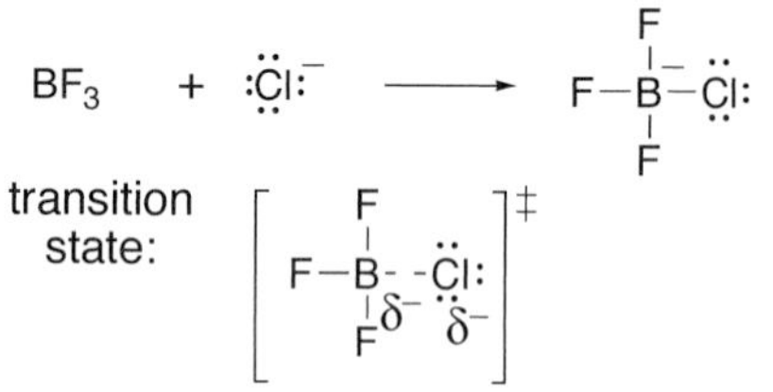

d.

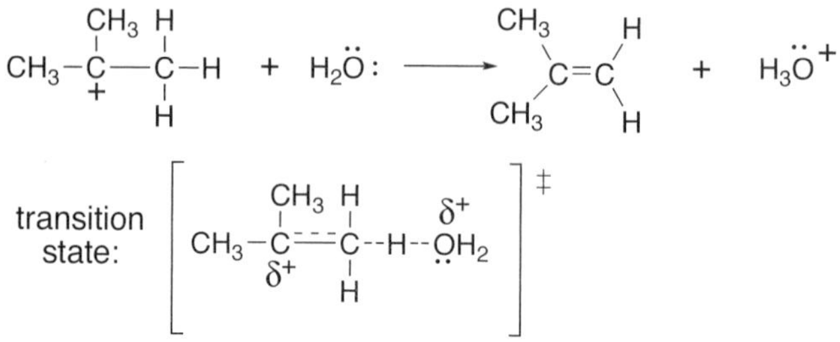

transition
state:

**6.39**

a.

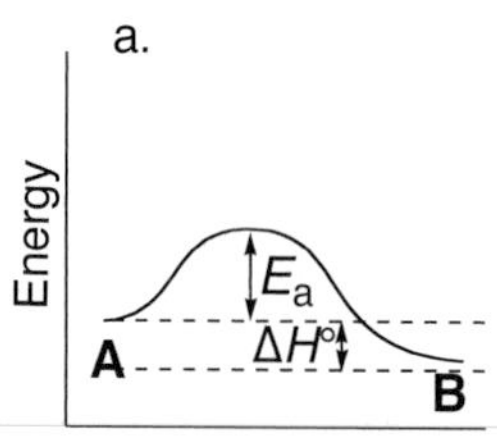

Reaction coordinate
• one step **A → B**
• exothermic since
  **B** lower than **A**
• low $E_a$
  (small energy barrier)

b.

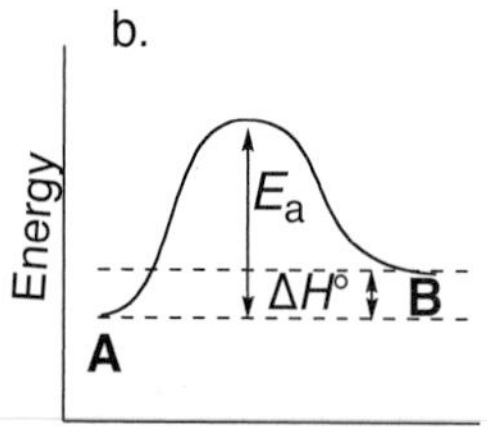

Reaction coordinate
• one step **A → B**
• endothermic since
  **B** higher than **A**
• high $E_a$
  (large energy barrier)

c.

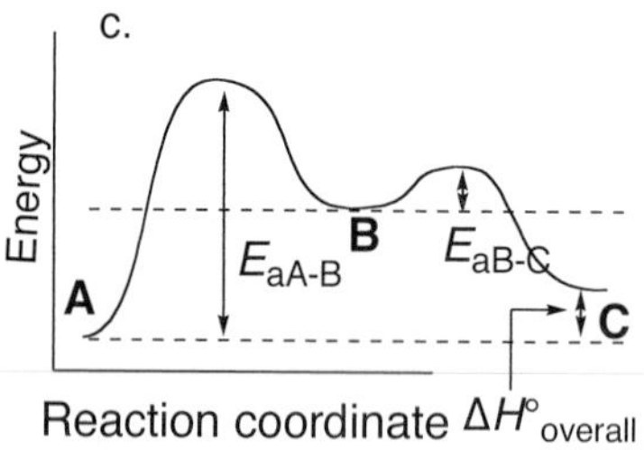

Reaction coordinate $\Delta H^{\circ}_{overall}$
• two steps
• **A** lowest energy
• **B** highest energy
• $E_{a(A-B)}$ **is rate-**
  **determining**,
  since the transtion
  state for Step [1] is
  higher in energy.

d.

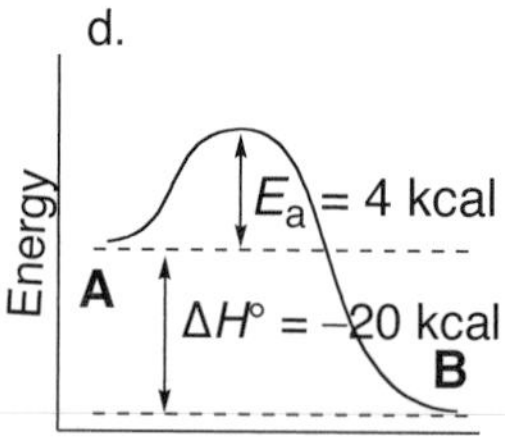

Reaction coordinate
• one step **A → B**
• exothermic since
  **B** lower than **A**

$E_a$ = 4 kcal
$\Delta H^{\circ}$ = –20 kcal

**6.40**

a.   CH$_3$—H  +  ·Cl  ⟶  ·CH$_3$  +  HCl

b.  ·Cl + CH$_4$  ⟶  ·CH$_3$  +  HCl

| **[1] Bonds broken** | | **[2] Bonds formed** | | **[3] Overall $\Delta H^{\circ}$ =** |
|---|---|---|---|---|
| | $\Delta H^{\circ}$ (kcal/mol) | | $\Delta H^{\circ}$ (kcal/mol) | + 104  kcal/mol |
| CH$_3$—H | + 104  kcal/mol | H—Cl | – 103  kcal/mol | – 103  kcal/mol |
| | | | | ANSWER: + 1  kcal/mol |

c.

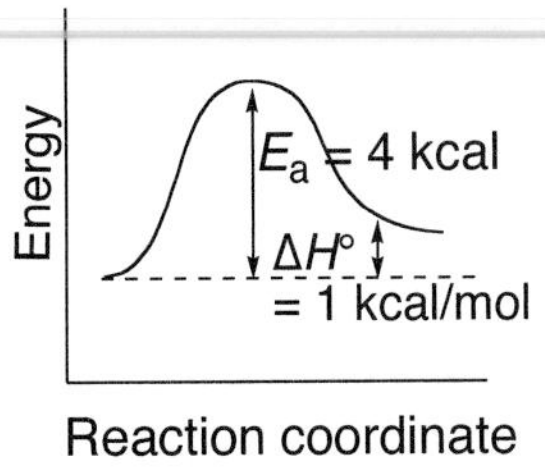

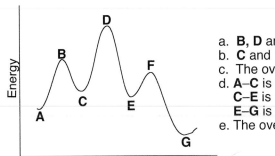

Energy / Reaction coordinate

$E_a = 4$ kcal

$\Delta H° = 1$ kcal/mol

d. The $E_a$ for the reverse reaction is the difference in energy between the products and the transition state, 3 kcal.

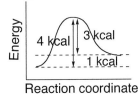

Energy / Reaction coordinate

4 kcal   3 kcal

1 kcal

**6.41**

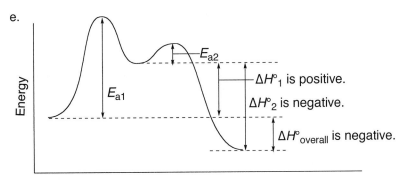

Energy / Reaction coordinate

a. **B, D** and **F** are transition states.
b. **C** and **E** are reactive intermediates.
c. The overall reaction has **three steps.**
d. **A–C** is endothermic.
   **C–E** is exothermic.
   **E–G** is exothermic.
e. The overall reaction is exothermic.

**6.42**

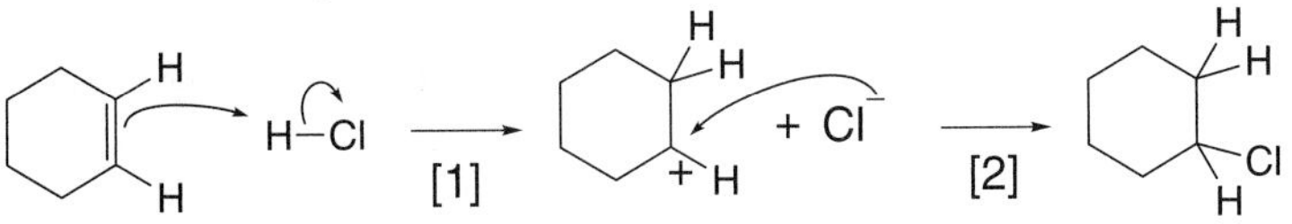

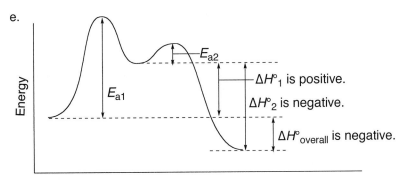

a. Step [1] breaks one π bond and the H–Cl bond, and one C–H bond is formed. The $\Delta H°$ for this step should be positive since more bonds are broken than formed.
b. Step [2] forms one bond. The $\Delta H°$ for this step should be negative since one bond is formed and none is broken.
c. Step [1] is rate-determining since it is more difficult.
d. Transition state for Step [1]:          Transition state for Step [2]:

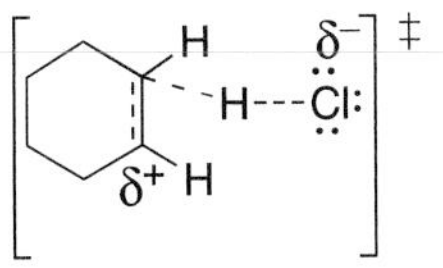

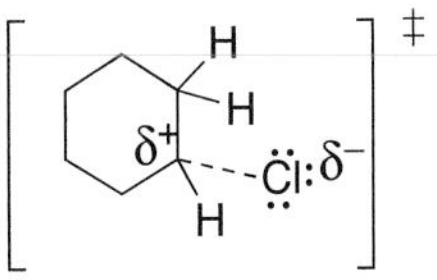

e.

Energy / Reaction coordinate

$E_{a2}$
$E_{a1}$
$\Delta H°_1$ is positive.
$\Delta H°_2$ is negative.
$\Delta H°_{overall}$ is negative.

**6.43**

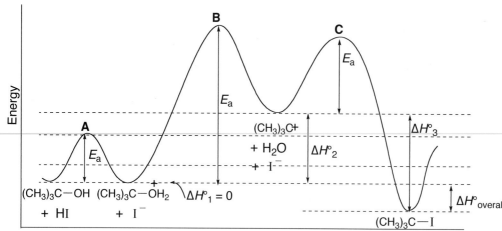

ß

a. The reaction has three steps, since there are three energy barriers.
b. See above.
c. Transition state **A** (see graph for location):

Transition state **B**:

Transition state **C**:

$$\left[ (CH_3)_3C \overset{\delta^+}{-}\overset{..}{\underset{|}{O}}H \atop H\text{----}\overset{..}{\underset{..}{I}}: \right]^{\ddagger} \qquad \left[ (CH_3)_3C\underset{\delta^+}{\text{----}}\overset{..}{O}H_2 \atop \delta^+ \right]^{\ddagger} \qquad \left[ (CH_3)_3\overset{\delta^+}{C}\text{---}\overset{\delta^-}{\underset{..}{I}}: \right]^{\ddagger}$$

d. Step 2 is rate-determining since this step has the highest energy transition state.

**6.44** $E_a$, concentration, catalysts, rate constant and temperature affect reaction rate so c, d, e, g, and h affect rate.

**6.45**

a. **rate = $k$[CH₃Br][ NaCN]**
b. Double [CH₃Br] = **rate doubles.**
c. Halve [NaCN] = **rate halved.**
d. Increase both [CH₃Br] and [NaCN] by factor of 5 = [5][5] = **rate increases by a factor of 25.**

**6.46**

acetyl chloride

$$CH_3-\overset{\overset{..}{O}:}{\underset{}{\overset{\|}{C}}}-Cl \quad \overset{[1]}{\underset{slow}{\longrightarrow}} \quad CH_3-\overset{:\overset{..}{O}\overset{-}{:}}{\underset{\overset{|}{CH_3\overset{..}{O}:}}{\overset{|}{C}}}-Cl \quad \overset{[2]}{\underset{fast}{\longrightarrow}} \quad CH_3\overset{\overset{..}{O}:}{\underset{}{\overset{\|}{C}}}\overset{..}{O}CH_3 + Cl^-$$

methyl acetate

a. Only the slow step is included in the rate equation: **Rate = $k$[CH₃O⁻][CH₃COCl]**
b. CH₃O⁻ is in the rate equation. Increasing its concentration by 10 times would increase the rate by **10 times.**
c. When both reactant concentrations are increased by 10 times, the rate increases by **100 times (10 x 10 = 100).**
d. This is a **substitution reaction** (OCH₃ substitutes for Cl).

**6.47**

    a. **True**: Increasing temperature increases reaction rate.

    b. **True**: If a reaction is fast, it has a large rate constant.

    c. **False: Corrected** - There is no relationship between $\Delta G°$ and reaction rate.

    d. **False: Corrected** - When the $E_a$ is large, *the rate constant is small.*

    e. **False: Corrected** - There is no relationship between $K_{eq}$ and reaction rate.

    f. **False: Corrected** - Increasing the concentration of a reactant increases the rate of a reaction *only if the reactant appears in the rate equation.*

**6.48**

    a. The first mechanism has one step: **Rate = $k[(CH_3)_3CI][^-OH]$**

    b. The second mechanism has two steps, but only the first step would be in the rate equation since it is slow and therefore rate-determining: **Rate = $k[(CH_3)_3CI]$**

    c. Possibility [1] is second order; possibility [2] is first order.

    d. These rate equations can be used to show which mechanism is plausible by changing the concentration of [$^-$OH]. If this affects the rate, possibility [1] is reasonable. If it does not affect the rate, possibility [2] is reasonable.

**6.49** The difference in both the acidity and the bond dissociation energy of $CH_3CH_3$ versus $HC\equiv CH$ is due to the same factor: percent *s*-character. The difference results because one process is based on homolysis and one is based on heterolysis.
Bond dissociation energy:

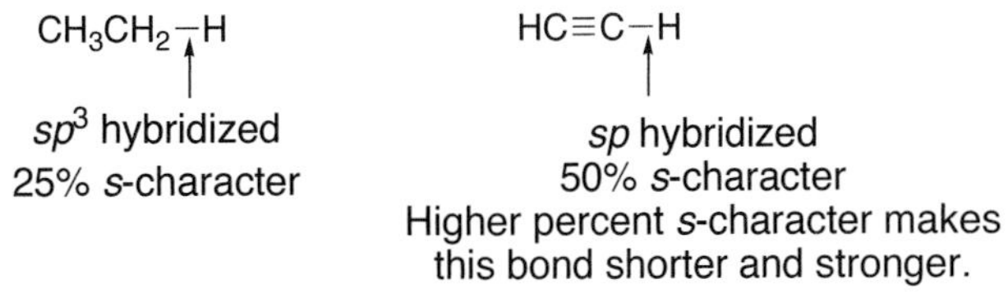

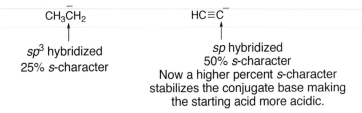

    Acidity. To compare acidity, we must compare the stability of the conjugate bases:

**6.50** In Reaction [1], the number of molecules of reactants and products stays the same, so entropy is not a factor. In Reaction [2], a single molecule of starting material forms two molecules of products, so entropy increases. This makes $\Delta G°$ more favorable, thus increasing $K_{eq}$.

**6.51**

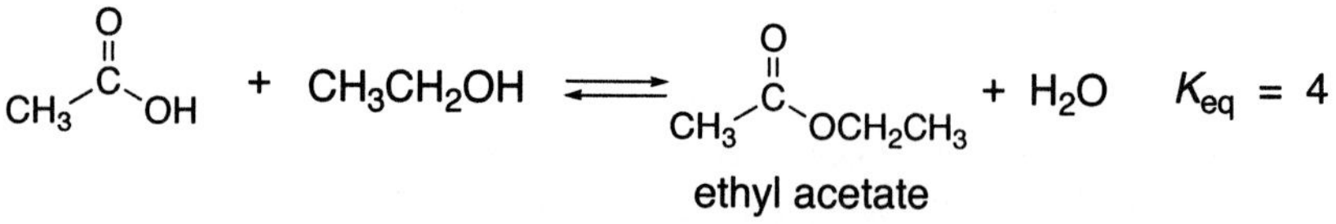

ethyl acetate

To increase the yield of ethyl acetate, $H_2O$ can be removed from the reaction mixture, or there can be a large excess of one of the starting materials.

**6.52** Since O atoms are more electronegative than all other atoms except F, each O atom in the O–O bond pulls electron density towards itself. The O atoms are less "willing" to share electron density in a two-electron bond, and this weakens the bond.

Chapter 7: Alkyl Halides and Nucleophilic Substitution

## ♦ General facts about alkyl halides

- Alkyl halides contain a halogen atom X bonded to an $sp^3$ hybridized carbon (7.1).
- Alkyl halides are named as halo alkanes, with the halogen as a substituent (7.2).
- Alkyl halides have a polar C–X bond, so they exhibit dipole–dipole interactions but are incapable of intermolecular hydrogen bonding (7.3).
- The polar C–X bond containing an electrophilic carbon makes alkyl halides reactive towards nucleophiles and bases (7.5).

## ♦ The central theme (7.6)

- Nucleophilic substitution is one of the two main reactions of alkyl halides. A nucleophile replaces a leaving group on an $sp^3$ hybridized carbon.

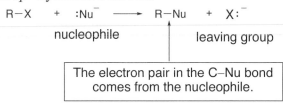

- One σ bond is broken and one σ bond is formed.
- There are two possible mechanisms: $S_N1$ and $S_N2$.

## ♦ $S_N1$ and $S_N2$ mechanisms compared

|  |  | **$S_N2$ mechanism** | **$S_N1$ mechanism** |
|---|---|---|---|
| [1] Mechanism | • | One step (7.11B) | • Two steps (7.13B) |
| [2] Alkyl halide | • | Order of reactivity: $CH_3X >$ $RCH_2X > R_2CHX > R_3CX$ (7.11D) | • Order of reactivity: $R_3CX >$ $R_2CHX > RCH_2X > CH_3X$ (7.13D) |
| [3] Rate equation | • | rate = $k[RX][:Nu^-]$ | • rate = $k[RX]$ |
|  | • | second order kinetics (7.11A) | • first order kinetics (7.13A) |
| [4] Stereochemistry | • | backside attack of the nucleophile (7.11C) | • trigonal planar carbocation intermediate (7.13C) |
|  | • | inversion of configuration at a stereogenic center | • racemization at a stereogenic center |
| [5] Nucleophile | • | favored by stronger nucleophiles (7.17B) | • favored by weaker nucleophiles (7.17B) |
| [6] Leaving group | • | better leaving group → faster reaction (7.17C) | • better leaving group → faster reaction (7.17C) |
| [7] Solvent | • | favored by polar aprotic solvents (7.17D) | • favored by polar protic solvents (7.17D) |

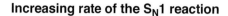

**Increasing rate of the S$_N$1 reaction**

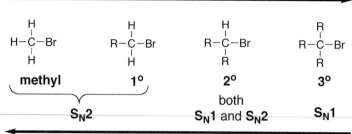

**Increasing rate of an S$_N$2 reaction**

## ♦ Important trends

- The best leaving group is the weakest base. Leaving group ability increases across a row and down a column of the periodic table (7.7).

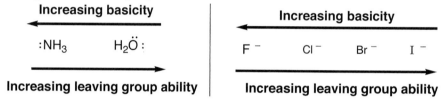

- Nucleophilicity decreases across a row of the periodic table (7.8A).

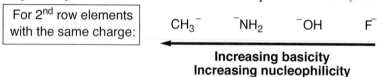

- Nucleophilicity decreases down a column of the periodic table in polar aprotic solvents (7.8C).

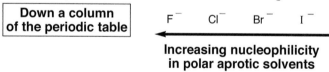

- Nucleophilicity increases down a column of the periodic table in polar protic solvents (7.8C).

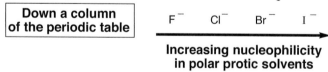

- The stability of a carbocation increases as the number of R groups bonded to the positively charged carbon increases (7.14).

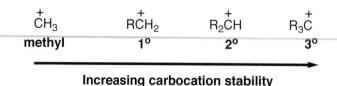

♦ Important principles

| Principle | Example |
|---|---|
| • Electron-donating groups (such as R groups) stabilize a positive charge (7.14A). | • 3° Carbocations ($R_3C^+$) are more stable than 2° carbocations ($R_2CH^+$), which are more stable than 1° carbocations ($RCH_2^+$). |
| • Steric hindrance decreases nucleophilicity but not basicity (7.8B). | • $(CH_3)_3CO^-$ is a stronger base but a weaker nucleophile than $CH_3CH_2O^-$. |
| • Hammond postulate: In an endothermic reaction, the more stable product is formed faster. In an exothermic reaction, this fact is not necessarily true (7.15). | • $S_N1$ reactions are faster when more stable (more substituted) carbocations are formed, because the rate-determining step is endothermic. |
| • Planar, $sp^2$ hybridized atoms react with reagents from both sides of the plane (7.13C). | • A trigonal planar carbocation reacts with nucleophiles from both sides of the plane. |

## Chapter 7: Answers to Problems

**7.1** Classify the alkyl halide as 1°, 2°, or 3° **by counting the number of carbons directly bonded to the carbon bonded to the halogen.**

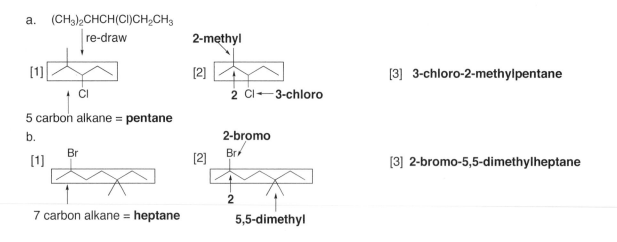

C bonded to 1 C
**1° alkyl halide**

C bonded to 2 C's
**2° alkyl halide**

a. CH₃CH₂CH₂CH₂CH₂—Br

b.

c. CH₃—C—CHCH₃

d.

C bonded to 3 C's
**3° alkyl halide**

C bonded to 3 C's
**3° alkyl halide**

**7.2** Use the directions from Answer 7.1.

$sp^3$ C's = **alkyl halides**
C's bonded to 2 C's
**2° alkyl halides**

telfairine

$sp^3$ C = **alkyl halide**
C bonded to 3 C's
**3° alkyl halide**

$sp^2$ C =
**vinyl halide**

$sp^3$ C's = **alkyl halides**
C's bonded to 3 C's
**3° alkyl halides**

This 3° alkyl halide is also **allylic.**

halomon

$sp^3$ C = **alkyl halide**
C bonded to 1 C
**1° alkyl halide**

$sp^3$ C = **alkyl halide**
C bonded to 2 C's
**2° alkyl halide**

$sp^2$ C =
**vinyl halide**

**7.3** To name a compound with the IUPAC system:
[1] **Name the parent** chain by finding the longest carbon chain.
[2] **Number the chain** so the first substituent gets the lower number. Then **name and number all substituents**, giving like substituents a prefix (di, tri, etc.). **To name the halogen substituent, change the -ine ending to -o.**
[3] **Combine all parts**, alphabetizing substituents, and ignoring all prefixes except iso.

a. (CH₃)₂CHCH(Cl)CH₂CH₃

re-draw

**2-methyl**

[1]

[2]

[3] **3-chloro-2-methylpentane**

Cl

2 Cl ← **3-chloro**

5 carbon alkane = **pentane**

b.

[1] Br

**2-bromo**

[2] Br

[3] **2-bromo-5,5-dimethylheptane**

2

7 carbon alkane = **heptane**

**5,5-dimethyl**

c.

[1] 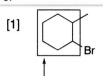 Br

↑
6 carbon cycloalkane =
**cyclohexane**

[2] 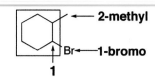 ← **2-methyl**

Br ← **1-bromo**
↑
**1**

[3] **1-bromo-2-methylcyclohexane**

d.

[1]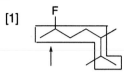

7 carbon alkane = **heptane**

[2]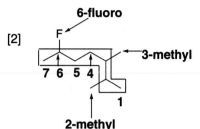

**6-fluoro**
F

**7 6 5 4** ← **3-methyl**

**1**
↑
**2-methyl**

[3] **6-fluoro-2,3-dimethylheptane**

## 7.4 To work backwards from a name to a structure:

[1] Find the parent name and draw that number of carbons. Use the suffix to identify the functional group. (**-ane = alkane**)

[2] Arbitrarily number the carbons in the chain. Add the substituents to the appropriate carbon.

a. 3-chloro-2-methyl**hexane**

[1]    6 carbon alkane

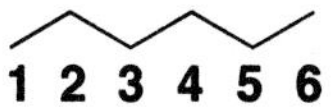

1 2 3 4 5 6

[2]    ┌─**methyl at C2**

Cl ← **chloro at C3**

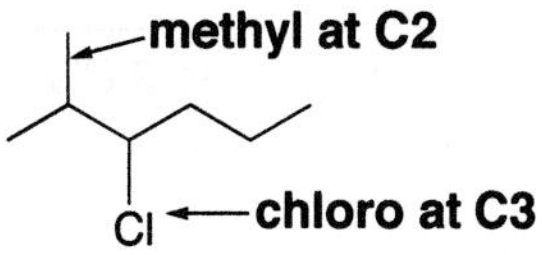

b. 4-ethyl-5-iodo-2,2-dimethyl**octane**

[1]   8 carbon alkane

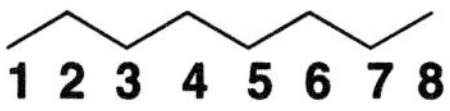

1 2 3 4 5 6 7 8

[2]

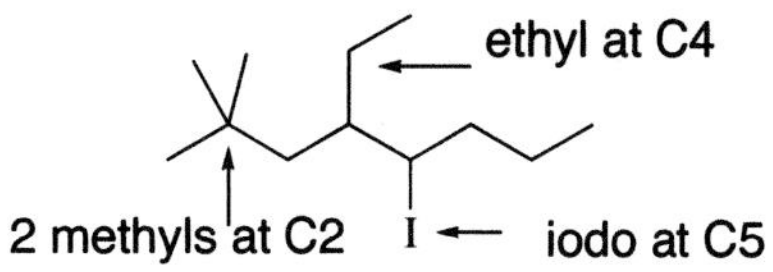

ethyl at C4

2 methyls at C2    I ← iodo at C5

c. *cis*-1,3-dichloro**cyclopentane**

[1]    5 carbon cycloalkane

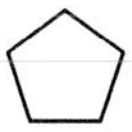

[2]   chloro groups at C1 and C3, both on the same side

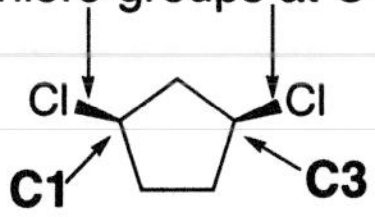

Cl         Cl
**C1**         **C3**

d. 1,1,3-tribromo**cyclohexane**

[1]    6 carbon cycloalkane

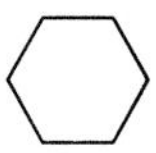

[2]                  3 Br groups
Br
**C3** →          Br
Br
**C1**

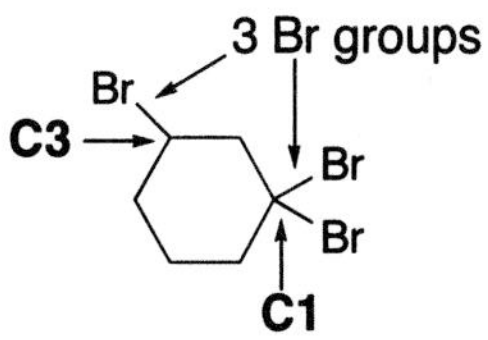

e. **propyl** chloride

[1] 3 carbon alkyl group

CH₃CH₂CH₂—

[2] **chloride on end**

CH₃CH₂CH₂–Cl

f. **sec-butyl** bromide

[1]  4 carbon alkyl group          [2]  **bromide**

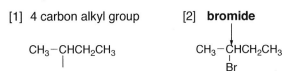

$CH_3-CHCH_2CH_3$          $CH_3-CHCH_2CH_3$
                                                            |
                                                            Br

**7.5**  Boiling points of alkyl halides increase as the size (and polarizability) of X increases.  Remember: **stronger intermolecular forces = higher boiling point.**

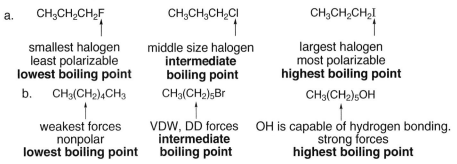

a.      $CH_3CH_2CH_2F$                    $CH_3CH_3CH_2Cl$                    $CH_3CH_2CH_2I$

smallest halogen            middle size halogen            largest halogen
least polarizable              **intermediate**                most polarizable
**lowest boiling point**      **boiling point**          **highest boiling point**

b.      $CH_3(CH_2)_4CH_3$            $CH_3(CH_2)_5Br$                    $CH_3(CH_2)_5OH$

weakest forces          VDW, DD forces        OH is capable of hydrogen bonding.
nonpolar                  **intermediate**                strong forces
**lowest boiling point**      **boiling point**          **highest boiling point**

**7.6**  a. Because an $sp^2$ hybridized C has a higher percent s-character than an $sp^3$ hybridized C, it holds electron density closer to C.  This pulls a little more electron density towards C, away from Cl, and thus a $C_{sp^2}$–Cl bond is less polar than a $C_{sp^3}$–Cl bond.

b.

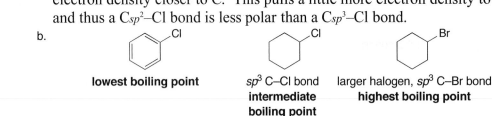

**lowest boiling point**          $sp^3$ C–Cl bond          larger halogen, $sp^3$ C–Br bond
                                          **intermediate**              **highest boiling point**
                                          **boiling point**

**7.7**  Since more polar molecules are more water soluble, look for polarity differences between methoxychlor and DDT.

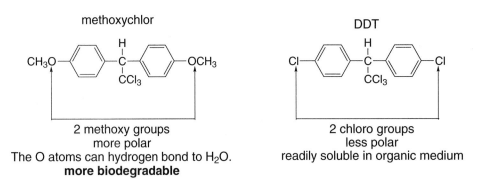

methoxychlor                                                        DDT

2 methoxy groups                                  2 chloro groups
more polar                                          less polar
The O atoms can hydrogen bond to $H_2O$.        readily soluble in organic medium
**more biodegradable**

**7.8**  To draw the products of a nucleophilic substitution reaction:
[1] **Find the $sp^3$ hybridized electrophilic carbon** with a leaving group.
[2] **Find the nucleophile** with lone pairs or electrons in π bonds.
[3] **Substitute the nucleophile for the leaving group** on the electrophilic carbon.

a. [structure with Br] **leaving group**  +  ⁻OCH₂CH₃ (nonbonded e⁻ pairs **nucleophile**)  ⟶  [structure with OCH₂CH₃]  +  Br⁻

b. [cyclohexane with Cl] **leaving group**  +  Na⁺ ⁻OH (nonbonded e⁻ pairs **nucleophile**)  ⟶  [cyclohexane with OH]  +  Na⁺Cl⁻

c. [structure with I] **leaving group**  +  N₃⁻ (nonbonded e⁻ pairs **nucleophile**)  ⟶  [structure with N₃]  +  I⁻

d. [cyclohexane with Br] **leaving group**  +  Na⁺ ⁻CN (nonbonded e⁻ pairs **nucleophile**)  ⟶  [cyclohexane with CN]  +  Na⁺Br⁻

**7.9** Use the steps from Answer 7.8 and then draw the proton transfer reaction.

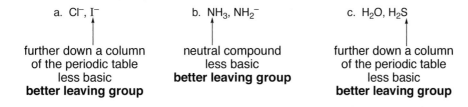

a. [structure with Br] **leaving group**  +  :N(CH₂CH₃)₃ **nucleophile**  —substitution→  [structure with N⁺(CH₂CH₃)₃]  +  Br⁻

b. (CH₃)₃C–Cl **leaving group**  +  H₂Ö: **nucleophile**  —substitution→  (CH₃)₃C–⁺Ö–H with H  +  Cl⁻  —proton transfer→  (CH₃)₃C–Ö–H  +  HCl

c. [cyclohexane with CH₃ and Br] **leaving group**  +  CH₃–Ö–H **nucleophile**  —substitution→  [cyclohexane with CH₃ and ⁺Ö–CH₃ with H]  +  Br⁻  —proton transfer→  [cyclohexane with CH₃ and Ö–CH₃]  +  HBr

**7.10** Compare the compounds based on these leaving group trends:
- Better leaving groups are weaker bases.
- A neutral leaving group is always better than its conjugate base.

a. Cl⁻, I⁻

further down a column
of the periodic table
less basic
**better leaving group**

b. NH₃, NH₂⁻

neutral compound
less basic
**better leaving group**

c. H₂O, H₂S

further down a column
of the periodic table
less basic
**better leaving group**

**7.11** Good leaving groups include $Cl^-$, $Br^-$, $I^-$, $H_2O$.

a. $CH_3CH_2CH_2 \boxed{Br}$

$Br^-$ is a **good leaving group.**

b. $CH_3CH_2CH_2OH$

No good leaving group. $^-OH$ is too strong a base.

c. $CH_3CH_2CH_2 \boxed{\overset{+}{O}H_2}$

$H_2O$ is a **good leaving group.**

d. $CH_3CH_3$

No good leaving group. $H^-$ is too strong a base.

**7.12** To decide whether the equilibrium favors the starting material or the products, **compare the nucleophile and the leaving group.** The reaction proceeds towards the weaker base.

a. $CH_3CH_2-NH_2$ + $Br^-$ $\longrightarrow$ $CH_3CH_2-Br$ + $^-NH_2$

nucleophile
better leaving group
weaker base
$pK_a$ (HBr) = –9

leaving group
$pK_a$ (NH$_3$) = 38

| **Reaction favors starting material.** |

b. (structure) + $^-CN$ $\longrightarrow$ (structure) $CN$ + $I^-$

nucleophile
$pK_a$ (HCN) = 9.1

leaving group
better leaving group
weaker base
$pK_a$ (HI) = –10

| **Reaction favors product.** |

**7.13** Use these three rules to find the stronger nucleophile in each pair:

[1] Comparing two nucleophiles having the *same attacking atom*, **the stronger base is a stronger nucleophile.**

[2] **Negatively charged nucleophiles** are always **stronger than their conjugate acids.**

[3] **Across a row of the periodic table, nucleophilicity decreases** when comparing a species of similar charge.

a. $NH_3$, $NH_2^-$

A negatively charged nucleophile is stronger than its conjugate acid. **stronger nucleophile**

b. $CH_3^-$, $HO^-$

Across a row of the periodic table, nucleophilicity decreases with species of same charge. **stronger nucleophile**

c. $CH_3NH_2$, $CH_3OH$

Across a row of the periodic table, nucleophilicity decreases with species of same charge. **stronger nucleophile**

d. (structure) $CH_3CH_2O^-$

same attacking atom (O) stronger base **stronger nucleophile**

**7.14** *Polar protic solvents* **are capable of H-bonding**, and therefore must contain a **H bonded to an electronegative O or N.** *Polar aprotic solvents* **are incapable of H-bonding**, and therefore do not contain any O–H or N–H bonds.

a. $HOCH_2CH_2OH$

contains 2 O–H bonds **polar protic**

b. $CH_3CH_2OCH_2CH_3$

no O–H bonds **polar aprotic**

c. $CH_3COOCH_2CH_3$

no O–H bonds **polar aprotic**

**7.15**
- In *polar protic solvents,* **the trend in nucleophilicity is opposite to the trend in basicity** down a column of the periodic table so that nucleophilicity increases.
- In *polar aprotic solvents,* **the trend is identical to basicity** so that nucleophilicity decreases down a column.

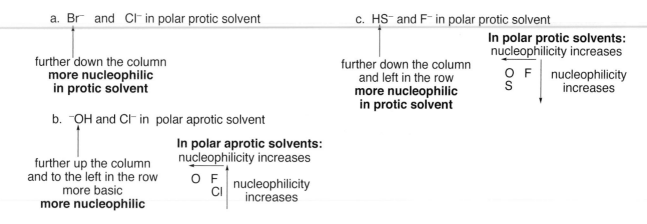

a. Br⁻ and Cl⁻ in polar protic solvent

further down the column
**more nucleophilic
in protic solvent**

c. HS⁻ and F⁻ in polar protic solvent

**In polar protic solvents:**
nucleophilicity increases ←

further down the column
and left in the row
**more nucleophilic
in protic solvent**

| | | |
|---|---|---|
| O | F | nucleophilicity |
| S | | increases ↓ |

b. ⁻OH and Cl⁻ in polar aprotic solvent

further up the column
and to the left in the row
**more basic
more nucleophilic**

**In polar aprotic solvents:**
nucleophilicity increases ←

| | | |
|---|---|---|
| O | F | nucleophilicity |
| | Cl | increases ↓ |

**7.16** The stronger base is the stronger nucleophile except in polar protic solvents when nucleophilicity increases down a column. For other rules, see Answers 7.13 and 7.15.

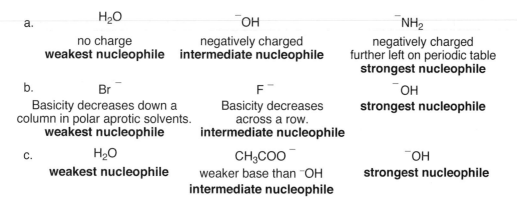

a.    $H_2O$                      ⁻OH                      ⁻NH₂

no charge            negatively charged          negatively charged
**weakest nucleophile    intermediate nucleophile**    further left on periodic table
                                                **strongest nucleophile**

b.    Br⁻                      F⁻                      ⁻OH

Basicity decreases down a        Basicity decreases        **strongest nucleophile**
column in polar aprotic solvents.    across a row.
**weakest nucleophile    intermediate nucleophile**

c.    $H_2O$                  $CH_3COO⁻$                ⁻OH

**weakest nucleophile**    weaker base than ⁻OH    **strongest nucleophile**
                    **intermediate nucleophile**

**7.17** To determine what nucleophile is needed to carry out each reaction, look at the product to see what has replaced the leaving group.

a. $(CH_3)_2CHCH_2CH_2$—Br ⟶ $(CH_3)_2CHCH_2CH_2$—SH
   SH replaces Br.
   **HS⁻ is needed.**

c. $(CH_3)_2CHCH_2CH_2$—Br ⟶ $(CH_3)_2CHCH_2CH_2$—OCOCH₃
   OCOCH₃ replaces Br.
   **CH₃COO⁻ is needed.**

b. $(CH_3)_2CHCH_2CH_2$—Br ⟶ $(CH_3)_2CHCH_2CH_2$—OCH₂CH₃
   OCH₂CH₃ replaces Br.
   **CH₃CH₂O⁻ is needed.**

d. $(CH_3)_2CHCH_2CH_2$—Br ⟶ $(CH_3)_2CHCH_2CH_2$—C≡CH
   C≡CH replaces Br.
   **HC≡C⁻ is needed.**

**7.18** The general rate equation for an $S_N2$ reaction is rate = $k[RX][:Nu⁻]$

a. [RX] is tripled, and [:Nu⁻] stays the same: **rate triples.**
b. Both [RX] and [:Nu⁻] are tripled: **rate increases by a factor of 9 (3 x 3 = 9).**
c. [RX] is halved, and [:Nu⁻] stays the same: **rate halved.**
d. [RX] is halved, and [:Nu⁻] is doubled: **rate stays the same (1/2 x 2 = 1).**

**7.19** The transition state in an $S_N2$ reaction has **dashed bonds to both the leaving group and the nucleophile**, and must contain partial charges.

a. $CH_3CH_2CH_2-Cl$ + $^-OCH_3$ ⟶ $CH_3CH_2CH_2-OCH_3$ + $Cl^-$ $\left[\begin{array}{c} CH_3CH_2CH_2 \text{---} Cl\ \delta^- \\ | \\ CH_3\overset{}{O}\ \delta^- \end{array}\right]^{\ddagger}$

b. 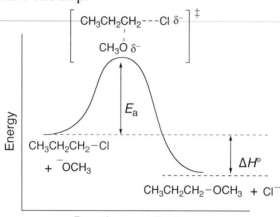 (propyl) Br + $^-SH$ ⟶ (propyl) SH + $Br^-$ $\left[\begin{array}{c} \overset{\delta^-}{SH} \\ \text{(butyl fragment)} \\ Br\ \delta^- \end{array}\right]^{\ddagger}$

**7.20** All $S_N2$ reactions have one step.

$\left[\begin{array}{c} CH_3CH_2CH_2 \text{---} Cl\ \delta^- \\ | \\ CH_3\overset{}{O}\ \delta^- \end{array}\right]^{\ddagger}$

Energy (y-axis) vs Reaction coordinate (x-axis)

$E_a$

$CH_3CH_2CH_2-Cl$
$+ \ ^-OCH_3$

$\Delta H°$

$CH_3CH_2CH_2-OCH_3$ + $Cl^-$

**7.21** To draw the products of $S_N2$ reactions, **replace the leaving group by the nucleophile, and then draw the stereochemistry with *inversion* at the stereogenic center**.

a. $CH_3CH_2$ / D / C—Br / H + $^-OCH_2CH_3$ ⟶ $CH_3CH_2O-C$ / D / $CH_2CH_3$ / H

b. (cyclopentane)—I + $^-CN$ ⟶ (cyclopentane)···CN

**7.22** *Increasing* the number of R groups *increases* crowding of the transition state and *decreases* the rate of an $S_N2$ reaction.

a. $CH_3CH_2-Cl$  or  $CH_3-Cl$

1° alkyl halide    methyl halide
**faster reaction**

b. (2° alkyl halide with Cl)  or  (1° alkyl halide with Cl)

2° alkyl halide    1° alkyl halide
**faster reaction**

c. (cyclohexane)—Br  or  (cyclohexane with Br)

2° alkyl halide    3° alkyl halide
**faster reaction**

**7.23**

| These three methyl groups make the alkyl halide sterically hindered. This slows the rate of an $S_N2$ reaction even though it is a 1° alkyl halide. |

$CH_3-\overset{\overset{\displaystyle CH_3}{|}}{\underset{\underset{\displaystyle CH_3}{|}}{C}}-CH_2Br$

**7.24**

(pyridine-pyrrolidine **A**) + $CH_3-SR_2^+$ ⟶ (N-methylated intermediate) + $SR_2$ → loss of a proton → nicotine

**7.25** In a first-order reaction, **the rate changes with any change in [RX]**. The rate is independent of any change in [nucleophile].

    a. [RX] is tripled, and [:Nu⁻] stays the same: **rate triples.**

    b. Both [RX] and [:Nu⁻] are tripled: **rate triples.**

    c. [RX]  is halved, and [:Nu⁻] stays the same: **rate halved.**

    d. [RX]  is halved, and [:Nu⁻] is doubled: **rate halved.**

**7.26** In $S_N1$ reactions, **racemization always occurs at a stereogenic center**. Draw two products, with the two possible configurations at the stereogenic center.

**7.27** **Carbocations are classified by the number of R groups bonded to the carbon**: 0 R groups = methyl, 1 R group = 1°, 2 R groups = 2°, and 3 R groups = 3°.

a. $CH_3\overset{+}{C}HCH_2CH_3$

    2 R groups
    **2° carbocation**

b. (structure)

    2 R groups
    **2° carbocation**

c. $(CH_3)_3C\overset{+}{C}H_2$

    1 R group
    **1° carbocation**

d. (structure)

    3 R groups
    **3° carbocation**

e. (structure)

    2 R groups
    **2° carbocation**

**7.28** For carbocations: Increasing number of R groups = Increasing stability.

a. $(CH_3)_2CHCH_2\overset{+}{C}H_2$

    1° carbocation
    **least stable**

$(CH_3)_2CH\overset{+}{C}HCH_3$

    2° carbocation
    **intermediate stability**

$(CH_3)_2\overset{+}{C}CH_2CH_3$

    3° carbocation
    **most stable**

b. (structure)

    1° carbocation
    **least stable**

(structure)

    2° carbocation
    **intermediate stability**

(structure)

    3° carbocation
    **most stable**

**7.29**

3 Cl groups -
**electron-*withdrawing***
destabilizing
**less stable**

methyl group
without added Cl's
**more stable**

In $Cl_3CCH_2^+$, the three electron-withdrawing Cl atoms place a partial positive charge on the carbon adjacent to the carbocation, destabilizing it.

**7.30 The rate of an $S_N1$ reaction increases with increasing alkyl substitution.**

a.　　(CH₃)₃CBr　　and　　(CH₃)₃CCH₂Br

　　　　3° alkyl halide　　　　　1° alkyl halide
　　　**faster $S_N1$ reaction　slower $S_N1$ reaction**

c. and

　　　3° alkyl halide　　　　2° alkyl halide
　　**faster $S_N1$ reaction　slower $S_N1$ reaction**

b. and

　　2° alkyl halide　　　　　　3° alkyl halide
　**slower $S_N1$ reaction　　faster $S_N1$ reaction**

**7.31** • For **methyl and 1° alkyl halides**, only $S_N2$ will occur.
　　• For **2° alkyl halides**, $S_N1$ and $S_N2$ will occur.
　　• For **3° alkyl halides**, only $S_N1$ will occur.

a.
$$CH_3-\overset{\overset{\displaystyle CH_3}{|}}{\underset{\underset{\displaystyle CH_3}{|}}{C}}-\overset{\overset{\displaystyle H}{|}}{\underset{\underset{\displaystyle CH_3}{|}}{C}}-Br$$
　2° alkyl halide
　**$S_N1$ and $S_N2$**

b.
　1° alkyl halide
　**$S_N2$**

c.
　2° alkyl halide
　**$S_N1$ and $S_N2$**

d.
　3° alkyl halide
　**$S_N1$**

**7.32** • Draw the product of nucleophilic substitution for each reaction.
　　• For **methyl and 1° alkyl halides**, only $S_N2$ will occur.
　　• For **2° alkyl halides**, $S_N1$ and $S_N2$ will occur and other factors determine which mechanism operates.
　　• For **3° alkyl halides**, only $S_N1$ will occur.

a.
　3° alkyl halide
　**only $S_N1$**

b.
　1° alkyl halide
　**only $S_N2$**

c. 
Strong nucleophile
**favors $S_N2$.**
　2° alkyl halide
　**Both $S_N1$ and $S_N2$
　are possible.**

d.
Weak nucleophile
**favors $S_N1$.**
　2° alkyl halide
　**Both $S_N1$ and $S_N2$
　are possible.**

**7.33** First decide whether the reaction will proceed via an $S_N1$ or $S_N2$ mechanism.  Then draw the products with stereochemistry.

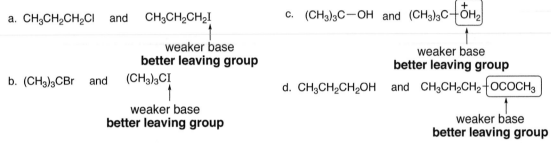

a.

2° alkyl halide
$S_N1$ and $S_N2$

weak nucleophile
**favors $S_N1$**

**enantiomers**

$S_N1$ = racemization at the
stereogenic C

b.

1° alkyl halide
$S_N2$ only

$S_N2$ = inversion at the stereogenic C

**7.34** Compounds with better leaving groups react faster.  Weaker bases are better leaving groups.

a. $CH_3CH_2CH_2Cl$ and $CH_3CH_2CH_2I$

weaker base
**better leaving group**

c. $(CH_3)_3C-OH$ and $(CH_3)_3C-\overset{+}{O}H_2$

weaker base
**better leaving group**

b. $(CH_3)_3CBr$ and $(CH_3)_3Cl$

weaker base
**better leaving group**

d. $CH_3CH_2CH_2OH$ and $CH_3CH_2CH_2-OCOCH_3$

weaker base
**better leaving group**

**7.35**  • **Polar protic solvents** favor the $S_N1$ mechanism by solvating the intermediate carbocation and halide.
  • **Polar aprotic solvents** favor the $S_N2$ mechanism by making the nucleophile stronger.

a. $CH_3CH_2OH$
*polar protic solvent*
contains an O–H bond
**favors $S_N1$**

b. $CH_3CN$
*polar aprotic solvent*
no O–H or N–H bond
**favors $S_N2$**

c. $CH_3COOH$
*polar protic solvent*
contains an O–H bond
**favors $S_N1$**

d. $CH_3CH_2OCH_2CH_3$
*polar aprotic solvent*
no O–H or N–H bond
**favors $S_N2$**

**7.36** Compare the solvents in the reactions below.  **For the solvent to increase the reaction rate of an $S_N1$ reaction, the solvent must be *polar protic*.**

a. $(CH_3)_3CBr$ + $H_2O$ $\xrightarrow[\text{or}]{H_2O}$ $(CH_3)_3COH$ + HBr

3° RX - $S_N1$ reaction $(CH_3)_2C=O$

$H_2O$
**Polar protic solvent**
increases the rate of an
$S_N1$ reaction.

b.

Cl + $CH_3OH$ $\xrightarrow[\text{DMSO}]{CH_3OH}$

OCH$_3$ + HCl

3° RX - $S_N1$ reaction

$CH_3OH$
**Polar protic solvent**
increases the rate of an
$S_N1$ reaction.

c.

Br + $^-$OH $\xrightarrow[\text{DMF}]{H_2O}$ OH + Br$^-$

1° RX - $S_N2$ reaction

DMF [$HCON(CH_3)_2$]
**Polar aprotic solvent**
increases the rate of an
$S_N2$ reaction.

**d.** [structure: 2° RX] + CH₃O⁻ →(CH₃OH or HMPA)→ [product] + Cl⁻

2° RX    strong nucleophile
S$_N$2 reaction

HMPA [(CH₃)₂N]₃P=O
**Polar aprotic solvent**
increases the rate of an
S$_N$2 reaction.

**7.37** To predict whether the reaction follows an S$_N$1 or S$_N$2 mechanism:
  [1] **Classify RX as a methyl, 1°, 2°, or 3° halide.** (methyl, 1° = S$_N$2; 3° = S$_N$1; 2° = either)
  [2] **Classify the nucleophile as strong or weak.** (strong favors S$_N$2, weak favors S$_N$1)
  [3] **Classify the solvent as polar protic or polar aprotic.** (polar protic favors S$_N$1, polar aprotic favors S$_N$2).

**a.** [structure]—CH₂Br + CH₃CH₂O⁻ ⟶ [structure]—CH₂OCH₂CH₃ + Br⁻     S$_N$2 reaction

1° alkyl halide
**S$_N$2**

**b.** [structure with Br] + N₃⁻ ⟶ [structure with N₃] + Br⁻     S$_N$2 reaction = *inversion* at the stereogenic center
The leaving group was "up."
The nucleophile attacks from below.

2° alkyl halide      Strong nucleophile
**S$_N$1 or S$_N$2**    **favors S$_N$2.**

**c.** [structure with I] + CH₃OH ⟶ [structure with OCH₃] + HI     S$_N$1 reaction

3° alkyl halide    Weak nucleophile
**S$_N$1**      **favors S$_N$1.**

**d.** [structure with Cl] + H₂O ⟶ [structure with OH] + [structure with HO] + HCl     S$_N$1 reaction
forms **two enantiomers.**

3° alkyl halide    Weak nucleophile
**S$_N$1**      **favors S$_N$1.**

**7.38** Vinyl carbocations are even less stable than 1° carbocations.

CH₃CH₂CH₂CH₂CH=ĊH          CH₃CH₂CH₂CH₂CH₂ĊH₂          CH₃CH₂CH₂CH₂ĊHCH₃

vinyl carbocation          1° carbocation          2° carbocation
**least stable**          **intermediate stability**          **most stable**

**7.39**

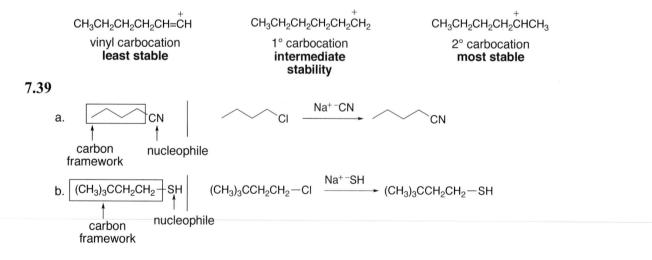

**a.** [framework with CN]  |  [structure with Cl] →(Na⁺ ⁻CN)→ [structure with CN]

carbon framework    nucleophile

**b.** (CH₃)₃CCH₂CH₂—SH  |  (CH₃)₃CCH₂CH₂—Cl →(Na⁺ ⁻SH)→ (CH₃)₃CCH₂CH₂—SH

carbon framework    nucleophile

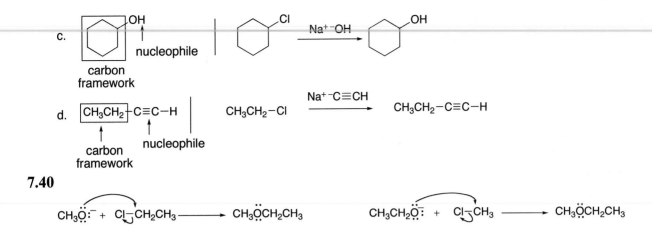

c.

carbon
framework

d.

carbon
framework

**7.40**

$CH_3\ddot{O}{:}^-$ + $Cl{-}CH_2CH_3$ ⟶ $CH_3\ddot{O}CH_2CH_3$      $CH_3CH_2\ddot{O}{:}^-$ + $Cl{-}CH_3$ ⟶ $CH_3\ddot{O}CH_2CH_3$

**7.41** Use the directions from Answer 7.3 to name the compounds.

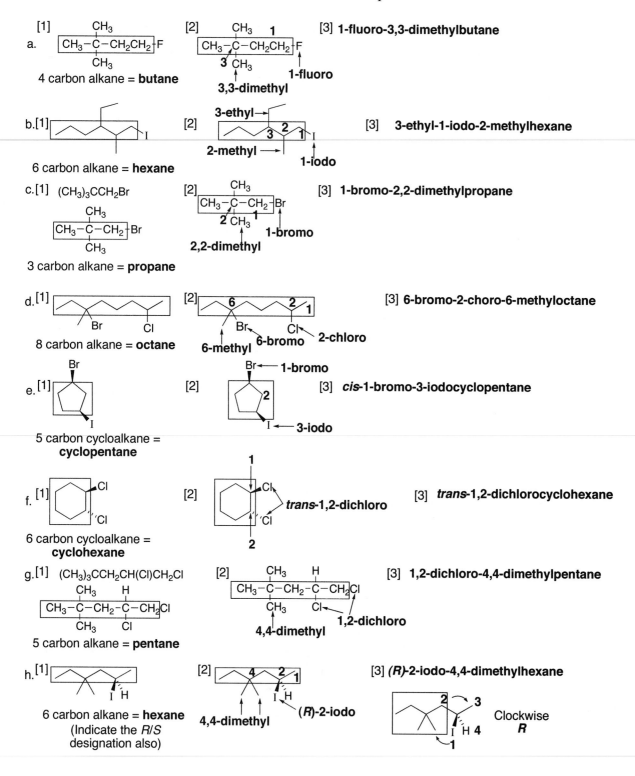

[1] CH₃ | CH₃−C−CH₂CH₂⌐F | CH₃
a. 4 carbon alkane = **butane**

[2] CH₃ 1 | CH₃−C−CH₂CH₂⌐F | 3 CH₃ | 1-fluoro | 3,3-dimethyl

[3] **1-fluoro-3,3-dimethylbutane**

b.[1] 6 carbon alkane = **hexane**

[2] 3-ethyl→ | 2-methyl → | 1-iodo

[3] **3-ethyl-1-iodo-2-methylhexane**

c.[1] (CH₃)₃CCH₂Br | CH₃−C−CH₂⌐Br | CH₃
3 carbon alkane = **propane**

[2] CH₃ | CH₃−C−CH₂⌐Br | 2 CH₃ 1 | 2,2-dimethyl | 1-bromo

[3] **1-bromo-2,2-dimethylpropane**

d.[1] Br Cl | 8 carbon alkane = **octane**

[2] 6 2 1 | Br Cl | 6-methyl 6-bromo 2-chloro

[3] **6-bromo-2-choro-6-methyloctane**

e.[1] Br | I | 5 carbon cycloalkane = **cyclopentane**

[2] Br← 1-bromo | 2 | I ← 3-iodo

[3] *cis*-**1-bromo-3-iodocyclopentane**

f.[1] Cl | Cl | 6 carbon cycloalkane = **cyclohexane**

[2] 1 | Cl | Cl | 2 | *trans*-1,2-dichloro

[3] *trans*-**1,2-dichlorocyclohexane**

g.[1] (CH₃)₃CCH₂CH(Cl)CH₂Cl | CH₃ H | CH₃−C−CH₂−C−CH₂Cl | CH₃ Cl | 5 carbon alkane = **pentane**

[2] CH₃ H | CH₃−C−CH₂−C−CH₂Cl | CH₃ Cl | 4,4-dimethyl | 1,2-dichloro

[3] **1,2-dichloro-4,4-dimethylpentane**

h.[1] I H | 6 carbon alkane = **hexane** | (Indicate the *R/S* designation also)

[2] 4 2 1 | I H | 4,4-dimethyl (*R*)-2-iodo

[3] (*R*)-**2-iodo-4,4-dimethylhexane** | 2 3 | I H 4 | 1 | Clockwise *R*

**7.42** To work backwards to a structure, use the directions in Answer 7.4.

a. **isopropyl** bromide

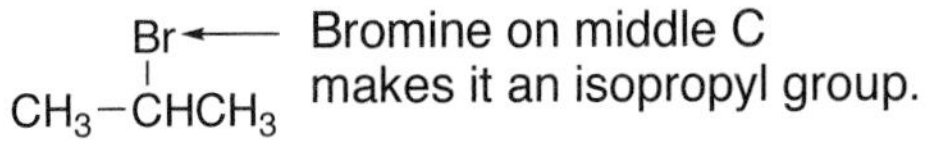

Br ◄— Bromine on middle C
|
$CH_3$–$CHCH_3$   makes it an isopropyl group.

b. 3-bromo-4-ethyl**heptane**

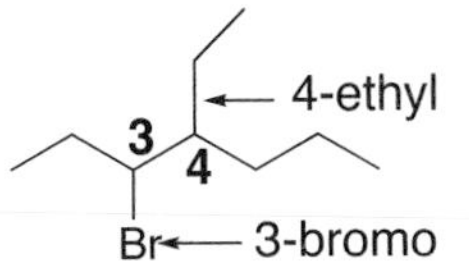

◄— 4-ethyl
**3** **4**

Br ◄— 3-bromo

c. 1,1-dichloro-2-methyl**cyclohexane**

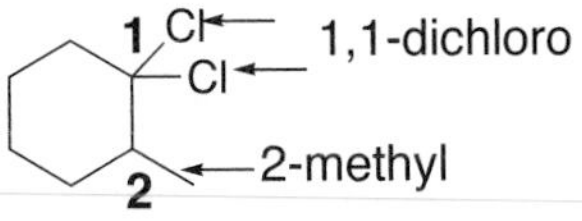

**1** Cl ◄— 1,1-dichloro
Cl ◄—

◄—2-methyl
**2**

d. *trans*-1-chloro-3-iodo**cyclobutane**

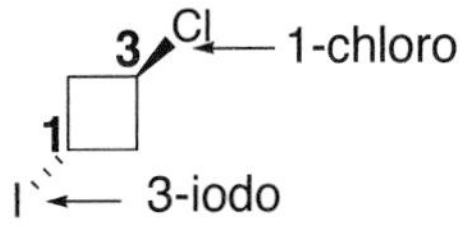

**3** Cl — 1-chloro
**1**
I ◄— 3-iodo

e. 1-bromo-4-ethyl-3-fluoro**octane**

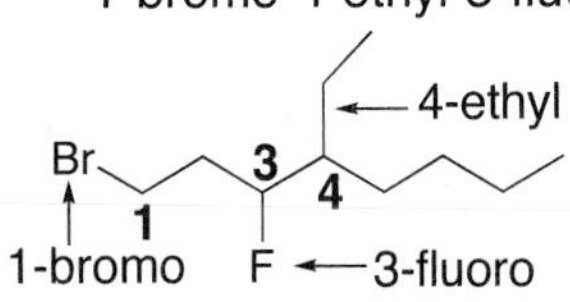

◄— 4-ethyl
Br **3** **4**
**1**
1-bromo   F ◄—3-fluoro

**7.43**

a.
$CH_3$ 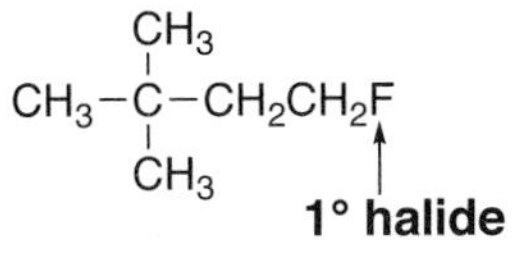
|
$CH_3$–C–$CH_2CH_2F$
|
$CH_3$
↑
**1° halide**

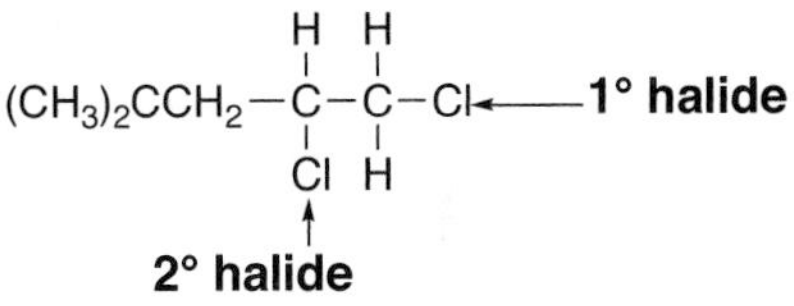

b.
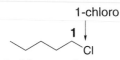
I
↑
**1° halide**

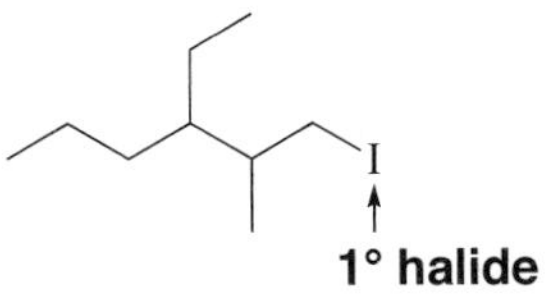

c. $(CH_3)_3CCH_2Br$
↑
**1° halide**

d.
Br
↑      Cl
↑
**3° halide**   **2° halide**

e.
Br ◄— **2° halide**

I ◄— **2° halide**

f.
Cl ◄—
Both are
**2° halides.**
Cl ◄—

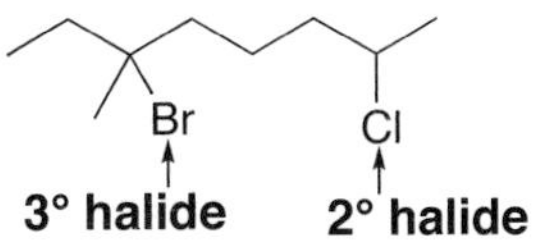

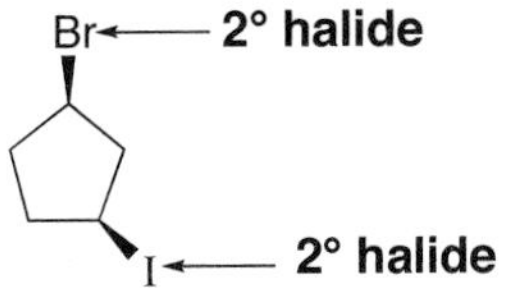

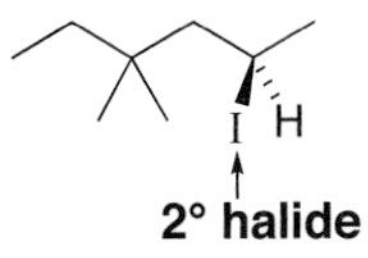

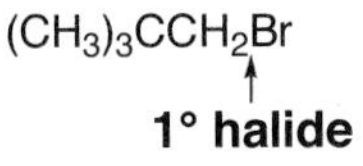

g. $(CH_3)_2CCH_2$–C–C–Cl ◄—**1° halide**
H H
| |
C–C–Cl
| |
Cl H
↑
**2° halide**

h.
I
↑ H
**2° halide**

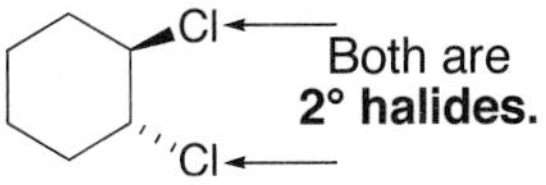

**7.44**

1-chloro
**1**
Cl
**1-chloropentane**

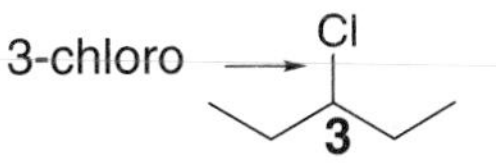

1-chloro
Cl
**2** **1**
**1-chloro-2,2-dimethylpropane**

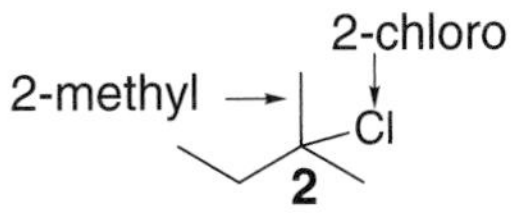

3-chloro — Cl
**3**
**3-chloropentane**

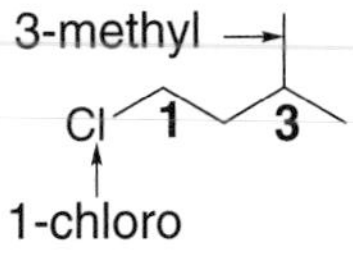

2-methyl — 2-chloro
Cl
**2**
**2-chloro-2-methylbutane**

3-methyl —
Cl **1** **3**
↑
1-chloro
**1-chloro-3-methylbutane**

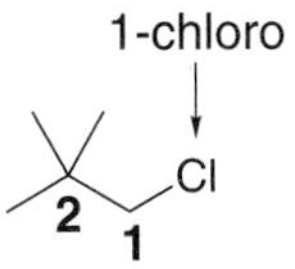

**Two stereoisomers**

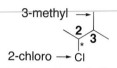

2-chloro → Cl

**2-chloropentane**

[* denotes stereogenic center]

Clockwise
"4" in back =
**R**

Clockwise
"4" in *front* =
**S**

**Two stereoisomers**

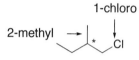

3-methyl →

2-chloro → Cl

**2-chloro-3-methylbutane**

[* denotes stereogenic center]

Counterclockwise
"4" in front =
**R**

Counterclockwise
"4" in *back* =
**S**

**Two stereoisomers**

1-chloro

2-methyl →

Cl

**1-chloro-2-methylbutane**

[* denotes stereogenic center]

Clockwise
"4" in *front* =
**S**

Clockwise
"4" in *back* =
**R**

**7.45** Use the directions from Answer 7.5.

a.  $(CH_3)_3CBr$  or  $CH_3CH_2CH_2CH_2Br$

larger surface area =
stronger intermolecular forces =
**higher boiling point**

b.  [structure] I  or  [structure] Br

larger halide = more polarizable =
**higher boiling point**

c.  [cyclohexane]  or  [bromocyclohexane] Br

nonpolar
only VDW forces

more polar =
**higher boiling point**

**7.46**

a.  $CH_3CH_2CH_2CH_2-Br$  +  $^-OH$  ⟶  $CH_3CH_2CH_2CH_2OH$  +  $Br^-$

b.  $CH_3CH_2CH_2CH_2-Br$  +  $^-SH$  ⟶  $CH_3CH_2CH_2CH_2SH$  +  $Br^-$

c.  $CH_3CH_2CH_2CH_2-Br$  +  $^-CN$  ⟶  $CH_3CH_2CH_2CH_2CN$  +  $Br^-$

d.  $CH_3CH_2CH_2CH_2-Br$  +  $^-OCH(CH_3)_2$  ⟶  $CH_3CH_2CH_2CH_2OCH(CH_3)_2$  +  $Br^-$

e.  $CH_3CH_2CH_2CH_2-Br$  +  $^-C{\equiv}CH$  ⟶  $CH_3CH_2CH_2CH_2C{\equiv}CH$  +  $Br^-$

f.  $CH_3CH_2CH_2CH_2-Br$  +  $H_2\ddot{O}:$  ⟶  $CH_3CH_2CH_2CH_2\overset{+}{\underset{..}{O}}H_2$  +  $Br^-$  ⟶  $CH_3CH_2CH_2CH_2OH$  +  $HBr$

g.  $CH_3CH_2CH_2CH_2-Br$  +  $\ddot{N}H_3$  ⟶  $CH_3CH_2CH_2CH_2\overset{+}{N}H_3$  +  $Br^-$  ⟶  $CH_3CH_2CH_2CH_2NH_2$  +  $HBr$

h.  $CH_3CH_2CH_2CH_2-Br$  +  $Na^+I^-$  ⟶  $CH_3CH_2CH_2CH_2I$  +  $Na^+Br^-$

i.  $CH_3CH_2CH_2CH_2-Br$  +  $Na^+N_3^-$  ⟶  $CH_3CH_2CH_2CH_2N_3$  +  $Na^+Br^-$

**7.47** Use the steps from Answer 7.8 and then draw the proton transfer reaction, when necessary.

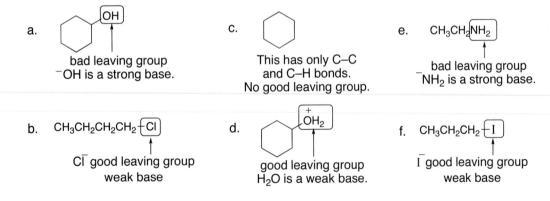

a. [structure with Cl leaving group] + CH₃C(=O)O⁻ nucleophile ⟶ [ester product] + Cl⁻

    **leaving group**     **nucleophile**

b. [pentyl–I] + Na⁺ ⁻CN ⟶ [pentyl–CN] + NaI

    **leaving group**    **nucleophile**

c. [cyclohexyl with ethyl and I] + H₂Ö: ⟶ [cyclohexyl with ethyl and OH] + HI

    **leaving group**    **nucleophile**

d. [cyclopentyl with methyl and Cl] + CH₃CH₂ÖH ⟶ [cyclopentyl with methyl and OCH₂CH₃] + HCl

    **leaving group**    **nucleophile**

e. [cyclohexyl–Br] + Na⁺ ⁻OCH₃ ⟶ [cyclohexyl–OCH₃] + NaBr

    **leaving group**   **nucleophile**

f. [propyl–Cl] + CH₃S̈CH₃ ⟶ [propyl–S⁺(CH₃)CH₃] + Cl⁻

    **leaving group**    **nucleophile**

**7.48** A good leaving group is a weak base.

a. [cyclohexyl–OH]

    bad leaving group
    ⁻OH is a strong base.

c. [cyclohexane]

    This has only C–C
    and C–H bonds.
    No good leaving group.

e. CH₃CH₂NH₂

    bad leaving group
    ⁻NH₂ is a strong base.

b. CH₃CH₂CH₂CH₂—Cl

    Cl⁻ good leaving group
    weak base

d. [cyclohexyl–⁺OH₂]

    good leaving group
    H₂O is a weak base.

f. CH₃CH₂CH₂—I

    I⁻ good leaving group
    weak base

**7.49** Use the rules from Answer 7.10.

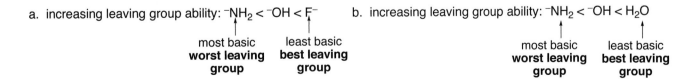

a. increasing leaving group ability: ⁻NH₂ < ⁻OH < F⁻

          most basic     least basic
          **worst leaving**   **best leaving**
          **group**         **group**

b. increasing leaving group ability: ⁻NH₂ < ⁻OH < H₂O

          most basic     least basic
          **worst leaving**   **best leaving**
          **group**         **group**

c. increasing leaving group ability: Cl⁻ < Br⁻ < I⁻    d. increasing leaving group ability: $NH_3 < H_2O < H_2S$

　　　　　　　　　↑　　　　　↑　　　　　　　　　　　　　　　　　　　　　　　　↑　　　　　↑
　　　　　　most basic    least basic　　　　　　　　　　　　　　　　　most basic    least basic
　　　　**worst leaving   best leaving**　　　　　　　　　　　　　**worst leaving   best leaving**
　　　　　　**group　　　　group**　　　　　　　　　　　　　　　　　　**group　　　　group**

**7.50** Compare the nucleophile and the leaving group in each reaction. The reaction will occur if it proceeds towards the weaker base. Remember that the stronger the acid (lower $pK_a$), the weaker the conjugate base.

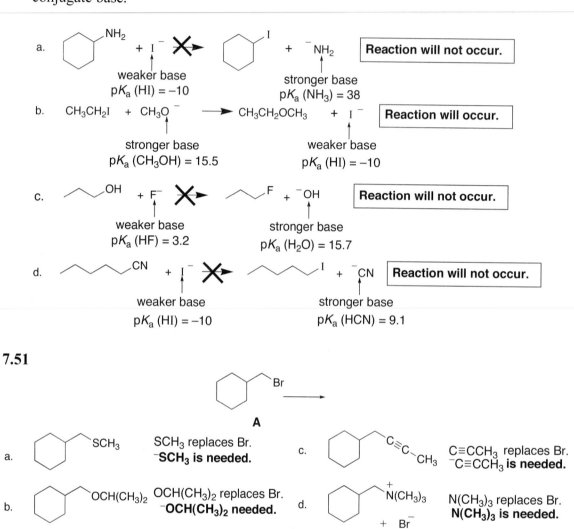

**7.51**

**7.52** Use the directions in Answer 7.16.

a. Across a row of the periodic table nucleophilicity decreases.

$$^-OH < {}^-NH_2 < CH_3{}^-$$

b. • In a **polar protic solvent** ($CH_3OH$), nucleophilicity *increases down a column* of the periodic table, so: $^-SH$ is more nucleophilic than $^-OH$.
   • *Negatively charged species* are more nucleophilic than neutral species so $^-OH$ is more nucleophilic than $H_2O$.

$$H_2O < {}^-OH < {}^-SH$$

c. • In a **polar protic solvent** ($CH_3OH$), nucleophilicity *increases down a column* of the periodic table, so: $CH_3CH_2S^-$ is more nucleophilic than $CH_3CH_2O^-$.
   • For two species with same attacking atom the more basic is more nucleophilic so $CH_3CH_2O^-$ is more nucleophilic than $CH_3COO^-$.

$$CH_3COO^- < CH_3CH_2O^- < CH_3CH_2S^-$$

d. Compare the nucleophilicity of N, S and O. In a polar aprotic solvent (acetone), nucleophilicity parallels basicity.

$$CH_3SH < CH_3OH < CH_3NH_2$$

e. In a **polar aprotic solvent** (acetone), nucleophilicity parallels basicity. Across a row and down a column of the periodic table nucleophilicity decreases.

$$Cl^- < F^- < {}^-OH$$

f. Nucleophilicity decreases across a row so $^-SH$ is more nucleophilic than $Cl^-$. In a **polar protic solvent** ($CH_3OH$), nucleophilicity increases down a column so $Cl^-$ is more nucleophilic than $F^-$.

$$F^- < Cl^- < {}^-SH$$

**7.53** Halide ions in the gas phase would experience no solvent effects. Therefore, the trends would be the same as trends in basicity—a stronger base is a stronger nucleophile. Thus **$F^- > Cl^- > Br^- > I^-$.**

**7.54** *Polar protic solvents* **are capable of hydrogen bonding**, and therefore must contain a H bonded to an electronegative O or N. *Polar aprotic solvents* **are incapable of hydrogen bonding**, and therefore do not contain any O–H or N–H bonds.

a. $(CH_3)_2CHOH$

   contains O–H bond
   **protic**

b. $CH_3NO_2$

   no O–H or N–H bond
   **aprotic**

c. $CH_2Cl_2$

   no O–H or N–H bond
   **aprotic**

d. $NH_3$

   contains N–H bond
   **protic**

e. $N(CH_3)_3$

   no O–H or N–H bond
   **aprotic**

f. $HCONH_2$

   contains an N–H bond
   **protic**

**7.55**

a. Mechanism:

1° alkyl halide
**$S_N2$ reaction**

b. Energy diagram:

c. Transition state:

d. Rate equation: one step reaction with both nucleophile and alkyl halide in the only step:

rate = $k$[R–Br][⁻CN]

e. [1] The leaving group is changed from Br⁻ to I⁻:

**Leaving group becomes less basic → a better leaving group → faster reaction.**

[2] The solvent is changed from acetone to $CH_3CH_2OH$:

**Solvent changed to polar protic → decreases reaction rate.**

[3] The alkyl halide is changed from $CH_3(CH_2)_4Br$ to $CH_3CH_2CH_2CH(Br)CH_3$:

**Changed from 1° to 2° alkyl halide → the alkyl halide gets more crowded and the reaction rate decreases.**

[4] The concentration of ⁻CN is increased by a factor of 5.

**Reaction rate will increase by a factor of five.**

[5] The concentration of both the alkyl halide and ⁻CN are increased by a factor of 5:

**Reaction rate will increase by a factor of 25 (5 x 5 = 25).**

**7.56** Use the directions for Answer 7.22.

a.

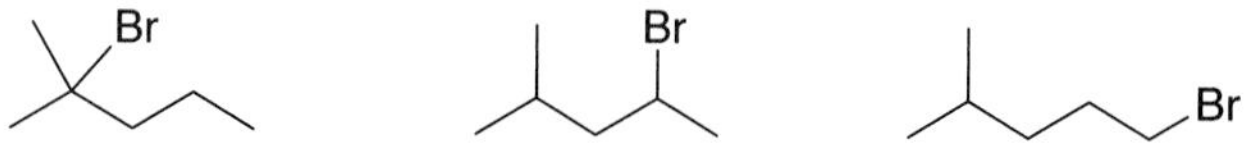

3° alkyl halide
**least reactive**

2° alkyl halide
**intermediate
reactivity**

1° alkyl halide
**most reactive**

b.

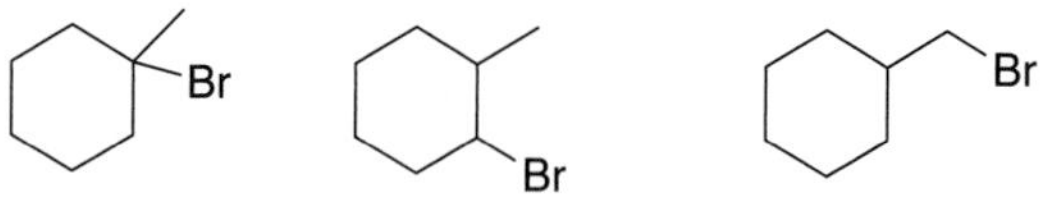

3° alkyl halide
**least reactive**

2° alkyl halide
**intermediate
reactivity**

1° alkyl halide
**most reactive**

c.

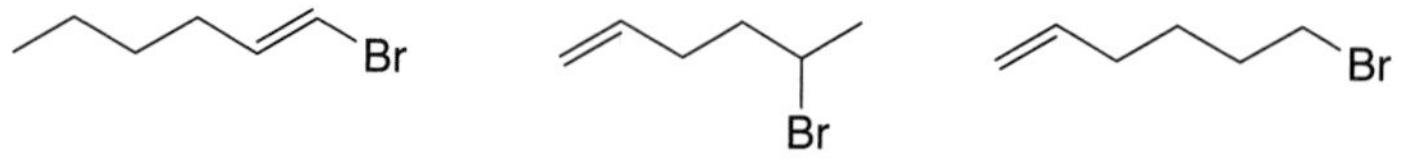

vinyl halide
**least reactivity**

2° alkyl halide
**intermediate
reactivity**

1° alkyl halide
**most reactive**

**7.57**

better leaving group
↓

a. $CH_3CH_2Br$ + ⁻OH ⟶ **faster reaction**

$CH_3CH_2Cl$ + ⁻OH ⟶

stronger nucleophile
↓

b.  Br + ⁻OH ⟶ **faster reaction**

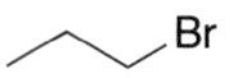

Br + $H_2O$ ⟶

stronger nucleophile
↓

c. 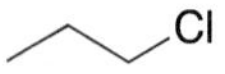 Cl + NaOH ⟶ **faster reaction**

Cl + $NaOCOCH_3$ ⟶

d. 
$$\text{CH}_3\text{CH}_2\text{CH}_2\text{I} + {}^-\text{OCH}_3 \xrightarrow[\text{CH}_3\text{OH}]{}$$

$$\text{CH}_3\text{CH}_2\text{CH}_2\text{I} + {}^-\text{OCH}_3 \xrightarrow[\substack{\text{DMSO} \\ \text{polar aprotic} \\ \text{solvent}}]{} \textbf{faster reaction}$$

less steric hinderance

e. 
$$\text{Br} + {}^-\text{OCH}_2\text{CH}_3 \longrightarrow \textbf{faster reaction}$$

$$\text{Br} + {}^-\text{OCH}_2\text{CH}_3 \longrightarrow$$

**7.58** All $S_N2$ reactions proceed with backside attack of the nucleophile. When nucleophilic attack occurs at a stereogenic center, inversion of configuration occurs.

a. [structure] $+ {}^-\text{OCH}_3 \longrightarrow$ [structure] $+ \text{Cl}^-$   inversion of configuration

b. [structure] $+ {}^-\text{OH} \longrightarrow$ [structure with OH] $+ \text{I}^-$

c. [structure with CI] $+ {}^-\text{OCH}_2\text{CH}_3 \longrightarrow$ [structure with OCH₂CH₃] $+ \text{Cl}^-$   

[ No bond to the stereogenic center is broken, since the leaving group is not bonded to the stereogenic center. ]

d. [cyclopentane structure with Br] $+ {}^-\text{CN} \longrightarrow$ [cyclopentane structure with CN] $+ \text{Br}^-$   inversion of configuration

[* denotes a stereogenic center]

**7.59** Follow the definitions from Answer 7.27.

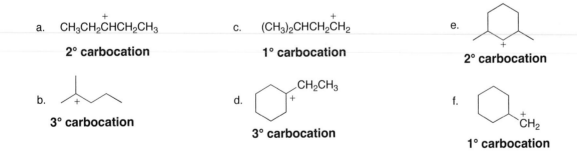

a.   $\text{CH}_3\text{CH}_2\overset{+}{\text{C}}\text{HCH}_2\text{CH}_3$
     **2° carbocation**

c.   $(\text{CH}_3)_2\text{CH}\overset{+}{\text{C}}\text{H}_2\text{CH}_2$
     **1° carbocation**

e.   **2° carbocation**

b.   **3° carbocation**

d.   $\text{CH}_2\text{CH}_3$ **3° carbocation**

f.   $\overset{+}{\text{C}}\text{H}_2$ **1° carbocation**

**7.60** For carbocations: **Increasing number of R groups = Increasing stability**.

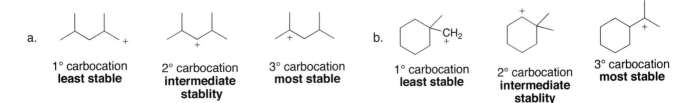

a.   **1° carbocation** **least stable**   **2° carbocation** **intermediate stablity**   **3° carbocation** **most stable**

b.   **1° carbocation** **least stable**   **2° carbocation** **intermediate stablity**   **3° carbocation** **most stable**

**7.61**

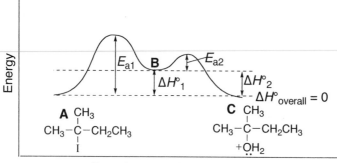

a. Mechanism:
**S$_N$1 only**

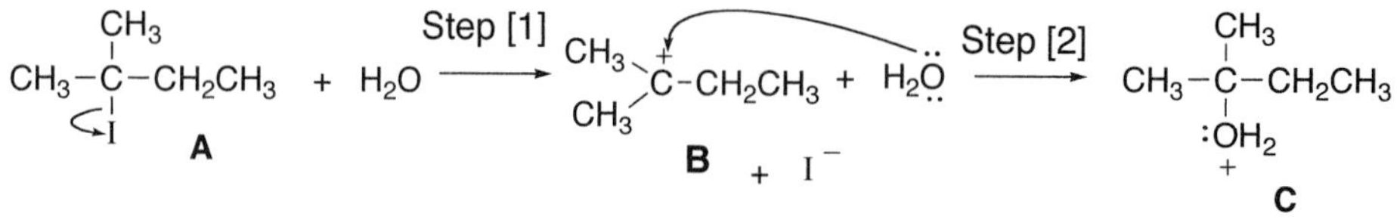

b. Energy diagram:

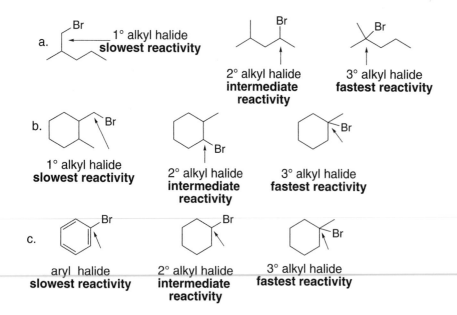

c. Transition states:

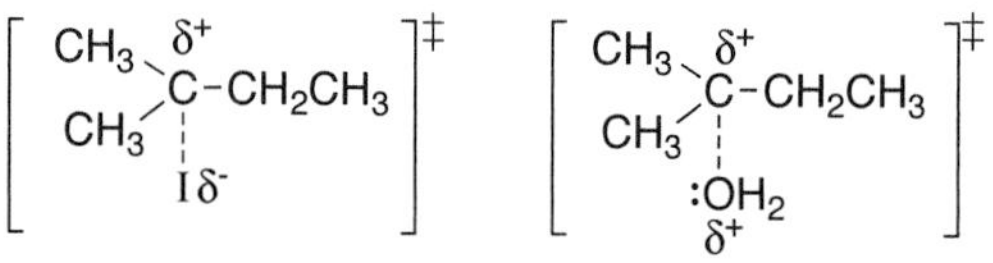

$$\left[ \begin{array}{c} CH_3 \\ CH_3 \end{array} \overset{\delta+}{C}-CH_2CH_3 \\ I\delta- \right]^{\ddagger} \qquad \left[ \begin{array}{c} CH_3 \\ CH_3 \end{array} \overset{\delta+}{C}-CH_2CH_3 \\ :OH_2 \\ \delta+ \right]^{\ddagger}$$

d. rate equation: **rate = $k$[(CH$_3$)$_2$CICH$_2$CH$_3$]**

e. [1] Leaving group changed from I$^-$ to Cl$^-$: **rate decreases** since I$^-$ is a better leaving group.
[2] Solvent changed from H$_2$O (polar protic) to DMF (polar aprotic):
   **rate decreases** since polar protic solvent favors S$_N$1.
[3] Alkyl halide changed from 3° to 2°: **rate decreases** since 2° carbocations are less stable.
[4] [H$_2$O] increased by factor of five: **no change in rate** since H$_2$O is not in rate equation.
[5] [R–X] and [H$_2$O] increased by factor of five: **rate increases** by a factor of five. (Only the concentration of R–X affects the rate.)

**7.62 The rate of S$_N$1 reaction increases with increasing alkyl substitution.**

**7.63** The rate of $S_N1$ reaction increases with increasing alkyl substitution, polar protic solvents, and better leaving groups.

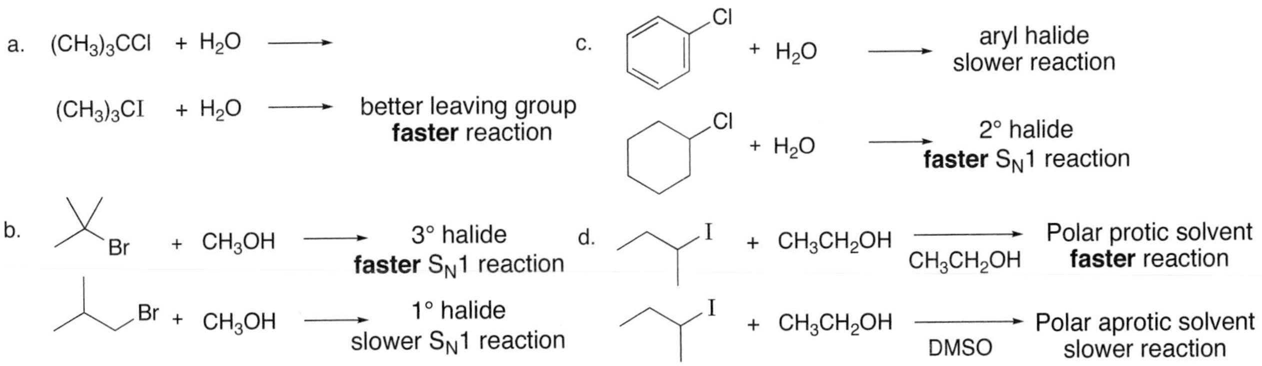

a. $(CH_3)_3CCl + H_2O \longrightarrow$

$(CH_3)_3CI + H_2O \longrightarrow$ **better leaving group** **faster** reaction

c. $+ H_2O \longrightarrow$ aryl halide slower reaction

$+ H_2O \longrightarrow$ $2°$ halide **faster** $S_N1$ reaction

b. $+ CH_3OH \longrightarrow$ $3°$ halide **faster** $S_N1$ reaction

$+ CH_3OH \longrightarrow$ $1°$ halide slower $S_N1$ reaction

d. $+ CH_3CH_2OH \xrightarrow{CH_3CH_2OH}$ Polar protic solvent **faster** reaction

$+ CH_3CH_2OH \xrightarrow{DMSO}$ Polar aprotic solvent slower reaction

**7.64**

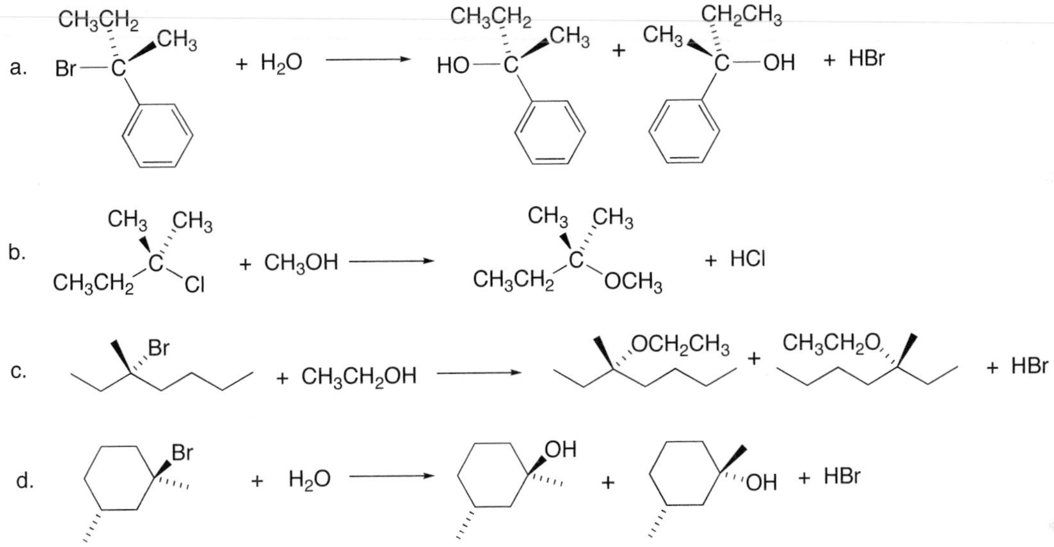

a.

b.

c.

d.

**7.65** The $1°$ alkyl halide is also allylic, so it forms a resonance-stabilized carbocation. Increasing the stability of the carbocation by resonance, increases the rate of the $S_N1$ reaction.

Use each resonance structure individually to continue the mechanism:

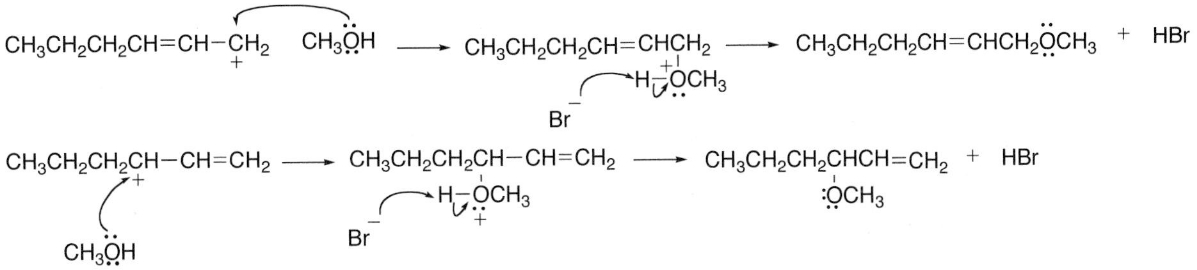

**7.66** More polar solvents favor $S_N1$ reactions by stabilizing the carbocation intermediate. Decreasing the polarity of a solvent would decrease the rate of an $S_N1$ reaction by making the $E_a$ higher.

**7.67** First decide whether the reaction will proceed via an $S_N1$ or $S_N2$ mechanism (Answer 7.37), and then draw the mechanism.

a.

3° alkyl halide
$S_N1$ only

can attack from
above or below

+ HBr

b.

1° alkyl halide
$S_N2$ only

$CH_3CH_2CH_2CH_2\overset{..}{N}H_2$ + $CH_3-I$ (excess)

$CH_3CH_2CH_2CH_2-\overset{+}{\underset{H}{\overset{CH_3}{N}}}-H$ + $I^-$ + $CO_3^{2-}$

$CH_3CH_2CH_2CH_2-\overset{CH_3}{\underset{..}{N}}-H$ + $HCO_3^-$

$CH_3CH_2CH_2CH_2-\overset{+}{\underset{H}{\overset{CH_3}{N}}}-CH_3$ + $I^-$ + $CO_3^{2-}$

$CH_3CH_2CH_2CH_2-\overset{CH_3}{\underset{..}{N}}-CH_3$

$CH_3CH_2CH_2CH_2\overset{+}{N}(CH_3)_3$ + $I^-$

+ $HCO_3^-$

**7.68**

a.

1° alkyl halide
$S_N2$ only

+ $^-CN$ $\xrightarrow{\text{acetone}}$ + $Br^-$

b.

2° alkyl halide
$S_N1$ and $S_N2$

+ $^-OCH_3$ $\xrightarrow{\text{DMSO}}$ + $Br^-$

strong nucleophile
polar aprotic solvent
**Both favor $S_N2$.**

reaction at a stereogenic center
**inversion of configuration**

c.

3° alkyl halide
$S_N1$ only

+ $CH_3OH$ ⟶ + HBr

d.

$\underset{H}{\overset{CH_2CH_3}{\underset{CH_3}{C}}}I$ + $CH_3COOH$ ⟶ $\underset{H}{\overset{CH_2CH_3}{\underset{CH_3}{C}}}OOCCH_3$ + $CH_3COO\underset{H}{\overset{CH_2CH_3}{\underset{CH_3}{C}}}$ + HI

2° alkyl halide  Weak nucleophile
$S_N1$ and $S_N2$  favors $S_N1$.

reaction at a stereogenic center
**racemization of product**

e.

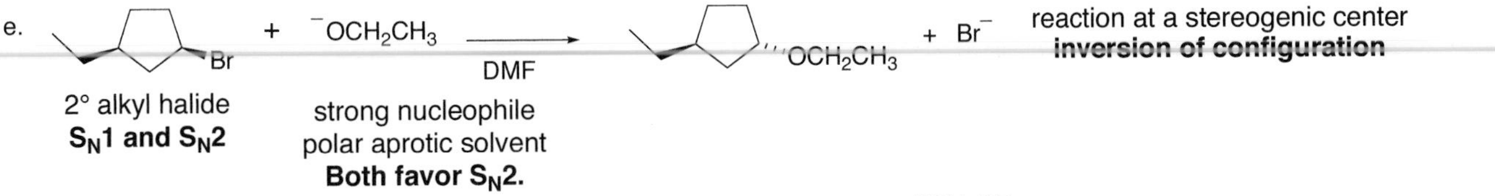
Br
$^{-}OCH_2CH_3$
DMF
$OCH_2CH_3$
+ $Br^-$
reaction at a stereogenic center
inversion of configuration
2° alkyl halide
$S_N1$ and $S_N2$
strong nucleophile
polar aprotic solvent
Both favor $S_N2$.

f.

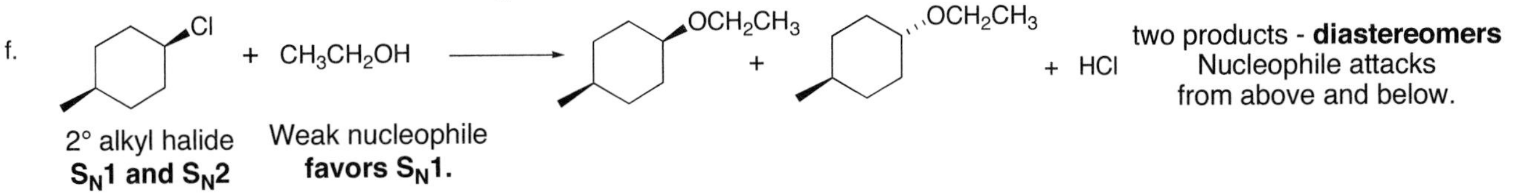
Cl
+ $CH_3CH_2OH$
$OCH_2CH_3$
+
$OCH_2CH_3$
+ HCl
two products - diastereomers
Nucleophile attacks
from above and below.
2° alkyl halide
$S_N1$ and $S_N2$
Weak nucleophile
favors $S_N1$.

**7.69**

a.

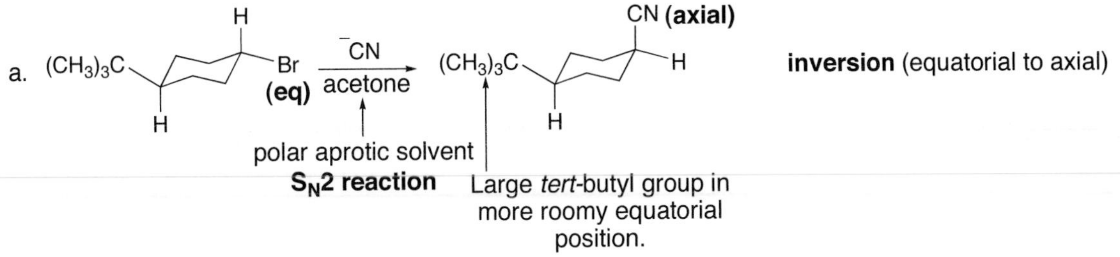
$(CH_3)_3C$
H
Br
(eq)
$^{-}CN$
acetone
CN (axial)
H
inversion (equatorial to axial)
polar aprotic solvent
$S_N2$ reaction
Large tert-butyl group in
more roomy equatorial
position.

b.

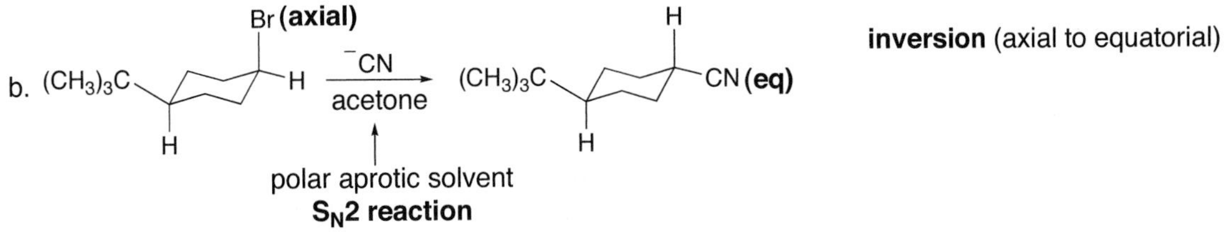
$(CH_3)_3C$
Br (axial)
H
$^{-}CN$
acetone
CN (eq)
inversion (axial to equatorial)
polar aprotic solvent
$S_N2$ reaction

**7.70**

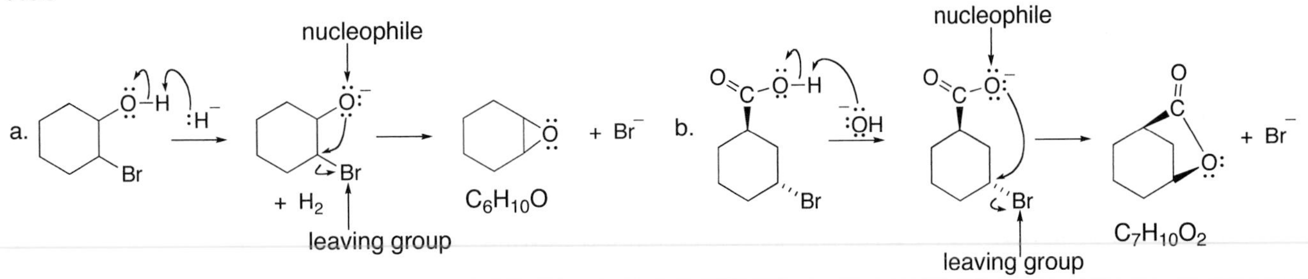
a.
Br
:H⁻
nucleophile
+ $H_2$
leaving group
$C_6H_{10}O$
+ $Br^-$
b.
:ÖH
nucleophile
leaving group
$C_7H_{10}O_2$
+ $Br^-$

**7.71**

a. Hexane is nonpolar and therefore few nucleophiles will dissolve in it.

b. $(CH_3)_3CO^-$ is a stronger base than $CH_3CH_2O^-$:

The three electron-donating $CH_3$ groups add electron density to the negative charge of the conjugate base, destabilizing it and making it a stronger base.

c. By the Hammond postulate, the $S_N1$ reaction is faster with RX that form more stable carbocations.

$(CH_3)_3C+$ — 3° Carbocation is stabilized by three electron-donor $CH_3$ groups.

$(CH_3)_2C+$ with $CF_3$ — Although this carbocation is also 3°, the three electron-withdrawing F atoms destabilize the positive charge. Since the carbocation is less stable, the reaction to form it is slower.

d. The identity of the nucleophile does not affect the rate of $S_N1$ reactions since the nucleophile does not appear in the rate-determining step.

e.

**Polar aprotic solvent favors $S_N2$ reaction.**

2° alkyl halide
$S_N1$ or $S_N2$

acetone    +    Br⁻

(**R**)-2-bromobutane
optically active

(**S**)

strong nucleophile
favors $S_N2$ reaction

This compound reacts with Br⁻ until a 50:50 mixture results, making the mixture optically inactive. Then either compound can react with Br⁻ and the mixture remains optically inactive.

**7.72**

1° Alkyl halides react by
$S_N2$ reactions.
$H_2O$ is a weak nucleophile
and favors $S_N1$ reactions.
This makes the reactions slow.

$CH_3Cl \xrightarrow[\text{slow}]{H_2O} CH_3OH$

$CH_2Cl_2 \xrightarrow[\text{extremely slow}]{H_2O} ClCH_2OH$

$H_2O$ is a weak nucleophile. Since $CH_3Cl$ must react by $S_N2$, the weak nucleophile means a slower reaction. Adding more Cl's adds steric hindrance, decreasing the rate even more.

**7.73**

a.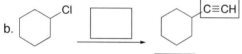

The nucleophile has replaced the leaving group.
Missing reagent:

b.

C≡CH

The nucleophile has replaced the leaving group.
Missing reagent: ⁻C≡CH

c.

N₃

The nucleophile has replaced the halide.
Starting material: Cl

d.

SH

The nucleophile has replaced the halide.
Starting material:

Cl

The leaving must have the opposite orientation to the position of the nucleophile in the product.

**7.74** To devise a synthesis, look for the carbon framework and the functional group in the product. **The carbon framework is from the alkyl halide and the functional group is from the nucleophile.**

a.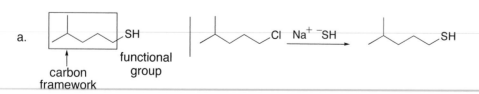

SH

functional
group

carbon
framework

Cl    Na⁺ ⁻SH

SH

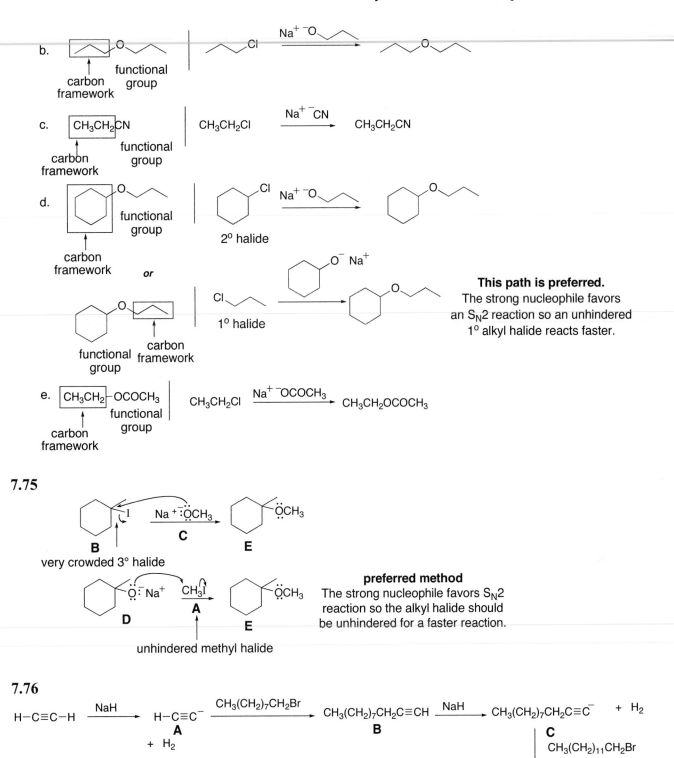

b.

carbon framework → functional group

c.

CH₃CH₂CN

carbon framework → functional group

$CH_3CH_2Cl \xrightarrow{Na^+ \; ^-CN} CH_3CH_2CN$

d.

carbon framework → functional group

2° halide

or

functional group / carbon framework

1° halide

**This path is preferred.**
The strong nucleophile favors
an S$_N$2 reaction so an unhindered
1° alkyl halide reacts faster.

e.

CH₃CH₂—OCOCH₃
carbon framework → functional group

$CH_3CH_2Cl \xrightarrow{Na^+ \; ^-OCOCH_3} CH_3CH_2OCOCH_3$

## 7.75

**B**
very crowded 3° halide

**C**

**E**

**D**

**A**
unhindered methyl halide

**E**

**preferred method**
The strong nucleophile favors S$_N$2
reaction so the alkyl halide should
be unhindered for a faster reaction.

## 7.76

$H-C\equiv C-H \xrightarrow{NaH} H-C\equiv C^-$ **A** $+ H_2$

$\xrightarrow{CH_3(CH_2)_7CH_2Br} CH_3(CH_2)_7CH_2C\equiv CH$ **B** $\xrightarrow{NaH} CH_3(CH_2)_7CH_2C\equiv C^-$ **C** $+ H_2$

$\downarrow CH_3(CH_2)_{11}CH_2Br$

$CH_3(CH_2)_7CH_2C\equiv CCH_2(CH_2)_{11}CH_3$ **D**

$\xleftarrow[\text{(1 equiv)}]{\text{addition of } H_2}$

muscalure

**7.77**

a. **A** and **B** can't react by an S$_N$2 mechanism because the backside attack of the nucleophile is blocked:

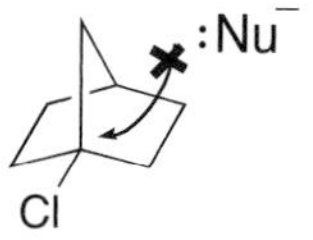

The S$_N$1 reaction would require a planar carbocation, and geometry doesn't allow this to occur.  The resulting carbocation cannot adopt the needed  trigonal planar geometry and thus it does not form.

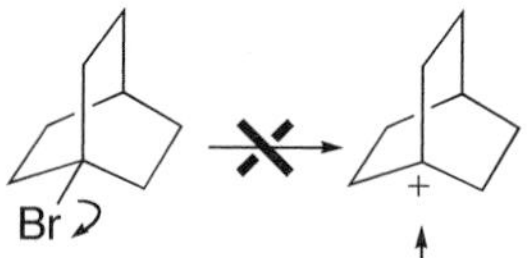

This carbocation cannot adopt a trigonal planar geometry.

b.

quinuclidine

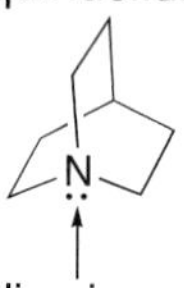

triethylamine

The three alkyl groups are "tied back" in a ring, making the electron pair more available.

This electron pair is more hindered by the three CH$_2$CH$_3$ groups.

These bulky groups around the N cause steric hindrance and this decreases nucleophilicity.

This electron pair on quinuclidine is much more available than the one on triethylamine.

less steric hindrance
**more nucleophilic**

**7.78**

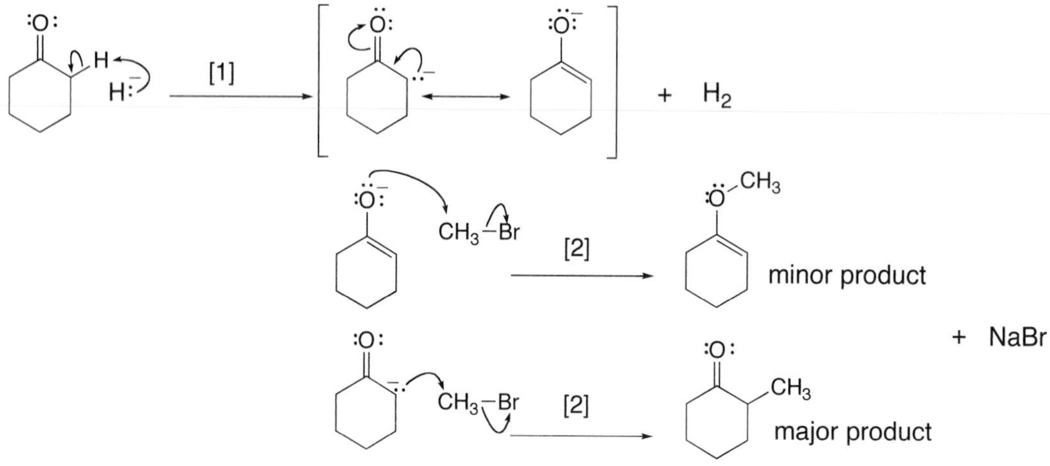

[1]

+ H$_2$

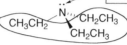

[2]

minor product

+ NaBr

[2]

major product

**7.79** CH$_3$CH$_2$OCH$_2$Cl affords a resonance-stabilized carbocation, making an S$_N$1 reaction possible even though the alkyl halide is 1°.

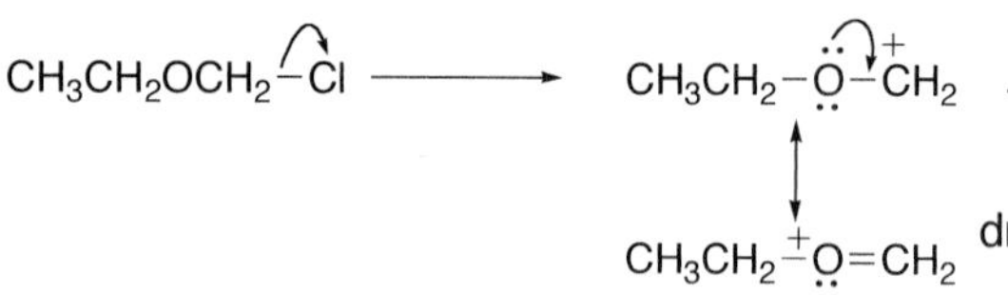

CH$_3$CH$_2$OCH$_2$—Cl  $\longrightarrow$  CH$_3$CH$_2$—O—CH$_2$ + Cl$^-$

CH$_3$CH$_2$—O=CH$_2$

Two resonance structures can be drawn for the carbocation, stabilizing it.

The carbocation can then continue the reaction:

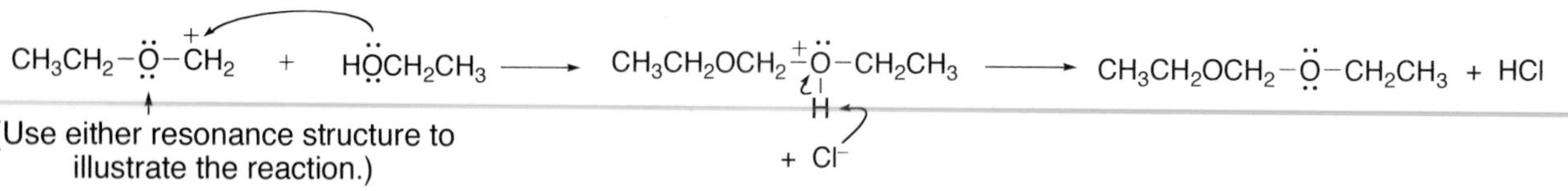

CH$_3$CH$_2$—O—CH$_2$  +  HOCH$_2$CH$_3$  $\longrightarrow$  CH$_3$CH$_2$OCH$_2$—O—CH$_2$CH$_3$  $\longrightarrow$  CH$_3$CH$_2$OCH$_2$—O—CH$_2$CH$_3$ + HCl

(Use either resonance structure to illustrate the reaction.)

+ Cl$^-$

Chapter 8: Alkyl Halides and Elimination Reactions

## ♦ A comparison between nucleophilic substitution and β-elimination

**Nucleophilic substitution**—A nucleophile attacks a carbon atom (7.6).

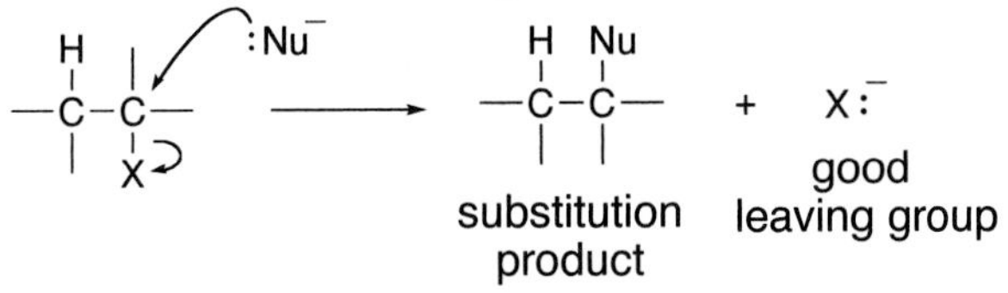

substitution product

good leaving group

**β-Elimination**—A base attacks a proton (8.1).

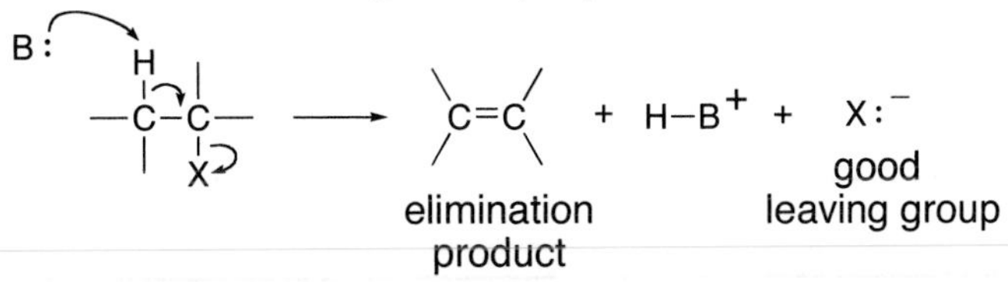

elimination product

good leaving group

| Similarities | Differences |
|---|---|
| • In both reactions RX acts as an electrophile, reacting with an electron rich reagent. <br> • Both reactions require a **good leaving group X:⁻** willing to accept the electron density in the C–X bond. | • In substitution, a nucleophile attacks a single carbon atom. <br> • In elimination, a Brønsted–Lowry base removes a proton to form a π bond, and two carbons are involved in the reaction. |

## ♦ The importance of the base in E2 and E1 reactions (8.9)

The strength of the base determines the mechanism of elimination.
- Strong bases favor E2 reactions.
- Weak bases favor E1 reactions.

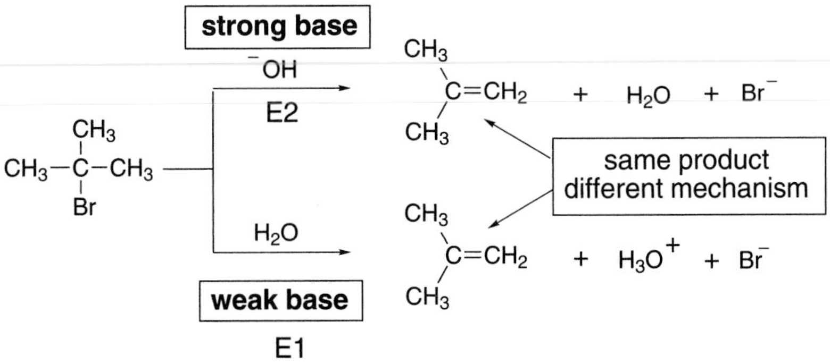

♦ E1 and E2 mechanisms compared

| | E2 mechanism | E1 mechanism |
|---|---|---|
| [1] Mechanism | • One step (8.4B) | • Two steps (8.6B) |
| [2] Alkyl halide | • rate: $R_3CX > R_2CHX >$ $RCH_2X$ (8.4C) | • rate: $R_3CX > R_2CHX >$ $RCH_2X$ (8.6C) |
| [3] Rate equation | • rate = $k$[RX][B:] <br> • second order kinetics (8.4A) | • rate = $k$[RX] <br> • first order kinetics (8.6A) |
| [4] Stereochemistry | • anti periplanar arrangement of H and X (8.8) | • trigonal planar carbocation intermediate (8.6B) |
| [5] Base | • favored by strong bases (8.4B) | • favored by weak bases (8.6C) |
| [6] Leaving group | • better leaving group $\rightarrow$ faster reaction (8.4B) | • better leaving group $\rightarrow$ faster reaction (Table 8.4) |
| [7] Solvents | • favored by polar aprotic solvents (8.4B) | • favored by polar protic solvents (Table 8.4) |
| [8] Product | • more substituted alkene favored (Zaitsev rule, 8.5) | • more substituted alkene favored (Zaitsev rule, 8.6C) |

♦ Summary chart on the four mechanisms: $S_N1$, $S_N2$, E1, or E2

| Alkyl halide type | Conditions | Mechanism |
|---|---|---|
| $1^o$ $RCH_2X$ | strong nucleophile | $S_N2$ |
| | strong bulky base | E2 |
| $2^o$ $R_2CHX$ | strong base and nucleophile | $S_N2$ + E2 |
| | strong bulky base | E2 |
| | weak base and nucleophile | $S_N1$ + E1 |
| $3^o$ $R_3CX$ | weak base and nucleophile | $S_N1$ + E1 |
| | strong base | E2 |

♦ Zaitsev rule

• β-Elimination affords the more stable product having the more substituted double bond.
• Zaitsev products predominate in E2 reactions except when a cyclohexane ring prevents trans diaxial arrangement.

## Chapter 8: Answers to Problems

**8.1** • The carbon bonded to the leaving group is the **α carbon**. Any carbon bonded to it is a **β carbon**.

   • **To draw the products of an elimination reaction:** Remove the leaving group from the α carbon and a H from the β carbon and form a π bond.

a. $\overset{\beta}{C}H_3CH_2CH_2\overset{\alpha}{C}H_2CH_2-Cl$ $\xrightarrow{K^+ {}^-OC(CH_3)_3}$ $CH_3CH_2CH_2CH=CH_2$

c. $\xrightarrow{K^+ {}^-OH}$ $(CH_3CH_2)_2C=CH_2$
   +
   $CH_3CH=C(CH_3)CH_2CH_3$

b. $\xrightarrow{Na^+ {}^-OCH_2CH_3}$ $CH_3CH_2CH_2CH=CHCH_3$
   +
   $CH_3CH_2CH_2CH_2CH=CH_2$

d. $\xrightarrow{K^+ {}^-OH}$

**8.2** **Alkenes are classified by the number of carbon atoms bonded to the double bond.** A monosubstituted alkene has one carbon atom bonded to the double bond, a disubstituted alkene has two carbon atoms bonded to the double bond, etc.

a.
4 C's bonded to C=C
**tetrasubstituted**
2 C's bonded to each C=C
**disubstituted**
3 C's bonded to each C=C
**trisubstituted**
vitamin A

b.
vitamin D₃
3 C's bonded to each C=C
**trisubstituted**
2 C's bonded to the C=C
**disubstituted**

**8.3** To have stereoisomers at a C=C, the two groups on each end of the double bond must be different from each other.

a.
two CH₃ groups
no stereoisomers
possible

b. $CH_3CH_2CH=CHCH_3$
Two different groups
(CH₃CH₂ and H)
Two different groups
(H and CH₃)
stereoisomers possible

c. $CH_2=CHCH_2CH_2CH_3$
two H's
no stereoisomers
possible

d.
Two different groups
(cyclohexyl and H)
Two different groups
(cyclohexyl and H)
stereoisomers possible

**8.4** Two definitions:
- **Constitutional isomers** differ in the connectivity of the atoms.
- **Stereoisomers** differ only in the 3-D arrangement of atoms in space.

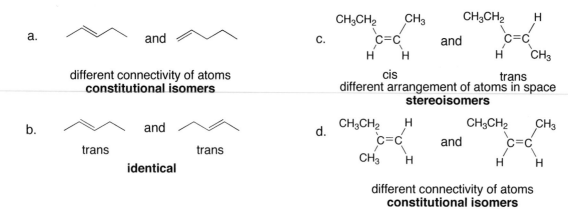

**8.5** Two rules to predict the relative stability of alkenes:
[1] Trans alkenes are more stable than cis alkenes.
[2] The stability of an alkene increases as the number of R groups on the C=C bond increases.

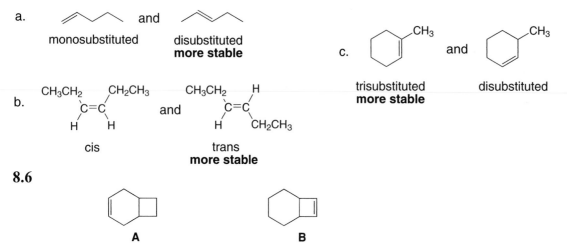

**8.6**

A      B

Alkene **A** is more stable than alkene **B** because the double bond in **A** is in a six-membered ring. The double bond in **B** is in a four-membered ring, which has considerable angle strain due to the small ring size.

**8.7** In an E2 mechanism, four bonds are involved in the single step. Use curved arrows to show these simultaneous actions:
[1] The base attacks a hydrogen on a β carbon.
[2] A π bond forms.
[3] The leaving group comes off.

**8.8** For E2 elimination to occur there must be at least one hydrogen on a β carbon.

no H's on β carbon
**inert to E2 elimination**

**8.9**

better leaving group
**faster reaction**

a. $CH_3CH_2-Br$ + $^-OH$ ⟶

b. $CH_3CH_2-Br$ + $^-OC(CH_3)_3$ ⟶

$CH_3CH_2-Br$ + $^-OC(CH_3)_3$ ⟶

$CH_3CH_2-Cl$ + $^-OC(CH_3)_3$ ⟶

stronger base
**faster reaction**

**8.10** As the number of R groups on the carbon with the leaving group increases, the rate of an E2 reaction increases.

a. $(CH_3)_2CHCH_2CH_2CH_2Br$     $(CH_3)_2CHCH_2CH(Br)CH_3$     $(CH_3)_2C(Br)CH_2CH_2CH_3$

1° alkyl halide       2° alkyl halide       3° alkyl halide
**least reactive**    **intermediate reactivity**    **most reactive**

b.

1° alkyl halide       2° alkyl halide       3° alkyl halide
**least reactive**    **intermediate reactivity**    **most reactive**

**8.11** Use the following characteristics of an E2 reaction to answer the questions:
     [1] E2 reactions are second order and one step.
     [2] More substituted halides react faster.
     [3] Reactions with strong bases or better leaving groups are faster.
     [4] Reactions with polar aprotic solvents are faster.

**Rate equation: rate = $k$[RX][Base]**
     a. tripling the concentration of the alkyl halide = **rate triples**
     b. halving the concentration of the base = **rate halved**
     c. changing the solvent from $CH_3OH$ to DMSO = **rate increases** (Polar aprotic solvent is better for E2.)
     d. changing the leaving group from $I^-$ to $Br^-$ = **rate decreases** ($I^-$ is a better leaving group.)
     e. changing the base from $^-OH$ to $H_2O$ = **rate decreases** (weaker base)
     f. changing the alkyl halide from $CH_3CH_2Br$ to $(CH_3)_2CHBr$ = **rate increases** (More substituted halide reacts faster.)

**8.12** The Zaitsev rule states: In a β-elimination reaction, the major product has the more substituted double bond.

a.

$$CH_3-\underset{\underset{H}{|}}{\overset{\overset{CH_3}{|}}{C}}-\underset{\underset{Br}{|}}{\overset{\overset{H}{|}\alpha}{C}}-CH_2CH_3 \xrightarrow{\text{loss of H and Br}}$$ (CH_3)_2C=CHCH_2CH_3  +  (CH_3)_2CHCH=CHCH_3

                                                        trisubstituted           disubstituted

                                                     **major product**       **minor product**

b.

trisubstituted       tetrasubstituted      disubstituted
**minor product**       **major product**      **minor product**

c.

                                   +  CH_3CH_2CH_2CH_2CH=CHCH_3

                             monosubstituted     disubstituted
                             **minor product**      **major product**

d.

                                            trisubstituted
                                         **ONLY product**

**8.13** An E1 mechanism has two steps:
   [1] The leaving group comes off, creating a carbocation.
   [2] A base pulls off a proton from a β carbon, and a π bond forms.

**8.14** The Zaitsev rule states: In a β-elimination reaction, the major product has the more substituted double bond.

a.

                                      trisubstituted            disubstituted
                                       **major product**

b.

                                tetrasubstituted     disubstituted     trisubstituted
                                **major product**

**8.15** Use the following characteristics of an **E1 reaction** to answer the questions:
   [1] E1 reactions are first order and two steps.
   [2] More substituted halides react faster.
   [3] Weaker bases are preferred.
   [4] Reactions with better leaving groups are faster.
   [5] Reactions in polar protic solvents are faster.

**Rate equation:  rate = $k$[RX]  The base doesn't affect rate.**
   a. doubling the concentration of the alkyl halide = **rate doubles**
   b. doubling the concentration of the base = **no change** (Base is not in the rate equation.)
   c. changing the alkyl halide from $(CH_3)_3CBr$ to $CH_3CH_2CH_2Br$ = **rate decreases** (More substituted halides react faster.)
   d. changing the leaving group from $Cl^-$ to $Br^-$ = **rate increases** (better leaving group)
   e. changing the solvent from DMSO to $CH_3OH$ = **rate increases** (Polar protic solvent favors E1.)

**8.16** Both $S_N1$ and E1 reactions occur by forming a carbocation. To draw the products:
   [1] **For the $S_N1$ reaction,** substitute the nucleophile for the leaving group.
   [2] **For the E1 reaction,** remove a proton from a β carbon and create a new π bond.

a.

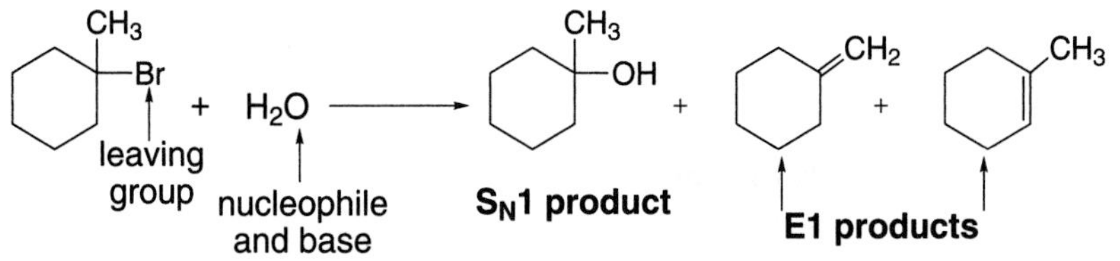

b.

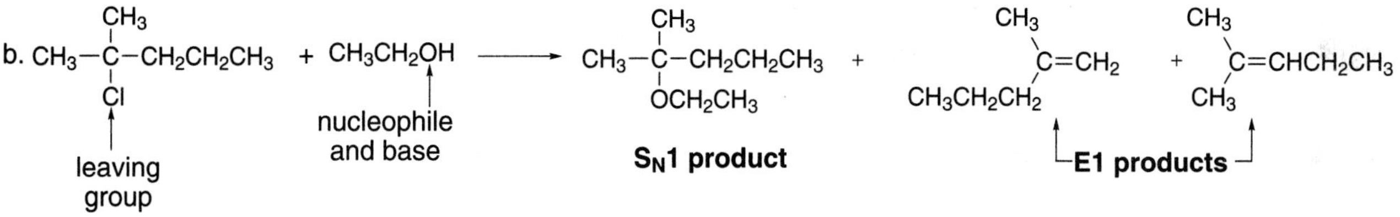

**8.17** E2 reactions occur with anti periplanar geometry. **The anti periplanar arrangement uses a staggered conformation and has the H and X on opposite sides of the C–C bond.**

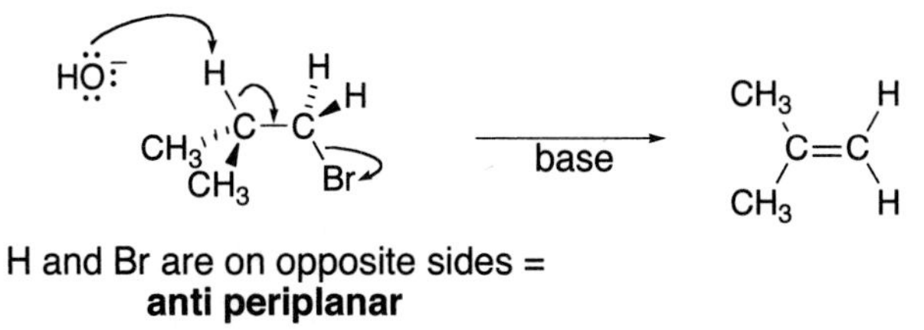

H and Br are on opposite sides =
**anti periplanar**

**8.18** The E2 elimination reactions will occur in the anti periplanar orientation as drawn. To draw the product of elimination, maintain the orientation of the remaining groups around the C=C.

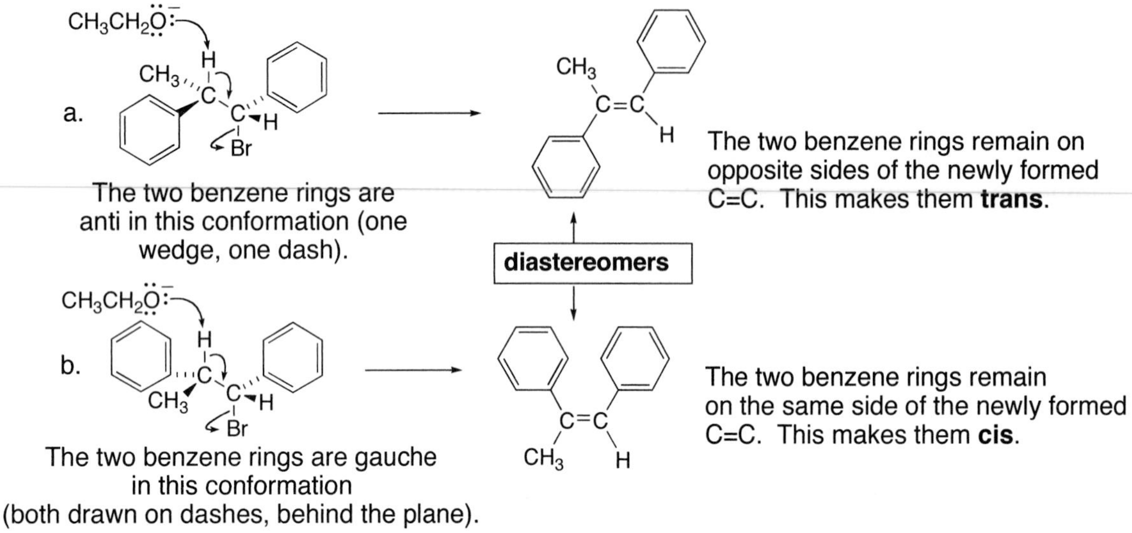

a.

The two benzene rings are anti in this conformation (one wedge, one dash).

The two benzene rings remain on opposite sides of the newly formed C=C. This makes them **trans**.

diastereomers

b.

The two benzene rings are gauche in this conformation (both drawn on dashes, behind the plane).

The two benzene rings remain on the same side of the newly formed C=C. This makes them **cis**.

**8.19** **Note:** The Zaitsev products predominate in E2 elimination *except* when substituents on a cyclohexane ring prevent a **trans diaxial** arrangement of H and X.

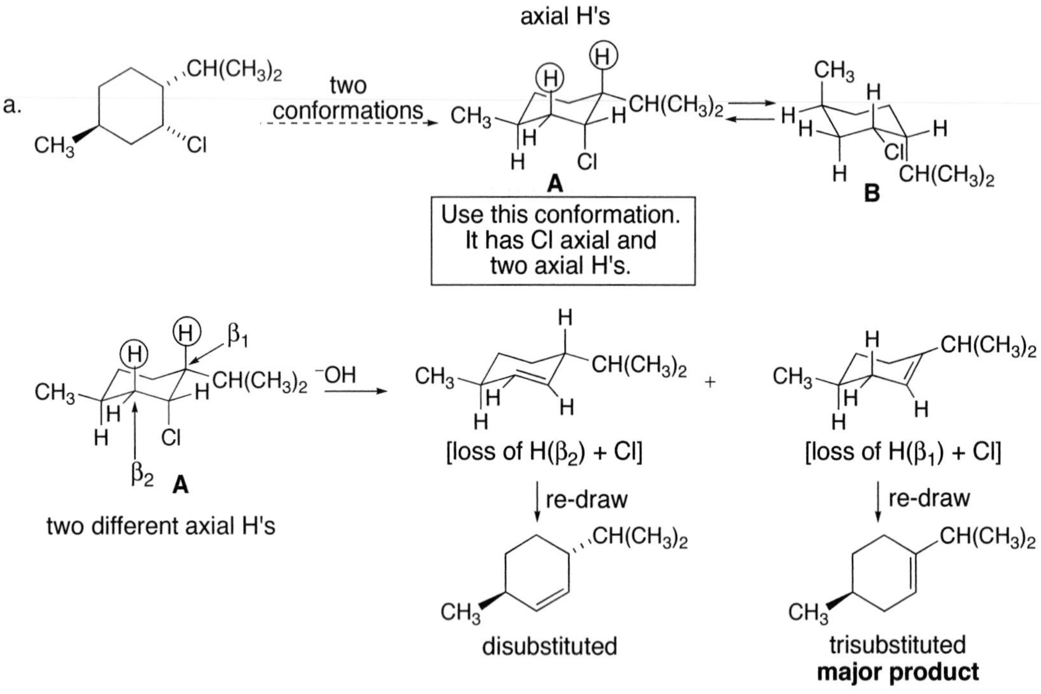

axial H's

a.

two conformations

Use this conformation. It has Cl axial and two axial H's.

two different axial H's

[loss of H($\beta_2$) + Cl]

re-draw

disubstituted

[loss of H($\beta_1$) + Cl]

re-draw

trisubstituted
**major product**

b.

CH(CH₃)₂   two
conformations

**A**

**B**

Use this conformation.
It has Cl axial and
one axial H.

β₂

⁻OH

**B**

only one axial H
on a β carbon

[loss of H(β₁) + Cl]

disubstituted
**only product**

**8.20** Draw the chair conformations of *cis*-1-chloro-2-methylcyclohexane and its trans isomer.  For E2
elimination reactions to occur, **there must be a H and X trans diaxial to each other.**

| Two conformations of the cis isomer: | Two conformations of the trans isomer: |
|---|---|

**A**
reacting conformation (axial Cl)

**B**
reacting conformation (axial Cl)

This reacting conformation has only one
group axial, making it more stable and
present in a higher concentration than **B**.
This makes a **faster elimination reaction
with the cis isomer.**

This conformation is less stable than **A**,
since both CH₃ and Cl are axial.
**This slows the rate of elimination
from the trans isomer.**

**8.21** **E2 reactions are favored by strong negatively charged bases** and occur with 1°, 2°, and 3°
halides, with 3° being the most reactive.
**E1 reactions are favored by weaker neutral bases** and do not occur with 1° halides since they
would have to form highly unstable carbocations.

a.   $CH_3-\overset{\overset{\displaystyle CH_3}{|}}{\underset{\underset{\displaystyle Cl}{|}}{C}}-CH_3$   +   ⁻OCH₃   →

strong negatively
charged base
**E2**

b.   + H₂O   →

weak neutral
base
**E1**

c.   + CH₃OH   →

weak neutral
base
**E1**

d.   CH₃CH₂Br   +   ⁻OC(CH₃)₃   →

strong negatively
charged base
**E2**

**8.22** Draw the alkynes that result from removal of two equivalents of HX.

a. [cyclohexyl-C(Cl)(H)-C(Cl)(H)-CH₂CH₃] $\xrightarrow{\ ^-NH_2\ }$ [cyclohexyl-C≡C-CH₂CH₃]

c. $CH_3-\underset{Br}{\overset{Br}{C}}-CH_2CH_3$ $\xrightarrow{\ ^-NH_2\ }$ $CH_3C≡CCH_3$

$+\ HC≡CCH_2CH_3$

b. $CH_3CH_2CH_2CHCl_2$ $\xrightarrow[\text{DMSO}]{\text{KOC(CH}_3)_3}$ $CH_3CH_2C≡CH$

d. [PhCH(Br)-CH(Br)Ph] $\xrightarrow{\ ^-NH_2\ }$ [Ph-C≡C-Ph]

**8.23**

a. [straight chain with Cl] $\xrightarrow{\ K^+\ ^-OC(CH_3)_3\ }$ [alkene]

1° halide
$S_N2$ or E2

strong sterically
hindered base
E2

b. $CH_3-\underset{Cl}{\overset{H}{C}}-CH_2CH_3$ $\xrightarrow{\ ^-OH\ }$ $CH_3-\underset{OH}{\overset{H}{C}}-CH_2CH_3$ $+$ $CH_3-CH=CHCH_3$ $+$ $CH_2=CH-CH_2CH_3$

2° halide
any mechanism

strong base
$S_N2$ and E2

$S_N2$ product

disubstituted
major E2 product

monosubstituted
minor E2 product

c. [cyclohexane with CH₂CH₃ and I] $\xrightarrow{\ CH_3CH_2OH\ }$ [cyclohexane with CH₂CH₃ and OCH₂CH₃] $+$ [cyclohexane with CHCH₃] $+$ [cyclohexene with CH₂CH₃]

3° halide
no $S_N2$

weak base
$S_N1$ and E1

$S_N1$ product

E1 product

E1 product

d. [branched chain with Cl] $\xrightarrow[\text{CH}_3CH_2OH]{\begin{array}{c}\text{strong base}\\ CH_3CH_2O^-\end{array}}$ [alkene] $+$ [alkene]

3° halide
no $S_N2$

E2

major E2 product

minor E2 product

**8.24**

3° halide          weak base
no $S_N2$          $S_N1$ and E1

[cyclohexane with CH₃, Br, CH₃] $\xrightarrow[\begin{array}{c}\textbf{overall}\\ \textbf{reaction}\end{array}]{CH_3OH}$ [cyclohexane with CH₃, ÖCH₃, CH₃] $+$ [cyclohexene with CH₃, CH₃] $+$ HBr

The steps: ↓

$S_N1$ [cyclohexane cation with +CH₃, CH₃] $\xrightarrow{CH_3\ddot{O}H}$ [cyclohexane with CH₃, Ö⁺-CH₃, H, CH₃]
$+\ Br^-$                    $Br^-$

or

E1 [cyclohexane cation with +CH₃, H, CH₃] $Br^-$

**8.25**

a. CH₃CH₂CH₂CH₂CH₂CH₂Br ⟶ CH₃CH₂CH₂CH₂CH=CH₂

b. [structure with Br] ⟶ CH₃CH₂CH₂CH₂CH=CHCH₂CH₃ + CH₃CH₂CH₂CH₂CH₂CH=CHCH₃

c. CH₃CH₂CHCHCH₃ (with CH₃ and Cl substituents) ⟶ CH₃CH₂CH=C(CH₃)₂ + CH₃CH=CHCH(CH₃)₂

d. [cyclohexane with CH₂I] ⟶ [cyclohexane with =CH₂]

**8.26** To give only one product in an elimination reaction, **the starting alkyl halide must have only one type of β carbon.**

a. CH₂=CHCH₂CH₂CH₃ ⟵ $\overset{\alpha}{CH_2}-\overset{\beta}{CH_2}CH_2CH_2CH_3$ (with Cl on α)

d. [cyclohexene with CH₃] ⟵ [cyclohexane with β, α Cl, CH₃, β]  Two β carbons are identical.

b. (CH₃)₂CHCH=CH₂ ⟵ $(CH_3)_2CH\overset{\beta}{CH_2}-\overset{\alpha}{CH_2}Cl$

c. [cyclohexane with =CH₂] ⟵ [cyclohexane with α CH₂Cl, β]

e. [cyclopentene with C(CH₃)₃] ⟵ [cyclopentane with β Cl, C(CH₃)₃, α, β]  Two β carbons are identical.

**8.27** To have stereoisomers, the two groups on each end of the double bond must be different from each other.

farnesene

(CH₃)₂C=CHCH₂CH₂ $\overset{CH_3}{\underset{}{C}}$=CHCH₂CH₂CH(CH₃)CH=CH₂

two methyl groups — no stereoisomers

two different groups at each end **can have stereoisomers**

2 H's — no stereoisomers

**8.28** Use the definitions in Answer 8.4.

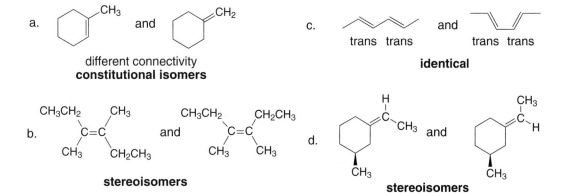

a. [cyclohexene with CH₃] and [cyclohexane with CH₂]

different connectivity
**constitutional isomers**

c. [trans trans diene] and [trans trans diene]

**identical**

b. [alkene: CH₃CH₂, CH₃ / C=C / CH₃, CH₂CH₃] and [alkene: CH₃CH₂, CH₂CH₃ / C=C / CH₃, CH₃]

**stereoisomers**

d. [cyclohexane with H, C, CH₃ and CH₃] and [cyclohexane with CH₃, C, H and CH₃]

**stereoisomers**

**8.29** There are three different isomers. Cis and trans isomers are diastereomers.

**A**
**constitutional isomer**
**of B and C**

**B**     **C**

**diastereomers**

**8.30**

Double bond can be cis or trans.

PGF$_{2\alpha}$

—CH$_2$CH=CH(CH$_2$)$_3$COOH

HO

CH=CHCH(OH)(CH$_2$)$_4$CH$_3$

$sp^3$ stereogenic center
Double bond can be cis or trans.

a. five $sp^3$ stereogenic centers (four circled, one labeled)
b. Two double bonds can both be cis or trans.
c. $2^7 = 128$ stereoisomers possible

**8.31** Use the rules from Answer 8.5 to rank the alkenes.

a. CH$_2$=CHCH$_2$CH$_2$CH$_3$

| monosubstituted | disubstituted cis | disubstituted trans |
|---|---|---|
| **least stable** | **intermediate stability** | **most stable** |

b. CH$_2$=CHCH(CH$_3$)$_2$     CH$_2$=C(CH$_3$)CH$_2$CH$_3$     (CH$_3$)$_2$C=CHCH$_3$

| monosubstituted | disubstituted | trisubstituted |
|---|---|---|
| **least stable** | **intermediate stability** | **most stable** |

**8.32 A larger negative value for $\Delta H°$ means the reaction is more exothermic.** Since both 1-butene and *cis*-2-butene form the same product (butane), these data show that 1-butene was higher in energy to begin with, **since more energy is released in the hydrogenation reaction**.

CH$_2$=CHCH$_2$CH$_3$   +   H$_2$   ⟶   CH$_3$CH$_2$CH$_2$CH$_3$
1-butene
$\Delta H° = -30.3$ kcal/mol

+   H$_2$   ⟶   CH$_3$CH$_2$CH$_2$CH$_3$
$\Delta H° = -28.6$ kcal/mol
*cis*-2-butene

1-butene
*cis*-2-butene
larger $\Delta H°$ for 1-butene
**higher in energy**

Energy

butane

smaller $\Delta H°$ for *cis*-2-butene
**lower in energy, more stable**

**8.33**

a.

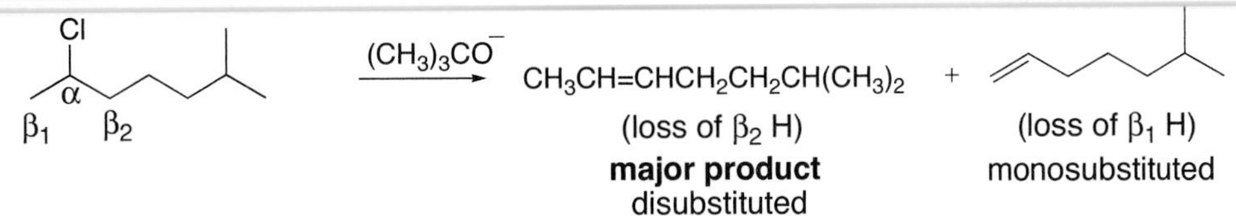
Cl
α
β1 β2
(CH3)3CO−
CH3CH=CHCH2CH2CH(CH3)2
(loss of β2 H)
major product
disubstituted
+
(loss of β1 H)
monosubstituted

b.

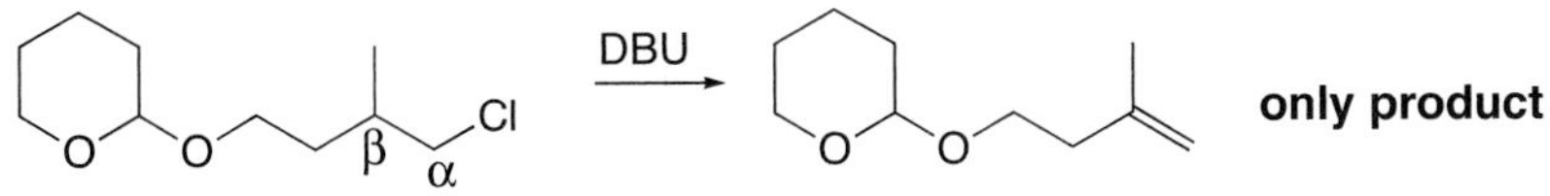
O
O
β
α
Cl
DBU
only product

c.

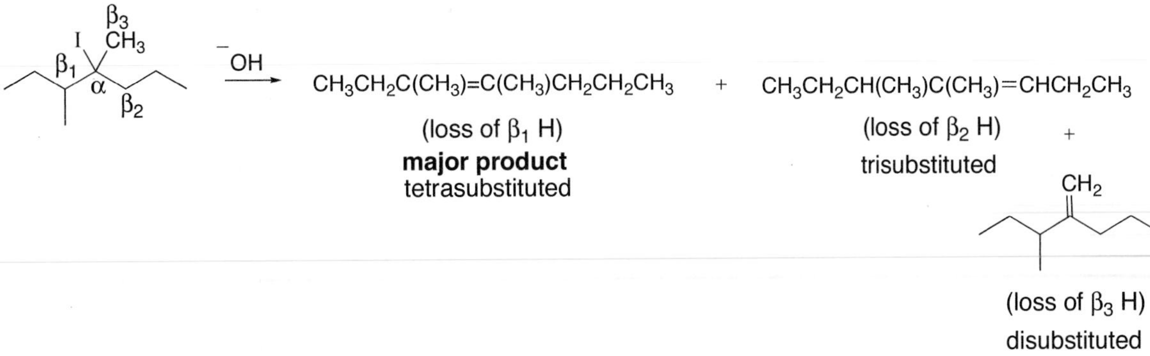
β3
I CH3
β1
α
β2
−OH
CH3CH2C(CH3)=C(CH3)CH2CH2CH3
(loss of β1 H)
major product
tetrasubstituted
+
CH3CH2CH(CH3)C(CH3)=CHCH2CH3
(loss of β2 H)
trisubstituted
+
CH2
(loss of β3 H)
disubstituted

d.

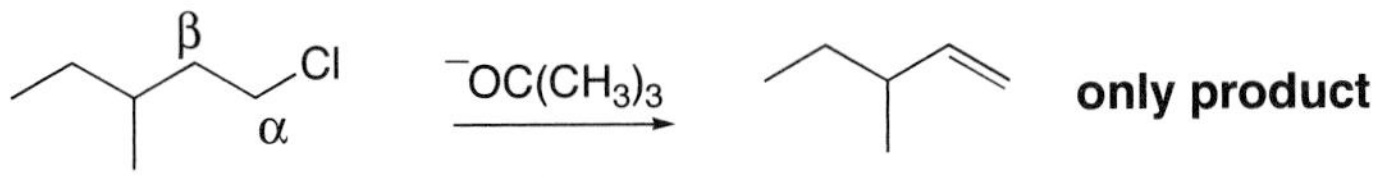
β
Cl
α
−OC(CH3)3
only product

e.

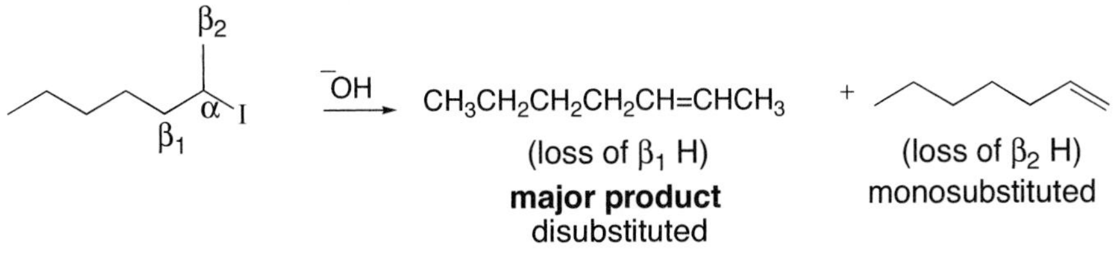
β2
α I
β1
−OH
CH3CH2CH2CH2CH=CHCH3
(loss of β1 H)
major product
disubstituted
+
(loss of β2 H)
monosubstituted

f.

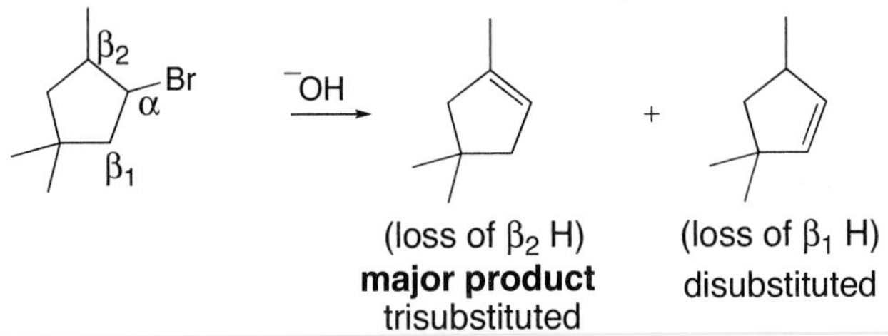
β2
Br
α
β1
−OH
+
(loss of β2 H)
major product
trisubstituted
(loss of β1 H)
disubstituted

**8.34** To give only one alkene as the product of elimination, the alkyl halide must have either:

- Only one β carbon with a hydrogen atom.
- All identical β carbons so the resulting elimination products are identical.

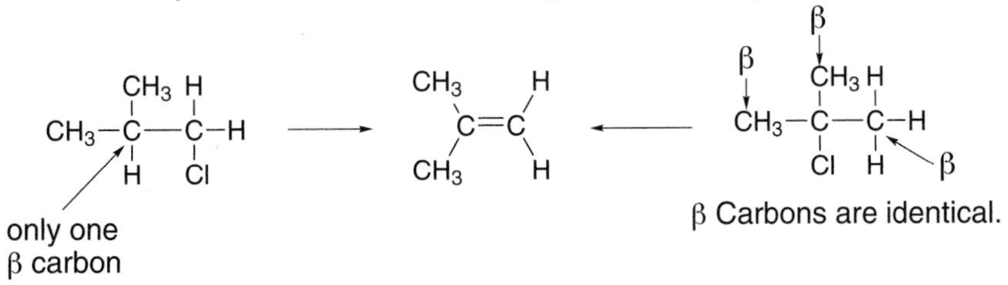
CH3 H
CH3–C–C–H
H Cl
only one
β carbon
CH3 H
C=C
CH3 H
β
β
CH3 H
CH3–C–C–H
Cl H
β
β Carbons are identical.

**8.35** Draw the products of the E2 reaction and compare the number of C's bonded to the C=C.

**major product**
trisubstituted

+ $(CH_3)_2CHCH=CHCH_3$
disubstituted

**major product**
disubstituted

$(CH_3)_2CHCH=CHCH_3$

+ monosubstituted

A yields a trisubstituted alkene as the major product and a disubstituted alkene as minor product. **B** yields a disubstituted alkene as the major product and a monosubstituted alkene as minor product. Since the major and minor products formed from **A** have more alkyl groups (making them more stable) than those formed from **B**, **A** reacts faster in an elimination reaction.

**8.36**

   a. Mechanism:

by-products

   b. Rate = $k$[R–Br][⁻OC(CH₃)₃]

     [1] Solvent changed to DMF (polar aprotic) = **rate increases**
     [2] [⁻OC(CH₃)₃] decreased = **rate decreases**
     [3] Base changed to ⁻OH = **rate decreases** (weaker base)
     [4] Halide changed to 2° = **rate increases** (More substituted RX reacts faster.)
     [5] Leaving group changed to I⁻ = **rate increases** (better leaving group)

**8.37**

1-chloro-1-methyl-
cyclopropane

K⁺ ⁻OC(CH₃)₃

A      +      B

The dehydrohalogenation of an alkyl halide usually forms the more stable alkene. In this case **A** is more stable than **B** even though **A** contains a disubstituted C=C whereas **B** contains a trisubstituted C=C. The double bond in **B** is part of a three-membered ring, and is less stable than **A** because of severe angle strain around both C's of the double bond.

**8.38**

CH₃CH₂O⁻    21%      79%

(CH₃)₃CO⁻    73%      27%

With a less sterically hindered base, more of the more stable product is formed.

As the base gets bigger, the more accessible proton is removed more easily.

**Removal of the less accessible 2° H gives the more substituted, more stable alkene:**

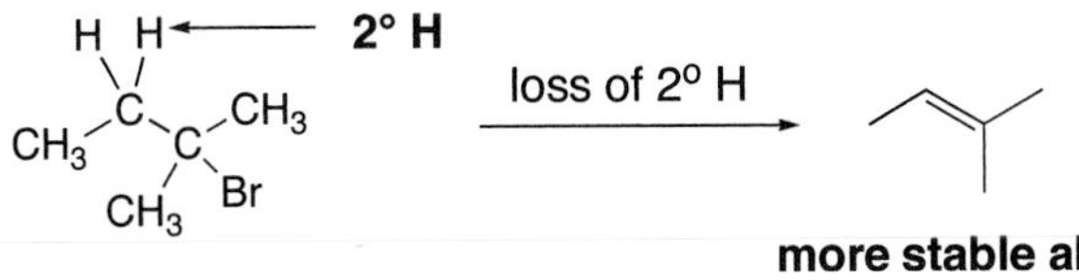

loss of 2° H

This pathway is usually **favored**, as is the case with CH₃CH₂O⁻ as base.

**more stable alkene**

**Removal of the more accessible 1° H gives the less substituted, less stable alkene:**

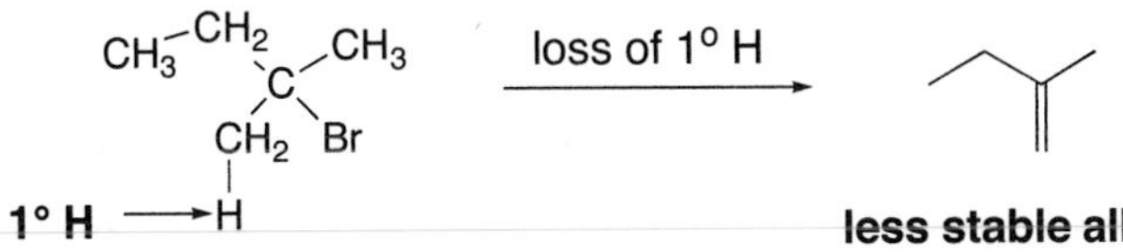

loss of 1° H

**less stable alkene**

As the base gets **bulkier**, the more accessible proton is removed faster; thus, the 1° H is removed faster than the 2° H, and the less stable alkene predominates.

1° H ⟶ H

The 1° H is more accessible, less sterically hindered. With a bulkier base, this proton is more readily removed.

**Explanation: In this example, 1° H's are more easily removed than 2° H's with sterically hindered bases.**

**8.39**

a. 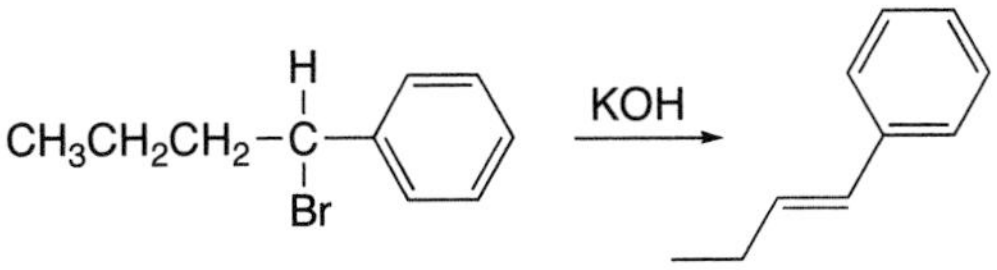 CH₃CH₂CH₂–C(H)(Br)–C₆H₅   KOH ⟶

**trans isomer more stable**
**major product**

b. 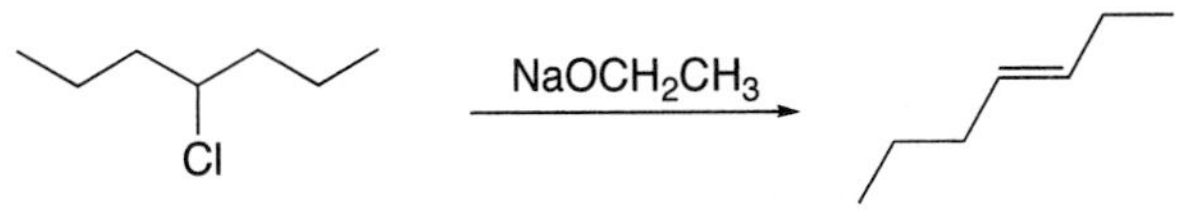 NaOCH₂CH₃ ⟶

**trans isomer more stable**
**major product**

**8.40**

a.

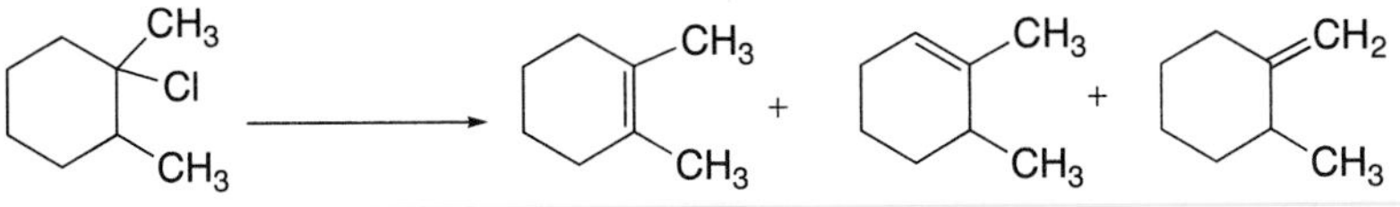

tetrasubstituted    trisubstituted    disubstituted
**major product**

b.

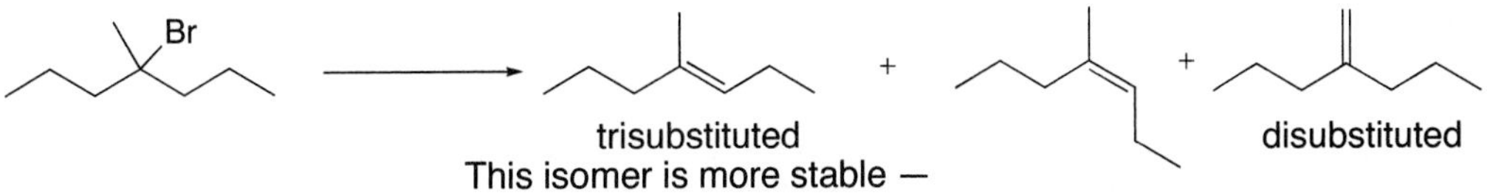

trisubstituted
This isomer is more stable —
large groups further away.
**major product**

trisubstituted

disubstituted

c.

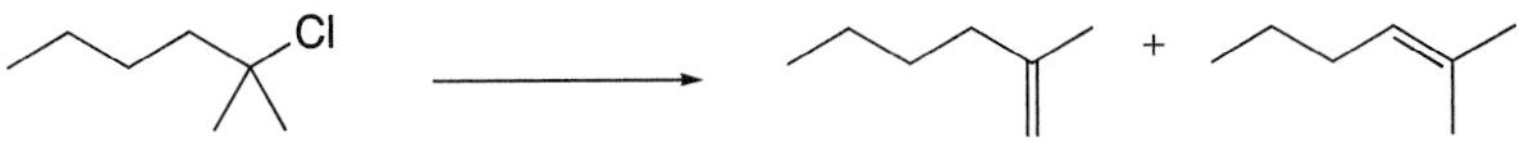

disubstituted

trisubstituted
**major product**

**8.41** Use the rules from Answer 8.21.

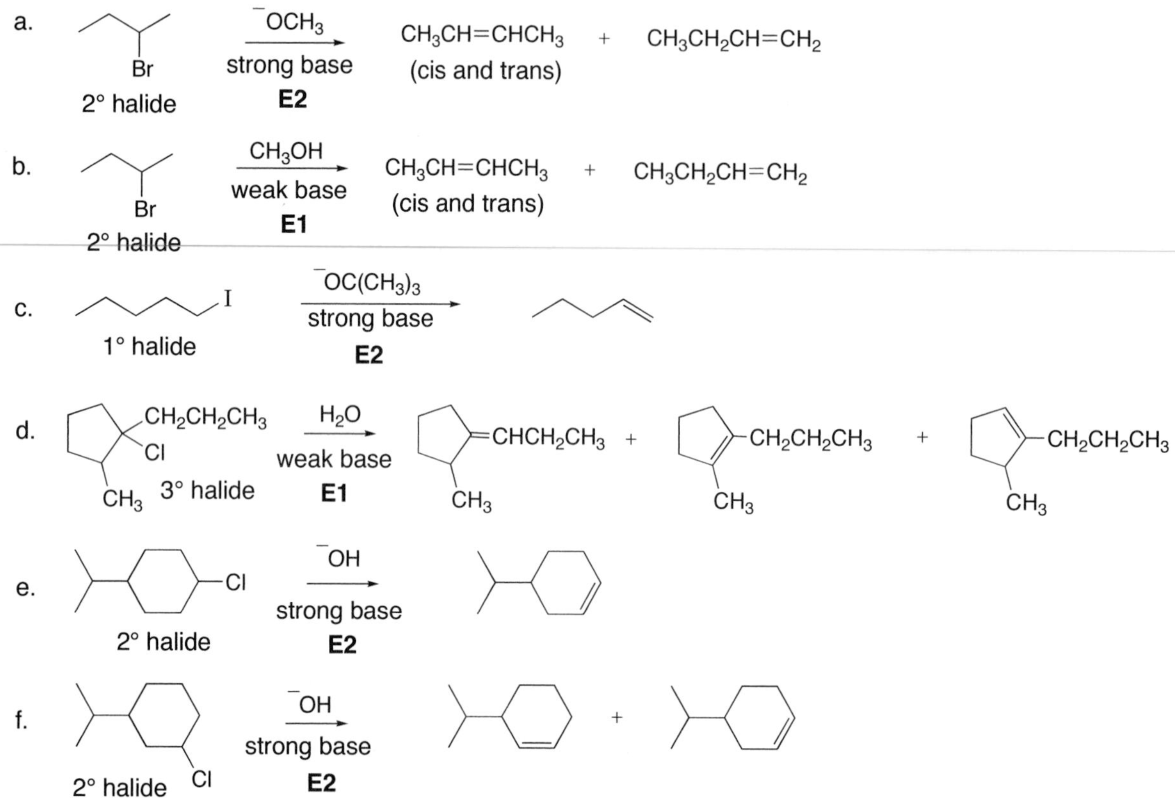

a.

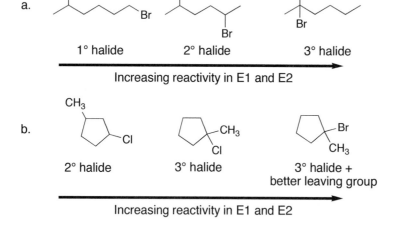

$CH_3CH=CHCH_3$  +  $CH_3CH_2CH=CH_2$

(cis and trans)

2° halide | strong base | E2 | $^-OCH_3$

b.

$^-$
$CH_3OH$
weak base
E1

$CH_3CH=CHCH_3$  +  $CH_3CH_2CH=CH_2$

(cis and trans)

Br
2° halide

c.

I
1° halide

$^-OC(CH_3)_3$
strong base
E2

d.

$CH_2CH_2CH_3$
$Cl$
$CH_3$  3° halide

$H_2O$
weak base
E1

$=CHCH_2CH_3$  +
$CH_3$

$-CH_2CH_2CH_3$  +
$CH_3$

$-CH_2CH_2CH_3$
$CH_3$

e.

$Cl$
2° halide

$^-OH$
strong base
E2

f.

$OH$
strong base
E2

$Cl$
2° halide

+

**8.42** The order of reactivity is the same for both E2 and E1: 1° < 2° < 3°

a.

Br
1° halide

Br
2° halide

Br
3° halide

Increasing reactivity in E1 and E2

b.

$CH_3$
$Cl$
2° halide

$CH_3$
$Cl$
3° halide

$Br$
$CH_3$
3° halide +
better leaving group

Increasing reactivity in E1 and E2

**8.43**

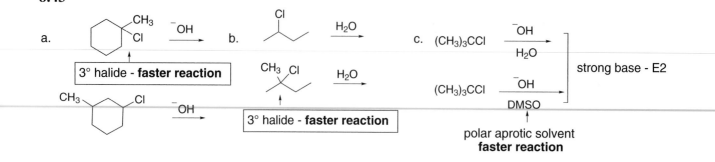

a.

$CH_3$
$Cl$
$^-OH$

3° halide - **faster reaction**

$CH_3$
$Cl$
$^-OH$

b.

$Cl$
$H_2O$

$CH_3$  $Cl$
$H_2O$

3° halide - **faster reaction**

c.  $(CH_3)_3CCl$
$^-OH$
$H_2O$

$(CH_3)_3CCl$
$^-OH$
DMSO

strong base - E2

polar aprotic solvent
**faster reaction**

**8.44**

a.

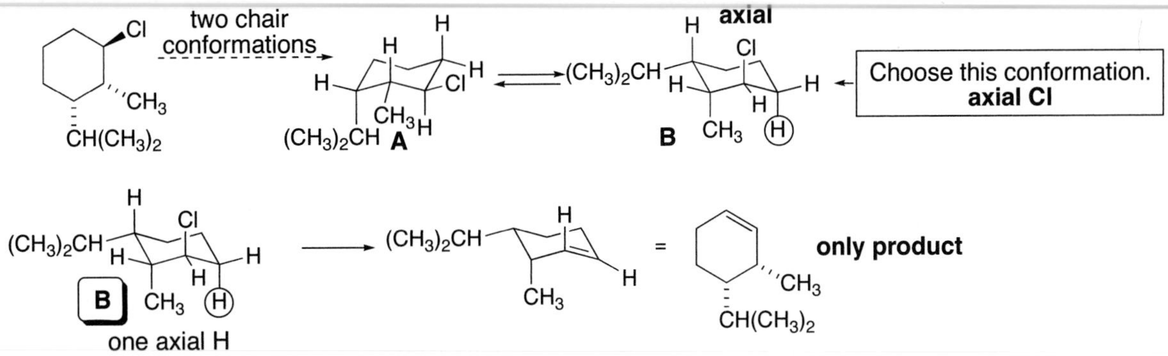

two chair conformations

**Choose this conformation.**
**axial Cl**

(CH₃)₂CH **A**

**B**

only product

**B**
one axial H

= **only product**

b. two chair conformations

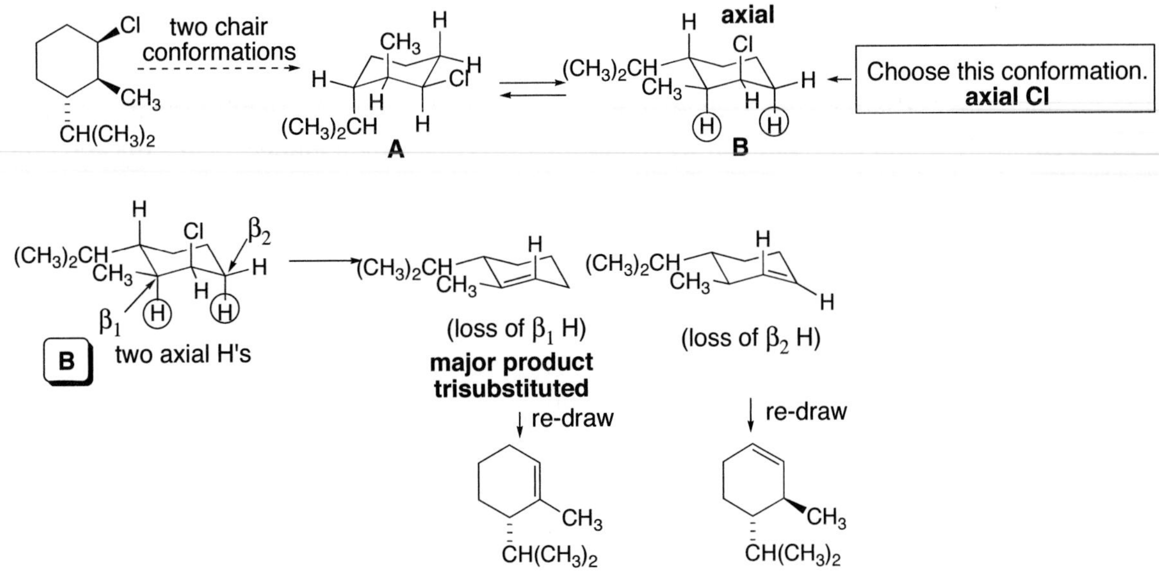

**Choose this conformation.**
**axial Cl**

(CH₃)₂CH **A**

**B**

**B** two axial H's

(loss of β₁ H)
**major product**
**trisubstituted**

(loss of β₂ H)

re-draw

re-draw

(loss of β₂ H)

c.

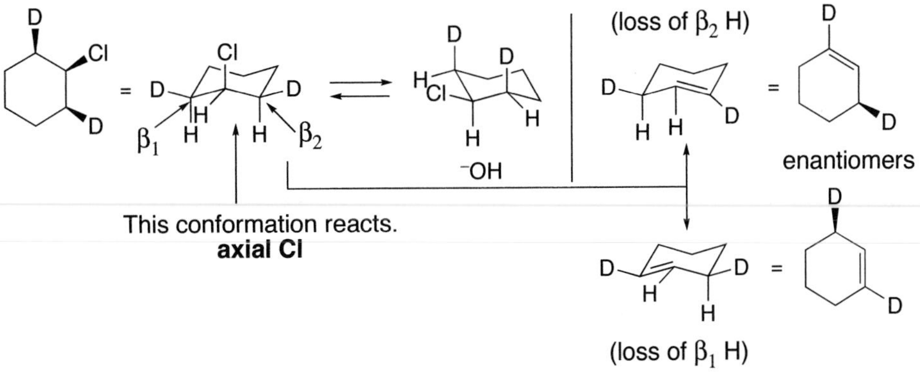

= 

This conformation reacts.
**axial Cl**

⁻OH

enantiomers

(loss of β₁ H)

(loss of β₂ D)

d.

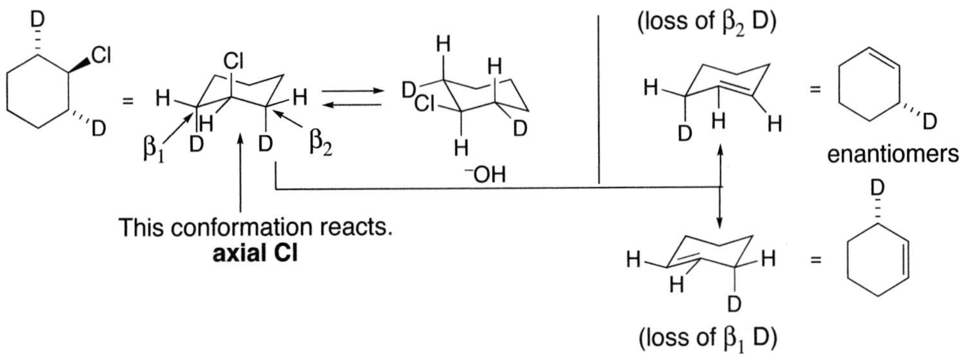

=

This conformation reacts.
**axial Cl**

⁻OH

enantiomers

(loss of β₁ D)

**8.45**

a.

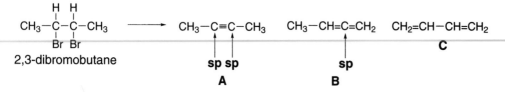

2-chloro-3-methylpentane

H and Cl are arranged anti in each stereoisomer, for anti periplanar elimination.

enantiomers

enantiomers

**A**   **B**   **C**   **D**

–HCl   –HCl   –HCl   –HCl

identical   identical

b. Two different alkenes are formed as products.

c. The products are diastereomers: Two enantiomers (**A** and **B**) give identical products. **A** and **B** are diastereomers of **C** and **D**. Each pair of enantiomers gives a single alkene. Thus diastereomers give diastereomeric products.

**8.46**

a.   $\overset{\quad CH_2CHCl_2}{\bigcirc}$  $\xrightarrow[\text{(2 equiv)}]{\quad ^-NH_2 \quad}$  $\overset{\quad C \equiv CH}{\bigcirc}$

b.   $CH_3CH_2-\overset{\overset{\displaystyle CH_3}{|}}{\underset{\underset{\displaystyle CH_3}{|}}{C}}-CHCH_2Br$   $\xrightarrow[\text{(2 equiv)}]{\quad ^-NH_2 \quad}$   $CH_3CH_2-\overset{\overset{\displaystyle CH_3}{|}}{\underset{\underset{\displaystyle CH_3}{|}}{C}}-C \equiv CH$

c.   $CH_3-\overset{\overset{\displaystyle Cl}{|}}{\underset{\underset{\displaystyle Cl}{|}}{C}}-CH_2CH_3$   $\xrightarrow[\text{(excess)}]{\quad ^-NH_2 \quad}$   $HC \equiv C-CH_2CH_3$   +   $CH_3-C \equiv C-CH_3$

d.   $\bigcirc-\overset{\overset{\displaystyle H}{|}}{\underset{\underset{\displaystyle Cl}{|}}{C}}-\overset{\overset{\displaystyle H}{|}}{\underset{\underset{\displaystyle Cl}{|}}{C}}-\bigcirc$   $\xrightarrow[\text{(2 equiv)}]{\quad ^-NH_2 \quad}$   $\bigcirc-C \equiv C-\bigcirc$

**8.47**

a.   $CH_3C \equiv CCH_3$   |   $CH_3-\overset{\overset{\displaystyle Br}{|}}{\underset{\underset{\displaystyle Br}{|}}{C}}-CH_2CH_3$   or   $CH_3-\overset{\overset{\displaystyle H}{|}}{\underset{\underset{\displaystyle Br}{|}}{C}}-\overset{\overset{\displaystyle H}{|}}{\underset{\underset{\displaystyle Br}{|}}{C}}-CH_3$

b.   $CH_3-\overset{\overset{\displaystyle CH_3}{|}}{\underset{\underset{\displaystyle CH_3}{|}}{C}}-C \equiv CH$   |   $CH_3-\overset{\overset{\displaystyle CH_3}{|}}{\underset{\underset{\displaystyle CH_3}{|}}{C}}-\overset{\overset{\displaystyle Br}{|}}{\underset{\underset{\displaystyle Br}{|}}{C}}-CH_3$   or   $CH_3-\overset{\overset{\displaystyle CH_3}{|}}{\underset{\underset{\displaystyle CH_3}{|}}{C}}-\overset{\overset{\displaystyle Br}{|}}{\underset{}{C}}H-CH_2Br$   or   $CH_3-\overset{\overset{\displaystyle CH_3}{|}}{\underset{\underset{\displaystyle CH_3}{|}}{C}}-CH_2CHBr_2$

c.   $\bigcirc-C \equiv C-\bigcirc$   |   $\bigcirc-\overset{\overset{\displaystyle Br}{|}\ \overset{\displaystyle H}{|}}{\underset{\underset{\displaystyle Br}{|}\ \underset{\displaystyle H}{|}}{C\ C}}-\bigcirc$   or   $\bigcirc-\overset{\overset{\displaystyle Br}{|}\ \overset{\displaystyle Br}{|}}{\underset{\underset{\displaystyle H}{|}\ \underset{\displaystyle H}{|}}{C\ C}}-\bigcirc$

**8.48**

$CH_3-\overset{\overset{\displaystyle H}{|}}{\underset{\underset{\displaystyle Br}{|}}{C}}-\overset{\overset{\displaystyle H}{|}}{\underset{\underset{\displaystyle Br}{|}}{C}}-CH_3$   $\longrightarrow$   $CH_3-C \equiv C-CH_3$   $CH_3-CH=C=CH_2$   $CH_2=CH-CH=CH_2$

2,3-dibromobutane

**sp sp**            **sp**            **C**

**A**            **B**

**8.49**

a. [1° halide structure with Br] ⁻OC(CH₃)₃ / sterically hindered base → [structure] **E2**

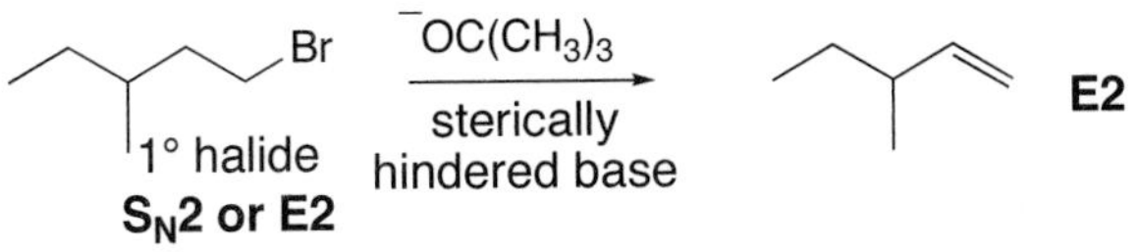

1° halide
**S_N2 or E2**

b. [1° halide structure with I] ⁻OCH₂CH₃ / strong nucleophile → [structure with OCH₂CH₃] **S_N2**

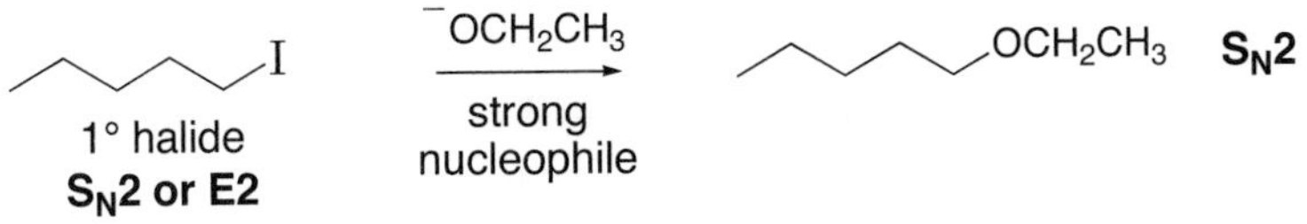

1° halide
**S_N2 or E2**

c. CH₃–C(Cl)(Cl)–CH₃ (dihalide) ⁻NH₂ (2 equiv) / strong base → HC≡C–CH₃

dihalide

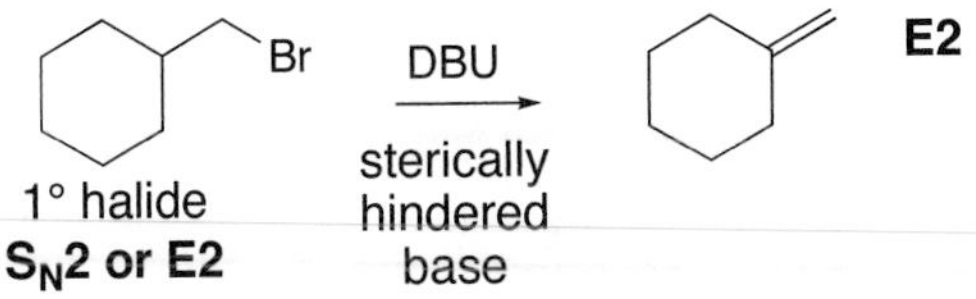

d. [cyclohexyl-CH₂Br structure] DBU / sterically hindered base → [methylenecyclohexane structure] **E2**

1° halide
**S_N2 or E2**

e. [cyclohexane with CH₂CH₃ and Br] ⁻OC(CH₃)₃ / sterically hindered base → [cyclohexene with CH₂CH₃] + [cyclohexene with CH₂CH₃] **E2**

**major product**

2° halide
**S_N1, S_N2, E1, E2**

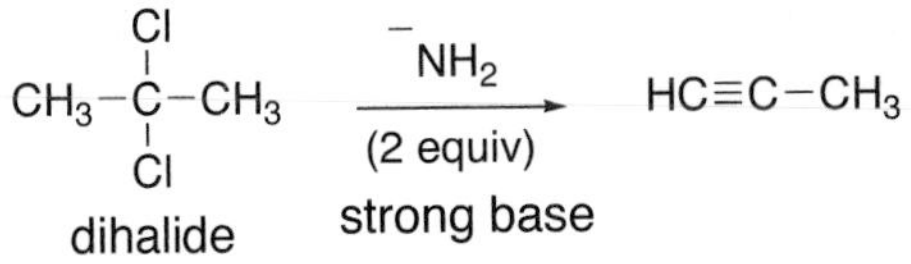

f. [cyclohexane with Br and CH₂CH₃] CH₃CH₂OH / weak base → [cyclohexane with OCH₂CH₃ and CH₂CH₃] + [cyclohexene with CH₂CH₃] + [cyclohexane with =CHCH₃]

3° halide
**no S_N2**

**S_N1 product**    **E1 products**

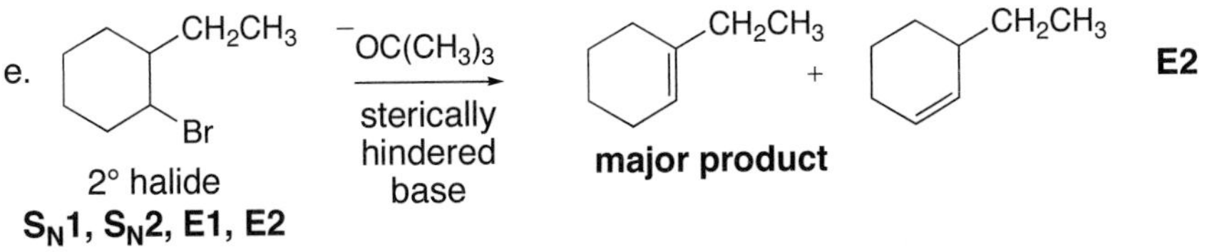

g. (CH₃)₂CH–CHCH₂Br (with Br below) dihalide    2 NaNH₂ → (CH₃)₂CH–C≡CH

dihalide  Br

h. [structure with Cl Cl] KOC(CH₃)₃ (2 equiv) DMSO → CH₃–C(CH₃)(CH₃)–C≡CH

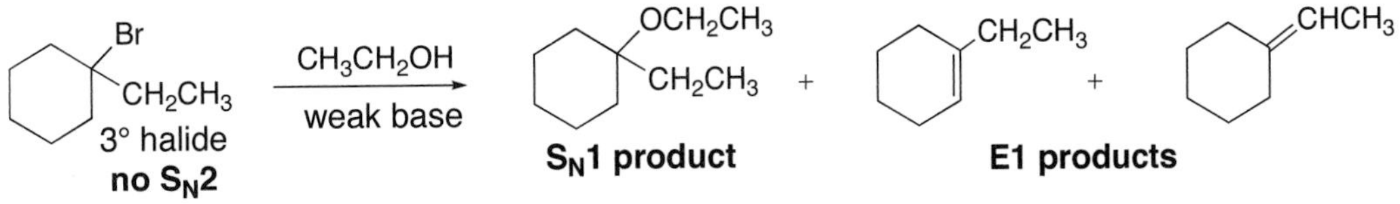

dihalide

i. [2° halide with I] CH₃CH₂OH / weak base → [structure with OCH₂CH₃] + [1-butene structure] + CH₃CH=CHCH₃ (cis and trans) **E1 product**

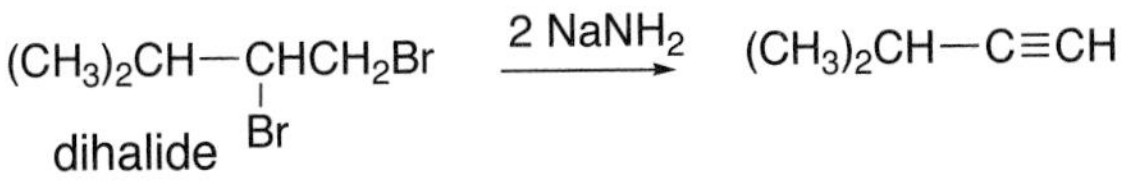

2° halide
**S_N1, S_N2, E1, E2**

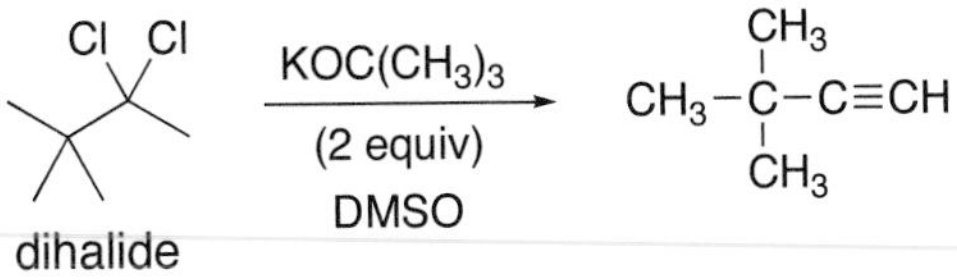

**S_N1 product**    **E1 product**

j. [3° halide with Cl] H₂O / weak base → [structure with OH] + CH₃CH₂C(CH₃)=CHCH₃ (cis and trans) + [structure] 

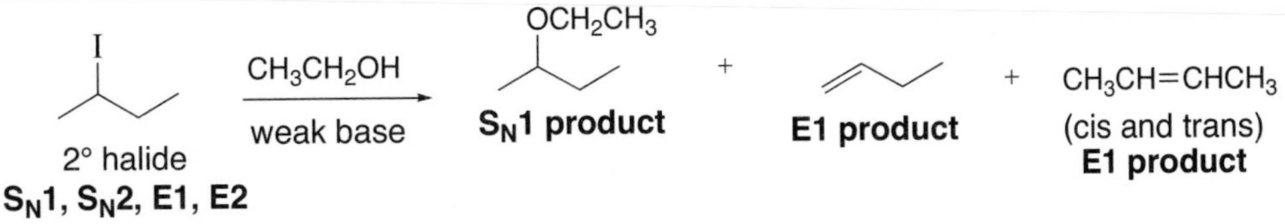

3° halide
**no S_N2**

**S_N1 product**    **E1 product**    **E1 product**

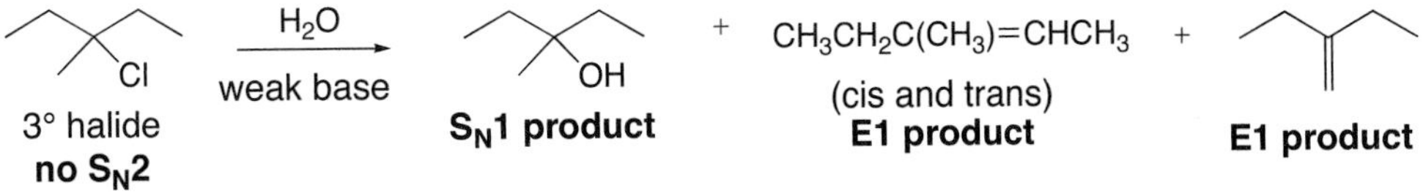

**8.50**

a. two enantiomers:

A        B

b. The bulky *tert*-butyl group anchors the cyclohexane ring and occupies the more roomy equatorial position. The cis isomer has the Br atom axial, while the trans isomer has the Br atom equatorial. For dehydrohalogenation to occur on a halo cyclohexane, the halogen must be axial to afford trans diaxial elimination of H and X. The cis isomer readily reacts since the Br atom is axial. The only way for the trans isomer to react is for the six-membered ring to flip into a highly unstable conformation having both $(CH_3)_3C$ and Br axial. Thus, the trans isomer reacts much more slowly.

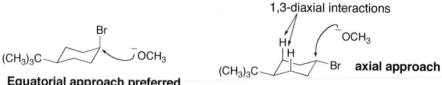

*cis*-1-bromo-4-*tert*-butylcyclohexane                 *trans*-1-bromo-4-*tert*-butylcyclohexane

c. two products:

C = $(CH_3)_3C$–OCH₃     D = $(CH_3)_3C$–OCH₃

d. *cis*-1-Bromo-4-*tert*-butylcyclohexane reacts faster. With the strong nucleophile ⁻OCH₃, backside attack occurs by an S_N2 reaction, and with the cis isomer, the nucleophile can approach from the equatorial direction, avoiding 1,3-diaxial interactions.

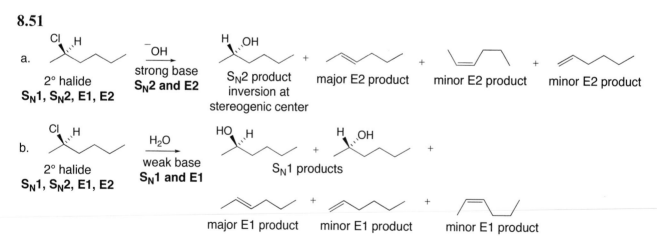

Equatorial approach preferred.

*cis*-1-bromo-4-*tert*-butylcyclohexane       *trans*-1-bromo-4-*tert*-butylcyclohexane

e. The bulky base ⁻$OC(CH_3)_3$ favors elimination by an E2 mechanism, affording a mixture of two enantiomers **A** and **B**. The strong nucleophile ⁻OCH₃ favors nucleophilic substitution by an S_N2 mechanism. Inversion of configuration results from backside attack of the nucleophile.

**8.51**

a.

2° halide
S_N1, S_N2, E1, E2

$\xrightarrow[\text{S_N2 and E2}]{\text{⁻OH strong base}}$

S_N2 product
inversion at
stereogenic center

+ major E2 product + minor E2 product + minor E2 product

b.

2° halide
S_N1, S_N2, E1, E2

$\xrightarrow[\text{S_N1 and E1}]{\text{H₂O weak base}}$

S_N1 products

+

major E1 product    minor E1 product    minor E1 product

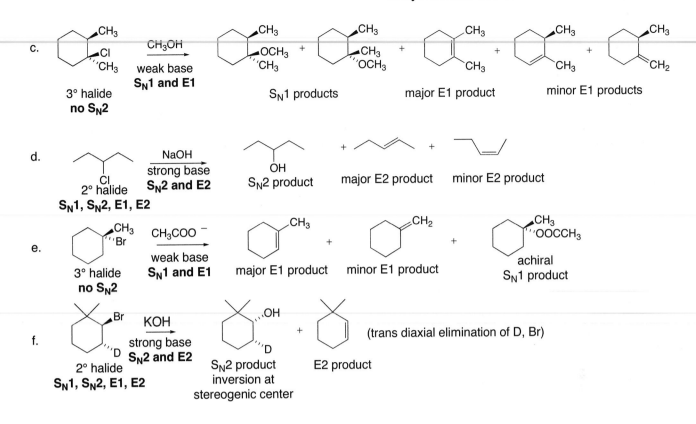

c. 3° halide
no S$_N$2
CH$_3$OH
weak base
S$_N$1 and E1

S$_N$1 products

major E1 product

minor E1 products

d. 2° halide
S$_N$1, S$_N$2, E1, E2
NaOH
strong base
S$_N$2 and E2

S$_N$2 product

+ major E2 product + minor E2 product

e. 3° halide
no S$_N$2
CH$_3$COO$^-$
weak base
S$_N$1 and E1

major E1 product + minor E1 product +

achiral
S$_N$1 product

f. 2° halide
S$_N$1, S$_N$2, E1, E2
KOH
strong base
S$_N$2 and E2

S$_N$2 product
inversion at
stereogenic center

E2 product

(trans diaxial elimination of D, Br)

**8.52**

a. 3° halide
$^-$OC(CH$_3$)$_3$
strong
bulky base
E2

**major product**
more substituted alkene

+

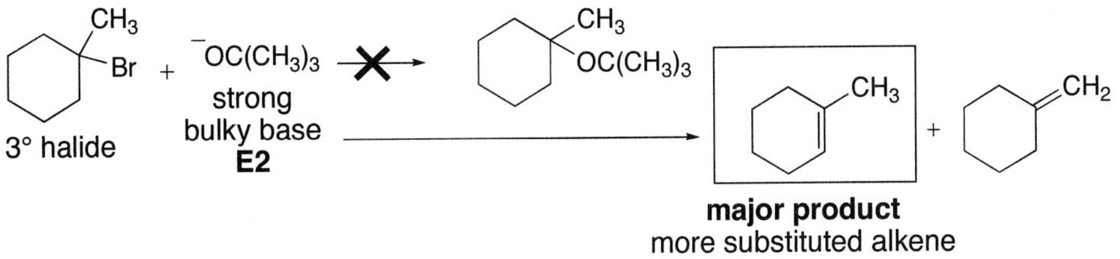

No substitution occurs with a strong bulky base and a 3° RX. The C with the leaving group is too crowded for an S$_N$2 substitution to occur. Elimination occurs instead by an E2 mechanism.

b. 1° halide
$^-$OCH$_3$
strong nucleophile
S$_N$2

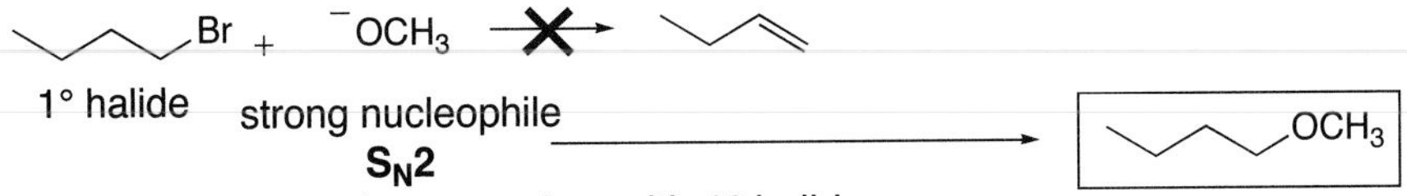

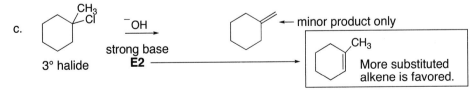

All elimination reactions are slow with 1° halides.
The strong nucleophile reacts by an S$_N$2 mechanism instead.

c. 3° halide
$^-$OH
strong base
E2

← minor product only

More substituted
alkene is favored.

d.

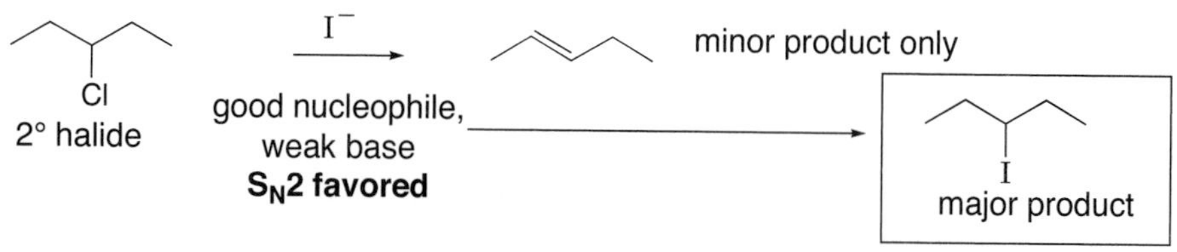

I⁻
good nucleophile,
weak base
**S_N2 favored**

minor product only

major product

The 2° halide can react by an E2 or S_N2 reaction with a negatively charged nucleophile or base. Since I⁻ is a weak base, substitution by an S_N2 mechanism is favored.

**8.53**

2° halide, weak base:
**S_N1 and E1**

a.

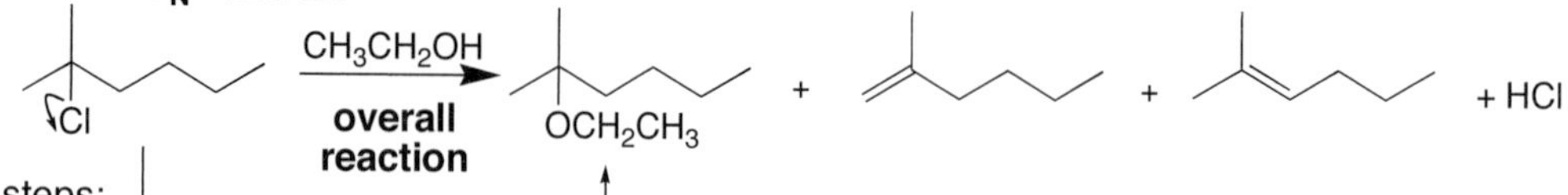

$CH_3CH_2OH$

**overall reaction**

$OCH_2CH_3$

+ ⟶ + ⟶ + HCl

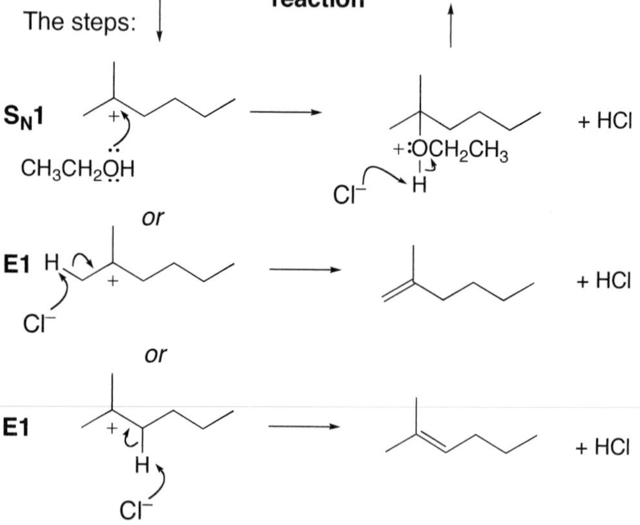

The steps:

**S_N1**

$CH_3CH_2\overset{..}{O}H$

⟶ + HCl

+:$OCH_2CH_3$
Cl⁻  H

Any base (such as $CH_3CH_2OH$ or Cl⁻) can be used to remove a proton to form an alkene. If Cl⁻ is used, HCl is formed as a reaction by-product. If $CH_3CH_2OH$ is used, $(CH_3CH_2OH_2)^+$ is formed instead.

or

**E1** H

Cl⁻

⟶ + HCl

or

**E1**

H
Cl⁻

⟶ + HCl

b.

$CH_3$
Cl

⁻OH
**overall reaction**

$CH_3$ + $CH_2$ + $H_2O$ + Cl⁻

3° halide
strong base
**E2**

Each product:

$CH_3$
Cl
H
⁻:$\overset{..}{O}$H

one step ⟶

$CH_3$

or

+ $H_2\overset{..}{\underset{..}{O}}$ + Cl⁻

$CH_2$─H   ⁻:$\overset{..}{O}$H
Cl
H

one step ⟶

$CH_2$

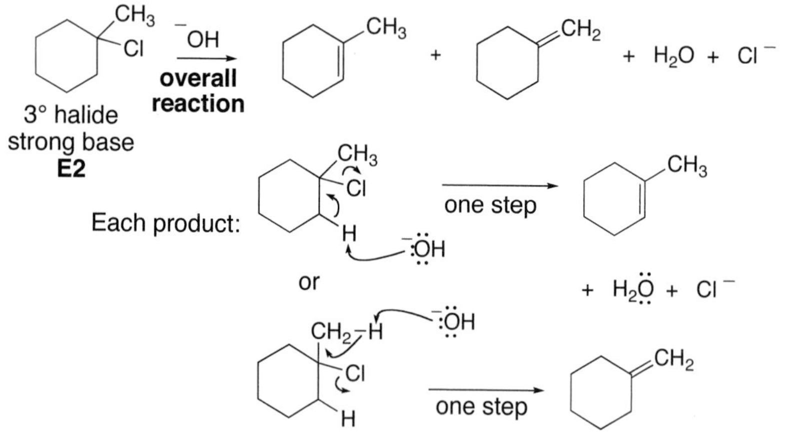

**8.54**

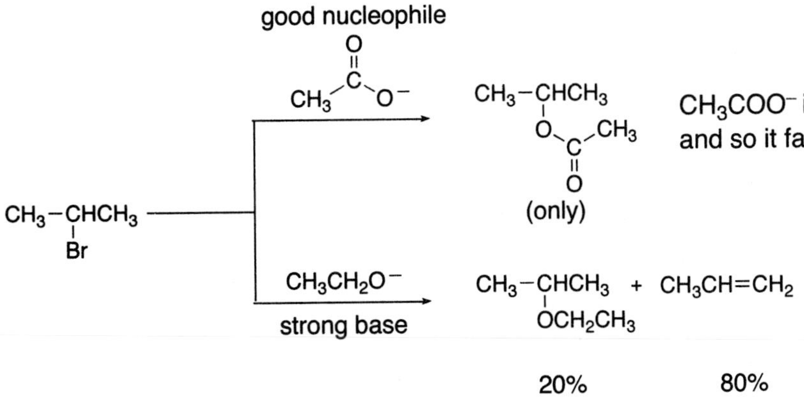

good nucleophile

$CH_3COO^-$ is a good nucleophile and a weak base and so it favors substitution by $S_N2$.

(only)

strong base

20%        80%

The strong base gives both $S_N2$ and E2 products, but since the 2° RX is somewhat hindered to substitution, the E2 product is favored.

**8.55**

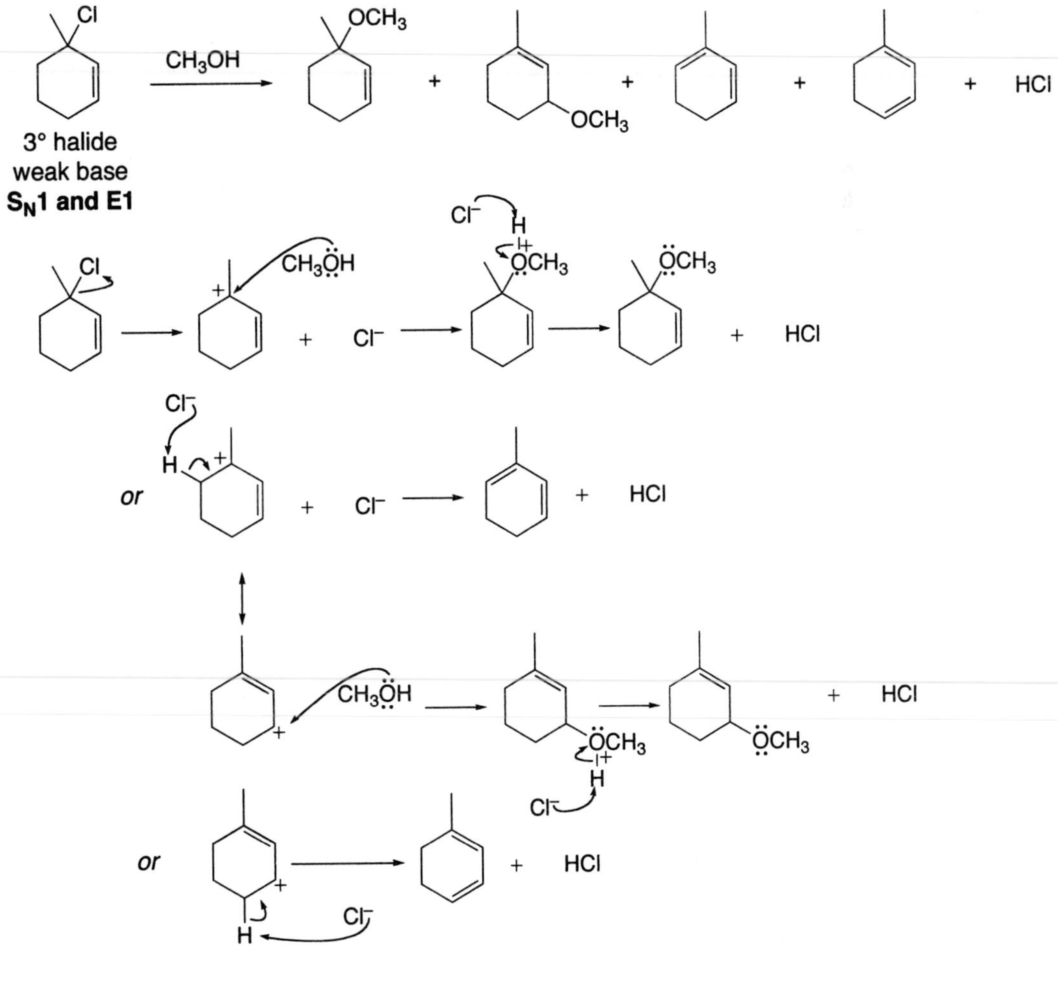

$CH_3OH$

+ HCl

3° halide
weak base
**$S_N1$ and E1**

+ Cl⁻ ⟶ + HCl

or

+ Cl⁻ ⟶ + HCl

$CH_3OH$ ⟶ + HCl

or ⟶ + HCl

**8.56** E2 elimination needs a leaving group and a hydrogen in the **trans diaxial** position.

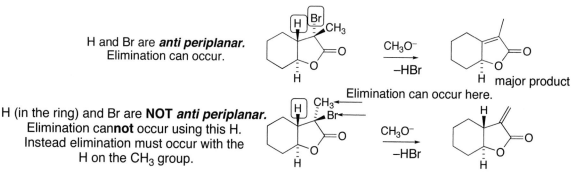

Two different conformations:

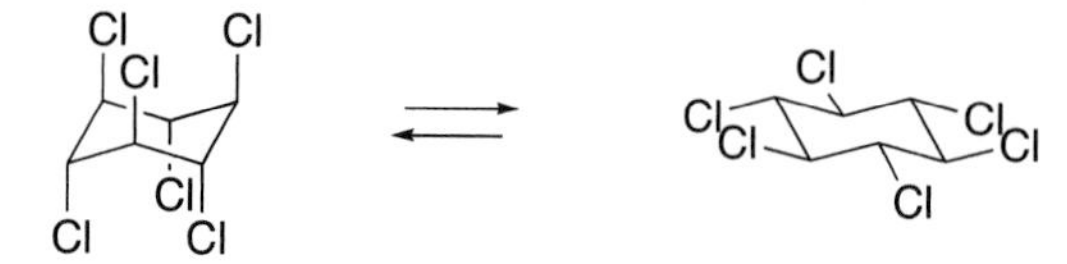

This conformation has Cl's axial, but no H's axial.     This conformation has no Cl's axial.

For elimination to occur, a cyclohexane must have a H and Cl in the trans diaxial arrangement. Neither conformation of this isomer has both atoms—H and Cl—axial; thus, this isomer only slowly loses HCl by elimination.

**8.57**

H and Br are *anti periplanar.*
Elimination can occur.

$CH_3O^-$

−HBr

major product

Elimination can occur here.

H (in the ring) and Br are **NOT** *anti periplanar.*
Elimination can**not** occur using this H.
Instead elimination must occur with the
H on the $CH_3$ group.

$CH_3O^-$

−HBr

Elimination cannot occur in the ring
because the required anti periplanar geometry is not present.

**8.58**

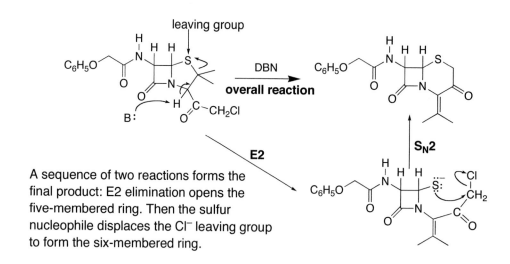

leaving group

$C_6H_5O$

DBN

**overall reaction**

$C_6H_5O$

**E2**

**S$_N$2**

A sequence of two reactions forms the
final product: E2 elimination opens the
five-membered ring. Then the sulfur
nucleophile displaces the Cl⁻ leaving group
to form the six-membered ring.

$C_6H_5O$

## Chapter 9: Alcohols, Ethers, and Epoxides

### ◆ General facts about ROH, ROR, and epoxides

- All three compounds contain an O atom that is $sp^3$ hybridized and tetrahedral (9.2).

$$CH_3 \overset{\ddot{O}}{\diagdown} H \qquad CH_3 \overset{\ddot{O}}{\diagdown} CH_3 \qquad \overset{60°}{\underset{H \quad H}{\triangle}} \overset{\ddot{O}}{\underset{H \quad H}{}}$$

$$109° \qquad\qquad 111°$$

**an alcohol**      **an ether**      **an epoxide**

- All three compounds have polar C–O bonds, but only alcohols have an O–H bond for intermolecular hydrogen bonding (9.4).

hydrogen bond

- Alcohols and ethers do not contain a good leaving group. Nucleophilic substitution can occur only after the OH (or OR) group is converted to a better leaving group (9.7A).

$$R-\ddot{O}H \; + \; H-Cl \; \rightleftharpoons \; R-\overset{+}{\underset{\ddot{.}}{O}H_2} \; + \; Cl^-$$

strong acid

weak base
**good leaving group**

- Epoxides have a leaving group located in a strained three-membered ring, making them reactive to strong nucleophiles and acids HZ that contain a nucleophilic atom Z (9.15).

leaving group

With strong nucleophiles,
:Nu⁻

$$\overset{O}{\underset{:Nu^-}{C-C}} \xrightarrow{[1]} \overset{O^-}{\underset{Nu}{C-C}} \xrightarrow[\text{[2]}]{H-OH} \overset{OH}{\underset{Nu}{C-C}} + \; ^-OH$$

### ◆ A new reaction of carbocations (9.9)

- Less stable carbocations rearrange to more stable carbocations by shift of a hydrogen atom or an alkyl group. Besides rearrangement, carbocations also react with nucleophiles (7.13) and bases (8.6).

$$-\overset{|}{\underset{R}{C}}-\overset{|}{\underset{+}{C}}- \xrightarrow{\text{1,2-shift}} -\overset{|}{\underset{+}{C}}-\overset{|}{\underset{R}{C}}-$$

(or H)          (or H)

### ◆ Preparation of alcohols, ethers, and epoxides (9.6)

#### [1] Preparation of alcohols

$$R-X \; + \; \boxed{^-OH} \longrightarrow R-\boxed{OH} \; + \; X^-$$

- The mechanism is $S_N2$.
- The reaction works best for $CH_3X$ and 1° RX.

## [2] Preparation of alkoxides (a Brønsted–Lowry acid–base reaction)

R—O—H + Na⁺H⁻ ⟶ $\boxed{R-O^{-}}$ Na⁺ + H₂

alkoxide

## [3] Preparation of ethers (Williamson ether synthesis)

R—X + $\boxed{^{-}OR'}$ ⟶ R—$\boxed{OR'}$ + X⁻

- The mechanism is S$_N$2.
- The reaction works best for CH₃X and 1° RX.

## [4] Preparation of epoxides (Intramolecular S$_N$2 reaction)

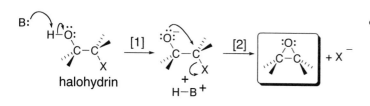

halohydrin

- A two-step reaction sequence:
  [1] Removal of a proton with base forms an alkoxide.
  [2] Intramolecular **S$_N$2** reaction forms the epoxide.

## ◆ Reactions of alcohols

### [1] Dehydration to form alkenes

#### [a] Using strong acid (9.8, 9.9)

- Order of reactivity: R₃COH > R₂CHOH > RCH₂OH.
- The mechanism for 2° and 3° ROH is E1; carbocations are intermediates and rearrangements occur.
- The mechanism for 1° ROH is E2.
- The Zaitsev rule is followed.

#### [b] Using POCl₃ and pyridine (9.10)

- The mechanism is E2.
- No carbocation rearrangements occur.

## [2] Reaction with HX to form RX (9.11)

R—OH + H—X ⟶ $\boxed{R-X}$ + H₂O

- Order of reactivity: R₃COH > R₂CHOH > RCH₂OH.
- The mechanism for 2° and 3° ROH is S$_N$1; carbocations are intermediates and rearrangements occur.
- The mechanism for CH₃OH and 1° ROH is S$_N$2.

## [3] Reaction with other reagents to form RX (9.12)

R–OH + SOCl$_2$ $\xrightarrow{\text{pyridine}}$ R–Cl

R–OH + PBr$_3$ $\longrightarrow$ R–Br

- Reactions occur with $CH_3OH$ and 1° and 2° ROH.
- The reactions follow an $S_N2$ mechanism.

## [4] Reaction with tosyl chloride to form tosylates (9.13A)

R–OH + Cl–S(=O)(=O)–C$_6$H$_4$–CH$_3$ $\xrightarrow{\text{pyridine}}$ R–O–S(=O)(=O)–C$_6$H$_4$–CH$_3$

R–OTs

- The C–O bond is not broken so the configuration at a stereogenic center is retained.

## ♦ Reactions of tosylates

Tosylates undergo either substitution or elimination depending on the reagent (9.13B).

- Substitution is carried out with strong :Nu⁻ so the mechanism is $S_N2$.
- Elimination is carried out with strong bases so the mechanism is E2.

## ♦ Reactions of ethers

Only one reaction is useful: Cleavage with strong acids (9.14)

R–O–R' + H–X $\longrightarrow$ R–X + R'–X + H$_2$O

(X = Br or I)

- With 2° and 3° R groups, the mechanism is $S_N1$.
- With $CH_3$ and 1° R groups the mechanism is $S_N2$.

## ♦ Reactions of epoxides

Epoxide rings are opened with nucleophiles :Nu⁻ and acids HZ (9.15).

[1] :Nu⁻ [2] H$_2$O or HZ → product (Z)

- The reaction occurs with backside attack, resulting in trans or anti products.
- With :Nu⁻, the mechanism is $S_N2$, and nucleophilic attack occurs at the less substituted C.
- With HZ, the mechanism is between $S_N1$ and $S_N2$, and attack of Z⁻ occurs at the more substituted C.

## Chapter 9: Answers to Problems

**9.1**   • **Alcohols** are classified as 1°, 2°, or 3°, depending on the number of carbon atoms bonded to the carbon with the OH group.
   • **Symmetrical ethers** have two identical R groups, and **unsymmetrical ethers** have R groups that are different.

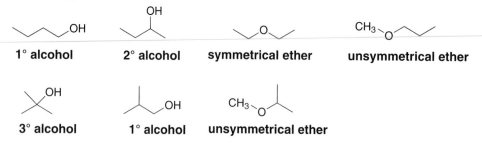

| 1° alcohol | 2° alcohol | symmetrical ether | unsymmetrical ether |

3° alcohol      1° alcohol      unsymmetrical ether

**9.2**  To name an alcohol:
   [1] **Find the longest chain that has the OH group as a substituent.**  Name the molecule as a derivative of that number of carbons by changing the **-e** ending of the alkane to the suffix **-ol**.
   [2] **Number the carbon chain to give the OH group the lower number.**  When the OH group is bonded to a ring, the ring is numbered beginning with the OH group, and the "1" is usually omitted.
   [3] Apply the other rules of nomenclature to complete the name.

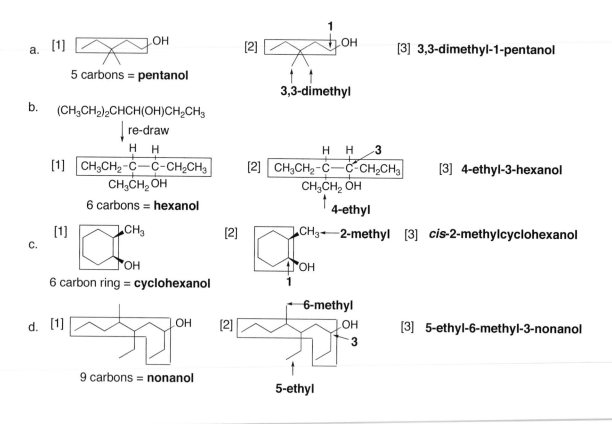

a.  [1]                              [2]                              [3] **3,3-dimethyl-1-pentanol**
      5 carbons = **pentanol**           **3,3-dimethyl**

b.   $(CH_3CH_2)_2CHCH(OH)CH_2CH_3$
                      re-draw
      [1]  $CH_3CH_2-C-C-CH_2CH_3$      [2]  $CH_3CH_2-C-C-CH_2CH_3$      [3]  **4-ethyl-3-hexanol**
           $CH_3CH_2\,OH$                    $CH_3CH_2\,OH$
      6 carbons = **hexanol**                **4-ethyl**

c.   [1]                              [2]      **2-methyl**      [3] *cis*-**2-methylcyclohexanol**
      6 carbon ring = **cyclohexanol**       1

d.   [1]                              [2]      **6-methyl**      [3]  **5-ethyl-6-methyl-3-nonanol**
                                                 3
      9 carbons = **nonanol**              **5-ethyl**

**9.3** To work backwards from a structure to a name:
  [1] Find the parent name and draw its structure.
  [2] Add the substituents to the long chain.

a. 7,7-dimethyl-4-**octanol**

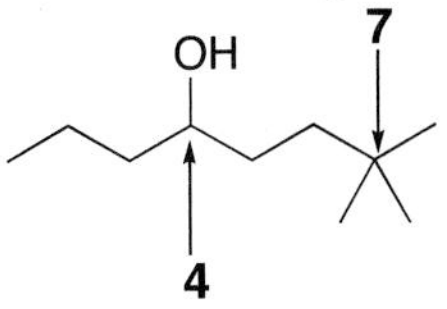

c. 2-*tert*-butyl-3-methyl**cyclohexanol**

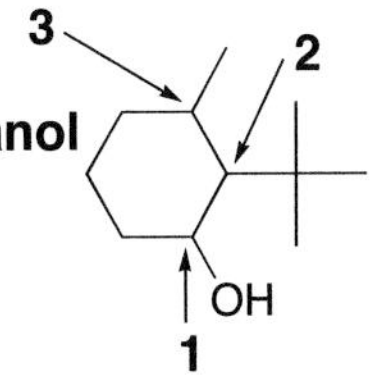

b. 5-methyl-4-propyl-3-**heptanol**

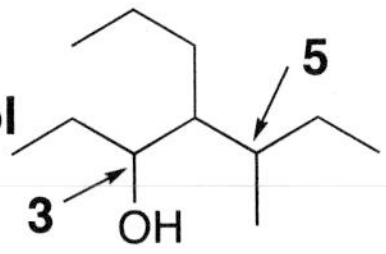

d. *trans*-1,2-**cyclohexanediol**

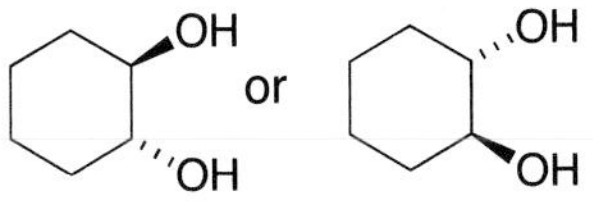

**9.4 To name simple ethers:**
  [1] Name both alkyl groups bonded to the oxygen.
  [2] Arrange these names alphabetically and add the word ***ether***. For symmetrical ethers, name the alkyl group and add the prefix ***di***.

**To name ethers using the IUPAC system:**
  [1] Find the two alkyl groups bonded to the ether oxygen. The smaller chain becomes the substituent, named as an alkoxy group.
  [2] Number the chain to give the lower number to the first substituent.

a. **common name:**

$CH_3-O-CH_2CH_2CH_2CH_3$

methyl          butyl

**butyl methyl ether**

**IUPAC name:**

$CH_3-O\text{-}CH_2CH_2CH_2CH_3$ ◄─larger group – 4 C's
                                      butane

substituent:
methoxy

**1-methoxybutane**

b. **common name:**

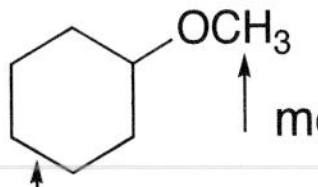

$OCH_3$
methyl

cyclohexyl

**cyclohexyl methyl ether**

**IUPAC name:**

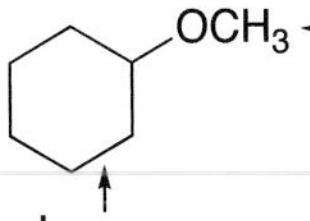

$OCH_3$ ◄─substituent –
          methoxy

larger group – 6 C's
cyclohexane

**methoxycyclohexane**

c. **common name:**

$CH_3CH_2CH_2-O-CH_2CH_2CH_3$

propyl          propyl

**dipropyl ether**

**IUPAC name:**

$CH_3CH_2CH_2-O\text{-}CH_2CH_2CH_3$

propoxy          propane

**1-propoxypropane**

**9.5** **Three ways to name epoxides**:

[1] Epoxides are named as derivations of oxirane, the simplest epoxide.

[2] Epoxides can be named by considering the oxygen a substituent called an **epoxy** group, bonded to a hydrocarbon chain or ring. Use two numbers to designate which two atoms the oxygen is bonded to.

[3] Epoxides can be named as **alkene oxides** by mentally replacing the epoxide oxygen by a double bond. Name the alkene (Chapter 10) and add the word *oxide*.

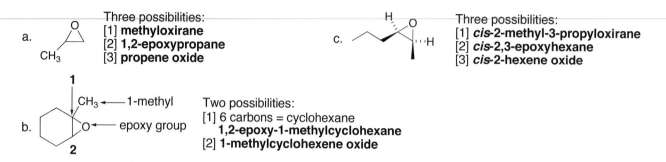

a.
Three possibilities:
[1] **methyloxirane**
[2] **1,2-epoxypropane**
[3] **propene oxide**

c.
Three possibilities:
[1] *cis*-**2-methyl-3-propyloxirane**
[2] *cis*-**2,3-epoxyhexane**
[3] *cis*-**2-hexene oxide**

b.
CH₃ ← 1-methyl
O ← epoxy group

Two possibilities:
[1] 6 carbons = cyclohexane
 **1,2-epoxy-1-methylcyclohexane**
[2] **1-methylcyclohexene oxide**

**9.6** Two rules for boiling point:

[1] **The stronger the forces the higher the bp.**

[2] **Bp increases as the extent of the hydrogen bonding increases.** For alcohols with the same number of carbon atoms: hydrogen bonding and bp's increase: $3° ROH < 2° ROH < 1° ROH$.

a.
VDW
**lowest bp**

VDW
DD
**intermediate bp**

VDW
DD
hydrogen
bonding
**highest bp**

b.
$3°$ ROH
**lowest bp**

$2°$ ROH
**intermediate bp**

$1°$ ROH
**highest bp**

**9.7** Draw dimethyl ether and ethanol and analyze their intermolecular forces to explain the observed trend.

**dimethyl ether**

$CH_3$ O $CH_3$

VDW
DD
**no HB**
**much lower bp**

**ethanol**

$CH_3CH_2OH$

VDW
DD
**HB**
Two molecules of $CH_3CH_2OH$
can hydrogen bond to each other.
**stronger forces =**
**much higher bp**

Both molecules contain an O atom
and can hydrogen bond with water. They
have fewer than 5 C's and are
therefore **water soluble**.

H O H
$CH_3$ O $CH_3$

H O H
$CH_3CH_2$ O H

**9.8** Melting points depend on intermolecular forces and symmetry. $(CH_3)_2CHCH_2OH$ has a lower melting point than $CH_3CH_2CH_2CH_2OH$ because branching decreases surface area and makes $(CH_3)_2CHCH_2OH$ less symmetrical so it packs less well. Although $(CH_3)_3COH$ has the most branching and least surface area, it is the most symmetrical so it packs best in a crystalline lattice, giving it the highest melting point.

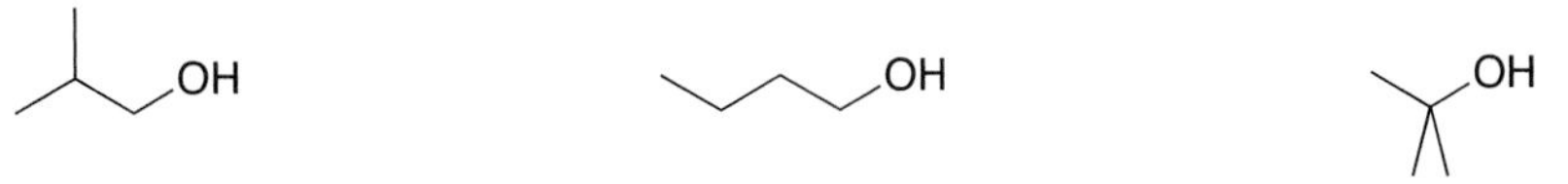

| | | |
|---|---|---|
| −108 °C | −90 °C | 26 °C |
| lowest melting point | intermediate melting point | highest melting point |

**9.9** Strong nucleophiles (like ⁻CN) favor $S_N2$ reactions. The use of crown ethers in nonpolar solvents increases the nucleophilicity of the anion, and this increases the rate of the $S_N2$ reaction. The nucleophile does not appear in the rate equation for the $S_N1$ reaction. Nonpolar solvents cannot solvate carbocations so this disfavors $S_N1$ reactions as well.

**9.10** Although epothilone B contains 27 C's, each of the labeled functional groups can hydrogen bond to $H_2O$, thus making it somewhat water soluble.

**epothilone B**
The nitrogen- and oxygen-containing functional groups increase the water solubility of epothilone B.

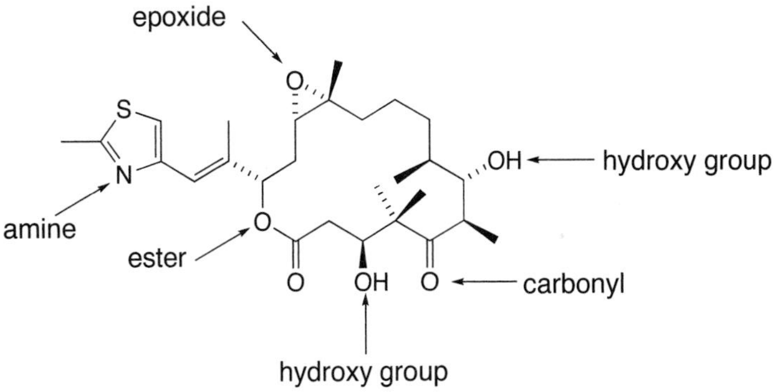

**9.11** Draw the products of substitution in the following reactions by substituting ⁻OH or ⁻OR for X in the starting material.

a. $CH_3CH_2CH_2CH_2-Br$ + ⁻OH ⟶ $CH_3CH_2CH_2CH_2-OH$ + Br⁻ **alcohol**

b. 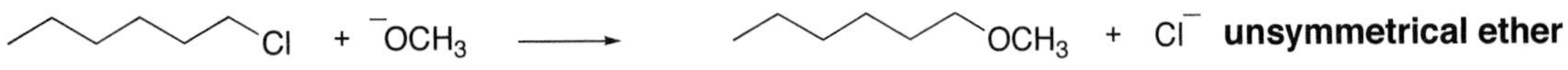 Cl + ⁻OCH₃ ⟶ OCH₃ + Cl⁻ **unsymmetrical ether**

c. 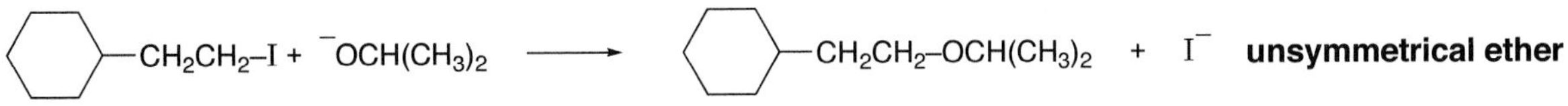 —CH₂CH₂–I + ⁻OCH(CH₃)₂ ⟶ —CH₂CH₂–OCH(CH₃)₂ + I⁻ **unsymmetrical ether**

d. 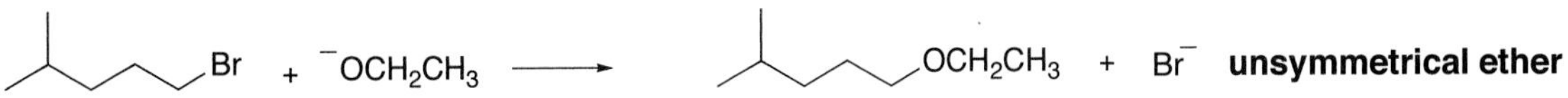 Br + ⁻OCH₂CH₃ ⟶ OCH₂CH₃ + Br⁻ **unsymmetrical ether**

**9.12** To synthesize an ether using a Williamson ether synthesis:
[1] First find the two possible alkoxides and alkyl halides needed for nucleophilic substitution.
[2] Classify the alkyl halides as 1°, 2°, or 3°. The favored path has the less hindered halide.

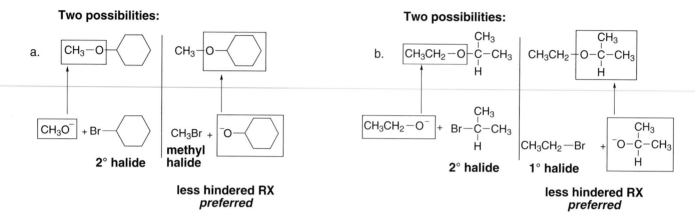

**9.13** **NaH and NaNH₂ are strong bases that will remove a proton from an alcohol**, creating a nucleophile.

a. $CH_3CH_2CH_2-\overset{..}{\underset{..}{O}}-H$ + $Na^+H:^-$ ⟶ $CH_3CH_2CH_2-\overset{..}{\underset{..}{O}}:^-$ $Na^+$ + $H_2$

b. (cyclopentane)$\overset{CH_3}{\underset{H}{C}}-\overset{..}{\underset{..}{O}}-H$ + $Na^+:\overset{..}{N}H_2$ ⟶ (cyclopentane)$\overset{CH_3}{\underset{H}{C}}-\overset{..}{\underset{..}{O}}:^-$ + $Na^+$ + $\overset{..}{N}H_3$

c. (chain)$\overset{H}{\underset{..}{:O}}$ $Na^+H:^-$ ⟶ (chain)$\overset{..}{\underset{..}{O}}:^-$ $CH_3CH_2CH_2-Br$ + $Na^+$ + $H_2$

(chain)$\overset{..}{\underset{..}{O}}-CH_2CH_2CH_3$ + $Br^-$

d. (cyclohexane with O⁻H and Br) $Na^+H:^-$ ⟶ (cyclohexane with :Ö:⁻ and Br) + $Na^+$ + $H_2$ ⟶ (cyclohexane epoxide) + $Br^-$

$C_6H_{10}O$

**9.14** **Dehydration follows the Zaitsev rule**, so the more stable, more substituted alkene is the major product.

a. $CH_3-\overset{H}{\underset{OH}{C}}-CH_3$ $\xrightarrow{H_2SO_4}$ $CH_2=CH-CH_3$ + $H_2O$

b. (branched alcohol) $\xrightarrow{TsOH}$ $\overset{CH_3CH_2}{\underset{CH_3}{C}}=CHCH_3$ + $\overset{}{\underset{CH_2}{}}$(alkene) + $H_2O$

trisubstituted          disubstituted
**major product**    **minor product**

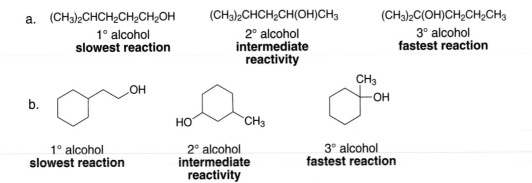

**c.**

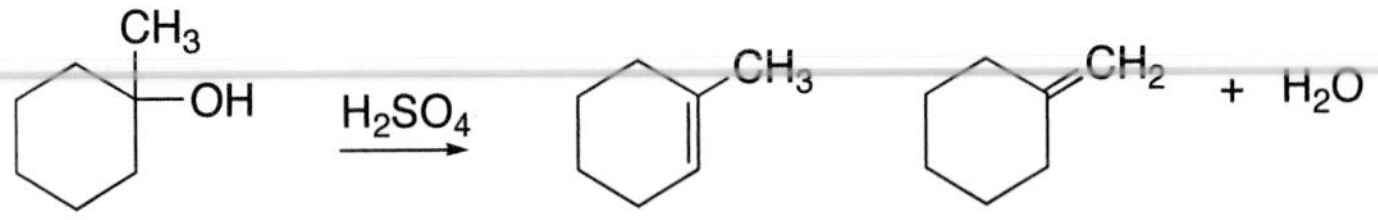

trisubstituted     disubstituted
**major product   minor product**

**9.15** The rate of dehydration increases as the number of R groups increases.

**a.**  (CH₃)₂CHCH₂CH₂CH₂OH          (CH₃)₂CHCH₂CH(OH)CH₃          (CH₃)₂C(OH)CH₂CH₂CH₃
$(CH_3)_2CHCH_2CH_2CH_2OH$          $(CH_3)_2CHCH_2CH(OH)CH_3$          $(CH_3)_2C(OH)CH_2CH_2CH_3$

1° alcohol                          2° alcohol                          3° alcohol
**slowest reaction**                **intermediate**                    **fastest reaction**
                                    **reactivity**

**b.**

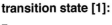

1° alcohol                          2° alcohol                          3° alcohol
**slowest reaction**                **intermediate**                    **fastest reaction**
                                    **reactivity**

**9.16** There are three steps in the E1 mechanism for dehydration of alcohols, and three transition states.

**transition state [1]:**          **transition state [2]:**          **transition state [3]:**

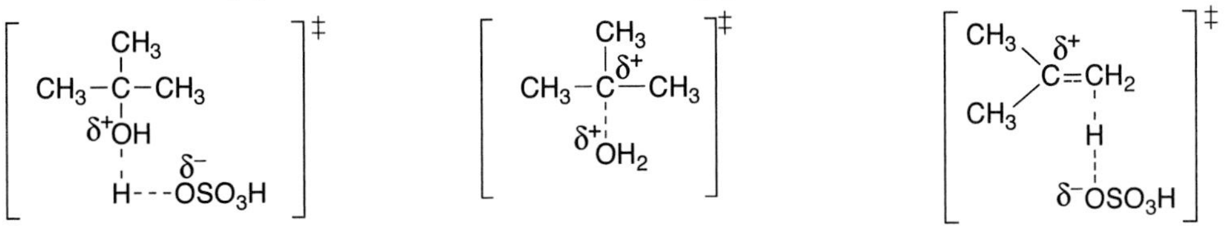

**9.17**

**transition state [1]:**          **transition state [2]:**

**9.18**

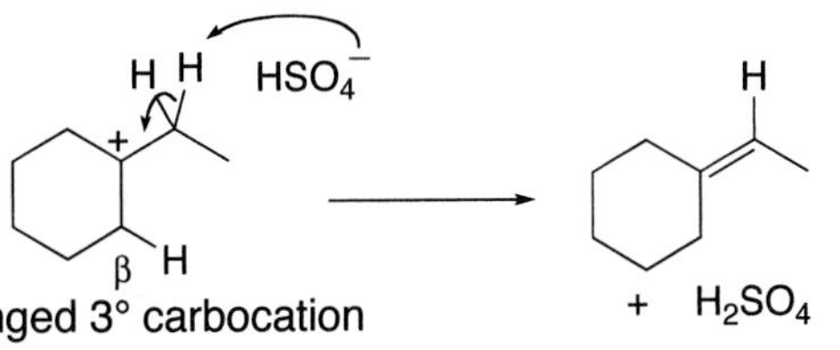

HSO₄⁻

rearranged 3° carbocation          + H₂SO₄

This alkene is also formed in addition to **Y** from the rearranged carbocation.

The initially formed 2° carbocation gives two alkenes:

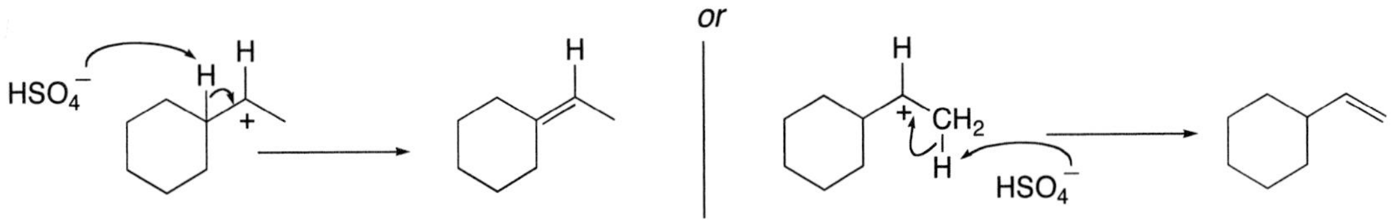

*or*

HSO₄⁻          HSO₄⁻

**9.19**

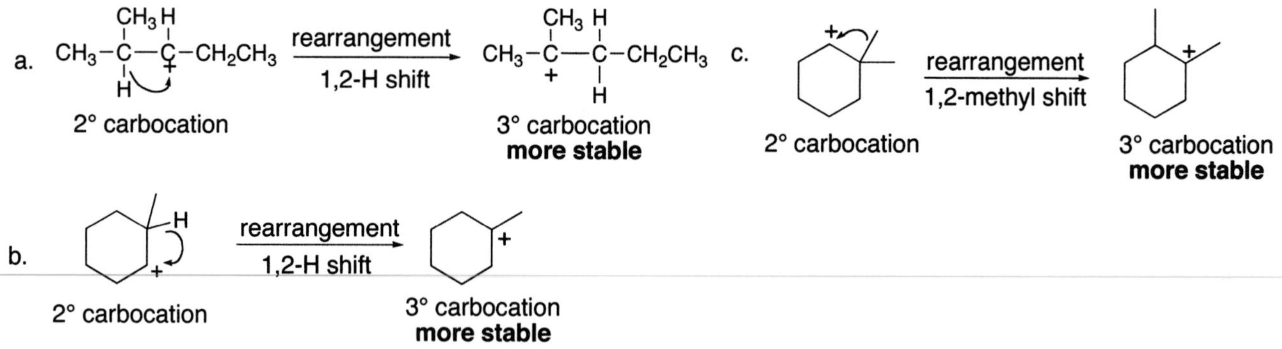

a.
rearrangement
1,2-H shift
2° carbocation
3° carbocation
more stable
b.
rearrangement
1,2-H shift
2° carbocation
3° carbocation
more stable
c.
rearrangement
1,2-methyl shift
2° carbocation
3° carbocation
more stable


**9.20**

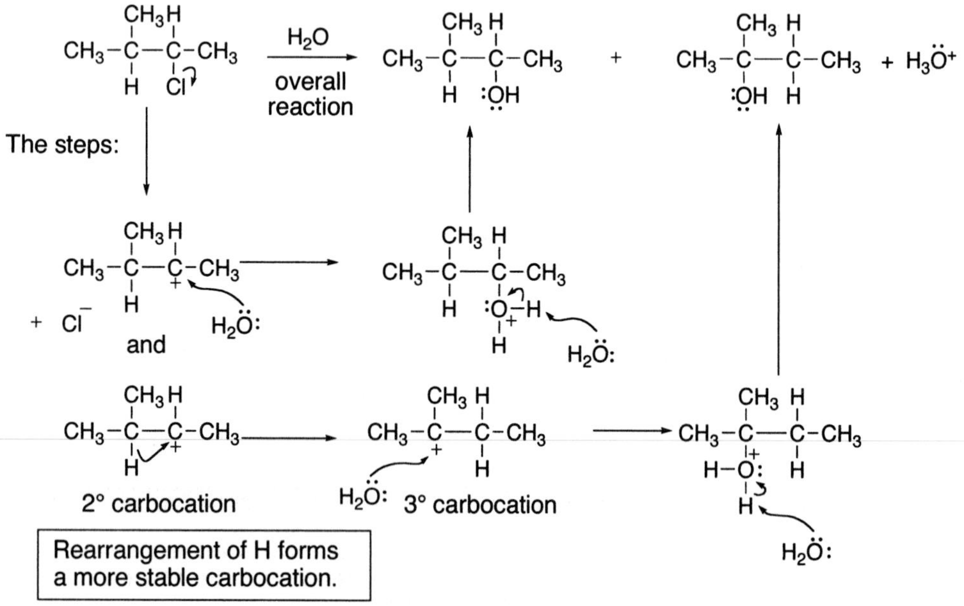

overall
reaction
The steps:
and
2° carbocation
3° carbocation
Rearrangement of H forms
a more stable carbocation.


**9.21**

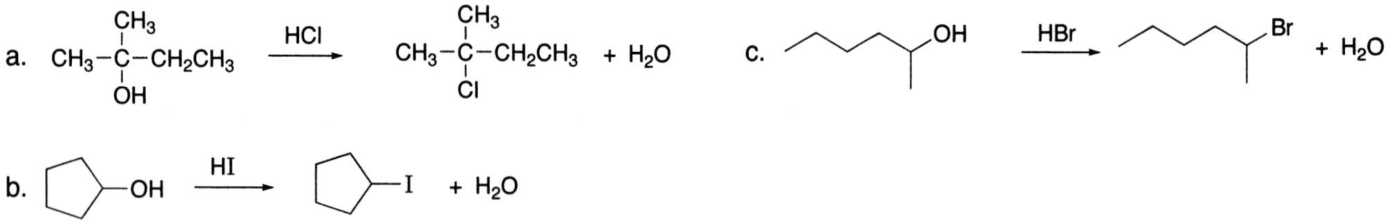

a.
b.
c.


**9.22** • **$CH_3OH$ and 1° alcohols** follow an $S_N2$ mechanism, which results in inversion of configuration.
- **Secondary (2°) and 3° alcohols** follow an $S_N1$ mechanism, which results in racemization at a stereogenic center.

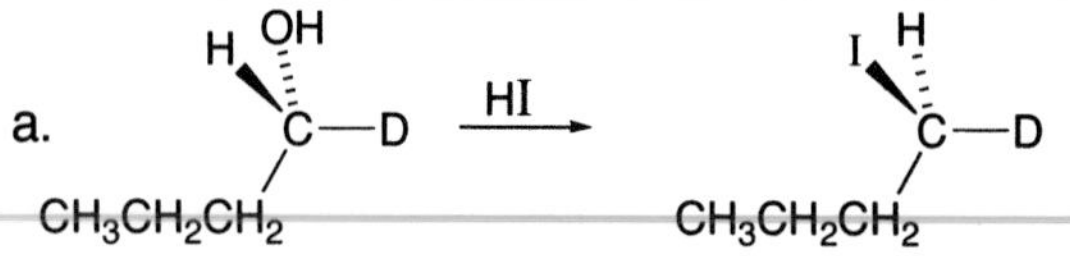

a.


1° alcohol so **inversion of configuration**

b.

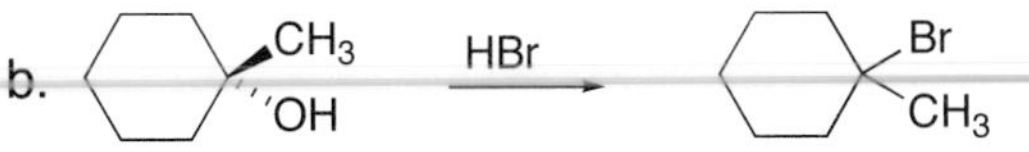

HBr

3° alcohol, so Br⁻ attacks from above and below.
The product is achiral.

achiral starting material          achiral product

c.

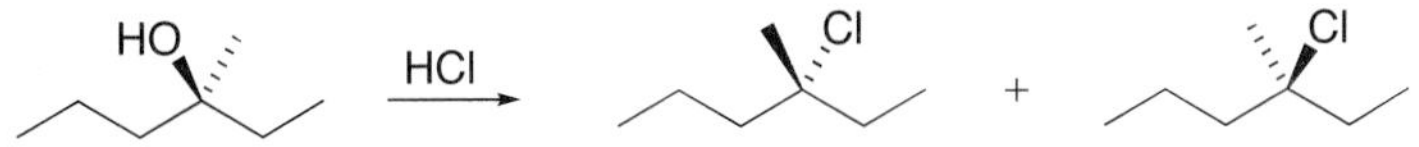

HCl

+

3° alcohol = **racemization**

**9.23**

a.

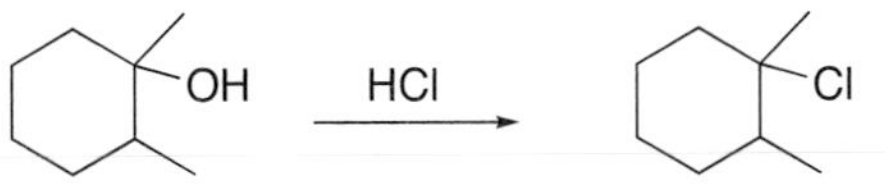

HCl

c.

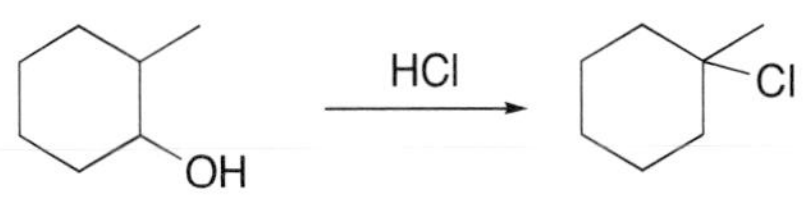

HCl

(product formed after a 1,2-H shift)

b.

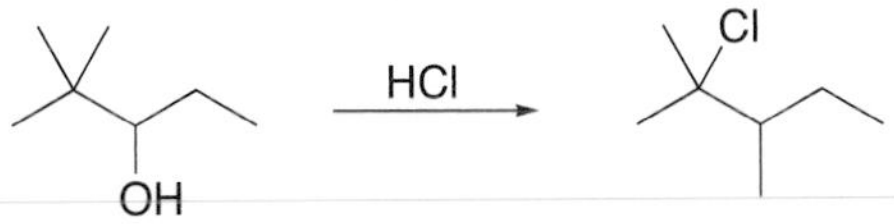

HCl

(product formed after a 1,2-CH₃ shift)

**9.24** Substitution reactions of alcohols using SOCl₂ proceed by an S_N2 mechanism. Therefore, there is **inversion of configuration** at a stereogenic center.

SOCl₂ / pyridine

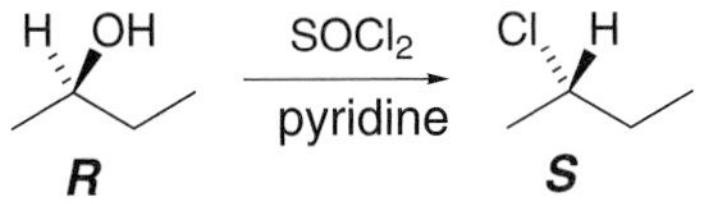

Reactions using SOCl₂
proceed by an S_N2 mechanism =
**inversion of configuration.**

**9.25** Substitution reactions of alcohols using PBr₃ proceed by an S_N2 mechanism. Therefore, there is inversion of configuration at a stereogenic center.

PBr₃

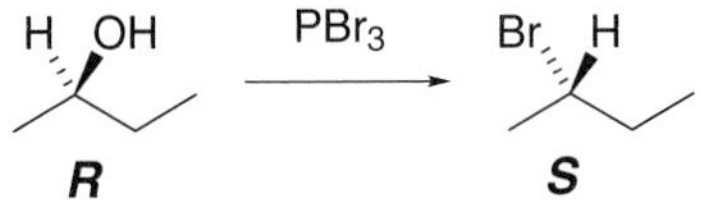

Reactions using PBr₃
proceed by an S_N2 mechanism =
**inversion of configuration.**

**9.26** Stereochemistry for conversion of ROH to RX by reagent:
   [1] **HX**—with 1°, S_N2, so inversion of configuration; with 2° and 3°, S_N1, so racemization.
   [2] **SOCl₂**—S_N2, so inversion of configuration.
   [3] **PBr₃**—S_N2, so inversion of configuration.

a.

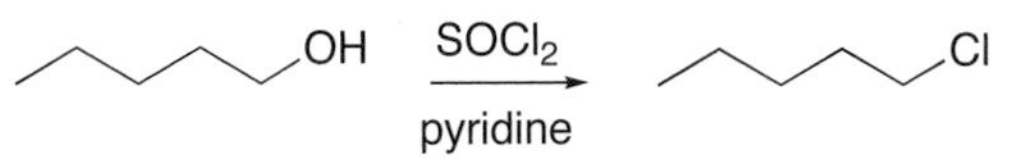

SOCl₂ / pyridine

c.

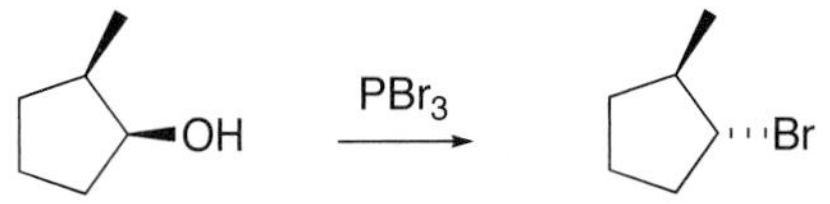

PBr₃

S_N2 = **inversion**

b.

HI     +

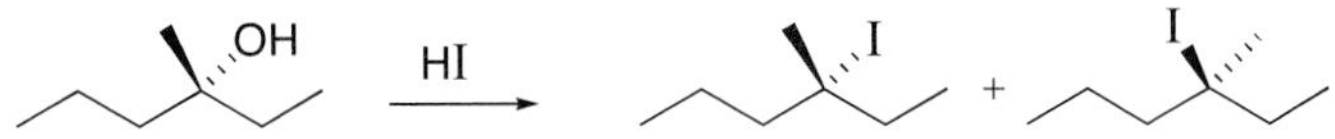

3° alcohol, S_N1 = **racemization**

**9.27** To do a two-step synthesis with this starting material:
[1] Convert the OH group into a good leaving group (by using either PBr$_3$ or SOCl$_2$).
[2] Add the nucleophile for the S$_N$2 reaction.

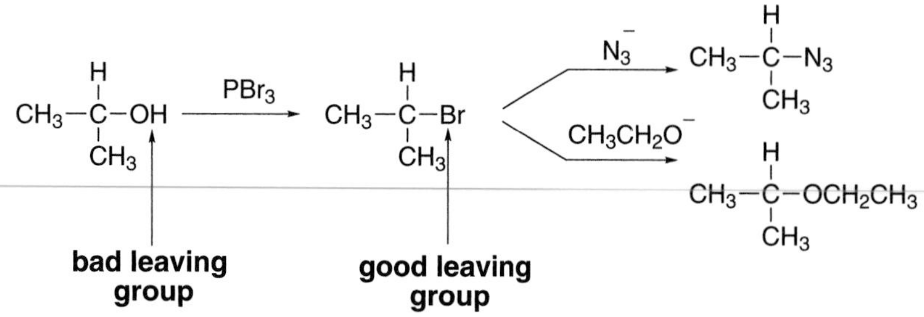

**9.28**

a. 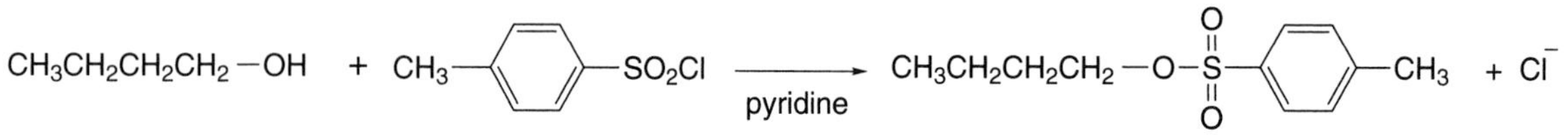 $CH_3CH_2CH_2CH_2-OH$ + $CH_3-$⟨benzene⟩$-SO_2Cl$ →(pyridine) $CH_3CH_2CH_2CH_2-O-S(=O)_2-$⟨benzene⟩$-CH_3$ + $Cl^-$

b.

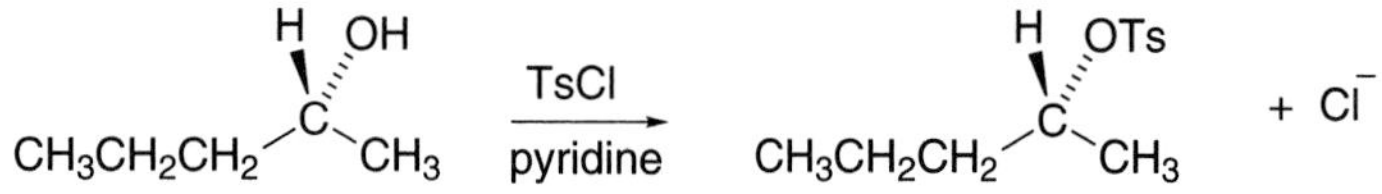

**9.29**

a.

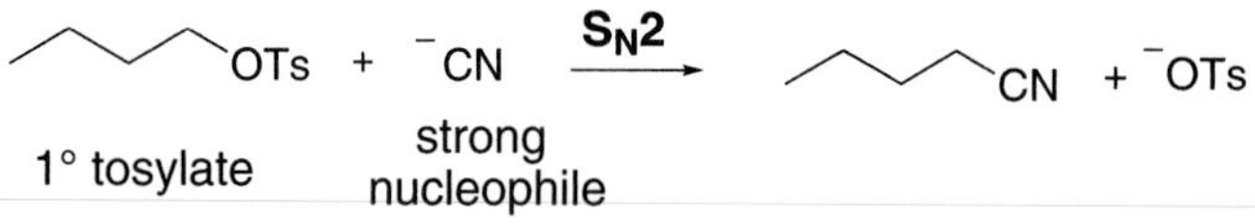

b. 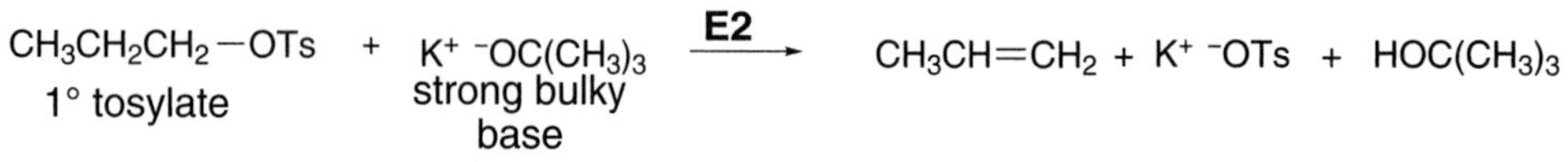 $CH_3CH_2CH_2-OTs$ + $K^+ \, ^-OC(CH_3)_3$ →(E2) $CH_3CH=CH_2$ + $K^+ \, ^-OTs$ + $HOC(CH_3)_3$
1° tosylate · strong bulky base

c.

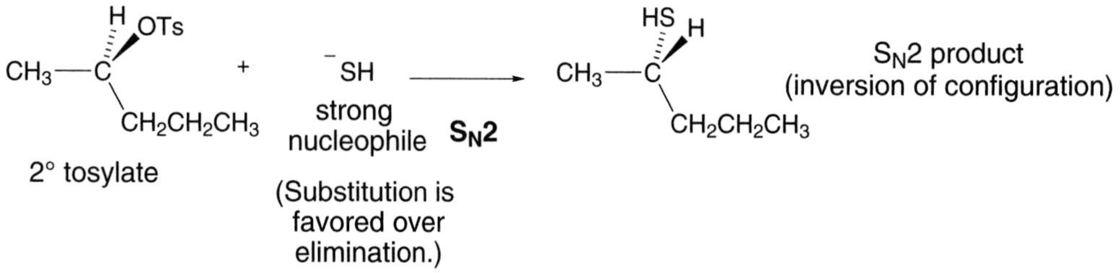

**9.30**

One inversion from starting material to product.

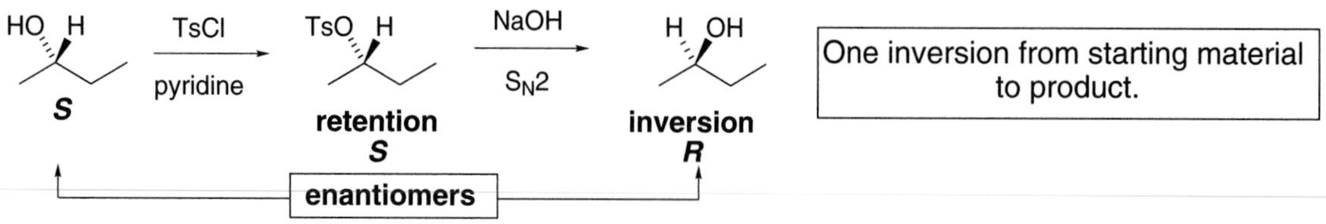

**9.31** These reagents can be classified as:

[1] SOCl₂, PBr₃, HCl, and HBr replace OH with X by a substitution reaction.
[2] Tosyl chloride (TsCl) makes OH a better leaving group by converting it to OTs.
[3] Strong acids (H₂SO₄) and POCl₃ (pyridine) result in elimination by dehydration.

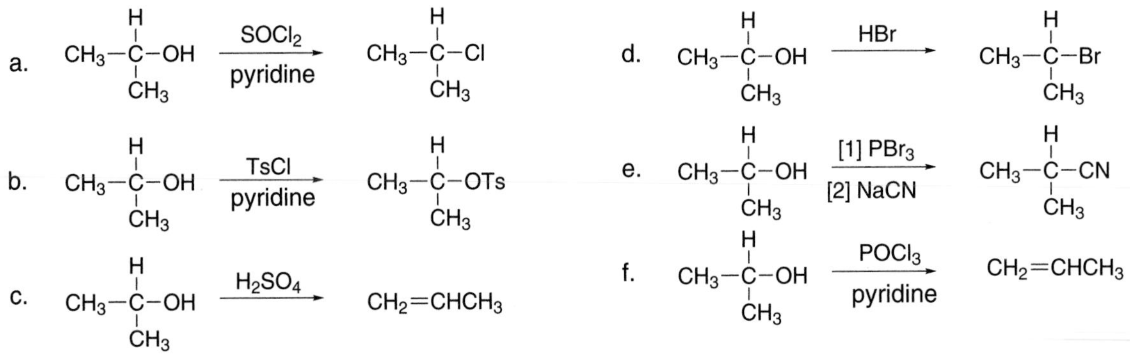

a. CH₃–C(H)(CH₃)–OH  —SOCl₂/pyridine→  CH₃–C(H)(CH₃)–Cl

d. CH₃–C(H)(CH₃)–OH  —HBr→  CH₃–C(H)(CH₃)–Br

b. CH₃–C(H)(CH₃)–OH  —TsCl/pyridine→  CH₃–C(H)(CH₃)–OTs

e. CH₃–C(H)(CH₃)–OH  —[1] PBr₃ [2] NaCN→  CH₃–C(H)(CH₃)–CN

c. CH₃–C(H)(CH₃)–OH  —H₂SO₄→  CH₂=CHCH₃

f. CH₃–C(H)(CH₃)–OH  —POCl₃/pyridine→  CH₂=CHCH₃

**9.32**

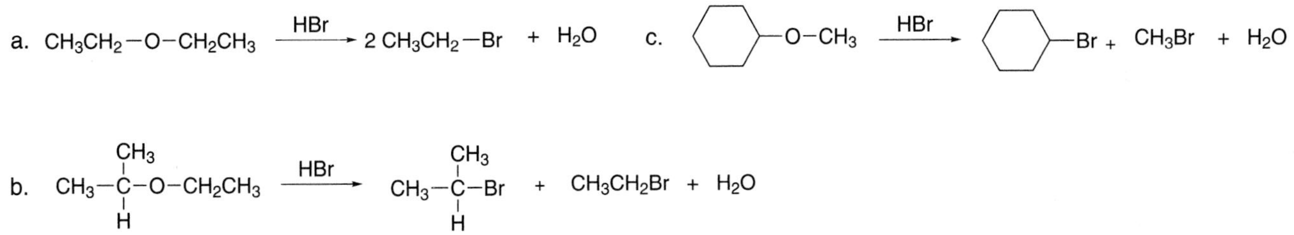

a. CH₃CH₂–O–CH₂CH₃  —HBr→  2 CH₃CH₂–Br + H₂O

c. [cyclohexyl]–O–CH₃  —HBr→  [cyclohexyl]–Br + CH₃Br + H₂O

b. CH₃–C(CH₃)(H)–O–CH₂CH₃  —HBr→  CH₃–C(CH₃)(H)–Br + CH₃CH₂Br + H₂O

**9.33** Ether cleavage can occur by either an S_N1 or S_N2 mechanism, but neither mechanism can occur when the ether O atom is bonded to an aromatic ring. An S_N1 reaction would require formation of a highly unstable carbocation on a benzene ring, a process that does not occur. An S_N2 reaction would require backside attack through the plane of the aromatic ring, which is also not possible. Thus cleavage of the Ph–OCH₃ bond does not occur.

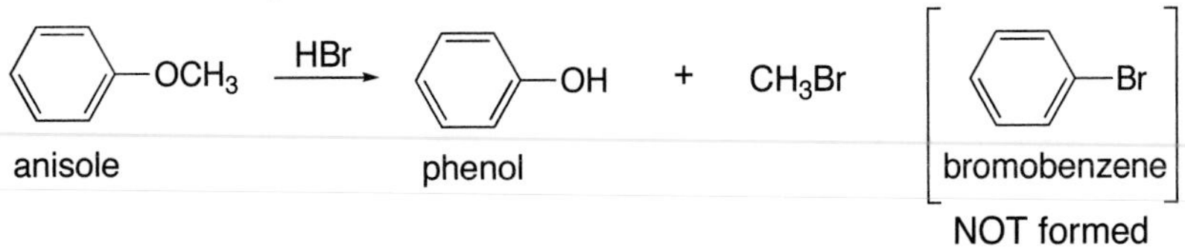

[Ph]–OCH₃  —HBr→  [Ph]–OH + CH₃Br   [ [Ph]–Br  bromobenzene NOT formed ]

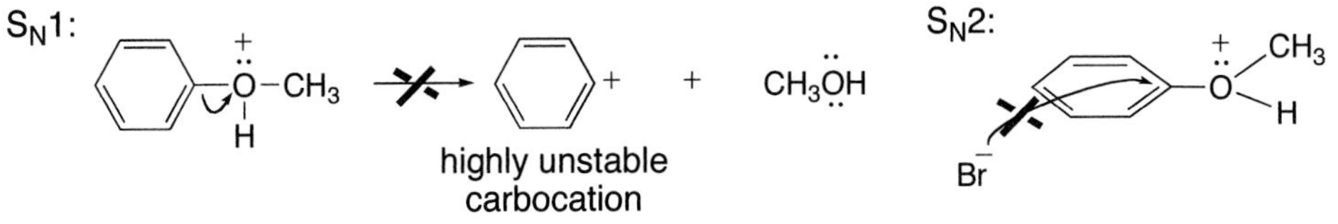

anisole → phenol

S_N1: [Ph]–O(CH₃)(H)⁺ ⇸ [Ph]⁺ highly unstable carbocation + CH₃OH

S_N2: Br attacking [Ph]–O⁺(CH₃)(H)

**9.34** Compare epoxides and cyclopropane. For a compound to be reactive towards nucleophiles, it must be electrophilic.

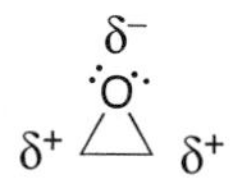

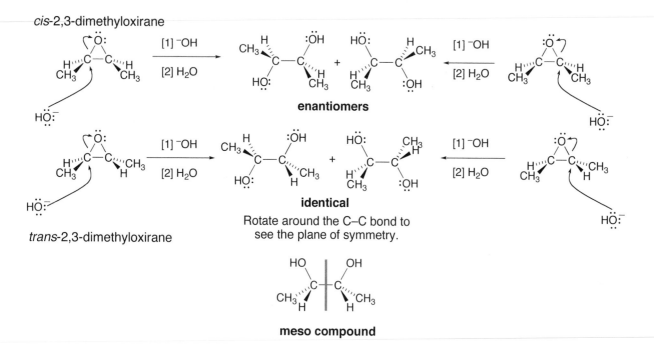

**epoxide**

O is electronegative and pulls electron density away from C's. This makes them electrophilic and reactive with nucleophiles.

**cyclopropane**

Cyclopropane has all C's and H's, so all nonpolar bonds. There are no electrophilic C's so it will not react with nucleophiles.

**9.35** Two rules for reaction of an epoxide:
  [1] Nucleophiles attack from the **back side** of the epoxide.
  [2] Negatively charged nucleophiles attack at the **less substituted carbon**.

a.

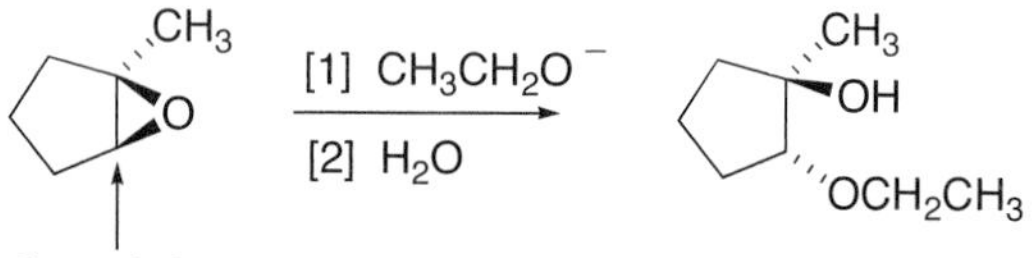

**Attack here:**
less substituted C
backside attack

b.

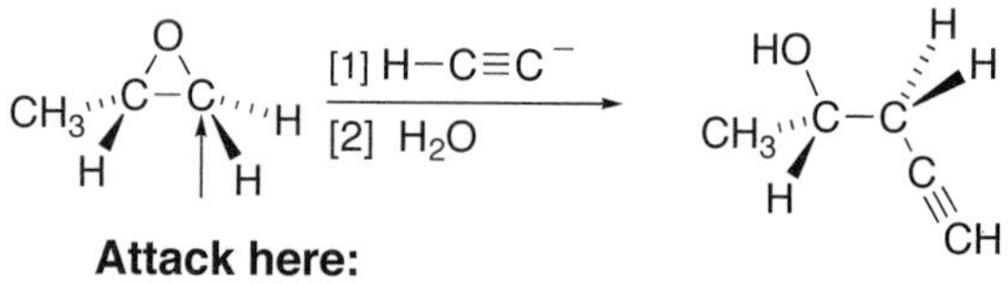

**Attack here:**
less substituted C
backside attack

**9.36** In both isomers, ⁻OH attacks from the back side at either C–O bond.

*cis*-2,3-dimethyloxirane

**enantiomers**

*trans*-2,3-dimethyloxirane

**identical**

Rotate around the C–C bond to see the plane of symmetry.

**meso compound**

**9.37** Remember the difference between negatively charged nucleophiles and neutral nucleophiles:

- **Negatively charged nucleophiles attack first,** followed by protonation, and the nucleophile attacks at the **less substituted carbon.**
- **Neutral nucleophiles have protonation first,** followed by nucleophilic attack at the **more substituted carbon.**

BUT – **trans or anti products are always formed regardless of the nucleophile.**

a.

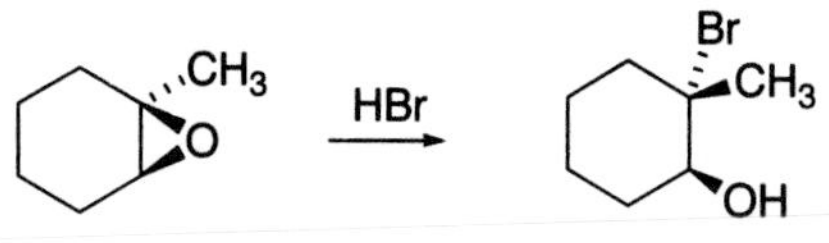

neutral nucleophile:
attack at **more**
substituted C

b.

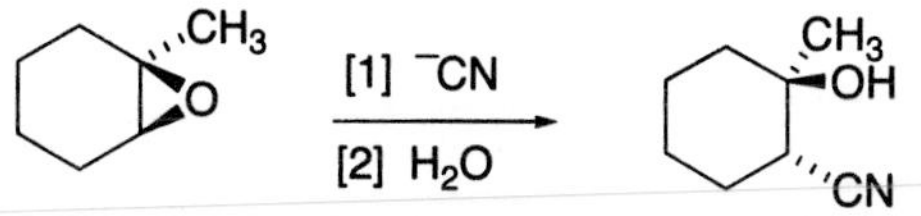

negatively charged
nucleophile:
attack at **less**
substituted C

c.

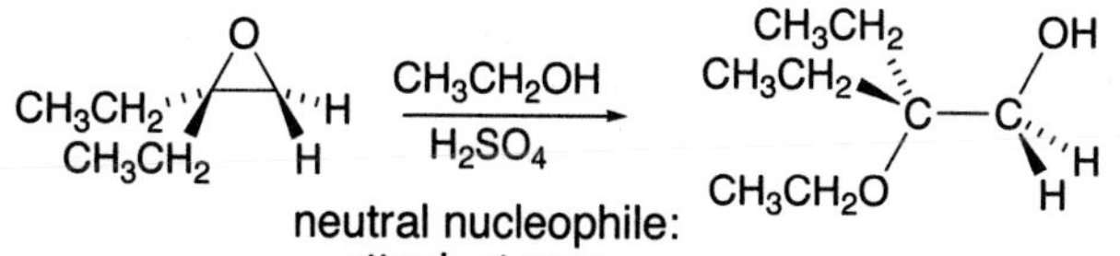

neutral nucleophile:
attack at **more**
substituted C

d.

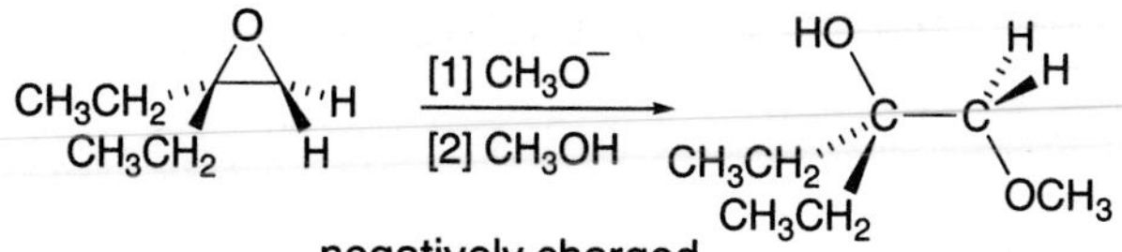

negatively charged
nucleophile:
attack at **less**
substituted C

Chapter 9–16

**9.38** Use the directions from Answer 9.2.

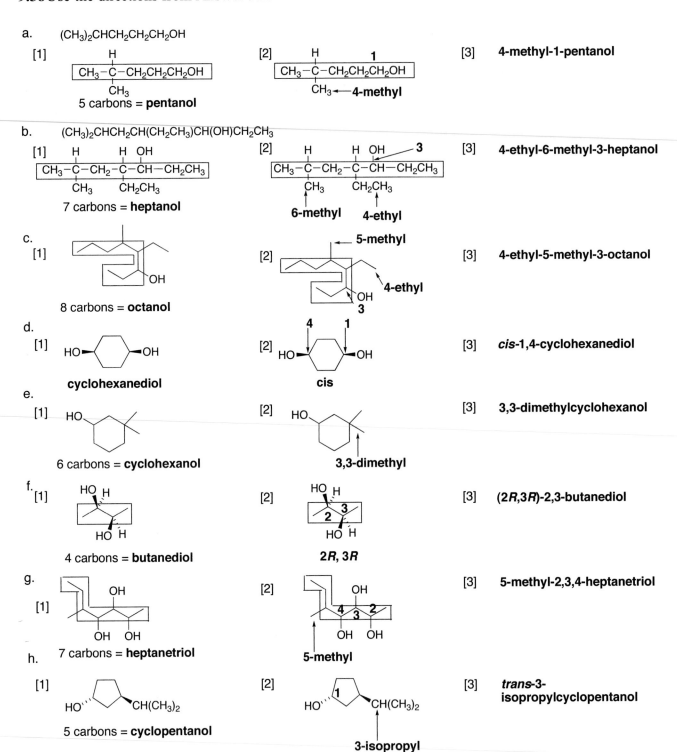

a. (CH₃)₂CHCH₂CH₂CH₂OH

[1] 
CH₃–C–CH₂CH₂CH₂OH 
CH₃ 
5 carbons = **pentanol**

[2] 
CH₃–C–CH₂CH₂CH₂OH 
CH₃ ← **4-methyl**

[3] **4-methyl-1-pentanol**

b. (CH₃)₂CHCH₂CH(CH₂CH₃)CH(OH)CH₂CH₃

[1] 
CH₃–C–CH₂–C–CH–CH₂CH₃ 
CH₃   CH₂CH₃ 
7 carbons = **heptanol**

[2] 
CH₃–C–CH₂–C–CH–CH₂CH₃ 
CH₃   CH₂CH₃ 
**6-methyl  4-ethyl**

[3] **4-ethyl-6-methyl-3-heptanol**

c. [1] 8 carbons = **octanol**

[2] **5-methyl  4-ethyl**

[3] **4-ethyl-5-methyl-3-octanol**

d. [1] HO—⬡—OH **cyclohexanediol**

[2] HO—⬡—OH **cis**

[3] *cis*-1,4-cyclohexanediol

e. [1] 6 carbons = **cyclohexanol**

[2] **3,3-dimethyl**

[3] **3,3-dimethylcyclohexanol**

f. [1] 4 carbons = **butanediol**

[2] **2R, 3R**

[3] **(2R,3R)-2,3-butanediol**

g. [1] 7 carbons = **heptanetriol**

[2] **5-methyl**

[3] **5-methyl-2,3,4-heptanetriol**

h. [1] 5 carbons = **cyclopentanol**

[2] **3-isopropyl**

[3] *trans*-3-isopropylcyclopentanol

**9.39** Use the rules from Answers 9.4 and 9.5.

a. **dicyclohexyl ether**

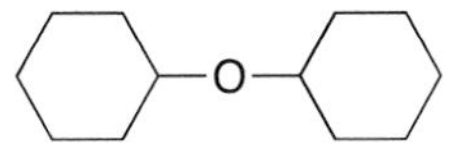

b. **4,4-dimethyl**

longest chain =
**heptane**

OCH₂CH₂CH₃

substituent =
**3-propoxy**

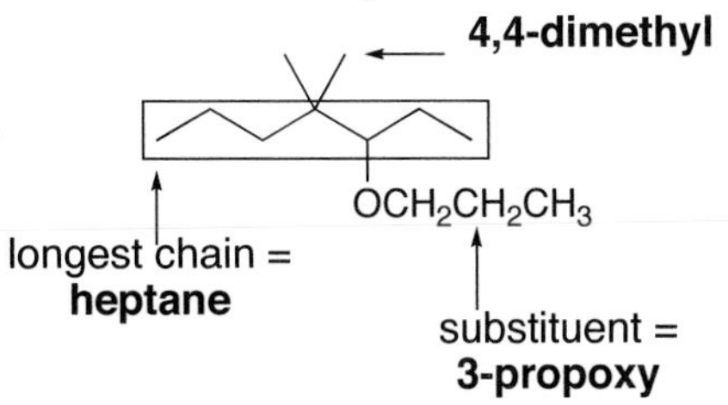

**4,4-dimethyl-3-propoxyheptane**

c. **ethyl isobutyl ether**
or **1-ethoxy-2-methylpropane**

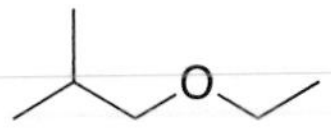

d. **1,2-epoxy-2-methylhexane**
or **2-butyl-2-methyloxirane**
or **2-methylhexene oxide**

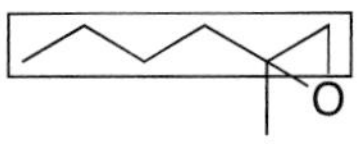

e. CH₂CH₃

**2**   **epoxy**

5 carbons =
**cyclopentane**

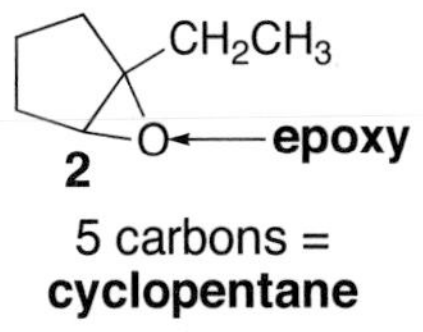

**1,2-epoxy-1-ethylcyclopentane**
or **1-ethylcyclopentene oxide**

f.   CH₃ CH₃
CH₃−C−O−C−CH₃
CH₃ CH₃

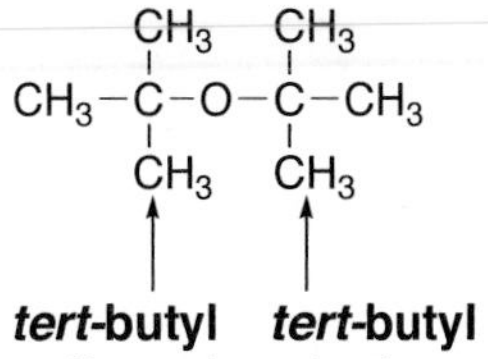

*tert*-butyl   *tert*-butyl
**di-*tert*-butyl ether**

**9.40** Use the directions from Answer 9.3.

a. 4-ethyl-3-**heptanol**

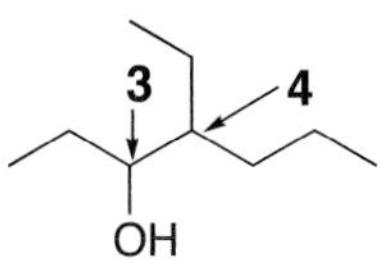

b. *trans*-2-methyl**cyclohexanol**

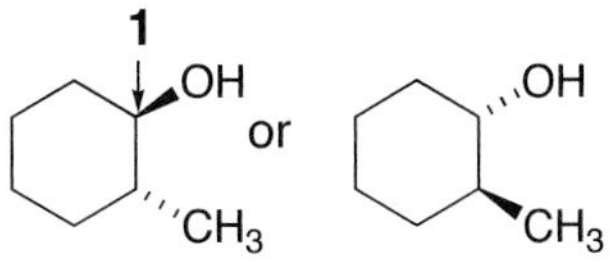

**1** OH

or

OH

CH₃   CH₃

c. 2,3,3-trimethyl-2-**butanol**

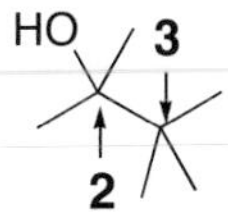

HO   **3**

**2**

d. 6-*sec*-butyl-7,7-diethyl-4-**decanol**

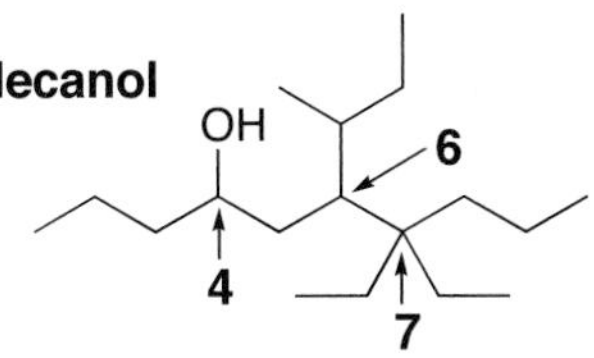

OH   **6**

**4**

**7**

e. 3-chloro-1,2-**propanediol**

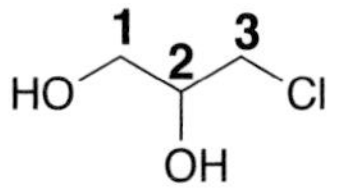

**1**   **3**
HO   **2**   Cl
OH

f. diisobutyl ether

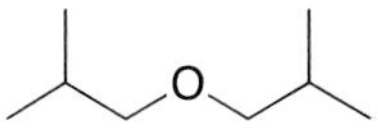

g. 1,2-epoxy-1,3,3-trimethyl**cyclohexane**

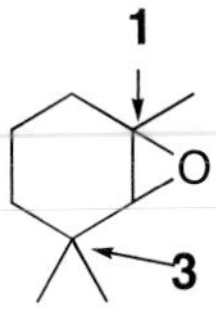

**1**

O

**3**

h. 1-ethoxy-3-ethyl**heptane**

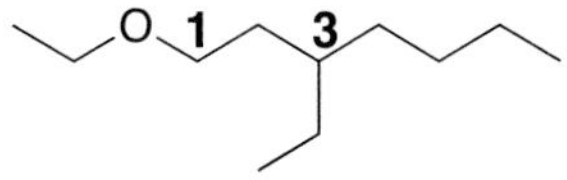

O **1**   **3**

**9.41**

Eight constitutional isomers of molecular formula $C_5H_{12}O$ containing an OH group:

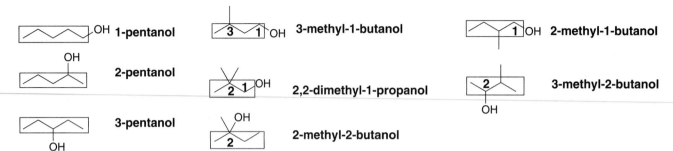

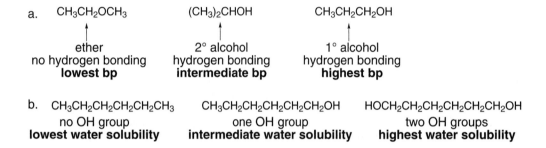

**9.42** Use the boiling point rules from Answer 9.6.

a.　$CH_3CH_2OCH_3$　　　$(CH_3)_2CHOH$　　　$CH_3CH_2CH_2OH$

ether
no hydrogen bonding
**lowest bp**

2° alcohol
hydrogen bonding
**intermediate bp**

1° alcohol
hydrogen bonding
**highest bp**

b.　$CH_3CH_2CH_2CH_2CH_2CH_3$　　$CH_3CH_2CH_2CH_2CH_2CH_2OH$　　$HOCH_2CH_2CH_2CH_2CH_2CH_2OH$

no OH group
**lowest water solubility**

one OH group
**intermediate water solubility**

two OH groups
**highest water solubility**

**9.43**

a.　$CH_3CH_2CH_2OH$　$\xrightarrow{H_2SO_4}$　$CH_3CH=CH_2$　+　$H_2O$

b.　$CH_3CH_2CH_2OH$　$\xrightarrow{NaH}$　$CH_3CH_2CH_2O^-\ Na^+$　+　$H_2$

c.　$CH_3CH_2CH_2OH$　$\xrightarrow[ZnCl_2]{HCl}$　$CH_3CH_2CH_2Cl$　+　$H_2O$

d.　$CH_3CH_2CH_2OH$　$\xrightarrow{HBr}$　$CH_3CH_2CH_2Br$　+　$H_2O$

e.　$CH_3CH_2CH_2OH$　$\xrightarrow[pyridine]{SOCl_2}$　$CH_3CH_2CH_2Cl$

f.　$CH_3CH_2CH_2OH$　$\xrightarrow{PBr_3}$　$CH_3CH_2CH_2Br$

g.　$CH_3CH_2CH_2OH$　$\xrightarrow[pyridine]{TsCl}$　$CH_3CH_2CH_2OTs$

h.　$CH_3CH_2CH_2OH$　$\xrightarrow{[1]\ NaH}$　$CH_3CH_2CH_2O^-\ Na^+$　$\xrightarrow{[2]\ CH_3CH_2Br}$　$CH_3CH_2CH_2OCH_2CH_3$

i.　$CH_3CH_2CH_2OH$　$\xrightarrow{[1]\ TsCl}$　$CH_3CH_2CH_2OTs$　$\xrightarrow{[2]\ NaSH}$　$CH_3CH_2CH_2SH$

**9.44**

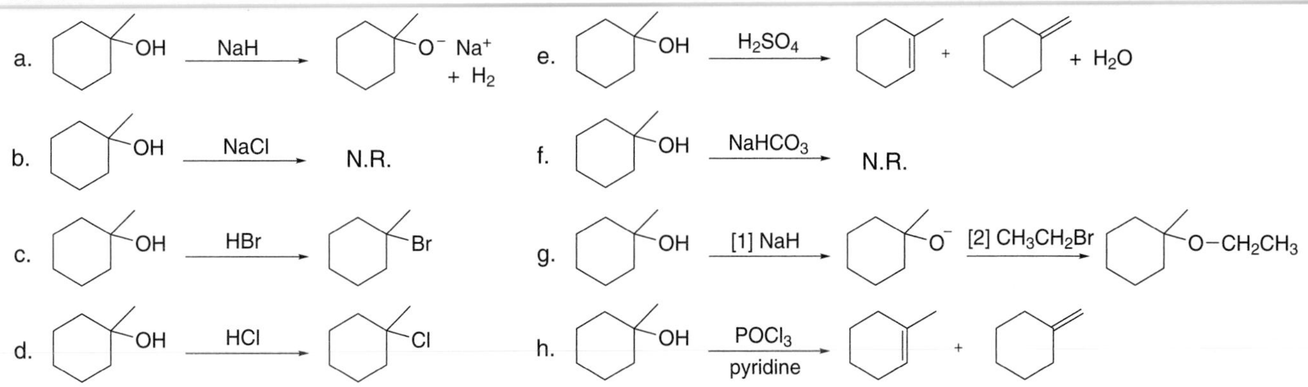

a. cyclohexanol + NaH → alkoxide O⁻ Na⁺ + H₂

e. + H₂SO₄ → + + H₂O

b. + NaCl → N.R.

f. + NaHCO₃ → N.R.

c. + HBr → Br

g. + [1] NaH → O⁻ [2] CH₃CH₂Br → O–CH₂CH₃

d. + HCl → Cl

h. + POCl₃ / pyridine → +

**9.45 Dehydration follows the Zaitsev rule**, so the more stable, more substituted alkene is the major product.

a. → TsOH →

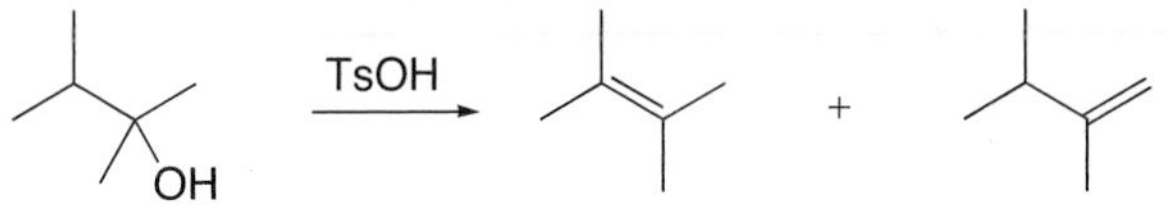

tetrasubstituted
**major product**

disubstituted

b. → TsOH →

CH₂CH₃ + CHCH₃

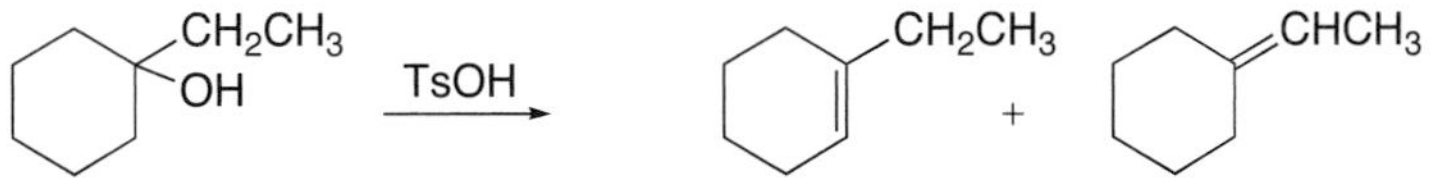

c. → TsOH →

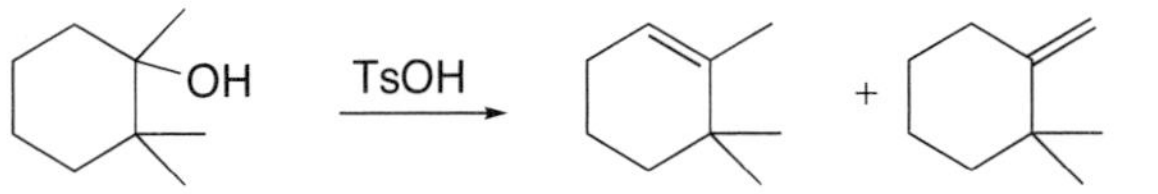

trisubstituted
**major product**

disubstituted

d. $CH_3CH_2CH_2CH_2OH$ → TsOH → $CH_3CH_2CH=CH_2$

e. → TsOH →

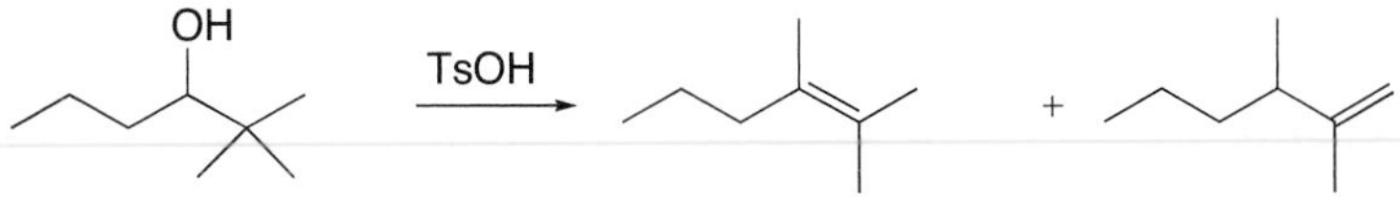

tetrasubstituted
**major product**

disubstituted

Two products formed
by carbocation rearrangement

**9.46 The more stable alkene is the major product.**

→ H₂SO₄ →

trans and disubstituted
**major product**

monosubstituted

cis and disubstituted

**9.47** OTs is a good leaving group and will easily be replaced by a nucleophile. Draw the products by substituting the nucleophile in the reagent for OTs in the starting material.

a. $CH_3CH_2CH_2CH_2{-}OTs \xrightarrow[S_N2]{CH_3SH} CH_3CH_2CH_2CH_2{-}SCH_3 + HOTs$

b. $CH_3CH_2CH_2CH_2{-}OTs \xrightarrow[S_N2]{NaOCH_2CH_3} CH_3CH_2CH_2CH_2{-}OCH_2CH_3 + Na^+ \ ^-OTs$

c. $CH_3CH_2CH_2CH_2{-}OTs \xrightarrow[S_N2]{NaOH} CH_3CH_2CH_2CH_2{-}OH + Na^+ \ ^-OTs$

d. $CH_3CH_2CH_2CH_2{-}OTs \xrightarrow[E2]{K^+ \ ^-OC(CH_3)_3} CH_3CH_2CH{=}CH_2 + (CH_3)_3COH + K^+ \ ^-OTs$

**9.48**

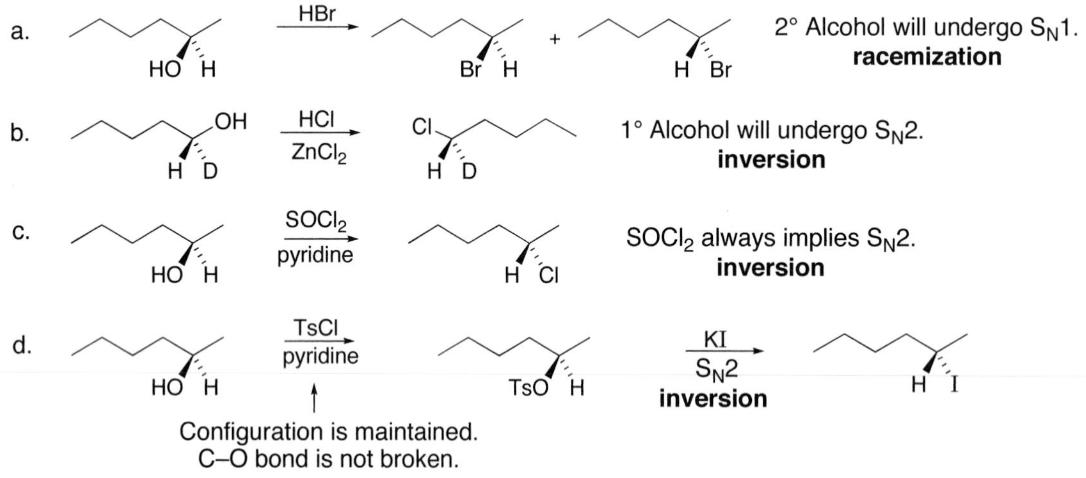

a.
HBr
HO H
Br H
+
H Br
2° Alcohol will undergo $S_N1$.
racemization
b.
OH
H D
HCl
$ZnCl_2$
Cl
H D
1° Alcohol will undergo $S_N2$.
inversion
c.
HO H
$SOCl_2$
pyridine
H Cl
$SOCl_2$ always implies $S_N2$.
inversion
d.
HO H
TsCl
pyridine
TsO H
KI
$S_N2$
inversion
H I
Configuration is maintained.
C–O bond is not broken.

**9.49**

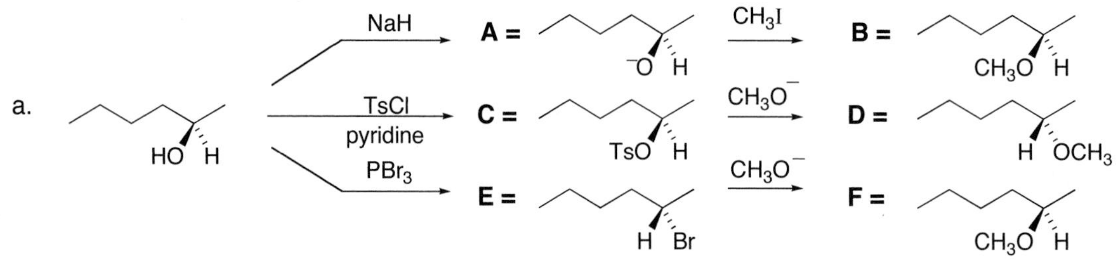

a.
HO H
NaH
A =
$^-$O H
$CH_3I$
B =
$CH_3O$ H
TsCl
pyridine
C =
TsO H
$CH_3O^-$
D =
H $OCH_3$
$PBr_3$
E =
H Br
$CH_3O^-$
F =
$CH_3O$ H

b. **B** and **D** are enantiomers.

c. **B** and **F** are identical.

**9.50** Acid-catalyzed dehydration follows an E1 mechanism for 2° and 3° ROH with an added step to make a good leaving group. The three steps are:

[1] Protonate the oxygen to make a good leaving group.

[2] Break the C–O bond to form a carbocation.

[3] Remove a β hydrogen to form the π bond.

a.

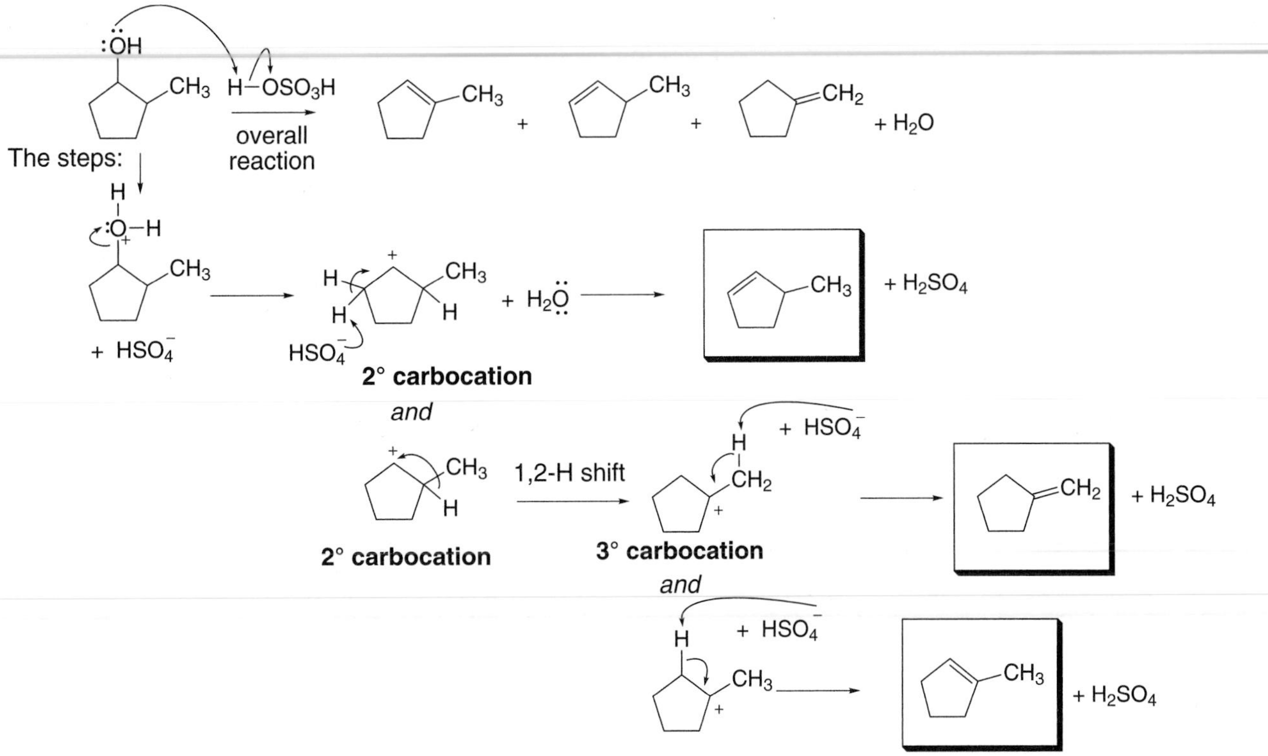

The steps:

+ HSO₄⁻          HSO₄⁻

**2° carbocation**

and

**2° carbocation**          **3° carbocation**

and

b.

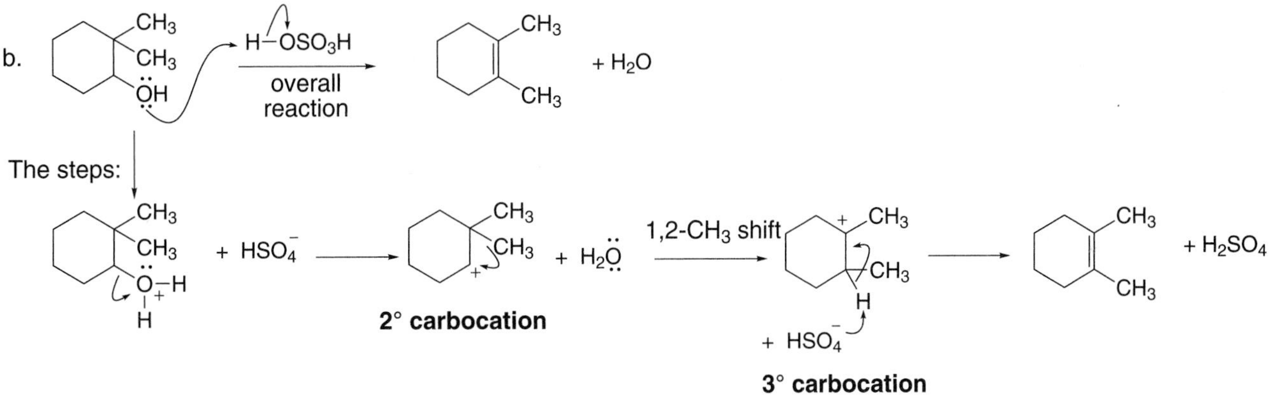

The steps:

**2° carbocation**          **3° carbocation**

**9.51**

a.

**A**          **B**          **C**          **D**

(E1 with acid)    (E2 with base)    (E2 with base)    (E1 with acid)

**3,3-dimethyl-cyclopentene**          major product

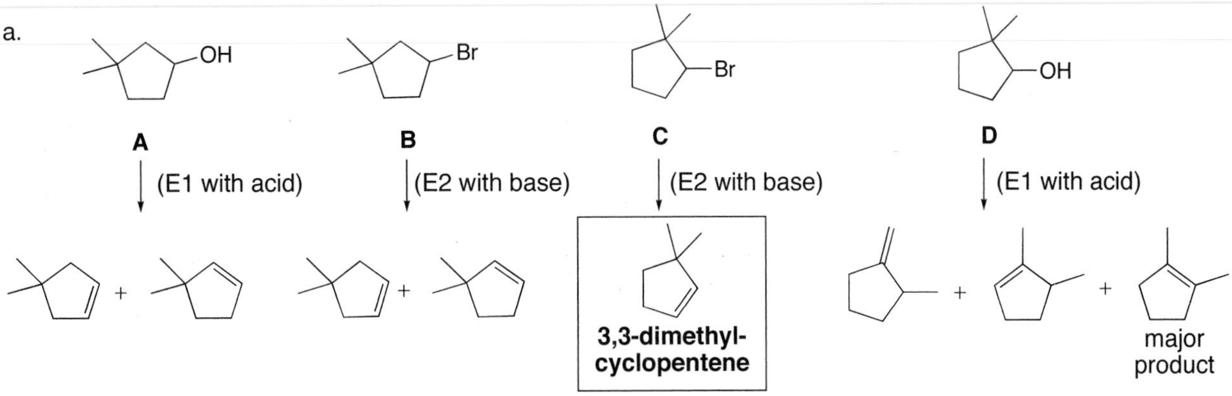

b. The best starting material to form 3,3-dimethylcyclopentene would be **C** since the alkene can be formed as a single product by an E2 elimination with base.

**9.52** To draw the mechanisms:
    [1] Protonate the oxygen to make a good leaving group.
    [2] Break the C–O bond to form a carbocation.
    [3] Look for possible rearrangements to make a more stable carbocation.
    [4] Remove a β hydrogen to form the π bond.

Dark and light circles are meant to show where the carbons in the starting material appear in the product.

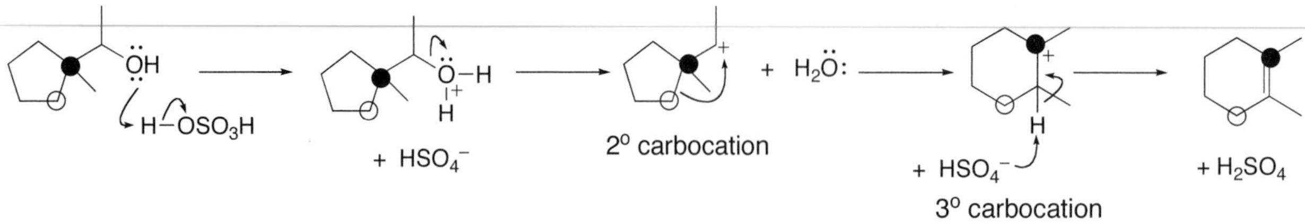

2° carbocation

3° carbocation

**9.53**

a. 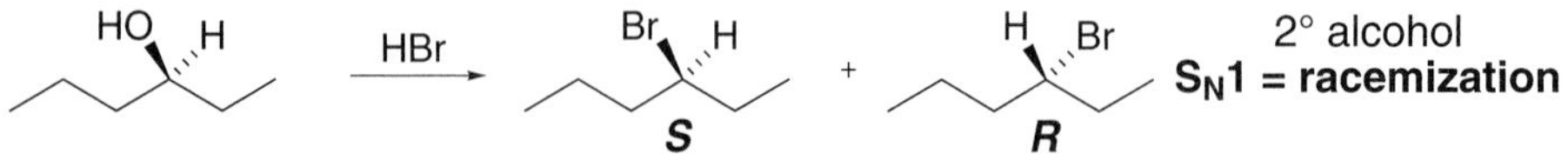  2° alcohol
    S$_N$1 = racemization

b. 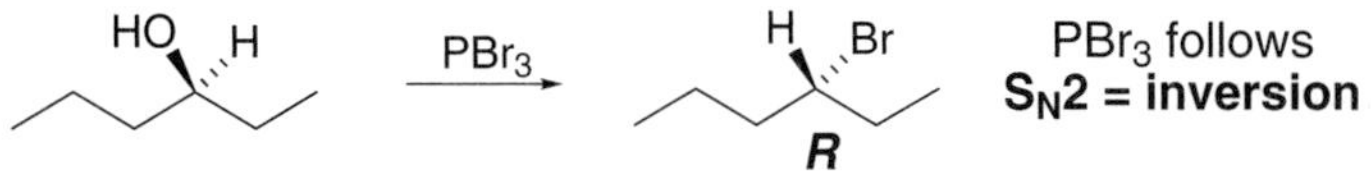  PBr$_3$ follows
    S$_N$2 = inversion

c. 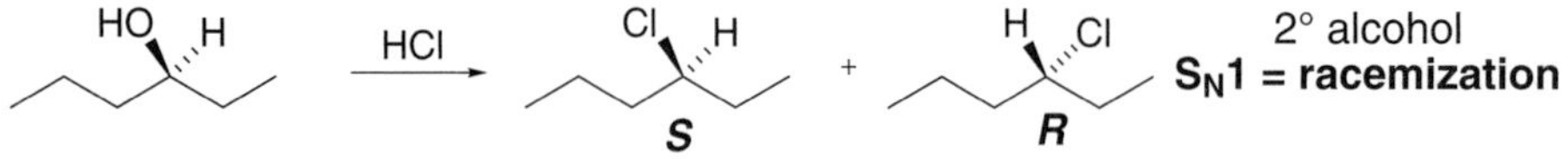  2° alcohol
    S$_N$1 = racemization

d. 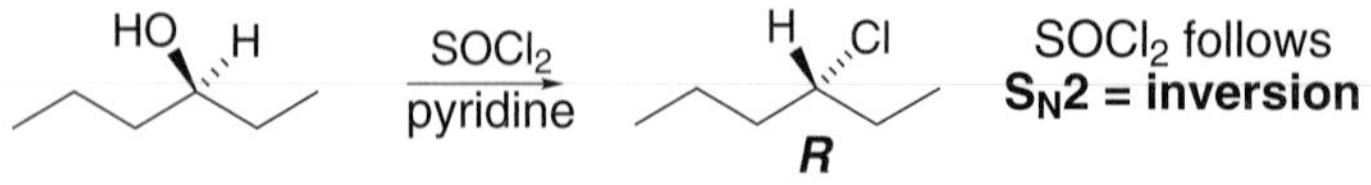  SOCl$_2$ follows
    S$_N$2 = inversion

**9.54**

**3-methyl-2-butanol**

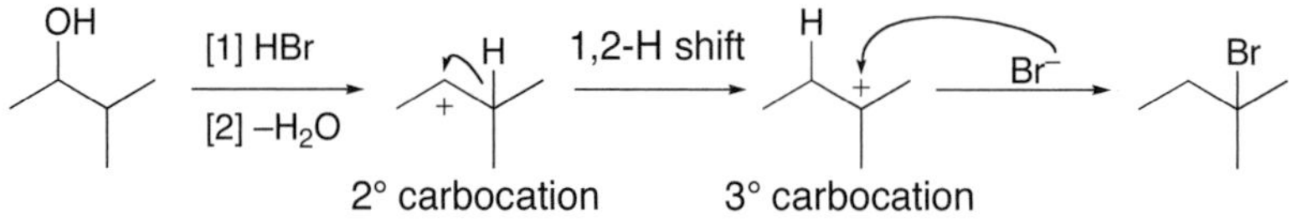

[1] HBr
[2] –H$_2$O

2° carbocation    1,2-H shift    3° carbocation

The 2° alcohol reacts by an S$_N$1 mechanism to form a carbocation that rearranges.

**2-methyl-1-propanol**

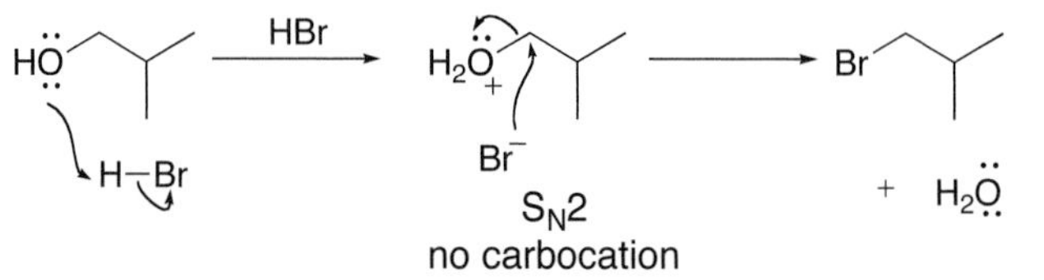

HBr

S$_N$2
no carbocation

The 1° alcohol reacts with HBr by an S$_N$2 mechanism.
**no carbocation intermediate = no rearrangement possible**

**9.55** Conversion of a 1° alcohol into a 1° alkyl chloride occurs by an S$_N$2 mechanism. S$_N$2 mechanisms occur more readily in polar aprotic solvents by making the nucleophile stronger. No added ZnCl$_2$ is necessary.

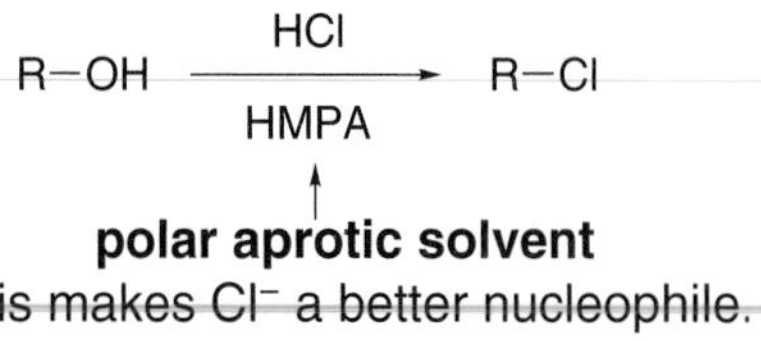

$$R-OH \xrightarrow[\text{HMPA}]{\text{HCl}} R-Cl$$

**polar aprotic solvent**
This makes Cl$^-$ a better nucleophile.

**9.56**

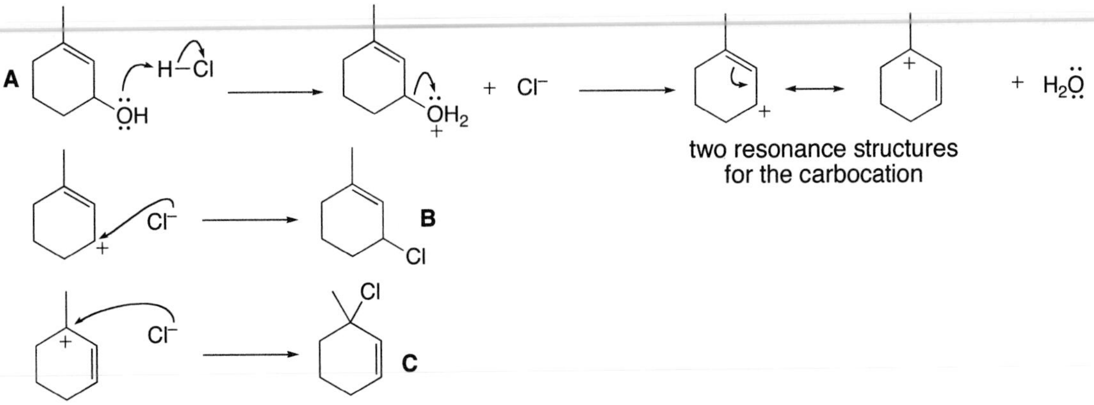

two resonance structures
for the carbocation

**B**

**C**

**9.57**

$$CH_3CH_2CH_2CH_2OH \xrightarrow[\text{overall reaction}]{H_2SO_4, \ NaBr} CH_3CH_2CH_2CH_2Br \ + \ CH_3CH_2CH=CH_2$$

$$+ \ CH_3CH_2CH_2CH_2OCH_2CH_2CH_2CH_3$$

**Step [1] for all products: Formation a good leaving group**

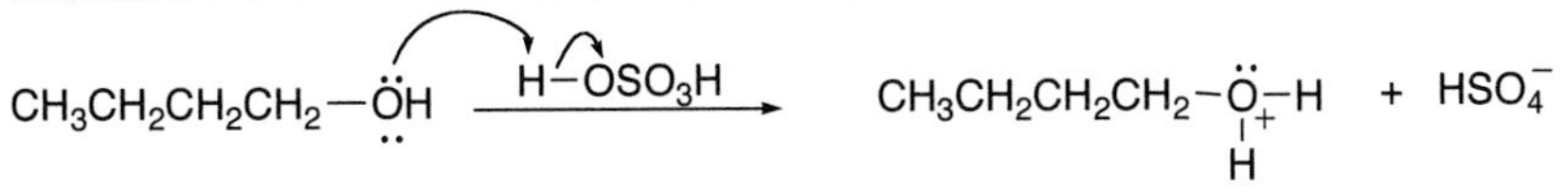

$$CH_3CH_2CH_2CH_2-\overset{..}{\underset{..}{O}}H \xrightarrow{H-OSO_3H} CH_3CH_2CH_2CH_2-\overset{+}{\underset{H}{O}}-H \ + \ HSO_4^-$$

**Formation of CH₃CH₂CH₂CH₂Br:**

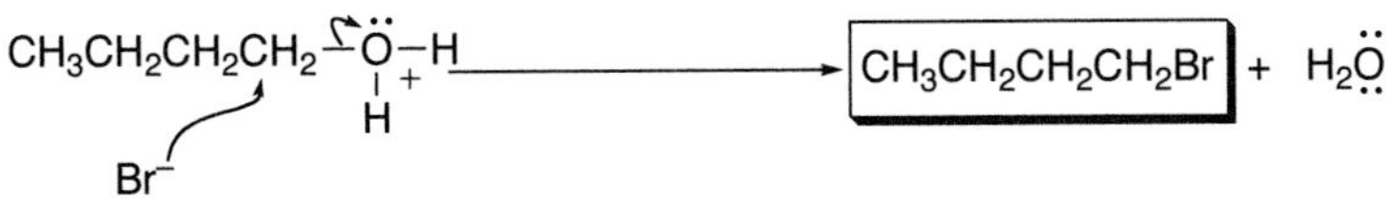

$$CH_3CH_2CH_2CH_2-\overset{+}{\underset{H}{O}}-H \longrightarrow \boxed{CH_3CH_2CH_2CH_2Br} \ + \ H_2\overset{..}{\underset{..}{O}}$$

$$Br^-$$

**Formation of CH₃CH₂CH=CH₂:**

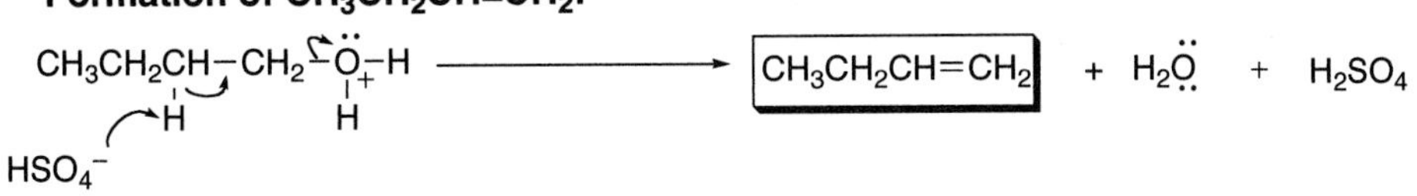

$$CH_3CH_2CH-CH_2-\overset{+}{\underset{H}{O}}-H \longrightarrow \boxed{CH_3CH_2CH=CH_2} \ + \ H_2\overset{..}{\underset{..}{O}} \ + \ H_2SO_4$$

$$HSO_4^-$$

**Ether forms (from the protonated alcohol):**

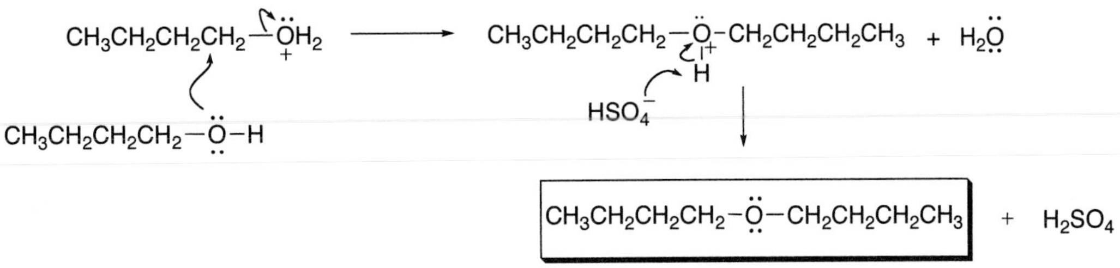

$$CH_3CH_2CH_2CH_2-\overset{+}{O}H_2 \longrightarrow CH_3CH_2CH_2CH_2-\overset{+}{\underset{H}{O}}-CH_2CH_2CH_2CH_3 \ + \ H_2\overset{..}{\underset{..}{O}}$$

$$CH_3CH_2CH_2CH_2-\overset{..}{\underset{..}{O}}-H$$

$$HSO_4^-$$

$$\boxed{CH_3CH_2CH_2CH_2-\overset{..}{\underset{..}{O}}-CH_2CH_2CH_2CH_3} \ + \ H_2SO_4$$

**9.58**

a.

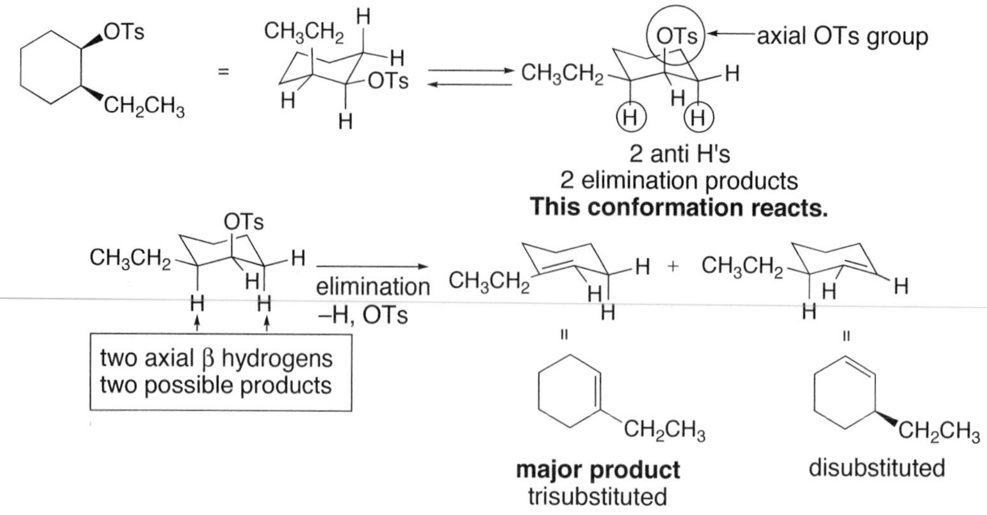

CH₃CH₂ ⎯ [structure] ⎯ ← axial OTs group

2 anti H's
2 elimination products
**This conformation reacts.**

elimination
−H, OTs

two axial β hydrogens
two possible products

CH₃CH₂ ⎯ [structure] + CH₃CH₂ ⎯ [structure]

**major product**
trisubstituted

disubstituted

b.

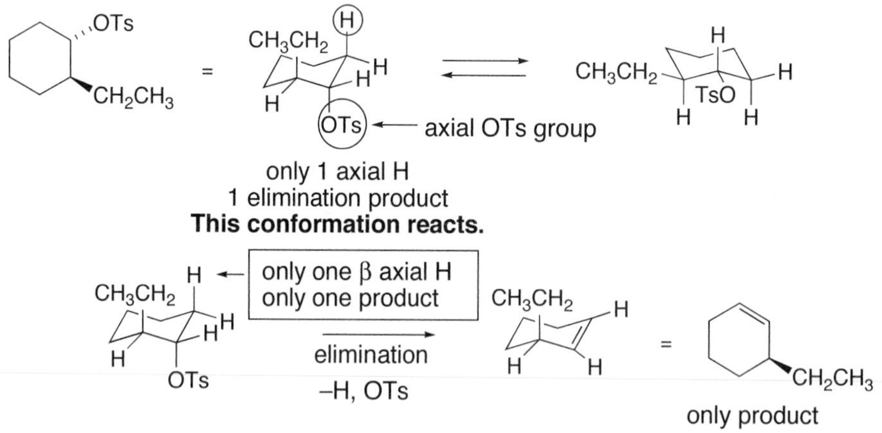

← axial OTs group

only 1 axial H
1 elimination product
**This conformation reacts.**

only one β axial H
only one product

elimination
−H, OTs

= [structure] CH₂CH₃

only product

**9.59**

a.

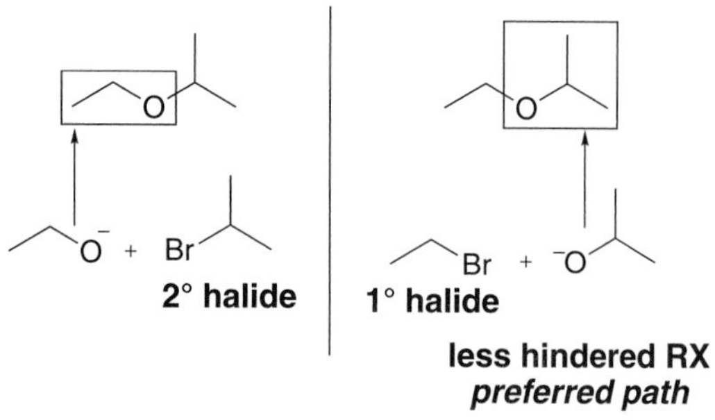

2° halide   1° halide

less hindered RX
*preferred path*

c.

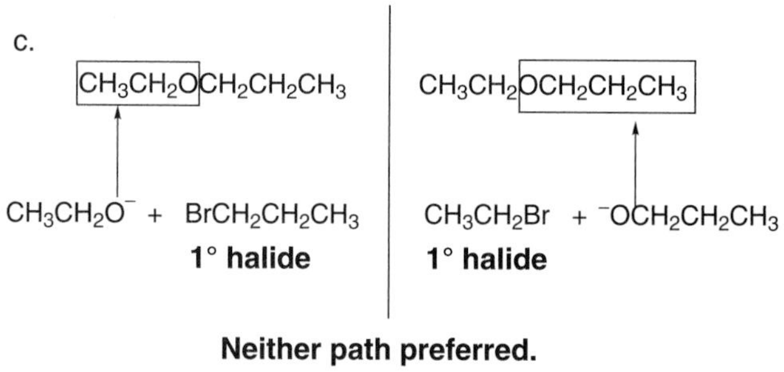

CH₃CH₂OCH₂CH₂CH₃    CH₃CH₂OCH₂CH₂CH₃

CH₃CH₂O⁻ + BrCH₂CH₂CH₃    CH₃CH₂Br + ⁻OCH₂CH₂CH₃
1° halide    1° halide

**Neither path preferred.**

b.

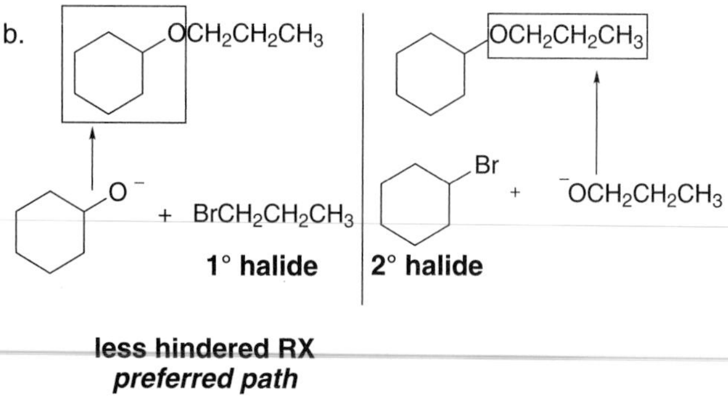

OCH₂CH₂CH₃    OCH₂CH₂CH₃

+ BrCH₂CH₂CH₃    + ⁻OCH₂CH₂CH₃
1° halide    2° halide

less hindered RX
*preferred path*

**9.60** A tertiary halide is too hindered and an aryl halide too unreactive to undergo a Williamson ether synthesis.

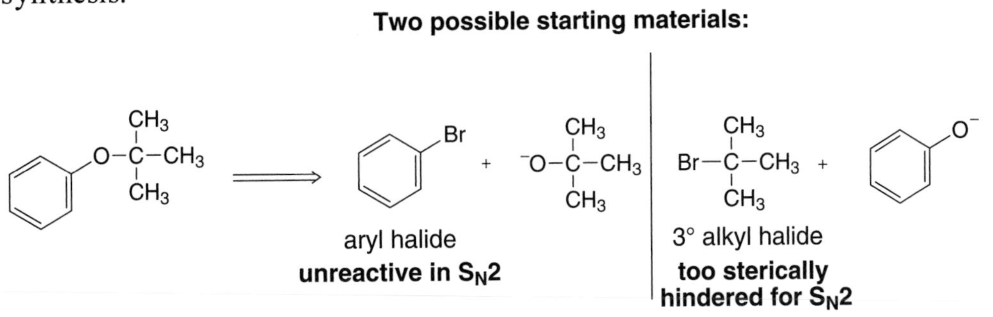

**Two possible starting materials:**

aryl halide
**unreactive in S_N2**

3° alkyl halide
**too sterically hindered for S_N2**

**9.61**

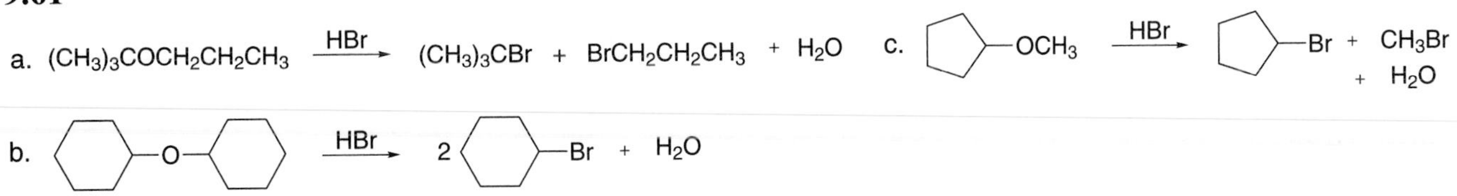

a. $(CH_3)_3COCH_2CH_2CH_3 \xrightarrow{HBr} (CH_3)_3CBr + BrCH_2CH_2CH_3 + H_2O$

b. $\xrightarrow{HBr}$ 2 —Br + $H_2O$

c. —OCH_3 $\xrightarrow{HBr}$ —Br + CH_3Br + H_2O

**9.62**

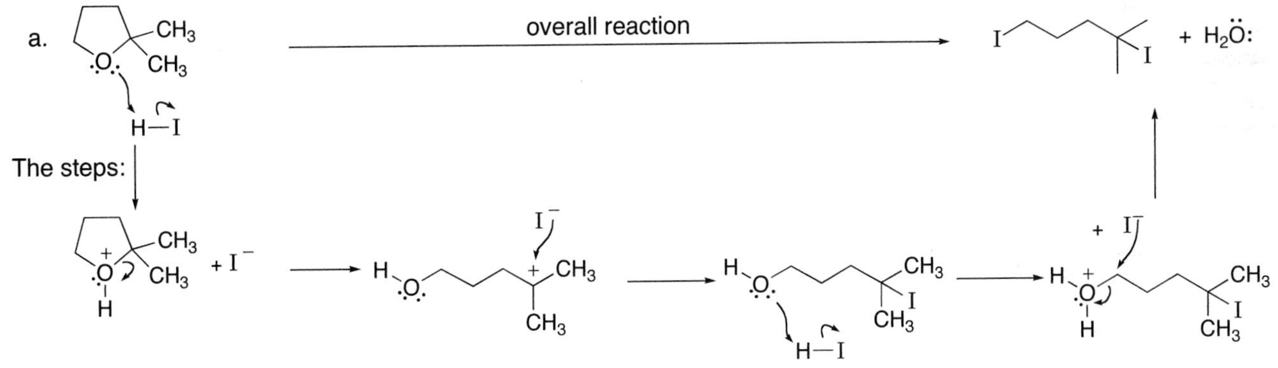

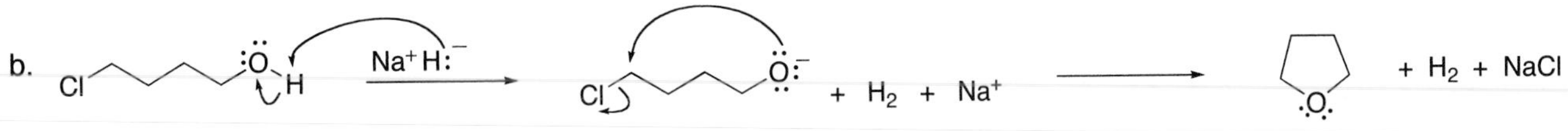

a.

overall reaction

The steps:

b.

**9.63**

a. $\xrightarrow{HBr}$ BrCH_2CH_2OH

b. $\xrightarrow[H_2SO_4]{H_2O}$ HOCH_2CH_2OH

c. $\xrightarrow[\text{[2] }H_2O]{\text{[1] }CH_3CH_2O^-}$ CH_3CH_2OCH_2CH_2OH

d. $\xrightarrow[\text{[2] }H_2O]{\text{[1] }HC\equiv C^-}$ HC$\equiv$C—CH_2—CH_2OH

e. $\xrightarrow[\text{[2] }H_2O]{\text{[1] }^-OH}$ HOCH_2CH_2OH

f. $\xrightarrow[\text{[2] }H_2O]{\text{[1] }CH_3S^-}$ CH_3SCH_2CH_2OH

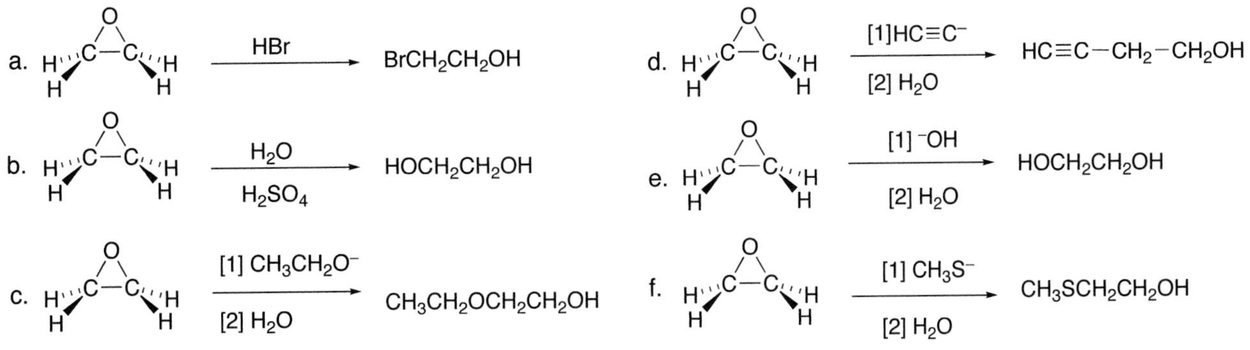

**9.64**

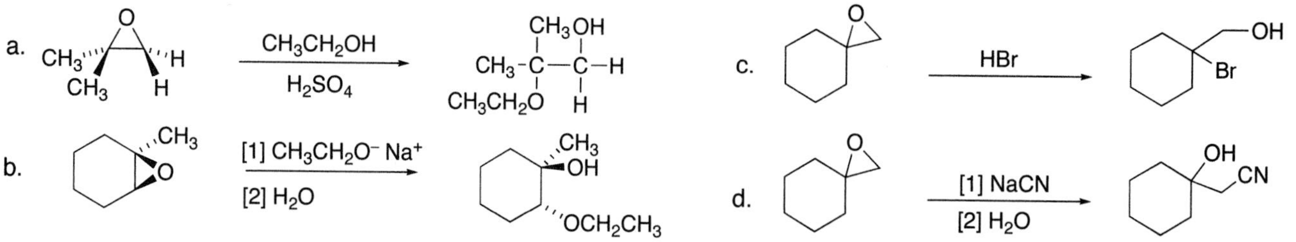

a.

CH$_3$CH$_2$OH / H$_2$SO$_4$

c. HBr

b. [1] CH$_3$CH$_2$O$^-$ Na$^+$ [2] H$_2$O

d. [1] NaCN [2] H$_2$O

**9.65**

a.

Na$^+$ H:$^-$

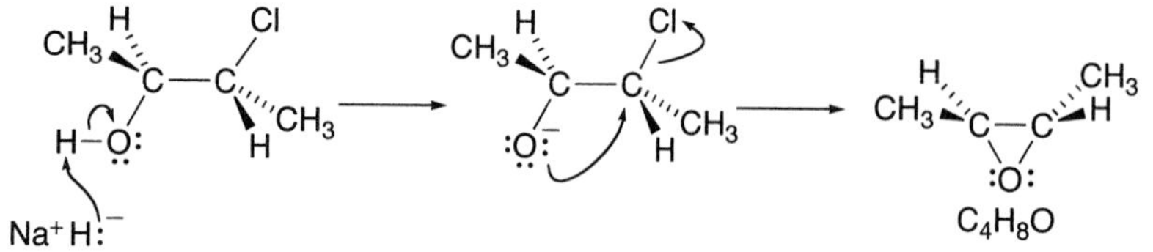

The 2 CH$_3$ groups are anti in the starting material, making them trans in the product.

C$_4$H$_8$O

b.

Na$^+$ H:$^-$

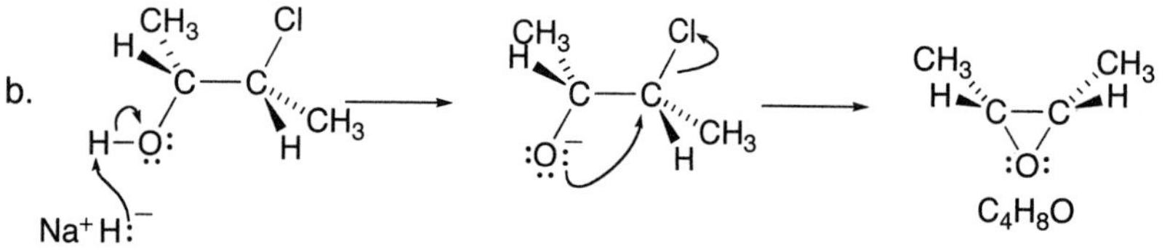

The 2 CH$_3$ groups are gauche in the starting material, making them cis in the product.

C$_4$H$_8$O

**9.66**

CH$_3$CH$_2$OCH$_2$ + Cl$^-$

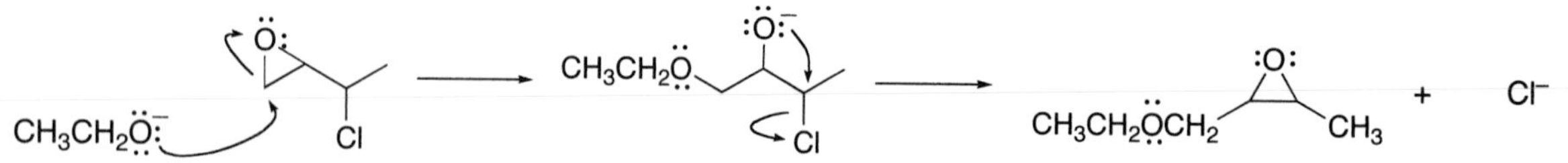

**9.67**

a. KOC(CH$_3$)$_3$

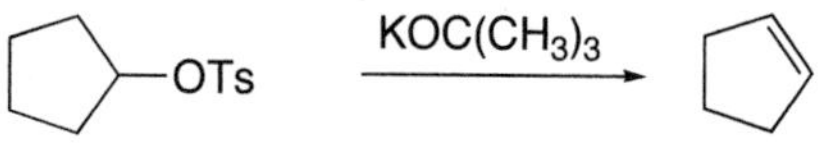

Bulky base favors E2.

b. HBr

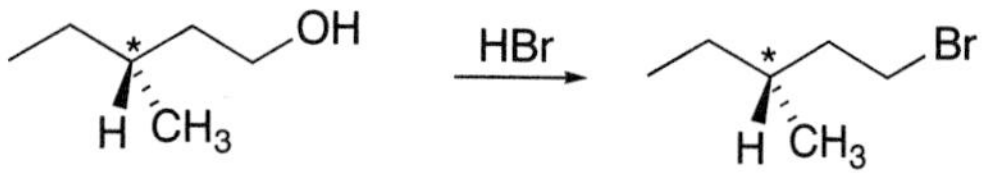

Keep the stereochemistry at the stereogenic center [*] the same here since no bond is broken to it.

c. [1] $^-$CN [2] H$_2$O

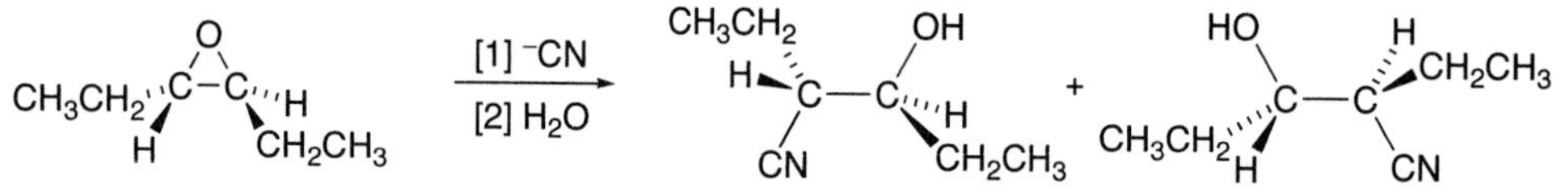

identical

d. KCN / S$_N$2 inversion

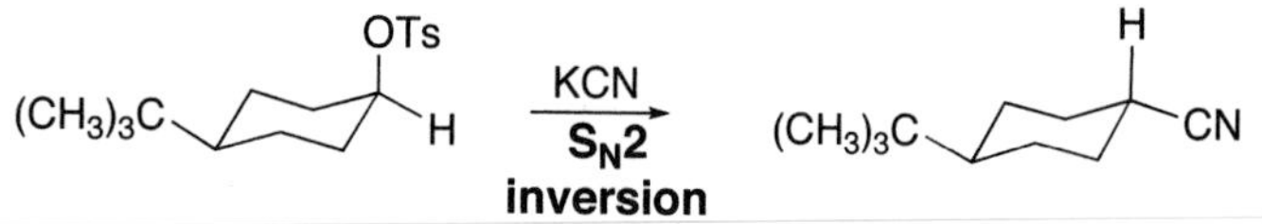

e. PBr$_3$

S$_N$2 inversion

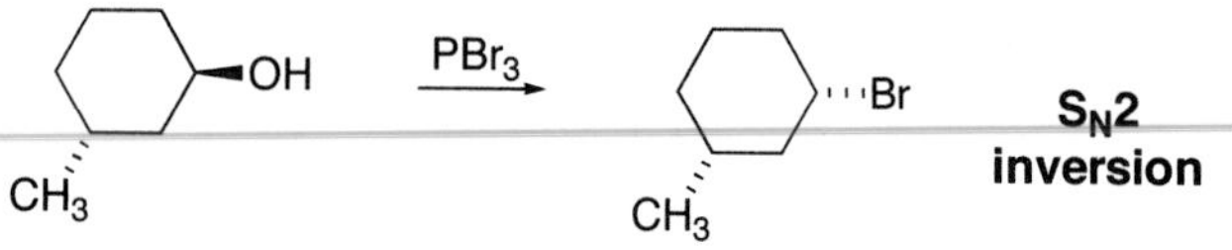

f.

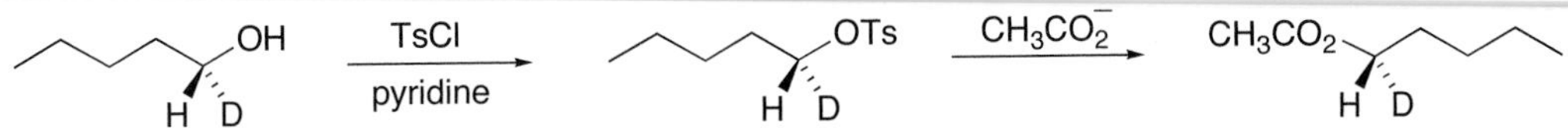

g.

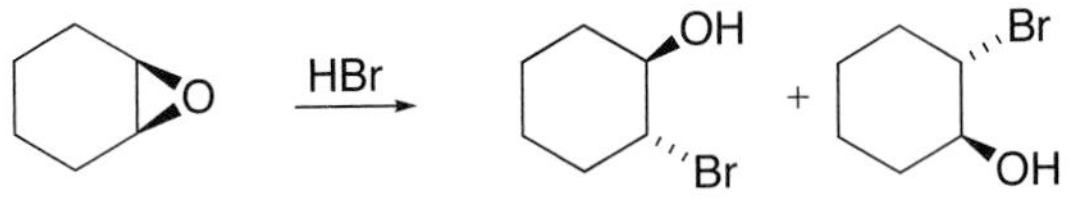

h.

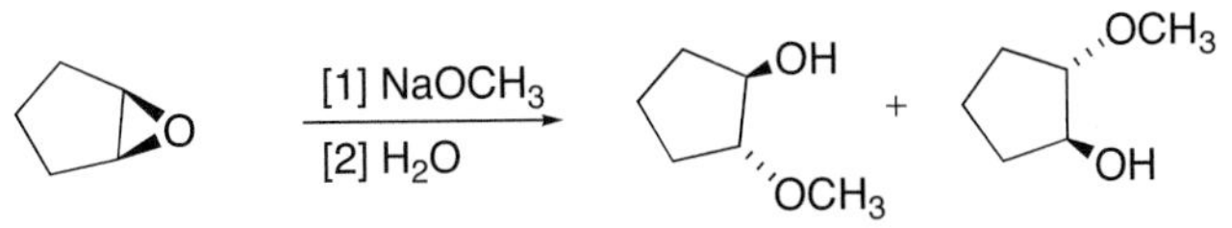

i.

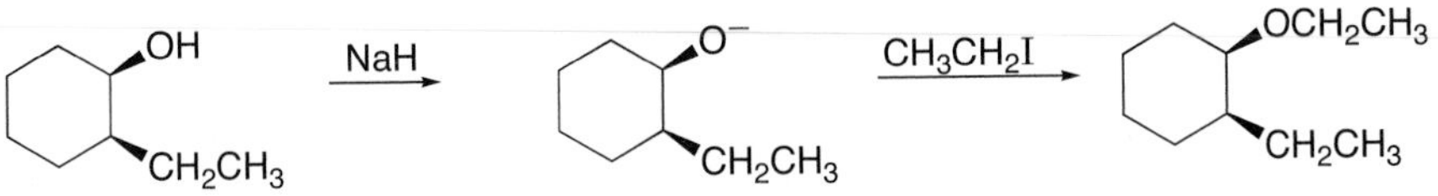

j.    $CH_3CH_2-\overset{\overset{\displaystyle CH_3}{|}}{\underset{\underset{\displaystyle CH_3}{|}}{C}}-O-CH_3 \xrightarrow{\ HI\ } CH_3CH_2-\overset{\overset{\displaystyle CH_3}{|}}{\underset{\underset{\displaystyle CH_3}{|}}{C}}-I \ + \ I-CH_3 \ + \ H_2O$

## 9.68

a.

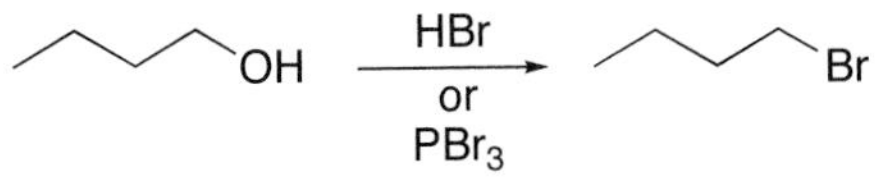

b.

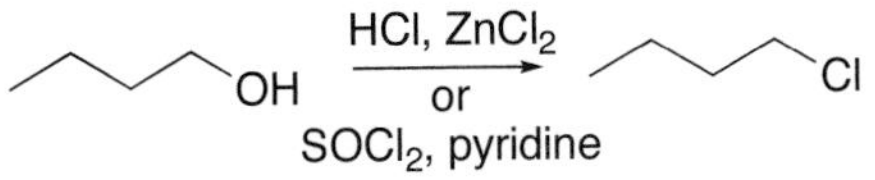

c.

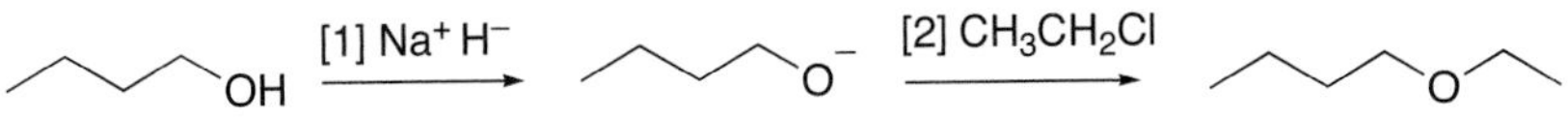

d.

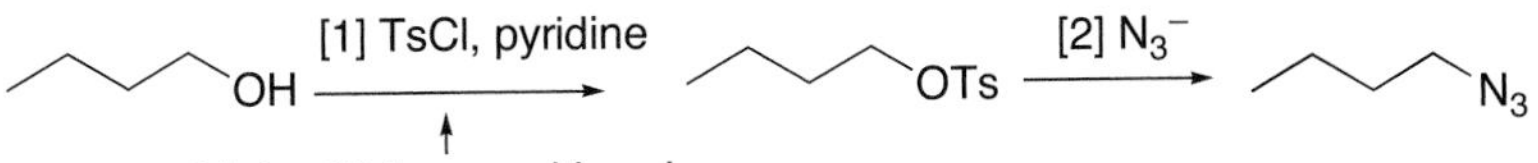

Make OH a good leaving
group (use TsCl), then add N$_3$⁻.

## 9.69

a.

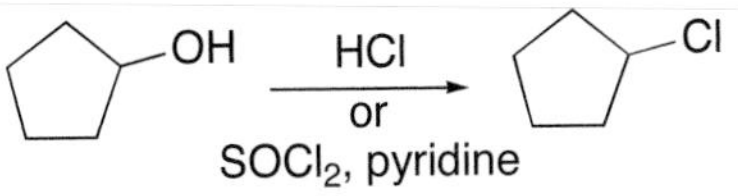

b.

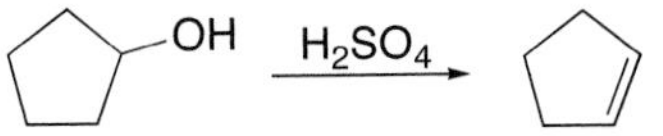

c.

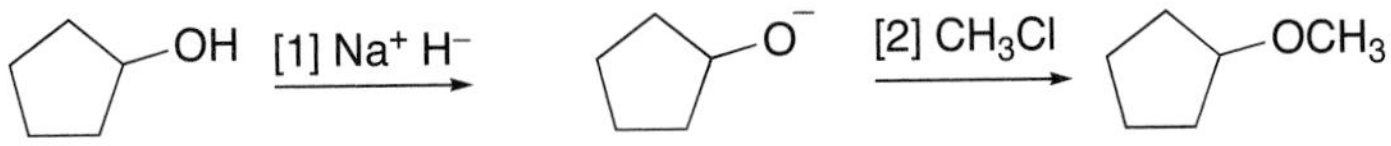

d.

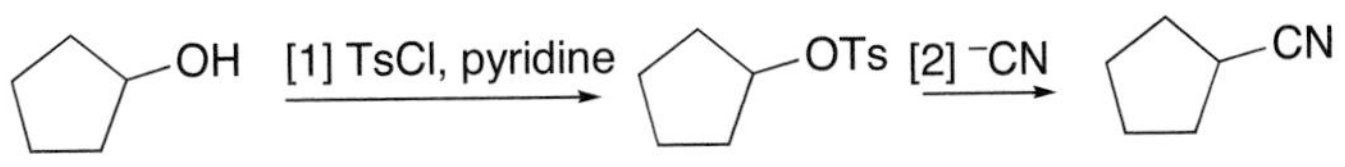

Make OH a good
leaving group (use TsCl),
then add ⁻CN.

**9.70**

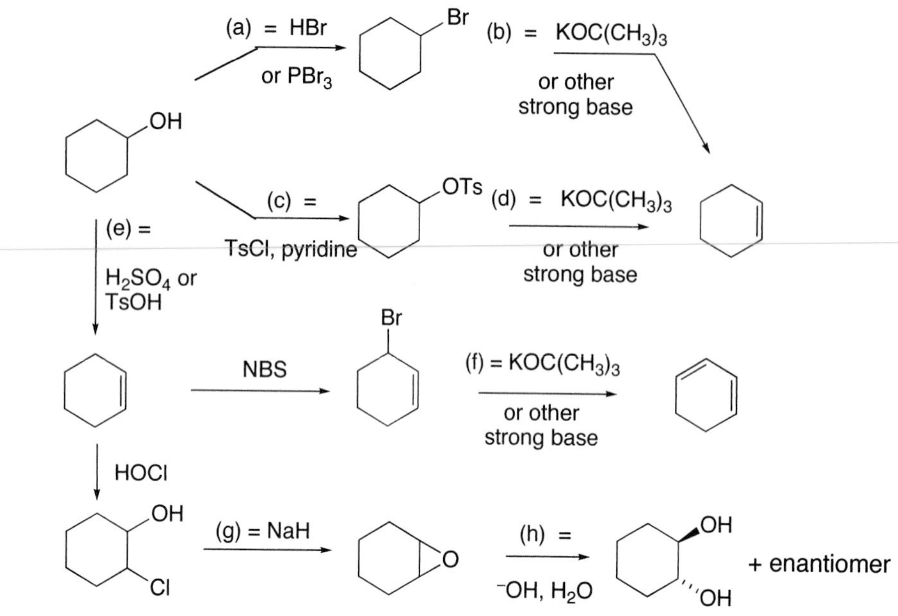

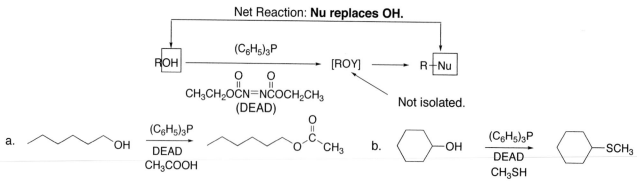

**9.71**

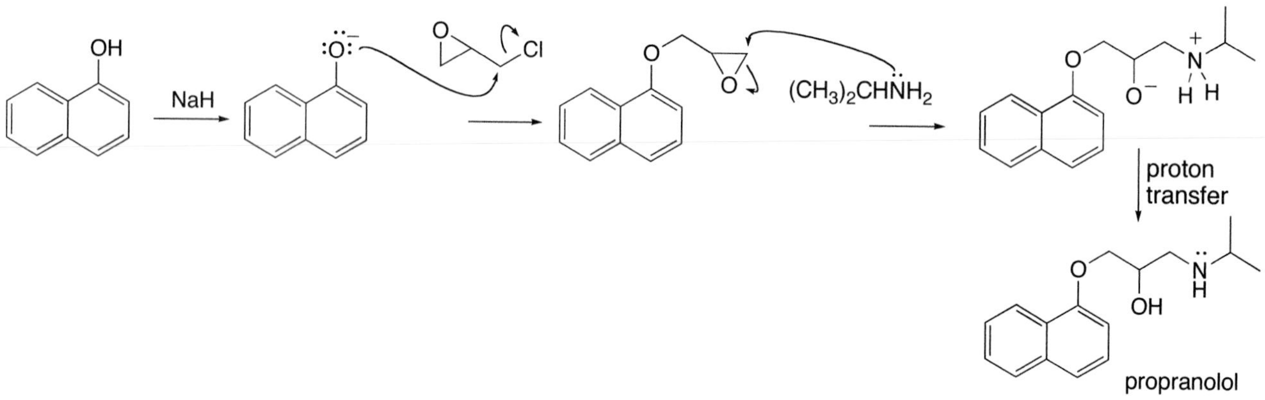

**9.72**

Net Reaction: **Nu replaces OH.**

**9.73**

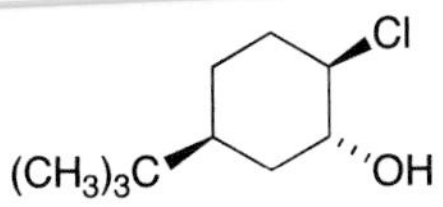

A

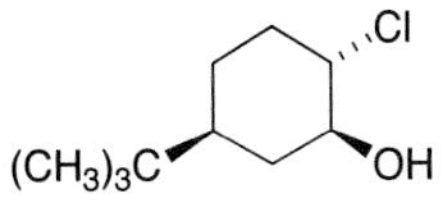

B

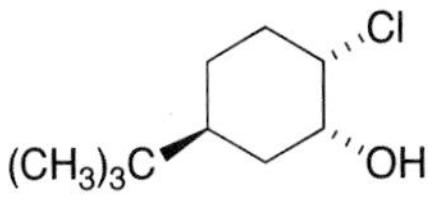

C

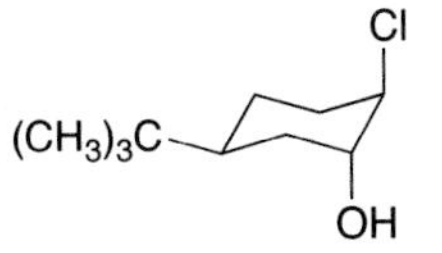

**rapid reaction**
This isomer can have the OH and Cl in the trans diaxial positions, while the larger group C(CH₃)₃ can be in the favorable equatorial position.

OH is in a favorable arrangement for backside attack of the nucleophile on the leaving group.

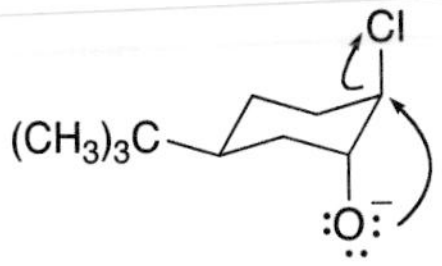

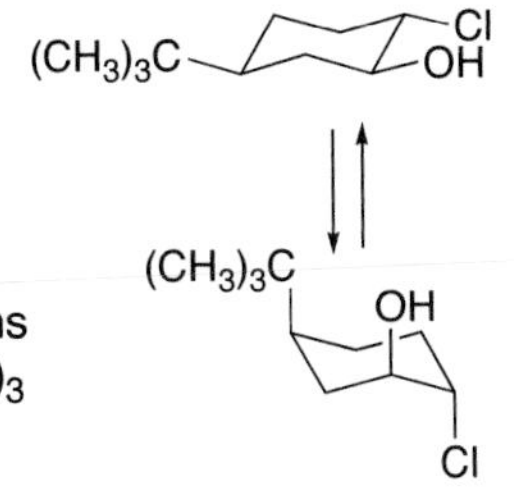

**intermediate reactivity**
This isomer can have the OH and Cl in the trans diaxial positions. But the larger group C(CH₃)₃ must be in the unfavorable axial position, making this reaction slower.

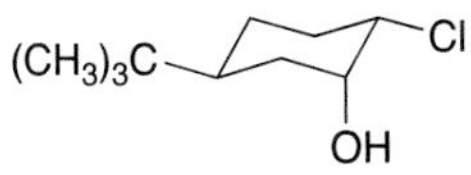

**no reaction**
This isomer has the OH and Cl in a cis position, prohibiting epoxide formation. The OH must be able to approach from the back side.

**9.74**

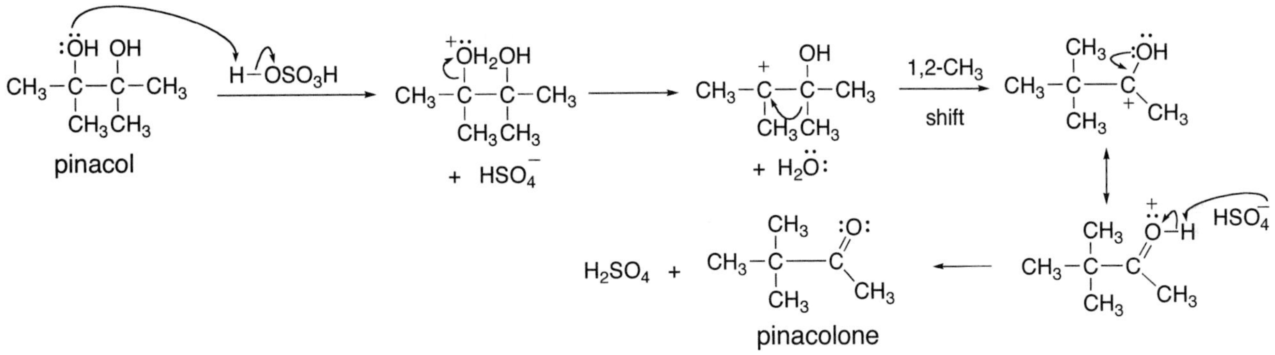

pinacol

+ HSO₄⁻

+ H₂Ö:

1,2-CH₃ shift

H₂SO₄ +

pinacolone

**9.75**

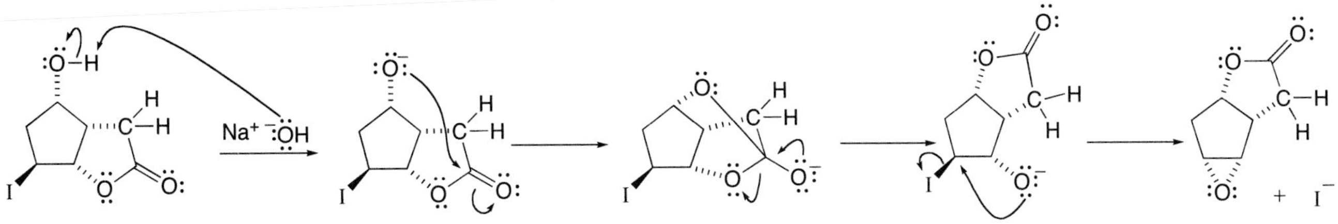

**9.76** You must draw the product that places the nucleophile and leaving group (O⁻) trans and diaxial.

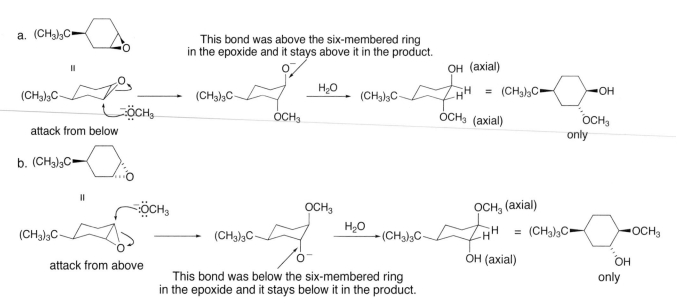

a. (CH₃)₃C—

This bond was above the six-membered ring
in the epoxide and it stays above it in the product.

(CH₃)₃C

attack from below

(CH₃)₃C

OCH₃

H₂O

(CH₃)₃C

OH (axial)
H
OCH₃ (axial)

= (CH₃)₃C

OH
OCH₃
only

b. (CH₃)₃C—

(CH₃)₃C

attack from above

This bond was below the six-membered ring
in the epoxide and it stays below it in the product.

(CH₃)₃C

OCH₃
O⁻

H₂O

(CH₃)₃C

OCH₃ (axial)
H
OH (axial)

= (CH₃)₃C

OCH₃
OH
only

The nucleophile must always approach by backside attack; i.e., if the epoxide is drawn "up" it must attack from below. Even though both ends of the epoxide are equally substituted, nucleophilic attack occurs at only one C–O bond, the one that gives trans diaxial products, as drawn.

## Chapter 10: Alkenes

### ♦ General facts about alkenes

- Alkenes contain a carbon–carbon double bond consisting of a stronger σ bond and a weaker π bond. Each carbon is $sp^2$ hybridized and trigonal planar (10.1).
- Alkenes are named using the suffix -**ene** (10.3).
- Alkenes with different groups on each end of the double bond exist as a pair of diastereomers, identified by the prefixes $E$ and $Z$ (10.3B).

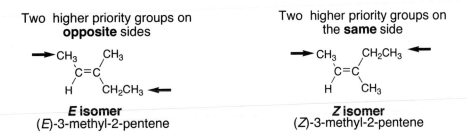

Two higher priority groups on **opposite** sides

**E isomer**
(E)-3-methyl-2-pentene

Two higher priority groups on the **same** side

**Z isomer**
(Z)-3-methyl-2-pentene

- Alkenes have weak intermolecular forces, giving them low mp's and bp's, and making them water insoluble. A cis alkene is more polar than a trans alkene, giving it a slightly higher boiling point (10.4).

**more polar isomer** →

cis-2-butene

a small net dipole
**higher bp**

trans-2-butene

← **less polar isomer**

**no net dipole**
**lower bp**

- Since a π bond is electron rich and much weaker than a σ bond, alkenes undergo addition reactions with electrophiles (10.8).

### ♦ Stereochemistry of alkene addition reactions (10.8)

A reagent XY adds to a double bond in one of three different ways:

- **Syn addition**—X and Y add from the same side.

  $$\text{C=C} \xrightarrow{\text{H−BH}_2} \text{C−C}$$

  - Syn addition occurs in **hydroboration.**

- **Anti addition**—X and Y add from opposite sides.

  $$\text{C=C} \xrightarrow[\substack{\text{or} \\ X_2, H_2O}]{X_2} \text{C−C}$$

  - Anti addition occurs in **halogenation** and **halohydrin formation.**

- **Both syn and anti addition** occur when carbocations are intermediates.

  $$\text{C=C} \xrightarrow[\substack{\text{or} \\ H_2O, H^+}]{H−X} \text{C−C} \quad \text{and} \quad \text{C−C}$$

  - Syn and anti addition occur in **hydrohalogenation** and **hydration.**

♦ Addition reactions of alkenes

## [1] Hydrohalogenation—Addition of HX (X = Cl, Br, I) (10.9–10.11)

$RCH{=}CH_2$ + H—X $\longrightarrow$

R—CH—CH$_2$
| |
X H
alkyl halide

- The mechanism has two steps.
- Carbocations are formed as intermediates.
- Carbocation rearrangements are possible.
- Markovnikov's rule is followed. H bonds to the less substituted C to form the more stable carbocation.
- Syn and anti addition occur.

## [2] Hydration and related reactions—Addition of $H_2O$ or ROH (10.12)

$RCH{=}CH_2$ + H—OH $\xrightarrow{H_2SO_4}$

R—CH—CH$_2$
| |
OH H
alcohol

$RCH{=}CH_2$ + H—OR $\xrightarrow{H_2SO_4}$

R—CH—CH$_2$
| |
OR H
ether

For both reactions:
- The mechanism has three steps.
- Carbocations are formed as intermediates.
- Carbocation rearrangements are possible.
- Markovnikov's rule is followed. H bonds to the less substituted C to form the more stable carbocation.
- Syn and anti addition occur.

## [3] Halogenation—Addition of $X_2$ (X = Cl or Br) (10.13–10.14)

$RCH{=}CH_2$ + X—X $\longrightarrow$

R—CH—CH$_2$
| |
X X
vicinal dihalide

- The mechanism has two steps.
- Bridged halonium ions are formed as intermediates.
- No rearrangements occur.
- Anti addition occurs.

## [4] Halohydrin formation—Addition of OH and X (X = Cl, Br) (10.15)

$RCH{=}CH_2$ + X—X $\xrightarrow{H_2O}$

R—CH—CH$_2$
| |
OH X
halohydrin

- The mechanism has three steps.
- Bridged halonium ions are formed as intermediates.
- No rearrangements occur.
- X bonds to the less substituted C.
- Anti addition occurs.
- NBS in DMSO and $H_2O$ adds Br and OH in the same fashion.

## [5] Hydroboration–oxidation—Addition of $H_2O$ (10.16)

$RCH{=}CH_2$ $\xrightarrow[\text{[2] } H_2O_2, \text{ HO}^-]{\text{[1] } BH_3 \text{ or 9-BBN}}$

R—CH—CH$_2$
| |
H OH
alcohol

- Hydroboration has a one-step mechanism.
- No rearrangements occur.
- OH bonds to the less substituted C.
- Syn addition of $H_2O$ results.

**10.1**

Six alkenes of molecular formula $C_5H_{10}$:

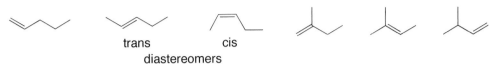

trans          cis

diastereomers

## 10.2 To determine the number of degrees of unsaturation:

[1] Calculate the maximum number of H's $(2n + 2)$.

[2] Subtract the actual number of H's from the maximum number.

[3] Divide by two.

a. $C_2H_2$

[1] maximum number of H's = $2n + 2 = 2(2) + 2 = 6$

[2] subtract actual from maximum = $6 - 2 = 4$

[3] divide by two = $4/2$ = **2 degrees of unsaturation**

b. $C_6H_6$

[1] maximum number of H's = $2n + 2 = 2(6) + 2 = 14$

[2] subtract actual from maximum = $14 - 6 = 8$

[3] divide by two = $8/2$ = **4 degrees of unsaturation**

c. $C_8H_{18}$

[1] maximum number of H's = $2n + 2 = 2(8) + 2 = 18$

[2] subtract actual from maximum = $18 - 18 = 0$

[3] divide by two = $0/2$ = **0 degrees of unsaturation**

d. $C_7H_8O$

Ignore the O.

[1] maximum number of H's = $2n + 2 = 2(7) + 2 = 16$

[2] subtract actual from maximum = $16 - 8 = 8$

[3] divide by two = $8/2$ = **4 degrees of unsaturation**

e. $C_7H_{11}Br$

Because of Br, add one more H (11 + 1 H = 12 H's).

[1] maximum number of H's = $2n + 2 = 2(7) + 2 = 16$

[2] subtract actual from maximum = $16 - 12 = 4$

[3] divide by two = $4/2$ = **2 degrees of unsaturation**

f. $C_5H_9N$

Because of N, subtract one H (9 − 1 H = 8 H's).

[1] maximum number of H's = $2n + 2 = 2(5) + 2 = 12$

[2] subtract actual from maximum = $12 - 8 = 4$

[3] divide by two = $4/2$ = **2 degrees of unsaturation**

**10.3** First determine the number of degrees of unsaturation in the compound. Then decide which combinations of rings and π bonds could exist.

$C_{10}H_{14}$

[1] maximum number of H's = $2n + 2 = 2(10) + 2 = 22$

[2] subtract actual from maximum = $22 - 14 = 8$

[3] divide by two = $8/2$ = **4 degrees of unsaturation**

possibilities:

4 π bonds

3 π bonds + 1 ring

2 π bonds + 2 rings

1 π bond + 3 rings

4 rings

## 10.4 To name an alkene:

[1] Find the longest chain that contains the double bond. Change the ending from *-ane* to *-ene*.

[2] Number the chain to give the double bond the lower number. The alkene is named by the first number.

[3] Apply all other rules of nomenclature.

### To name a cycloalkene:

[1] When a double bond is located in a ring, it is always located between C1 and C2. Omit the "1" in the name. Change the ending from *-ane* to *-ene*.

[2] Number the ring clockwise or counterclockwise to give the first substituent the lower number.

[3] Apply all other rules of nomenclature.

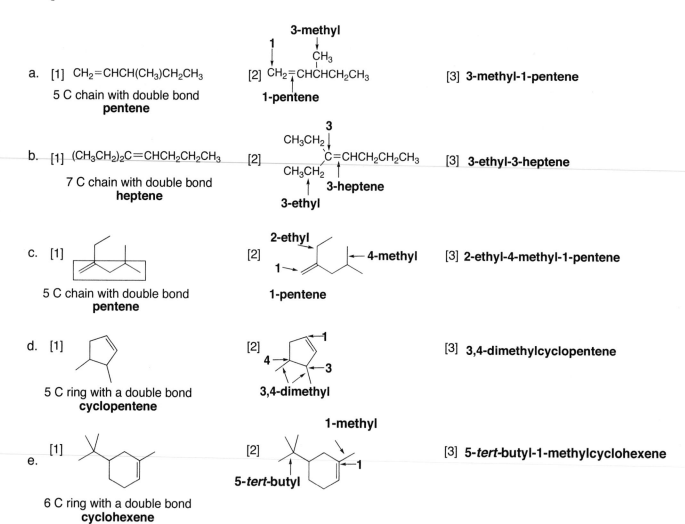

a. [1] CH$_2$=CHCH(CH$_3$)CH$_2$CH$_3$

5 C chain with double bond
**pentene**

[2] **3-methyl**
CH$_3$
CH$_2$=CHCHCH$_2$CH$_3$
**1-pentene**

[3] **3-methyl-1-pentene**

b. [1] (CH$_3$CH$_2$)$_2$C=CHCH$_2$CH$_2$CH$_3$

7 C chain with double bond
**heptene**

[2] **3**
CH$_3$CH$_2$
C=CHCH$_2$CH$_2$CH$_3$
CH$_3$CH$_2$   **3-heptene**
**3-ethyl**

[3] **3-ethyl-3-heptene**

c. [1]

5 C chain with double bond
**pentene**

[2] **2-ethyl**
1
**4-methyl**
**1-pentene**

[3] **2-ethyl-4-methyl-1-pentene**

d. [1]

5 C ring with a double bond
**cyclopentene**

[2] 1
4   3
**3,4-dimethyl**

[3] **3,4-dimethylcyclopentene**

e. [1]

6 C ring with a double bond
**cyclohexene**

[2] **1-methyl**
5-**tert**-butyl   1

[3] **5-tert-butyl-1-methylcyclohexene**

**10.5** Use the rules from Answer 10.4 to name the compounds. Enols are named to give the OH the lower number. Compounds with two C=C's are named with the suffix *–adiene*.

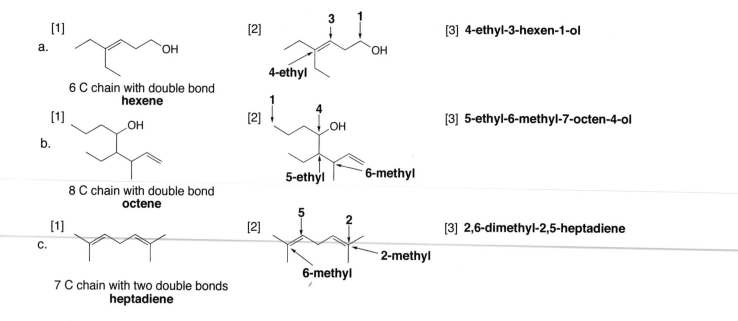

a. [1]
OH

6 C chain with double bond
**hexene**

[2] 3   1
OH
**4-ethyl**

[3] **4-ethyl-3-hexen-1-ol**

b. [1]
OH

8 C chain with double bond
**octene**

[2] 1   4
OH
**5-ethyl**   **6-methyl**

[3] **5-ethyl-6-methyl-7-octen-4-ol**

c. [1]

7 C chain with two double bonds
**heptadiene**

[2] 5   2
**6-methyl**   **2-methyl**

[3] **2,6-dimethyl-2,5-heptadiene**

**10.6** To label an alkene as *E* or *Z*:

    [1] **Assign priorities** to the two substituents *on each end* using the rules for *R,S* nomenclature.

    [2] **Assign *E* or *Z*** depending on the location of the two higher priority groups.

- The *E* prefix is used when the two higher priority groups are on **opposite sides**.
- The *Z* prefix is used when the two higher priority groups are on the **same side** of the double bond.

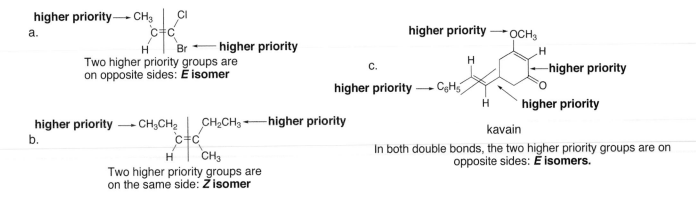

**10.7** **To name an alkene:** First follow the rules from Answer 10.4. Then, when necessary, assign an *E* or *Z* prefix based on priority, as in 10.6.

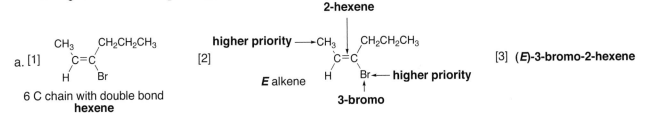

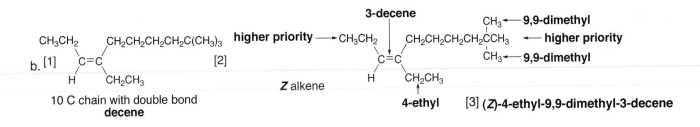

**10.8 To work backwards from a name to a structure:**
   [1] Find the parent name and functional group and draw, remembering that the double bond is between C1 and C2 for cycloalkenes.
   [2] Add the substituents to the appropriate carbon.

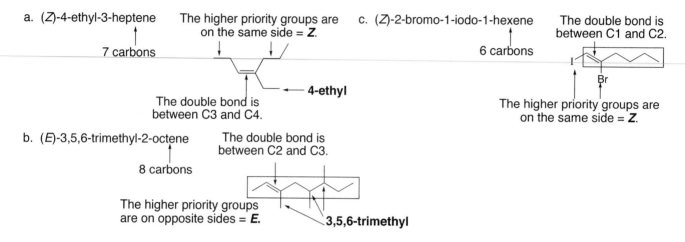

a. (*Z*)-4-ethyl-3-heptene

The higher priority groups are on the same side = **Z**.

7 carbons

The double bond is between C3 and C4.

4-ethyl

c. (*Z*)-2-bromo-1-iodo-1-hexene

The double bond is between C1 and C2.

6 carbons

The higher priority groups are on the same side = **Z**.

b. (*E*)-3,5,6-trimethyl-2-octene

The double bond is between C2 and C3.

8 carbons

The higher priority groups are on opposite sides = **E**.

3,5,6-trimethyl

**10.9 To rank the isomers by increasing boiling point:**
   **Look for polarity differences:** *small net dipoles* make an alkene more polar giving it a higher boiling point than an alkene with *no net dipole*. Cis isomers have a higher boiling point than their trans isomers.

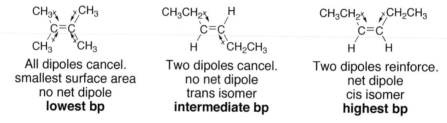

All dipoles cancel.
smallest surface area
no net dipole
**lowest bp**

Two dipoles cancel.
no net dipole
trans isomer
**intermediate bp**

Two dipoles reinforce.
net dipole
cis isomer
**highest bp**

**10.10**   Increasing number of double bonds = decreasing melting point

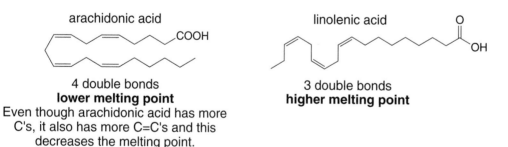

arachidonic acid

COOH

4 double bonds
**lower melting point**
Even though arachidonic acid has more
C's, it also has more C=C's and this
decreases the melting point.

linolenic acid

3 double bonds
**higher melting point**

**10.11**

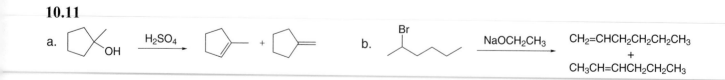

a.

OH   $\xrightarrow{\text{H}_2\text{SO}_4}$   +

b.

Br   $\xrightarrow{\text{NaOCH}_2\text{CH}_3}$   $CH_2=CHCH_2CH_2CH_2CH_3$
+
$CH_3CH=CHCH_2CH_2CH_3$

**10.12** To draw the products of an addition reaction:

[1] Locate the two bonds that will be broken in the reaction. Always break the π bond.

[2] Draw the product by forming two new σ bonds.

a.  [cyclopentene structure]  +  HI  ⟶  [cyclopentane with H and I added]  two new σ bonds

b.  [dimethylcyclohexene with CH₃ groups]  + HCl  ⟶  [cyclohexane with H, CH₃, CH₃, Cl]  two new σ bonds

**10.13  Addition reactions of HX occur in two steps:**

[1] The double bond attacks the H atom of HX to form a carbocation.

[2] X⁻ attacks the carbocation to form a C–X bond.

[reaction mechanism scheme showing cyclohexene + H–Cl → carbocation intermediate → chlorocyclohexane, with transition state step [1] and transition state step [2] shown in brackets]

transition state step [1]:  [structure with -H---Cl, δ⁻, δ⁺ H]

transition state step [2]:  [structure with δ⁺--Cl, δ⁻]

**10.14** Addition to alkenes follows Markovnikov's rule: When HX adds to an unsymmetrical alkene, the H bonds to the C that has more H's to begin with.

a.  no H's / Cl adds here.  [methylcyclohexene structure with CH₃ and H, "one H / H adds here."]  HCl ⟶  [product: cyclohexane with CH₃, Cl, H, H]

b.  no H's / Cl adds here  CH₃—C(CH₃)=CH₂  "2 H's / H adds here."  HCl ⟶  Cl–C(CH₃)(CH₃)–CH₂–H  [CH₃ and H labels]

c.  2 H's / H adds here.  [alkene structure]  "no H's / Cl adds here."  HCl ⟶  [product: CH₂—H, C—Cl, CH₃]

**10.15** To determine which alkene in each pair will react faster, draw the carbocation that forms in the rate-determining step. The more stable, more substituted carbocation, the lower the $E_a$ to form it and the faster the reaction.

[left reaction: (CH₃)₂C=CH₂ type alkene + H–X ⟶ ⁺C(CH₃)(CH₃)–CH₂–H]  3° carbocation  **faster** reaction

[right reaction: CH₃–CH=CH₂ + H–X ⟶ CH₃CHCH₂ with ⁺ and H]  2° carbocation  **slower** reaction

**10.16** Look for rearrangements of a carbocation intermediate to explain these results.

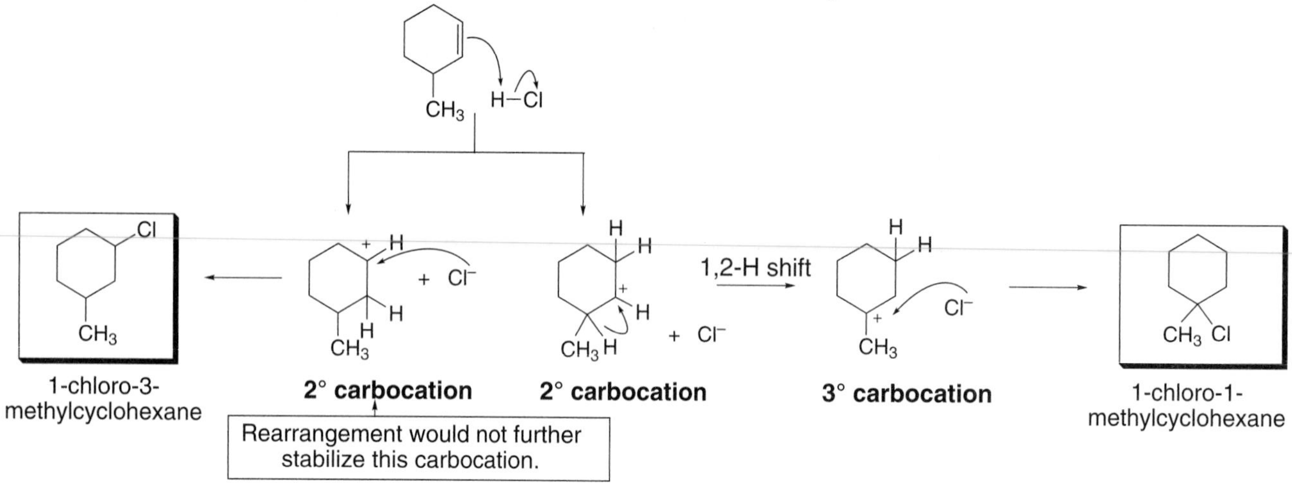

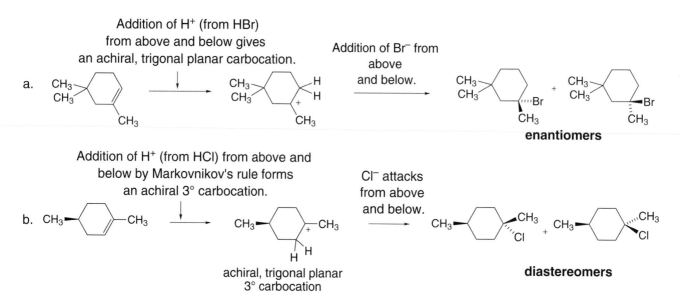

**10.17** To draw the products, remember that addition of HBr proceeds via a carbocation intermediate.

a.

Addition of H⁺ (from HBr) from above and below gives an achiral, trigonal planar carbocation.

Addition of Br⁻ from above and below.

**enantiomers**

b.

Addition of H⁺ (from HCl) from above and below by Markovnikov's rule forms an achiral 3° carbocation.

Cl⁻ attacks from above and below.

achiral, trigonal planar 3° carbocation

**diastereomers**

**10.18** The product of syn addition will have H and Cl both up or down (both on wedges or both dashes), while the product of anti addition will have one up and one down (one wedge, one dash).

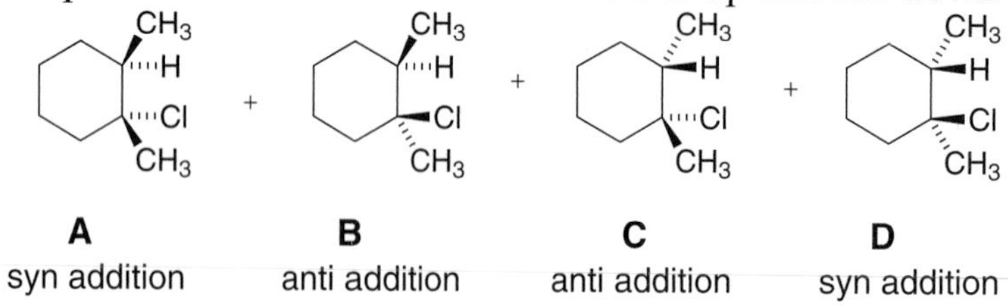

| A | B | C | D |
|---|---|---|---|
| syn addition | anti addition | anti addition | syn addition |

**10.19**

a.

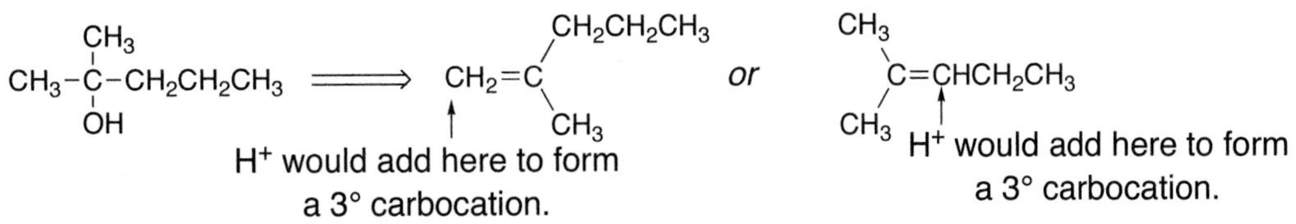

$$CH_3-\underset{\underset{OH}{|}}{\overset{\overset{CH_3}{|}}{C}}-CH_2CH_2CH_3 \Longrightarrow$$

$$CH_2=\underset{\underset{CH_3}{|}}{\overset{\overset{CH_2CH_2CH_3}{|}}{C}}$$

H⁺ would add here to form
a 3° carbocation.

*or*

$$\underset{\underset{CH_3}{|}}{\overset{\overset{CH_3}{|}}{C}}=CHCH_2CH_3$$

H⁺ would add here to form
a 3° carbocation.

b.

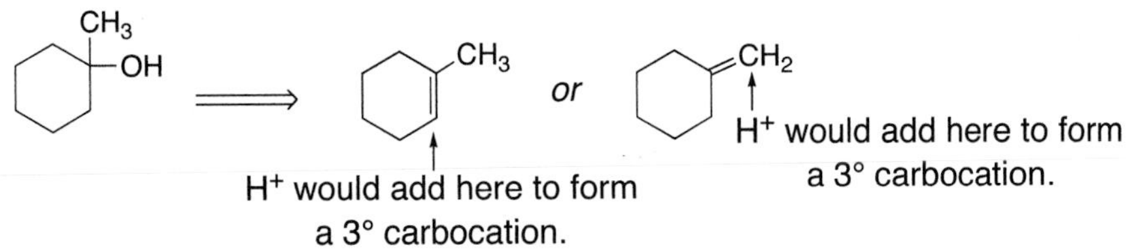

H⁺ would add here to form
a 3° carbocation.

*or*

H⁺ would add here to form
a 3° carbocation.

c.

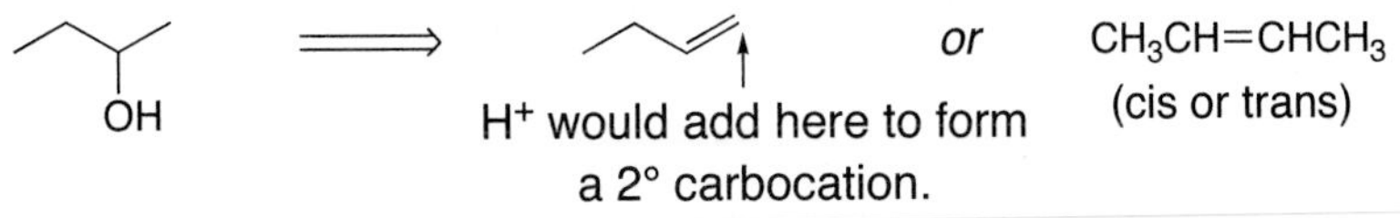

H⁺ would add here to form
a 2° carbocation.

*or* $CH_3CH=CHCH_3$
(cis or trans)

**10.20**

1-pentene $\xrightarrow[H_2SO_4]{H_2O}$

**enantiomers**

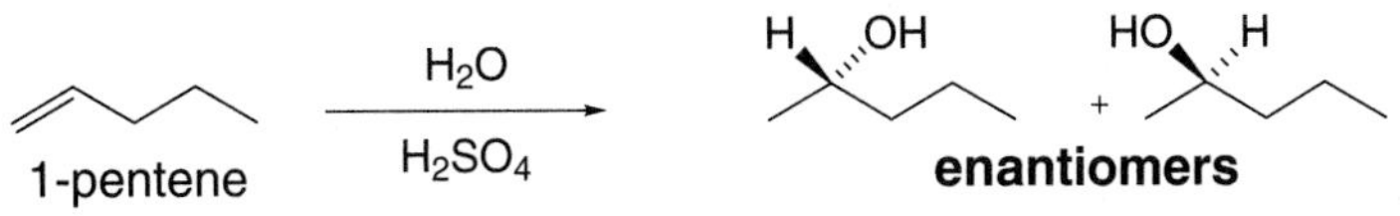

**10.21** **The two steps in the mechanism for the halogenation of an alkene are:**
[1] Addition of X⁺ to the alkene to form a bridged halonium ion.
[2] Nucleophilic attack by X⁻.

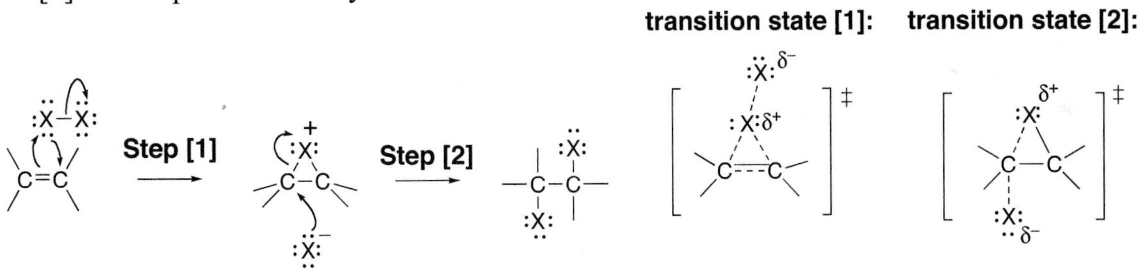

**transition state [1]:**   **transition state [2]:**

**10.22** Halogenation of an alkene adds two elements of X in an anti fashion.

a.

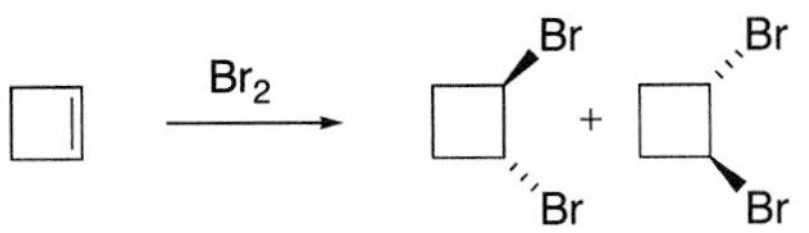

$\xrightarrow{Br_2}$

b.

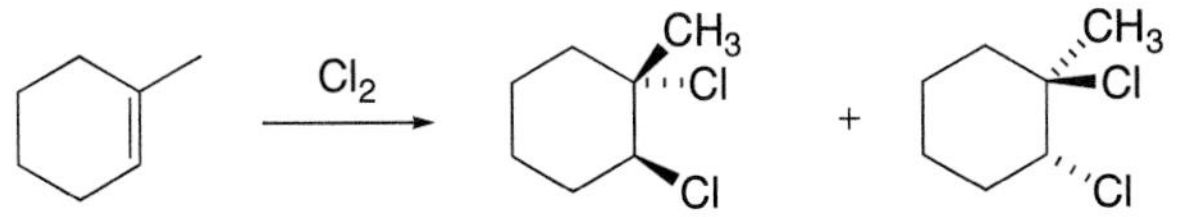

$\xrightarrow{Cl_2}$

**10.23** To draw the products of halogenation of an alkene, remember that the halogen adds to both ends of the double bond but only anti addition occurs.

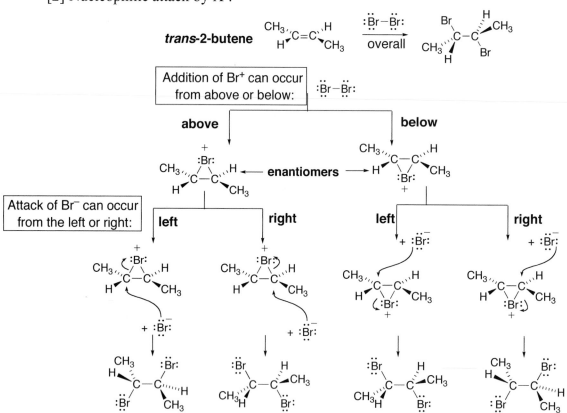

**10.24** **The two steps in the mechanism for the halogenation of an alkene are:**
[1] Addition of X$^+$ to the alkene to form a bridged halonium ion.
[2] Nucleophilic attack by X$^-$.

All four compounds are identical—an achiral meso compound.

**10.25** Halohydrin formation adds the elements of X and OH across the double bond in an anti fashion. The reaction is regioselective so X ends up on the carbon that had more H's to begin with.

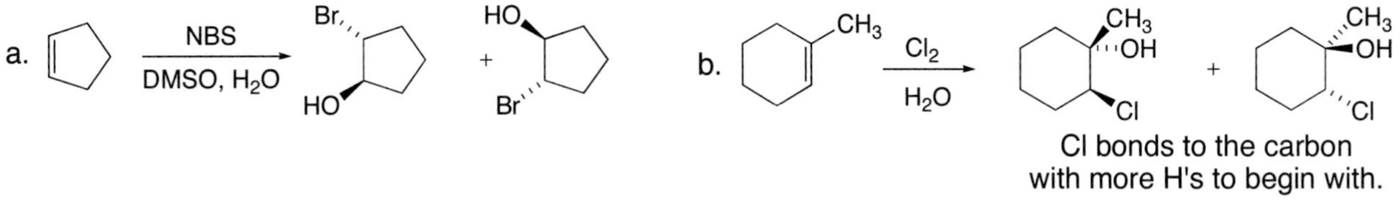

a. $\xrightarrow[\text{DMSO, H}_2\text{O}]{\text{NBS}}$ +

b. $\xrightarrow[\text{H}_2\text{O}]{\text{Cl}_2}$ +

Cl bonds to the carbon with more H's to begin with.

**10.26**

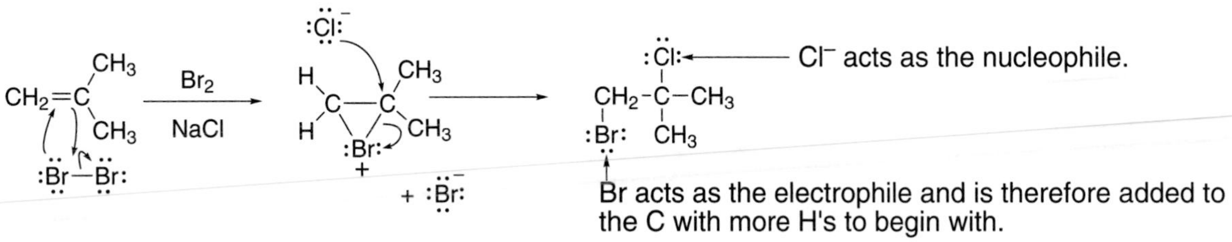

Cl⁻ acts as the nucleophile.

Br acts as the electrophile and is therefore added to the C with more H's to begin with.

**10.27**

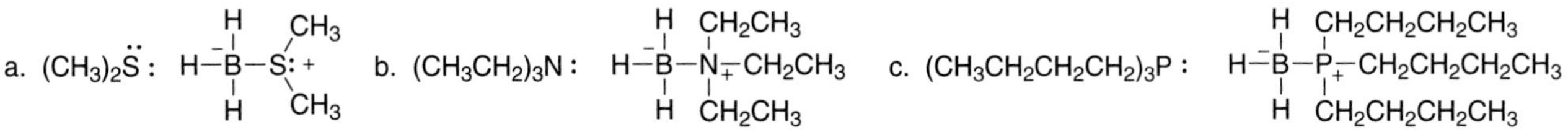

a. $(CH_3)_2\ddot{S}:$   b. $(CH_3CH_2)_3N:$   c. $(CH_3CH_2CH_2CH_2)_3P:$

**10.28** In hydroboration the boron atom is the electrophile and becomes bonded to the carbon atom that had more H's to begin with.

a. $\xrightarrow{BH_3}$

C with more H's. B will add here.

c. $\xrightarrow{BH_3}$

C with more H's. B will add here.

b. $\xrightarrow{BH_3}$

C with more H's. B will add here.

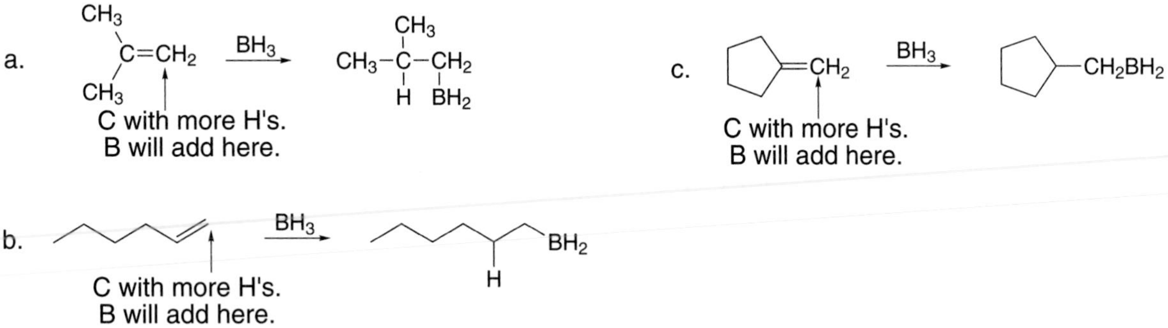

**10.29** The hydroboration–oxidation reaction occurs in two steps:
[1] Syn addition of BH₃, with the boron on the less substituted carbon atom.
[2] OH replaces the BH₂ with retention of configuration.

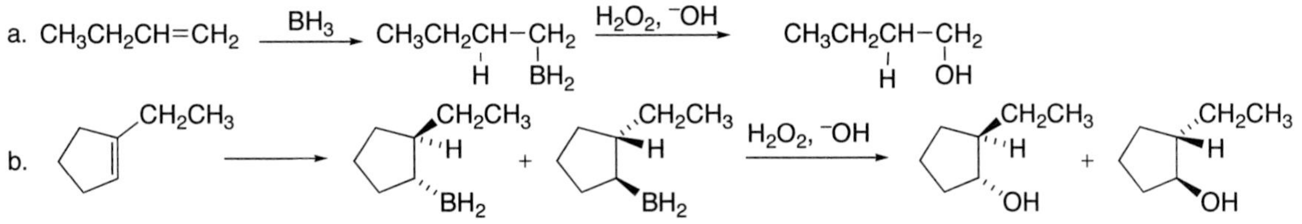

a. $CH_3CH_2CH{=}CH_2$ $\xrightarrow{BH_3}$ $\xrightarrow{H_2O_2,\ ^-OH}$

b. $\longrightarrow$ + $\xrightarrow{H_2O_2,\ ^-OH}$ +

c. CH₃—(ring)—CH₃  →BH₃→  CH₃—(ring with ,,,CH₃, H, BH₂)  +  CH₃—(ring with ,CH₃, H, BH₂)

$H_2O_2$, ⁻OH  ↓

CH₃—(ring with ,,,CH₃, H, OH)  +  CH₃—(ring with ,CH₃, H, OH)

**10.30**  Remember that hydroboration results in addition of OH on the less substituted C.

a.  (HO, cyclopentane with CH₃)  ←  (methylcyclopentene)

c.  (cyclohexane with CH(OH)CH₃)  ←  (ethylidenecyclohexane)

b.  (chain with OH)  ←  (branched alkene)      (*E* or *Z* isomer can be used.)

**10.31**

a.  (cyclopentane)=CH₂  →$\frac{H_2O}{H_2SO_4}$→  (cyclopentane with OH, CH₃)   Hydration places the OH on the more substituted carbon.

b.  (cyclopentane)=CH₂  →[1] BH₃ / [2] H₂O₂, HO⁻→  (cyclopentane)—CH₂OH   Hydroboration–oxidation places the OH on the less substituted carbon.

c.  (cyclohexane with isopropenyl)  →$\frac{H_2O}{H_2SO_4}$→  (cyclohexane with C(OH)(CH₃)₂)   Hydration places the OH on the more substituted carbon.

d.  (cyclohexane with isopropenyl)  →[1] BH₃ / [2] H₂O₂, HO⁻→  (cyclohexane with CH(CH₃)CH₂OH)   Hydroboration–oxidation places the OH on the less substituted carbon.

**10.32**  There are always two steps in this kind of question:
  [1] **Identify the functional group and decide what types of reactions it undergoes** (for example, substitution, elimination, or addition).
  [2] **Look at the reagent and determine if it is an electrophile, nucleophile, acid, or base.**

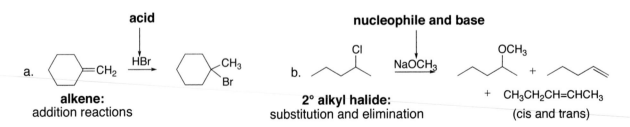

a.
**acid**
(cyclohexane)=CH₂  →HBr→  (cyclohexane with CH₃, Br)
**alkene:**
addition reactions

b.
**nucleophile and base**
(chain with Cl)  →NaOCH₃→  (chain with OCH₃)  +  (alkene)
  +  CH₃CH₂CH=CHCH₃
**2° alkyl halide:**
substitution and elimination
(cis and trans)

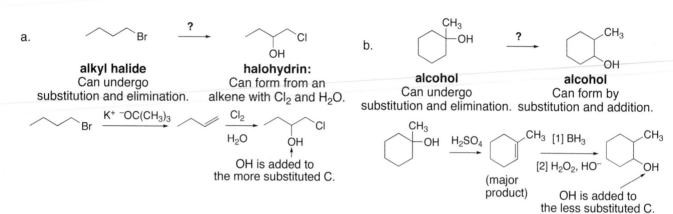

c.

**acid:**
catalyzes loss of H₂O

**alcohol:**
substitution and elimination

**10.33** To devise a synthesis:
[1] Look at the starting material and decide what reactions it can undergo.
[2] Look at the product and decide what reactions could make it.

a.

**alkyl halide**
Can undergo
substitution and elimination.

**halohydrin:**
Can form from an
alkene with Cl₂ and H₂O.

b.

**alcohol**
Can undergo
substitution and elimination.

**alcohol**
Can form by
substitution and addition.

OH is added to
the more substituted C.

(major
product)

OH is added to
the less substituted C.

**10.34** Name the alkenes using the rules in Answers 10.4 and 10.6.

a. CH₂=CHCH₂CH(CH₃)CH₂CH₃

    6 C chain with a double bond =
    **hexene**

$$\underset{\text{1-hexene}}{CH_2=CHCH_2} \overset{CH_3}{\underset{\text{4-methyl}}{CHCH_2CH_3}}$$

**4-methyl-1-hexene**

b.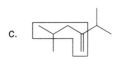

    8 C chain with a double bond =
    **octene**

5-ethyl    2-methyl→    **2-octene**

**5-ethyl-2-methyl-2-octene**

c.

    5 C chain with a double bond =
    **pentene**

2-isopropyl    1-pentene    4-methyl

**2-isopropyl-4-methyl-1-pentene**

d.
$$\underset{H}{\overset{CH_3}{\phantom{}}} C=C \underset{CH_2CH(CH_3)_2}{\overset{CH_3}{\phantom{}}}$$

    6 C chain with a double bond =
    **hexene**

CH₃   C=C   CH₃ ←3-methyl   CH₃←5-methyl
H   CH₂CHCH₃   **2-hexene**

higher priority→CH₃   C=C   CH₃   **(E)-3,5-dimethyl-2-hexene**
H   CH₂CH(CH₃)₂← higher priority

Higher priority groups
are on opposite sides =
***E* alkene.**

e.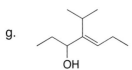

    6 C ring with a double bond =
    **cyclohexene**

1-ethyl    5-isopropyl

**1-ethyl-5-isopropylcyclohexene**

f.

    5 C ring with a double bond =
    **cyclopentene**

2-methyl→    1-*sec*-butyl

**1-*sec*-butyl-2-methylcyclopentene**

g.

    7 C chain with a double bond =
    **heptene**

4-isopropyl →    *E* **double bond** (higher priority groups on opposite sides with bold bonds)

3-ol → OH    4-heptene

**(E)-4-isopropyl-4-hepten-3-ol**

h.

    6 C ring with a double bond =
    **cyclohexene**

2-cyclohexene    1-ol    5-*sec*-butyl

**5-*sec*-butyl-2-cyclohexenol**

**10.35** Use the directions from Answer 10.8.

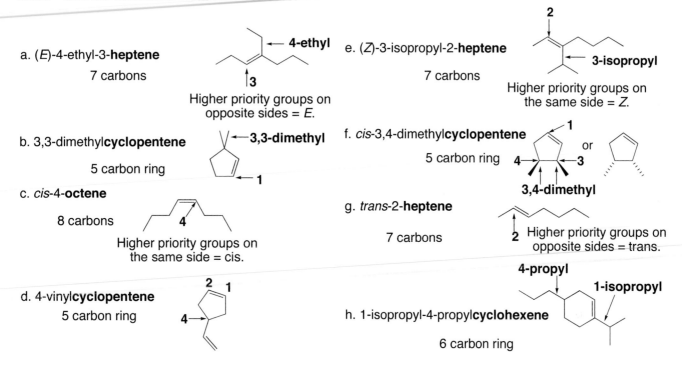

a. (*E*)-4-ethyl-3-**heptene**
7 carbons

← **4-ethyl**
↑3
Higher priority groups on opposite sides = *E*.

b. 3,3-dimethyl**cyclopentene**
5 carbon ring

← **3,3-dimethyl**
← 1

c. *cis*-4-**octene**
8 carbons

4
Higher priority groups on the same side = cis.

d. 4-vinyl**cyclopentene**
5 carbon ring

2 1
4→

e. (*Z*)-3-isopropyl-2-**heptene**
7 carbons

2
← **3-isopropyl**
Higher priority groups on the same side = *Z*.

f. *cis*-3,4-dimethyl**cyclopentene**
5 carbon ring

1
4→ ←3
**3,4-dimethyl**
or

g. *trans*-2-**heptene**
7 carbons

2 Higher priority groups on opposite sides = trans.

h. 1-isopropyl-4-propyl**cyclohexene**
6 carbon ring

**4-propyl**
**1-isopropyl**

**10.36** Name the alkene from which the epoxide can be derived and add the word *oxide*.

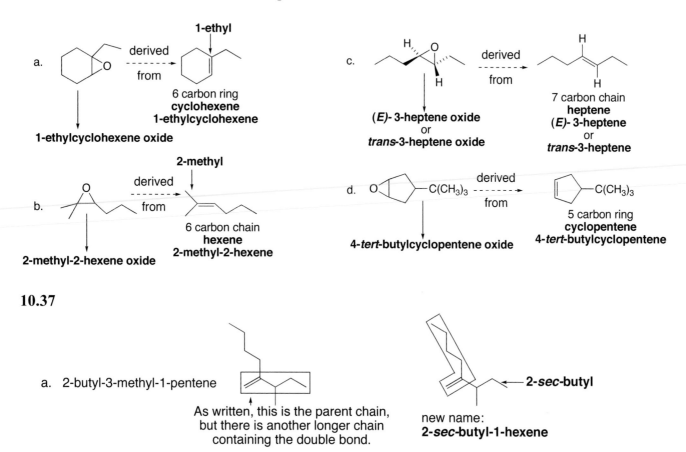

a.

derived
from

**1-ethyl**

6 carbon ring
**cyclohexene**
**1-ethylcyclohexene**

**1-ethylcyclohexene oxide**

b.

derived
from

**2-methyl**

6 carbon chain
**hexene**
**2-methyl-2-hexene**

**2-methyl-2-hexene oxide**

c.

H O
H

derived
from

H
H

7 carbon chain
**heptene**
(*E*)- **3-heptene**
or
*trans*-3-heptene

(*E*)- 3-heptene oxide
or
*trans*-3-heptene oxide

d. O —C(CH₃)₃

derived
from

—C(CH₃)₃

5 carbon ring
**cyclopentene**
**4-*tert*-butylcyclopentene**

**4-*tert*-butylcyclopentene oxide**

**10.37**

a. 2-butyl-3-methyl-1-pentene

As written, this is the parent chain, but there is another longer chain containing the double bond.

← **2-*sec*-butyl**

new name:
**2-*sec*-butyl-1-hexene**

b.  (*Z*)-2-methyl-2-hexene

new name:
**2-methyl-2-hexene**

Two groups on one end of the C=C
are the same (2 CH₃'s), so no *E* and *Z* isomers are possible.

c.  (*E*)-1-isopropyl-1-butene

As written, this is the parent chain,
but there is another longer chain
containing the double bond.

new name:
**(*E*)-2-methyl-3-hexene**

d.  5-methylcyclohexene

As written the methyl
is at C5. Re-number
to put it at C4.

new name:
**4-methylcyclohexene**

e.  4-isobutyl-2-methylcylohexene

As written this methyl
is at C2. Re-number
to put it at C1.

new name:
**5-isobutyl-1-methylcyclohexene**

f.  1-*sec*-butyl-2-cyclopentene

This has the double bond between
C2 and C3. Cycloalkenes must
have the double bond between
C1 and C2. Re-number.

new name:
**3-*sec*-butylcyclopentene**

g.  1-cyclohexen-4-ol

The numbering is incorrect. When a compound
contains both a double bond and a OH group,
number the C skeleton to give the OH group the
lower number.

**3-cyclohexenol** (The "1" can be omitted.)

h.  3-ethyl-3-octen-5-ol

The numbering is incorrect. When a compound
contains both a double bond and a OH group,
number the C skeleton to give the OH group the
lower number.

**6-ethyl-5-octen-4-ol**

**10.38**

a. and b.

bongkrekic acid

c. Since there are 7 double bonds
and 2 tetrahedral stereogenic
centers, $2^9 = 512$ possible stereoisomers.

## 10.39

2-methyl-1-pentene   (E)-4-methyl-2-pentene   4-methyl-1-pentene   (E)-3-methyl-2-pentene   (R)-3-methyl-1-pentene

2-methyl-2-pentene   (Z)-4-methyl-2-pentene   2-ethyl-1-butene   (Z)-3-methyl-2-pentene   (S)-3-methyl-1-pentene

**10.40**   Use the directions from Answer 10.2 to calculate degrees of unsaturation.

a. $C_3H_4$
  [1] maximum number of H's = $2n + 2 = 2(3) + 2 = 8$
  [2] subtract actual from maximum = $8 - 4 = 4$
  [3] divide by 2 = $4/2 =$ **2 degrees of unsaturation**

b. $C_6H_8$
  [1] maximum number of H's = $2n + 2 = 2(6) + 2 = 14$
  [2] subtract actual from maximum = $14 - 8 = 6$
  [3] divide by 2 = $6/2 =$ **3 degrees of unsaturation**

c. $C_{40}H_{56}$
  [1] maximum number of H's = $2n + 2 = 2(40) + 2 = 82$
  [2] subtract actual from maximum = $82 - 56 = 26$
  [3] divide by 2 = $26/2 =$ **13 degrees of unsaturation**

d. $C_8H_8O$
  Ignore the O.
  [1] maximum number of H's = $2n + 2 = 2(8) + 2 = 18$
  [2] subtract actual from maximum = $18 - 8 = 10$
  [3] divide by 2 = $10/2 =$ **5 degrees of unsaturation**

e. $C_{10}H_{16}O_2$
  Ignore both O's.
  [1] maximum number of H's = $2n + 2 = 2(10) + 2 = 22$
  [2] subtract actual from maximum = $22 - 16 = 6$
  [3] divide by 2 = $6/2 =$ **3 degrees of unsaturation**

f. $C_8H_9Br$
  Because of Br, add one H ($9 + 1 = 10$ H's).
  [1] maximum number of H's = $2n + 2 = 2(8) + 2 = 18$
  [2] subtract actual from maximum = $18 - 10 = 8$
  [3] divide by 2 = $8/2 =$ **4 degrees of unsaturation**

g. $C_8H_9ClO$
  Ignore the O; count Cl as one more H ($9 + 1 = 10$ H's).
  [1] maximum number of H's = $2n + 2 = 2(8) + 2 = 18$
  [2] subtract actual from maximum = $18 - 10 = 8$
  [3] divide by 2 = $8/2 =$ **4 degrees of unsaturation**

h. $C_7H_9Br$
  Because of Br, add one H ($9 + 1 = 10$ H's).
  [1] maximum number of H's = $2n + 2 = 2(7) + 2 = 16$
  [2] subtract actual from maximum = $16 - 10 = 6$
  [3] divide by 2 = $6/2 =$ **3 degrees of unsaturation**

i. $C_7H_{11}N$
  Because of N, subtract one H ($11 - 1 = 10$ H's).
  [1] maximum number of H's = $2n + 2 = 2(7) + 2 = 16$
  [2] subtract actual from maximum = $16 - 10 = 6$
  [3] divide by 2 = $6/2 =$ **3 degrees of unsaturation**

j. $C_4H_8BrN$
  Because of Br, add one H, but subtract one for N
  ($8 + 1 - 1 = 8$ H's).
  [1] maximum number of H's = $2n + 2 = 2(4) + 2 = 10$
  [2] subtract actual from maximum = $10 - 8 = 2$
  [3] divide by 2 = $2/2 =$ **1 degree of unsaturation**

## 10.41

**One possibility for $C_6H_{10}$:**

a. a compound that has 2 π bonds

c. a compound with 2 rings

b. a compound that has 1 ring and 1 π bond

d. compound with 1 triple bond

**10.42**

stearic acid 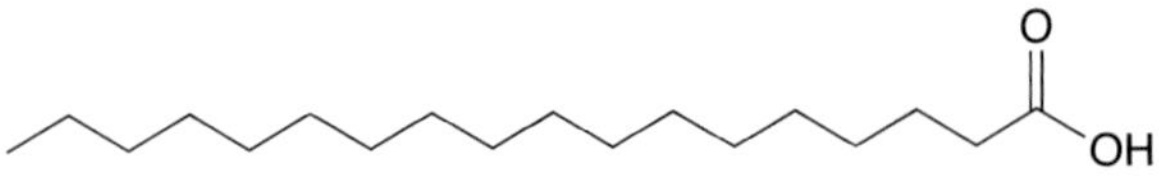
**highest melting point**
no double bonds

elaidic acid 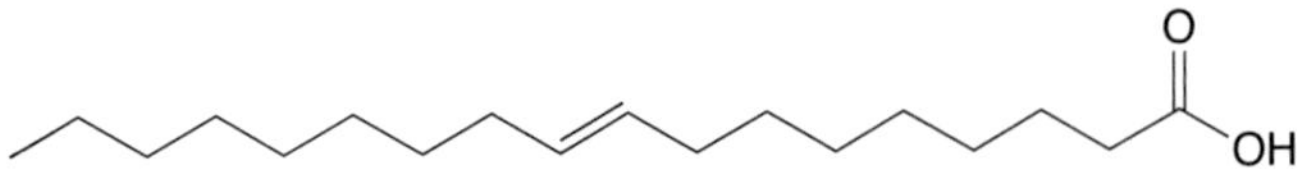
**intermediate melting point**
one *E* double bond

oleic acid 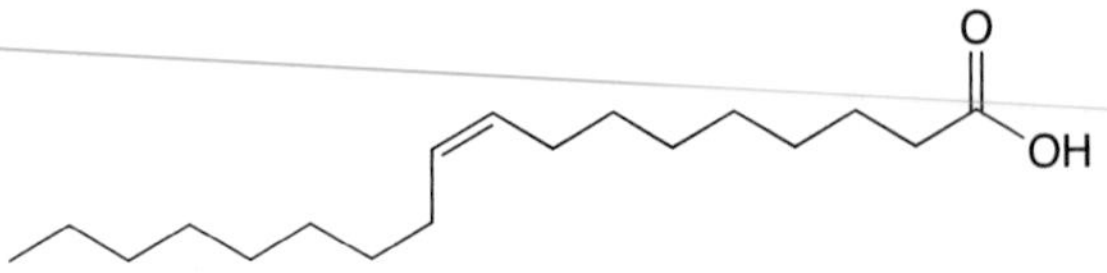
**lowest melting point**
one *Z* double bond

**10.43**

a.

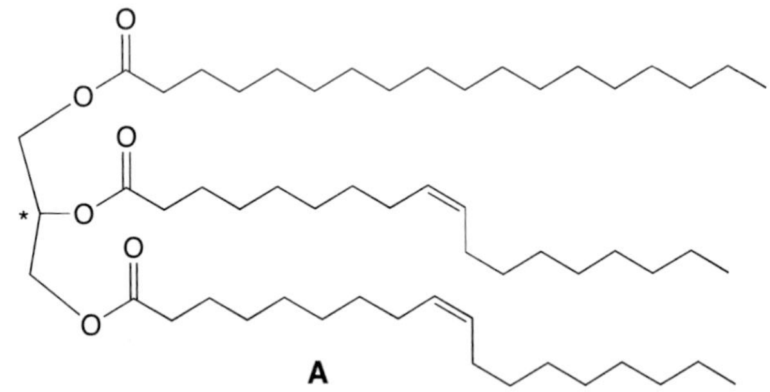

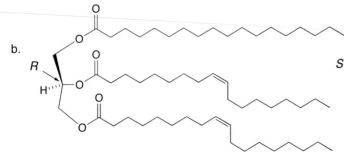

**A**
**A** has one tetrahedral stereogenic center,
labeled with an asterisk [*].

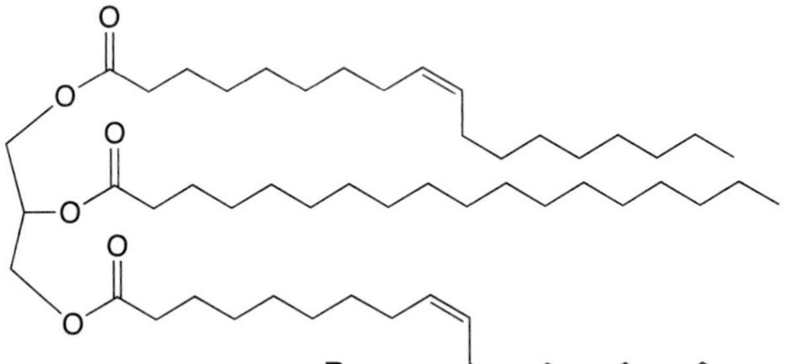

**B**

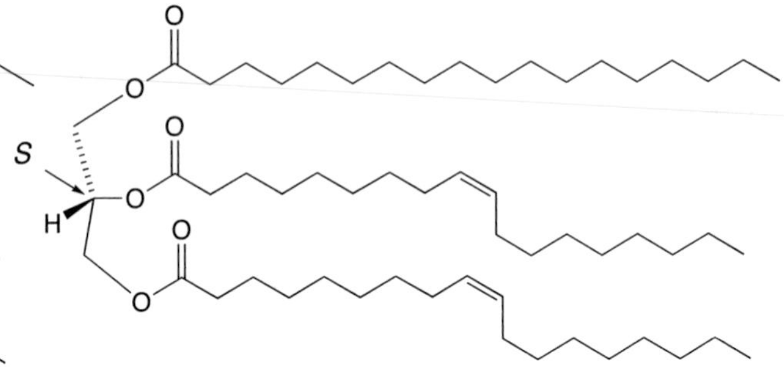

b.

*R*

*S*

**10.44**

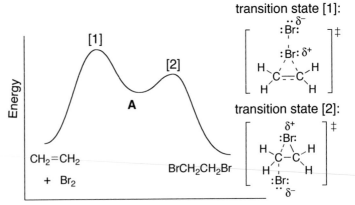

transition state [1]:

transition state [2]:

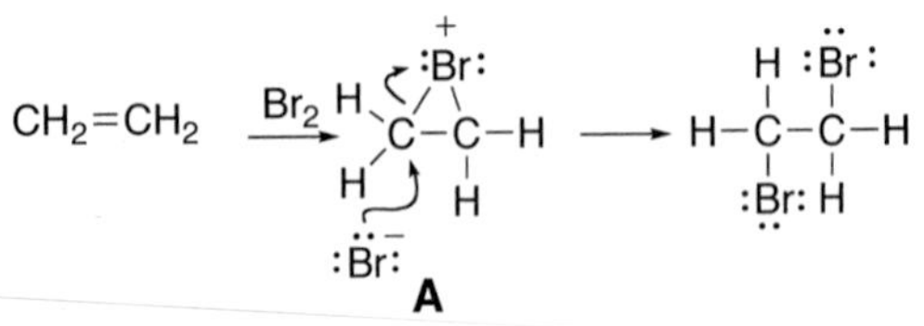

**10.45**  The more negative the $\Delta H°$, the larger the $K_{eq}$ assuming entropy changes are comparable. Calculate the $\Delta H°$ for each reaction and compare.

$CH_2=CH_2$  +  HI  $\longrightarrow$  $CH_3CH_2-I$

| **[1] Bonds broken** | | **[2] Bonds formed** | | **[3] Overall $\Delta H°$ =** |
|---|---|---|---|---|

| | $\Delta H°$ (kcal/mol) | | $\Delta H°$ (kcal/mol) |
|---|---|---|---|
| C–C π bond | + 64 | $CH_2ICH_2-H$ | – 98 |
| H–I | + 71 | C–I | – 53 |
| Total | + 135  kcal/mol | Total | – 151  kcal/mol |

[3] Overall $\Delta H°$ =

sum in Step [1] + sum in Step [2]

+ 135  kcal/mol
– 151  kcal/mol

**– 16 kcal/mol**

$CH_2=CH_2$  +  HCl  $\longrightarrow$  $CH_3CH_2-Cl$

| **[1] Bonds broken** | | **[2] Bonds formed** | |
|---|---|---|---|

| | $\Delta H°$ (kcal/mol) | | $\Delta H°$ (kcal/mol) |
|---|---|---|---|
| C–C π bond | + 64 | $CH_2ClCH_2-H$ | – 98 |
| H–Cl | + 103 | C–Cl | – 81 |
| Total | + 167  kcal/mol | Total | – 179  kcal/mol |

**[3] Overall $\Delta H°$ =**

sum in Step [1] + sum in Step [2]

+ 167  kcal/mol
– 179  kcal/mol

**– 12 kcal/mol**

Compare the $\Delta H°$:
Addition of  HI: **–16 kcal/mol** more negative $\Delta H°$, larger $K_{eq}$.
Addition of HCl: **–12 kcal/mol**

**10.46**

a.  [cyclohexene] — HBr → [cyclohexane with H and Br]

b.  [cyclohexene] — HI → [cyclohexane with H and I]

c.  [cyclohexene] — $H_2O$ / $H_2SO_4$ → [cyclohexanol with H and OH]

d.  [cyclohexene] — $CH_3CH_2OH$ / $H_2SO_4$ → [cyclohexane with H and $OCH_2CH_3$]

e.  [cyclohexene] — $Cl_2$ → [cyclohexane with Cl, Cl] + [cyclohexane with Cl, Cl]

f.  [cyclohexene] — $Br_2$, $H_2O$ → [cyclohexane with Br and OH] + [cyclohexane with Br and OH]

g.  [cyclohexene] — NBS / aqueous DMSO → [cyclohexane with Br and OH] + [cyclohexane with Br and OH]

h.  [cyclohexene] — [1] $BH_3$ / [2] $H_2O_2$, HO⁻ → [cyclohexane with H and OH]

i.  [cyclohexene] — [1] 9-BBN / [2] $H_2O_2$, HO⁻ → [cyclohexane with H and OH]

**10.47**

a. $\underset{CH_3}{\overset{CH_3}{C}}=CH_2 \xrightarrow{\text{HBr}} CH_3-\underset{Br}{\overset{CH_3}{C}}-CH_3$

b. $\underset{CH_3}{\overset{CH_3}{C}}=CH_2 \xrightarrow{\text{HI}} CH_3-\underset{I}{\overset{CH_3}{C}}-CH_3$

c. $\underset{CH_3}{\overset{CH_3}{C}}=CH_2 \xrightarrow[\text{H}_2\text{SO}_4]{\text{H}_2\text{O}} CH_3-\underset{OH}{\overset{CH_3}{C}}-CH_3$

d. $\underset{CH_3}{\overset{CH_3}{C}}=CH_2 \xrightarrow[\text{H}_2\text{SO}_4]{\text{CH}_3\text{CH}_2\text{OH}} CH_3-\underset{OCH_2CH_3}{\overset{CH_3}{C}}-CH_3$

e. $\underset{CH_3}{\overset{CH_3}{C}}=CH_2 \xrightarrow{\text{Cl}_2} CH_3-\underset{Cl}{\overset{CH_3}{C}}-CH_2Cl$

f. $\underset{CH_3}{\overset{CH_3}{C}}=CH_2 \xrightarrow{\text{Br}_2,\ \text{H}_2\text{O}} CH_3-\underset{OH}{\overset{CH_3}{C}}-CH_2Br$

g. $\underset{CH_3}{\overset{CH_3}{C}}=CH_2 \xrightarrow[\text{aqueous DMSO}]{\text{NBS}} CH_3-\underset{OH}{\overset{CH_3}{C}}-CH_2Br$

h. $\underset{CH_3}{\overset{CH_3}{C}}=CH_2 \xrightarrow[\text{[2] H}_2\text{O}_2,\ \text{HO}^-]{\text{[1] BH}_3} CH_3-\underset{H}{\overset{CH_3}{C}}-CH_2OH$

i. $\underset{CH_3}{\overset{CH_3}{C}}=CH_2 \xrightarrow[\text{[2] H}_2\text{O}_2,\ \text{HO}^-]{\text{[1] 9-BBN}} CH_3-\underset{H}{\overset{CH_3}{C}}-CH_2OH$

**10.48**

a. $\xrightarrow{\text{Br}_2}$ Br–CH(Br)... **Halogenation**

b. $\xrightarrow[\text{H}_2\text{O}]{\text{Br}_2}$ Br...OH **Halohydrin formation**: Br adds to the C that had more H's to begin with.

c. $\xrightarrow[\text{CH}_3\text{OH}]{\text{Br}_2}$ Br...OCH$_3$ Same as halohydrin formation, except CH$_3$OH in place of H$_2$O.

**10.49**

a. cyclopentyl-Br ← cyclopentene

b. cyclohexane(Cl)(Cl) ← cyclohexene

c. cyclohexyl-C(CH$_3$)$_2$Br ← $\underset{\text{cyclohexyl}}{\overset{CH_3}{C}}=CH_2$

d. CH$_3$CH$_2$CH$_2$-CH(Br)-CH(Br)-CH$_3$ ← CH$_3$CH$_2$CH$_2$CH=CHCH$_3$

e. cyclopentane(CH$_3$)(Cl) ← cyclopentylidene=CH$_2$ or cyclopentene–CH$_3$

f. (CH$_3$CH$_2$)$_3$CBr ← $\underset{CH_3CH_2}{\overset{CH_3CH_2}{C}}=CHCH_3$

**10.50**

a.  $(CH_3CH_2)_2C=CHCH_2CH_3$

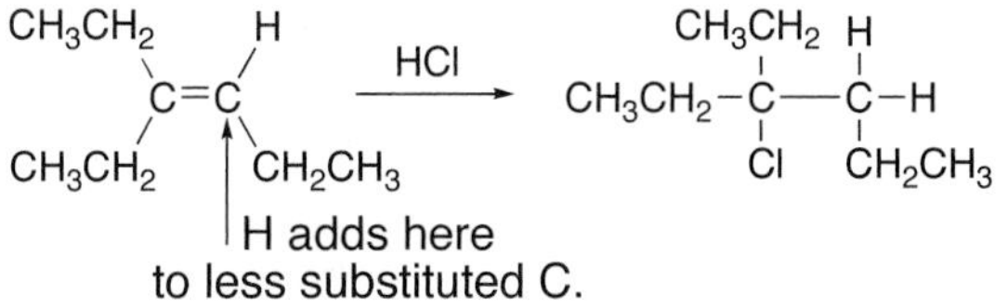

HCl

H adds here
to less substituted C.

b.  $(CH_3CH_2)_2C=CH_2$

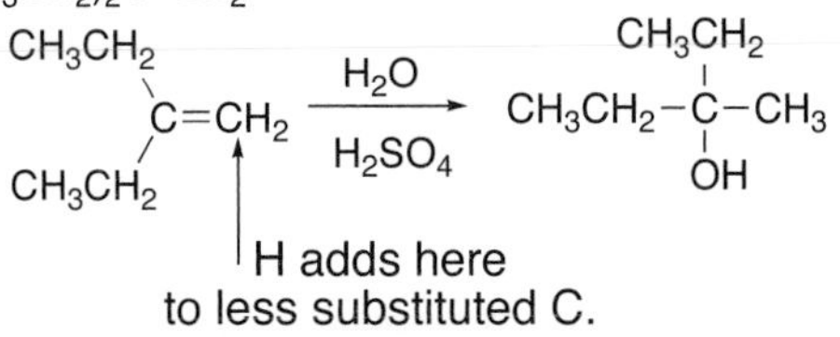

$H_2O$
$H_2SO_4$

H adds here
to less substituted C.

c.  $(CH_3)_2C=CHCH_3$

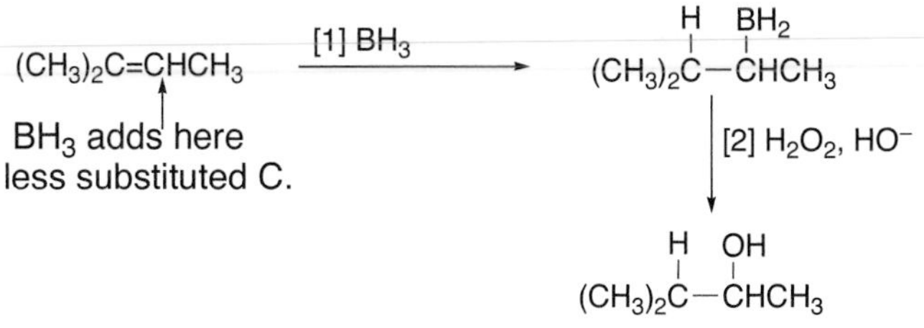

[1] BH$_3$

BH$_3$ adds here
to less substituted C.

[2] H$_2$O$_2$, HO$^-$

d.

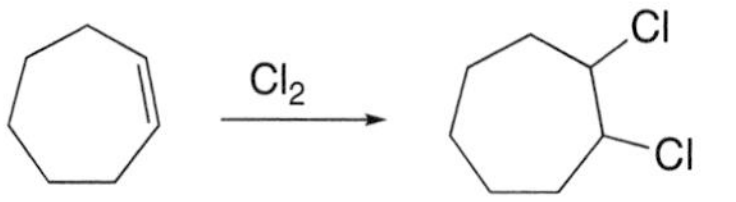

Cl$_2$

e.

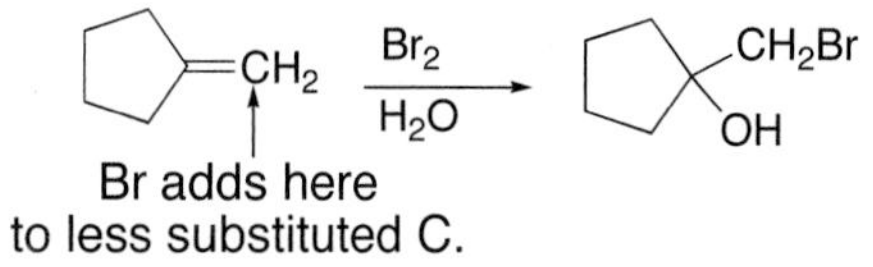

$=CH_2$

Br$_2$
H$_2$O

Br adds here
to less substituted C.

f.

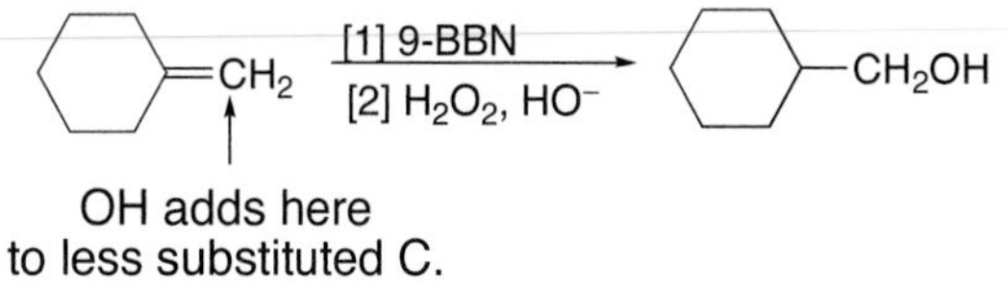

$=CH_2$

[1] 9-BBN
[2] H$_2$O$_2$, HO$^-$

OH adds here
to less substituted C.

g.

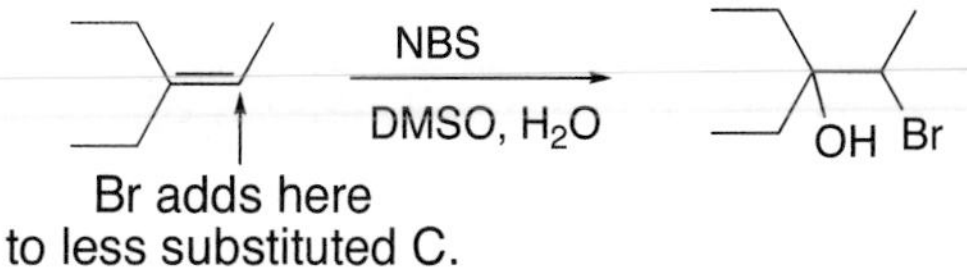

NBS
DMSO, H$_2$O

Br adds here
to less substituted C.

h.

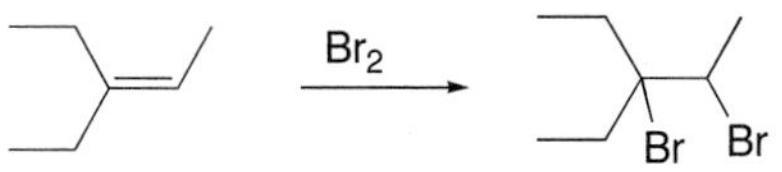

Br$_2$

**10.51**

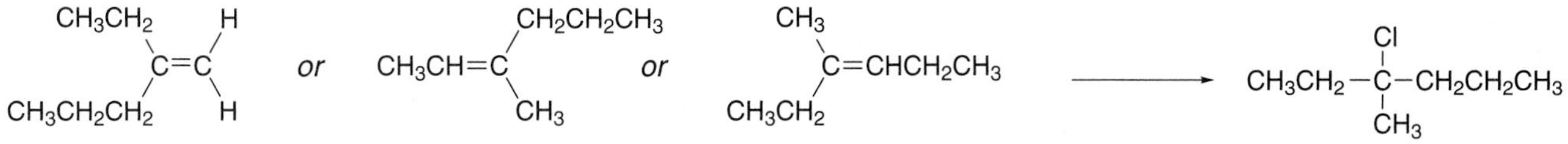

or

or

**10.52**

a.

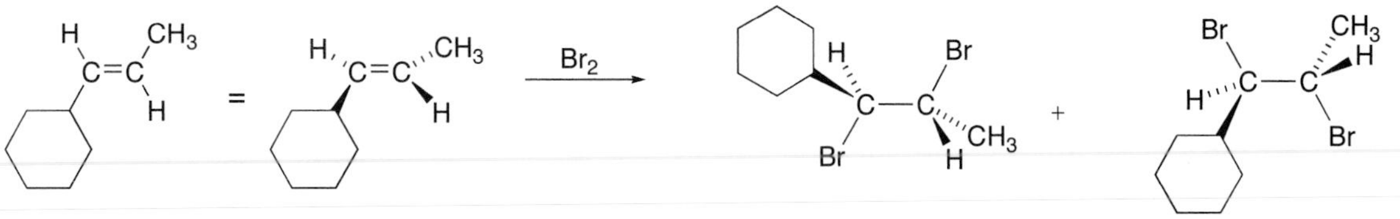

Br$_2$

b.

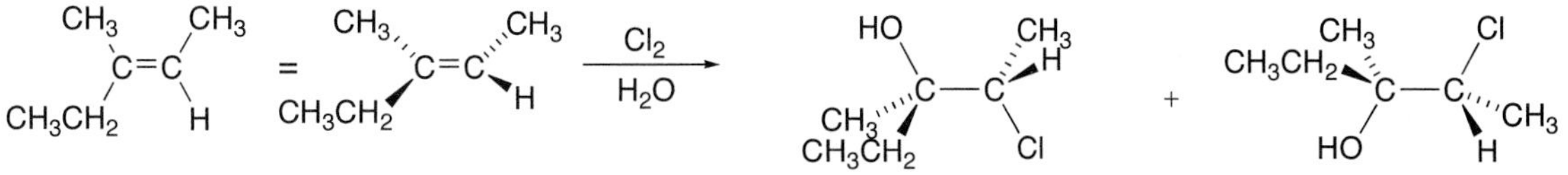

Cl$_2$
H$_2$O

c.

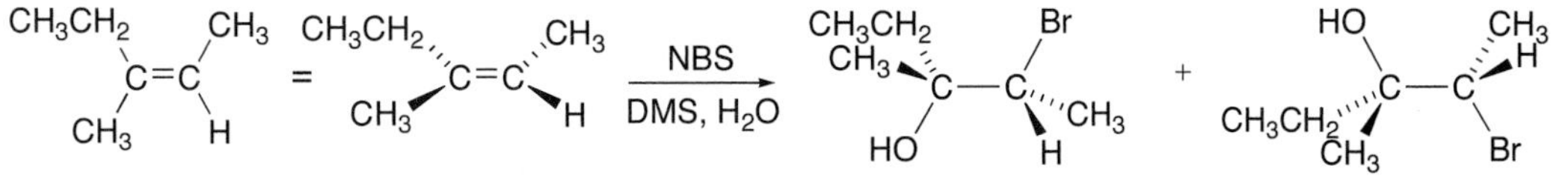

NBS
DMS, H$_2$O

**10.53**

a.  $(CH_3)_3C$—⬡$=CH_2$

$H_2O$
$H_2SO_4$

$(CH_3)_3C$—⬡ CH$_3$ OH + $(CH_3)_3C$—⬡ CH$_3$ OH

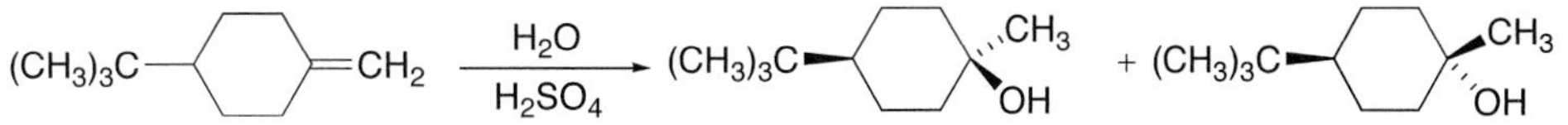

b.

HI

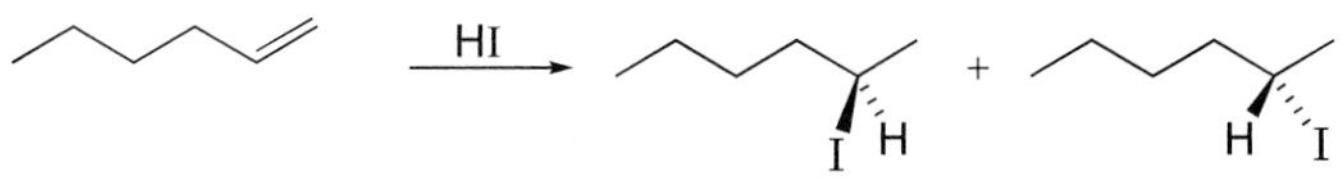

c.

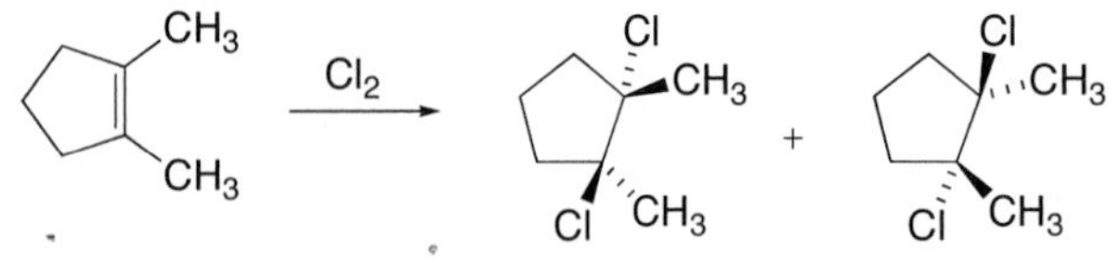

only anti addition

d.

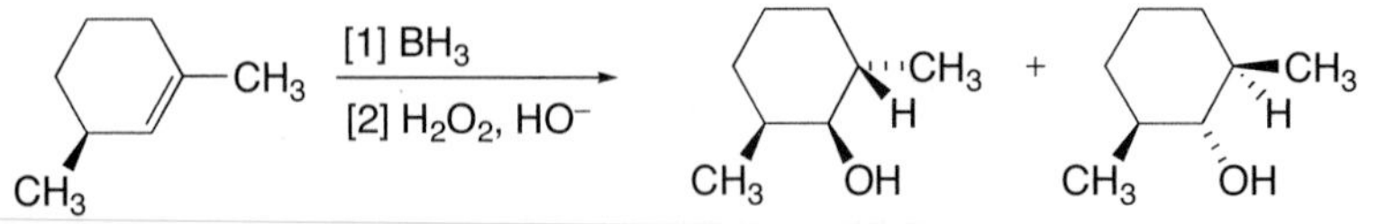

[1] BH₃
[2] H₂O₂, HO⁻

only syn addition

e.

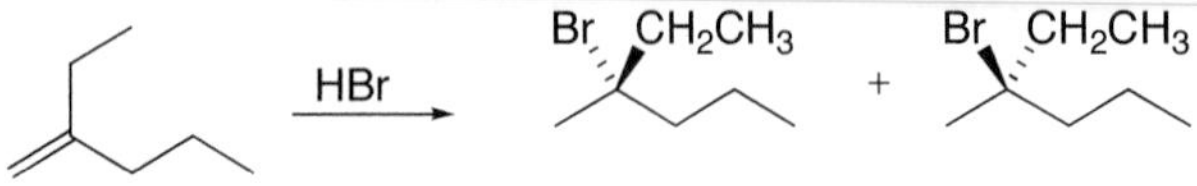

HBr

f.

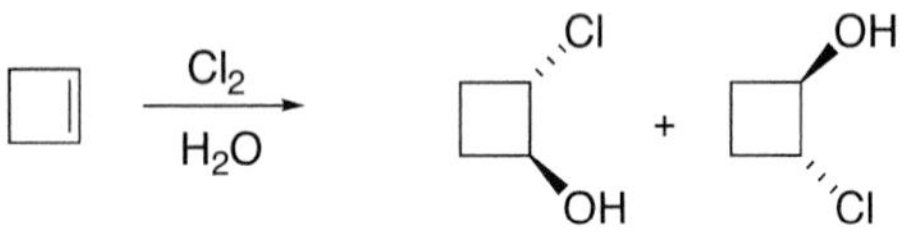

Cl₂
H₂O

only anti addition

g.

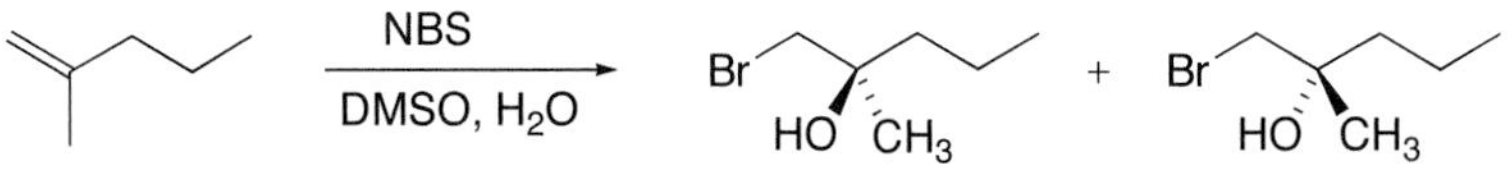

NBS
DMSO, H₂O

h. 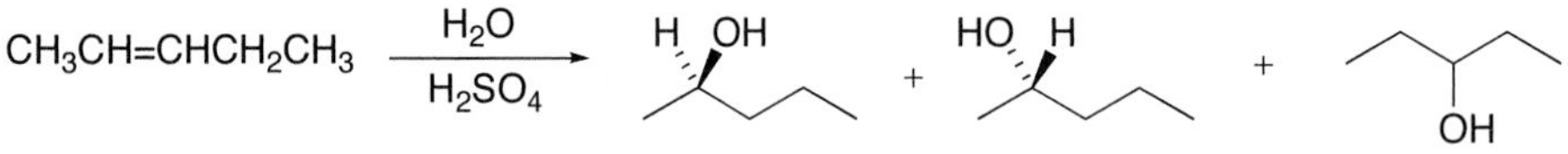  $CH_3CH=CHCH_2CH_3$  $\xrightarrow[H_2SO_4]{H_2O}$

## 10.54

a.   $CH_3CH=CHCH_3 + CH_3SCl \longrightarrow$  $CH_3-\underset{\boxed{Cl}}{CH}-\underset{CH_3}{CH}-\boxed{SCH_3}$

b. 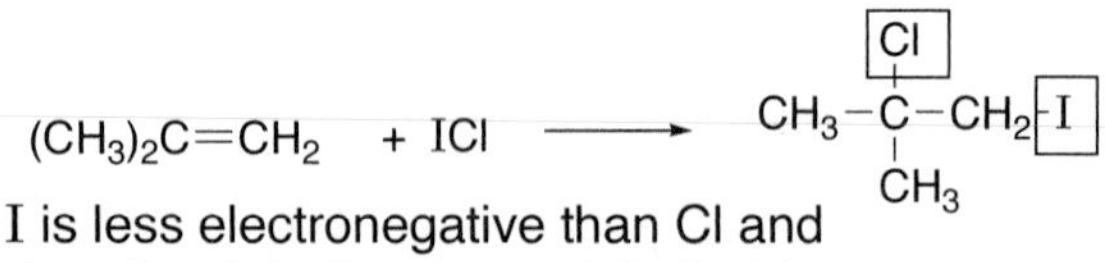  $(CH_3)_2C=CH_2 + ICl \longrightarrow$  $CH_3-\underset{CH_3}{\overset{\boxed{Cl}}{C}}-CH_2\boxed{-I}$

I is less electronegative than Cl and therefore it is the electrophile in this reaction. It ends up on the less substituted C.

## 10.55

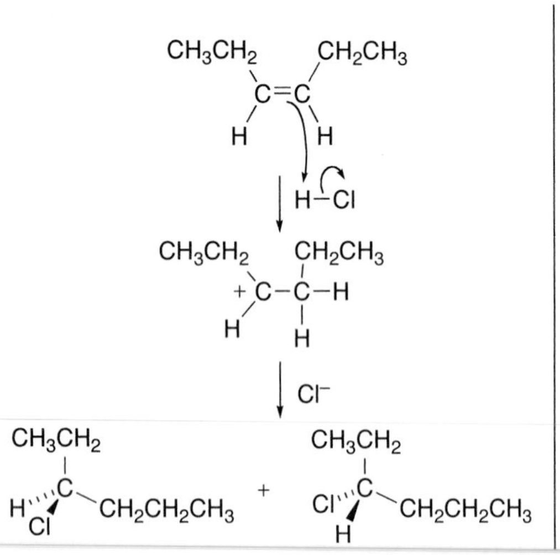

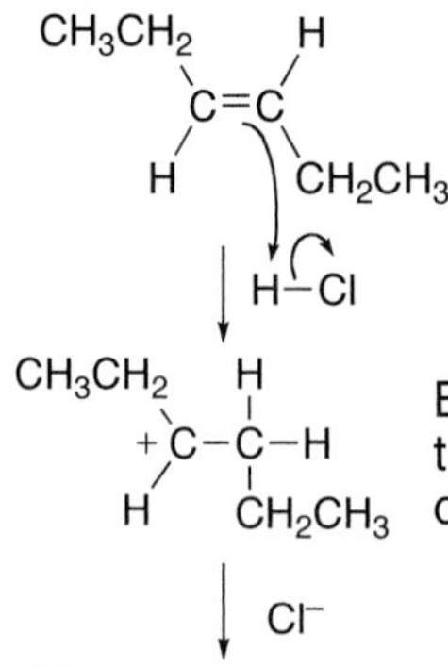

By protonation of the alkene, the cis and trans isomers produce **identical** carbocation intermediates.

Both *cis*- and *trans*-3-hexene give the same racemic mixture of products, so the reaction is not stereospecific.

**10.56** The alkene that forms the more stable carbocation reacts faster, according to the Hammond postulate.

a.

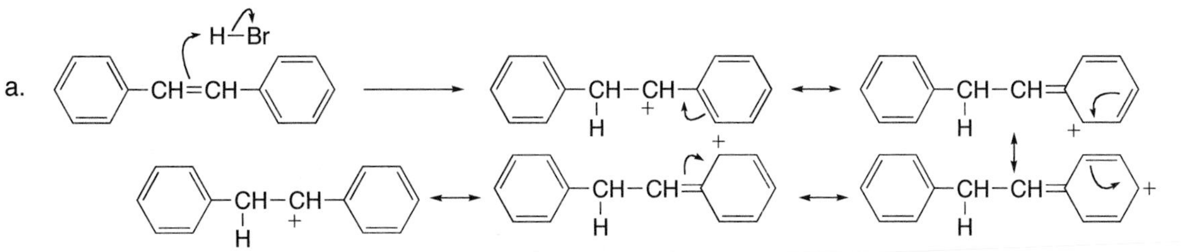

This **2° carbocation is resonance stabilized**, making it more stable, so the starting alkene reacts **faster** with HBr.

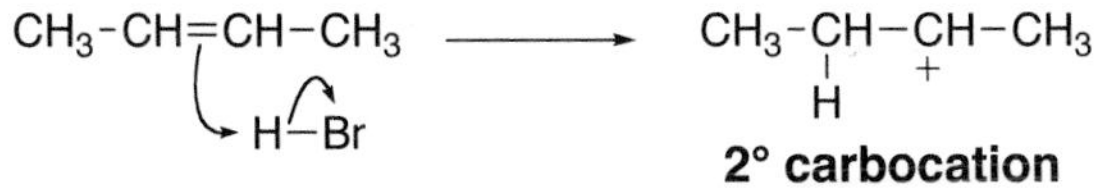

$$CH_3-CH=CH-CH_3 \xrightarrow{\quad H-Br \quad} CH_3-\overset{+}{CH}-\underset{H}{CH}-CH_3$$

**2° carbocation**

b.

$$CH_2=C\overset{CH_3}{\underset{CH_3}{\diagdown}} \xrightarrow[\text{addition}]{\text{Markovnikov}} \overset{+}{CH_2}-C(CH_3)_2 \quad \textbf{faster}$$
$$\underset{H}{} \quad \textbf{3° carbocation}$$

$$CH_2=C\overset{CH_3}{\underset{CH_2OCH_3}{\diagdown}} \xrightarrow[\text{addition}]{\text{Markovnikov}} CH_2-\overset{CH_3}{\underset{CH_2-OCH_3}{\overset{+}{C}}}$$

**3° carbocation**

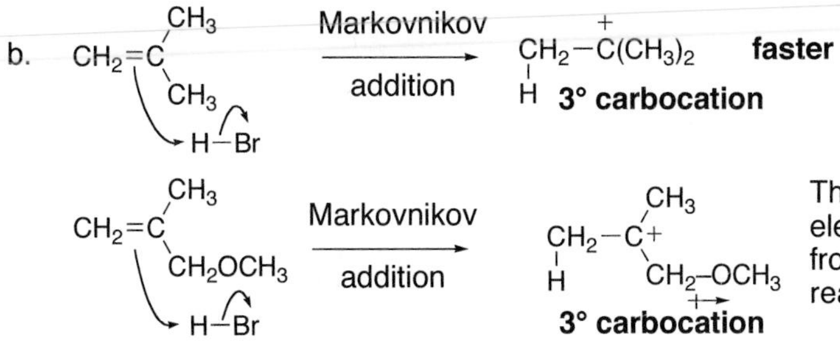

This carbocation is still 3°, but the nearby electronegative O atom withdraws electron density from the carbocation, destabilizing it. Thus, the reaction to form this carbocation occurs more slowly.

**10.57**

a.

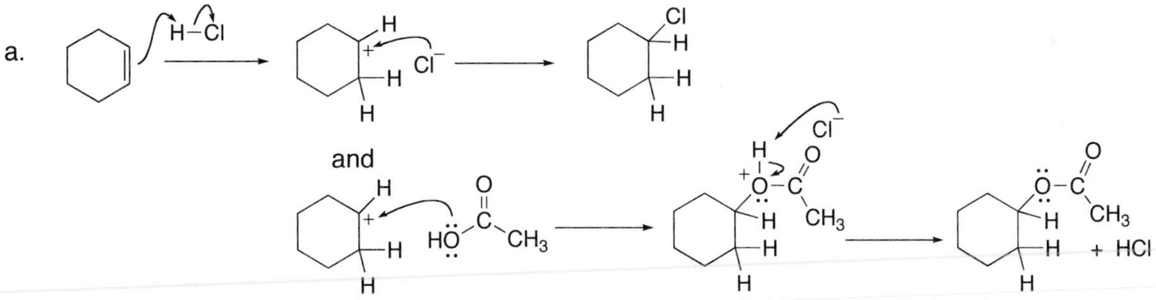

and

**b.**

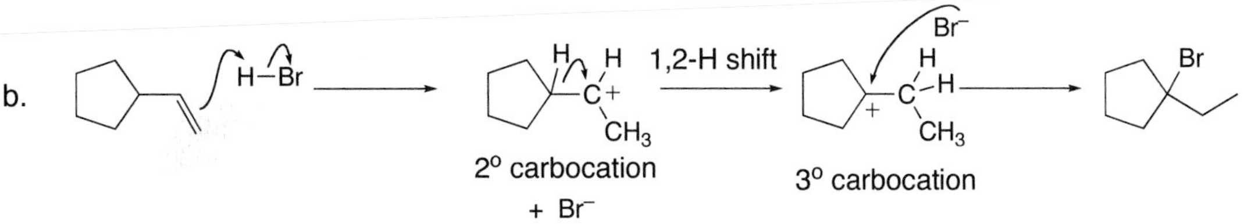

2° carbocation
+ Br⁻

1,2-H shift

3° carbocation

**10.58**

a.

$H_2O$ / $H_2SO_4$

2° carbocation
+ HSO₄⁻

1,2-shift

3° carbocation

+ $H_2SO_4$

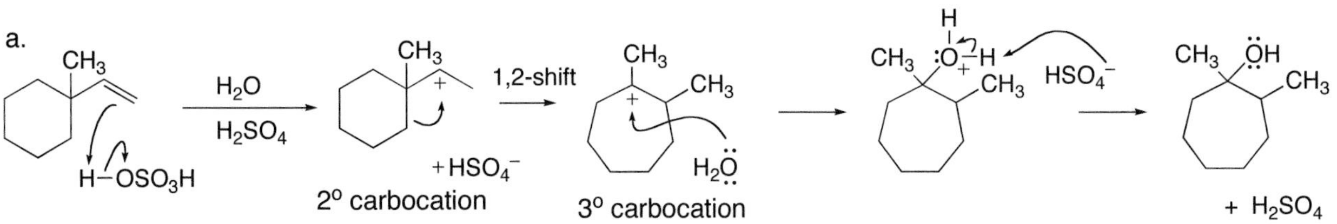

b.

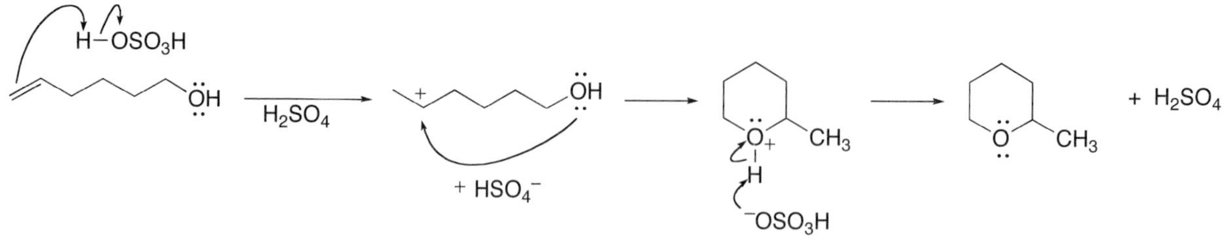

**10.59** The isomerization reaction occurs by protonation and deprotonation.

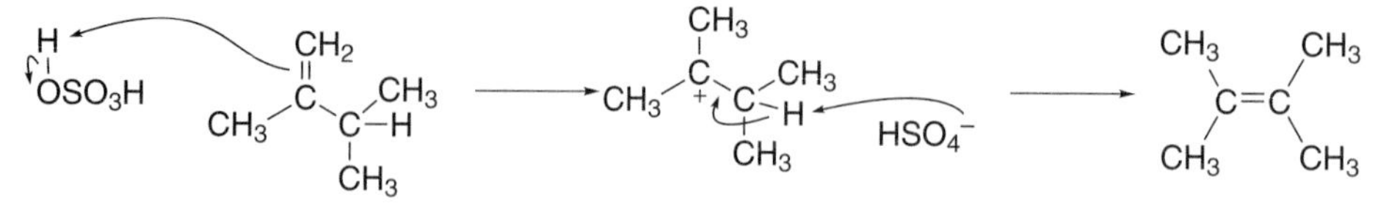

**10.60**

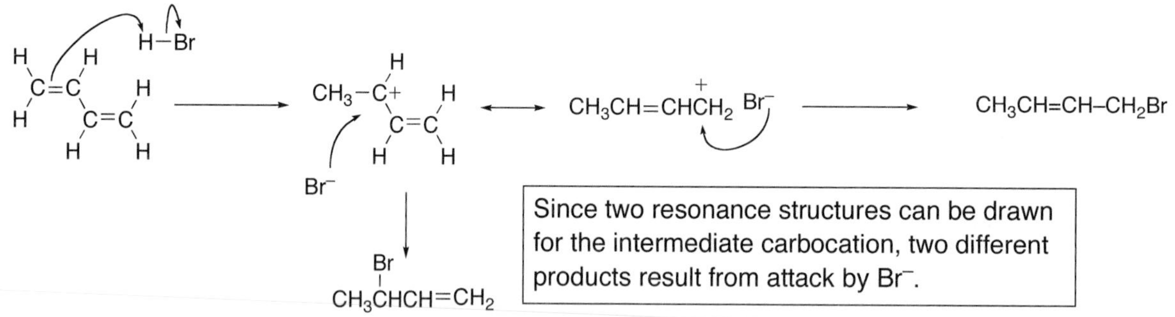

Since two resonance structures can be drawn for the intermediate carbocation, two different products result from attack by Br⁻.

**10.61**

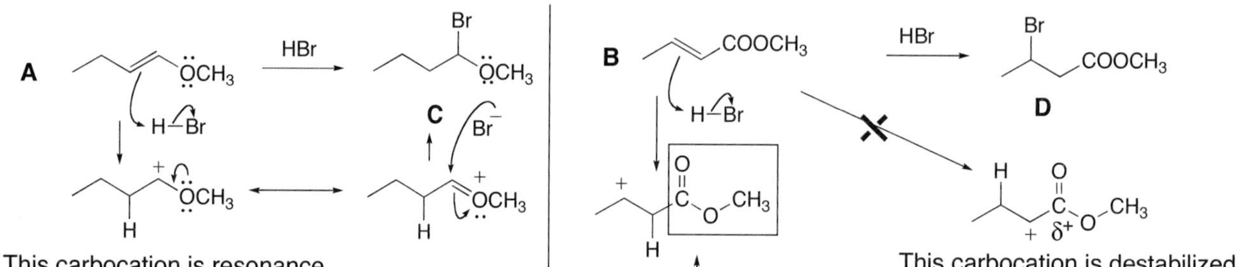

**A**

This carbocation is resonance stabilized by the O atom, and therefore preferentially forms and results in **C**.

This carbocation is formed preferentially and results in product **D**. It is not destabilized by an adjacent electron-withdrawing COOCH₃ group.

This carbocation is destabilized by the δ⁺ on the adjacent C, so it does not form.

**10.62**

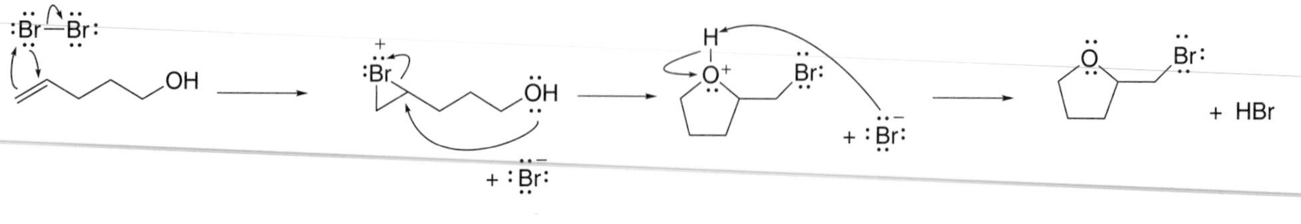

**10.63**

a.

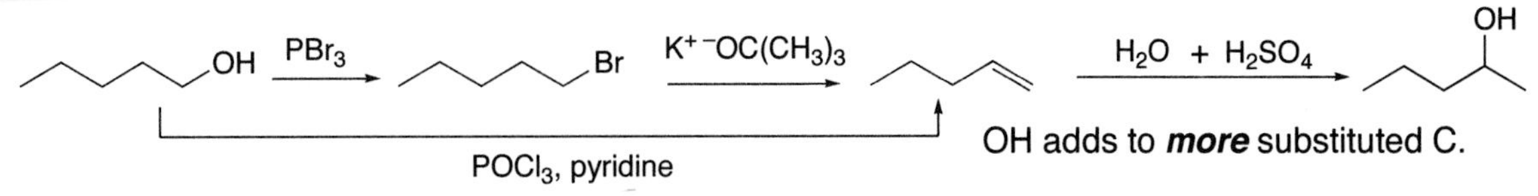

OH adds to **more** substituted C.

b.

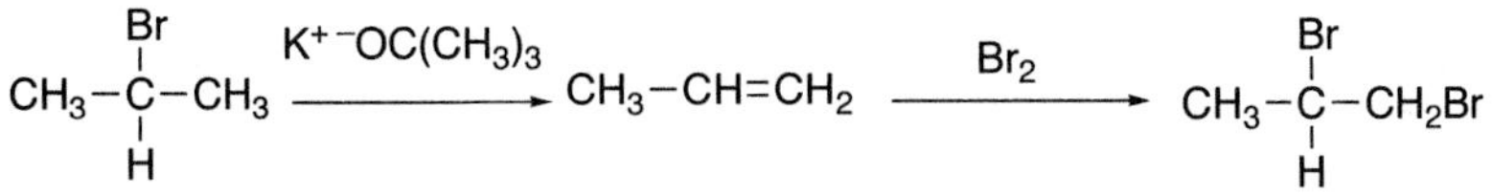

c.

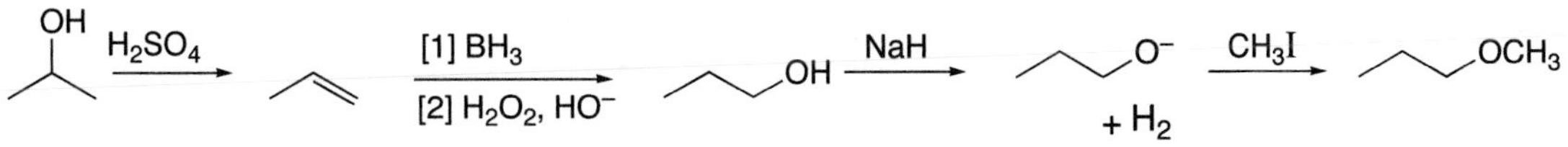

+ H₂

d.

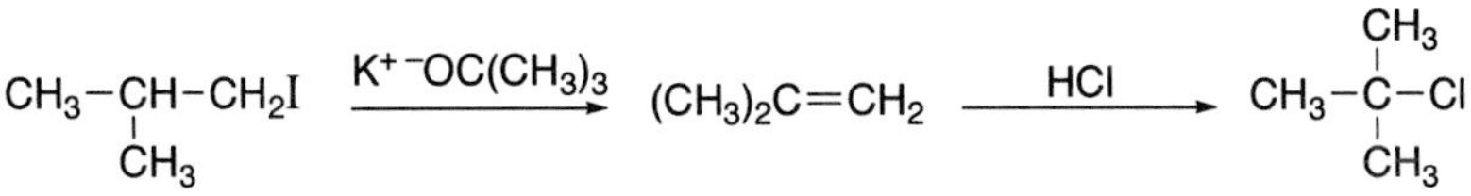

e.

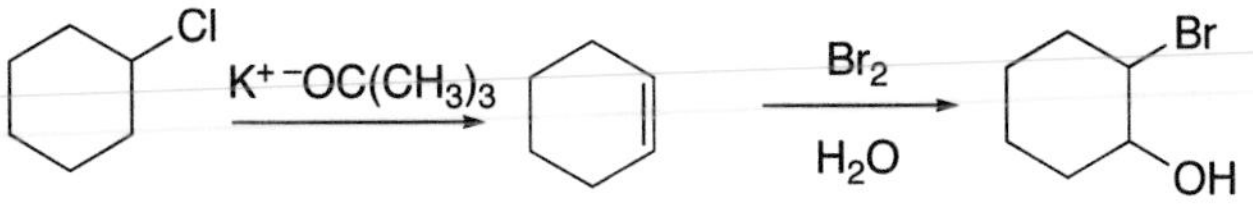

f.   CH₃CH=CH₂ →(Br₂) CH₃CH–CH₂ (Br Br) →(NaNH₂, (2 equiv)) CH₃C≡CH

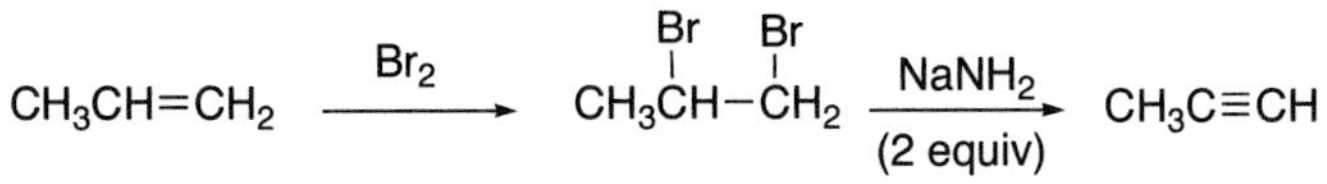

**10.64**

a.

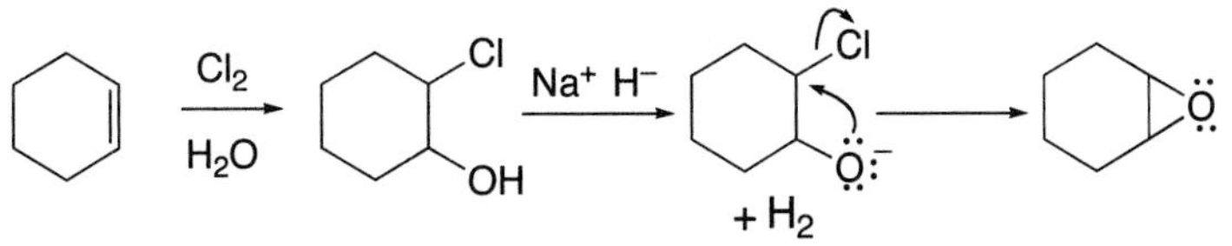

+ H₂

b.

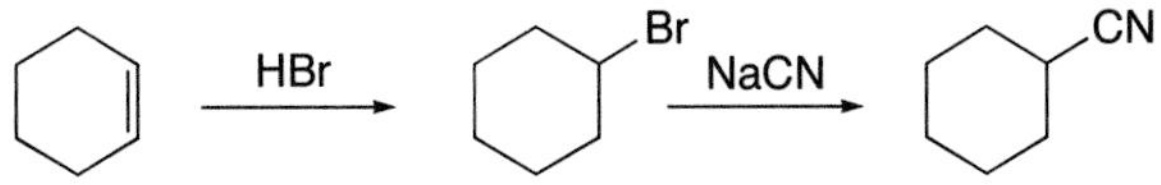

c.

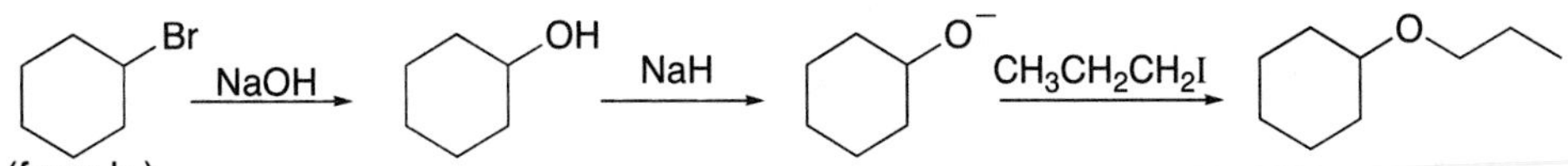

(from b.)

d.

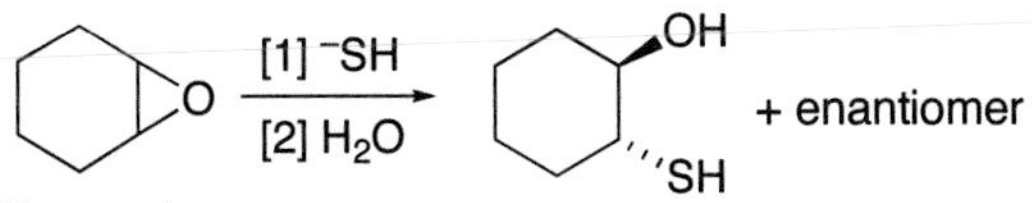

+ enantiomer

(from a.)

e.

+ enantiomer

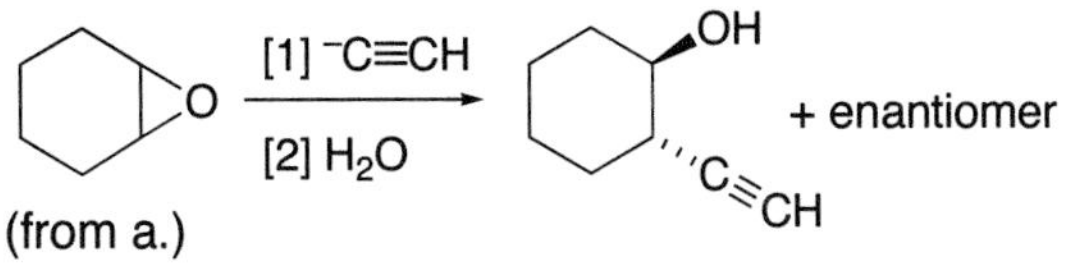

(from a.)

**10.65**

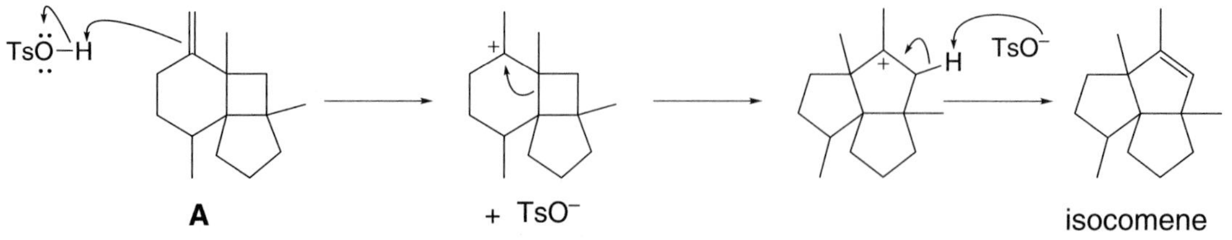
TsÖ–H
+ TsO⁻
TsO⁻
A
isocomene

**10.66**

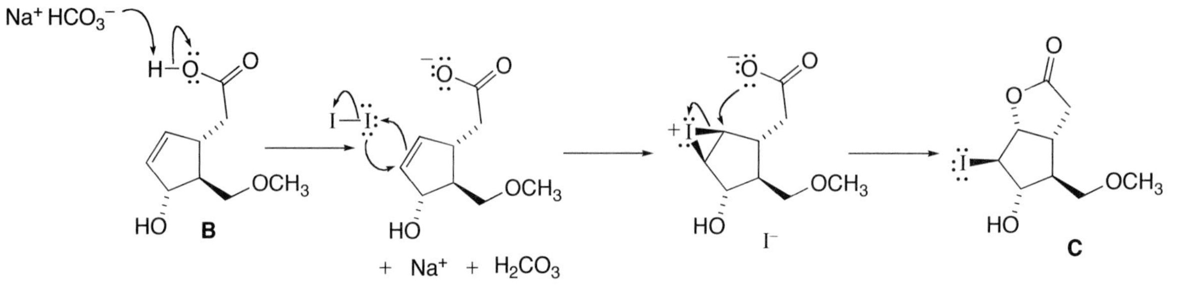
Na⁺ HCO₃⁻
OCH₃
HO
B
+ Na⁺ + H₂CO₃
I⁻
C

**10.67**

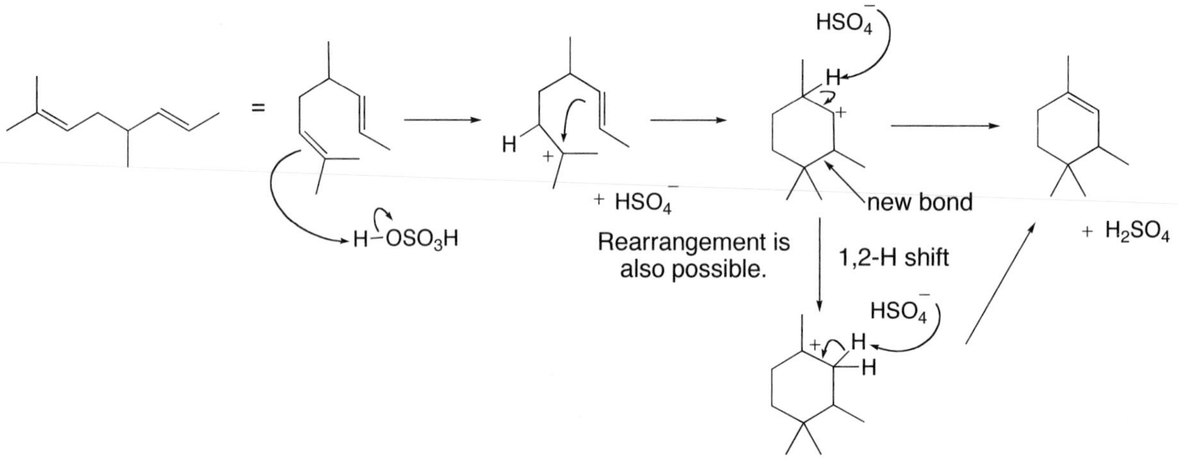
H–OSO₃H
+ HSO₄⁻
HSO₄⁻
new bond
Rearrangement is also possible.
1,2-H shift
+ H₂SO₄

# Chapter 11: Alkynes

## ♦ General facts about alkynes

- Alkynes contain a carbon–carbon triple bond consisting of a strong σ bond and two weak π bonds. Each carbon is *sp* hybridized and linear (11.1).

$$H-C\equiv C-H$$
acetylene

180°

*sp* hybridized

- Alkynes are named using the suffix **-yne** (11.2).
- Alkynes have weak intermolecular forces, giving them low mp's and low bp's, and making them water insoluble (11.3).
- Since its weaker π bonds make an alkyne electron rich, alkynes undergo addition reactions with electrophiles (11.6).

## ♦ Addition reactions of alkynes

### [1] Hydrohalogenation—Addition of HX (X = Cl, Br, I) (11.7)

$$R-C\equiv C-H \xrightarrow[\text{(2 equiv)}]{H-X}$$

X H
|   |
R−C−C−H
|   |
X H
geminal dihalide

- Markovnikov's rule is followed. H bonds to the less substituted C in order to form the more stable carbocation.

### [2] Halogenation—Addition of X₂ (X = Cl or Br) (11.8)

$$R-C\equiv C-H \xrightarrow[\text{(2 equiv)}]{X-X}$$

X X
|   |
R−C−C−H
|   |
X X
tetrahalide

- Bridged halonium ions are formed as intermediates.
- Anti addition of X₂ occurs.

### [3] Hydration—Addition of H₂O (11.9)

$$R-C\equiv C-H \xrightarrow[\substack{H_2SO_4 \\ HgSO_4}]{H_2O}$$

R   H
 \ /
  C=C
 /   \
HO    H
enol

O
||
R−C−CH₃
ketone

- Markovnikov's rule is followed. H bonds to the less substituted C in order to form the more stable carbocation.
- The unstable enol that is first formed rearranges to a carbonyl group.

---

**[4] Hydroboration–oxidation—Addition of $H_2O$ (11.10)**

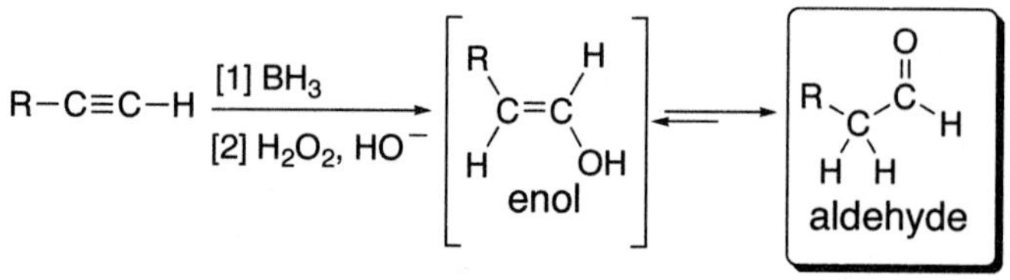

- The unstable enol, first formed after oxidation, rearranges to a carbonyl group.

---

### ♦ Reactions involving acetylide anions

**[1] Formation of acetylide anions from terminal alkynes (11.6B)**

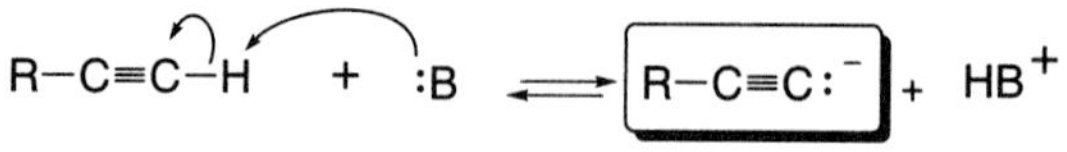

- Typical bases used for the reaction are $NaNH_2$ and $NaH$.

---

**[2] Reaction of acetylide anions with alkyl halides (11.11A)**

- The reaction follows an $S_N2$ mechanism.
- The reaction works best with $CH_3X$ and $RCH_2X$.

---

**[3] Reaction of acetylide anions with epoxides (11.11B)**

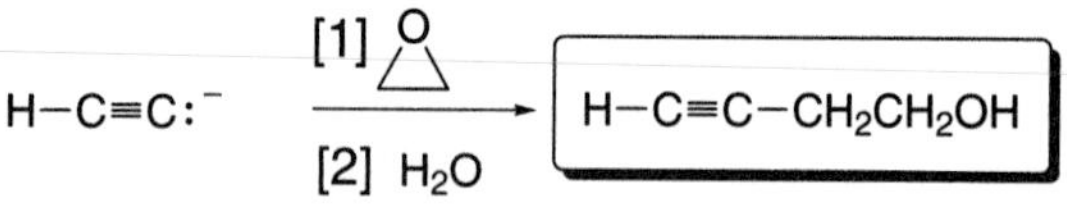

- The reaction follows an $S_N2$ mechanism.
- Ring opening occurs from the back side at the less substituted end of the epoxide.

**11.1**• An *internal alkyne* has the triple bond somewhere in the *middle* of the carbon chain.
  • A *terminal alkyne* has the triple bond at the *end* of the carbon chain.

$HC{\equiv}C-CH_2CH_2CH_3$    $CH_3-C{\equiv}C-CH_2CH_3$    $HC{\equiv}C-\underset{\underset{CH_3}{|}}{CH}-CH_3$

terminal alkyne          internal alkyne          terminal alkyne

**11.2**

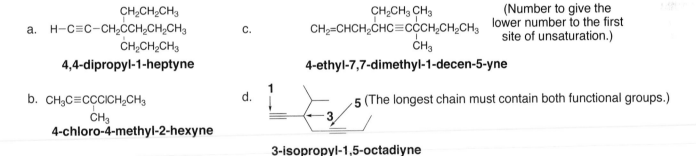

Increasing bond strength: a < c < b

**11.3** Like alkenes, the larger the number of alkyl groups bonded to the *sp* hybridized C, the more stable the alkyne. This makes internal alkynes more stable than terminal alkynes.

**11.4** To name an alkyne:
  [1] Find the longest chain that contains both atoms of the triple bond, and number the chain to give the first carbon of the triple bond the lower number.
  [2] Name all substituents following the other rules of nomenclature.

a.  $H-C{\equiv}C-CH_2\underset{\underset{CH_2CH_2CH_3}{|}}{\overset{\overset{CH_2CH_2CH_3}{|}}{C}}CH_2CH_2CH_3$

**4,4-dipropyl-1-heptyne**

c.  $CH_2{=}CHCH_2\underset{\underset{CH_3}{|}}{\overset{\overset{CH_2CH_3}{|}}{CH}}C{\equiv}C\overset{\overset{CH_3}{|}}{C}CH_2CH_2CH_3$    (Number to give the lower number to the first site of unsaturation.)

**4-ethyl-7,7-dimethyl-1-decen-5-yne**

b.  $CH_3C{\equiv}C\underset{\underset{CH_3}{|}}{C}ClCH_2CH_3$

**4-chloro-4-methyl-2-hexyne**

d.  (The longest chain must contain both functional groups.)

**3-isopropyl-1,5-octadiyne**

**11.5 To work backwards from a name to a structure:**
  [1] Find the parent name and the functional group.
  [2] Add the substituents to the appropriate carbon.

a.  *trans*-2-ethynyl**cyclopentanol**
    5 C ring with OH at C1

b.  4-*tert*-butyl-5-**decyne**
    10 C chain with
    a triple bond

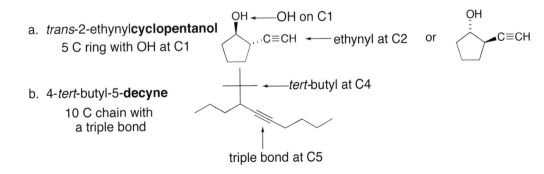

OH ← OH on C1
C≡CH ← ethynyl at C2    or
tert-butyl at C4
triple bond at C5

c. 3-methyl**cyclononyne**
9 C ring with a
triple bond at C1

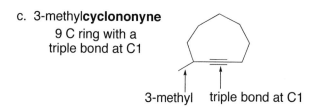

3-methyl    triple bond at C1

**11.6** Two factors cause the boiling point increase. The linear *sp* hybridized C's of the alkyne allow for more van der Waals attraction between alkyne molecules. Also, since a triple bond is more polarizable than a double bond, this increases the van der Waals forces between two molecules as well.

**11.7 To convert an alkene to an alkyne:**
[1] Make a vicinal dihalide from the alkene by addition of $X_2$.
[2] Add base to remove two equivalents of HX and form the alkyne.

a. $Br_2CH(CH_2)_4CH_3$ $\xrightarrow{\text{Na}^+ \ ^-NH_2}$ $\left[ BrCH{=}CHCH_2CH_2CH_2CH_3 \right]$ $\xrightarrow{\text{Na}^+ \ ^-NH_2}$ $HC{\equiv}CCH_2CH_2CH_2CH_3$
not isolated

b. $CH_2{=}CCl(CH_2)_3CH_3$ $\xrightarrow{\text{Na}^+ \ ^-NH_2}$ $HC{\equiv}CCH_2CH_2CH_2CH_3$

c. $CH_2{=}CH(CH_2)_3CH_3$ $\xrightarrow{Cl_2}$ $\underset{\underset{Cl \ \ \ Cl}{|\ \ \ |}}{CH_2CHCH_2CH_2CH_2CH_3}$ $\xrightarrow[\text{(2 equiv)}]{\text{Na}^+ \ ^-NH_2}$ $HC{\equiv}CCH_2CH_2CH_2CH_3$

**11.8**

$CH_3CH_2{-}\overset{\overset{Br \ H}{|\ \ |}}{C}{-}\overset{}{C}{-}H$     $\overset{CH_3CH_2}{\underset{Br \ \ \ H}{>}}C{=}C\overset{H}{<}$ :NH_2     $\boxed{CH_3CH_2{-}C{\equiv}C{-}H}$
$\underset{Br \ H}{|\ \ |}$
:NH_2     H_2N:                                            1-butyne

$CH_3{-}\overset{\overset{H \ \ \ Br}{|\ \ \ |}}{C}{-}\overset{}{C}{-}CH_3$     $\overset{H}{\underset{CH_3}{>}}C{=}C\overset{CH_3}{<}$ Br     $\boxed{CH_3{-}C{\equiv}C{-}CH_3}$     $\boxed{\begin{array}{c}\textbf{major product}\\ \text{more substituted alkyne}\end{array}}$
$\underset{H \ \ \ Br}{|\ \ \ |}$
:NH_2                       (*E* and *Z* isomers possible)     2-butyne

$CH_3CH_2{-}\overset{\overset{Br \ H}{|\ \ |}}{C}{-}\overset{}{C}{-}H$     :NH_2 $\overset{H}{\underset{CH_3{-}C{-}C}{>}}\overset{H}{<}$ C{-}H     $\boxed{CH_3CH{=}C{=}CH_2}$
$\underset{Br \ H}{|\ \ |}$                                   $\underset{H \ \ \ Br}{|\ \ \ |}$
:NH_2                                              1,2-butadiene

Reaction by-products:
$2 \ \ddot{N}H_3 \ + \ 2 \ Br^-$

**11.9** Acetylene has a p$K_a$ of 25, so **bases having a conjugate acid with a p$K_a$ *above* 25 will be able to** deprotonate it.

a. $CH_3NH^-$ [p$K_a$ $(CH_3NH_2)$= 40]
   p$K_a$ > 25 = **Can deprotonate acetylene.**
b. $CO_3^{2-}$ [p$K_a$ $(HCO_3^-)$= 10.2]
   p$K_a$ < 25 = **Cannot deprotonate acetylene.**

c. $CH_2{=}CH^-$ [p$K_a$ $(CH_2{=}CH_2)$ = 44]
   p$K_a$ > 25 = **Can deprotonate acetylene.**
d. $(CH_3)_3CO^-$ {p$K_a$ [$(CH_3)_3COH$]= 18}
   p$K_a$ < 25 = **Cannot deprotonate acetylene.**

**11.10** To draw the products of reactions with HX:
  • Add two moles of HX to the triple bond, following Markovnikov's rule.
  • Both X's end up on the more substituted C.

a. $CH_3CH_2CH_2CH_2-C\equiv C-H$ $\xrightarrow{\text{2 HBr}}$ $CH_3CH_2CH_2CH_2-\overset{\overset{\displaystyle Br}{|}}{\underset{\underset{\displaystyle Br}{|}}{C}}-CH_3$

b. $CH_3-C\equiv C-CH_2CH_3$ $\xrightarrow{\text{2 HBr}}$ $CH_3-CH_2-\overset{\overset{\displaystyle Br}{|}}{\underset{\underset{\displaystyle Br}{|}}{C}}-CH_2CH_3$ + $CH_3-\overset{\overset{\displaystyle Br}{|}}{\underset{\underset{\displaystyle Br}{|}}{C}}-CH_2-CH_2CH_3$

c. $-C\equiv CH$ $\xrightarrow{\text{2 HBr}}$ $-\overset{\overset{\displaystyle Br}{|}}{\underset{\underset{\displaystyle Br}{|}}{C}}-CH_3$

**11.11**

a.

b. $CH_3-\overset{..}{\underset{..}{O}}\overset{\curvearrowright}{-}\overset{+}{C}H_2$ $\longleftrightarrow$ $CH_3-\overset{+}{\underset{..}{O}}=CH_2$

c.

**11.12** Addition of one equivalent of $X_2$ to alkynes forms a trans dihalide.
  Addition of two equivalents of $X_2$ to alkynes forms tetrahalides.

$CH_3CH_2-C\equiv C-CH_2CH_3$ $\xrightarrow{\text{2 Br}_2}$ $CH_3CH_2-\overset{\overset{\displaystyle Br}{|}}{\underset{\underset{\displaystyle Br}{|}}{C}}-\overset{\overset{\displaystyle Br}{|}}{\underset{\underset{\displaystyle Br}{|}}{C}}-CH_2CH_3$

$CH_3CH_2-C\equiv C-CH_2CH_3$ $\xrightarrow{\text{Cl}_2}$ 
$\underset{CH_3CH_2}{\overset{Cl}{>}}C=C\underset{Cl}{\overset{CH_2CH_3}{<}}$ **trans dihalide**

**11.13**

$CH_3-C\equiv C-CH_3$ $\xrightarrow{\text{Cl}_2}$ $\underset{CH_3}{\overset{Cl}{>}}C=C\underset{Cl}{\overset{CH_3}{<}}$ The two Cl atoms are electron withdrawing, making the π bond less electron rich and therefore less reactive with an electrophile.

**11.14 To draw the keto form of each enol:**
  [1] Change the C–OH to a C=O at one end of the double bond.
  [2] At the other end of the double bond, add a proton.

a. $\longleftrightarrow$
new C–H bond

c. $\longrightarrow$
new C–H bond

b. $\longleftrightarrow$
new C–H bond

**11.15** Treatment of alkynes with $H_2O$, $H_2SO_4$, and $HgSO_4$ yields ketones.

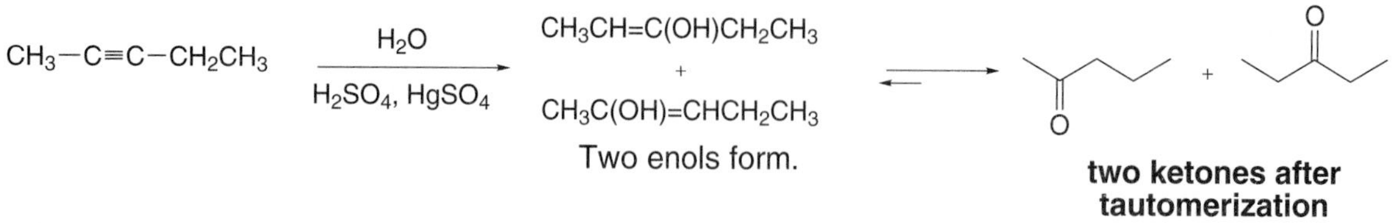

$CH_3-C\equiv C-CH_2CH_3$ $\xrightarrow[H_2SO_4,\ HgSO_4]{H_2O}$ $CH_3CH=C(OH)CH_2CH_3$

+

$CH_3C(OH)=CHCH_2CH_3$

Two enols form.

**two ketones after
tautomerization**

**11.16** To determine what two alkynes could yield the given ketone, work backwards by drawing the enols and then the alkynes.

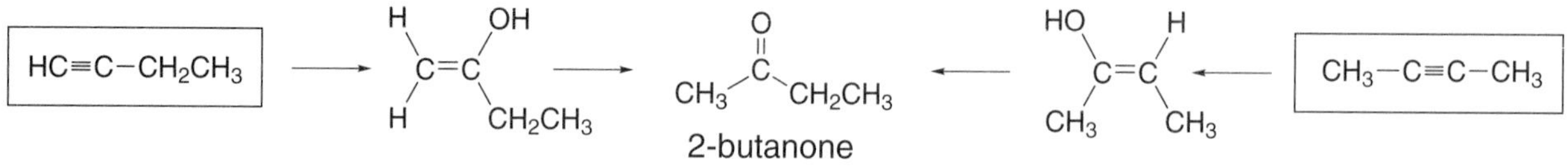

2-butanone

**11.17** Reaction with $H_2O$, $H_2SO_4$, and $HgSO_4$ adds the oxygen to the *more* substituted carbon. Reaction with [1] $BH_3$, [2] $H_2O_2$, $^-OH$ adds the oxygen to the *less* substituted carbon.

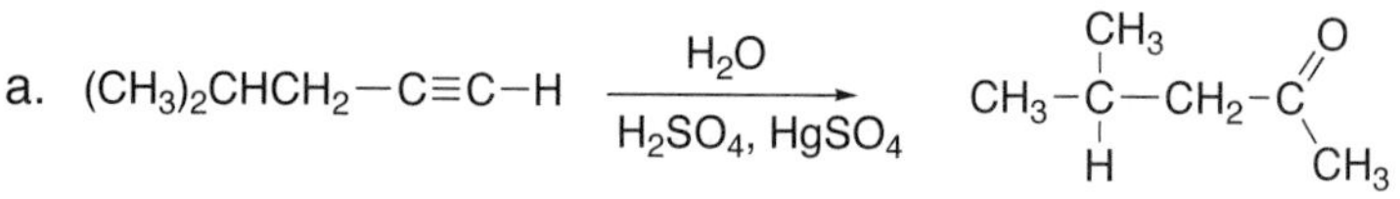

a. $(CH_3)_2CHCH_2-C\equiv C-H$ $\xrightarrow[H_2SO_4,\ HgSO_4]{H_2O}$ Forms a **ketone.** $H_2O$ is added with the O atom on the *more* substituted carbon.

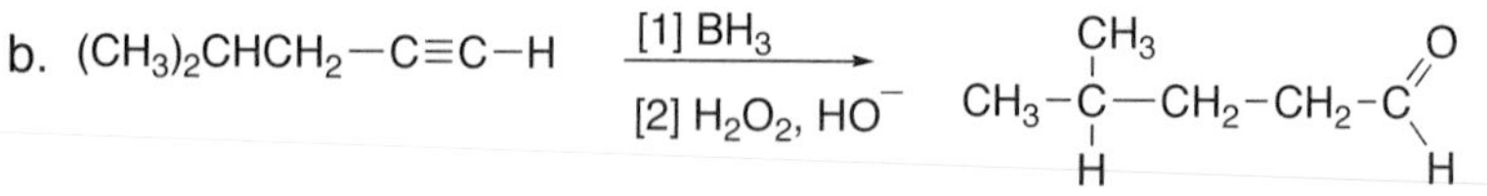

b. $(CH_3)_2CHCH_2-C\equiv C-H$ $\xrightarrow[[2]\ H_2O_2,\ HO^-]{[1]\ BH_3}$ Forms an **aldehyde.** $H_2O$ is added with the O atom on the *less* substituted carbon.

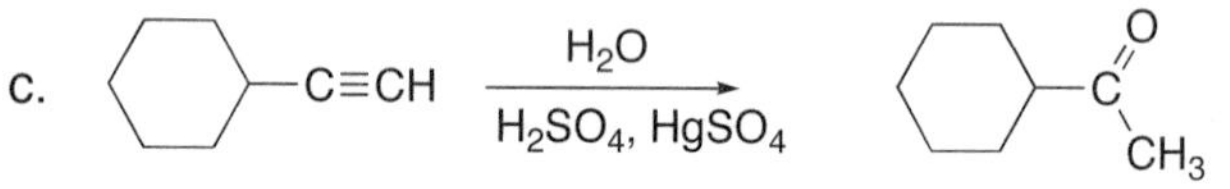

c. $\xrightarrow[H_2SO_4,\ HgSO_4]{H_2O}$ Forms a **ketone.** $H_2O$ is added with the O atom on the *more* substituted carbon.

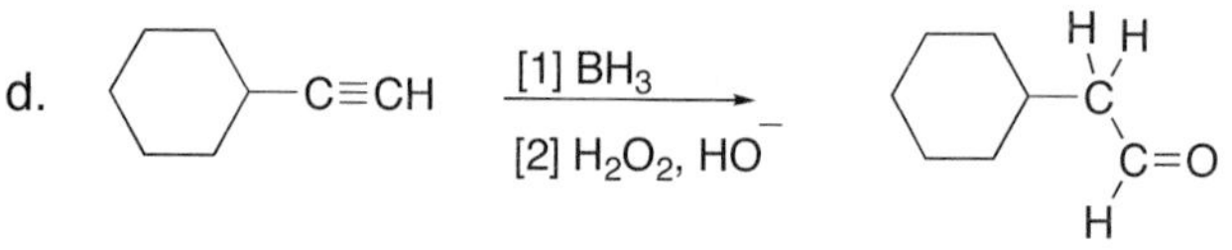

d. $\xrightarrow[[2]\ H_2O_2,\ HO^-]{[1]\ BH_3}$ Forms an **aldehyde.** $H_2O$ is added with the O atom on the *less* substituted carbon.

**11.18**

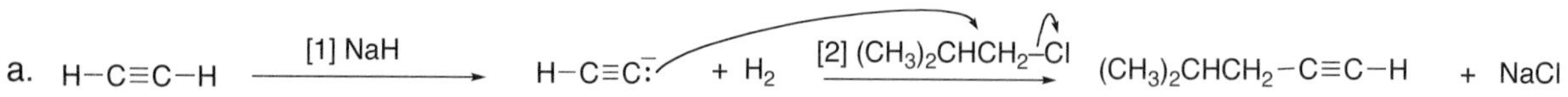

a. $H-C\equiv C-H$ $\xrightarrow{[1]\ NaH}$ $H-C\equiv C:^-$ $+\ H_2$ $\xrightarrow{[2]\ (CH_3)_2CHCH_2-Cl}$ $(CH_3)_2CHCH_2-C\equiv C-H$ $+\ NaCl$

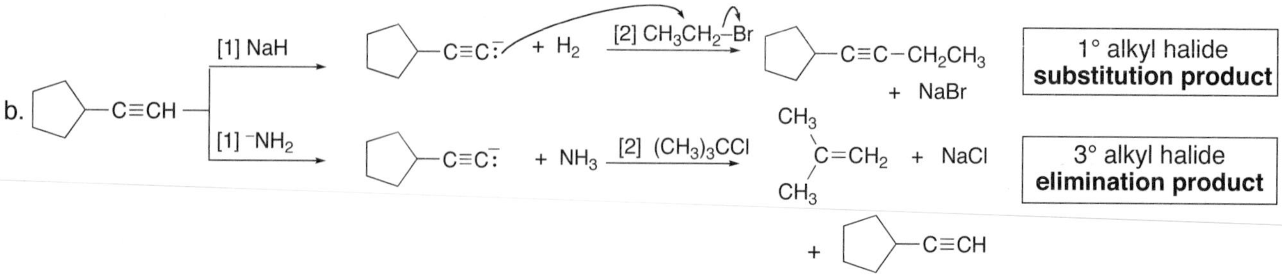

b.

**11.19**

a.  $(CH_3)_2CHCH_2C{\equiv}CH$

$CH_3-\overset{\underset{|}{CH_3}}{\underset{|}{H}}{C}-CH_2-\{-C{\equiv}C-H \implies CH_3-\overset{\underset{|}{CH_3}}{\underset{|}{H}}{C}-CH_2Cl + {}^-C{\equiv}C-H$

1° RX

**terminal alkyne**
only one possibility

b. $CH_3C{\equiv}CCH_2CH_2CH_2CH_3$

$CH_3\cdot\{-C{\equiv}C-\}-CH_2CH_2CH_2CH_3 \implies$
[1]  [2]

[1]  $CH_3Cl + {}^-C{\equiv}C-CH_2CH_2CH_2CH_3$

[2]  $CH_3-C{\equiv}C^- + Cl-CH_2CH_2CH_2CH_3$

1° RX

**internal alkyne**
two possibilities

c.  $(CH_3)_3CC{\equiv}CCH_2CH_3$

$CH_3-\overset{\underset{|}{CH_3}}{\underset{|}{CH_3}}{C}-C{\equiv}C-\}-CH_2CH_3 \implies CH_3-\overset{\underset{|}{CH_3}}{\underset{|}{CH_3}}{C}-C{\equiv}C^- + Cl-CH_2CH_3$

1° RX

**internal alkyne**
only one possibility

$\overset{\times}{\implies} CH_3-\overset{\underset{|}{CH_3}}{\underset{|}{CH_3}}{C}-Cl + {}^-C{\equiv}CCH_2CH_3$

3° RX
too crowded for S$_N$2 reaction

The 3° alkyl halide
would undergo elimination.

**11.20**

$HC{\equiv}C-H \xrightarrow{Na^+H:^-} HC{\equiv}C:^- \xrightarrow{CH_3CH_2-Br} H-C{\equiv}C-CH_2CH_3 \xrightarrow{Na^+H:^-} {}^-:C{\equiv}C-CH_2CH_3 + H_2$

+ H$_2$     + Na$^+$        + Na$^+$Br$^-$

$(CH_3)_2CHCH_2\overset{\underset{|}{H}}{\underset{|}{H}}{C}-Br$

$(CH_3)_2CHCH_2CH_2-C{\equiv}C-CH_2CH_3 + Br^-$

**11.21**

$CH_3-\overset{\underset{|}{CH_3}}{\underset{|}{CH_3}}{C}-C{\equiv}C-\overset{\underset{|}{CH_3}}{\underset{|}{CH_3}}{C}-CH_3 \implies CH_3-\overset{\underset{|}{CH_3}}{\underset{|}{CH_3}}{C}-C{\equiv}C^- + X-\overset{\underset{|}{CH_3}}{\underset{|}{CH_3}}{C}-CH_3$

2,2,5,5-tetramethyl-3-hexyne

The 3° alkyl halide is too crowded for nucleophilic substitution.  Instead, it would undergo elimination with the acetylide anion.

**11.22**

a.  $\xrightarrow[\text{[2] }H_2O]{\text{[1] }{}^-:C{\equiv}C-H}$

Epoxide is drawn up, so the
**acetylide anion attacks from *below*
at *less* substituted C.**

b.  $\xrightarrow[\text{[2] }H_2O]{\text{[1] }{}^-:C{\equiv}C-H}$ +

enantiomers

Backside attack of the nucleophile
($^-C{\equiv}CH$) at either C since both
ends are equally substituted.

**11.23** To use a retrosynthetic analysis:

[1] **Count the number of carbon atoms** in the starting material and product.

[2] **Look at the functional groups** in the starting material and product.

Determine what types of reactions can form the product.

Determine what types of reactions the starting material can undergo.

[3] **Work backwards** from the product to make the starting material.

[4] Write out the synthesis in synthetic direction.

$$CH_3CH_2C\equiv CCH_2CH_3 \xrightarrow{\ ?\ } HC\equiv CH$$
6 C's                              2 C's

$$CH_3CH_2C\equiv CCH_2CH_3 \Longrightarrow CH_3CH_2C\equiv C^- + CH_3CH_2Br \Longrightarrow HC\equiv C^- + CH_3CH_2Br$$

$$HC\equiv C-H \xrightarrow{Na^+ H:^-} HC\equiv C:^- \xrightarrow{CH_3CH_2-Br} CH_3CH_2C\equiv C-H \xrightarrow{Na^+ H:^-} CH_3CH_2C\equiv C:^- \xrightarrow{CH_3CH_2-Br} CH_3CH_2C\equiv CCH_2CH_3$$

**11.24**

$$CH_3CH_2CH_2-C\!\!\begin{array}{c}O\\ \diagup\!\diagdown\\ H\end{array} \Longrightarrow HC\equiv CH$$

product:
**4 carbons, aldehyde functional group**
(can be made by hydroboration–oxidation of
a terminal alkyne)

starting material:
**2 carbons, C≡C functional group**
(can form an acetylide
anion by reaction with NaH)

Retrosynthetic analysis:  $CH_3CH_2CH_2-C\!\!\begin{array}{c}O\\ \diagup\!\diagdown\\ H\end{array} \Longrightarrow CH_3CH_2C\equiv C-H \Longrightarrow H-C\equiv C-H$

Forward direction: $H-C\equiv C-H \xrightarrow{Na^+ H:^-} {}^-:C\equiv C-H \xrightarrow{CH_3CH_2-Br} CH_3CH_2C\equiv C-H \xrightarrow[{[2]\ H_2O_2,\ HO^-}]{[1]\ BH_3} CH_3CH_2CH_2-C\!\!\begin{array}{c}O\\ \diagup\!\diagdown\\ H\end{array}$

**11.25** Use the rules from Answer 11.4 to name the alkynes.

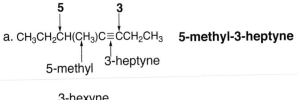

1-hexyne          3-hexyne          3-methyl-1-pentyne          3,3-dimethyl-1-butyne

2-hexyne     4-methyl-1-pentyne     4-methyl-2-pentyne

**11.26** Use the rules from Answer 11.4 to name the alkynes.

a. $\overset{5}{\downarrow}$ $\overset{3}{\downarrow}$

a. $CH_3CH_2CH(CH_3)C\equiv CCH_2CH_3$  **5-methyl-3-heptyne**

        5-methyl $\overset{\uparrow}{}$ 3-heptyne $\overset{\uparrow}{}$

$CH_3 \leftarrow$ 3-methyl

e. $CH_3CH_2\!\!\boxed{\begin{array}{c}C-C\equiv CH \\ CH_2CH_2CH_3\end{array}}$  **3-ethyl-3-methyl-1-hexyne**

     3-ethyl $\uparrow$

3-hexyne $\downarrow$

b. $CH_3CHC\equiv CCHCH_3$  **2,5-dimethyl-3-hexyne**

     $CH_3$     $CH_3$

      2,5-dimethyl $\uparrow$

f. $CH_3CH_2C\equiv CCH_2C\equiv CCH_3$  **2,5-octadiyne**

        **5**      **2**

4-nonyne $\downarrow$   $CH_3 \leftarrow$ 7-methyl

c. $CH_3CH_2CHC\equiv CCHCHCH_2CH_3$

     $CH_3CH_2$      $CH_2CH_3$

           **3,6-diethyl-7-methyl-4-nonyne**

      3,6-diethyl $\uparrow$

g. **2**      $\leftarrow$ 5-ethyl

      **6**   **(E)-4,5-diethyl-2-decen-6-yne**

    $\leftarrow$ 4-ethyl

6-methyl $\searrow$     1

h. $\equiv \leftarrow$ ethynyl

     **1-ethynyl-6-methylcyclohexene**

d. $HC\equiv C-CH(CH_2CH_3)CH_2CH_2CH_3$  **3-ethyl-1-hexyne**

  1-hexyne $\uparrow$    3-ethyl $\uparrow$

**11.27** Use the directions from Answer 11.5 to draw each structure.

a. 5,6-dimethyl-2-**heptyne**

2-heptyne $\longrightarrow$

b. 5-*tert*-butyl-6,6-dimethyl-3-**nonyne**

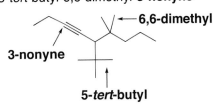

           $\leftarrow$ **6,6-dimethyl**

3-nonyne $\nearrow$

        5-*tert*-**butyl** $\uparrow$

c. (*S*)-4-chloro-2-**pentyne**

Cl ← **4-chloro**
···H *S* stereochemistry

**2-pentyne**

e. 3,4-dimethyl-1,5-**octadiyne**

**3,4-dimethyl**

**diyne**    **diyne**

d. *cis*-1-ethynyl-2-methyl**cyclopentane**

**1-ethynyl**

⫶C≡CH    or    ···C≡CH

← **2-methyl**

f. (*Z*)-6-methyl-6-**octen-1-yne**

← **6-methyl**

**Z**

**1**

**11.28**  Keto–enol tautomers are constitutional isomers in equilibrium that differ in the location of a double bond and a hydrogen.  The OH in an enol must be bonded to a C=C.

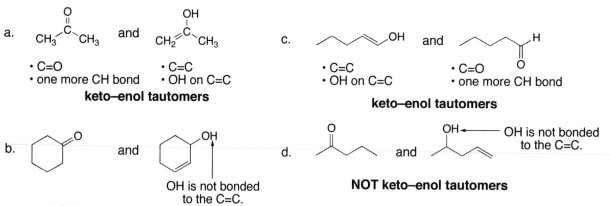

a.

CH₃–C(=O)–CH₃    and    CH₂=C(OH)–CH₃

• C=O                • C=C
• one more CH bond    • OH on C=C
**keto–enol tautomers**

c.

~~~OH    and    ~~~CHO (H, O)

• C=C                • C=O
• OH on C=C          • one more CH bond
**keto–enol tautomers**

b.

(cyclohexanone) and (cyclohexenol with OH)

OH is not bonded
to the C=C.
**NOT keto–enol tautomers**

d.

(ketone) and (enol OH)    OH ← OH is not bonded
to the C=C.

**NOT keto–enol tautomers**

**11.29**  **To draw the enol form of each keto form:**
[1] Change the C=O to a C–OH.
[2] Change one single C–C bond to a double bond, making sure the OH group is bonded to the C=C.

a.  ⟶  [E/Z isomers possible]

c.  (cyclohexanone with CH₃) ⟶ (enol OH, CH₃) + (enol OH, CH₃)

b.  CH₃CH₂CHO =  ⟶  [E/Z isomers possible]

**11.30**  Use the directions from Answer 11.14 to draw each keto form.

a.  ⟶

b.  OH ⟶

c.  (cyclohexenol OH) ⟶ (cyclohexanone O)

**11.31**

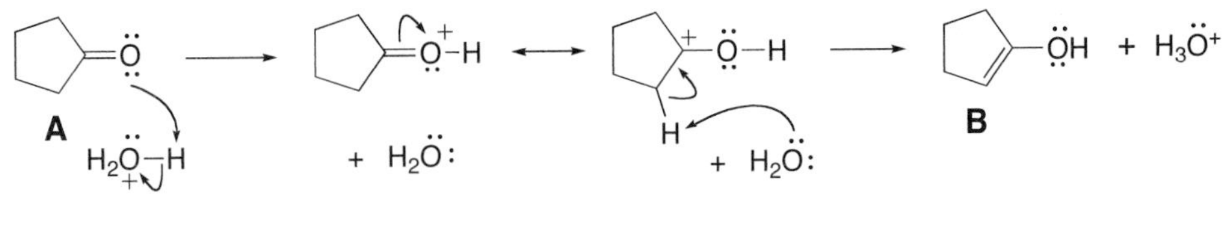

**11.32**

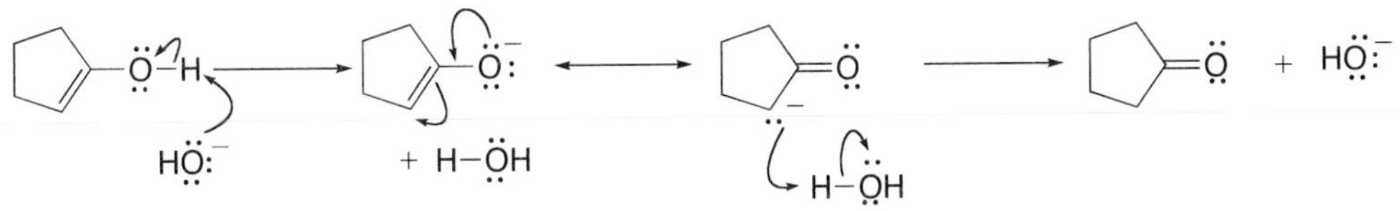

**11.33**

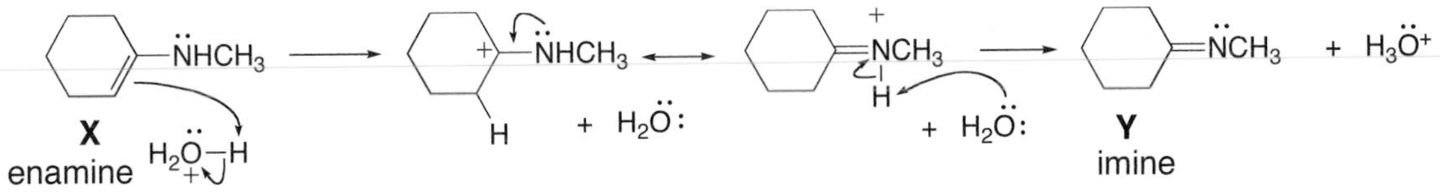

**11.34** The equilibrium always favors the formation of the weaker acid and the weaker base.

a. $HC\equiv C^- + CH_3OH \rightleftharpoons HC\equiv CH + CH_3O^-$

    $pK_a = 15.5$      $pK_a = 25$
                    weaker acid
**Equilibrium favors products.**

c. $HC\equiv CH + Na^+Br^- \rightleftharpoons HC\equiv C^- Na^+ + HBr$

    $pK_a = 25$                         $pK_a = -9$
   weaker acid
**Equilibrium favors
starting materials.**

b. $CH_3C\equiv CH + CH_3^- \rightleftharpoons CH_3C\equiv C^- + CH_4$

  $pK_a = 25$                         $pK_a = 50$
                             weaker acid
**Equilibrium favors products.**

d. $CH_3CH_2C\equiv C^- + CH_3COOH \rightleftharpoons CH_3CH_2C\equiv CH + CH_3COO^-$

                 $pK_a = 4.8$              $pK_a = 25$
                                weaker acid
**Equilibrium favors products.**

**11.35**

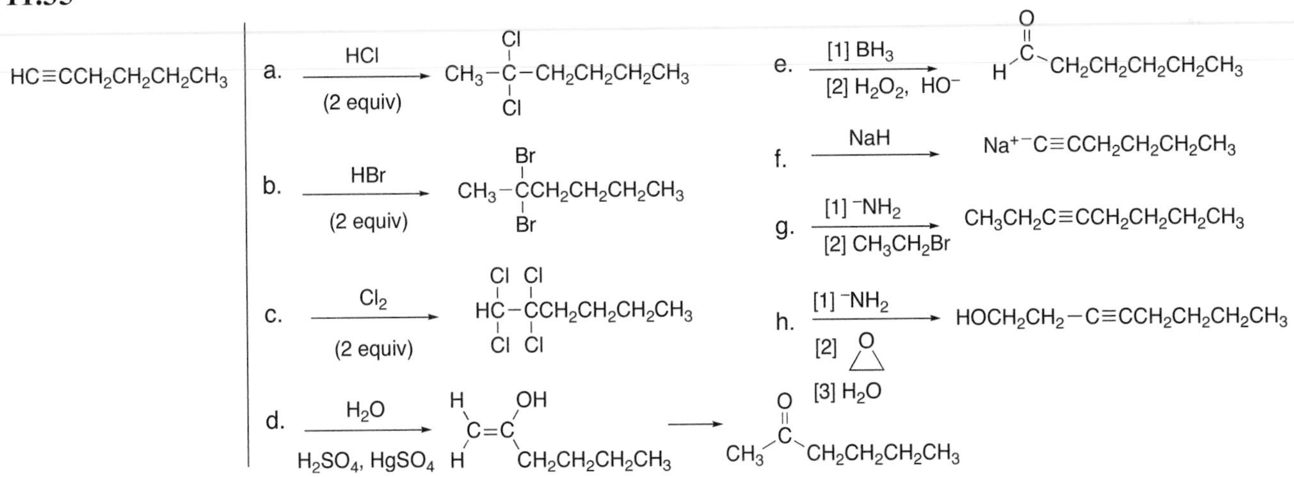

**11.36**

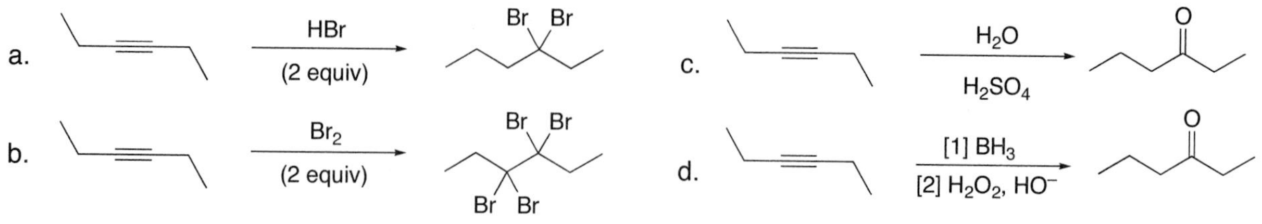

a. HBr (2 equiv) →

c. H₂O / H₂SO₄ →

b. Br₂ (2 equiv) →

d. [1] BH₃ / [2] H₂O₂, HO⁻ →

**11.37**

a. $(CH_3CH_2)_3C-C{\equiv}CH$

- $\xrightarrow[\text{H}_2\text{SO}_4,\ \text{HgSO}_4]{\text{H}_2\text{O}}$ $(CH_3CH_2)_3C-\overset{\overset{\displaystyle O}{\|}}{C}-CH_3$
- $\xrightarrow[\text{[2] H}_2\text{O}_2,\ \text{HO}^-]{\text{[1] BH}_3}$ $(CH_3CH_2)_3C-CH_2CHO$
- $\xrightarrow[\text{[2] CH}_3\text{Br}]{\text{[1] NaH}}$ $(CH_3CH_2)_3C-C{\equiv}CCH_3$

b. $CH_3CH_2CH_2CH(CH_3)CH_2CHBr_2$ $\xrightarrow{2\ ^-\text{NH}_2}$ $CH_3CH_2CH_2CH(CH_3)C{\equiv}CH$

c. $CH_3CH_2CH{=}CHCH_2CH_3$ $\xrightarrow{\text{Br}_2}$ $CH_3CH_2CH{-}CHCH_2CH_3$ (Br Br) $\xrightarrow{2\ ^-\text{NH}_2}$ $CH_3CH_2C{\equiv}CCH_2CH_3$

**11.38**

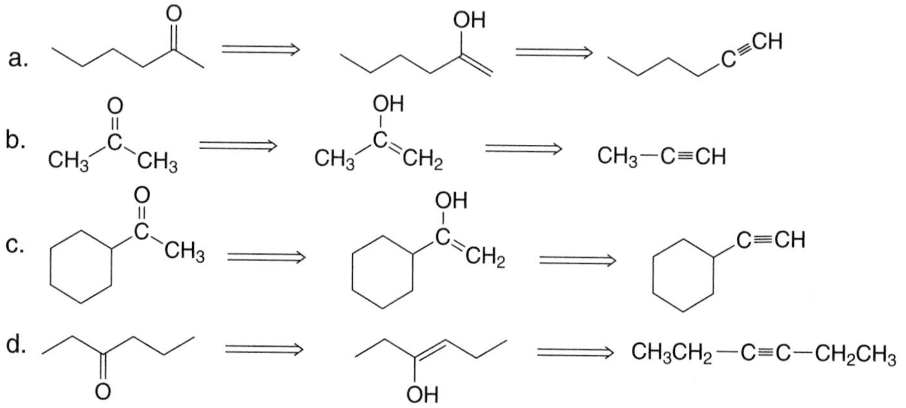

a. ⟹ ⟹

b. ⟹ ⟹ $CH_3-C{\equiv}CH$

c. ⟹ ⟹

d. ⟹ ⟹ $CH_3CH_2-C{\equiv}C-CH_2CH_3$

**11.39**

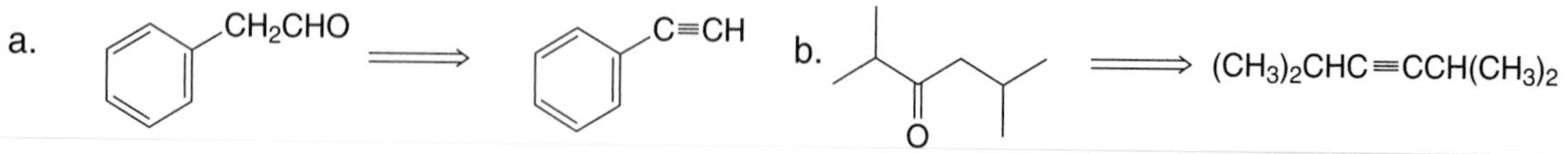

a. ⟹

b. ⟹ $(CH_3)_2CHC{\equiv}CCH(CH_3)_2$

**11.40**

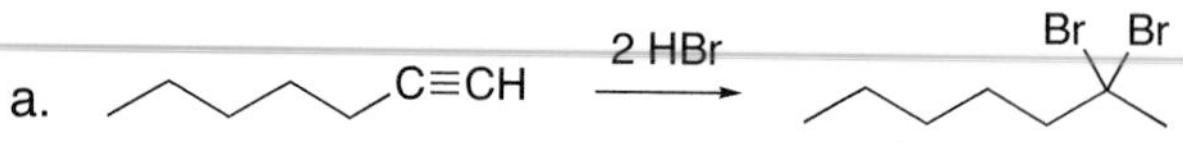

a. C≡CH $\xrightarrow{\text{2 HBr}}$

b. $(CH_3)_3CC{\equiv}CH$ $\xrightarrow{\text{2 Cl}_2}$ $(CH_3)_3C\overset{\overset{\displaystyle Cl}{|}}{\underset{\underset{\displaystyle Cl}{|}}{C}}-CHCl_2$

c. [structure: PhCH=CHPh] $\xrightarrow{\text{[1] Cl}_2}$ [structure: Ph–CHCl–CHCl–Ph] $\xrightarrow[\text{(2 equiv)}]{\text{[2] NaNH}_2}$ [structure: Ph–C≡C–Ph]

d. [structure: CH₃CH₂CH₂CH₂CH₂C≡CH] $\xrightarrow[\text{[2] H}_2\text{O}_2, \text{ HO}^-]{\text{[1] BH}_3}$ [structure: aldehyde CH₃CH₂CH₂CH₂CH₂CH₂CHO]

e. [structure: cyclopentyl–C≡CH] $\xrightarrow[\text{(1 equiv)}]{\text{HCl}}$ [structure: cyclopentyl–C(Cl)=CH₂]

f. $HC≡C^- + D_2O \longrightarrow HC≡CD + DO^-$

g. [structure: cyclopentyl–C≡C–CH₃] $\xrightarrow[\text{H}_2\text{SO}_4]{\text{H}_2\text{O}}$ [structure: cyclopentyl–CH₂–CO–CH₃] + [structure: cyclopentyl–CO–CH₂CH₃]

h. $CH_3CH_2C≡C^- + CH_3CH_2CH_2-OTs \longrightarrow CH_3CH_2C≡CCH_2CH_2CH_3 + {}^-OTs$

i. [structure: epoxide with CH₃] $\xrightarrow{\text{[1] HC≡C}^-}$ [structure: cyclohexane with CH₃, O⁻, C≡CH] $\xrightarrow{\text{[2] HO–H}}$ [structure: cyclohexane with CH₃, OH, C≡CH]

j. $CH_3C≡C-H \xrightarrow{\text{[1] Na}^+ \text{H}^-} CH_3C≡C^-$ $\xrightarrow[\substack{2° \text{ halide} \\ \text{E2}}]{\text{[2] cyclopentyl-Br}}$ [structure: cyclopentene] $+ CH_3C≡C-H$

k. $CH_3CH_2C≡C-H \xrightarrow{\text{[1] Na}^+ {}^-\text{NH}_2} CH_3CH_2C≡C^-$ $\xrightarrow{\text{[2] cyclohexyl-CH}_2-\text{I}}$ $CH_3CH_2C≡C-CH_2$cyclohexyl

l. [structure: Ph–C≡C–H] $\xrightarrow{\text{[1] Na}^+ \text{H}^-}$ [structure: Ph–C≡C⁻] $\xrightarrow{\text{[2] epoxide}}$ [structure: Ph–C≡CCH₂CH₂O⁻] $\xrightarrow{\text{[3] HO–H}}$ [structure: Ph–C≡C–CH₂CH₂OH]

## 11.41

[structure: cyclohexyl–CH₂CH₂Br] $\xrightarrow{\text{KOC(CH}_3)_3}$ [structure: cyclohexyl–CH=CH₂] $\xrightarrow{\text{Br}_2}$ [structure: cyclohexyl–CHBr–CHBr–H] $\xrightarrow[\substack{\text{DMSO} \\ \text{(2 equiv)}}]{\text{KOC(CH}_3)_3}$ [structure: cyclohexyl–C≡CH]

**A**      **B**      **C**      **D** $\downarrow$ NaNH₂

[structure: cyclohexyl–C≡CCH₃] $\xleftarrow{\text{CH}_3\text{I}}$ [structure: cyclohexyl–C≡C⁻]

     **E**

**11.42**

most acidic H

H−C≡C−CH₂CH₂CH₂OH  →✗→  CH₃−C≡C−CH₂CH₂CH₂OH

**A**

1) NaNH₂
2) CH₃I

NaNH₂ will remove the proton from the OH since it is more acidic.

H−C≡C−CH₂CH₂CH₂Ö:⁻  ⟶  H−C≡C−CH₂CH₂CH₂ÖCH₃  = **B**

CH₃−I

**11.43**

Stereogenic center at the site of reaction.

a.

[structure] Cl  HC≡C⁻  →  [structure] C≡CH

H D → D H

inversion

identical

c.

[epoxide structure] [1] HC≡C⁻ / [2] H₂O → HO CH₃ ... + CH₃ ... OH

Stereogenic center NOT at the site of reaction.

Configuration is retained.

b.

[structure] Cl  HC≡C⁻  →  [structure] C≡CH

CH₃ H → CH₃ H

d.

[epoxide structure] [1] HC≡C⁻ / [2] H₂O → HO ... + CH₃ ... OH

enantiomers

**11.44**

a.

[structure] OH  PBr₃  →  [structure] Br  :C≡CCH₂OR'  →  [structure] C≡CCH₂OR'

[structure] OCOR  →  [structure] OCOR  →  [structure] OCOR

**A**  **B**  **C**

H−C≡CCH₂OR'

⁻CH₃ Li⁺

new C−C bond

b.

[epoxide] OR  [1] **D = HC≡C⁻** / [2] H₂O  →  H−[C≡C]−[structure] OR  →  H−C≡C−[structure] OH / OH

These 2 C's are added.

**E = TsCl, pyridine**

[structure] OCH₂CH₃  [1] NaH / [2] CH₃I  ←  H−C≡C−[structure] OCH₂CH₃  [1] NaH / [2] **F = CH₃CH₂Br**  ←  H−C≡C−[structure] OH  ←  H−C≡C−[structure] OH / OTs

CH₃−C≡C−[structure]

**G**

new C−C bond

**11.45** Draw two diagrams to show σ and π bonds.

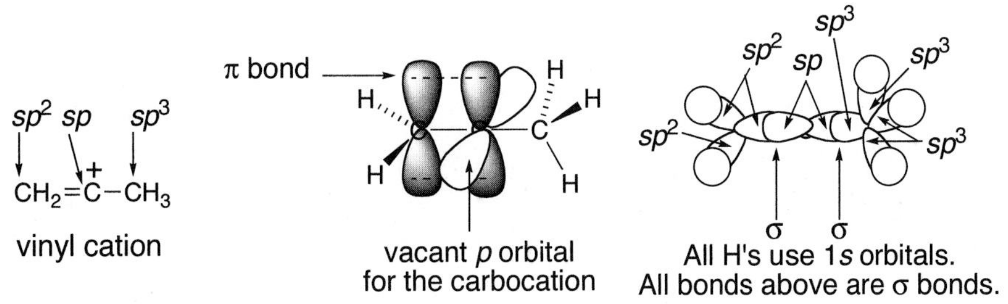

$sp^2\ sp\ \ sp^3$

$CH_2=\overset{+}{C}-CH_3$

vinyl cation

π bond

vacant $p$ orbital
for the carbocation

All H's use $1s$ orbitals.
All bonds above are σ bonds.

The positive charge in a vinyl carbocation resides on a carbon that is $sp$ hybridized, while in $(CH_3)_2CH^+$, the positive charge is located on an $sp^2$ hybridized carbon. The higher percent $s$-character on carbon destabilizes the positive charge in the vinyl cation. Moreover, the positively charged carbocation is now bonded to an $sp^2$ hybridized carbon, which donates electrons less readily than an $sp^3$ hybridized carbon.

**11.46**

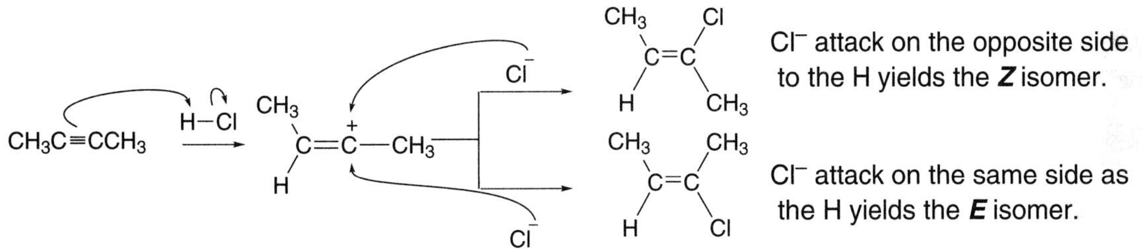

$Cl^-$ attack on the opposite side to the H yields the **Z** isomer.

$Cl^-$ attack on the same side as the H yields the **E** isomer.

**11.47**

a.

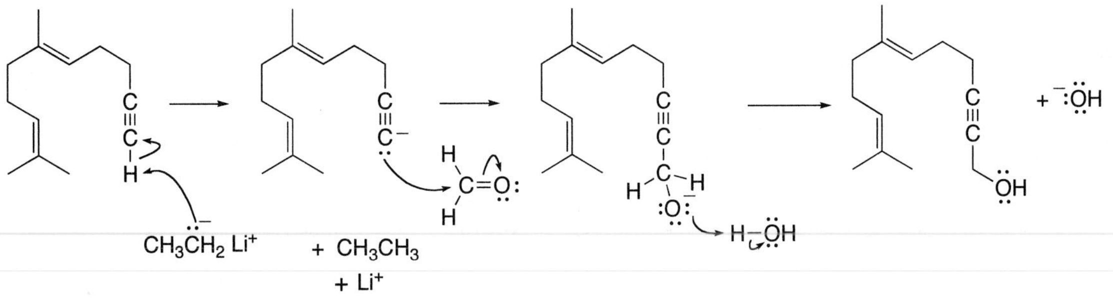

b.

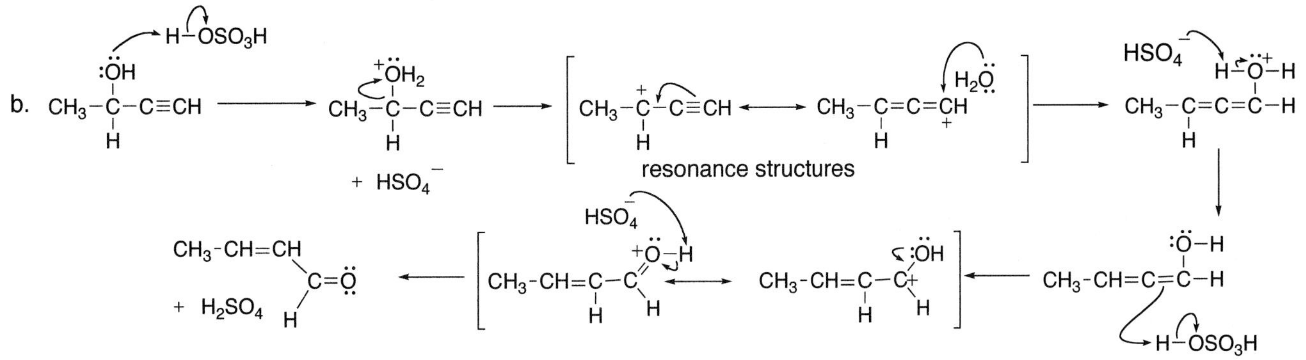

**11.48**

a.

b. A more stable internal alkyne can be isomerized to a less stable terminal alkyne under these reaction conditions because when CH$_3$CH$_2$C≡CH is first formed, it contains an *sp* hybridized C–H bond, which is more acidic than any proton in CH$_3$–C≡C–CH$_3$. Under the reaction conditions, this proton is removed with base. Formation of the resulting acetylide anion drives the equilibrium to favor its formation. Protonation of this acetylide anion gives the less stable terminal alkyne.

**11.49** The alkyl halides must be methyl or 1°.

a. HC≡C–$\xi$–CH$_2$CH$_2$CH(CH$_3$)$_2$ ⟹ HC≡C: $^-$ + Cl–CH$_2$CH$_2$CH(CH$_3$)$_2$  $\boxed{1° RX}$

b. CH$_3$–$\xi$–C≡C–C(CH$_3$)(CH$_3$)–CH$_2$CH$_3$ ⟹ CH$_3$–Cl + $^-$:C≡C–C(CH$_3$)(CH$_3$)–CH$_2$CH$_3$

c. (cyclohexyl)–C≡C–$\xi$–CH$_2$CH$_2$CH$_3$ ⟹ (cyclohexyl)–C≡C: $^-$ + Cl–CH$_2$CH$_2$CH$_3$  $\boxed{1° RX}$

**11.50**

a. → 2 $^-$NH$_2$ →

b. → Cl$_2$ → → KOC(CH$_3$)$_3$ (2 equiv) DMSO →

c. → POCl$_3$ / pyridine → → Cl$_2$ → → KOC(CH$_3$)$_3$ (2 equiv) DMSO →

**11.51**

a. HC≡C–H → Na$^+$ H$^-$ → HC≡C$^-$ → (CH$_3$)$_2$CHCH$_2$–Cl → (CH$_3$)$_2$CHCH$_2$C≡CH

b. HC≡C–H → Na$^+$ H$^-$ → HC≡C$^-$ → CH$_3$CH$_2$CH$_2$–Cl → $\boxed{\text{CH}_3\text{CH}_2\text{CH}_2\text{C≡CH}}$ → NaH → CH$_3$CH$_2$CH$_2$C≡C$^-$

CH$_3$CH$_2$CH$_2$–Cl ↓

CH$_3$CH$_2$CH$_2$C≡CCH$_2$CH$_2$CH$_3$

c. CH$_3$CH$_2$CH$_2$C≡CH (from b.) → [1] BH$_3$ [2] H$_2$O$_2$, HO$^-$ → CH$_3$CH$_2$CH$_2$CH$_2$CHO

d. $CH_3CH_2CH_2C\equiv CH$ $\xrightarrow[\substack{H_2SO_4 \\ HgSO_4}]{H_2O}$ $CH_3CH_2CH_2\overset{\displaystyle O}{\overset{\|}{C}}CH_3$

(from b.)

e. $CH_3CH_2CH_2C\equiv CH$ $\xrightarrow{2\ HCl}$ $CH_3CH_2CH_2CCl_2CH_3$

(from b.)

f. $CH_3CH_2CH_2C\equiv CCH_2CH_2CH_3$ $\xrightarrow[H_2SO_4,\ HgSO_4]{H_2O}$ $CH_3CH_2CH_2\overset{\displaystyle O}{\overset{\|}{C}}CH_2CH_2CH_2CH_3$

(from b.)

## 11.52

a. $CH_3CH_2CH=CH_2$ $\xrightarrow{Cl_2}$ $CH_3CH_2\overset{\displaystyle Cl}{\overset{|}{CH}}-\overset{\displaystyle Cl}{\overset{|}{CH_2}}$ $\xrightarrow{2\ ^-NH_2}$ $CH_3CH_2C\equiv CH$

b. $CH_3CH_2C\equiv CH$ $\xrightarrow[\text{(2 equiv)}]{HBr}$ $CH_3CH_2CBr_2CH_3$
(from a.)

c. $CH_3CH_2C\equiv CH$ $\xrightarrow[\text{(2 equiv)}]{Cl_2}$ $CH_3CH_2CCl_2CHCl_2$
(from a.)

d. $CH_3CH_2CH=CH_2$ $\xrightarrow{Br_2}$ $CH_3CH_2CHBrCH_2Br$

e. $CH_3CH_2C\equiv CH$ $\xrightarrow{NaH}$ $CH_3CH_2C\equiv C^-$ $\xrightarrow{CH_3CH_2CH_2-Br}$ $CH_3CH_2C\equiv C-CH_2CH_2CH_3$
(from a.)

f. $CH_3CH_2C\equiv C^-$ $\xrightarrow[\text{[2] } H_2O]{\text{[1] } \triangle O}$ $CH_3CH_2C\equiv C-CH_2CH_2OH$
(from e.)

g. $CH_3CH_2C\equiv C^-$ $\xrightarrow[\text{[2] } H_2O]{\text{[1]}}$
(from e.)

(+ enantiomer)

## 11.53

a. $HC\equiv CH$ $\xrightarrow{NaH}$ $HC\equiv C^-$ $\xrightarrow{CH_3CH_2CH_2CH_2CH_2CH_2Br}$ $CH_3CH_2CH_2CH_2CH_2CH_2C\equiv CH$

b. $CH_3CH_2CH_2CH_2CH_2CH_2C\equiv CH$ $\xrightarrow{NaH}$ $CH_3CH_2CH_2CH_2CH_2CH_2C\equiv C^-$ $\xrightarrow{CH_3CH_2Br}$ $CH_3CH_2CH_2CH_2CH_2CH_2C\equiv CCH_2CH_3$
(from a.)

c. $CH_3CH_2CH_2CH_2CH_2CH_2C\equiv C^-$ $\xrightarrow[\text{[2] } H_2O]{\text{[1] } \triangle O}$ $CH_3CH_2CH_2CH_2CH_2CH_2C\equiv CCH_2CH_2OH$
(from b.)

d. $CH_3CH_2CH_2CH_2CH_2CH_2C\equiv CCH_2CH_2OH$ $\xrightarrow[\text{[2] } CH_3CH_2Br]{\text{[1] NaH}}$ $CH_3CH_2CH_2CH_2CH_2CH_2C\equiv CCH_2CH_2OCH_2CH_3$
(from c.)

## 11.54

a. $\xrightarrow{H_2SO_4}$ $\xrightarrow{Br_2}$ $\xrightarrow{2\ ^-NH_2}$ $CH_3-C\equiv CH$ $\xrightarrow{NaH}$ $CH_3-C\equiv C^-$ $\longrightarrow$ $CH_3-C\equiv CCH_2CH_2CH_3$

$\xrightarrow{SOCl_2}$

b.

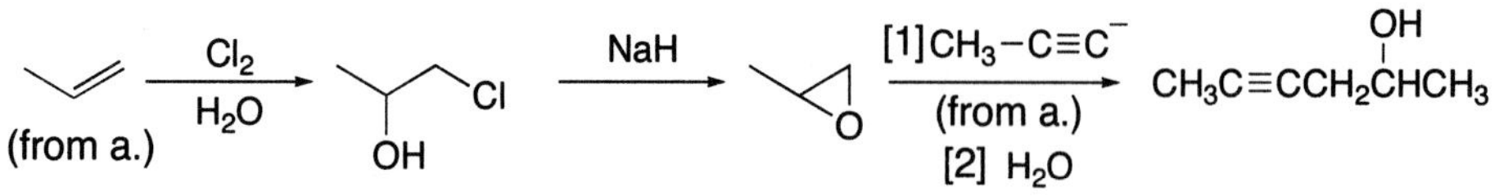

(from a.)
$Cl_2$
$H_2O$
OH
Cl
NaH
O
[1] $CH_3-C\equiv C^-$
(from a.)
[2] $H_2O$
OH
$CH_3C\equiv CCH_2CHCH_3$


**11.55**

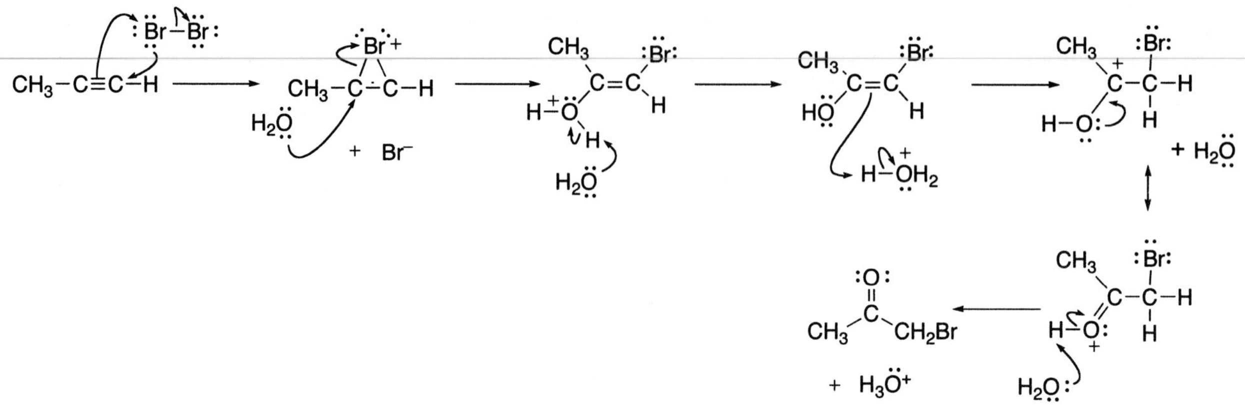

+ Br⁻
+ $H_2O$
+ $H_3O^+$


**11.56**

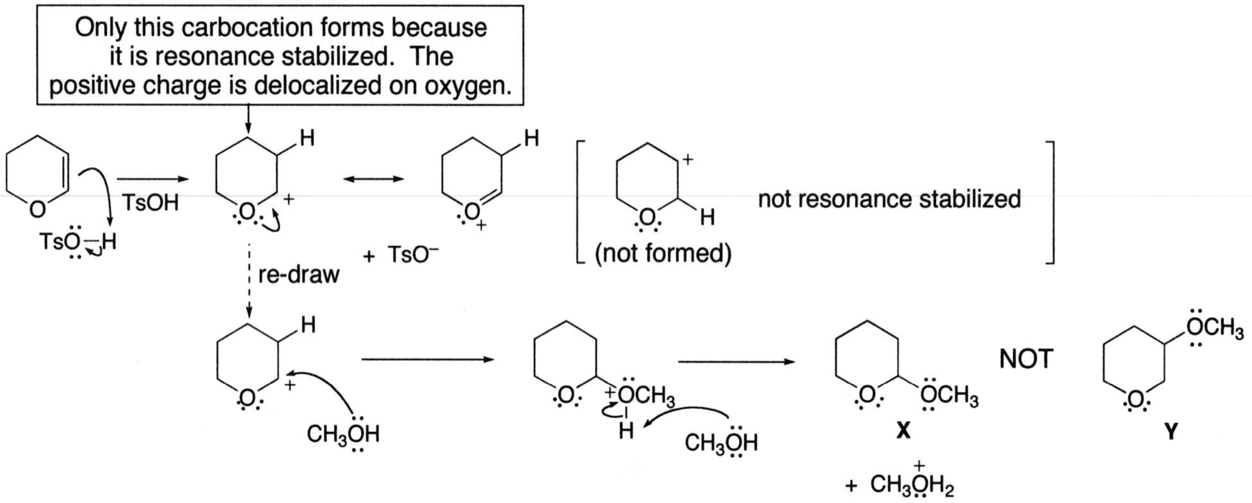

Only this carbocation forms because it is resonance stabilized. The positive charge is delocalized on oxygen.
TsOH
+ TsO⁻
re-draw
not resonance stabilized
(not formed)
NOT
X
Y

**11.57**

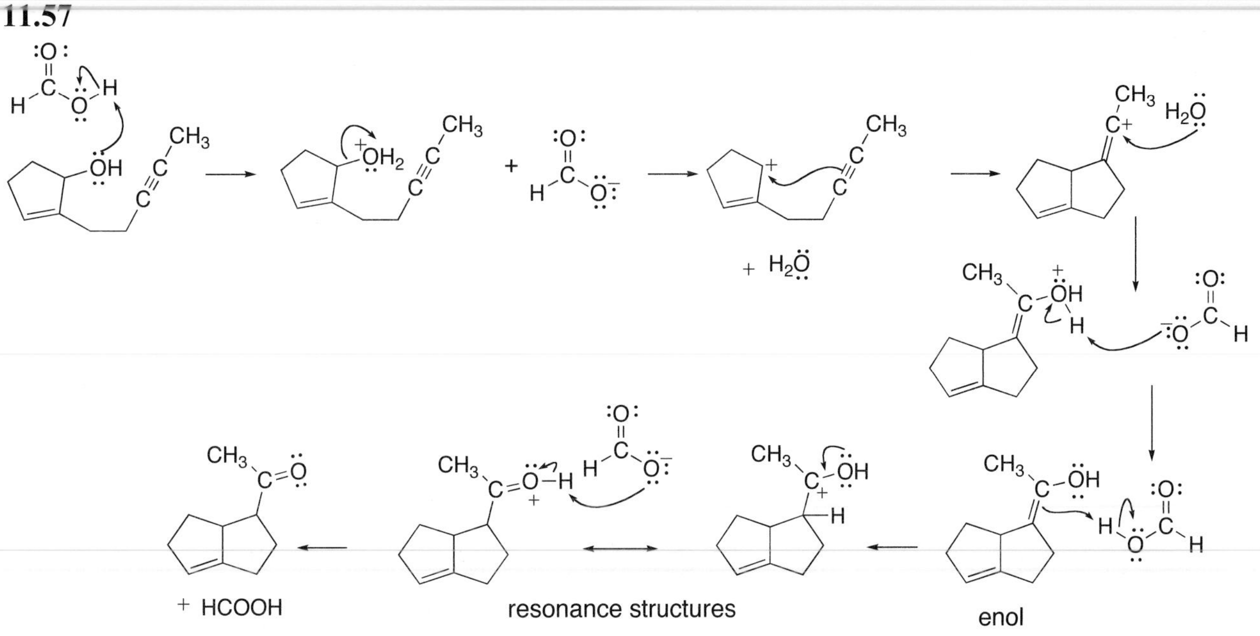

+ HCOOH            resonance structures                          enol

Chapter 12: Oxidation and Reduction

♦ Summary: Terms that describe reaction selectivity

- A **regioselective reaction** forms predominately or exclusively one constitutional isomer (Section 8.5).

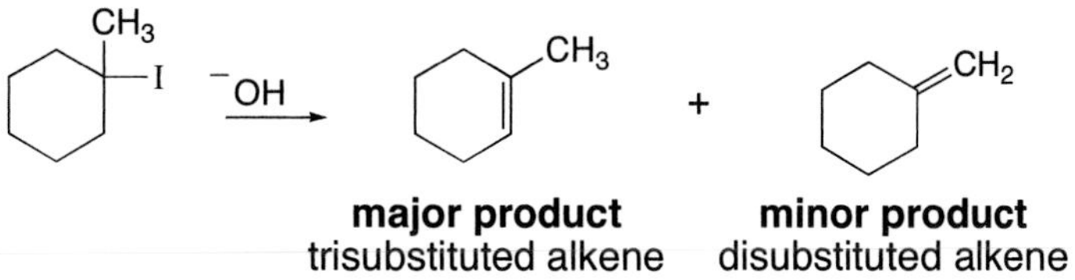

**major product**
trisubstituted alkene

**minor product**
disubstituted alkene

- A **stereoselective reaction** forms predominately or exclusively one stereoisomer (Section 8.5).

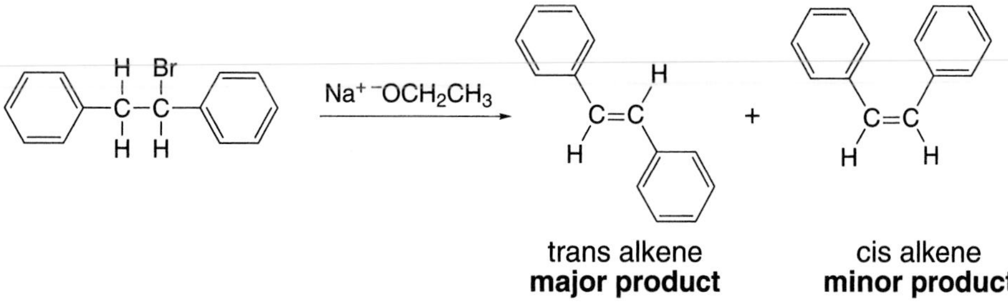

trans alkene
**major product**

cis alkene
**minor product**

- An **enantioselective reaction** forms predominately or exclusively one enantiomer (Section 12.15).

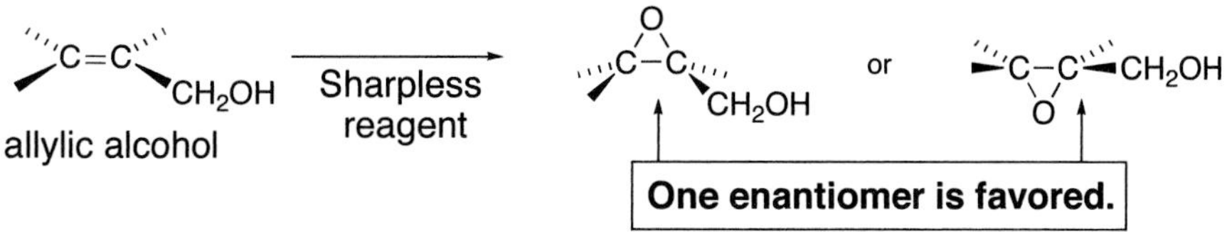

allylic alcohol

Sharpless reagent

or

**One enantiomer is favored.**

♦ Definitions of oxidation and reduction

| **Oxidation** reactions result in: | **Reduction** reactions result in: |
|---|---|
| - an increase in the number of C–Z bonds, *or* <br> - a decrease in the number of C–H bonds. | - a decrease in the number of C–Z bonds, *or* <br> - an increase in the number of C–H bonds. <br><br> [Z = an element more electronegative than C] |

♦ Reduction reactions

[1] Reduction of alkenes—Catalytic hydrogenation (12.3)

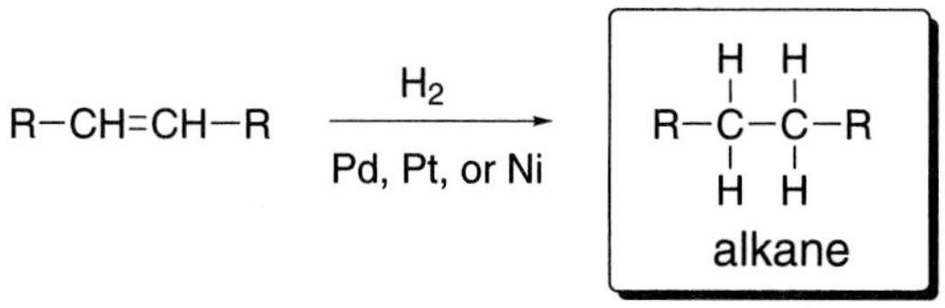

R–CH=CH–R  $\xrightarrow[\text{Pd, Pt, or Ni}]{\text{H}_2}$

alkane

- **Syn addition** of $H_2$ occurs.
- Increasing alkyl substitution on the C=C decreases the rate of reaction.

## [2] Reduction of alkynes

[a] $R-C\equiv C-R$ $\xrightarrow{\text{2 H}_2}{\text{Pd-C}}$

H H
R–C–C–R
H H
**alkane**

- Two equivalents of $H_2$ are added and four new C–H bonds are formed (12.5A).

[b] $R-C\equiv C-R$ $\xrightarrow{\text{H}_2}{\text{Lindlar catalyst}}$

R     R
  C=C
H     H
**cis alkene**

- **Syn addition** of $H_2$ occurs, forming a **cis** alkene (12.5B).
- The Lindlar catalyst is deactivated so that reaction stops after one equivalent of $H_2$ has added.

[c] $R-C\equiv C-R$ $\xrightarrow{\text{Na}}{\text{NH}_3}$

R     H
  C=C
H     R
**trans alkene**

- **Anti addition** of $H_2$ occurs, forming a **trans** alkene (12.5C).

## [3] Reduction of alkyl halides (12.6)

$R-X$ $\xrightarrow{\text{[1] LiAlH}_4}{\text{[2] H}_2\text{O}}$

R–H
**alkane**

- The reaction follows an $S_N2$ mechanism.
- $CH_3X$ and $RCH_2X$ react faster than more substituted RX.

## [4] Reduction of epoxides (12.6)

O
C–C $\xrightarrow{\text{[1] LiAlH}_4}{\text{[2] H}_2\text{O}}$

OH
C–C
H
**alcohol**

- The reaction follows an $S_N2$ mechanism.
- In unsymmetrical epoxides, $H^-$ (from $LiAlH_4$) attacks at the less substituted carbon.

## ◆ Oxidation reactions

## [1] Oxidation of alkenes

### [a] Epoxidation (12.8)

$C=C$ + $RCO_3H$ $\longrightarrow$

O
C–C
**epoxide**

- The mechanism has **one step**.
- **Syn addition** of an O atom occurs.
- The reaction is stereospecific.

### [b] Anti dihydroxylation (12.9A)

$C=C$ $\xrightarrow{\text{[1] RCO}_3\text{H}}{\text{[2] H}_2\text{O ( H}^+ \text{ or HO}^-)}$

HO
C–C
OH
**1,2-diol**

- Ring opening of an epoxide intermediate with $^-OH$ or $H_2O$ forms a 1,2-diol with two OH groups added in an **anti** fashion.

[c] Syn dihydroxylation (12.9B)

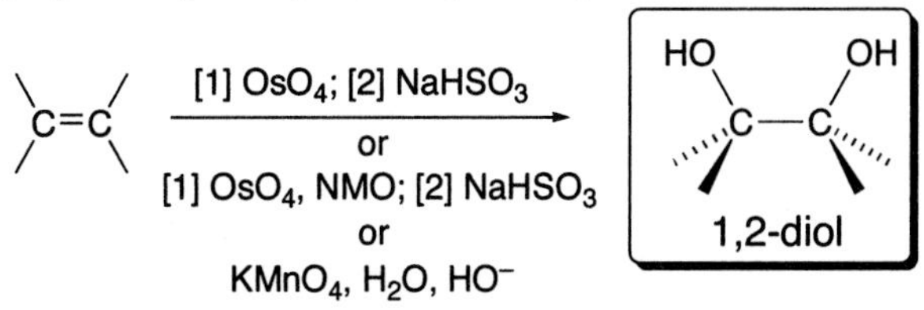

- Each reagent adds two new C–O bonds to the C=C in a **syn** fashion.

[d] Oxidative cleavage (12.10)

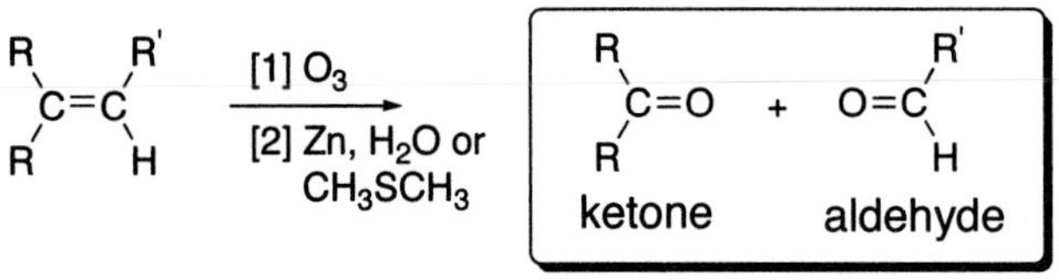

- Both the σ and π bonds of the alkene are cleaved to form two carbonyl groups.

[2] Oxidative cleavage of alkynes (12.11)

[a]

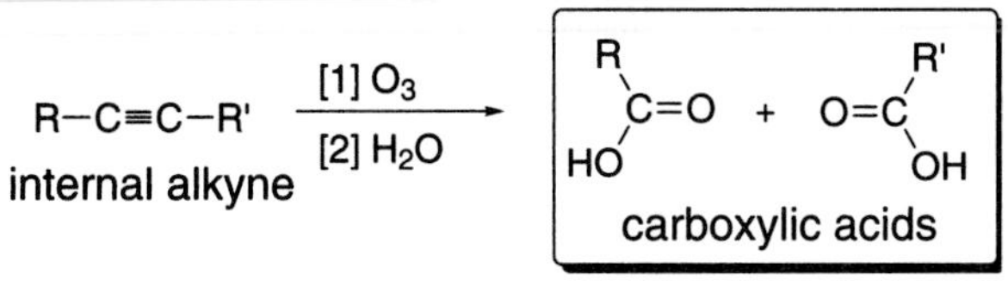

- The σ bond and both π bonds of the alkyne are cleaved.

[b]

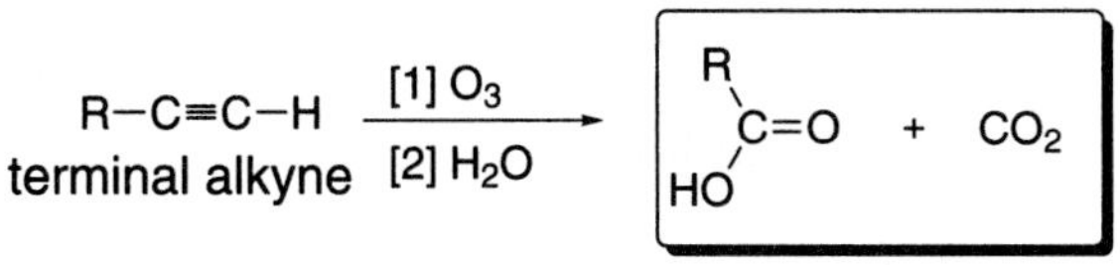

[3] Oxidation of alcohols (12.12, 12.13)

[a]

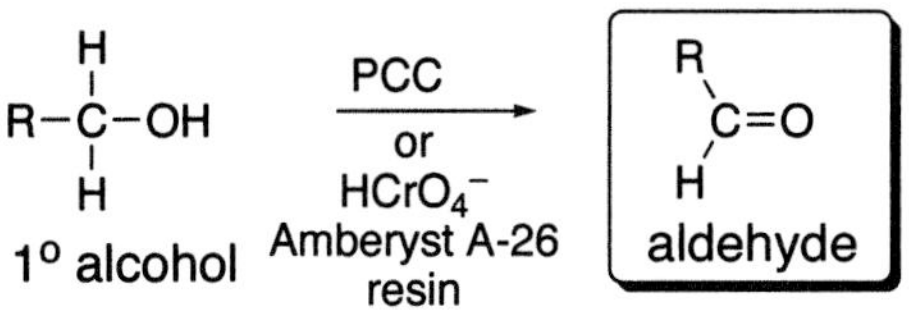

- Oxidation of a 1° alcohol with PCC or $HCrO_4^-$ (Amberlyst A-26 resin) stops at the aldehyde stage. Only one C–H bond is replaced by a C–O bond.

[b]

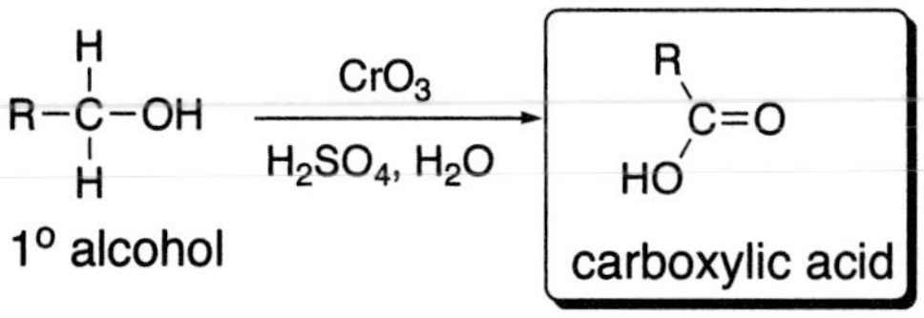

- Oxidation of a 1° alcohol under harsher reaction conditions—$CrO_3$ (or $Na_2Cr_2O_7$ or $K_2Cr_2O_7$) + $H_2O$ + $H_2SO_4$—affords a RCOOH. Two C–H bonds are replaced by two C–O bonds.

[c]

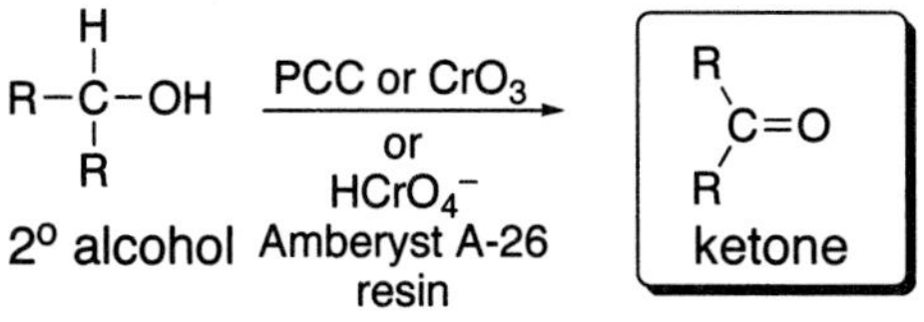

- Since a 2° alcohol has only one C–H bond on the carbon bearing the OH group, all $Cr^{6+}$ reagents—PCC, $CrO_3$, $Na_2Cr_2O_7$, $K_2Cr_2O_7$, or $HCrO_4^-$ (Amberlyst A-26 resin)—oxidize a 2° alcohol to a ketone.

[4] Asymmetric epoxidation of allylic alcohols (12.15)

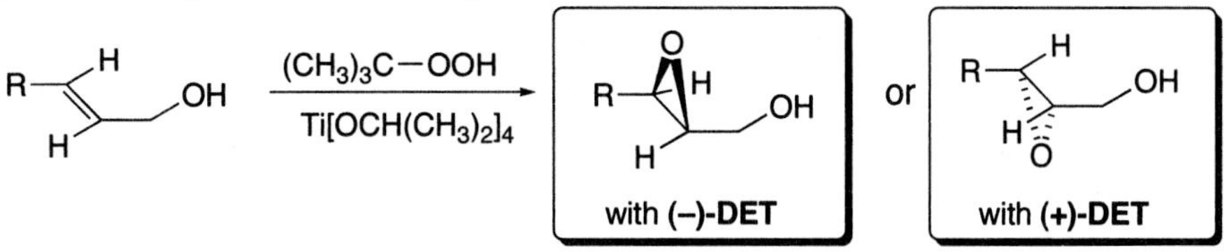

## Chapter 12: Answers to Problems

**12.1** *Oxidation* results in an *increase* in the number of C–Z bonds (usually C–O bonds) *or* a *decrease* in the number of C–H bonds.
*Reduction* results in a *decrease* in the number of C–Z bonds (usually C–O bonds) *or* an *increase* in the number of C–H bonds.

a. [structure] $\longrightarrow$ [structure] **oxidation**

b. [benzene] $\longrightarrow$ [cyclohexane] **reduction**

c. $CH_3-C(=O)-CH_3 \longrightarrow CH_3-C(=O)-OCH_3$ **oxidation**

d. $CH_2{=}CH_2 \longrightarrow CH_3CH_2Cl$ **neither**
1 new C–H bond
and 1 new C–Cl bond

**12.2** **Hydrogenation** is the addition of hydrogen. When alkenes are hydrogenated, they are *reduced* by the addition of $H_2$ to the π bond. To draw the alkane product, add a H to each C of the double bond.

a. [structure with $CH_3$, $CH_2CH(CH_3)_2$, $C{=}C$, $CH_3$, H] $\longrightarrow$ $CH_3-\underset{\underset{H}{|}}{C}-\underset{\underset{H}{|}}{\overset{\overset{CH_3\,CH_2CH(CH_3)_2}{|}}{C}}-H$

c. [methylcyclohexene] $\longrightarrow$ [methylcyclohexane]

b. [structure] $\longrightarrow$ [structure]

**12.3** Cis alkenes are less stable than trans alkenes, so they have larger heats of hydrogenation. Increasing alkyl substitution increases the stability of a C=C, decreasing the heat of hydrogenation.

a. [cis structure $CH_3CH_2$, $CH_2CH_3$, $C{=}C$, H, H] **and** [trans structure $CH_3CH_2$, H, $C{=}C$, H, $CH_2CH_3$]

  cis alkane            trans alkane
  less stable
  *larger* **heat of hydrogenation**

b. [trisubstituted cyclohexene] **and** [disubstituted cyclohexene]

  trisubstituted            disubstituted
                            less stable
            *larger* **heat of hydrogenation**

**12.4** Hydrogenation products must be identical to use hydrogenation data to evaluate the relative stability of the starting materials.

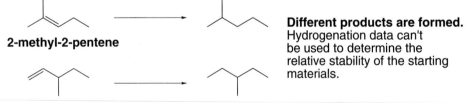

**2-methyl-2-pentene**

**3-methyl-1-pentene**

**Different products are formed.** Hydrogenation data can't be used to determine the relative stability of the starting materials.

**12.5** Increasing alkyl substitution on the C=C decreases the rate of hydrogenation.

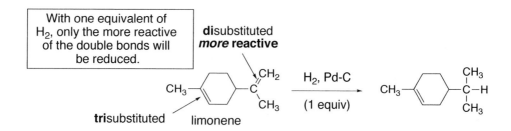

With one equivalent of H₂, only the more reactive of the double bonds will be reduced.

**di**substituted **more reactive**

**tri**substituted    limonene

H₂, Pd-C (1 equiv)

**12.6**

new stereogenic center

Two enantiomers are formed in equal amounts:

a.

b.

diastereomers

c.

diastereomers

**12.7**

| Compound | Molecular formula before hydrogenation | Molecular formula after hydrogenation | Number of rings | Number of π bonds |
|---|---|---|---|---|
| **A** | $C_{10}H_{12}$ | $C_{10}H_{16}$ | 3 | 2 |
| **B** | $C_4H_8$ | $C_4H_{10}$ | 0 | 1 |
| **C** | $C_6H_8$ | $C_6H_{12}$ | 1 | 2 |

**12.8**

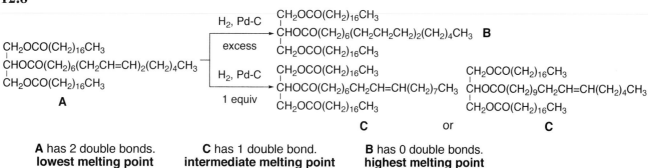

CH₂OCO(CH₂)₁₆CH₃
CHOCO(CH₂)₆(CH₂CH=CH)₂(CH₂)₄CH₃
CH₂OCO(CH₂)₁₆CH₃
**A**

H₂, Pd-C
excess

CH₂OCO(CH₂)₁₆CH₃
CHOCO(CH₂)₆(CH₂CH₂CH₂)₂(CH₂)₄CH₃  **B**
CH₂OCO(CH₂)₁₆CH₃

H₂, Pd-C
1 equiv

CH₂OCO(CH₂)₁₆CH₃
CHOCO(CH₂)₆CH₂CH=CH(CH₂)₇CH₃
CH₂OCO(CH₂)₁₆CH₃
**C**

or

CH₂OCO(CH₂)₁₆CH₃
CHOCO(CH₂)₉CH₂CH=CH(CH₂)₄CH₃
CH₂OCO(CH₂)₁₆CH₃
**C**

**A** has 2 double bonds.
**lowest melting point**

**C** has 1 double bond.
**intermediate melting point**

**B** has 0 double bonds.
**highest melting point**

**12.9** Hydrogenation of HC≡CCH₂CH₂CH₃ and CH₃C≡CCH₂CH₃ yields the same compound. The heat of hydrogenation is larger for HC≡CCH₂CH₂CH₃ than for CH₃C≡CCH₂CH₃ because internal alkynes are more stable (lower in energy) than terminal alkynes.

**12.10**

$CH_2-C\equiv C-CH_2CH_3$

A

$H_2$
Lindlar catalyst →

*cis*-jasmone
(perfume component
isolated from jasmine flowers)

**12.11** To draw the products of catalytic hydrogenation remember:
- $H_2$ (excess), Pd will reduce **alkenes and alkynes to alkanes**.
- $H_2$ (excess), Lindlar catalyst will reduce *only* **alkynes to cis alkenes**.

a.  $CH_2=CHCH_2CH_2-C\equiv C-CH_3$ $\xrightarrow[\text{Pd-C}]{H_2\ \text{(excess)}}$ $CH_3CH_2CH_2CH_2CH_2CH_2CH_3$

b.  $CH_2=CHCH_2CH_2-C\equiv C-CH_3$ $\xrightarrow[\substack{\text{Lindlar}\\ \text{catalyst}}]{H_2\ \text{(excess)}}$ $CH_2=CH-CH_2-CH_2\ \ CH_3$ / $C=C$ / H H

**12.12** Use the directions from Answer 12.11.

a.  $CH_3OCH_2CH_2C\equiv CCH_2CH(CH_3)_2$ $\xrightarrow[\text{Pd-C}]{H_2\ \text{(excess)}}$ $CH_3OCH_2CH_2CH_2CH_2CH_2CH(CH_3)_2$

b.  $CH_3OCH_2CH_2C\equiv CCH_2CH(CH_3)_2$ $\xrightarrow[\text{Lindlar catalyst}]{H_2\ \text{(1 equiv)}}$ $CH_3OCH_2CH_2\ \ CH_2CH(CH_3)_2$ / $C=C$ / H H

c.  $CH_3OCH_2CH_2C\equiv CCH_2CH(CH_3)_2$ $\xrightarrow[\text{Lindlar catalyst}]{H_2\ \text{(excess)}}$ $CH_3OCH_2CH_2\ \ CH_2CH(CH_3)_2$ / $C=C$ / H H

d.  $CH_3OCH_2CH_2C\equiv CCH_2CH(CH_3)_2$ $\xrightarrow{Na,\ NH_3}$ $CH_3OCH_2CH_2\ \ H$ / $C=C$ / H $CH_2CH(CH_3)_2$

**12.13**

$CH_3-C\equiv C-CH_2CH_2CH_3$

$\xrightarrow[\text{Pd-C}]{D_2}$ $CH_3CD_2CD_2CH_2CH_2CH_3$

$\xrightarrow[\substack{\text{Lindlar}\\ \text{catalyst}}]{D_2}$ $CH_3\ \ CH_2CH_2CH_3$ / $C=C$ / D D

$\xrightarrow[ND_3]{Na}$ $CH_3\ \ D$ / $C=C$ / D $CH_2CH_2CH_3$

**12.14**

$\xrightarrow[\substack{\text{Lindlar}\\ \text{catalyst}}]{H_2}$

A → B

**12.15** LiAlH₄ reduces alkyl halides to alkanes and epoxides to alcohols.

a.

$$\text{(structure: alkyl chloride)} \xrightarrow[\text{[2] H}_2\text{O}]{\text{[1] LiAlH}_4} \text{(structure: alkane with H)}$$

H replaces Cl.

b.

$$\xrightarrow[\text{[2] H}_2\text{O}]{\text{[1] LiAlH}_4}$$

**12.16** To draw the product, add an O atom across the π bond of the C=C.

a. $(CH_3)_2C{=}CH_2 \xrightarrow{\text{mCPBA}} (CH_3)_2C\overset{O}{-}CH_2$

c. $\text{(cyclohexane)}{=}CH_2 \xrightarrow{\text{mCPBA}} \text{(epoxide)}CH_2$

b. $(CH_3)_2C{=}C(CH_3)_2 \xrightarrow{\text{mCPBA}} (CH_3)_2C\overset{O}{-}C(CH_3)_2$

**12.17** For epoxidation reactions:
- There are two possible products: O adds from above and below the double bond.
- Substituents on the C=C retain their original configuration in the products.

a.

$$\underset{H}{\overset{CH_3}{C}}{=}\underset{H}{\overset{H}{C}} \xrightarrow{\text{mCPBA}} CH_3\overset{O}{\underset{H}{C}}{-}\underset{H}{C}H \quad + \quad H\overset{}{\underset{O}{C}}{-}\underset{H}{C}H \quad \text{enantiomers}$$

b.

$$\underset{H}{\overset{CH_3CH_2}{C}}{=}\underset{H}{\overset{CH_2CH_3}{C}} \xrightarrow{\text{mCPBA}} CH_3CH_2\overset{O}{\underset{H}{C}}{-}\underset{H}{C}CH_2CH_3 \quad + \quad \underset{H}{\overset{CH_3CH_2}{C}}\overset{}{\underset{O}{}}\underset{H}{\overset{CH_2CH_3}{C}}$$

identical

c.

$$\text{(cyclohexene with CH}_3) \xrightarrow{\text{mCPBA}} \text{(epoxide with CH}_3, H) \quad + \quad \text{(epoxide with CH}_3, H) \quad \text{enantiomers}$$

**12.18** Treatment of an alkene with a peroxyacid followed by H₂O, HO⁻ adds two hydroxy groups in an **anti** fashion. *cis*-2-Butene and *trans*-2-butene yield different products of dihydroxylation. *cis*-2-Butene gives a mixture of two enantiomers and *trans*-2-butene gives a meso compound. The reaction is stereospecific because two stereoisomeric starting materials give different products that are also stereoisomers of each other.

$$\underset{H}{\overset{CH_3}{C}}{=}\underset{H}{\overset{CH_3}{C}} \xrightarrow[\text{[2] H}_2\text{O, HO}^-]{\text{[1] RCO}_3\text{H}} \text{(diol)} \quad + \quad \text{(diol)}$$

**cis-2-butene**     **enantiomers**

$$\underset{H}{\overset{CH_3}{C}}{=}\underset{CH_3}{\overset{H}{C}} \xrightarrow[\text{[2] H}_2\text{O, HO}^-]{\text{[1] RCO}_3\text{H}} \text{(diol)} \quad + \quad \text{(diol)}$$

**trans-2-butene**     **identical**
**meso compound**

**12.19** Treatment of an alkene with $OsO_4$ adds two hydroxy groups in a **syn** fashion. *cis*-2-Butene and *trans*-2-butene yield different stereoisomers in this dihydroxylation, so the reaction is stereospecific.

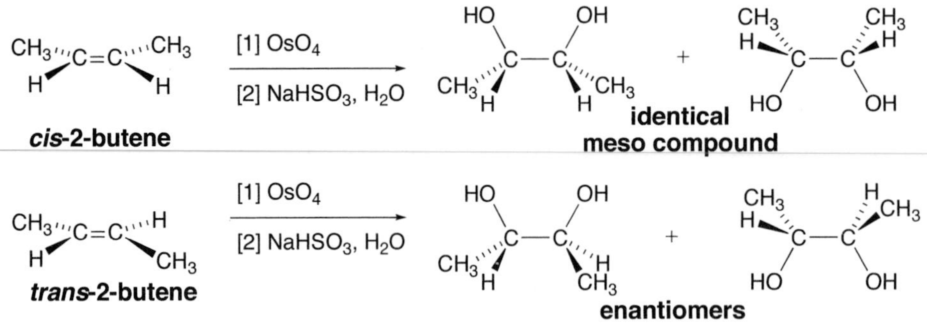

**12.20** To draw the oxidative cleavage products:

- **Locate all the π bonds** in the molecule.
- **Replace all C=C's with *two* C=O's.**

Replace this π bond with two C=O's.

a. $(CH_3)_2C=CHCH_2CH_2CH_2CH_3$ —[1] $O_3$; [2] Zn, $H_2O$→ $(CH_3)_2C=O$ (ketone) + $O=CHCH_2CH_2CH_2CH_3$ (aldehyde)

One **ketone** and one **aldehyde** are formed.

b. 


Two **aldehydes** are formed.

c. 


A **dicarbonyl** compound is formed.

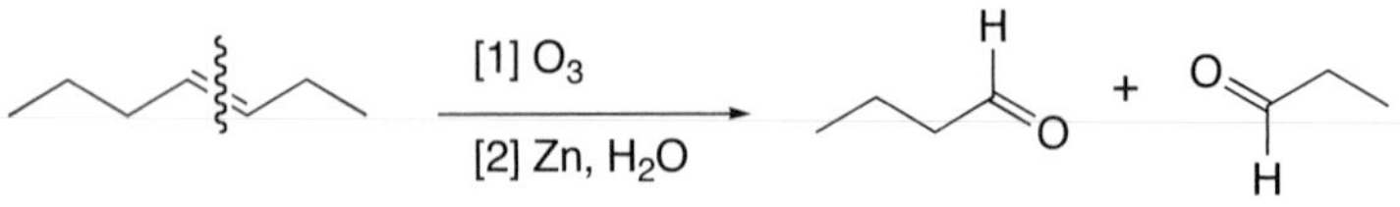

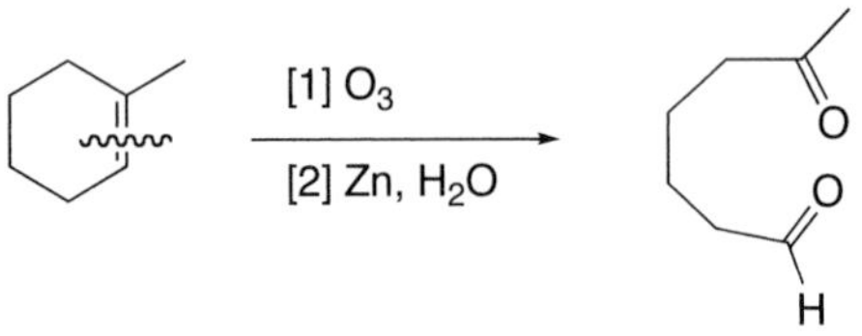

**12.21** To find the alkene that yields the oxidative cleavage products:

- **Find the two carbonyl groups** in the products.
- **Join the two carbonyl carbons** together with a double bond. This is the double bond that was broken during ozonolysis.

a. $(CH_3)_2C=O$ + $(CH_3CH_2)_2C=O$ ⟹ $(CH_3)_2C=C(CH_2CH_3)_2$

Join these two C's.

c. 


With only one product, the alkene must be symmetrical around the double bond. Join this C to the same C in another identical molecule.

b. 


Join these two C's.

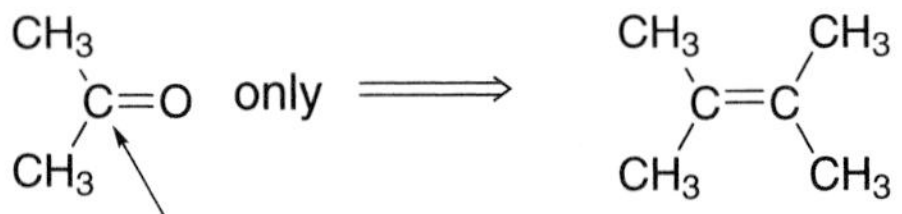

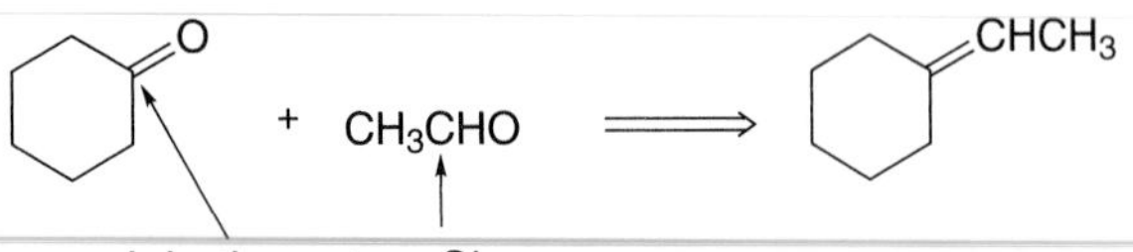

**12.22**

a. [1] $O_3$ / [2] $CH_3SCH_3$

c. [1] $O_3$ / [2] $CH_3SCH_3$

b. [1] $O_3$ / [2] $CH_3SCH_3$

d. [1] $O_3$ / [2] $CH_3SCH_3$

**12.23** To draw the products of oxidative cleavage of alkynes:
- **Locate the triple bond**.
- For internal alkynes, **convert the *sp* hybridized C to COOH.**
- For terminal alkynes, the *sp* hybridized C–H becomes $CO_2$.

a. $CH_3CH_2-C{\equiv}C-CH_2CH_2CH_3$ $\xrightarrow[\text{[2] }H_2O]{\text{[1] }O_3}$ $CH_3CH_2\overset{O}{\underset{}{C}}OH$ + $HO\overset{O}{\underset{}{C}}CH_2CH_2CH_3$

internal alkyne

b. internal alkyne $\xrightarrow[\text{[2] }H_2O]{\text{[1] }O_3}$ + 

identical compounds

c. $H-C{\equiv}C-CH_2-CH_2-C{\equiv}C-CH_3$ $\xrightarrow[\text{[2] }H_2O]{\text{[1] }O_3}$ $CO_2$ + $HO\overset{O}{\underset{O}{C}}...\overset{O}{\underset{}{C}}OH$ + $HO\overset{O}{\underset{}{C}}CH_3$

terminal alkyne   internal alkyne

**12.24**

a. $CO_2$ + $CH_3(CH_2)_8CO_2H$
$\Downarrow$
$CH_3(CH_2)_8-C{\equiv}CH$

b. $CH_3CH_2CH(CH_3)CO_2H$
$\Downarrow$
$CH_3CH_2CH-C{\equiv}C-CHCH_2CH_3$
$\quad\ \ \overset{|}{CH_3}\qquad\qquad\overset{|}{CH_3}$

c. $CH_3CH_2CO_2H$, $HO_2CCH_2CO_2H$, $CH_3CO_2H$
$\Downarrow$
$CH_3CH_2-C{\equiv}C-CH_2-C{\equiv}C-CH_3$

d. $HO_2C(CH_2)_{14}CO_2H \Rightarrow$

**12.25** For **oxidation of alcohols**, remember:
- **1° Alcohols** are oxidized to aldehydes with PCC.
- **1° Alcohols** are oxidized to carboxylic acids with oxidizing agents like $CrO_3$ or $Na_2Cr_2O_7$.
- **2° Alcohols** are oxidized to ketones with all $Cr^{6+}$ reagents.

a. $\xrightarrow{\text{PCC}}$

c. $\xrightarrow[H_2SO_4,\ H_2O]{CrO_3}$

b. $\xrightarrow{\text{PCC}}$

d. $\xrightarrow[H_2SO_4,\ H_2O]{CrO_3}$

**12.26** Upon treatment with Amberlyst A-26 resin–$HCrO_4^-$:
- **1° Alcohols** are oxidized to aldehydes.
- **2° Alcohols** are oxidized to ketones.

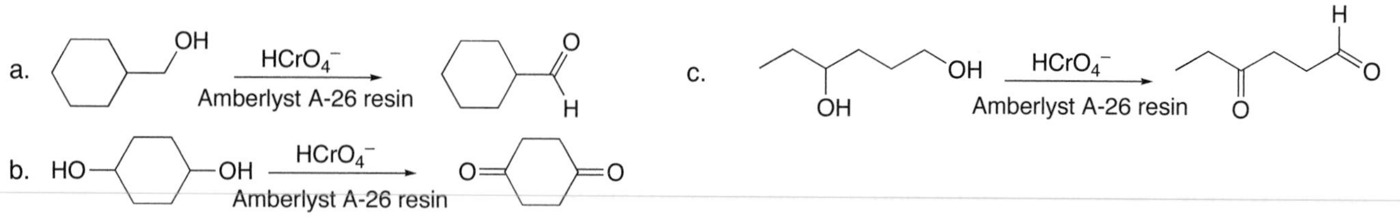

a.

c.

b.

**12.27**

a.

The by-products of the reaction with sodium hypochlorite are water and table salt (NaCl), as opposed to the by-products with Amberlyst A-26 resin, which contain carcinogenic $Cr^{3+}$ metal.

b. Oxidation with NaOCl has at least two advantages over oxidation with $CrO_3$, $H_2SO_4$ and $H_2O$. Since no $Cr^{6+}$ is used as oxidant, there are no Cr by-products that must be disposed of. Also, $CrO_3$ oxidation is carried out in corrosive inorganic acids ($H_2SO_4$) and oxidation with NaOCl avoids this.

**12.28**

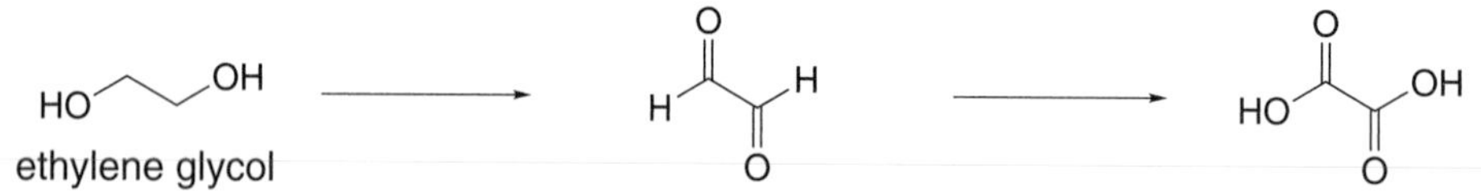

ethylene glycol

**12.29** To draw the products of a **Sharpless epoxidation**:
- With the C=C vertical, draw the allylic alcohol with the OH on the **bottom right** of the alkene.
- Add the new oxygen **above** the plane if (–)-DET is used and **below** the plane if (+)-DET is used.

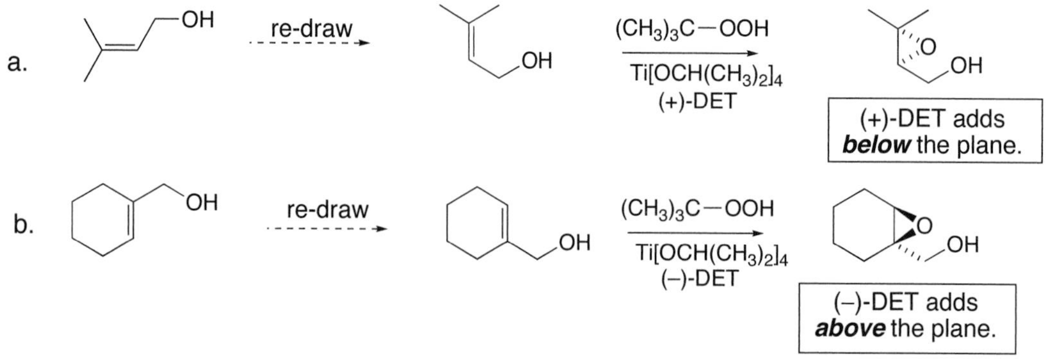

a.

(+)-DET adds **below** the plane.

b.

(–)-DET adds **above** the plane.

**12.30**  Sharpless epoxidation needs an *allylic alcohol* as the starting material.  Alkenes with no allylic OH group will not undergo reaction with the Sharpless reagent.

This alkene is part of an **allylic alcohol**
and will be epoxidized.

geraniol

This alkene is **not** part of an
allylic alcohol and will not be epoxidized.

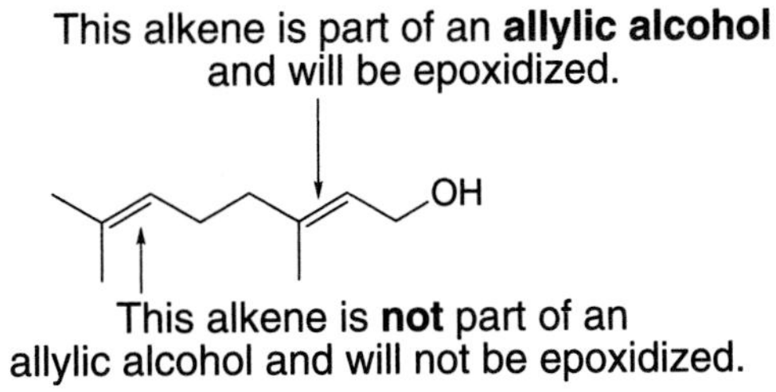

**12.31** Use the rules from Answer 12.1.

a. [cyclohexyl]—C≡CH ⟶ [cyclohexyl]—CH₂CH₃ **reduction**

b. [CH₃CH₂CH₂CH₂OH] ⟶ [CH₃CH₂CH₂COOH] **oxidation**

c. CH₃CH₂Br ⟶ CH₂=CH₂ **neither**
1 C–H and 1 C–Br
bond are removed.

d. [epoxide on cyclohexane] ⟶ [cyclohexanol] **reduction**

e. CH₂=CH₂ ⟶ ClCH₂CH₂Cl **oxidation**
(2 new C–Cl bonds)

f. HO—⟨benzene⟩—OH ⟶ O=⟨quinone⟩=O **oxidation**

**12.32** Use the principles from Answer 12.2 and draw the products of syn addition of H₂ from above and below the C=C.

a. [CH₂=CHCH₂CH₂CH₂CH₃] $\xrightarrow[\text{Pd-C}]{\text{H}_2}$ [CH₃CH₂CH₂CH₂CH₂CH₃]

b. [1,4-dimethylcyclohexene with CH₃] $\xrightarrow[\text{Pd-C}]{\text{H}_2}$ [cis-dimethylcyclohexane with CH₃, CH₃] + [trans-dimethylcyclohexane with CH₃, CH₃]

c. [1-ethyl-2-methylcyclohexene with CH₂CH₃, CH₃] $\xrightarrow[\text{Pd-C}]{\text{H}_2}$ [cyclohexane with CH₂CH₃, CH₃] + [cyclohexane with CH₂CH₃, CH₃]

d. [CH₃CH₂ and (CH₃)₂CH]C=CH₂ $\xrightarrow[\text{Pd-C}]{\text{H}_2}$ [H⋯C(CH₂CH₃)(CH₃) with (CH₃)₂CH] + [(CH₃)₂CH⋯C(CH₂CH₃)(CH₃) with H]

**12.33** Increasing alkyl substitution increases alkene stability, decreasing the heat of hydrogenation.

[2-methyl-2-butene structure]     [2-methyl-1-butene structure]     [3-methyl-1-butene structure]

**2-methyl-2-butene**            **2-methyl-1-butene**            **3-methyl-1-butene**

**tri**substituted              **di**substituted               **mono**substituted

**smallest** $\Delta H^{\circ} = -26.9$ kcal/mol    **intermediate** $\Delta H^{\circ} = -28.5$ kcal/mol    **largest** $\Delta H^{\circ} = -30.3$ kcal/mol

**12.34**

A possible structure:

a. Compound **A**: molecular formula $C_5H_8$: hydrogenated to $C_5H_{10}$.
2 degrees of unsaturation, 1 is hydrogenated.
**1 ring and 1 π bond** ──────────────────⟶

b. Compound **B**: molecular formula $C_{10}H_{16}$: hydrogenated to $C_{10}H_{18}$.
3 degrees of unsaturation, 1 is hydrogenated.
**2 rings and 1 π bond** ──────────────⟶

c. Compound **C**: molecular formula $C_8H_8$: hydrogenated to $C_8H_{16}$.
5 degrees of unsaturation, 4 are hydrogenated.
**1 ring and 4 π bonds** ──────────────⟶

**12.35**

**A**

a. **mono**substituted
   **largest** heat of hydrogenation
b. **fastest** reaction rate

c.
$$\xrightarrow[\text{[2] Zn, H}_2\text{O}]{\text{[1] O}_3}$$

(products shown)

**B**

a. **tetra**substituted
   **smallest** heat of hydrogenation
b. **slowest** reaction rate

c.
$$\xrightarrow[\text{[2] Zn, H}_2\text{O}]{\text{[1] O}_3}$$

identical

**C**

a. **tri**substituted
   **intermediate** heat of hydrogenation
b. **intermediate** reaction rate

c.
$$\xrightarrow[\text{[2] Zn, H}_2\text{O}]{\text{[1] O}_3}$$

(products shown)

**12.36** Work backwards to find the alkene that will be hydrogenated to form 3-methylpentane.

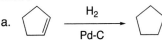

3-methylpentane

2 possible enantiomers:

*R* isomer    or    *S* isomer

**12.37**

a.
$$\xrightarrow[\text{Pd-C}]{\text{H}_2}$$
(product)

b.
$$\xrightarrow[\text{Lindlar catalyst}]{\text{H}_2}$$
no reaction

c.
$$\xrightarrow[\text{NH}_3]{\text{Na}}$$
no reaction

d.
$$\xrightarrow{\text{CH}_3\text{CO}_3\text{H}}$$
(epoxide)

e.
$$\xrightarrow[\text{[2] H}_2\text{O, HO}^-]{\text{[1] CH}_3\text{CO}_3\text{H}}$$
(diol) + (diol)    anti addition

f.
$$\xrightarrow[\text{[2] NaHSO}_3]{\text{[1] OsO}_4 + \text{NMO}}$$
(diol)    syn addition

g.
$$\xrightarrow[\text{H}_2\text{O, HO}^-]{\text{KMnO}_4}$$
(diol)    syn addition

h.
$$\xrightarrow[\text{[2] H}_2\text{O}]{\text{[1] LiAlH}_4}$$
no reaction

i.
$$\xrightarrow[\text{[2] CH}_3\text{SCH}_3]{\text{[1] O}_3}$$
(dialdehyde)

j.
$$\xrightarrow[\substack{\text{Ti[OCH(CH}_3\text{)}_2]_4 \\ (-)\text{-DET}}]{\text{(CH}_3)_3\text{COOH}}$$
no reaction

k.
$$\xrightarrow{\text{mCPBA}}$$
(epoxide)

l.
$$\xrightarrow[\text{[2] H}_2\text{O}]{\text{[1] LiAlH}_4}$$
(alcohol)

**12.38**

a. $\text{CH}_3\text{CH}_2\text{CH}_2-\text{C}\equiv\text{C}-\text{CH}_2\text{CH}_2\text{CH}_3$
$$\xrightarrow[\text{Pd-C}]{\text{H}_2 \text{ (excess)}}$$
(product)

b. $CH_3CH_2CH_2-C\equiv C-CH_2CH_2CH_3$ $\xrightarrow[\text{Lindlar catalyst}]{H_2}$ [cis alkene structure] cis alkene

c. $CH_3CH_2CH_2-C\equiv C-CH_2CH_2CH_3$ $\xrightarrow[NH_3]{Na}$ [trans alkene structure] trans alkene

d. $CH_3CH_2CH_2-C\equiv C-CH_2CH_2CH_3$ $\xrightarrow[[2]\ H_2O]{[1]\ O_3}$ $CH_3CH_2CH_2-\overset{O}{\underset{}{C}}-OH$ + $CH_3CH_2CH_2-\overset{O}{\underset{}{C}}-OH$

identical

## 12.39

a. [alkene structure] $\xrightarrow[Pd-C]{H_2}$ $H-\overset{CH_3CH_2}{\underset{CH_3}{\overset{*}{C}}}-\overset{H}{\underset{H}{C}}-CH_2OH$   [* = new stereogenic center.]   [two structures] + Two enantiomers are formed.

b. [alkene structure] $\xrightarrow{mCPBA}$ [epoxide structure] + [epoxide structure]

c. [alkene structure] $\xrightarrow{PCC}$ [structure with CHO]

d. [alkene structure] $\xrightarrow[H_2SO_4,\ H_2O]{CrO_3}$ [structure with COOH]

e. [alkene structure] $\xrightarrow[\substack{Ti[OCH(CH_3)_2]_4 \\ (+)\text{-DET}}]{(CH_3)_3COOH}$ [epoxide structure]

f. [alkene structure] $\xrightarrow[\substack{Ti[OCH(CH_3)_2]_4 \\ (-)\text{-DET}}]{(CH_3)_3COOH}$ [epoxide structure]

g. [alkene structure] $\xrightarrow{[1]\ PBr_3}$ [structure with CH₂Br] $\xrightarrow[[3]\ H_2O]{[2]\ LiAlH_4}$ [alkene structure]

h. [alkene structure] $\xrightarrow{\substack{\text{Amberlyst} \\ \text{A-26–HCrO}_4^-}}$ [structure with CHO]

## 12.40

a. [cyclohexene-CH₂OH structure] $\xrightarrow{\substack{H_2 \\ Pd-C}}$ [cyclohexane-CH₂OH structure]

b. [cyclohexene-CH₂OH structure] $\xrightarrow[H_2SO_4,\ H_2O]{Na_2Cr_2O_7}$ [cyclohexene-COOH structure]

c. [cyclohexene-CH₂OH structure] $\xrightarrow{PCC}$ [cyclohexene-CHO structure]

d. [cyclohexene-CH₂OH structure] $\xrightarrow{CF_3CO_3H}$ [epoxide structure] + [epoxide structure]

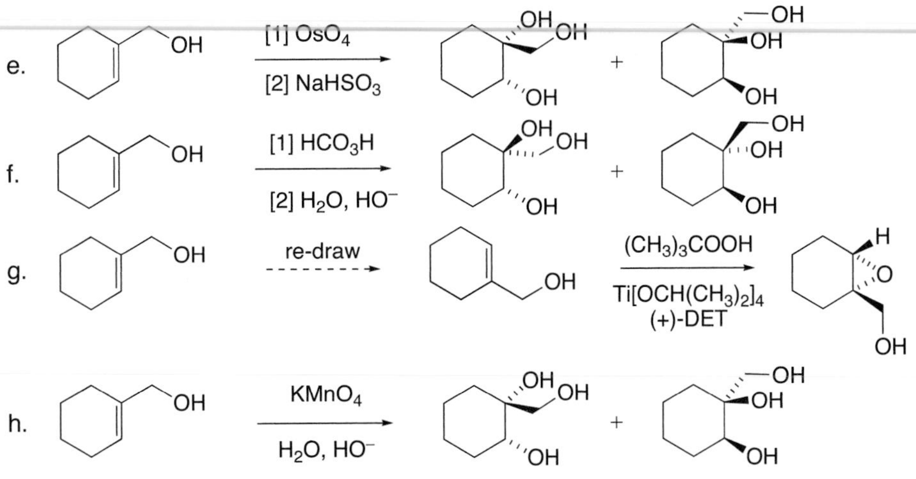

e.

[1] OsO₄

[2] NaHSO₃

f.

[1] HCO₃H

[2] H₂O, HO⁻

g.

re-draw

(CH₃)₃COOH

Ti[OCH(CH₃)₂]₄

(+)-DET

h.

KMnO₄

H₂O, HO⁻

## 12.41

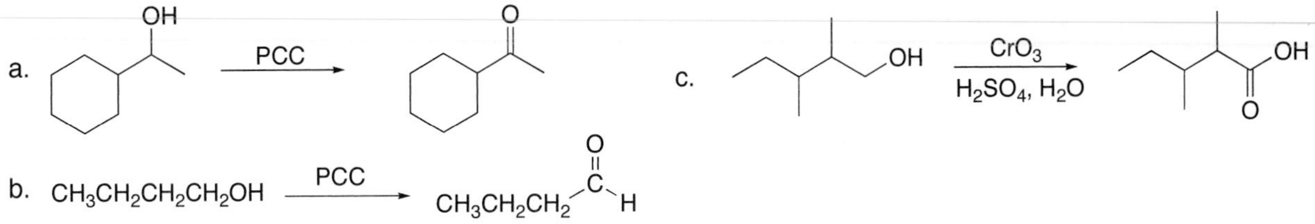

a.

PCC

c.

CrO₃

H₂SO₄, H₂O

b. CH₃CH₂CH₂CH₂OH

PCC

CH₃CH₂CH₂ C—H

## 12.42

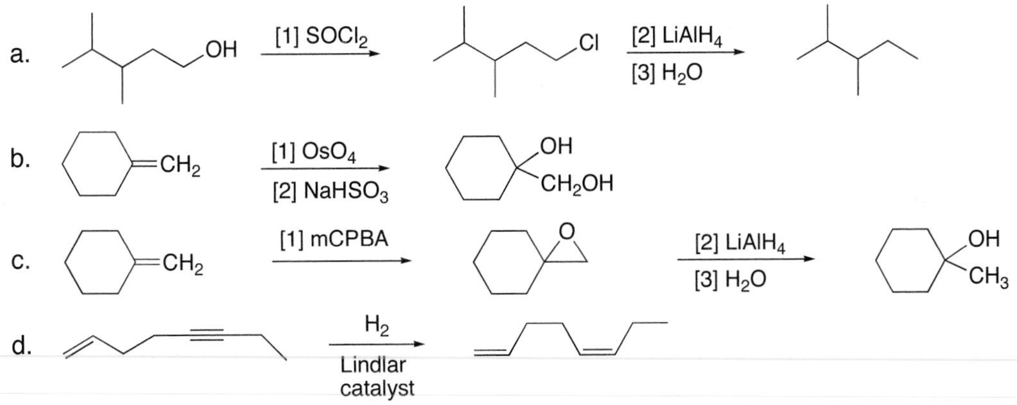

a.

[1] SOCl₂

[2] LiAlH₄

[3] H₂O

b.

[1] OsO₄

[2] NaHSO₃

c.

[1] mCPBA

[2] LiAlH₄

[3] H₂O

d.

H₂

Lindlar
catalyst

## 12.43

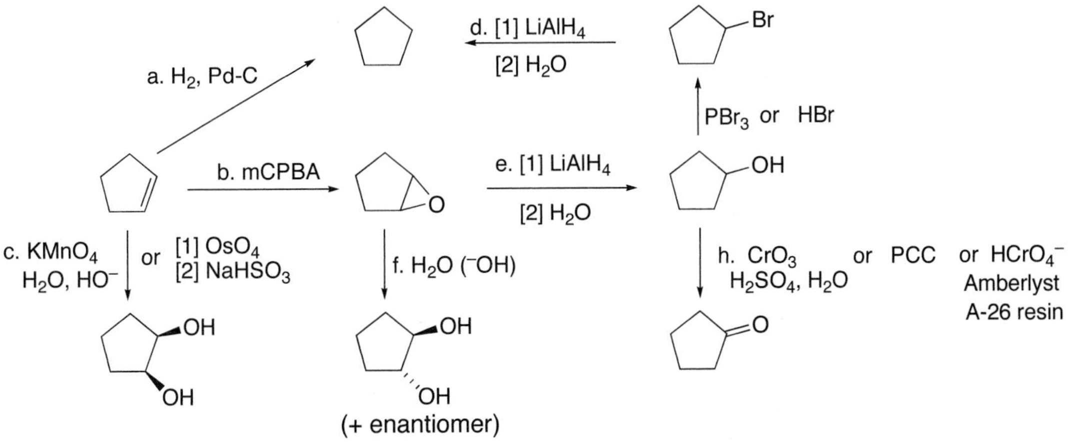

a. H₂, Pd-C

d. [1] LiAlH₄

[2] H₂O

PBr₃ or HBr

b. mCPBA

e. [1] LiAlH₄

[2] H₂O

c. KMnO₄     or   [1] OsO₄
   H₂O, HO⁻        [2] NaHSO₃

f. H₂O (⁻OH)

h. CrO₃     or   PCC   or  HCrO₄⁻
   H₂SO₄, H₂O              Amberlyst
                           A-26 resin

(+ enantiomer)

**12.44**

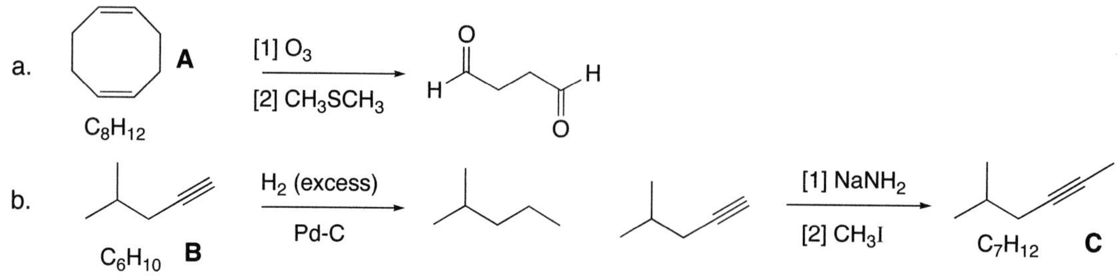

a. **A** $\xrightarrow[\text{[2] CH}_3\text{SCH}_3]{\text{[1] O}_3}$

$C_8H_{12}$

b. $\xrightarrow[\text{Pd-C}]{\text{H}_2 \text{ (excess)}}$ $\xrightarrow[\text{[2] CH}_3\text{I}]{\text{[1] NaNH}_2}$ **C**

$C_6H_{10}$ **B** $C_7H_{12}$

**12.45** Use the directions from Answer 12.20.

a. $(CH_3CH_2)_2C=CHCH_2CH_3$ $\xrightarrow[\text{[2] CH}_3\text{SCH}_3]{\text{[1] O}_3}$ $(CH_3CH_2)_2C=O$ + $O=CHCH_2CH_3$

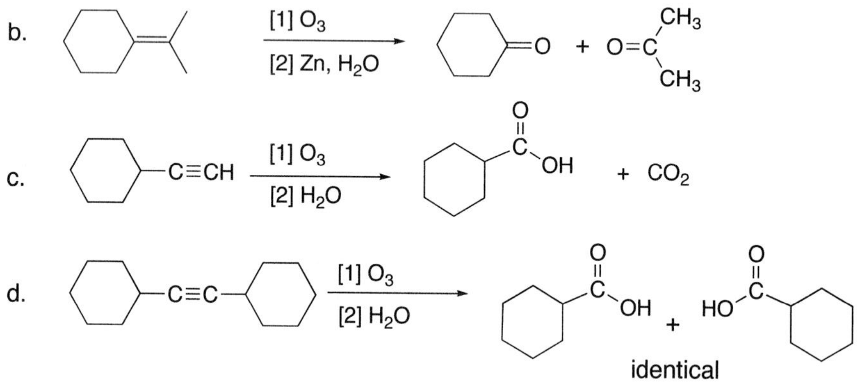

b. $\xrightarrow[\text{[2] Zn, H}_2\text{O}]{\text{[1] O}_3}$

c. $\xrightarrow[\text{[2] H}_2\text{O}]{\text{[1] O}_3}$ + $CO_2$

d. $\xrightarrow[\text{[2] H}_2\text{O}]{\text{[1] O}_3}$

identical

**12.46**

a. $(CH_3)_2C=O$ and $CH_2=O$ $\Longrightarrow$ $(CH_3)_2C=CH_2$

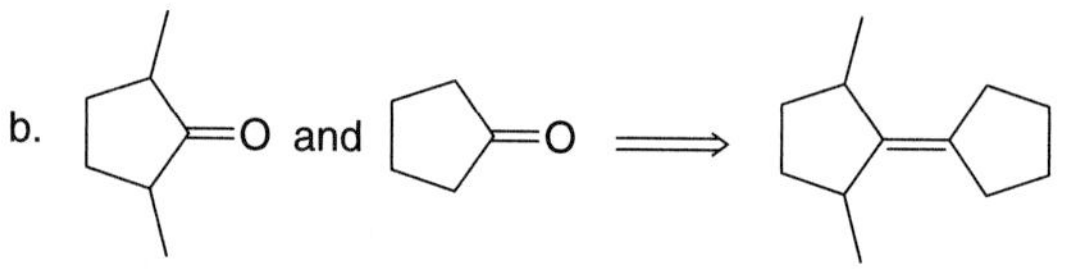

b. $=O$ and $=O$ $\Longrightarrow$

c. $CH_3CH_2CH_2CHO$ only $\Longrightarrow$ $CH_3CH_2CH_2CH=CHCH_2CH_2CH_3$

Join this C to the same C
in another identical molecule.

d. and 2 equivalents of $CH_2=O$ $\Longrightarrow$

Join both of these C's
to a C from formaldehyde.

formaldehyde C

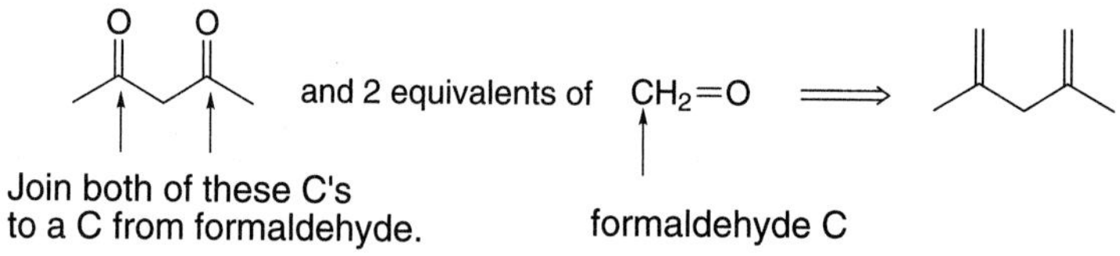

**12.47** Use the directions from Answer 12.21.

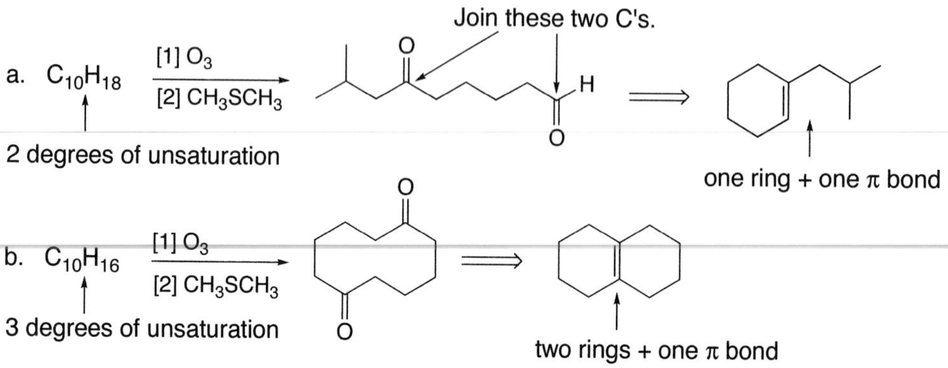

a. $C_{10}H_{18}$ $\xrightarrow[\text{[2] CH}_3\text{SCH}_3]{\text{[1] O}_3}$ $\Longrightarrow$

2 degrees of unsaturation

Join these two C's.

one ring + one π bond

b. $C_{10}H_{16}$ $\xrightarrow[\text{[2] CH}_3\text{SCH}_3]{\text{[1] O}_3}$ $\Longrightarrow$

3 degrees of unsaturation

two rings + one π bond

**12.48**

Join these two C's.

a. $CH_3CH_2CH_2CH_2COOH$ and $CO_2 \implies CH_3CH_2CH_2CH_2C \equiv CH$

Join these two C's.

c. [benzene ring]COOH and $CH_3COOH \implies$ [benzene ring]$C \equiv CCH_3$

Join these two C's.

b. $CH_3CH_2COOH$ and $CH_3CH_2CH_2COOH \implies CH_3CH_2C \equiv CCH_2CH_2CH_3$

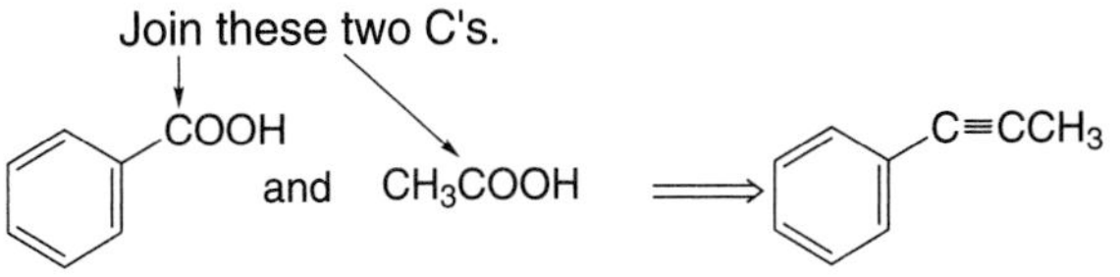

**12.49**

a.

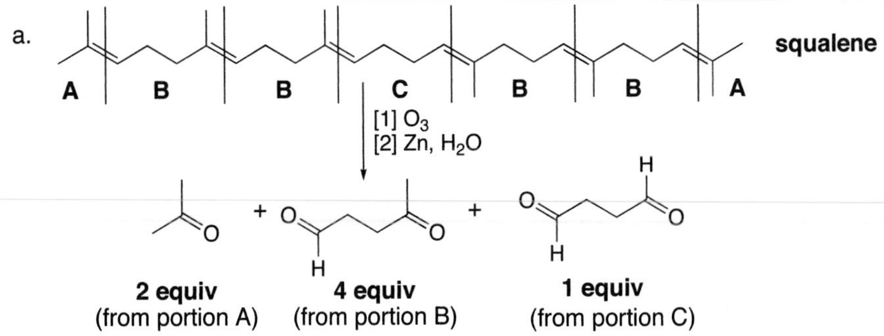

squalene

A | B | B | C | B | B | A

[1] $O_3$
[2] Zn, $H_2O$

**2 equiv**    **4 equiv**    **1 equiv**
(from portion A)   (from portion B)   (from portion C)

b.

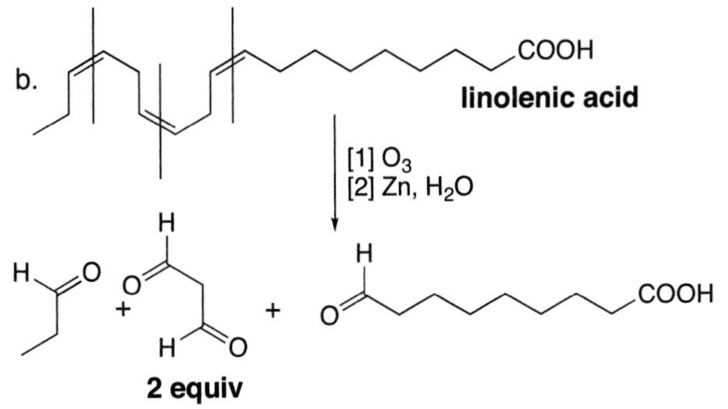

**linolenic acid**

[1] $O_3$
[2] Zn, $H_2O$

**2 equiv**

c.

**zingiberene**

[1] $O_3$
[2] Zn, $H_2O$

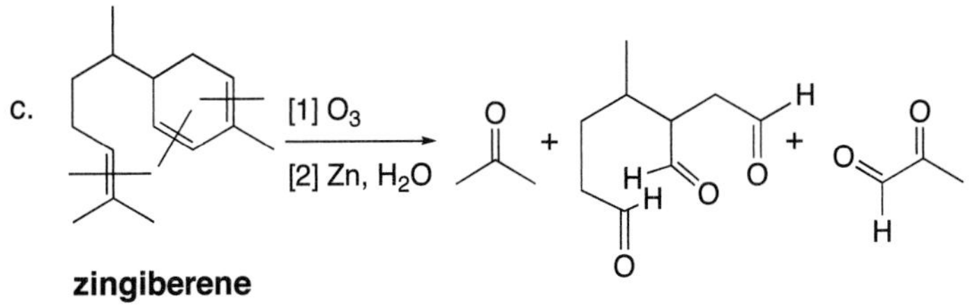

**12.50**

$C_{10}H_{16} \xrightarrow[\text{Pd-C}]{H_2}$

2,6-dimethyloctane

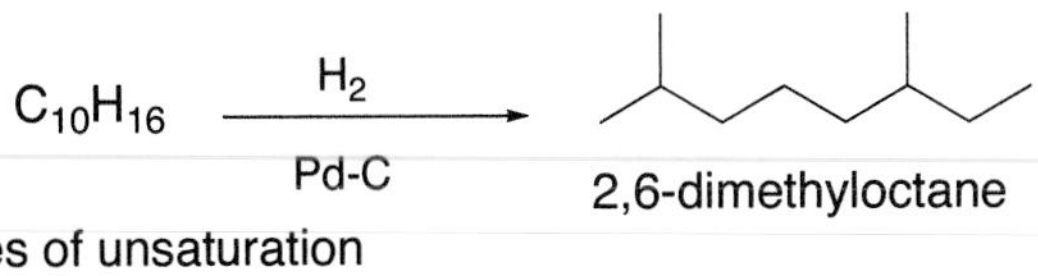

3 degrees of unsaturation

The hydrogenation reaction tells you that both oximene and myrcene have 3 π bonds (and no rings). Use this carbon backbone and add in the double bonds based on the oxidative cleavage products.

**Oximene:**   $(CH_3)_2C=O$   $CH_2=O$   $CH_2(CHO)_2$   $CH_3-\overset{O}{\underset{}{C}}-CHO$ $\implies$

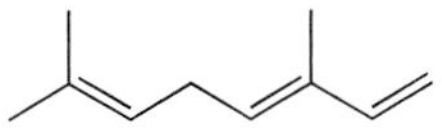

**Myrcene:**   $(CH_3)_2C=O$   $CH_2=O$   $H-\overset{O}{\underset{}{C}}-CH_2CH_2-\overset{O}{\underset{}{C}}-CHO$ $\implies$
            (2 equiv)

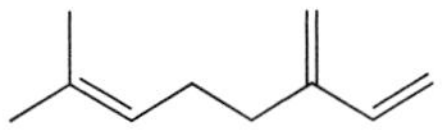

## 12.51

a.

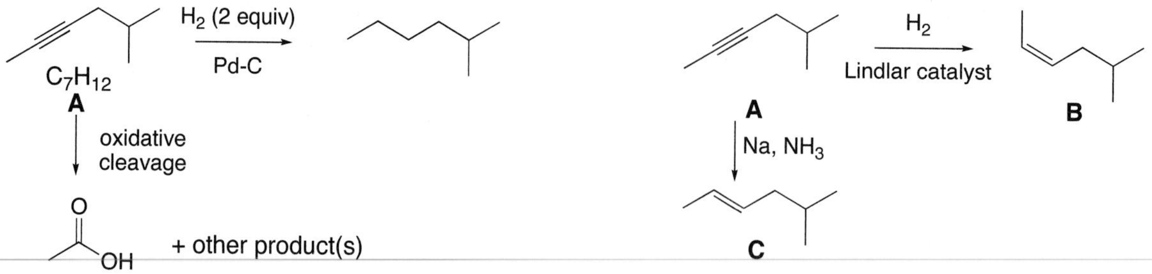
H2 (2 equiv)
Pd-C
C7H12
A
oxidative cleavage
OH
+ other product(s)
A
H2
Lindlar catalyst
B
Na, NH3
C

b. **A** does not react with NaH because it is not a terminal alkyne.

## 12.52

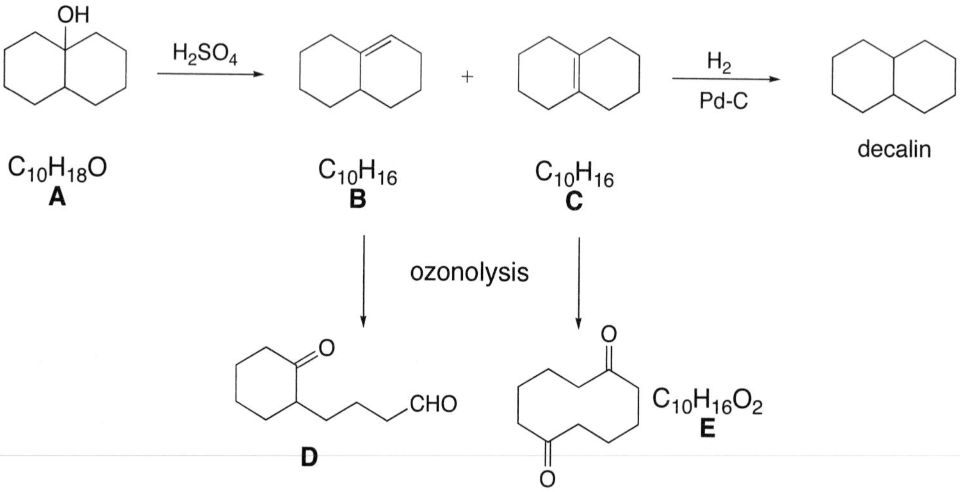
OH
H2SO4
+
H2
Pd-C
decalin
C10H18O
A
C10H16
B
C10H16
C
ozonolysis
CHO
D
C10H16O2
E

## 12.53

a.

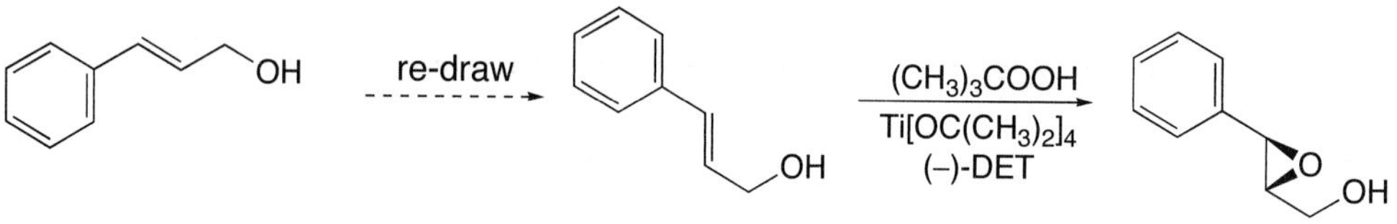
OH
re-draw
OH
(CH3)3COOH
Ti[OC(CH3)2]4
(–)-DET
OH

b.

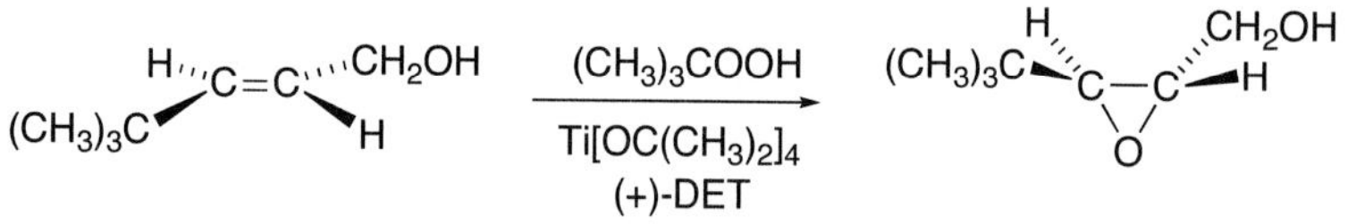
H
C=C
CH2OH
(CH3)3C
H
(CH3)3COOH
Ti[OC(CH3)2]4
(+)-DET
H
CH2OH
(CH3)3C
C—C
H
O

## 12.54

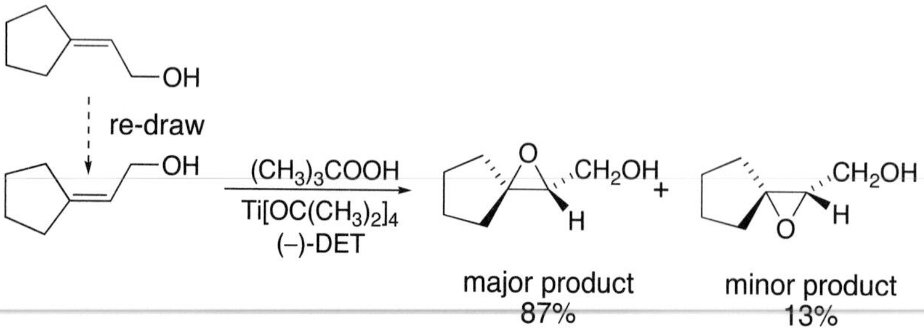
OH
re-draw
OH
(CH3)3COOH
Ti[OC(CH3)2]4
(–)-DET
CH2OH
H
+
CH2OH
H
major product
87%
minor product
13%

**enantiomeric excess** =

% one enantiomer – % second enantiomer

***ee*** **=** 87% – 13% = **74%**

**12.55**

a.
Replace this O to make an alkene.
(–)-DET

b.
Replace this O to make an alkene.
(–)-DET

c.
Replace this O to make an alkene.
(+)-DET

**12.56**  Use retrosynthetic analysis to devise a synthesis of each hydrocarbon from acetylene.

a.  $CH_3CH_2CH=CH_2 \Longrightarrow CH_3CH_2CH\equiv CH \Longrightarrow {}^-C\equiv CH \Longrightarrow HC\equiv CH$

$HC\equiv CH \xrightarrow{\text{NaH}} {}^-C\equiv CH \xrightarrow{CH_3CH_2Cl} CH_3CH_2C\equiv CH \xrightarrow[\text{Lindlar catalyst}]{H_2} CH_3CH_2CH=CH_2$

b. $\Longrightarrow CH_3-C\equiv C-CH_3 \Longrightarrow CH_3-C\equiv CH \Longrightarrow {}^-C\equiv CH \Longrightarrow HC\equiv CH$

$HC\equiv CH \xrightarrow{\text{NaH}} {}^-C\equiv CH \xrightarrow{CH_3Cl} CH_3-C\equiv CH \xrightarrow{\text{NaH}} CH_3-C\equiv C^- \xrightarrow{CH_3Cl} CH_3-C\equiv C-CH_3 \xrightarrow[\text{Lindlar catalyst}]{H_2}$

c. $\Longrightarrow CH_3-C\equiv C-CH_3 \Longrightarrow CH_3-C\equiv CH \Longrightarrow {}^-C\equiv CH \Longrightarrow HC\equiv CH$

$HC\equiv CH \xrightarrow{\text{NaH}} {}^-C\equiv CH \xrightarrow{CH_3Cl} CH_3-C\equiv CH \xrightarrow{\text{NaH}} CH_3-C\equiv C^- \xrightarrow{CH_3Cl} CH_3-C\equiv C-CH_3 \xrightarrow[\text{NH}_3]{\text{Na}}$

d.  $(CH_3)_2CHCH_2CH_2CH_2CH_2CH(CH_3)_2 \Longrightarrow HC\equiv CCH_2CH(CH_3)_2 \Longrightarrow {}^-C\equiv CH \Longrightarrow HC\equiv CH$

$HC\equiv CH \xrightarrow{\text{NaH}} {}^-C\equiv CH \xrightarrow{\text{Cl}\diagup\diagdown} HC\equiv C\diagup \xrightarrow{\text{NaH}} {}^-C\equiv C\diagup \xrightarrow{\text{Cl}\diagup} \diagdown C\equiv C\diagup \xrightarrow[\text{Pd-C}]{H_2 \text{ (2 equiv)}}$

**12.57**

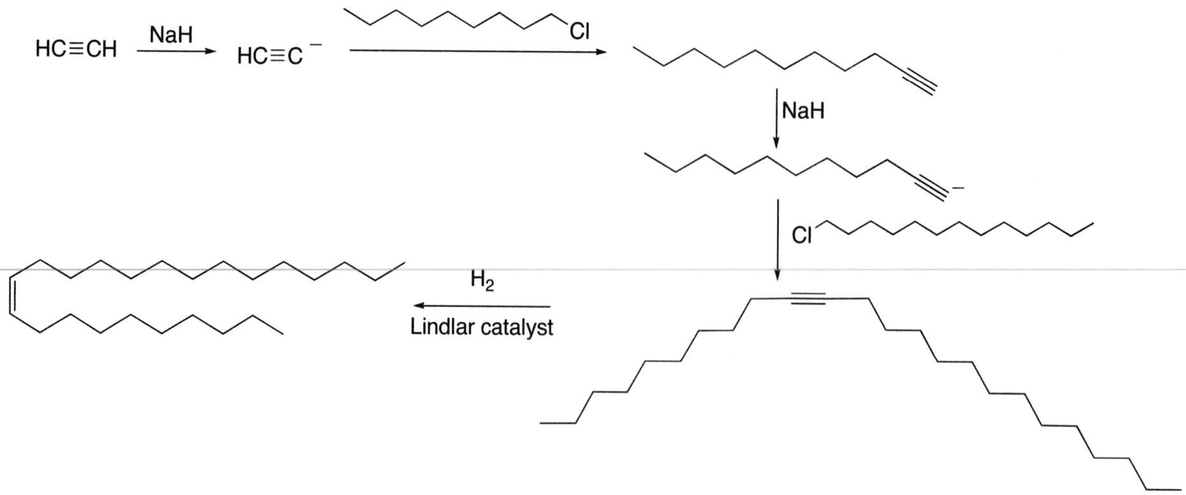

HC≡CH  —NaH→  HC≡C⁻  ———————→ (to terminal alkyne)

|NaH

(internal) ⁻

| Cl———————

(long alkyne)

H₂ / Lindlar catalyst → (cis alkene)

**12.58**

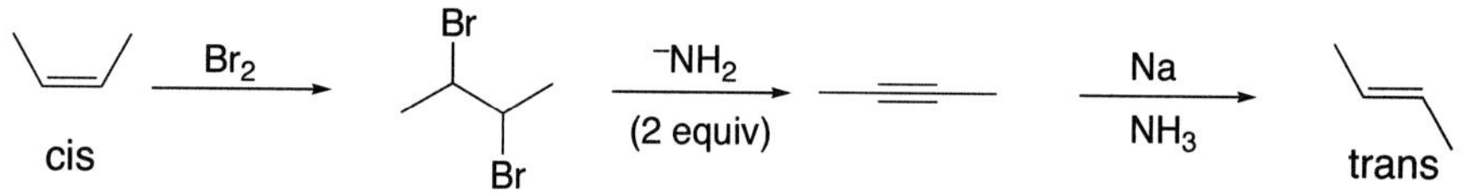

cis  —Br₂→  Br / Br  —⁻NH₂ (2 equiv)→  (alkyne)  —Na / NH₃→  trans

**12.59**

a.

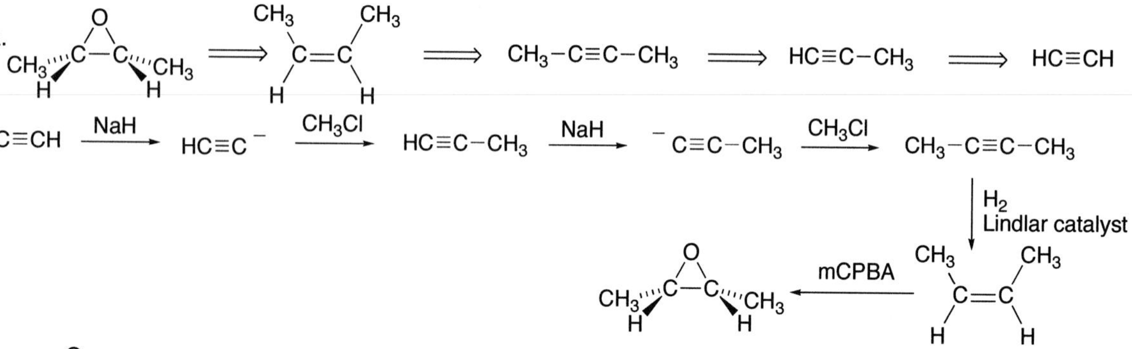

$CH_3$ $C=C$ $CH_3$  ⟹  $C=C$  ⟹  $CH_3-C≡C-CH_3$  ⟹  $HC≡C-CH_3$  ⟹  $HC≡CH$

HC≡CH  —NaH→  HC≡C⁻  —CH₃Cl→  HC≡C-CH₃  —NaH→  ⁻C≡C-CH₃  —CH₃Cl→  $CH_3-C≡C-CH_3$

|H₂ / Lindlar catalyst

$CH_3$ $C-C$ $CH_3$  ←mCPBA—  $CH_3$ $C=C$ $CH_3$

b.

H $C-C$ $CH_3$ (+ enantiomer)  ⟹  H / $CH_3$ $C=C$ $CH_3$ / H  ⟹  $CH_3-C≡C-CH_3$  ⟹  $HC≡C-CH_3$  ⟹  $HC≡CH$

$CH_3-C≡C-CH_3$ (from a.)  —Na, NH₃→  H / $CH_3$ $C=C$ $CH_3$ / H  —mCPBA→  (epoxide) + enantiomer

c.

HO / OH $C-C$ $CH_3$ $CH_3$  ⟹  H / H $C=C$ $CH_3$ $CH_3$  ⟹  $CH_3-C≡C-CH_3$  ⟹  $HC≡C-CH_3$  ⟹  $HC≡CH$

H / H $C=C$ $CH_3$ $CH_3$ (from a.)  —KMnO₄ / H₂O, HO⁻→  HO / OH $C-C$ H CH₃ CH₃

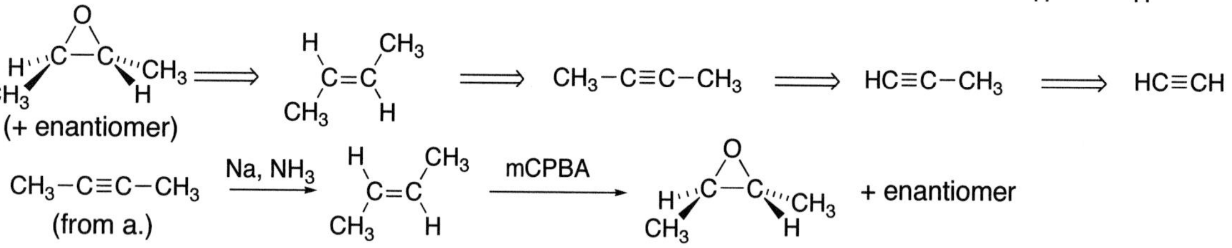

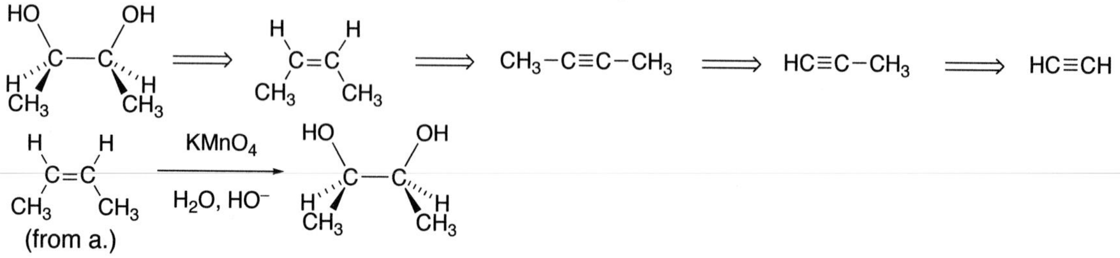

d.

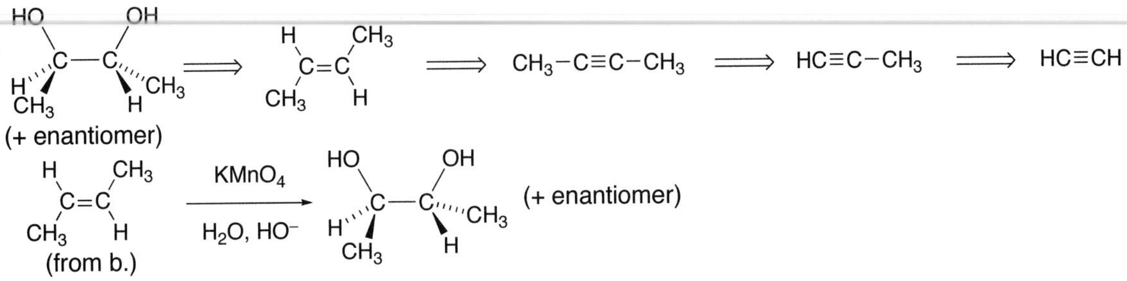
HO
OH
CH3
H
(+ enantiomer)
CH3−C≡C−CH3
HC≡C−CH3
HC≡CH
KMnO4
H2O, HO−
(from b.)

**12.60**

a.

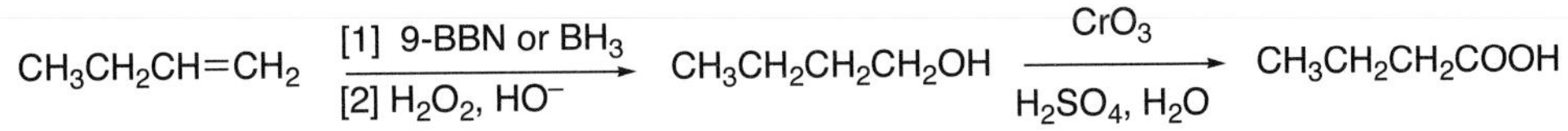
CH3CH2CH=CH2
[1] 9-BBN or BH3
[2] H2O2, HO−
CH3CH2CH2CH2OH
CrO3
H2SO4, H2O
CH3CH2CH2COOH

b.

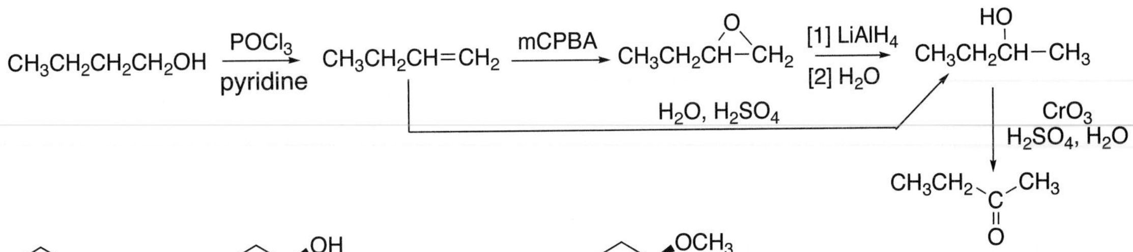
CH3CH2CH2CH2OH
POCl3
pyridine
CH3CH2CH=CH2
mCPBA
[1] LiAlH4
[2] H2O
H2O, H2SO4
CrO3
H2SO4, H2O

c.

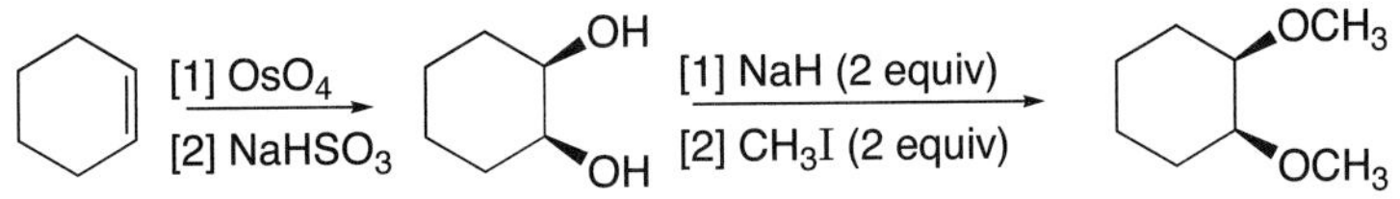
[1] OsO4
[2] NaHSO3
OH
[1] NaH (2 equiv)
[2] CH3I (2 equiv)
OCH3

d.

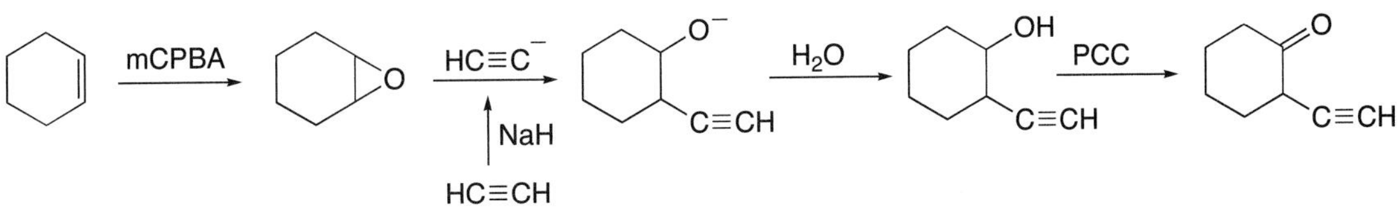
mCPBA
HC≡C−
NaH
HC≡CH
C≡CH
H2O
OH
PCC

**12.61**

a.

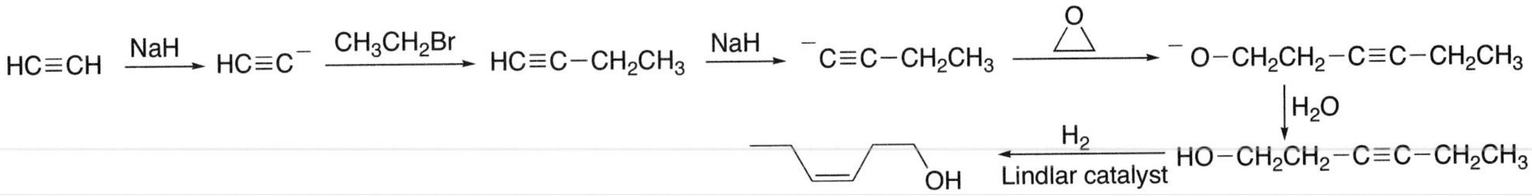
HC≡CH
NaH
HC≡C−
CH3CH2Br
HC≡C−CH2CH3
−C≡C−CH2CH3
−O−CH2CH2−C≡C−CH2CH3
H2O
HO−CH2CH2−C≡C−CH2CH3
H2
Lindlar catalyst
OH

b.

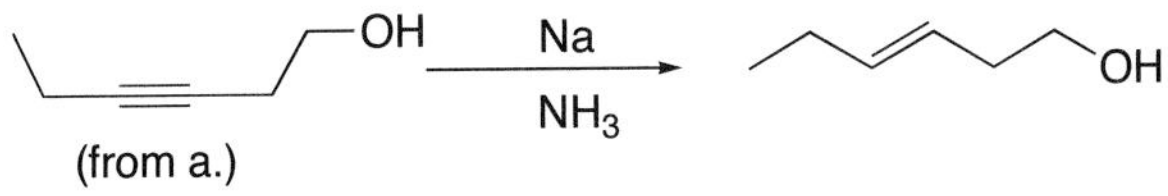
OH
Na
NH3
(from a.)

c.

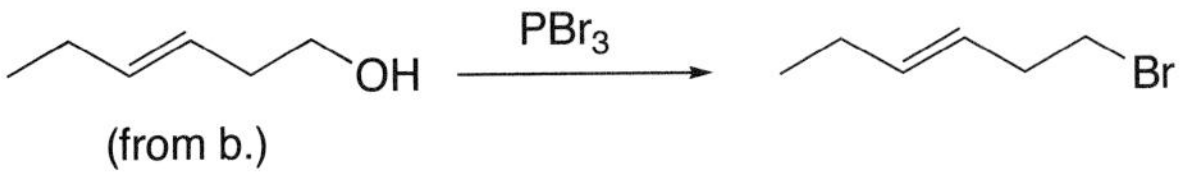
OH
PBr3
Br
(from b.)

d.

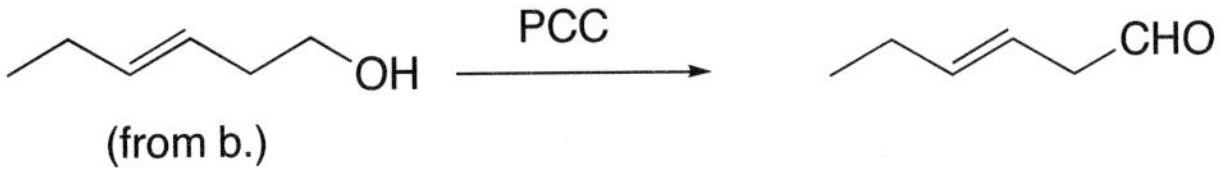
OH
PCC
CHO
(from b.)

**12.62**

a.

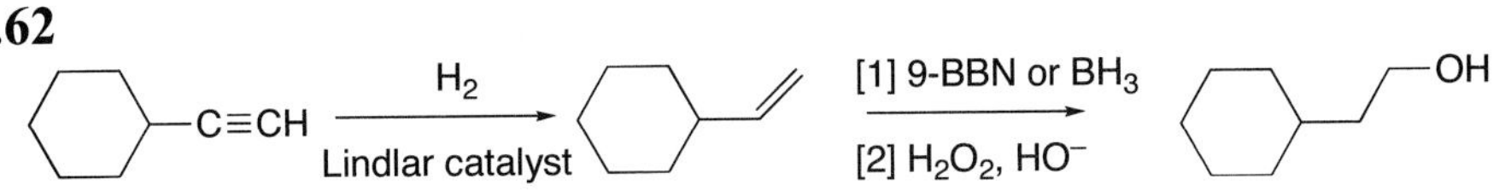
C≡CH
H2
Lindlar catalyst
[1] 9-BBN or BH3
[2] H2O2, HO−
OH

b.

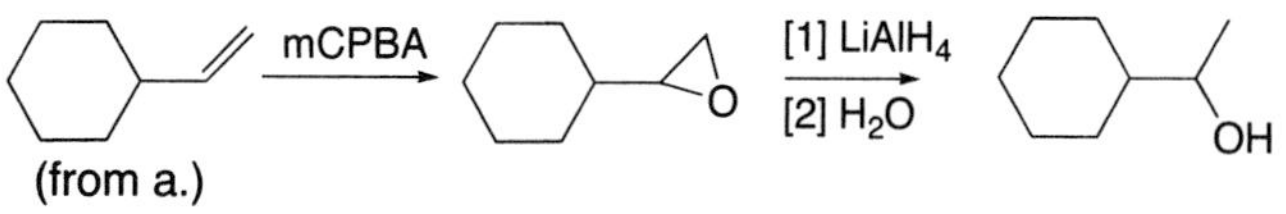

(from a.)

c.

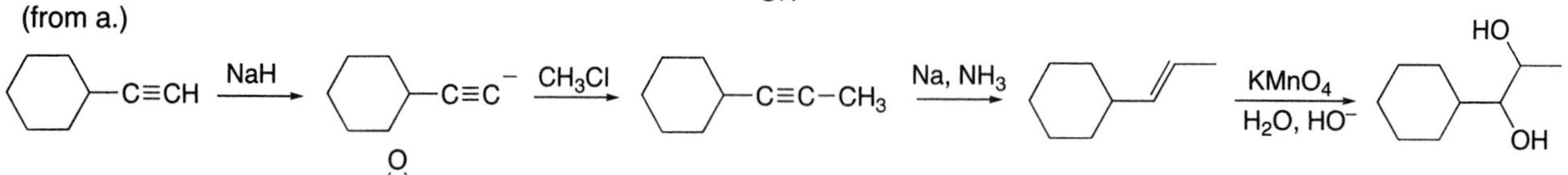

d.

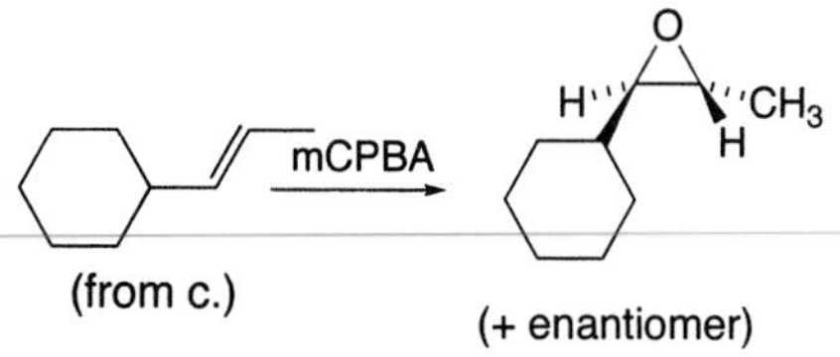

(from c.)

(+ enantiomer)

**12.63**

**disparlure**

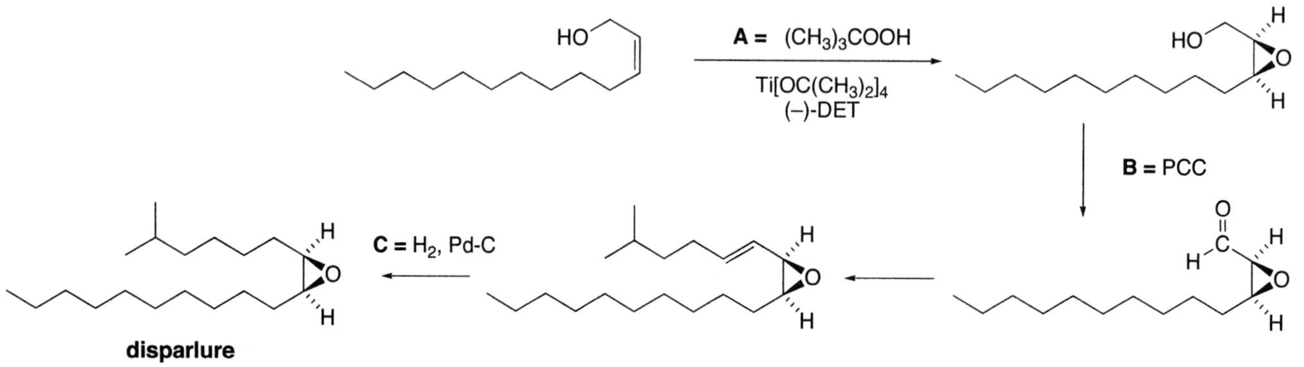

**12.64**

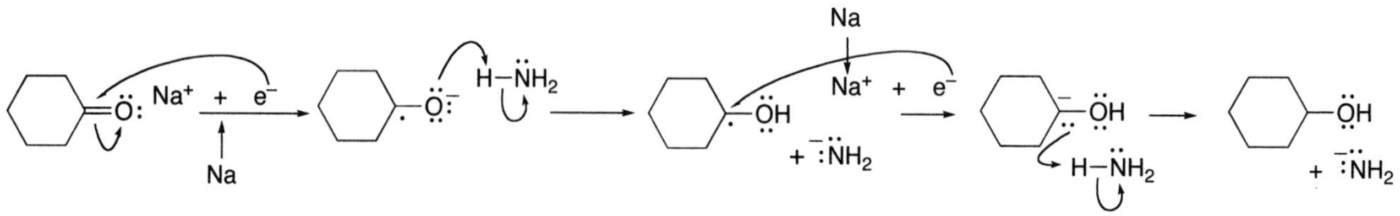

**12.65**

The favored conformation
for both molecules places
the *tert*-butyl group equatorial.

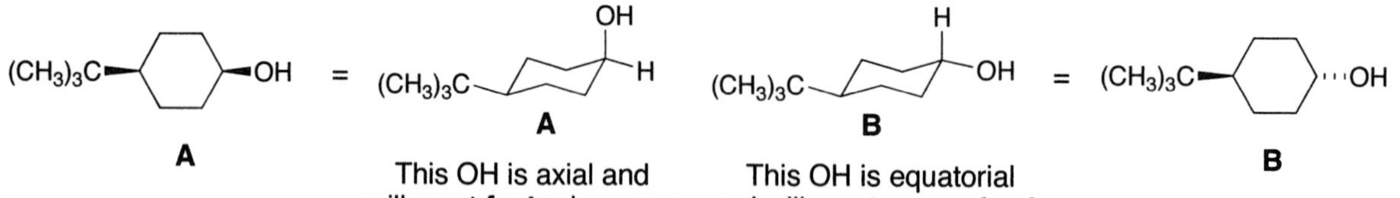

A

A

This OH is axial and
will react **faster** because
the OH group is more
hindered.

B

This OH is equatorial
and will react **more slowly**
because the OH group
is less hindered.

B

**12.66**

## Chapter 13: Mass Spectrometry and Infrared Spectroscopy

### ◆ Mass spectrometry (MS)

- Mass spectrometry measures the molecular weight of a compound (13.1A).
- The mass of the molecular ion (**M**) = the molecular weight of a compound. Except for isotope peaks at M + 1 and M + 2, the molecular ion has the highest mass in a mass spectrum (13.1A).
- The base peak is the tallest peak in a mass spectrum (13.1A).
- A compound with an odd number of N atoms gives an odd molecular ion. A compound with an even number of N atoms (including zero) gives an even molecular ion (13.1B).
- Organic chlorides show two peaks for the molecular ion (M and M + 2) in a 3:1 ratio (13.2).
- Organic bromides show two peaks for the molecular ion (M and M + 2) in a 1:1 ratio (13.2).
- High-resolution mass spectrometry gives the molecular formula of a compound (13.3A).

### ◆ Electromagnetic radiation

- The wavelength and frequency of electromagnetic radiation are *inversely* related by the following equations: $\lambda = c/\nu$ or $\nu = c/\lambda$ (13.4).
- The energy of a photon is proportional to its frequency; the higher the frequency the higher the energy: $E = h\nu$ (13.4).

### ◆ Infrared spectroscopy (IR, 13.5 and 13.6)

- Infrared spectroscopy identifies functional groups.
- IR absorptions are reported in wavenumbers:

$$\boxed{\text{wavenumber} = \tilde{\nu} = 1/\lambda}$$

- The functional group region from **4000–1500 cm$^{-1}$** is the most useful region of an IR spectrum.
- C–H, O–H, and N–H bonds absorb at high frequency, $\geq 2500$ cm$^{-1}$.
- As bond strength increases, the wavenumber of an absorption increases; thus triple bonds absorb at higher wavenumber than double bonds.

$$
\begin{array}{cc}
C=C & C\equiv C \\
\sim 1650 \text{ cm}^{-1} & \sim 2250 \text{ cm}^{-1}
\end{array}
$$

**Increasing bond strength**
**Increasing $\tilde{\nu}$**

- The higher the percent *s*-character, the stronger the bond, and the higher the wavenumber of an IR absorption.

$$
\begin{array}{ccc}
\overset{|}{\underset{|}{-C}}-H & =C\overset{/}{\underset{\backslash H}{}} & \equiv C-H \\
C_{sp^3}-H & C_{sp^2}-H & C_{sp}-H \\
25\% \text{ } s\text{-character} & 33\% \text{ } s\text{-character} & 50\% \text{ } s\text{-character} \\
3000\text{–}2850 \text{ cm}^{-1} & 3150\text{–}3000 \text{ cm}^{-1} & 3300 \text{ cm}^{-1}
\end{array}
$$

**Increasing percent *s*-character**
**Increasing $\tilde{\nu}$**

## Chapter 13: Answers to Problems

**13.1** The molecular ion formed from each compound is equal to its molecular weight.

a. $C_3H_6O$
molecular weight = **58**
molecular ion ($m/z$) = **58**

b. $C_{10}H_{20}$
molecular weight = **140**
molecular ion ($m/z$) = **140**

c. $C_8H_8O_2$
molecular weight = **136**
molecular ion ($m/z$) = **136**

d. $C_{10}H_{15}N$
molecular weight = **149**
molecular ion ($m/z$) = **149**

**13.2** Some possible formulas for each molecular ion:
  a. Molecular ion at 72: $C_5H_{12}$, $C_4H_8O$, $C_3H_4O_2$
  b. Molecular ion at 100: $C_8H_4$, $C_7H_{16}$, $C_6H_{12}O$, $C_5H_8O_2$
  c. Molecular ion at 73: $C_4H_{11}N$, $C_2H_7N_3$

**13.3** To calculate the molecular ions you would expect for compounds with Cl, calculate the molecular weight using each of the two most common isotopes of Cl ($^{35}Cl$ and $^{37}Cl$). Do the same for Br, using $^{79}Br$ and $^{81}Br$.

a. $C_4H_9{}^{35}Cl$ = **92**
$C_4H_9{}^{37}Cl$ = **94**
Two peaks in 3:1 ratio at $m/z$ 92 and 94.
b. $C_3H_7F$ = **62**
One peak at $m/z$ 62
c. $C_6H_{11}{}^{79}Br$ = **162**
$C_6H_{11}{}^{81}Br$ = **164**
Two peaks in a 1:1 ratio at $m/z$ 162 and 164.

d. $C_4H_{11}N$ = **73**
One peak at $m/z$ 73.
e. $C_4H_4N_2$ = **80**
One peak at $m/z$ 80.

**13.4** Use the exact values given in Table 13.1 to calculate the exact mass of each compound.

$C_7H_5NO_3$
mass: 151.0270

$C_8H_9NO_2$
mass: 151.0634
compound **X**

$C_{10}H_{17}N$
mass: 151.1362

**13.5**

benzene
$C_6H_6$ $m/z$ = 78

toluene
$C_7H_8$ $m/z$ = 92

$p$-xylene
$C_8H_{10}$ $m/z$ = 106

**GC–MS analysis:**
Three peaks in the gas chromatogram.
Order of peaks: benzene, toluene, $p$-xylene, in order of increasing bp.
Molecular ions observed in the three mass spectra: 78, 92, 106.

**13.6** **Wavelength and frequency are inversely proportional.** The higher frequency light will have a shorter wavelength.
  a. Light having a $\lambda$ of $10^2$ nm has a higher $\nu$ than light with a $\lambda$ of $10^4$ nm.
  b. Light having a $\lambda$ of 100 nm has a higher $\nu$ than light with a $\lambda$ of 100 μm.
  c. Blue light has a higher $\nu$ than red light.

**13.7** The **energy of a photon** is *proportional* to its **frequency**, and inversely proportional to its wavelength.
  a. Light having a $v$ of $10^8$ Hz is of higher energy than light having a $v$ of $10^4$ Hz.
  b. Light having a $\lambda$ of 10 nm is of higher energy than light having a $\lambda$ of 1000 nm.
  c. Blue light is of higher energy than red light.

**13.8** The larger the energy difference between two states, the higher the frequency needed for absorption. The 100 kcal/mol transition requires a higher $v$ of radiation than a 5 kcal/mol transition.

**13.9** Higher wavenumbers are proportional to higher frequencies and higher energies.
  a. IR light with a wavenumber of 3000 $cm^{-1}$ is higher in energy than IR light with a wavenumber of 1500 $cm^{-1}$.
  b. IR light having a $\lambda$ of 10 μm is higher in energy than IR light having a $\lambda$ of 20 μm.

**13.10** Stronger bonds absorb at a higher wavenumber. Bonds to lighter atoms (H versus D) absorb at higher wavenumber.

  a. $CH_3-C \equiv C-CH_2CH_3$ or $CH_2 = C(CH_3)_2$      b. $CH_3-H$ or $CH_3-D$

  stronger bond                                              lighter atom H
  higher wavenumber                                          higher wavenumber

**13.11** Cyclopentane and 1-pentene are both composed of C–C and C–H bonds, but 1-pentene also has a C=C bond. This difference will give the IR of 1-pentene an additional peak at 1650 $cm^{-1}$ (for the C=C). 1-Pentene will also show C–H absorptions for $sp^2$ hybridized C–H bonds at 3150–3000 $cm^{-1}$.

**13.12** Look at the functional groups in each compound below to explain how each IR is different.

  A: $CH_3-\overset{\overset{O}{\|}}{C}-CH_3$      B: $CH_3OCH=CH_2$      C: ▷—OH

  C=O peak at ~1700 $cm^{-1}$      C=C peak at 1650 $cm^{-1}$      O–H peak at 3200–3600 $cm^{-1}$
                                   $Csp^2$–H at 3150–3000 $cm^{-1}$

**13.13**
  a. Compound **A** has peaks at ~3150 ($sp^2$ hybridized C–H), 3000–2850 ($sp^3$ hybridized C–H), and 1650 (C=C) $cm^{-1}$.
  b. Compound **B** has a peak at 3000–2850 ($sp^3$ hybridized C–H) $cm^{-1}$.

**13.14** All compounds show an absorption at 3000–2850 $cm^{-1}$ due to the $sp^3$ hybridized C–H bonds. Additional peaks in the functional group region for each compound are shown.

  a. [structure]      b. [structure]—OH      c. [structure]      d. [structure]=O

  no additional peaks      O–H bond at 3600–3200 $cm^{-1}$      $Csp^2$–H at 3150–3000 $cm^{-1}$      C=O bond at ~1700 $cm^{-1}$
                                                                C=C bond at 1650 $cm^{-1}$

e.

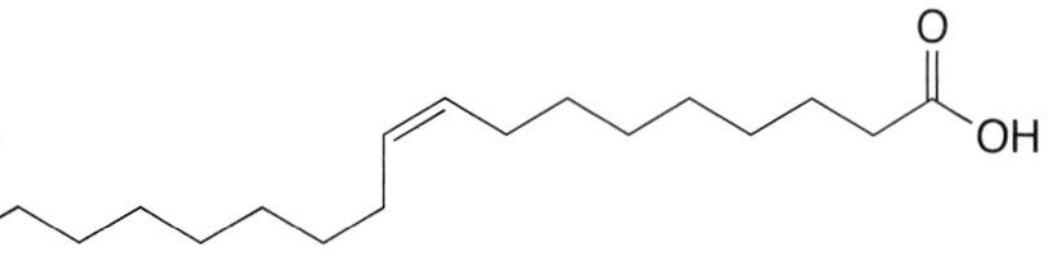

Csp²–H at 3150–3000 cm⁻¹
C=O bond at ~1700 cm⁻¹
C=C bond at 1650 cm⁻¹
O–H above 3000 cm⁻¹
[ The OH of a COOH is much broader
than the OH of an alcohol and occurs
at 3500–2500 cm⁻¹ (see Chapter 19). ]

f.

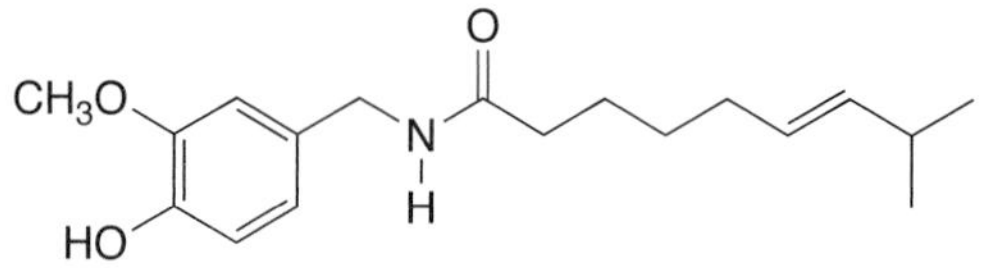

O–H bond at 3600–3200 cm⁻¹
N–H at 3500–3200 cm⁻¹
Csp²–H at 3150–3000 cm⁻¹
C=O bond at ~1700 cm⁻¹
C=C bond at 1650 cm⁻¹
aromatic ring at 1600, 1500 cm⁻¹

**13.15** Possible structures are (a) $CH_3COOCH_2CH_3$ and (c) $CH_3CH_2COOCH_3$. Compounds (b) and (d) also have an OH group that would give a strong absorption at ~3600–3200 cm⁻¹, which is absent in the IR spectrum of **X**, thus excluding them as possibilities.

**13.16**

a. Hydrocarbon with a molecular ion at $m/z = 68$
IR absorptions at 3310 cm⁻¹ = $Csp$–H bond
3000–2850 cm⁻¹ = $Csp^3$–H bonds
2120 cm⁻¹ = C≡C bond
Molecular formula: $C_5H_8$

H−C≡C−CH₂CH₂CH₃   or   H−C≡C−CHCH₃
                                      |
                                     CH₃

b. Compound with C, H, and O with a molecular ion at $m/z = 60$
IR absorptions at 3600–3200 cm⁻¹ = O–H bond
3000–2850 cm⁻¹ = $Csp^3$–H bonds
Molecular formula: $C_3H_8O$

CH₃CH₂CH₂−O−H   or   CH₃CH−O−H
                                  |
                                 CH₃

**13.17**

a.

molecular formula: $C_6H_6$
molecular ion (*m/z*): **78**

c.

molecular formula: $C_5H_{10}O$
molecular ion (*m/z*): **86**

e. $(CH_3)_3CCH(Br)CH(CH_3)_2$

molecular formula: $C_8H_{17}Br$
molecular ions (*m/z*): **192, 194**

b.

molecular formula: $C_{10}H_{16}$
molecular ion (*m/z*): **136**

d.  Cl

molecular formula: $C_5H_{11}Cl$
molecular ions (*m/z*): **106, 108**

**13.18**

$CH_2CH_2CH_3$

$C_9H_{12}$
molecular weight = 120

$\overset{O}{\overset{\|}{C}}-CH_2CH_3$

$C_9H_{10}O$
molecular weight = 134

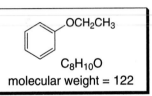  $OCH_2CH_3$

$C_8H_{10}O$
molecular weight = 122

**13.19**  Examples are given for each molecular ion.
a.  molecular ion 102: $C_8H_6$, $C_6H_{14}O$, $C_5H_{10}O_2$, $C_5H_{14}N_2$
b.  molecular ion 98: $C_8H_2$, $C_7H_{14}$, $C_6H_{10}O$, $C_5H_6O_2$
c.  molecular ion 119: $C_8H_9N$, $C_6H_5N_3$
d.  molecular ion 74: $C_6H_2$, $C_4H_{10}O$, $C_3H_6O_2$

**13.20**  Likely molecular formula, $C_8H_{16}$ (one degree of unsaturation—one ring or one π bond).

Four structures with *m/z* = 112

**13.21**

$CH_3-CH-CO_2CH_3$
         |
         Cl     **B**

$C_4H_7O_2Cl$
molecular weight: **122, 124**
should show 2 peaks for the
molecular ion with a **3:1 ratio**

Mass spectrum [1]

  CH_3

  OCH_3
**C**

$C_8H_{10}O$
molecular weight: **122**

Mass spectrum [2]

$CH_3CH_2CH_2Br$
**A**

$C_3H_7Br$
molecular weight: **122, 124**
should show 2 peaks for the
molecular ion with a **1:1 ratio**

Mass spectrum [3]

**13.22**

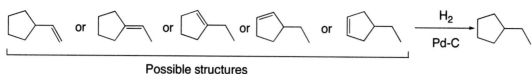

Possible structures
$C_7H_{12}$
(exact mass 96.0940)

**13.23** One possible structure is drawn for each set of data:

a. A compound that contains a benzene ring and has a molecular ion at $m/z = 107$.

$C_7H_9N$

c. A compound that contains a carbonyl group and gives a molecular ion at $m/z = 114$.

$$CH_3\overset{\overset{\displaystyle O}{\|}}{C}CH_2CH_2CH_2CH_2CH_3$$

$C_7H_{14}O$

b. A hydrocarbon that contains only $sp^3$ hybridized carbons and a molecular ion at $m/z = 84$.

$C_6H_{12}$

d. A compound that contains C, H, N, and O and has an exact mass for the molecular ion at 101.0841.

$$CH_3\overset{\overset{\displaystyle O}{\|}}{C}NHCH_2CH_2CH_3$$

$C_5H_{11}NO$

**13.24** Use the values given in Table 13.1 to calculate the exact mass of each compound. $C_8H_{11}NO_2$ (exact mass 153.0790) is the correct molecular formula.

**13.25** Molecules with an odd number of N's have an odd number of H's, making the molecular ion odd as well.

**13.26** Two isomers such as $CH_2=CHCH_2CH_2CH_2CH_3$ and $(CH_3)_2C=CHCH_2CH_3$ have the same molecular formulas and therefore give the same exact mass, so they are not distinguishable by their exact mass spectra.

**13.27**

a. $(CH_3)_2C=O$  or  $(CH_3)_2CH{-}OH$

stronger bond
higher $\bar{\nu}$ absorption

b. $(CH_3)_2C=NCH_3$  or  $(CH_3)_2CH{-}NCH_3$

stronger bond
higher $\bar{\nu}$ absorption

c.

stronger bond
higher $\bar{\nu}$ absorption

**13.28** Locate the functional groups in each compound. Use Table 13.2 to determine what IR absorptions each would have.

a.  $C sp^3$–H at 2850–3000 cm$^{-1}$

b.  $-C{\equiv}CH$  $C sp$–H at 3300 cm$^{-1}$
$C sp^3$–H at 2850–3000 cm$^{-1}$
C–C triple bond at 2250 cm$^{-1}$

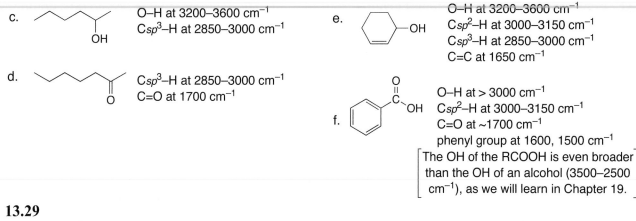

c. O–H at 3200–3600 cm$^{-1}$
Csp$^3$–H at 2850–3000 cm$^{-1}$

d. Csp$^3$–H at 2850–3000 cm$^{-1}$
C=O at 1700 cm$^{-1}$

e. O–H at 3200–3600 cm$^{-1}$
Csp$^2$–H at 3000–3150 cm$^{-1}$
Csp$^3$–H at 2850–3000 cm$^{-1}$
C=C at 1650 cm$^{-1}$

f. O–H at > 3000 cm$^{-1}$
Csp$^2$–H at 3000–3150 cm$^{-1}$
C=O at ~1700 cm$^{-1}$
phenyl group at 1600, 1500 cm$^{-1}$
[The OH of the RCOOH is even broader than the OH of an alcohol (3500–2500 cm$^{-1}$), as we will learn in Chapter 19.]

**13.29**

a. C=C bond
1650 cm$^{-1}$
Csp$^2$–H at 3150–3000 cm$^{-1}$

and HC≡CCH$_2$CH$_2$CH$_3$
C≡C bond
2250 cm$^{-1}$
Csp–H at 3300 cm$^{-1}$

d. no C=O bond

and CH$_3$(CH$_2$)$_5$ ... OCH$_3$
C=O bond
~1700 cm$^{-1}$

b. CH$_3$CH$_2$ ... OH
O–H bond
> 3000 cm$^{-1}$
[See note on OH in Answer 13.28f.]

and CH$_3$ ... OCH$_3$
no O–H bond

e. CH$_3$C≡CCH$_3$
no C≡C absorption
due to symmetry

and CH$_3$CH$_2$C≡CH
Csp–H bond
3300 cm$^{-1}$
C≡C bond at ~2250 cm$^{-1}$

c. CH$_3$CH$_2$ ... CH$_3$
C=O bond
1700 cm$^{-1}$

and CH$_3$CH=CHCH$_2$OH
O–H bond
3200–3600 cm$^{-1}$
Csp$^2$–H at 3150–3000 cm$^{-1}$
C=C bond at 1650 cm$^{-1}$

f. HC≡CCH$_2$N(CH$_2$CH$_3$)$_2$
Csp–H bond
3300 cm$^{-1}$

and CH$_3$(CH$_2$)$_5$C≡N

**13.30** The IR absorptions above 1500 cm$^{-1}$ are different for each of the narcotics.

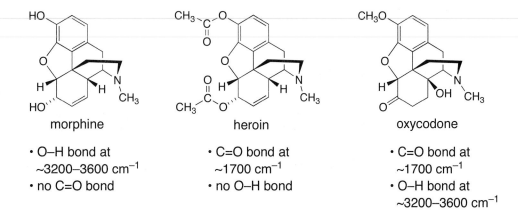

morphine
• O–H bond at ~3200–3600 cm$^{-1}$
• no C=O bond

heroin
• C=O bond at ~1700 cm$^{-1}$
• no O–H bond

oxycodone
• C=O bond at ~1700 cm$^{-1}$
• O–H bond at ~3200–3600 cm$^{-1}$

**13.31** Look for a **change in functional groups** from starting material to product to see how IR could be used to determine when the reaction is complete.

a. 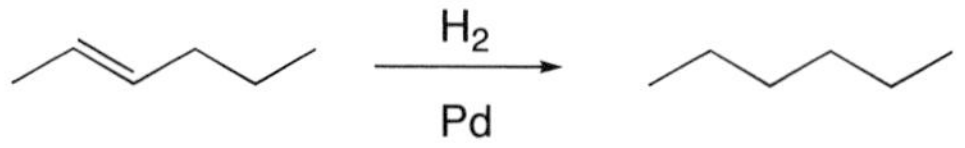  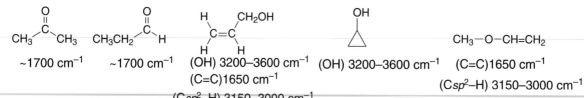  $\xrightarrow[\text{Pd}]{\text{H}_2}$  Loss of the C=C will be visible in the IR by disappearance of the peak at 1650 cm$^{-1}$.

b. 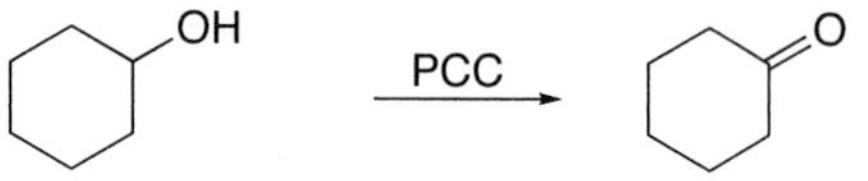  $\xrightarrow{\text{PCC}}$  Loss of the O–H group will be visible in the IR by disappearance of the peak at 3200–3600 cm$^{-1}$ and appearance of the C=O at ~1700 cm$^{-1}$.

c. 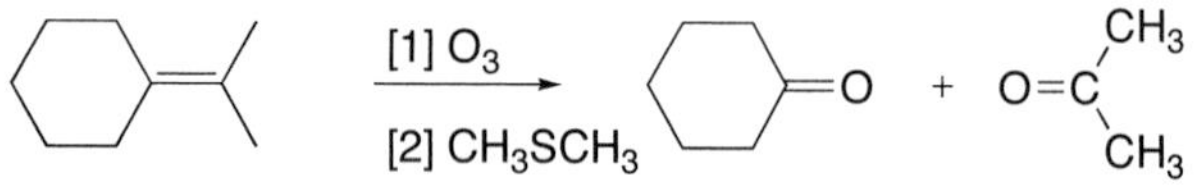  $\xrightarrow[\text{[2] CH}_3\text{SCH}_3]{\text{[1] O}_3}$  $=$O  +  O$=$C$\begin{smallmatrix}\text{CH}_3\\\text{CH}_3\end{smallmatrix}$  Loss of the C=C will be visible in the IR by disappearance of the peak at 1650 cm$^{-1}$ and appearance of the C=O at ~1700 cm$^{-1}$.

d. 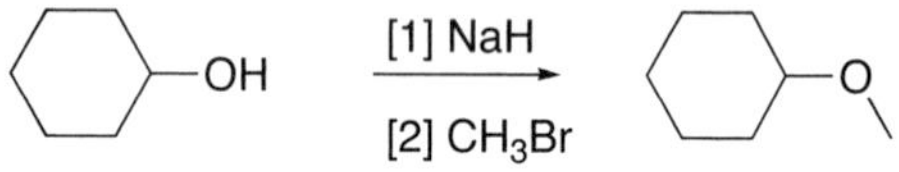  —OH  $\xrightarrow[\text{[2] CH}_3\text{Br}]{\text{[1] NaH}}$  —O  Loss of the O–H will be visible in the IR by disappearance of the peak at 3200–3600 cm$^{-1}$.

**13.32** In addition to C$sp^3$–H at ~3000–2850 cm$^{-1}$:

**Spectrum [1]:**

$CH_2=C(CH_3)CH_2CH_2CH_2CH_3$ (**B**)
    C=C peak at 1650 cm$^{-1}$
    C$sp^2$–H at ~3150 cm$^{-1}$

**Spectrum [2]:**

$(CH_3CH_2)_3COH$ (**F**)
    OH at 3600–3200 cm$^{-1}$

**Spectrum [3]:**

$(CH_3)_2CHOCH(CH_3)_2$ (**D**)
    No other peaks above 1500 cm$^{-1}$.

**Spectrum [4]:**

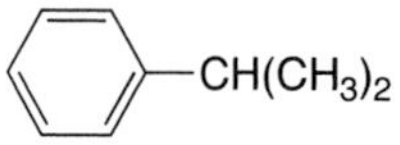

—CH(CH$_3$)$_2$  (**C**)
    C$sp^2$–H at ~3150 cm$^{-1}$
    Phenyl peaks at 1600 and 1500 cm$^{-1}$

**Spectrum [5]:**

$CH_3CH_2CH_2CH_2COOH$ (**A**)
    OH at ~3500–2500 cm$^{-1}$
    C=O at ~1700 cm$^{-1}$

**Spectrum [6]:**

$CH_3COOC(CH_3)_3$ (**E**)
    C=O at ~1700 cm$^{-1}$

**13.33** In addition to C$sp^3$–H at ~3000–2850 cm$^{-1}$:

$CH_3$–C(=O)–$CH_3$
~1700 cm$^{-1}$

$CH_3CH_2$–C(=O)–H
~1700 cm$^{-1}$

H, CH$_2$OH / C=C / H, H
(OH) 3200–3600 cm$^{-1}$
(C=C) 1650 cm$^{-1}$
(C$sp^2$–H) 3150–3000 cm$^{-1}$

OH (cyclopropane with OH)
(OH) 3200–3600 cm$^{-1}$

$CH_3$–O–CH=CH$_2$
(C=C) 1650 cm$^{-1}$
(C$sp^2$–H) 3150–3000 cm$^{-1}$

No enols (such as CH$_3$CH=CHOH) are drawn since these compounds are not stable.

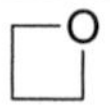

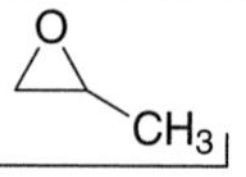

(oxetane) (methyl oxirane with CH$_3$)
No additional peaks above 1500 cm$^{-1}$.

**13.34**

a. Compound with a molecular ion at $m/z = 72$
   IR absorption at 1725 cm$^{-1}$ = C=O bond
   Molecular formula: $C_4H_8O$

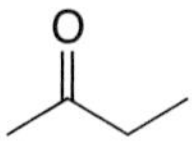

c. Compound with a molecular ion at $m/z = 74$
   IR absorption at 3600–3200 cm$^{-1}$ = O–H bond
   Molecular formula: $C_4H_{10}O$

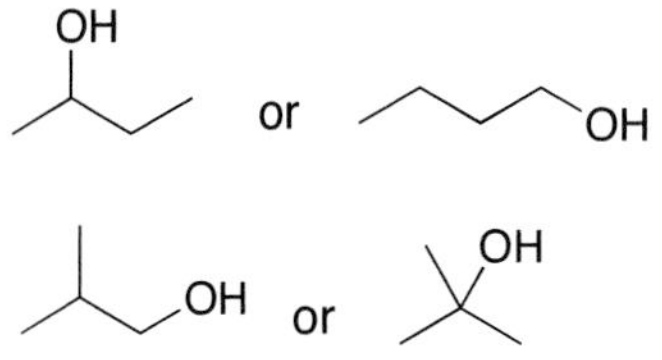

b. Compound with a molecular ion at $m/z = 55$
   The odd molecular ion means an odd number
   of N's present. Molecular formula: $C_3H_5N$
   IR absorption at 2250 cm$^{-1}$ = C≡N bond

   $CH_3CH_2C≡N$

**13.35**

Chiral hydrocarbon with a molecular ion at $m/z = 82$
Molecular formula: $C_6H_{10}$

   IR absorptions at 3300 cm$^{-1}$ = C$sp$–H bond
                     3000–2850 cm$^{-1}$ = C$sp^3$–H bonds
                     2250 cm$^{-1}$ = C≡C bond

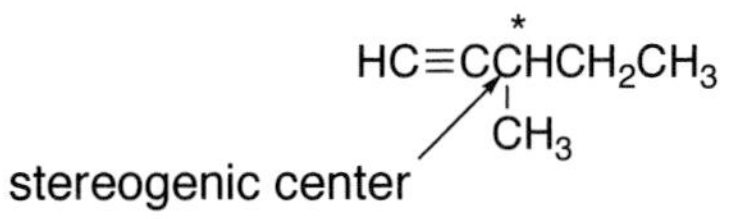

   $HC≡C\overset{*}{C}HCH_2CH_3$
      $\quad CH_3$
   stereogenic center

   Two possible enantiomers:

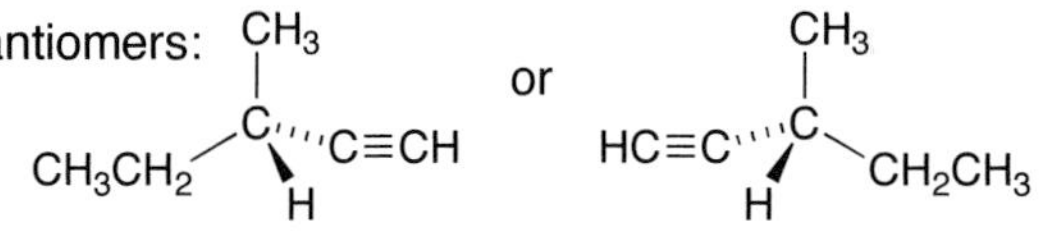

**13.36** The chiral compound **Y** has a strong absorption at 2970–2840 cm$^{-1}$ in its IR spectrum due to $sp^3$
hybridized C–H bonds. The two peaks of equal intensity at 136 and 138 indicate the presence of
a Br atom. The molecular formula is $C_4H_9Br$. Only one alkyl bromide of this molecular formula
has a stereogenic center:

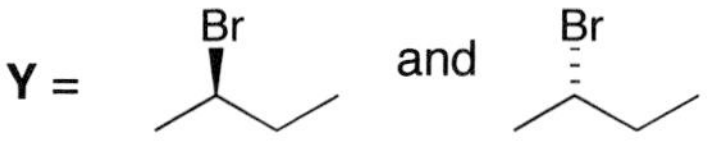

   **Y** =            and

**13.37**

               Zn(Hg)
   ────────→
               HCl

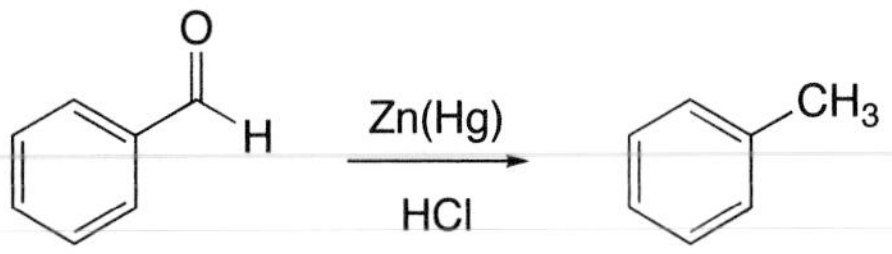

   **Z** =
   $m/z = 92$; molecular formula $C_7H_8$
   IR absorptions at:
   3150–2950 cm$^{-1}$ = C$sp^3$–H and C$sp^2$–H bonds
   1605 cm$^{-1}$ and 1496 cm$^{-1}$ due to phenyl group

**13.38**

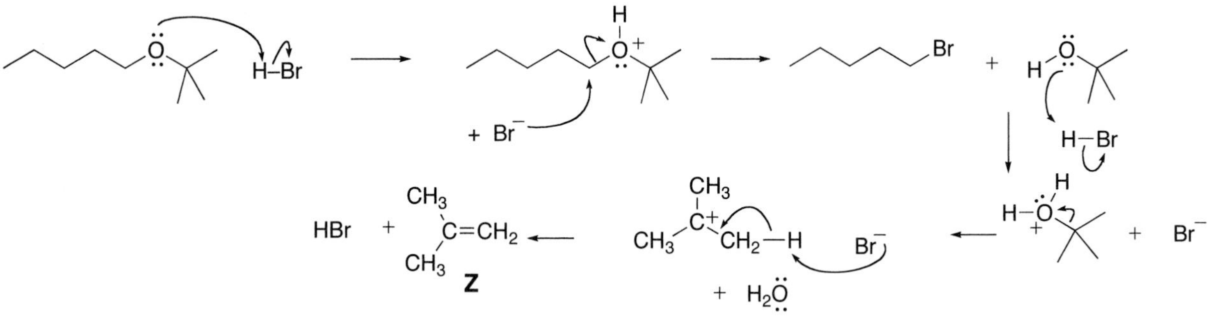

   HBr  +   $\underset{CH_3}{\overset{CH_3}{C}}$=CH$_2$
              **Z**

**13.39** The mass spectrum has a molecular ion at 71. The odd mass suggests the presence of an odd number of N atoms;  likely formula, $C_4H_9N$.  The IR absorption at ~3300 $cm^{-1}$ is due to N–H and the 3000–2850 $cm^{-1}$ is due to $sp^3$ hybridized C–H bonds.

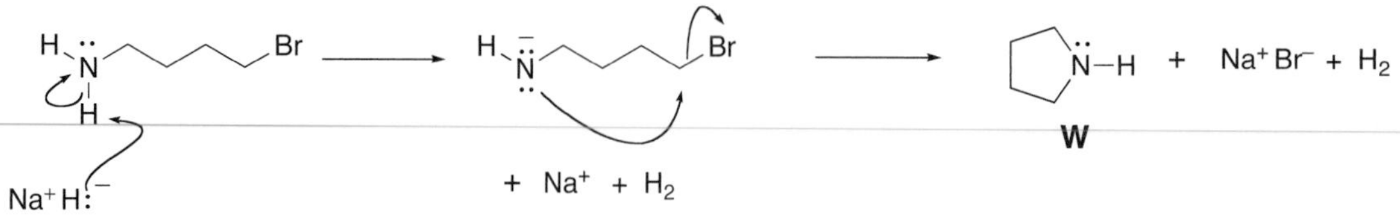

**13.40** The $\alpha,\beta$-unsaturated carbonyl compound has three resonance structures, two of which place a single bond between the C and O atoms.  This means that the C–O bond has partial single bond character, making it weaker than a regular C=O bond, and moving the absorption to lower wavenumber.

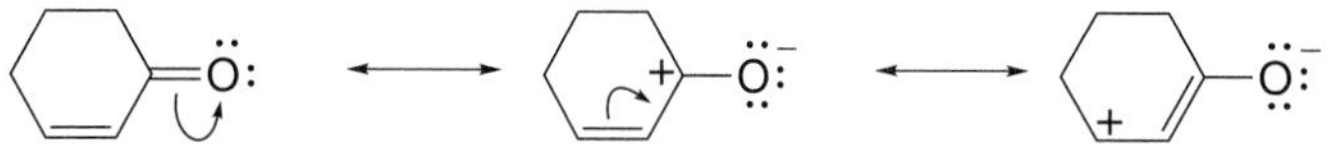

three resonance structures for 2-cyclohexenone

**13.41**

4,4'-dichlorobiphenyl
(a common PCB)

Three peaks would be seen for the molecular ions:

$m/z$

$C_{12}H_8Cl_2{}^{35}$: 222 (Both $^{35}Cl$ isotopes)
$C_{12}H_8Cl_2{}^{35,37}$: 224 (One $^{35}Cl$ and one $^{37}Cl$ isotope)
$C_{12}H_8Cl_2{}^{37}$: 226 (Both $^{37}Cl$ isotopes)

**13.42**

a. and b.

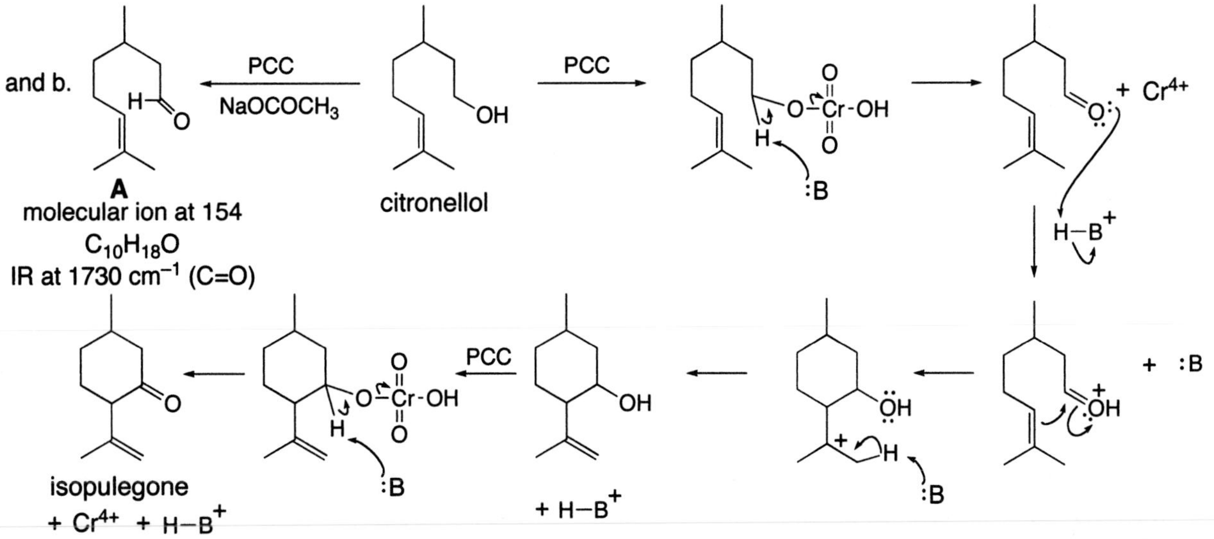
PCC
NaOCOCH3
PCC
OH
+ Cr4+
A
molecular ion at 154
C10H18O
IR at 1730 cm−1 (C=O)
citronellol
H–B+
+ :B
PCC
isopulegone
+ Cr4+ + H–B+
:B
+ H–B+
:B

a. and c.

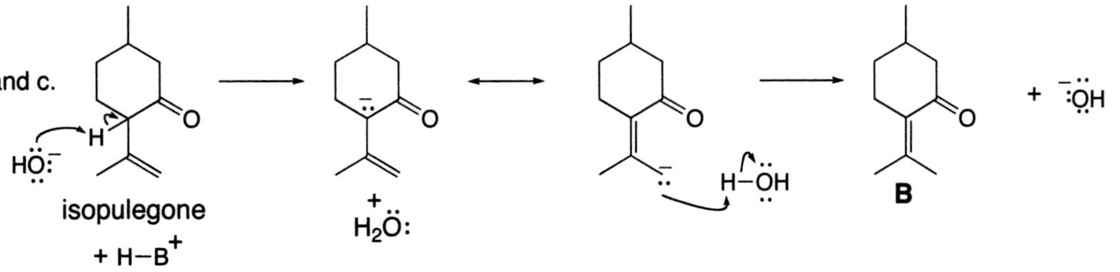
isopulegone
+ H–B+
H2O:
H–OH
B
+ :OH

## Chapter 14: Nuclear Magnetic Resonance Spectroscopy

♦ $^1$H NMR spectroscopy

[1] The **number of signals** equals the number of different types of protons (14.2).

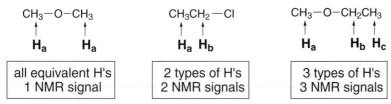

[2] The **position of a signal** (its chemical shift) is determined by shielding and deshielding effects.
- Shielding shifts an absorption upfield; deshielding shifts an absorption downfield.
- Electronegative atoms withdraw electron density, deshield a nucleus, and shift an absorption downfield (14.3).

| | This proton is shielded. Its absorption is upfield, 0.9–2 ppm. | | This proton is deshielded. Its absorption is further downfield, 2.5–4 ppm. |

- Loosely held π electrons can either shield or deshield a nucleus. Protons on benzene rings and double bonds are deshielded and absorb downfield, whereas protons on triple bonds are shielded and absorb upfield (14.4).

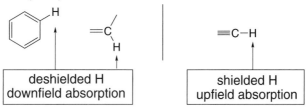

[3] The **area under an NMR signal** is proportional to the number of absorbing protons (14.5).

[4] **Spin-spin splitting** tells about nearby nonequivalent protons (14.6–14.8).
- Equivalent protons do not split each other's signals.
- A set of $n$ nonequivalent protons on the same carbon or adjacent carbons split an NMR signal into $n + 1$ peaks.
- OH and NH protons do not cause splitting (14.9).
- When an absorbing proton has two sets of nearby nonequivalent protons that are equivalent to each other, use the $n + 1$ rule to determine splitting.
- When an absorbing proton has two sets of nearby nonequivalent protons that are not equivalent to each other, the number of peaks in the NMR signal = $(n + 1)(m + 1)$.

♦ $^{13}$C NMR spectroscopy (14.11)

[1] The number of signals equals the number of different types of carbon atoms. All signals are single lines.
[2] The relative position of $^{13}$C signals is determined by shielding and deshielding effects.
- Carbons that are $sp^3$ hybridized are shielded and absorb upfield.
- Electronegative elements (N, O, and X) shift absorptions downfield.
- The carbons of alkenes and benzene rings absorb downfield.
- Carbonyl carbons are highly deshielded, and absorb further downfield than other carbon types.

## Chapter 14: Answers to Problems

**14.1** To calculate chemical shift, use the following formula:

$$\delta = [\text{observed chemical shift (Hz)}] / \nu \text{ of the NMR (MHz)}]$$

a. $\delta = [60 \text{ Hz}] / [60 \text{ MHz}]$
= **1 ppm**

b. $\delta = [1600 \text{ Hz}] / [200 \text{ MHz}]$
= **8 ppm**

**14.2** Use the formula from Answer 14.1 to calculate the chemical shifts.

a. **CH₃ protons:**
$\delta = [1715 \text{ Hz}] / [500 \text{ MHz}]$
= **3.43 ppm**

**OH proton:**
$\delta = [1830 \text{ Hz}] / [500 \text{ MHz}]$
= **3.66 ppm**

b. The positive direction of the δ scale is downfield from TMS. The $CH_3$ protons absorb upfield from the OH proton.

**14.3** To determine if two H's are equivalent replace each by an atom X. If this yields the same compound or mirror images, the two H's are equivalent. Each kind of H will give one NMR signal.

a. $CH_3CH_3$
1 kind of H
1 NMR signal

c. $CH_3CH_2CH_2CH_3$
2 kinds of H's
2 NMR signals

e. $CH_3CH_2CO_2CH_2CH_3$
4 kinds of H's
4 NMR signals

g. $CH_3CH_2OCH_2CH_3$
2 kinds of H's
2 NMR signals

b. $CH_3CH_2CH_3$
2 kinds of H's
2 NMR signals

d. $(CH_3)_2CHCH(CH_3)_2$
2 kinds of H's
2 NMR signals

f. $CH_3OCH_2CH(CH_3)_2$
4 kinds of H's
4 NMR signals

h. $CH_3CH_2CH_2OH$
4 kinds of H's
4 NMR signals

**14.4**

$CH_3CH_2CH_2CH_2CH_2CH_2CH_2CH_2Cl$

Each C is a different distance from the Cl. This makes each C different, and each set of H's different. There are 8 different kinds of protons.

**14.5** Draw in all of the H's and compare them. If two H's are cis and trans to the same group, they are equivalent.

a. 4 identical H's

**2 NMR signals**

b.

**4 NMR signals**

c.

**3 NMR signals**

**14.6** The two protons of a $CH_2$ group are different from each other if a compound has one stereogenic center. Replace one of the protons of one enantiomer with X and compare the products.

stereogenic center

**2-chlorobutane**

**diastereomers**

This means the two H's on the $CH_2$ group are diastereotopic protons. They are not equivalent and give different signals.

**5 NMR signals altogether**

**14.7** If replacement of H with X yields enantiomers, the protons are **enantiotopic**.
If replacement of H with X yields diastereomers, the protons are **diastereotopic**. In general, if the compound has **one stereogenic center**, the protons in a $CH_2$ group are **diastereotopic**.

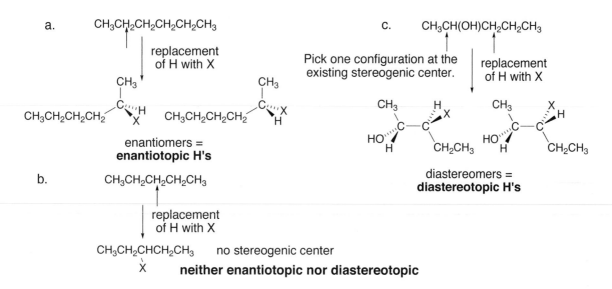

a.

b.

c.

enantiomers =
**enantiotopic H's**

diastereomers =
**diastereotopic H's**

$CH_3CH_2CHCH_2CH_3$     no stereogenic center
$\overset{|}{X}$          **neither enantiotopic nor diastereotopic**

**14.8** The two protons of a $CH_2$ group are different from each other if the compound has one stereogenic center. Replace one proton with X and compare the products.

a. The stereogenic center makes the H's in the $CH_2$ group diastereotopic and therefore different from each other.

**5 NMR signals**

b. stereogenic center

**7 NMR signals**

**14.9** Decreased electron density deshields a nucleus and the absorption goes downfield. Absorption also shifts downfield with increasing alkyl substitution.

a. $FC\underline{H}_2CH_2C\underline{H}_2Cl$
F is more electronegative than Cl. The $CH_2$ group adjacent to the F is more deshielded and the H's will absorb further downfield.

b. $CH_3C\underline{H}_2CH_2C\underline{H}_2OCH_3$
The $CH_2$ group adjacent to the O will absorb further downfield because it is closer to the electronegative O atom.

c. $C\underline{H}_3OC(CH_3)_3$
The $CH_3$ group bonded to the O atom will absorb further downfield.

**14.10**

a. $ClCH_2CH_2CH_2Br$
      $H_a$ $H_b$ $H_c$
3 types of protons:
$H_b < H_c < H_a$

b. $CH_3OCH_2OC(CH_3)_3$
      $H_a$ $H_b$ $H_c$
3 types of protons:
$H_c < H_a < H_b$

c. $CH_3\overset{\overset{O}{\|}}{C}CH_2CH_3$
      $H_a$ $H_b$ $H_c$
3 types of protons:
$H_c < H_a < H_b$

**14.11**

a. $CH_3-C{\equiv}C-H$    $CH_3CH{=}CH_2$    $CH_3CH_2CH_3$
         ↑                        ↑                    ↑
        $H_a$                    $H_b$                $H_c$

**$H_c$** protons are shielded because they are bonded to an $sp^3$ C.
**$H_a$** is shielded because it is bonded to an $sp$ C.
**$H_b$** protons are deshielded because they are bonded to an $sp^2$ C.

$$H_c < H_a < H_b$$

b.

$$CH_3\overset{\overset{\displaystyle O}{\|}}{C}OCH_2CH_3$$
    ↑              ↑   ↑
   $H_a$          $H_b$ $H_c$

**$H_c$** protons are shielded because they are bonded to an $sp^3$ C.
**$H_a$** protons are deshielded slightly because the $CH_3$ group is bonded to a C=O.
**$H_b$** protons are deshielded because the $CH_2$ group is bonded to an O atom.

$$H_c < H_a < H_b$$

**14.12** An integration ratio of 2:3 means that there are two types of hydrogens in the compound, and that the ratio of one type to another type is 2:3.

a. $CH_3CH_2Cl$
2 types of H's
3:2 - YES

b. $CH_3CH_2CH_3$
2 types of H's
6:2 or 3:1 - no

c. $CH_3CH_2OCH_2CH_3$
2 types of H's
6:4 or 3:2 - YES

d. $CH_3OCH_2CH_2OCH_3$
2 types of H's
6:4 or 3:2 - YES

**14.13** To determine how many protons give rise to each signal:
- Divide the total number of integration units by the total number of protons to find the number of units per H.
- Divide each integration value by this value and round to the nearest whole number.

$C_8H_{14}O_2$
total number of integration units = 14 + 12 + 44 = 70 units
total number of protons = 14H's
70 units/14H's = **5 units per H**

Signal [A] = 14/5 = **3H**
Signal [B] = 12/5 = **2H**
Signal [C] = 44/5 = **9H**

**14.14** To determine the **splitting pattern** for a molecule:
- Determine the number of different kinds of protons.
- Nonequivalent protons on the same C or adjacent C's split each other
- Apply the $n + 1$ rule.

a.

$$CH_3CH_2\overset{\overset{\displaystyle O}{\|}}{C}Cl$$
      ↑   ↑
     $H_a$ $H_b$

$H_a$: 3 peaks - triplet
$H_b$: 4 peaks - quartet

c.

$$CH_3\overset{\overset{\displaystyle O}{\|}}{C}CH_2CH_2Br$$
   ↑              ↑   ↑
  $H_a$          $H_b$ $H_c$

$H_a$: 1 peak - singlet
$H_b$: 3 peaks - triplet
$H_c$: 3 peaks - triplet

e.

$CH_3CH_2$          H ← $H_a$
         \C=C/
$CH_3$          H ← $H_b$

$H_a$: 2 peaks - doublet
$H_b$: 2 peaks - doublet

b.

H ← $H_b$
|
$CH_3-C-Br$
  ↑   |
 Br
$H_a$

$H_a$: 2 peaks - doublet
$H_b$: 4 peaks - quartet

d.

$H_a$ → H    Cl
     \C=C/
  Br    H ← $H_b$

$H_a$: 2 peaks - doublet
$H_b$: 2 peaks - doublet

f. $ClCH_2CH(OCH_3)_2$
          ↑   ↑
         $H_a$ $H_b$

$H_a$: 2 peaks - doublet
$H_b$: 3 peaks - triplet

**14.15**  Use the directions from Answer 14.14.

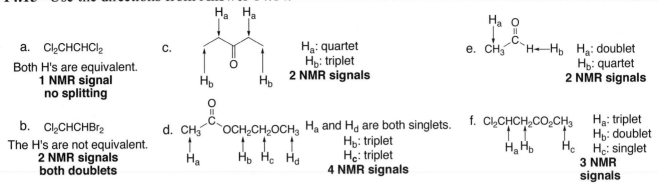

a.  Cl₂CHCHCl₂

Both H's are equivalent.
**1 NMR signal**
**no splitting**

b.  Cl₂CHCHBr₂

The H's are not equivalent.
**2 NMR signals**
**both doublets**

c.

Hₐ: quartet
H_b: triplet
**2 NMR signals**

d.  CH₃—C(=O)—OCH₂CH₂OCH₃

Hₐ and H_d are both singlets.
H_b: triplet
H_c: triplet
**4 NMR signals**

e.  CH₃—CH(=O)—H←H_b

Hₐ: doublet
H_b: quartet
**2 NMR signals**

f.  Cl₂CHCH₂CO₂CH₃

Hₐ: triplet
H_b: doublet
H_c: singlet
**3 NMR signals**

**14.16**  CH₃CH₂Cl

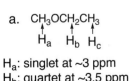

**2 units**

**3 units**

3      1

chemical shift (ppm)

There are two kinds of protons, and they can split each other. The CH₃ protons will be split by the CH₂ protons into 2+1 = 3 peaks. It will be upfield from the CH₂ protons since it is further from the Cl. The CH₂ protons will be split by the CH₃ protons into 3+1 = 4 peaks. It will be downfield from the CH₃ protons since the CH₂ protons are closer to the Cl. The ratio of integration units will be 3:2.

**14.17**

a.  (CH₃)₂CHCO₂CH₃

split by 6 equivalent H's
(6+1) = **7 peaks**

b.  CH₃CH₂CH₂CH₂CH₃
        Hₐ H_b H_c

Hₐ: split by 2 H's
**3 peaks**
H_b: split by 2 sets of H's
(3+1)(2+1) = **12 peaks**
H_c: split by 4 equivalent H's
**5 peaks**

c.   Cl      CH₂Br
       C=C
   Hₐ→H      H ← H_b

Hₐ: split by 1 H
**2 peaks**
H_b: split by 2 sets of H's
(1+1)(2+1) = **6 peaks**

d.  Hₐ→H    H← H_b
       C=C   (all H's)
    Br    H←H_c

Hₐ: split by 2 different H's
(1+1)(1+1) = **4 peaks**
H_b: split by 2 different H's
(1+1)(1+1) = **4 peaks**
H_c: split by 2 different H's
(1+1)(1+1) = **4 peaks**

**14.18**

a.  CH₃OCH₂CH₃
        Hₐ H_b H_c

Hₐ: singlet at ~3 ppm
H_b: quartet at ~3.5 ppm
H_c: triplet at ~1 ppm

b.  CH₃CH₂—C(=O)—OCH(CH₃)₂
        Hₐ H_b        H_c H_d

Hₐ: triplet at ~1 ppm
H_b: quartet at ~2 ppm
H_c: septet at ~3.5 ppm
H_d: doublet at ~1 ppm

c.  CH₃OCH₂CH₂CH₂OCH₃
       Hₐ  H_b H_c H_b Hₐ

Hₐ: singlet at ~3 ppm
H_b: triplet at ~3.5 ppm
H_c: quintet at ~1.5 ppm

d.  CH₃CH₂     H_b     CH₂CH₃
       Hₐ  C=C    Hₐ
         H      H
           H_c

Hₐ: triplet at ~1 ppm
H_b: multiplet (8 peaks) at ~2.5 ppm
H_c: triplet at ~5 ppm

**14.19**

a. two singlets

b. singlet in the 3–4 ppm region

c. septet in the 3–4 ppm region

d. doublet in the 1–2 ppm region

e. six signals

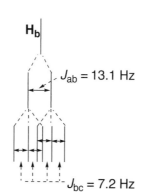

This stereogenic center makes the 2 H's of the $CH_2$ group not equivalent to each other.

**14.20**

trans-1,3-dichloropropene

2 $H_c$ protons

$J_{ab}$ = 13.1 Hz

$J_{bc}$ = 7.2 Hz

**Splitting diagram for $H_b$**

1 **trans** $H_a$ proton splits $H_b$ into
**1 + 1 = 2 peaks**
**a doublet**

2 $H_c$ protons split $H_b$ into
**2 + 1 = 3 peaks**
**Now it's a doublet of triplets.**

**14.21**

**A**

$H_a$: 1.75 ppm, doublet, 3 H, $J$ = 6.9 Hz
$H_b$: 5.89 ppm, quartet, 1 H, $J$ = 6.9 Hz

$C_3H_4Cl_2$

**B**

$ClCH_2$ H ← doublet
singlet Cl H ← doublet

signal at 4.16 ppm, singlet, 2 H
signal at 5.42 ppm, doublet, 1 H, $J$ = 1.9 Hz
signal at 5.59 ppm, doublet, 1 H, $J$ = 1.9 Hz

**14.22** Remember that **OH (or NH) protons** do not split other signals, and are not split by adjacent protons.

singlet

a. $(CH_3)_3CCH_2OH$

singlet   singlet

**3 NMR signals**

triplet   triplet

b. $CH_3CH_2CH_2OH$

12 peaks   singlet

**4 NMR signals**

**14.23**

$H_d$

5 H's on benzene ring

H ← $H_c$
C–$CH_3$ ← $H_a$
OH ← $H_b$

**A**

$H_a$: doublet at ~1.4 due to the $CH_3$ group, split into two peaks by one adjacent nonequivalent H ($H_c$).

$H_b$: singlet at ~2.7 due to the OH group. OH protons are not split by nor do they split adjacent protons.

$H_c$: quartet at ~4.7 due to the CH group, split into four peaks by the adjacent $CH_3$ group.

$H_d$: Five protons on the benzene ring.

**14.24** Use these steps to propose a structure consistent with the molecular formula, IR, and NMR data.

- Calculate the **degrees of unsaturation**.
- Use the IR data to determine what types of **functional groups** are present.
- Determine the number of different **types of protons**.
- Calculate the **number of H's** giving rise to each signal.
- Analyze the **splitting pattern** and put together a molecule.
- Use the **chemical shift** information to check the structure.

- Molecular formula $C_7H_{14}O_2$

  $2n + 2 = 2(7) + 2 = 16$
  $16 - 14 = 2/2 = \textbf{1 degree of unsaturation}$
  **1 $\pi$ bond or one ring**

- IR peak at 1740 cm$^{-1}$

  **C=O** absorption is around 1700 cm$^{-1}$ (causes the degree of unsaturation)
  no signal at 3200–3600 cm$^{-1}$ means there is no O–H bond

- NMR data:

  | absorption | ppm | integration |
  |---|---|---|
  | singlet | 1.2 | 26 - - - - - - → 26/3 = **9 H's** |
  | triplet | 1.3 | 10 - - - - - - → 10/3 = **3 H's** (probably a $CH_3$ group) |
  | quartet | 4.1 | 6 - - - - - - → 6/3 = **2 H's** (probably a $CH_2$ group) |

  - 3 kinds of H's
  - number of H's per signal
    total integration units:  26 + 10 + 6 = 42 units
    42 units / 14 H's = 3 units per H
  - look at the splitting pattern
    the singlet (9H) is likely from a *tert*-butyl group:

    $$-\overset{\overset{\displaystyle CH_3}{|}}{\underset{\underset{\displaystyle CH_3}{|}}{C}}-CH_3$$

    the $CH_3$ and $CH_2$ groups split each other:  $CH_3-CH_2-$

- Join the pieces together.

  Pick this structure due to the chemical shift data.
  The $CH_2$ group is shifted downfield  (4 ppm), so it
  is close to the electron-withdrawing O.

**14.25**

- Molecular formula: $C_3H_8O$

  ➢ Calculate degrees of unsaturation
    $2n + 2 = 2(3) + 2 = 8$
    $8 - 8 = \textbf{0 degrees of unsaturation}$

- IR peak at 3200–3600 cm$^{-1}$

  ➢ Peak at 3200–3600 cm$^{-1}$ is due to an **O–H bond**.

- NMR data:
  - Doublet at ~1.2 (6H)
  - Singlet at ~2.2 (1H)
  - Septet at ~4 (1H)

  3 types of H's
  **septet** from 1 H ⟵——— split by 6 H's
  **singlet** from 1 H
  **doublet** from 6 H's ⟵——— split by 1 H
  from the O–H proton

➢ Put information together:

**14.26**

a.

| | | |
|---|---|---|
| Absorption [**A**]: | Singlet at ~3.8 ppm | CH₃O– |
| Absorption [**B**]: | Multiplet at ~3.6 ppm | CH₂N |
| Absorption [**C**]: | Triplet at ~2.9 ppm | CH₂ adjacent to five-membered ring |
| Absorption [**D**]: | Singlet at ~1.9 ppm | CH₃C=O |

b.

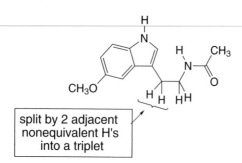

split by 2 adjacent
nonequivalent H's
into a triplet

**14.27** Identify each compound from the ¹H NMR data.

a.

CH₂=CHCOCH₃  $\xrightarrow{\text{HCl}}$

CH₃ ... Cl

triplet at 3.6

singlet

**A** triplet at 3.05

b.

(CH₃)₂C=O  $\xrightarrow[\text{H₂O}]{\text{base}}$

singlet at 2.5

singlet at 1.3

OH — singlet at 3.8

singlet at 2.2

**B**

**14.28** Each different kind of carbon atom will give a different ¹³C NMR signal.

a. CH₃CH₂CH₂CH₃
Cₐ C_b C_b Cₐ
2 kinds of C's
**2 ¹³C NMR signals**

b.
CH₃CH₂—C(=O)—OCH₃
Each C is different.
4 kinds of C's
**4 ¹³C NMR signals**

c. CH₃CH₂CH₂—O—CH₂CH₂CH₃
Cₐ C_b C_c      C_c C_b Cₐ
same groups on both sides of O
3 kinds of C's
**3 ¹³C NMR signals**

d.
CH₃CH₂ ... H
  C=C
H      H
Each C is different.
4 kinds of C's
**4 ¹³C NMR signals**

**14.29** To give only one peak in both its ¹H and ¹³C NMR spectra, a compound must have only one type of C and H.

CH₃O—C(=O)—OCH₃

△

CH₃—C(CH₃)(CH₃)—O—C(CH₃)(CH₃)—CH₃

Br₂CHCHBr₂

| | | | |
|---|---|---|---|
| 1 ¹H NMR signal | **1 ¹H NMR signal** | 1 ¹H NMR signal | **1 ¹H NMR signal** |
| 2 ¹³C signals | **1 ¹³C signal** | 2 ¹³C signals | **1 ¹³C signal** |

**14.30** Electronegative elements shift absorptions downfield. The carbons of alkenes and benzene rings, and carbonyl carbons are also shifted downfield.

a. $CH_3CH_2OCH_2CH_3$
↑  ↑

b. $BrCH_2CHBr_2$
↑  ↑

c.

d. $CH_3CH=CH_2$
↑       ↑

The CH$_2$ group is closer to the electronegative O and will be further downfield.

The C of the CHBr$_2$ group has two bonds to electronegative Br atoms and will be further downfield.

The carbonyl carbon is highly deshielded and will be further downfield.

The CH$_2$ group is part of a double bond and will be further downfield.

**14.31**

a. In order of lowest to highest chemical shift:

$C_a$  $C_b$  $C_c$  $C_d$
↑    ↑    ↑    ↑
$CH_3CHCH_2CH_3$
       |
       OH

$C_d < C_a < C_c < C_b$

b. In order of lowest to highest chemical shift:

$(CH_3CH_2)_2C=O$
↑      ↑      ↑
$C_a$  $C_b$   $C_c$

$C_a < C_b < C_c$

**14.32**

- molecular formula $C_4H_8O_2$

  $2n + 2 = 2(4) + 2 = 10$

  $10 - 8 = 2/2 =$ **1 degree of unsaturation**

- no IR peaks at 3200–3600 or 1700 cm$^{-1}$

  no O–H or C=O

- $^1H$ NMR spectrum at 3.69 ppm

  only one kind of proton

- $^{13}C$ NMR spectrum at 67 ppm

  only one kind of carbon

This structure satisfies all the data. One ring is one degree of unsaturation. All carbons and protons are identical.

**14.33** Use the directions from Answer 14.3.

a. (CH$_3$)$_3$CH
**2 kinds** of H's

b. (CH$_3$)$_3$CC(CH$_3$)$_3$
**1 kind** of H

c. CH$_3$CH$_2$OCH$_2$CH$_2$CH$_2$CH$_2$CH$_3$
**7 kinds** of H's

d. 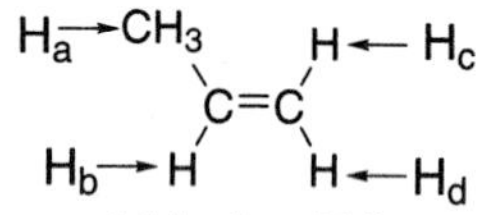
**4 kinds** of H's

e. 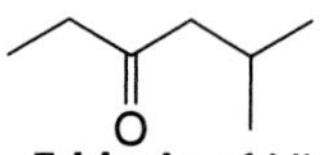
**5 kinds** of H's

f. 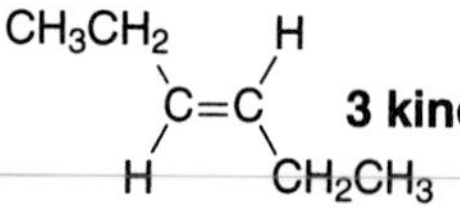
**3 kinds** of H's

g. CH$_3$CH$_2$CH$_2$OCH$_2$CH$_2$CH$_3$
**3 kinds** of H's

h. 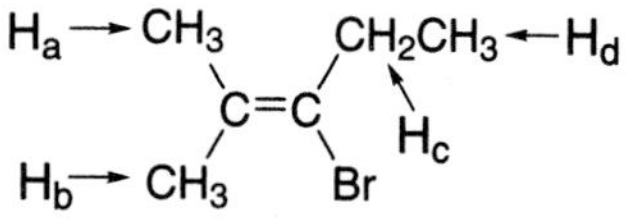
**4 kinds** of H's

i. 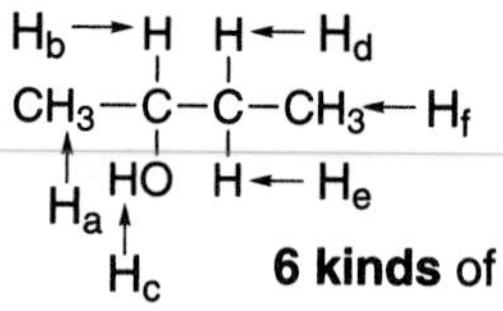
**6 kinds** of H's

j. 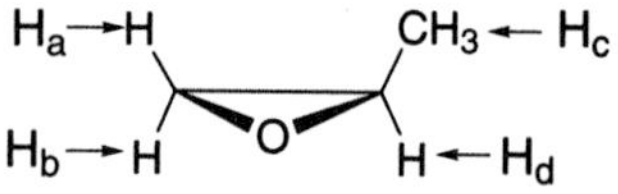
**4 kinds** of H's

**14.34**

**2 kinds** of protons

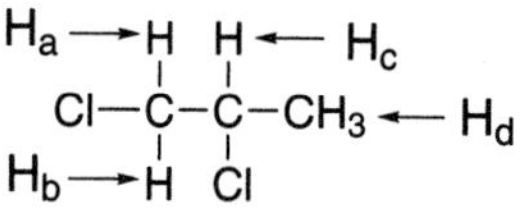

**4 kinds** of protons

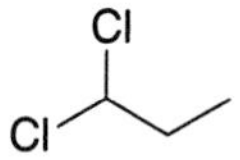

**3 kinds** of protons

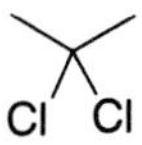

**1 kind** of proton

**14.35**

a. 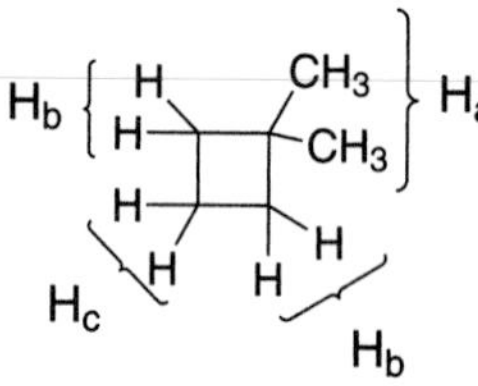
**3 kinds** of protons

b.
**7 kinds** of protons

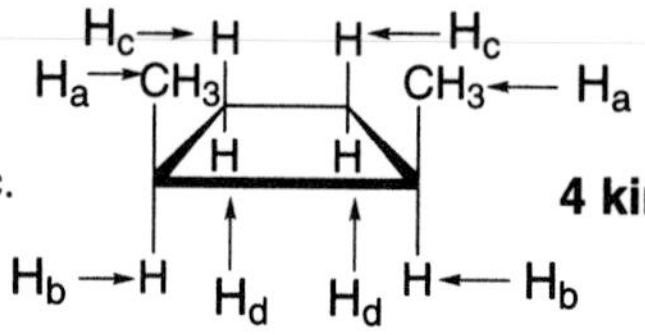

c.
**4 kinds** of protons

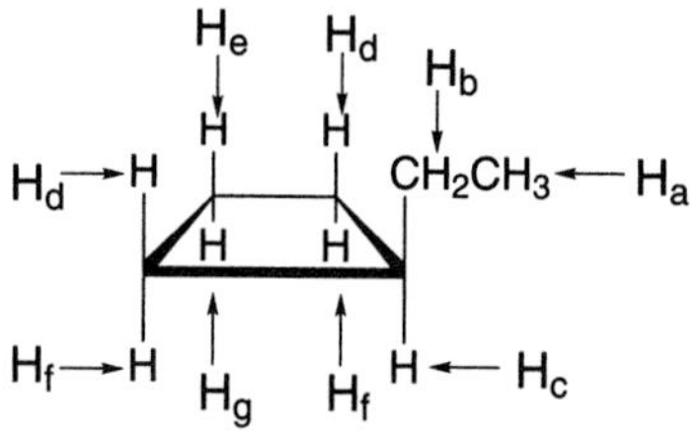

d. 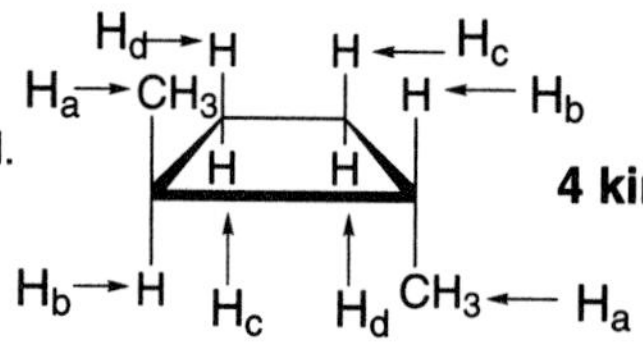
**4 kinds** of protons

**14.36**

a. 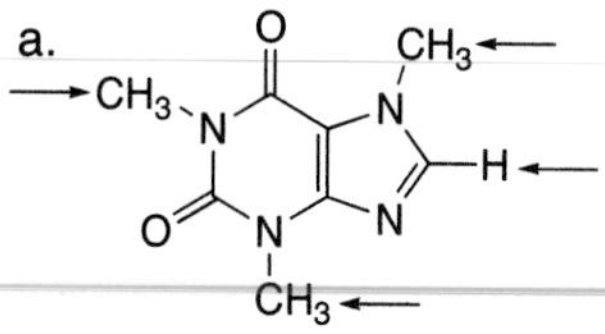
caffeine
**4 NMR signals**

b. 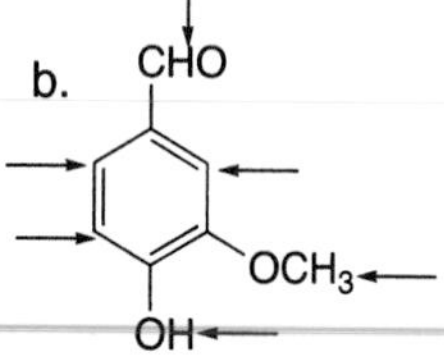
vanillin
**6 NMR signals**

c. 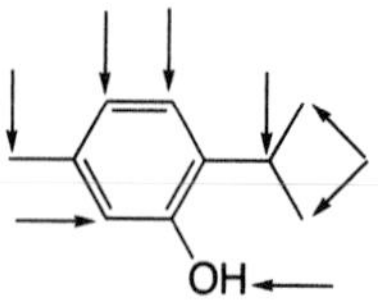
thymol
**7 NMR signals**

d. 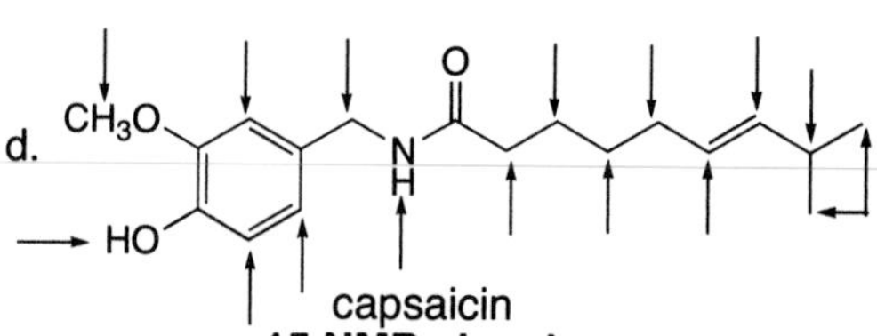
capsaicin
**15 NMR signals**

**14.37**

$$\delta = [\text{observed chemical shift (Hz)}] / \nu \text{ of the NMR (MHz)}]$$

a. $2.5 = $ observed chemical shift/300 MHz
   **observed chemical shift = 750 Hz**

b. ppm = 1200 Hz/300 Hz
   **4 ppm**

c. $2.0 = $ Hz/300 MHz
   **600 Hz**

**14.38** Use the directions from Answer 14.9.

a. $CH_3CH_2CH_2CH_2CH_3$  or  $CH_3CH_2CH_2OCH_3$
   ↑
   Adjacent O deshields the H's.
   **further downfield**

b. $CH_3CH_2CH_2I$  or  $CH_3CH_2CH_2F$
   ↑                        ↑
   More electronegative F
   deshields the H's.
   **further downfield**

c. $CH_3OCH_2CH_3$
   ↑ or ↑
   Increasing alkyl substitution
   **further downfield**

d. $CH_3CH_2CHBr_2$  or  $CH_3CH_2CH_2Br$
   ↑                          ↑
   Two electronegative
   Br's deshield the H.
   **further downfield**

**14.39** Use the directions from Answer 14.13.

[total number of integration units] / [total number of protons]  signal of 13 units is from **1 H**
$[13 + 33 + 73] / 10 = {\sim}12$ units per proton                signal of 33 units is from **3 H's**
                                                                 signal of 73 units is from **6 H's**

**14.40** The following compounds give one singlet in a $^1$H NMR spectrum:

$CH_3CH_3$    $CH_3{-}C{\equiv}C{-}CH_3$       Cl–C(Br)(Br)–Cl with $CH_2Cl$ groups    $(CH_3)(CH_3)C{=}C(CH_3)(CH_3)$    $(CH_3)_3C{-}\underset{O}{\overset{\|}{C}}{-}C(CH_3)_3$

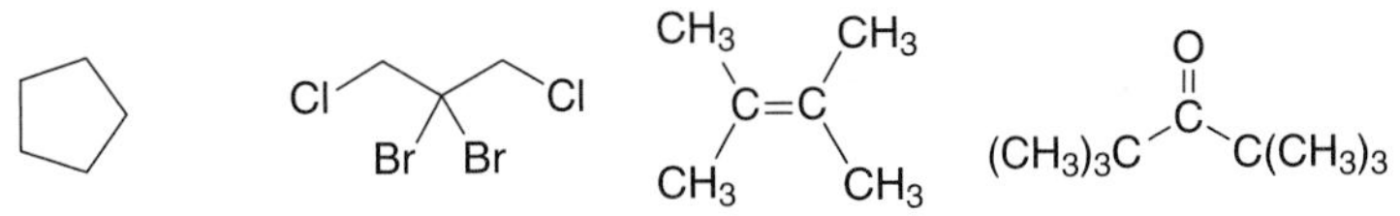

**14.41**

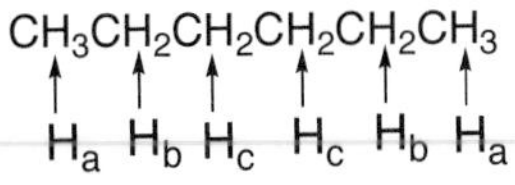

$CH_3CH_2CH_2CH_2CH_2CH_3$
  ↑   ↑   ↑   ↑   ↑   ↑
  $H_a$  $H_b$  $H_c$  $H_c$  $H_b$  $H_a$

**3 signals:**
$H_a$: split by 2 $H_b$ protons - triplet
$H_b$: split by 3 $H_a$ + 2 $H_c$ protons - 12 peaks
$H_c$: split by 2 $H_b$ protons - triplet

$CH_3 \leftarrow H_a$
$CH_3CHCH_2CH_2CH_3$
  ↑   ↑   ↑   ↑   ↑
  $H_a$  $H_b$  $H_c$  $H_d$  $H_e$

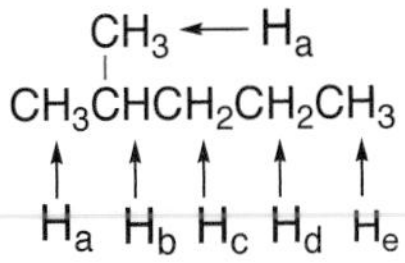

**5 signals:**
$H_a$: split by 1 $H_b$ proton - doublet
$H_b$: split by 6 $H_a$ + 2 $H_c$ protons - 21 peaks
$H_c$: split by 1 $H_b$ + 2 $H_d$ protons - 6 peaks
$H_d$: split by 2 $H_c$ + 3 $H_e$ protons - 12 peaks
$H_e$: split by 2 $H_d$ protons - triplet

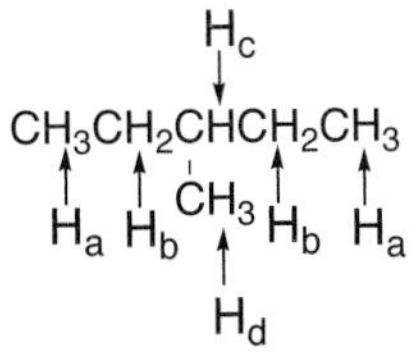

$H_c$
 ↓
$CH_3CH_2CHCH_2CH_3$
  ↑   ↑  $|$  ↑   ↑
  $H_a$  $H_b$ $CH_3$ $H_b$  $H_a$
            ↑
            $H_d$

**4 signals:**
$H_a$: split by 2 $H_b$ protons - triplet
$H_b$: split by 3 $H_a$ + 1 $H_c$ protons - 8 peaks
$H_c$: split by 4 $H_b$ + 3 $H_d$ protons - 20 peaks
$H_d$: split by 1 $H_c$ proton - doublet

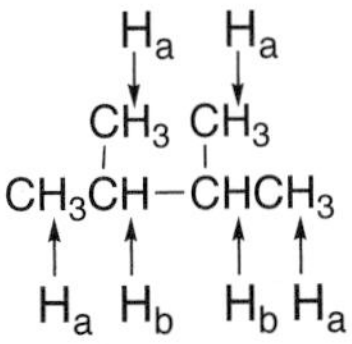

   $H_a$   $H_a$
   ↓       ↓
  $CH_3$  $CH_3$
$CH_3CH{-}CHCH_3$
  ↑   ↑   ↑   ↑
  $H_a$  $H_b$  $H_b$  $H_a$

**2 signals:**
$H_a$: split by 1 $H_b$ proton - doublet
$H_b$: split by 6 $H_a$ protons - septet

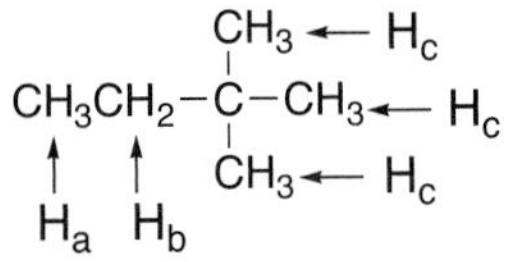

              $CH_3 \leftarrow H_c$
$CH_3CH_2{-}C{-}CH_3 \leftarrow H_c$
  ↑   ↑    $|$
  $H_a$  $H_b$  $CH_3 \leftarrow H_c$

**3 signals:**
$H_a$: split by 2 $H_b$ protons - triplet
$H_b$: split by 3 $H_a$ protons - quartet
$H_c$: no splitting - singlet

**14.42**

a. $CH_3CH(OCH_3)_2$

$CH_3$ protons split by 1 H = **doublet**
CH proton split by 3 H's = **quartet**

e. $(CH_3)_2CH$—C(=O)—$OCH_2CH_3$
      $H_a$   $H_b$     $H_c$   $H_d$

$H_a$ protons split by 1 H = **doublet**
$H_b$ proton split by 6 H's = **septet**
$H_c$ protons split by 3 H's = **quartet**
$H_d$ protons split by 2 H's = **triplet**

i.   $CH_3CH_2$—C(=O)—H
           $H_a$    $H_b$

$H_a$: split by $CH_3$ group + $H_b$
    = **8 peaks**
$H_b$: split by 2 H's = **triplet**

b.   $CH_3OCH_2CH_2$—C(=O)—$OCH_3$

both $CH_2$ groups split
each other = **triplets**

f.   $HOCH_2CH_2CH_2OH$
            $H_a$   $H_b$

$H_a$ protons split by 2 $CH_2$ groups =
**quintet**
$H_b$ protons split by 2 H's = **triplet**

j.
   $CH_3$     H—$H_a$
        C=C
  $CH_3CH_2$    H—$H_b$

$H_a$: split by 1 H = **doublet**
$H_b$: split by 1 H = **doublet**

c.   [phenyl]—$CH_2CH_3$

$CH_3$ protons split by 2 H's = **triplet**
$CH_2$ protons split by 3 H's = **quartet**

g.   $CH_3CH_2CH_2CH_2OH$
        $H_a$   $H_b$   $H_c$   $H_d$

$H_a$ protons split by 2 H's = **triplet**
$H_b$ protons split by $CH_3$ + $CH_2$ protons = **12 peaks**
$H_c$ protons split by 2 different $CH_2$ groups = **9 peaks**
$H_d$ protons split by 2 H's = **triplet**

k.
   $CH_3$     H—$H_a$
        C=C
   Br      H—$H_b$

$H_a$: split by 1 H = **doublet**
$H_b$: split by 1 H = **doublet**

d.   $CH_3OCH_2CHCl_2$

$CH_2$ protons split by 1 H = **doublet**
CH proton split by 2 H's = **triplet**

h.   $CH_3CH_2CH_2$—C(=O)—OH
       $H_a$   $H_b$   $H_c$

$H_a$ protons split by 2 H's = **triplet**
$H_b$ protons split by $CH_3$ + $CH_2$ protons = **12 peaks**
$H_c$ protons split by 2 H's = **triplet**

l.
   $CH_3$     H—$H_a$
        C=C
  $H_c$—H    H—$H_b$

$H_a$: split by $H_b$ + $H_c$ -
**doublet of doublets** (4 peaks)
$H_b$: split by $H_a$ + $H_c$ -
**doublet of doublets** (4 peaks)
$H_c$: split by $CH_3$, $H_a$ + $H_b$ - **16 peaks**

**14.43**

$H_b$, $H_a$
  C=C
$H_c$, CN

$J_{ab}$ = 11.8 Hz
$J_{bc}$ = 0.9 Hz
$J_{ac}$ = 18 Hz

$H_a$: doublet of doublets at 5.7 ppm. Two large $J$ values are seen for the H's cis ($J_{ab}$ = 11.8 Hz) and trans ($J_{ac}$ = 18 Hz) to $H_a$.
$H_b$: doublet of doublets at ~6.2 ppm. One large $J$ value is seen for the cis H ($J_{ab}$ = 11.8 Hz). The geminal coupling ($J_{bc}$ = 0.9 Hz) is hard to see.
$H_c$: doublet of doublets at ~6.6 ppm. One large $J$ value is seen for the trans H ($J_{ac}$ = 18 Hz). The geminal coupling ($J_{bc}$ = 0.9 Hz) is hard to see.

| Splitting diagram for $H_a$ | $H_a$ |
|---|---|

1 **trans** $H_c$ proton splits $H_a$ into
**1 + 1 = 2 peaks**
**a doublet**

$J_{ac}$ = the coupling constant between $H_a$ and $H_c$

1 **cis** $H_b$ proton splits $H_a$ into
**1 + 1 = 2 peaks**
**Now it's a doublet of doublets.**

$J_{ab}$ = the coupling constant between $H_a$ and $H_b$

**14.44** Use the directions from Answer 14.24.

a. $C_4H_8Br_2$: 0 degrees of unsaturation
   **IR** peak at 3000–2850 $cm^{-1}$: $Csp^3$–H bonds
   **NMR**: singlet at 1.87 ppm (6H) (2 $CH_3$ groups)
   singlet at 3.86 ppm (2H) ($CH_2$ group)

$$CH_3-\underset{\underset{Br}{|}}{\overset{\overset{CH_3}{|}}{C}}-CH_2Br$$

b. $C_3H_6Br_2$: 0 degrees of unsaturation
   **IR** peak at 3000–2850 $cm^{-1}$: $Csp^3$–H bonds
   **NMR**: quintet at 2.4 ppm (split by 2 $CH_2$ groups)
   triplet at 3.5 ppm (split by 2 H's)

Br⌒⌒Br

c. $C_5H_{10}O_2$: 1 degree of unsaturation
   **IR** peak at 1740 $cm^{-1}$: **C=O**
   **NMR**: triplet at 1.15 ppm (3H) ($CH_3$ split by 2 H's)
   triplet at 1.25 ppm (3H) ($CH_3$ split by 2 H's)
   quartet at 2.30 ppm (2H) ($CH_2$ split by 3 H's)
   quartet at 4.72 ppm (2H) ($CH_2$ split by 3 H's)

$$CH_3CH_2\overset{\overset{O}{\|}}{C}O-CH_2CH_3$$

d. $C_6H_{14}O$: 0 degrees of unsaturation
   **IR** peak at 3600–3200 $cm^{-1}$: **O–H**
   **NMR**: triplet at 0.8 ppm (6H) (2 $CH_3$ groups split by $CH_2$ groups)
   singlet at 1.0 ppm (3H) ($CH_3$)
   quartet at 1.5 ppm (4H) (2 $CH_2$ groups split by $CH_3$ groups)
   singlet at 1.6 ppm (1H) (O–H proton)

$$CH_3CH_2-\underset{\underset{OH}{|}}{\overset{\overset{CH_3}{|}}{C}}-CH_2CH_3$$

e. $C_6H_{14}O$: 0 degrees of unsaturation
   **IR** peak at 3000–2850 $cm^{-1}$: $Csp^3$–H bonds
   **NMR**: doublet at 1.10 ppm (integration = 30 units) (from 12 H's)
   septet at 3.60 ppm (integration = 5 units) (from 2 H's)

$$CH_3-\underset{\underset{CH_3}{|}}{\overset{\overset{H}{|}}{C}}-O-\underset{\underset{CH_3}{|}}{\overset{\overset{H}{|}}{C}}-CH_3$$

f. $C_3H_6O$: 1 degree of unsaturation
   **IR** peak at 1730 $cm^{-1}$: **C=O**
   **NMR**: triplet at 1.11 ppm
   multiplet at 2.46 ppm
   triplet at 9.79 ppm

$$CH_3CH_2\overset{\overset{O}{\|}}{C}H$$

## 14.45

**Two isomers of $C_9H_{10}O$: 5 degrees of unsaturation (benzene ring likely)**

**Compound A:**
   **IR** absorption at 1742 $cm^{-1}$: **C=O**
   **NMR** data:
   Absorptions:
   singlet at 2.15 (3H) ($CH_3$ group)
   singlet at 3.70 (2H) ($CH_2$ group)
   broad singlet at 7.20 (5H)
     (likely a monosubstituted benzene ring)

$$\text{C}_6\text{H}_5-CH_2\overset{\overset{O}{\|}}{C}CH_3$$

**Compound B:**
   **IR** absorption at 1688 $cm^{-1}$: **C=O**
   **NMR** data:
   Absorptions:
   triplet at 1.22 (3H) ($CH_3$ group split by 2 H's)
   quartet at 2.98 (2H) ($CH_2$ group split by 3 H's)
   multiplet at 7.28–7.95 (5H)
     (likely a monosubstituted benzene ring)

$$\text{C}_6\text{H}_5\overset{\overset{O}{\|}}{C}CH_2CH_3$$

**14.46**

In addition to protons for the benzene rings, the following absorptions are observed:

a. [benzene ring]—OCH₃  and  [benzene ring]—CH₂OH

**one singlet** for the CH₃ protons

**two singlets** for the O–H proton and CH₂ protons

b. [benzene ring]—CH₂CH₃  and  [benzene ring]—OCH₂CH₃

The CH₂ group is adjacent to the benzene ring, but not an O atom, and so absorbs around 2.5 ppm.

The CH₂ group is adjacent to an O atom, and so it is more deshielded and absorbs around 3.5 ppm.

---

**14.47**

Compound [C]:
**C₄H₈O₂:**
   **1 degree of unsaturation**
**IR** absorption at 1743 cm⁻¹: **C=O**
**NMR data:**

   total integration units/# H's
   (23 + 29 + 30)/8 = ~10 units per H

   Hₐ: quartet at 4.1 (23 units - **2H**)
   H_b: singlet at 2.0 (29 units - **3H**)
   H_c: triplet at 1.4 (30 units - **3H**)

Compound [D]:
**C₄H₈O₂:**
   **1 degree of unsaturation**
**IR** absorption at 1730 cm⁻¹: **C=O**
**NMR data:**

   total integration units/# H's
   (18 + 30 + 31)/8 = ~10 units per H

   Hₐ: singlet at 4.1 (18 units - **2H**)
   H_b: singlet at 3.4 (30 units - **3H**)
   H_c: singlet at 2.1 (31 units - **3H**)

---

**14.48**

a. **C₉H₁₀O₂:**
   **5 degrees of unsaturation**
**IR** absorption at 1718 cm⁻¹: **C=O**
**NMR data:**
   multiplet at 7.8 ppm, **5H** on a benzene ring
   quartet at 4.4 ppm, **2H**, split by 3 H's
   triplet at 1.3 ppm, **3H**, split by 2 H's

downfield due to the O atom

b. **C₉H₁₂:**
   **4 degrees of unsaturation**
**IR** absorption at 2850–3150 cm⁻¹:
   **C–H bonds**
**NMR data:**
   singlet at 7.2 ppm, **5H**, benzene
   septet at 2.8 ppm, **1H**, split by 6 H's
   doublet at 1.3 ppm, **6H**, split by 1 H

---

**14.49**

a. Compound **A** has a molecular ion at 72: molecular formula **C₄H₈O**
   **1 degree of unsaturation**
   **IR** spectrum at 1710 cm⁻¹: **C=O**
   ¹H NMR data (ppm):
      1.0 (triplet, 3H), split by 2 H's
      2.1 (singlet, 3H)
      2.4 (quartet, 2H), split by 3 H's

b. Compound **B** has a molecular ion at 88: molecular formula $C_5H_{12}O$

  **0 degrees if unsaturation**

  IR spectrum at 3600–3200 cm$^{-1}$: **O–H bond**

  $^1$H NMR data (ppm):

    0.9 (triplet, 3H), split by 2 H's

    1.2 (singlet, 6H), due to 2 CH$_3$ groups

    1.5 (quartet, 2H), split by 3 H's

    1.6 (singlet, 1H), due to the OH proton

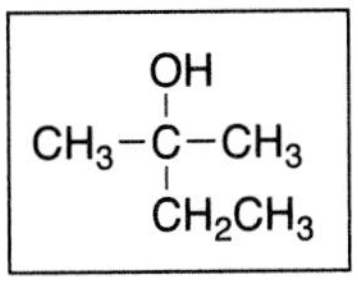

## 14.50

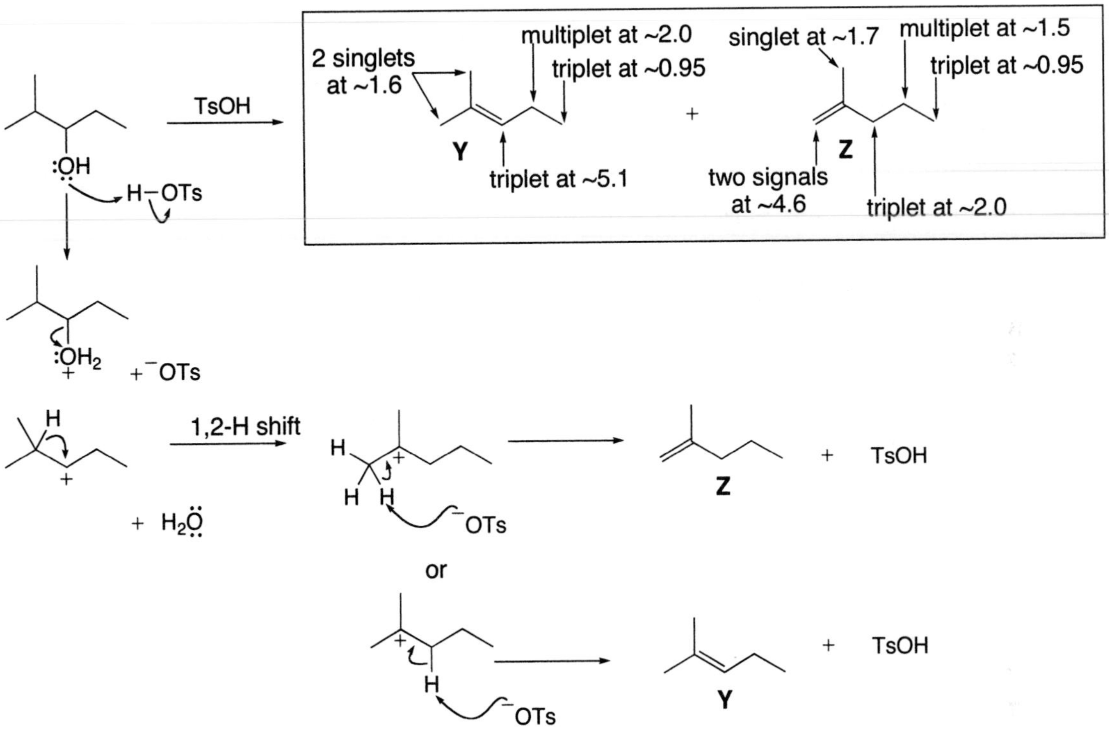

## 14.51

**Compound W**: Molecular ion at 86.

Molecular formula: $C_5H_{10}O$:

  **1 degree of unsaturation**

**IR** absorption at ~1700 cm$^{-1}$: **C=O**

**NMR data:**

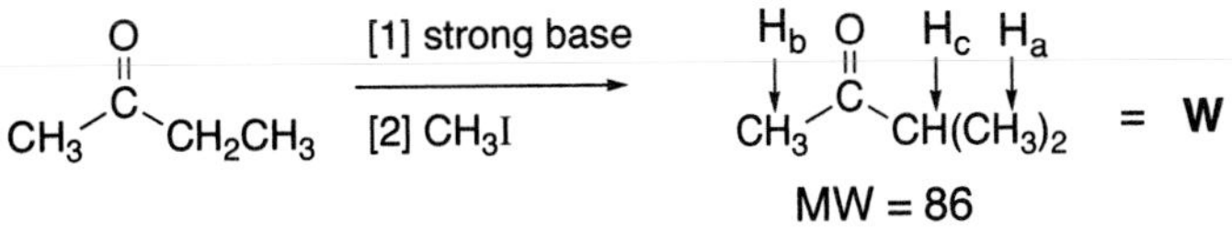

  H$_a$: doublet at 1.1 ppm, 2 CH$_3$ groups split by 1 H

  H$_b$: singlet at 2.1 ppm, CH$_3$ group

  H$_c$: septet at 2.6 ppm, 1 H split by 6 H's

**14.52**

a. Compound **A**, the odor of banana: $C_7H_{14}O_2$
   **1 degree of unsaturation**
   $^1H$ NMR (ppm):

   $H_a$: 0.93 (doublet, 6H)
   $H_b$: 1.52 (multiplet, 2H)
   $H_c$: 1.69 (multiplet, 1H)
   $H_d$: 2.04 (singlet, 3H)
   $H_e$: 4.10 (triplet, 2H)

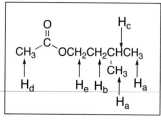

b. Compound **B**, the odor of rum: $C_7H_{14}O_2$
   **1 degree of unsaturation**
   $^1H$ NMR (ppm):

   $H_a$: 0.94 (doublet, 6H)
   $H_b$: 1.15 (triplet, 3H)
   $H_c$: 1.91 (multiplet, 1H)
   $H_d$: 2.33 (quartet, 2H)
   $H_e$: 3.86 (doublet, 2H)

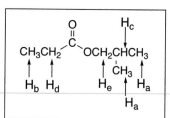

**14.53**

**$C_6H_{12}$:**
**1 degree of unsaturation**

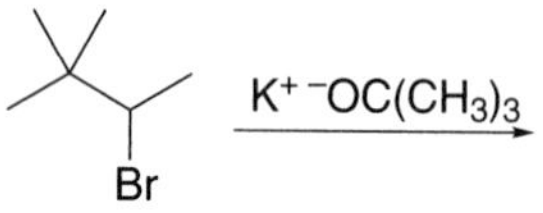

$\xrightarrow{K^+ \ ^-OC(CH_3)_3}$ **A**

$^1H$ NMR of **A** (ppm):

   $H_a$: 1.01 (singlet, 9H)
   $H_b$: 4.82 (doublet of doublets, 1H, $J$ = 10, 1.7 Hz)
   $H_c$: 4.93 (doublet of doublets, 1H, $J$ = 18, 1.7 Hz)
   $H_d$: 5.83 (doublet of doublets, 1H, $J$ = 18, 10 Hz)

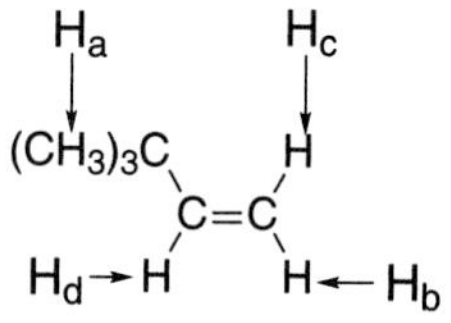

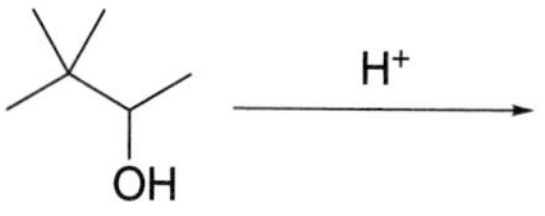

$\xrightarrow{H^+}$ **B**

$^1H$ NMR of **B**: 1.60 (singlet) ppm.

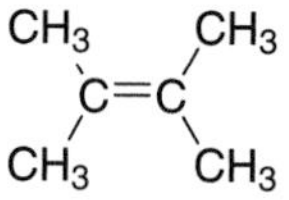

All H's are identical, so there is only one singlet in the NMR.

**14.54** Both **A** and **B** have the same molecular ion—since they are isomers— and show a C=O peak in their IR spectra. $^1H$ NMR spectroscopy is the best way to distinguish the two compounds.

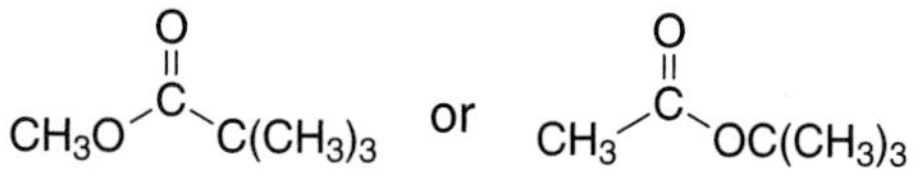

$CH_3O-\overset{\overset{O}{\|}}{C}-C(CH_3)_3$ **A**   or   $CH_3-\overset{\overset{O}{\|}}{C}-OC(CH_3)_3$ **B**

Both **A** and **B** have two singlets in a 3:1 ratio in their $^1H$ NMR spectra. But **A** has a peak at ~3 ppm due to the deshielded $CH_3$ group bonded to the O atom. **B** has no proton that is so deshielded. Both of its singlets are in the 1–2.5 ppm region.

**14.55**

Four constitutional isomers of **$C_4H_9Br$**:

4 different C's          4 different C's          2 different C's          3 different C's

**14.56** Only two compounds in Problem 14.40 give one signal in their $^{13}C$ NMR spectrum:

$CH_3CH_3$

**14.57**

a. HC(CH₃)₃
**2 signals**

b.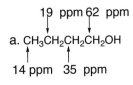
**5 signals**

c. CH₃OCH(CH₃)₂
**3 signals**

d.
**7 signals**

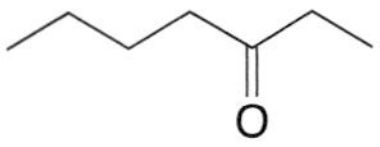

e. 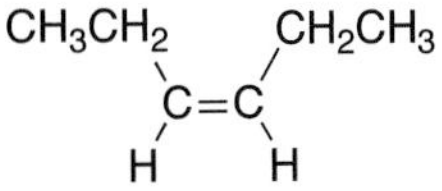 CH₃CH₂     CH₂CH₃
     C=C
     H    H
**3 signals**

f.  ~~~~~~OH
**7 signals**

g. 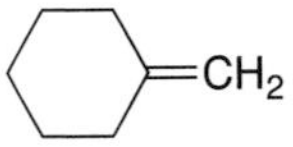 ⬡=CH₂
**5 signals**

h. 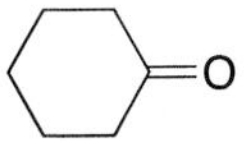 ⬡=O
**4 signals**

i. 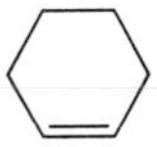 ⬡
**3 signals**

**14.58**

a. 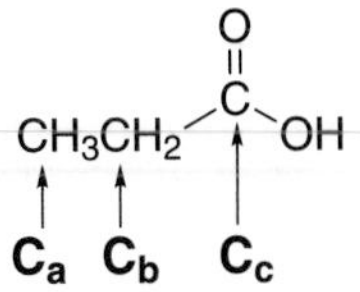 CH₃CH₂—C(=O)—OH
   Cₐ   C_b   C_c
   Cₐ < C_b < C_c

b. 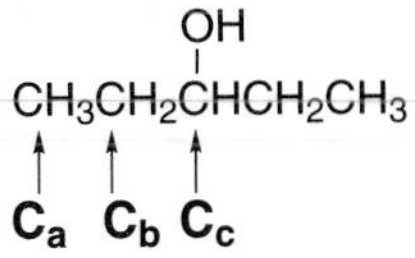 CH₃CH₂CH(OH)CH₂CH₃
   Cₐ   C_b   C_c
   Cₐ < C_b < C_c

c. Ph—C(=O)—CH₂CH₃
   Cₐ    C_b   C_c
   C_c < C_b < Cₐ

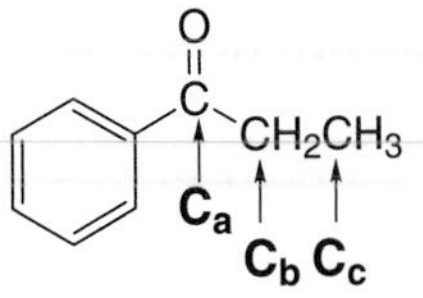

d. 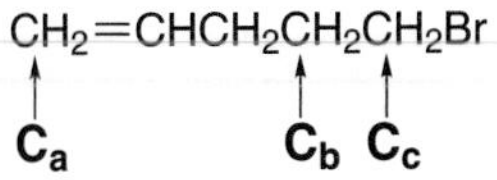 CH₂=CHCH₂CH₂CH₂Br
   Cₐ     C_b   C_c
   C_b < C_c < Cₐ

**14.59**

a. CH₃CH₂CH₂CH₂OH
   19 ppm   62 ppm (top)
   14 ppm   35 ppm (bottom)

b. (CH₃)₂CHCHO
   16 ppm   205 ppm (top)
   41 ppm (bottom)

c. CH₂=CHCH(OH)CH₃
   143 ppm   23 ppm (top)
   113 ppm   69 ppm (bottom)

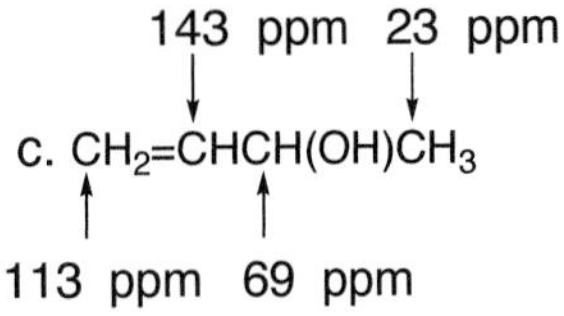

**14.60**

a.

**C₆H₁₂O₂:**
    **1 degree of unsaturation**
**IR** peak at 1740 cm⁻¹: **C=O**
**¹H NMR** 2 signals: 2 types of H's
**¹³C NMR:** 4 signals: 4 kinds of C's,
including one at ~170 ppm due a C=O

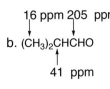

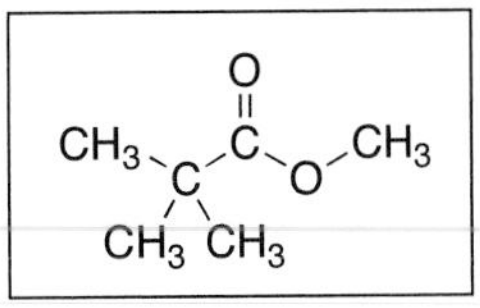

b.

**C₆H₁₀:**
    **2 degrees of unsaturation**
**IR** peak at 3000 cm⁻¹: **C_{sp³}–H bonds**
    peak at 3300 cm⁻¹: **C_{sp}–H bond**
    peak at ~2150 cm⁻¹: **C≡C bond**
**¹³C NMR:** 4 signals: 4 kinds of C's

HC≡C—C(CH₃)(CH₃)—CH₃

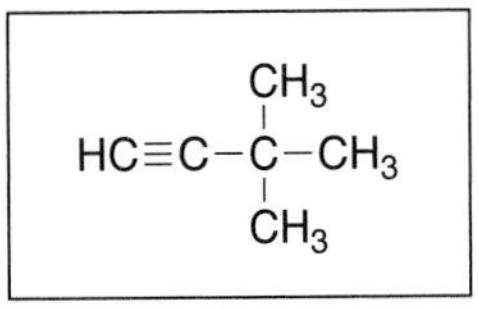

**14.61**

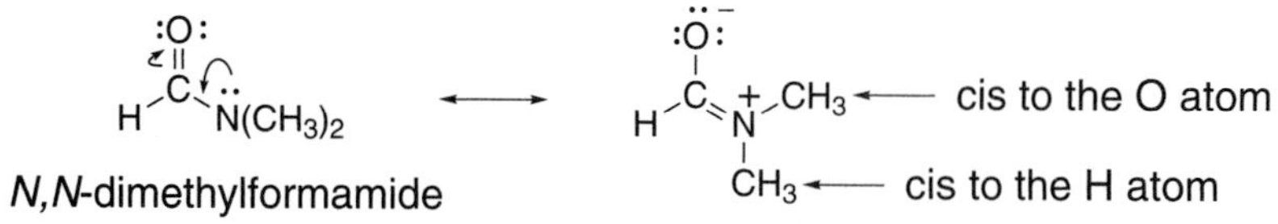

N,N-dimethylformamide

cis to the O atom
cis to the H atom

A second resonance structure for *N,N*-dimethylformamide places the two $CH_3$ groups in different environments. One $CH_3$ group is cis to the O atom, and one is cis to the H atom. This gives rise to two different absorptions for the $CH_3$ groups.

**14.62**

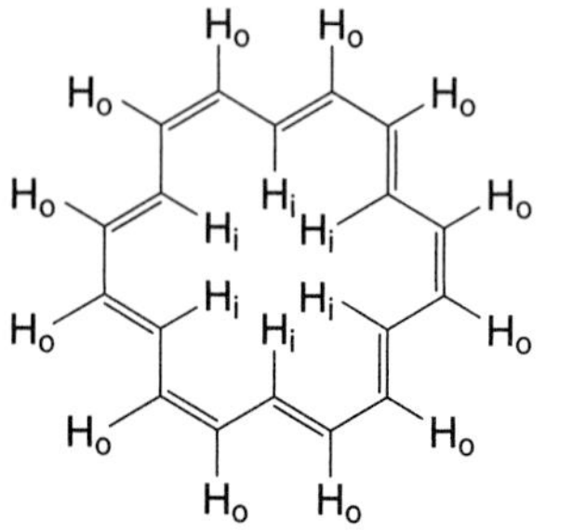

18-Annulene has 18 π electrons that create an **induced magnetic field** similar to the 6 π electrons of benzene. 18-Annulene has 12 protons that are oriented on the outside of the ring (labeled $H_o$), and 6 protons that are oriented inside the ring (labeled $H_i$). The induced magnetic field reinforces the external field in the vicinity of the protons on the outside of the ring. These protons are deshielded and so they absorb downfield (8.9 ppm). In contrast, the induced magnetic field is opposite in direction to the applied magnetic field in the vicinity of the protons on the inside of the ring. This shields the protons and the absorption is therefore very far upfield, even higher than TMS (–1.8 ppm).

**14.63**

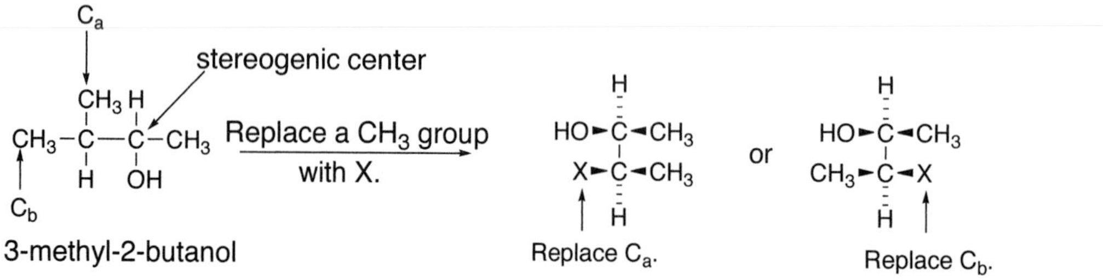

stereogenic center

3-methyl-2-butanol

Replace a $CH_3$ group with X.

Replace $C_a$.

or

Replace $C_b$.

The $CH_3$ groups are not equivalent to each other, since replacement of each by X forms two diastereomers.

Thus, every C in this compound is different and there are five $^{13}C$ signals.

Chapter 15: Radical Reactions

♦ General features of radicals

- A radical is a reactive intermediate with an unpaired electron (15.1).
- A carbon radical is $sp^2$ hybridized and trigonal planar (15.1).
- The stability of a radical increases as the number of C's bonded to the radical carbon increases (15.1).

| least stable | | | most stable |
|---|---|---|---|
| $\overset{\cdot}{C}H_3$ | $R\overset{\cdot}{C}H_2$ | $R_2\overset{\cdot}{C}H$ | $R_3\overset{\cdot}{C}$ |
| | 1° | 2° | 3° |

**Increasing alkyl substitution**
**Increasing radical stability**

- Allylic radicals are stabilized by resonance, making them more stable than 3° radicals (15.10).

$$CH_2{=}CH{-}\overset{\cdot}{C}H_2 \longleftrightarrow \overset{\cdot}{C}H_2{-}CH{=}CH_2$$

two resonance structures for the allyl radical

♦ Radical reactions

[1] Halogenation of alkanes (15.4)

$$R{-}H \xrightarrow[\text{hv or } \Delta]{X_2} \boxed{R{-}X \\ \text{alkyl halide}}$$

X = Cl or Br

- The reaction follows a radical chain mechanism.
- The weaker the C–H bond, the more readily the H is replaced by X.
- Chlorination is faster and less selective than bromination (15.6).
- Radical substitution results in racemization at a stereogenic center (15.8).

[2] Allylic halogenation (15.10)

$$CH_2{=}CH{-}CH_3 \xrightarrow[\text{hv or ROOR}]{NBS} \boxed{CH_2{=}CHCH_2Br \\ \text{allylic halide}}$$

- The reaction follows a radical chain mechanism.

[3] Radical addition of HBr to an alkene (15.13)

$$RCH{=}CH_2 \xrightarrow[\substack{\text{hv, } \Delta, \text{ or} \\ \text{ROOR}}]{HBr} \boxed{\underset{\text{alkyl bromide}}{R{-}\overset{\overset{H}{|}}{\underset{\underset{H}{|}}{C}}{-}\overset{\overset{H}{|}}{\underset{\underset{Br}{|}}{C}}{-}H}}$$

- A radical addition mechanism is followed.
- Br bonds to the less substituted carbon atom to form the more substituted, more stable radical.

[4] Radical polymerization of alkenes (15.14)

$$CH_2{=}CHZ \xrightarrow{ROOR} \boxed{\text{polymer}}$$

- A radical addition mechanism is followed.

## Chapter 15: Answers to Problems

**15.1** 1° Radicals are on C's bonded to one other C; 2° radicals are on C's bonded to two other C's; 3° radicals are on C's bonded to three other C's.

a. $CH_3CH_2-\overset{\cdot}{C}HCH_2CH_3$     b.     c.     d.

 **2° radical**          **3° radical**          **2° radical**          **1° radical**

**15.2** The stability of a radical increases as the number of alkyl groups bonded to the radical carbon increases.

a. $(CH_3)_2CH\overset{\cdot}{C}H_2$     $CH_3CH_2\overset{\cdot}{C}HCH_3$     $(CH_3)_3C\cdot$     b.

| 1° radical | 2° radical | 3° radical | 1° radical | 2° radical | 3° radical |
|---|---|---|---|---|---|
| least stable | intermediate stability | most stable | least stable | intermediate stability | most stable |

**15.3** Reaction of a radical with:
- an alkane abstracts a hydrogen atom and creates a new carbon radical.
- an alkene generates a new bond to one carbon, and a new carbon radical.
- another radical forms a bond.

a. $CH_3-CH_3 \xrightarrow{:\overset{\cdot\cdot}{\underset{\cdot\cdot}{C}}l\cdot} CH_3-\overset{\cdot}{C}H_2 + H-\overset{\cdot\cdot}{\underset{\cdot\cdot}{C}}l:$     c. $:\overset{\cdot\cdot}{\underset{\cdot\cdot}{C}}l\cdot \xrightarrow{:\overset{\cdot\cdot}{\underset{\cdot\cdot}{C}}l\cdot} :\overset{\cdot\cdot}{\underset{\cdot\cdot}{C}}l-\overset{\cdot\cdot}{\underset{\cdot\cdot}{C}}l:$

b. $CH_2=CH_2 \xrightarrow{:\overset{\cdot\cdot}{\underset{\cdot\cdot}{C}}l\cdot} \overset{\cdot}{C}H_2-CH_2$     d. $:\overset{\cdot\cdot}{\underset{\cdot\cdot}{C}}l\cdot + \cdot\overset{\cdot\cdot}{\underset{\cdot\cdot}{O}}-\overset{\cdot\cdot}{\underset{\cdot\cdot}{O}}\cdot \longrightarrow :\overset{\cdot\cdot}{\underset{\cdot\cdot}{C}}l-\overset{\cdot\cdot}{\underset{\cdot\cdot}{O}}-\overset{\cdot\cdot}{\underset{\cdot\cdot}{O}}\cdot$

**15.4 Monochlorination** is a radical substitution reaction in which a Cl replaces a H generating an alkyl halide.

a. $\xrightarrow[\Delta]{Cl_2}$

b. $CH_3CH_2CH_2CH_2CH_2CH_3 \xrightarrow[\Delta]{Cl_2} ClCH_2CH_2CH_2CH_2CH_2CH_3 + CH_3-\overset{\overset{H}{|}}{\underset{\underset{Cl}{|}}{C}}-CH_2CH_2CH_2CH_3 + CH_3CH_2-\overset{\overset{H}{|}}{\underset{\underset{Cl}{|}}{C}}-CH_2CH_2CH_3$

c. $(CH_3)_3CH \xrightarrow[\Delta]{Cl_2} CH_3-\overset{\overset{Cl}{|}}{\underset{\underset{CH_3}{|}}{C}}-CH_3 + CH_3-\overset{\overset{H}{|}}{\underset{\underset{CH_3}{|}}{C}}-CH_2Cl$

**15.5**

**A** $\xrightarrow[\Delta]{Cl_2}$

**B** $\xrightarrow[\Delta]{Cl_2}$ + +

**15.6**

**Initiation:** $:\!\ddot{B}r\!-\!\ddot{B}r\!:$ $\xrightarrow[\text{or } \Delta]{h\nu}$ $:\!\ddot{B}r\!\cdot$ + $\cdot\ddot{B}r\!:$

**Propagation:** $CH_3\!-\!H$ + $\cdot\ddot{B}r\!:$ $\longrightarrow$ $\dot{C}H_3$ + $H\!-\!\ddot{B}r\!:$

$\dot{C}H_3$ + $:\!\ddot{B}r\!-\!\ddot{B}r\!:$ $\longrightarrow$ $CH_3\!-\!\ddot{B}r\!:$ + $\cdot\ddot{B}r\!:$

**Termination:** $:\!\ddot{B}r\!\cdot$ + $\cdot\ddot{B}r\!:$ $\longrightarrow$ $:\!\ddot{B}r\!-\!\ddot{B}r\!:$

or

$\dot{C}H_3$ + $\dot{C}H_3$ $\longrightarrow$ $CH_3\!-\!CH_3$

or

$\dot{C}H_3$ + $\cdot\ddot{B}r\!:$ $\longrightarrow$ $CH_3\!-\!\ddot{B}r\!:$

**15.7**

**Initiation:** $:\!\ddot{C}l\!-\!\ddot{C}l\!:$ $\xrightarrow[\text{or } \Delta]{h\nu}$ $:\!\ddot{C}l\!\cdot$ + $\cdot\ddot{C}l\!:$

**Propagation:** $CH_3\!-\!H$ + $\cdot\ddot{C}l\!:$ $\longrightarrow$ $\dot{C}H_3$ + $H\!-\!\ddot{C}l\!:$

$\dot{C}H_3$ + $:\!\ddot{C}l\!-\!\ddot{C}l\!:$ $\longrightarrow$ $CH_3\!-\!\ddot{C}l\!:$ + $\cdot\ddot{C}l\!:$

**One possibility for termination:** $\dot{C}H_3$ + $\dot{C}H_3$ $\longrightarrow$ $\boxed{CH_3\!-\!CH_3}$

$\downarrow$

$CH_3\!-\!CH_2\!-\!H$ + $\cdot\ddot{C}l\!:$ $\longrightarrow$ $CH_3\dot{C}H_2$ + $H\!-\!\ddot{C}l\!:$

$CH_3\dot{C}H_2$ + $:\!\ddot{C}l\!-\!\ddot{C}l\!:$ $\longrightarrow$ $\boxed{CH_3CH_2\!-\!Cl}$ + $\cdot\ddot{C}l\!:$

**15.8** Transition states are hypothetical structures at the "peaks" on an energy diagram. They are drawn with dashed lines to indicate partially broken and partially formed bonds.

Transition state 1:

$$\left[ :\!\overset{\delta\cdot}{\ddot{C}l}\text{----}H\text{---}\overset{\delta\cdot}{CH_2CH_3} \right]^{\ddagger}$$

Transition state 2:

$$\left[ :\!\overset{\delta\cdot}{\ddot{C}l}\text{----}\ddot{C}l\text{---}\overset{\delta\cdot}{CH_2CH_3} \right]^{\ddagger}$$

**15.9**

**Step 1:** $CH_3\!-\!H$ + $\cdot\ddot{B}r\!:$ $\longrightarrow$ $\dot{C}H_3$ + $H\!-\!\ddot{B}r\!:$ $\qquad \Delta H^{\circ} = +16$ kcal/mol

1 bond broken
+104 kcal/mol

1 bond formed
−88 kcal/mol

**Step 2:** $\dot{C}H_3$ + $:\!\ddot{B}r\!-\!\ddot{B}r\!:$ $\longrightarrow$ $CH_3\!-\!Br$ + $\cdot\ddot{B}r\!:$ $\qquad \Delta H^{\circ} = -24$ kcal/mol

1 bond broken
+46 kcal/mol

1 bond formed
−70 kcal/mol

**15.10** The rate-determining step for halogenation reactions is formation of $CH_3\cdot + HX$.

$$CH_3-H \ + \ \cdot\ddot{I}: \longrightarrow \ \cdot CH_3 \ + \ H-\ddot{I}:$$

1 bond broken
+104 kcal/mol

1 bond formed
−71 kcal/mol

$\Delta H° = +33$ kcal/mol

This is reaction is more endothermic and has a higher $E_a$ than a similar reaction with $Cl_2$ or $Br_2$.

**15.11** The **weakest C–H bond** in each alkane is the **most readily cleaved** during radical halogenation.

a.

3°
**most reactive**

b.

.H

3°
**most reactive**

c. $CH_3CHCH_2CH_3$

H

2°
**most reactive**

**15.12** To draw the product of bromination:
- Draw out the starting material and find the most reactive C–H bond (on the most substituted C).
- The major product is formed by **cleavage of the** *weakest* **C–H bond**.

a.

$\xrightarrow[\Delta]{Br_2}$

Br

c.

$\xrightarrow[\Delta]{Br_2}$

Br

b.

$\xrightarrow[\Delta]{Br_2}$

Br

d.

$\xrightarrow[\Delta]{Br_2}$

Br

**15.13**

$\xrightarrow[h\nu]{Cl_2}$

Cl

+

Cl

+

Cl

+

Cl

+

Cl

This is the desired product,
**1-chloro-1-methylcyclohexane**,
but many other products are formed.

**15.14** If 1° C–H and 3° C–H bonds were equally reactive there would be nine times as much $(CH_3)_2CHCH_2Cl$ as $(CH_3)_3CCl$ since the ratio of 1° to 3° H's is 9:1. The fact that the ratio is only 63:37 shows that the 1° C–H bond is less reactive than the 3° C–H bond. $(CH_3)_2CHCH_2Cl$ is still the major product, though, because there are nine 1° C–H bonds and only one 3° C–H bond.

**15.15**

a. $CH_3-\overset{\overset{\displaystyle CH_3}{|}}{\underset{\underset{\displaystyle CH_3}{|}}{C}}-H$ $\xrightarrow[\Delta]{Br_2}$ $CH_3-\overset{\overset{\displaystyle CH_3}{|}}{\underset{\underset{\displaystyle CH_3}{|}}{C}}-Br$

c. $\overset{\displaystyle CH_3}{\underset{\displaystyle CH_3}{C}}=CH_2$ $\xrightarrow[H_2SO_4]{H_2O}$ $CH_3-\overset{\overset{\displaystyle CH_3}{|}}{\underset{\underset{\displaystyle CH_3}{|}}{C}}-OH$

(from b.)

b. $CH_3-\overset{\overset{\displaystyle CH_3}{|}}{\underset{\underset{\displaystyle CH_3}{|}}{C}}-Br$ $\xrightarrow{(CH_3)_3CO^-K^+}$ $\overset{\displaystyle CH_3}{\underset{\displaystyle CH_3}{C}}=CH_2$

d. $\overset{\displaystyle CH_3}{\underset{\displaystyle CH_3}{C}}=CH_2$ $\xrightarrow[CCl_4]{Cl_2}$ $CH_3-\overset{\overset{\displaystyle CH_2Cl}{|}}{\underset{\underset{\displaystyle CH_3}{|}}{C}}-Cl$

(from a.)

(from b.)

**15.16**

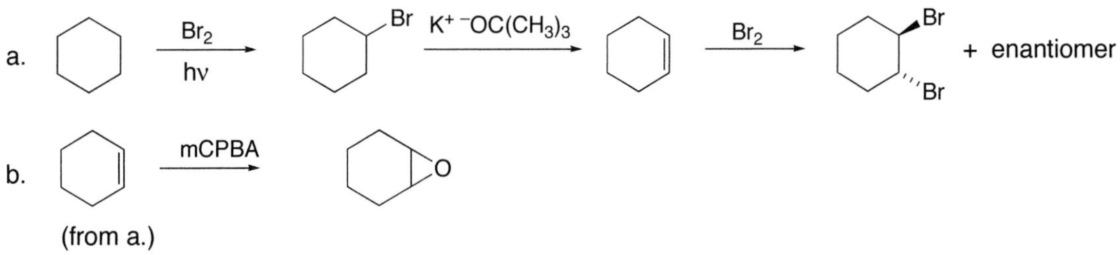

a.

b.

(from a.)

**15.17** Since the reaction does not occur at the stereogenic center, leave it as is.

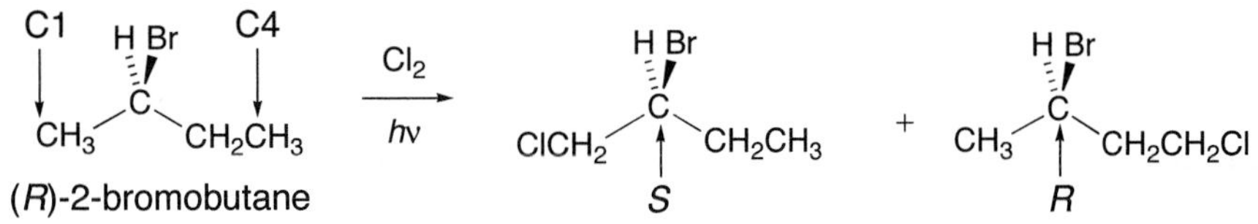

**15.18**

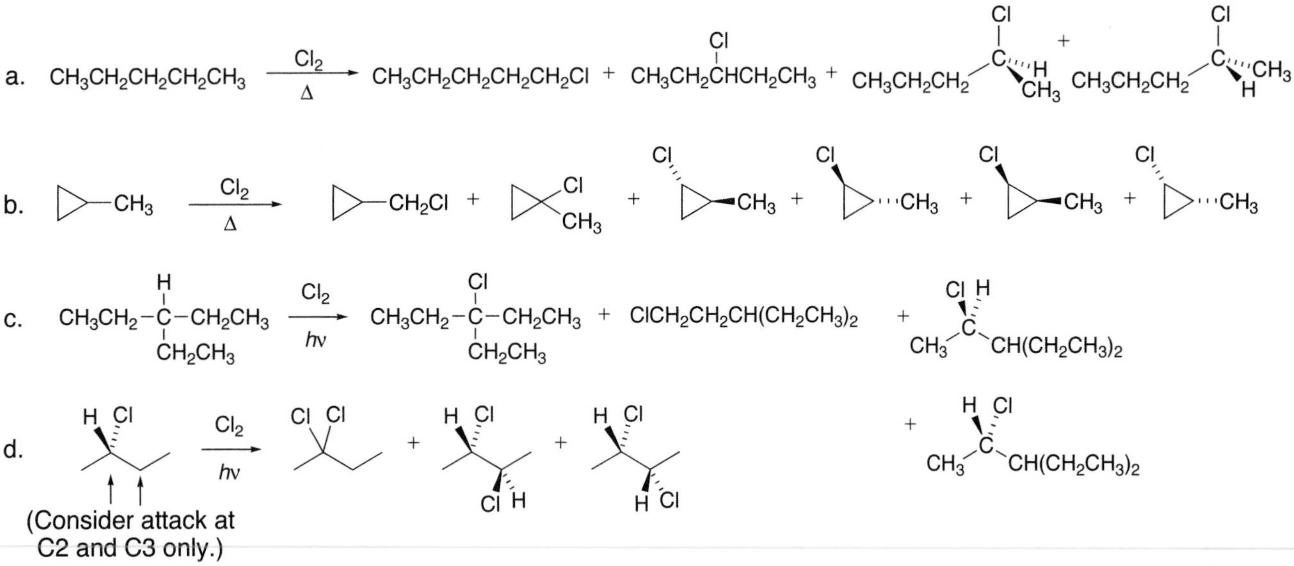

a. $CH_3CH_2CH_2CH_2CH_3$ →($Cl_2$, Δ)→ $CH_3CH_2CH_2CH_2CH_2Cl$ + $CH_3CH_2CHClCH_2CH_3$ + (stereoisomers as drawn)

b.

c.

d.

(Consider attack at C2 and C3 only.)

**15.19**

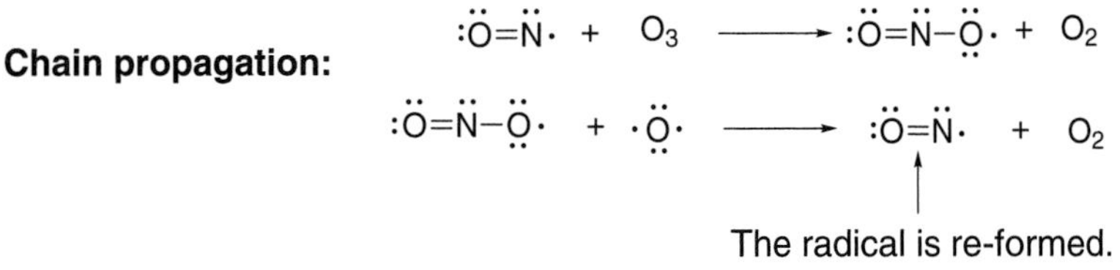

**Chain propagation:**

$:\ddot{O}=\ddot{N}\cdot$ + $O_3$ ⟶ $:\ddot{O}=\ddot{N}-\ddot{O}\cdot$ + $O_2$

$:\ddot{O}=\ddot{N}-\ddot{O}\cdot$ + $\cdot\ddot{O}\cdot$ ⟶ $:\ddot{O}=\ddot{N}\cdot$ + $O_2$

The radical is re-formed.

**15.20** Draw the resonance structure by moving the π bond and the unpaired electron. The hybrid is drawn with dashed lines for bonds that are in one resonance structure but not another. The symbol δ• is used on any atom that has an unpaired electron in any resonance structure.

a. $CH_3-CH{=}CH{-}CH_2$ ⟷ $CH_3-\overset{\cdot}{C}H-CH{=}CH_2$

hybrid: $CH_3-\overset{\delta\cdot}{CH}{=}CH{-}\overset{\delta\cdot}{CH_2}$

b. 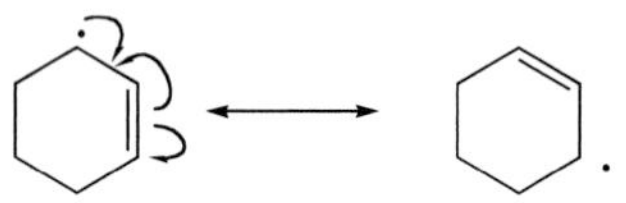 ⟷

hybrid:

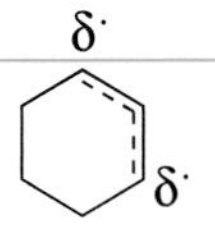

c.  ⟷

hybrid:

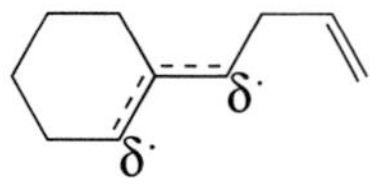

d. 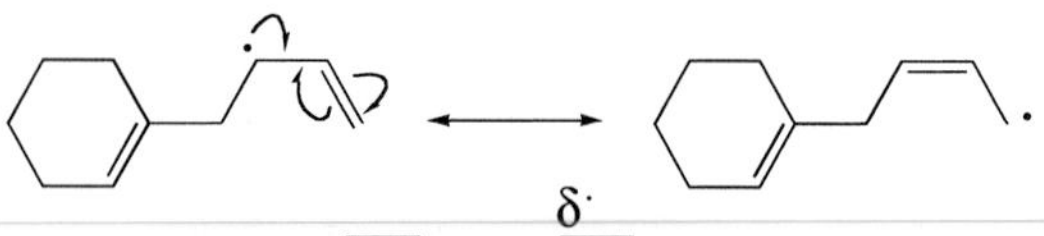 ⟷

hybrid:

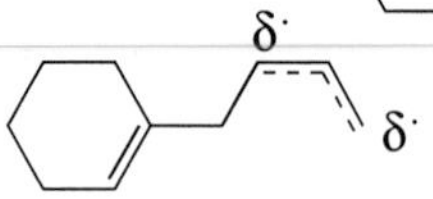

**15.21** Reaction of an alkene with NBS or $Br_2 + h\nu$ yields allylic substitution products.

a.

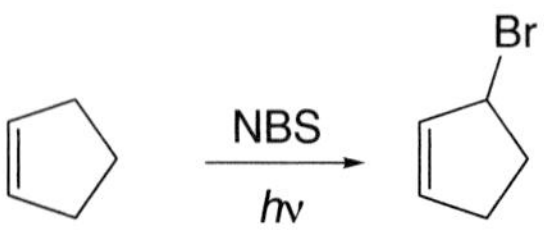

c. $CH_2{=}CH{-}CH_3 \xrightarrow{Br_2} \underset{\underset{Br\ \ Br}{|\ \ \ |}}{CH_2CHCH_3}$

b. $CH_2{=}CH{-}CH_3 \xrightarrow[h\nu]{NBS} CH_2{=}CH{-}CH_2Br$

**15.22**

a. $CH_3-CH{=}CH{-}CH_3 \xrightarrow[h\nu]{NBS} CH_3{-}CH{=}CH{-}CH_2Br$
+
$CH_3CH(Br)CH{=}CH_2$

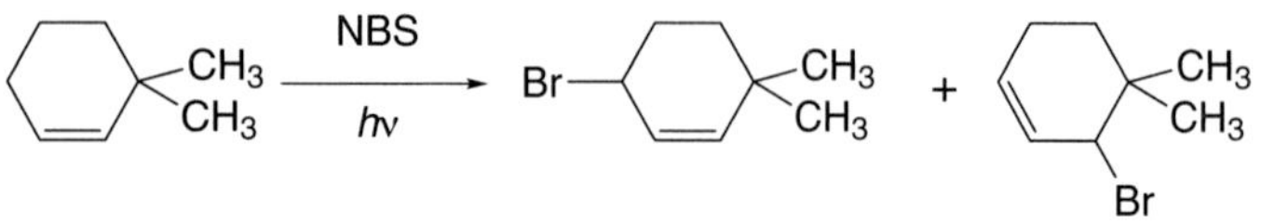

c. $CH_2{=}C(CH_2CH_3)_2 \xrightarrow[h\nu]{NBS} CH_2{=}C\overset{CHBrCH_3}{\underset{CH_2CH_3}{\big<}}$
+
$BrCH_2C(CH_2CH_3){=}CHCH_3$

b. $\xrightarrow[h\nu]{NBS}$ +

**15.23**

$\xrightarrow[h\nu]{NBS}$ + + +

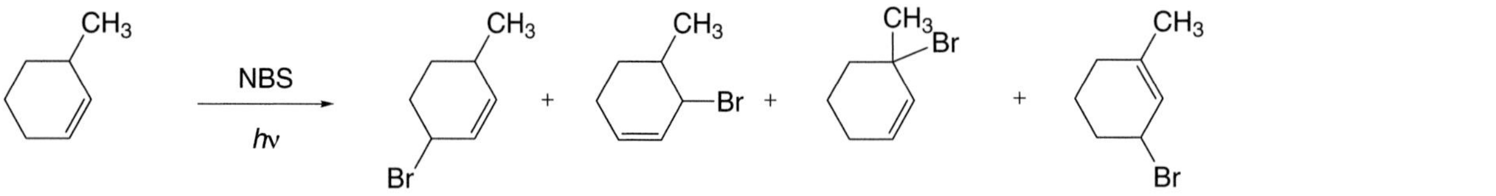

**15.24** Reaction of an alkene with NBS + *hv* yields allylic substitution products.

a. 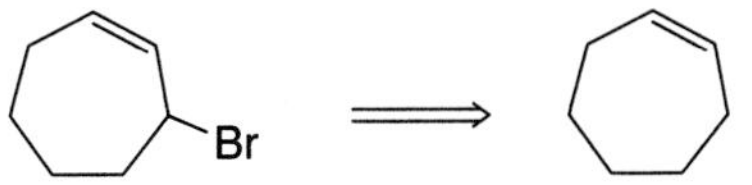       one possible product: **high yield**

b.    CH₃CH₂CH=CHCH₂Br    c.

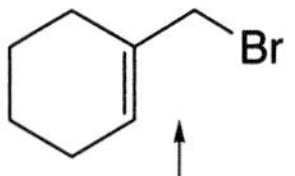

Cannot be made in high yield by allylic halogenation.
Any alkene starting material would yield a mixture of allylic halides.

**15.25**

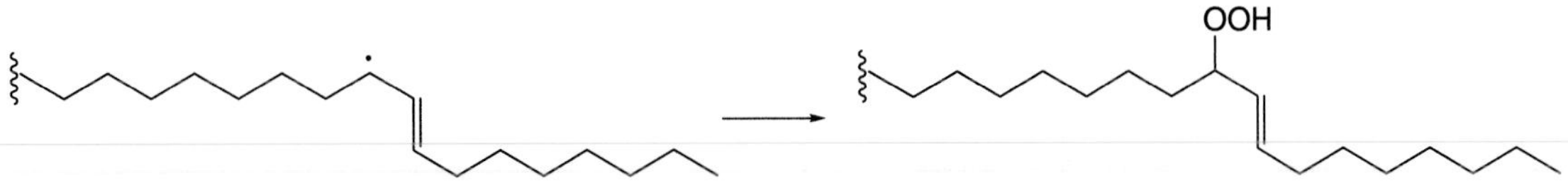

second resonance structure

**15.26** The weakest C–H bond is most readily cleaved. To draw the hydroperoxide products, add OOH to each carbon that bears a radical in one of the resonance structures.

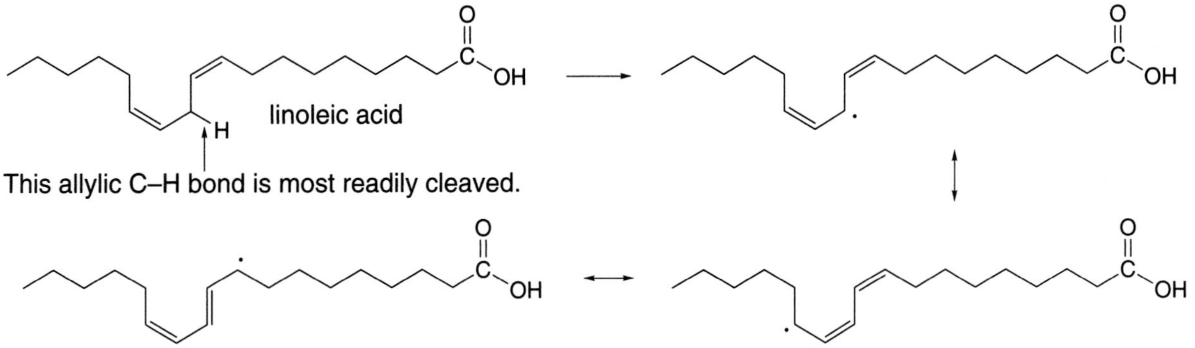

linoleic acid

This allylic C–H bond is most readily cleaved.

**hydroperoxide products:**

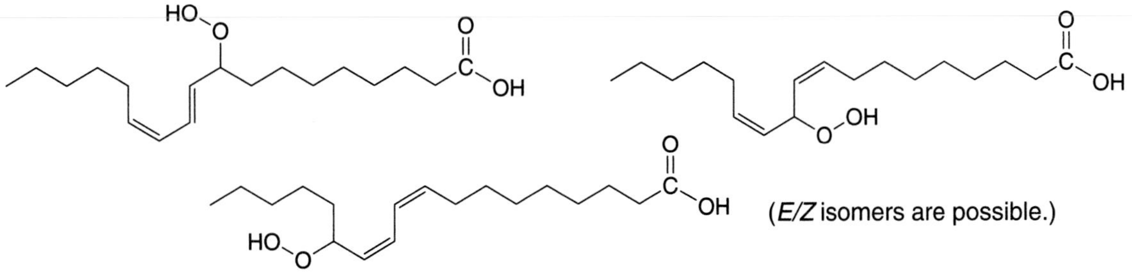

(*E/Z* isomers are possible.)

**15.27**

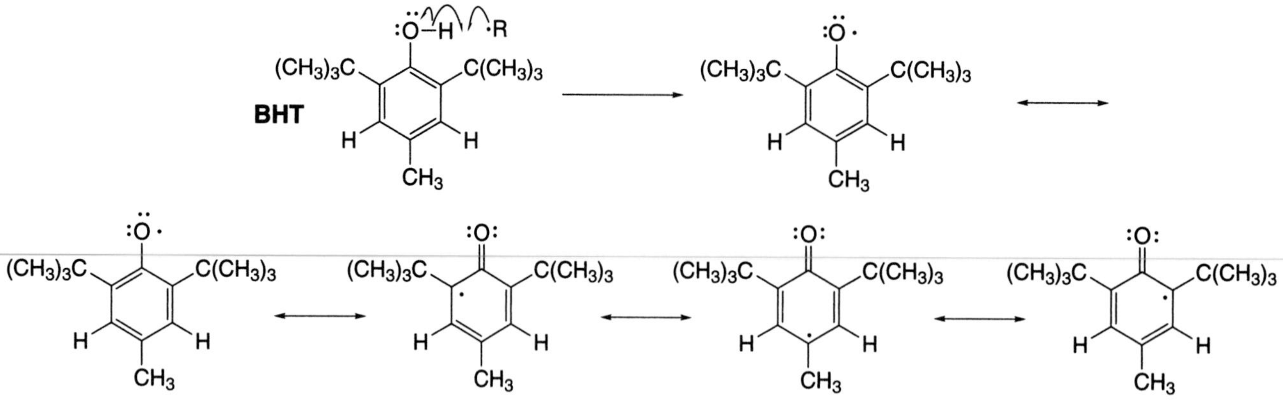

**15.28**

a.  $CH_2=CHCH_2CH_2CH_2CH_3$ $\xrightarrow{HBr}$ $CH_3CHCH_2CH_2CH_2CH_3$
$\quad\quad\quad$ | $\quad\quad\quad\quad\quad\quad\quad\quad\quad\quad\quad\quad\quad\quad\quad\quad$ Br

b. 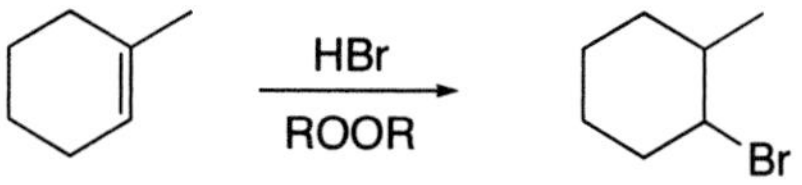 $\xrightarrow[ROOR]{HBr}$

c.  $(CH_3)_2C=CHCH_3$ $\xrightarrow{HBr}$ $(CH_3)_2\overset{\overset{\displaystyle Br}{|}}{C}-CH_2CH_3$

d.  $CH_3CH=CHCH_2CH_2CH_3$ $\xrightarrow[ROOR]{HBr}$ $CH_3\underset{\underset{\displaystyle Br}{|}}{CH}-CH_2CH_2CH_2CH_3$ + $CH_3CH_2-\underset{\underset{\displaystyle Br}{|}}{CH}CH_2CH_2CH_3$

**15.29** In addition of HBr under radical conditions:
- Br• adds first to form the more stable radical.
- Then H• is added to the carbon radical.

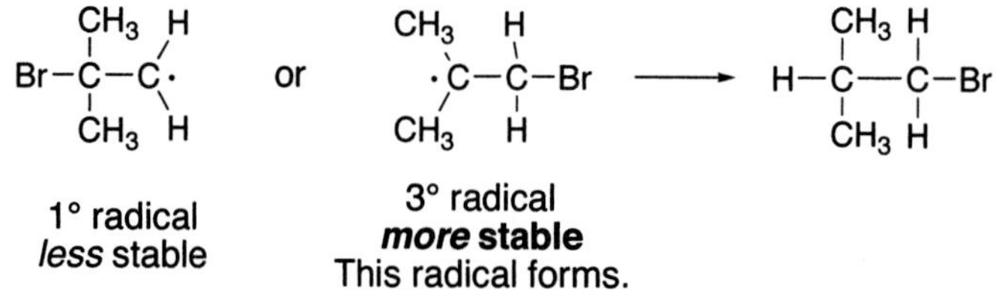

**15.30**

a.  $\xrightarrow[ROOR]{HBr}$

c.  $\xrightarrow{Br_2}$

b. 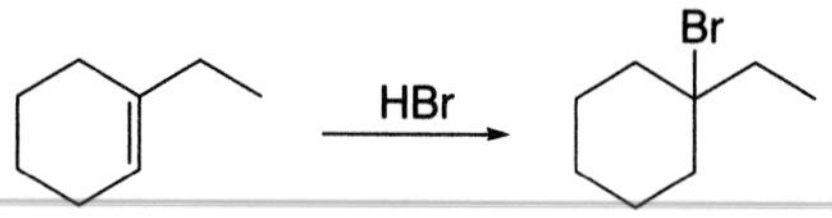 $\xrightarrow{HBr}$

**15.31**

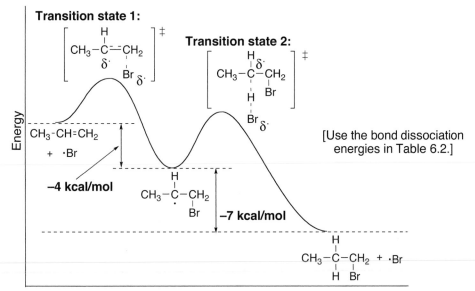

**Transition state 1:**

$$\left[ \begin{array}{c} H \\ CH_3-C\!=\!CH_2 \\ \delta \cdot \quad | \\ Br \; \delta \cdot \end{array} \right]^{\ddagger}$$

**Transition state 2:**

$$\left[ \begin{array}{c} H \; \delta \cdot \\ CH_3-C-CH_2 \\ | \qquad | \\ H \qquad Br \\ | \\ Br \; \delta \cdot \end{array} \right]^{\ddagger}$$

[Use the bond dissociation energies in Table 6.2.]

$CH_3-CH\!=\!CH_2$
$+ \; \cdot Br$

−4 kcal/mol

$$CH_3-\overset{H}{\underset{\underset{Br}{|}}{C}}-CH_2$$

−7 kcal/mol

$$CH_3-\overset{H}{\underset{\underset{H \;\; Br}{| \;\; |}}{C}}-CH_2 \; + \; \cdot Br$$

Energy

Reaction coordinate

**15.32**

**Step 1:**  $CH_3-CH\!=\!CH_2 \; + \; \cdot Cl \longrightarrow CH_3-\overset{H}{\underset{\underset{Cl}{|}}{\underset{\cdot}{C}}}-CH_2 \qquad \Delta H_1° = -17$ kcal/mol

1 bond broken
+64 kcal/mol

1 bond formed
−81 kcal/mol

**Step 2:**  $CH_3-\overset{H}{\underset{\underset{Cl}{|}}{\underset{\cdot}{C}}}-CH_2 \; + \; H\!-\!Cl \longrightarrow CH_3-\overset{H}{\underset{\underset{H \;\; Cl}{| }}{C}}-CH_2 \qquad \Delta H_2° = +8$ kcal/mol

This step of propagation is endothermic. It prohibits chain propagation from occurring over and over.

1 bond broken
+103 kcal/mol

1 bond formed
−95 kcal/mol

**15.33**

a.  $CH_2\!=\!\overset{CH_3}{\underset{CH_3}{\overset{|}{\underset{|}{C}}}} \longrightarrow \{CH_2-\overset{CH_3}{\underset{\underset{CH_3}{|}}{C}}-CH_2-\overset{CH_3}{\underset{\underset{CH_3}{|}}{C}}-CH_2-\overset{CH_3}{\underset{\underset{CH_3}{|}}{C}}-\}$

b.  $\{$ ... $\} \Longrightarrow CH_2\!=\!CH-OCOCH_3$

$$\underset{COCH_3 \quad COCH_3 \quad COCH_3}{\underset{O \qquad\quad O \qquad\quad O}{|\qquad\quad |\qquad\quad |}}$$

poly(vinyl acetate)

**15.34**

Initiation:

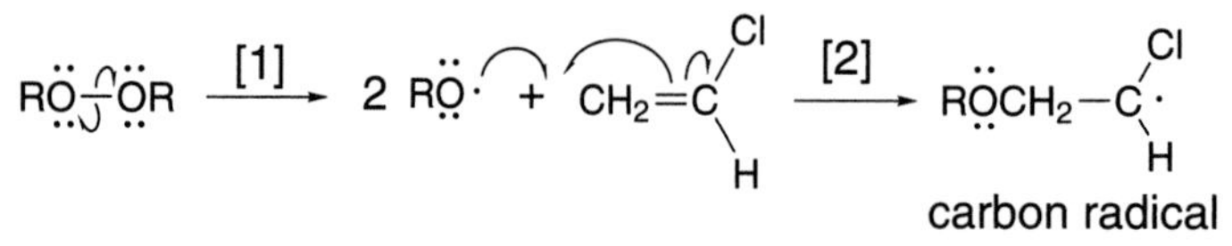

$$RO{:}-{:}OR \xrightarrow{[1]} 2\ RO{\cdot} \ + \ CH_2{=}C\!\!\begin{array}{c} Cl \\ \\ H \end{array} \xrightarrow{[2]} ROCH_2{-}\overset{Cl}{\underset{H}{C}}{\cdot}$$

carbon radical

Propagation:

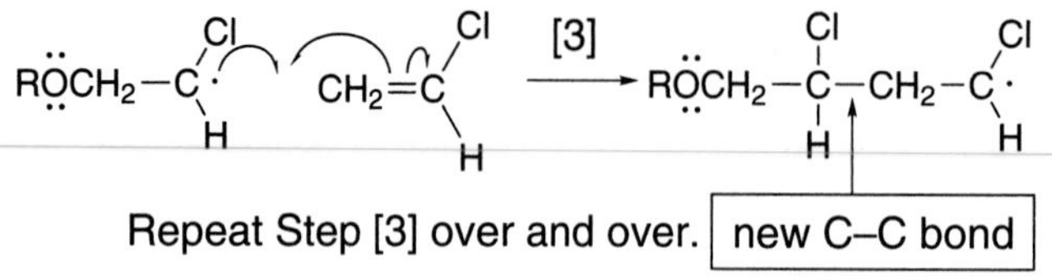

$$ROCH_2{-}\overset{Cl}{\underset{H}{C}}{\cdot} \quad CH_2{=}C\!\!\begin{array}{c} Cl \\ \\ H \end{array} \xrightarrow{[3]} ROCH_2{-}\overset{Cl}{\underset{H}{C}}{-}CH_2{-}\overset{Cl}{\underset{H}{C}}{\cdot}$$

Repeat Step [3] over and over. | new C–C bond |

Termination:

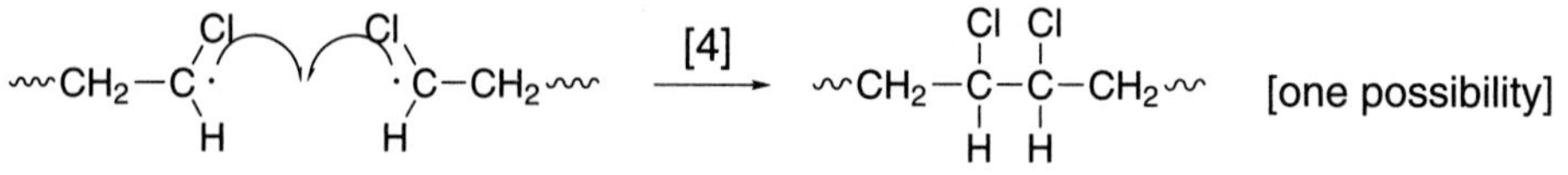

$$\sim\!\!CH_2{-}\overset{Cl}{\underset{H}{C}}{\cdot} \quad {\cdot}\overset{Cl}{\underset{H}{C}}{-}CH_2\!\!\sim \xrightarrow{[4]} \sim\!\!CH_2{-}\overset{Cl}{\underset{H}{C}}{-}\overset{Cl}{\underset{H}{C}}{-}CH_2\!\!\sim \quad \text{[one possibility]}$$

**15.35**

a. increasing bond strength: b < c < a

b. and c.

| · CH$_2$-C(CH$_3$)(H)-CHCH$_3$(H) | H-CH$_2$-C(CH$_3$)(H)-CHCH$_3$(·) | H-CH$_2$-C(CH$_3$)(·)-CHCH$_3$(H) |
|---|---|---|
| **1° radical**<br>**least stable** | **2° radical**<br>**intermediate stability** | **3° radical**<br>**most stable** |

d. increasing ease of H abstraction: a < c < b

**15.36** Use the directions from Answer 15.2 to rank the radicals.

a.

| (CH$_3$)$_2$CHCH$_2$CH(CH$_3$)ĊH$_2$ | (CH$_3$)$_2$CHĊHCH(CH$_3$)$_2$ | (CH$_3$)$_2$ĊCH$_2$CH(CH$_3$)$_2$ |
|---|---|---|
| **1° radical**<br>**least stable** | **2° radical**<br>**intermediate stability** | **3° radical**<br>**most stable** |

b.

| ⬡–ĊHCH$_3$ | ⬡–ĊCH$_2$CH$_3$ | ⬡=CHĊH$_2$ |
|---|---|---|
| **2° radical**<br>**least stable** | **3° radical**<br>**intermediate stability** | **allylic radical**<br>**most stable** |

**15.37** Draw the radical formed by cleavage of the benzylic C–H bond. Then draw all of the resonance structures. Having more resonance structures (five in this case) makes the radical more stable, and the benzylic C–H bond weaker.

benzylic C–H bond
bond dissociation energy = 85 kcal/mol

**15.38**

$H_a$ = bonded to an $sp^3$ 3° carbon
$H_b$ = bonded to an allylic carbon
$H_c$ = bonded to an $sp^3$ 1° carbon
$H_d$ = bonded to an $sp^3$ 2° carbon

| Increasing ease of abstraction: |
|---|
| $H_c < H_d < H_a < H_b$ |

**15.39**

a.

b. (CH$_3$)$_3$CCH$_2$CH$_2$CH$_2$CH$_3$ ⟶ (CH$_3$)$_3$CCH$_2$CH$_2$CH$_2$CH$_2$Cl + (CH$_3$)$_3$CCH$_2$CH$_2$CHClCH$_3$ + (CH$_3$)$_3$CCH$_2$CHClCH$_2$CH$_3$

+ (CH$_3$)$_2$(CH$_2$Cl)CCH$_2$CH$_2$CH$_2$CH$_3$ + (CH$_3$)$_3$CCHClCH$_2$CH$_2$CH$_3$

c.

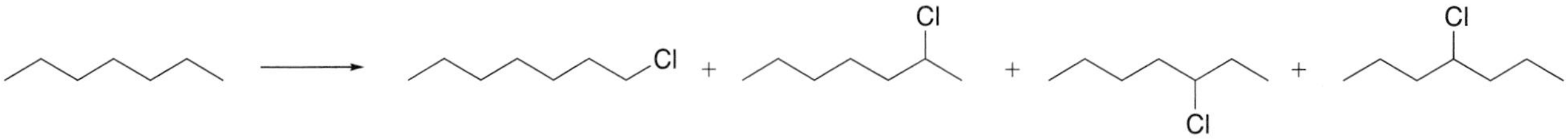

d.

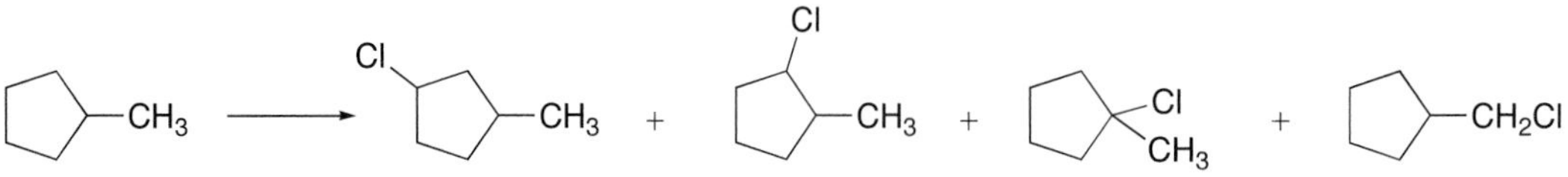

**15.40** To draw the product of bromination:
- Draw out the starting material and find the most reactive C–H bond (on the most substituted C).
- The major product is formed by **cleavage of the *weakest* C–H bond**.

a.

c.

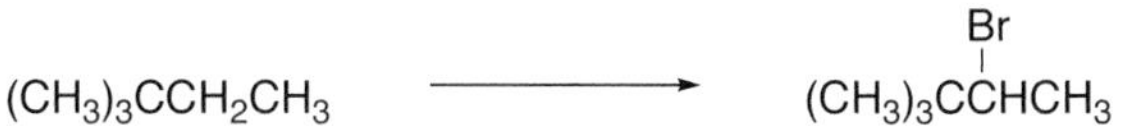

b. 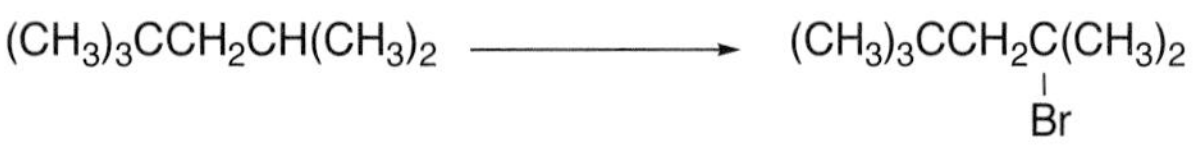  $(CH_3)_3CCH_2CH(CH_3)_2$ ⟶ $(CH_3)_3CCH_2C(CH_3)_2$ with Br

d.  $(CH_3)_3CCH_2CH_3$ ⟶ $(CH_3)_3CCHCH_3$ with Br

**15.41** Draw all of the alkane isomers of $C_6H_{14}$ and their products on chlorination. Then determine which letter corresponds to which alkane.

**A**

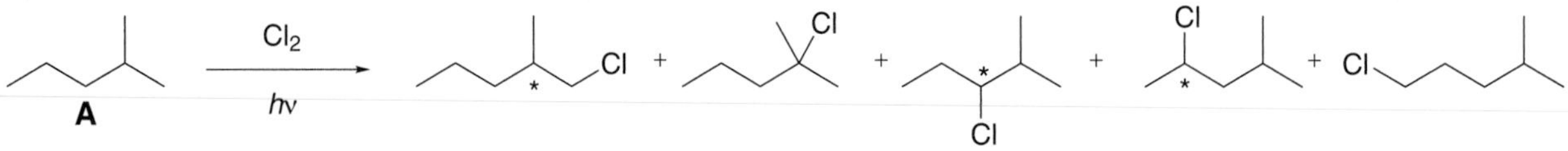

**B**

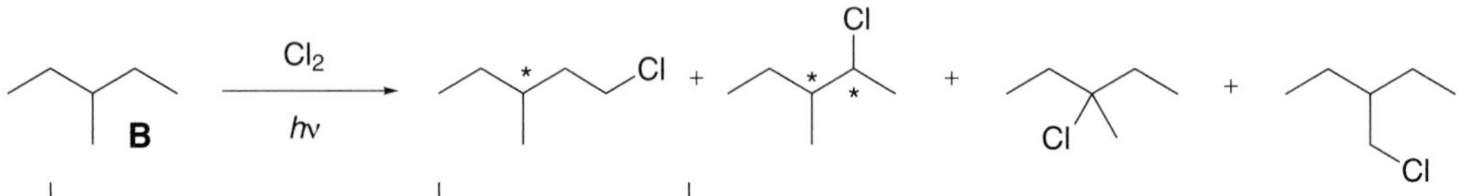

**C**

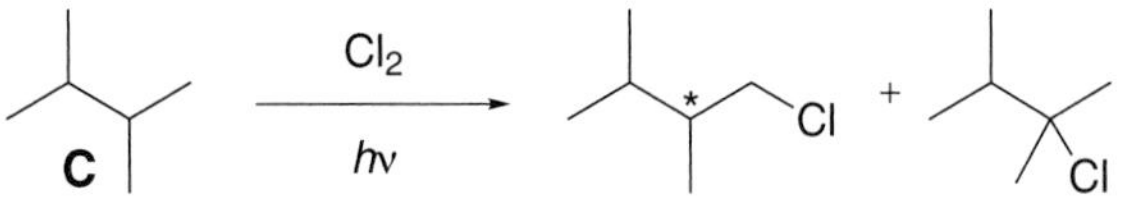

**D**

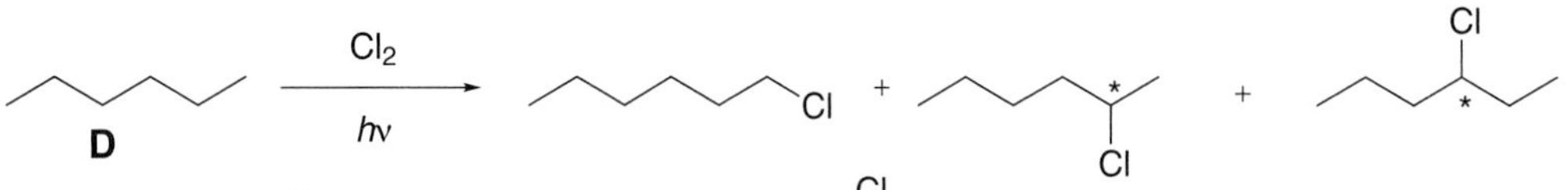

**E**

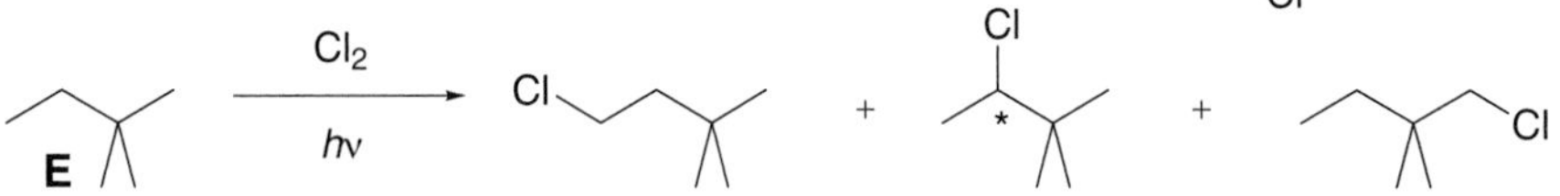

[* = stereogenic center]

**15.42** Halogenation replaces a C–H bond with a C–X bond. To find the alkane needed to make each of the alkyl halides, replace the X with a H.

a.

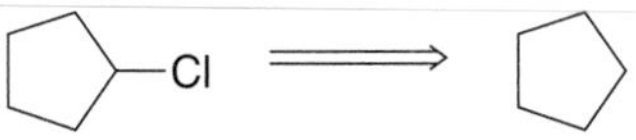

c.

b.

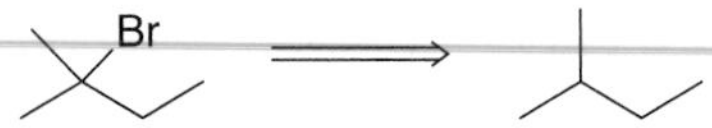

d.  $(CH_3)_3CCH_2Cl$ ⟹ $(CH_3)_3CCH_3$

**15.43** For an alkane to yield one major product on monohalogenation with $Cl_2$, all of the hydrogens must be identical in the starting material. For an alkane to yield one major product on bromination, it must have a more substituted carbon in the starting material.

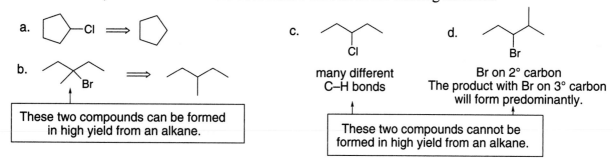

a.

These two compounds can be formed in high yield from an alkane.

c. many different C–H bonds

d. Br on 2° carbon The product with Br on 3° carbon will form predominantly.

These two compounds cannot be formed in high yield from an alkane.

**15.44** Chlorination with two equivalents of $Cl_2$ yields a variety of products.

The desired product is only one of four products formed.

**15.45** Draw the resonance structures by moving the π bonds and the radical.

a.

b.

c.

**15.46** Reaction of an alkene with NBS + $h\nu$ yields allylic substitution products.

a.

b. $CH_3CH_2CH=CHCH_2CH_3 \xrightarrow[h\nu]{NBS} CH_3CH_2CH=CHCHCH_3$ (Br) $+$ $CH_3CH_2CHCH=CHCH_3$ (Br)

c. $(CH_3)_2C=CHCH_3 \xrightarrow[h\nu]{NBS} (CH_3)_2C=CHCH_2Br + BrCH_2C(CH_3)=CHCH_3 + CH_2=C(CH_3)CH(Br)CH_3$
$+$
$(CH_3)_2C(Br)CH=CH_2$

d.

**15.47**

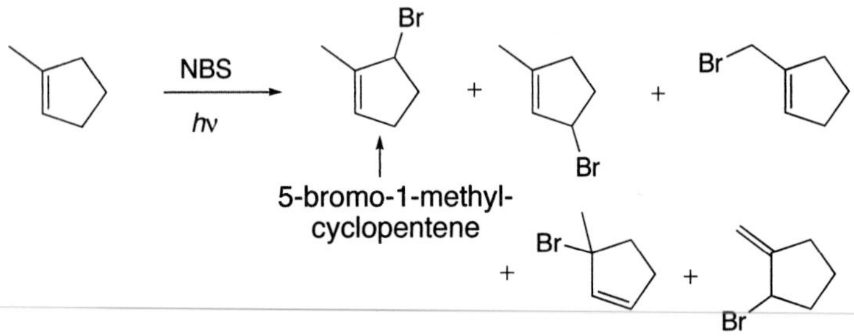

It is not possible to form 5-bromo-1-methylcyclopentene in good yield by allylic bromination because several other products are formed. 1-Methylcyclopentene has three different types of allylic hydrogens (labeled with *), all of which can be removed during radical bromination.

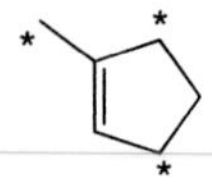

**15.48**

a.

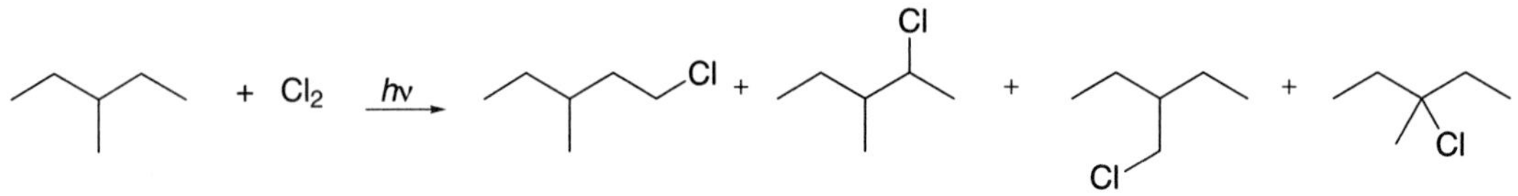

b.

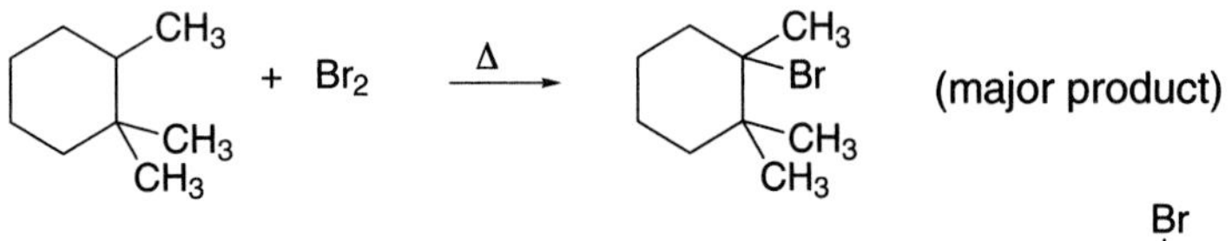

(major product)

c.

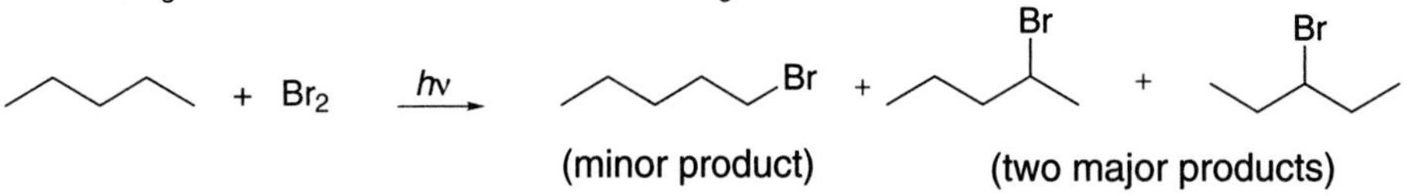

(minor product)     (two major products)

d.

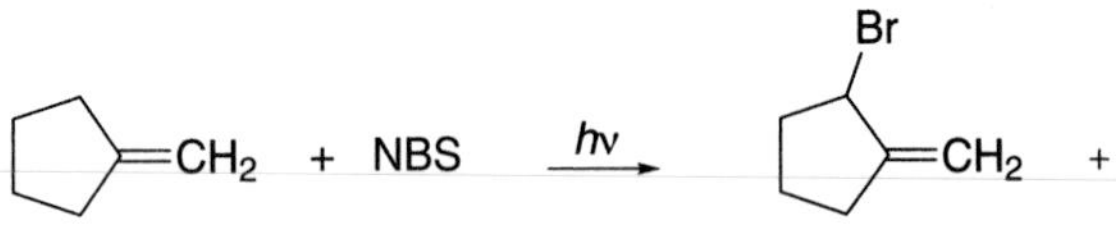

e.

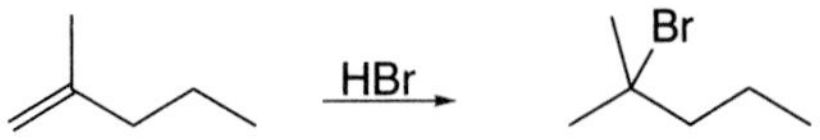

f.

g.

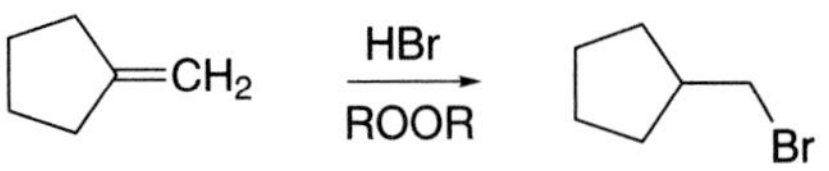

h.

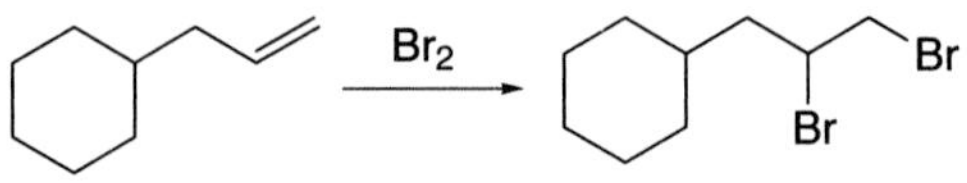

i.

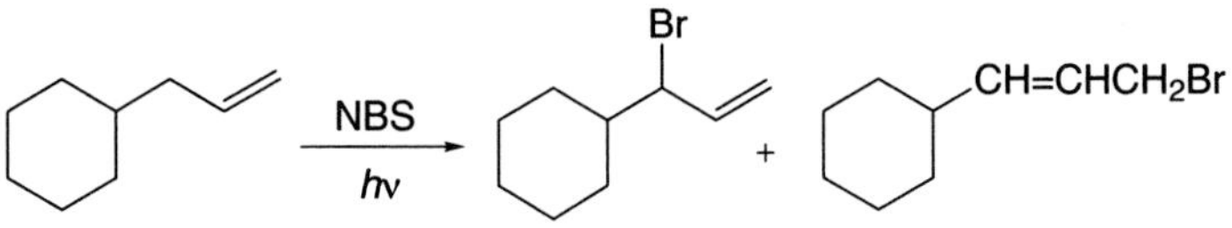

**15.49**

a.

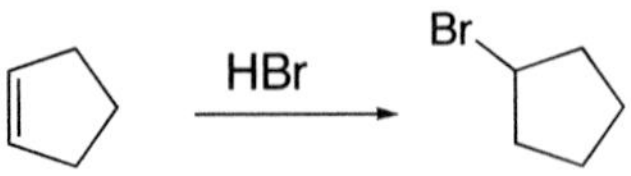

b.

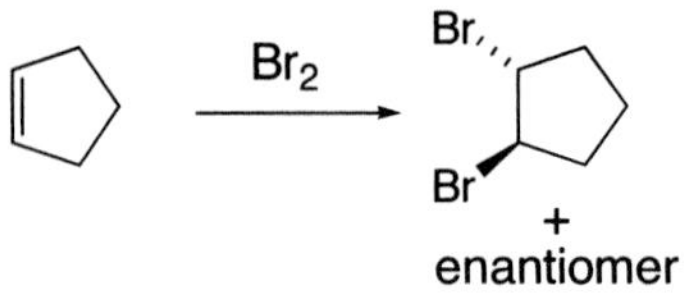

+
enantiomer

c.

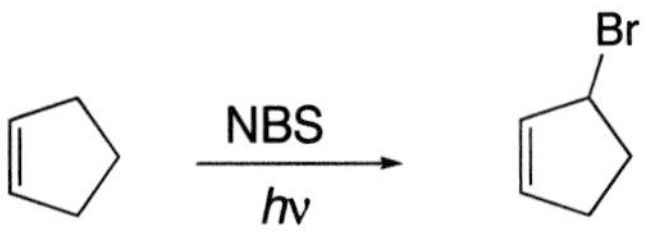

**15.50**

cyclohexanone     acetone

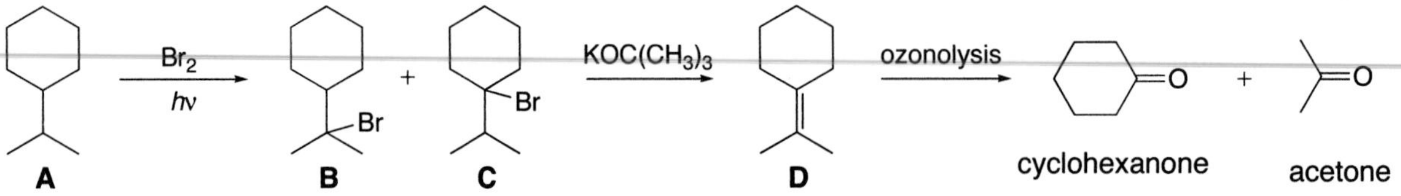

**15.51**

a. 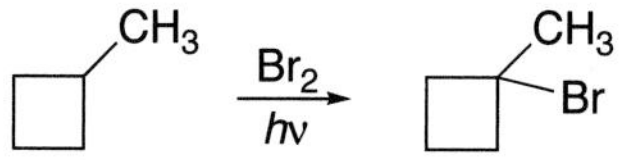

b. 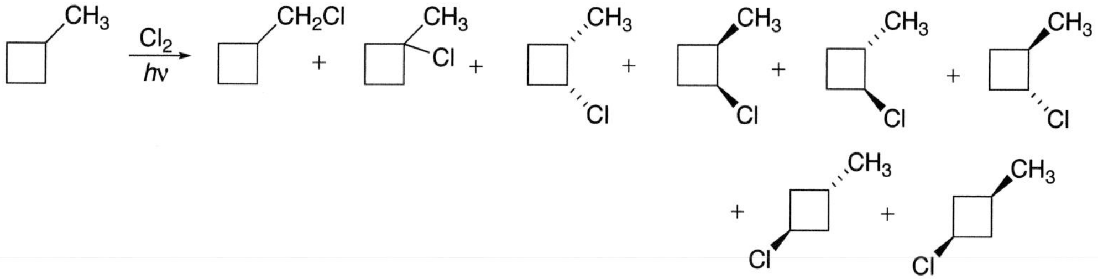

c. 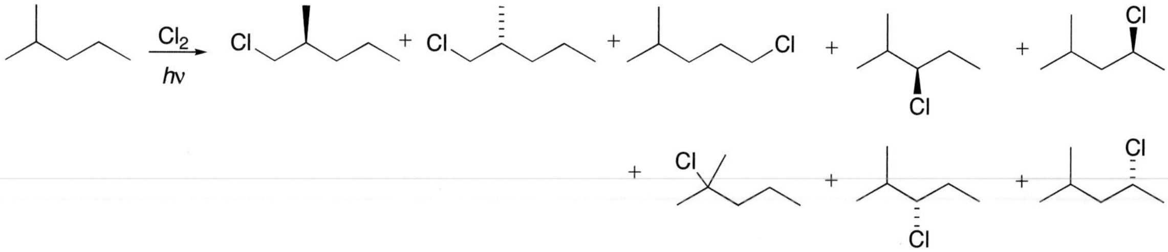

d. 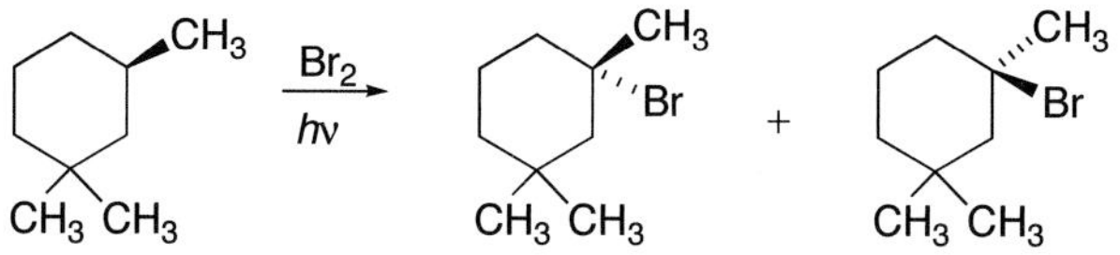

e. 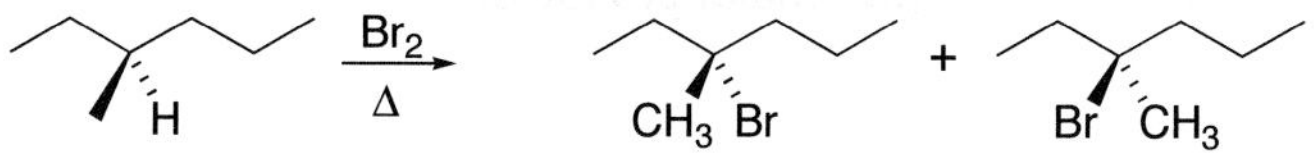

f. 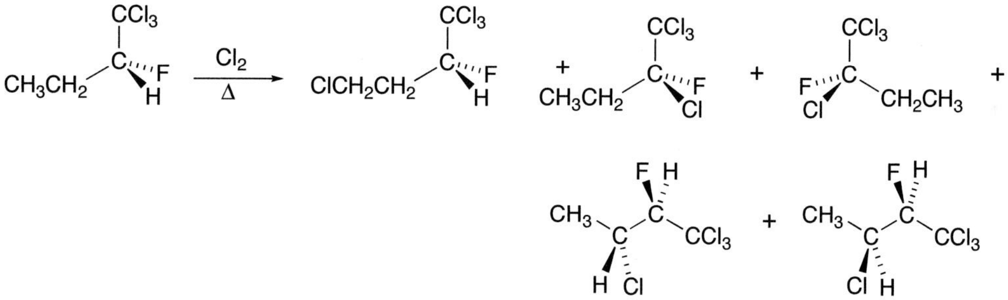

**15.52**

a. 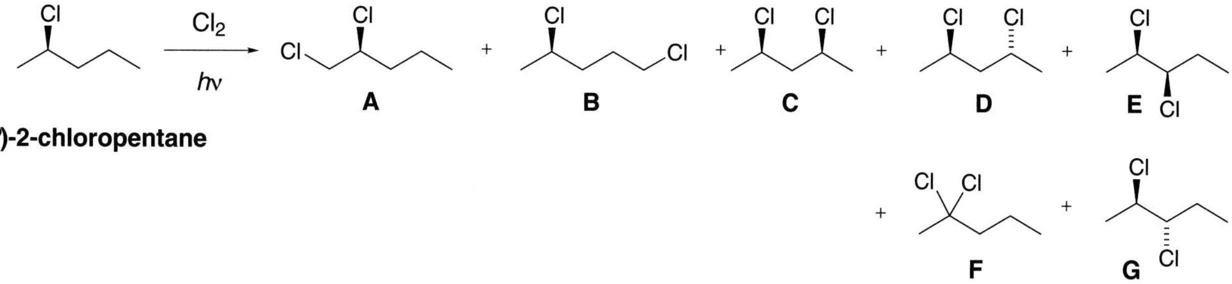
A
B
C
D
E
F
G
(R)-2-chloropentane

b. There would be seven fractions, since each molecule drawn has different physical properties.

c. Fractions **A**, **B**, **D**, **E**, and **G** would show optical activity.

**15.53**

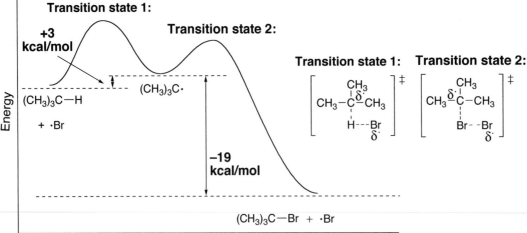

**15.54**

a.

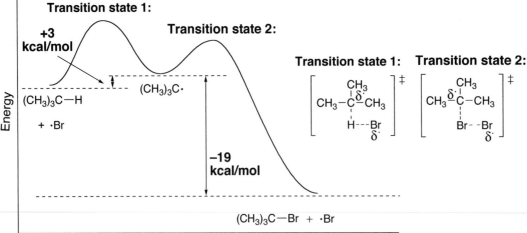

C–H bond broken    Br–Br bond broken    C–Br bond formed    H–Br bond formed
  +91 kcal/mol        +46 kcal/mol          –65 kcal/mol         –88 kcal/mol

total bonds broken = +137 kcal/mol        total bonds formed = –153 kcal/mol        $\Delta H° = -16$ kcal/mol

b.  Initiation:    $:Br\!-\!Br:$ $\xrightarrow[\text{or } \Delta]{h\nu}$ $:Br\cdot$ + $\cdot Br:$

Propagation:    $(CH_3)_3C\!-\!H$ + $\cdot Br:$ $\longrightarrow$ $(CH_3)_3C\cdot$ + $H\!-\!Br:$

c. $\Delta H° =$ (bonds broken) – (bonds formed)
   = (+91 kcal/mol) + (–88 kcal/mol)
   = +3 kcal/mol

$(CH_3)_3C\cdot$ + $:Br\!-\!Br:$ $\longrightarrow$ $(CH_3)_3C\!-\!Br$ + $\cdot Br:$

$\Delta H° =$ (bonds broken) – (bonds formed)
   = (+46 kcal/mol) + (–65 kcal/mol)
   = –19 kcal/mol

Termination:    $:Br\cdot$ + $\cdot Br:$ $\longrightarrow$ $:Br\!-\!Br:$
(one possibility)

d. and e.

**Transition state 1:**

+3 kcal/mol

**Transition state 2:**

Energy

$(CH_3)_3C\!-\!H$
+ $\cdot Br$

$(CH_3)_3C\cdot$

–19 kcal/mol

$(CH_3)_3C\!-\!Br$ + $\cdot Br$

Reaction coordinate

**Transition state 1:**

$\left[ CH_3\!-\!\overset{CH_3}{\underset{H\text{---}Br}{\overset{\delta\cdot}{C}}}\!CH_3 \right]^{\ddagger}$

$\delta\cdot$

**Transition state 2:**

$\left[ CH_3\!-\!\overset{CH_3}{\underset{Br\text{--}Br}{C}}\!CH_3 \right]^{\ddagger}$

$\delta\cdot$

**15.55**

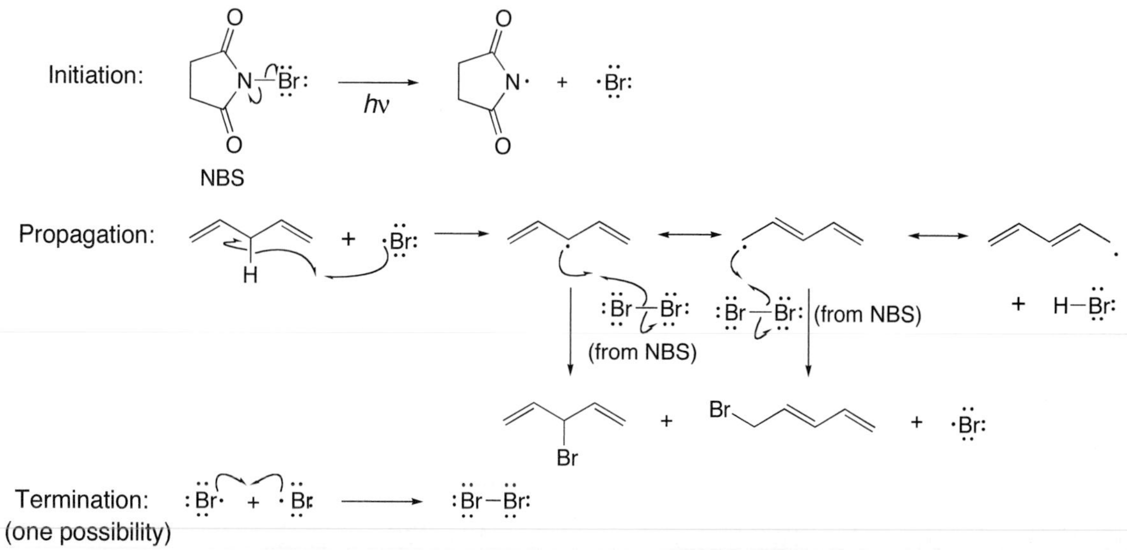

Initiation:

Propagation:

Termination: : $\ddot{B}r\cdot$ + $\cdot\ddot{B}r$ ⟶ : $\ddot{B}r$—$\ddot{B}r$ :
(one possibility)

**15.56** Calculate the $\Delta H°$ for the propagation steps of the reaction of $CH_4$ with $I_2$ to show why it does not occur at an appreciable rate.

$CH_3$—H + $\cdot\ddot{I}$ : ⟶ $\dot{C}H_3$ + H—$\ddot{I}$ : $\Delta H° = +33$ kcal/mol
+104 kcal/mol                              −71 kcal/mol

| This step is highly endothermic, making it difficult for chain propagation to occur over and over again. |

$\dot{C}H_3$ + : $\ddot{I}$—$\ddot{I}$ : ⟶ $CH_3$—I + $\cdot\ddot{I}$ : $\Delta H° = -20$ kcal/mol
      +36 kcal/mol   −56 kcal/mol

**15.57** Calculate $\Delta H°$ for each of these steps, and use these values to explain why this alternate mechanism is unlikely.

[1]   $CH_4$ + : $\ddot{C}l\cdot$ ⟶ $CH_3Cl$ + H $\cdot$
C–H bond broken       C–Cl bond formed   $\Delta H° = +20$ kcal/mol
+104 kcal/mol          −84 kcal/mol

| The $\Delta H°$ for Step [1] is very endothermic, making this mechanism unlikely. |

[2]   H $\cdot$ + $Cl_2$ ⟶ HCl + : $\ddot{C}l\cdot$
Cl–Cl bond broken      H–Cl bond formed   $\Delta H° = -45$ kcal/mol
+58 kcal/mol            −103 kcal/mol

**15.58**

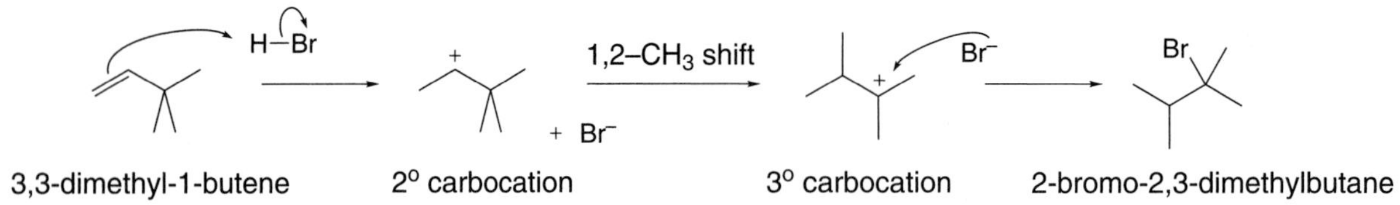

3,3-dimethyl-1-butene → H–Br → 2° carbocation → 1,2–CH₃ shift (+ Br⁻) → 3° carbocation → Br⁻ → 2-bromo-2,3-dimethylbutane

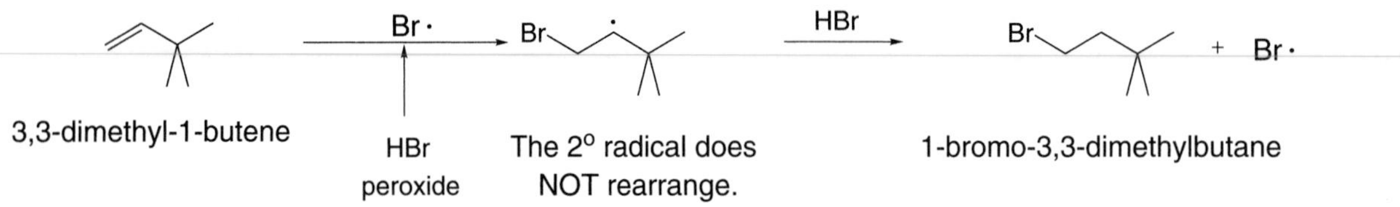

3,3-dimethyl-1-butene → Br· (HBr peroxide) → Br (radical) The 2° radical does NOT rearrange. → HBr → 1-bromo-3,3-dimethylbutane + Br·

Addition of HBr without added peroxide occurs by an ionic mechanism and forms a 2° carbocation, which rearranges to a more stable 3° carbocation. The addition of H⁺ occurs first, followed by Br⁻. Addition of HBr with added peroxide occurs by a radical mechanism and forms a 2° radical that does not rearrange. In the radical mechanism Br• adds first, followed by H•.

**15.59**

a. 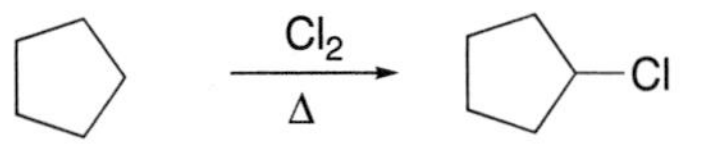 cyclopentane → $Cl_2$, Δ → chlorocyclopentane

b. 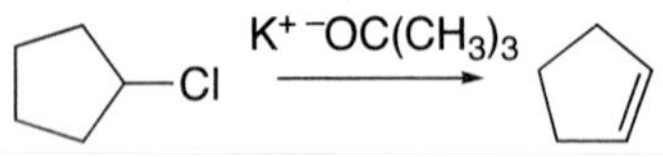 chlorocyclopentane (from a.) → $K^+ \, {}^-OC(CH_3)_3$ → cyclopentene

c. 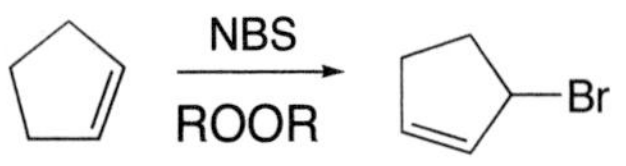 cyclopentene (from b.) → NBS, ROOR → bromocyclopentene

d. 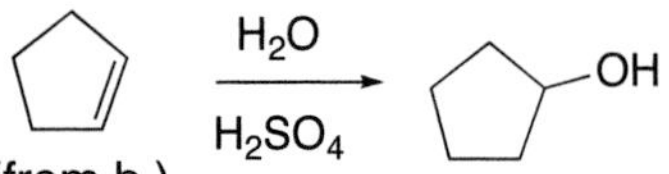 cyclopentene (from b.) → $H_2O$, $H_2SO_4$ → cyclopentanol

e. 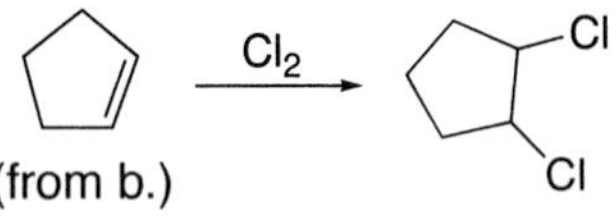 cyclopentene (from b.) → $Cl_2$ → 1,2-dichlorocyclopentane

f. 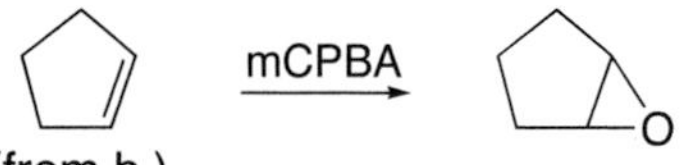 cyclopentene (from b.) → mCPBA → epoxide

g. 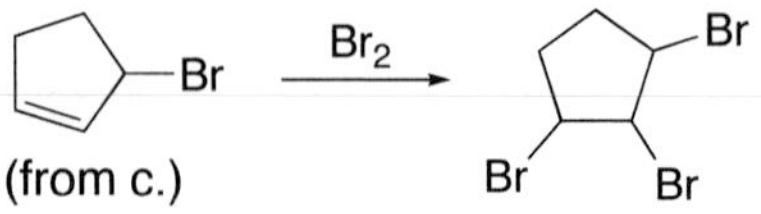 bromocyclopentene (from c.) → $Br_2$ → tribromocyclopentane

h. 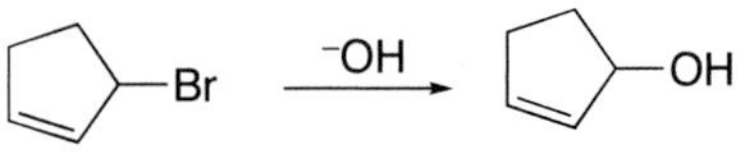 bromocyclopentene (from c.) → ⁻OH → hydroxycyclopentene

i. 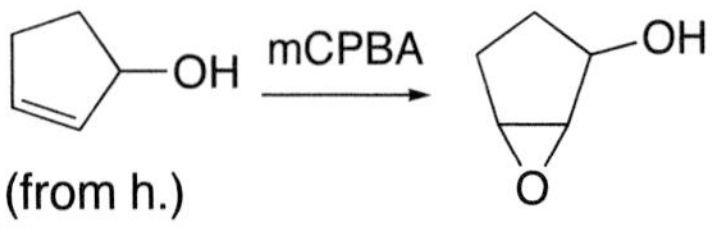 hydroxycyclopentene (from h.) → mCPBA → epoxy alcohol

j. 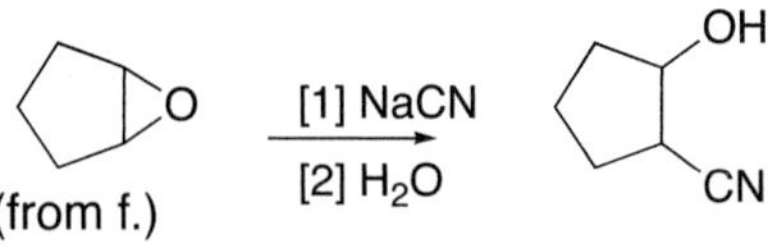 epoxide (from f.) → [1] NaCN [2] $H_2O$ → trans-2-hydroxycyclopentane-CN

## 15.60

a. $CH_3CH_2CH_3$ $\xrightarrow{\text{Br}_2}{h\nu}$ $CH_3\overset{\displaystyle |}{\underset{\displaystyle Br}{C}}HCH_3$ $\xrightarrow{K^+ {}^-OC(CH_3)_3}$ $CH_3CH=CH_2$ $\xrightarrow{\text{Br}_2}$ $CH_3-\overset{Br}{\underset{H}{C}}-\overset{Br}{\underset{H}{C}}-H$ $\xrightarrow[\substack{(2\ \text{equiv})\\ DMSO}]{K^+ {}^-OC(CH_3)_3}$ $CH_3C\equiv CH$

b. (isopropyl bromide) $\xrightarrow{K^+ {}^-OC(CH_3)_3}$ (propene) $\xrightarrow[\text{ROOR}]{\text{HBr}}$ (1-bromopropane)

c. $HC\equiv CH$ $\xrightarrow{NaH}$ $HC\equiv C^-$ $\xrightarrow{CH_3CH_2Br}$ (1-butyne) $\xrightarrow[\text{Lindlar catalyst}]{H_2}$ (1-butene) $\xrightarrow[\text{ROOR}]{\text{HBr}}$ (1-bromobutane)

d. (methylcyclohexane) $\xrightarrow{\text{Br}_2}{h\nu}$ (1-bromo-1-methylcyclohexane) $\xrightarrow{K^+ {}^-OC(CH_3)_3}$ (1-methylcyclohexene) $\xrightarrow[\text{H}_2\text{SO}_4]{H_2O}$ (1-methylcyclohexanol, OH)

major product

$\uparrow H_2O$

## 15.61

a. $CH_3-CH_3$ $\xrightarrow{\text{Br}_2}{h\nu}$ $CH_3-CH_2Br$ $\xrightarrow{K^+ {}^-OC(CH_3)_3}$ $CH_2=CH_2$ $\xrightarrow{\text{Br}_2}$ $BrCH_2-CH_2Br$ $\xrightarrow{2\ NaNH_2}$ $HC\equiv CH$

b. $HC\equiv CH$ $\xrightarrow{NaH}$ $HC\equiv C^-$ $\xrightarrow[\text{(from a.)}]{CH_3-CH_2Br}$ $HC\equiv CCH_2CH_3$
   (from a.)

c. $CH_2=CH_2$ $\xrightarrow{mCPBA}$ (epoxide) $\xrightarrow[\substack{\text{(from b.)}\\ [2]\ H_2O}]{[1]\ HC\equiv C^-}$ $HC\equiv CCH_2CH_2OH$
   (from a.)

d. $HC\equiv CCH_2CH_3$ $\xrightarrow{NaH}$ $^-C\equiv CCH_2CH_3$ $\xrightarrow[\text{(from a.)}]{CH_3-CH_2Br}$ $CH_3CH_2-C\equiv C-CH_2CH_3$ $\xrightarrow[\text{NH}_3]{Na}$ (trans-3-hexene)
   (from b.)

e. $CH_3CH_2-C\equiv C-CH_2CH_3$ $\xrightarrow[\substack{H_2SO_4\\ HgSO_4}]{H_2O}$ (3-hexanone)
   (from d.)

## 15.62

O$_2$ abstracts a H here.

arachidonic acid

$\cdot \ddot{O}-\ddot{O} \cdot$

+ $H\ddot{O}\ddot{O}\cdot$

$\cdot \ddot{O}-\ddot{O} \cdot$

$R-H$

another molecule of arachidonic acid

5-HPETE

This process is repeated.

**15.63**

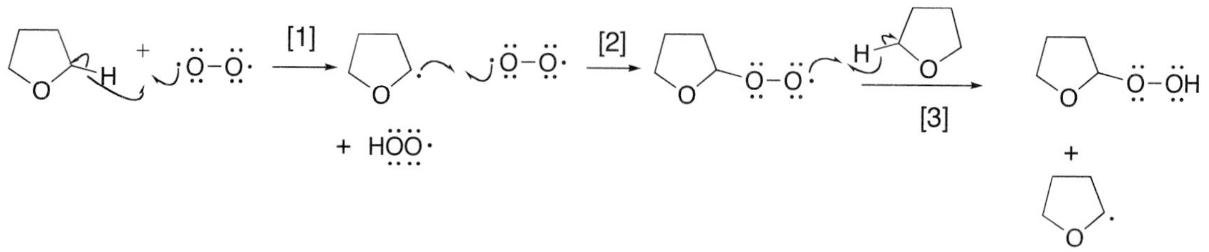

+ HÖÖ·

Then, repeat
Steps [2] and [3].

**15.64**

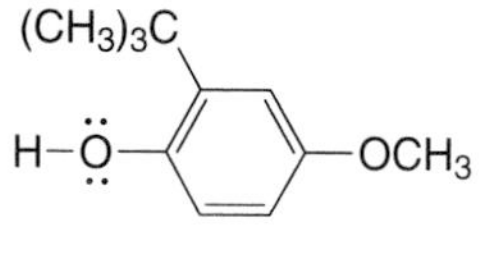

BHA

Abstraction of the phenol H
produces a resonance-stabilized radical.

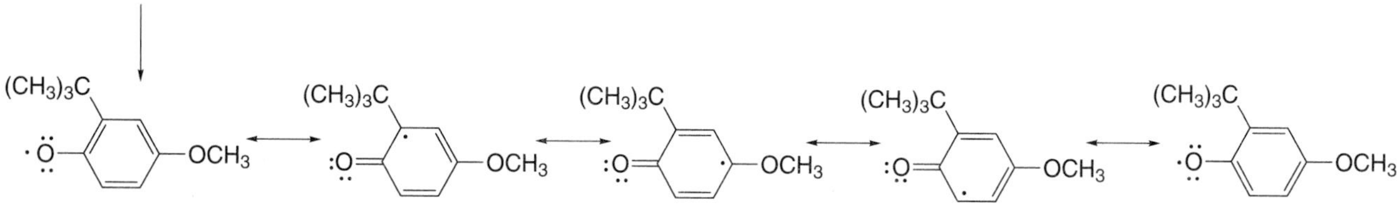

**15.65** Abstraction of the labeled H forms a highly resonance-stabilized radical. Four of the possible resonance structures are drawn.

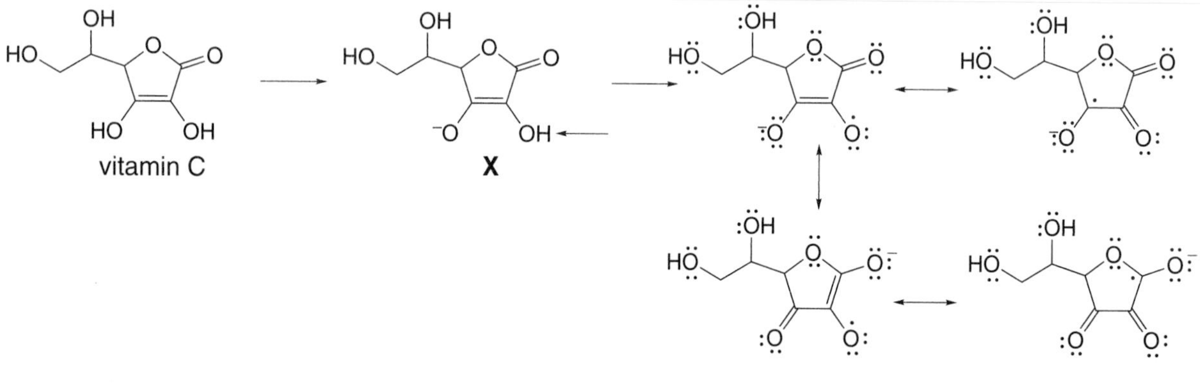

vitamin C

**X**

**15.66** The monomers used in polymerization always contain double bonds.

a.

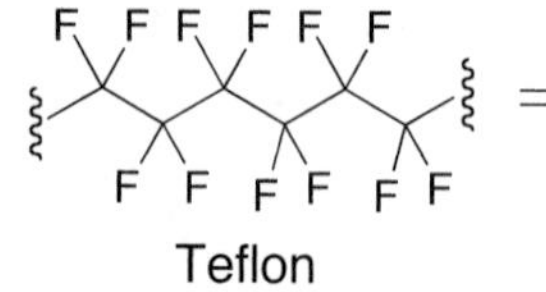

$CF_2=CF_2$

Teflon
(nonstick surface coating)

b.

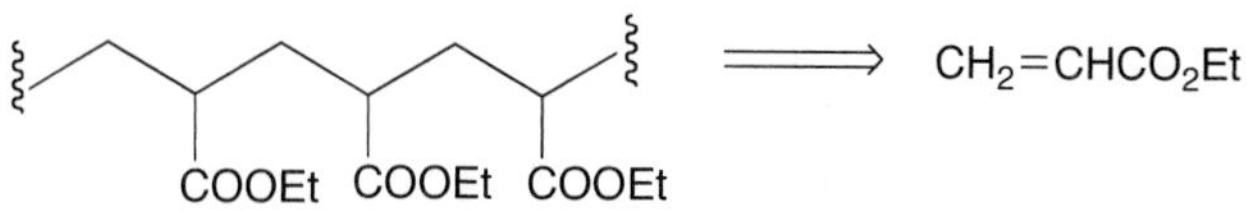

$CH_2=CHCO_2Et$

poly(ethyl acrylate)
(used in Latex paints)    Et = $CH_2CH_3$

**15.67**

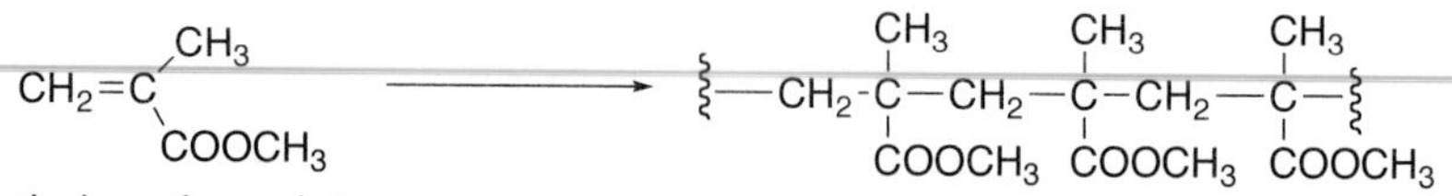

methyl methacrylate

## 15.68

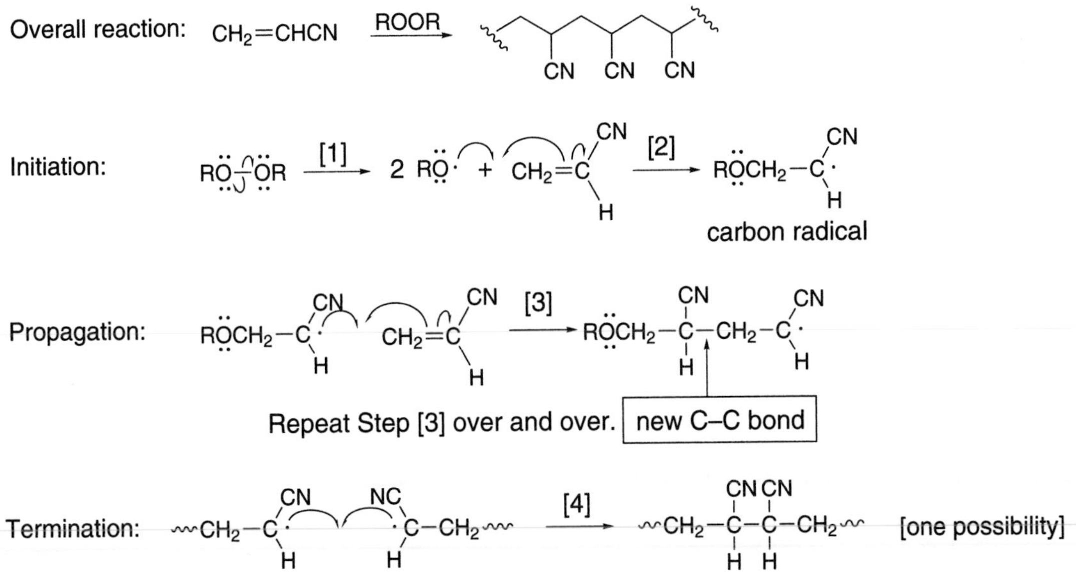

Overall reaction: $CH_2=CHCN$ $\xrightarrow{\text{ROOR}}$

Initiation:

carbon radical

Propagation:

Repeat Step [3] over and over. | new C–C bond |

Termination: [one possibility]

## 15.69

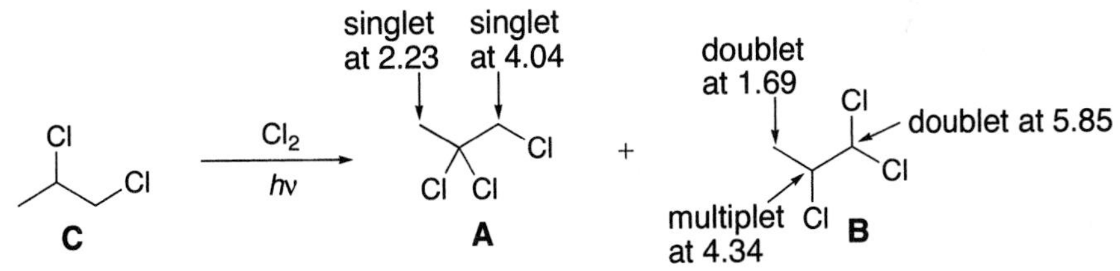

singlet at 2.23  singlet at 4.04

doublet at 1.69

doublet at 5.85

multiplet at 4.34

**C**   **A**   **B**

## 15.70

a. $CH_3CH_3$ $\xrightarrow[h\nu]{Cl_2}$

equivalent H's singlet
$ClCH_2CH_2Cl$
**X**

+

doublet  quartet
$CH_3CHCl_2$
**Y**

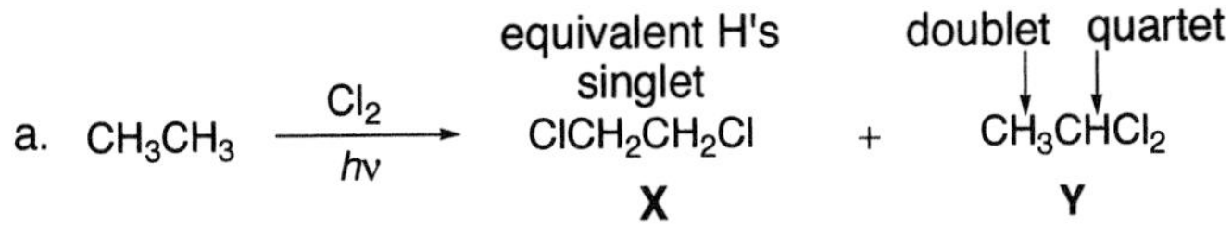

b. Initiation:

$$:\!\overset{..}{Cl}\!-\!\overset{..}{Cl}\!: \xrightarrow[\text{or }\Delta]{h\nu} :\!\overset{..}{Cl}\!\cdot \; + \; \cdot\overset{..}{Cl}\!:$$

Propagation:

$$CH_3\!-\!\underset{H}{\overset{H}{C}}\!-\!H \; + \; \cdot\overset{..}{Cl}\!: \longrightarrow CH_3\!-\!\underset{H}{\overset{H}{C}}\!\cdot \; + \; H\!-\!\overset{..}{Cl}\!:$$

$$CH_3\!-\!\underset{H}{\overset{H}{C}}\!\cdot \; + \; :\!\overset{..}{Cl}\!-\!\overset{..}{Cl}\!: \longrightarrow CH_3\!-\!\underset{H}{\overset{H}{C}}\!-\!Cl \; + \; \cdot\overset{..}{Cl}\!:$$

Formation of **Y**:

$$CH_3\!-\!\overset{H}{CH}\!-\!Cl \quad \overset{..}{Cl}: \longrightarrow CH_3\!-\!\overset{\cdot}{C}H\!-\!Cl \; + \; H\!-\!\overset{..}{Cl}\!:$$

$$CH_3\!-\!\overset{\cdot}{C}H\!-\!Cl \quad :\!\overset{..}{Cl}\!-\!\overset{..}{Cl}\!: \longrightarrow CH_3CHCl_2 \; + \; \cdot\overset{..}{Cl}\!:$$
$$\mathbf{Y}$$

Formation of **X**:

$$H\!-\!CH_2\!-\!CH_2Cl \quad \overset{..}{Cl}: \longrightarrow \cdot CH_2\!-\!CH_2Cl \; + \; H\!-\!\overset{..}{Cl}\!:$$

$$\overset{\cdot}{C}H_2\!-\!CH_2Cl \quad :\!\overset{..}{Cl}\!-\!\overset{..}{Cl}\!: \longrightarrow ClCH_2CH_2Cl \; + \; \cdot\overset{..}{Cl}$$
$$\mathbf{X}$$

Termination:

$$:\!\overset{..}{Cl}\!\cdot \; + \; \cdot\overset{..}{Cl}\!: \longrightarrow :\!\overset{..}{Cl}\!-\!\overset{..}{Cl}\!:$$

## 15.71

$$RO\!-\!OR \longrightarrow R\overset{..}{O}\!\cdot \; + \; \cdot\overset{..}{O}R$$

$$+ \; H\!-\!\overset{..}{O}R$$

(Repeat Steps [2] and [3])

## 15.72

Initiation:

$$R_3SnH \; + \; \cdot Z \longrightarrow R_3Sn\cdot \; + \; HZ$$

Propagation:

## Chapter 16: Conjugation, Resonance, and Dienes

### ♦ Conjugation and delocalization of electron density

- The overlap of $p$ orbitals on three or more adjacent atoms allows electron density to delocalize, thus adding stability (16.1).
- An allyl carbocation ($CH_2=CHCH_2^+$) is more stable than a 1° carbocation because of $p$ orbital overlap (16.2).
- In any system X=Y–Z:, Z is $sp^2$ hybridized to allow the lone pair to occupy a $p$ orbital, making the system conjugated (16.5).

### ♦ Four common examples of resonance (16.3)

[1] The three atom "allyl" system:

$$X=Y-Z \longleftrightarrow X-Y=Z \qquad \boxed{* = +, -, \cdot, \text{ or } \cdot\cdot}$$

[2] Conjugated double bonds:

[3] Cations having a positive charge adjacent to a lone pair:

$$\ddot{X}-\overset{+}{Y} \longleftrightarrow \overset{+}{X}=Y$$

[4] Double bonds having one atom more electronegative than the other:

$$X=\overset{\frown}{Y} \longleftrightarrow \overset{+}{X}-\overset{-}{Y}: \qquad \boxed{\text{Electronegativity of Y} > \text{X}}$$

### ♦ Rules on evaluating the relative "stability" of resonance structures (16.4)

[1] Structures with more bonds and fewer charges are more stable.

all neutral atoms
one more bond

charge separation

[2] Structures in which every atom has an octet are more stable.

All 2$^{nd}$ row elements have an octet.

[3] Structures that place a negative charge on a more electronegative element are more stable.

The (–) charge is on the more electronegative O atom.

## ♦ The unusual properties of conjugated dienes

[1] The C–C σ bond joining the two double bonds is unusually short (16.8).

[2] Conjugated dienes are more stable than similar isolated dienes. $\Delta H°$ of hydrogenation is smaller for a conjugated diene than for an isolated diene converted to the same product (16.9).

[3] The reactions are unusual:
- Electrophilic addition affords products of 1,2-addition and 1,4-addition (16.10, 16.11).
- Conjugated dienes undergo the Diels–Alder reaction, a reaction that does not occur with isolated dienes (16.12–16.14).

[4] Conjugated dienes absorb UV light in the 200–400 nm region. As the number of conjugated π bonds increases, the absorption shifts to longer wavelength (16.15).

## ♦ Reactions of conjugated dienes

[1] Electrophilic addition of HX (X = halogen) (16.10–16.11)

- The mechanism has two steps.
- Markovnikov's rule is followed. Addition of $H^+$ forms the more stable allylic carbocation.
- The 1,2-product is the kinetic product. When $H^+$ adds to the double bond, $X^-$ adds to the end of the allylic carbocation to which it is closer (C2 not C4). The kinetic product is formed faster at low temperature.
- The thermodynamic product has the more substituted, more stable double bond. The thermodynamic product predominates at equilibrium. With 1,3-butadiene, the thermodynamic product is the 1,4-product.

[2] Diels–Alder reaction (16.12–16.14)

**1,3-diene   dienophile**

- The reaction forms two σ and one π bond in a six-membered ring.
- The reaction is initiated by heat.
- The mechanism is concerted: all bonds are broken and formed in a single step.
- The diene must react in the s-cis conformation (16.13A).
- Electron-withdrawing groups in the dienophile increase the reaction rate (16.13B).
- The stereochemistry of the dienophile is retained in the product (16.13C).
- Endo products are preferred (16.13D).

Chapter 16: Answers to Problems

**16.1** **Isolated dienes** have two double bonds separated by two or more σ bonds.
**Conjugated dienes** have two double bonds separated by only one σ bond.

a.

b.

c.

d.

One σ bond separates
two double bonds =
**conjugated diene**

Two σ bonds separate
two double bonds =
**isolated diene**

One σ bond separates
two double bonds =
**conjugated diene**

Four σ bonds separate
two double bonds =
**isolated diene**

**16.2 Conjugation** occurs when there are overlapping *p* orbitals on three or more adjacent atoms. Double
bonds separated by 2 σ bonds are not conjugated.

a. 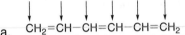 CH₂=CH−CH=CH−CH=CH₂

All of the carbon atoms are *sp²*
hybridized. Each π bond is
separated by only one σ bond.
**conjugated**

c.

The two π bonds are
separated by only one σ bond.
**conjugated**

e.

This carbon is not *sp²*
hybridized.
**NOT conjugated**

b.

The two π bonds are
separated by three σ bonds.
**NOT conjugated**

d.

Three adjacent carbon atoms are *sp²* hybridized
and have an unhybridized *p* orbital.
**conjugated**

**16.3** Two resonance structures differ only in the placement of electrons. All σ bonds stay in the same
place. Nonbonded electrons and π bonds can be moved. To draw the hybrid:
  • Use a dashed line between atoms that have a π bond in one resonance structure and not the other.
  • Use a δ symbol for atoms with a charge or radical in one structure but not the other.

a.

resonance hybrid:

δ⁺     δ⁺

| The + charge is delocalized |
| on two carbons. |

b.

resonance hybrid:

---- δ⁺

δ⁺

| The + charge is delocalized |
| on two carbons. |

c.

resonance hybrid:

δ⁺     δ⁺

| The + charge is delocalized |
| on two carbons. |

**16.4** Each different kind of carbon atom will give a different $^{13}C$ signal. When a carbocation is delocalized as in structure **B**, carbons become equivalent.

The two $\delta^+$ carbons are identical.

**A**
5 different kinds of C
**5 $^{13}C$ NMR signals**

**B**
4 different kinds of C
**4 $^{13}C$ NMR signals**

**16.5** $S_N1$ reactions proceed via a carbocation intermediate. Draw the carbocation formed on loss of Cl and compare. The more stable the carbocation, the faster the $S_N1$ reaction.

3-chloro-1-propene  $CH_2=CHCH_2Cl$
more reactive

resonance-stabilized carbocation

Two resonance structures delocalize the positive charge on 2 C's making 3-chloro-1-propene more reactive.

$CH_3CH_2CH_2Cl$ is a 1° halide, which does not react by an $S_N1$ reaction because cleavage of the C–Cl bond forms a highly unstable 1° carbocation.

1-chloropropane  $CH_3CH_2CH_2Cl$
less reactive

$CH_3CH_2\overset{+}{CH_2}$

only one Lewis structure
**very unstable**

**16.6**

a.

Move the charge and the double bond.

b.

Move the charge and the double bond.

c.

Move the lone pair.

d.

Move the charge and the double bond.

**16.7** To compare the resonance structures remember:

- Resonance structures with **more bonds** are better.
- Resonance structures in which **every atom has an octet** are better.
- Resonance structures with **neutral atoms** are better than those with charge separation.
- Resonance structures that place a **negative charge on a more electronegative atom** are better.

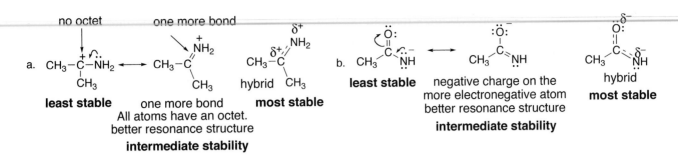

a.
**least stable**      **one more bond**      **most stable**
All atoms have an octet.
better resonance structure
**intermediate stability**

b.   **least stable**   negative charge on the   **most stable**
more electronegative atom
better resonance structure
**intermediate stability**

**16.8**

a.
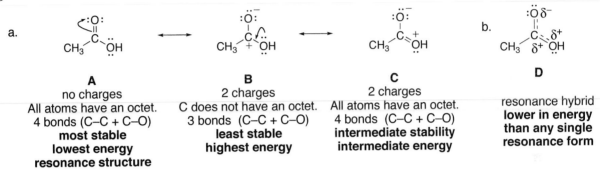

| **A** | **B** | **C** | **D** |
|---|---|---|---|
| no charges | 2 charges | 2 charges | resonance hybrid |
| All atoms have an octet. | C does not have an octet. | All atoms have an octet. | **lower in energy** |
| 4 bonds (C–C + C–O) | 3 bonds (C–C + C–O) | 4 bonds (C–C + C–O) | **than any single** |
| **most stable** | **least stable** | **intermediate stability** | **resonance form** |
| **lowest energy** | **highest energy** | **intermediate energy** | |
| resonance structure | | | |

c.  In order of increasing energy: **D < A < C < B**

**16.9** Remember that in any allyl system, there must be *p* orbitals to delocalize the lone pair.

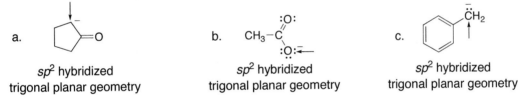

a.
*sp²* hybridized
trigonal planar geometry

b.
*sp²* hybridized
trigonal planar geometry

c.
*sp²* hybridized
trigonal planar geometry

**16.10**   The **s-cis** conformation has two double bonds on the **same side** of the single bond.
The **s-trans** conformation has two double bonds on **opposite sides** of the single bond.

a. (2*E*,4*E*)-2,4-**octadiene** in the *s*-trans conformation

double bonds on opposite sides
**s-trans**

b. (3*E*,5*Z*)- 3,5-**nonadiene** in the *s*-cis conformation

*Z*
*E*
double bonds on the same side
**s-cis**

c. (3*Z*,5*Z*)-4,5-dimethyl-3,5-**decadiene** in
both the *s*-cis and *s*-trans conformation

*s*-cis   →
*Z*
*Z*

*s*-trans   →
*Z*
*Z*

**16.11**  Bond length depends on hybridization and percent *s*-character.  Bonds with a higher percent *s*-character have smaller orbitals and are shorter.

HC≡C–C≡CH                CH₂=CH–CH=CH₂                CH₃–CH₃

*sp* hybridized carbons        *sp²* hybridized carbons        *sp³* hybridized carbons
50% *s*-character             33% *s*-character             25% *s*-character
**shortest bond**           **intermediate length bond**      **longest bond**

**16.12**  Two equivalent resonance structures delocalize the π bond and the negative charge.

$$\left[ CH_3-C\begin{matrix}\ddot{O}: \\ \ddot{O}:^- \end{matrix} \longleftrightarrow CH_3-C\begin{matrix}:\ddot{O}:^- \\ \ddot{O}: \end{matrix} \right]$$

hybrid:  CH₃–C with ⟶ These bond lengths are equal because they are identical.

**16.13**  The **less stable** (higher energy) **diene** has the **larger heat of hydrogenation.**  Isolated dienes are higher in energy than conjugated dienes, so they will have a larger heat of hydrogenation.

a.

or

Double bonds separated by
one σ bond = **conjugated diene**
**smaller** heat of hydrogenation

Double bonds separated by
two σ bonds = **isolated diene**
**larger** heat of hydrogenation

b.

or

Double bonds separated by
one σ bond = **conjugated diene**
**smaller** heat of hydrogenation

Double bonds separated by
two σ bonds = **isolated diene**
**larger** heat of hydrogenation

**16.14**  Isolated dienes are higher in energy than conjugated dienes.  Compare the location of the double bonds in the compounds below.

0 conjugated double bonds
**least stable**

2 conjugated double bonds
**intermediate stability**

3 conjugated double bonds
**most stable**

**16.15**  Conjugated dienes react with HX to form 1,2- and 1,4-products.

a.   CH₃–CH=CH–CH=CH–CH₃   →(HCl)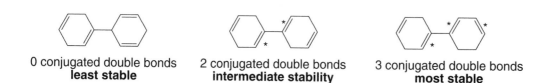

CH₃–CH–CH–CH=CH–CH₃   +   CH₃–CH–CH=CH–CH–CH₃
     H  Cl                          H                   Cl

**1,2-product**                    **1,4-product**

b.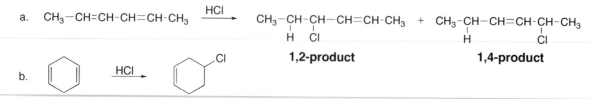

**isolated diene**

c.

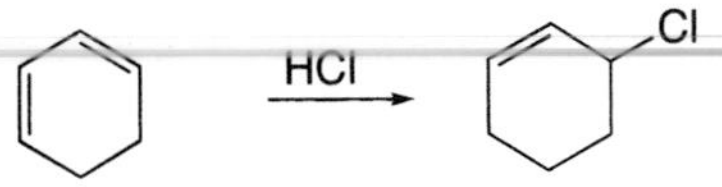

d.

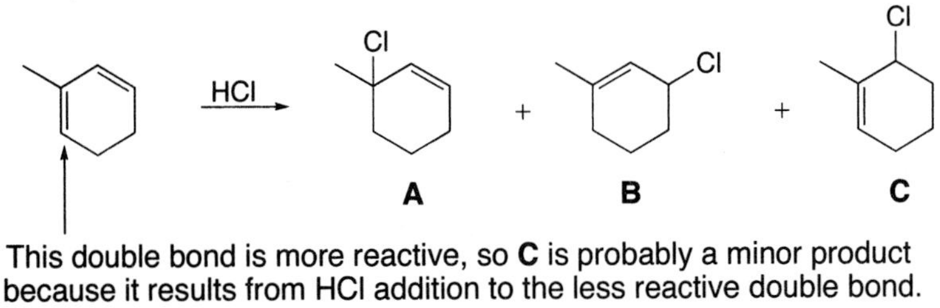

This double bond is more reactive, so **C** is probably a minor product because it results from HCl addition to the less reactive double bond.

**16.16** The mechanism for addition of DCl has two steps:
[1] **Addition of D$^+$** forms a resonance-stabilized carbocation.
[2] **Nucleophilic attack of Cl$^-$** forms 1,2- and 1,4-products.

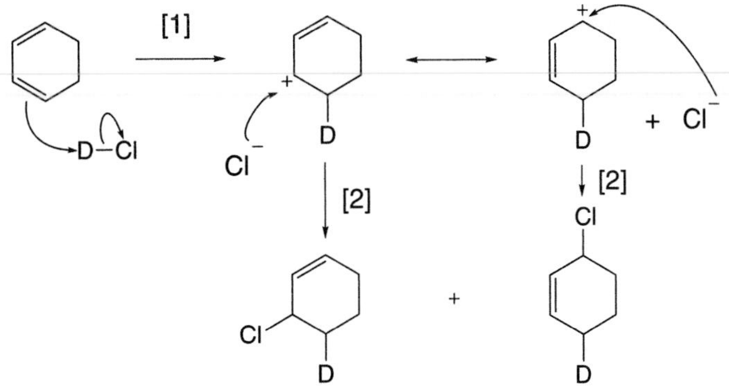

**16.17** Label the products as 1,2- or 1,4-products. The 1,2-product is the kinetic product, and the 1,4-product, which has the more substituted double bond, is the thermodynamic product.

This is C1.

This is C4.

The H added here. The H added here.

1,2-product
kinetic product

1,4-product
thermodynamic product

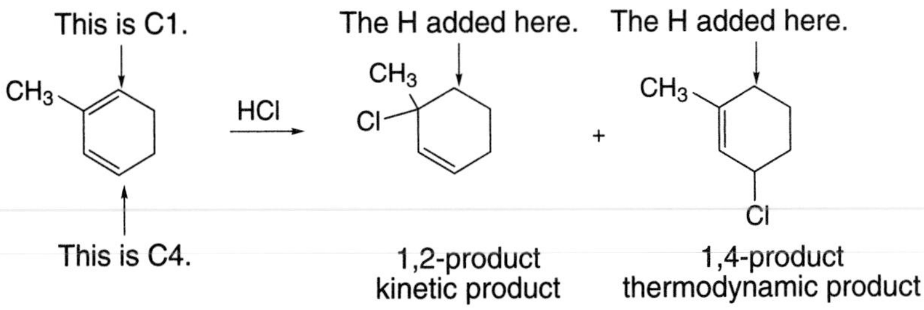

**16.18** To draw the products of a Diels–Alder reaction:
[1] Find the 1,3-diene and the dienophile.
[2] Arrange them so the diene is on the left and the dienophile is on the right.
[3] Cleave three bonds and use arrows to show where the new bonds will be formed.

a.

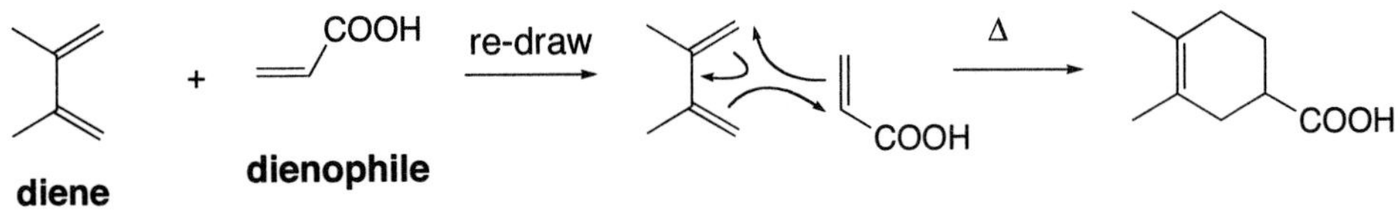

diene      dienophile

b.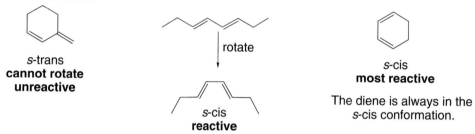

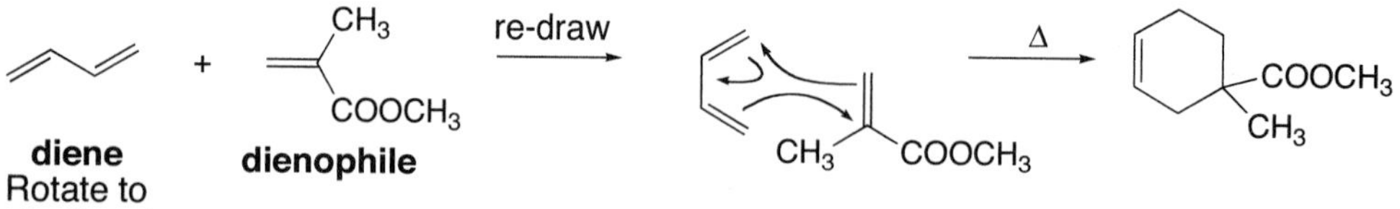

**diene**
Rotate to
make it *s*-cis.

**dienophile**

c.

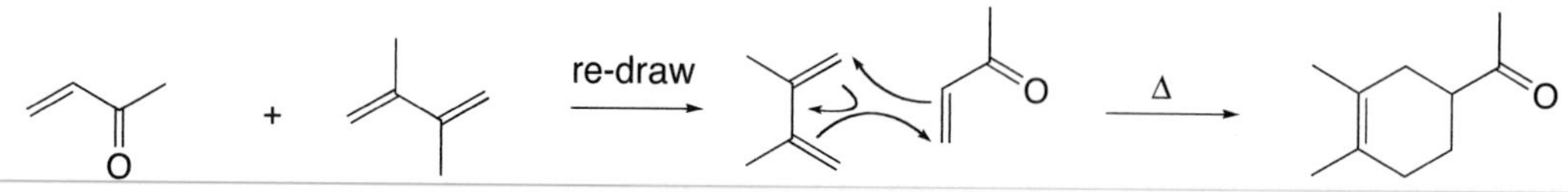

**dienophile**

**diene**
Rotate to
make it *s*-cis.

**16.19** For a diene to be reactive in a Diels–Alder reaction, **a diene must be able to adopt an *s*-cis conformation.**

*s*-trans
**cannot rotate
unreactive**

| rotate

*s*-cis
**reactive**

*s*-cis
**most reactive**

The diene is always in the
*s*-cis conformation.

**16.20**

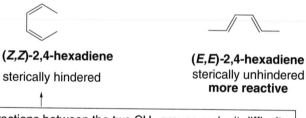

**(Z,Z)-2,4-hexadiene**
sterically hindered

**(E,E)-2,4-hexadiene**
sterically unhindered
**more reactive**

Steric interactions between the two $CH_3$ groups make it difficult
for the diene to adopt the needed *s*-cis conformation.

**16.21** Electron-withdrawing substituents in the dienophile increase the reaction rate.

$CH_2=CH_2$

$CH_2=C$ H / COOH

C=C, HOOC / COOH, H / H

no electron-withdrawing groups
**least reactive**

one electron-withdrawing group
**intermediate reactivity**

two electron-withdrawing groups
**most reactive**

**16.22** A cis dienophile forms a cis-substituted cyclohexene.
A trans dienophile forms a trans-substituted cyclohexene.

a.

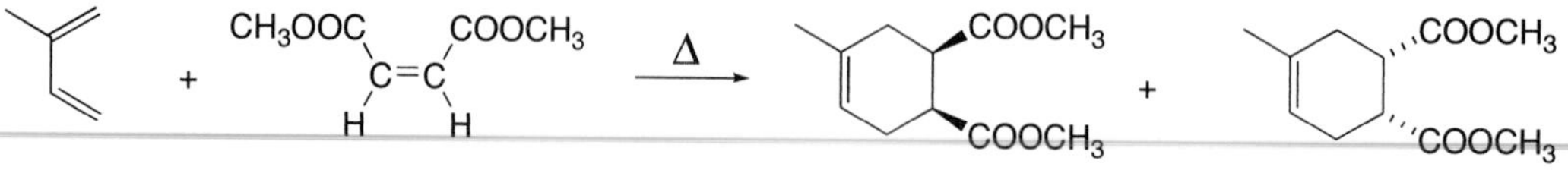

$CH_3OOC$  $COOCH_3$
C=C
H   H

cis dienophile

COOCH₃

+

COOCH₃

cis-substituted products

b.

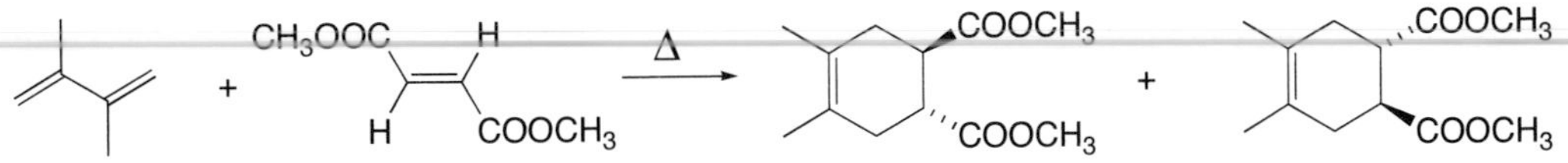

trans dienophile          trans-substituted products

c.

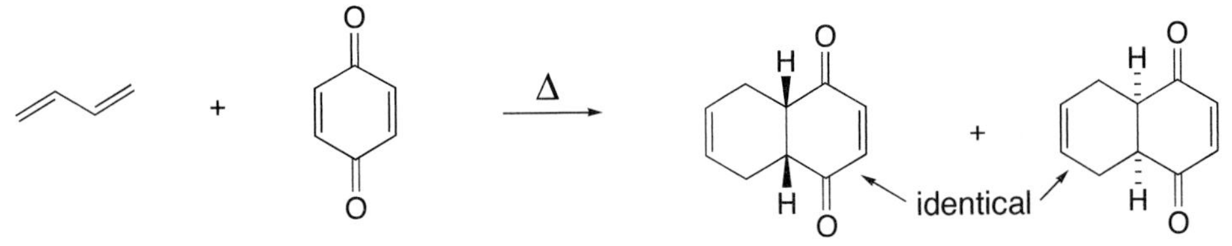

cis dienophile          cis-substituted product

**16.23** The **endo product** (with the substituents under the plane of the new six-membered ring) is the preferred product.

a.

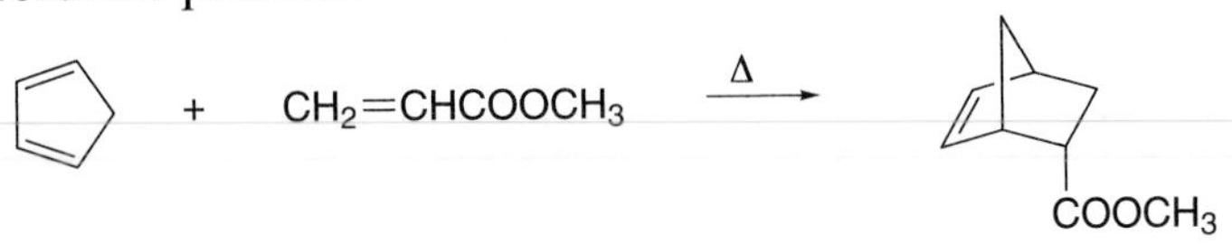

COOCH₃   **endo** substituent

b.

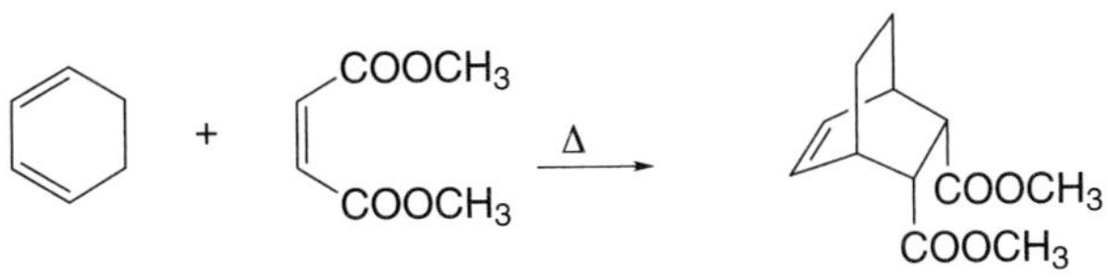

COOCH₃  both groups **endo**
COOCH₃

**16.24** To find the diene and dienophile needed to make each of the products:
[1] Find the six-membered ring with a C–C double bond.
[2] Draw three arrows to work backwards.
[3] Follow the arrows to show the diene and dienophile.

a.

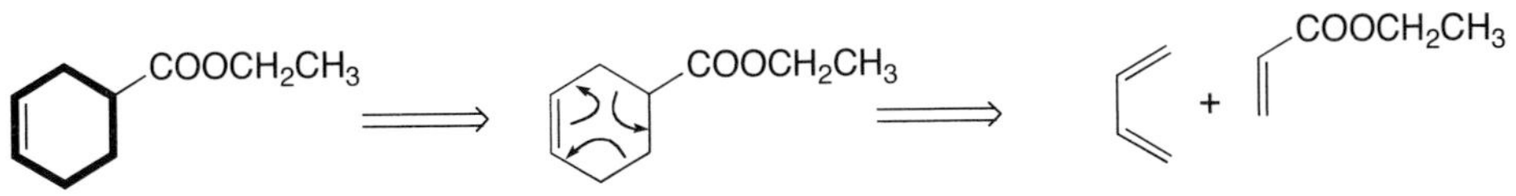

b.

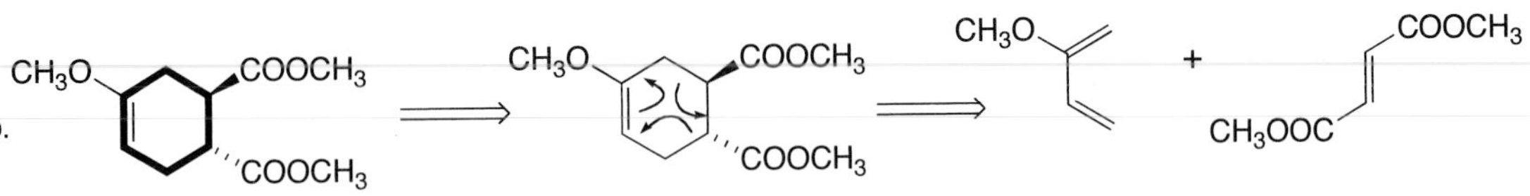

c.

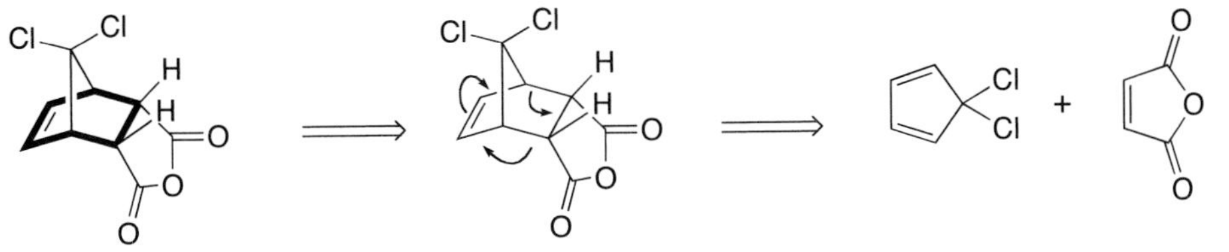

**16.25**

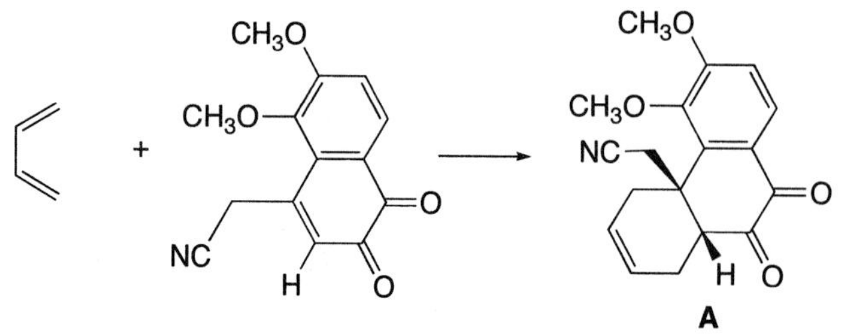

**16.26** Conjugated molecules absorb light at a longer wavelength than molecules that are not conjugated.

a.

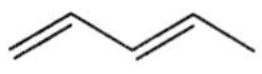

and

conjugated
longer wavelength            not conjugated

b.

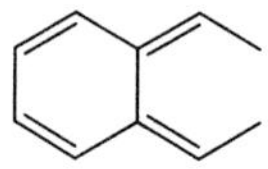

and

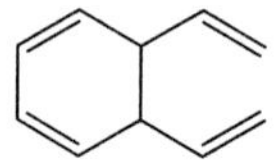

all double bonds conjugated
longer wavelength            one set of
conjugated dienes

**16.27** Sunscreens contain conjugated systems to absorb UV radiation from sunlight.  Look for conjugated systems in the compounds below.

a.

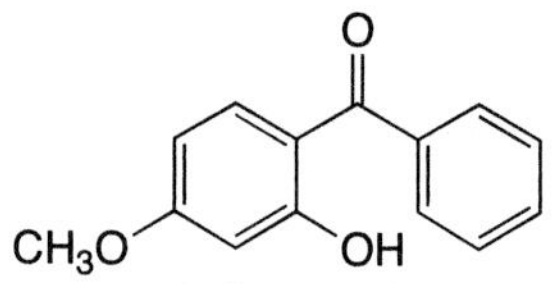

**conjugated system**
could be a sunscreen

b.

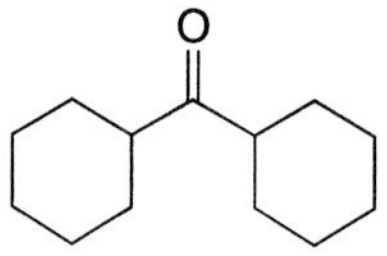

not a conjugated system

c.

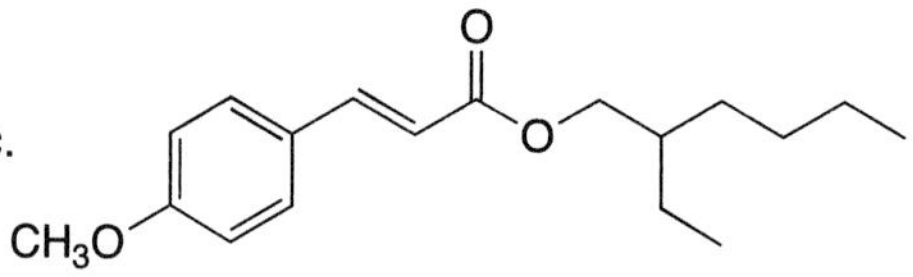

**conjugated system**
could be a sunscreen

**16.28** Use the definition from Answer 16.1.

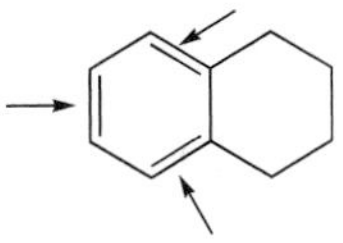

3 π bonds with only
1 σ bond between
**conjugated**

CH₂=CHC≡N

2 multiple bonds with only
1 σ bond between
**conjugated**

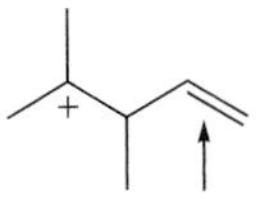

1 π bond with
no adjacent *sp²* hybridized atoms
**NOT conjugated**

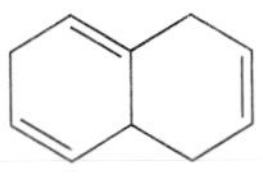

3 π bonds with 2 or
more σ bonds between
**NOT conjugated**

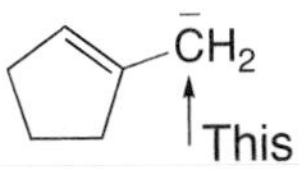

⁻CH₂

This C is *sp²*.

1 π bond with
an adjacent *sp²* hybridized atom

The lone pair occupies a *p* orbital,
so there are *p* orbitals on
three adjacent atoms.
**conjugated**

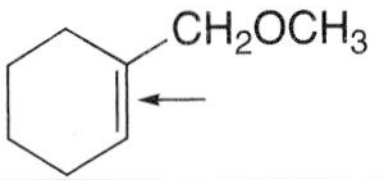

CH₂OCH₃

1 π bond with
no adjacent *sp²* hybridized atoms
**NOT conjugated**

**16.29** Use the definition from Answer 16.1.

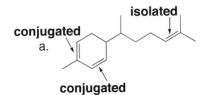

**conjugated**
a.
**conjugated**
**isolated**

b. 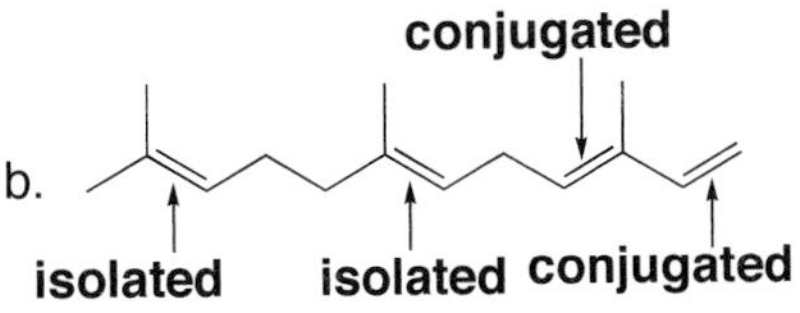
**conjugated**
**isolated** **isolated** **conjugated**

c. 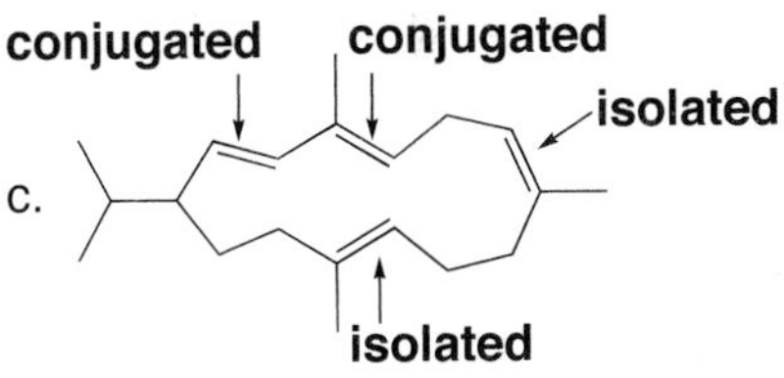
**conjugated** **conjugated**
**isolated**
**isolated**

**16.30**

a. (CH₃)₂C⁺—CH=CH₂ ⟷ (CH₃)₂C=CH—C⁺H₂

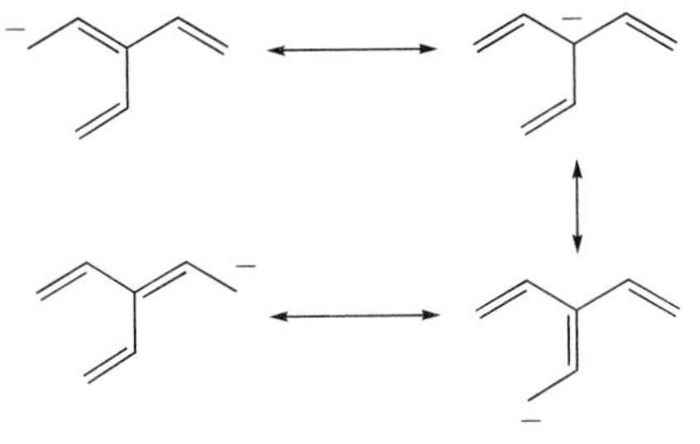

e. CH₃Ö—CH=CH—C⁺H₂ ⟷ CH₃Ö⁺—CH—CH=CH₂

CH₃Ö⁺=CH—CH=CH₂

b.

c.  ⁺CH₂ ⟷ =CH₂

f. 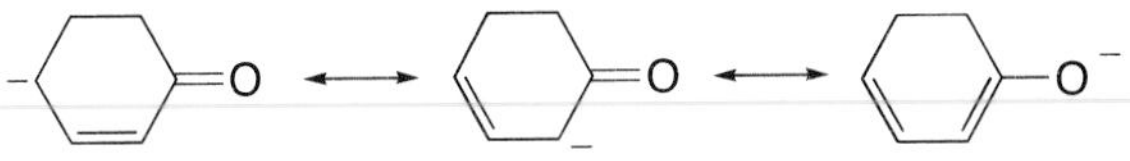 ⟷ ⟷

d. ⟷

⟷

g. 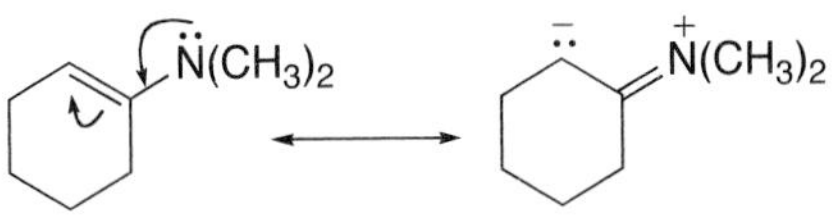 N̈(CH₃)₂ ⟷ N⁺(CH₃)₂

h. 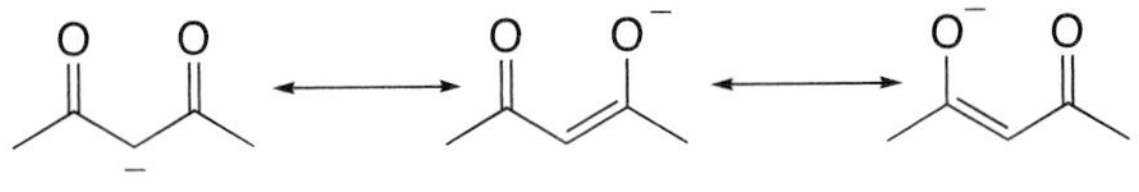 ⟷ ⟷

**16.31**

a.

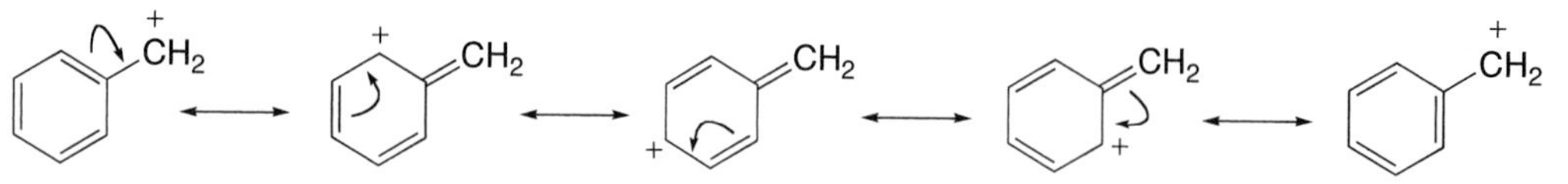

b.

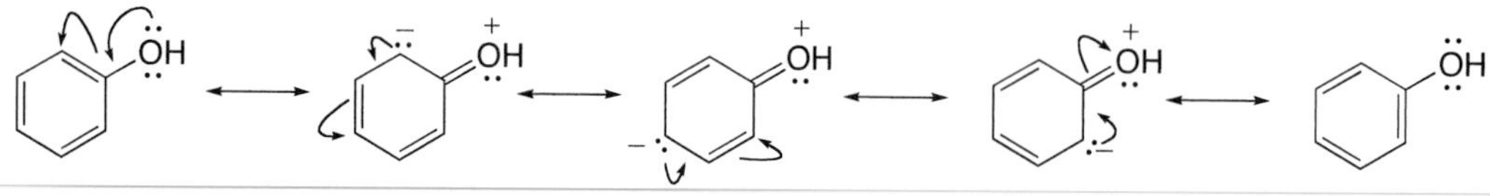

c.

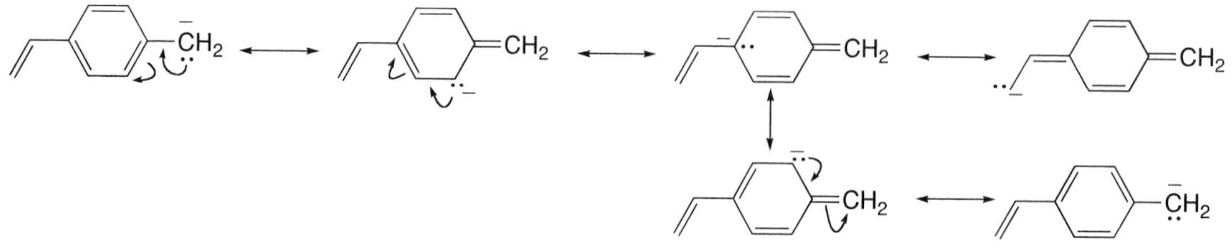

d.

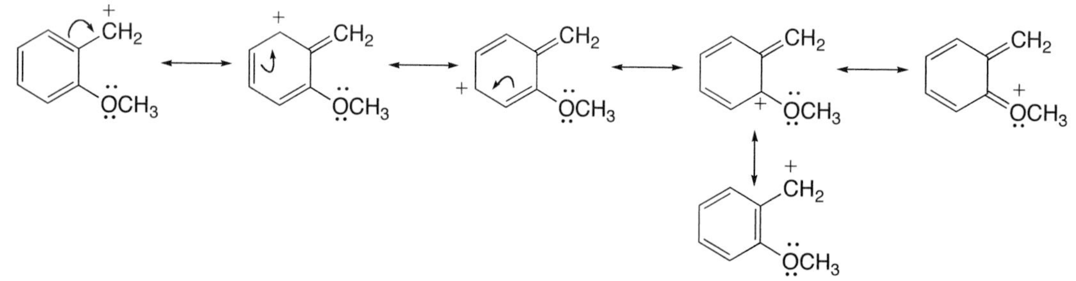

**16.32**

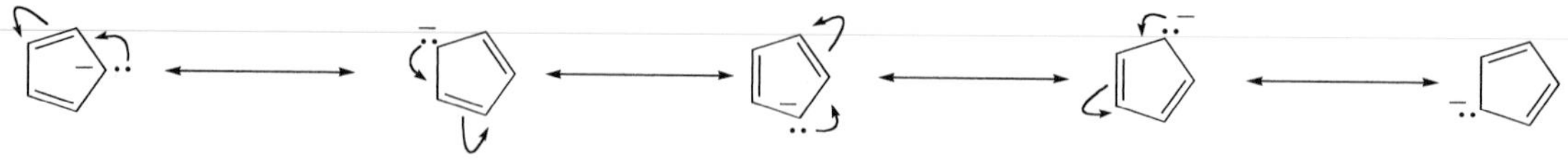

resonance hybrid:

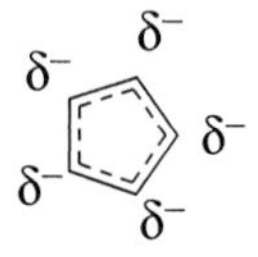

Five resonance structures delocalize the negative charge
on five C's making them all equivalent.

All of the carbons are identical in the anion.

**16.33**

a.

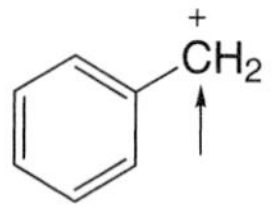

The benzylic C is $sp^2$ hybridized.
**trigonal planar**

b.

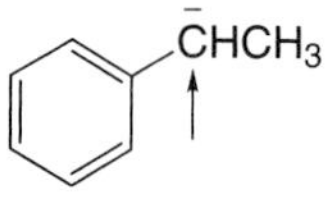

The benzylic C is $sp^2$ hybridized.
**trigonal planar**

c.

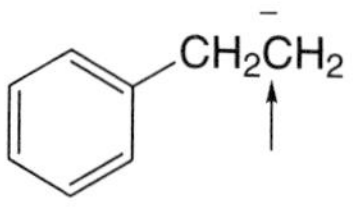

This carbon is NOT part of
a conjugated system.
4 groups = $sp^3$ hybridized
**tetrahedral**

**16.34**

a.

$$CH_2=CH-CH_2-H$$

↓ more acidic

$$\overset{..}{\underset{..}{C}}H_2-CH=CH_2 \longleftrightarrow CH_2=CH-\overset{..}{\underset{..}{C}}H_2$$

Resonance stabilization delocalizes the
negative charge on 2 C's after loss of a proton.
This makes propene more acidic than propane.

$$CH_3CH_2CH_2-H$$

↓ less acidic

$$CH_3CH_2\overset{..}{\underset{..}{C}}H_2$$

only one Lewis structure

b. Draw the products of cleavage of the bond.

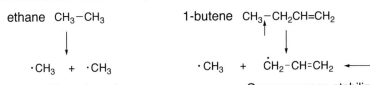

ethane  $CH_3-CH_3$

$\downarrow$

$\cdot CH_3$  +  $\cdot CH_3$

Two unstable radicals form.

1-butene  $CH_3-CH_2CH=CH_2$

$\downarrow$

$\cdot CH_3$  +  $\dot{C}H_2-CH=CH_2$  ⟷  $CH_2=CH-\dot{C}H_2$

One resonance-stabilized radical forms.
This makes the bond dissociation
energy lower because a more stable radical is formed.

**16.35**

$CH_3CH=CHCH_2\ddot{O}H$ ⟶ $CH_3CH=CHCH_2-\overset{+}{\underset{\raise4pt|}{\ddot{O}}}H_2$  +  $Br^-$ ⟶ $CH_3CH=CH\overset{+}{C}H_2$  +  $H_2\ddot{O}$ ⟶ $CH_3CH=CHCH_2Br$

$\overset{\frown}{H-Br}$

two resonance
structures:

+ Br

$CH_3\overset{+}{C}HCH=CH_2$ ⟶ $CH_3CHCH=CH_2$

$\underset{Br}{|}$

+ Br

**16.36** Use the directions from Answer 16.10.

a. (3*Z*)-1,3-pentadiene in the *s*-trans conformation

double bonds on opposite sides
**s-trans**

b. (2*E*,4*Z*)-1-bromo-3-methyl-2,4-hexadiene

c. (2*E*,4*E*,6*E*)-2,4,6-octatriene

d. (*E*,*E*)-3-methyl-2,4-hexadiene in the *s*-cis conformation

**s-cis**

**16.37**

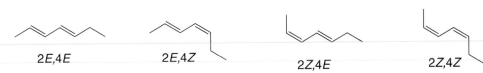

2*E*,4*E*                    2*E*,4*Z*                    2*Z*,4*E*                    2*Z*,4*Z*

**16.38**

a.           and

(3*E*)-1,3,5-hexatriene    (3*E*)-1,3,5-hexatriene
both *s*-cis              both *s*-trans

**different conformations**

c.           and

(3*E*)-1,3,5-hexatriene    (3*Z*)-1,3,5-hexatriene

**different stereoisomers**

b.          and

(3*Z*)-1,3,5-hexatriene    (3*Z*)-1,3,5-hexatriene
both *s*-cis              both *s*-trans

**different conformations**

**16.39** Use the directions from Answer 16.13 and recall that more substituted double bonds are more stable.

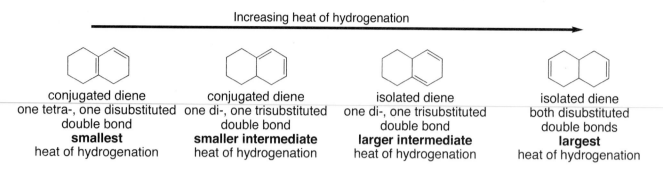

Increasing heat of hydrogenation

| conjugated diene | conjugated diene | isolated diene | isolated diene |
|---|---|---|---|
| one tetra-, one disubstituted double bond | one di-, one trisubstituted double bond | one di-, one trisubstituted double bond | both disubstituted double bonds |
| **smallest** heat of hydrogenation | **smaller intermediate** heat of hydrogenation | **larger intermediate** heat of hydrogenation | **largest** heat of hydrogenation |

**16.40** Conjugated dienes react with HX to form 1,2- and 1,4-products.

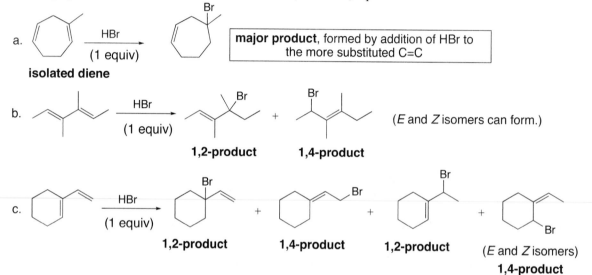

a.
HBr
(1 equiv)

**isolated diene**

**major product**, formed by addition of HBr to the more substituted C=C

b.
HBr
(1 equiv)

**1,2-product**     **1,4-product**     (*E* and *Z* isomers can form.)

c.
HBr
(1 equiv)

**1,2-product**     **1,4-product**     **1,2-product**     (*E* and *Z* isomers) **1,4-product**

**16.41**

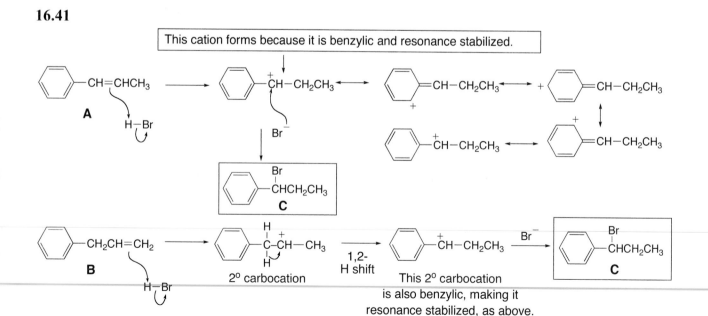

This cation forms because it is benzylic and resonance stabilized.

—CH=CHCH₃

**A**

H—Br

—ĊH—CH₂CH₃

Br⁻

Br
|
—CHCH₂CH₃

**C**

—CH₂CH=CH₂

**B**

H—Br

—Ċ—CH—CH₃
H

**2° carbocation**

1,2-
H shift

—ĊH—CH₂CH₃

This 2° carbocation is also benzylic, making it resonance stabilized, as above.

Br⁻

Br
|
—CHCH₂CH₃

**C**

**16.42** To draw the mechanism for reaction of a diene with HBr and ROOR, recall from Chapter 15 that when an alkene is treated with HBr under these radical conditions, the Br ends up on the carbon with more H's to begin with.

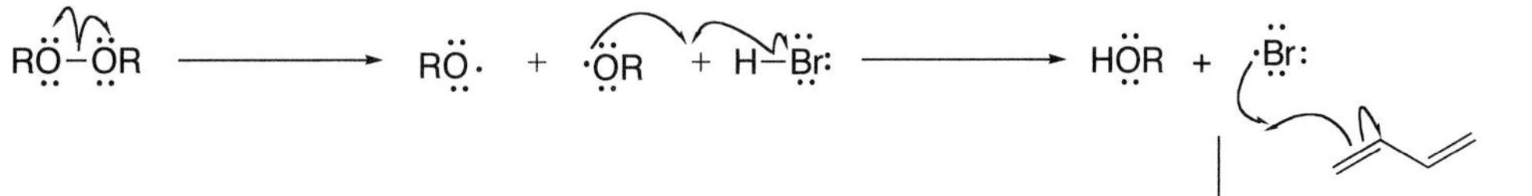

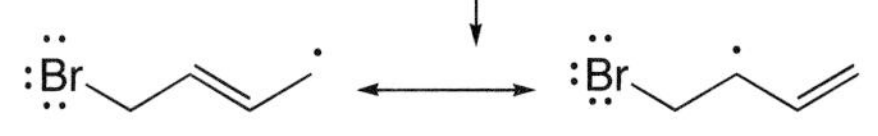

Use each resonance structure to react with HBr.

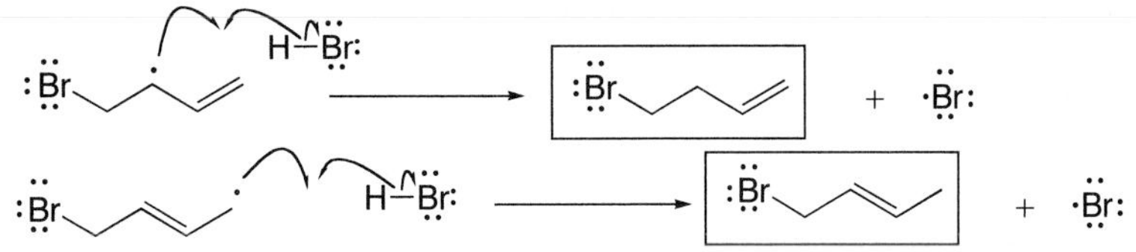

**16.43**

a. and b.

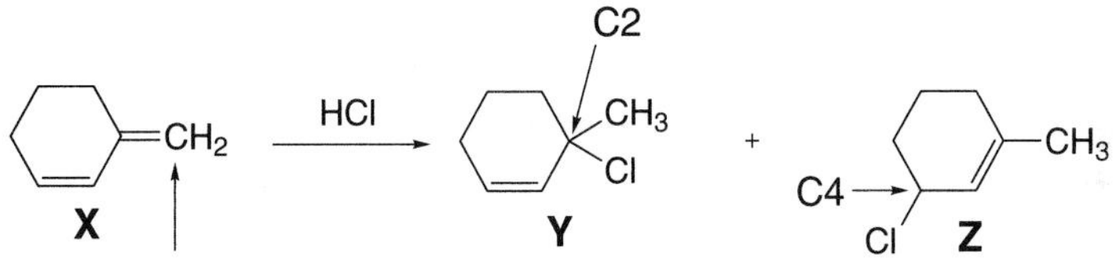

H adds here at C1.   Cl added at C2.   Cl added at C4.
**1,2-product**   **1,4-product**
*kinetic product*   *thermodynamic product*

**Y** is the kinetic product because of the proximity effect. H and Cl add across two adjacent atoms.
**Z** is the thermodynamic product because it has a more stable trisubstituted double bond.

Addition occurs at the labeled double bond due to the stability of the carbocation intermediate.

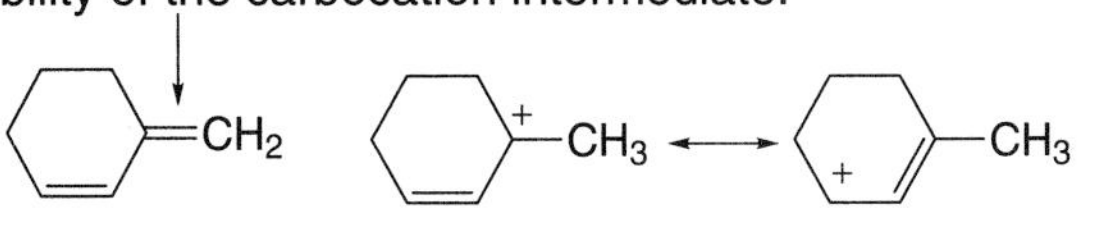

If addition occurred at the other C=C, the following allylic carbocation would form:

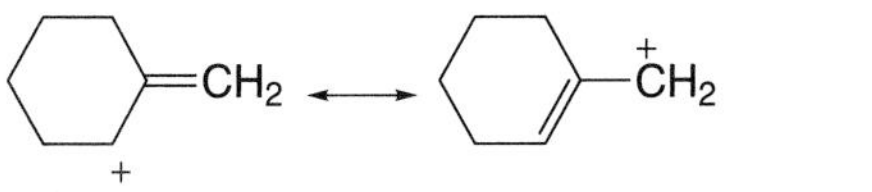

c.

The two resonance structures for this allylic cation are 3° and 2° carbocations.

| more stable intermediate Addition occurs here. |

The two resonance structures for this allylic cation are 1° and 2° carbocations.

**less stable**

**16.44** Addition of HCl at the terminal double bond forms a carbocation that is highly resonance stabilized since it is both allylic and benzylic. Such stabilization does not occur when HCl is added to the other double bond. This gives rise to two products of electrophilic addition.

H–Cl
Cl
Cl
Cl
Cl
+
+

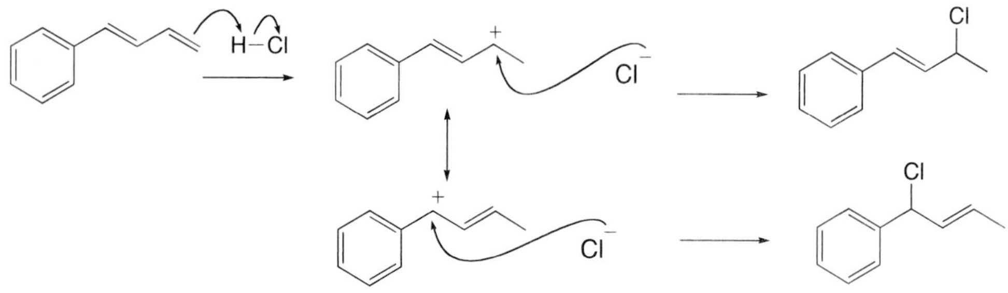

1,2-product

1,4-product

(+ three more resonance structures that delocalize the positive charge onto the benzene ring)

**16.45** There are two possible products:

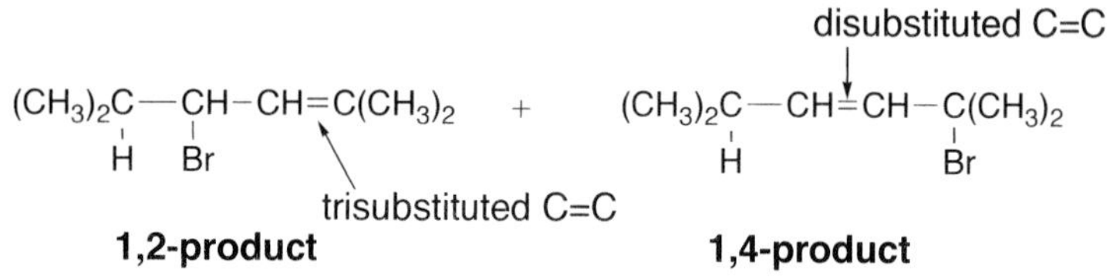

disubstituted C=C
(CH3)2C—CH–CH=C(CH3)2
H Br
+
(CH3)2C—CH=CH–C(CH3)2
H Br
trisubstituted C=C
1,2-product
1,4-product

The 1,2-product is always the kinetic product because of the proximity effect. In this case, it is also the thermodynamic (more stable) product because it contains a more highly substituted C=C (trisubstituted) than the 1,4-product (disubstituted). Thus, the 1,2-product is the major product at high and low temperature.

**16.46** The electron pairs on O can be donated to the double bond through resonance. This increases the electron density of the double bond, making it less electrophilic and therefore less reactive in a Diels–Alder reaction.

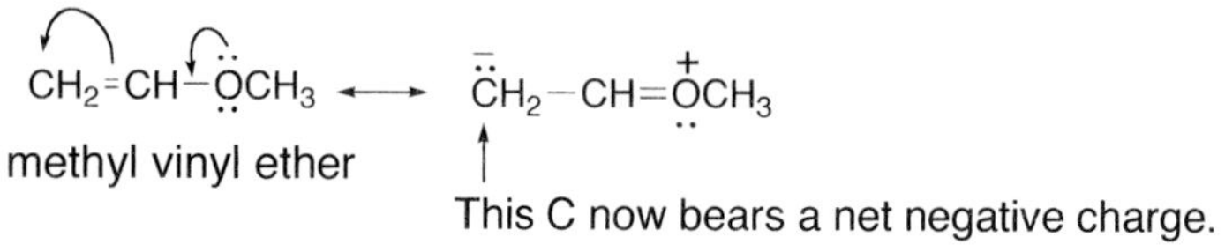

CH2=CH–OCH3
methyl vinyl ether
CH2–CH=OCH3
This C now bears a net negative charge.

**16.47** Use the directions from Answer 16.18.

a. 
+
re-draw
Δ

diene dienophile

b. 
COOCH3
Cl
Δ
COOCH3
Cl
+
COOCH3
Cl

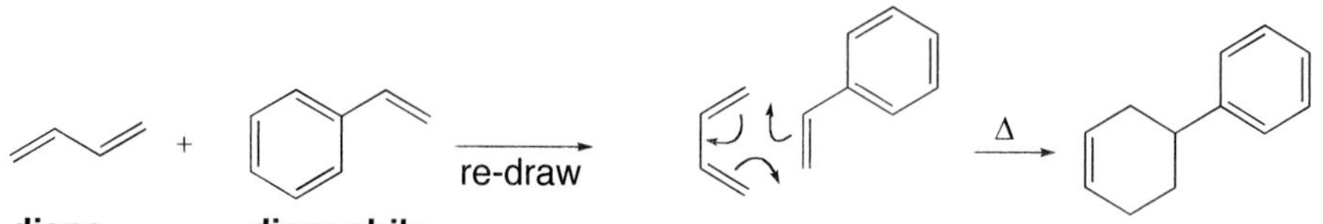

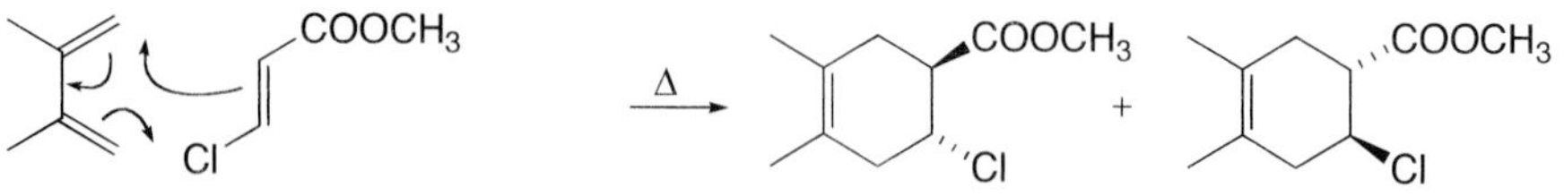

diene trans dienophile trans-substituted products

c. 
COOCH3
Cl
Δ
COOCH3
Cl
+
COOCH3
Cl

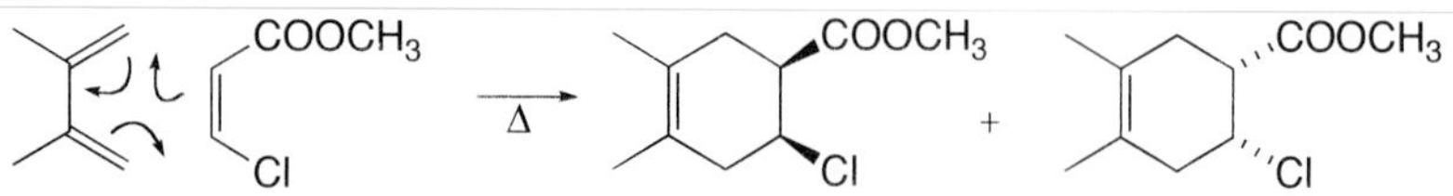

diene cis dienophile cis-substituted products

d. 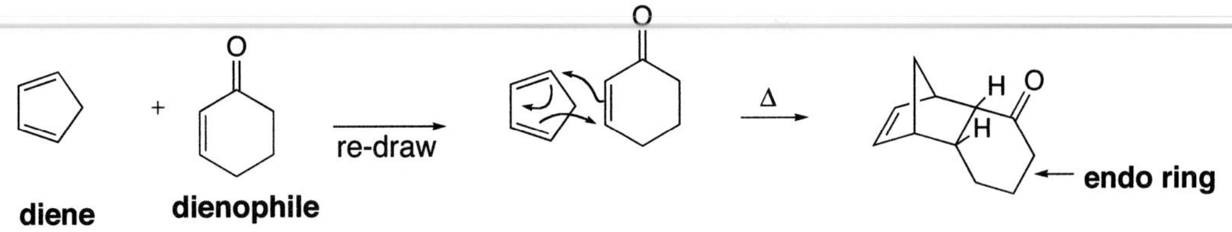 diene + dienophile → re-draw → → Δ → ← endo ring

e. 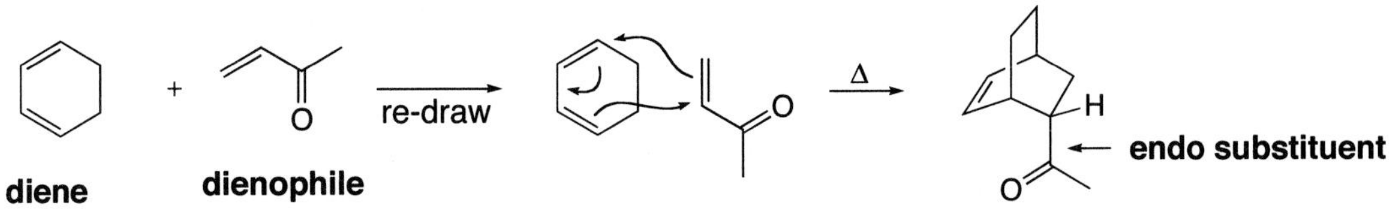 diene + dienophile → re-draw → → Δ → ← endo substituent

f. 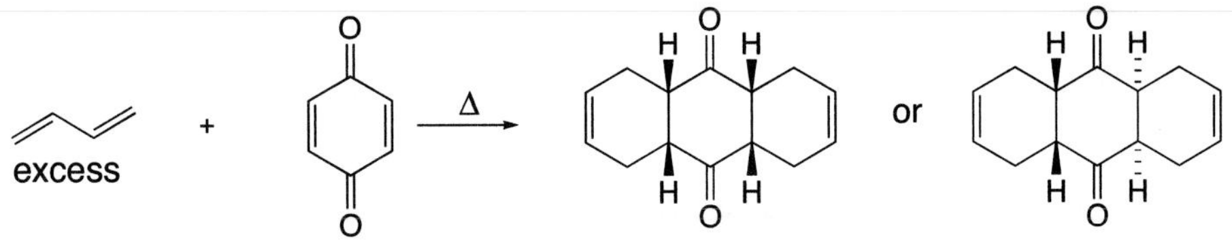 excess + → Δ → or

**16.48** Use the directions from 16.24.

a.

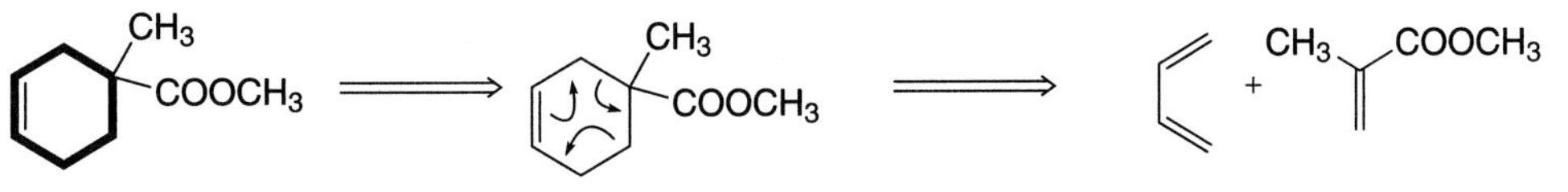

b.

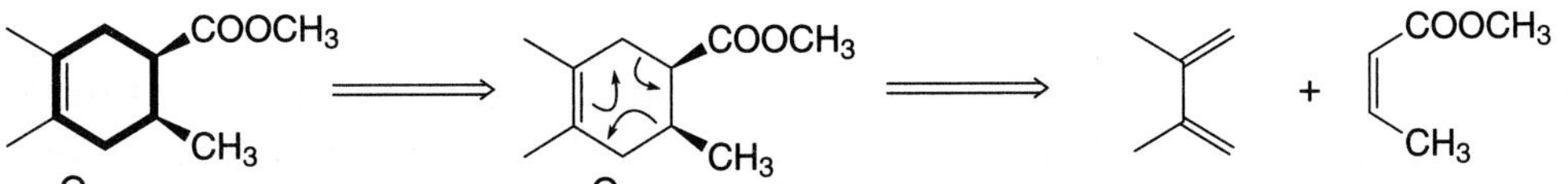

c.

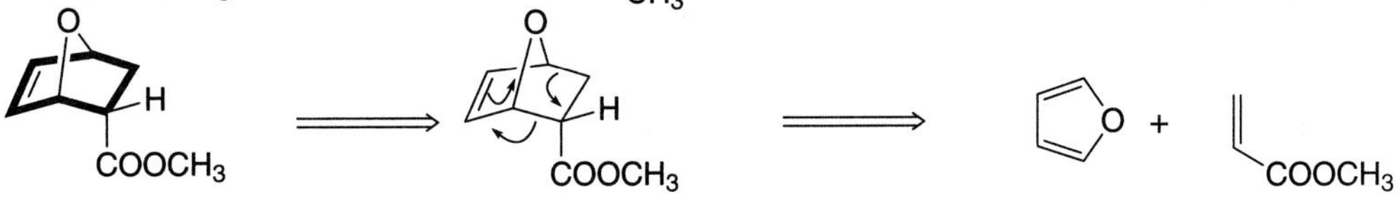

d.

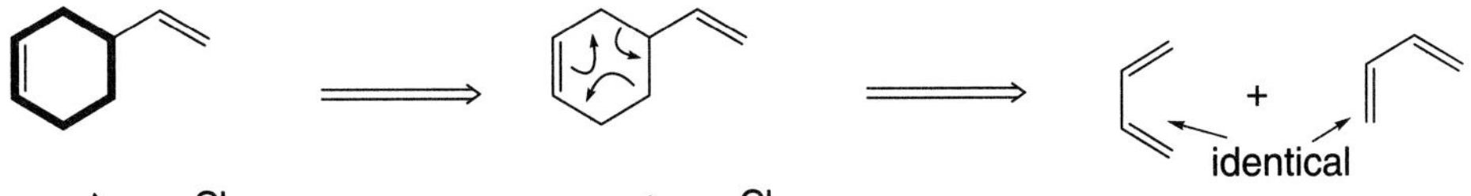

e.

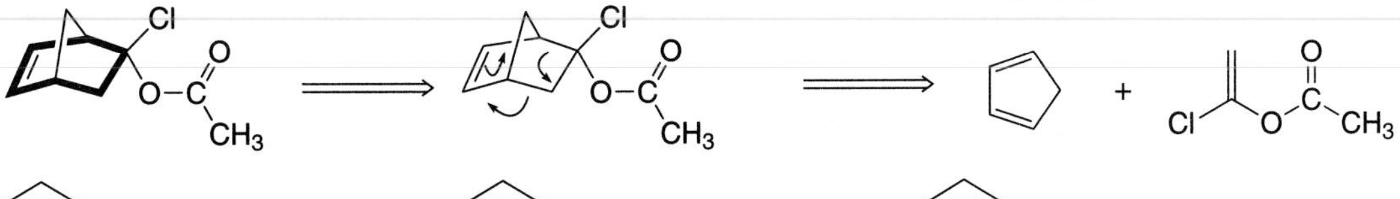

f.

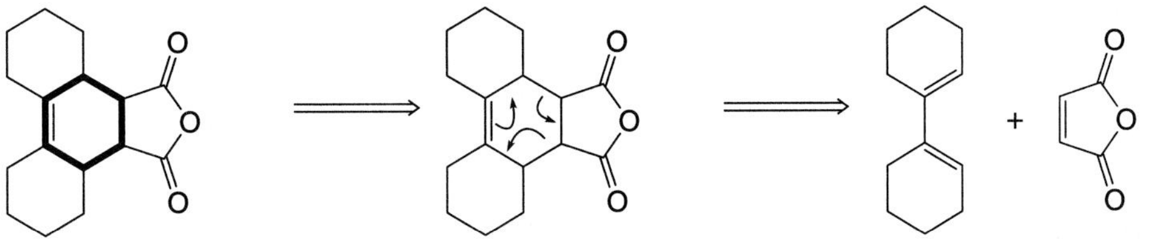

**16.49**

CH2
CH2
+

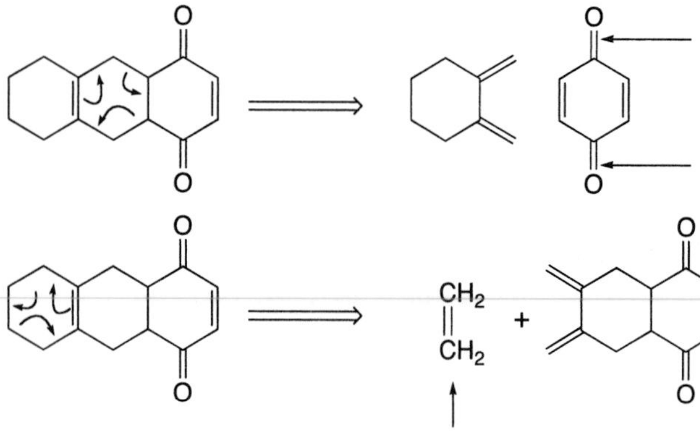

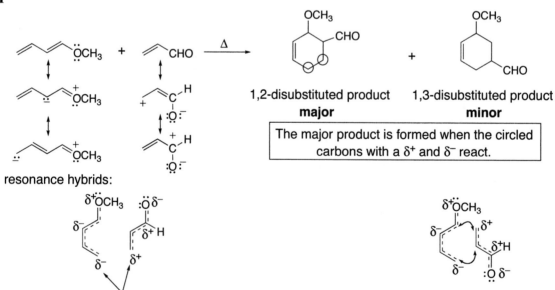

This pathway is **preferred** because the dienophile has electron-withdrawing C=O groups that make it more reactive.

no electron-withdrawing groups
**less reactive**

**16.50**

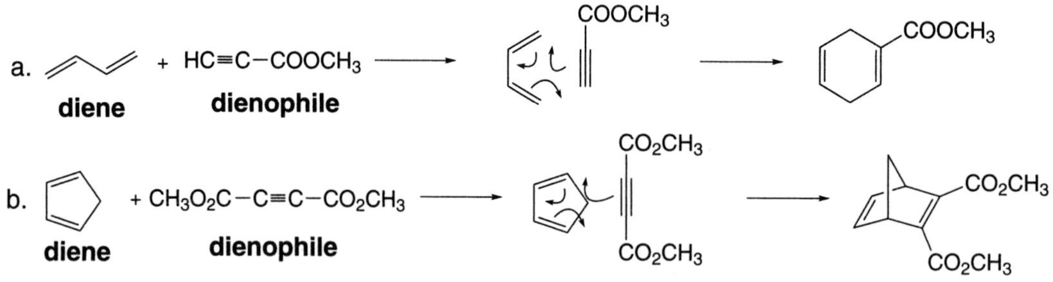

a.
+ HC≡C−COOCH3
diene
dienophile
COOCH3
COOCH3
b.
+ CH3O2C−C≡C−CO2CH3
diene
dienophile
CO2CH3
CO2CH3
CO2CH3
CO2CH3

**16.51**

OCH3
+
CHO
Δ
OCH3
CHO
+
OCH3
CHO
1,2-disubstituted product
major
1,3-disubstituted product
minor
resonance hybrids:

The major product is formed when the circled carbons with a $\delta^+$ and $\delta^-$ react.

δ+OCH3
:O δ−
δ−
δ+ H
δ−
δ+

δ+OCH3
δ−
δ+
δ+H
δ−
:O δ−

For the 1,2-product, carbons with unlike charges would react. This is favored because the electron-rich and the electron-poor C's can bond.

For the 1,3-product, there are no partial charges of opposite sign on reacting carbons. This arrangement is less attractive.

**16.52**

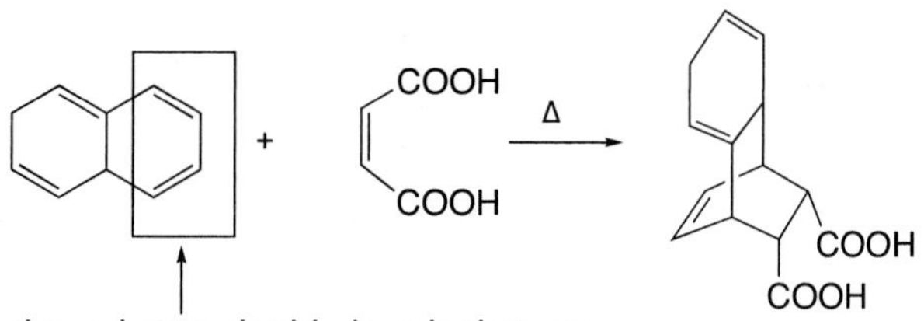

These are the only two double bonds that are
conjugated and have the *s*-cis conformation
needed for a Diels–Alder reaction.

**16.53**

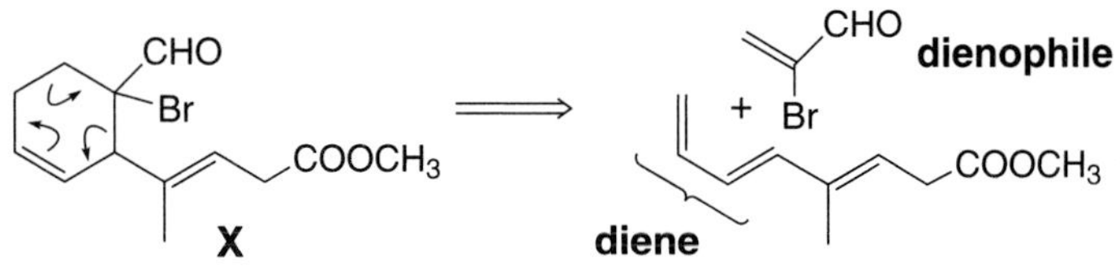

**dienophile**

**diene**

**X**

**16.54**

aldrin

dieldrin

**Z**

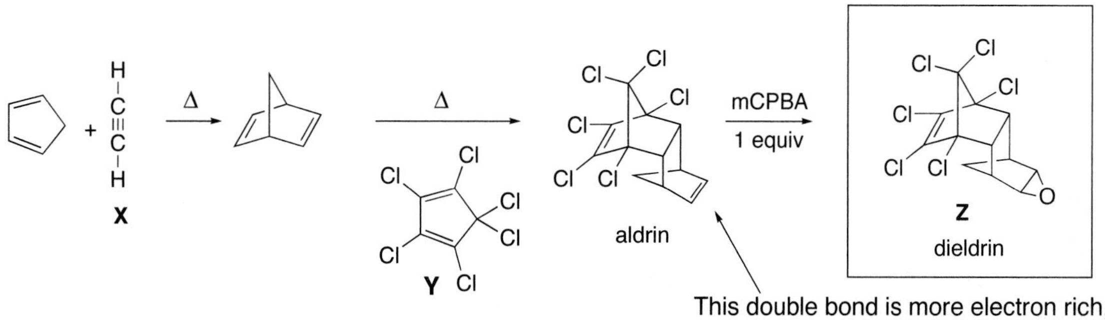

This double bond is more electron rich,
so it is epoxidized more readily.

**16.55**

In each problem, the synthesis must
begin with the preparation of
cyclopentadiene from dicyclopentadiene.

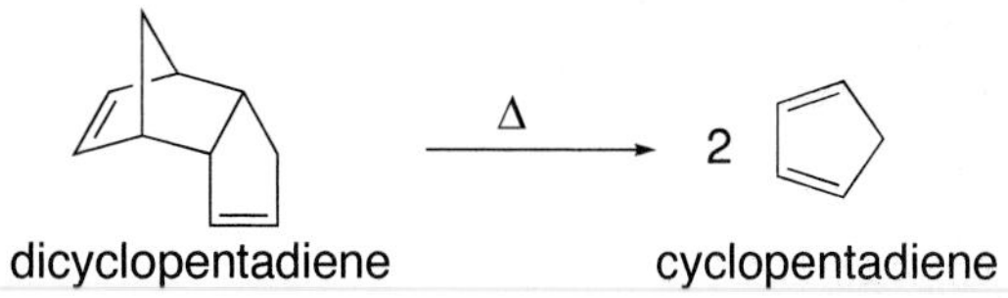

dicyclopentadiene

2  cyclopentadiene

a.

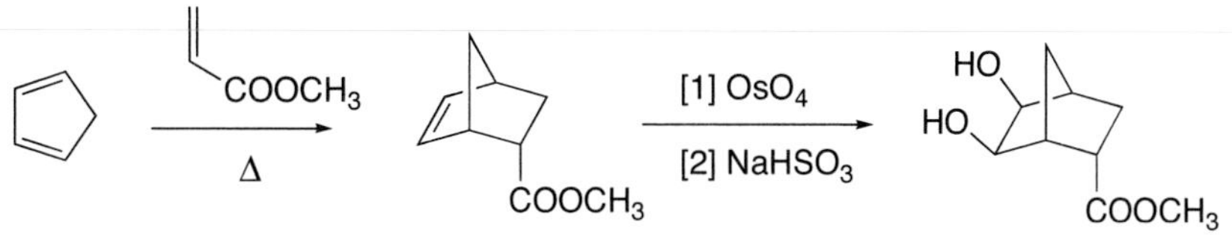

[1] OsO₄

[2] NaHSO₃

b.

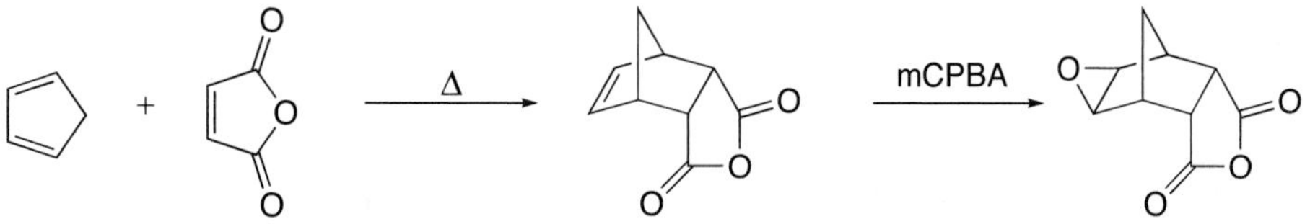

mCPBA

c.

H₂ (excess)

Pd-C

[1] NaH

[2] CH₃CH₂CH₂CH₂Br

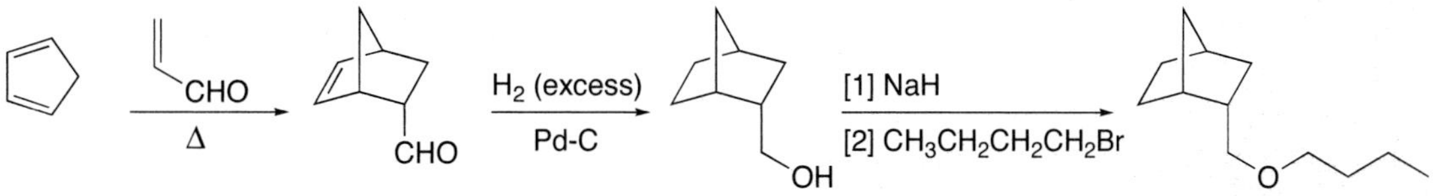

## 16.56

a. 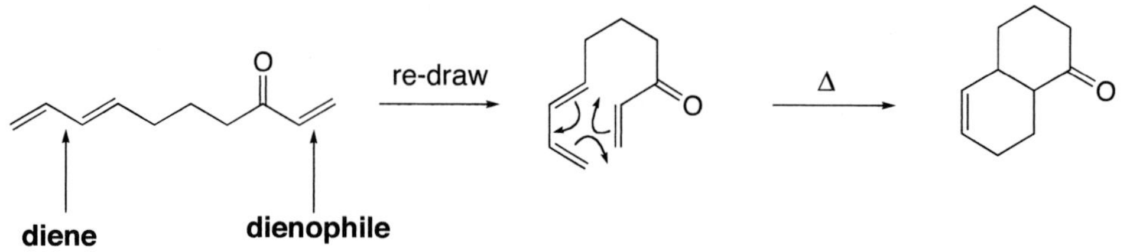
re-draw
Δ
diene
dienophile

b. 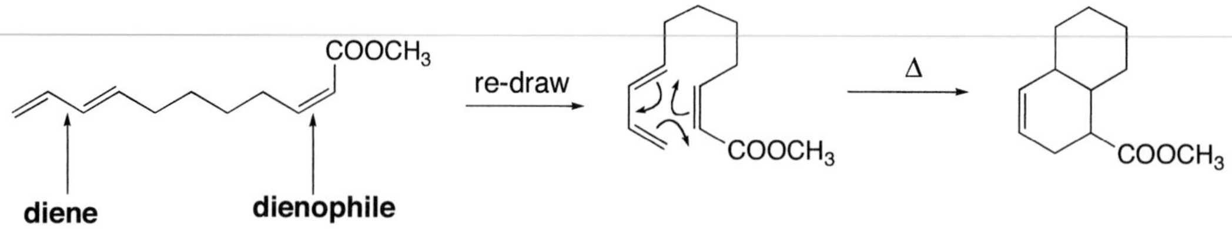
COOCH3
re-draw
Δ
COOCH3
COOCH3
diene
dienophile

## 16.57

a. $(CH_3)_2C{=}CHCH_2CH_2CH{=}CH_2$ **isolated diene** $\xrightarrow{HCl}$ $(CH_3)_2C(Cl)CH_2CH_2CH_2CH{=}CH_2$ **major product** + $(CH_3)_2C{=}CHCH_2CH_2CHClCH_3$ **minor product**

b. 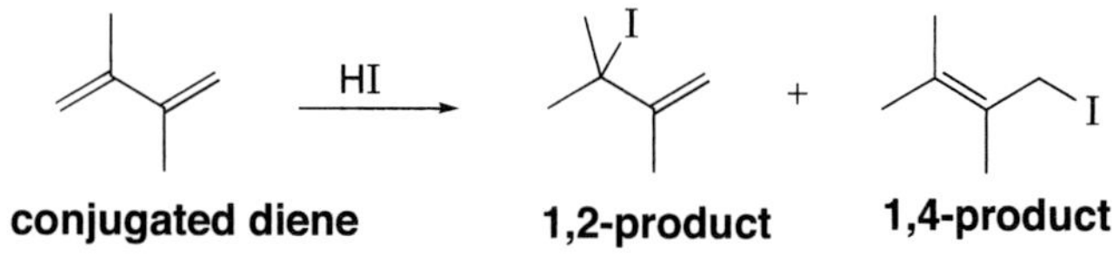
HI
I
+
I
conjugated diene
1,2-product
1,4-product

c. 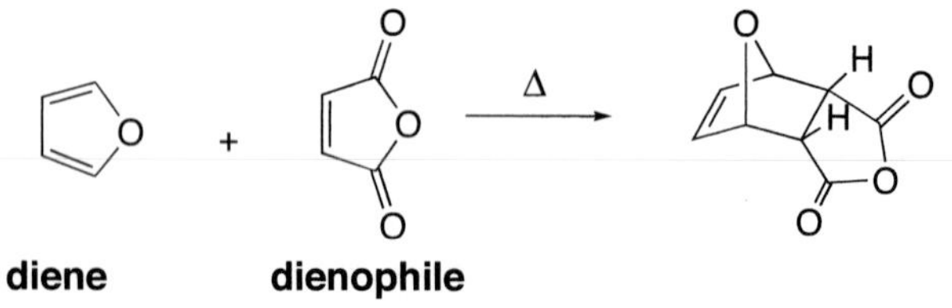
+
Δ
diene
dienophile

d. 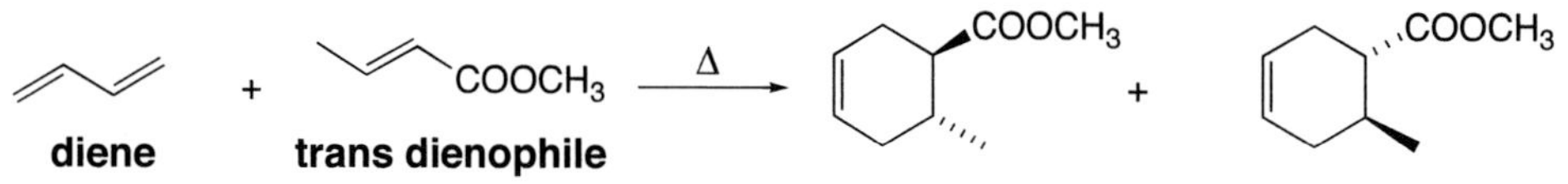
+
COOCH3
Δ
COOCH3
+
COOCH3
diene
trans dienophile

e. 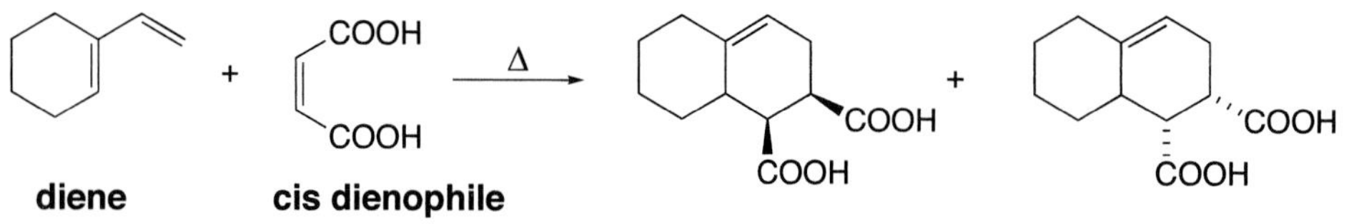
+
COOH
COOH
Δ
COOH
COOH
+
COOH
COOH
diene
cis dienophile

f. 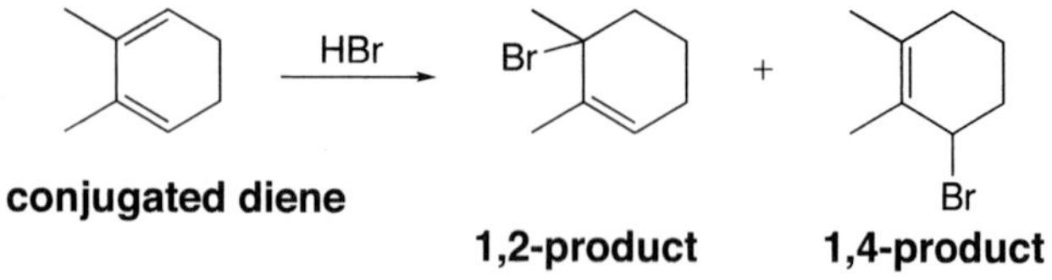
HBr
Br
+
Br
conjugated diene
1,2-product
1,4-product

## 16.58

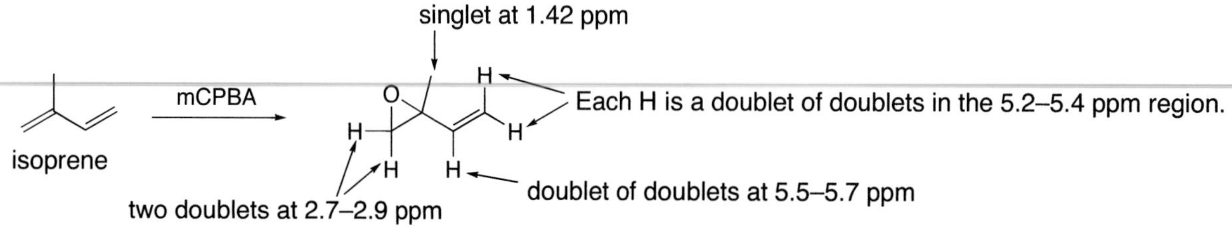
singlet at 1.42 ppm
mCPBA
isoprene
Each H is a doublet of doublets in the 5.2–5.4 ppm region.
doublet of doublets at 5.5–5.7 ppm
two doublets at 2.7–2.9 ppm

**16.59**

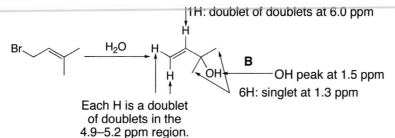

1H: doublet of doublets at 6.0 ppm
Br
H₂O
B
OH peak at 1.5 ppm
6H: singlet at 1.3 ppm
Each H is a doublet of doublets in the 4.9–5.2 ppm region.

The IR shows an OH absorption at 3200–3600 $cm^{-1}$.

**16.60**

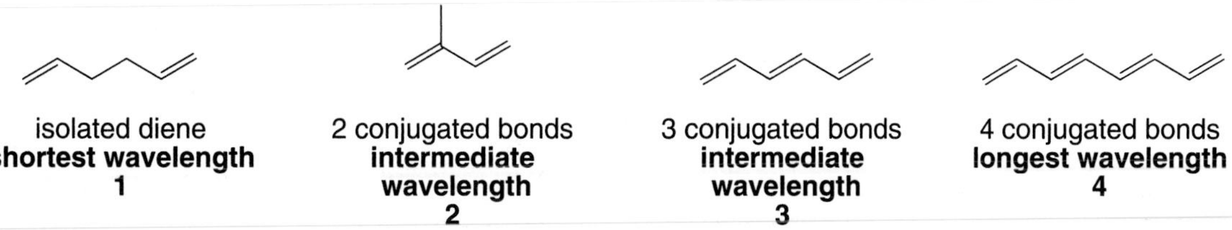

isolated diene
shortest wavelength
1
2 conjugated bonds
intermediate wavelength
2
3 conjugated bonds
intermediate wavelength
3
4 conjugated bonds
longest wavelength
4

**16.61**

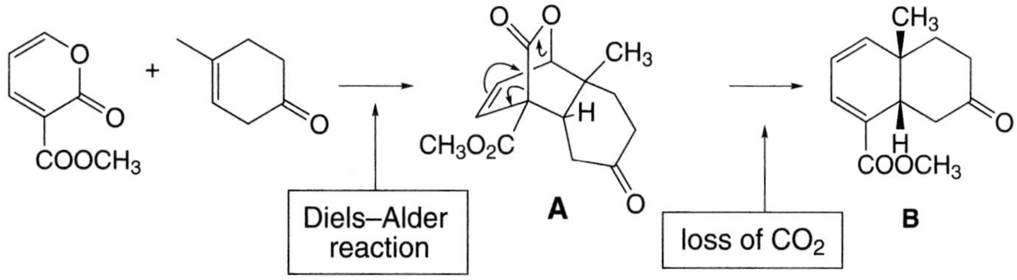

COOCH3
CH3
CH3O2C
A
Diels–Alder reaction
loss of CO2
B
O=C=O

**16.62**

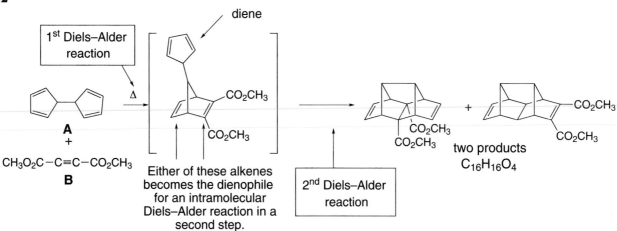

1st Diels–Alder reaction
diene
Δ
A
+
CH3O2C–C≡C–CO2CH3
B
CO2CH3
Either of these alkenes becomes the dienophile for an intramolecular Diels–Alder reaction in a second step.
2nd Diels–Alder reaction
two products
C16H16O4

## Chapter 17: Benzene and Aromatic Compounds

### ♦ Comparing aromatic, antiaromatic, and nonaromatic compounds (17.7)

- **Aromatic compound**
  - A cyclic, planar, completely conjugated compound that contains $4n + 2$ $\pi$ electrons ($n = 0, 1, 2, 3$, and so forth).
  - An aromatic compound is more stable than a similar acyclic compound having the same number of $\pi$ electrons.

- **Antiaromatic compound**
  - A cyclic, planar, completely conjugated compound that contains $4n$ $\pi$ electrons ($n = 0, 1, 2, 3$, and so forth).
  - An antiaromatic compound is less stable than a similar acyclic compound having the same number of $\pi$ electrons.

- **A compound that is not aromatic**
  - A compound that lacks one (or more) of the requirements to be aromatic or antiaromatic.

### ♦ Properties of aromatic compounds

- Every carbon has a *p* orbital to delocalize electron density (17.2).
- They are unusually stable. $\Delta H^\circ$ for hydrogenation is much less than expected, given the number of degrees of unsaturation (17.6).
- They do not undergo the usual addition reactions of alkenes (17.6).
- $^1$H NMR spectra show highly deshielded protons because of ring currents (17.4).

### ♦ Examples of aromatic compounds with 6 π electrons (17.8)

benzene        pyridine        pyrrole    cyclopentadienyl    tropylium
                                            anion               cation

### ♦ Examples of compounds that are not aromatic (17.8)

not cyclic           not planar          not completely
                                          conjugated

## Chapter 17: Answers to Problems

**17.1** Move the electrons in the π bonds to draw all major resonance structures.

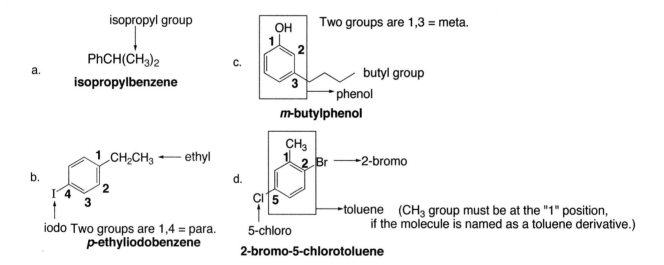

**17.2** Look at the hybridization of the atoms involved in each bond.  Carbons in a benzene ring are surrounded by three groups and are $sp^2$ hybridized.

a.
$Csp^2–Csp^3$
$Csp^2–H1s$
$Csp^2–Csp^2$
$Cp–Cp$
**shortest** of all the indicated bonds in (a) and (b).

b.
$Csp^2–Csp^2$
$Csp^2–Csp^2$
$Cp–Cp$

**17.3**

- To name a benzene ring with **one substituent**, name the substituent and add the word *benzene*.
- To name a **disubstituted ring**, select the correct prefix (ortho = 1,2; meta = 1,3; para = 1,4) and alphabetize the substituents.  Use a common name if it is a derivative of that monosubstituted benzene.
- To name a **polysubstituted ring**, number the ring to give the lowest possible numbers and then follow other rules of nomenclature.

a.
isopropyl group
$PhCH(CH_3)_2$
**isopropylbenzene**

c.
OH
Two groups are 1,3 = meta.
butyl group
phenol
*m*-butylphenol

b.
$CH_2CH_3$ ← ethyl
iodo Two groups are 1,4 = para.
*p*-ethyliodobenzene

d.
$CH_3$
Br → 2-bromo
toluene (CH₃ group must be at the "1" position, if the molecule is named as a toluene derivative.)
5-chloro
**2-bromo-5-chlorotoluene**

**17.4** Work backwards to draw the structure from the names.

a. isobutylbenzene

isobutyl
group

b. *o*-dichlorobenzene

c. *cis*-1,2-diphenylcyclohexane

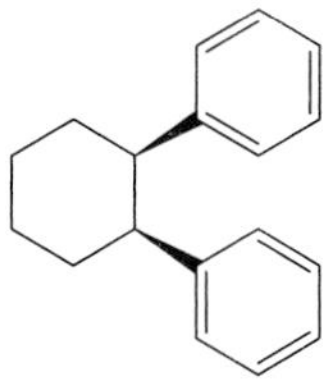

d. *m*-bromoaniline

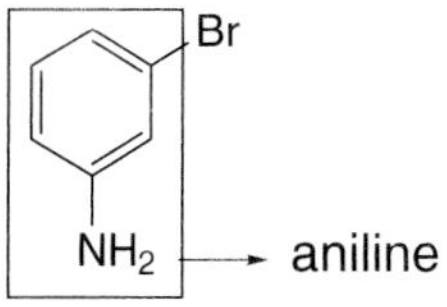

aniline

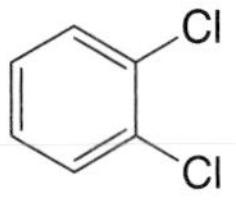

e. 4-chloro-1,2-diethylbenzene

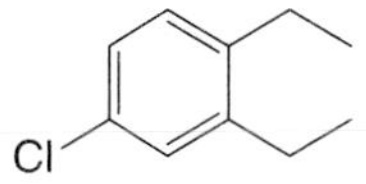

**17.5**

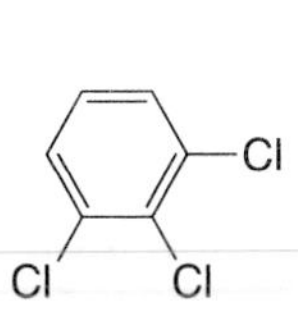

1,2,3-trichlorobenzene 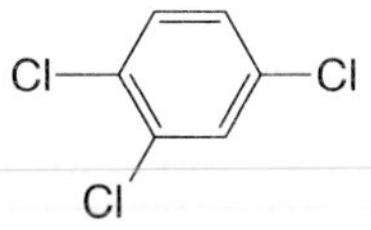    1,2,4-trichlorobenzene 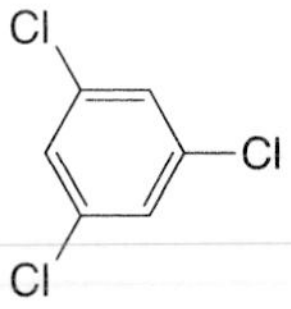    1,3,5-trichlorobenzene

**17.6**

Molecular formula $C_{10}H_{14}O_2$: 4 degrees of unsaturation
IR absorption at 3150–2850 cm$^{-1}$: $sp^2$ and $sp^3$ hybridized C–H bonds
NMR absorptions (ppm):
   1.4 (triplet, 6H)
   4.0 (quartet, 4H)
   6.8 (singlet, 4H)

**17.7** Count the different types of carbons to determine the number of $^{13}C$ NMR signals.

a.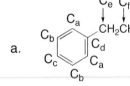

4 types of C's in the benzene ring
6 peaks

b.

All C's are different.
7 peaks

c.

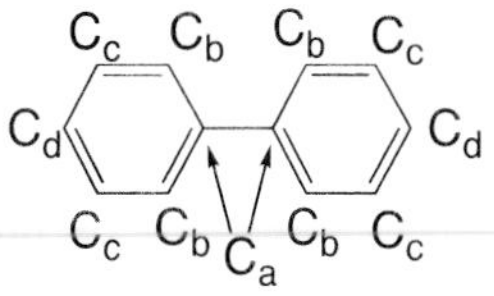

4 peaks

**17.8** The less stable compound has a larger heat of hydrogenation.

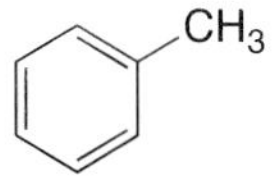

**A**
benzene ring, more stable
smaller $\Delta H°$

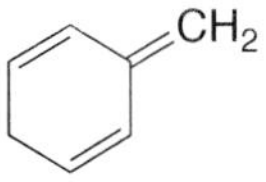

**B**
no benzene ring, less stable
larger $\Delta H°$

**17.9** The protons on $sp^2$ hybridized carbons in aromatic hydrocarbons are highly deshielded and absorb at 6.5–8 ppm whereas hydrocarbons that are not aromatic show an absorption at 4.5–6 ppm, typical of protons bonded to the C=C of an alkene.

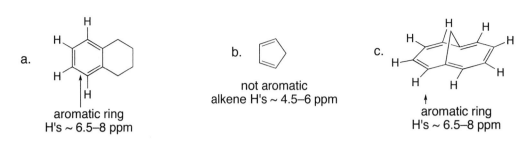

a.

aromatic ring
H's ~ 6.5–8 ppm

b.

not aromatic
alkene H's ~ 4.5–6 ppm

c.

aromatic ring
H's ~ 6.5–8 ppm

**17.10** To be aromatic, a ring must have $4n + 2$ π electrons.

| 16 π e⁻ | 20 π e⁻ | 22 π e⁻ |
|---|---|---|
| $4n$ | $4n$ | $4n + 2$ |
| $4(4) = 16$ | $4(5) = 20$ | $4(5) + 2 = 22$ |
| **antiaromatic** | **antiaromatic** | **aromatic** |

**17.11** Annulenes have alternating double and single bonds. An odd number of carbon atoms in the ring would mean there would be two adjacent single bonds. Therefore an annulene having an odd number of carbon atoms cannot exist.

**17.12**

**17.13** In determining if a heterocycle is aromatic, count a nonbonded electron pair if it makes the ring aromatic in calculating $4n + 2$. Lone pairs on atoms already part of a multiple bond cannot be delocalized in a ring, and so they are never counted in determining aromaticity.

a.

Count one lone pair from O.
$4n + 2 = 4(1) + 2 = 6$
**aromatic**

b.

no lone pair from O
$4n + 2 = 4(1) + 2 = 6$
**aromatic**

c.

With one lone
pair from each O
there would be 8 electrons.
If O's are $sp^3$ hybridized,
the ring is not completely
conjugated.
**not aromatic**

d.

Both N atoms are part of
a double bond, so the lone
pairs cannot be counted:
there are 6 electrons.
$4n + 2 = 4(1) + 2 = 6$
**aromatic**

**17.14**

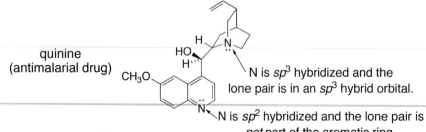

quinine
(antimalarial drug)

N is $sp^3$ hybridized and the
lone pair is in an $sp^3$ hybrid orbital.

N is $sp^2$ hybridized and the lone pair is
*not* part of the aromatic ring.
This means it occupies an $sp^2$ hybrid orbital.

**17.15**

**17.16** Compare the conjugate base of 1,3,5-cycloheptatriene with the conjugate base of cyclopentadiene. Remember that the compound with the more stable conjugate base will have a lower p$K_a$.

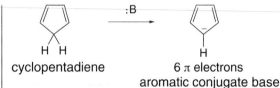

1,3,5-cycloheptatriene
p$K_a$ = 39

8 π Electrons make this conjugate base especially unstable (**antiaromatic**).

Since the conjugate base is unstable, the p$K_a$ of 1,3,5-cycloheptatriene is **high**.

cyclopentadiene

6 π electrons aromatic conjugate base **very stable anion**

Since the conjugate base is very stable, the p$K_a$ of cyclopentadiene is much **lower**.

**17.17** The compound with the most stable conjugate base is the most acidic.

Conjugate bases:

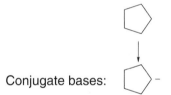

no resonance delocalization

Most unstable base so **least acidic acid.**

2 resonance structures

The acid is **intermediate in acidity.**

**aromatic conjugate base** most stable

The acid is the **most acidic.**

**17.18**

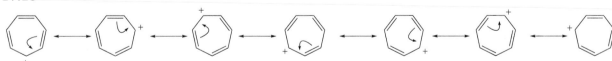

**17.19** To be aromatic, the ions must have 4$n$ + 2 π electrons.

a.

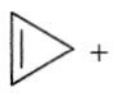

2 π electrons
4(0) + 2 = 2
**aromatic**

d.

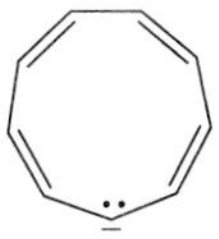

10 π electrons
4(2) + 2 = 10
**aromatic**

**17.20**

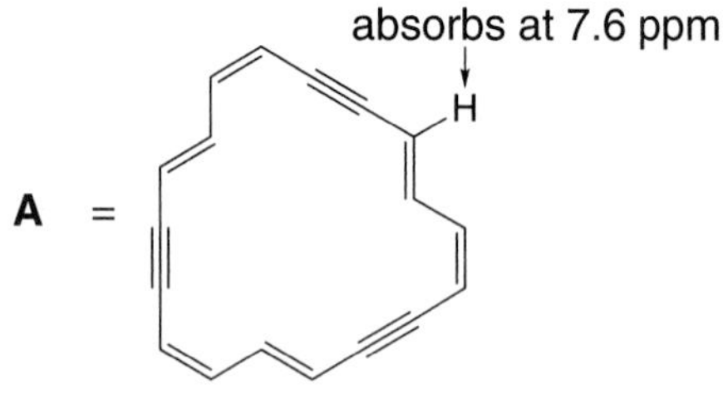

absorbs at 7.6 ppm

**A** =

The NMR indicates that **A** is aromatic. The C's of the triple bond are *sp* hybridized. Each triple bond has one set of electrons in *p* orbitals that overlap with other *p* orbitals on adjacent atoms in the ring. This overlap allows electrons to delocalize. Each C of the triple bonds also has a *p* orbital in the plane of the ring. The electrons in these *p* orbitals are localized between the C's of the triple bond, and not delocalized in the ring. Although **A** has 24 π e⁻ electrons total, only 18 e⁻ are delocalized around the ring.

**17.21**   In using the inscribed polygon method, always draw the vertex pointing down.

2 antibonding MOs

1 bonding MO

2 π electrons
All bonding MOs are filled.
**aromatic**

**17.22**   Draw the inscribed pentagons with the vertex pointing down. Then draw the molecular orbitals (MOs) and add the electrons.

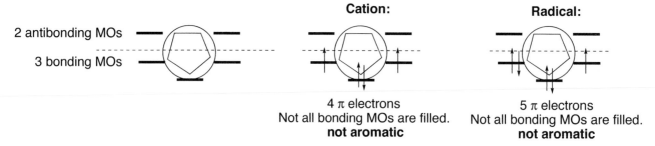

2 antibonding MOs

3 bonding MOs

**Cation:**

4 π electrons
Not all bonding MOs are filled.
**not aromatic**

**Radical:**

5 π electrons
Not all bonding MOs are filled.
**not aromatic**

**17.23**   $C_{60}$ would exhibit only one $^{13}C$ NMR signal because all the carbons are identical.

**17.24**

a. If benzene could be described by a single Kekulé structure, only one product would form in Reaction [1], but there would be four (not three) dibromobenzenes (**A**–**D**), because adjacent C–C bonds are different—one is single and one is double. Thus, compounds **A** and **B** would *not* be identical. **A** has two Br's bonded to the same double bond, but **B** has two Br's on different double bonds.

b. In the resonance description, only one product would form in Reaction [1], since all C's are identical, but only three dibromobenzenes (ortho, meta, and para isomers) are possible. **A** and **B** are identical because each C–C bond is identical and intermediate in bond length between a C–C single and C–C double bond.

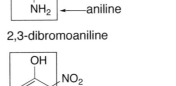

**A**     **B**     **C**     **D**

**17.25** To name the compounds use the directions from Answer 17.3.

a.

*sec*-butylbenzene

b.

CH₂CH₃

*m*-chloroethylbenzene

c.

CH₃—⟨ ⟩—Cl

toluene
*p*-chlorotoluene

d.

NH₂ ← aniline

Cl

*o*-chloroaniline

e.

Br

Br

NH₂ ←aniline

2,3-dibromoaniline

f.

OH

NO₂

NO₂ ←phenol (OH at C1)

2,5-dinitrophenol

g.

CH₃CH₂        CH₂CH₂CH₃

CH(CH₃)₂

1-ethyl-3-isopropyl-5-propylbenzene

h.

Br

Ph

*cis*-1-bromo-2-phenylcyclohexane

**17.26**

a. *p*-dichlorobenzene

Cl

Cl

b. *m*-chlorophenol

Cl

OH

c. *p*-iodoaniline

I

H₂N

d. *o*-bromonitrobenzene

Br

NO₂

e. 2,6-dimethoxytoluene

OCH₃

CH₃

OCH₃

f. 2-phenyl-1-butene

g. 2-phenyl-2-propen-1-ol

OH

h. *trans*-1-benzyl-3-phenylcyclopentane

or

**17.27** Count the electrons in the π bonds. Each π bond holds two electrons.

a.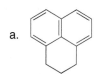

b.

c.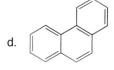

d.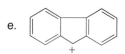

e.

10 π electrons     7 π electrons     10 π electrons     14 π electrons     12 π electrons

**17.28** To be aromatic, the compounds must be cyclic, planar, completely conjugated, and have $4n + 2$ π electrons.

a.    Circled C's are not $sp^2$.
not completely conjugated
**not aromatic**

d.     Circled C is not $sp^2$.
not completely conjugated
**not aromatic**

b.    14 π electrons in outer ring
**aromatic**

e.    12 π electrons
does **not** have $4n + 2$
π electrons
**not aromatic**

c.    4 benzene rings
joined together
**aromatic**

f.    12 π electrons
does **not** have $4n + 2$
π electrons
**not aromatic**

**17.29** In determining if a heterocycle is aromatic, count a nonbonded electron pair if it makes the ring aromatic in calculating $4n +2$. Lone pairs on atoms already part of a multiple bond cannot be delocalized in a ring, and so they are never counted in determining aromaticity.

a.    6 π electrons
counting a lone pair from S
$4(1) + 2 = 6$
**aromatic**

c.    **not aromatic**

e.    **not aromatic**

g.     N is not $sp^2$ (no p orbital)
**not aromatic**

b.    6 π electrons
counting a lone pair from O
$4(1) + 2 = 6$
**aromatic**

d.    10 π electrons
$4(2) + 2 = 10$
**aromatic**

f.    6 π electrons,
counting a lone pair from O
$4(1) + 2 = 6$
**aromatic**

h.    Count
these $2e^-$.

10 π electrons
$4(2) + 2 = 10$
**aromatic**    These lone
pairs are on
doubly
bonded N
atoms, so they
can't be
counted.

**17.30**

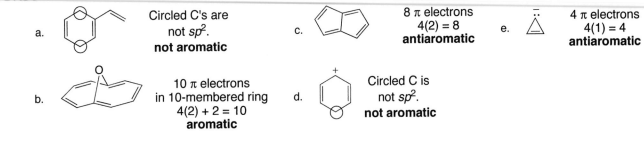

a. Circled C's are not $sp^2$. **not aromatic**

c. 8 π electrons 4(2) = 8 **antiaromatic**

e. 4 π electrons 4(1) = 4 **antiaromatic**

b. 10 π electrons in 10-membered ring 4(2) + 2 = 10 **aromatic**

d. Circled C is not $sp^2$. **not aromatic**

**17.31**

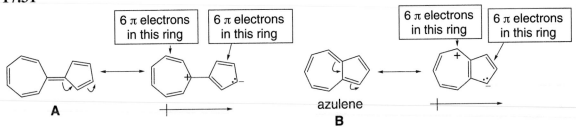

6 π electrons in this ring

6 π electrons in this ring

6 π electrons in this ring

6 π electrons in this ring

**A**

azulene

**B**

In both **A** and **B**, resonance structures can be drawn that place a negative charge in the five-membered ring and a positive charge in the seven-membered ring. These resonance structures show that each ring has 6 π electrons, making it aromatic. Each molecule possesses a dipole such that the seven-membered ring is electron deficient and the five-membered ring is electron rich.

**17.32** Benzene has C–C bonds of equal length, intermediate between a C–C double and single bond. Cyclooctatetraene is not planar and not aromatic so its double bonds are localized.

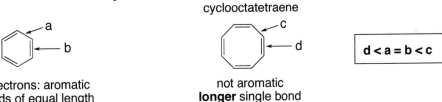

cyclooctatetraene

a
b

c
d

$d < a = b < c$

6 π electrons: aromatic
all bonds of equal length
**intermediate**

not aromatic
**longer** single bond
localized double bond: **shorter**

**17.33**

purine

$sp^2$ hybridized but with lone pair in $p$ orbital.

a. Each N atom is $sp^2$ hybridized.

b. The three unlabeled N atoms are $sp^2$ hybridized with lone pairs in one of the $sp^2$ hybrid orbitals. The labeled N has its lone pair in a $p$ orbital.

c. 10 π electrons

d. Purine is cyclic, planar, completely conjugated, and has 10 π electrons [4(2) + 2] so it is aromatic.

**17.34**

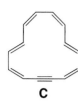

**C**

a. 16 total π electrons
b. 14 π electrons delocalized in the ring. [Note: Two of the electrons in the triple bond are localized between two C's, perpendicular to the π electrons delocalized in the ring.]
c. By having two of the *p* orbitals of the C–C triple bond co-planar with the *p* orbitals of all the C=C's, the total number of π electrons delocalized in the ring is 14. 4(3) + 2 = 14, so the ring is **aromatic.**

**17.35** The rate of an $S_N1$ reaction increases with increasing stability of the intermediate carbocation.

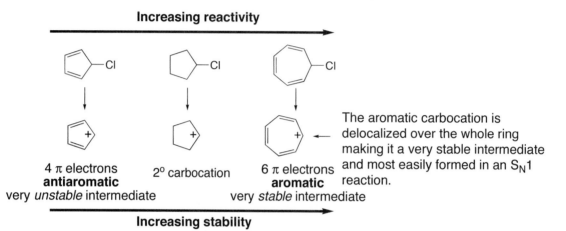

**Increasing reactivity**

4 π electrons
**antiaromatic**
very *unstable* intermediate

2° carbocation

6 π electrons
**aromatic**
very *stable* intermediate

The aromatic carbocation is delocalized over the whole ring making it a very stable intermediate and most easily formed in an $S_N1$ reaction.

**Increasing stability**

**17.36**

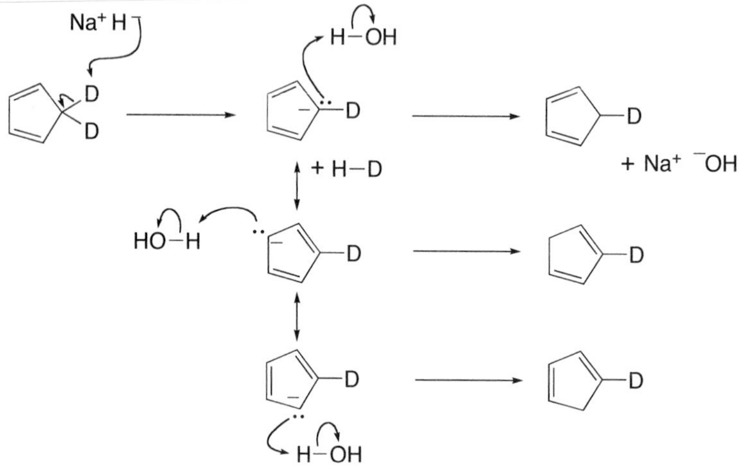

**17.37** α-Pyrone reacts like benzene because it is aromatic. A second resonance structure can be drawn showing how the ring has six π electrons. Thus, α-pyrone undergoes reactions characteristic of aromatic compounds—that is, substitution rather than addition.

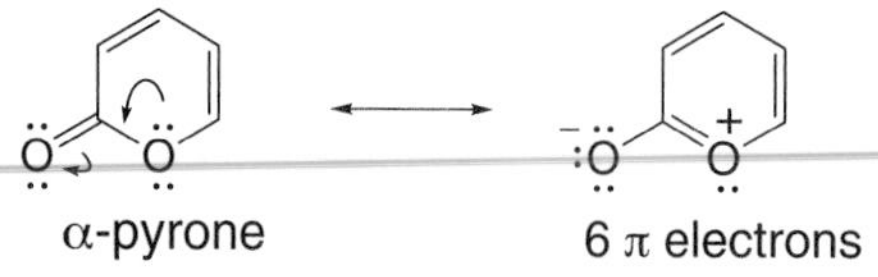

α-pyrone

6 π electrons

**17.38**

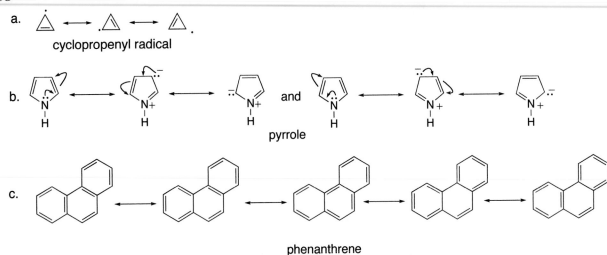

a.

cyclopropenyl radical

b.

and

pyrrole

c.

phenanthrene

**17.39**

Naphthalene can be drawn as three resonance structures:

(a)

(b)

(a)

(b)

(a)

(b)

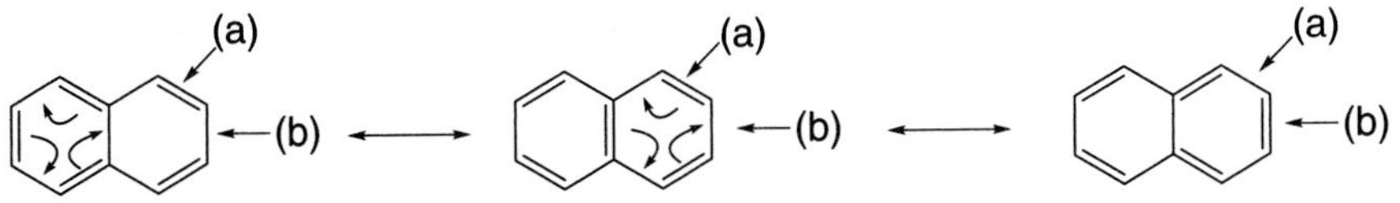

In two of the resonance structures bond (a) is a double bond, and
bond (b) is a single bond. Therefore, bond (b) has more single
bond character, making it longer.

**17.40** The compound with the more stable conjugate base is the stronger acid. Draw and compare the
conjugate bases of each pair of compounds.

conjugate bases

a.  and

**more acidic**

and

resonance-stabilized
but not aromatic

6 π electrons, aromatic
more stable conjugate base
**Its acid is more acidic.**

b. 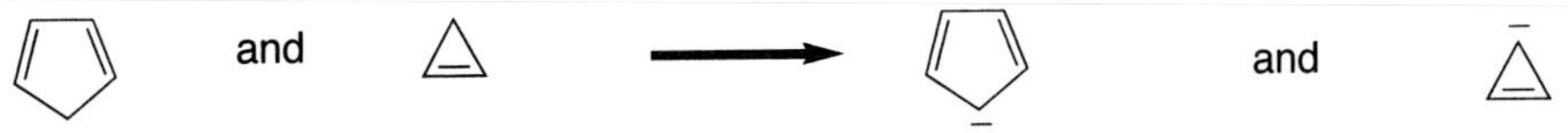 and

**more acidic**

and

6 π electrons, aromatic
more stable conjugate base
**Its acid is more acidic.**

antiaromatic
highly destabilized
conjugate base

**17.41**

indene + NaNH₂ ⟶ [indenyl anion] Na⁺ + NH₃

The conjugate base of indene has 10 π electrons making it aromatic
and very stable. Therefore, indene is more acidic than many hydrocarbons.

**17.42**

pyrrole ⟶ conjugate base

Both pyrrole and the conjugate base of pyrrole have 6 π
electrons in the ring, making them both aromatic. Thus,
deprotonation of pyrrole does not result in a gain of aromaticity
since the starting material is aromatic to begin with.

cyclopentadiene
more acidic ⟶ conjugate base

Cyclopentadiene is not aromatic, but the conjugate base has 6 π
electrons and is therefore aromatic. This makes the C–H bond in
cyclopentadiene more acidic than the N–H bond in pyrrole, since
deprotonation of cyclopentadiene forms an aromatic conjugate base.

**17.43**

a.

pyrrole

**A** ⟷ ⟷

**B**

Protonation at C2 forms conjugate acid
**A** because the positive charge can be
delocalized by resonance. There is no
resonance stabilization of the positive
charge in **B**.

b.

**A**
p$K_a$ = 0.4

Loss of a proton from **A** (which is not
aromatic) gives two electrons to N, and forms
pyrrole, which has six π electrons that can
then delocalize in the five-membered ring,
making it aromatic. This makes deprotonation
a highly favorable process, and **A** more acidic.

p$K_a$ = 5.2
**C**

Both **C** and its conjugate base pyridine are
aromatic. Since **C** has six π electrons, it is
already aromatic to begin with, so there is less
to be gained by deprotonation, and **C** is thus
less acidic than **A**.

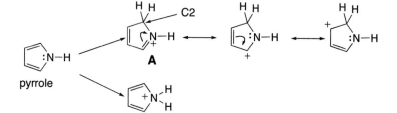

**17.44**

a.

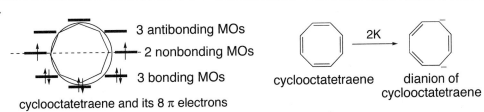

3 antibonding MOs
2 nonbonding MOs
3 bonding MOs

cyclooctatetraene and its 8 π electrons

cyclooctatetraene    dianion of cyclooctatetraene

b. Even if cyclooctatetraene were flat, it has two unpaired electrons in its HOMOs (nonbonding MOs) so it cannot be aromatic.

c. The dianion has 10 π electrons.

d. The two additional electrons fill the nonbonding MOs; that is, all the bonding and nonbonding MOs are filled with electrons in the dianion.

e. The dianion is aromatic since its HOMOs are completely filled, and it has no electrons in antibonding MOs.

**17.45**

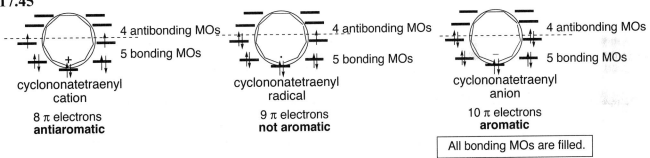

4 antibonding MOs
5 bonding MOs

cyclononatetraenyl cation

8 π electrons
**antiaromatic**

4 antibonding MOs
5 bonding MOs

cyclononatetraenyl radical

9 π electrons
**not aromatic**

4 antibonding MOs
5 bonding MOs

cyclononatetraenyl anion

10 π electrons
**aromatic**

All bonding MOs are filled.

**17.46** Use the directions from Answer 17.7.

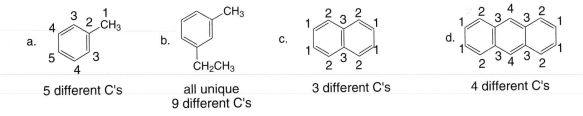

a. 5 different C's

b. all unique 9 different C's

c. 3 different C's

d. 4 different C's

**17.47** Draw the three isomers and count the different types of carbon in each. Then match the structures with the data.

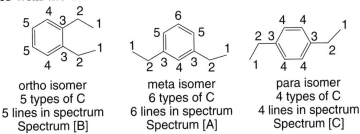

ortho isomer
5 types of C
5 lines in spectrum
Spectrum [B]

meta isomer
6 types of C
6 lines in spectrum
Spectrum [A]

para isomer
4 types of C
4 lines in spectrum
Spectrum [C]

**17.48**

a. $C_{10}H_{14}$: IR absorptions at 3150–2850 ($sp^2$ and $sp^3$ hybridized C–H), 1600, and 1500 (due to a benzene ring) $cm^{-1}$.

$^1H$ NMR data:

| Absorption | ppm | # of H's | Explanation |
|---|---|---|---|
| doublet | 1.2 | 6 | 6H's adjacent to 1H |
| singlet | 2.3 | 3 | $CH_3$ |
| septet | 3.1 | 1 | 1H adjacent to 6H's |
| multiplet | 7–7.4 | 4 | a disubstituted benzene ring |

$(CH_3)_2CH$ group

You can't tell from these data where the two groups are on the benzene ring. They are not para. That usually gives two sets of distinct peaks (resembling two doublets) so there are two possible structures— ortho and meta isomers.

or

b. $C_9H_{12}$: $^{13}C$ NMR signals at 21, 127, and 138 ppm → means three different types of C's.
$^1H$ NMR shows 2 types of H's: 9H's probably means 3 $CH_3$ groups; the other 3H's are very deshielded so they are bonded to a benzene ring.
Only one possible structure fits:

c. $C_8H_{10}$: IR absorptions at 3108–2875 ($sp^2$ and $sp^3$ hybridized C–H), 1606, and 1496 (due to a benzene ring) $cm^{-1}$.

$^1H$ NMR data:

| Absorption | ppm | # of H's | Explanation | Structure: |
|---|---|---|---|---|
| triplet | 1.3 | 3 | 3H's adjacent to 2H's | |
| quartet | 2.7 | 2 | 2H's adjacent to 3H's | $CH_2CH_3$ |
| multiplet | 7.3 | 5 | a monosubstituted benzene ring | |

**17.49**

a. Compound **A**: Molecular formula $C_8H_{10}O$.
IR absorption at 3150–2850 ($sp^2$ and $sp^3$ hybridized C–H) $cm^{-1}$.

$^1H$ NMR data:

| Absorption | ppm | # of H's | Explanation | Structure: |
|---|---|---|---|---|
| triplet | 1.4 | 3 | 3H's adjacent to 2H's | |
| quartet | 3.95 | 2 | 2H's adjacent to 3H's | $OCH_2CH_3$ |
| multiplet | 6.8–7.3 | 5 | a monosubstituted benzene ring | |

b. Compound **B**: Molecular formula $C_9H_{10}O_2$.
   IR absorption at 1669 (C=O) cm$^{-1}$.
   $^1$H NMR data:

| Absorption | ppm | # of H's | Explanation |
|---|---|---|---|
| singlet | 2.5 | 3 | CH$_3$ group |
| singlet | 3.8 | 3 | CH$_3$ group |
| doublet | 6.9 | 2 | 2H's on a benzene ring |
| doublet | 7.9 | 2 | 2H's on a benzene ring |

**Structure:**

It would be hard to distinguish
these two compounds.

**17.50**

3 equivalent H's
singlet at ~2.3 ppm

**thymol**

* 3 other H's on benzene ring
(arrows with *)
at ~6.9 ppm

1 H
septet at ~3.2 ppm

6 equivalent H's
doublet at ~1.2 ppm

IR absorptions:
3500–3200 cm$^{-1}$ (O–H)
3150–2850 cm$^{-1}$ (C–H bonds)
1621 and 1585 cm$^{-1}$ (benzene ring)

basic structure of
thymol

Thymol must have this basic structure given the NMR and IR data since it is a
trisubstituted benzene ring with one singlet and two doublets in the NMR at
~6.9 ppm. However, which group [OH, CH$_3$, or CH(CH$_3$)$_2$] corresponds to X, Y,
and Z is not readily distinguished. The correct structure for thymol is given.

**17.51**

The induced magnetic field by the circulating π electrons
opposes the applied field in this vicinity, shifting the absorption upfield to
a lower chemical shift than other $sp^3$ C–H protons.
In this case the protons absorb upfield from TMS, an unusual
phenomenon for C–H protons.

Induced magnetic field

H H

2.84 ppm → H

–0.6 ppm
(upfield from TMS)

H

Induced magnetic field

$B_o$

The induced magnetic field by the circulating π electrons
reinforces the applied field in this vicinity, shifting the absorption
downfield to a somewhat higher chemical shift.

**17.52**

a.

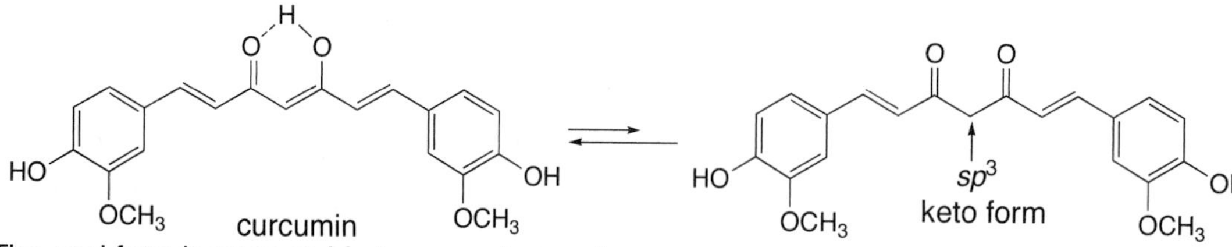

curcumin      keto form

The enol form is more stable because the enol
double bond makes a highly conjugated
system. The enol OH can also intramolecularly
hydrogen bond to the nearby carbonyl O atom.

b.

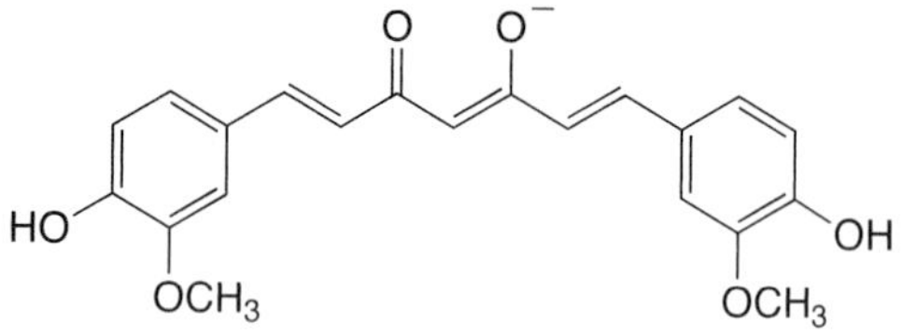

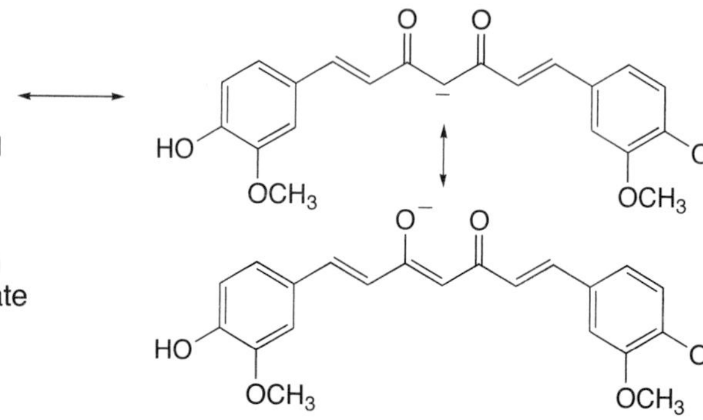

The O–H proton is more acidic than an
alcohol O–H proton because the conjugate
base is resonance stabilized.

c. Curcumin is colored because it has many conjugated π electrons, which shift absorption of light from the UV to the visible region.

d. Curcumin is an antioxidant because it contains a phenol. Homolytic cleavage affords a resonance-stabilized phenoxy radical, which can inhibit oxidation from occurring, much like vitamin E and BHT in Chapter 15.

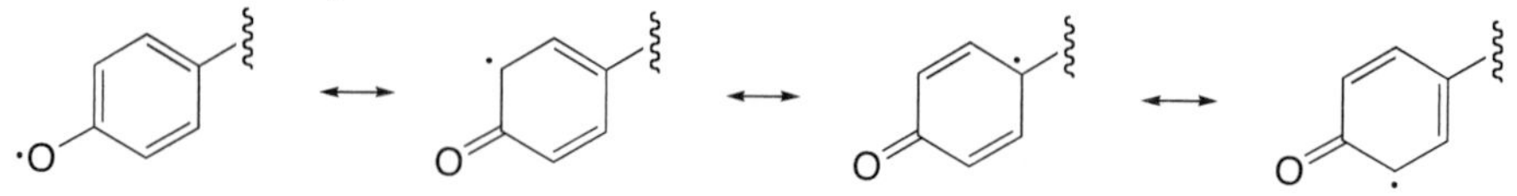

(+ other resonance structures)

phenoxy radical
Resonance delocalizes the radical on the ring and C chain of curcumin.

**17.53** A second resonance structure for **A** shows that the ring is completely conjugated and has 6 π electrons, making it aromatic and especially stable. A similar charge-separated resonance structure for **B** makes the ring completely conjugated, but gives the ring 4 π electrons, making it antiaromatic and especially unstable.

**A**      6 π electrons
**aromatic**
**stable**

**B**      4 π electrons
**antiaromatic**
**not stable**

**17.54**  Resonance structures for triphenylene:

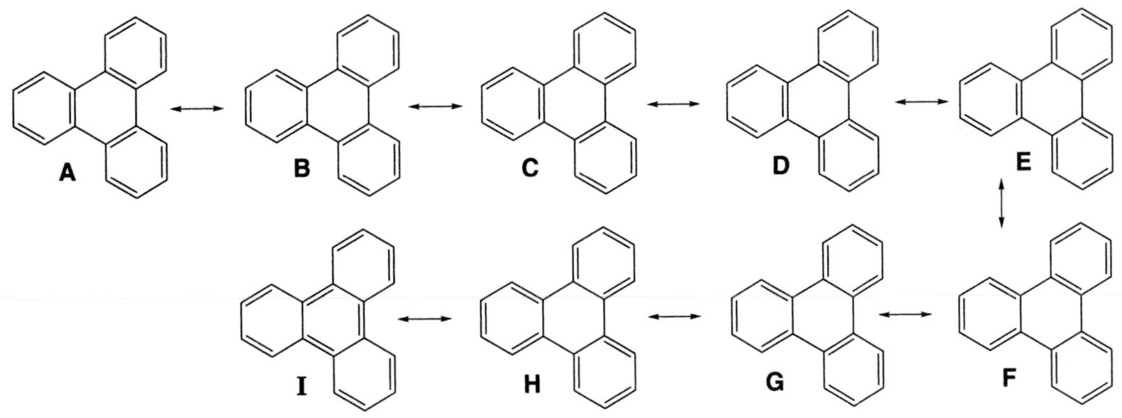

Resonance structures **A–H** all keep three double and three single bonds in the three six-membered rings on the periphery of the molecule.  This means that each ring behaves like an isolated benzene ring undergoing substitution rather than addition because the π electron density is delocalized within each six-membered ring.  Only resonance structure **I** does not have this form.  Each C–C bond of triphenylene has four (or five) resonance structures in which it is a single bond and four (or five) resonance structures in which it is a double bond.

Resonance structures for phenanthrene:

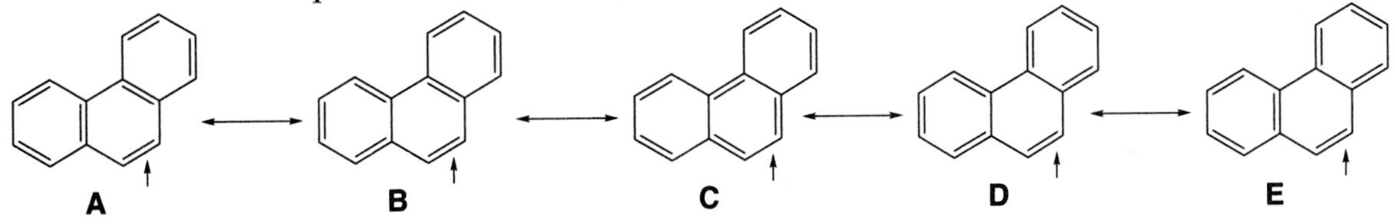

With phenanthrene, however, four of the five resonance structures keep a double bond at the labeled C's.  (Only **C** does not.)  This means that these two C's have more double bond character than other  C–C bonds in phenanthrene, making them more susceptible to addition rather than substitution.

**17.55**

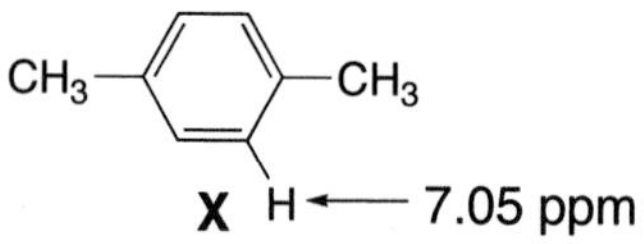

CH$_3$—⬡—CH$_3$

**X**  H ⟵ 7.05 ppm

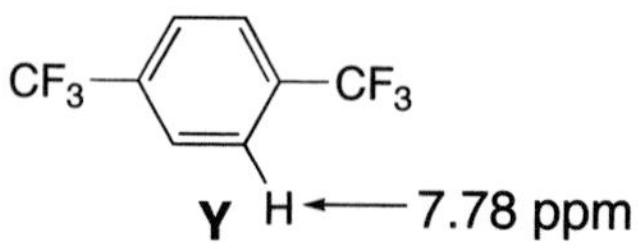

CH$_3$ groups are electron donating. Increasing the electron density of the ring shields the protons and shifts the absorption slightly upfield.

CF$_3$—⬡—CF$_3$

**Y**  H ⟵ 7.78 ppm

The 3 electronegative F's make CF$_3$ an electron-withdrawing group. Decreasing the electron density of the ring deshields the protons and the absorption goes slightly downfield.

When F's are directly bonded to the benzene ring, two conflicting factors come into play. Since F is very electronegative, it withdraws electron density from the ring. But, F atoms also contain lone pairs of electrons that can be donated to the ring by resonance. For example:

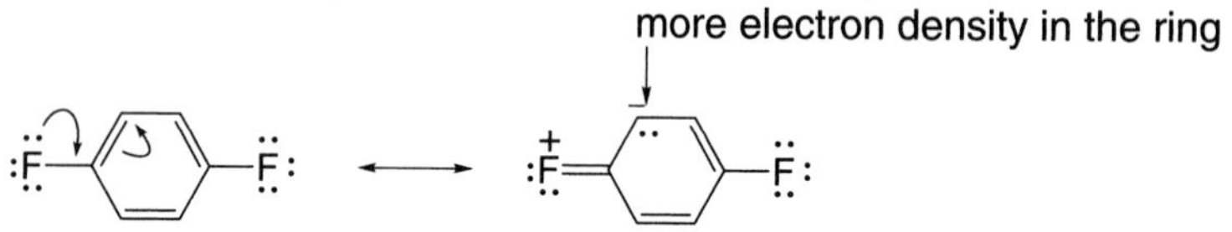

more electron density in the ring

:F—⬡—F:  ⟷  :F⁺=⬡—F:

This increases the electron density of the ring. On balance, these factors just about cancel, so the absorption occurs at ~7 ppm.

## Chapter 18:  Electrophilic Aromatic Substitution

### ◆ Mechanism of electrophilic aromatic substitution (18.2)

- Electrophilic aromatic substitution follows a two-step mechanism.  Reaction of the aromatic ring with an electrophile forms a carbocation, and loss of a proton regenerates the aromatic ring.
- The first step is rate-determining.
- The intermediate carbocation is stabilized by resonance;  a minimum of three resonance structures can be drawn.  The positive charge is always located ortho or para to the new C–E bond.

| (+) ortho to E | (+) para to E | (+) ortho to E |

### ◆ Three rules describing the reactivity and directing effects of common substituents (18.7–18.9)

[1]  All ortho, para directors except the halogens activate the benzene ring.
[2]  All meta directors deactivate the benzene ring.
[3]  The halogens deactivate the benzene ring.

### ◆ Summary of substituent effects in electrophilic aromatic substitution (18.6–18.9)

| | Substituent | Inductive effect | Resonance effect | Reactivity | Directing effect |
|---|---|---|---|---|---|
| [1] | R = alkyl | donating | none | activating | ortho, para |
| [2] | Z = N or O | withdrawing | donating | activating | ortho, para |
| [3] | X = halogen | withdrawing | donating | deactivating | ortho, para |
| [4] | Y (δ⁺ or +) | withdrawing | withdrawing | deactivating | meta |

### ◆ Five examples of electrophilic aromatic substitution

### [1]  Halogenation–Replacement of H by Cl or Br (18.3)

aryl chloride  or  aryl bromide

[X = Cl,  Br]

- Polyhalogenation occurs on benzene rings substituted by OH and NH$_2$ (and related substituents) (18.10A).

## [2]  Nitration–Replacement of H by NO₂ (18.4)

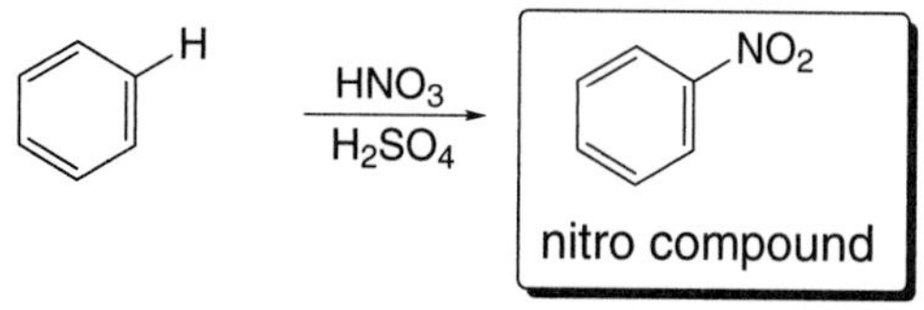

benzene $\xrightarrow[\text{H}_2\text{SO}_4]{\text{HNO}_3}$ nitro compound (NO₂)

## [3]  Sulfonation–Replacement of H by SO₃H (18.4)

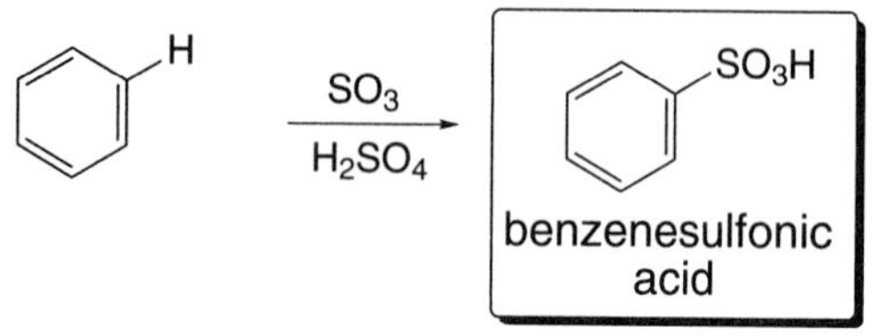

benzene $\xrightarrow[\text{H}_2\text{SO}_4]{\text{SO}_3}$ benzenesulfonic acid (SO₃H)

## [4]  Friedel–Crafts alkylation–Replacement of H by R (18.5)

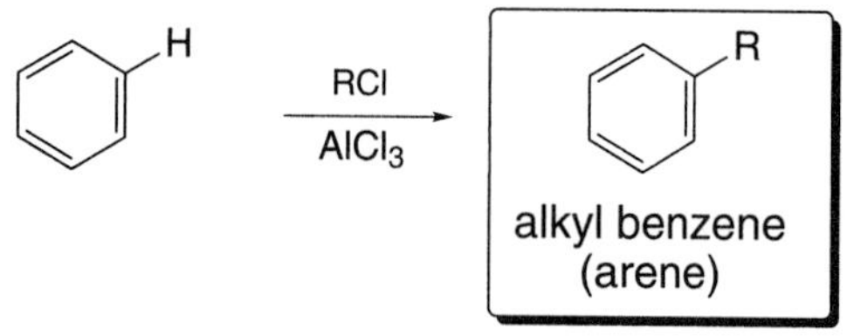

benzene $\xrightarrow[\text{AlCl}_3]{\text{RCl}}$ alkyl benzene (arene) (R)

- Rearrangements can occur.
- Vinyl halides and aryl halides are unreactive.
- The reaction does not occur on benzene rings substituted by meta deactivating groups or NH₂ groups (18.10B).
- Polyalkylation can occur.

**Variations:**

[1] with alcohols

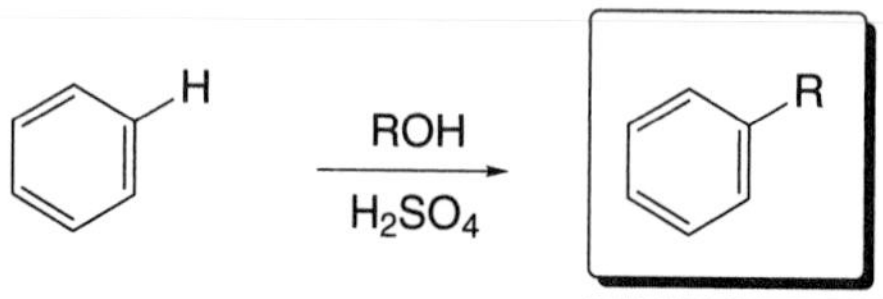

benzene $\xrightarrow[\text{H}_2\text{SO}_4]{\text{ROH}}$ (R)

[2] with alkenes

benzene $\xrightarrow[\text{H}_2\text{SO}_4]{\text{CH}_2=\text{CHR}}$ (R, CH₃)

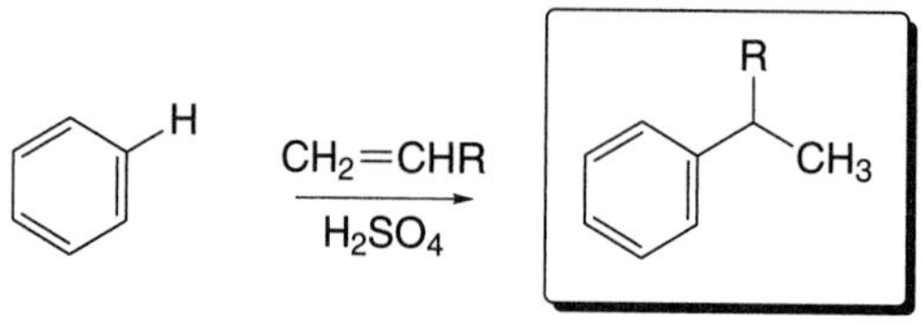

## [5]  Friedel–Crafts acylation–Replacement of H by RCO (18.5)

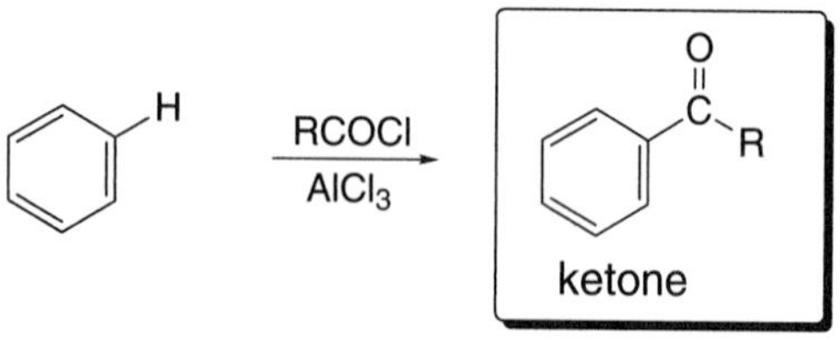

benzene $\xrightarrow[\text{AlCl}_3]{\text{RCOCl}}$ ketone

- The reaction does not occur on benzene rings substituted by meta deactivating groups or NH₂ groups (18.10B).

◆ Other reactions of benzene derivatives

## [1] Benzylic halogenation (18.13)

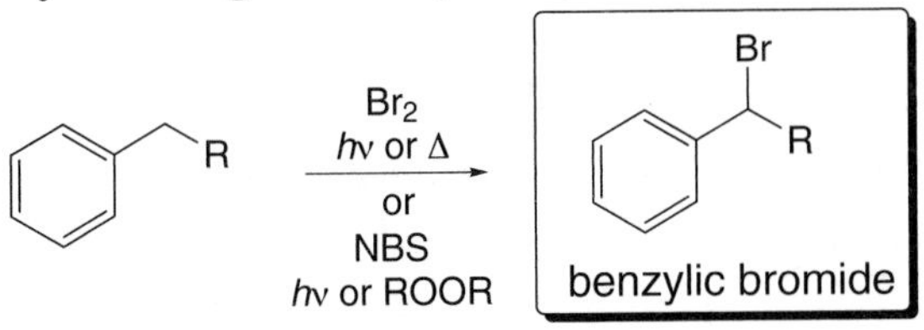
R
$Br_2$
hν or Δ
or
NBS
hν or ROOR
Br
R
benzylic bromide

## [2] Oxidation of alkyl benzenes (18.14A)

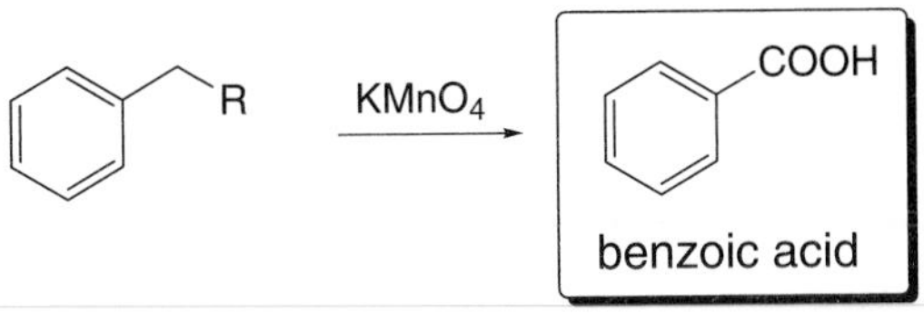
R
$KMnO_4$
COOH
benzoic acid

- A benzylic C–H bond is needed for reaction.

## [3] Reduction of ketones to alkyl benzenes (18.14B)

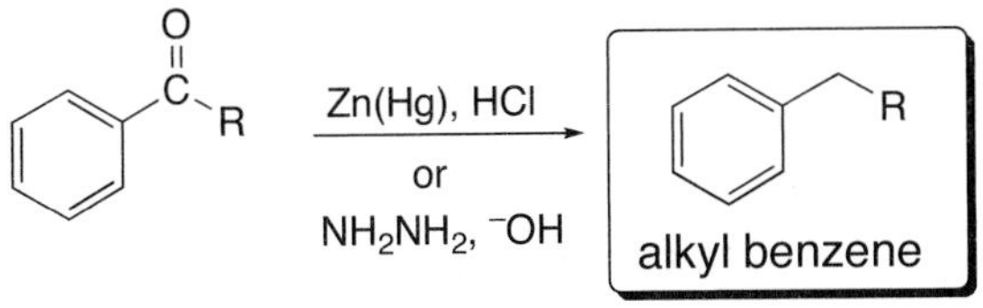
O
C
R
Zn(Hg), HCl
or
$NH_2NH_2$, $^{-}OH$
R
alkyl benzene

## [4] Reduction of nitro groups to amino groups (18.14C)

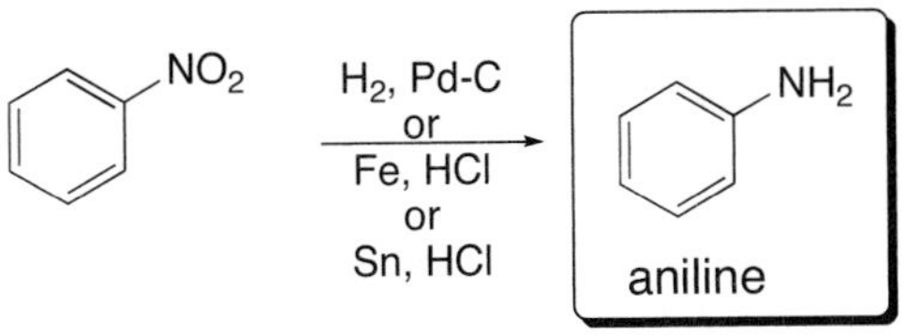
$NO_2$
$H_2$, Pd-C
or
Fe, HCl
or
Sn, HCl
$NH_2$
aniline

## Chapter 18: Answers to Problems

**18.1** The π electrons of benzene are delocalized over the six atoms of the ring, increasing benzene's stability and making them less available for electron donation. With an alkene, the two π electrons are localized between the two C's making them more nucleophilic and thus more reactive with an electrophile than the delocalized electrons in benzene.

**18.2**

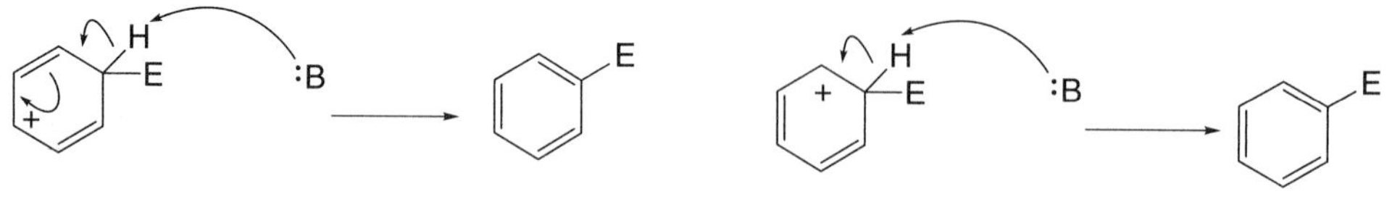

**18.3** Addition of $Cl_2$ with $FeCl_3$ as the catalyst occurs in two parts. First is the formation of an electrophile, followed by a two-step substitution reaction.

[1]

[2]

[3]

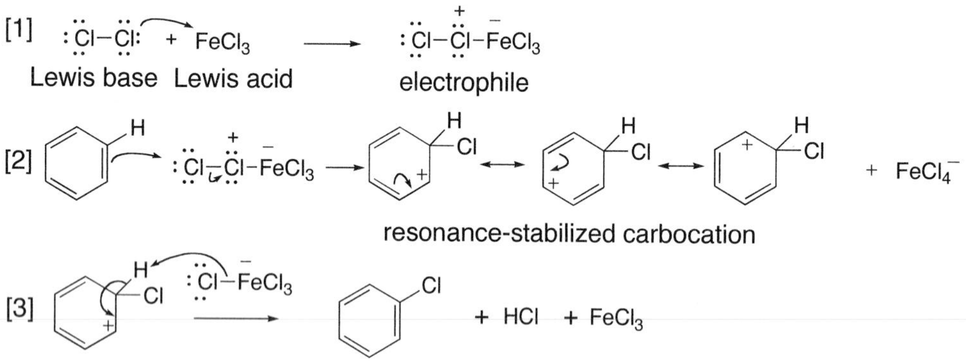

**18.4** There are two parts in the mechanism. The first part is formation of an electrophile. The second part is a two-step substitution reaction.

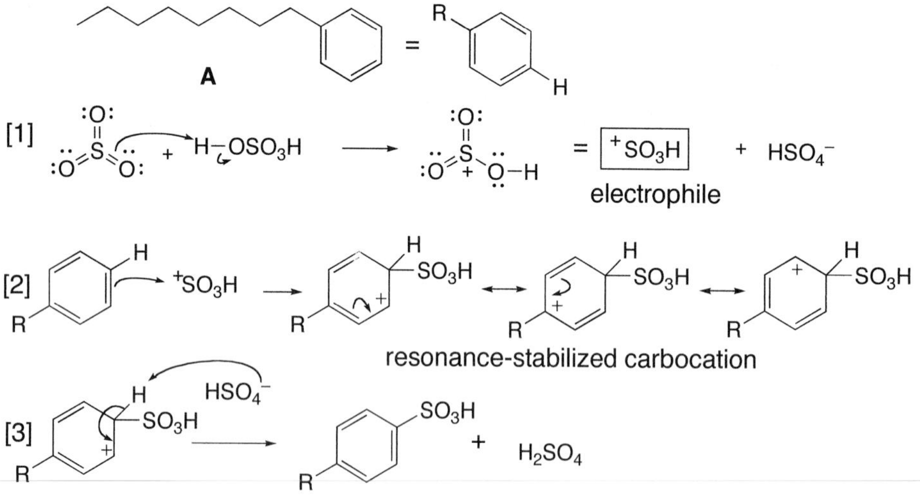

[1]

[2]

[3]

**18.5** Friedel–Crafts alkylation results in the transfer of an alkyl group from a halogen to a benzene ring. In Friedel–Crafts acylation an acyl group is transferred from a halogen to a benzene ring.

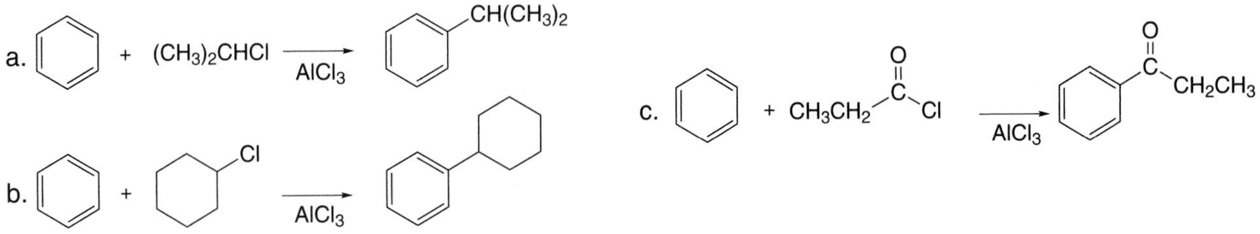
a.
(CH3)2CHCl
AlCl3
CH(CH3)2
b.
Cl
AlCl3
c.
CH3CH2
Cl
AlCl3
CH2CH3

**18.6** Remember that an acyl group is transferred from a Cl atom to a benzene ring. To draw the acid chloride, substitute a Cl for the benzene ring.

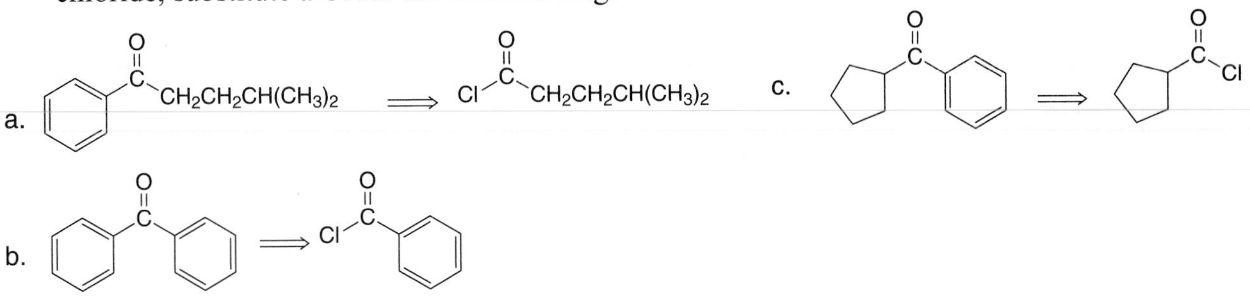
a.
CH2CH2CH(CH3)2
Cl
CH2CH2CH(CH3)2
b.
Cl
c.
Cl

**18.7**

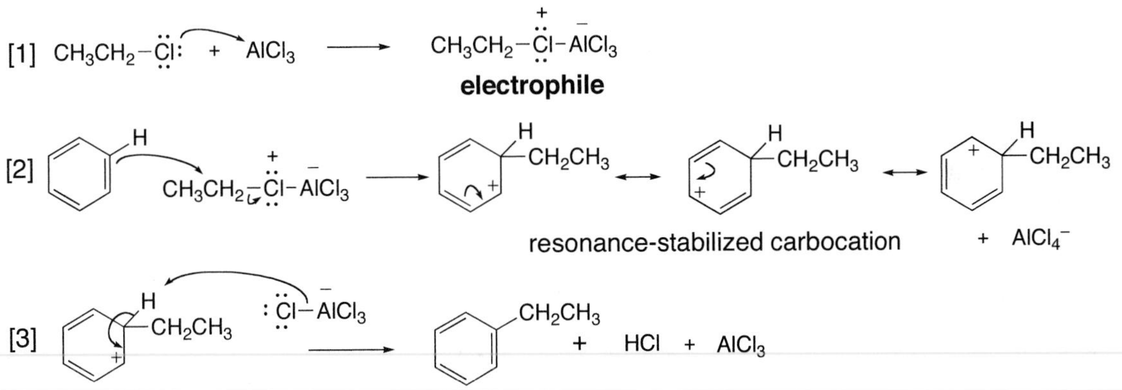
[1]
CH3CH2–Cl:
AlCl3
CH3CH2–Cl–AlCl3
electrophile
[2]
resonance-stabilized carbocation
+ AlCl4–
[3]
CH2CH3
HCl + AlCl3

**18.8** To be reactive in a Friedel–Crafts alkylation reaction, the X must be bonded to an $sp^3$ hybridized carbon atom.

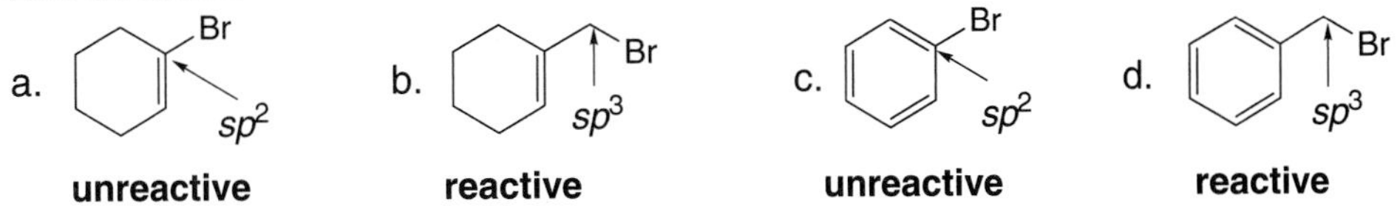
a.
Br
sp2
unreactive
b.
Br
sp3
reactive
c.
Br
sp2
unreactive
d.
Br
sp3
reactive

**18.9** The product has an “unexpected” carbon skeleton, so rearrangement must have occurred.

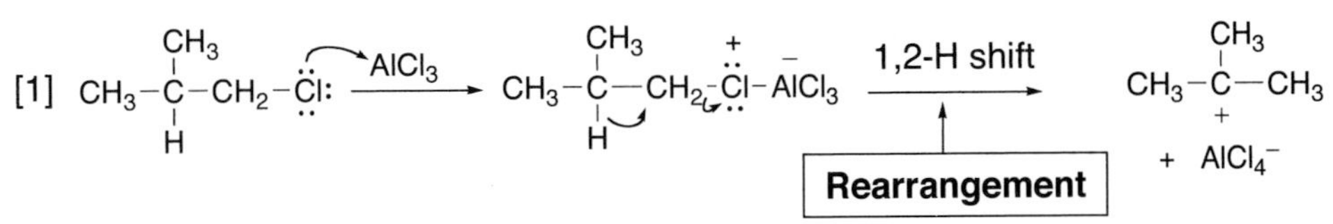
[1]
AlCl3
1,2-H shift
Rearrangement
+ AlCl4–

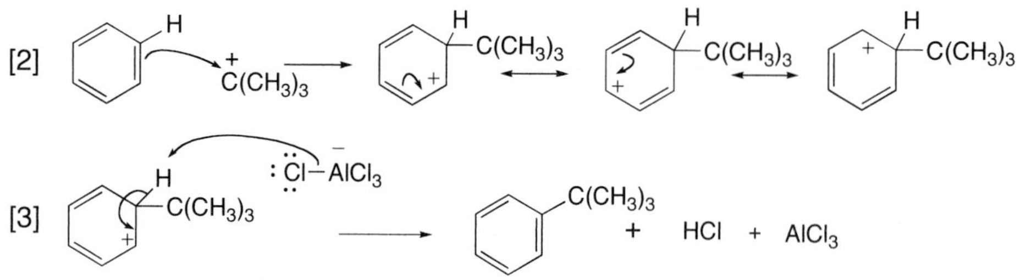

**18.10** Both alkenes and alcohols can form carbocations for Friedel–Crafts alkylation reactions.

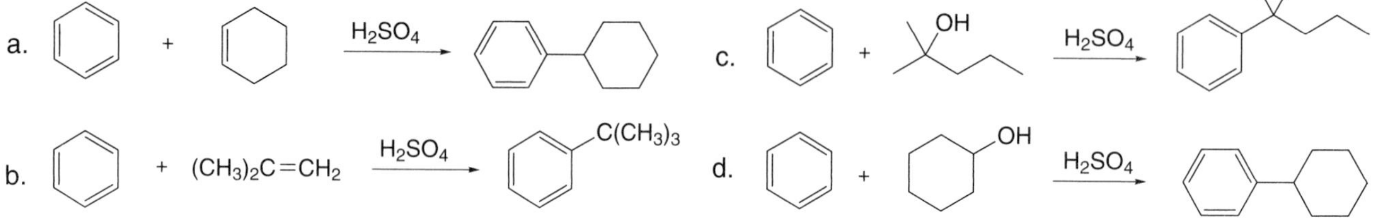

**18.11**

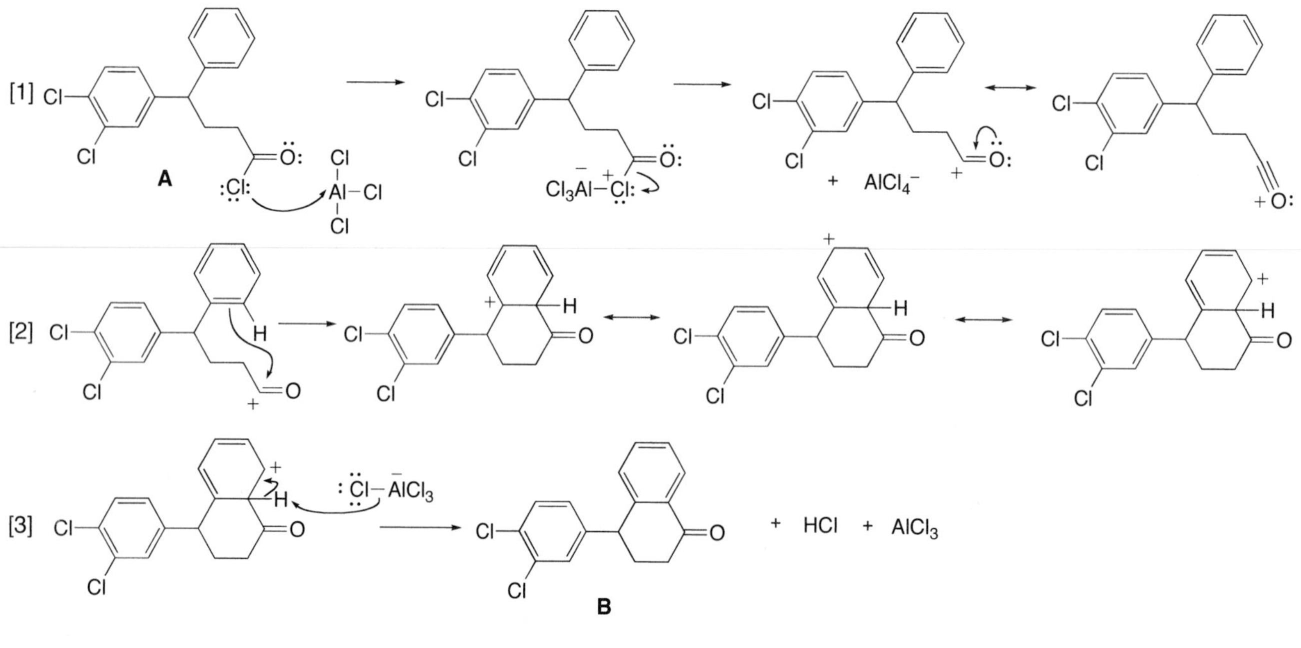

**18.12**

a. —CH$_2$CH$_2$CH$_2$CH$_3$
   alkyl group
   **electron donating**

b. —Br
   halide
   **electron withdrawing**

c. —OCH$_2$CH$_3$
   electronegative O
   **electron withdrawing**

**18.13** Electron-donating groups place a negative charge in the benzene ring. Draw the resonance structures to show how —OCH$_3$ puts a negative charge in the ring. Electron-withdrawing groups place a positive charge in the benzene ring. Draw the resonance structures to show how —COCH$_3$ puts a positive charge in the ring.

a.

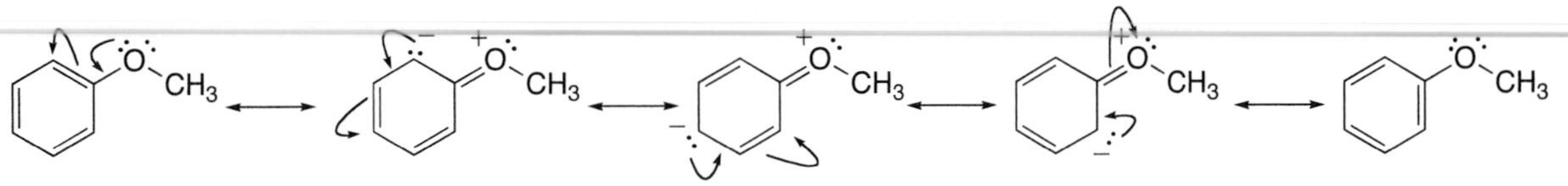

b.

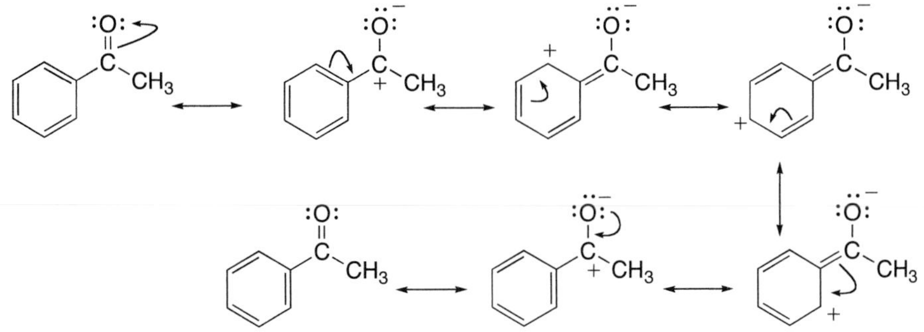

**18.14** To classify each substituent, look at the atom directly bonded to the benzene ring. All R groups and Z groups (except halogens) are electron donating. All groups with a positive charge, $\delta^+$, or halogens are electron withdrawing.

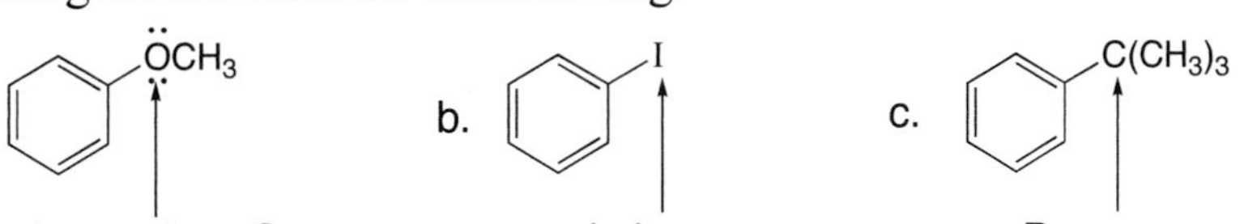

a.

OCH₃

lone pair on O
**electron donating**

b.

I

halogen
**electron withdrawing**

c.

C(CH₃)₃

R group
**electron donating**

**18.15** First classify the starting material as: ortho, para activating, ortho, para deactivating, or meta deactivating. Then draw the products.

a.

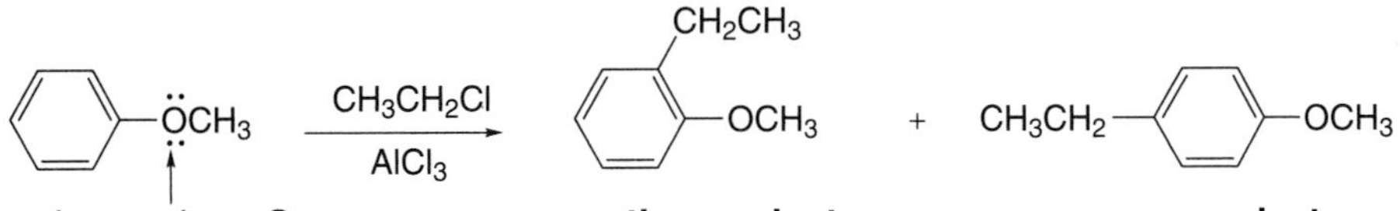

—ÖCH₃    $\xrightarrow[\text{AlCl}_3]{\text{CH}_3\text{CH}_2\text{Cl}}$    CH₂CH₃ / —OCH₃    +    CH₃CH₂— —OCH₃

lone pair on O
**o,p activating**            **ortho product**                    **para product**

b.

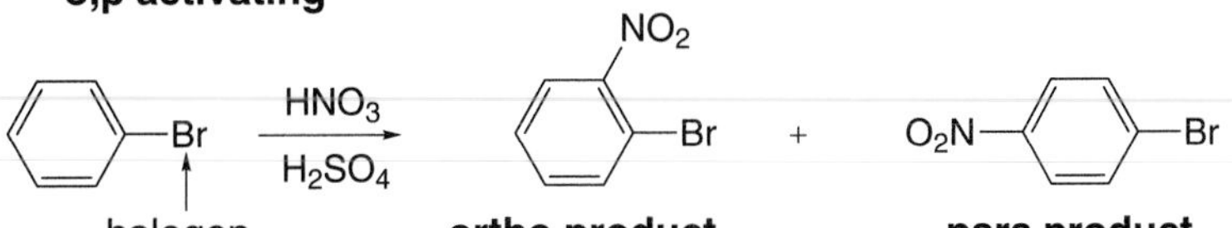

—Br    $\xrightarrow[\text{H}_2\text{SO}_4]{\text{HNO}_3}$    NO₂ / —Br    +    O₂N— —Br

halogen
**o,p deactivating**        **ortho product**            **para product**

c.

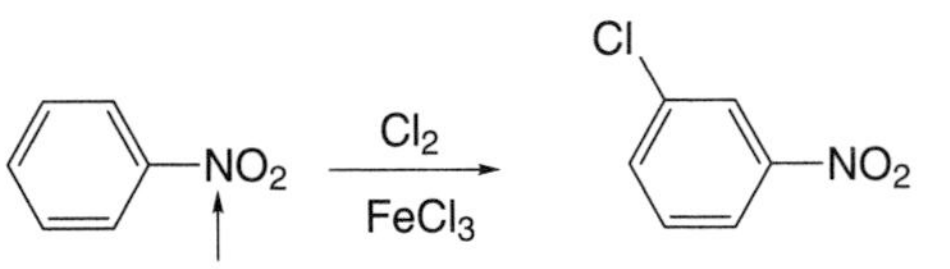

—NO₂    $\xrightarrow[\text{FeCl}_3]{\text{Cl}_2}$    Cl / —NO₂

**meta deactivating**        **meta product**

**18.16** **Electron-donating groups** make the compound ***react faster*** than benzene in electrophilic aromatic substitution. **Electron-withdrawing groups** make the compound ***react more slowly*** than benzene in electrophilic aromatic substitution.

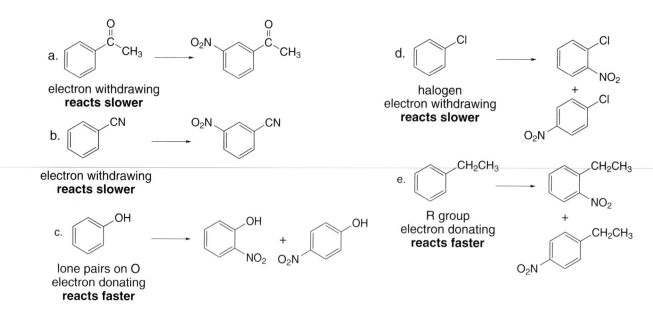

**18.17** **Electron-donating groups** make the compound ***more reactive*** than benzene in electrophilic aromatic substitution. **Electron-withdrawing groups** make the compound ***less reactive*** than benzene in electrophilic aromatic substitution.

a. 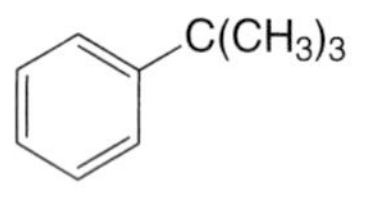
R group
electron donating
**more reactive**

b. 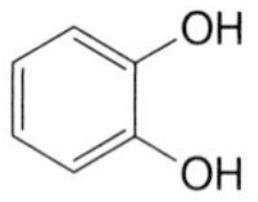
two OH's
electron donating
**more reactive**

c. 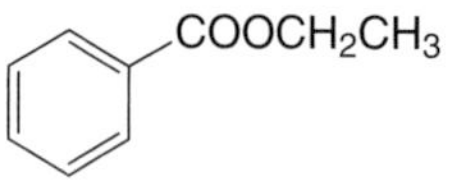
C with 2 electronegative O's
electron withdrawing
**less reactive**

d. 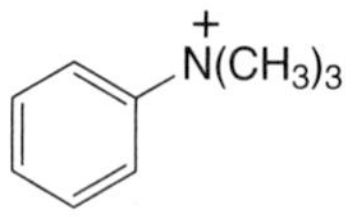
electron withdrawing
**less reactive**

**18.18**

a. 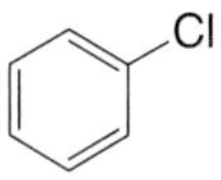
halogen
electron withdrawing
**least reactive**

intermediate
**reactivity**

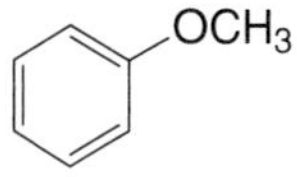

lone pairs on O
electron donating
**most reactive**

b. 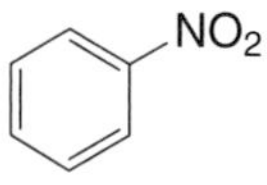
electron withdrawing
**least reactive**

intermediate
**reactivity**

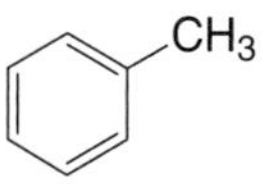

R group
electron donating
**most reactive**

**18.19** Especially stable resonance structures have all atoms with an octet. Carbocations with additional electron donor R groups are also more stable structures. Especially unstable resonance structures have adjacent like charges.

a.

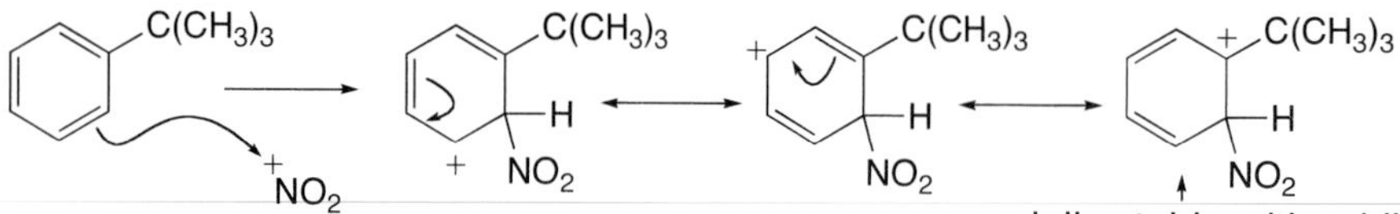

especially stable with additional R group
**stabilized carbocation**

b.

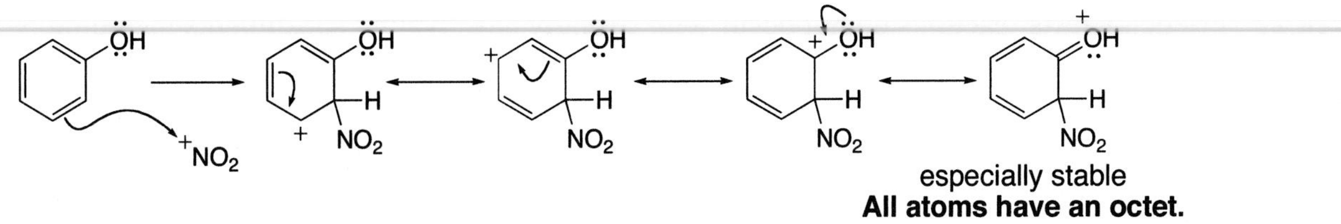

especially stable
**All atoms have an octet.**

c.

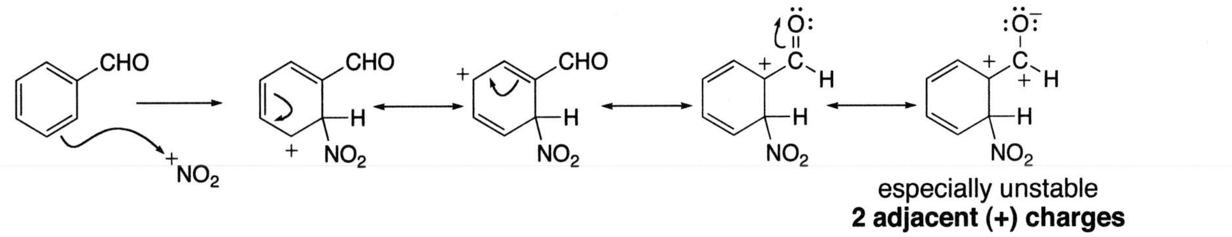

especially unstable
**2 adjacent (+) charges**

**18.20**

ortho
attack

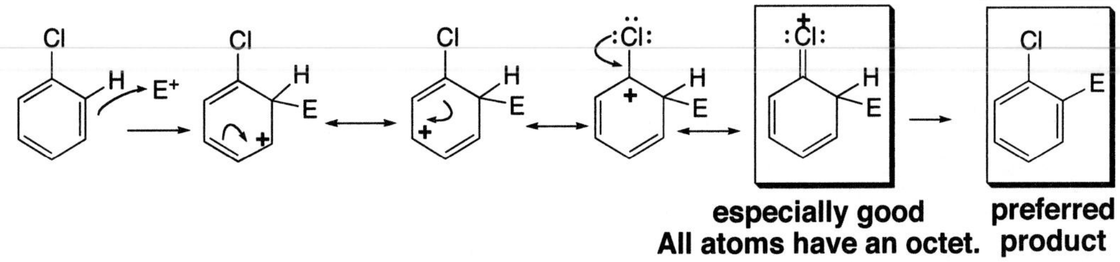

especially good
**All atoms have an octet.**    preferred
product

meta
attack

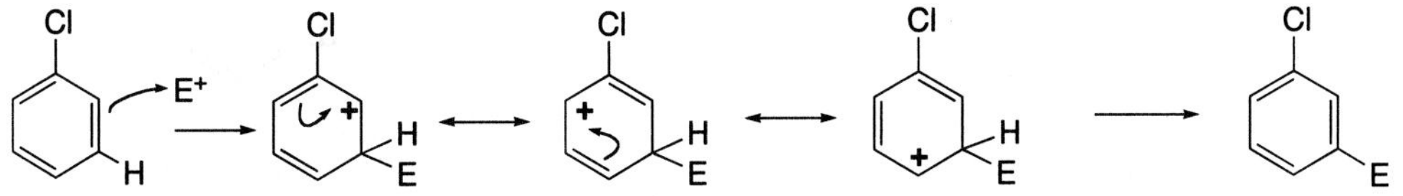

para
attack

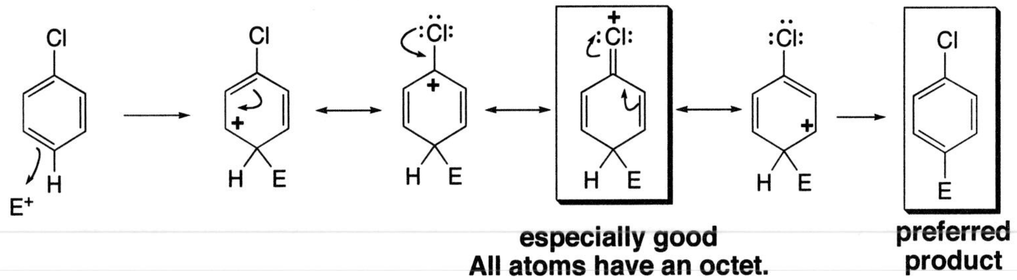

especially good
**All atoms have an octet.**    preferred
product

**18.21**   Polyhalogenation occurs with highly activated benzene rings containing OH, $NH_2$, and related groups with a catalyst.

a.

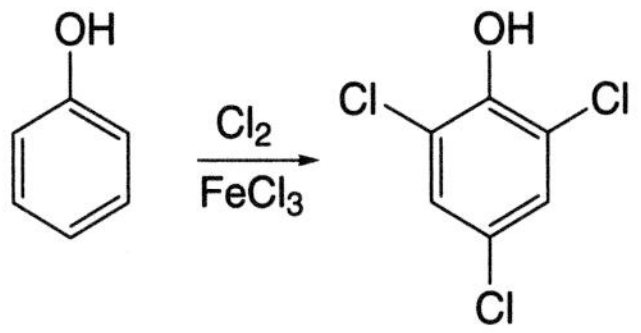

b.

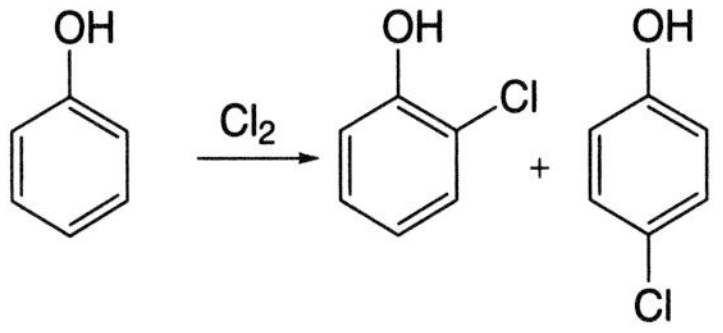

c.

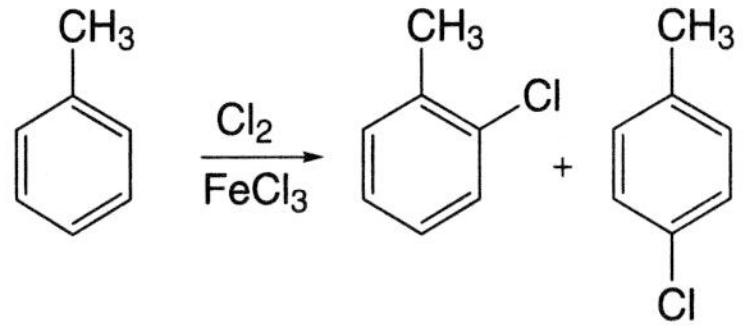

**18.22** Friedel–Crafts reactions do not occur with strongly deactivating substituents including $NO_2$, or with $NH_2$, $NR_2$, or $NHR$ groups.

a. [benzene]—$SO_3H$ $\xrightarrow[AlCl_3]{CH_3Cl}$ **no reaction**

strongly deactivating

b. [benzene]—$Cl$ $\xrightarrow[AlCl_3]{CH_3Cl}$ [toluene with $CH_3$ top, $Cl$]  +  $CH_3$—[benzene]—$Cl$

Cl is an o,p director.

c. [benzene]—$N(CH_3)_2$ $\xrightarrow[AlCl_3]{CH_3Cl}$ **no Friedel–Crafts reaction**

d. [benzene]—$NHCOCH_3$ $\xrightarrow[AlCl_3]{CH_3Cl}$ [benzene with $CH_3$ and $NHCOCH_3$]  +  $CH_3$—[benzene]—$NHCOCH_3$

**18.23** To draw the product of reaction with these disubstituted benzene derivatives and $HNO_3$, $H_2SO_4$ remember:

- If the two directing effects reinforce each other, the new substituent will be on the position reinforced by both.
- If the directing effects oppose each other, the stronger activator wins.
- No substitution occurs between two meta substituents.

a. o,p ↓ $OCH_3$ / meta ↑ $COOCH_3$ $\xrightarrow[H_2SO_4]{HNO_3}$ [product with $OCH_3$, $NO_2$, $COOCH_3$]

b. $OCH_3$ / $Br$ (o,p strong) (o,p oppose) $\xrightarrow[H_2SO_4]{HNO_3}$ [$O_2N$, $OCH_3$, $Br$ product]  +  [$OCH_3$, $Br$, $NO_2$ product]

products due to $OCH_3$ directing effects

c. o,p ↓ $CH_3$ / meta $NO_2$ $\xrightarrow[H_2SO_4]{HNO_3}$ [$O_2N$, $CH_3$, $NO_2$ product]  +  [$CH_3$, $NO_2$, $NO_2$ product]

d. o,p ↓ $Cl$ / o,p ↑ $Br$ $\xrightarrow[H_2SO_4]{HNO_3}$ [$O_2N$, $Cl$, $Br$ product]  +  [$Cl$, $Br$, $NO_2$ product]

**18.24**

a. [benzene] $\xrightarrow[\ ]{SO_3, H_2SO_4}$ [benzene]—$SO_3H$ $\xrightarrow[\ ]{Cl_2, FeCl_3}$ [$Cl$, benzene, $SO_3H$ product]

Put meta director on first.

c. [benzene]—$OH$ $\xrightarrow[\ ]{CH_3Cl, AlCl_3}$ [$OH$, $CH_3$ product] $\xrightarrow[\ ]{Br_2}$ [$OH$, $Br$, $CH_3$ product]

(+ ortho isomer)  Br goes ortho to the stronger activator.

b. [benzene] $\xrightarrow[AlCl_3]{ClCOCH_3}$ [phenyl—C(=O)—$CH_3$] $\xrightarrow[\ ]{HNO_3, H_2SO_4}$ [$O_2N$—phenyl—C(=O)—$CH_3$]

**18.25** This reaction proceeds via a radical bromination mechanism and two radicals are possible: **A** (2°
and benzylic) and **B** (1°). Since **B** (which leads to C$_6$H$_5$CH$_2$CH$_2$Br) is much less stable, this
radical is not formed so only C$_6$H$_5$CH(Br)CH$_3$ is formed as product.

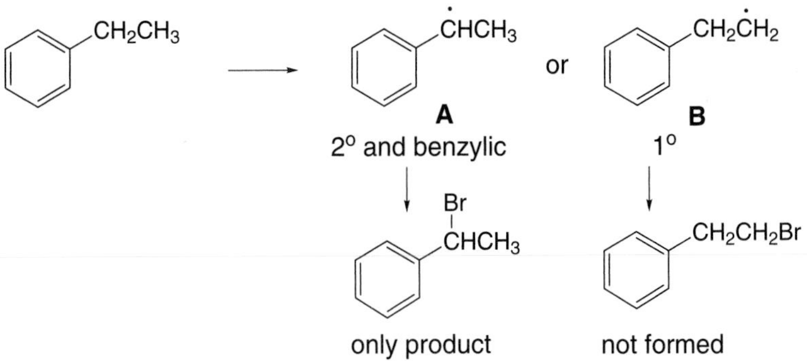

**A**
2° and benzylic

or

**B**
1°

only product

not formed

**18.26** **Radical substitution** occurs at the carbon adjacent to the benzene ring (at the **benzylic
position**).

a.

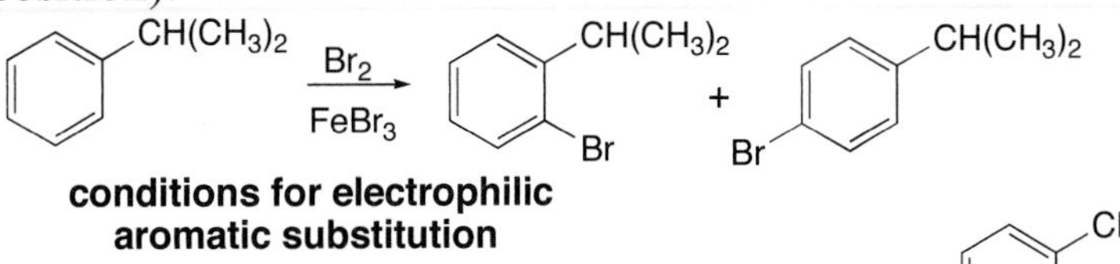

**conditions for electrophilic
aromatic substitution**

b.

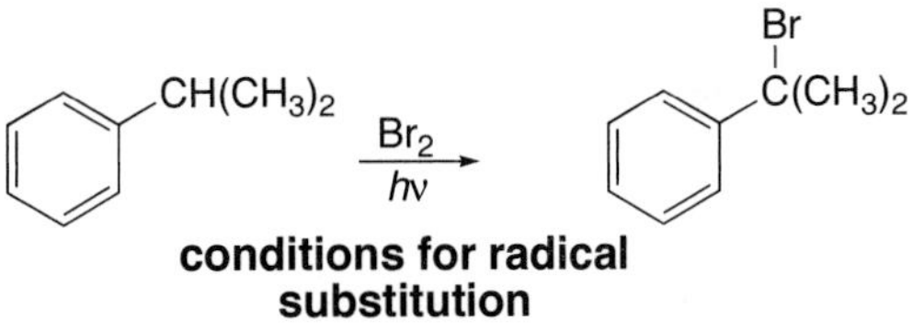

**conditions for radical
substitution**

c.

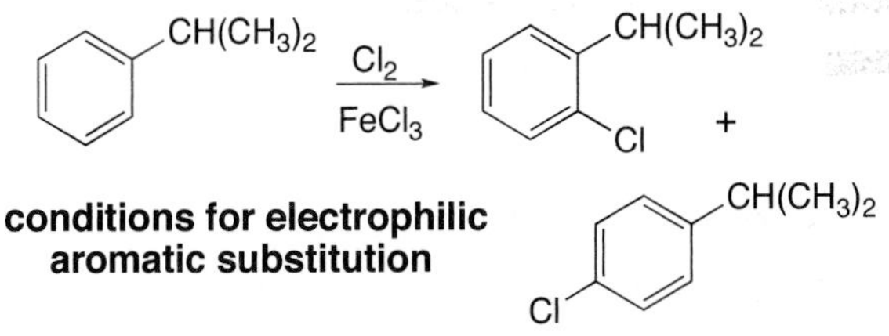

**conditions for electrophilic
aromatic substitution**

**18.27**

a.

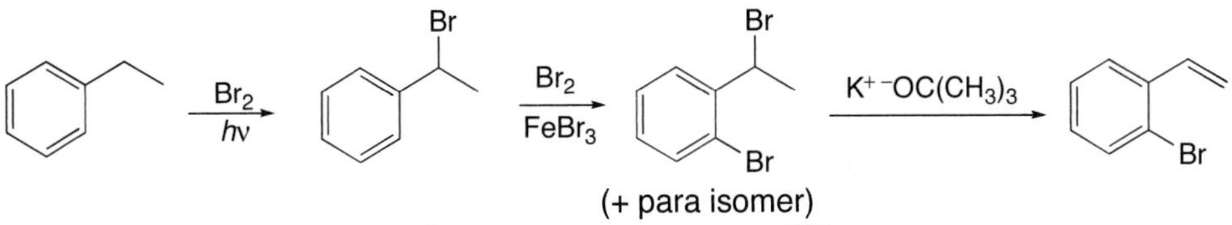

(+ para isomer)

b.

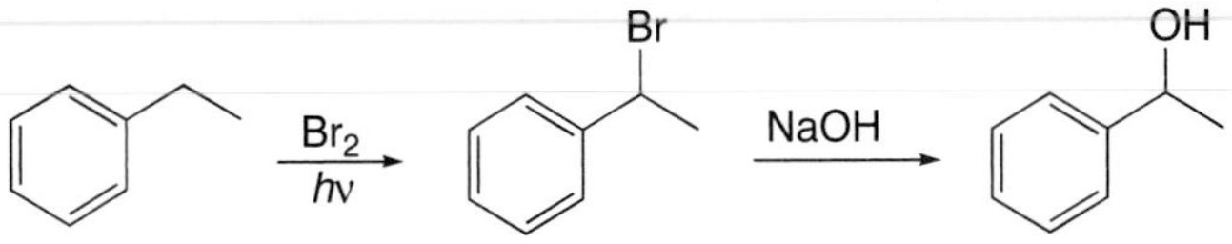

c.

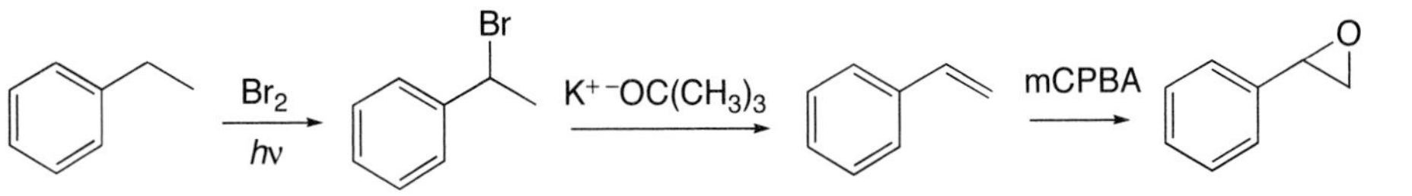

d.

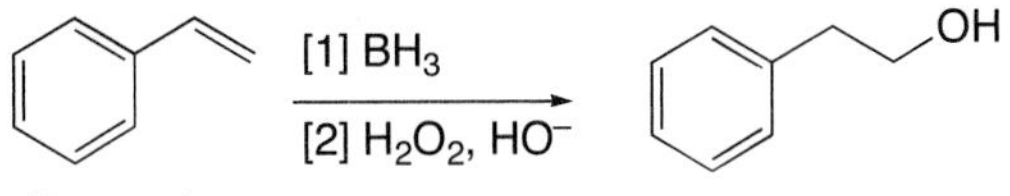

(from c.)

**18.28** First use an acylation reaction, and then reduce the carbonyl group to form the alkyl benzenes.

a.

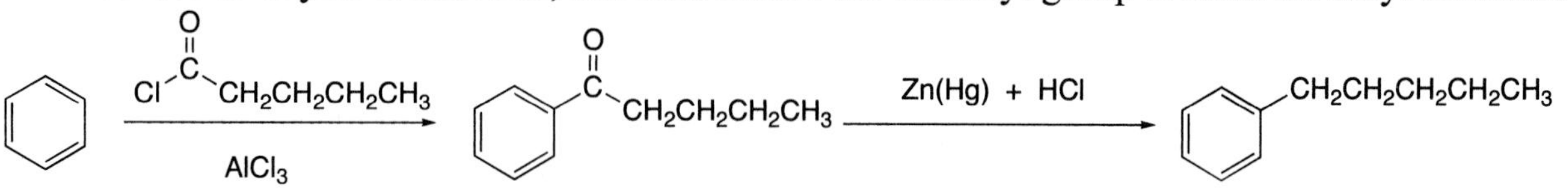

b.

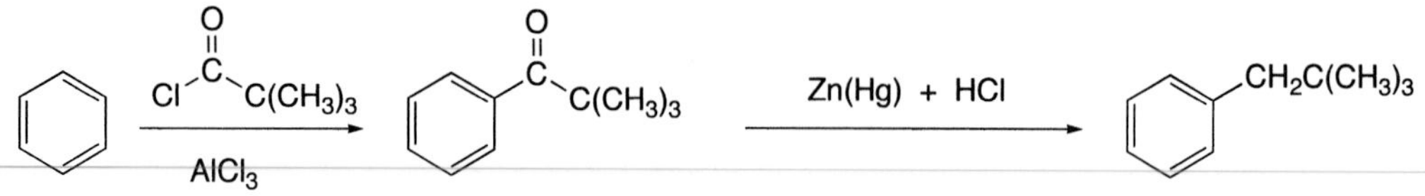

**18.29**

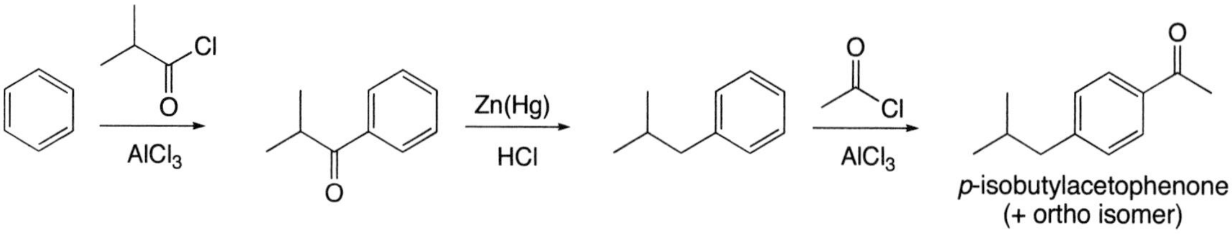

p-isobutylacetophenone
(+ ortho isomer)

**18.30**

a.

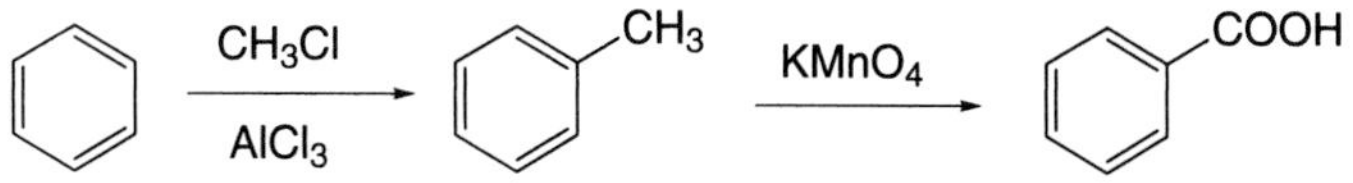

b.

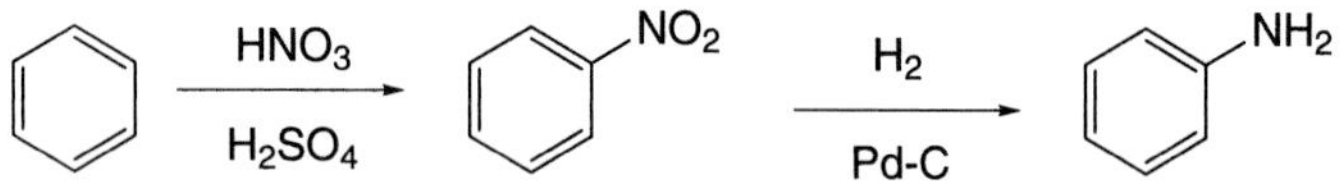

c.

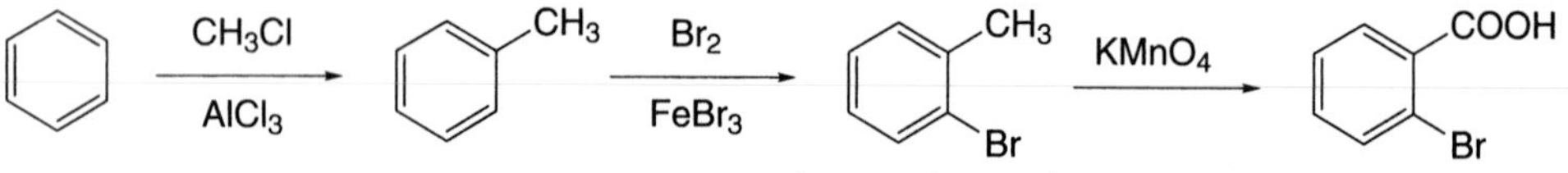

(+ para isomer)

**18.31**

a.

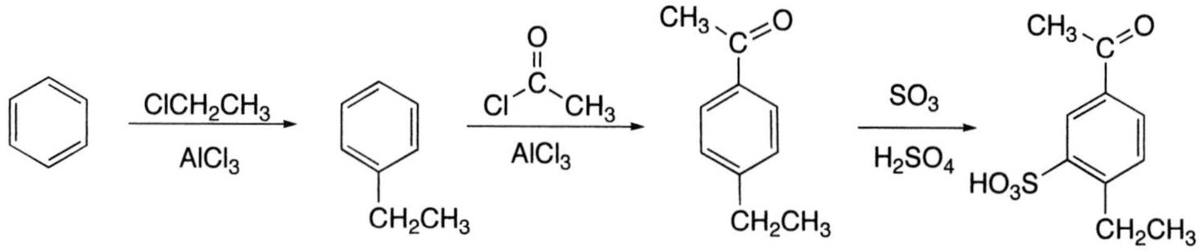

(+ ortho isomer)

b.

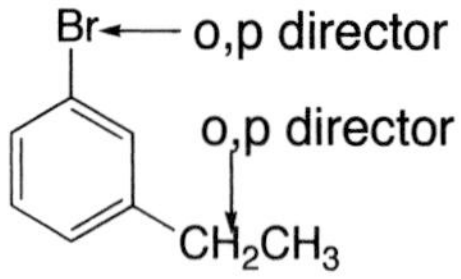

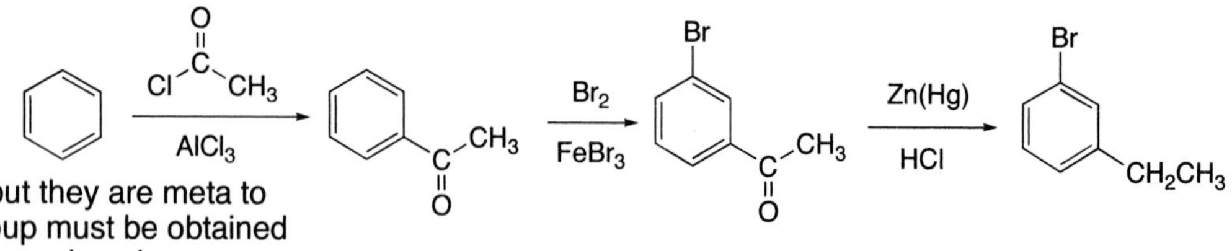

Both are o,p directors, but they are meta to
each other. The alkyl group must be obtained
by reduction of a carbonyl.

c.

**18.32** OH is an ortho, para director.

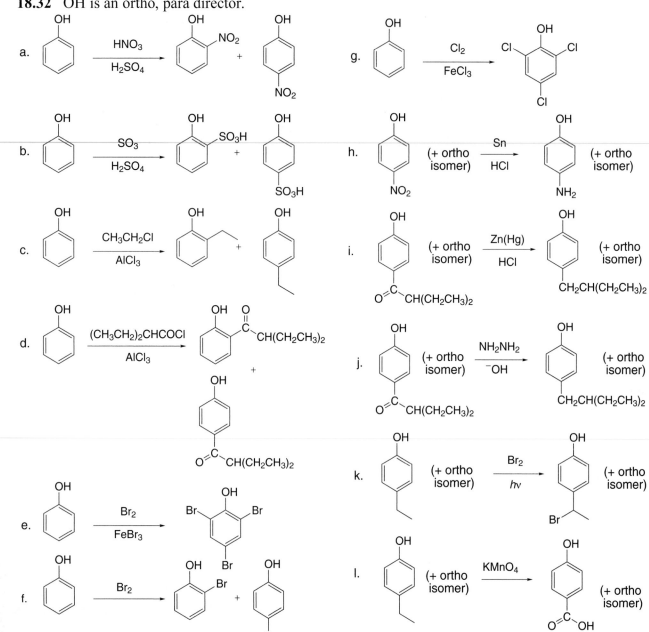

**18.33** CN is a meta director that deactivates the benzene ring.

**18.34**

a.

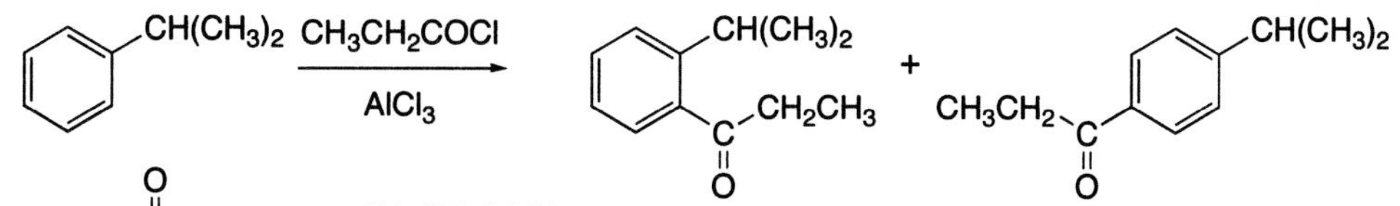

b.

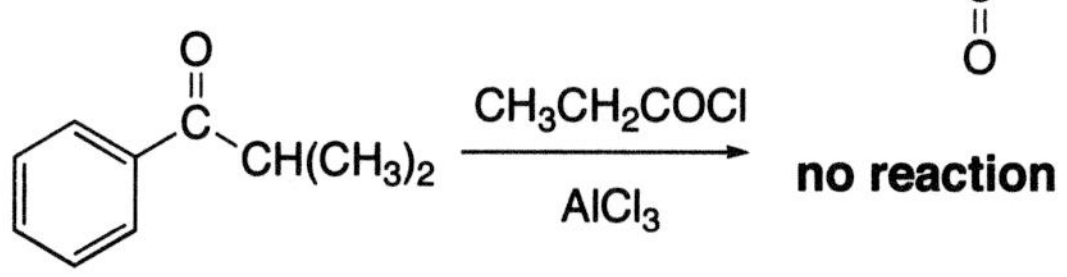

c.

d.

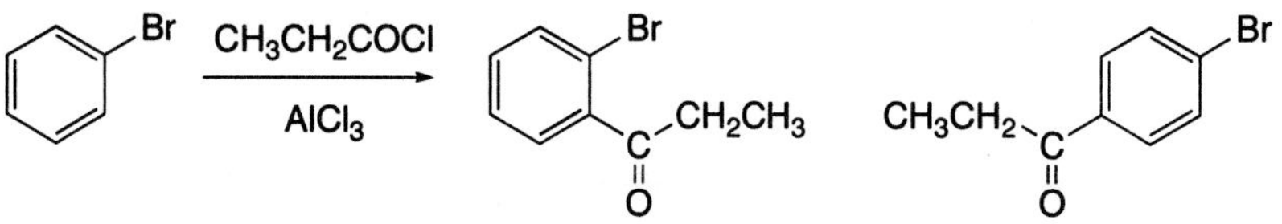

e.

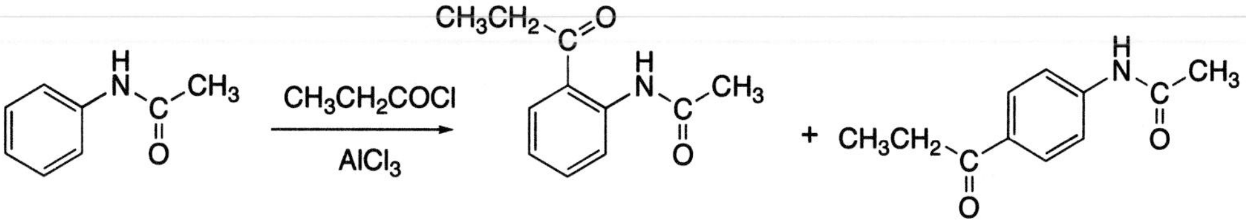

**18.35**

a.

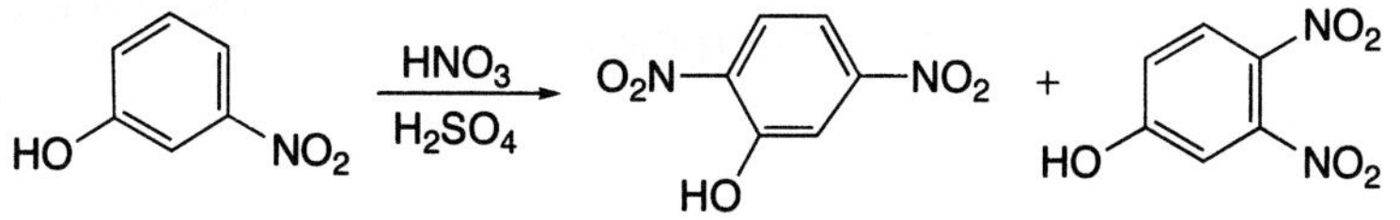

b.

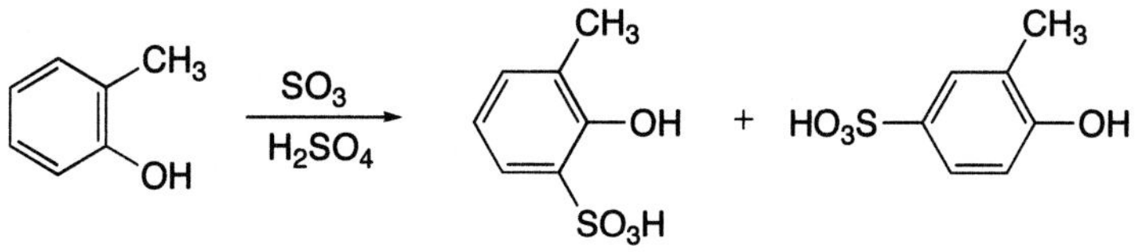

c.

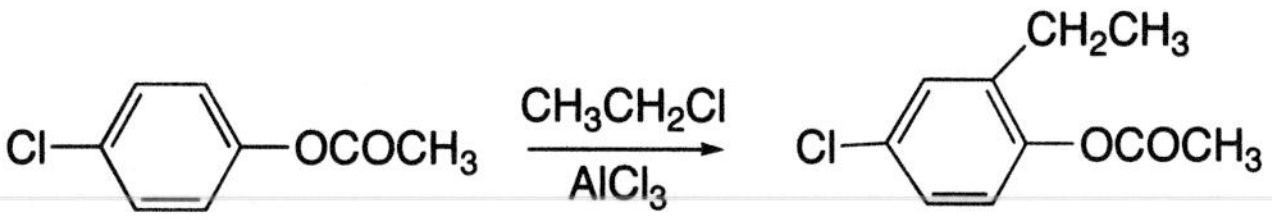

d.

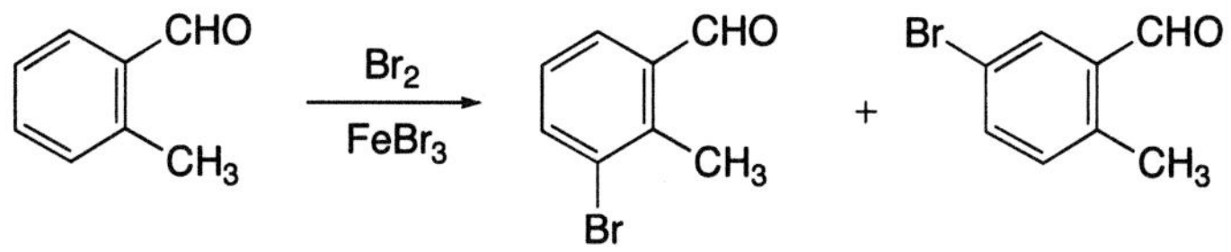

e.

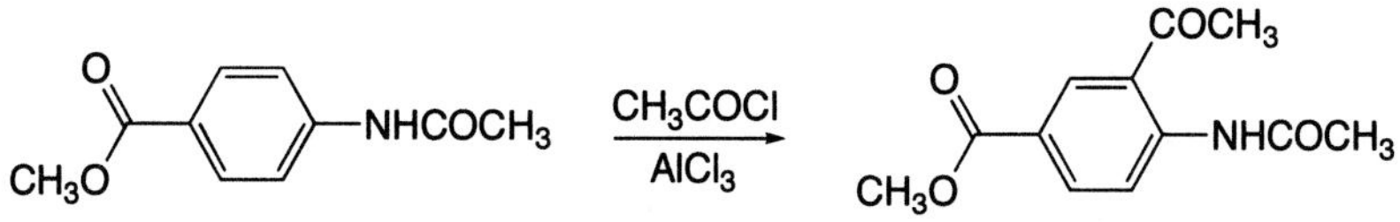

f.

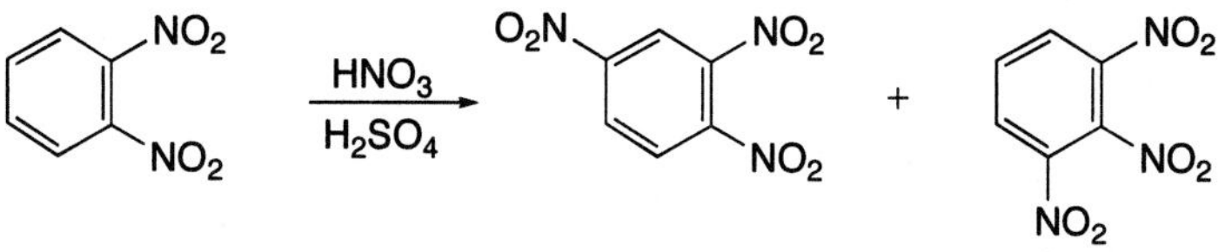

g.

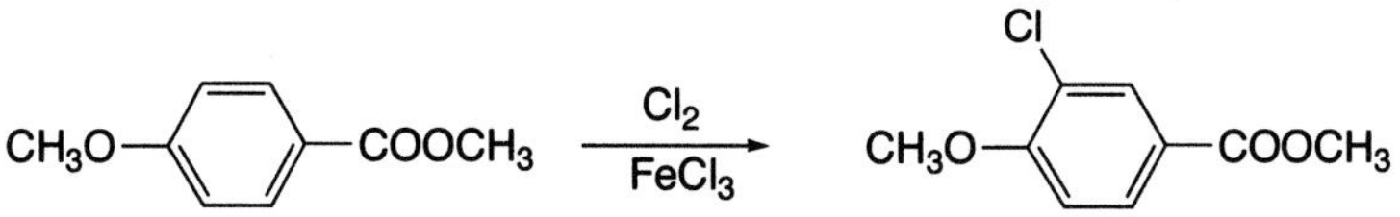

h.

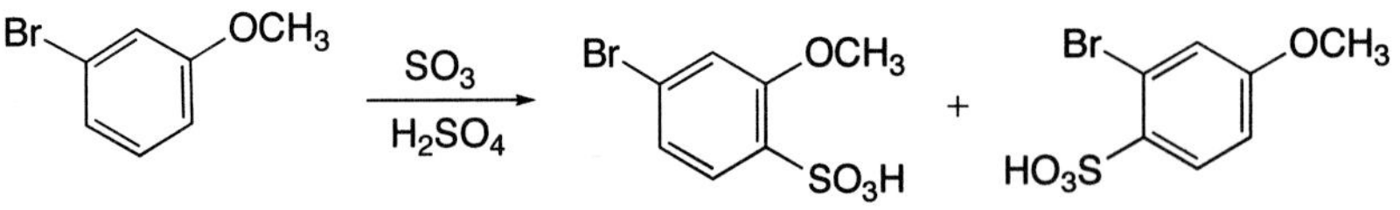

**18.36** Watch out for rearrangements.

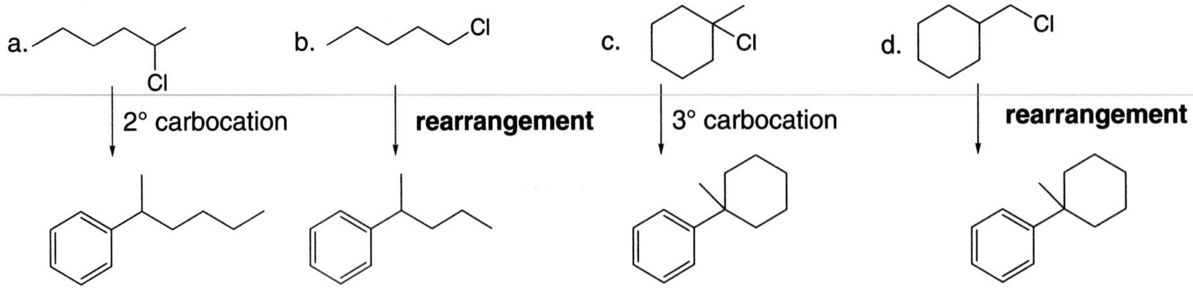

a.   2° carbocation

b.   **rearrangement**

c.   3° carbocation

d.   **rearrangement**

**18.37**

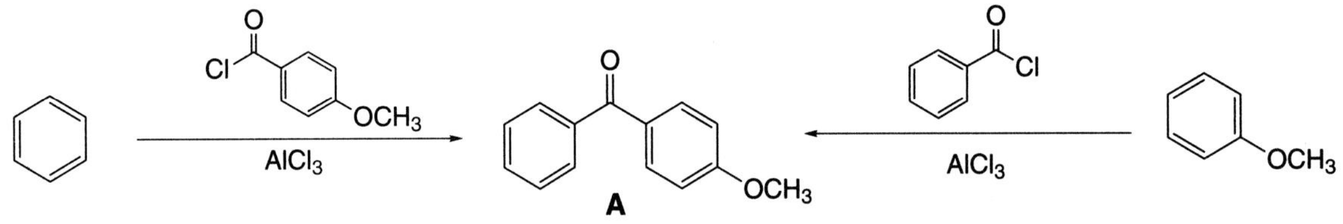

**18.38**

a.

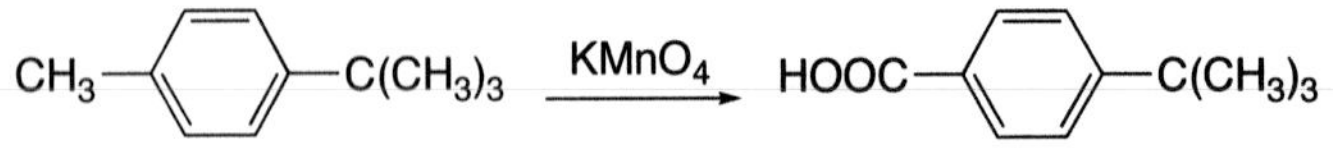

b.

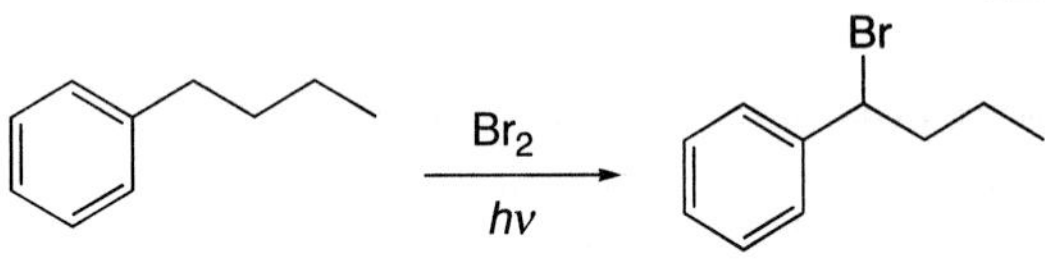

c.

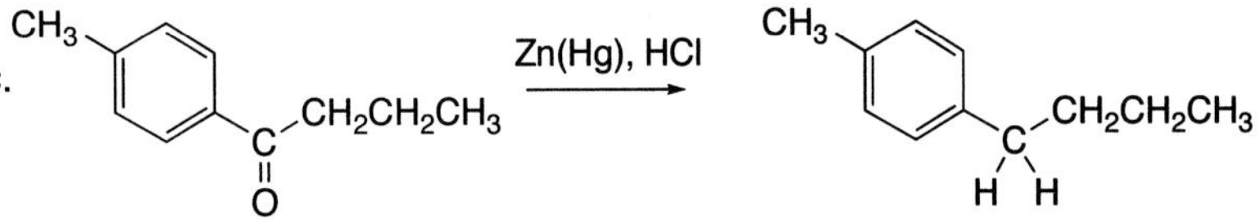

d.

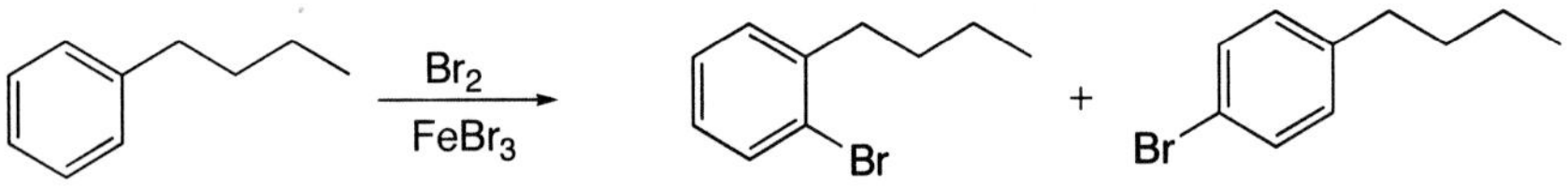

e.

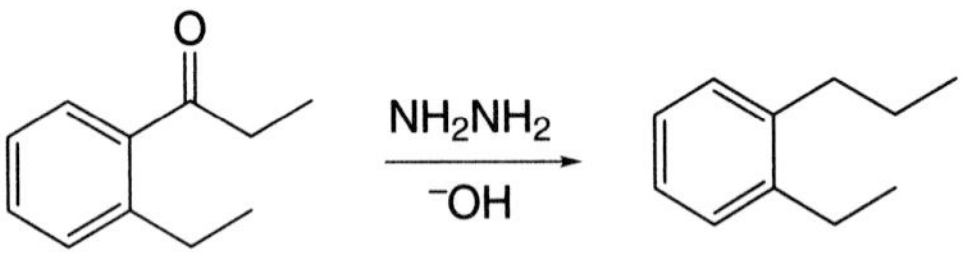

f.

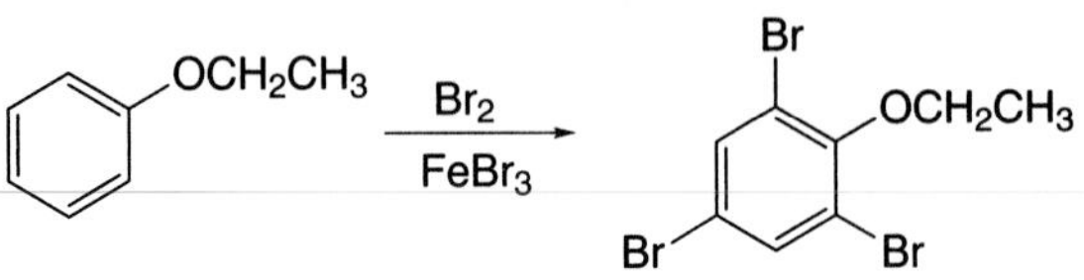

**18.39**

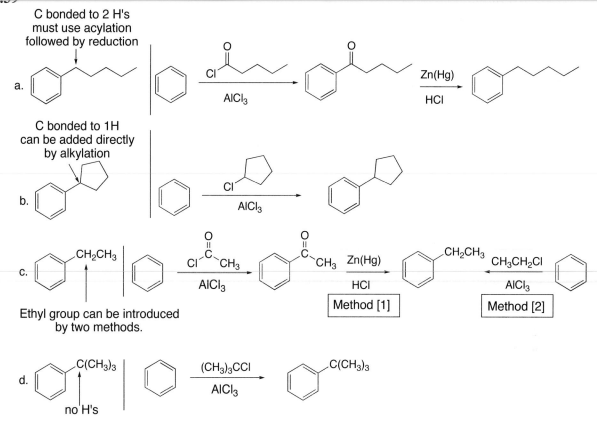

a.

C bonded to 2 H's must use acylation followed by reduction
Cl
O
AlCl3
O
Zn(Hg)
HCl

b.

C bonded to 1H can be added directly by alkylation
Cl
AlCl3

c.

CH2CH3
Ethyl group can be introduced by two methods.
O
Cl C CH3
AlCl3
O
C CH3
Zn(Hg)
HCl
Method [1]
CH2CH3
CH3CH2Cl
AlCl3
Method [2]

d.

C(CH3)3
no H's
(CH3)3CCl
AlCl3
C(CH3)3

**18.40**

a.

SO3H
[1] CH3COCl, AlCl3
[2] Cl2, FeCl3
SO3H
= A
Cl
O
CH3

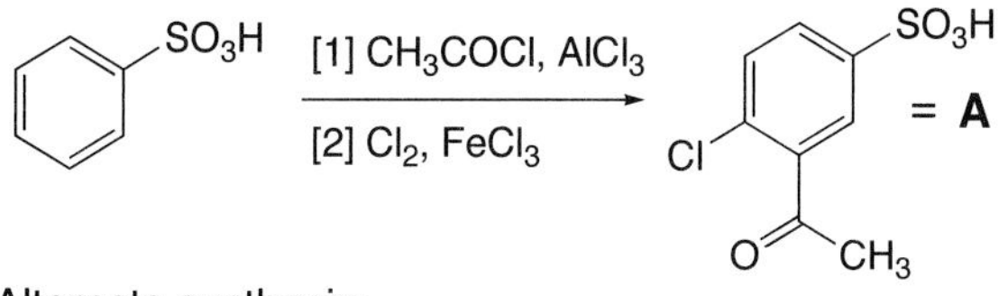

Step [1] won't work because a Friedel–Crafts reaction can't be done on a deactivated benzene ring, as is the case with the $SO_3H$ substituent. Even if Step [1] did work, the second step would introduce Cl meta to $SO_3H$, not para as drawn.

Alternate synthesis:

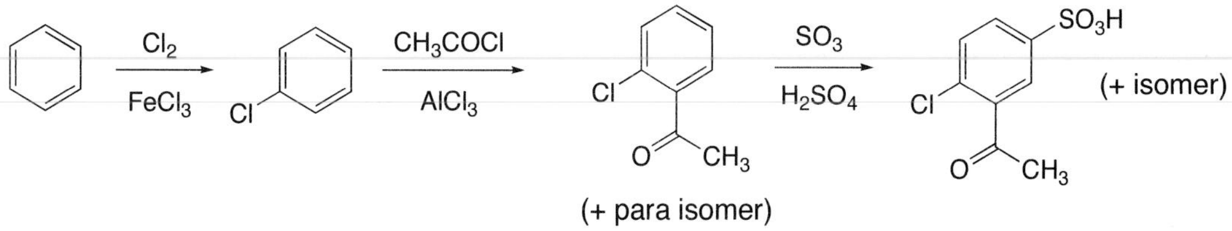

Cl2
FeCl3
Cl
CH3COCl
AlCl3
Cl
O
CH3
(+ para isomer)
SO3
H2SO4
SO3H
Cl
O
CH3
(+ isomer)

b.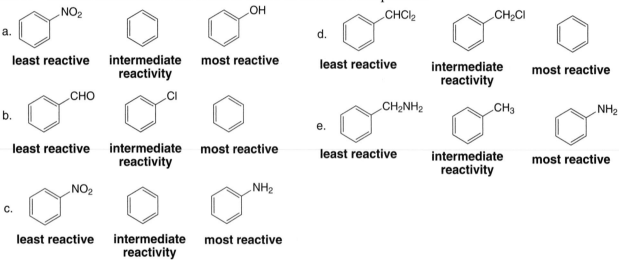

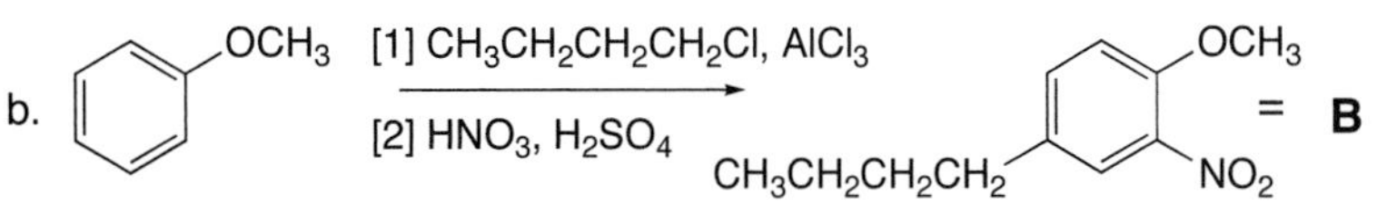

Step [1] involves a Friedel–Crafts alkylation using a 1° alkyl halide that will undergo rearrangement, so that a butyl group will not be introduced as a side chain.

Alternate synthesis:

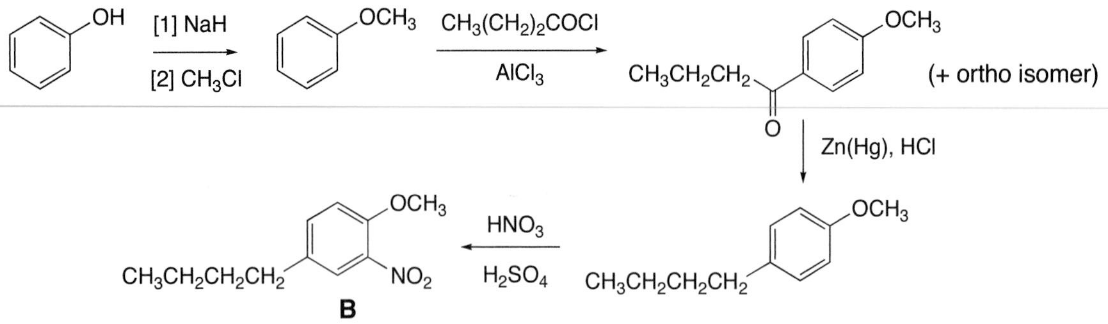

**18.41** Use the directions from Answer 18.17 to rank the compounds.

a. 

NO₂ / (benzene) / OH

least reactive    intermediate reactivity    most reactive

d.

CHCl₂ / CH₂Cl / (benzene)

least reactive    intermediate reactivity    most reactive

b.

CHO / Cl / (benzene)

least reactive    intermediate reactivity    most reactive

e.

CH₂NH₂ / CH₃ / NH₂

least reactive    intermediate reactivity    most reactive

c.

NO₂ / (benzene) / NH₂

least reactive    intermediate reactivity    most reactive

**18.42** Electron-withdrawing groups place a positive charge in the benzene ring. Draw the resonance structures to show how NO₂ puts a positive charge in the ring, giving it an electron-withdrawing resonance effect. Electron-donating groups place a negative charge in the benzene ring. Draw the resonance structures to show how F puts a negative charge in the ring, giving it an electron-donating resonance effect.

a.

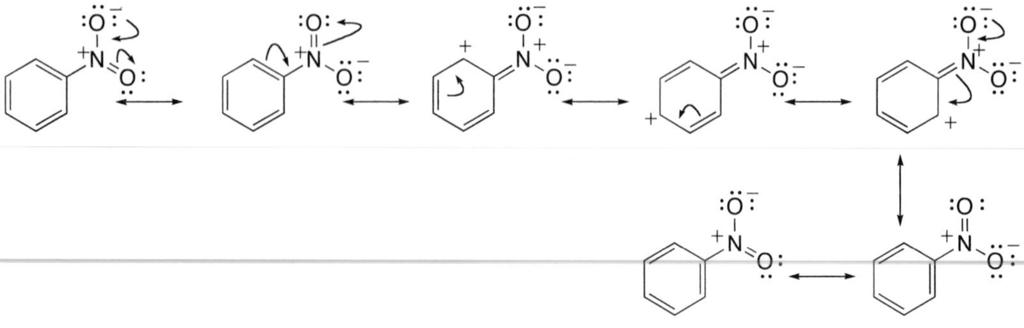

b.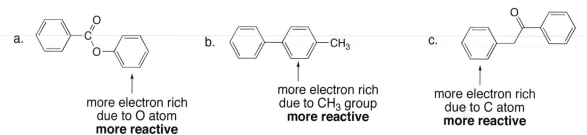

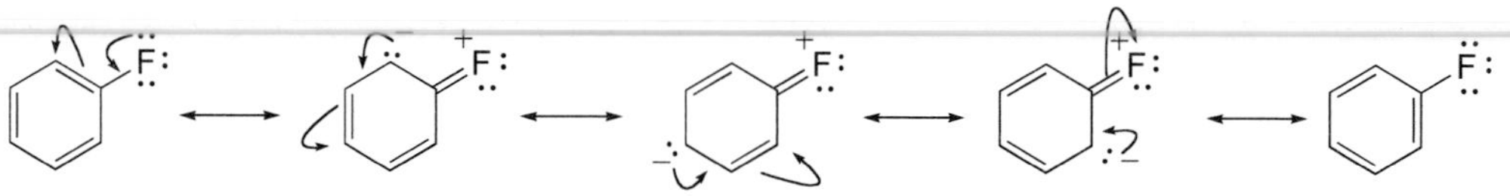

## 18.43

[1] (benzene with Br)

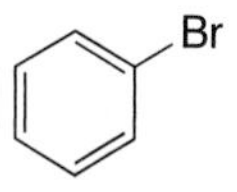

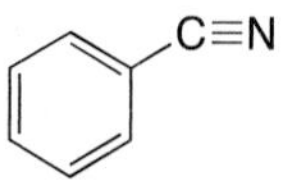

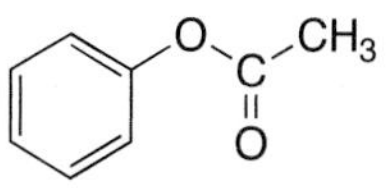

a. withdraw
b. donate
c. less
d. deactivate

[2] (benzene with C≡N)

a. withdraw
b. withdraw
c. less
d. deactivate

[3] (benzene with O–C(=O)CH₃)

a. withdraw
b. donate
c. more
d. activate

## 18.44

a. (phenyl benzoate structure)

more electron rich
due to O atom
**more reactive**

b. (biphenyl with CH₃ structure)

more electron rich
due to CH₃ group
**more reactive**

c. (phenyl benzyl ketone structure)

more electron rich
due to C atom
**more reactive**

## 18.45

a.

**ortho
attack**

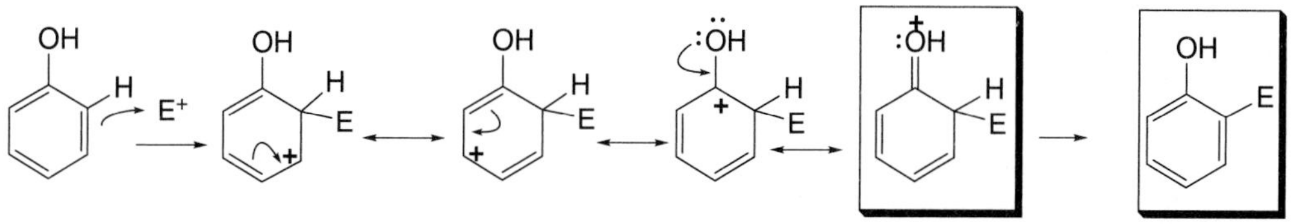

(mechanism with resonance structures)

especially good
All atoms have an octet.

preferred
product

**meta
attack**

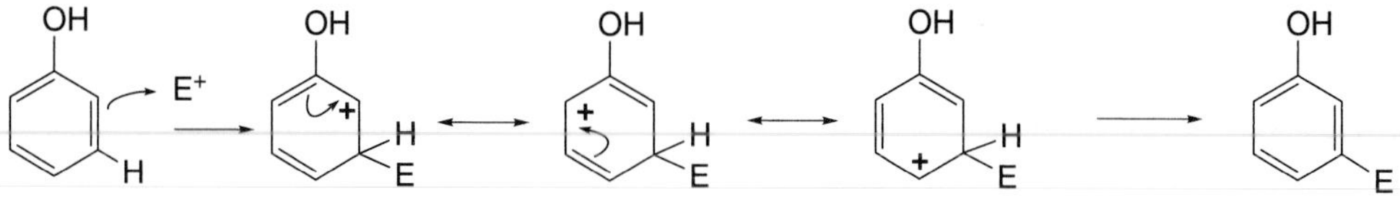

(mechanism with resonance structures)

**para
attack**

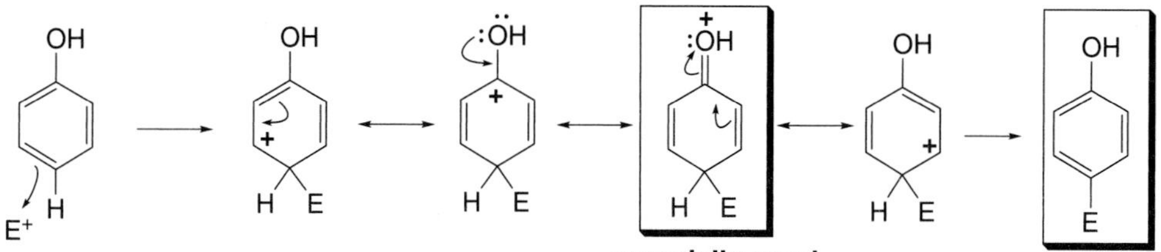

(mechanism with resonance structures)

especially good
All atoms have an octet.

preferred
product

b.

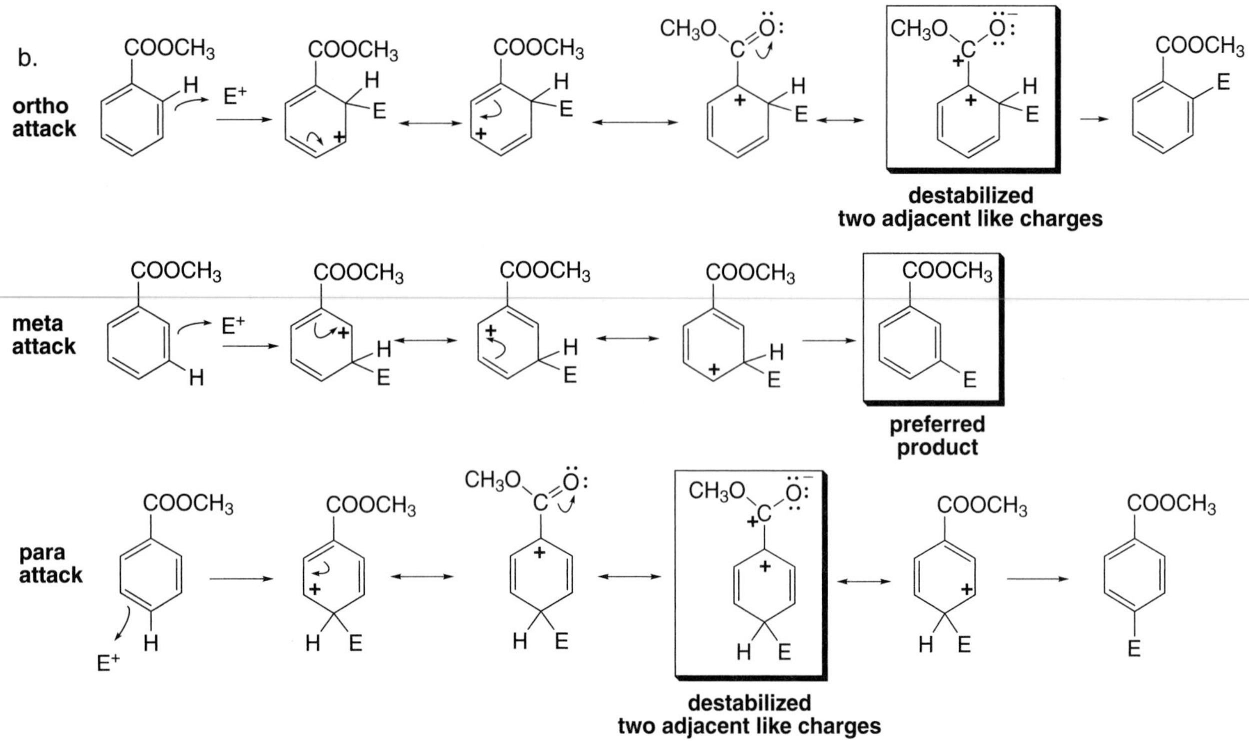

**ortho attack**

**destabilized
two adjacent like charges**

**meta attack**

**preferred
product**

**para attack**

**destabilized
two adjacent like charges**

**18.46**

a.

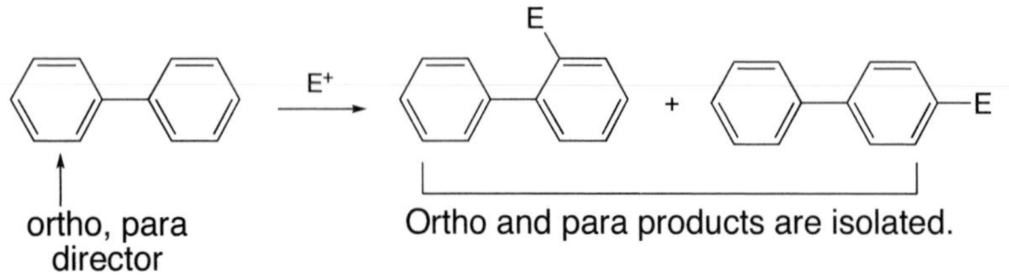

ortho, para
director

Ortho and para products are isolated.

A benzene ring is an electron-rich substituent that stabilizes an intermediate positive charge by an electron-donating resonance effect. As a result, it activates a benzene ring toward reaction with electrophiles.

With ortho and para attack there is additional resonance stabilization that delocalizes the positive charge onto the second benzene ring. Such additional stabilization is not possible with meta attack.

Ortho attack:

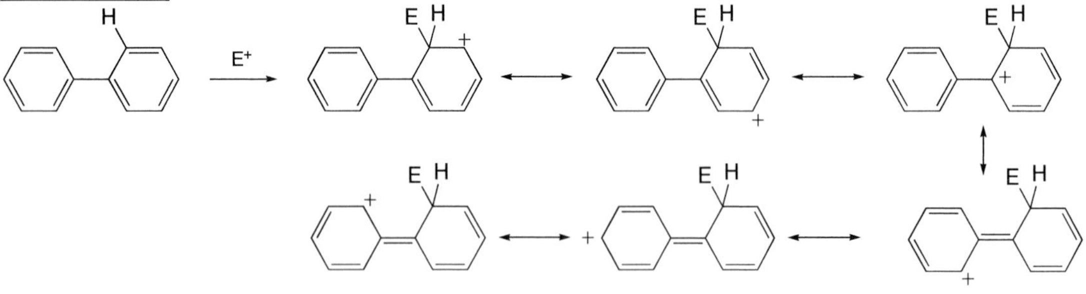

Meta attack:

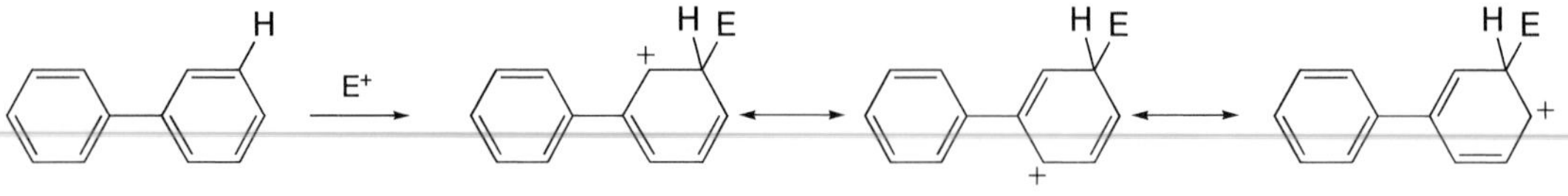

**Para attack:**

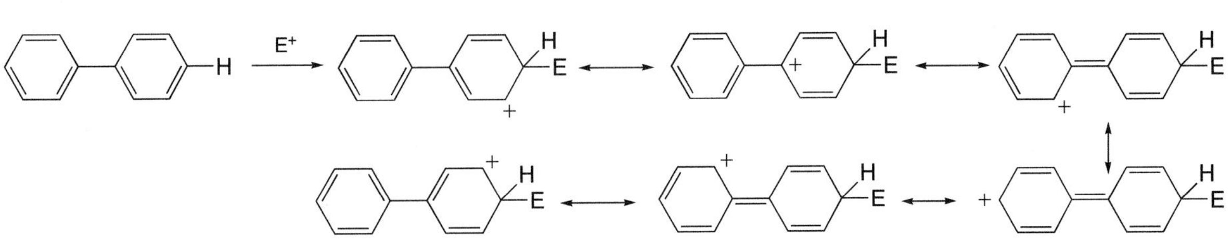

b.

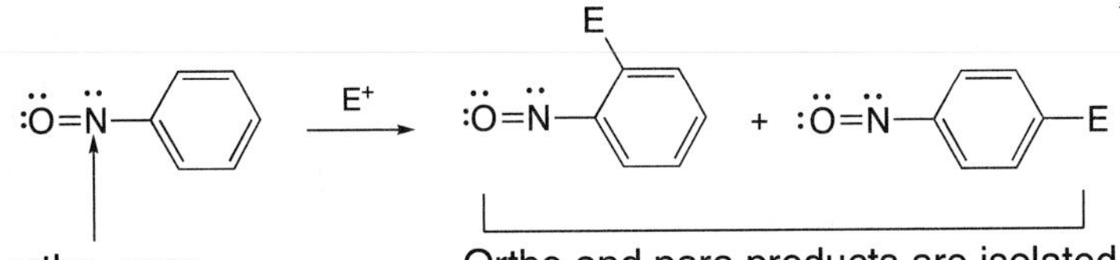

ortho, para director

Ortho and para products are isolated.

With ortho and para attack there is additional resonance stabilization that delocalizes the positive charge onto the nitroso group. Such additional stabilization is not possible with meta attack. This makes –NO an ortho, para director. Since the N atom bears a partial (+) charge (because it is bonded to a more electronegative O atom), the –NO group inductively withdraws electron density, thus deactivating the benzene ring towards electrophilic attack. In this way, the –NO group resembles the halogens. Thus, the electron-donating resonance effect makes –NO an o,p director, but the electron-withdrawing inductive effect makes it a deactivator.

**Ortho attack:**

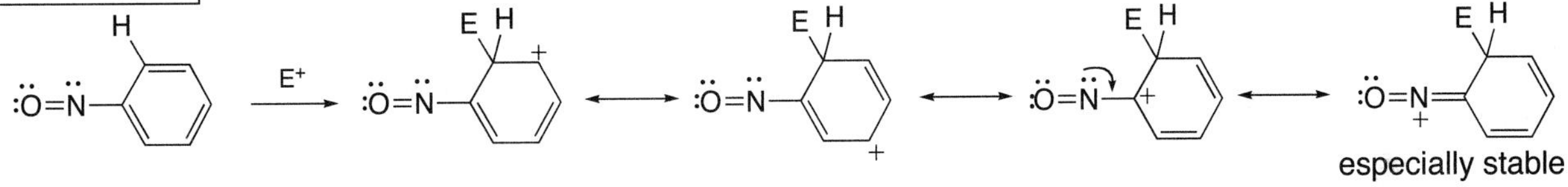

especially stable

**Meta attack:**

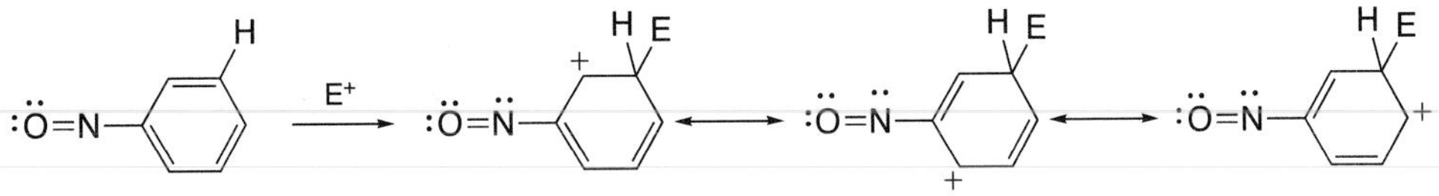

**Para attack:**

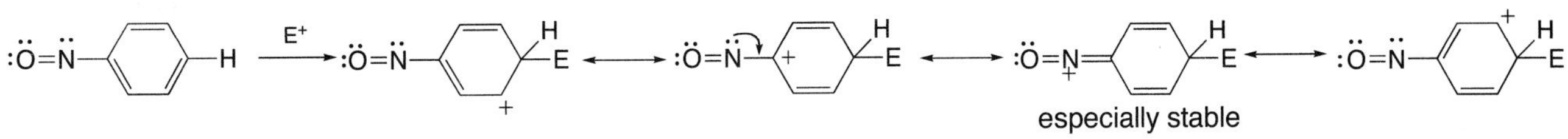

especially stable

**18.47** Under the acidic conditions of nitration, the N atom of the starting material gets protonated, so the atom directly bonded to the benzene ring bears a (+) charge. This makes it a meta director, so the new $NO_2$ group is introduced meta to it.

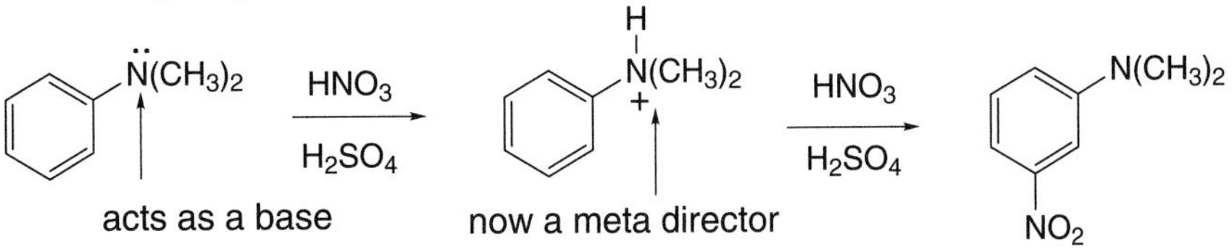

acts as a base      now a meta director      $NO_2$

**18.48**

a.

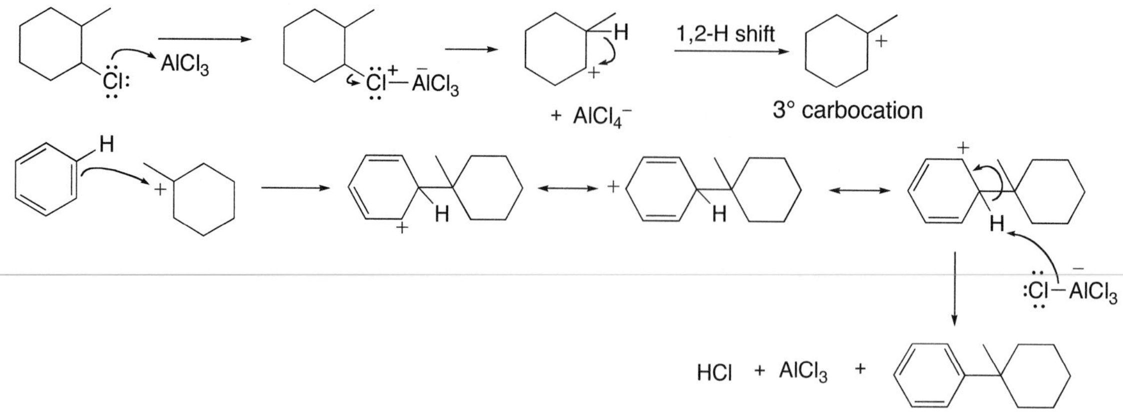
AlCl3
1,2-H shift
+ AlCl4−
3° carbocation
HCl + AlCl3 +
:Cl−AlCl3

b.

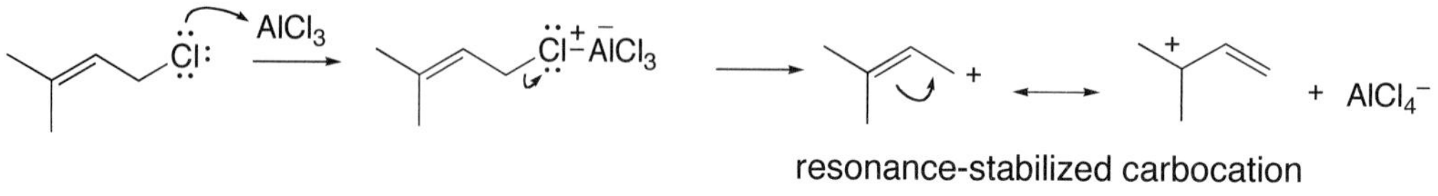
AlCl3
+ AlCl4−
resonance-stabilized carbocation

Use both resonance forms to show how two products are formed.

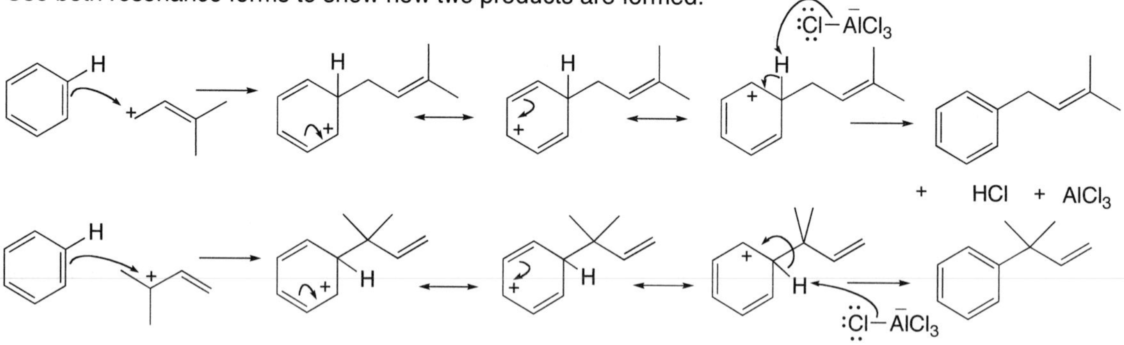
:Cl−AlCl3
+ HCl + AlCl3

c.

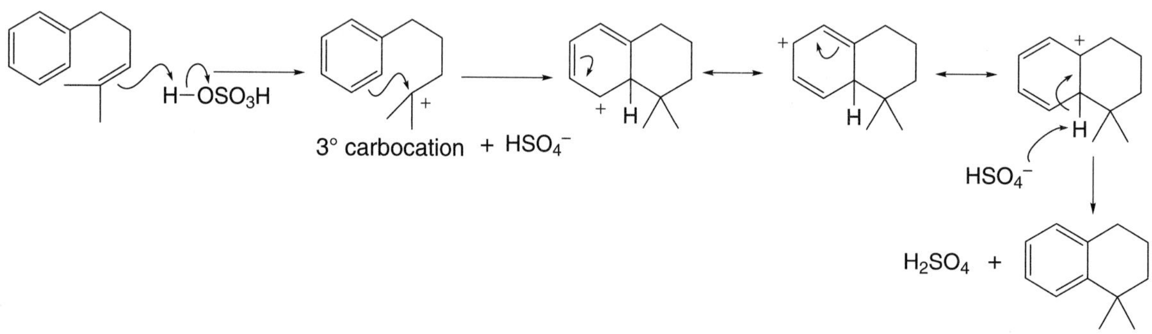
H−OSO3H
3° carbocation + HSO4−
HSO4−
H2SO4 +

**18.49**

a. The product has one stereogenic center.

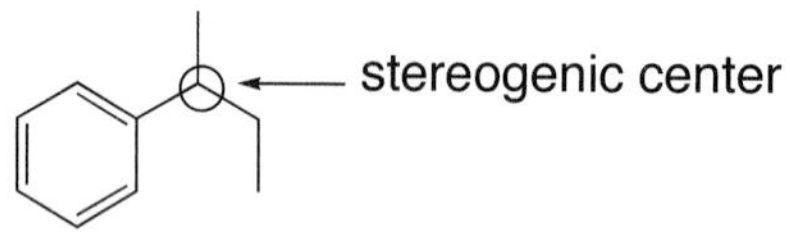
stereogenic center

b. The mechanism for Friedel–Crafts alkylation with this 2° halide involves formation of a trigonal planar carbocation. Since the carbocation is achiral, it reacts with benzene with equal probability from two possible directions (above and below) to afford an optically inactive, racemic mixture of two products.

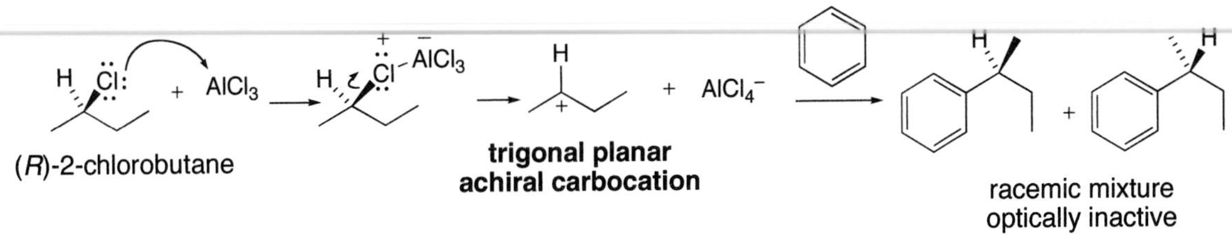
+ AlCl3
AlCl3
H
+ AlCl4−
H
H
(R)-2-chlorobutane
trigonal planar
achiral carbocation
racemic mixture
optically inactive

**18.50**

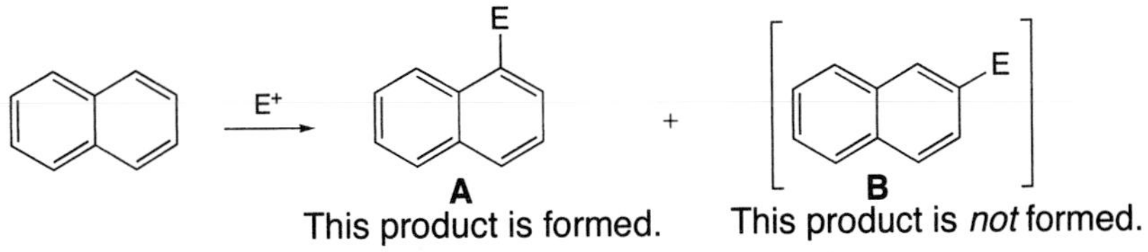
E+
E
E
A
This product is formed.
B
This product is *not* formed.

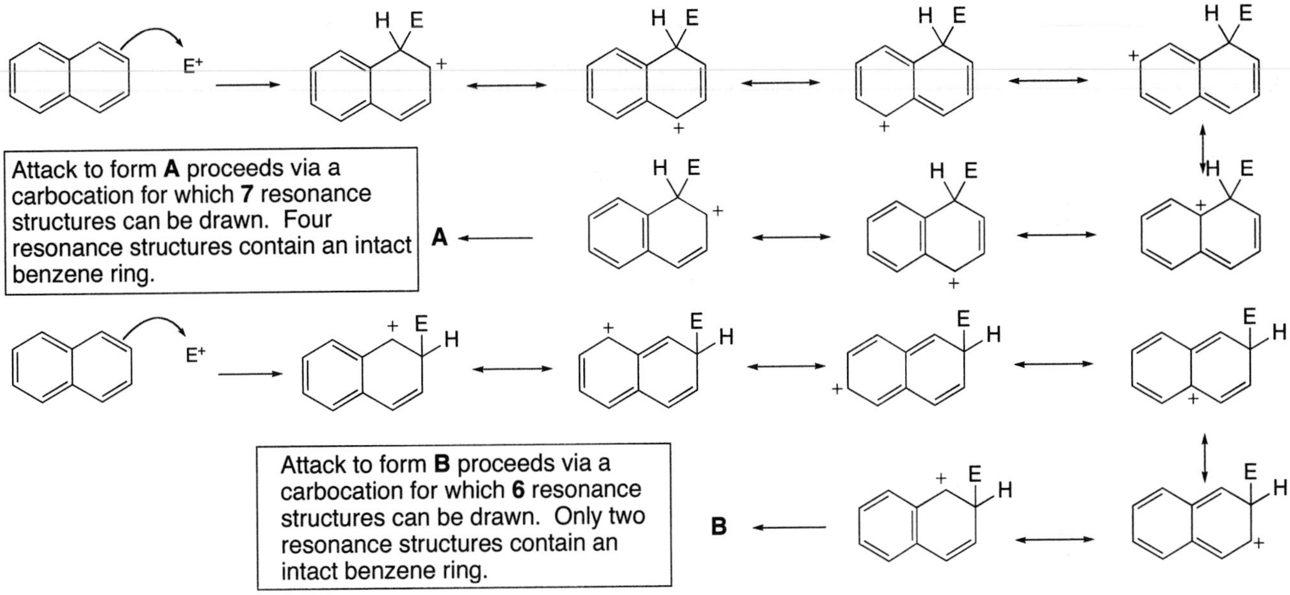
E+
H E
Attack to form A proceeds via a carbocation for which 7 resonance structures can be drawn. Four resonance structures contain an intact benzene ring.
A
E H
Attack to form B proceeds via a carbocation for which 6 resonance structures can be drawn. Only two resonance structures contain an intact benzene ring.
B

A reaction that occurs by way of the more stable carbocation is preferred so product **A** is formed.

**18.51**

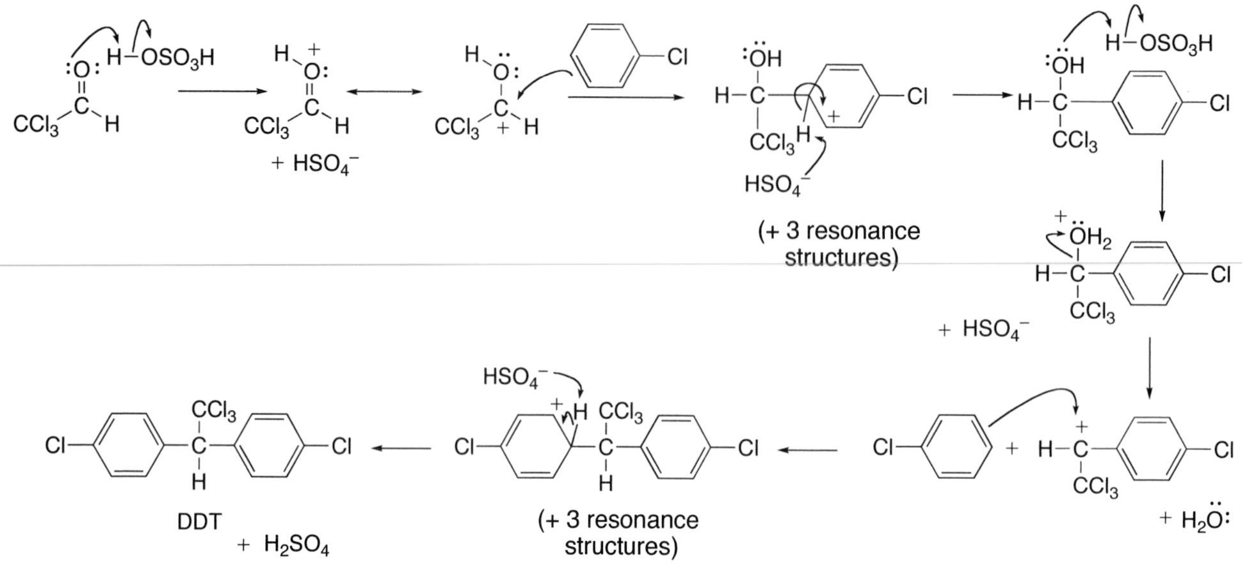

**18.52** Benzyl bromide forms a resonance-stabilized intermediate that allows it to react rapidly under $S_N1$ conditions.

Formation of a resonance-stabilized carbocation:

benzyl bromide

resonance-stabilized carbocation

benzyl methyl ether

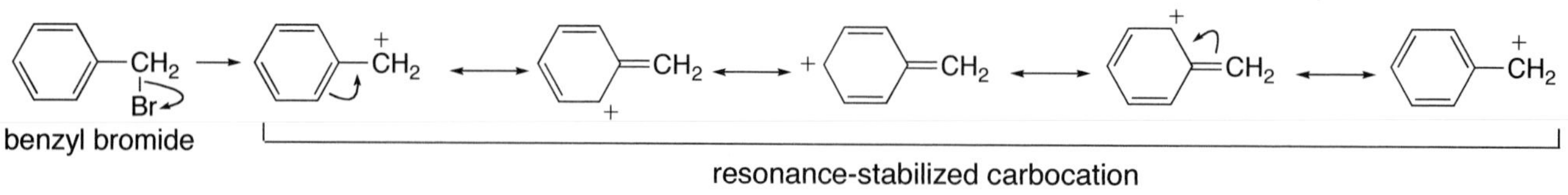

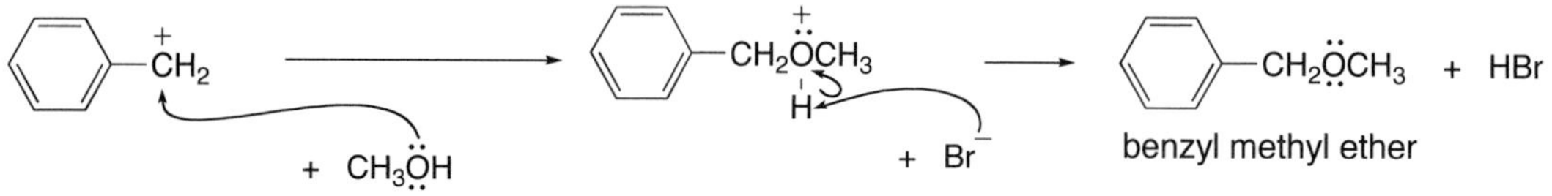

**18.53** Addition of HBr will afford only one alkyl bromide because the intermediate carbocation leading to its formation is resonance stabilized.

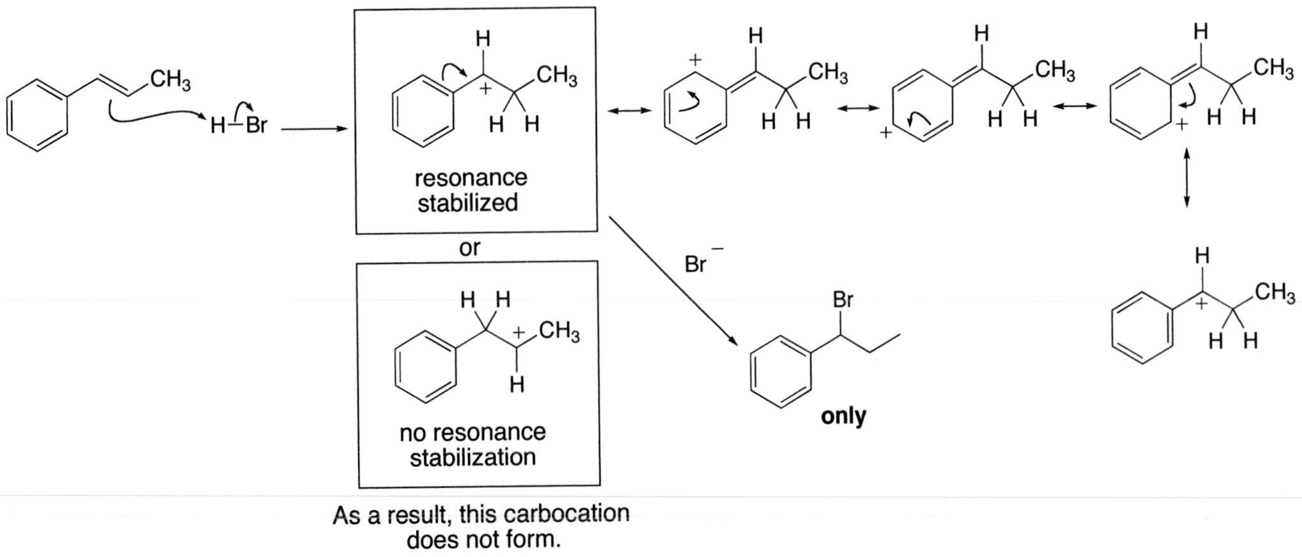

resonance stabilized

or

no resonance stabilization

As a result, this carbocation does not form.

only

**18.54**

a.

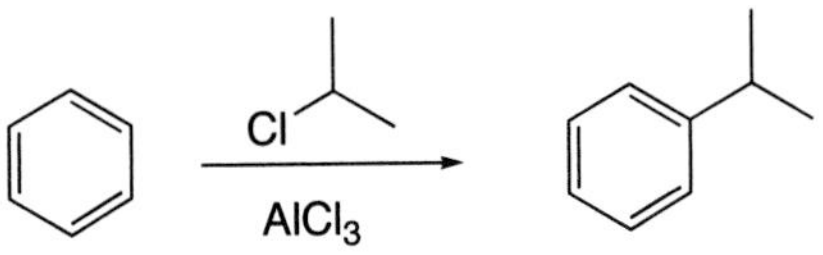

b.

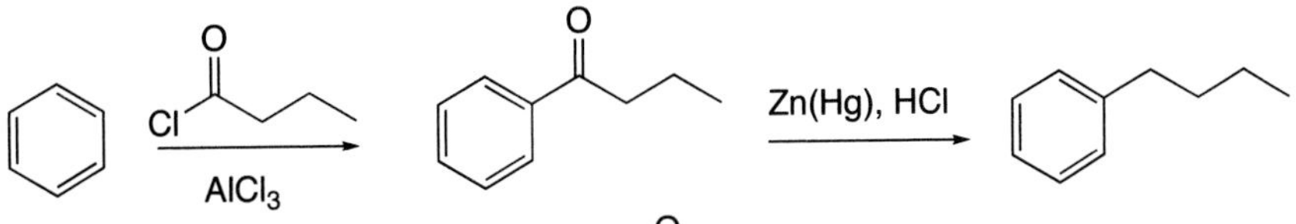

c. (+ para isomer)

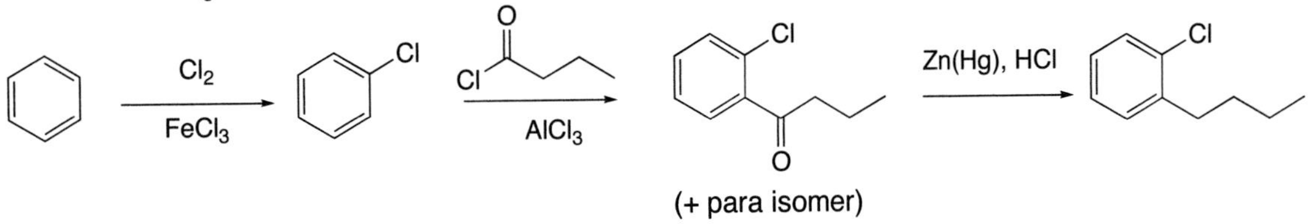

d.

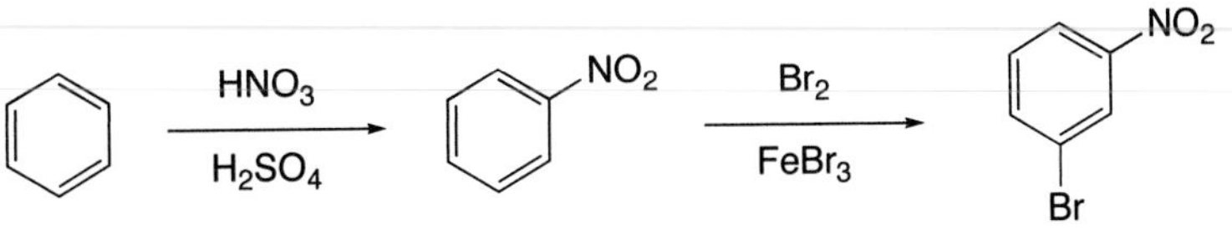

e. (+ para isomer)

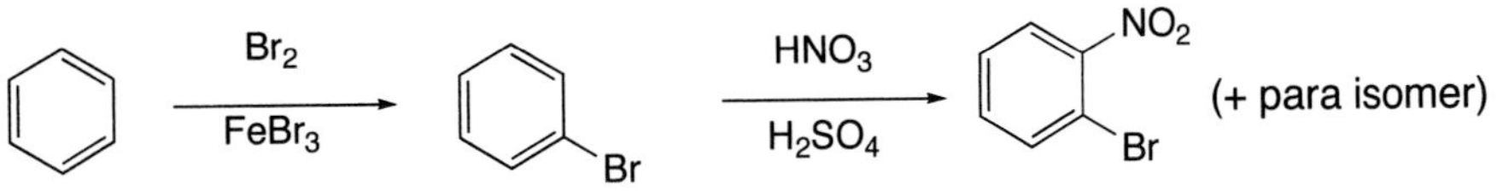

f. (+ para isomer)

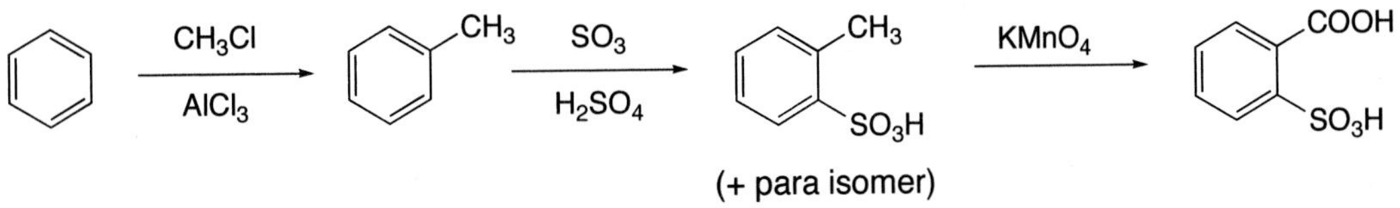

g.

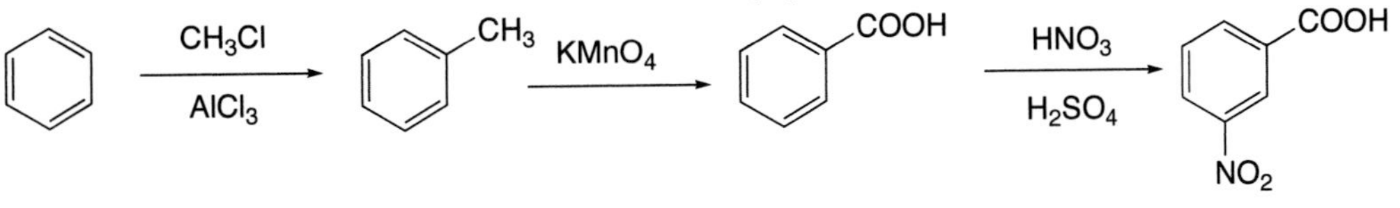

h.    CH₃Cl / AlCl₃ → CH₃ → Br₂ / FeBr₃ → Br—CH₃ (+ ortho isomer) → KMnO₄ → Br—COOH

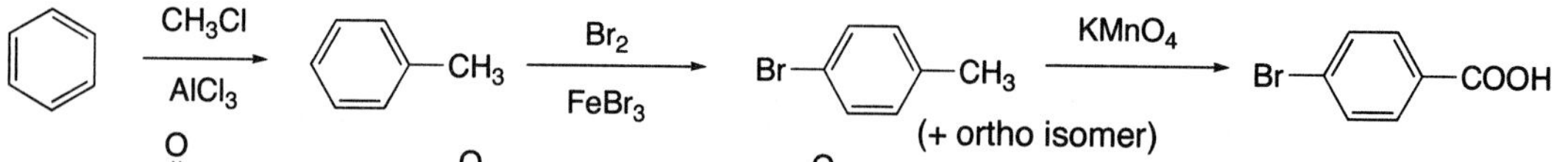

i.    Cl—C(=O)—CH₂CH₃ / AlCl₃ → (=O)CH₂CH₃ → Br₂ / FeBr₃ → (=O)CH₂CH₃, Br → Zn(Hg), HCl → CH₂CH₂CH₃, Br → HNO₃ / H₂SO₄ → NO₂, CH₂CH₂CH₃, Br (+ isomer)

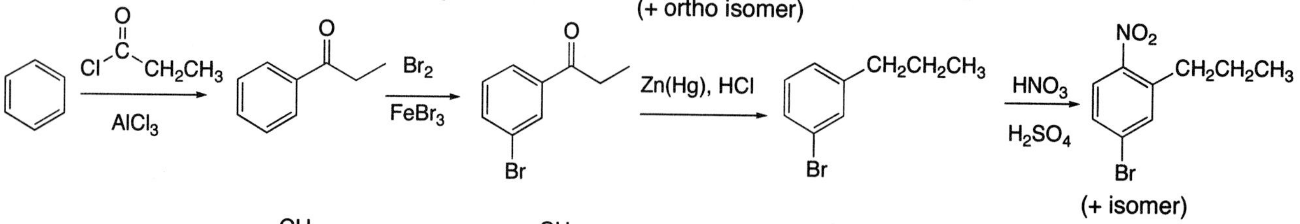

j.    CH₃Cl / AlCl₃ → CH₃ → SO₃ / H₂SO₄ → CH₃, SO₃H (+ ortho isomer) → Cl₂ (excess) / FeCl₃ → Cl, CH₃, Cl, SO₃H → KMnO₄ → Cl, COOH, Cl, SO₃H

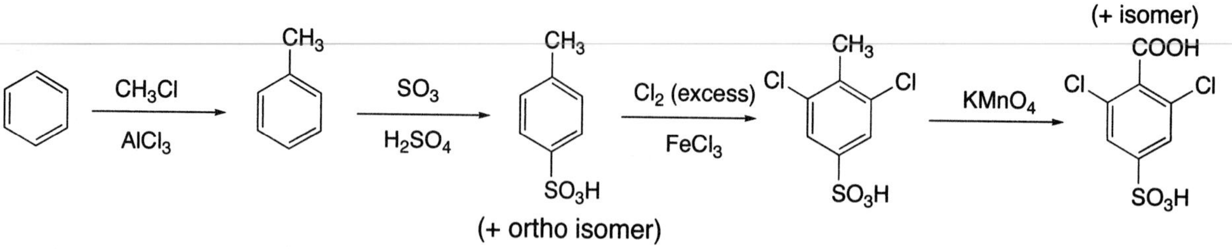

k.    (CH₃)₂CHCl / AlCl₃ → CH(CH₃)₂ → HNO₃ / H₂SO₄ → CH(CH₃)₂, NO₂ (+ para isomer) → Cl₂ / FeCl₃ → Cl, CH(CH₃)₂, NO₂ (+ isomer) → Br₂ / hν → Cl, C(CH₃)₂Br, NO₂ → K⁺ ⁻OC(CH₃)₃ → Cl, C(=CH₂)CH₃, NO₂

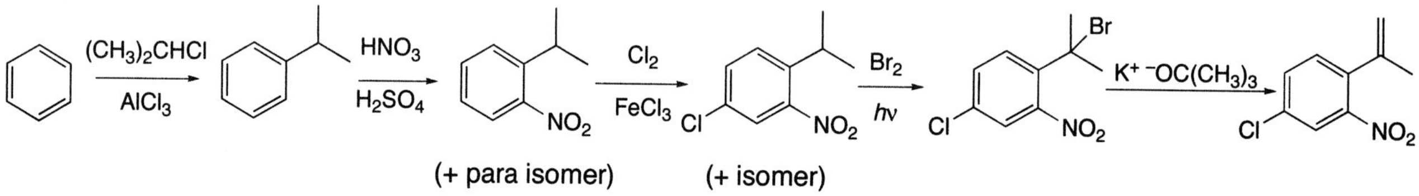

**18.55**

a.    Br₂ / FeBr₃ → Br → HNO₃ / H₂SO₄ → NO₂, Br (+ ortho isomer) → Sn / HCl → NH₂, Br

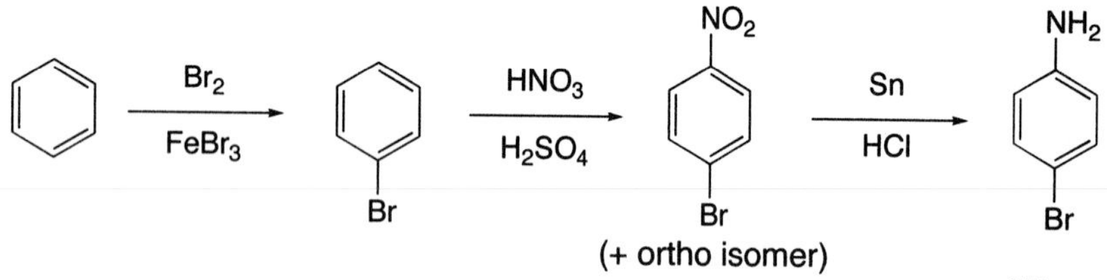

b.    CH₃Cl / AlCl₃ → CH₃ → HNO₃ / H₂SO₄ → CH₃, NO₂ (+ ortho isomer) → Br₂ / FeBr₃ → CH₃, Br, NO₂

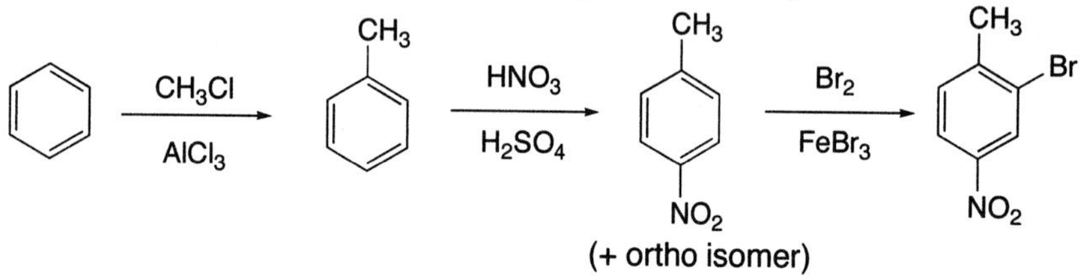

c.    Cl—CH(CH₃)₂ / AlCl₃ → CH(CH₃)₂ → CH₃C(=O)Cl / AlCl₃ → C(=O)CH₃, CH(CH₃)₂ (+ para isomer) → HNO₃ / H₂SO₄ → NO₂, C(=O)CH₃, CH(CH₃)₂ (+ isomer)

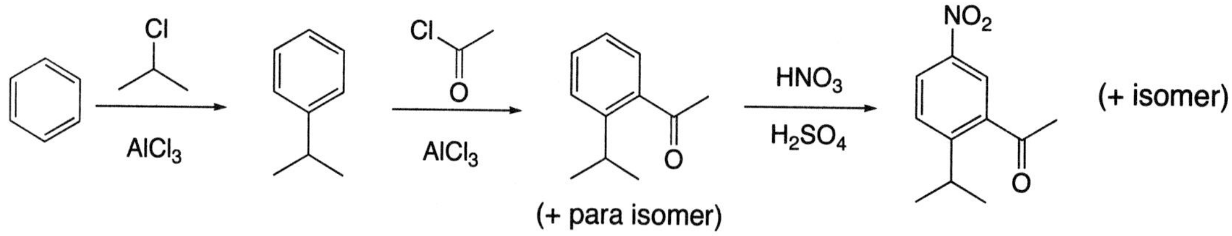

d.    CH₃Cl / AlCl₃ → CH₃ → HNO₃ / H₂SO₄ → CH₃, NO₂ (+ ortho isomer) → KMnO₄ → HOOC, NO₂ → Sn / HCl → HOOC, NH₂

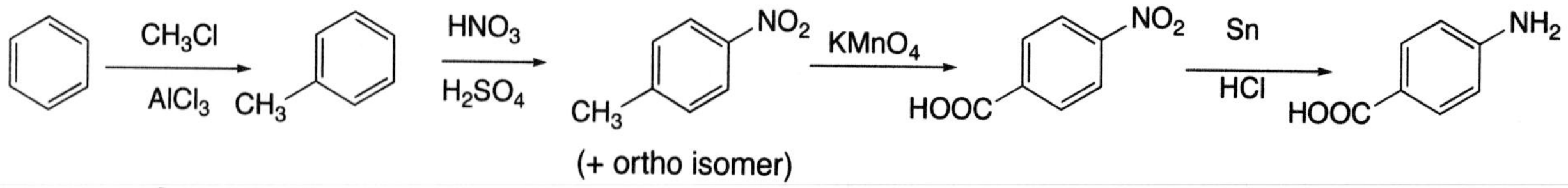

e.    CH₃CH₂C(=O)Cl / AlCl₃ → C(=O)CH₂CH₃ → Zn(Hg), HCl → CH₂CH₂CH₃ → SO₃ / H₂SO₄ → CH₂CH₂CH₃, SO₃H (+ ortho isomer) → NaOH → CH₂CH₂CH₃, SO₃⁻ Na⁺

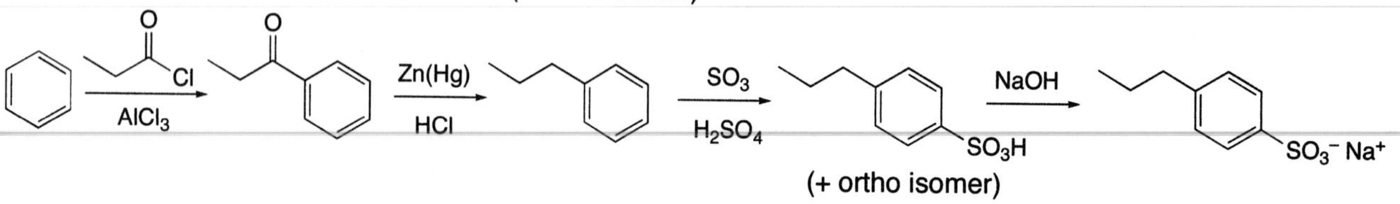

**f.**

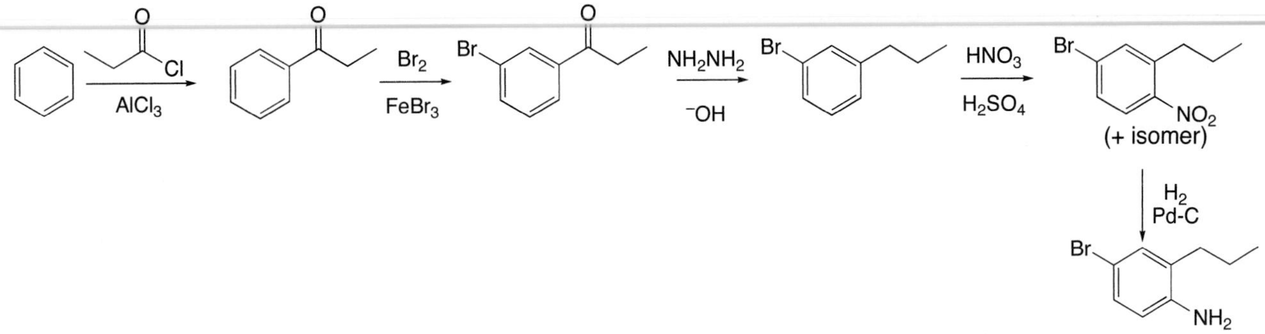

**g.**

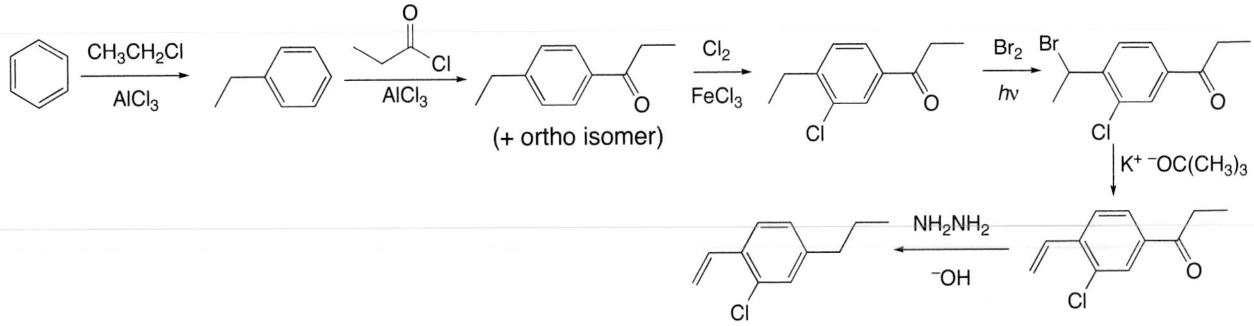

## 18.56

**a.**

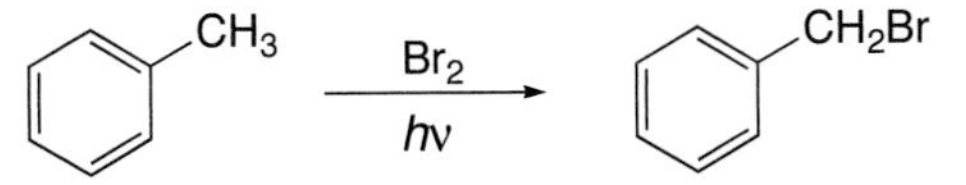

**b.** product in (a) →(NaOH)→

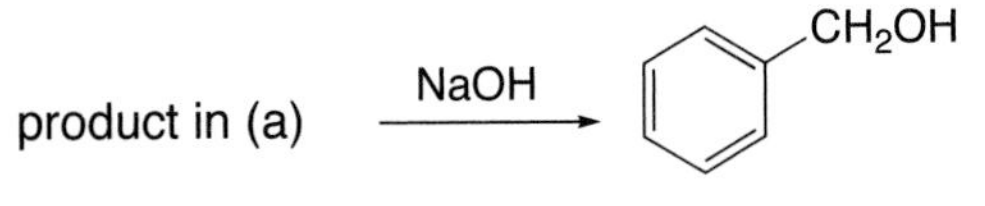

**c.** product in (a) →(⁻OC(CH₃)₃)→

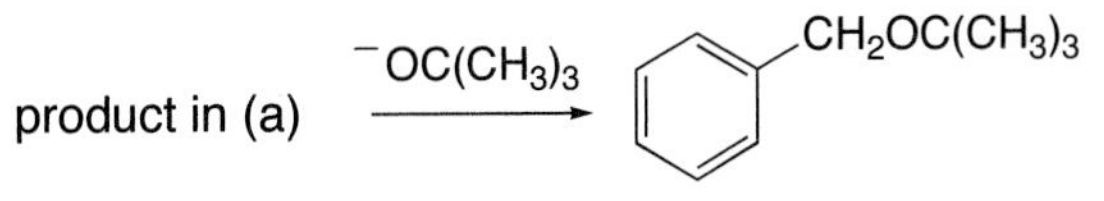

**d.** product in (b) →(PCC)→

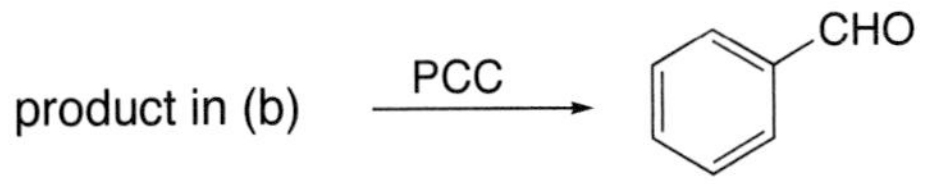

**e.**

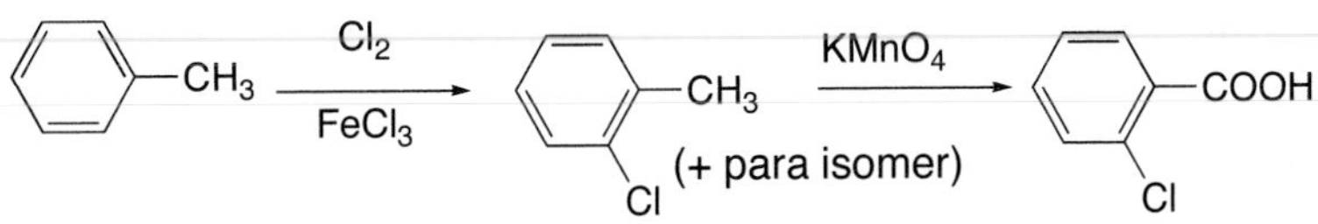

**f.**

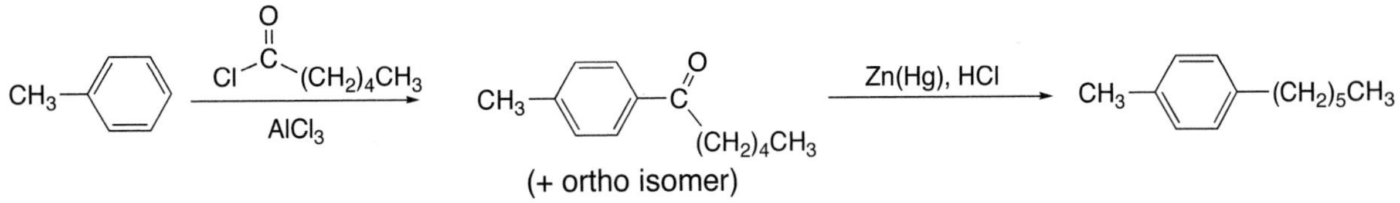

**g.**

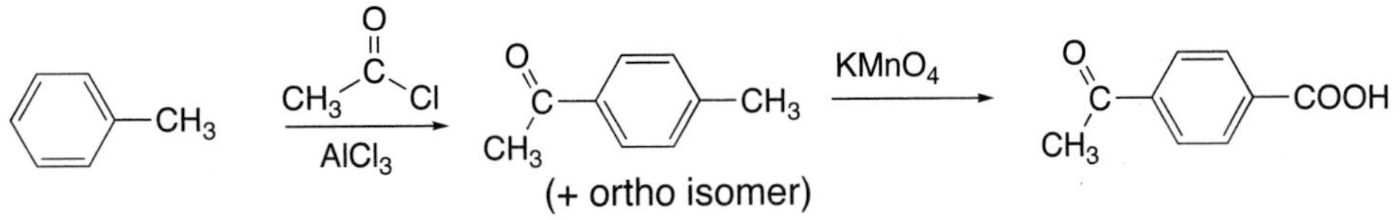

h.

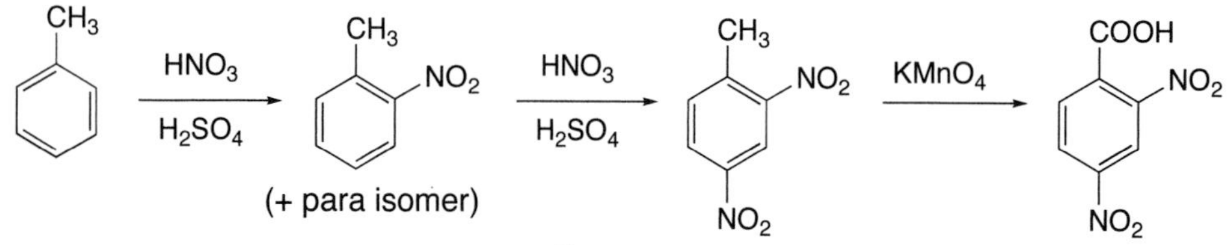

CH₃ → HNO₃ / H₂SO₄ → NO₂ (+ para isomer) → HNO₃ / H₂SO₄ → NO₂, NO₂ → KMnO₄ → COOH, NO₂, NO₂

i.

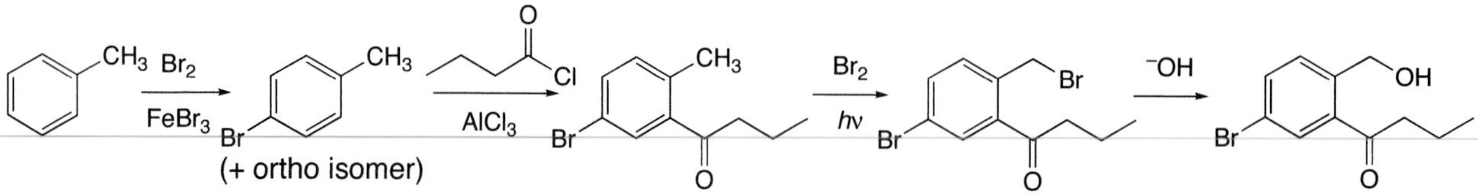

CH₃, Br₂ / FeBr₃ → Br, CH₃ (+ ortho isomer) → (acyl chloride) / AlCl₃ → Br, CH₃, O → Br₂ / hv → Br, CH₂Br, O → ⁻OH → Br, CH₂OH, O

j.

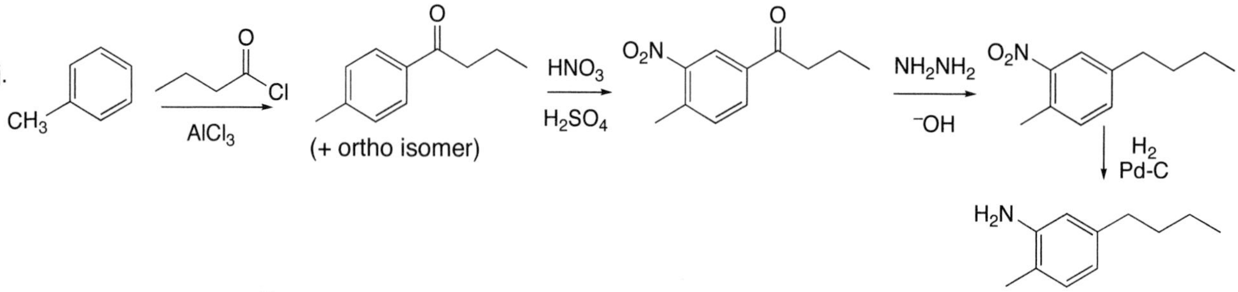

CH₃, (acyl chloride) / AlCl₃ → O (+ ortho isomer) → HNO₃ / H₂SO₄ → O₂N, O → NH₂NH₂ / ⁻OH → O₂N → H₂ / Pd-C → H₂N

k.

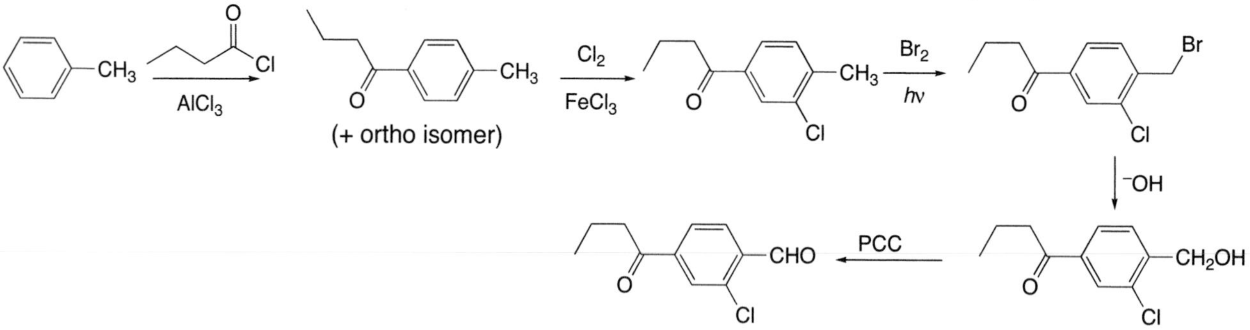

CH₃, (acyl chloride) / AlCl₃ → CH₃, O (+ ortho isomer) → Cl₂ / FeCl₃ → CH₃, O, Cl → Br₂ / hv → CH₂Br, O, Cl → ⁻OH → CH₂OH, O, Cl → PCC → CHO, O, Cl

**18.57**

a.

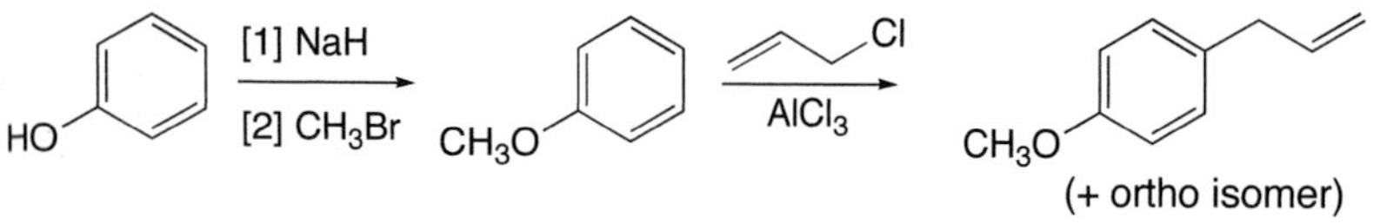

HO → [1] NaH [2] CH₃Br → CH₃O → (allyl chloride) / AlCl₃ → CH₃O, allyl (+ ortho isomer)

b.

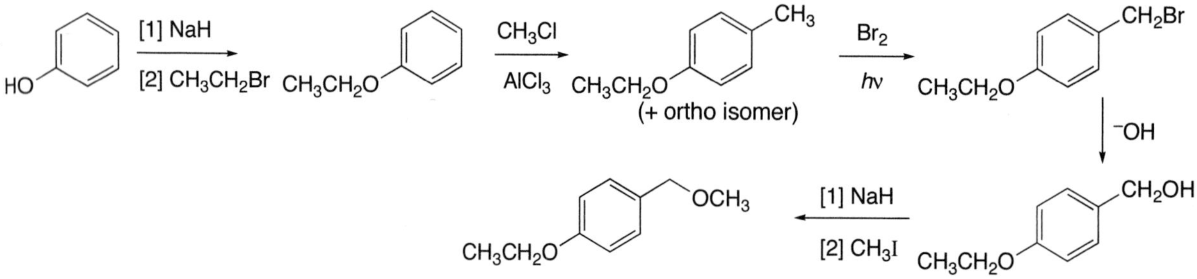

HO → [1] NaH [2] CH₃CH₂Br → CH₃CH₂O → CH₃Cl / AlCl₃ → CH₃CH₂O, CH₃ (+ ortho isomer) → Br₂ / hv → CH₃CH₂O, CH₂Br → ⁻OH → CH₃CH₂O, CH₂OH → [1] NaH [2] CH₃I → CH₃CH₂O, CH₂OCH₃

c.

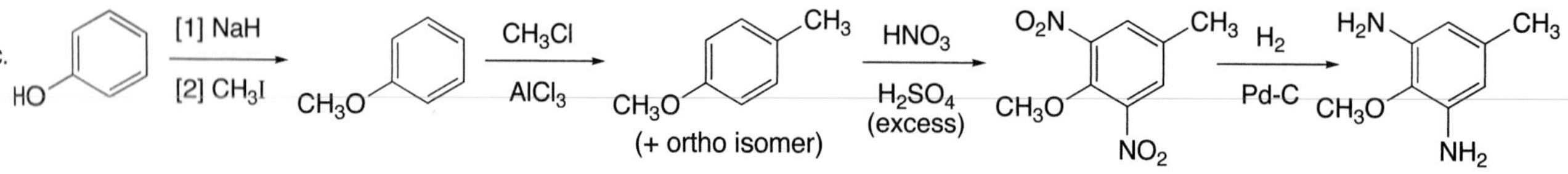

HO → [1] NaH [2] CH₃I → CH₃O → CH₃Cl / AlCl₃ → CH₃O, CH₃ (+ ortho isomer) → HNO₃ / H₂SO₄ (excess) → CH₃O, O₂N, CH₃, NO₂ → H₂ / Pd-C → CH₃O, H₂N, CH₃, NH₂

**18.58**

Cl—/ AlCl₃ → (ethylbenzene) → Br₂ / hv → Br (benzylic) → K⁺ ⁻OC(CH₃)₃ → (styrene) → Br₂ → Br, Br → 2 ⁻NH₂ → C≡CH

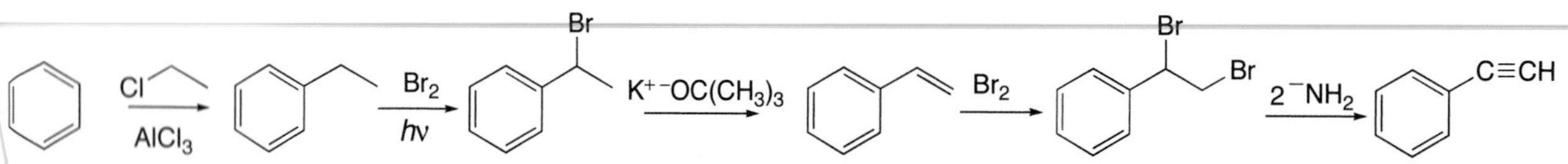

b.

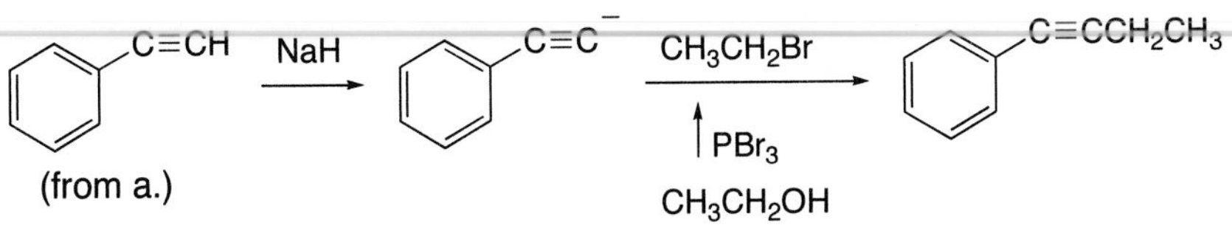

(from a.)

c.

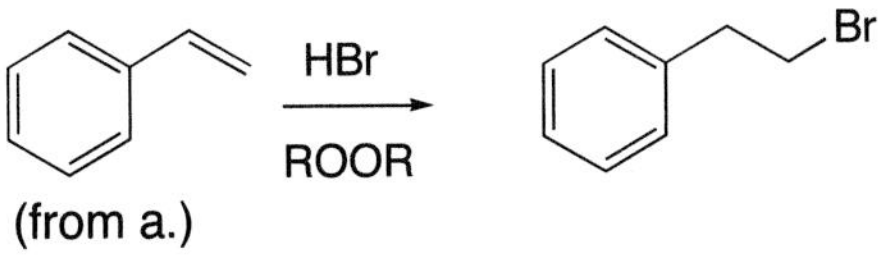

(from a.)

d.

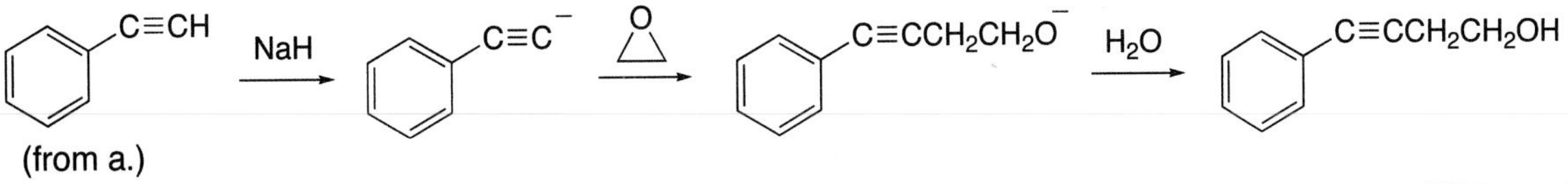

(from a.)

e.

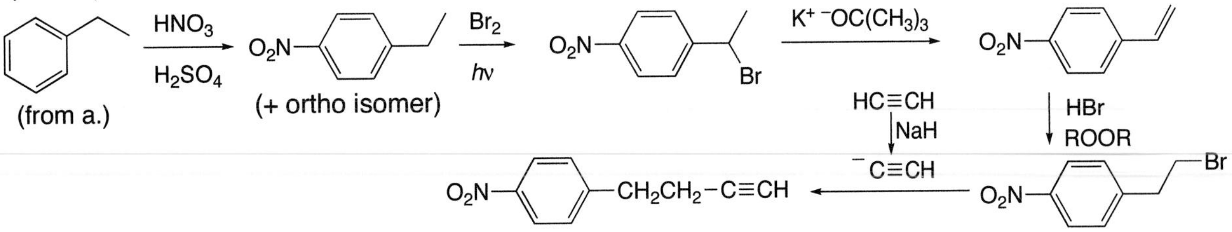

(from a.) (+ ortho isomer)

f.

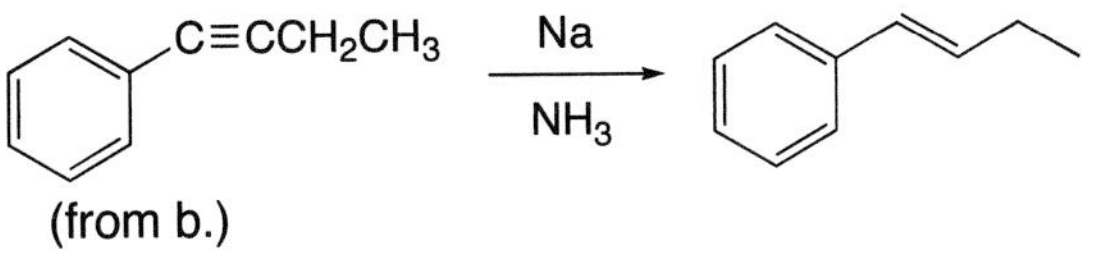

(from b.)

g.

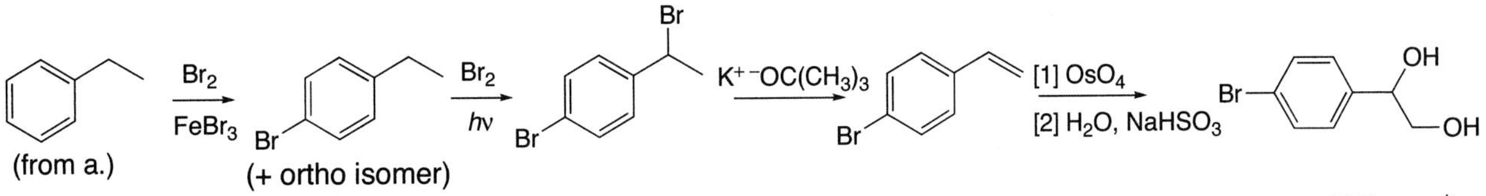

(from a.) (+ ortho isomer)

h.

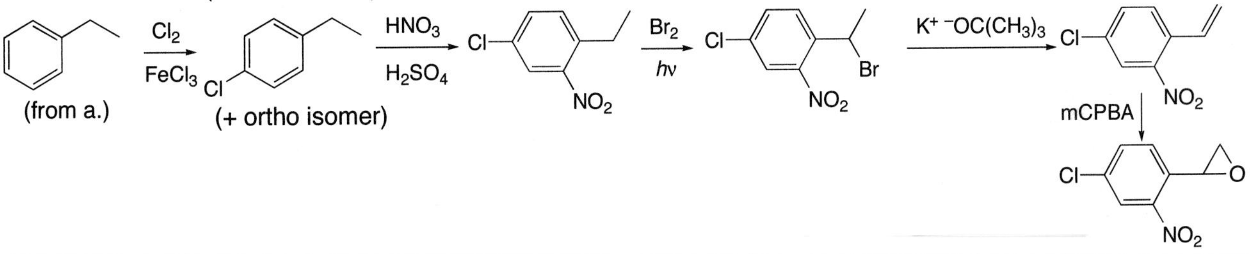

(from a.) (+ ortho isomer)

**18.59**

(+ ortho isomer)

**X**

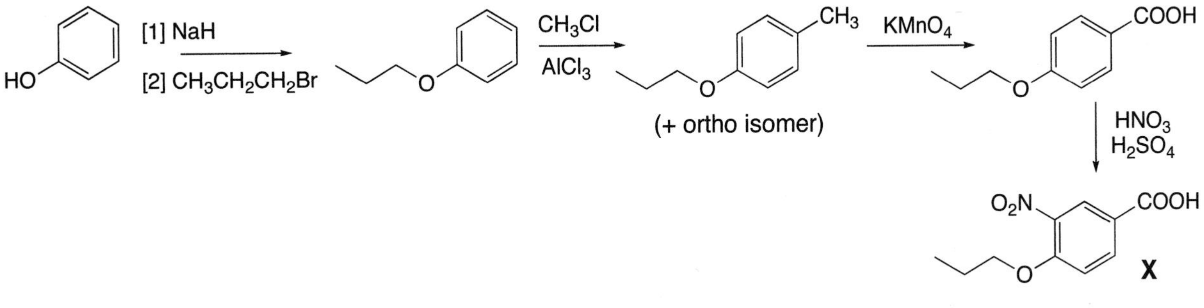

**18.60**

[1]H NMR data Compound **X** ($C_{10}H_{12}O$):

| absorption | ppm | # of H's | Explanation |
|---|---|---|---|
| doublet | 1.3 | 6 | 6H's adjacent to 1H |
| septet | 3.5 | 1 | 1H adjacent to 6H's |
| multiplet | 7.4–8.1 | 5 | monosubstituted benzene |

**Structure:**

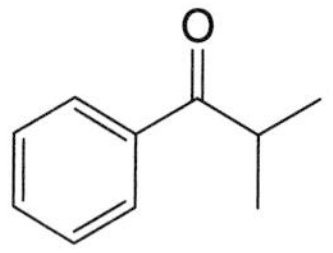

[1]H NMR data Compound **Y** ($C_{10}H_{14}$):

| absorption | ppm | # of H's | Explanation |
|---|---|---|---|
| doublet | 0.9 | 6 | 6H's adjacent to 1H |
| multiplet | 1.8 | 1 | 1H adjacent to many H's |
| doublet | 2.5 | 2 | 2H's adjacent to 1H |
| multiplet | 7.1–7.3 | 5 | monosubstituted benzene |

**Structure:**

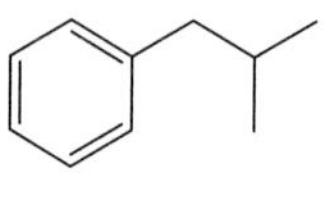

**18.61**

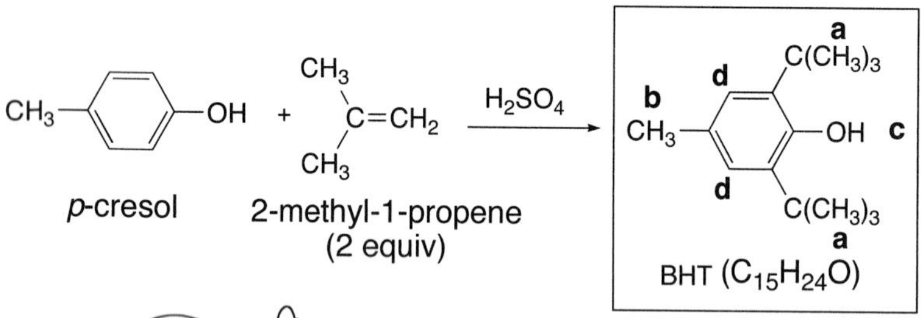


[1]H NMR spectral data:
1.4 (singlet, 18H) (a)
2.27 (singlet, 3H) (b)
5.0 (singlet, 1H) (c)
7.0 (singlet, 2H) (d) ppm

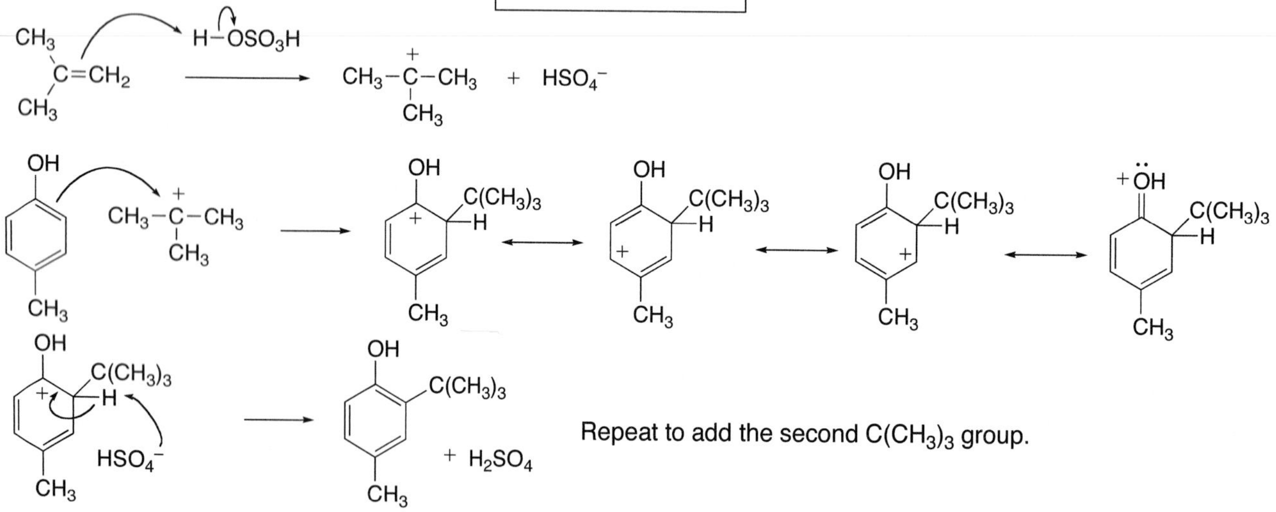


Repeat to add the second $C(CH_3)_3$ group.

**18.62**

Molecular formula (**Z**): $C_9H_9ClO$

IR absorption at 1683 $cm^{-1}$: C=O

[1]H NMR spectral data:

| absorption | ppm | # of H's | Explanation |
|---|---|---|---|
| triplet | 1.2 | 3 | 3H's adjacent to 2H's |
| quartet | 2.9 | 2 | 2H's adjacent to 3H's |
| multiplet | 7.2–8.0 | 4 | disubstituted benzene |

**Structure:**

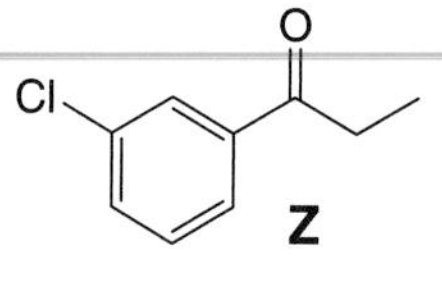

**18.63** Five resonance structures can be drawn for phenol, three of which place a negative charge on the ortho and para carbons. These illustrate that the electron density at these positions is increased, thus shielding the protons at these positions, and shifting the absorptions to lower chemical shift. Similar resonance structures cannot be drawn with a negative charge at the meta position, so it is more deshielded and absorbs further downfield, at higher chemical shift.

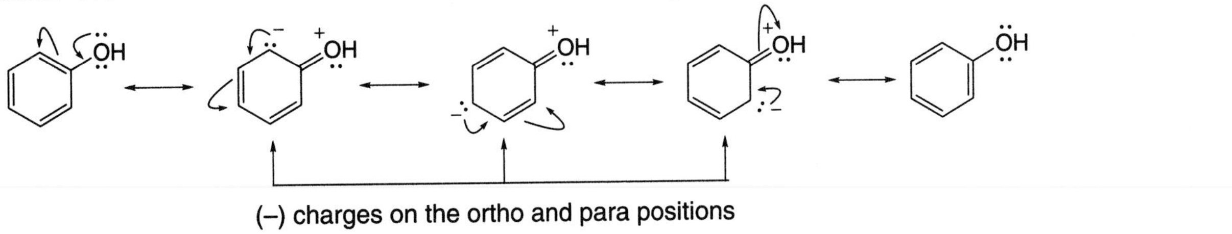

(–) charges on the ortho and para positions

**18.64** Attack at the 2-position is favored because the resulting carbocation is more highly resonance stabilized than the carbocation that results from attack at the 3-position.

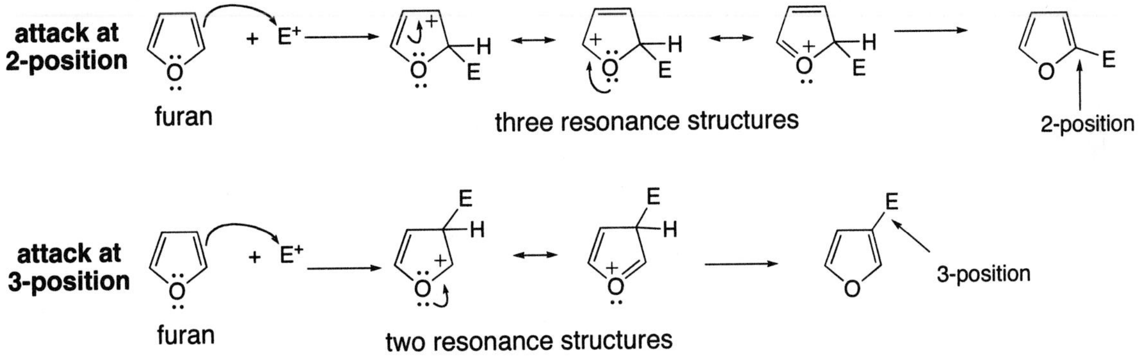

**18.65** Draw a stepwise mechanism for the following intramolecular reaction, which was used in the synthesis of the female sex hormone estrone.

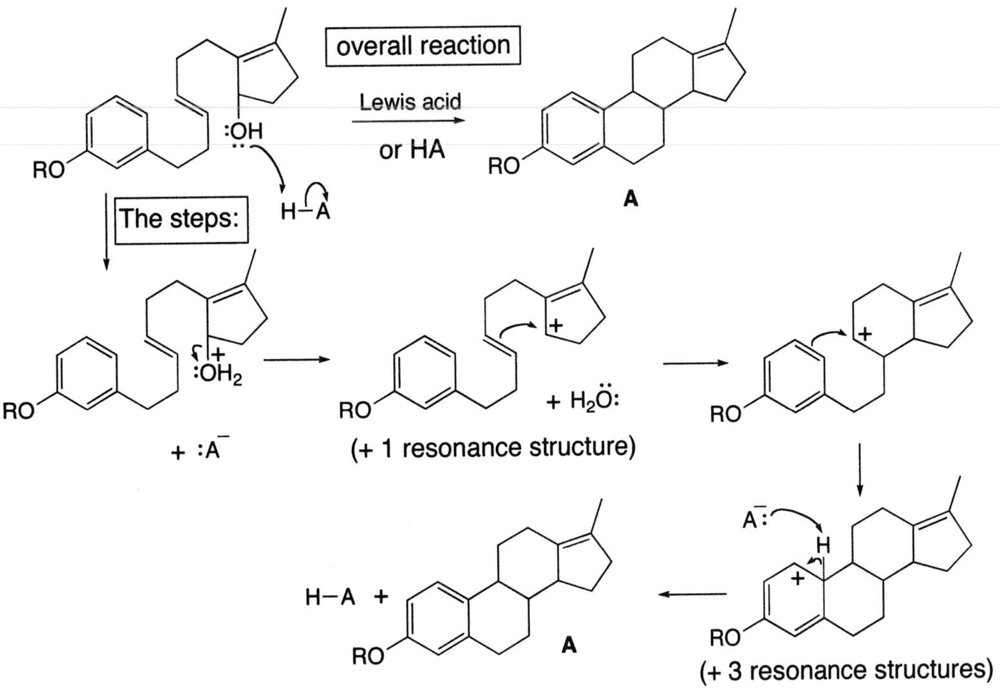

**18.66**

a. The reaction could follow a two-step mechanism: [1] addition of the nucleophile to form a carbanion, followed by [2] elimination of the leaving group.

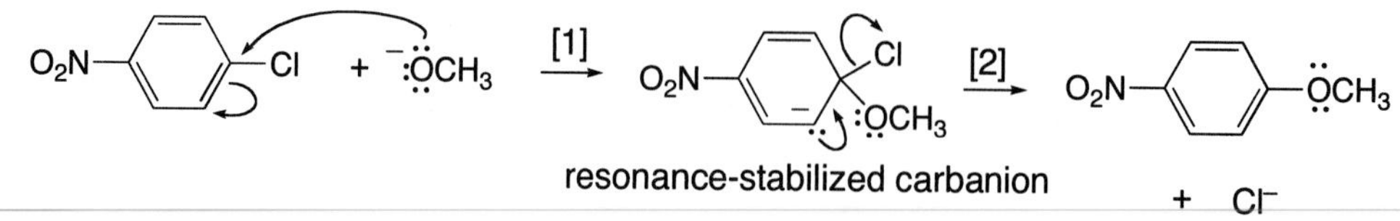

resonance-stabilized carbanion

b. The NO$_2$ group stabilizes the negatively charged intermediate by an electron-withdrawing inductive effect and by resonance.

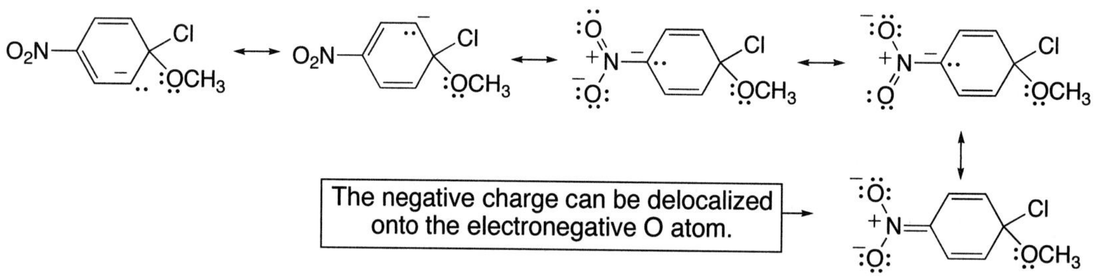

The negative charge can be delocalized onto the electronegative O atom.

c. *m*-Chloronitrobenzene does not undergo this reaction because no resonance structure can be drawn that delocalizes the negative charge of the reactive intermediate onto the O atom of the NO$_2$ group.

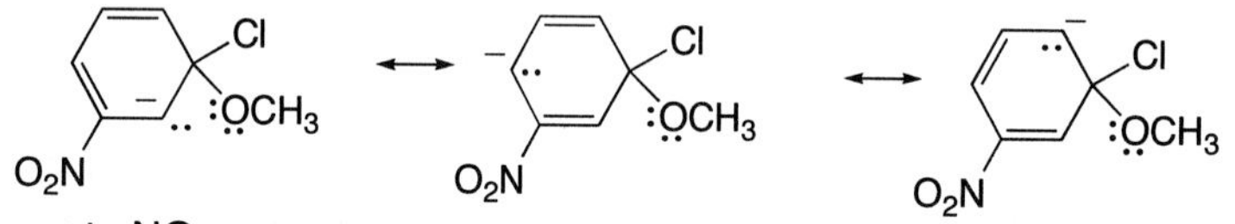

meta NO$_2$ group

## Chapter 19: Carboxylic Acids and the Acidity of the O–H Bond

### ◆ General facts

- Carboxylic acids contain a carboxy group (COOH). The central carbon is $sp^2$ hybridized and trigonal planar (19.1).
- Carboxylic acids are identified by the suffixes -*oic acid*, *carboxylic acid*, or -*ic acid* (19.2).
- Carboxylic acids are polar compounds that exhibit hydrogen bonding interactions (19.3).

### ◆ Summary of spectroscopic absorptions (19.4)

| | | |
|---|---|---|
| **IR absorptions** | C=O | $\sim$1710 cm$^{-1}$ |
| | O–H | 3500–2500 cm$^{-1}$ (very broad and strong) |
| **$^1$H NMR absorptions** | O–H | 10–12 ppm (highly deshielded proton) |
| | C–H α to COOH | 2–2.5 ppm (somewhat deshielded C$sp^3$–H) |
| **$^{13}$C NMR absorption** | C=O | 170–210 ppm (highly deshielded carbon) |

### ◆ General acid–base reaction of carboxylic acids (19.9)

$pK_a \approx 5$    carboxylate anion

- Carboxylic acids are especially acidic because carboxylate anions are resonance stabilized.
- For equilibrium to favor the products, the base must have a conjugate acid with a $pK_a > 5$. Common bases are listed in Table 19.3.

### ◆ Factors that affect acidity

**Resonance effects.** A carboxylic acid is more acidic than an alcohol or phenol because its conjugate base is more effectively stabilized by resonance (19.9).

ROH          
$pK_a = 16–18$    $pK_a = 10$    $pK_a \approx 5$

**Increasing acidity**

**Inductive effects.** Acidity increases with the presence of electron-withdrawing groups (like the electronegative halogens) and decreases with the presence of electron-donating groups (like polarizable alkyl groups) (19.10).

## Substituted benzoic acids.

- Electron-donor groups (D) make a substituted benzoic acid less acidic than benzoic acid.
- Electron-withdrawing groups (W) make a substituted benzoic acid more acidic than benzoic acid.

D →

less acidic
higher p$K_a$
p$K_a$ > 4.2

p$K_a$ = 4.2

W ←

more acidic
lower p$K_a$
p$K_a$ < 4.2

**Increasing acidity**

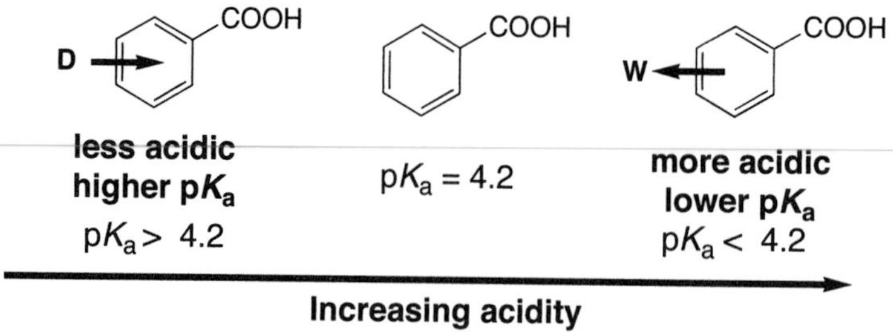

♦ **Other facts**

- Extraction is a useful technique for separating compounds having different solubility properties. Carboxylic acids can be separated from other organic compounds by extraction, because aqueous base converts a carboxylic acid into a water-soluble carboxylate anion (19.12).
- A sulfonic acid ($RSO_3H$) is a strong acid because it forms a weak, resonance-stabilized conjugate base on deprotonation (19.13).
- Amino acids have an amino group on the $\alpha$ carbon to the carboxy group [$RCH(NH_2)COOH$]. Amino acids exist as zwitterions at pH $\approx$ 7. Adding acid forms a species with a net (+1) charge [$RCH(NH_3)COOH$]$^+$. Adding base forms a species with a net (−1) charge [$RCH(NH_2)COO$]$^-$ (19.14).

**19.1** To name a carboxylic acid:

[1] Find the longest chain containing the COOH group and change the -*e* ending to -*oic acid*.

[2] Number the chain to put the COOH carbon at C1, but omit the number from the name.

[3] Follow all other rules of nomenclature.

a.

$$3 \quad CH_3$$
$$CH_3CH_2CH_2-\overset{|}{\underset{\underset{1}{\overset{|}{CH_3}}}{C}}-CH_2COOH$$

Number the chain to put COOH at C1.
6 carbon chain = **hexanoic acid**
**3,3-dimethylhexanoic acid**

c.

$$CH_3CH_2-\overset{4}{\underset{CH_3CH_2}{\overset{H}{\underset{|}{C}}}}-CH_2-\overset{2}{\underset{CH_2CH_3}{\overset{H}{\underset{|}{C}}}}-\overset{1}{COOH}$$

Number the chain to put COOH at C1.
6 carbon chain = **hexanoic acid**
**2,4-diethylhexanoic acid**

b.

$$CH_3-\overset{4}{\underset{\underset{1}{\overset{H}{\underset{|}{Cl}}}}{C}}-CH_2CH_2COOH$$

Number the chain to put COOH at C1.
5 carbon chain = **pentanoic acid**
**4-chloropentanoic acid**

d.

Number the chain to put COOH at C1.
9 carbon chain = **nonanoic acid**
**4-isopropyl-6,8-dimethylnonanoic acid**

**19.2**

a. 2-bromo**butanoic acid**

c. 3,3,4-trimethyl**heptanoic acid**

e. 3,4-diethyl**cyclohexanecarboxylic acid**

b. 2,3-dimethyl**pentanoic acid**

d. 2-*sec*-butyl-4,4-diethyl**nonanoic acid**

f. 1-isopropyl**cyclobutane-carboxylic acid**

**19.3**

a. α-methoxy**valeric acid**

c. α,β-dimethyl**caproic acid**

b. β-phenyl**propionic acid**

d. α-chloro-β-methyl**butyric acid**

**19.4**

a.

6 carbon chain = **hexanoic acid** *or* **caproic acid**

**4-bromo-4-chlorohexanoic acid**
*or*
**γ-bromo-γ-chlorocaproic acid**

b.

4 carbon chain = **butanoic acid** *or* **butyric acid**

**2-iodo-2-phenylbutanoic acid**
*or*
**α-iodo-α-phenylbutyric acid**

**19.5**

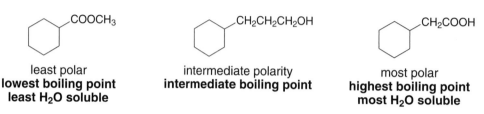

a. lithium benzoate

b. sodium formate
or
sodium methanoate

c. potassium 2-methylpropanoate

d. sodium 4-bromo-6-ethyl-octanoate

**19.6** More polar molecules have a higher boiling point and are more water soluble.

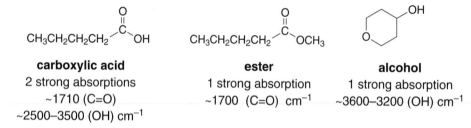

COOCH$_3$

least polar
**lowest boiling point
least H$_2$O soluble**

CH$_2$CH$_2$CH$_2$OH

intermediate polarity
**intermediate boiling point**

CH$_2$COOH

most polar
**highest boiling point
most H$_2$O soluble**

**19.7** Look for functional group differences to distinguish the compounds by IR. Besides $sp^3$ hybridized C–H bonds at 3000–2850 cm$^{-1}$ (which all three compounds have), the following functional group absorptions are seen:

CH$_3$CH$_2$CH$_2$CH$_2$—C(=O)—OH

**carboxylic acid**
2 strong absorptions
~1710 (C=O)
~2500–3500 (OH) cm$^{-1}$

CH$_3$CH$_2$CH$_2$CH$_2$—C(=O)—OCH$_3$

**ester**
1 strong absorption
~1700 (C=O) cm$^{-1}$

OH (tetrahydropyran with OH)

**alcohol**
1 strong absorption
~3600–3200 (OH) cm$^{-1}$

**19.8**

Molecular formula: C$_4$H$_8$O$_2$
one degree of unsaturation

$^1$H NMR data (ppm):
0.95 (triplet, 3H)
1.65 (multiplet, 2H)
2.30 (triplet, 2H)
11.8 (singlet, 1H)

HO—C(=O)—CH$_2$CH$_2$CH$_3$

**19.9**

HO ... COOH

PGF$_{2\alpha}$
a prostaglandin

HO ... COOH

enantiomer

There are five tetrahedral stereogenic centers. Both double bonds can exhibit cis–trans isomerism. Therefore, there are $2^7 = 128$ stereoisomers.

**19.10** 1° Alcohols are converted to carboxylic acids by oxidation reactions.

a. ⟹

c. ⟹

b. ⟹ $(CH_3)_2CH-CH_2-OH$

**19.11**

a. $\xrightarrow{\text{Na}_2\text{Cr}_2\text{O}_7}{\text{H}_2\text{SO}_4, \text{H}_2\text{O}}$

**A**

c. $\xrightarrow{\text{KMnO}_4}$

**C**

(Any R group with benzylic H's can be present para to NO$_2$.)

b. $CH_3C{\equiv}CCH_3$ $\xrightarrow{[1] \text{O}_3}{[2] \text{H}_2\text{O}}$ $CH_3COOH$ (2 equiv)

**B**

d. OH ← 2° OH, 1° OH $\xrightarrow{\text{CrO}_3}{\text{H}_2\text{SO}_4, \text{H}_2\text{O}}$

**D**

**19.12**

a. $\xrightarrow{\text{NaOH}}$ + H$_2$O

c. $CH_3-\underset{\underset{CH_3}{|}}{\overset{\overset{CH_3}{|}}{C}}-OH$ $\xrightarrow{\text{NaH}}$ $CH_3-\underset{\underset{CH_3}{|}}{\overset{\overset{CH_3}{|}}{C}}-O^-$ Na$^+$ + H$_2$

b. $\xrightarrow{\text{NaOCH}_3}$ + HOCH$_3$

d. $\xrightarrow{\text{NaHCO}_3}$ + H$_2$CO$_3$

**19.13** $CH_3COOH$ has a p$K_a$ of 4.8. Any base with a conjugate acid with a p$K_a$ higher than 4.8 can deprotonate it.

a. F$^-$ p$K_a$ (HF) = 3.2 **not strong enough**
b. (CH$_3$)$_3$CO$^-$ p$K_a$ [(CH$_3$)$_3$COH]= 18 **strong enough**
c. CH$_3^-$ p$K_a$ (CH$_4$) = 50 **strong enough**
d. $^-$NH$_2$ p$K_a$ (NH$_3$) = 38 **strong enough**
e. Cl$^-$ p$K_a$ (HCl) = –7.0 **not strong enough**

**19.14**

**Increasing acidity: H$_a$ < H$_b$ < H$_c$**

mandelic acid

− H$_a$ → negative charge on C **unstable conjugate base**

− H$_b$ → negative charge on O **more stable conjugate base**

− H$_c$ → ↔ negative charge on O, resonance stabilized **most stable conjugate base**

**19.15** Electron-withdrawing groups make an acid more acidic, lowering its $pK_a$.

$CH_3CH_2-COOH$
least acidic
$pK_a = 4.9$

$ICH_2-COOH$
one electron-withdrawing group
intermediate acidity
$pK_a = 3.2$

$CF_3-COOH$
three electron-withdrawing F's
most acidic
$pK_a = 0.2$

**19.16** Acetic acid has an electron-donating methyl group bonded to the carboxy group. The $CH_3$ group both stabilizes the acid and destabilizes the nearby negative charge on the conjugate base, making $CH_3COOH$ less acidic (with a higher $pK_a$) than HCOOH.

electron-donating $CH_3$ group ⟶

acetic acid
$CH_3$ stabilizes the
partial positive charge.

conjugate base
$CH_3$ destabilizes the
negative charge.
a less stable conjugate base

**19.17**

a. $CH_3COOH$
**least acidic**

$HSCH_2COOH$
**intermediate
acidity**

$HOCH_2COOH$
**most acidic**

b. $ICH_2CH_2COOH$
**least acidic**

$ICH_2COOH$
**intermediate
acidity**

$I_2CHCOOH$
**most acidic**

**19.18**

a. $CH_3$—⟨ ⟩—COOH
**least acidic**

⟨ ⟩—COOH
**intermediate
acidity**

Cl—⟨ ⟩—COOH
**most acidic**

b. $CH_3O$—⟨ ⟩—COOH
**least acidic**

$CH_3$—⟨ ⟩—COOH
**intermediate
acidity**

⟨ ⟩—COOH (with acetyl group, $CH_3$)
**most acidic**

**19.19**

Phenol **A** has a higher $pK_a$ than phenol because of its substituents. Both the OH and $CH_3$ are electron-donating groups, which make the conjugate base less stable. Therefore, the acid is **less acidic.**

**19.20** To separate compounds by an extraction procedure, they must have different solubility properties.

a. $CH_3(CH_2)_6COOH$ and $CH_3CH_2CH_2CH_2CH=CH_2$: **YES.** The acid can be extracted into aqueous base, while the alkene will remain in an organic layer.

b. $CH_3CH_2CH_2CH_2CH=CH_2$ and $(CH_3CH_2CH_2)_2O$: **NO.** Both compounds are soluble in organic solvents and insoluble in water. Neither is acidic enough to be extracted into aqueous base.

c. $CH_3(CH_2)_6COOH$ and NaCl: one carboxylic acid, one salt: **YES.** The carboxylic acid is soluble in an organic solvent while the salt is soluble in water.

d. NaCl and KCl: two salts: **NO.**

**19.21** To separate compounds by an aqueous extraction technique, compounds must have different solubility properties. $CH_3CH_2COOH$ and $CH_3CH_2CH_2OH$ are low molecular weight organic compounds that can hydrogen bond to water, so they are water soluble. They also both dissolve in organic solvents. As a result, they are inseparable because of their similar solubility properties.

**19.22**

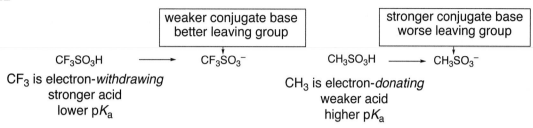

$CF_3SO_3H \longrightarrow CF_3SO_3^-$ (weaker conjugate base, better leaving group)

$CF_3$ is electron-*withdrawing*
stronger acid
lower p$K_a$

$CH_3SO_3H \longrightarrow CH_3SO_3^-$ (stronger conjugate base, worse leaving group)

$CH_3$ is electron-*donating*
weaker acid
higher p$K_a$

**19.23**

phenylalanine

*R* *S*

methionine

*R* *S*

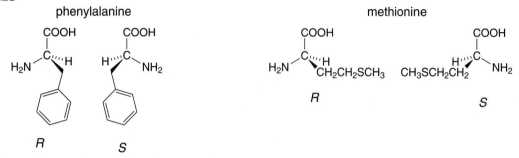

**19.24** Since amino acids exist as zwitterions (i.e., salts), they are too polar to be soluble in organic solvents like diethyl ether. Thus, they are soluble in water.

**19.25**

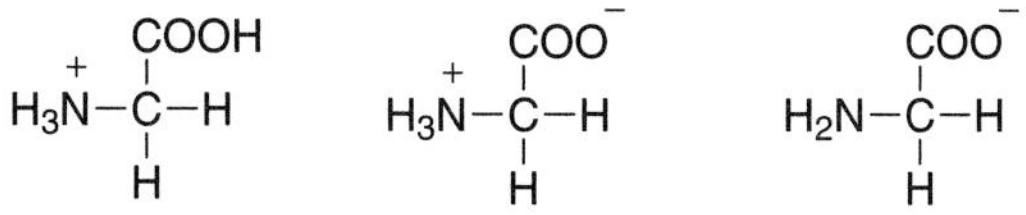

pH = 1 glycine neutral form pH = 11

**19.26**

$$pI = \frac{pK_a(\text{COOH}) + pK_a(\text{NH}_3^+)}{2} = \frac{(2.58) + (9.24)}{2} = 5.91$$

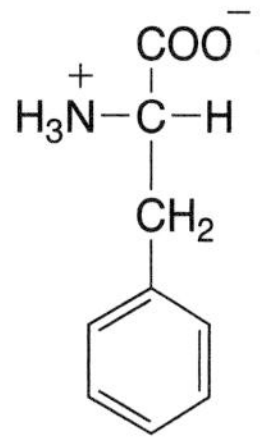

**19.27**

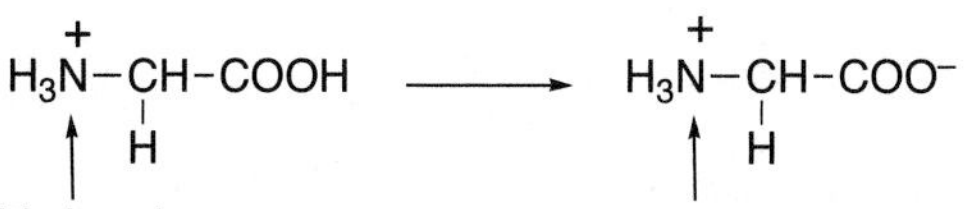

electron-withdrawing group

The nearby (+) stabilizes the conjugate base by an electron-withdrawing inductive effect, thus making the starting acid more acidic.

**19.28** Use the directions from Answer 19.1 to name the compounds.

a. $(CH_3)_2CHCH_2CH_2CO_2H$   4-methylpentanoic acid

b. $BrCH_2COOH$   2-bromoacetic acid
or 2-bromoethanoic acid

c.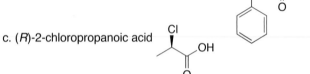
4,4,5,5-tetramethyloctanoic acid

d. $CH_3CH_2CH_2COO^-Li^+$   lithium butanoate

e. 1-ethylcyclopentanecarboxylic acid
COOH

f. 2,4-dimethylcyclohexanecarboxylic acid
COOH

g. COOH / Br   o-bromobenzoic acid

h. $CH_3CH_2$—◯—COOH  p-ethylbenzoic acid

i. O $O^-$ Na$^+$
sodium 2-methylhexanoate

j. 10 / COOH / 7 / 5 / 3
7-ethyl-5-isopropyl-3-methyldecanoic acid

**19.29**

a. 3,3-dimethylpentanoic acid   OH / O

b. 4-chloro-3-phenylheptanoic acid   Cl / OH / O

c. (R)-2-chloropropanoic acid   Cl / OH / O

d. β,β-dichloropropionic acid   Cl Cl / O / OH

e. m-hydroxybenzoic acid   HOOC / OH

f. o-chlorobenzoic acid   COOH / Cl

g. potassium acetate   O / CH$_3$—C—O$^-$ K$^+$

h. sodium α-bromobutyrate   O / O$^-$ Na$^+$ / Br

i. 2,2-dichloropentanedioic acid   O / O / HO / OH / Cl Cl

j. 4-isopropyl-2-methyloctanedioic acid   HO / O / O / OH

**19.30**

O / OH
pentanoic acid

O / OH
3-methylbutanoic acid

O / OH
2-methylbutanoic acid

O / OH
2,2-dimethylpropanoic acid

O / O$^-$ Na$^+$
sodium pentanoate

O / O$^-$ Na$^+$
sodium 3-methylbutanoate

O / O$^-$ Na$^+$
sodium 2-methylbutanoate

O / O$^-$ Na$^+$
sodium
2,2-dimethylpropanoate

### 19.31

a.

**lowest boiling point**    **intermediate boiling point**    **highest boiling point**

b.

**lowest boiling point**    **intermediate boiling point**    **highest boiling point**

### 19.32

a.

$$\xrightarrow[\text{H}_2\text{SO}_4,\ \text{H}_2\text{O}]{\text{CrO}_3}$$

c.

$$-\text{C}{\equiv}\text{C}-\text{H} \xrightarrow[\text{[2] H}_2\text{O}]{\text{[1] O}_3} \quad -\text{COOH} + \text{CO}_2$$

b. $(CH_3)_2CH$—⬡—$CH_3$ $\xrightarrow{\text{KMnO}_4}$ HOOC—⬡—COOH

d. $CH_3(CH_2)_6CH_2OH \xrightarrow[\text{H}_2\text{SO}_4,\ \text{H}_2\text{O}]{\text{Na}_2\text{Cr}_2\text{O}_7} CH_3(CH_2)_6COOH$

### 19.33

a. ⬡$=CH_2$ $\xrightarrow[\text{[2] H}_2\text{O}_2,\ ^-\text{OH}]{\text{[1] BH}_3}$ ⬡$-CH_2OH$ $\xrightarrow[\text{H}_2\text{SO}_4,\ \text{H}_2\text{O}]{\text{CrO}_3}$ ⬡$-COOH$

     **A**                     **B**

b. $HC{\equiv}CH$ $\xrightarrow[\text{[2] CH}_3\text{I}]{\text{[1] NaNH}_2}$ $HC{\equiv}CCH_3$ $\xrightarrow[\text{[2] CH}_3\text{CH}_2\text{I}]{\text{[1] NaNH}_2}$ $CH_3CH_2C{\equiv}CCH_3$ $\xrightarrow[\text{[2] H}_2\text{O}]{\text{[1] O}_3}$ $CH_3CH_2COOH$ + $CH_3COOH$

                       **C**                       **D**                       **E + F**

c. ⬡ $\xrightarrow[\text{AlCl}_3]{(CH_3)_2CHCl}$ ⬡$-CH(CH_3)_2$ $\xrightarrow{\text{KMnO}_4}$ ⬡$-COOH$

                     **G**                         **H**

### 19.34

Bases: [1] $^-$OH $pK_a$ ($H_2O$) = 15.7; [2] $CH_3CH_2^-$ $pK_a$ ($CH_3CH_3$) = 50; [3] $^-NH_2$ $pK_a$ ($NH_3$) = 38;
[4] $NH_3$ $pK_a$ ($NH_4^+$) = 9.4; [5] $HC{\equiv}C^-$ $pK_a$ ($HC{\equiv}CH$) = 25.

a. $CH_3$—⬡—COOH

$pK_a$ = 4.3
All of the bases
can deprotonate this.

b. $Cl$—⬡—OH

$pK_a$ = 9.4
$^-$OH, $CH_3CH_2^-$, $^-NH_2$ and $HC{\equiv}C^-$
can deprotonate this.

c. $(CH_3)_3COH$

$pK_a$ = 18
$CH_3CH_2^-$, $^-NH_2$ and $HC{\equiv}C^-$
can deprotonate this.

### 19.35

a. ⬡—COOH + $K^+$ $^-OC(CH_3)_3$ ⟶ ⬡—$COO^-$ $K^+$ + $HOC(CH_3)_3$

    $pK_a$ = 4.2                                          $pK_a$ = 18     | Reaction favors products. |

b. ⟍⟋⟍⟋OH + $NH_3$ ⟶ ⟍⟋⟍⟋$O^-$ + $NH_4^+$

    $pK_a \approx 16$                                  $pK_a$ = 9.4   | Reaction favors reactants. |

c. (benzene ring)—OH + NH$_2^-$ → (benzene ring)—O$^-$ + NH$_3$    | Reaction favors products. |

pK$_a$ = 10                                pK$_a$ = 38

d. (benzene ring)—COOH + CH$_3^-$Li$^+$ → (benzene ring)—COO$^-$ Li$^+$ + CH$_4$    | Reaction favors products. |

   CH$_3$                                    CH$_3$    pK$_a$ = 50

pK$_a$ ≈ 4

e. (structure)—OH + Na$^+$H$^-$ → (structure)—O$^-$ Na$^+$ + H$_2$    | Reaction favors products. |

pK$_a$ ≈ 16                                pK$_a$ = 35

f. CH$_3$—(benzene ring)—OH + Na$_2$CO$_3$ → CH$_3$—(benzene ring)—O$^-$ Na$^+$ + Na$^+$ HCO$_3^-$    | With the same pK$_a$ for the starting acid and the conjugate acid, an equal amount of starting materials and products is present. |

pK$_a$ = 10.2                              pK$_a$ = 10.2

**19.36**  The stronger acid has a lower pK$_a$ and a weaker conjugate base.

a. (benzene ring)—COOH    and    (cyclohexane)—CH$_2$OH

   carboxylic acid          alcohol
   stronger acid            weaker acid
   **lower pK$_a$**         higher pK$_a$
   weaker conjugate base    **stronger conjugate base**

c.  CH$_3$—(benzene ring)—COOH  and  Cl—(benzene ring)—COOH

    CH$_3$ is electron donating.    Cl is electron withdrawing.
    weaker acid                     stronger acid
    higher pK$_a$                   **lower pK$_a$**
    **stronger conjugate base**     weaker conjugate base

b.  ClCH$_2$COOH    and    FCH$_2$COOH

    weaker acid      F is more electronegative.
    higher pK$_a$    stronger acid
    **stronger conjugate base**    **lower pK$_a$**
                     weaker conjugate base

d.  NCCH$_2$COOH    and    CH$_3$COOH

    CN is electron withdrawing.    weaker acid
    stronger acid                  higher pK$_a$
    **lower pK$_a$**               **stronger conjugate base**
    weaker conjugate base

**19.37**

a.  (propyl)COOH          (with Br)COOH          (with Cl)COOH

    **least acidic**      Br is electronegative    Cl more electronegative
                          **intermediate acidity**  **most acidic**

b.  CH$_3$—(benzene ring)—OH    Cl—(benzene ring)—OH    O$_2$N—(benzene ring)—OH

    **least acidic**            **intermediate acidity**    **most acidic**

c.  CH$_3$—(benzene ring)—COOH    (benzene ring)—COOH    CF$_3$—(benzene ring)—COOH

    **least acidic**            **intermediate acidity**    **most acidic**

d.

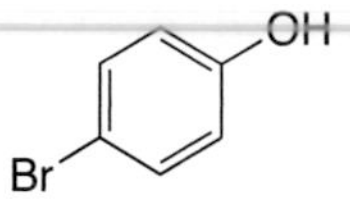

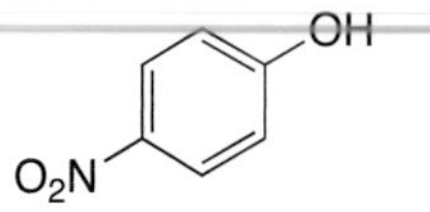

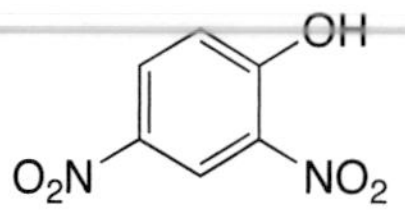

OH
Br
**least acidic**

OH
O₂N
**intermediate acidity**

OH
O₂N    NO₂
**most acidic**

## 19.38

a. $BrCH_2COO^-$     $BrCH_2CH_2COO^-$     $(CH_3)_3CCOO^-$

**weakest base**  **intermediate basicity**  **strongest base**

c.

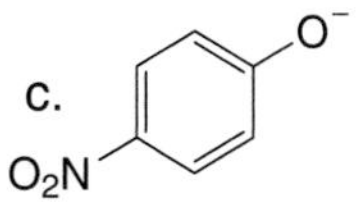

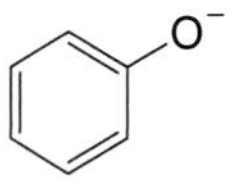

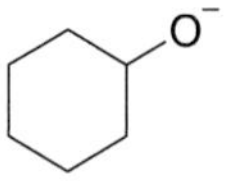

O⁻
O₂N

O⁻

O⁻

**weakest base**  **intermediate basicity**  **strongest base**

b.

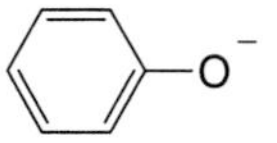

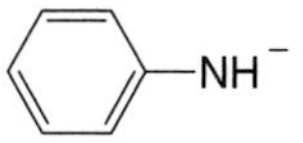

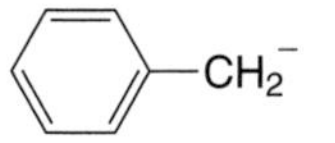

O⁻

NH⁻

CH₂⁻

**weakest base**  **intermediate basicity**  **strongest base**

## 19.39

Increasing acidity →

| | ICH₂COOH | BrCH₂COOH | FCH₂COOH | F₂CHCOOH | F₃CCOOH |
|---|---|---|---|---|---|
| p$K_a$ values | least acidic 3.12 | 2.86 | 2.66 | 1.24 | most acidic 0.28 |

## 19.40

a. The negative charge on the conjugate base of *p*-nitrophenol is delocalized on the NO₂ group, stabilizing the conjugate base, and making *p*-nitrophenol more acidic than phenol (where the negative charge is delocalized only around the benzene ring).

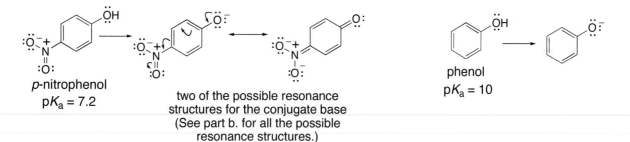

*p*-nitrophenol
p$K_a$ = 7.2

two of the possible resonance structures for the conjugate base (See part b. for all the possible resonance structures.)

phenol
p$K_a$ = 10

b. In the para isomer, the negative charge of the conjugate base is delocalized over both the benzene ring and onto the NO$_2$ group, whereas in the meta isomer it cannot be delocalized onto the NO$_2$ group. This makes the conjugate base from the para isomer more highly resonance stabilized, and the para substituted phenol more acidic than its meta isomer.

pK$_a$ = 7.2
p-nitrophenol

negative charge on
two O atoms
very good resonance structure
more stable conjugate base
stronger acid

pK$_a$ = 8.3
m-nitrophenol

**19.41** A CH$_3$O group has an electron-withdrawing inductive effect and an electron-donating resonance effect. In 2-methoxyacetic acid, the OCH$_3$ group is bonded to an $sp^3$ hybridized C, so there is no way to donate electron density by resonance. The CH$_3$O group withdraws electron density because of the electronegative O atom, stabilizing the conjugate base, and making CH$_3$OCH$_2$COOH a stronger acid than CH$_3$COOH.

more acidic acid                more stable conjugate base

In p-methoxybenzoic acid, the CH$_3$O group is bonded to an $sp^2$ hybridized C so it can donate electron density by a resonance effect. This destabilizes the conjugate base, making the starting material less acidic than C$_6$H$_5$COOH.

less acidic acid                like charges nearby
less stable conjugate base

**19.42** Both **A** and **B** have stabilizing groups for the conjugate base, making them more acidic than CH$_3$CH$_2$COOH.

The nearby partial positive charge on the C=O group stabilizes the conjugate base, so the starting acid is more acidic.

CH$_2$=CHCH$_2$COOH
**B**

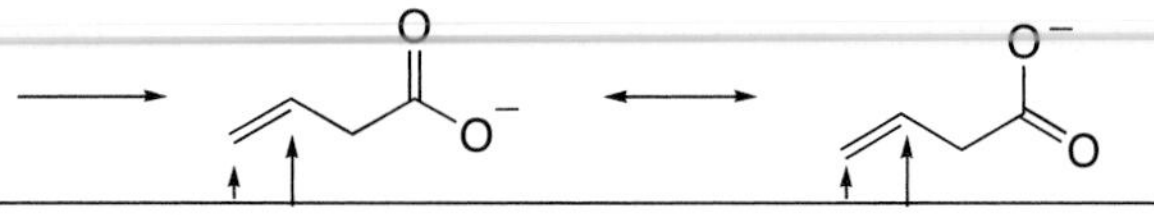

The *sp*$^2$ hybridized C's of the double bond have a higher percent *s*-character than an *sp*$^3$ hybridized C, so they pull more electron density toward them, stabilizing the conjugate base. This makes **B** more acidic than CH$_3$CH$_2$COOH.

**19.43**

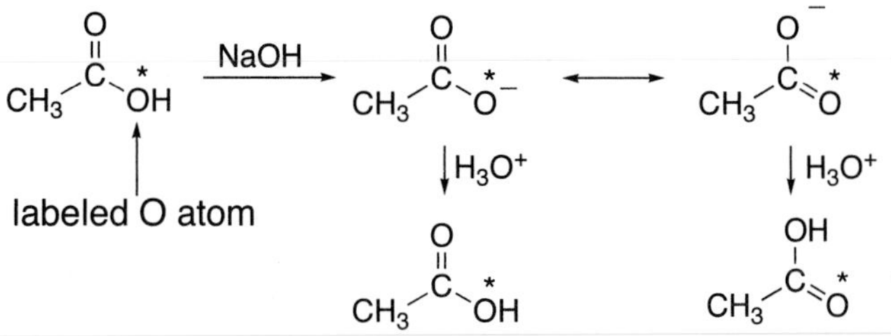

labeled O atom

The resonance-stabilized carboxylate anion can now be protonated on either O atom, the one with the label and the one without the label.

The label is now in two different locations.

**19.44**

a.

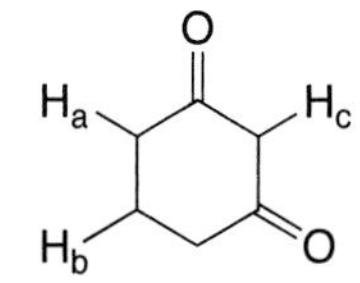

1,3-cyclohexanedione
**increasing acidity: H$_b$ < H$_a$ < H$_c$**

loss of H$_b$:

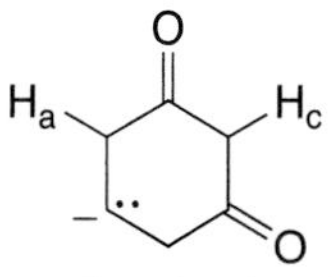

one Lewis structure
**least stable conjugate base**

The most acidic proton forms the most stable conjugate base.

loss of H$_a$:

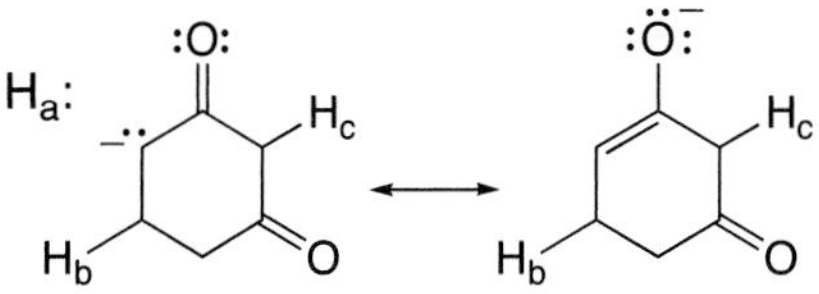

2 resonance structures
**intermediate stability**

loss of H$_c$:

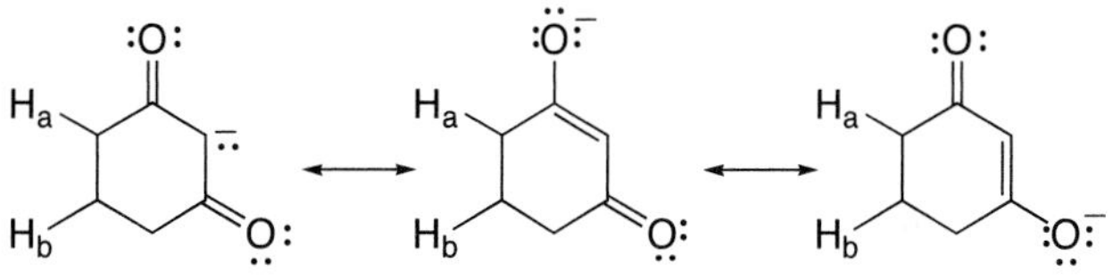

3 resonance structures
**most stable conjugate base**

b.

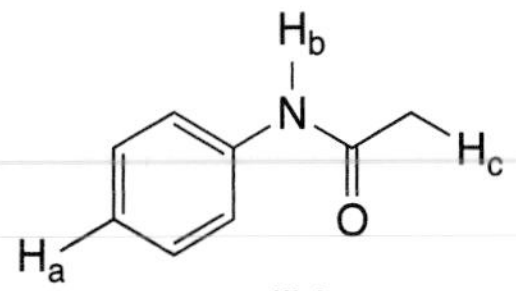

acetanilide
**increasing acidity: H$_a$ < H$_c$ < H$_b$**

loss of H$_b$:

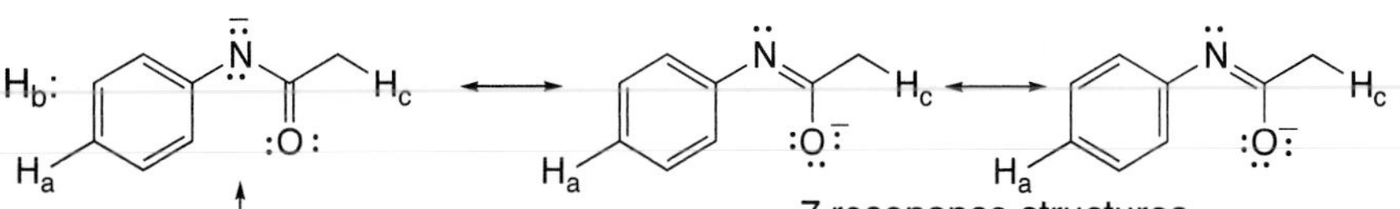

7 resonance structures
**most stable conjugate base**

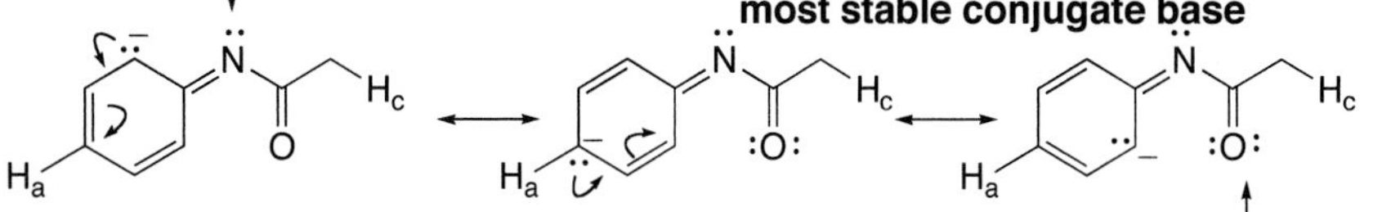

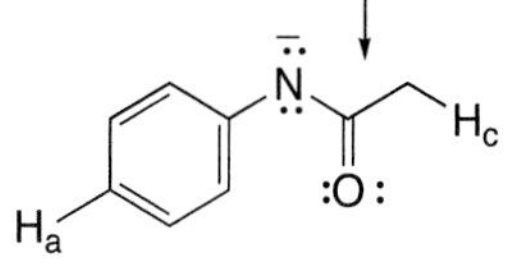

loss of H$_a$:

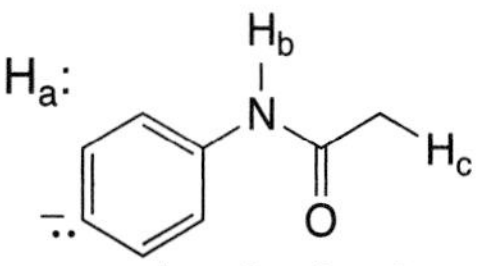

one Lewis structure
**least stable conjugate base**

loss of H$_c$:

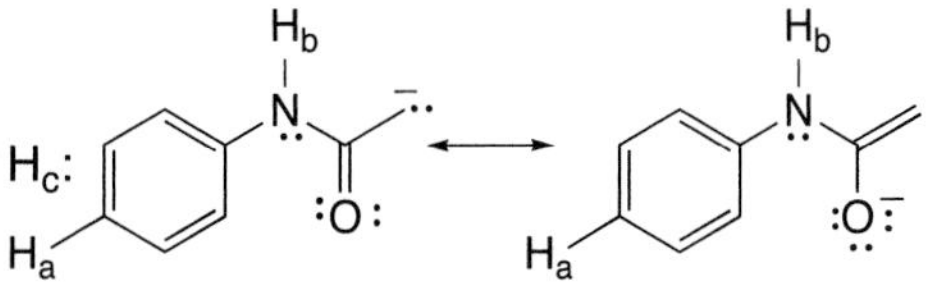

2 resonance structures
**intermediate stability**

**19.45**   As usual, compare the stability of the conjugate bases.  With RSO₃H, loss of a proton forms a conjugate base that has three resonance structures, all of which are equivalent and place a negative charge on a more electronegative O atom.  With the conjugate base of RCOOH, there are only two of these such resonance structures.  Thus, the conjugate base $RSO_3^-$ is more highly resonance stabilized than $RCOO^-$, so $RSO_3H$ is a stronger acid than RCOOH.

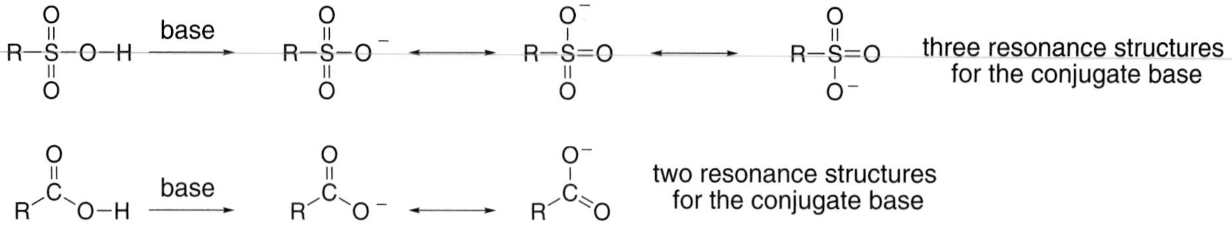

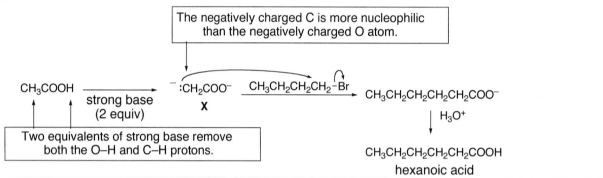

**19.46**

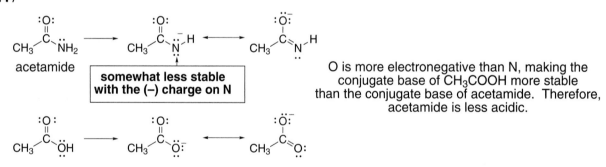

**19.47**

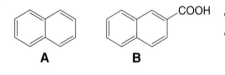

**19.48**

- Dissolve both compounds in $CH_2Cl_2$.
- Add 10% NaHCO₃ solution.  This makes a carboxylate anion ($C_{10}H_7COO^-$) from **B**, which dissolves in the aqueous layer.  The other compound (**A**) remains in the $CH_2Cl_2$.
- Separate the layers.

**19.49**

- Dissolve both compounds in $CH_2Cl_2$.
- Add 10% NaOH solution. This converts $C_6H_5OH$ into a phenoxide anion, $C_6H_5O^-$, which dissolves in the aqueous solution. The alcohol remains in the organic layer (neutral) since it is not acidic enough to be deprotonated to any significant extent by NaOH.
- Separate the layers.

**19.50** To separate two compounds in an aqueous extraction, one must be water soluble (or be able to be converted into a water-soluble ionic compound by an acid–base reaction), and the other insoluble. 1-Octanol has greater than 5 C's, making it insoluble in water. Octane is an alkane, also insoluble in water. Neither compound is acidic enough to be deprotonated by a base in aqueous solution. Since their solubility properties are similar, they cannot be separated by an extraction procedure.

**19.51**

a. Molecular formula: $C_3H_5ClO_2$ ⟶ one double bond or ring
IR: 3500–2500 cm$^{-1}$; 1714 cm$^{-1}$ ⟶ C=O and O–H
NMR data: 2.87 (triplet, 2H); 3.76 (triplet, 2H); 11.8 (singlet, 1H) ppm

$ClCH_2CH_2$ —C(=O)—OH

b. Molecular formula: $C_8H_8O_3$ ⟶ 5 double bonds or rings
IR: 3500–2500 cm$^{-1}$; 1688 cm$^{-1}$ ⟶ C=O and O–H
NMR data: 3.8 (singlet, 3H); 7.0 (doublet, 2H), 7.9 (doublet, 2H); 12.7 (singlet, 1H) ppm

$CH_3O$—⟨benzene⟩—COOH

| para disubstituted benzene ring |

c. Molecular formula: $C_8H_8O_3$ ⟶ 5 double bonds or rings
IR: 3500–2500 cm$^{-1}$; 1710 cm$^{-1}$ ⟶ C=O and O–H
NMR data: 4.7 (singlet, 2H); 6.9–7.3 (multiplet, 5H), 11.3 (singlet, 1H) ppm

⟨benzene⟩—$OCH_2COOH$

| monosubstituted benzene ring |

**19.52**
Compound A: Molecular formula $C_4H_8O_2$ (one degree of unsaturation)
IR absorptions at 3600–3200 (O–H), 3000–2800 (C–H), and 1700 (C=O) cm$^{-1}$.
$^1H$ NMR data:

| absorption | ppm | # of H's | Explanation | Structure: |
|---|---|---|---|---|
| singlet | 2.2 | 3 | a $CH_3$ group | |
| singlet | 2.55 | 1 | 1H adjacent to none or OH | |
| triplet | 2.7 | 2 | 2H's adjacent to 2H's | $CH_3$—C(=O)—$CH_2CH_2OH$  **A** |
| triplet | 3.9 | 2 | 2H's adjacent to 2H's | |

**Compound B:** Molecular formula $C_4H_8O_2$ (one degree of unsaturation)
IR absorptions at 3500–2500 (O–H) and 1700 (C=O) $cm^{-1}$.
$^1H$ NMR data:

| absorption | ppm | # of H's | Explanation | Structure: |
|---|---|---|---|---|
| doublet | 1.6 | 6 | 6H's adjacent to 1H | CH₃ |
| septet | 2.3 | 1 | 1H adjacent to 6H's | CH₃–C–COOH |
| singlet (very broad) | 10.7 | 1 | OH of RCOOH | H |

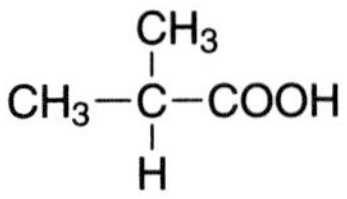

**19.53**

Molecular formula $C_6H_{12}O_2$  (1 double bond due to COOH)

$^1H$ NMR: 1.1 (singlet), 2.2 (singlet), 11.9 (singlet) ppm

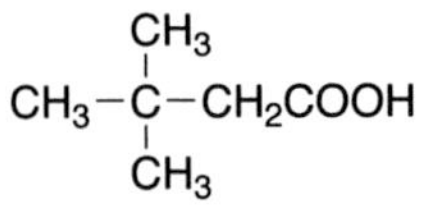

CH₃
|
CH₃–C–CH₂COOH
|
CH₃

**19.54**

Molecular formula: $C_8H_6O_4$: 6 degrees of unsaturation
IR 1692 $cm^{-1}$ (C=O)
$^1H$ NMR 8.2 and 10.0 ppm (singlets)

↑          ↑
aromatic H    COOH

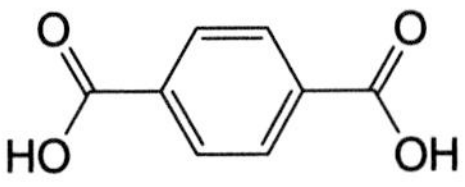

**19.55**

**A**   (CH₃)₃C–COOH   3 different C's
Spectrum [2]: peaks at 27, 39, 186 ppm

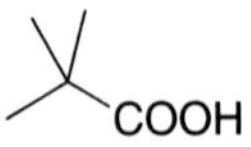

**B**   ~~~~COOH   5 different C's
Spectrum [1]: peaks at 14, 22, 27, 34, 181 ppm

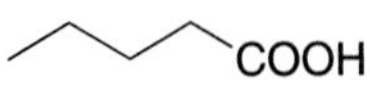

**C**   (structure)   4 different C's
Spectrum [3]: peaks at 22, 26, 43, 180 ppm

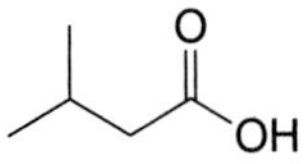

**19.56**

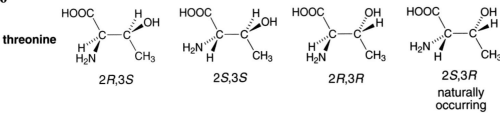

| threonine | | | |
|---|---|---|---|
| 2R,3S | 2S,3S | 2R,3R | 2S,3R naturally occurring |

**19.57**

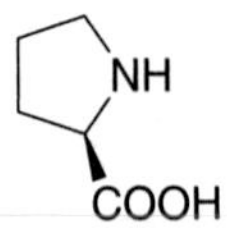

proline 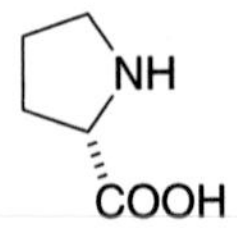        enantiomer 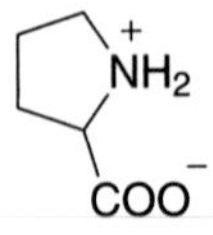        zwitterion

**19.58**

**a. methionine**

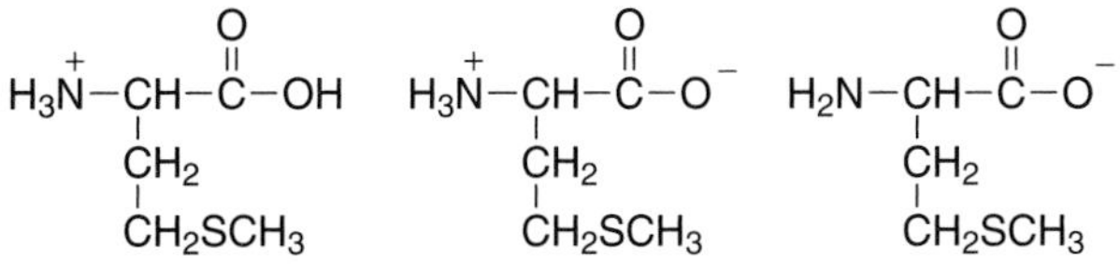

| $H_3\overset{+}{N}-CH-\overset{O}{\overset{\|}{C}}-OH$ | $H_3\overset{+}{N}-CH-\overset{O}{\overset{\|}{C}}-O^-$ | $H_2N-CH-\overset{O}{\overset{\|}{C}}-O^-$ |
| :---: | :---: | :---: |
| $CH_2$ | $CH_2$ | $CH_2$ |
| $CH_2SCH_3$ | $CH_2SCH_3$ | $CH_2SCH_3$ |
| pH = 1 | pH = 7 | pH = 11 |
| | form at isoelectric point | |

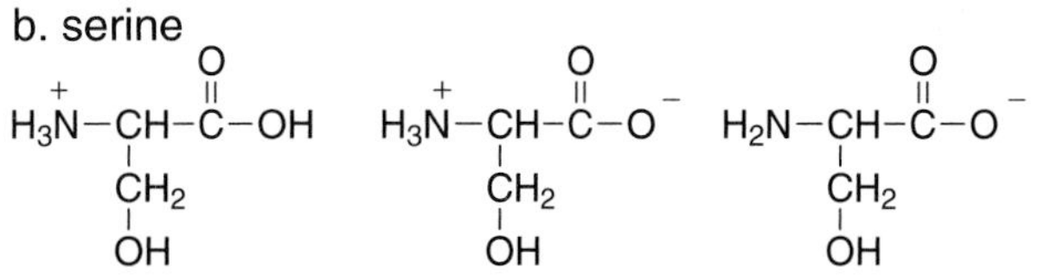

**b. serine**

| $H_3\overset{+}{N}-CH-\overset{O}{\overset{\|}{C}}-OH$ | $H_3\overset{+}{N}-CH-\overset{O}{\overset{\|}{C}}-O^-$ | $H_2N-CH-\overset{O}{\overset{\|}{C}}-O^-$ |
| :---: | :---: | :---: |
| $CH_2$ | $CH_2$ | $CH_2$ |
| $OH$ | $OH$ | $OH$ |
| pH = 1 | pH = 7 | pH = 11 |
| | form at isoelectric point | |

**19.59**

a. cysteine $pI = \dfrac{pK_a(COOH) + pK_a(NH_3^+)}{2} = (2.05) + (10.25)\,/\,2 = \mathbf{6.15}$

b. methionine $pI = \dfrac{pK_a(COOH) + pK_a(NH_3^+)}{2} = (2.28) + (9.21)\,/\,2 = \mathbf{5.75}$

**19.60** The first equivalent of $NH_3$ acts as a base to remove a proton from the carboxylic acid. A second equivalent then acts as a nucleophile to displace X to form the ammonium salt of the amino acid.

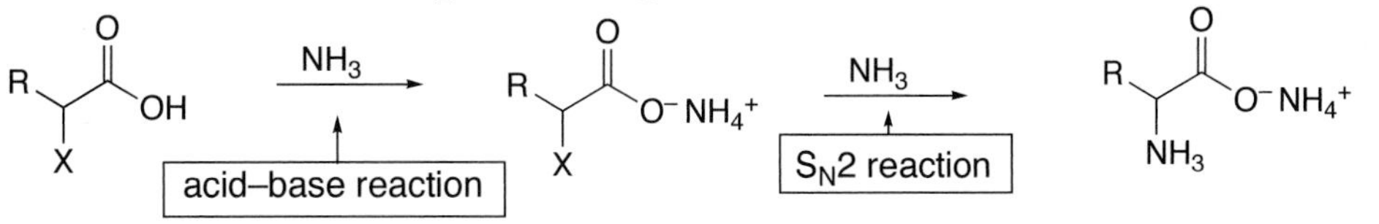

**19.61**

a. At pH = 1, the net charge is (+1).

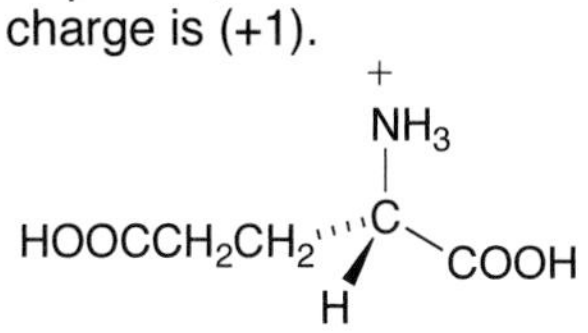

b. increasing pH: As base is added, the most acidic proton is removed first, then the next most acidic proton, and so forth.

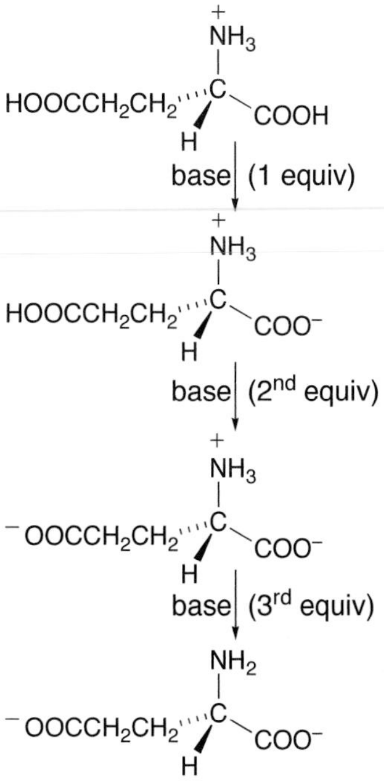

c. monosodium glutamate

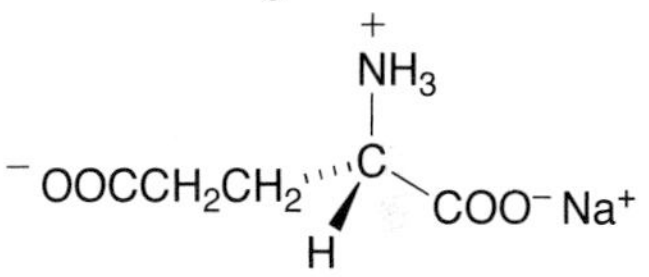

**19.62**   The first equivalent of NaH removes the most acidic proton; that is, the OH proton on the phenol.  The resulting phenoxide can then act as a nucleophile to displace I to form a substitution product.  With two equivalents, both OH protons are removed.  In this case the more nucleophilic O atom is the strongest base; that is, the alkoxide derived from the alcohol (not the phenoxide) so this negatively charged O atom reacts first in a nucleophilic substitution reaction.

**19.63**

p-hydroxybenzoic acid
**less acidic** than benzoic acid

like charges on nearby atoms
**destabilizing**

The OH group donates electron density by its resonance effect and this destabilizes the conjugate base, making the acid less acidic than benzoic acid.

o-hydroxybenzoic acid
**more acidic** than benzoic acid

Intramolecular hydrogen bonding stabilizes the conjugate base, making the acid more acidic than benzoic acid.

**19.64**

2-hydroxybutanedioic acid
increasing acidity:
$H_d < H_c < H_b < H_e < H_a$

$H_a$ and $H_e$ must be the two most acidic protons since they are part of carboxylic acids.  Loss of a proton forms a resonance-stabilized carboxylate anion that has the negative charge delocalized on two O atoms.  $H_a$ is more acidic than $H_e$ because the nearby OH group on the $\alpha$ carbon increases acidity by an electron-withdrawing inductive effect.  $H_b$ is the next most acidic proton because the conjugate base places a negative charge on the electronegative O atom, but it is not resonance stabilized.

The least acidic H's are $H_c$ and $H_d$ since these H's are bonded to C atoms.  The electronegative O atom further acidifies $H_c$ by an electron-withdrawing inductive effect.

## Chapter 20: Introduction to Carbonyl Chemistry

### ◆ Reduction reactions

### [1]  Reduction of aldehydes and ketones to 1° and 2° alcohols (20.4)

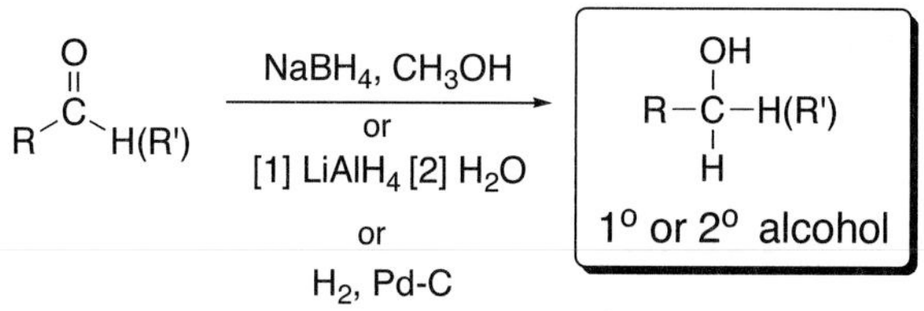

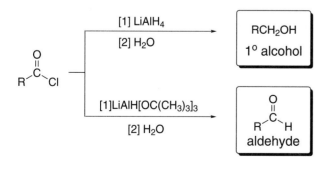

### [2]  Reduction of α,β-unsaturated aldehydes and ketones (20.4C)

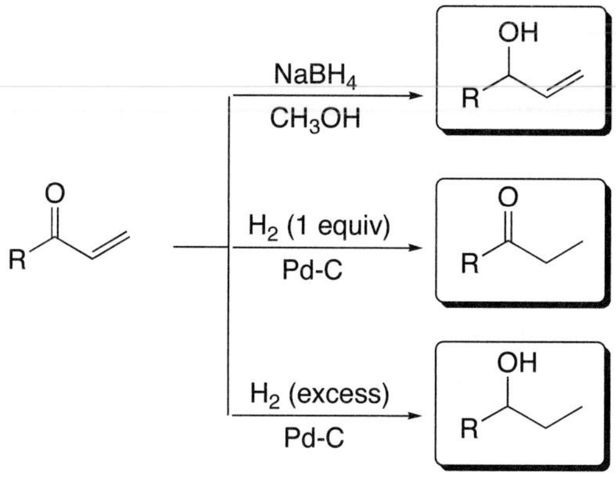

- reduction of the C=O only

- reduction of the C=C only

- reduction of both π bonds

### [3]  Enantioselective ketone reduction (20.6)

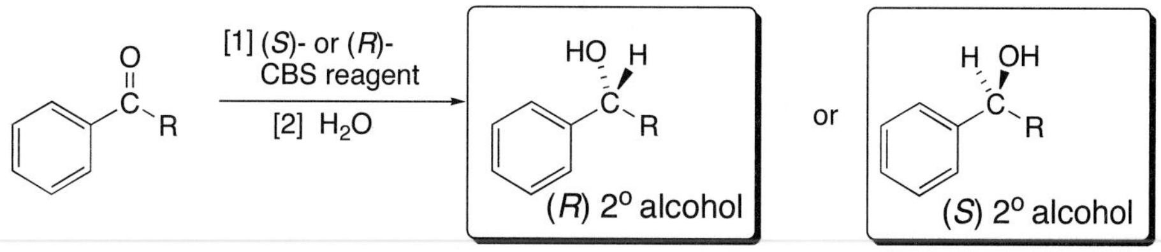

[1] (S)- or (R)-
CBS reagent

[2] H₂O

(R) 2° alcohol      or      (S) 2° alcohol

- A single enantiomer is formed.

### [4]  Reduction of acid chlorides (20.7A)

[1] LiAlH₄

[2] H₂O

RCH₂OH
1° alcohol

- LiAlH₄, a strong reducing agent, reduces an acid chloride all the way to a 1° alcohol.

[1] LiAlH[OC(CH₃)₃]₃

[2] H₂O

aldehyde

- With LiAlH[OC(CH₃)₃]₃, a milder reducing agent, reduction stops at the aldehyde stage.

## [5] Reduction of esters (20.7A)

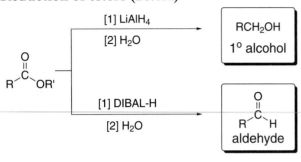

- LiAlH$_4$, a strong reducing agent, reduces an ester all the way to a 1° alcohol.

- With DIBAL-H, a milder reducing agent, reduction stops at the aldehyde stage.

## [6] Reduction of carboxylic acids to 1° alcohols (20.7B)

$$
R\text{–COOH} \xrightarrow[\text{[2] H}_2\text{O}]{\text{[1] LiAlH}_4} \boxed{\begin{array}{c} RCH_2OH \\ 1° \text{ alcohol} \end{array}}
$$

## [7] Reduction of amides to amines (20.7B)

$$
R\text{–C(=O)–N} \xrightarrow[\text{[2] H}_2\text{O}]{\text{[1] LiAlH}_4} \boxed{\begin{array}{c} RCH_2\text{–N–} \\ \text{amine} \end{array}}
$$

## ◆ Oxidation reactions

### Oxidation of aldehydes to carboxylic acids (20.8)

$$
R\text{–CHO} \xrightarrow[\text{Ag}_2\text{O, NH}_4\text{OH}]{\substack{\text{CrO}_3,\ \text{Na}_2\text{Cr}_2\text{O}_7,\ \text{K}_2\text{Cr}_2\text{O}_7,\ \text{KMnO}_4 \\ \text{or}}} \boxed{\begin{array}{c} R\text{–COOH} \\ \text{carboxylic acid} \end{array}}
$$

- All Cr$^{6+}$ reagents except PCC oxidize RCHO to RCOOH.
- Tollens reagent (Ag$_2$O + NH$_4$OH) oxidizes RCHO only. Primary (1°) and secondary (2°) alcohols do not react with Tollens reagent.

## ◆ Preparation of organometallic reagents (20.9)

**[1] Organolithium reagents:**

$$R\text{–X} + 2\,Li \longrightarrow \boxed{R\text{–Li}} + LiX$$

**[2] Grignard reagents:**

$$R\text{–X} + Mg \xrightarrow{(CH_3CH_2)_2O} \boxed{R\text{–Mg-X}}$$

**[3] Organocuprate reagents:**

$$R\text{–X} + 2\,Li \longrightarrow R\text{–Li} + LiX$$

$$2\,R\text{–Li} + CuI \longrightarrow \boxed{R_2Cu^-\ Li^+} + LiI$$

**[4] Lithium and sodium acetylides:**

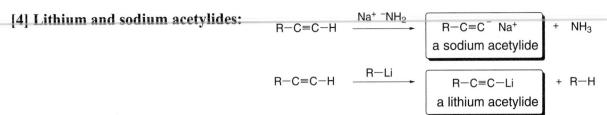

♦ **Reactions with organometallic reagents**

## [1] Reaction as a base (20.9C)

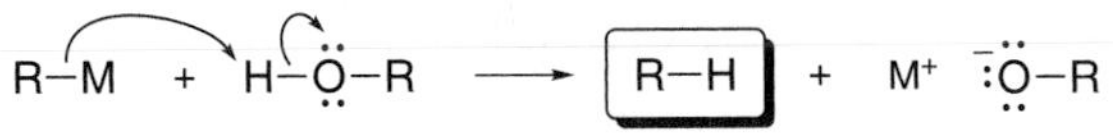

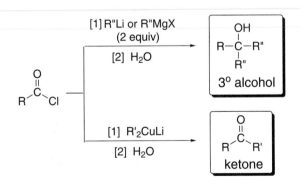

- RM = RLi, RMgX, R$_2$CuLi
- This acid–base reaction occurs with H$_2$O, ROH, RNH$_2$, R$_2$NH, RSH, RCOOH, RCONH$_2$, and RCONHR.

## [2] Reaction with aldehydes and ketones to form 1°, 2°, and 3° alcohols (20.10)

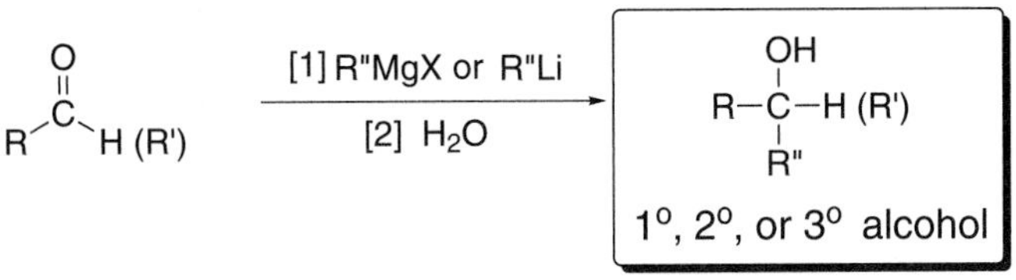

1°, 2°, or 3° alcohol

## [3] Reaction with esters to form 3° alcohols (20.13A)

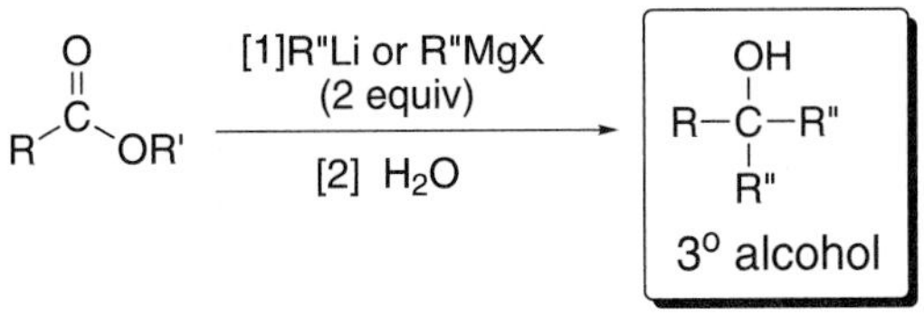

3° alcohol

## [4] Reaction with acid chlorides (20.13)

3° alcohol

ketone

- More reactive organometallic reagents—R″Li and R″MgX—add two equivalents of R″ to an acid chloride to form a 3° alcohol with two identical R″ groups.

- Less reactive organometallic reagents—R′$_2$CuLi—add only one equivalent of R′ to an acid chloride to form a ketone.

## [5] Reaction with carbon dioxide—Carboxylation (20.14A)

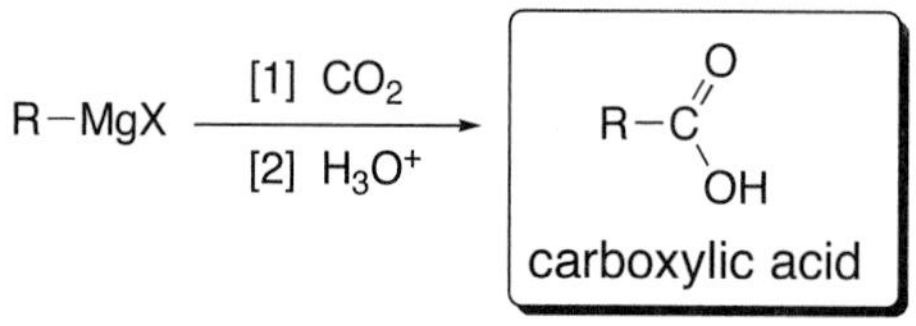

$$R-MgX \xrightarrow[\text{[2] } H_3O^+]{\text{[1] } CO_2} \quad R-\overset{\displaystyle O}{\underset{OH}{C}}$$

carboxylic acid

## [6] Reaction with epoxides (20.14B)

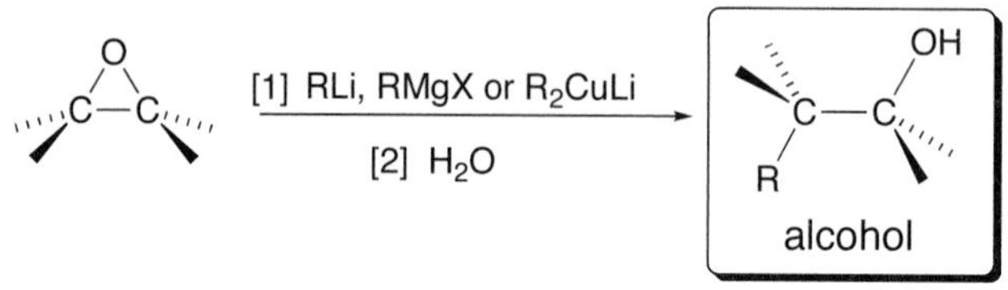

$$\text{(epoxide)} \xrightarrow[\text{[2] } H_2O]{\text{[1] RLi, RMgX or } R_2CuLi} \quad \text{(alcohol)}$$

alcohol

## [7] Reaction with α,β-unsaturated aldehydes and ketones (20.15B)

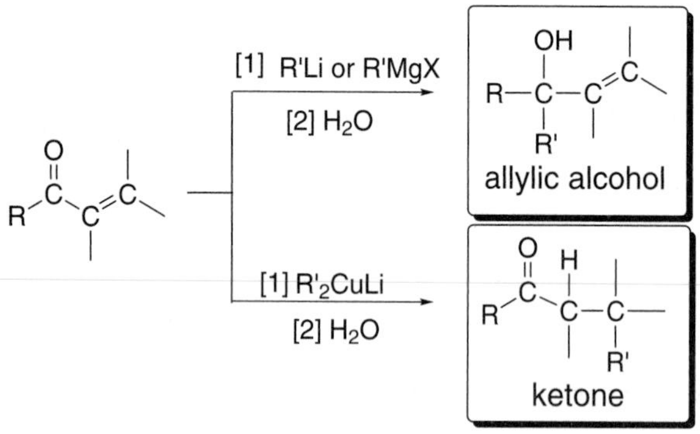

[1] R'Li or R'MgX
[2] $H_2O$

OH

allylic alcohol

- More reactive organometallic reagents—R'Li and R'MgX—react with α,β-unsaturated carbonyls by 1,2-addition.

[1] $R'_2CuLi$
[2] $H_2O$

ketone

- Less reactive organometallic reagents—$R'_2CuLi$— react with α,β-unsaturated carbonyls by 1,4-addition.

♦ Protecting groups (20.12)

## [1] Protecting an alcohol as a *tert*-butyldimethylsilyl ether

$$R-O-H \;+\; Cl-\overset{\displaystyle CH_3}{\underset{CH_3}{Si}}-C(CH_3)_3 \longrightarrow R-O-\overset{\displaystyle CH_3}{\underset{CH_3}{Si}}-C(CH_3)_3$$

[Cl—TBDMS]     [R—O—TBDMS]

*tert*-butyldimethylsilyl ether

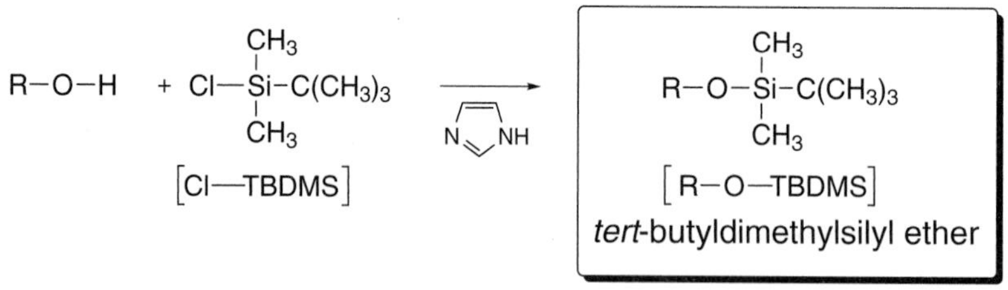

## [2] Deprotecting a *tert*-butyldimethylsilyl ether to re-form an alcohol

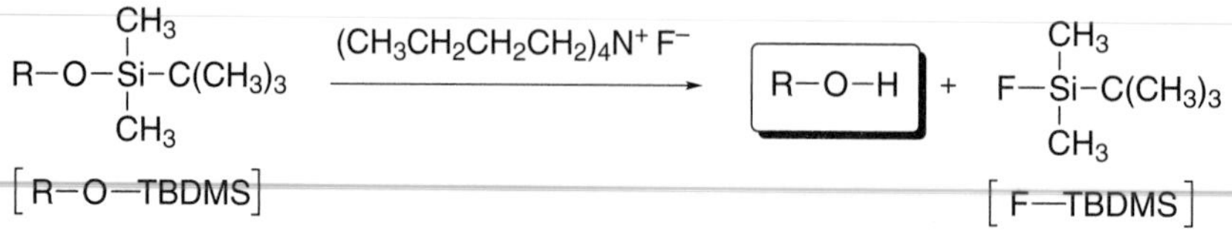

$$R-O-\overset{\displaystyle CH_3}{\underset{CH_3}{Si}}-C(CH_3)_3 \xrightarrow{(CH_3CH_2CH_2CH_2)_4N^+ F^-} R-O-H \;+\; F-\overset{\displaystyle CH_3}{\underset{CH_3}{Si}}-C(CH_3)_3$$

[R—O—TBDMS]     [F—TBDMS]

**20.1**

a.

(a) $C_{sp^3}$–$C_{sp^2}$

(b) σ: $C_{sp^2}$–$O_{sp^2}$
    π: $C_p$–$O_p$

(c) $C_{sp^3}$–$C_{sp^2}$

**A**

b. The O is $sp^2$ hybridized.
   Both lone pairs occupy $sp^2$ hybrid orbitals.

**20.2** A carbonyl compound with a reasonable leaving group undergoes substitution reactions. Those without good leaving groups undergo addition.

a.
$CH_3$–C(=O)–$CH_3$
no good leaving group
**addition reactions**

b.
$CH_3CH_2CH_2$–C(=O)–Cl
Cl–good leaving group
**substitution reactions**

c.
$CH_3$–C(=O)–$OCH_3$
$OCH_3$–reasonable leaving group
**substitution reactions**

d.
$C_6H_5$–C(=O)–H
no good leaving group
**addition reactions**

**20.3** Aldehydes are more reactive than ketones. In carbonyl compounds with leaving groups, the better the leaving group, the more reactive the carbonyl compound.

a. $CH_3CH_2CH_2$–C(=O)–H    and    $CH_3CH_2CH_2$–C(=O)–$CH_3$
   less hindered carbonyl
   **more reactive**

c. $CH_3CH_2$–C(=O)–Cl    and    $CH_3$–C(=O)–$OCH_3$
   better leaving group
   **more reactive**

b. $CH_3CH_2$–C(=O)–$CH_3$    and    $CH_3CH(CH_3)$–C(=O)–$CH_2CH_3$
   less hindered carbonyl
   **more reactive**

d. $CH_3$–C(=O)–$OCH_3$    and    $CH_3$–C(=O)–$NHCH_3$
   better leaving group
   **more reactive**

**20.4** NaBH$_4$ reduces aldehydes to 1° alcohols, and ketones to 2° alcohols.

a. $CH_3CH_2CH_2$–C(=O)–H  →(NaBH$_4$ / CH$_3$OH)→  $CH_3CH_2CH_2$–C(OH)(H)–H

c.  →(NaBH$_4$ / CH$_3$OH)→

b.  cyclohexanone =O  →(NaBH$_4$ / CH$_3$OH)→  cyclohexanol –OH

**20.5** 1° Alcohols are prepared from aldehydes and 2° alcohols are from ketones.

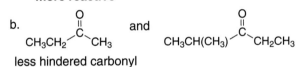

**20.6**

1-methylcyclohexanol, C(CH$_3$)(–OH)

3° Alcohols cannot be made by reduction of a carbonyl group, because they do not contain a H on the C with the OH.

**20.7**

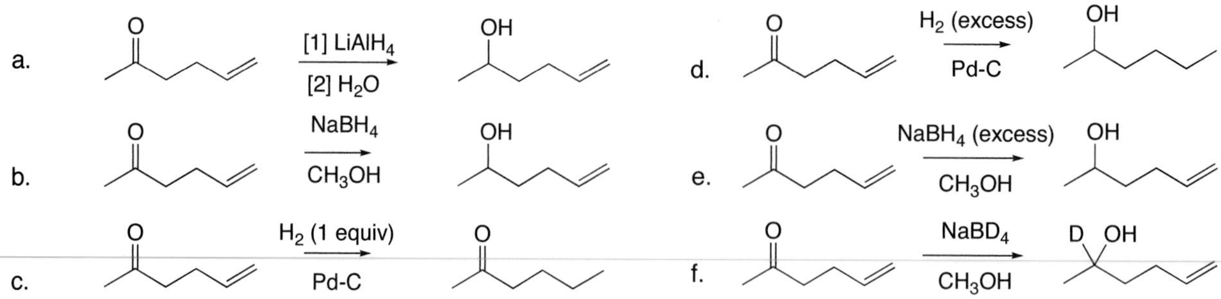

a. [1] LiAlH₄ / [2] H₂O →

d. H₂ (excess) / Pd-C →

b. NaBH₄ / CH₃OH →

e. NaBH₄ (excess) / CH₃OH →

c. H₂ (1 equiv) / Pd-C →

f. NaBD₄ / CH₃OH →

**20.8**

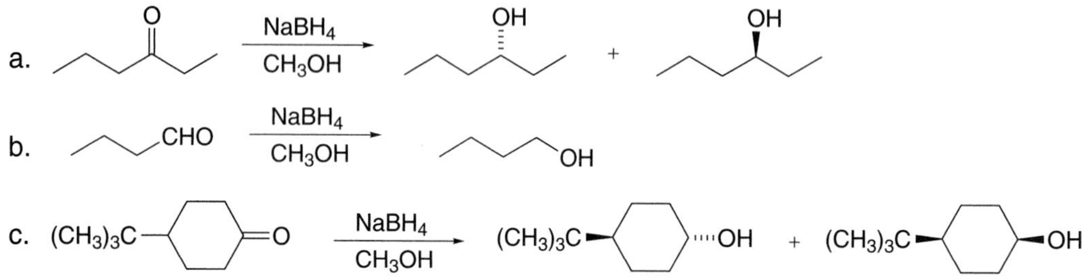

a. NaBH₄ / CH₃OH →

b. CHO → NaBH₄ / CH₃OH → OH

c. (CH₃)₃C—⬡=O → NaBH₄ / CH₃OH → (CH₃)₃C—⬡‴OH + (CH₃)₃C—⬡—OH

**20.9**

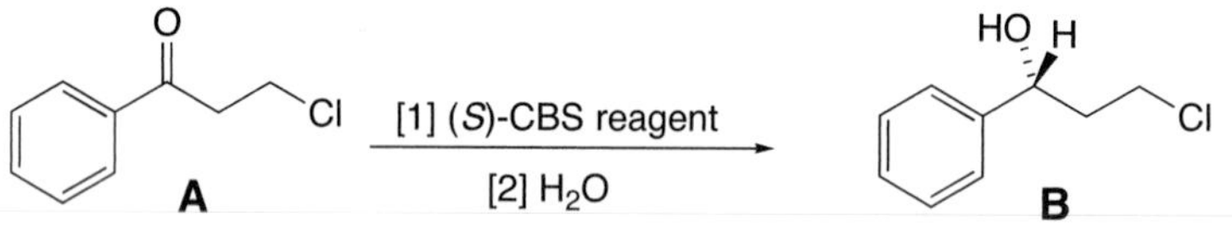

A → [1] (S)-CBS reagent / [2] H₂O → B

**20.10**

**Part [1]: Nucleophilic substitution of H for Cl**

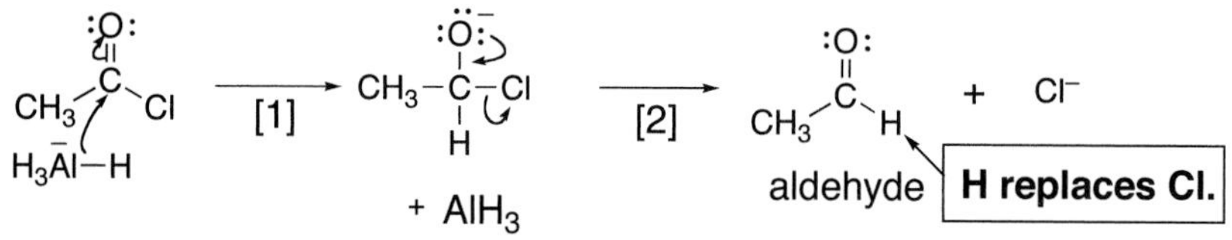

CH₃–C(=O)–Cl + H₃Al–H → [1] → CH₃–C(O⁻)(H)–Cl + AlH₃ → [2] → CH₃–C(=O)–H + Cl⁻

aldehyde | H replaces Cl.

**Part [2]: Nucleophilic addition of H⁻ to form an alcohol**

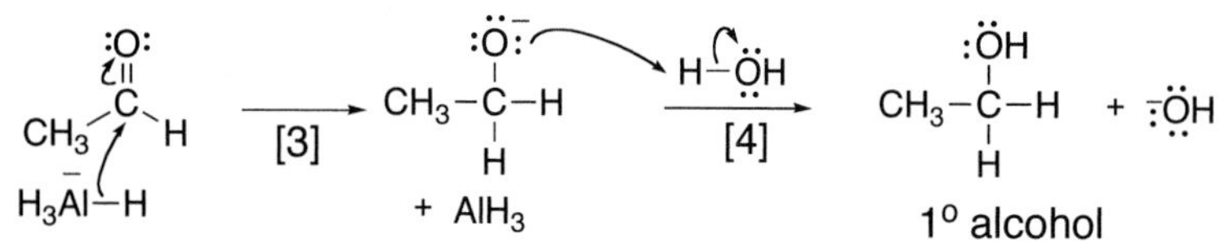

CH₃–C(=O)–H + H₃Al–H → [3] → CH₃–C(O⁻)(H)–H + AlH₃ → H–ÖH → [4] → CH₃–C(ÖH)(H)–H + :ÖH⁻

1° alcohol

**20.27**

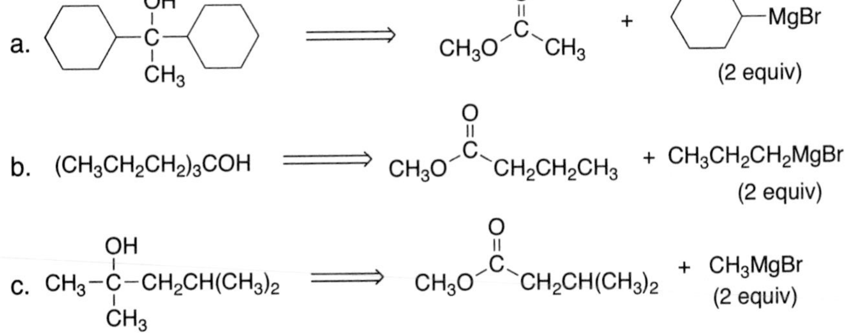

a.

b. (CH₃CH₂CH₂)₃COH ⟹ + CH₃CH₂CH₂MgBr (2 equiv)

c.

**20.28**

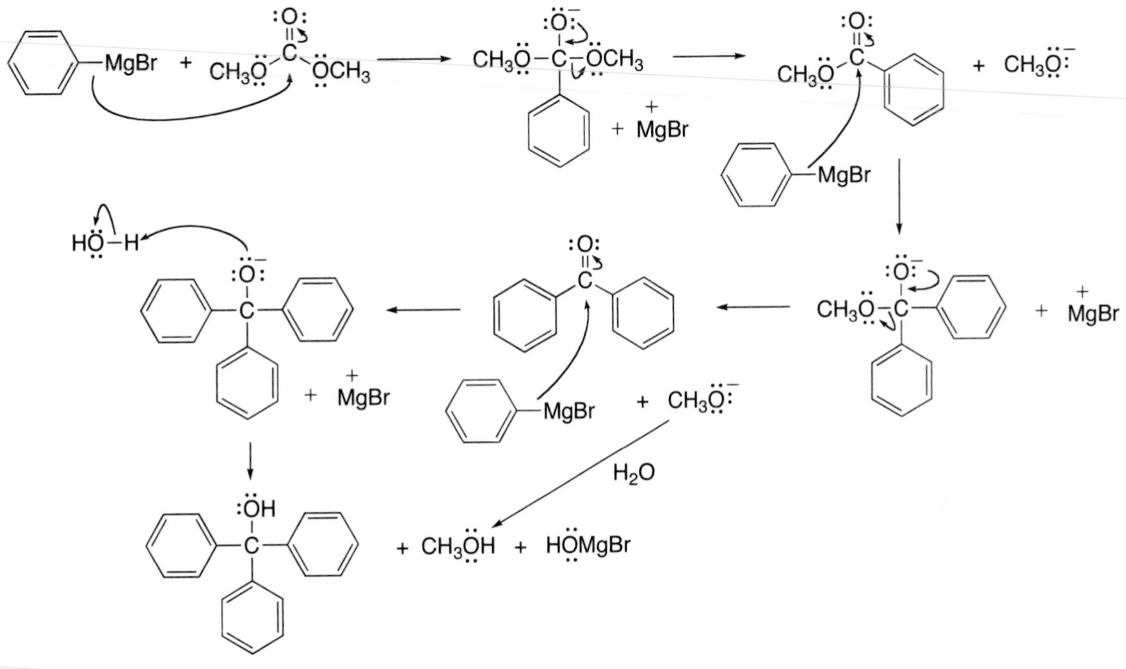

**20.29**  The R group of the organocuprate has replaced the Cl on the acid chloride.

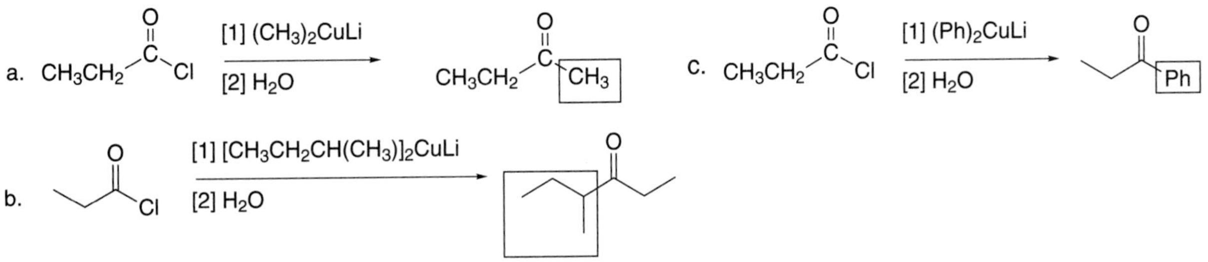

a.

b.

c.

## 20.30

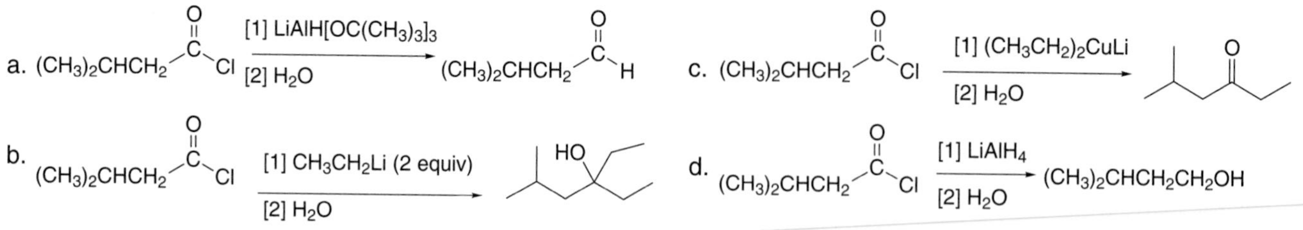

a.

c.

b.

d.

## 20.31

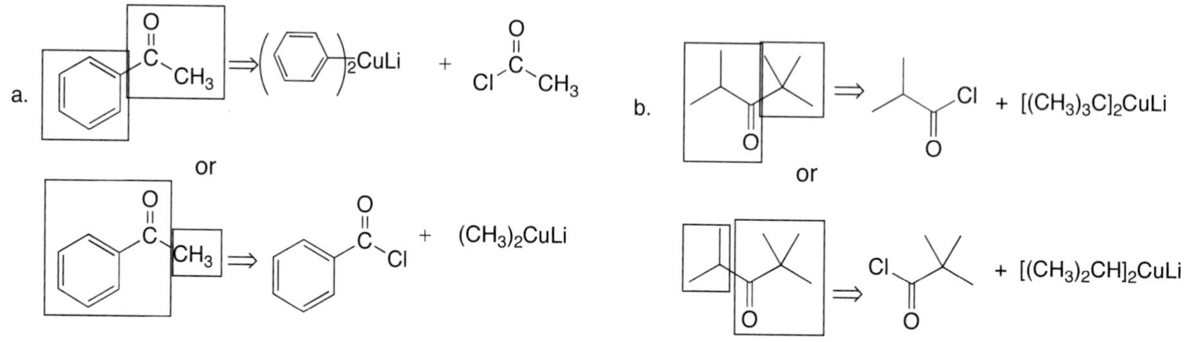

a.

or

b.

or

## 20.32

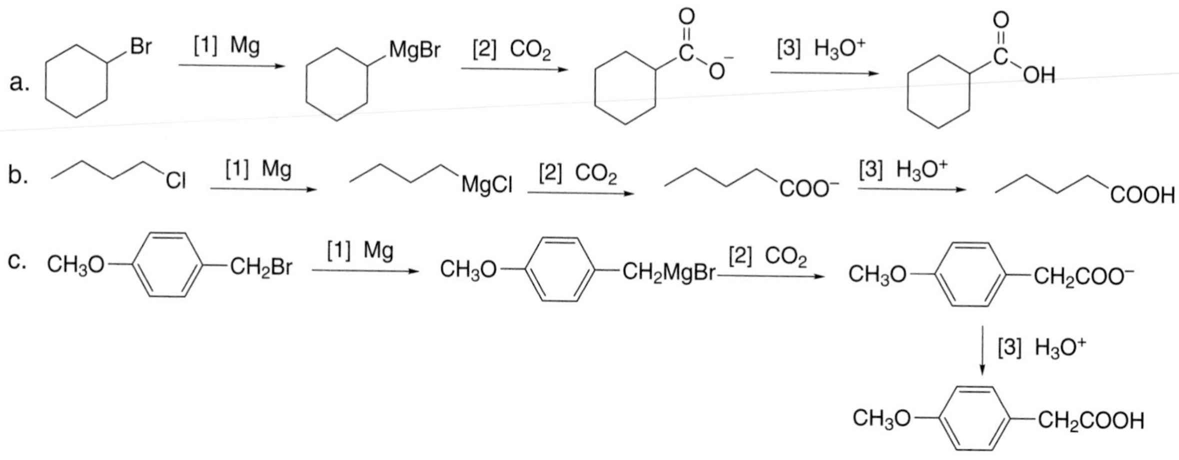

a.

b.

c.

## 20.33

a.

(+ enantiomer)

b.

c.

d.

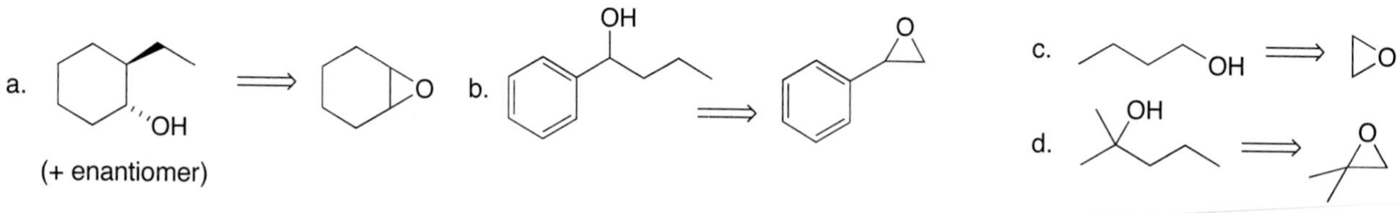

**20.34** The characteristic reaction of α,β-unsaturated carbonyl compounds is nucleophilic addition. Grignard and organolithium reagents react by 1,2-addition and organocuprate reagents react by 1,4-addition.

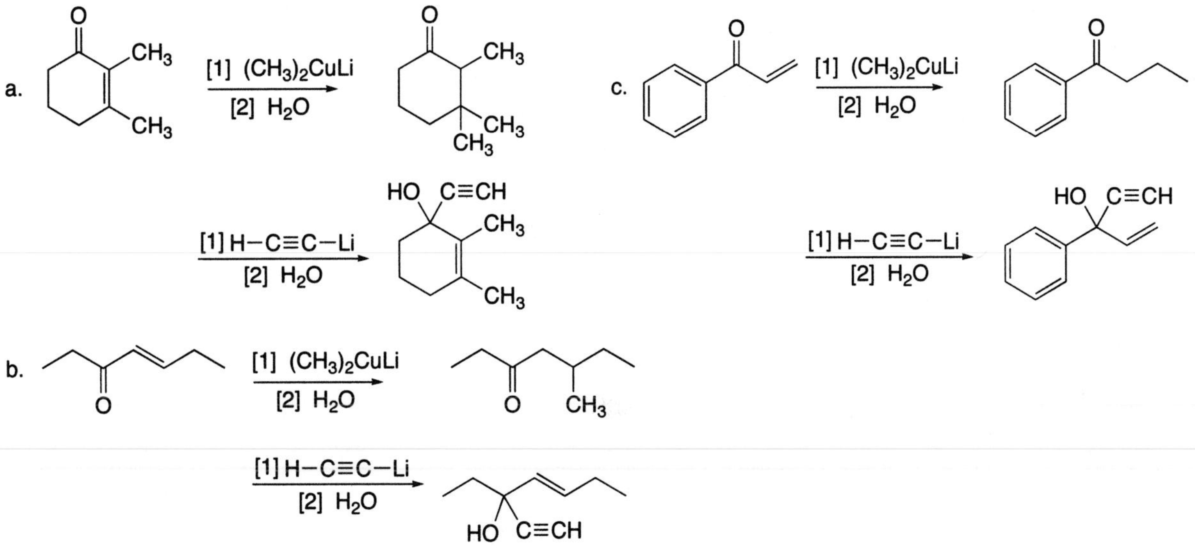

a.

[1] (CH₃)₂CuLi
[2] H₂O

[1] H−C≡C−Li
[2] H₂O

b.

[1] (CH₃)₂CuLi
[2] H₂O

[1] H−C≡C−Li
[2] H₂O

c.

[1] (CH₃)₂CuLi
[2] H₂O

[1] H−C≡C−Li
[2] H₂O

**20.35**

a.

[1] (CH₂=CH)₂CuLi
[2] H₂O

b.

[1] CH₂=CHLi
[2] H₂O

c.

(from a.)

[1] CH₂=CHLi
[2] H₂O

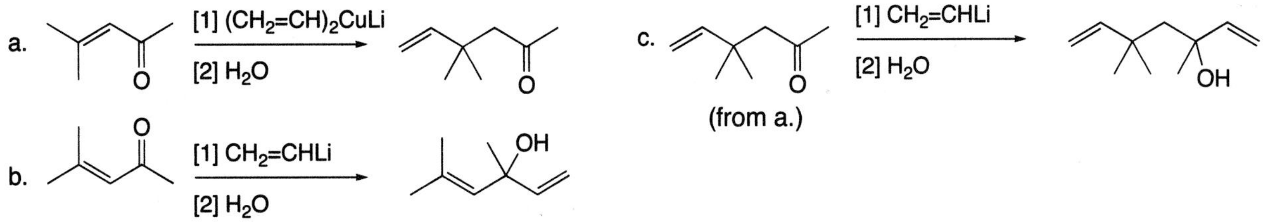

**20.36**

a. 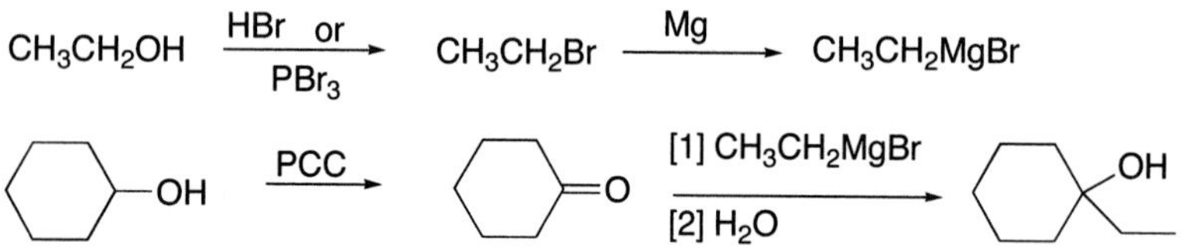 $CH_3CH_2OH$ $\xrightarrow[\text{PBr}_3]{\text{HBr or}}$ $CH_3CH_2Br$ $\xrightarrow{\text{Mg}}$ $CH_3CH_2MgBr$

—OH $\xrightarrow{\text{PCC}}$ =O $\xrightarrow[\text{[2] H}_2\text{O}]{\text{[1] CH}_3\text{CH}_2\text{MgBr}}$ OH

b. 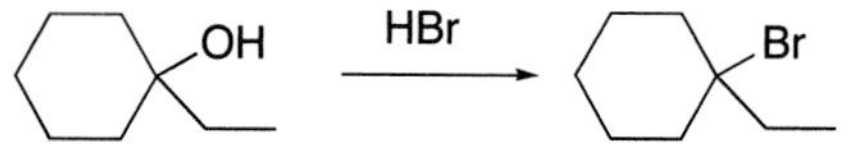 OH $\xrightarrow{\text{HBr}}$ Br

(from a.)

c. 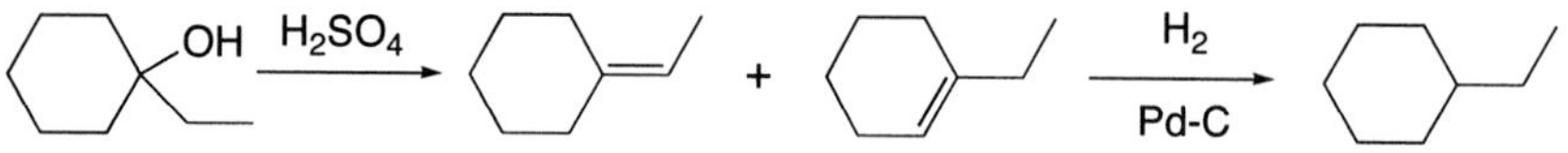 OH $\xrightarrow{\text{H}_2\text{SO}_4}$ + $\xrightarrow[\text{Pd-C}]{\text{H}_2}$

(from a.)

d. 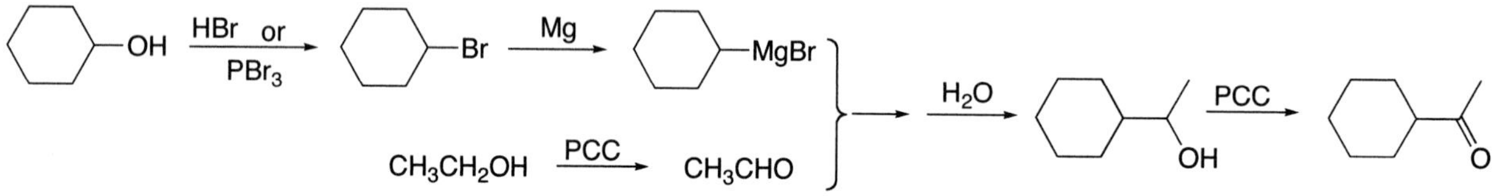 —OH $\xrightarrow[\text{PBr}_3]{\text{HBr or}}$ —Br $\xrightarrow{\text{Mg}}$ —MgBr

$CH_3CH_2OH$ $\xrightarrow{\text{PCC}}$ $CH_3CHO$

$\xrightarrow{\text{H}_2\text{O}}$ OH $\xrightarrow{\text{PCC}}$

e. 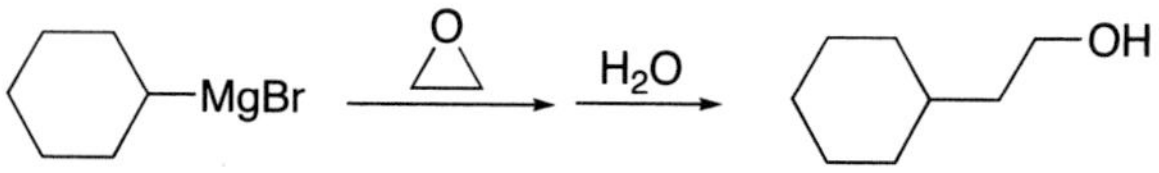 —MgBr $\xrightarrow{\triangle O}$ $\xrightarrow{\text{H}_2\text{O}}$ —OH

(from d.)

$CH_3CH_2OH$ $\xrightarrow{\text{H}_2\text{SO}_4}$ $CH_2$=$CH_2$ $\xrightarrow{\text{mCPBA}}$

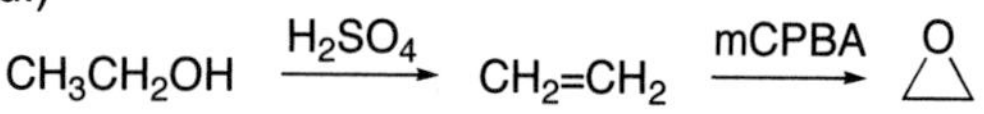

**20.37**

a. $\xrightarrow{\text{NaBH}_4}{\text{CH}_3\text{OH}}$

g. $\xrightarrow{\text{[1] CH}_3\text{MgBr}}{\text{[2] H}_2\text{O}}$

b. $\xrightarrow{\text{[1] LiAlH}_4}{\text{[2] H}_2\text{O}}$

h. $\xrightarrow{\text{[1] C}_6\text{H}_5\text{Li}}{\text{[2] H}_2\text{O}}$

c. $\xrightarrow{\text{H}_2}{\text{Pd-C}}$

i. $\xrightarrow{\text{[1] (CH}_3)_2\text{CuLi}}{\text{[2] H}_2\text{O}}$ **No reaction**

d. $\xrightarrow{\text{PCC}}$ **No reaction**

j. $\xrightarrow{\text{[1] HC}\equiv\text{CNa}}{\text{[2] H}_2\text{O}}$

e. $\xrightarrow{\text{Na}_2\text{Cr}_2\text{O}_7}{\text{H}_2\text{SO}_4, \text{H}_2\text{O}}$

k. $\xrightarrow{\text{[1] CH}_3\text{C}\equiv\text{CLi}}{\text{[2] H}_2\text{O}}$

f. $\xrightarrow{\text{Ag}_2\text{O}}{\text{NH}_4\text{OH}}$

l. $\xrightarrow{\text{TBDMSCl}}{\text{imidazole}}$ O–TBDMS

**20.38**

a. $\xrightarrow{\text{NaBH}_4}{\text{CH}_3\text{OH}}$

g. $\xrightarrow{\text{[1] CH}_3\text{MgBr}}{\text{[2] H}_2\text{O}}$

b. $\xrightarrow{\text{[1] LiAlH}_4}{\text{[2] H}_2\text{O}}$

h. $\xrightarrow{\text{[1] C}_6\text{H}_5\text{Li}}{\text{[2] H}_2\text{O}}$ $C_6H_5$

c. $\xrightarrow{\text{H}_2}{\text{Pd-C}}$

i. $\xrightarrow{\text{[1] (CH}_3)_2\text{CuLi}}{\text{[2] H}_2\text{O}}$ **No reaction**

d. $\xrightarrow{\text{PCC}}$ **No reaction**

j. $\xrightarrow{\text{[1] HC}\equiv\text{CNa}}{\text{[2] H}_2\text{O}}$

e. $\xrightarrow{\text{Na}_2\text{Cr}_2\text{O}_7}{\text{H}_2\text{SO}_4, \text{H}_2\text{O}}$ **No reaction**

k. $\xrightarrow{\text{[1] CH}_3\text{C}\equiv\text{CLi}}{\text{[2] H}_2\text{O}}$

f. $\xrightarrow{\text{Ag}_2\text{O}}{\text{NH}_4\text{OH}}$ **No reaction**

l. $\xrightarrow{\text{TBDMSCl}}{\text{imidazole}}$ O–TBDMS

**20.39**

a. $\xrightarrow{\text{Li (2 equiv)}}$ Li + LiBr

d. $\xrightarrow{\text{H}_2\text{O}}$ + LiOH

b. $\xrightarrow{\text{Mg}}$ MgBr

e. $\xrightarrow{\text{D}_2\text{O}}$ D + DOMgBr

c. $\xrightarrow{\text{[1] Li (2 equiv)}}{\text{[2] CuI (0.5 equiv)}}$ $(CH_3CH_2CH_2CH_2)_2CuLi$

f. $\xrightarrow{\text{CH}_3\text{C}\equiv\text{CH}}$ + LiC$\equiv$CCH$_3$

**20.40**

a. [structure] MgBr $\xrightarrow{CH_2=O \ \ H_2O}$ [structure] OH

g. [structure] MgBr $\xrightarrow{CH_3COOH}$ [structure] + $CH_3COO^-$

b. [structure] MgBr $\xrightarrow{\text{(cyclopentanone)} \ \ H_2O}$ [structure] HO

h. [structure] MgBr $\xrightarrow{HC\equiv CH}$ [structure] + $HC\equiv C^-$

c. [structure] MgBr $\xrightarrow{CH_3CH_2COCl \ \ H_2O}$ [structure] OH

i. [structure] MgBr $\xrightarrow{CO_2 \ \ H_3O^+}$ [structure] O OH

d. [structure] MgBr $\xrightarrow{CH_3CH_2COOCH_3 \ \ H_2O}$ [structure] OH

j. [structure] MgBr $\xrightarrow{\text{(epoxide)} \ \ H_2O}$ [structure] OH

e. [structure] MgBr $\xrightarrow{H_2O}$ [structure] + $^-OH$

k. [structure] MgBr $\xrightarrow{D_2O}$ [structure] D + $^-OD$

f. [structure] MgBr $\xrightarrow{CH_3CH_2OH}$ [structure] + $CH_3CH_2O^-$

l. [structure] MgBr $\xrightarrow{\text{(enone)} \ \ H_2O}$ [structure] OH

**20.41**

a. [structure] $\overset{O}{\underset{}{C}}$Cl $\xrightarrow{(CH_3CH_2CH_2CH_2)_2CuLi}$ [structure] $\overset{O}{\underset{}{C}}$CH_2CH_2CH_2CH_3

b. [structure] $\overset{O}{\underset{}{C}}$OCH_3 $\xrightarrow{(CH_3CH_2CH_2CH_2)_2CuLi}$ **No reaction**

c. [structure] CH_3 $\xrightarrow[\text{[2] } H_2O]{\text{[1] } (CH_3CH_2CH_2CH_2)_2CuLi}$ [structure] CH_3 CH_2CH_2CH_2CH_3

d. [structure] O CH_3 / CH_3 $\xrightarrow[\text{[2] } H_2O]{\text{[1] } (CH_3CH_2CH_2CH_2)_2CuLi}$ [structure] HO

**20.42** Arrange the larger group [$(CH_3)_3C-$] on the left side of the carbonyl.

[structure of ketone]

a. NaBH$_4$, CH$_3$OH $\longrightarrow$ [structure] H OH S + [structure] HO H R

b. [1] (S)-CBS reagent; [2] H$_2$O $\longrightarrow$ [structure] HO H R

c. [1] (R)-CBS reagent; [2] H$_2$O $\longrightarrow$ [structure] H OH S

**20.43**

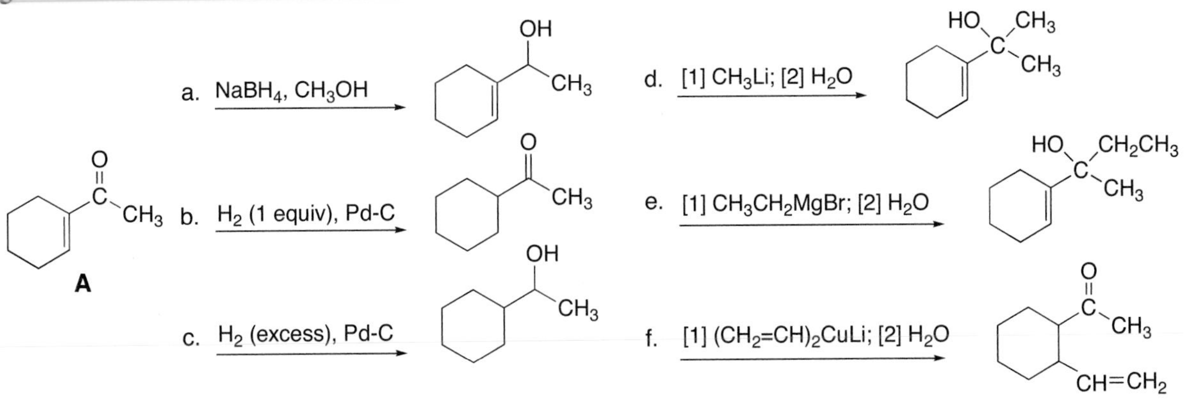

a. NaBH₄, CH₃OH

b. H₂ (1 equiv), Pd-C

c. H₂ (excess), Pd-C

d. [1] CH₃Li; [2] H₂O

e. [1] CH₃CH₂MgBr; [2] H₂O

f. [1] (CH₂=CH)₂CuLi; [2] H₂O

**A**

**20.44**

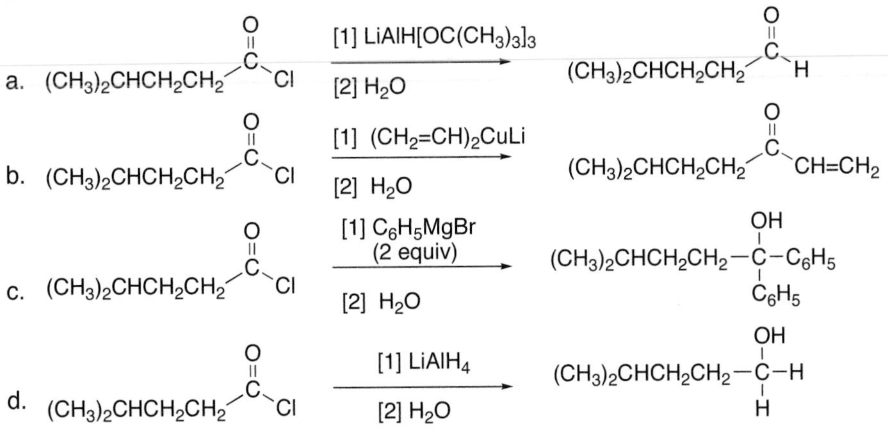

a. $(CH_3)_2CHCH_2CH_2$ —C(=O)—Cl
   [1] LiAlH[OC(CH₃)₃]₃
   [2] H₂O
   → $(CH_3)_2CHCH_2CH_2$ —C(=O)—H

b. $(CH_3)_2CHCH_2CH_2$ —C(=O)—Cl
   [1] (CH₂=CH)₂CuLi
   [2] H₂O
   → $(CH_3)_2CHCH_2CH_2$ —C(=O)—CH=CH₂

c. $(CH_3)_2CHCH_2CH_2$ —C(=O)—Cl
   [1] C₆H₅MgBr (2 equiv)
   [2] H₂O
   → $(CH_3)_2CHCH_2CH_2$ —C(OH)(C₆H₅)—C₆H₅

d. $(CH_3)_2CHCH_2CH_2$ —C(=O)—Cl
   [1] LiAlH₄
   [2] H₂O
   → $(CH_3)_2CHCH_2CH_2$ —C(OH)(H)—H

**20.45**

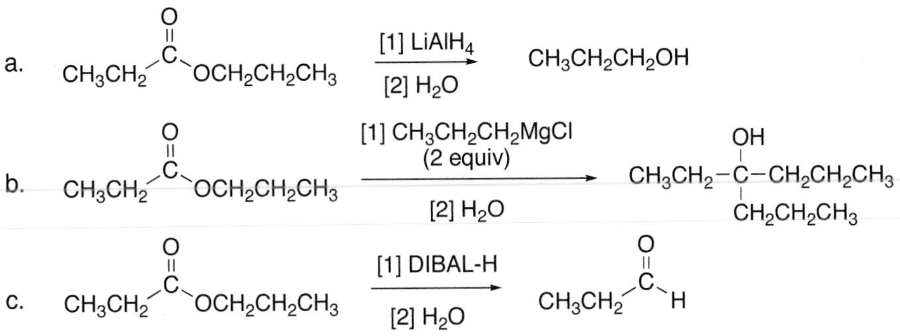

a. $CH_3CH_2$ —C(=O)—OCH₂CH₂CH₃
   [1] LiAlH₄
   [2] H₂O
   → CH₃CH₂CH₂OH

b. $CH_3CH_2$ —C(=O)—OCH₂CH₂CH₃
   [1] CH₃CH₂CH₂MgCl (2 equiv)
   [2] H₂O
   → CH₃CH₂—C(OH)(CH₂CH₂CH₃)—CH₂CH₂CH₃

c. $CH_3CH_2$ —C(=O)—OCH₂CH₂CH₃
   [1] DIBAL-H
   [2] H₂O
   → CH₃CH₂—C(=O)—H

**20.46**

a. HO—(CH₂)...—CHO
   CrO₃
   H₂SO₄, H₂O
   → HOOC—(CH₂)₃—COOH

b. HO—(CH₂)...—CHO
   PCC
   → OHC—(CH₂)₃—CHO

c. HO—(CH₂)...—CHO
   Ag₂O
   NH₄OH
   → HO—(CH₂)...—COOH

d. HO—(CH₂)...—CHO
   Na₂Cr₂O₇
   H₂SO₄, H₂O
   → HOOC—(CH₂)₃—COOH

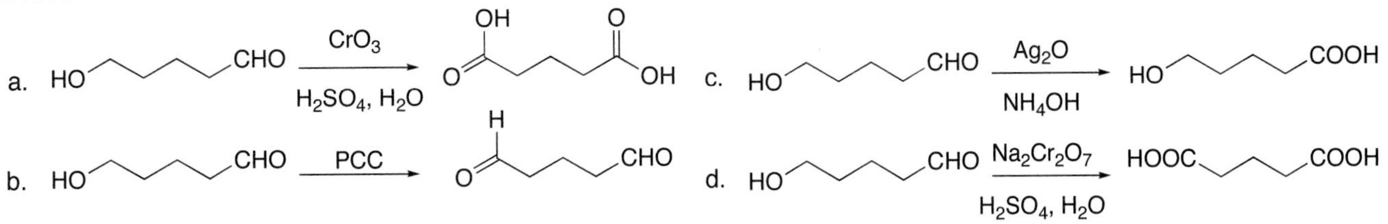

**20.47**

a.

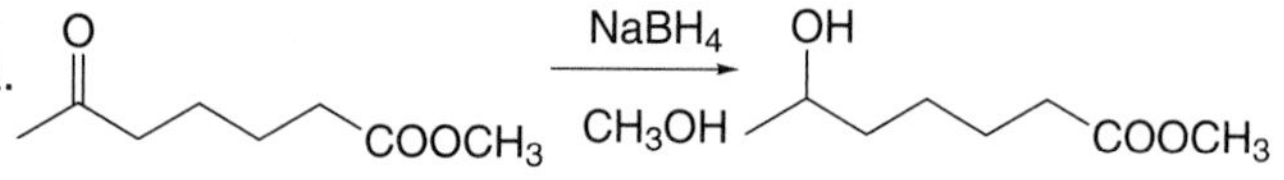

NaBH$_4$
CH$_3$OH

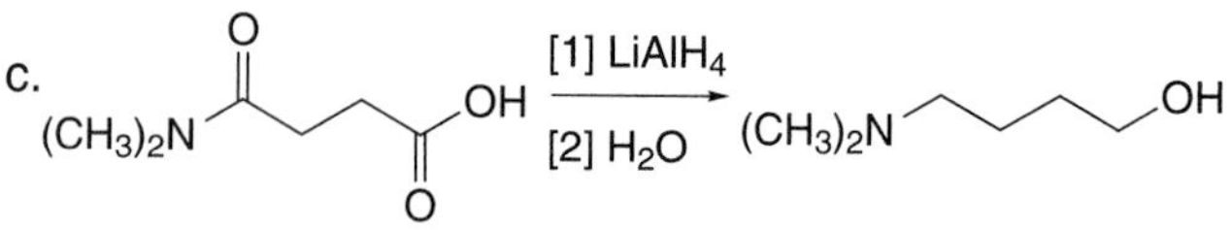

b.

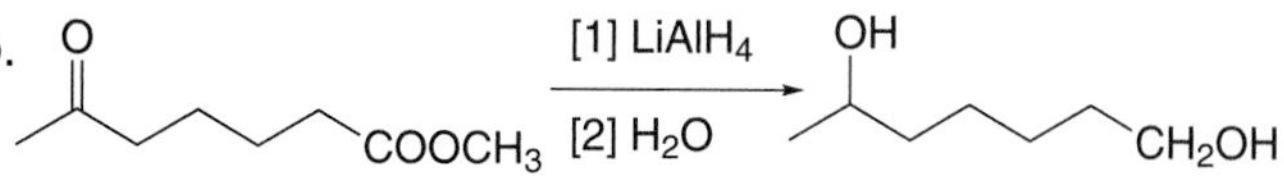

[1] LiAlH$_4$
[2] H$_2$O

c. (CH$_3$)$_2$N

[1] LiAlH$_4$
[2] H$_2$O

(CH$_3$)$_2$N

OH

d.

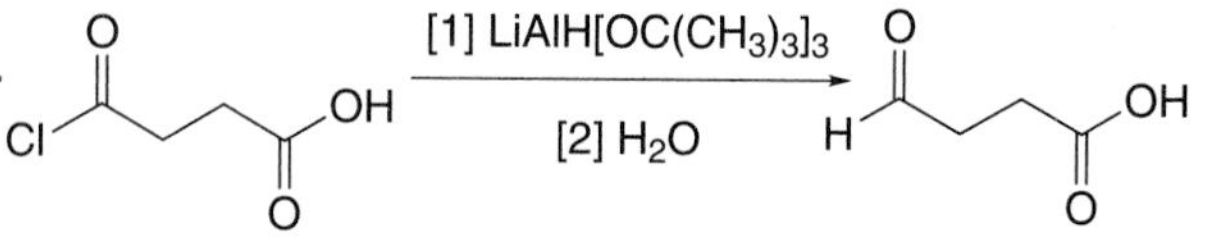

[1] LiAlH[OC(CH$_3$)$_3$]$_3$
[2] H$_2$O

**20.48**

a. 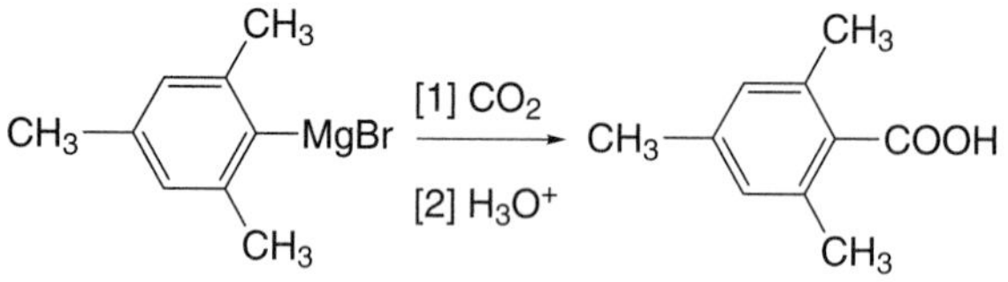 CH$_3$

MgBr

[1] CO$_2$
[2] H$_3$O$^+$

CH$_3$

COOH

f.

MgBr

[1] CH$_2$=O
[2] H$_2$O

CH$_2$OH

b.

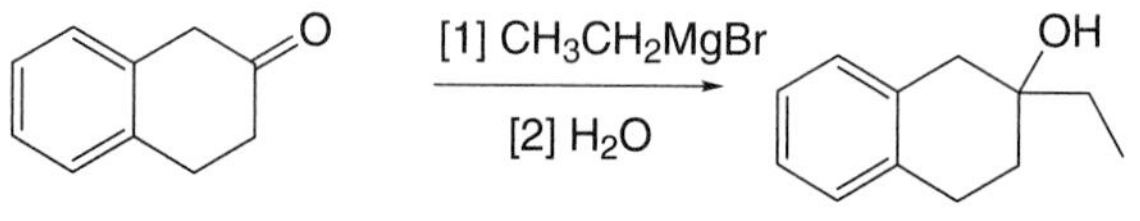

[1] CH$_3$CH$_2$MgBr
[2] H$_2$O

g.

[1] (CH$_3$)$_2$CuLi
[2] H$_2$O

c.

CHO

[1] C$_6$H$_5$Li
[2] H$_2$O

h.

[1] CH$_3$MgBr
[2] H$_2$O

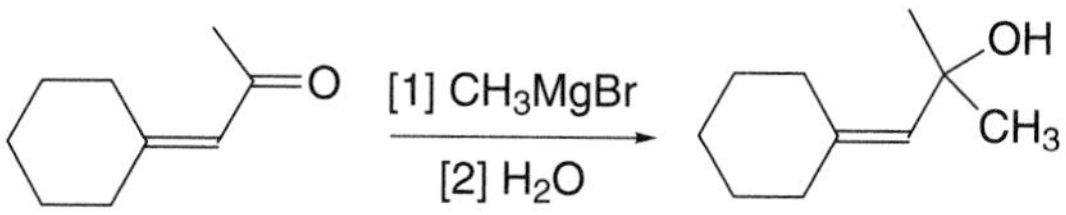

i.

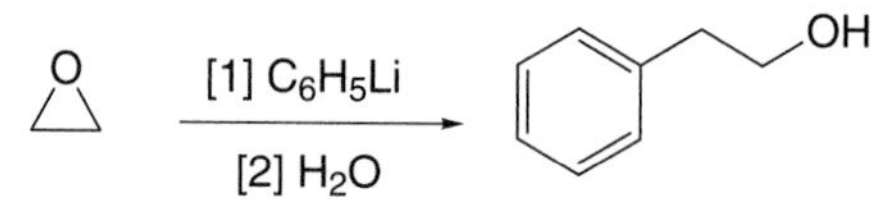

[1] C$_6$H$_5$Li
[2] H$_2$O

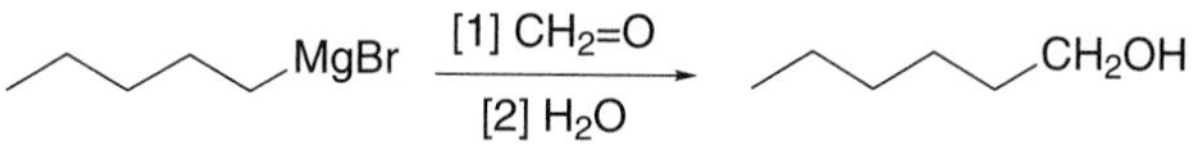

d.

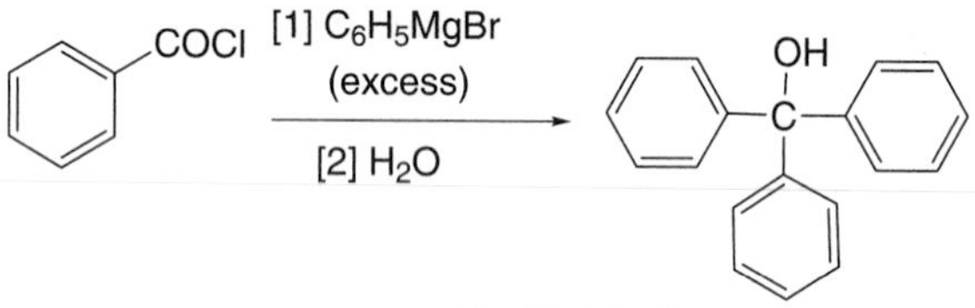

COCl

[1] C$_6$H$_5$MgBr
(excess)
[2] H$_2$O

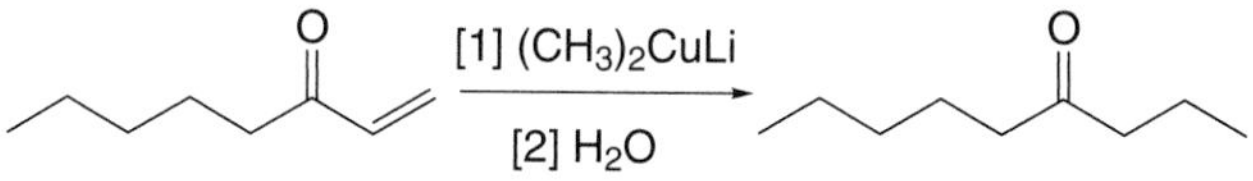

j. 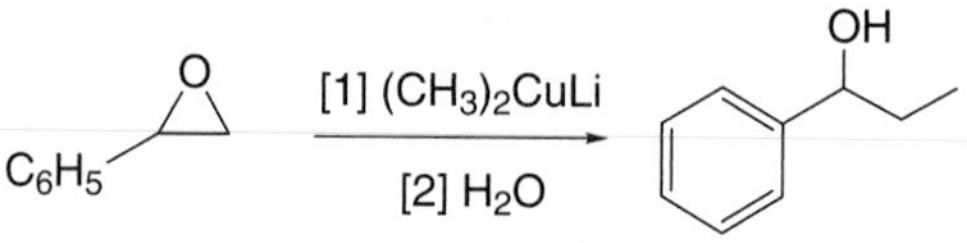 C$_6$H$_5$

[1] (CH$_3$)$_2$CuLi
[2] H$_2$O

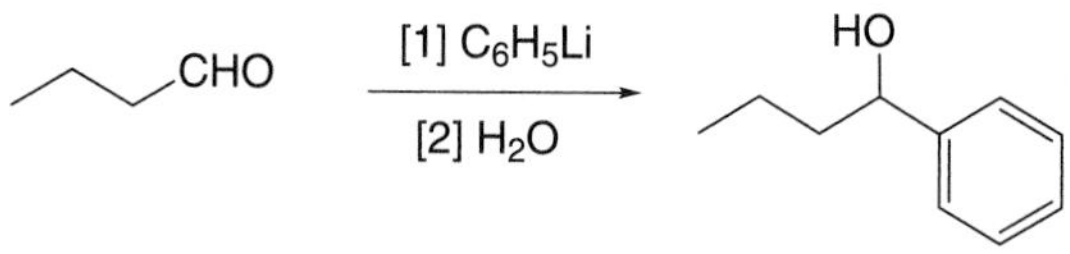

e.

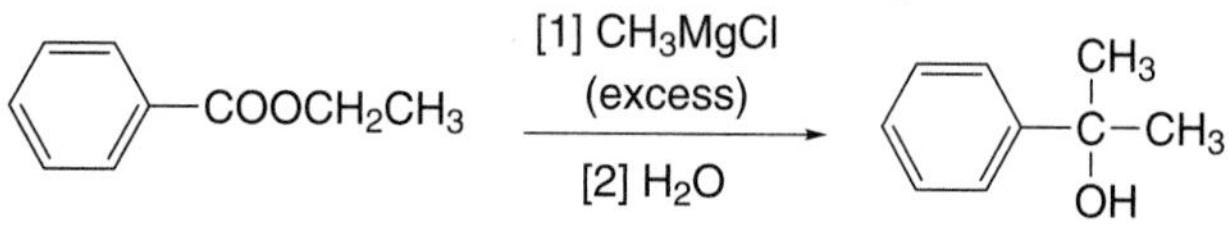

COOCH$_2$CH$_3$

[1] CH$_3$MgCl
(excess)
[2] H$_2$O

**20.49**

a.

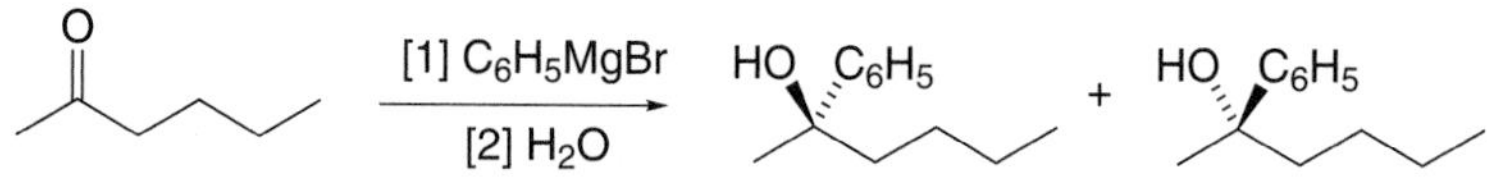

[1] C$_6$H$_5$MgBr
[2] H$_2$O

HO   C$_6$H$_5$

+

HO   C$_6$H$_5$

b. (CH$_3$)$_3$C

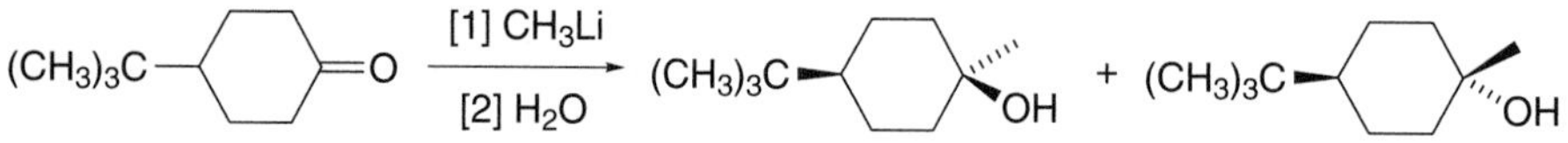

O

[1] CH$_3$Li
[2] H$_2$O

(CH$_3$)$_3$C

OH

+  (CH$_3$)$_3$C

OH

c.

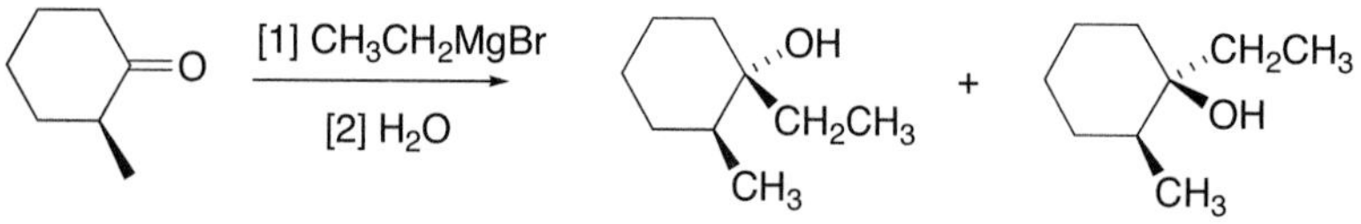

O

[1] CH$_3$CH$_2$MgBr
[2] H$_2$O

OH
CH$_2$CH$_3$
CH$_3$

+

CH$_2$CH$_3$
OH
CH$_3$

d.

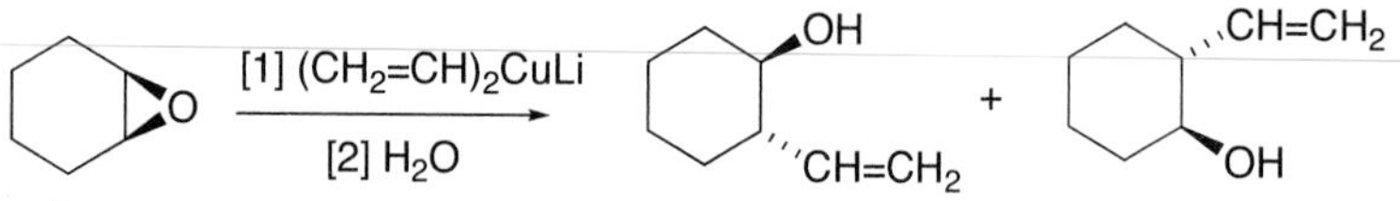

O

[1] (CH$_2$=CH)$_2$CuLi
[2] H$_2$O

OH

CH=CH$_2$

+

CH=CH$_2$

OH

e.

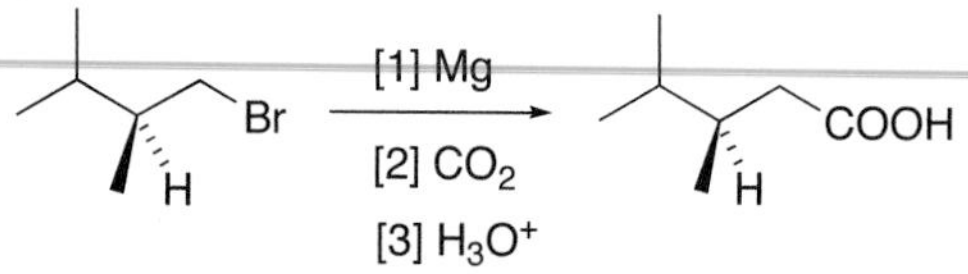

Br

[1] Mg
[2] CO$_2$
[3] H$_3$O$^+$

COOH

f.

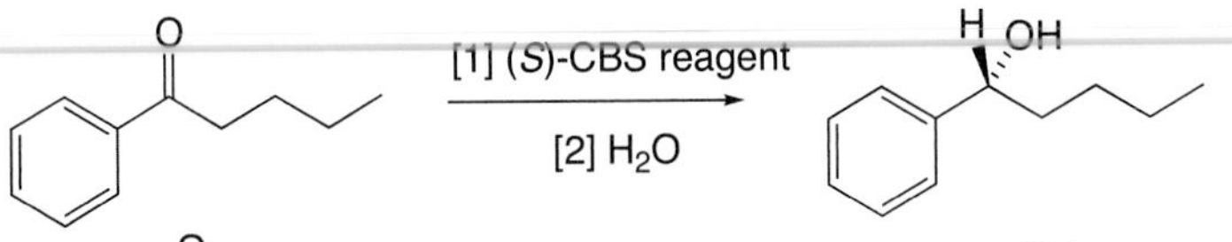

[1] (S)-CBS reagent

[2] H₂O

g.

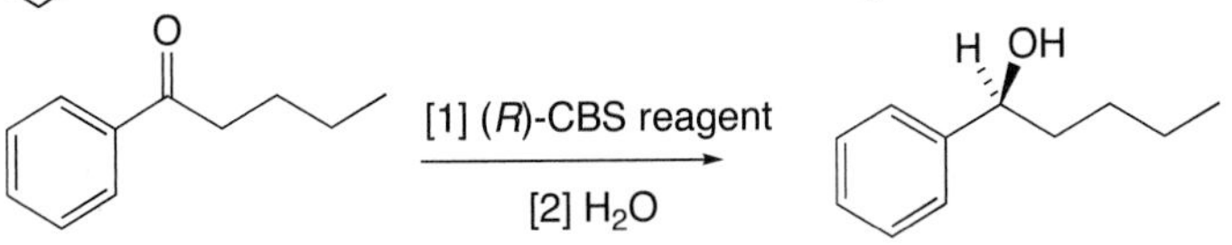

[1] (R)-CBS reagent

[2] H₂O

h.

[1] LiAlH₄

[2] H₂O

+ HOCH₂CH₃

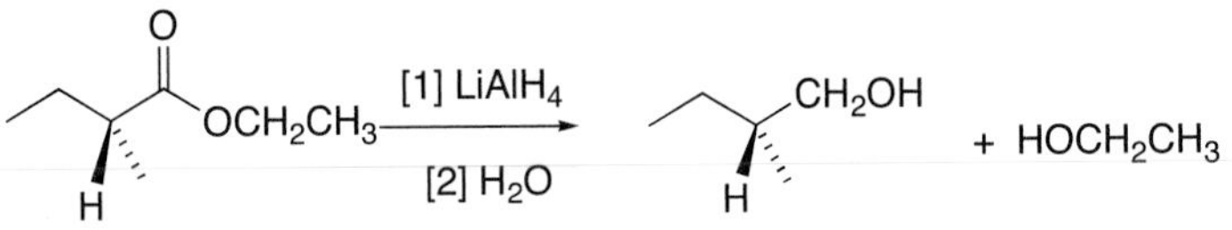

**20.50** Since a Grignard reagent contains a carbon atom with a partial negative charge, it acts as a base and reacts with the OH of the starting halide, $BrCH_2CH_2CH_2CH_2OH$. This acid–base reaction destroys the Grignard reagent so that addition cannot occur. To get around this problem, the OH group can be protected as a *tert*-butyldimethylsilyl ether, from which a Grignard reagent can be made.

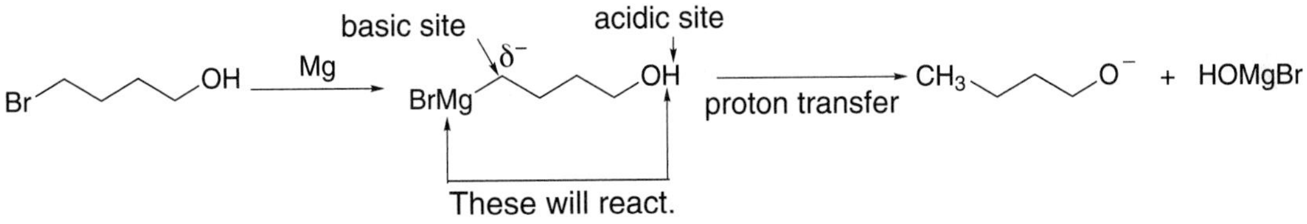

**INSTEAD: Use a protecting group.**

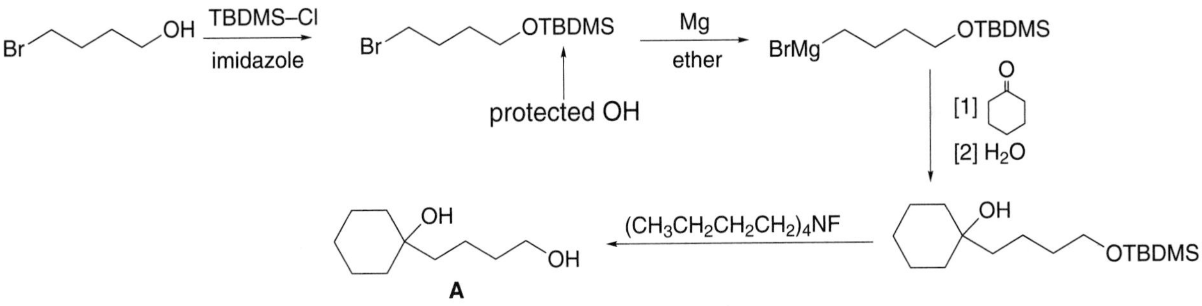

**20.51** Compounds **F, G,** and **K** are all alcohols with aromatic rings so there will be many similarities in their proton NMR spectra. These compounds will, however, show differences in absorptions due to the CH protons on the carbon bearing the OH group. **F** has a CH₂OH group, which will give a singlet in the 3–4 ppm region of the spectrum. **G** is a 3° alcohol that has no protons on the C bonded to the OH group so it will have no peak in the 3–4 ppm region of the spectrum. **K** is a 2° alcohol that will give a doublet in the 3–4 ppm region of the spectrum for the CH proton on the carbon with the OH group.

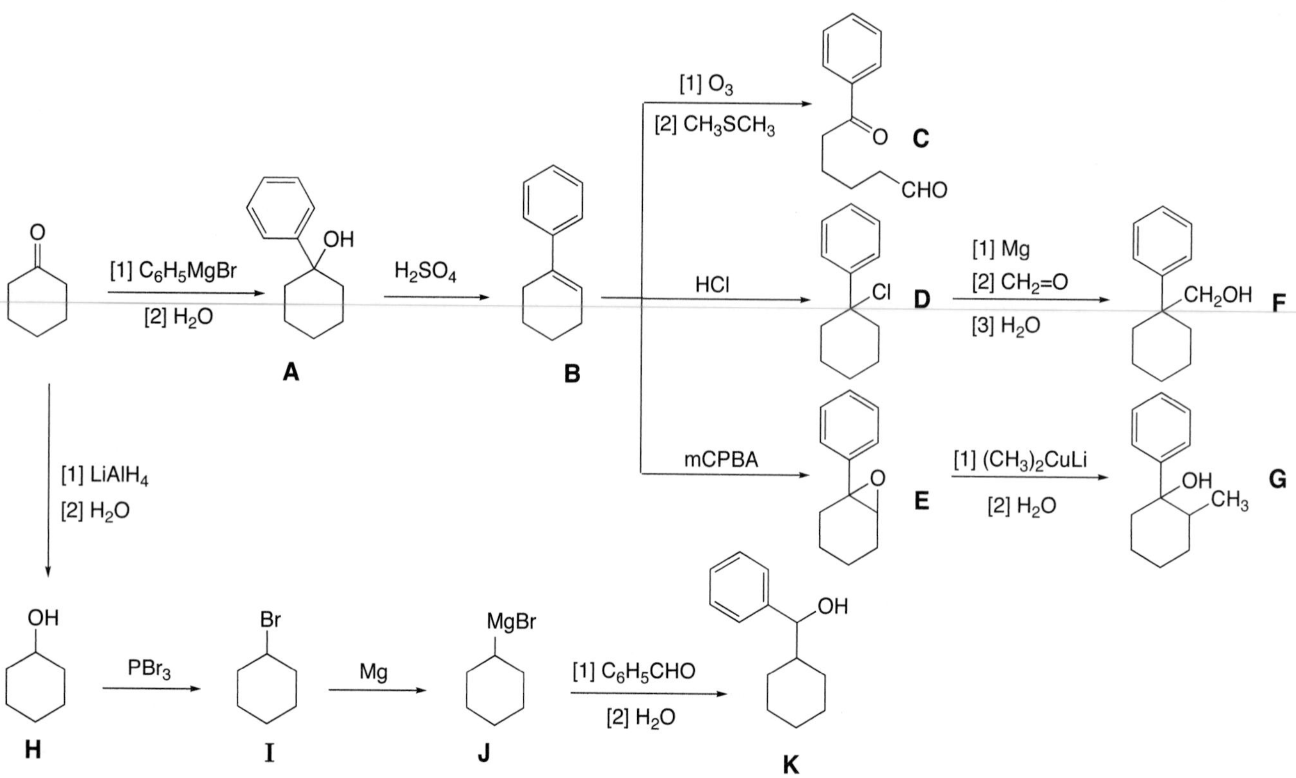

[1] C$_6$H$_5$MgBr  [2] H$_2$O → **A**

H$_2$SO$_4$ → **B**

[1] O$_3$  [2] CH$_3$SCH$_3$ → **C**

HCl → **D**

mCPBA → **E**

[1] Mg  [2] CH$_2$=O  [3] H$_2$O → **F**

[1] (CH$_3$)$_2$CuLi  [2] H$_2$O → **G**

[1] LiAlH$_4$  [2] H$_2$O → **H**

PBr$_3$ → **I**

Mg → **J**

[1] C$_6$H$_5$CHO  [2] H$_2$O → **K**

**20.52**

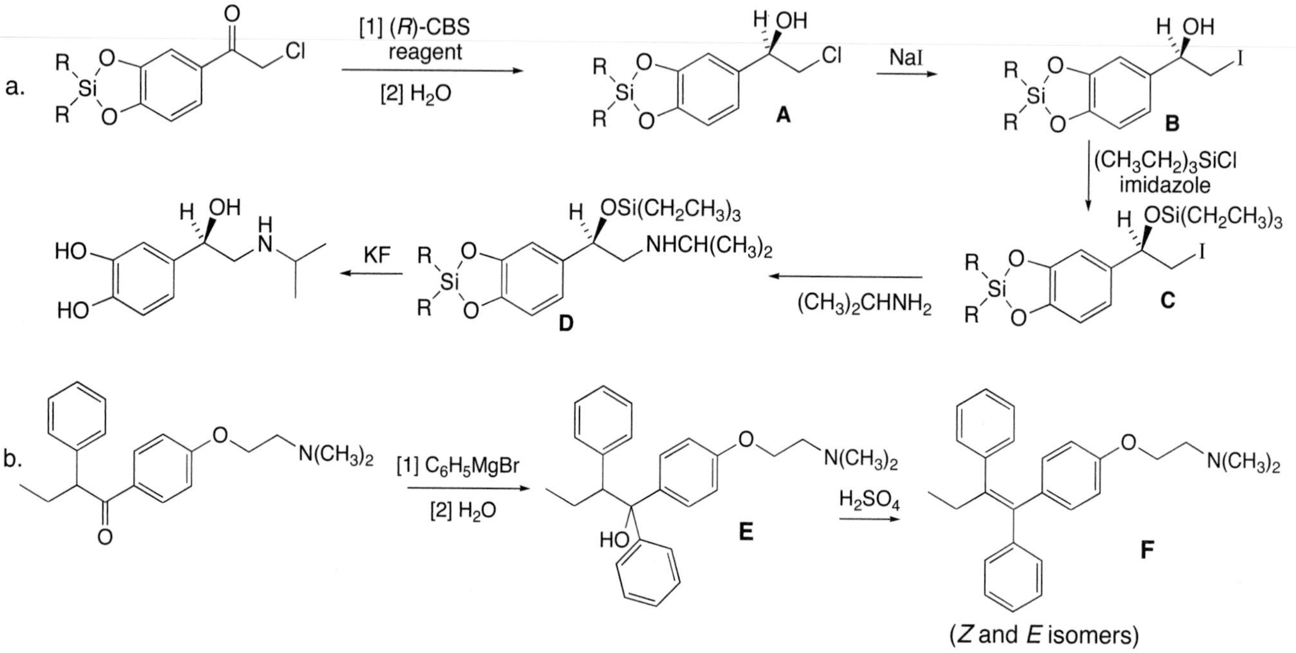

a.

[1] (R)-CBS reagent  [2] H$_2$O → **A**

NaI → **B**

(CH$_3$CH$_2$)$_3$SiCl  imidazole → **C**

(CH$_3$)$_2$CHNH$_2$ → **D**

KF →

b.

[1] C$_6$H$_5$MgBr  [2] H$_2$O → **E**

H$_2$SO$_4$ → **F**

(Z and E isomers)

c.

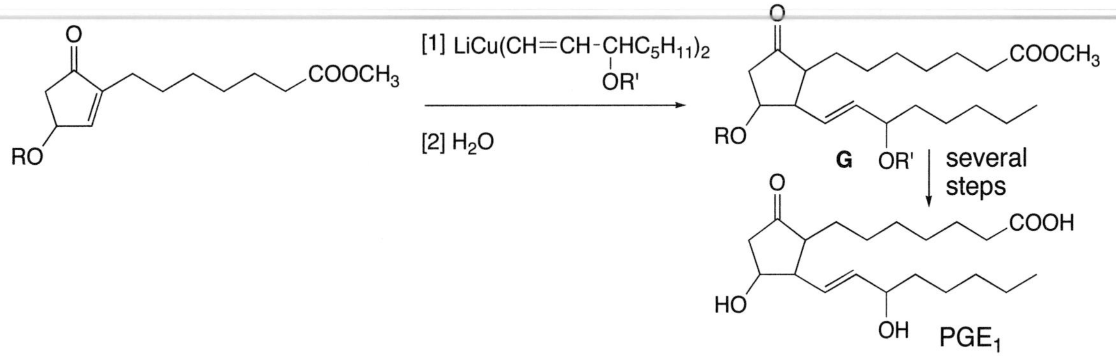

**20.53**

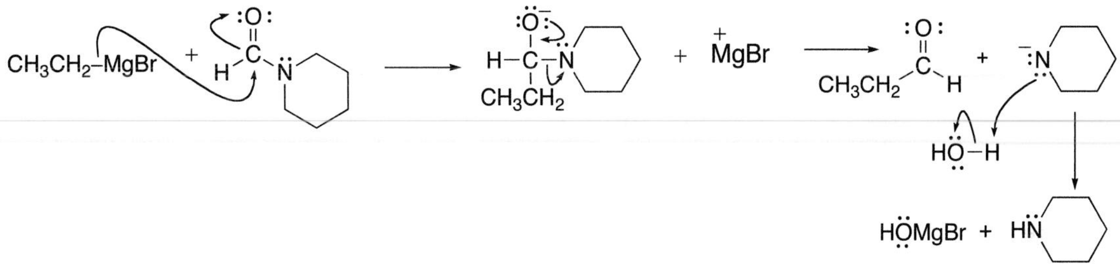

**20.54**

a.

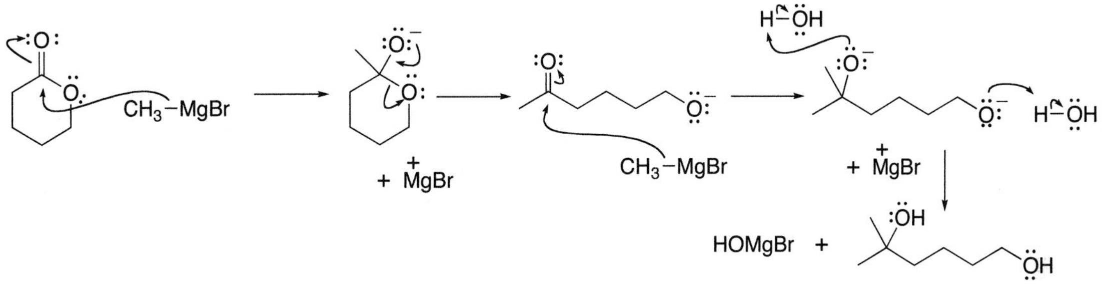

b.

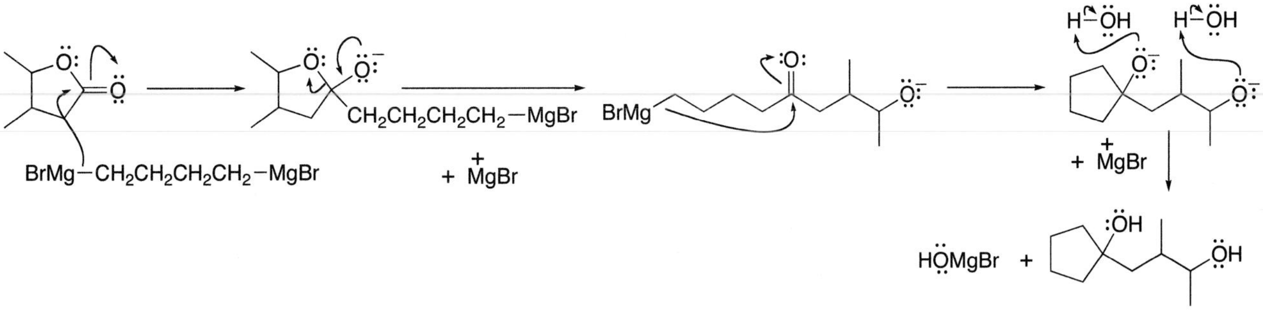

**20.55**

a.

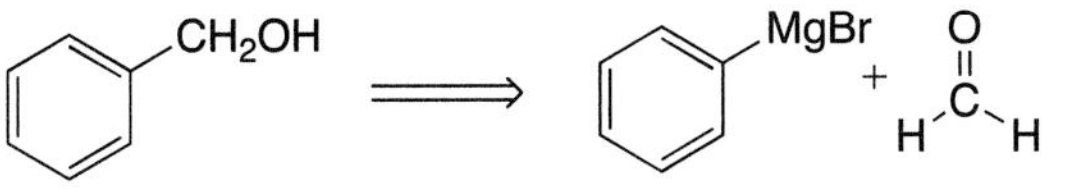

b.

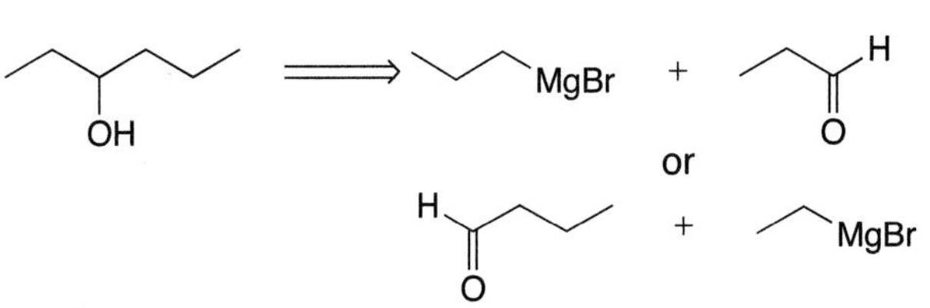

c. 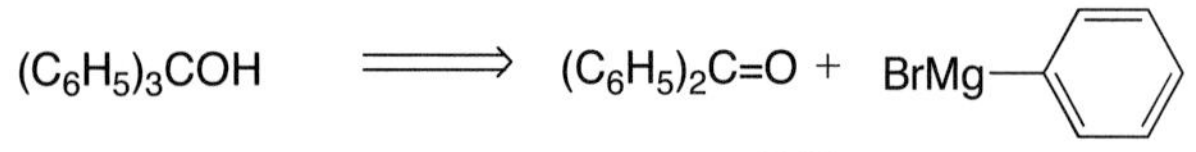 $(C_6H_5)_3COH \implies (C_6H_5)_2C=O +$ BrMg— (phenyl)

e.

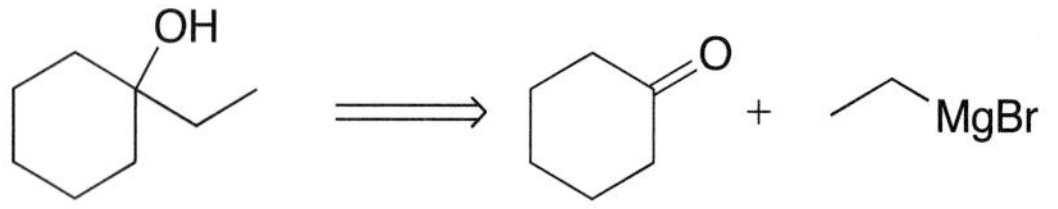

d.

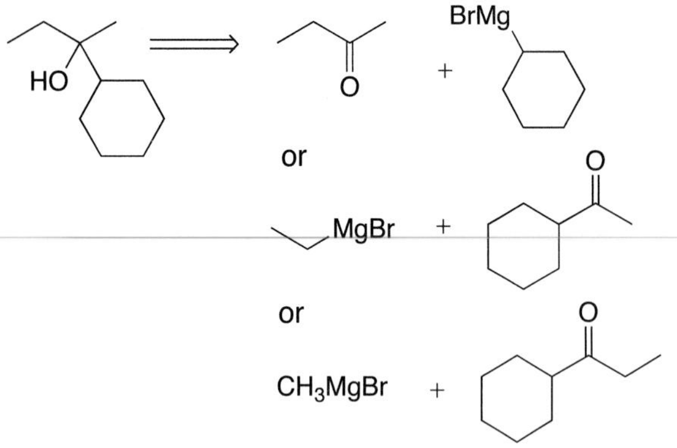

or

or

or

$CH_3MgBr$ +

**20.56**

a.

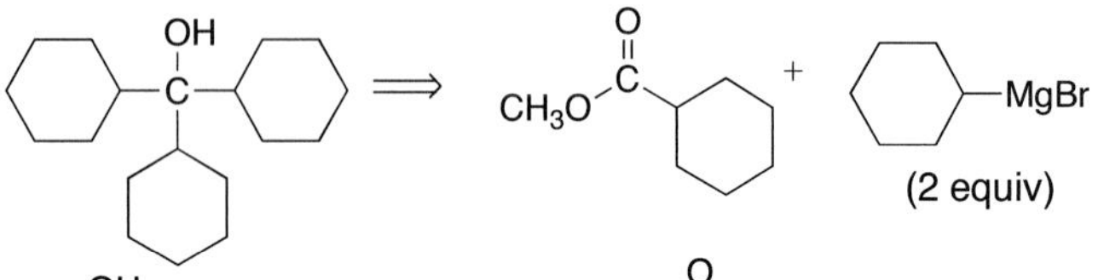

$\implies CH_3O-C(=O)-$ (cyclohexyl) + (cyclohexyl)—MgBr (2 equiv)

c. $(CH_3CH_2CH_2CH_2)_2C(OH)CH_3 \implies CH_3O-C(=O)-CH_3$ + BrMg— (butyl) (2 equiv)

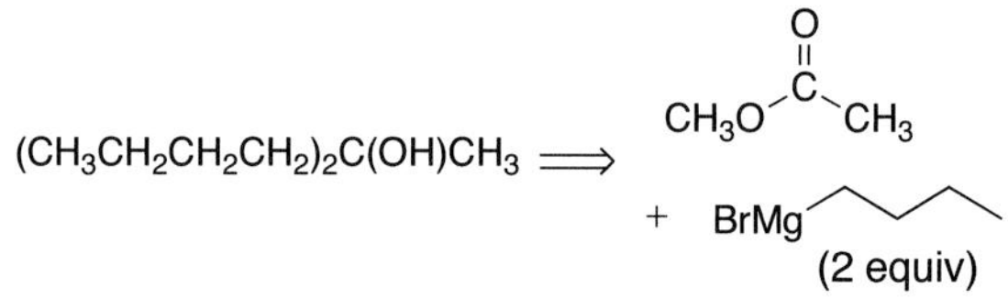

b. 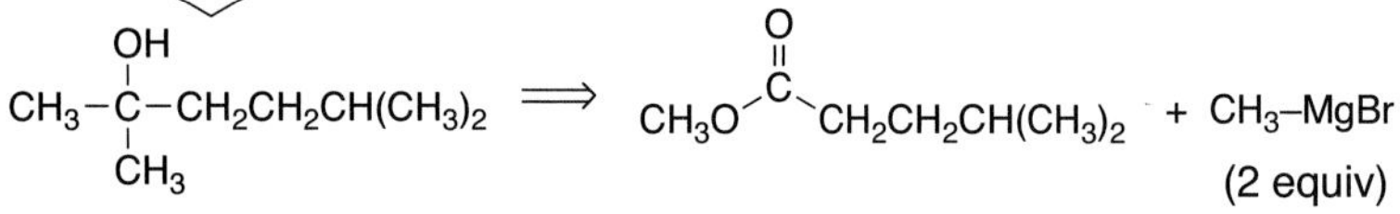 $CH_3-\overset{OH}{\underset{CH_3}{C}}-CH_2CH_2CH(CH_3)_2 \implies CH_3O-C(=O)-CH_2CH_2CH(CH_3)_2 + CH_3-MgBr$ (2 equiv)

**20.57**

a. 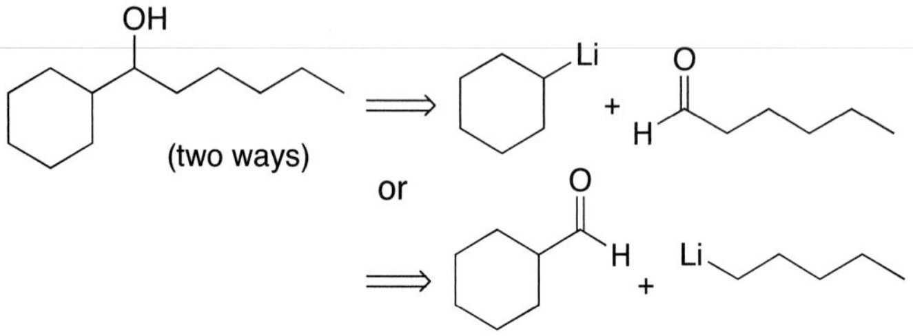 (two ways) $\implies$ (cyclohexyl)—Li + $H-C(=O)-$ (pentyl)

or

$\implies$ (cyclohexyl)—$CH(=O)$ H + Li— (butyl)

c. (three ways)

$\implies$

+ Li—$CH_2CH_3$

or

+ (cyclohexyl)—Li

or

$CH_3CH_2-C(=O)-OCH_3$ + (cyclohexyl)—Li (2 equiv)

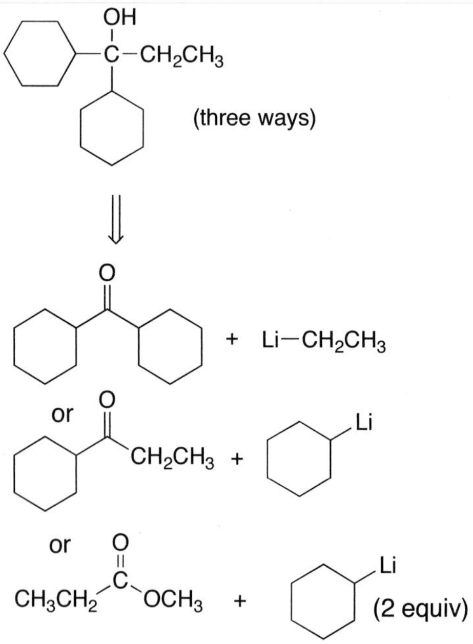

b. 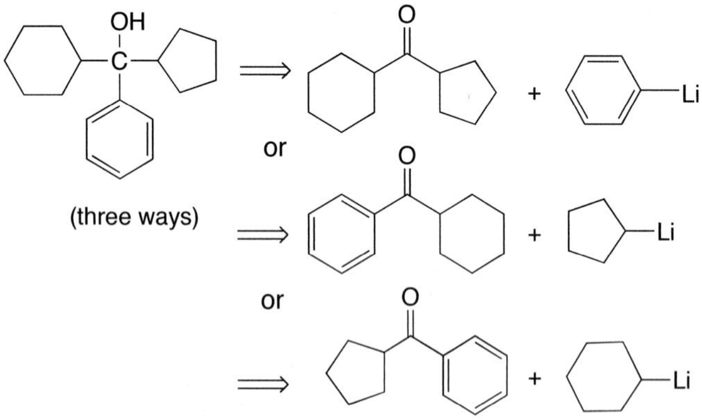 (three ways)

$\implies$ + (phenyl)—Li

or

$\implies$ + (cyclopentyl)—Li

or

$\implies$ + (cyclohexyl)—Li

**20.58**

a.

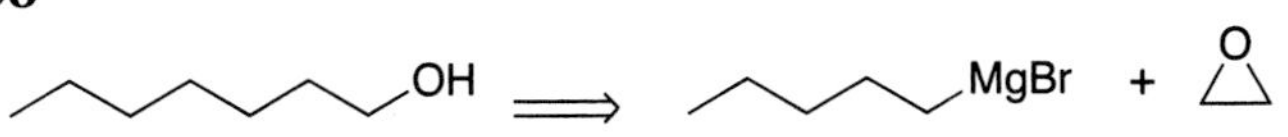

b.

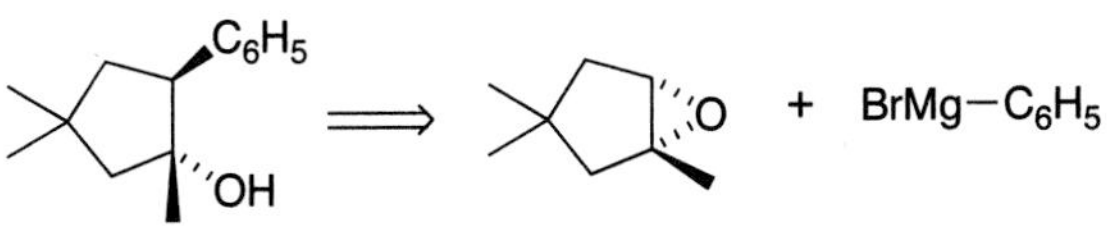

c.

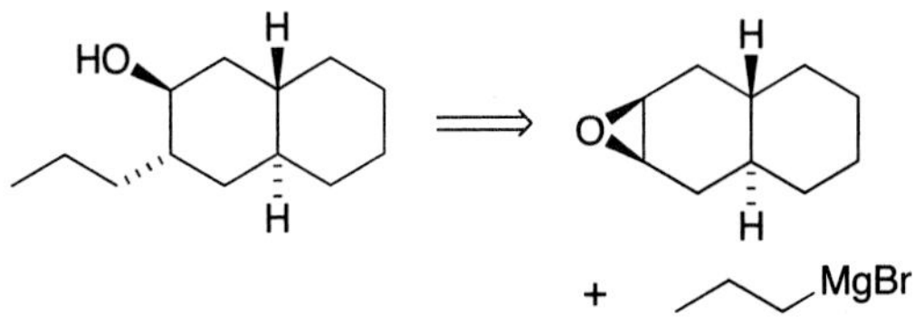

**20.59**

a.

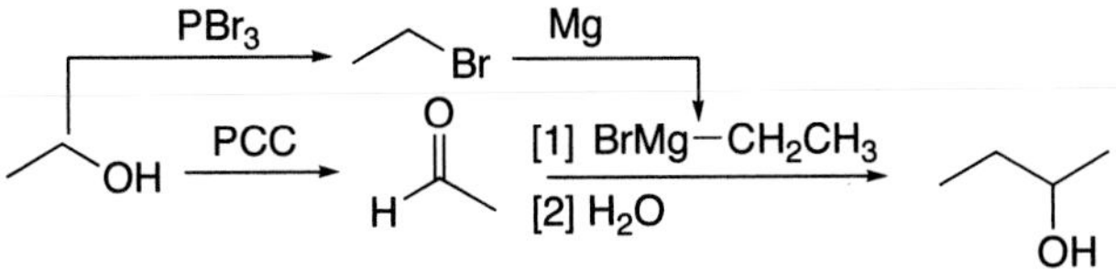

b.

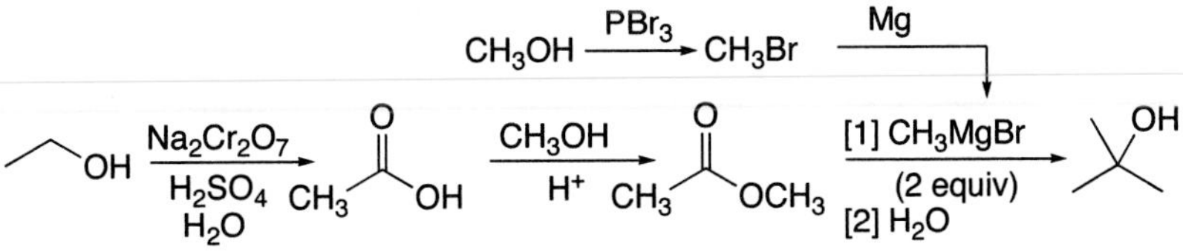

c.

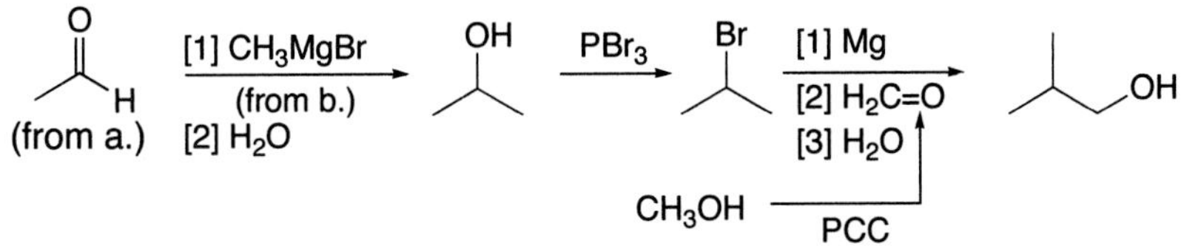

d.

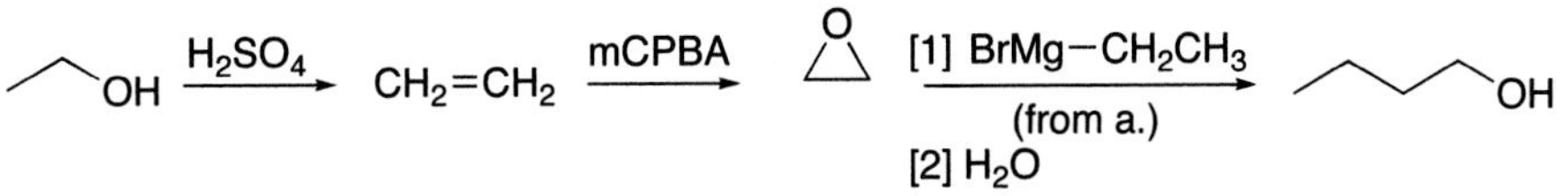

**20.60**

a.

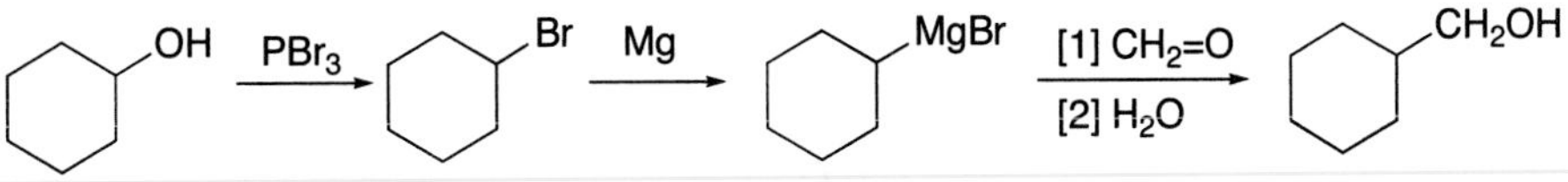

b.

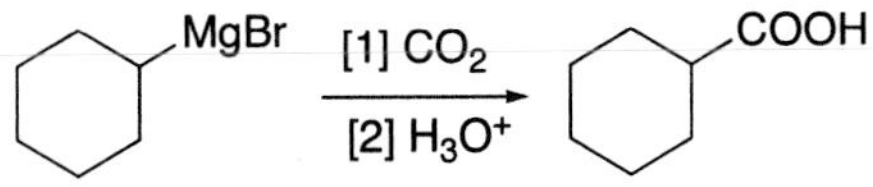

(from a.)

c.

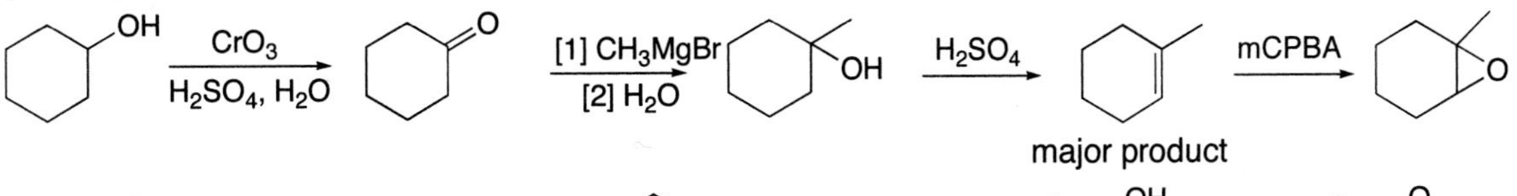

major product

d.

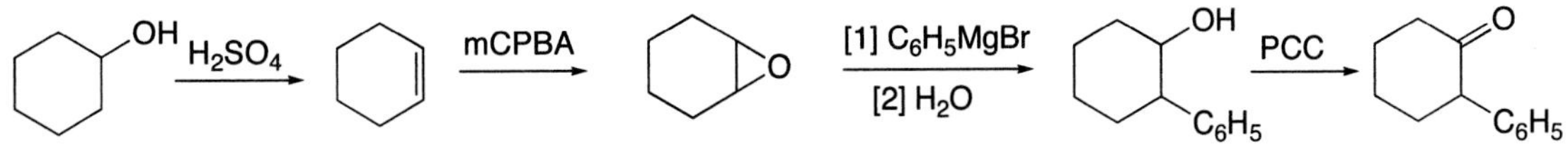

e.

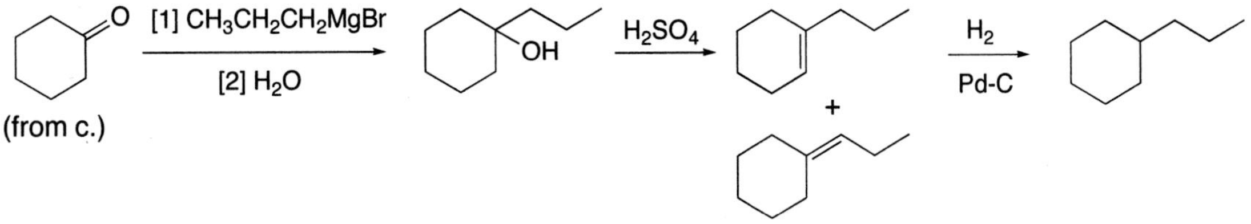

(from c.)

f. (from c.)

[1] HC≡CLi
[2] H₂O

H₂O
H₂SO₄
Hg₂SO₄

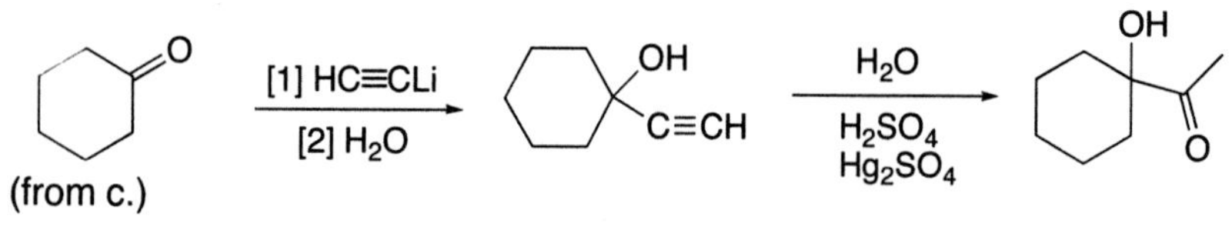

g. (from a.)

MgBr
[1] H₂CO
[2] H₂O

PCC

[1] MgBr
[2] H₂O

PCC

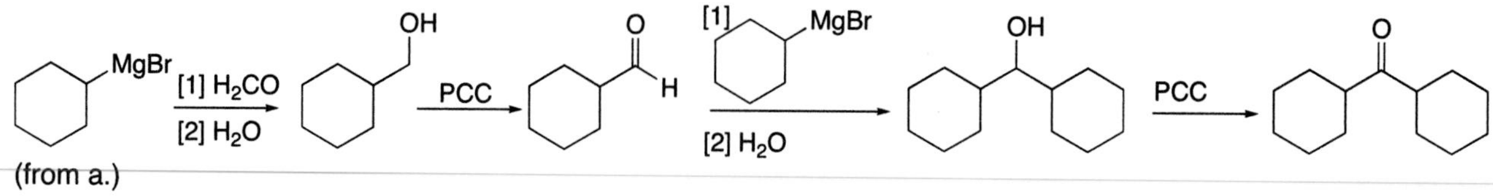

h. (from c.)

[1] MgBr
(from a.)
[2] H₂O

H₂SO₄

major product

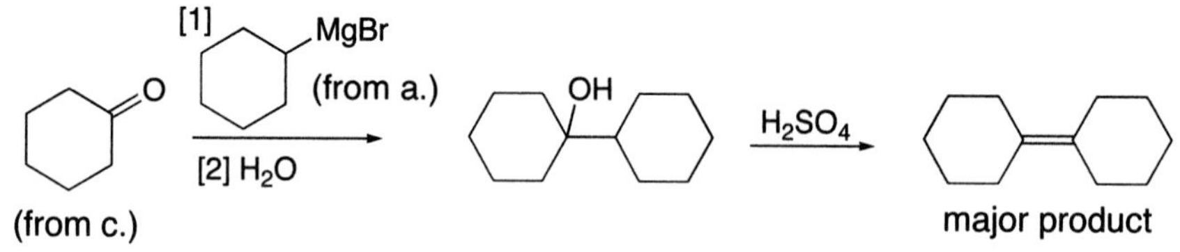

**20.61**

a.

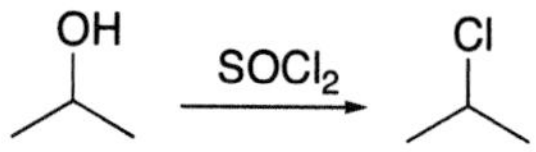

OH → SOCl₂ → Cl

b.

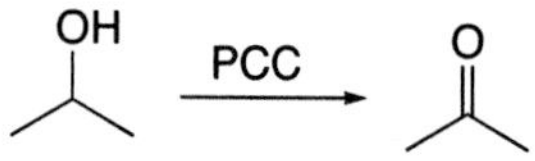

OH → PCC → O

c.

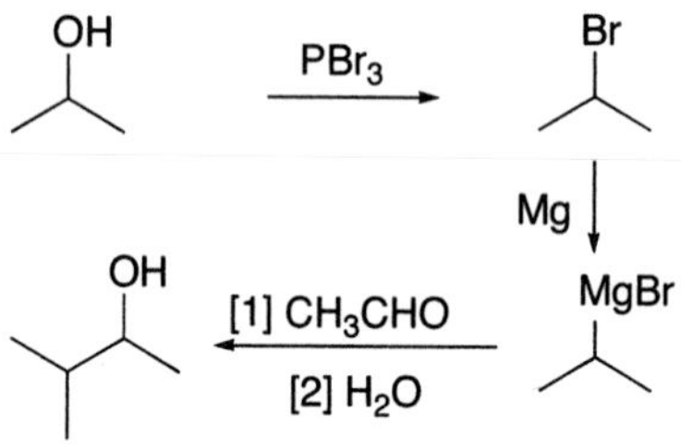

OH → PBr₃ → Br

Mg ↓

MgBr

[1] CH₃CHO
[2] H₂O

OH

d.

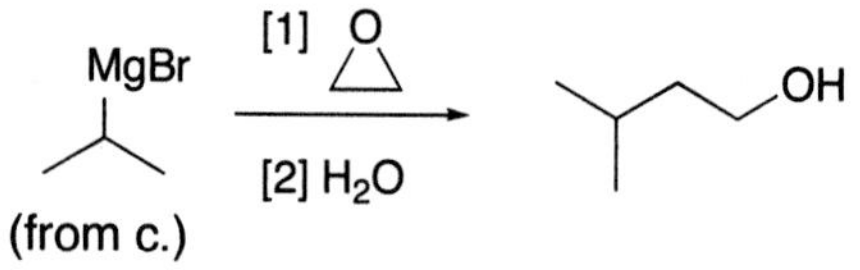

MgBr
(from c.)

[1] O (epoxide)
[2] H₂O

OH

e.

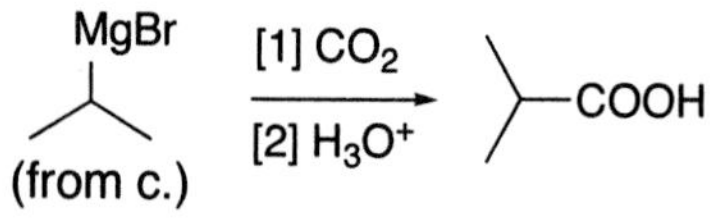

MgBr
(from c.)

[1] CO₂
[2] H₃O⁺

COOH

f.

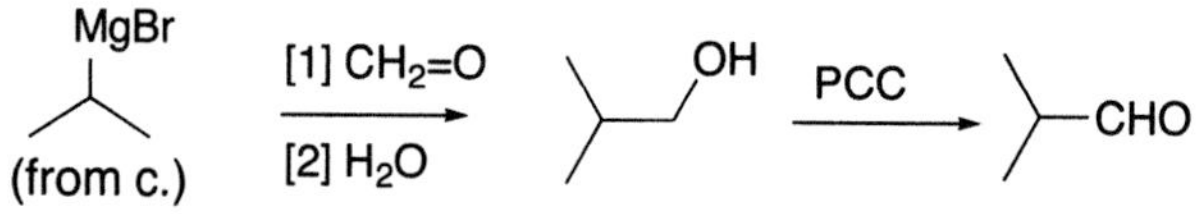

MgBr
(from c.)

[1] CH₂=O
[2] H₂O

OH

PCC

CHO

g.

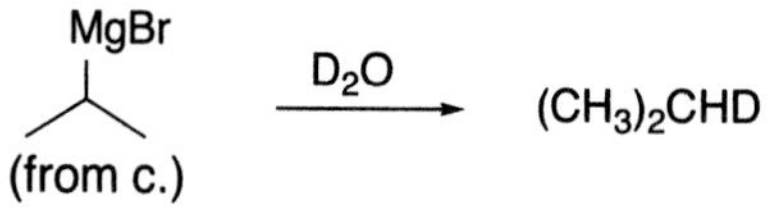

MgBr
(from c.)

D₂O

(CH₃)₂CHD

h.

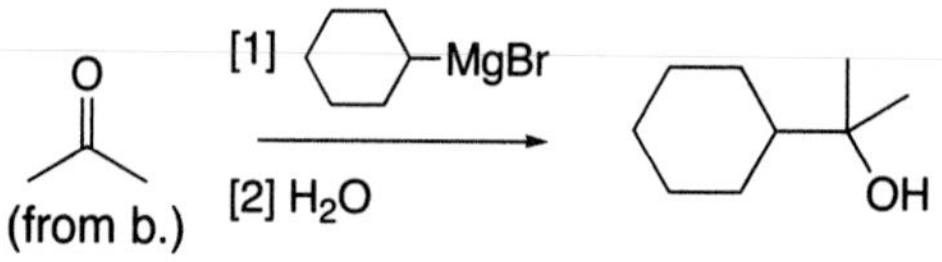

O

[1] MgBr (cyclohexyl)
(from b.) [2] H₂O

OH

i.

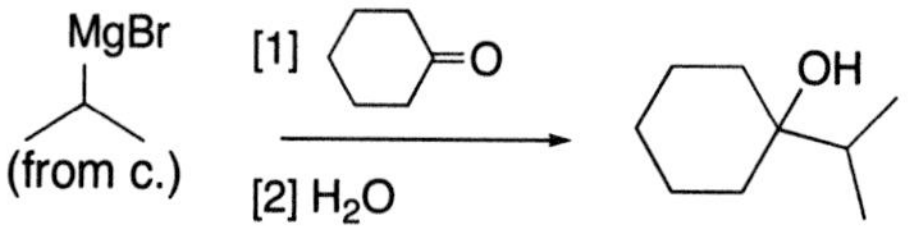

MgBr
(from c.)

[1] cyclohexanone =O
[2] H₂O

OH

j.

Br
(from c.)

[1] 2 Li
[2] CuI (0.5 equiv)

[(CH₃)₂CH]₂CuLi

[1] O (cyclopentenone)
[2] H₂O

O

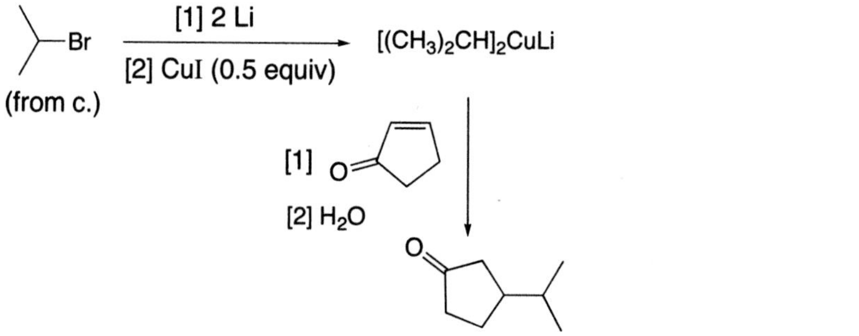

**20.62**  a.  Since estrone contains a OH group, the first equivalent of LiC≡CH removes the OH proton by an acid–base reaction to form a phenoxide.  Then, the second equivalent adds to the carbonyl group.

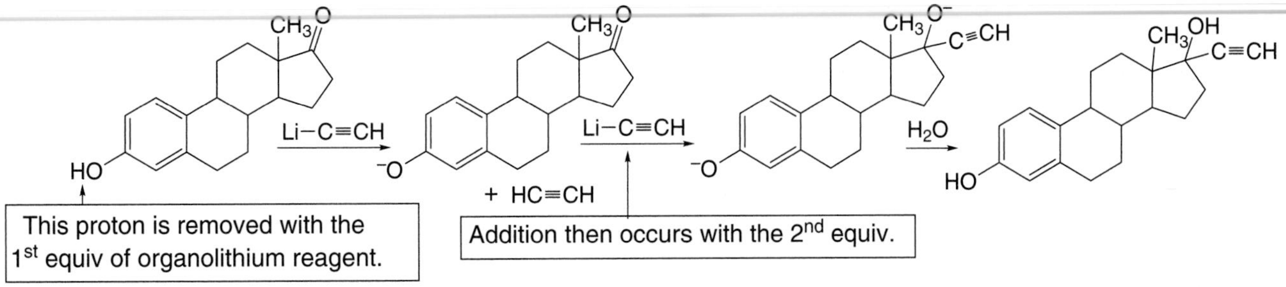

This proton is removed with the 1st equiv of organolithium reagent.

+ HC≡CH

Addition then occurs with the 2nd equiv.

b. **Alternate synthesis:**

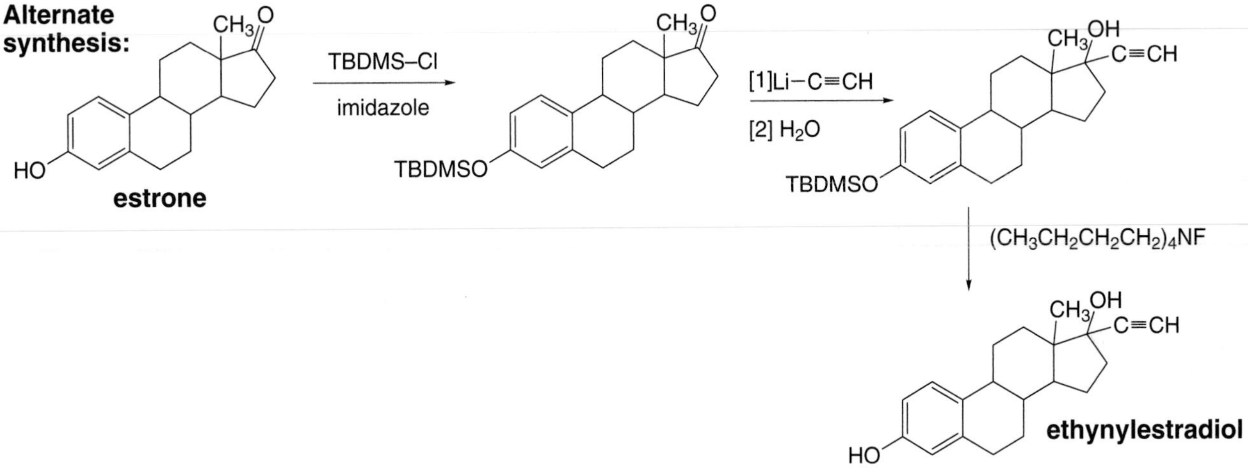

estrone

TBDMS–Cl

imidazole

[1] Li–C≡CH

[2] H₂O

(CH₃CH₂CH₂CH₂)₄NF

**ethynylestradiol**

**20.63**

a.

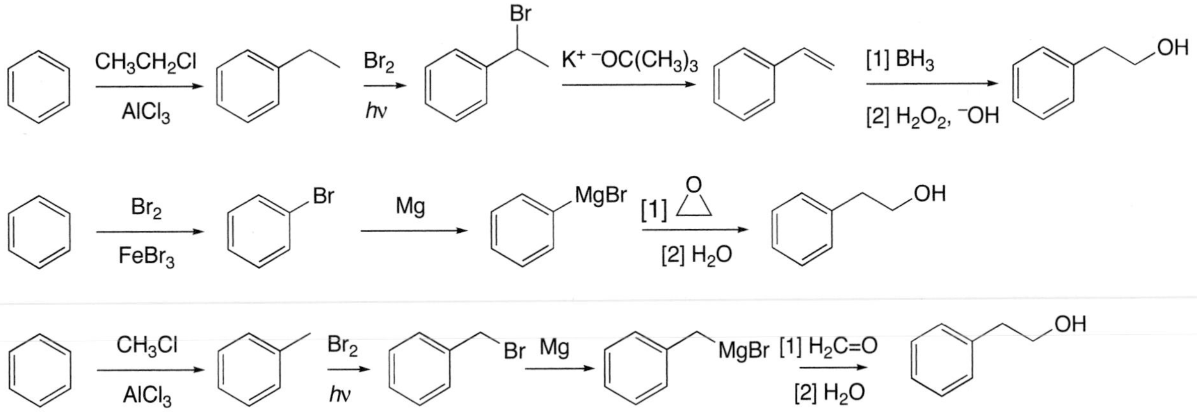

CH₃CH₂Cl / AlCl₃

Br₂ / hν

K⁺ ⁻OC(CH₃)₃

[1] BH₃

[2] H₂O₂, ⁻OH

Br₂ / FeBr₃

Mg

MgBr

[1] epoxide

[2] H₂O

CH₃Cl / AlCl₃

Br₂ / hν

Mg

MgBr

[1] H₂C=O

[2] H₂O

b.

MgBr  [1] CH₃CH=O

[2] H₂O

(from a.)

PCC

CH₃ Cl (acetyl chloride) / AlCl₃

Br  NaOH

OH  PCC

CH₃Br

Mg

[1] CH₃MgBr

[2] H₂O

PCC

(from a.)

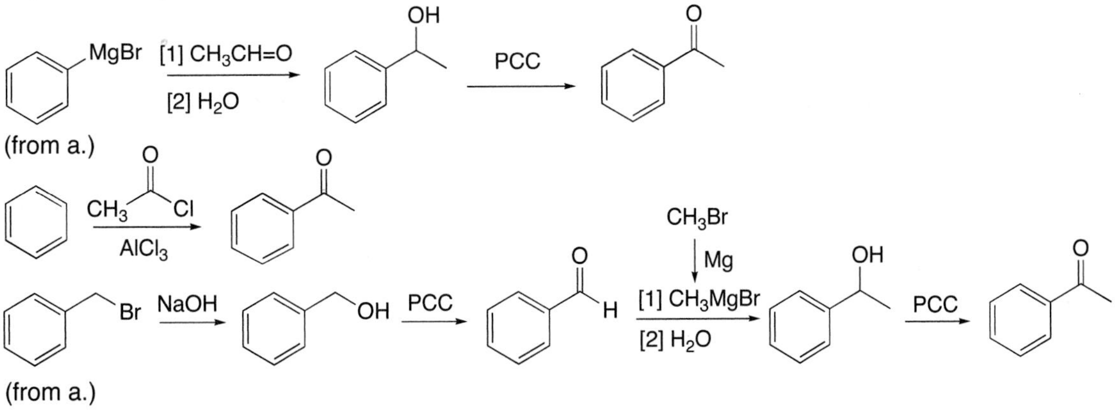

## 20.64

a.

$$\text{benzene} \xrightarrow[\text{FeBr}_3]{\text{Br}_2} \text{PhBr} \xrightarrow{\text{Mg}} \text{PhMgBr} \xrightarrow[\text{[2] H}_3\text{O}^+]{\text{[1] CO}_2} \text{PhCOOH}$$

b.

$$\text{PhMgBr (from a.)} \xrightarrow[\text{[2] H}_2\text{O}]{\text{[1] CH}_2=\text{O}} \text{PhCH}_2\text{OH} \xrightarrow{\text{PCC}} \text{PhCHO}$$

$$\text{CH}_3\text{OH} \xrightarrow{\text{PCC}} \text{CH}_2=\text{O}$$

c. (from b.)

$$\text{PhCHO} \xrightarrow[\text{[2] H}_2\text{O}]{\text{[1] PhMgBr (from a.)}} \text{Ph}_2\text{CHOH} \xrightarrow{\text{PCC}} \text{Ph}_2\text{C=O}$$

d.

$$\text{CH}_3\text{CH}_2\text{CH}_2\text{OH} \xrightarrow{\text{PBr}_3} \text{CH}_3\text{CH}_2\text{CH}_2\text{Br} \xrightarrow{\text{Mg}} \text{CH}_3\text{CH}_2\text{CH}_2\text{MgBr}$$

$$\text{CH}_3\text{CH}_2\text{CH}_2\text{OH} \xrightarrow{\text{PCC}} \text{CH}_3\text{CH}_2\text{CHO} \xrightarrow[\text{[2] H}_2\text{O}]{\text{[1] CH}_3\text{CH}_2\text{CH}_2\text{MgBr}} \cdots \xrightarrow{\text{PCC}} \cdots$$

$$\xrightarrow[\text{[2] H}_2\text{O}]{\text{[1] PhMgBr (from a.)}} \cdots$$

e.

$$\cdots\text{OH} \xrightarrow{\text{PCC}} \cdots \xrightarrow[]{} \xrightarrow{\text{H}_2\text{O}} \cdots \xrightarrow{\text{PCC}} \cdots \xrightarrow[\text{FeBr}_3]{\text{Br}_2} \cdots$$

(from a.)

## 20.65

a.

$$\text{HO}\cdots \xrightarrow{\text{PCC}} \cdots \xrightarrow[\text{[2] H}_2\text{O}]{\text{[1] CH}_3\text{CH}_2\text{MgBr}} \cdots \xrightarrow{\text{PCC}} \cdots$$

$$\text{CH}_3\text{CH}_2\text{OH} \xrightarrow{\text{PBr}_3} \text{CH}_3\text{CH}_2\text{Br} \xrightarrow{\text{Mg}} \text{CH}_3\text{CH}_2\text{MgBr}$$

b.

$$\cdots \xrightarrow{\text{PCC}} \cdots \xrightarrow[\text{[2] H}_2\text{O}]{\text{[1] CH}_3\text{CH}_2\text{CH}_2\text{MgBr}} \cdots$$

$$\text{CH}_3\text{CH}_2\text{CH}_2\text{OH} \xrightarrow{\text{PBr}_3} \text{CH}_3\text{CH}_2\text{CH}_2\text{Br} \xrightarrow{\text{Mg}} \text{CH}_3\text{CH}_2\text{CH}_2\text{MgBr}$$

c.

$$\cdots \xrightarrow{\text{PCC}} \cdots \xrightarrow[\text{[2] H}_2\text{O}]{\text{[1] CH}_3\text{CH}_2\text{CH}_2\text{CH}_2\text{MgBr}} \cdots \xrightarrow{\text{PCC}} \cdots$$

$$\text{CH}_3\text{CH}_2\text{CH}_2\text{CH}_2\text{OH} \xrightarrow{\text{PBr}_3} \text{CH}_3\text{CH}_2\text{CH}_2\text{CH}_2\text{Br} \xrightarrow{\text{Mg}} \text{CH}_3\text{CH}_2\text{CH}_2\text{CH}_2\text{MgBr}$$

d.

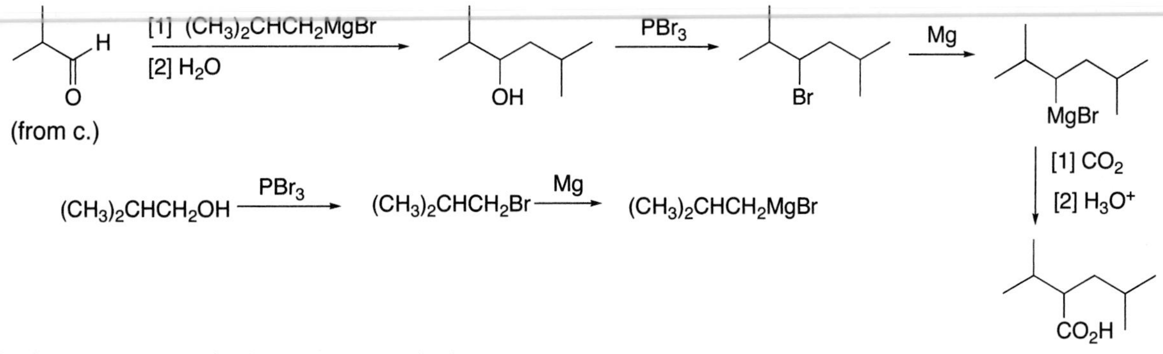

(from c.)

$(CH_3)_2CHCH_2OH \xrightarrow{PBr_3} (CH_3)_2CHCH_2Br \xrightarrow{Mg} (CH_3)_2CHCH_2MgBr$

e.

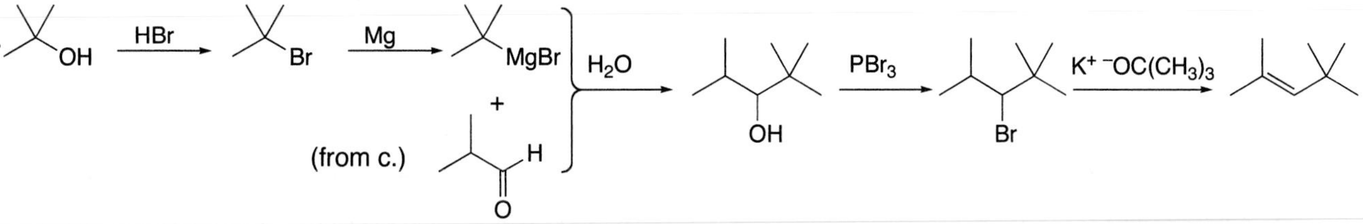

(from c.)

## 20.66

a.

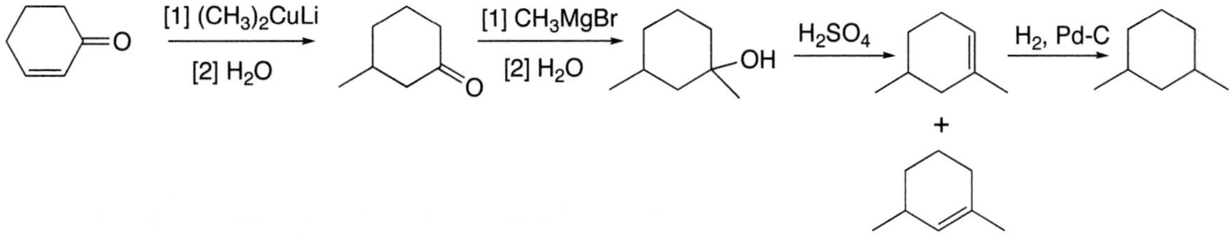

b.

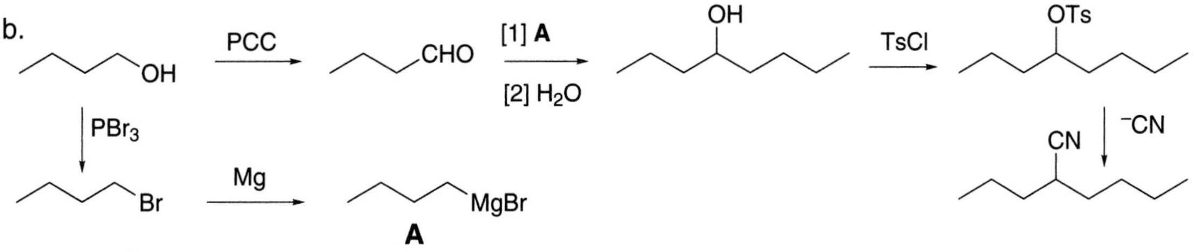

c.

(+ ortho isomer)

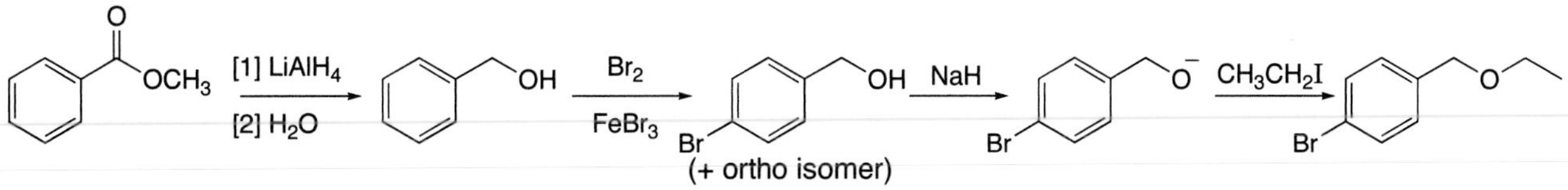

d.

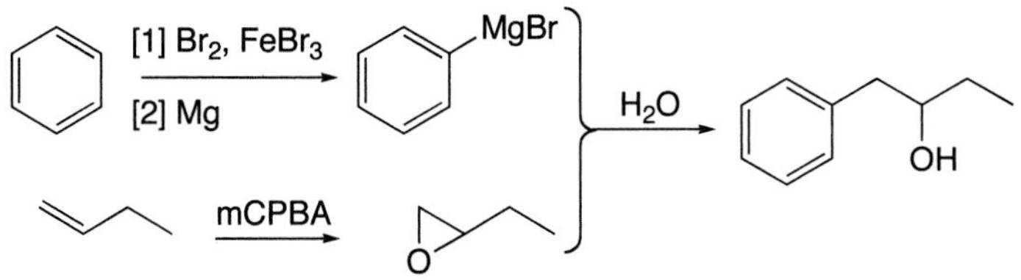

e.

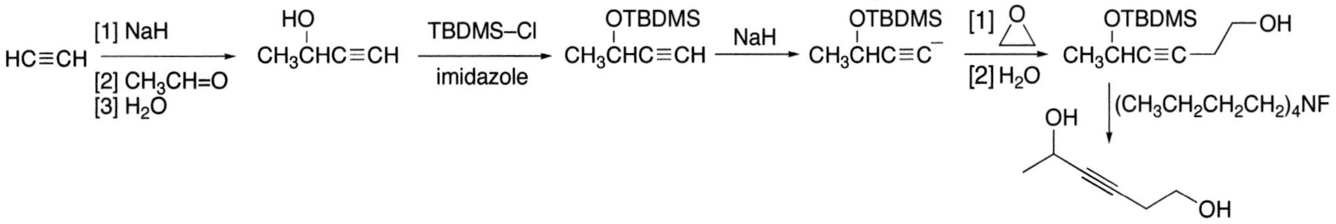

## 20.67

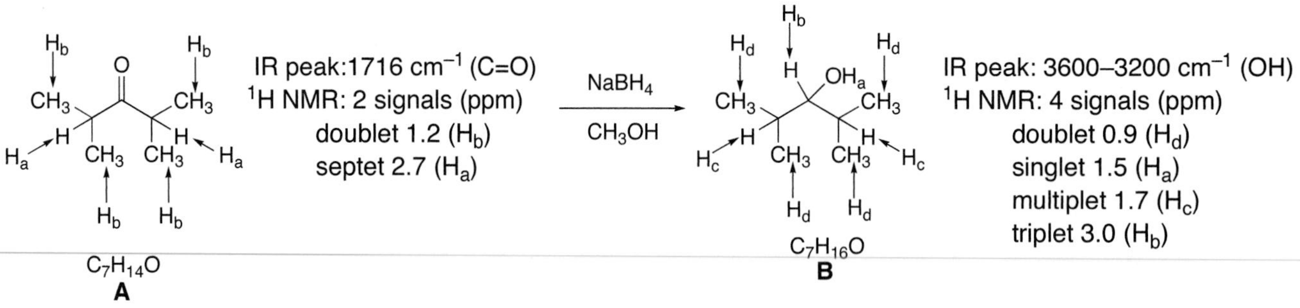

$H_b$ $H_b$ O $CH_3$ $CH_3$ $H_a$ H H $H_a$ $CH_3$ $CH_3$ $H_b$ $H_b$
$C_7H_{14}O$
A
IR peak:1716 $cm^{-1}$ (C=O)
$^1H$ NMR: 2 signals (ppm)
doublet 1.2 ($H_b$)
septet 2.7 ($H_a$)
$NaBH_4$
$CH_3OH$
$C_7H_{16}O$
B
IR peak: 3600–3200 $cm^{-1}$ (OH)
$^1H$ NMR: 4 signals (ppm)
doublet 0.9 ($H_d$)
singlet 1.5 ($H_a$)
multiplet 1.7 ($H_c$)
triplet 3.0 ($H_b$)


## 20.68

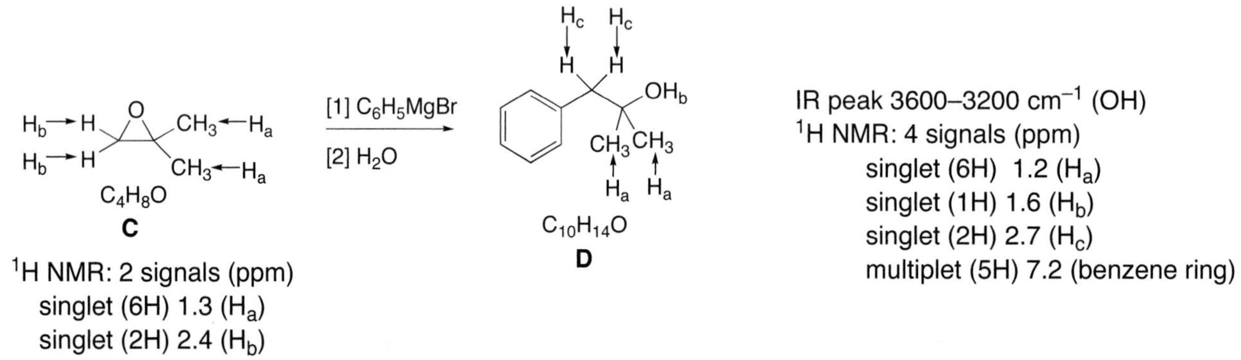

$C_4H_8O$
C
$^1H$ NMR: 2 signals (ppm)
singlet (6H) 1.3 ($H_a$)
singlet (2H) 2.4 ($H_b$)
[1] $C_6H_5MgBr$
[2] $H_2O$
$C_{10}H_{14}O$
D
IR peak 3600–3200 $cm^{-1}$ (OH)
$^1H$ NMR: 4 signals (ppm)
singlet (6H) 1.2 ($H_a$)
singlet (1H) 1.6 ($H_b$)
singlet (2H) 2.7 ($H_c$)
multiplet (5H) 7.2 (benzene ring)


## 20.69

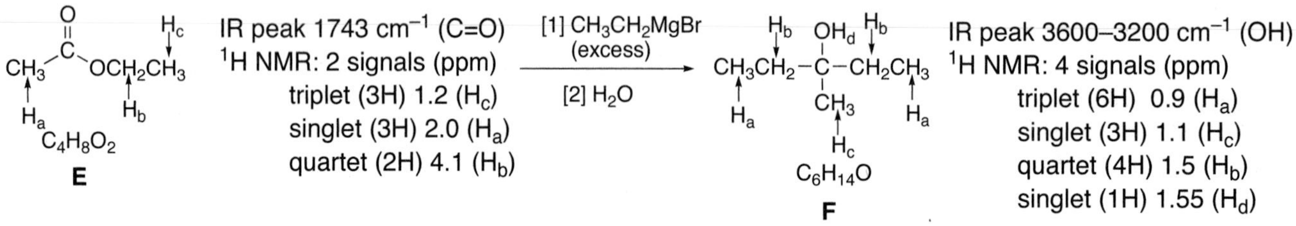

$C_4H_8O_2$
E
IR peak 1743 $cm^{-1}$ (C=O)
$^1H$ NMR: 2 signals (ppm)
triplet (3H) 1.2 ($H_c$)
singlet (3H) 2.0 ($H_a$)
quartet (2H) 4.1 ($H_b$)
[1] $CH_3CH_2MgBr$ (excess)
[2] $H_2O$
$C_6H_{14}O$
F
IR peak 3600–3200 $cm^{-1}$ (OH)
$^1H$ NMR: 4 signals (ppm)
triplet (6H) 0.9 ($H_a$)
singlet (3H) 1.1 ($H_c$)
quartet (4H) 1.5 ($H_b$)
singlet (1H) 1.55 ($H_d$)


## 20.70

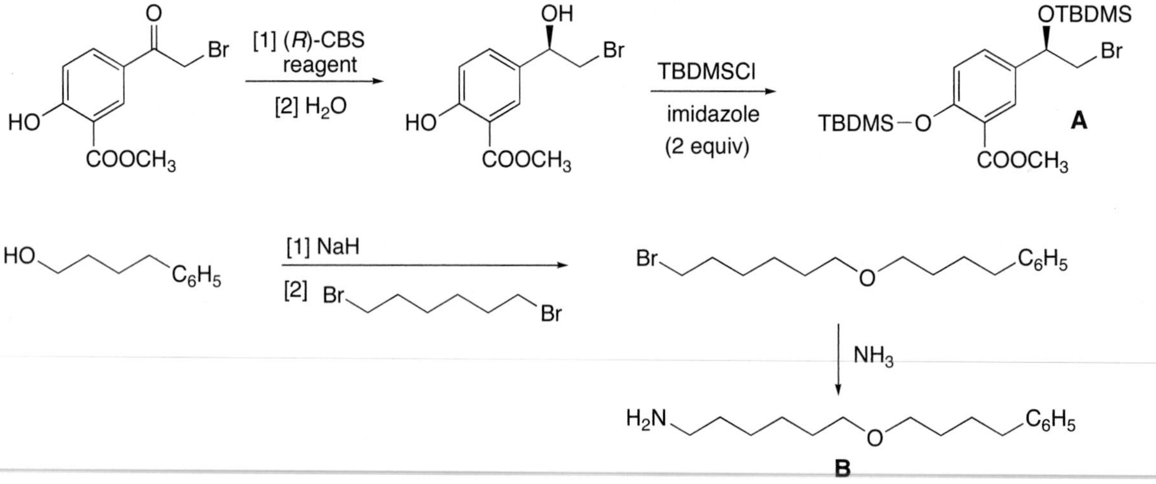

[1] (R)-CBS reagent
[2] $H_2O$
TBDMSCl
imidazole
(2 equiv)
A
[1] NaH
[2] Br⁀Br
$NH_3$
B

A + B $\longrightarrow$

[structure: aromatic compound with OTBDMS, H-N linkage to chain ending in C₆H₅, TBDMS-O and COOCH₃ substituents]

$\left|\begin{array}{l} \text{[1] LiAlH}_4 \\ \text{[2] H}_2\text{O} \end{array}\right.$

[structure: similar aromatic compound with OTBDMS, H-N chain to C₆H₅, TBDMS-O and CH₂OH substituents]

$\left| \text{(CH}_3\text{CH}_2\text{CH}_2\text{CH}_2)_4\text{NF} \right.$

(*R*)-salmeterol

## 20.71

4-*tert*-butylcyclohexanone $\xrightarrow[\text{[2] H}_2\text{O}]{\text{[1] L-selectride}}$ *cis*-4-*tert*-butylcyclohexanol
major product

= (CH₃)₃C— [chair structure] larger group equatorial / OH ← axial / equatorial

L-Selectride adds H⁻ to a C=O group. There are two possible reduction products—cis and trans isomers—but the cis isomer is favored. The key element is that the three *sec*-butyl groups make L-selectride a large, bulky reducing agent that attacks the carbonyl group from the less hindered direction.

(CH₃)₃C— [chair with O] R₃B̄—H $\longrightarrow$ (CH₃)₃C— [chair with O⁻, H equatorial] $\xrightarrow{\text{H}_2\text{O}}$ (CH₃)₃C— [chair OH axial, H equatorial] **cis product**

> When H⁻ adds from the equatorial direction, the product has an axial OH and a new equatorial H. Since the equatorial direction is less hindered, this mode of attack is favored with large bulky reducing agents like L-selectride. In this case, the product is cis.

Axial H's hinder axial attack. R₃B̄—H

(CH₃)₃C— [chair] $\longrightarrow$ (CH₃)₃C— [chair, H axial, O⁻] $\xrightarrow{\text{H}_2\text{O}}$ (CH₃)₃C— [chair, H axial, OH equatorial] **trans product**

> The axial H's hinder H⁻ attack from the axial direction. As a result, this mode of attack is more difficult with larger reducing agents. In this case the product is trans. This product is not formed to any appreciable extent.

**20.72** The β carbon of an α,β-unsaturated carbonyl compound absorbs further downfield in the $^{13}C$ NMR spectrum than the α carbon, because the β carbon is deshielded and bears a partial positive charge as a result of resonance. Since three resonance structures can be drawn for an α,β-unsaturated carbonyl compound, one of which places a positive charge on the β carbon, the decrease of electron density at this carbon deshields it, shifting the $^{13}C$ absorption downfield. This is not the case for the α carbon.

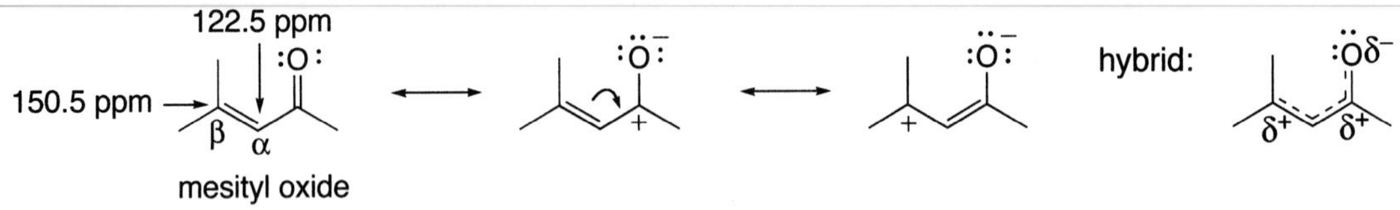

150.5 ppm 122.5 ppm
β α
mesityl oxide     hybrid:

# Chapter 21: Aldehydes and Ketones—Nucleophilic Addition

## ♦ General facts

- Aldehydes and ketones contain a carbonyl group bonded to only H atoms or R groups. The carbonyl carbon is $sp^2$ hybridized and trigonal planar (21.1).
- Aldehydes are identified by the suffix *-al*, while ketones are identified by the suffix *-one* (21.2).
- Aldehydes and ketones are polar compounds that exhibit dipole–dipole interactions (21.3).

## ♦ Summary of spectroscopic absorptions of RCHO and $R_2CO$ (21.4)

| | | |
|---|---|---|
| **IR absorptions** | C=O | ~1715 cm$^{-1}$ for ketones |
| | | • increasing frequency with decreasing ring size |
| | | ~1730 cm$^{-1}$ for aldehydes |
| | | • For both RCHO and $R_2CO$, the frequency decreases with conjugation. |
| | C$sp^2$–H of CHO | ~2700–2830 cm$^{-1}$ (one or two peaks) |
| **$^1$H NMR absorptions** | CHO | 9–10 ppm (highly deshielded proton) |
| | C–H α to C=O | 2–2.5 ppm (somewhat deshielded C$sp^3$–H) |
| **$^{13}$C NMR absorption** | C=O | 190–215 ppm |

## ♦ Nucleophilic addition reactions

### [1] Addition of hydride (H$^-$) (21.8)

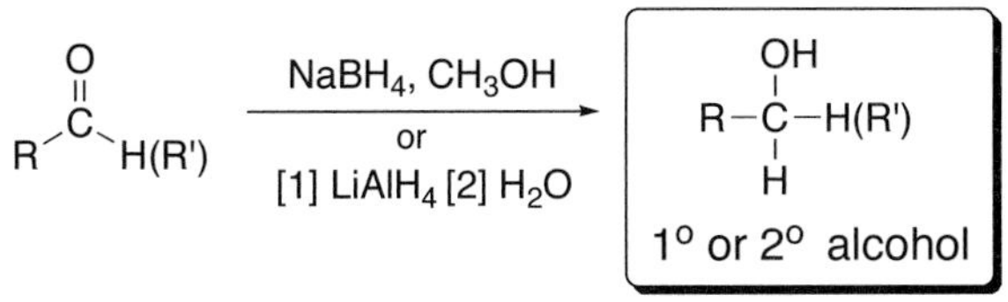

- The mechanism has two steps.
- H:$^-$ adds to the planar C=O from both sides.

### [2] Addition of organometallic reagents (R$^-$) (21.8)

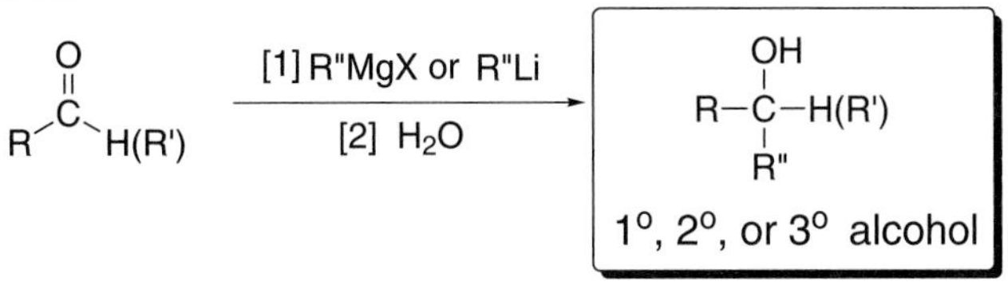

- The mechanism has two steps.
- R:$^-$ adds to the planar C=O from both sides.

### [3] Addition of cyanide ($^-$CN) (21.9)

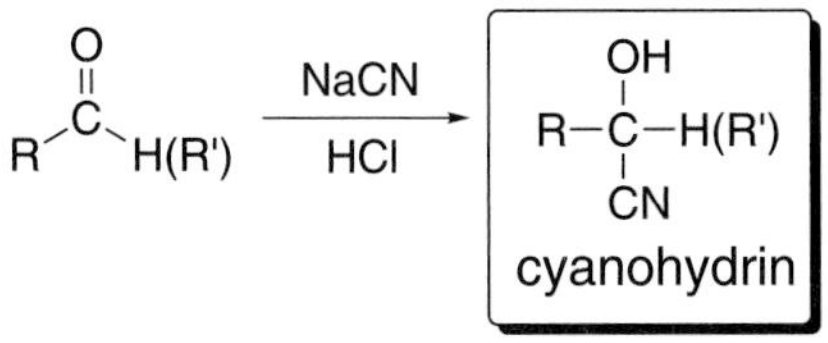

- The mechanism has two steps.
- $^-$CN adds to the planar C=O from both sides.

## [4] Wittig reaction (21.10)

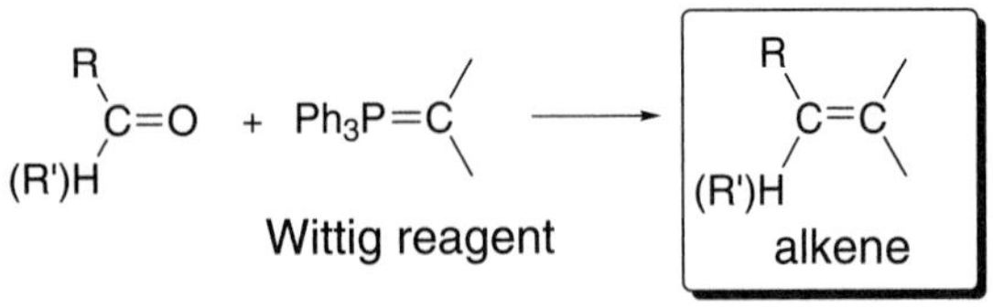

- The reaction forms a new C–C σ bond and a new C–C π bond.
- $Ph_3P=O$ is formed as by-product.

## [5] Addition of 1° amines (21.11)

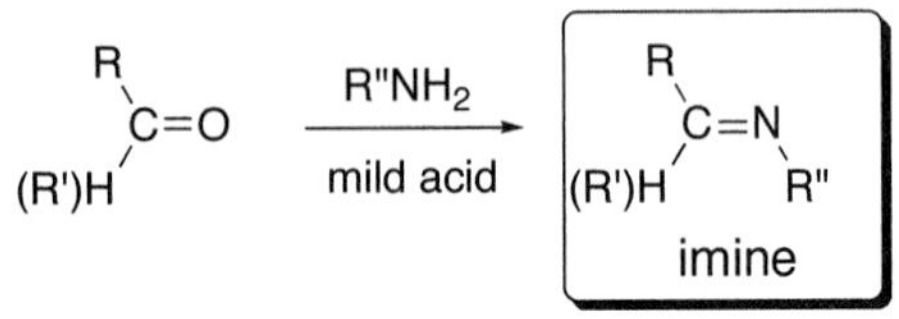

- The reaction is fastest at pH 4–5.
- The intermediate carbinolamine is unstable, and loses $H_2O$ to form the C=N.

## [6] Addition of 2° amines (21.12)

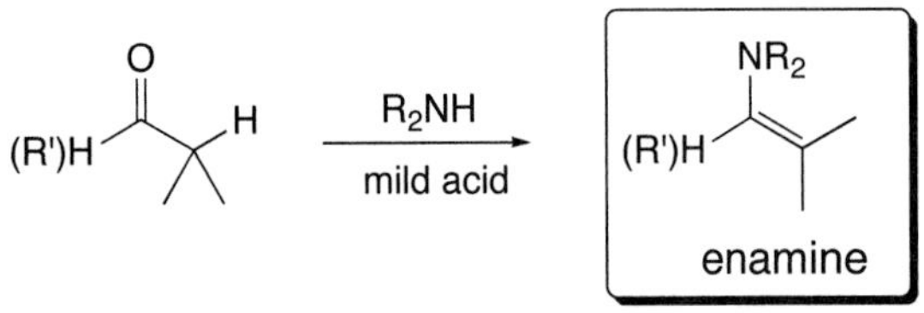

- The reaction is fastest at pH 4–5.
- The intermediate carbinolamine is unstable, and loses $H_2O$ to form the C=C.

## [7] Addition of H₂O—Hydration (21.13)

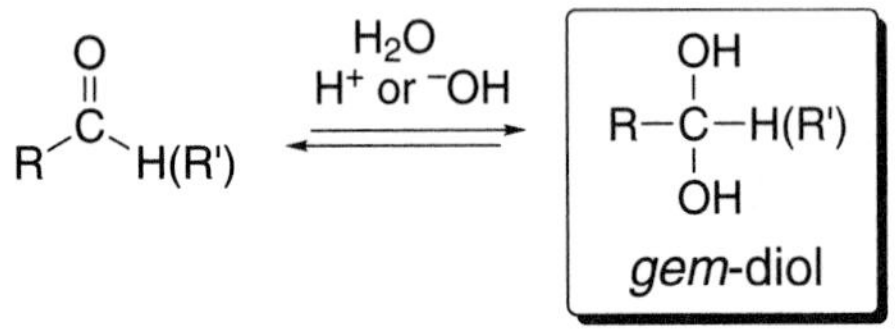

- The reaction is reversible. Equilibrium favors the product only with less stable carbonyl compounds (e.g., $H_2CO$ and $Cl_3CCHO$).
- The reaction is catalyzed with either $H^+$ or $^-OH$.

## [8] Addition of alcohols (21.14)

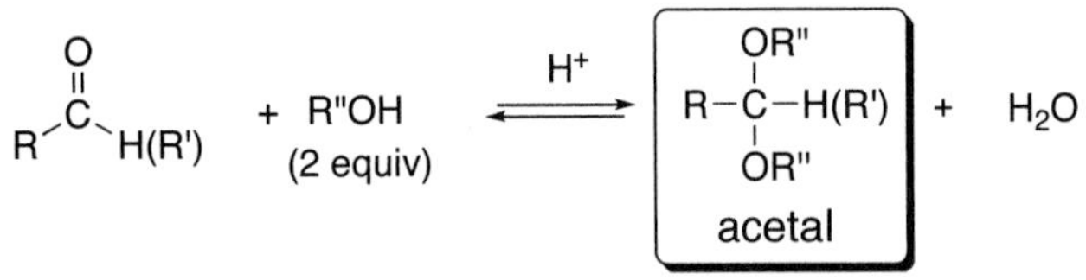

- The reaction is reversible.
- The reaction is catalyzed with acid.
- Removal of $H_2O$ drives the equilibrium to favor the products.

## ◆ Other reactions

## [1] Synthesis of Wittig reagents (21.10A)

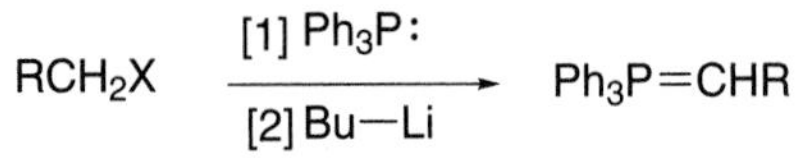

- Step [1] is best with $CH_3X$ and $RCH_2X$ since the reaction follows an $S_N2$ mechanism.
- A strong base is needed for proton removal in Step [2].

## [2] Conversion of cyanohydrins to aldehydes and ketones (21.9)

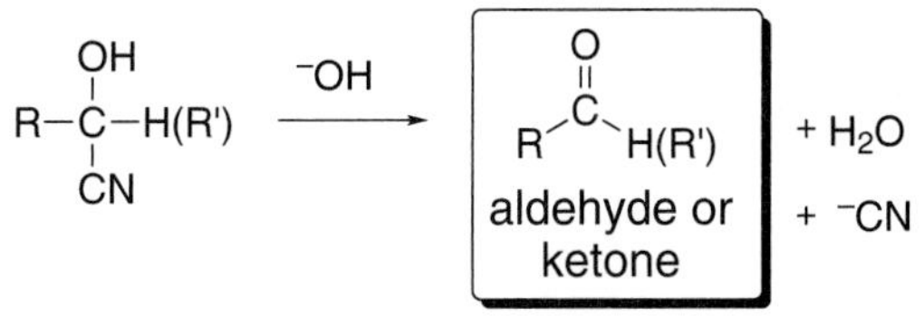

OH
|
R—C—H(R')    —OH→    [ aldehyde or ketone: R—C(=O)—H(R') ]   + $H_2O$    • This reaction is the reverse of cyanohydrin
|                                                               + $^-CN$       formation.
CN

## [3] Hydrolysis of nitriles (21.9)

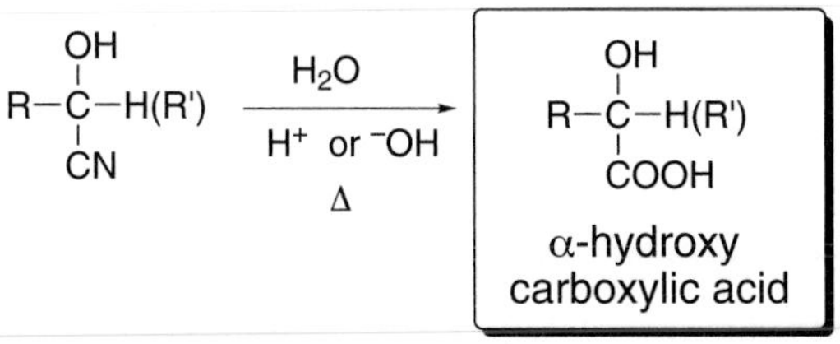

OH                         OH
|          $H_2O$          |
R—C—H(R')  ————————→   R—C—H(R')
|         $H^+$ or $^-OH$   |
CN            Δ            COOH

α-hydroxy
carboxylic acid

## [4] Hydrolysis of imines and enamines (21.12)

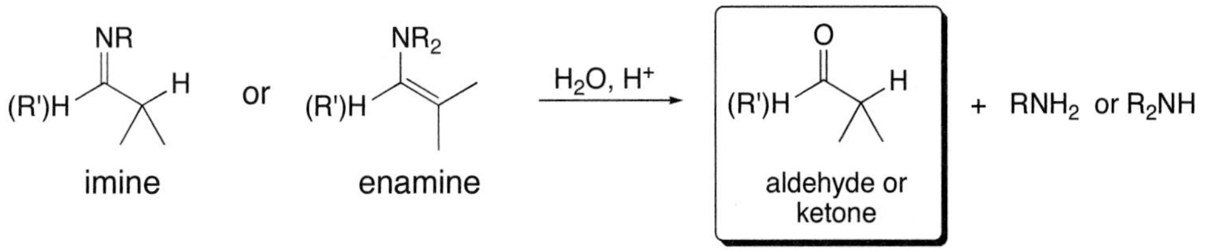

NR                    $NR_2$
‖                     |
(R')H—C—H    or   (R')H—C=C        $H_2O, H^+$ →   [ aldehyde or ketone: (R')H—C(=O)—H ]   +  $RNH_2$ or $R_2NH$
|                     |

imine                 enamine

## [5] Hydrolysis of acetals (21.14)

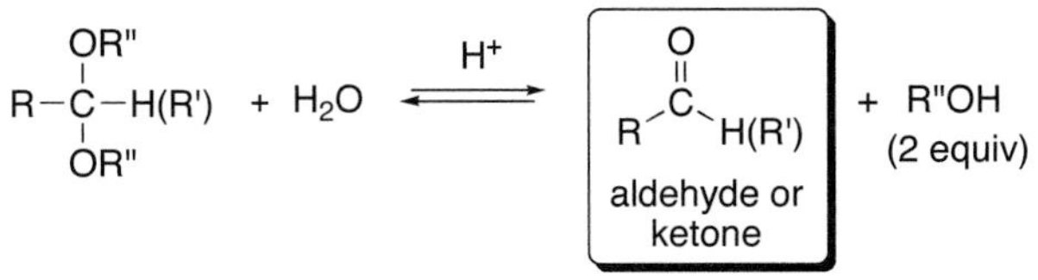

OR"
|                      $H^+$
R—C—H(R')  + $H_2O$  ⇌   [ aldehyde or ketone: R—C(=O)—H(R') ]   + R"OH      • The reaction is acid catalyzed and is
|                                                                 (2 equiv)    the reverse of acetal synthesis.
OR"                                                                          • A large excess of $H_2O$ drives the
                                                                              equilibrium to favor the products.

## Chapter 21: Answers to Problems

**21.1** As the number of R groups bonded to the carbonyl C increases, reactivity towards nucleophilic attack decreases.

a.  $(CH_3)_2C=O$     $CH_3CH=O$     $CH_2=O$      b.

         2 R groups     1 R group    0 R groups

**Increasing reactivity**
decreasing alkyl substitution

**Increasing reactivity**
decreasing steric hindrance

**21.2** More stable aldehydes are less reactive towards nucleophilic attack.

benzaldehyde
Several resonance structures delocalize the partial positive charge on the carbonyl carbon making it more stable, and less reactive towards nucleophilic attack.

cyclohexanecarbaldehyde
This aldehyde has no added resonance stabilization.

**21.3** • To name an aldehyde with a chain of atoms: [1] Find the longest chain with the CHO group and change the *-e* ending to *-al*. [2] Number the carbon chain to put the CHO at C1, but omit this number from the name. Apply all other nomenclature rules.
     • To name an aldehyde with the CHO bonded to a ring: [1] Name the ring and add the suffix *-carbaldehyde*. [2] Number the ring to put the CHO group at C1, but omit this number from the name. Apply all other nomenclature rules.

a.   $(CH_3)_3CC(CH_3)_2CH_2CHO$

5 C chain = pentanal   **3,3,4,4-tetramethylpentanal**

c.

4 C ring =
cyclobutanecarbaldehyde

**3,3-dichlorocyclobutane-
carbaldehyde**

b.

8 C chain = octanal      **2,5,6-trimethyloctanal**

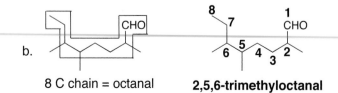

**21.4** Work backwards from the name to the structure, referring to the nomenclature rules in Answer 21.3.

a. 2-isobutyl-3-isopropyl**hexanal**

6 C chain

c. 1-methyl**cyclopropanecarbaldehyde**

3 carbon ring

b. *trans*-3-methyl**cyclopentanecarbaldehyde**

5 carbon ring

CHO          CHO

or

CH₃          CH₃

d. 3,6-diethyl**nonanal**

9 C chain

**21.5** • To name an acyclic ketone: [1] Find the longest chain with the carbonyl group and change the -*e* ending to -*one*. [2] Number the carbon chain to give the carbonyl C the lower number. Apply all other nomenclature rules.
• To name a cyclic ketone: [1] Name the ring and change the -*e* ending to -*one*. [2] Number the C's to put the carbonyl C at C1 and give the next substituent the lower number. Apply all other nomenclature rules.

a.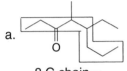

8 C chain =
octanone

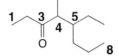

**5-ethyl-4-methyl-3-octanone**

c. (CH₃)₃CCOC(CH₃)₃

5 C chain =
pentanone

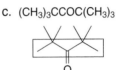

**2,2,4,4-tetramethyl-3-pentanone**

b.

(CH₃)₃C

CH₃

5 C ring =
cyclopentanone

(CH₃)₃C

CH₃

**3-*tert*-butyl-2-methylcyclopentanone**

**21.6** Most common names are formed by naming both alkyl groups on the carbonyl C, arranging them alphabetically, and adding the word ketone.

a. *sec*-butyl ethyl ketone

b. methyl vinyl ketone

c. *p*-ethylacetophenone

d. 2-benzyl-3-benzoylcyclopentanone

benzyl group:   benzoyl group:   5 C ketone

—CH₂—

e. 6,6-dimethyl-2-cyclohexenone

6 C ketone

f. 3-ethyl-5-hexenal

CHO

**21.7** Even though both compounds have polar C–O bonds, the electron pairs around the $sp^3$ hybridized O atom of diethyl ether are more crowded and less able to interact with electron-deficient sites in other diethyl ether molecules. The O atom of the carbonyl group of 2-butanone extends out from the carbon chain making it less crowded. The lone pairs of electrons on the O atom can more readily interact with the electron-deficient sites in the other molecules, resulting in stronger forces.

2-butanone          diethyl ether

**21.8** For cyclic ketones, the carbonyl absorption shifts to higher wavenumber as the size of the ring decreases and the ring strain increases. Conjugation of the carbonyl group with a C=C or a benzene ring shifts the absorption to lower wavenumber.

a.  ~CHO  and  ~CHO

conjugated C=O          **higher wavenumber**
lower wavenumber

b.  ⬠=O  and  ▷=O

smaller ring
**higher wavenumber**

**21.9**

cyclopropenone
(1640 cm$^{-1}$)

2-cyclohexenone
(1685 cm$^{-1}$)

These two resonance structures include an aromatic ring $(4n + 2) = 2\ \pi$ electrons. Although they are charge separated, the stabilized aromatic ring makes these two structures contribute to the hybrid more than usual. Since these two resonance contributors have a C–O single bond, the absorption is shifted to a lower wavenumber.

There are three resonance structures for 2-cyclohexenone, but the charge-separated resonance structures are not aromatic so they contribute less to the resonance hybrid. The C=O absorbs in the usual region for a conjugated carbonyl.

**21.10** Compound **A** shows three signals in its $^{13}$C NMR spectrum. Compound **B** has a signal at 9.8 ppm in its $^1$H NMR spectrum, due to the H of an aldehyde. Compound **C** has a singlet at 2.1 ppm in its $^1$H NMR spectrum, due to the protons on the α C, a methyl ketone.

$C_4H_8O$

3 types of C
aldehyde
**A**

4 types of C
aldehyde
**B**

4 types of C
methyl ketone
**C**

**21.11**

a. $CH_3CH_2CH_2COOCH_3$ $\xrightarrow[\text{[2] H}_2\text{O}]{\text{[1] DIBAL-H}}$ $CH_3CH_2CH_2CHO$

b. $CH_3CH_2CH_2CH_2OH$ $\xrightarrow{\text{PCC}}$ $CH_3CH_2CH_2CHO$

c. $HC{\equiv}CCH_2CH_3$ $\xrightarrow[\text{[2] H}_2\text{O}_2,\ \text{HO}^-]{\text{[1] BH}_3}$ $CH_3CH_2CH_2CHO$

d. $CH_3CH_2CH_2CH{=}CHCH_2CH_2CH_3$ $\xrightarrow[\text{[2] Zn, H}_2\text{O}]{\text{[1] O}_3}$ $CH_3CH_2CH_2CHO$

**21.12**

a.

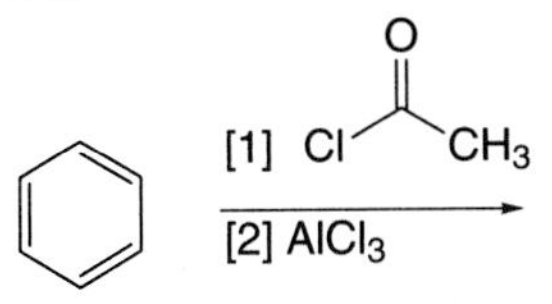

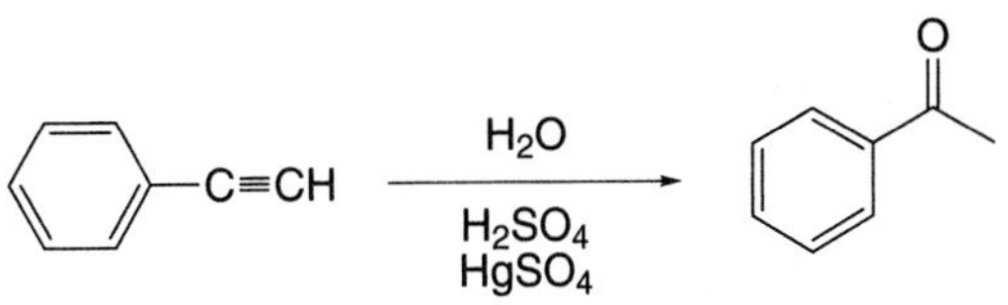

b.

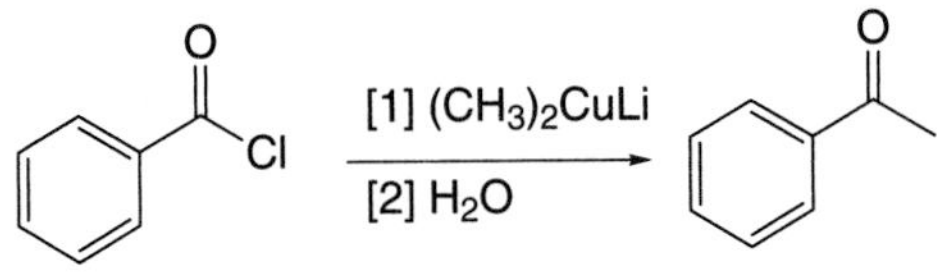

c.

**21.13**

O⁻ ← weaker base

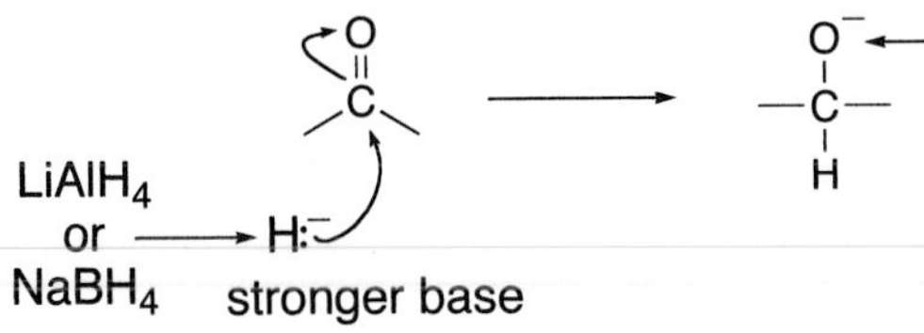

Equilibrium favors the weaker base.
The H⁻ nucleophile is a much stronger base than
the alkoxide product.

LiAlH₄
or ⟶ H:⁻
NaBH₄    stronger base

**21.14** Addition of hydride or R–M occurs at a planar carbonyl C, so two different configurations at a
new stereogenic center are possible.

a.

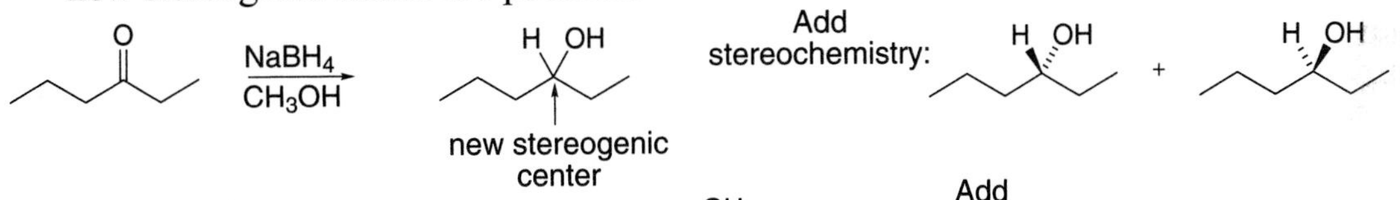

b.

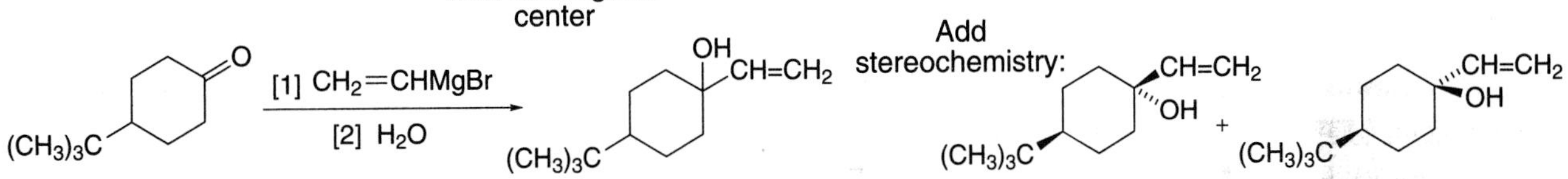

c.

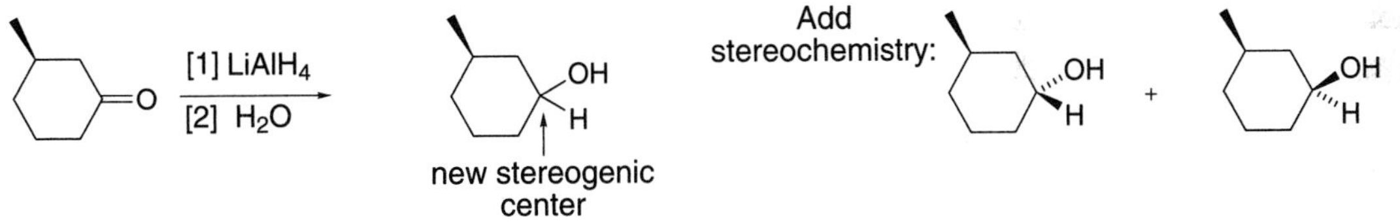

**21.15** Treatment of an aldehyde or ketone with NaCN, HCl adds HCN across the double bond. Cyano
groups are hydrolyzed by $H_3O^+$ to replace the 3 C–N bonds with 3 C–O bonds.

a.

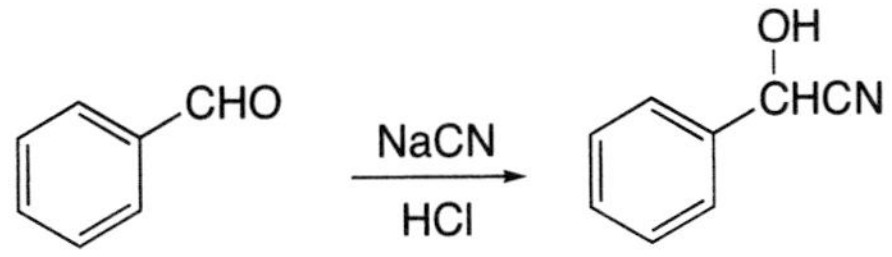

b.

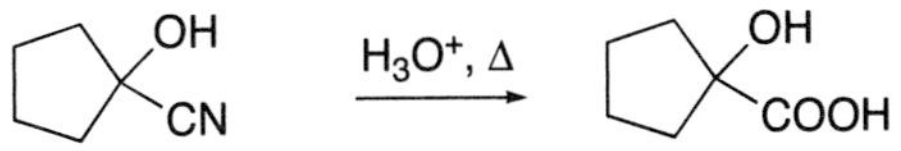

**21.16**

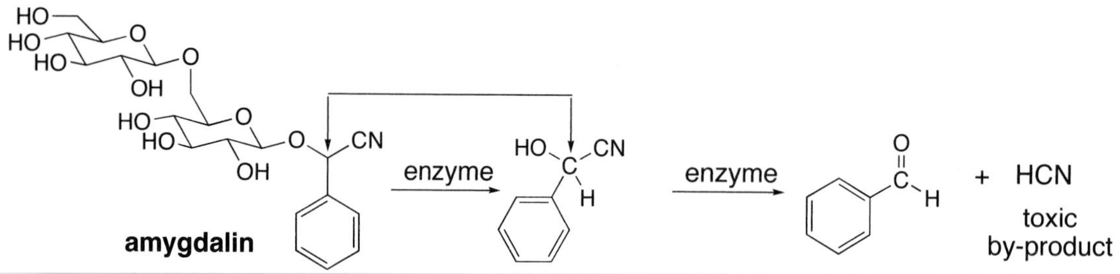

**21.17**

a.

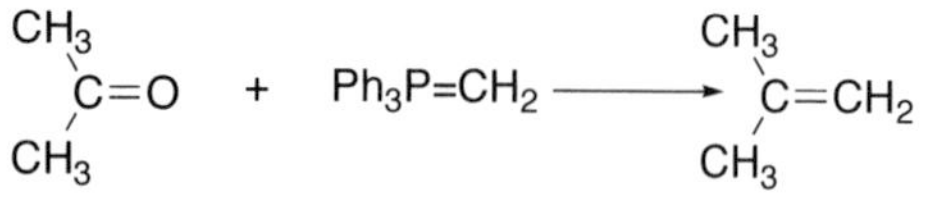

$$\begin{array}{c} CH_3 \\ C=O \\ CH_3 \end{array} + Ph_3P=CH_2 \longrightarrow \begin{array}{c} CH_3 \\ C=CH_2 \\ CH_3 \end{array}$$

b.

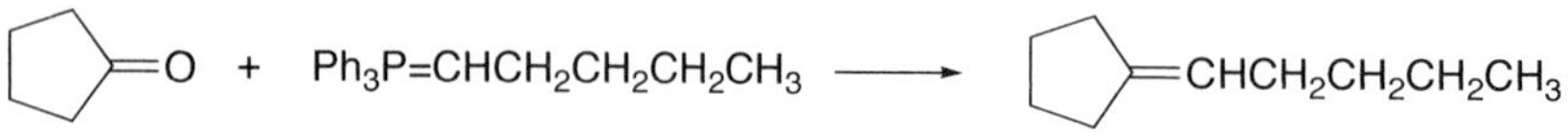

$O + Ph_3P=CHCH_2CH_2CH_2CH_3 \longrightarrow$ $=CHCH_2CH_2CH_2CH_3$

**21.18**

a.  $Ph_3P: + Br-CH_2CH_3 \longrightarrow \underset{Br^-}{\overset{+}{Ph_3P-CH_2CH_3}} \overset{BuLi}{\longrightarrow} Ph_3P=CHCH_3$

b.  $Ph_3P: + Br-CH(CH_3)_2 \longrightarrow \underset{Br^-}{\overset{+}{Ph_3P-CH(CH_3)_2}} \overset{BuLi}{\longrightarrow} Ph_3P=C(CH_3)_2$

c.  $Ph_3P: + Br-CH_2C_6H_5 \longrightarrow \underset{Br^-}{\overset{+}{Ph_3P-CH_2C_6H_5}} \overset{BuLi}{\longrightarrow} Ph_3P=CHC_6H_5$

**21.19**  You need to use a phosphorus reagent that has no H's on the C bonded to the P; otherwise more than one Wittig reagent is possible.  If $(CH_3CH_2)_3P$ is used, for example, a H from a $CH_2$ group of the reagent can be removed with base when the Wittig reagent is prepared.

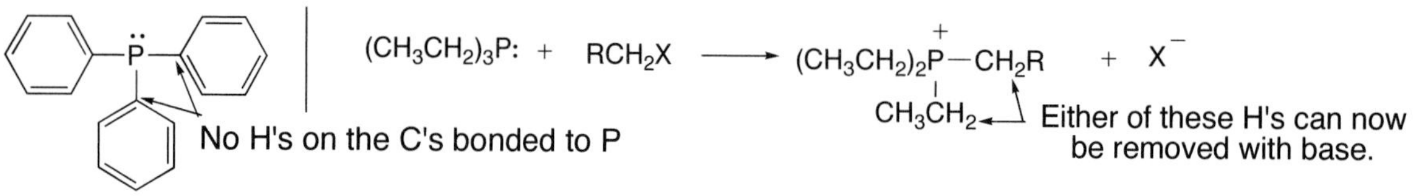

No H's on the C's bonded to P

$(CH_3CH_2)_3P: + RCH_2X \longrightarrow (CH_3CH_2)_2\overset{+}{P}-CH_2R + X^-$

$CH_3CH_2$ — Either of these H's can now be removed with base.

**21.20**

a.

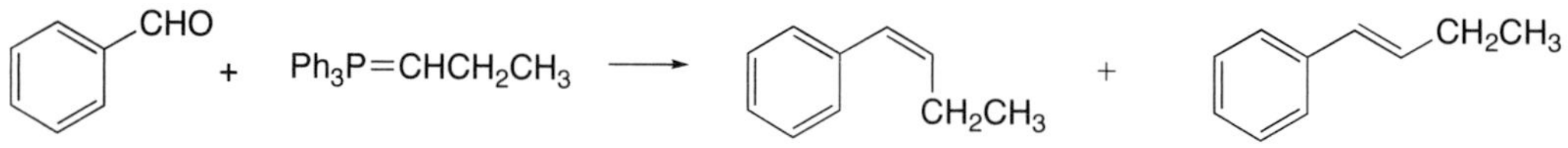

$+ Ph_3P=CHCH_2CH_3 \longrightarrow$ $+$

b.

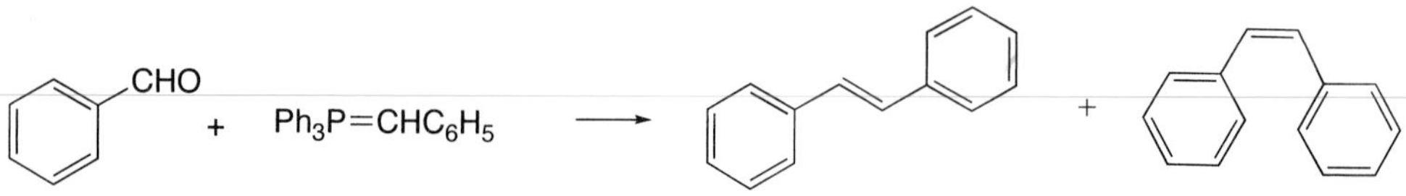

$+ Ph_3P=CHC_6H_5 \longrightarrow$ $+$

c.

$+ Ph_3P=CHCOOCH_3 \longrightarrow$ $+$

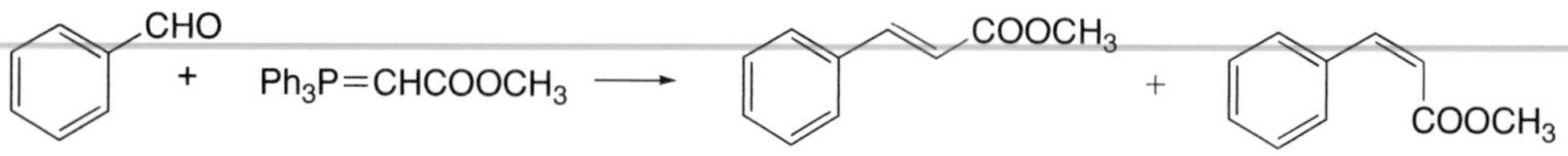

**21.21** To draw the starting materials of the Wittig reactions, find the C=C and cleave it. Replace it with a C=O in one half of the molecule and a C=PPh₃ in the other half. The preferred pathway uses a Wittig reagent derived from a less hindered alkyl halide.

a.

$$CH_3, CH_2CH_3 \atop CH_3 \quad H \; C\!=\!C \implies CH_3 \atop CH_3 \; C\!=\!PPh_3 \; + \; O\!=\!C \atop H \; CH_2CH_3 \quad \text{or} \quad CH_3 \atop CH_3 \; C\!=\!O \; + \; Ph_3P\!=\!C \atop H \; CH_2CH_3$$

2° halide precursor
(CH₃)₂CHX

1° halide precursor
XCH₂CH₂CH₃
**preferred pathway**

b.

$$CH_3CH_2, CH_2CH_3 \atop H \quad H \; C\!=\!C \implies CH_3CH_2 \atop H \; C\!=\!PPh_3 \; + \; O\!=\!C \atop H \; CH_2CH_3 \quad \text{(only one route possible)}$$

(cis or trans)

c.

$$C_6H_5, CH_3 \atop H \quad H \; C\!=\!C \implies C_6H_5 \atop H \; C\!=\!PPh_3 \; + \; O\!=\!C \atop H \; CH_3 \quad \text{or} \quad C_6H_5 \atop H \; C\!=\!O \; + \; Ph_3P\!=\!C \atop H \; CH_3$$

(cis or trans)

1° halide precursor
C₆H₅CH₂X

(both routes possible)

1° halide precursor
XCH₂CH₃

**21.22**

a. Two-step sequence:

[1] CH₃MgBr
[2] H₂O

H₂SO₄

One-step sequence:

Ph₃P=CH₂

**minor product**    tetrasubstituted **major product**    (*E* and *Z* isomers)

**only product**

b. Two-step sequence:

[1] C₆H₅CH₂MgBr
[2] H₂O

H₂SO₄

=CHC₆H₅    +    —CH₂C₆H₅

trisubstituted conjugated C=C    trisubstituted

One-step sequence:

Ph₃P=CHC₆H₅

=CHC₆H₅  **only product**

**21.23** When a 1° amine reacts with an aldehyde or ketone, the C=O is replaced by C=NR.

a.

—CHO    CH₃CH₂CH₂CH₂NH₂    —CH=NCH₂CH₂CH₂CH₃

b.

CH₃CH₂CH₂CH₂NH₂    =NCH₂CH₂CH₂CH₃

c.

=O    CH₃CH₂CH₂CH₂NH₂    =NCH₂CH₂CH₂CH₃

**21.24**  Remember that the C=NR is formed from a C=O and an NH₂ group of a 1° amine.

a. 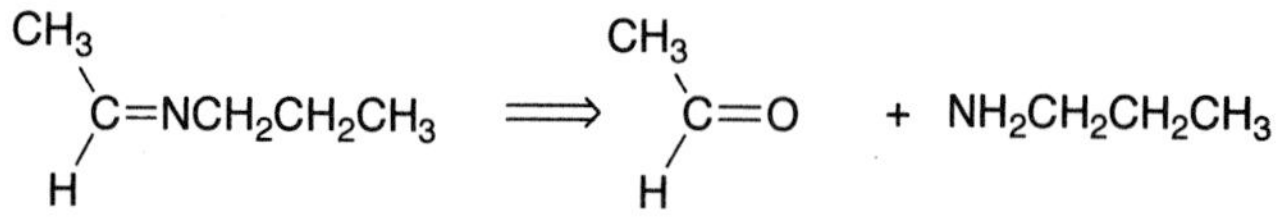 $\underset{H}{\overset{CH_3}{C}}=NCH_2CH_2CH_3 \implies \underset{H}{\overset{CH_3}{C}}=O + NH_2CH_2CH_2CH_3$

b. 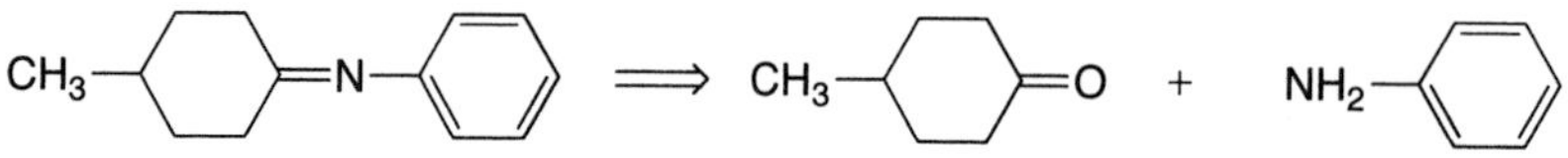 $CH_3-\bigcirc=N-\bigcirc \implies CH_3-\bigcirc=O + NH_2-\bigcirc$

**21.25**

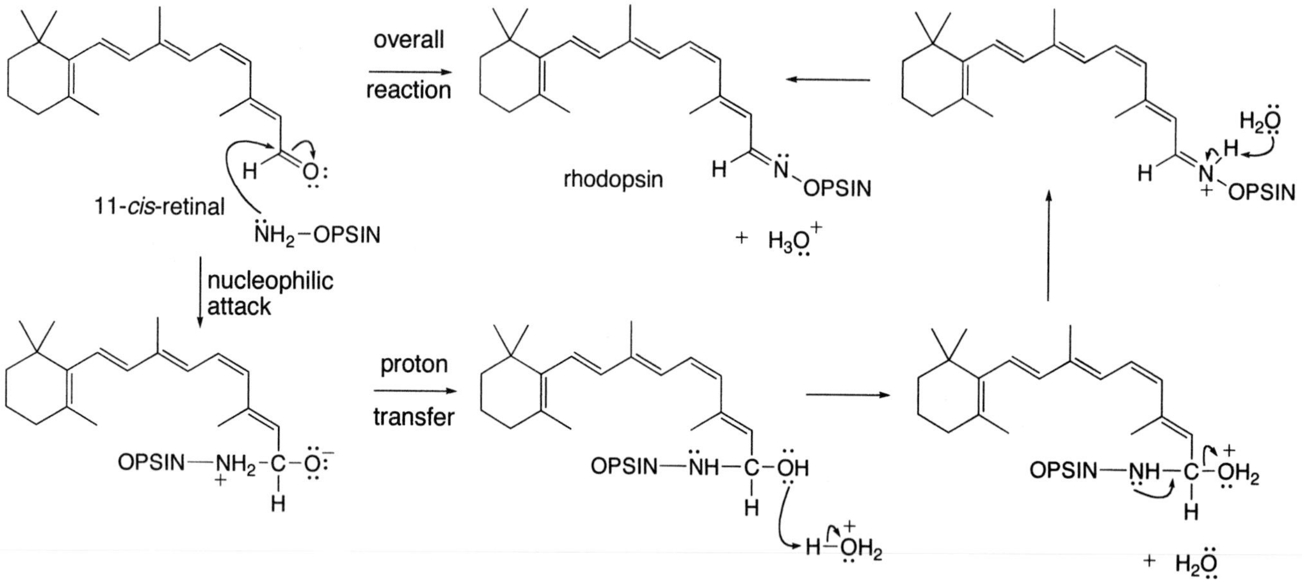

**21.26**

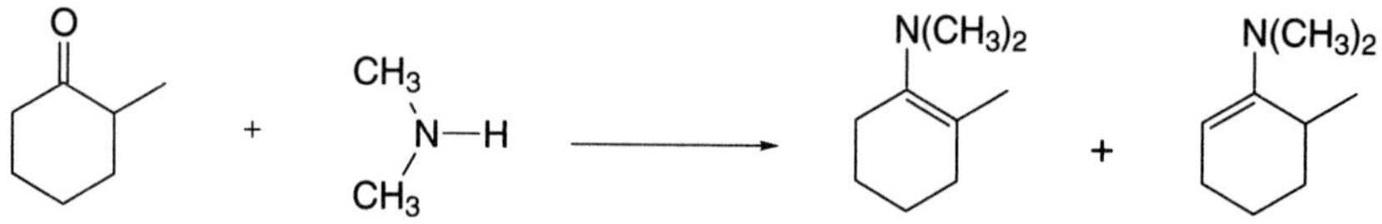

**21.27**

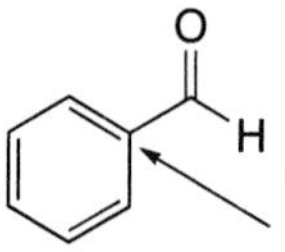

This carbon has four bonds to C's. To make an enamine, it needs a H atom, which is lost as H₂O when the enamine is formed.

**21.28**  • Imines are hydrolyzed to 1° amines and a carbonyl compound.
       • Enamines are hydrolyzed to 2° amines and a carbonyl compound.

a.

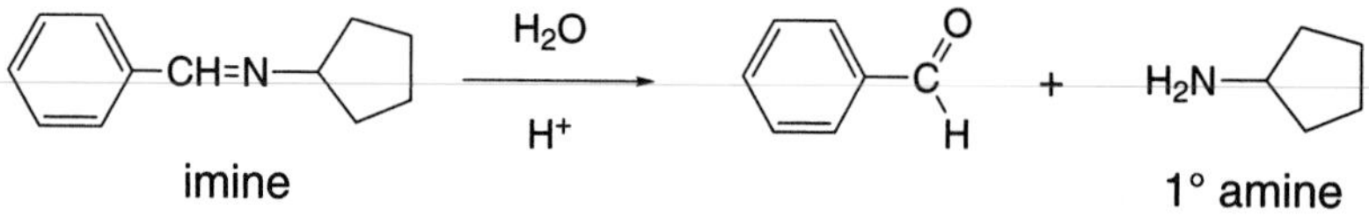

b.

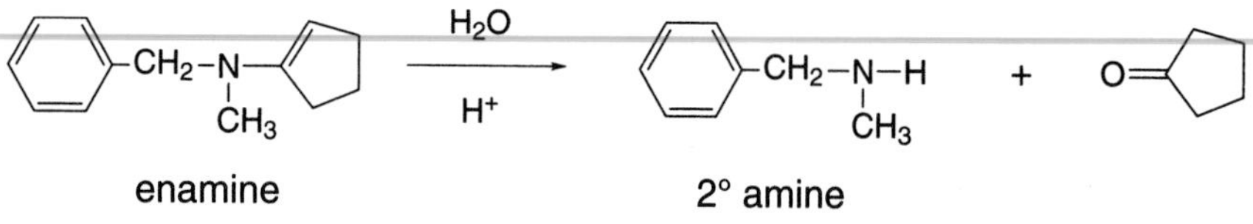

c.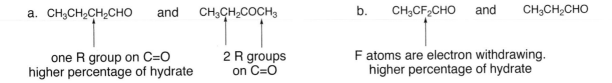

**21.29**
- A substituent that **donates** electron density to the carbonyl C stabilizes it, **decreasing** the percentage of hydrate at equilibrium.
- A substituent that **withdraws** electron density from the carbonyl C destabilizes it, **increasing** the percentage of hydrate at equilibrium.

a. $CH_3CH_2CH_2CHO$ and $CH_3CH_2COCH_3$

    one R group on C=O    2 R groups
 higher percentage of hydrate   on C=O

b.   $CH_3CF_2CHO$ and $CH_3CH_2CHO$

F atoms are electron withdrawing.
higher percentage of hydrate

**21.30** Electron-donating groups decrease the amount of hydrate at equilibrium by stabilizing the carbonyl starting material. Electron-withdrawing groups increase the amount of hydrate at equilibrium by destabilizing the carbonyl starting material.

$CH_3$—⟨benzene⟩—CHO     ⟨benzene⟩—CHO     NC—⟨benzene⟩—CHO

electron donating                      electron withdrawing

→ Increasing percentage of hydrate at equilibrium

**21.31**

⟨reaction mechanism scheme⟩

**21.32** Treatment of an aldehyde or ketone with two equivalents of alcohol results in the formation of an acetal (a C bonded to 2 OR groups).

a. ⟨cyclopentanone⟩=O + 2 CH₃OH →(TsOH)→ ⟨cyclopentane⟩ with OCH₃, OCH₃

b. ⟨ketone⟩ + HO–CH₂CH₂–OH →(TsOH)→ ⟨dioxolane⟩

**21.33**

a. ⟨cyclohexane with OCH₃, OCH₃ on different carbons⟩

2 OR groups
on different C's.
**2 ethers**

b. ⟨cyclohexane with OCH₃, OCH₃ on same carbon⟩

2 OR groups
on same C.
**acetal**

c. ⟨bicyclic dioxolane with CH₃, CH₃⟩

2 OR groups
on same C.
**acetal**

d. ⟨chain with OCH₃ and OH on same carbon⟩

1 OR group and
1 OH group
on same C.
**hemiacetal**

**21.34** The mechanism has two parts: [1] nucleophilic addition of ROH to form a hemiacetal; [2] conversion of the hemiacetal to an acetal.

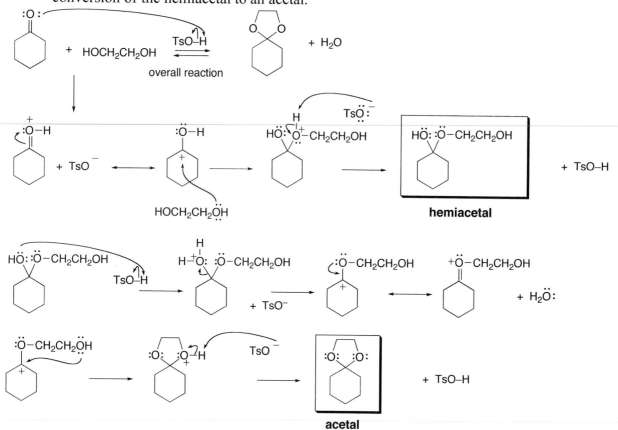

**21.35**

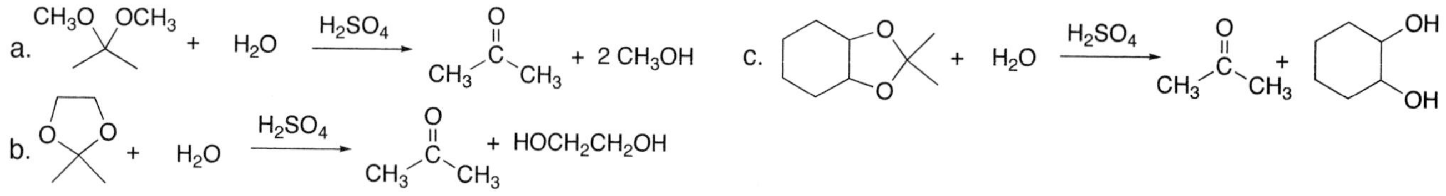

a. CH₃O  OCH₃  +  H₂O  →(H₂SO₄)→  CH₃–C(=O)–CH₃  + 2 CH₃OH

c. +  H₂O  →(H₂SO₄)→  CH₃–C(=O)–CH₃  +  cyclohexane-1,2-diol

b. +  H₂O  →(H₂SO₄)→  CH₃–C(=O)–CH₃  + HOCH₂CH₂OH

**21.36**

safrole  →(H₂O / H₂SO₄)→  formaldehyde (H–C(=O)–H)  +

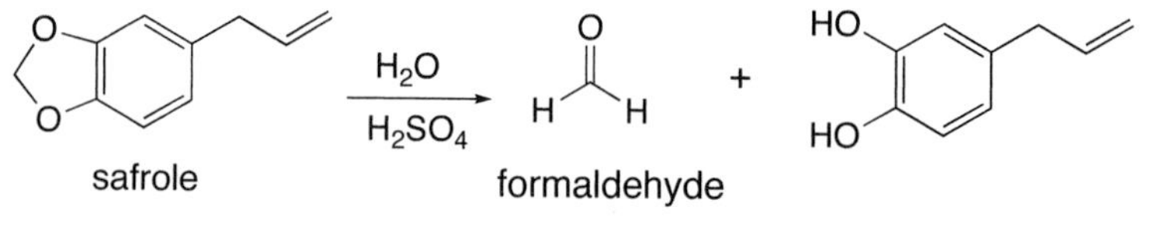

**21.37** Use an acetal protecting group to carry out the reaction.

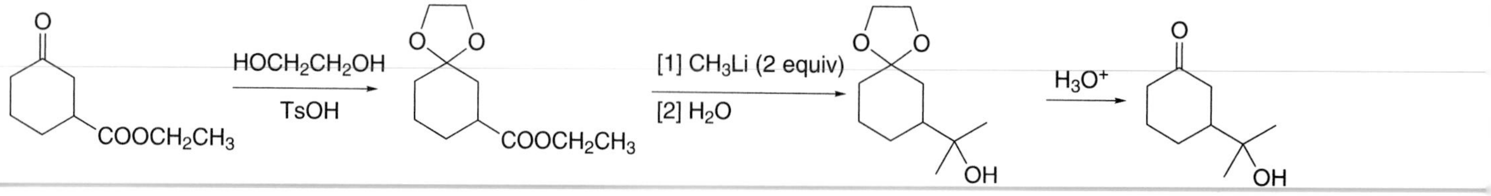

HOCH₂CH₂OH / TsOH  →  [1] CH₃Li (2 equiv) / [2] H₂O  →  H₃O⁺  →

**21.38**

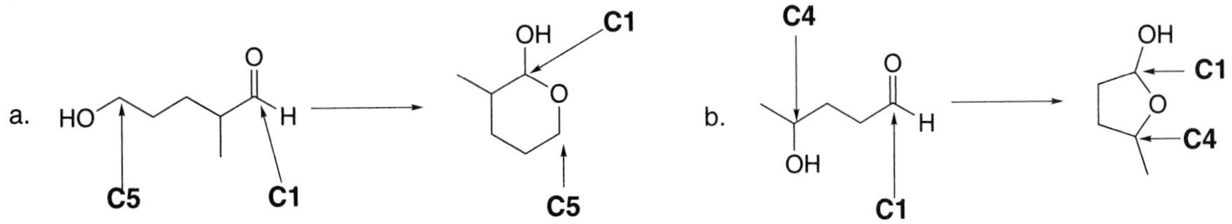

a.

b.

**21.39** The hemiacetal OH is replaced by an OR group to form an acetal.

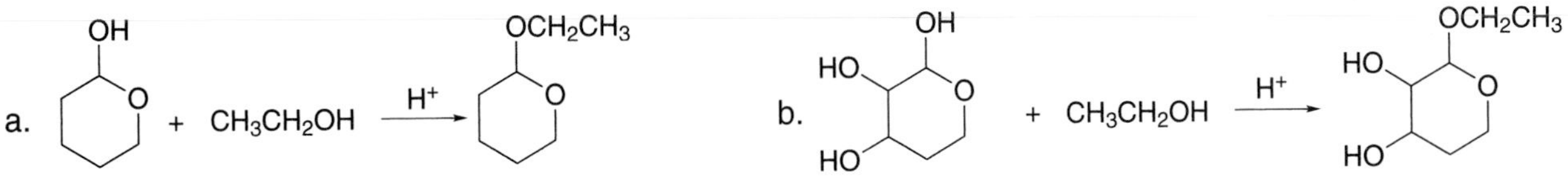

a.

b.

**21.40**

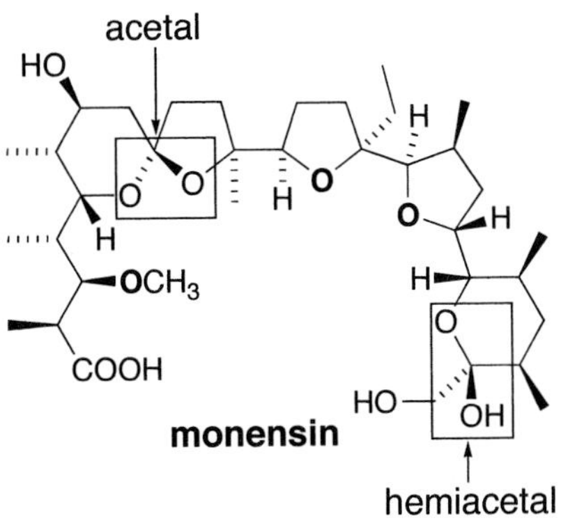

**monensin**

Ether **O** atoms are indicated in **bold.**

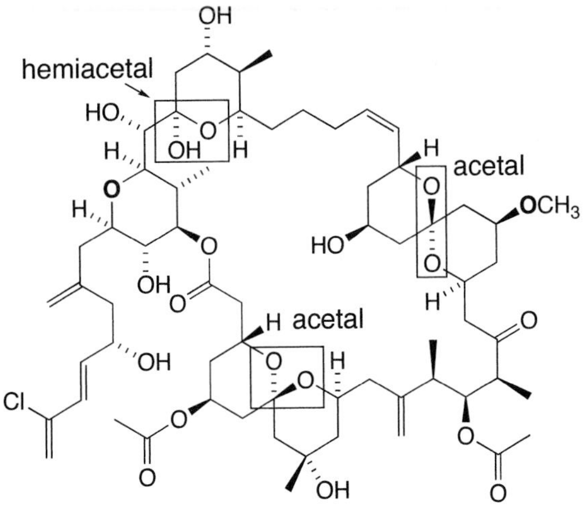

**spongistatin 1**

**21.41** β-D-Glucose has five stereogenic centers. α- and β-D-Glucose are stereoisomers (diastereomers). **A** and β-D-glucose are constitutional isomers.

**21.42**

a. 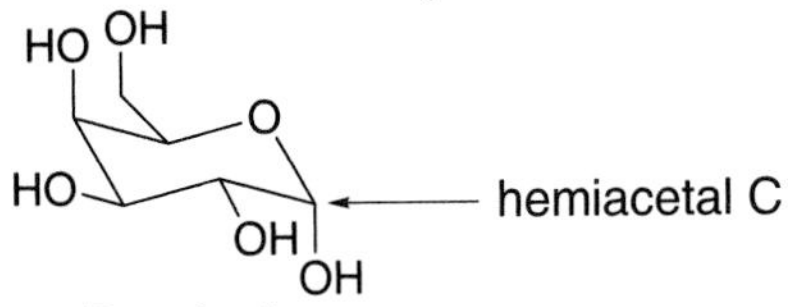
α-D-galactose — hemiacetal C

b. 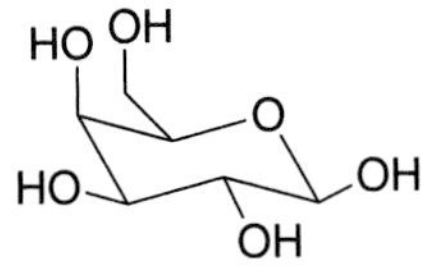
β-D-galactose

c.

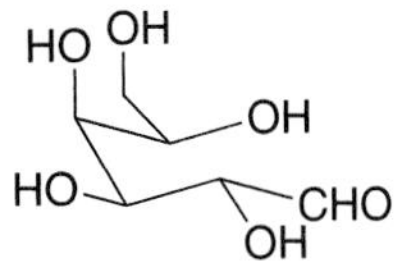

d.

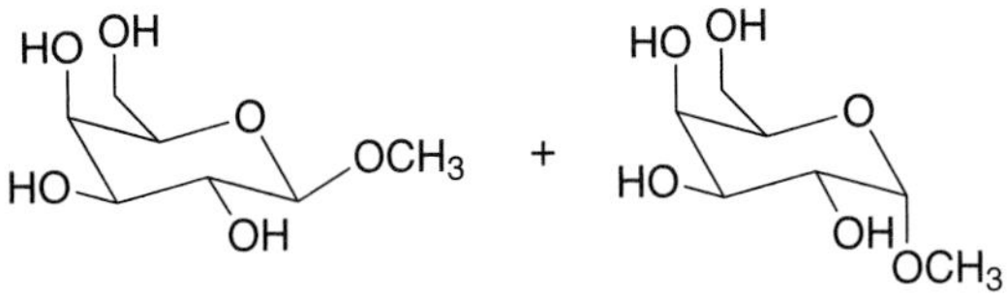

**21.43** Use the rules from Answer 21.3 and 21.5 to name the aldehydes and ketones.

a. $(CH_3)_3CCH_2CHO$ $\xrightarrow{\text{re-draw}}$

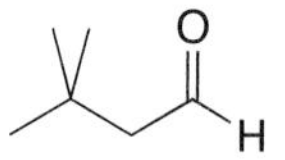

4 C = butanal
**3,3-dimethybutanal**

h. 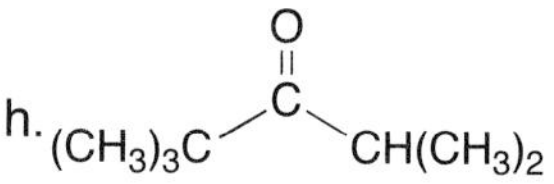 $(CH_3)_3C$—C(=O)—$CH(CH_3)_2$ $\xrightarrow{\text{re-draw}}$

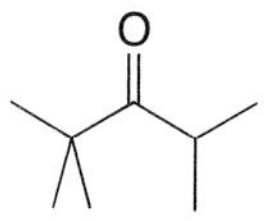

5 C = pentanone
**2,2,4-trimethyl-
3-pentanone**
(common name:
***tert*-butyl isopropyl ketone**)

b.

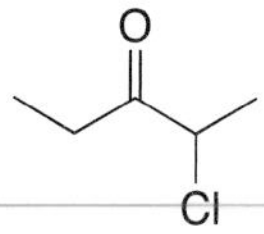

5 C = pentanone
**2-chloro-3-pentanone**

c. 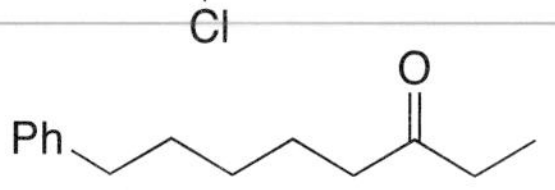 Ph

8 C = octanone
**8-phenyl-3-octanone**

d.

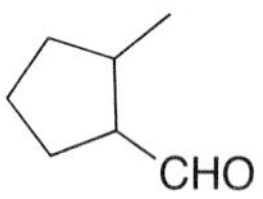

5 C ring
**2-methyl-
cyclopentanecarbaldehyde**

i.

***o*-nitroacetophenone**

e.

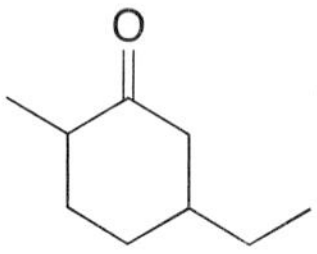

6 C ring = cyclohexanone
**5-ethyl-2-methyl-
cyclohexanone**

j.

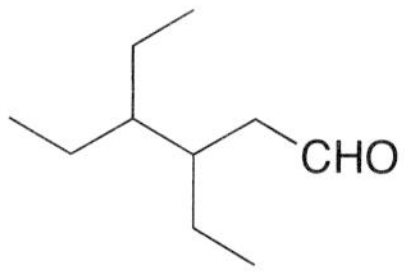

6 C = hexanal **3,4-diethylhexanal**

f. 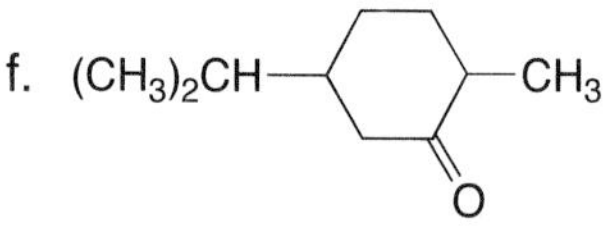 $(CH_3)_2CH$— —$CH_3$

6 C ring = cyclohexanone
**5-isopropyl-2-methyl-
cyclohexanone**

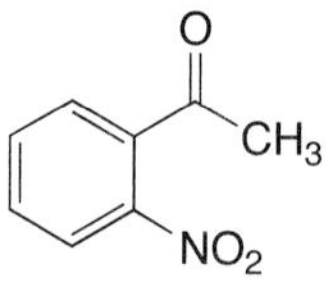

k.

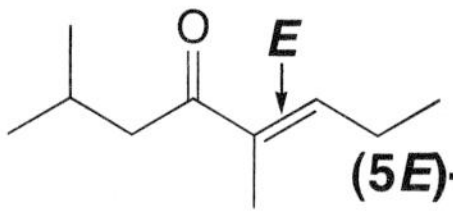

8 C = octenone
**(5*E*)-2,5-dimethyl-5-octen-4-one**

g.

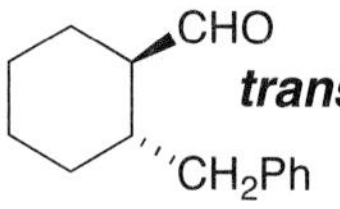

***trans*-2-benzylcyclohexanecarbaldehyde**

l.

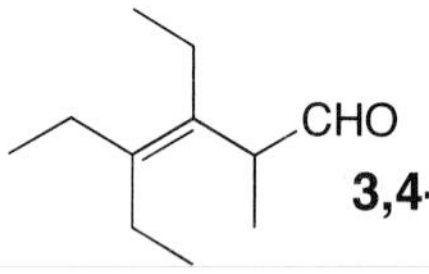

6 C = hexenal
**3,4-diethyl-2-methyl-3-hexenal**

**21.44**

a. 2-methyl-3-phenylbutanal

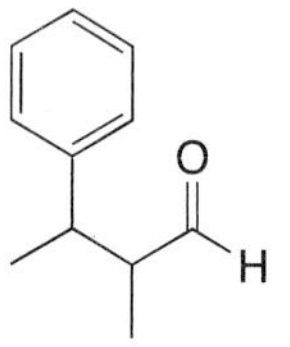

b. dipropyl ketone

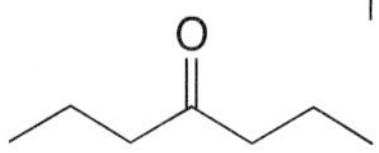

c. 3,3-dimethylcyclohexanecarbaldehyde

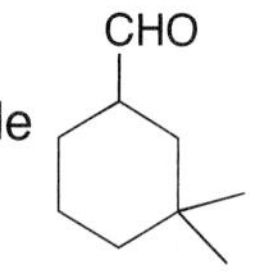

d. α-methoxypropionaldehyde

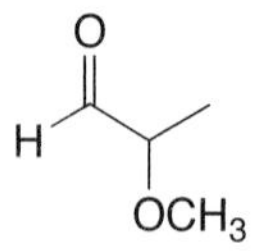

e. 3-benzoylcyclopentanone

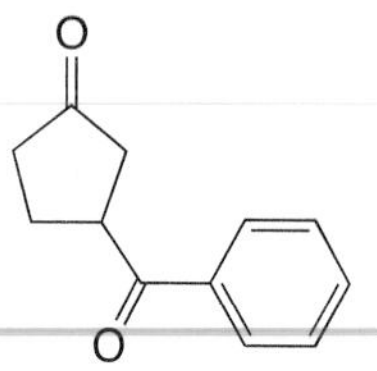

f. 2-formylcyclopentanone

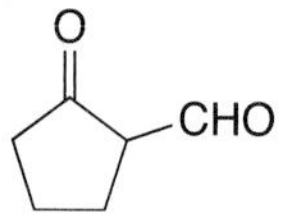

g. (*R*)-3-methyl-2-heptanone

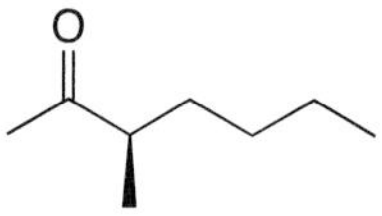

h. *m*-acetylbenzaldehyde

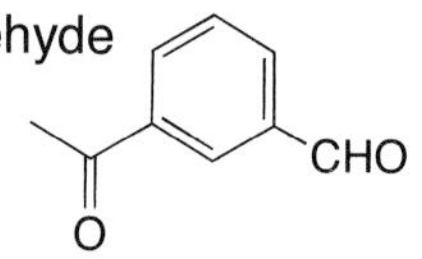

i. 2-*sec*-butyl-3-cyclopentenone

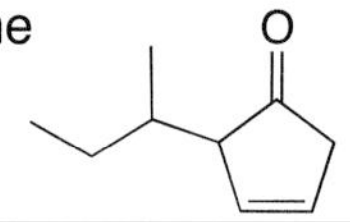

j. 5,6-dimethyl-1-cyclohexenecarbaldehyde

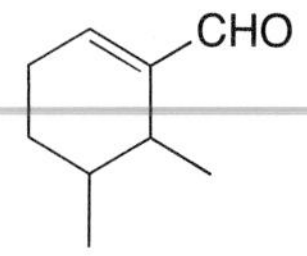

**21.45**

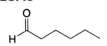

hexanal    2-methylpentanal    3-methylpentanal    4-methylpentanal    3,3-dimethylbutanal

2-ethylbutanal    2,2-dimethylbutanal    2,3-dimethylbutanal    3-hexanone    3,3-dimethyl-2-butanone

2-hexanone    3-methyl-2-pentanone    4-methyl-2-pentanone    2-methyl-3-pentanone

**21.46**

 = $C_6H_5CH_2CHO$    phenylacetaldehyde

a.  $\underrightarrow{NaBH_4,\ CH_3OH}$    $C_6H_5CH_2CH_2OH$

b.  $\underrightarrow{[1]\ LiAlH_4;\ [2]\ H_2O}$    $C_6H_5CH_2CH_2OH$

c.  $\underrightarrow{[1]\ CH_3MgBr;\ [2]\ H_2O}$    $C_6H_5CH_2CH(OH)CH_3$

d.  $\underrightarrow{NaCN,\ HCl}$    $C_6H_5CH_2CH(OH)CN$

e.  $Ph_3P=CHCH_3$    $C_6H_5CH_2CH=CHCH_3$
(E and Z isomers)

f.  $\underrightarrow{(CH_3)_2CHNH_2}$  mild acid

g.  $\underrightarrow{(CH_3CH_2)_2NH,\ mild\ acid}$    (E and Z isomers)

h.  $\underrightarrow{CH_3CH_2OH\ (excess),\ H^+}$

i.  $\underrightarrow{\text{piperidine NH}}$  mild acid    (E and Z isomers)

j.  $\underrightarrow{HO\text{—}OH,\ H^+}$

**21.47**

2-butanone

a.  $\underrightarrow{NaBH_4,\ CH_3OH}$

b.  $\underrightarrow{[1]\ LiAlH_4;\ [2]\ H_2O}$

c.  $\underrightarrow{[1]\ CH_3MgBr;\ [2]\ H_2O}$

d.  $\underrightarrow{NaCN,\ HCl}$

e.  $\underrightarrow{Ph_3P=CHCH_3}$    (E and Z isomers)

f.  $\underrightarrow{(CH_3)_2CHNH_2,\ mild\ H^+}$

g. $(CH_3CH_2)_2NH$, mild $H^+$ → $\overset{N(CH_2CH_3)_2}{\diagup}$ + $\overset{N(CH_2CH_3)_2}{\diagup}$

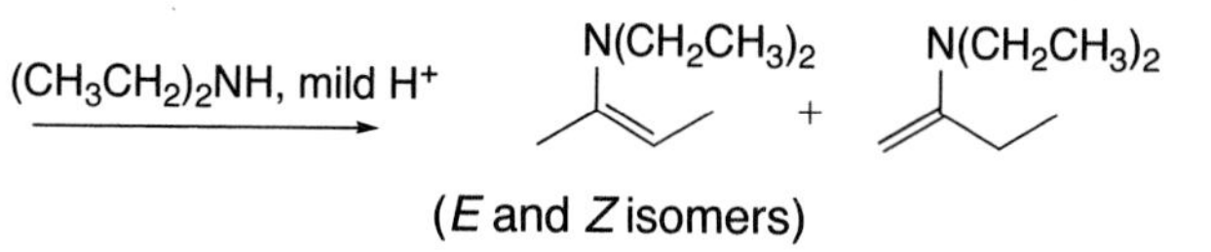

(*E* and *Z* isomers)

i. [piperidine NH], mild $H^+$ → [N-propenyl piperidine] + [N-butenyl piperidine]

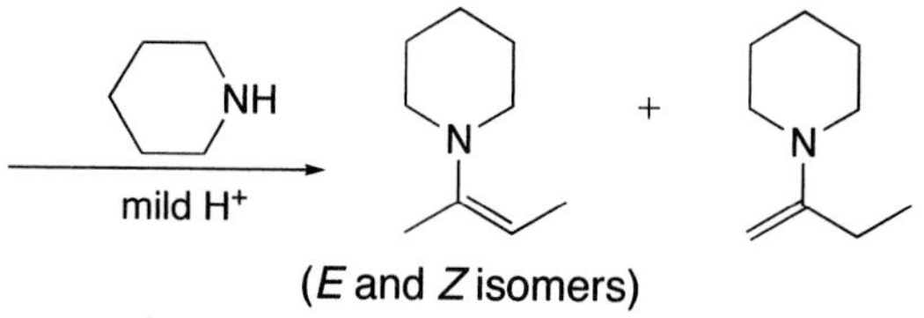

(*E* and *Z* isomers)

h. $CH_3CH_2OH$ (excess), $H^+$ → $CH_3CH_2O\;OCH_2CH_3$ [acetal structure]

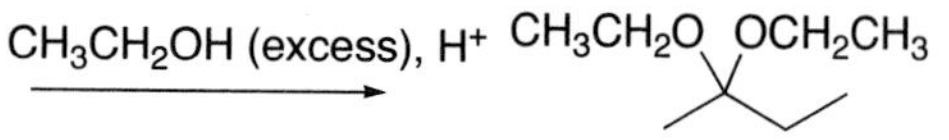

j. $HO\frown OH$, $H^+$ → [dioxolane structure]

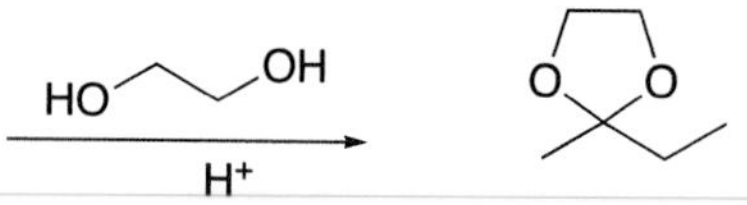

## 21.48

a. [cyclopentanone] $=O$ $\xrightarrow{Ph_3P=CHCH_2CH_3}$ [cyclopentylidene] $=CHCH_2CH_3$

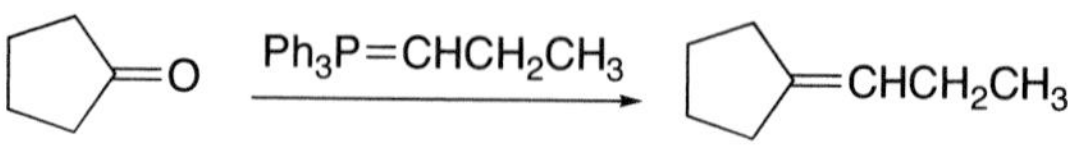

c. [cyclopentane–CHO] $\xrightarrow{Ph_3P=CHCOOCH_3}$ [(Z)-cyclopentyl-CH=CH-COOCH$_3$] + [(E)-cyclopentyl-CH=CH-COOCH$_3$]

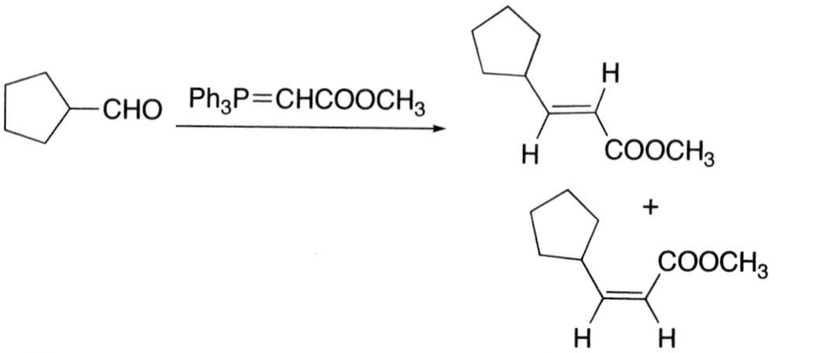

b. [cyclopentane–CHO] $\xrightarrow{Ph_3P=\text{cyclohexylidene}}$ [cyclopentyl–CH=cyclohexylidene]

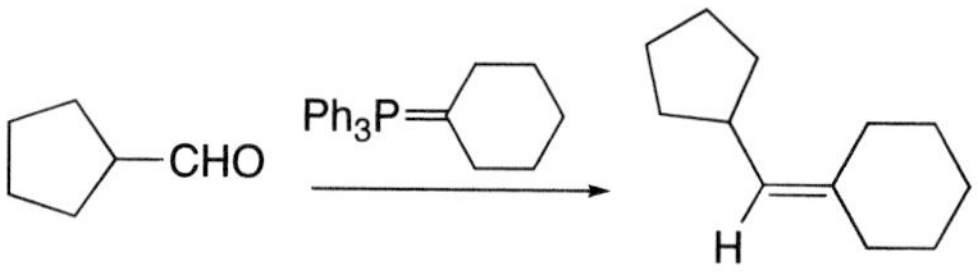

d. [cyclopentanone] $=O$ $\xrightarrow{Ph_3P=CH(CH_2)_5COOCH_3}$ [cyclopentylidene] $=CH(CH_2)_5COOCH_3$

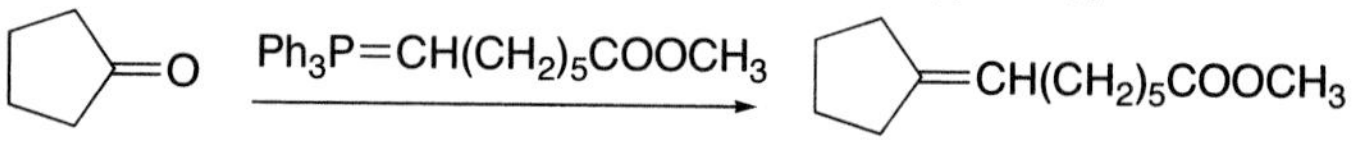

## 21.49

a. $CH_3CH_2Cl$ $\xrightarrow[\text{[2] BuLi}]{\text{[1] Ph}_3\text{P}}$ $CH_3CH=C(CH_3)_2$
[3] $(CH_3)_2C=O$

c. [cyclopentane–$CH_2Cl$] $\xrightarrow[\text{[2] BuLi}]{\text{[1] Ph}_3\text{P}}$ [cyclopentyl–CH=CHCH$_2$CH$_2$CH$_3$]
[3] $CH_3CH_2CH_2CHO$ (*E* and *Z* isomers)

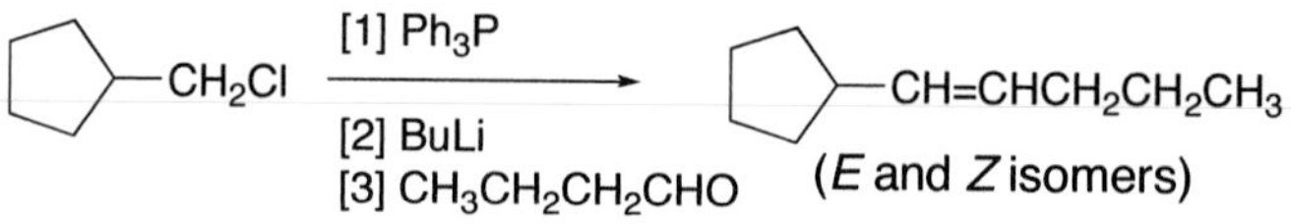

b. [benzene–$CH_2Br$] $\xrightarrow[\text{[2] BuLi}]{\text{[1]Ph}_3\text{P}}$ [phenyl–CH=CHCH$_2$CH$_2$C$_6$H$_5$]
[3] $C_6H_5CH_2CH_2CHO$ (*E* and *Z* isomers)

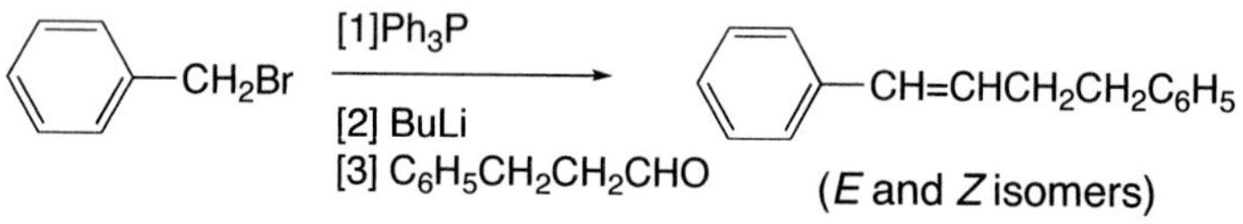

## 21.50

a. $Ph_3P=CHCH_2CH_2CH_3 \Longrightarrow BrCH_2CH_2CH_2CH_3$

c. $Ph_3P=CHCH=CH_2 \Longrightarrow BrCH_2CH=CH_2$

b. $Ph_3P=C(CH_2CH_2CH_3)_2 \Longrightarrow BrCH(CH_2CH_2CH_3)_2$

**21.51**

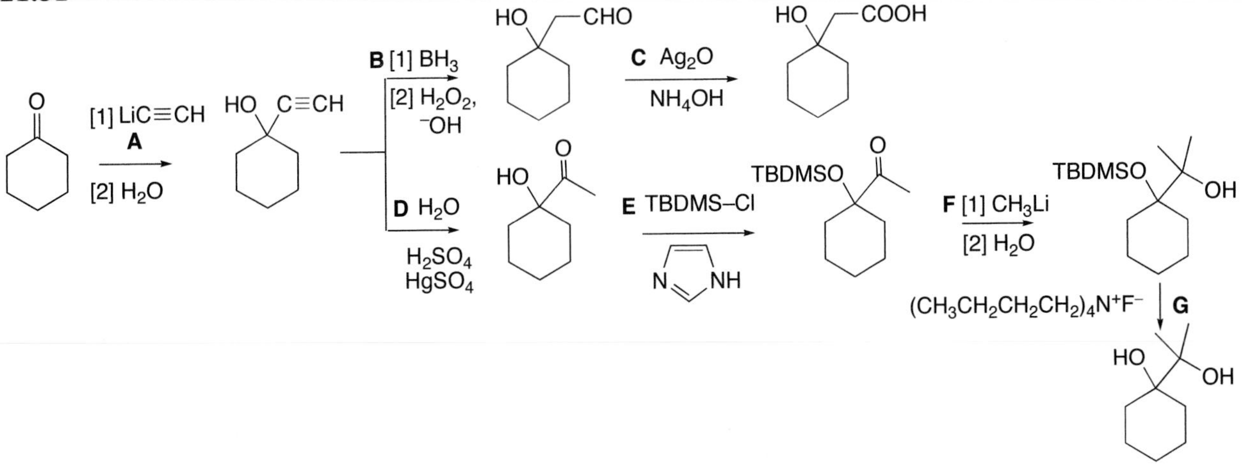

[1] LiC≡CH  **A**  [2] H₂O

**B** [1] BH₃  [2] H₂O₂, ⁻OH

**C** Ag₂O / NH₄OH

**D** H₂O / H₂SO₄ / HgSO₄

**E** TBDMS–Cl / imidazole (N≡NH)

**F** [1] CH₃Li  [2] H₂O

(CH₃CH₂CH₂CH₂)₄N⁺F⁻  **G**

**21.52**

a. 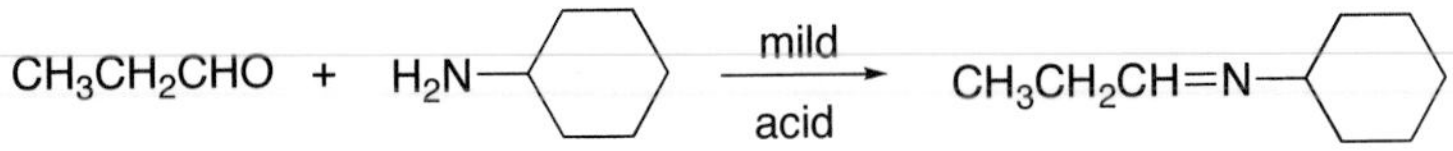  CH₃CH₂CHO + H₂N–cyclohexyl  →(mild acid)→  CH₃CH₂CH=N–cyclohexyl

b. 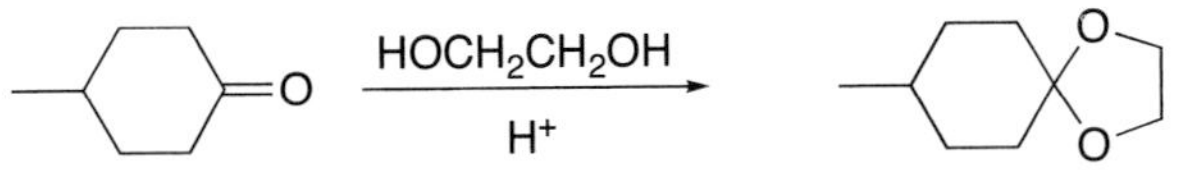  4-methylcyclohexanone + HOCH₂CH₂OH / H⁺ → dioxolane product

c. 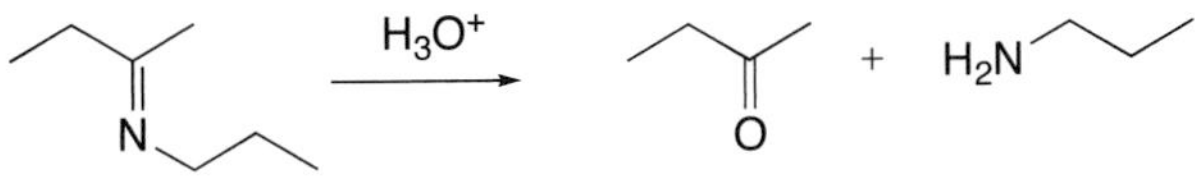  ketimine + H₃O⁺ → ketone + H₂N–CH₂CH₂CH₃

d. 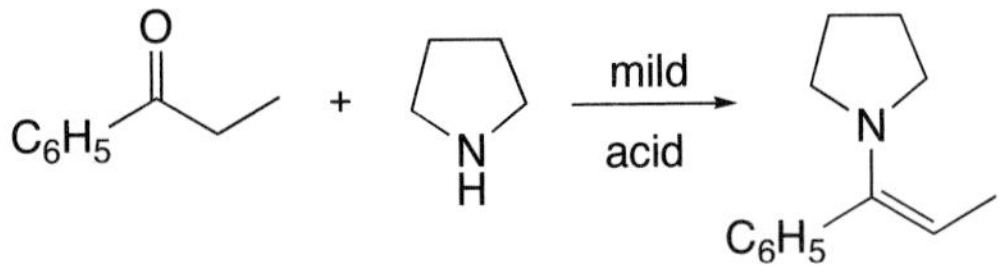  C₆H₅ ketone + pyrrolidine →(mild acid)→ enamine (E and Z isomers)

e. 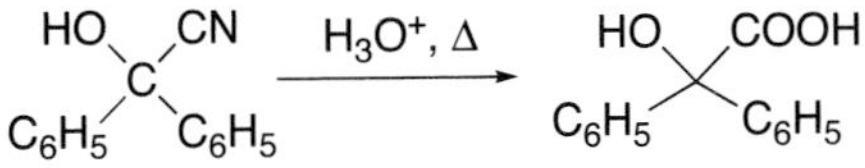  HO–C(CN)(C₆H₅)(C₆H₅)  →(H₃O⁺, Δ)→  HO–C(COOH)(C₆H₅)(C₆H₅)

f. 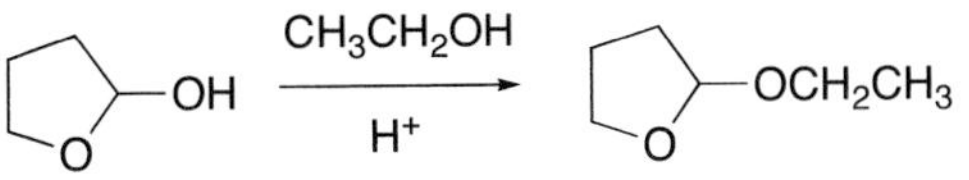  tetrahydrofuran-2-ol + CH₃CH₂OH / H⁺ → 2-OCH₂CH₃ tetrahydrofuran

g. 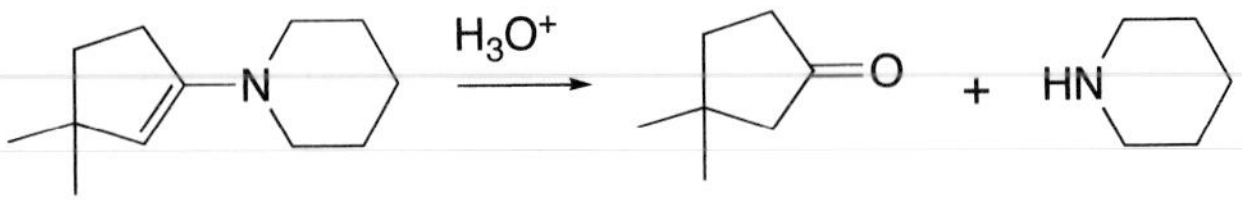  enamine + H₃O⁺ → cyclopentanone derivative + HN–piperidine

h.  CH₃O–cyclohexyl–(OCH₃)₂ + H₃O⁺ → CH₃O–cyclohexanone + HOCH₃

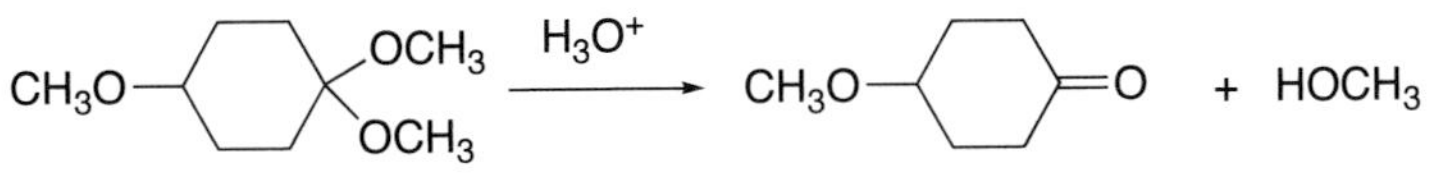

**21.53**

a. 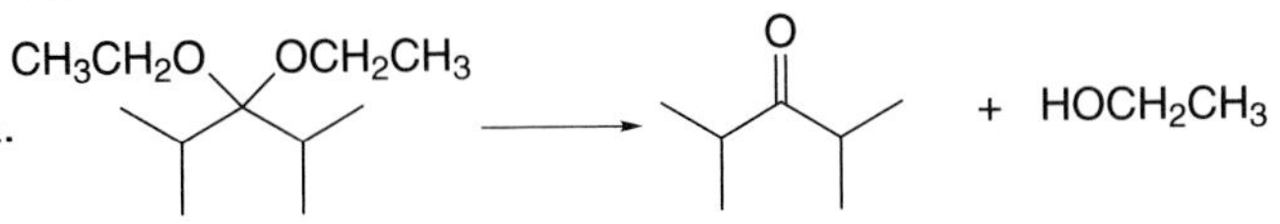  CH₃CH₂O–C(OCH₂CH₃)(iPr)(iPr) → ketone + HOCH₂CH₃

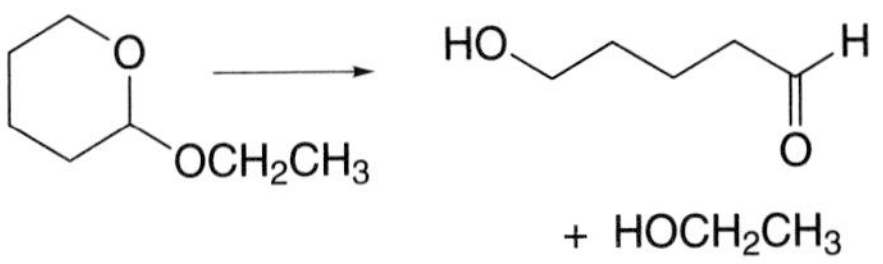

b. 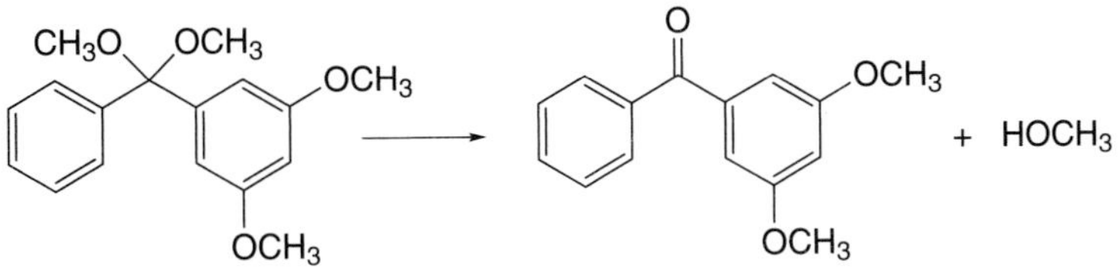  CH₃O–C(OCH₃)(C₆H₅)(aryl-OCH₃) → benzophenone derivative + HOCH₃

c.  tetrahydropyran-2-OCH₂CH₃ → HO–(CH₂)₃–CHO + HOCH₂CH₃

**21.54** Consider para product only, when an ortho, para mixture can result.

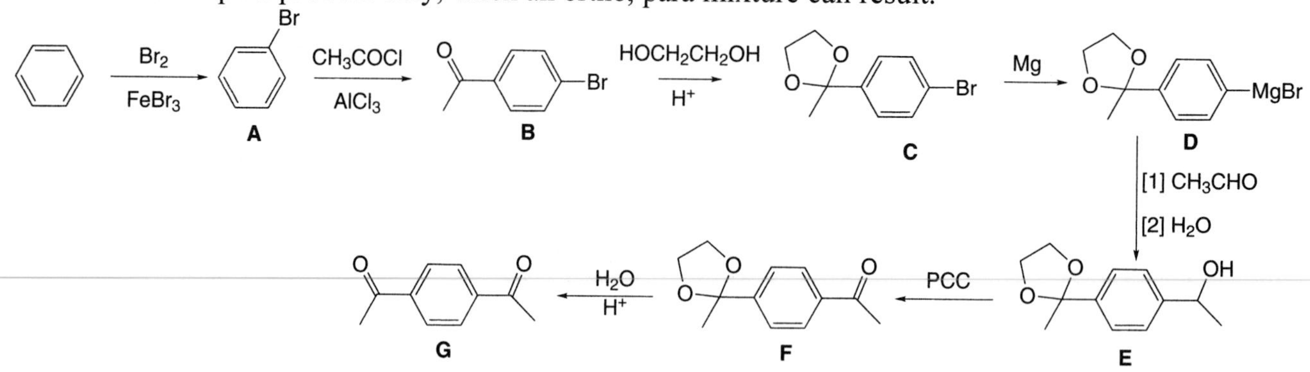
Br
Br2
FeBr3
A
CH3COCl
AlCl3
O
Br
B
HOCH2CH2OH
H+
O
O
Br
C
Mg
MgBr
D
[1] CH3CHO
[2] H2O
OH
E
PCC
F
H2O
H+
G

**21.55**

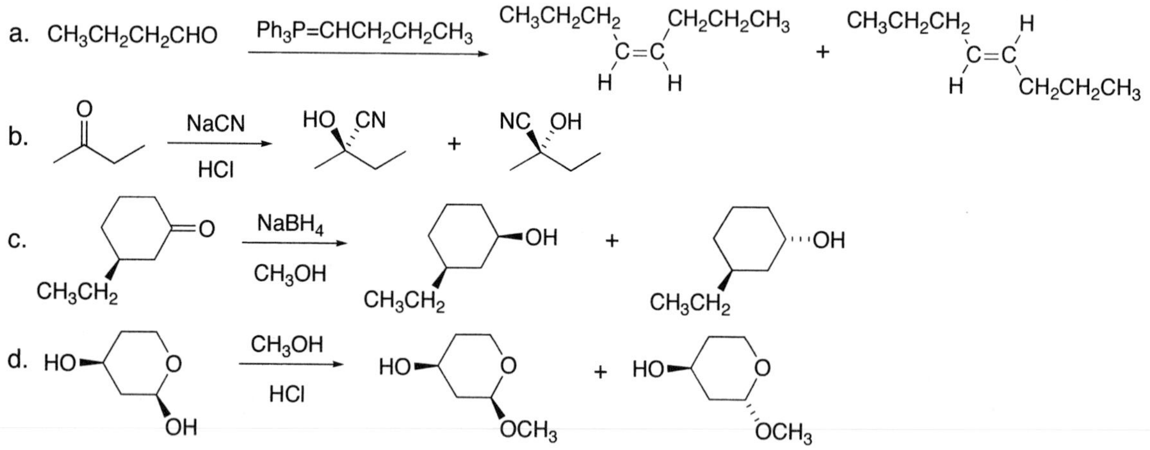
a. CH3CH2CH2CHO
Ph3P=CHCH2CH2CH3
CH3CH2CH2 CH2CH2CH3 C=C H H
+
CH3CH2CH2 H C=C H CH2CH2CH3
b.
NaCN
HCl
HO CN
+
NC OH
c.
NaBH4
CH3OH
CH3CH2
OH
+
OH
d. HO
OH
CH3OH
HCl
OCH3
+
OCH3

**21.56**

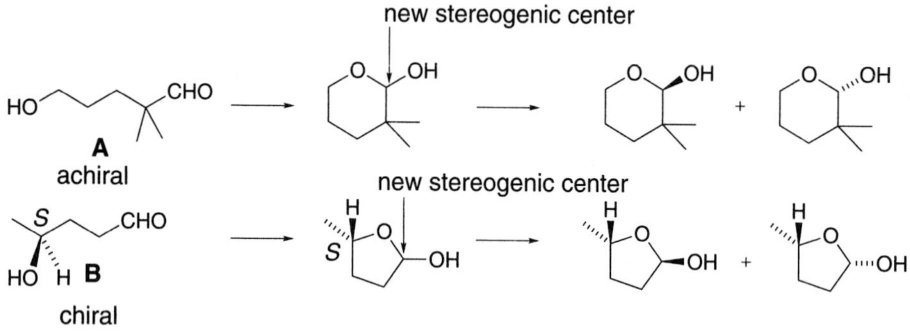
new stereogenic center
HO
CHO
A
achiral
OH
+
new stereogenic center
S
CHO
HO H B
chiral
H
S
OH
+

An equal mixture of enantiomers results, so the product is optically inactive.

A mixture of diastereomers results. Both compounds are chiral and they are not enantiomers, so the mixture is optically active.

**21.57**

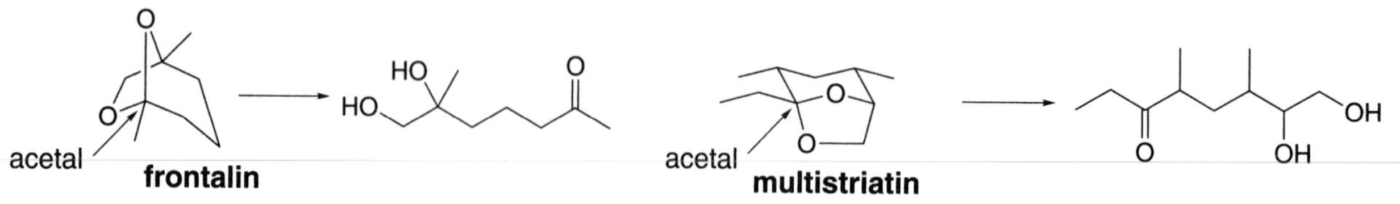
acetal
frontalin
HO
HO
O
acetal
multistriatin
O
OH
OH

**21.58** As the number of R groups bonded to the carbonyl C increases, stability also increases due to electron-donating effects. Then use the principles from Answer 21.29.

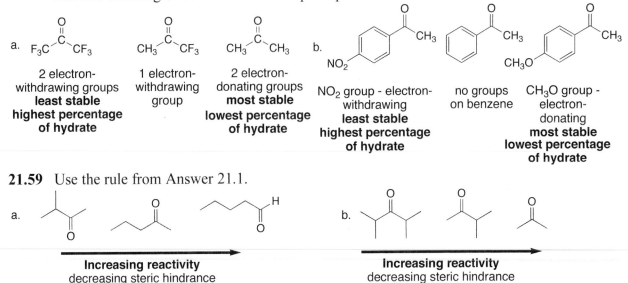

a.

F₃C–C(=O)–CF₃

2 electron-withdrawing groups
**least stable**
**highest percentage of hydrate**

CH₃–C(=O)–CF₃

1 electron-withdrawing group

CH₃–C(=O)–CH₃

2 electron-donating groups
**most stable**
**lowest percentage of hydrate**

b.

NO₂ group - electron-withdrawing
**least stable**
**highest percentage of hydrate**

no groups on benzene

CH₃O group - electron-donating
**most stable**
**lowest percentage of hydrate**

**21.59** Use the rule from Answer 21.1.

a.

→ **Increasing reactivity**
decreasing steric hindrance

b.

→ **Increasing reactivity**
decreasing steric hindrance

**21.60**

Less stable carbonyl compounds give a higher percentage of hydrate. Cyclopropanone is an unstable carbonyl compound because the bond angles around the carbonyl carbon deviate considerably from the desired angle. Since the carbonyl carbon is *sp²* hybridized, the optimum bond angle is 120°, but the three-membered ring makes the C–C–C bond angles only 60°. This destabilizes the ketone, giving a high concentration of hydrate when dissolved in H₂O.

**21.61** Use the principles from Answer 21.21.

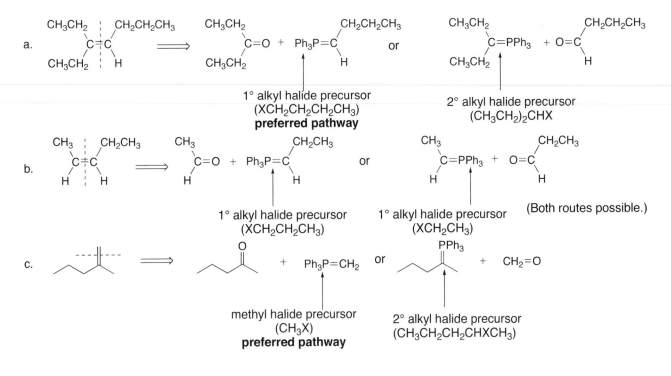

a.

CH₃CH₂    CH₂CH₂CH₃
    C=C
CH₃CH₂    H    ⟹

CH₃CH₂
    C=O  +  Ph₃P=C
CH₃CH₂           H    CH₂CH₂CH₃

or

CH₃CH₂
    C=PPh₃  +  O=C
CH₃CH₂           H    CH₂CH₂CH₃

1° alkyl halide precursor
(XCH₂CH₂CH₃)
**preferred pathway**

2° alkyl halide precursor
(CH₃CH₂)₂CHX

b.

CH₃    CH₂CH₃
   C=C
 H    H    ⟹

CH₃
   C=O  +  Ph₃P=C
 H           H    CH₂CH₃

or

CH₃
   C=PPh₃  +  O=C
 H           H    CH₂CH₃

1° alkyl halide precursor
(XCH₂CH₂CH₃)

1° alkyl halide precursor
(XCH₂CH₃)

(Both routes possible.)

c.

⟹

+  Ph₃P=CH₂

or

PPh₃

+  CH₂=O

methyl halide precursor
(CH₃X)
**preferred pathway**

2° alkyl halide precursor
(CH₃CH₂CH₂CHXCH₃)

d. 

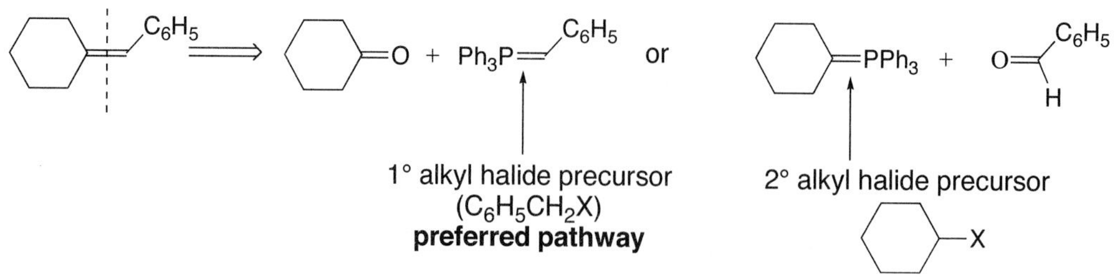

C6H5
O + Ph3P=
C6H5
or
=PPh3 +
O=
C6H5
H
1° alkyl halide precursor
(C6H5CH2X)
preferred pathway
2° alkyl halide precursor
X


## 21.62

a. $(CH_3)_2C{=}NCH_2CH_2CH_2CH_2CH_3 \Longrightarrow (CH_3)_2C{=}O$ + $H_2NCH_2CH_2CH_2CH_2CH_3$

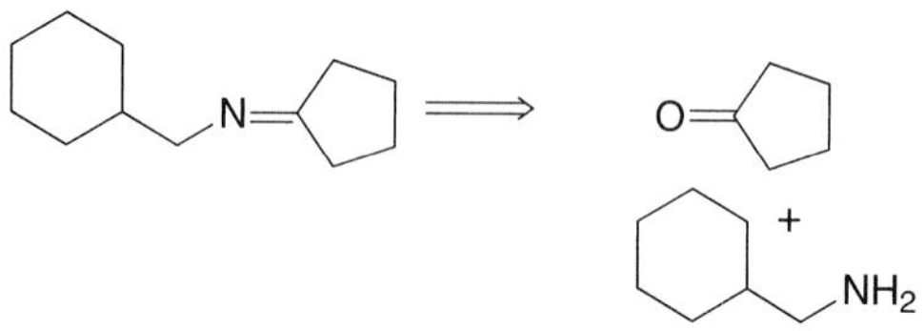

b. 

c. 

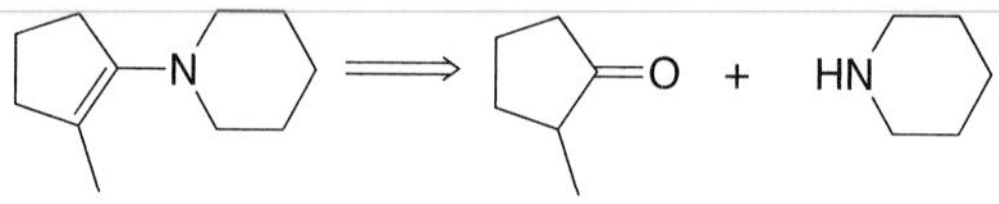

d. 

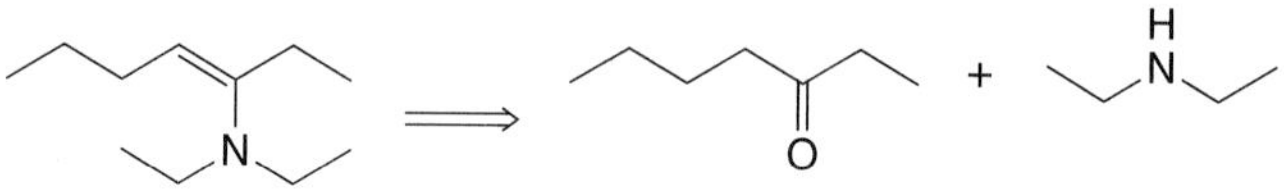

## 21.63

a. $(CH_3CH_2)_2C(OCH_2CH_3)_2 \Longrightarrow (CH_3CH_2)_2C{=}O$ + $HOCH_2CH_3$

b. 

OCH2CH2CH3
OCH2CH2CH3
O

+ $HOCH_2CH_2CH_3$

c. 
+ $HOCH_2CH_2OH$

d. 

CH3O
OCH3
OCH3
CH3O
H
O

+ $HOCH_3$

## 21.64

a. $C_6H_5{-}CH_2OH \xrightarrow{PCC} C_6H_5{-}CHO$

b. $C_6H_5{-}COCl \xrightarrow[\text{[2] } H_2O]{\text{[1] } LiAlH[OC(CH_3)_3]_3} C_6H_5{-}CHO$

c. $C_6H_5{-}COOCH_3 \xrightarrow[\text{[2] } H_2O]{\text{[1] DIBAL-H}} C_6H_5{-}CHO$

d. $C_6H_5{-}COOH \xrightarrow[\text{[2] } H_2O]{\text{[1] } LiAlH_4} C_6H_5{-}CH_2OH \xrightarrow{PCC} C_6H_5{-}CHO$

e. $C_6H_5{-}CH_3 \xrightarrow{KMnO_4} C_6H_5{-}COOH \xrightarrow[\text{[2] } H_2O]{\text{[1] } LiAlH_4} C_6H_5{-}CH_2OH \xrightarrow{PCC} C_6H_5{-}CHO$

f. $C_6H_5{-}CH{=}CH_2 \xrightarrow[\text{[2] Zn, } H_2O]{\text{[1] } O_3} C_6H_5{-}CHO$

g. $C_6H_5{-}CH{=}NCH_2CH_2CH_3 \xrightarrow[H^+]{H_2O} C_6H_5{-}CHO$

h. $C_6H_5{-}CH(OCH_2CH_3)_2 \xrightarrow[H^+]{H_2O} C_6H_5{-}CHO$

## 21.65

a. 

PCC
OH
O


b. 

Cl
O
(CH3)2CuLi
O


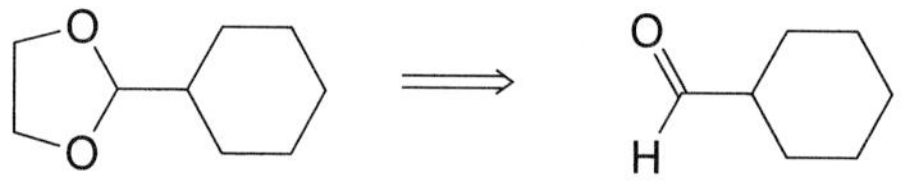

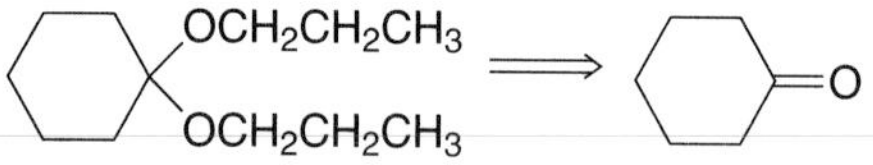

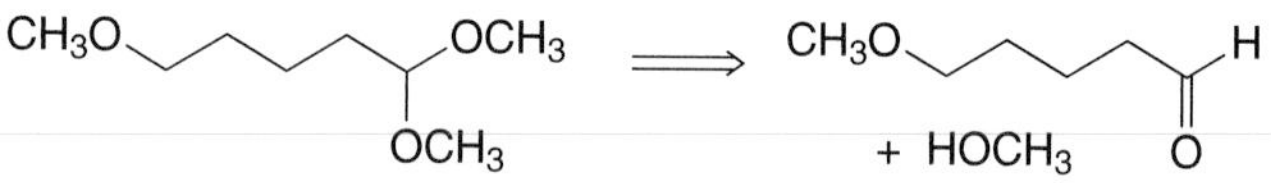

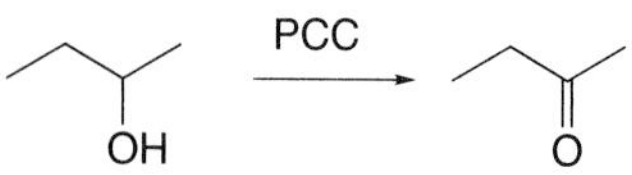

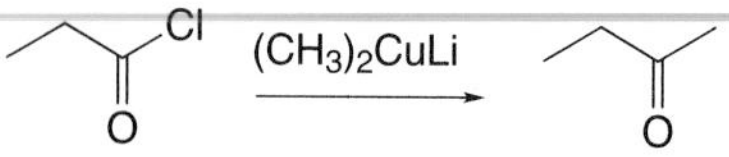

c. $CH_3COCl \xrightarrow{(CH_3CH_2)_2CuLi}$ 

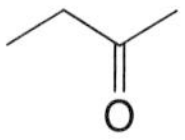

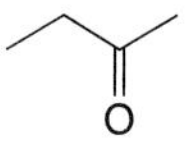

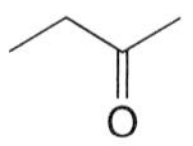

d. $CH_3CH_2C{\equiv}CH \xrightarrow[H_2SO_4,\ HgSO_4]{H_2O}$ 

e. $CH_3C{\equiv}CCH_3 \xrightarrow[H_2SO_4,\ HgSO_4]{H_2O}$

## 21.66

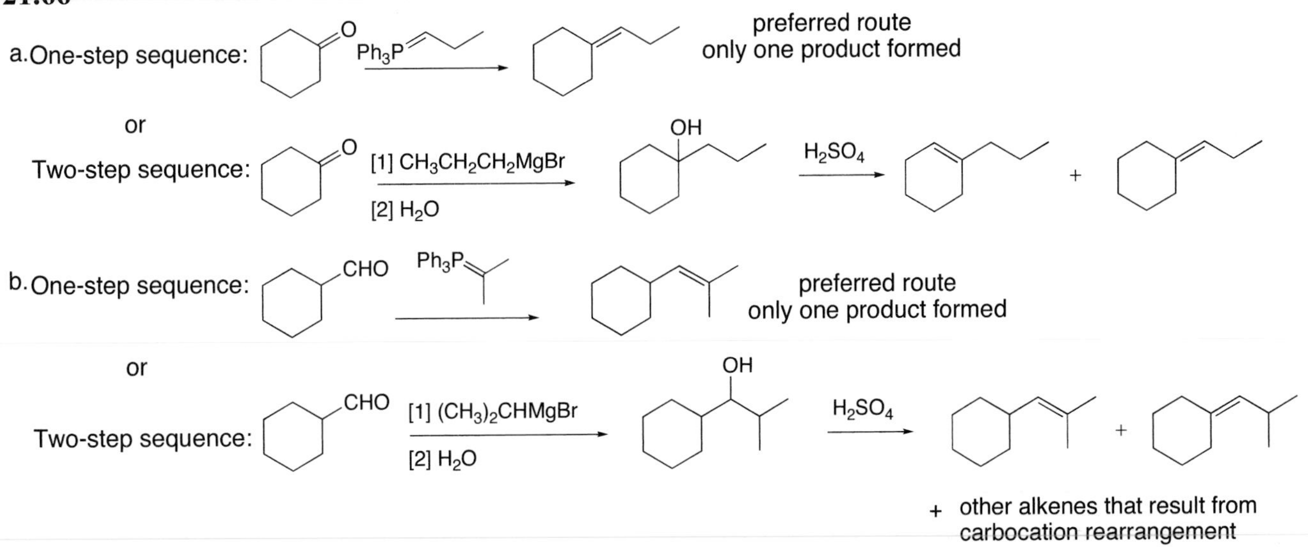
a. One-step sequence:
Ph3P
preferred route
only one product formed
or
Two-step sequence:
[1] CH3CH2CH2MgBr
[2] H2O
OH
H2SO4
+
b. One-step sequence:
CHO
Ph3P
preferred route
only one product formed
or
Two-step sequence:
CHO
[1] (CH3)2CHMgBr
[2] H2O
OH
H2SO4
+
+ other alkenes that result from
carbocation rearrangement

## 21.67

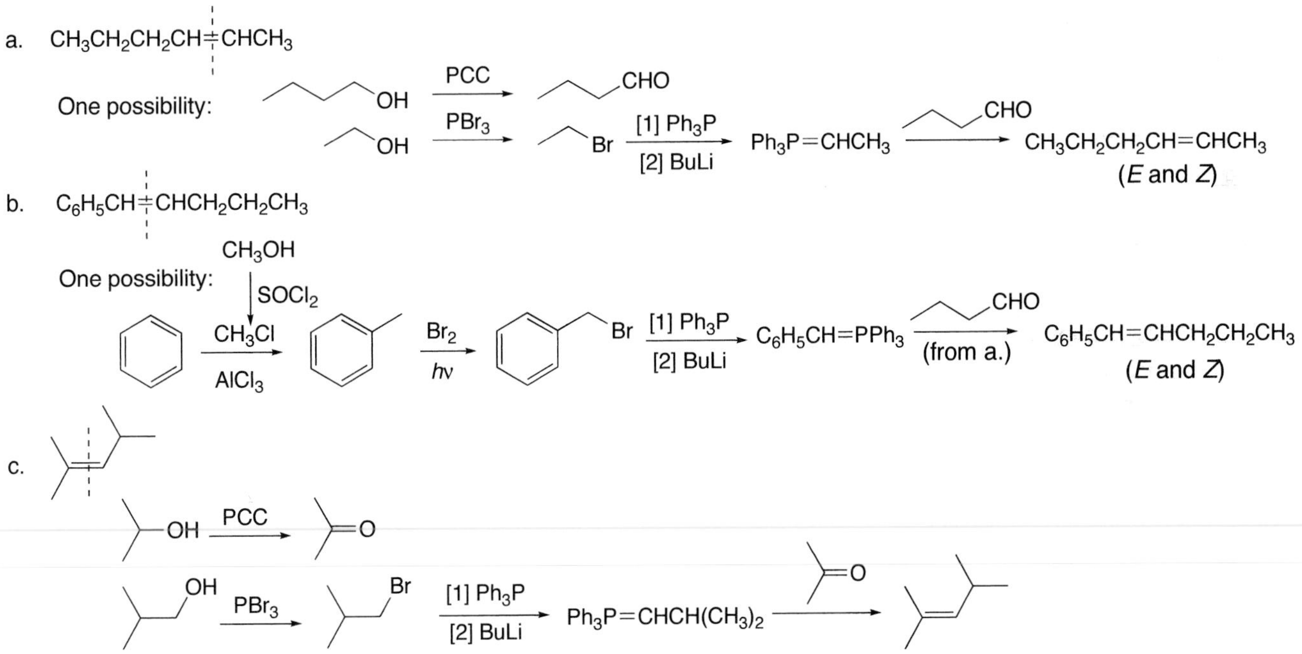
a. CH3CH2CH2CH=CHCH3
One possibility:
OH
PCC
CHO
OH
PBr3
Br
[1] Ph3P
[2] BuLi
Ph3P=CHCH3
CHO
CH3CH2CH2CH=CHCH3
(E and Z)
b. C6H5CH=CHCH2CH2CH3
One possibility:
CH3OH
SOCl2
CH3Cl
AlCl3
Br2
hν
Br
[1] Ph3P
[2] BuLi
C6H5CH=PPh3
CHO
(from a.)
C6H5CH=CHCH2CH2CH3
(E and Z)
c.
OH
PCC
O
OH
PBr3
Br
[1] Ph3P
[2] BuLi
Ph3P=CHCH(CH3)2
O

## 21.68

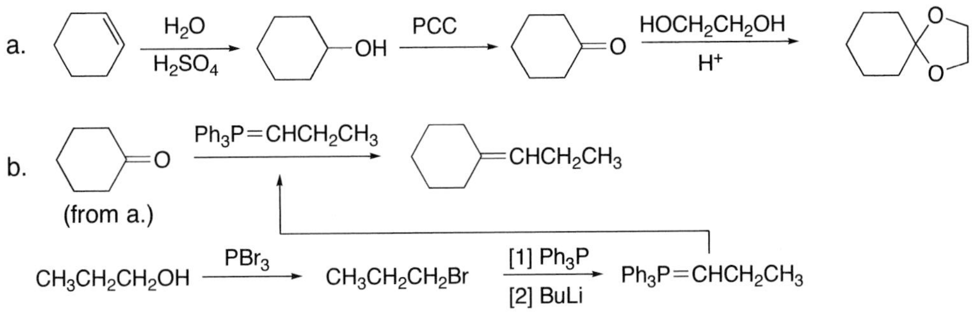
a.
H2O
H2SO4
OH
PCC
O
HOCH2CH2OH
H+
O
O
b.
O
(from a.)
Ph3P=CHCH2CH3
=CHCH2CH3
CH3CH2CH2OH
PBr3
CH3CH2CH2Br
[1] Ph3P
[2] BuLi
Ph3P=CHCH2CH3

c.

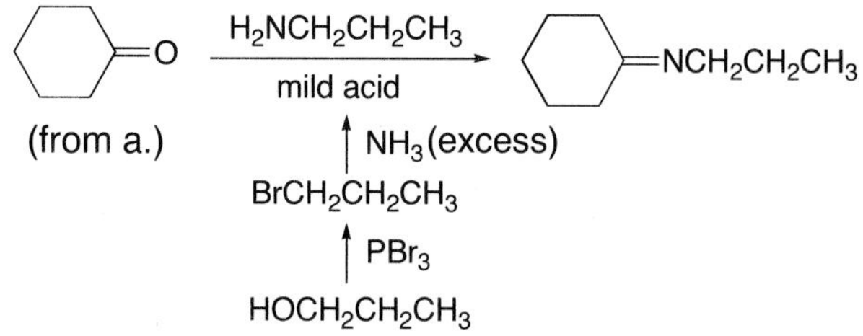

(from a.)

d.

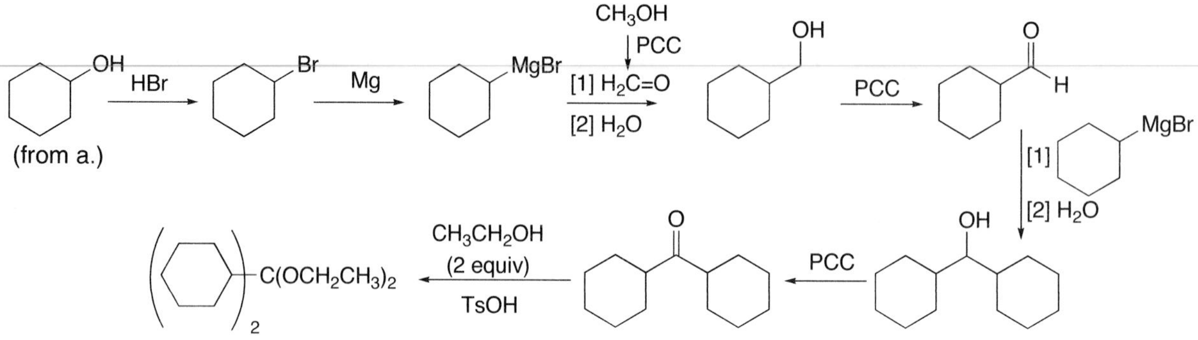

(from a.)

e.

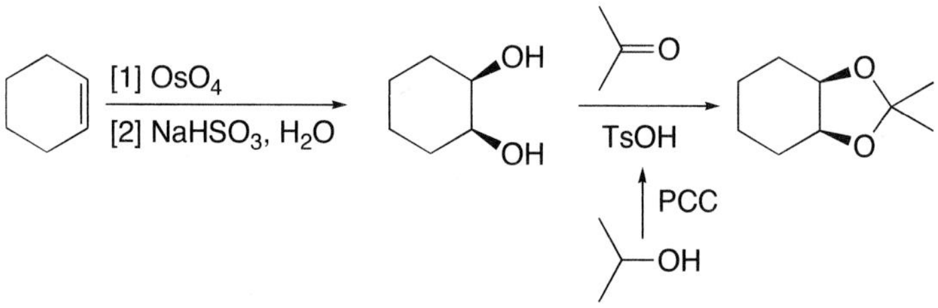

## 21.69

a.

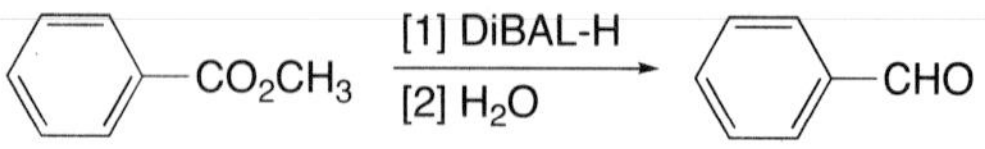

b.

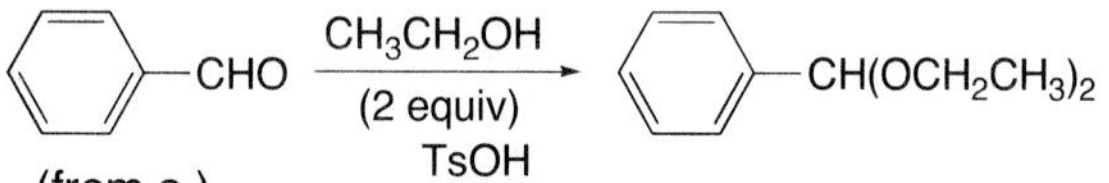

(from a.)

c.

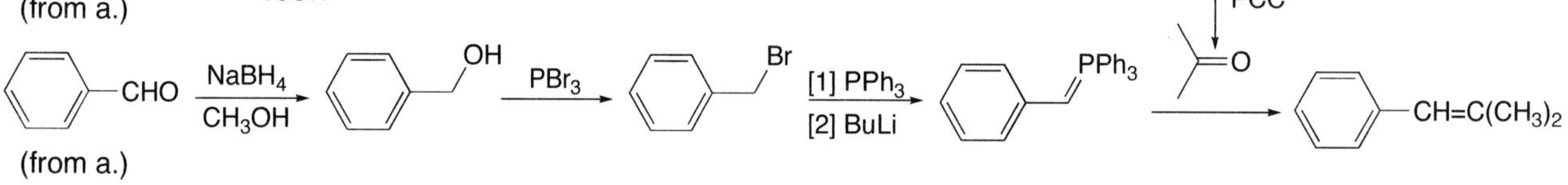

(from a.)

d.

(from c.)     (from a.)     (E and Z isomers)

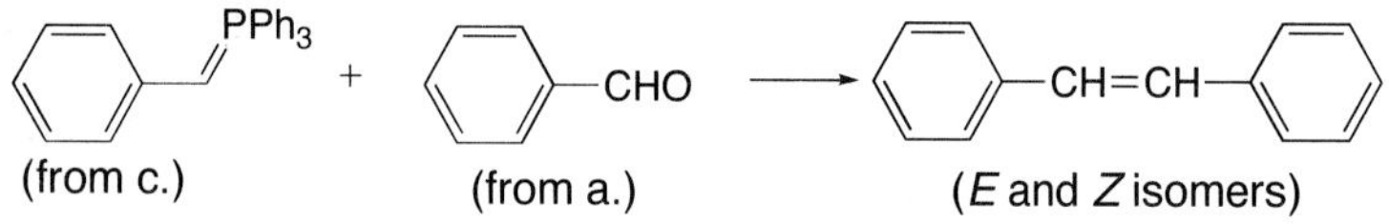

## 21.70

a.

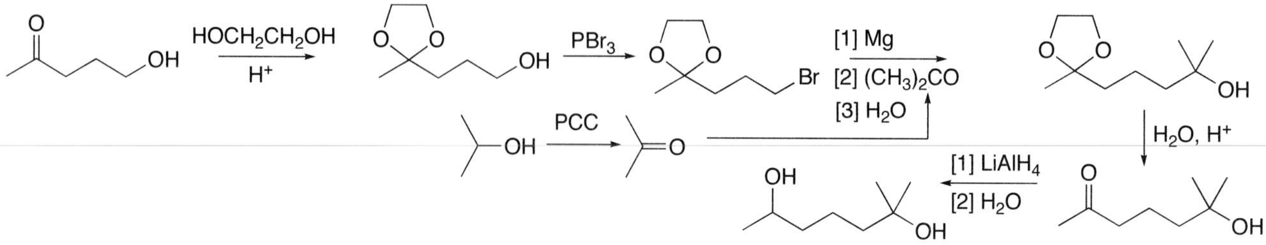

b.

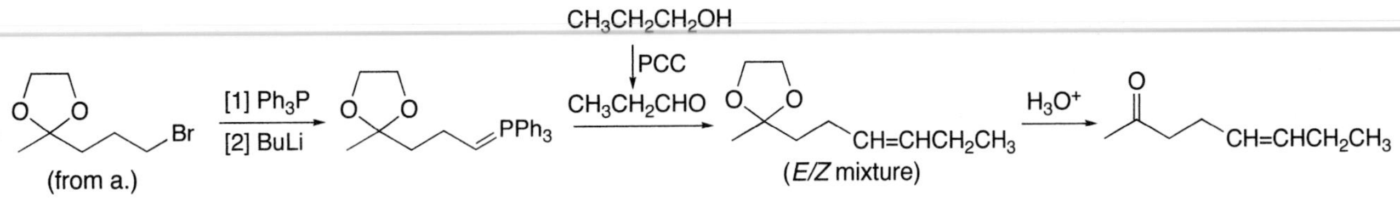

(from a.)

**21.71**

a.

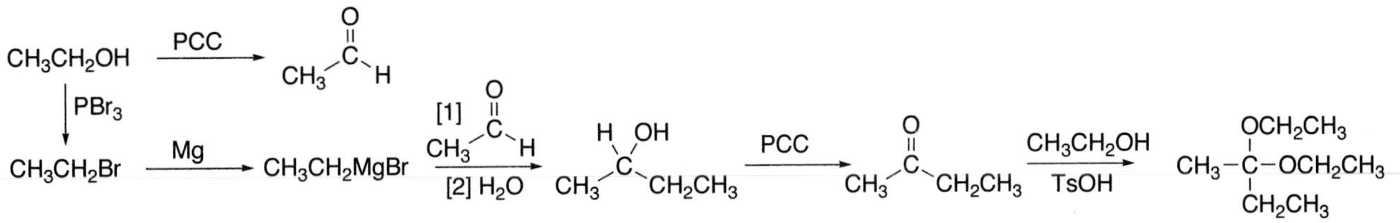

b.

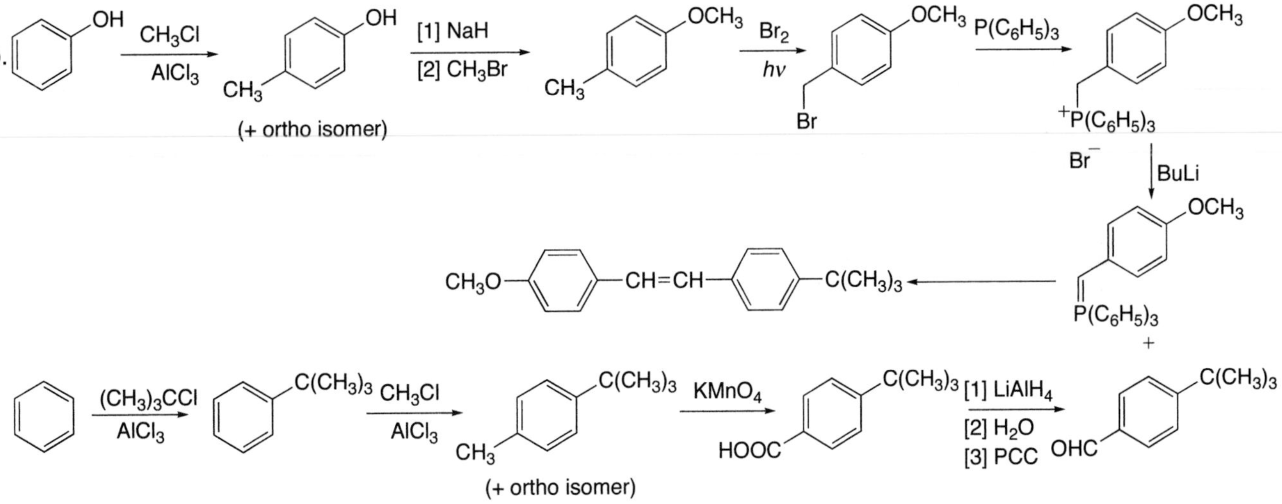

**21.72**

**A**

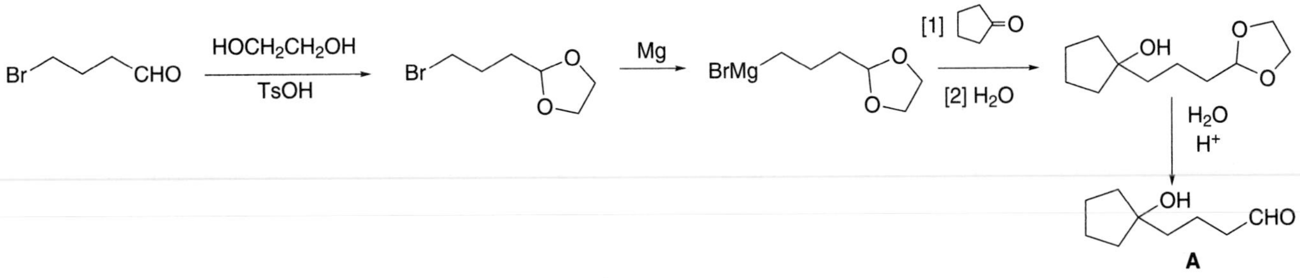

**21.73** a. The α,β-unsaturated carbonyl has a δ⁺ distributed over two carbons—the carbonyl carbon and the β carbon. As a result, there is less (+) charge on the carbonyl carbon, making it less electrophilic, and less reactive towards nucleophiles.

Three resonance structures:

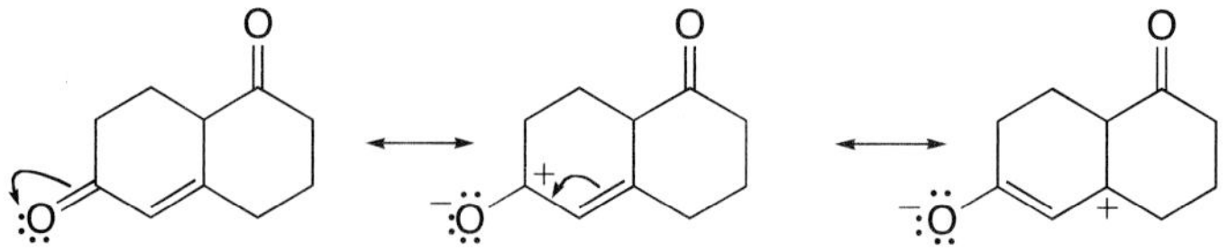

b.

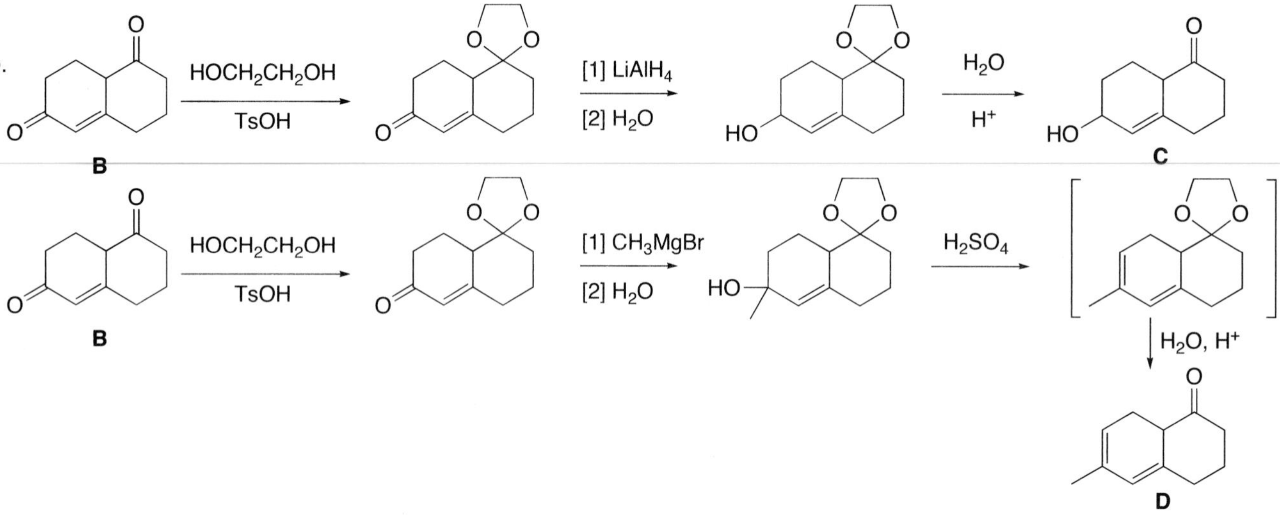

B → C

B → D

**21.74**

a.

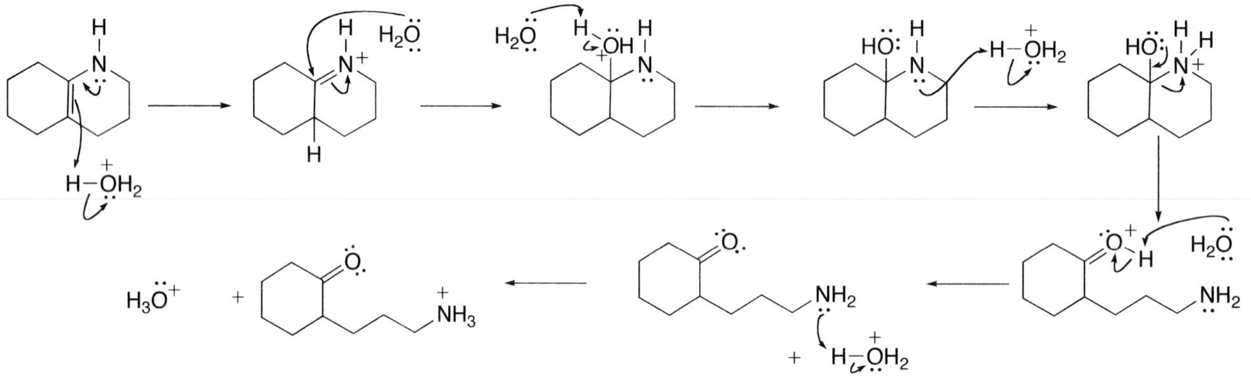

$H_3\overset{..}{O}{}^+$ +

b.

proton transfer

proton transfer

+ $H_2\overset{..}{O}$

+ $H_3\overset{..}{O}{}^+$

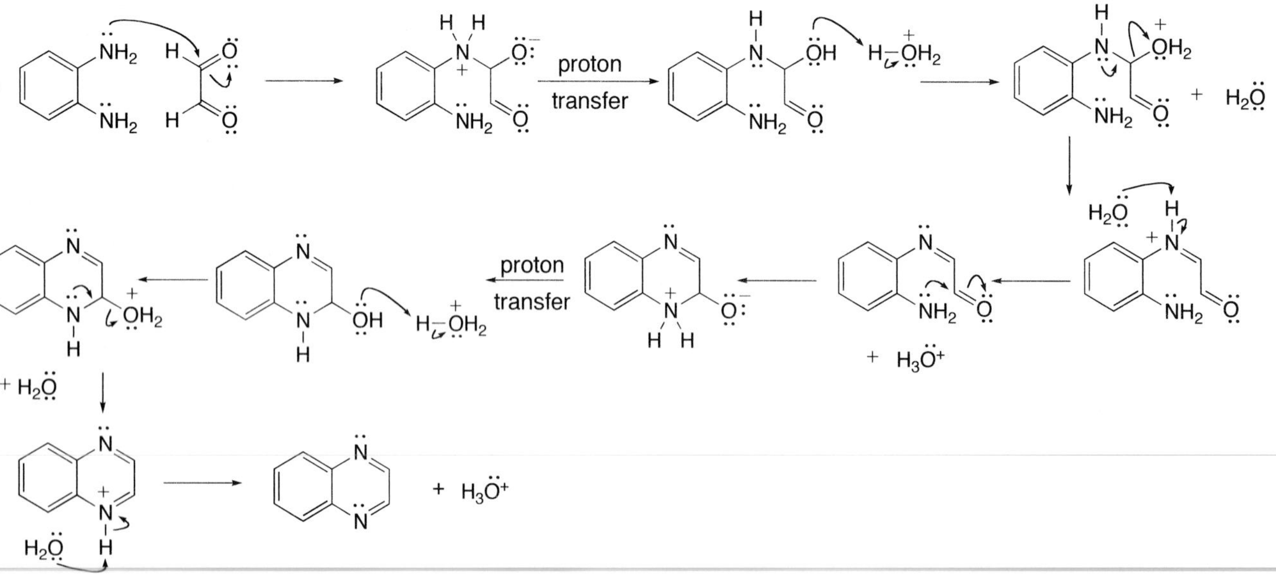

**21.75**

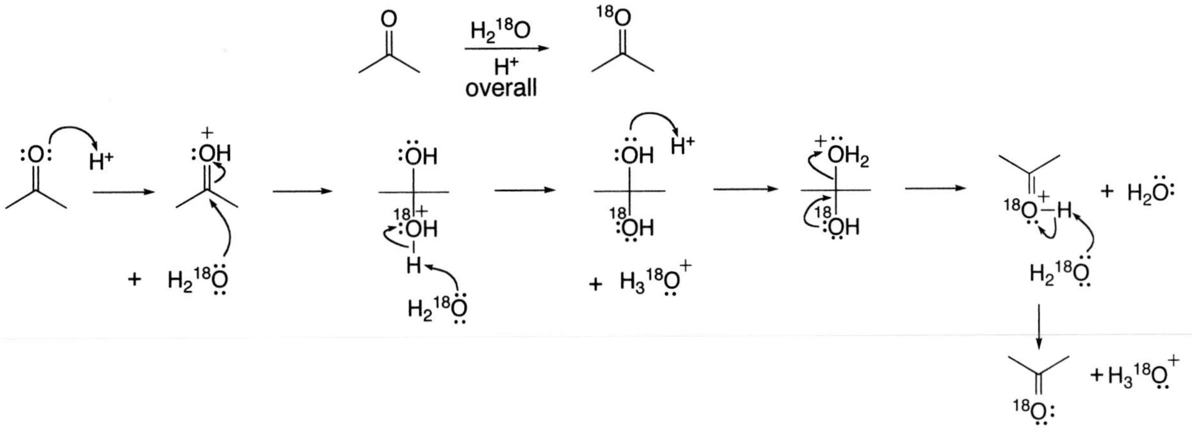

**21.76**

a.

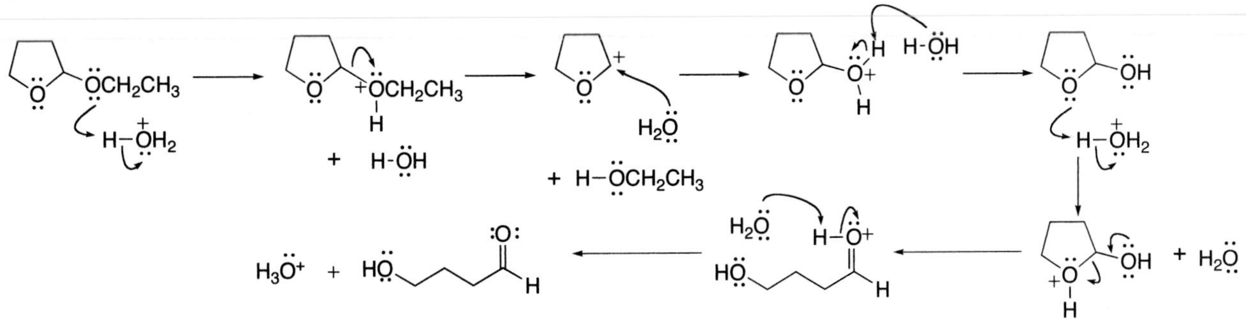

b.

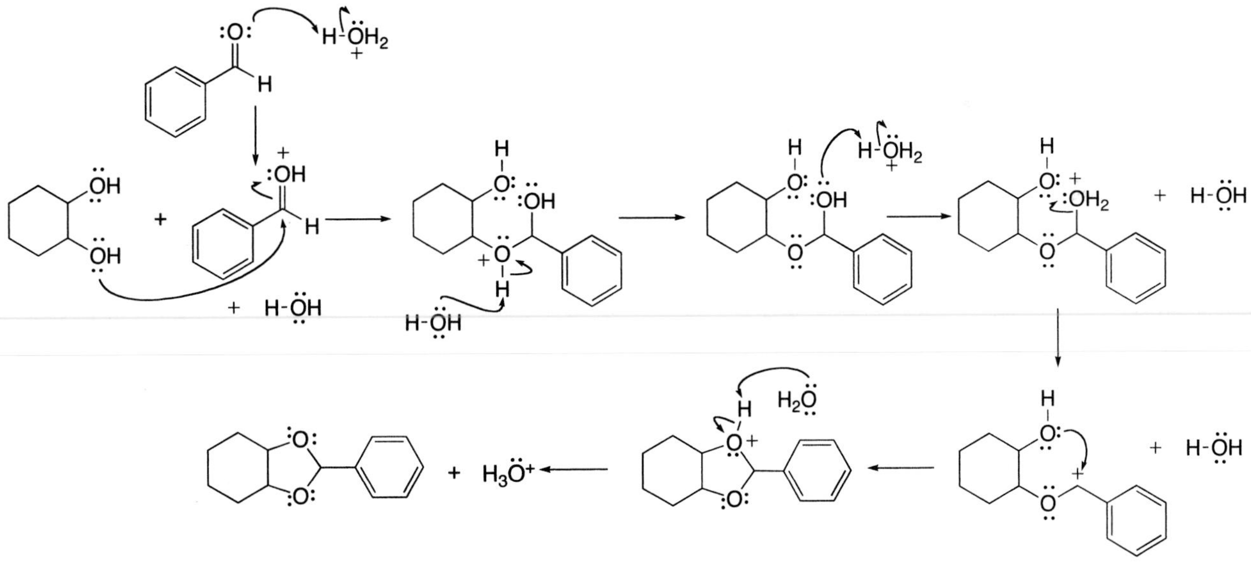

**21.77**

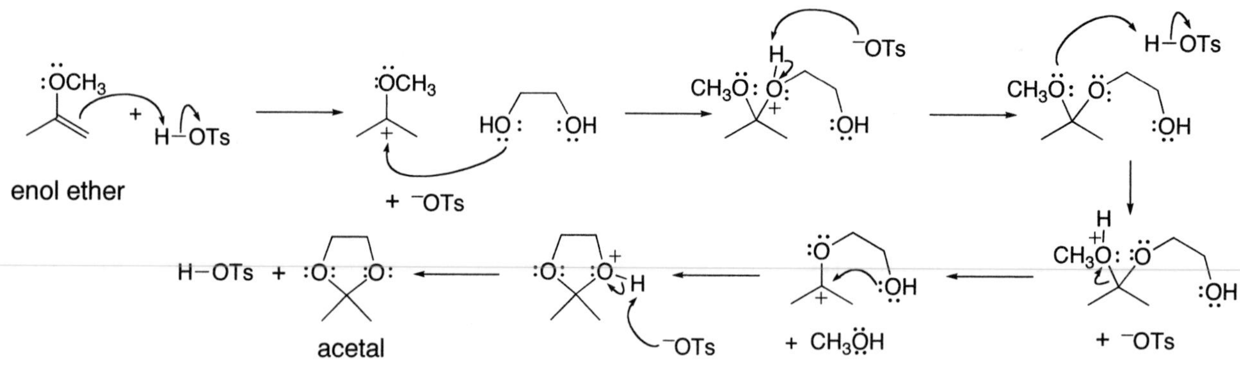

enol ether

acetal

**21.78**

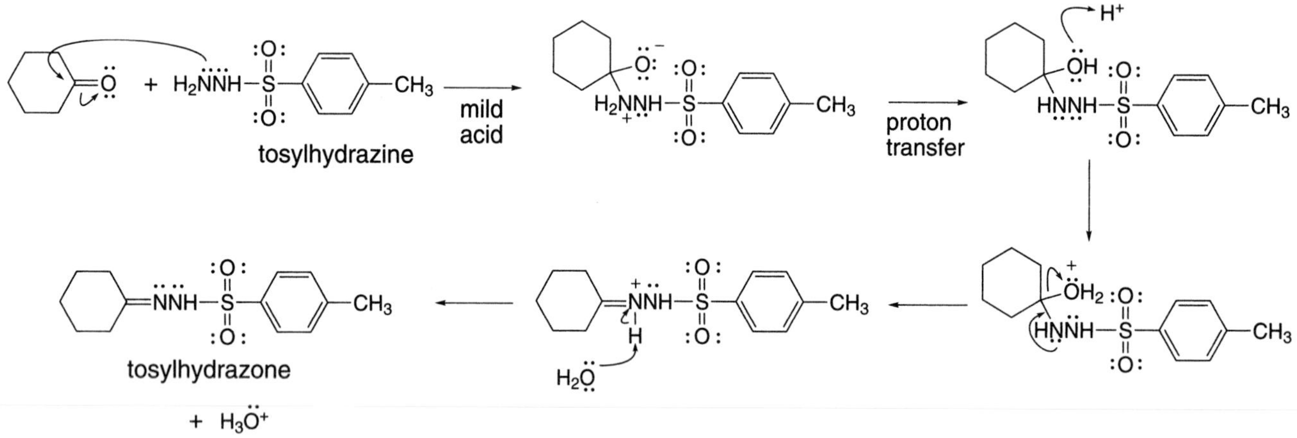

tosylhydrazine

mild acid

proton transfer

tosylhydrazone

+ $H_3\overset{..}{O}^+$

**21.79**

sulfonium salt

sulfur ylide

+ Bu−H + LiX

X

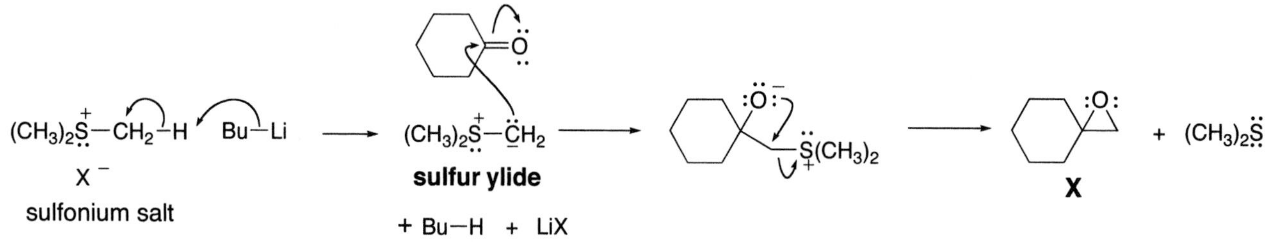

**21.80** Hemiacetal **A** is in equilibrium with its acyclic hydroxy aldehyde. The aldehyde is susceptible to hydride reduction and this forms 1,4-butanediol.

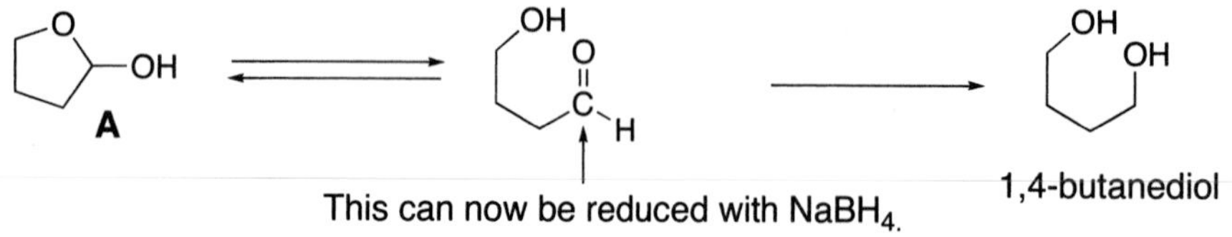

**A**

This can now be reduced with NaBH$_4$.

1,4-butanediol

**21.81**

a. and

      aldehyde            ketone

The $sp^2$ hybridized C–H bond of the aldehyde absorbs at 2700–2830 cm$^{-1}$.

c. and

                            smaller ring
                       **higher wavenumber for C=O**

b. and

  **higher wavenumber for C=O**      conjugated with a benzene ring lower wavenumber

**21.82**

**A.** Molecular formula $C_5H_{10}O$ ⟶ 1 degree of unsaturation
    IR absorptions at 1728, 2791, 2700 cm$^{-1}$ ⟶ C=O, CHO
    NMR data (ppm): singlet at 1.08 (9H) ⟶ 3 $CH_3$ groups
                singlet at 9.48 (1H) ppm ⟶ CHO

**B.** Molecular formula $C_5H_{10}O$ ⟶ 1 degree of unsaturation
    IR absorption at 1718 cm$^{-1}$ ⟶ C=O
    NMR data: doublet at 1.10 (6H) ⟶ 2 $CH_3$'s adjacent to H
           singlet at 2.14 (3H) ⟶ $CH_3$
           septet at 2.58 (1H) ppm ⟶ CH adjacent to 2 $CH_3$'s

**C.** Molecular formula $C_{10}H_{12}O$ ⟶ 5 degrees of unsaturation (4 due to a benzene ring)
    IR absorption at 1686 cm$^{-1}$ ⟶ C=O
    NMR data: triplet at 1.21 (3H) ⟶ $CH_3$ adjacent to 2 H's
          singlet at 2.39 (3H) ⟶ $CH_3$
          quartet at 2.95 (2H) ⟶ $CH_2$ adjacent to 3 H's
          doublet at 7.24 (2H) ⟶ 2 H's on benzene ring
          doublet at 7.85 (2H) ppm ⟶ 2 H's on benzene ring

**D.** Molecular formula $C_{10}H_{12}O$ ⟶ 5 degrees of unsaturation (4 due to a benzene ring)
    IR absorption at 1719 cm$^{-1}$ ⟶ C=O
    NMR data: triplet at 1.02 (3H) ⟶ $CH_3$ adjacent 2 H's
          quartet at 2.45 (2H) ⟶ 2 H's adjacent to 3 H's
          singlet at 3.67 (2H) ⟶ $CH_2$
          multiplet at 7.06–7.48 (5H) ppm ⟶ a monosubstituted benzene ring

**21.83**

$C_7H_{16}O_2$: 0 degrees of unsaturation
**IR: 3000 cm$^{-1}$: C–H bonds**
**NMR data (ppm):**
  $H_a$: quartet at 3.5 (**4H**), split by 3 H's
  $H_b$: singlet at 1.4 (**6H**)
  $H_c$: triplet at 1.2 (**6H**), split by 2 H's

$$CH_3CH_2-O-\underset{\underset{\underset{H_b}{|}}{CH_3}}{\overset{\overset{\overset{H_b}{|}}{CH_3}}{C}}-O-CH_2CH_3$$

       $H_c$  $H_a$             $H_a$  $H_c$

**21.84**

**A**. Molecular formula $C_9H_{10}O$
   5 degrees of unsaturation
   IR absorption at 1700 cm$^{-1}$ → C=O
   IR absorption at ~2700 cm$^{-1}$ → CH of RCHO
   NMR data (ppm):
   
   triplet at 1.2 (2H's adjacent)
   quartet at 2.7 (3H's adjacent)
   doublet at 7.3 (2H's on benzene)
   doublet at 7.7 (2H's on benzene)
   singlet at 9.9 (CHO)

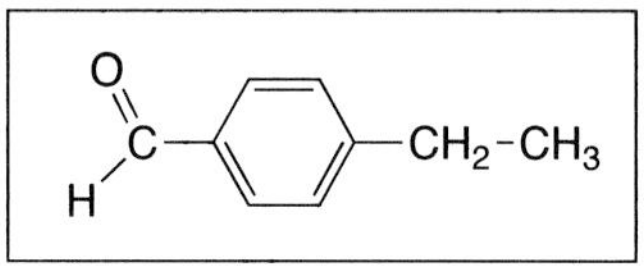

**B**. Molecular formula $C_9H_{10}O$
   5 degrees of unsaturation
   IR absorption at 1720 cm$^{-1}$ → C=O
   IR absorption at ~2700 cm$^{-1}$ → CH of RCHO
   NMR data (ppm):
   
   2 triplets at 2.85 and 2.95 (suggests –CH$_2$CH$_2$–)
   multiplet at 7.2 (benzene H's)
   signal at 9.8 (CHO)

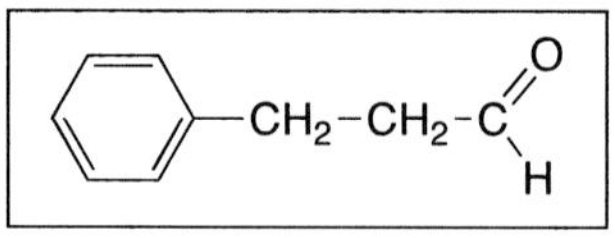

**21.85**

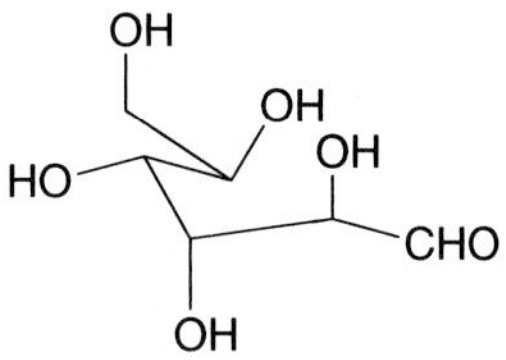

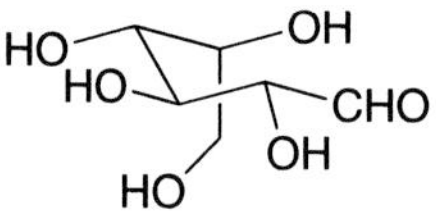

**21.86**

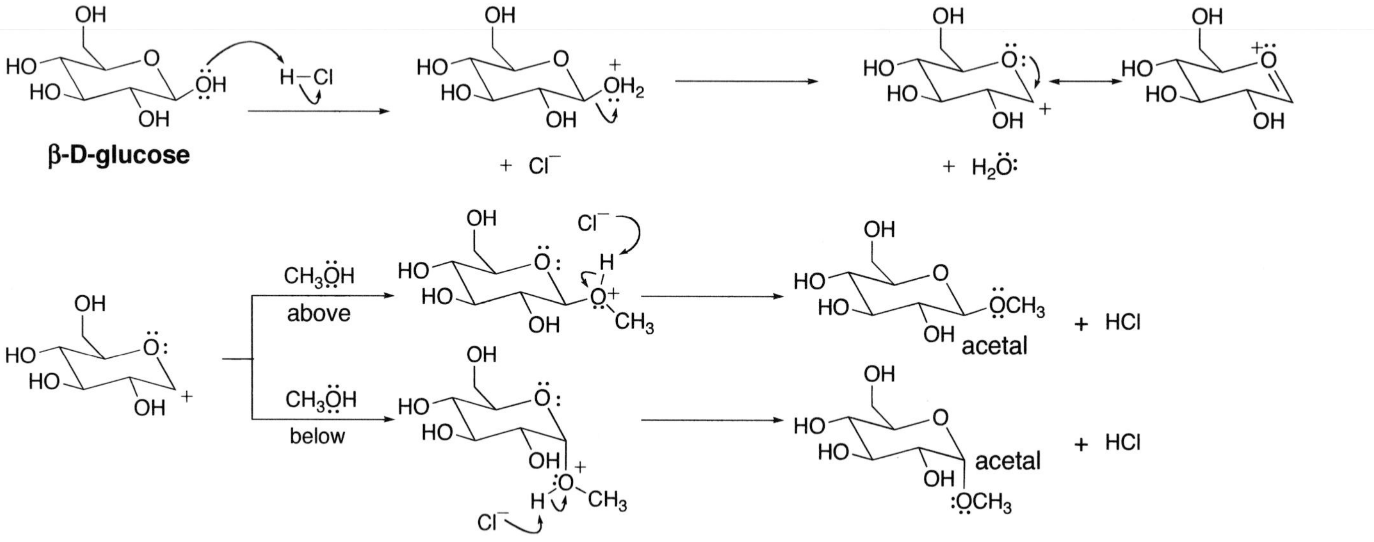

**21.87**

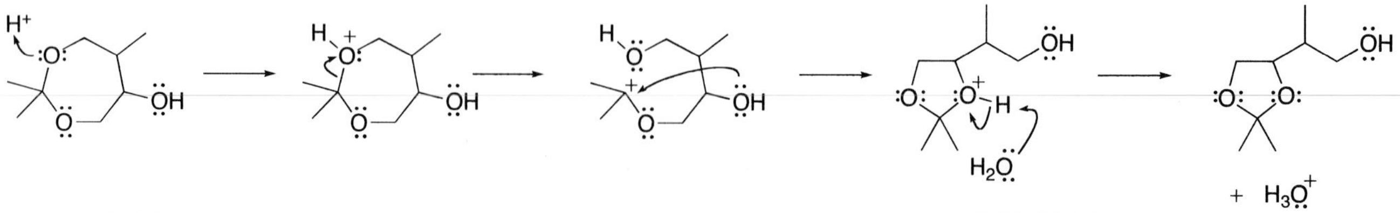

**21.88**

a.

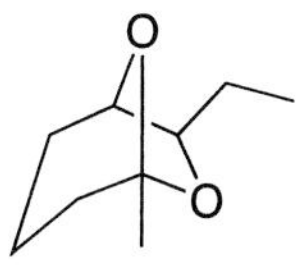

brevicomin

b.

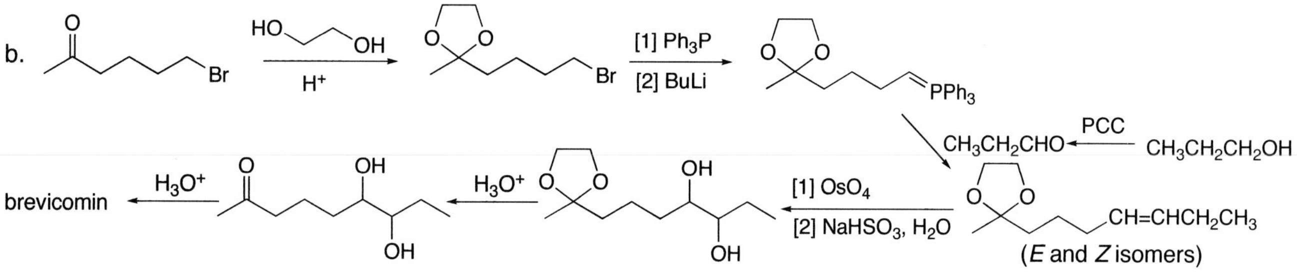
HO OH
H+
Br
[1] Ph3P
[2] BuLi
PPh3
PCC
CH3CH2CHO
CH3CH2CH2OH
[1] OsO4
[2] NaHSO3, H2O
CH=CHCH2CH3
(E and Z isomers)
H3O+
brevicomin

**21.89**

a.

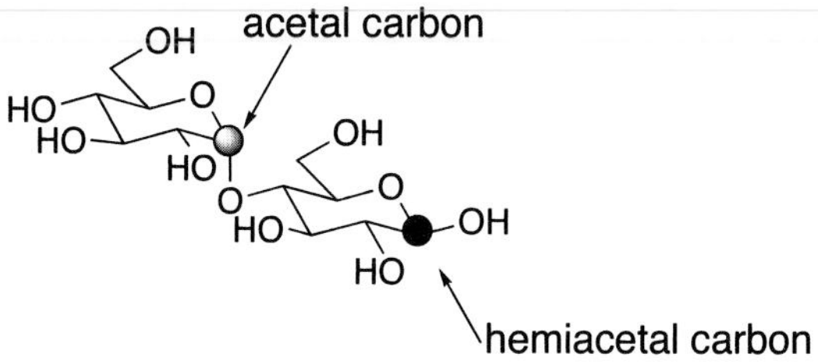
acetal carbon
hemiacetal carbon

c.

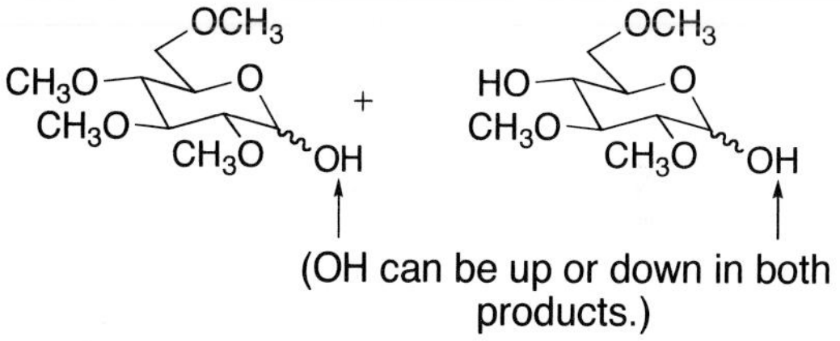
(OH can be up or down in both products.)

b.

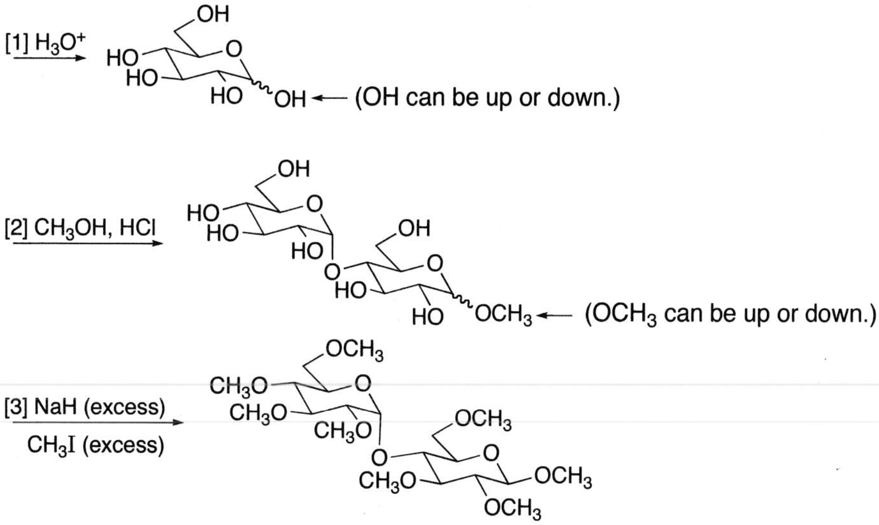
[1] H3O+
(OH can be up or down.)
[2] CH3OH, HCl
(OCH3 can be up or down.)
[3] NaH (excess)
CH3I (excess)

## Chapter 22: Carboxylic Acids and Their Derivatives—Nucleophilic Acyl Substitution

### ♦ Summary of spectroscopic absorptions of RCOZ (22.5)

| | |
|---|---|
| **IR absorptions** | • All ROCZ compounds have a C=O absorption in the region 1600–1850 $cm^{-1}$. |
| |    • RCOCl:  1800 $cm^{-1}$ |
| |    • $(RCO)_2O$: 1820 and 1760 $cm^{-1}$ (two peaks) |
| |    • RCOOR′: 1735–1745 $cm^{-1}$ |
| |    • $RCONR'_2$: 1630–1680 $cm^{-1}$ |
| | • Additional amide absorptions occur at 3200–3400 $cm^{-1}$ (N–H stretch) and 1640 $cm^{-1}$ (N–H bending). |
| | • Decreasing the ring size of a cyclic lactone, lactam, or anhydride increases the frequency of the C=O absorption. |
| | • Conjugation shifts the C=O to lower wavenumber. |
| **$^1$H NMR absorptions** | • C–H $\alpha$ to the C=O absorbs at 2–2.5 ppm. |
| | • N–H of an amide absorbs at 7.5–8.5 ppm. |
| **$^{13}$C NMR absorption** | • C=O absorbs at 160–180 ppm. |

### ♦ Summary of spectroscopic absorptions of RCN (22.5)

| | |
|---|---|
| **IR absorption** | • C≡N absorption at 2250 $cm^{-1}$ |
| **$^{13}$C NMR absorption** | • C≡N absorbs at 115–120 ppm. |

### ♦ Summary: The relationship between the basicity of Z⁻ and the properties of RCOZ

• **Increasing basicity of the leaving group** (22.2)
• **Increasing resonance stabilization** (22.2)

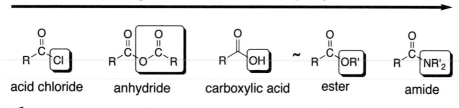

| acid chloride | anhydride | carboxylic acid | ester | amide |

• **Increasing leaving group ability** (22.7B)
• **Increasing reactivity** (22.7B)
• **Increasing frequency of the C=O absorption in the IR** (22.5)

### ♦ General features of nucleophilic acyl substitution

• The characteristic reaction of compounds having the general structure RCOZ is nucleophilic acyl substitution (22.1).
• The mechanism consists of two basic steps (22.7A):
  [1] Addition of a nucleophile to form a tetrahedral intermediate
  [2] Elimination of a leaving group
• More reactive acyl compounds can be used to prepare less reactive acyl compounds. The reverse is not necessarily true (22.7B).

## ♦ Nucleophilic acyl substitution reactions

### [1] Reactions that synthesize acid chlorides (RCOCl)

**[a] From RCOOH (22.10A):**

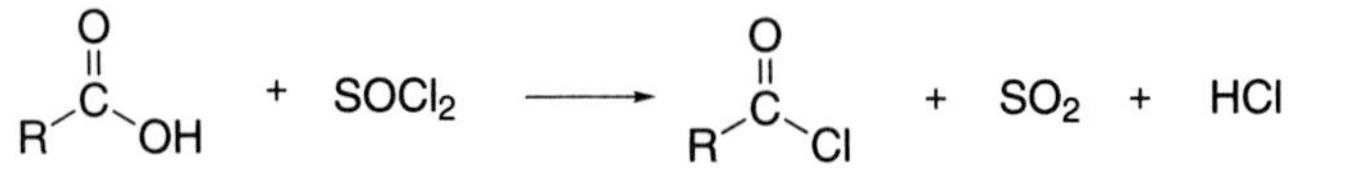

### [2] Reactions that synthesize anhydrides [(RCO)₂O]

**[a] From RCOCl (22.8):**

**[b] From dicarboxylic acids (22.10B):**

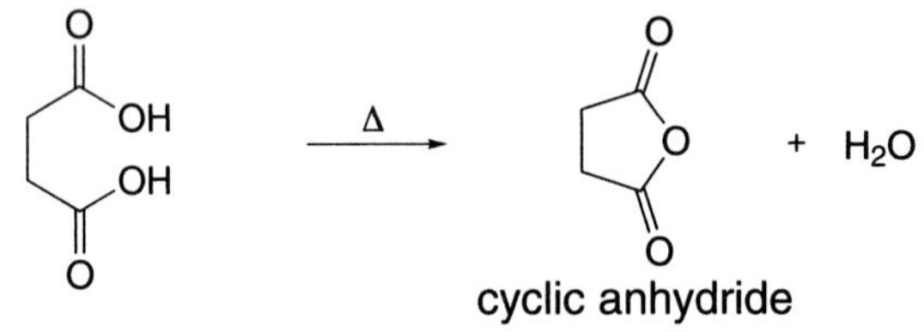

cyclic anhydride

### [3] Reactions that synthesize carboxylic acids (RCOOH)

**[a] From RCOCl (22.8):**

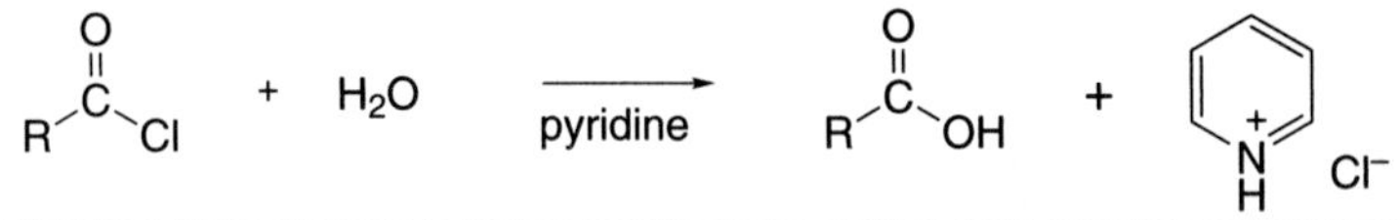

**[b] From (RCO)₂O (22.9):**

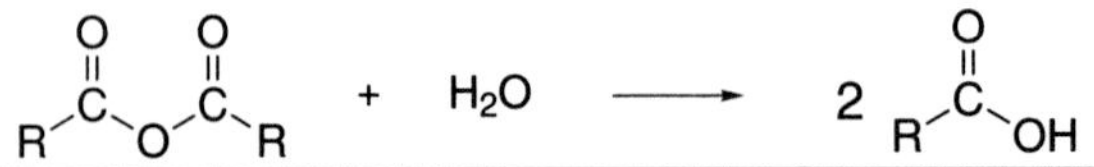

**[c] From RCOOR′ (22.11):**

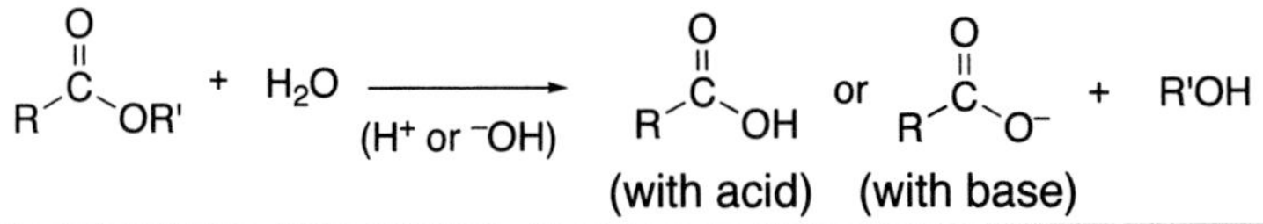

(with acid)   (with base)

**[d] From RCONR′₂ (R′ = H or alkyl, 22.13):**

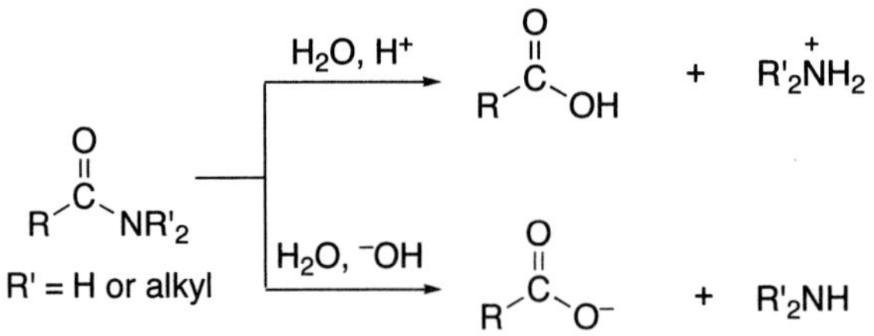

### [4] Reactions that synthesize esters (RCOOR′)

**[a] From RCOCl (22.8):**

**[b] From (RCO)₂O**
**(22.9):**

$$\underset{R}{\overset{O}{\underset{\|}{C}}}\text{—O—}\underset{R}{\overset{O}{\underset{\|}{C}}} \quad + \quad \text{R'OH} \longrightarrow \quad \underset{R}{\overset{O}{\underset{\|}{C}}}\text{—OR'} \quad + \quad \text{RCOOH}$$

**[c] From RCOOH**
**(22.10C):**

$$\underset{R}{\overset{O}{\underset{\|}{C}}}\text{—OH} \quad + \quad \text{R'OH} \xrightarrow{\text{H}_2\text{SO}_4} \quad \underset{R}{\overset{O}{\underset{\|}{C}}}\text{—OR'} \quad + \quad \text{H}_2\text{O}$$

**[5]** **Reactions that synthesize amides (RCONH₂)** [The reactions are written with NH₃ as the nucleophile to form RCONH₂.  Similar reactions occur with R'NH₂ to form RCONHR', and with R'₂NH to form RCONR'₂.]

**[a] From RCOCl**
**(22.8):**

$$\underset{R}{\overset{O}{\underset{\|}{C}}}\text{—Cl} \quad + \quad \underset{\text{(2 equiv)}}{\text{NH}_3} \longrightarrow \quad \underset{R}{\overset{O}{\underset{\|}{C}}}\text{—NH}_2 \quad + \quad \text{NH}_4{}^+\text{Cl}^-$$

**[b] From (RCO)₂O**
**(22.9):**

$$\underset{R}{\overset{O}{\underset{\|}{C}}}\text{—O—}\underset{R}{\overset{O}{\underset{\|}{C}}} \quad + \quad \underset{\text{(2 equiv)}}{\text{NH}_3} \longrightarrow \quad \underset{R}{\overset{O}{\underset{\|}{C}}}\text{—NH}_2 \quad + \quad \text{RCOO}^-\text{NH}_4{}^+$$

**[c] From RCOOH**
**(22.10D):**

$$\underset{R}{\overset{O}{\underset{\|}{C}}}\text{—OH} \xrightarrow[\text{[2] }\Delta]{\text{[1]NH}_3} \quad \underset{R}{\overset{O}{\underset{\|}{C}}}\text{—NH}_2 \quad + \quad \text{H}_2\text{O}$$

$$\underset{R}{\overset{O}{\underset{\|}{C}}}\text{—OH} \quad + \quad \text{R'NH}_2 \xrightarrow{\text{DCC}} \quad \underset{R}{\overset{O}{\underset{\|}{C}}}\text{—NHR'} \quad + \quad \text{H}_2\text{O}$$

**[d] From RCOOR'**
**(22.11):**

$$\underset{R}{\overset{O}{\underset{\|}{C}}}\text{—OR'} \quad + \quad \text{NH}_3 \longrightarrow \quad \underset{R}{\overset{O}{\underset{\|}{C}}}\text{—NH}_2 \quad + \quad \text{R'OH}$$

♦ **Nitrile synthesis (22.18)**

**Nitriles are prepared by S$_N$2 substitution using unhindered alkyl halides as starting materials.**

$$\text{R—X} \quad + \quad {}^-\text{CN} \xrightarrow{\text{S}_\text{N}2} \text{R—C}\equiv\text{N} \quad + \quad \text{X}^-$$

$$\text{R} = \text{CH}_3, 1°$$

♦ **Reactions of nitriles**

**[1] Hydrolysis (22.18A)**

$$\text{R—C}\equiv\text{N} \xrightarrow[\text{(H}^+ \text{ or } {}^-\text{OH)}]{\text{H}_2\text{O}} \quad \underset{R}{\overset{O}{\underset{\|}{C}}}\text{—OH} \quad \text{or} \quad \underset{R}{\overset{O}{\underset{\|}{C}}}\text{—O}^-$$

$$\phantom{xxxxxxxxxxxxxx}\text{(with acid)} \qquad \text{(with base)}$$

---

**[2]  Reduction (22.18B)**

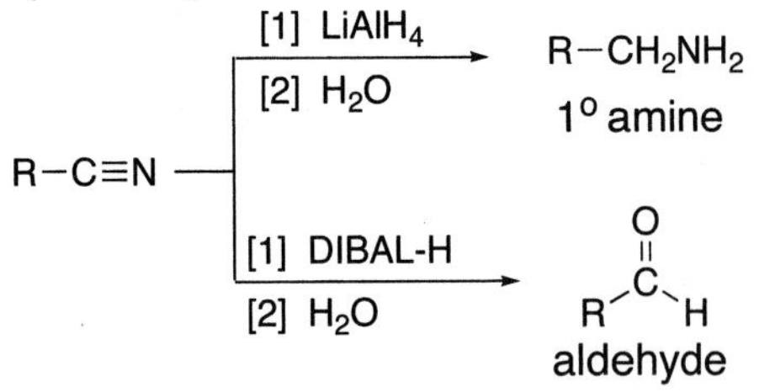

$$R-C\equiv N \quad \begin{cases} \xrightarrow[\text{[2] }H_2O]{\text{[1] LiAlH}_4} \quad R-CH_2NH_2 \\[2em] \xrightarrow[\text{[2] }H_2O]{\text{[1] DIBAL-H}} \end{cases}$$

R—CH₂NH₂
1° amine

$$\underset{\text{aldehyde}}{\overset{\displaystyle O}{\underset{R}{\parallel}}} \!\! \overset{}{C}\!\!_{H}$$

---

**[3]  Reaction with organometallic reagents (22.18C)**

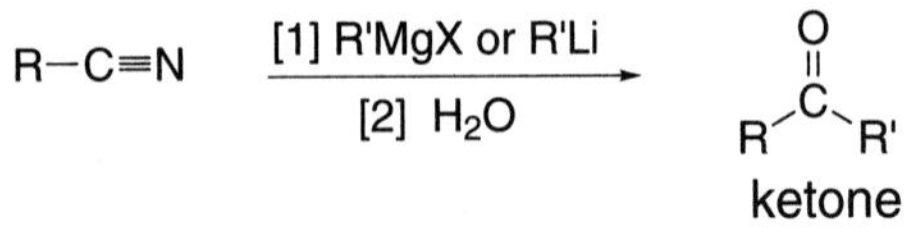

$$R-C\equiv N \xrightarrow[\text{[2] }H_2O]{\text{[1] R'MgX or R'Li}} \underset{\text{ketone}}{\overset{\displaystyle O}{\underset{R}{\parallel}}}\!\!\overset{}{C}\!\!_{R'}$$

**22.1** The number of C–N bonds determines the classification as a 1°, 2°, or 3° amide.

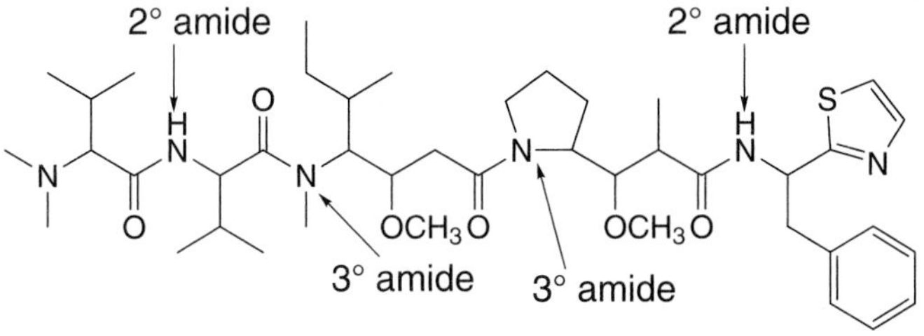

**22.2** As the basicity of Z increases, the stability of RCOZ increases because of added resonance stabilization.

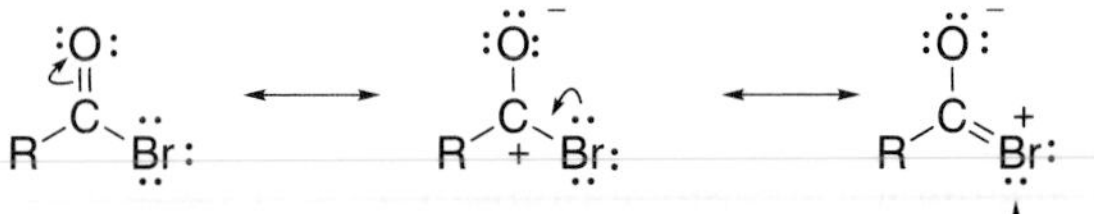

The **basicity of Z** determines how much this structure contributes to the hybrid. Br⁻ is less basic than ⁻OH, so RCOBr is less stable than RCOOH.

**22.3**

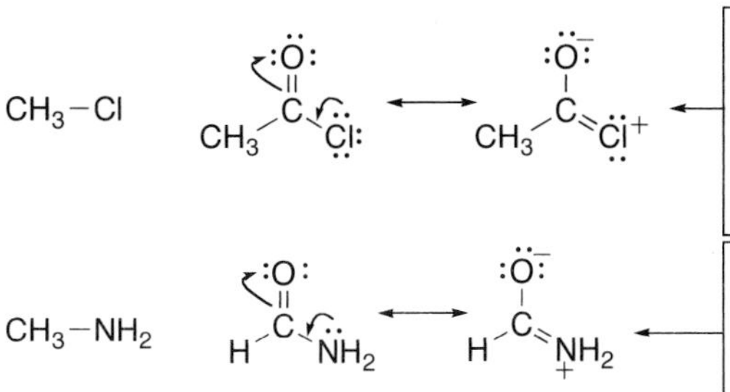

| This resonance structure contributes little to the hybrid since Cl⁻ is a weak base. Thus, the C–Cl bond has little double bond character, making it similar in length to the C–Cl bond in CH₃Cl. |

| This resonance structure contributes more to the hybrid since ⁻NH₂ is more basic. Thus, the C–N bond in HCONH₂ has more double bond character, making it shorter than the C–N bond in CH₃NH₂. |

**22.4**

a. $(CH_3CH_2)_2CH-COCl$

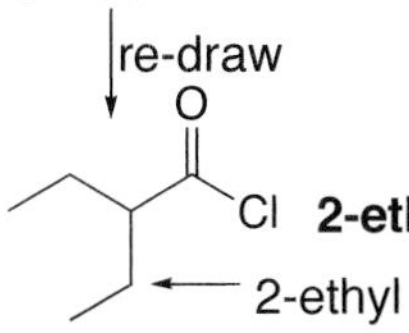

re-draw

Cl **2-ethylbutanoyl chloride**

⟵ 2-ethyl

b. $C_6H_5COOCH_3$

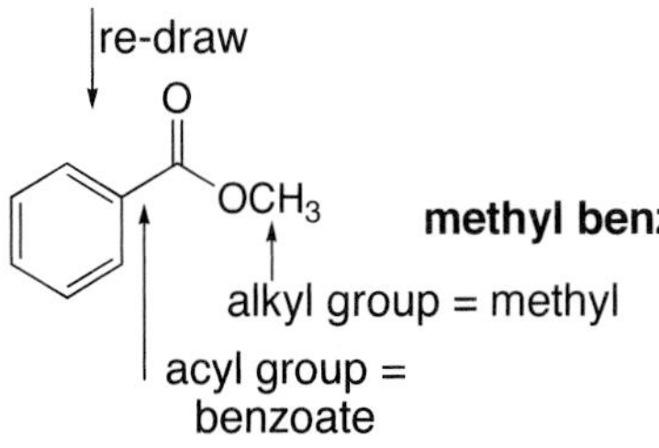

re-draw

OCH₃ **methyl benzoate**

alkyl group = methyl

acyl group = benzoate

c.  CH₃CH₂CON(CH₃)CH₂CH₃

|re-draw

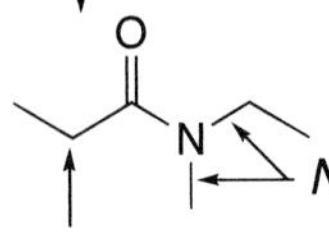

acyl group =
propanamide

*N*-ethyl-*N*-methyl

**N-ethyl-N-methylpropanamide**

e. 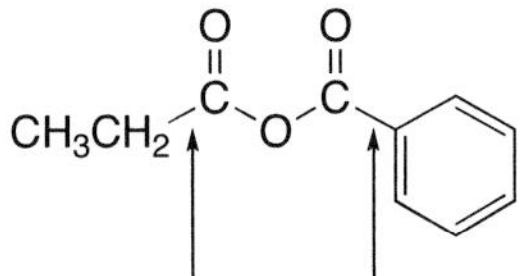  CH₃CH₂ —

**benzoic propanoic anhydride**

acyl group =        acyl group =
propanoic          benzoic

d.

H      OCH₂CH₃

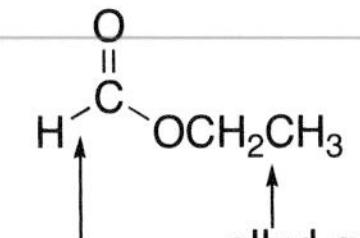

alkyl group = ethyl

acyl group =      **ethyl formate**
formate

f.

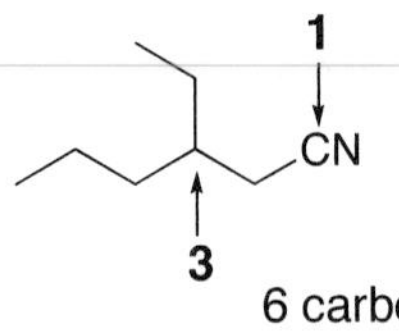

**3-ethylhexanenitrile**

6 carbon chain =
hexanenitrile

## 22.5

a.  5-methylheptanoyl chloride

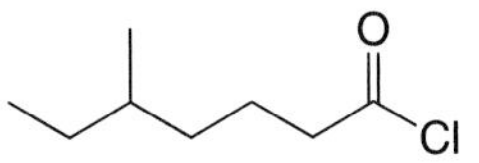

b.  isopropyl propanoate

c.  acetic formic anhydride

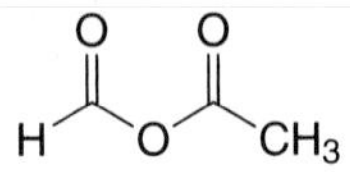

d.  *N*-isobutyl-*N*-methylbutanamide

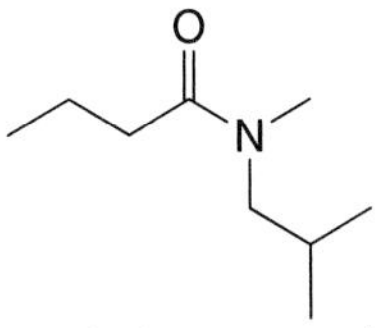

e.  3-methylpentanenitrile

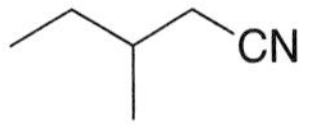

f.  *o*-cyanobenzoic acid

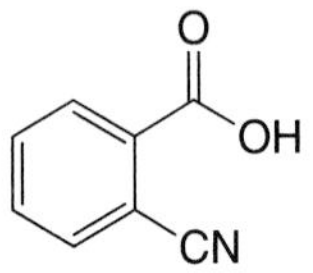

g.  *sec*-butyl 2-methylhexanoate

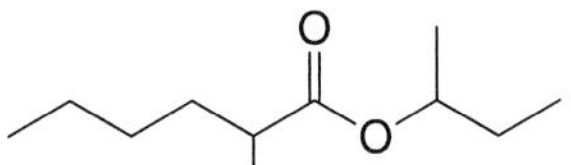

h.  *N*-ethylhexanamide

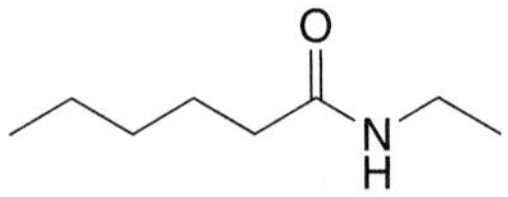

**22.6** CH₃CONH₂ has two H's bonded to N that can hydrogen bond. CH₃CON(CH₃)₂ does not have any H's capable of hydrogen bonding. This means CH₃CONH₂ has much stronger intermolecular forces, which leads to a higher boiling point.

## 22.7

a. 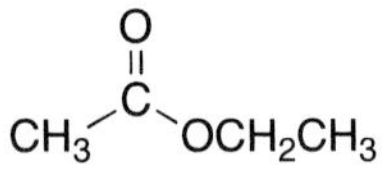  CH₃ — OCH₂CH₃  and  CH₃ — N(CH₂CH₃)₂

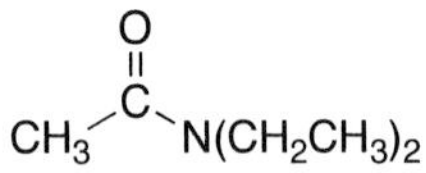

amide: C=O at
lower wavenumber

b.          and

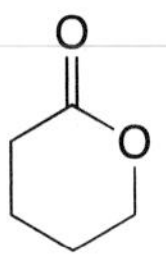

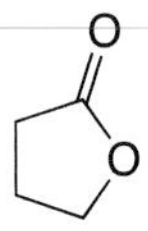

smaller ring:
C=O at a higher
wavenumber

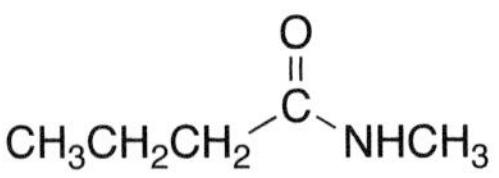

c.  CH₃CH₂CH₂ — NHCH₃  and  CH₃CH₂CH₂ — NH₂

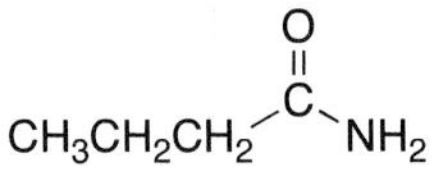

2° amide: 1 N–H
absorption at
3200–3400 cm⁻¹

1° amide: 2
N–H absorptions

d. 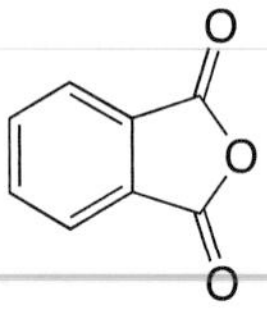          and

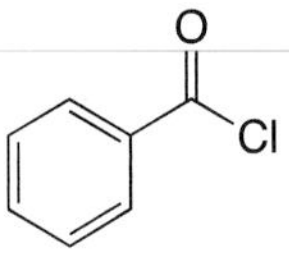

anhydride:
2 C=O peaks

**22.8**

$H_b \rightarrow$ H H O

$H_c$ signal from the 5 H's on the aromatic ring $\rightarrow$

**A**

$H_a$

CH₃

Molecular formula $C_9H_{10}O_2$
5 degrees of unsaturation
IR: 1743 cm$^{-1}$ from ester C=O
    3091–2895 cm$^{-1}$ from $sp^2$ and $sp^3$ C–H
$^1$H NMR:  $H_a$ = 2.06 ppm (singlet, 3 H) – CH₃
           $H_b$ = 5.08 ppm (singlet, 2 H) – CH₂
           $H_c$ = 7.33 ppm (broad singlet, 5 H)

$H_a \rightarrow$ H H O

$H_b \rightarrow$

**B**

$\leftarrow H_b$ signal from the 10 H's on the two aromatic rings

Molecular formula $C_{14}H_{12}O_2$
9 degrees of unsaturation
IR: 1718 cm$^{-1}$ from conjugated ester C=O
    3091–2953 cm$^{-1}$ from $sp^2$ and $sp^3$ C–H
$^1$H NMR: $H_a$ = 5.35 ppm (singlet, 2 H)
           $H_b$ = 7.26–8.15 ppm (multiplets, 10 H)

**22.9**

**amoxicillin**

a. 4 stereogenic centers
b. $2^4$ = 16 possible stereoisomers
c. enantiomer

**cephalexin**
(Trade name: Keflex)

a. 3 stereogenic centers
b. $2^3$ = 8 possible stereoisomers
c. enantiomer

**22.10**  To draw the products of these nucleophilic acyl substitution reactions, find the nucleophile and the leaving group. Then replace the leaving group with the nucleophile and draw a neutral product.

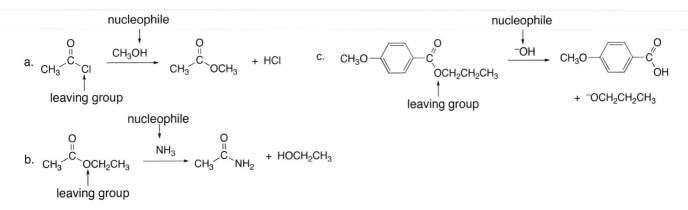

**22.11** More reactive acyl compounds can be converted to less reactive acyl compounds.

a.   $CH_3COCl$ $\longrightarrow$ $CH_3COOH$
more reactive    **YES**    less reactive

c.   $CH_3COOCH_3$ $\longrightarrow$ $CH_3COCl$
less reactive    **NO**    more reactive

b.   $CH_3CONHCH_3$ $\longrightarrow$ $CH_3COOCH_3$
less reactive    **NO**    more reactive

**22.12** The better the leaving group is, the more reactive the carboxylic acid derivative. The weakest base is the best leaving group.

a.

$^-NH_2$ stongest base
**least reactive**

$^-OCH_3$
intermediate

$^-Cl$ weakest base
**most reactive**

b.   $CH_3CH_2$ — C(=O) — $NHCH_3$     $CH_3CH_2$ — C(=O) — $OH$     $CH_3CH_2$ — C(=O) — O — C(=O) — $CH_2CH_3$

$^-NHCH_3$ stongest base
**least reactive**

$^-OH$
intermediate

$^-OOCCH_2CH_3$ weakest base
**most reactive**

**22.13**

$CH_3$ — C(=O) — O — C(=O) — $CH_3$
acetic anhydride

$Cl_3C$ — C(=O) — O — C(=O) — $CCl_3$
trichloroacetic anhydride

The Cl atoms are electron withdrawing, which makes the conjugate base (the leaving group, $CCl_3COO^-$) weaker and more stable.

**22.14**

a.   $\xrightarrow[\text{pyridine}]{H_2O}$     benzoic acid ($OH$)   +   pyridinium ($N^+H$ $Cl^-$)

b.   $\xrightarrow{CH_3COO^-}$     benzoic anhydride ( — O — C(=O) — $CH_3$)   +   $Cl^-$

c.   $\xrightarrow[\text{excess}]{NH_3}$     benzamide ($NH_2$)   +   $NH_4^+Cl^-$

d.   $\xrightarrow[\text{excess}]{(CH_3)_2NH}$     benzamide ($N(CH_3)_2$)   +   $(CH_3)_2\overset{+}{N}H_2$ $Cl^-$

**22.15** The mechanism has three steps: [1] nucleophilic attack by O; [2] proton transfer; [3] elimination of the $Cl^-$ leaving group to form the product.

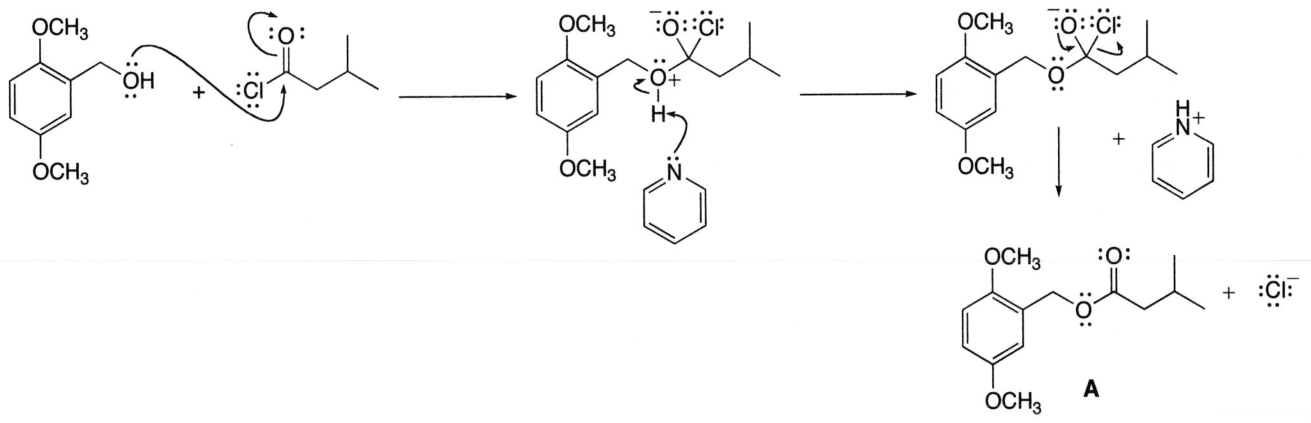

OCH3
OH
Cl
O+
H
N
N+
Cl−
A


**22.16**

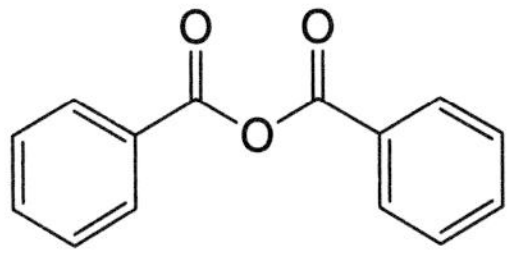

a. 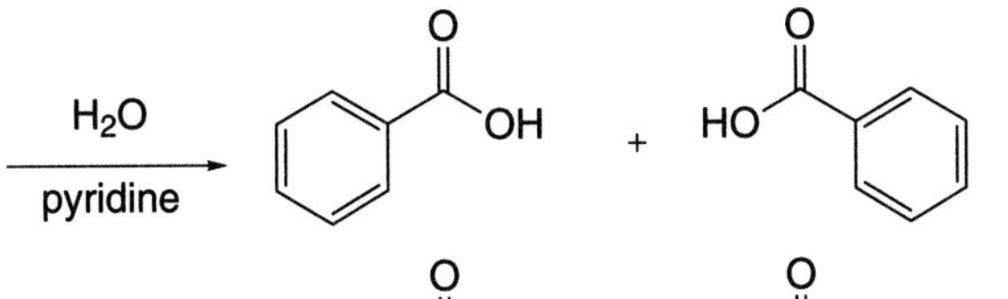

H2O
pyridine
OH
HO


b. 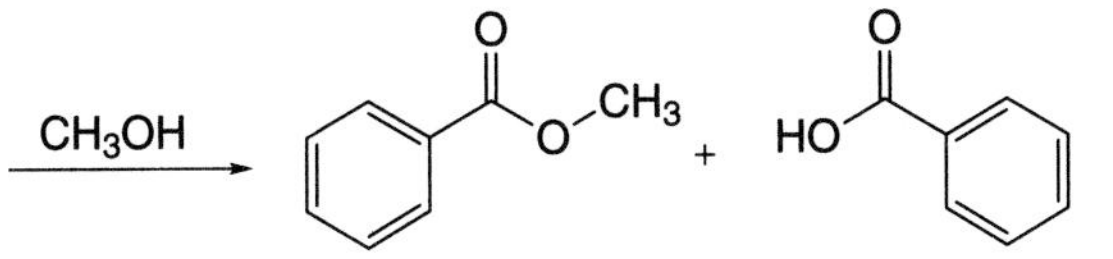

CH3OH
CH3
HO


c. 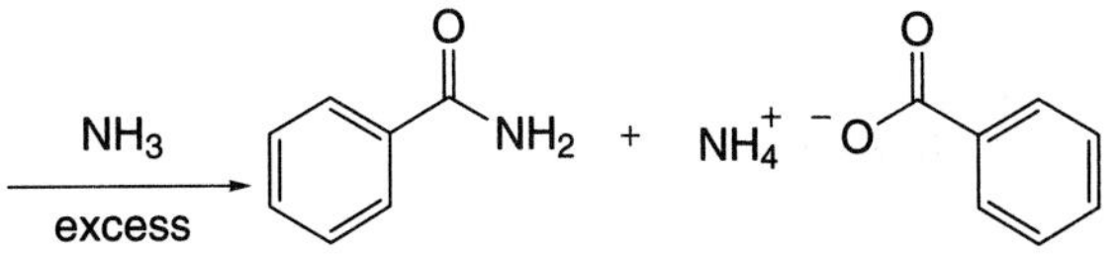

NH3
excess
NH2
NH4+


d. 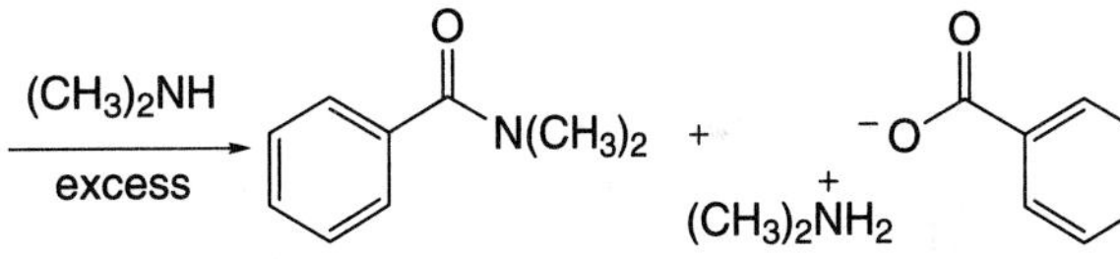

(CH3)2NH
excess
N(CH3)2
(CH3)2NH2+

**22.17**

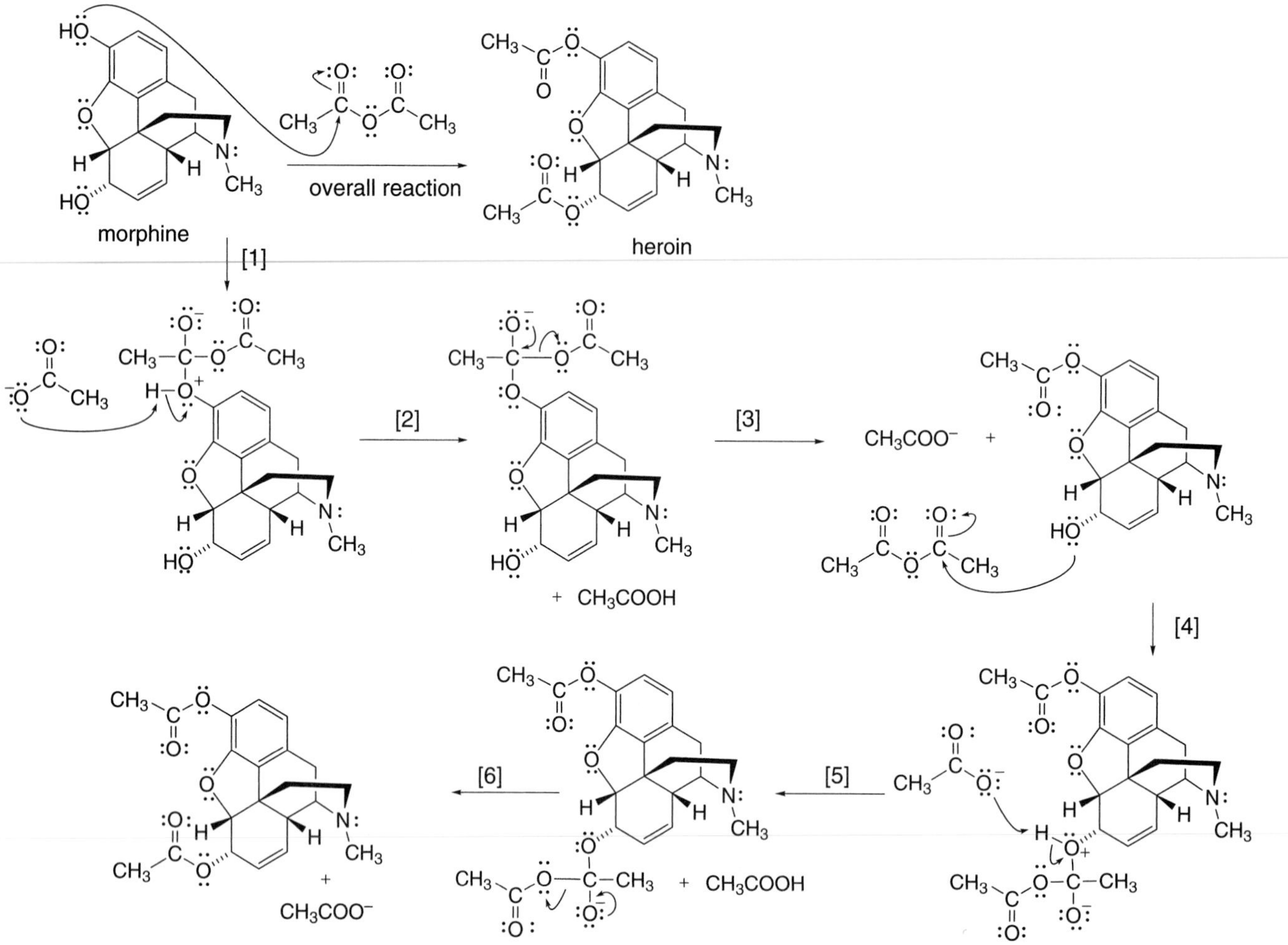
morphine
overall reaction
heroin
[1]
[2]
[3]
[4]
[5]
[6]
+ CH3COOH
CH3COO⁻ +
+ CH3COOH
CH3COO⁻

**22.18** Reaction of a carboxylic acid with thionyl chloride converts it to an acid chloride.

a. 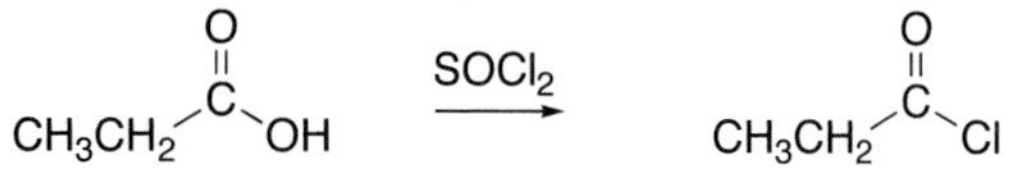
CH3CH2 OH
SOCl2
CH3CH2 Cl

b. 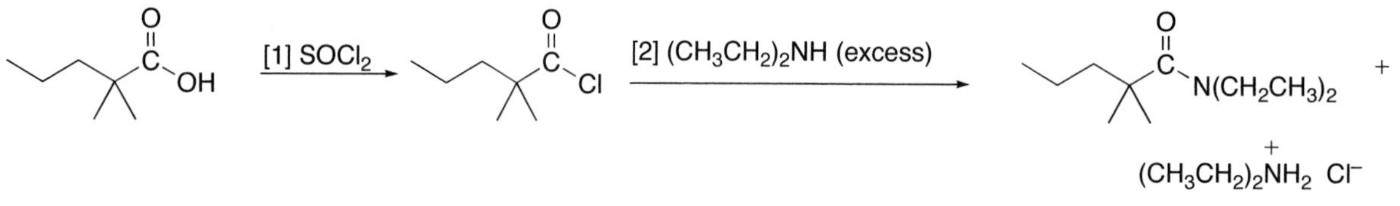
OH
[1] SOCl2
Cl
[2] (CH3CH2)2NH (excess)
N(CH2CH3)2
(CH3CH2)2NH2 Cl⁻

**22.19**

a. 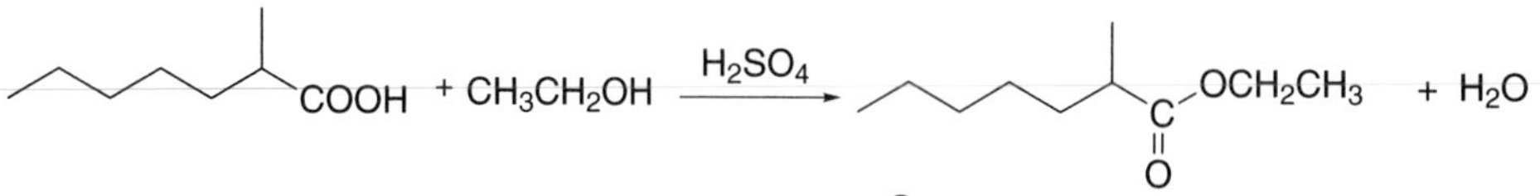
COOH
+ CH3CH2OH
H2SO4
OCH2CH3
+ H2O

b. 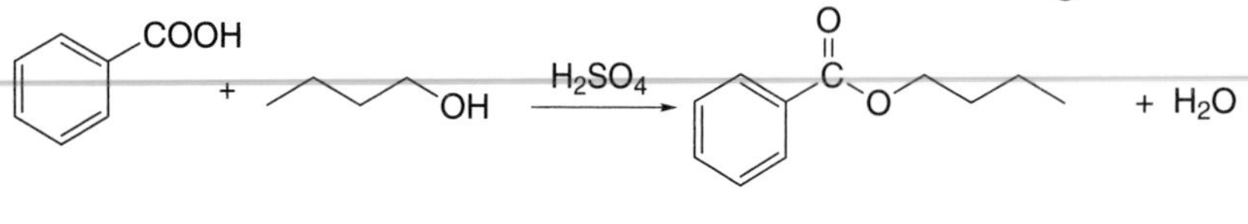
COOH
OH
H2SO4
+ H2O

c.

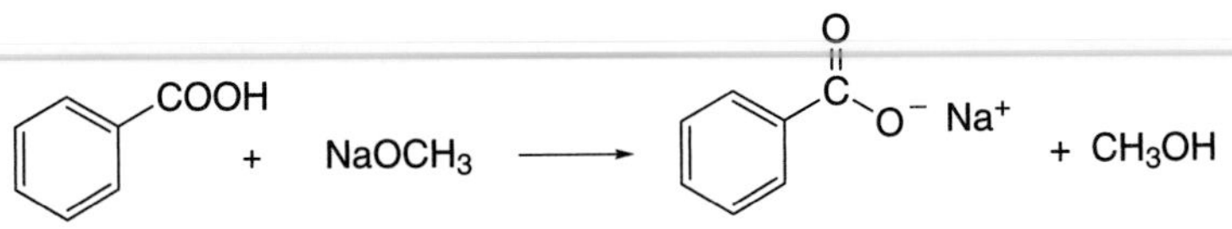

+ NaOCH$_3$ → benzoate + CH$_3$OH

d.

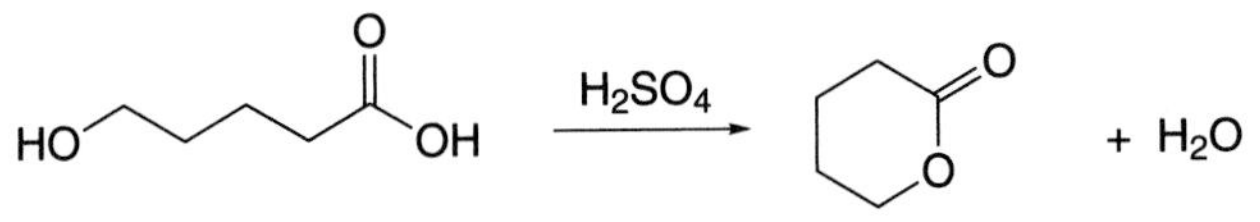

## 22.20

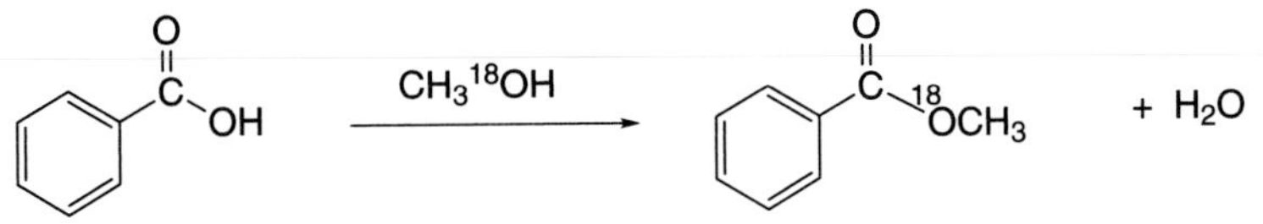

## 22.21

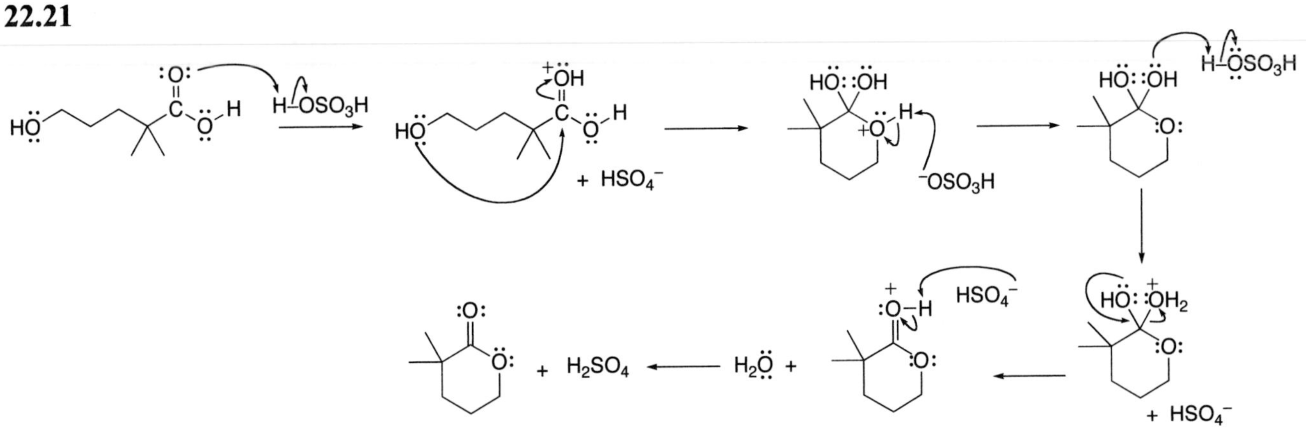

## 22.22

a.

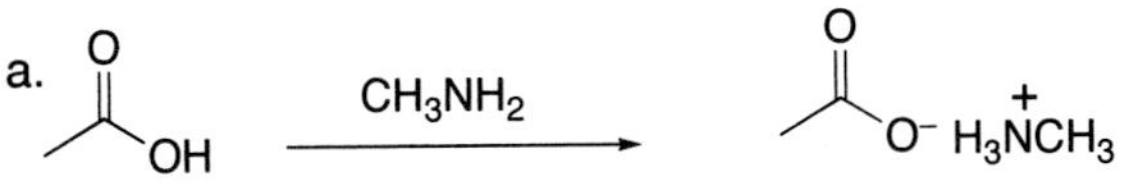

CH$_3$NH$_2$ →

c.

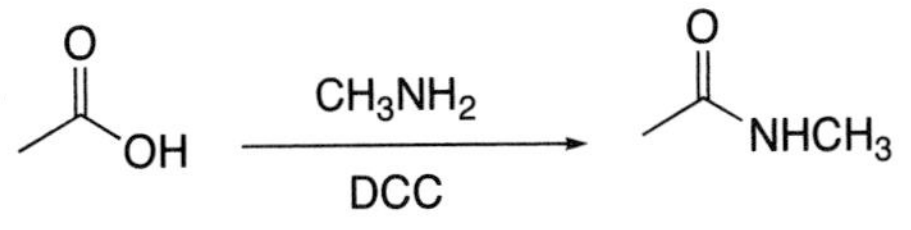

CH$_3$NH$_2$, DCC →

b.

[1] CH$_3$NH$_2$
[2] Δ

→ NHCH$_3$ + H$_2$O

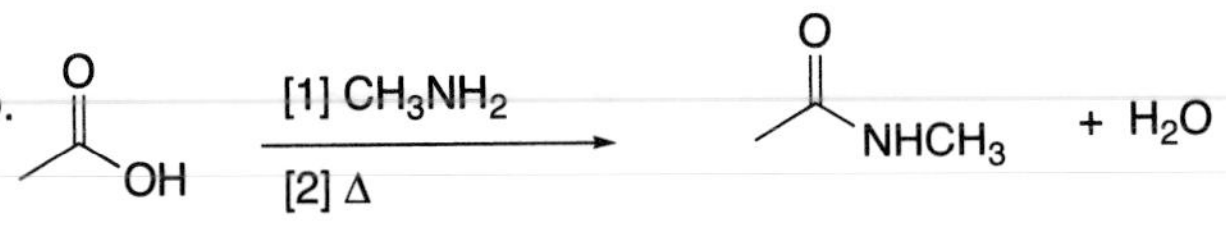

## 22.23

product of step [1]          product of step [5]

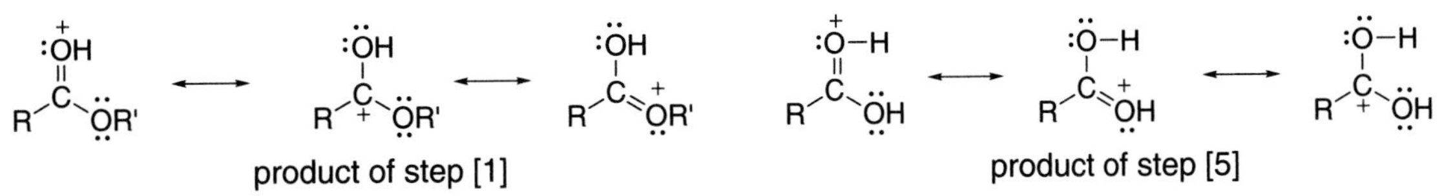

**22.24**

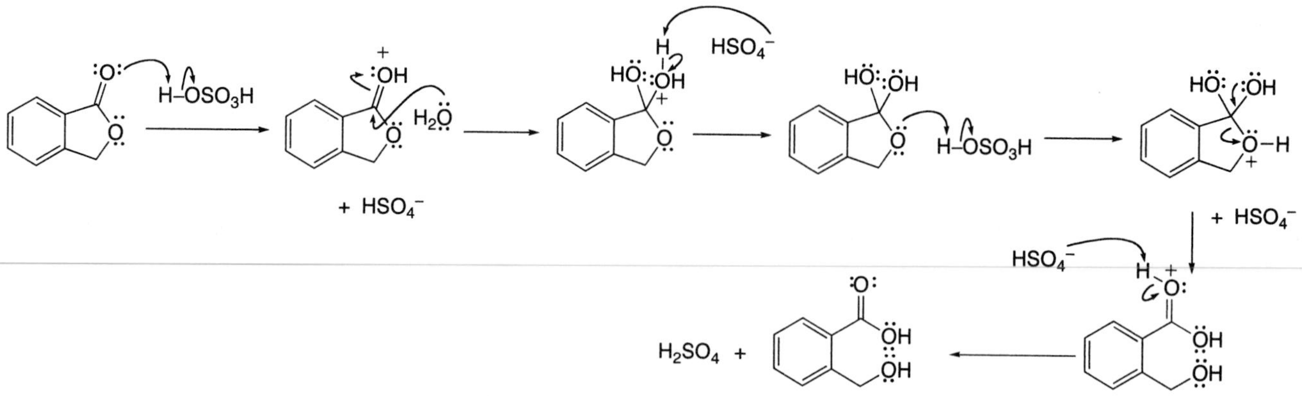

**22.25**  The ester is cleaved in these reactions to form a carboxylic acid and an alcohol.

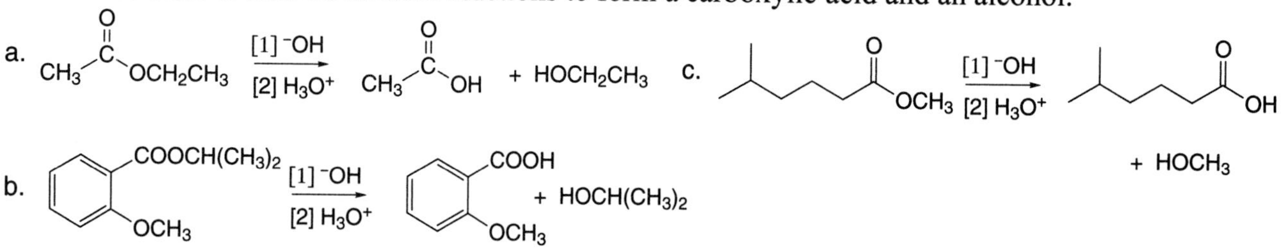

a.

b.

c.

**22.26**

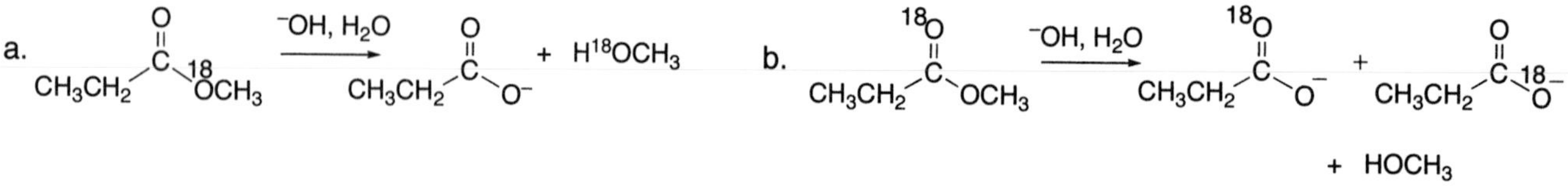

a.

b.

**22.27**

sucrose → olestra

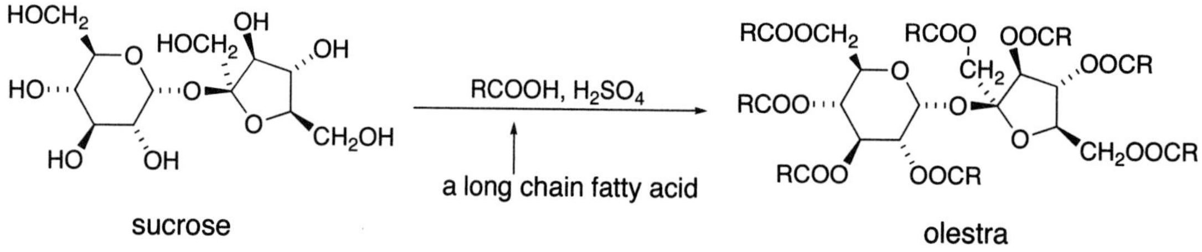

RCOOH, $H_2SO_4$

a long chain fatty acid

**22.28**

glycerol + soap

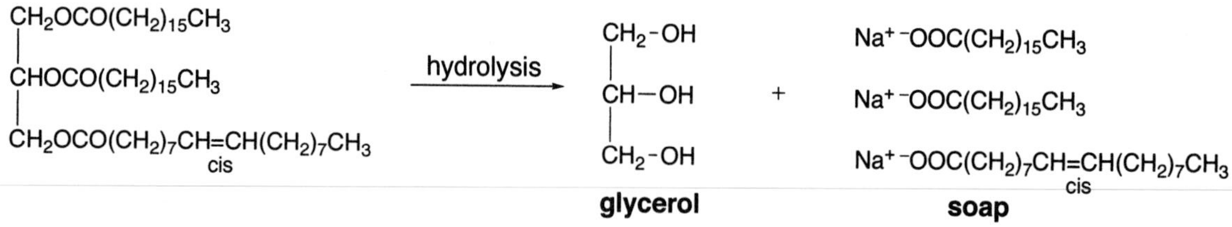

**22.29**

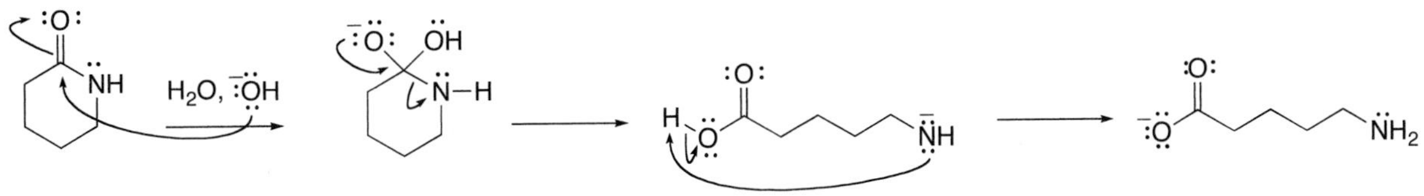

**22.30** Aspirin has an ester, a more reactive acyl group, but acetaminophen has an amide, a less reactive acyl group.

a. The ester makes aspirin more easily hydrolyzed with water from the air than acetaminophen. Therefore, Tylenol can be kept for many years, whereas aspirin decomposes.

b. Similarly, aspirin will be hydrolyzed and decompose in the aqueous medium of a liquid medication, but acetaminophen is stable due to the less reactive amide group, allowing it to remain unchanged while dissolved in $H_2O$.

acetylsalicylic acid          acetaminophen

**22.31**

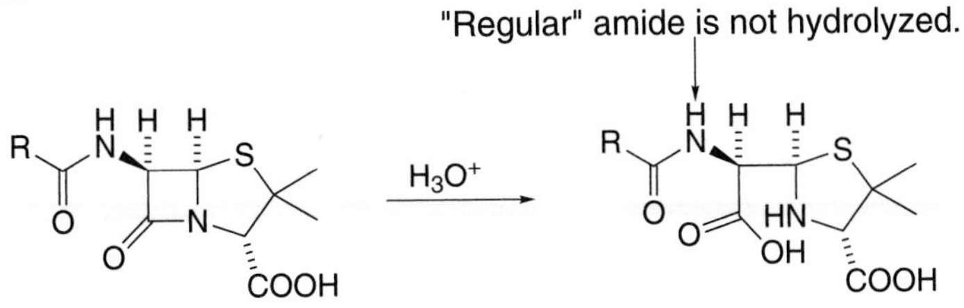

"Regular" amide is not hydrolyzed.

**22.32**

nylon 6,10

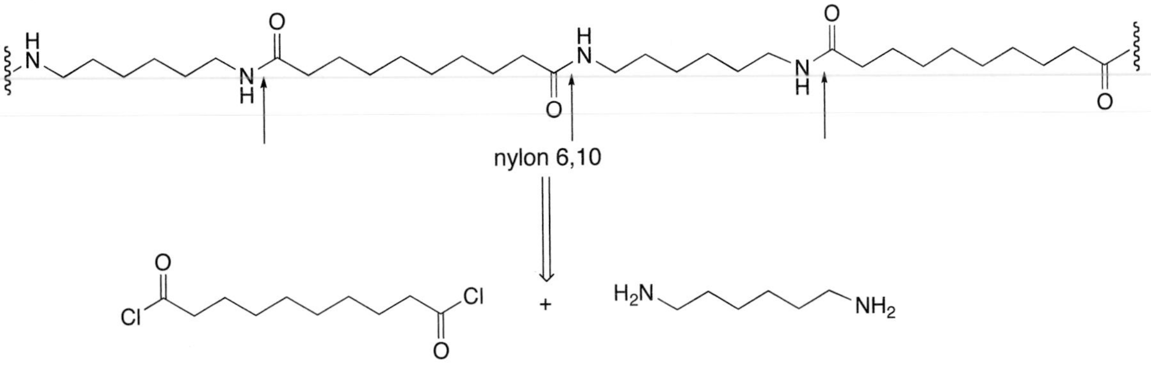

**22.33**

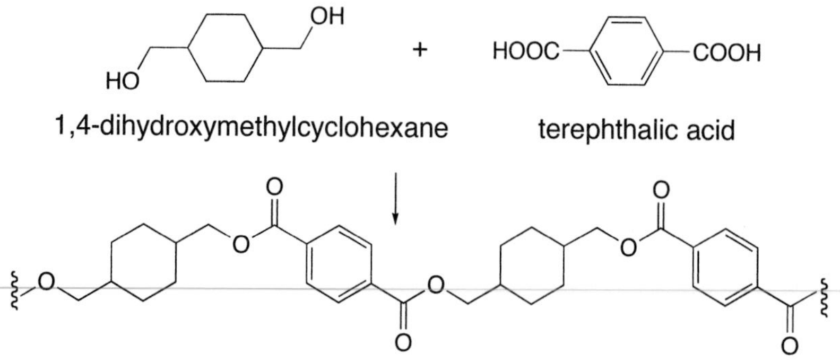

1,4-dihydroxymethylcyclohexane      terephthalic acid

In the polyester Kodel, most of the bonds in the polymer backbone are part of a ring, so there are fewer degrees of freedom. Fabrics made from Kodel are stiff and crease resistant, due to these less flexible polyester fibers.

Kodel

**22.34**

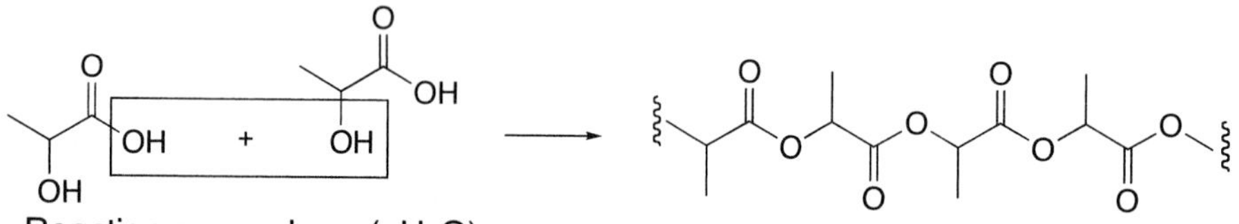

Reaction occurs here (–H$_2$O).

**PLA**
polylactic acid

**22.35**

a.

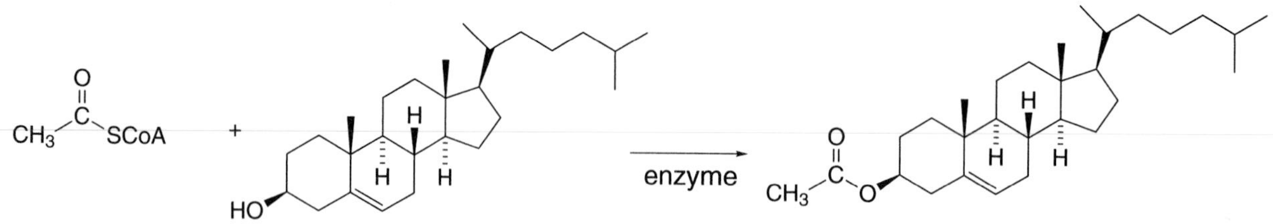

b.

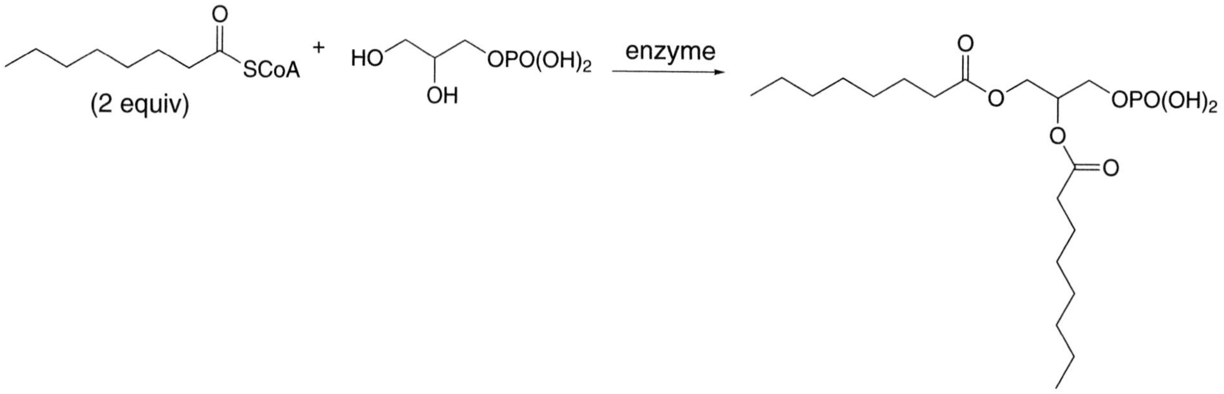

(2 equiv)

**22.36**

a.  CH$_3$CH$_2$CH$_2$–Br   $\xrightarrow{\text{NaCN}}$   CH$_3$CH$_2$CH$_2$–CN

c.

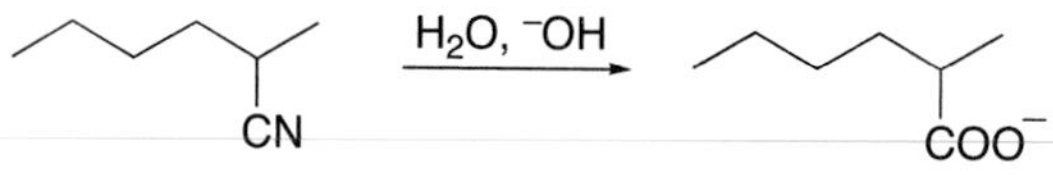

b.

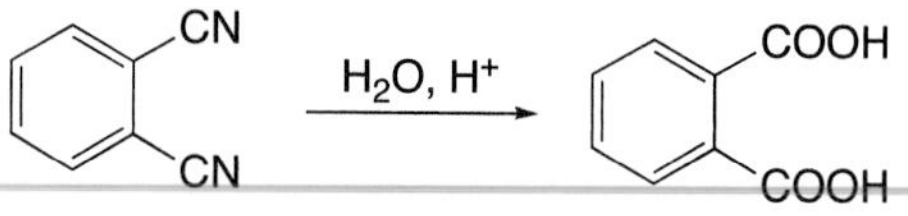

**22.37**

a.

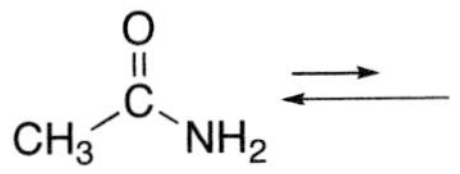

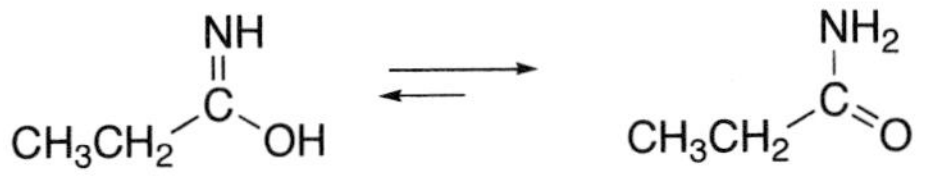

b.

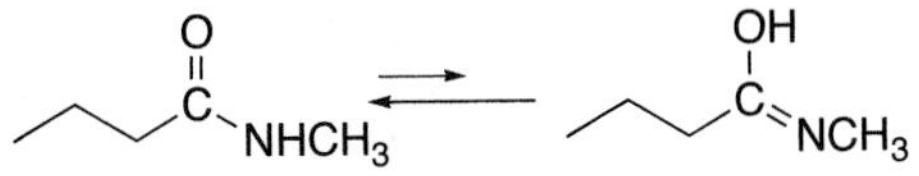

c.

**22.38**

a. 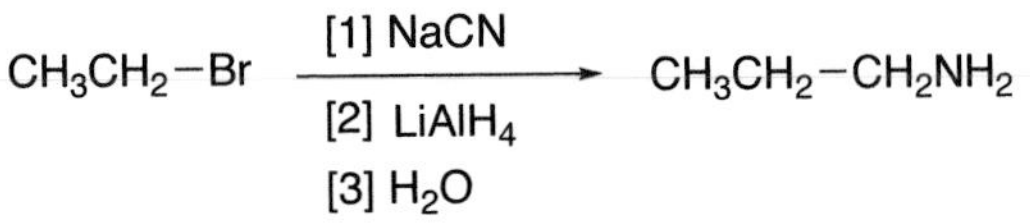 $CH_3CH_2-Br$ $\xrightarrow[\begin{array}{c}[2]\ LiAlH_4\\ [3]\ H_2O\end{array}]{[1]\ NaCN}$ $CH_3CH_2-CH_2NH_2$

b. 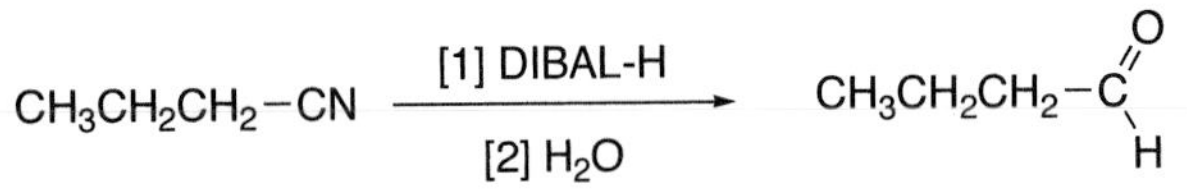 $CH_3CH_2CH_2-CN$ $\xrightarrow[\begin{array}{c}[2]\ H_2O\end{array}]{[1]\ DIBAL\text{-}H}$ $CH_3CH_2CH_2-\overset{\displaystyle O}{\underset{\displaystyle H}{C}}$

**22.39**

a.

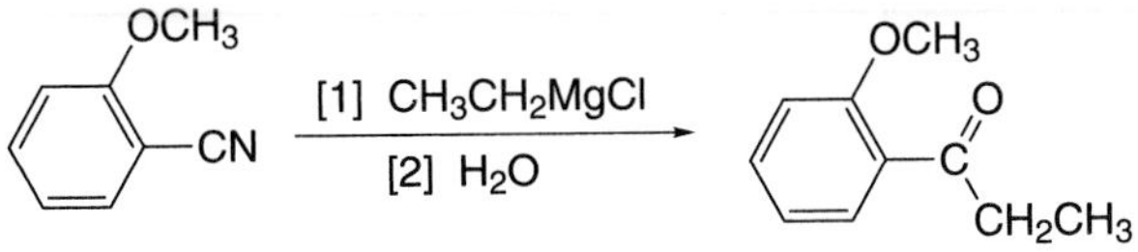

b.

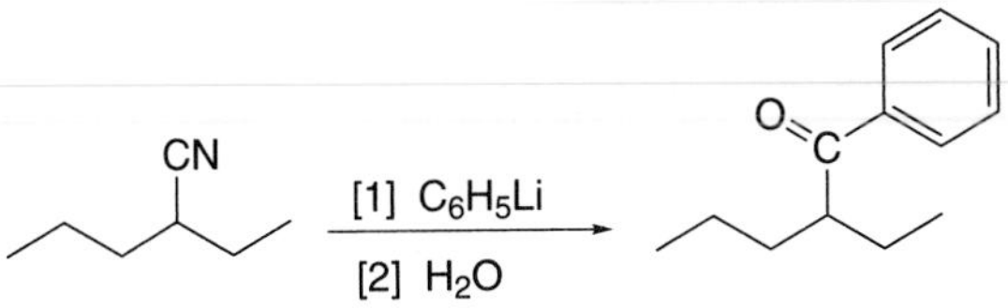

**22.40**

a.

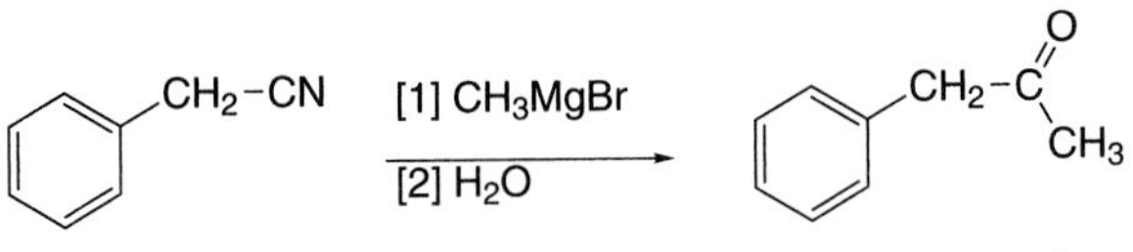

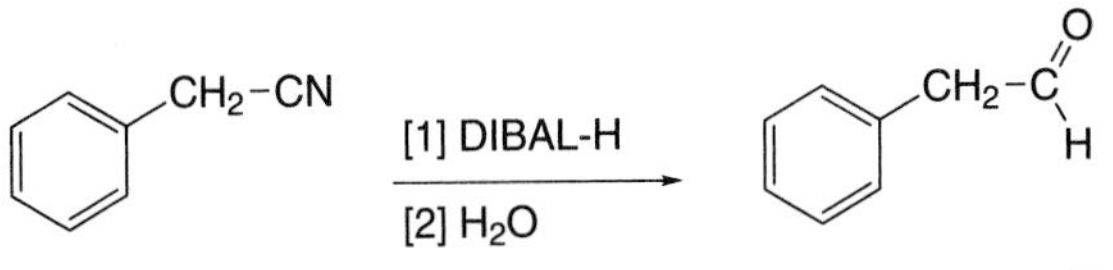

b.

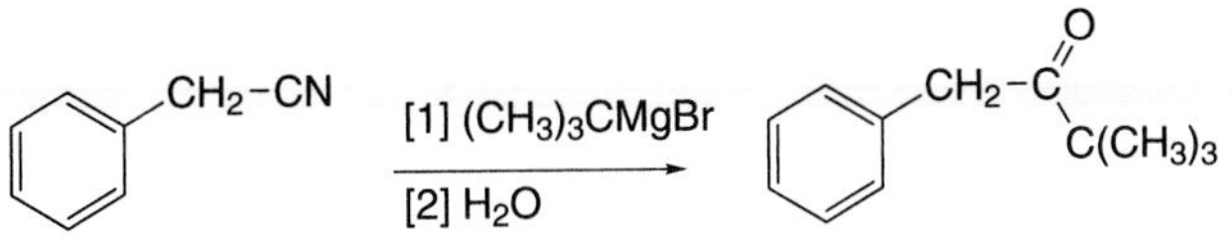

c.

d.

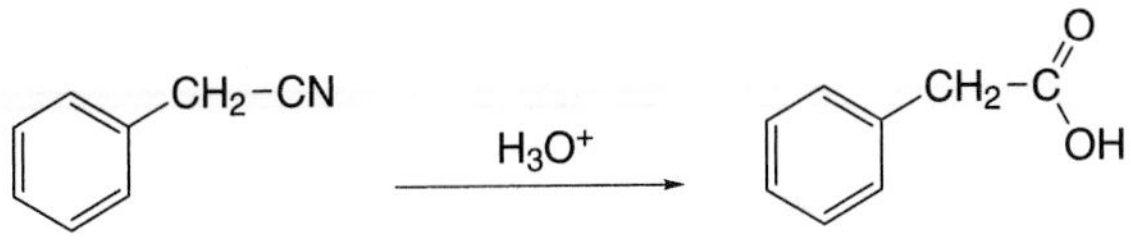

**22.41**

$CH_3-C\equiv N$ $\xrightarrow[\begin{array}{c}[2]\ H_2O\end{array}]{[1]\ CH_3CH_2MgBr}$

$CH_3CH_2-C\equiv N$ $\xrightarrow[\begin{array}{c}[2]\ H_2O\end{array}]{[1]\ CH_3MgBr}$

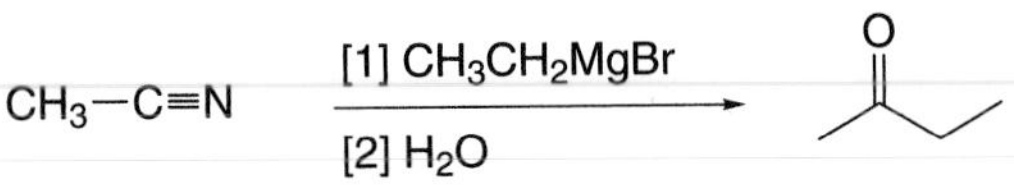

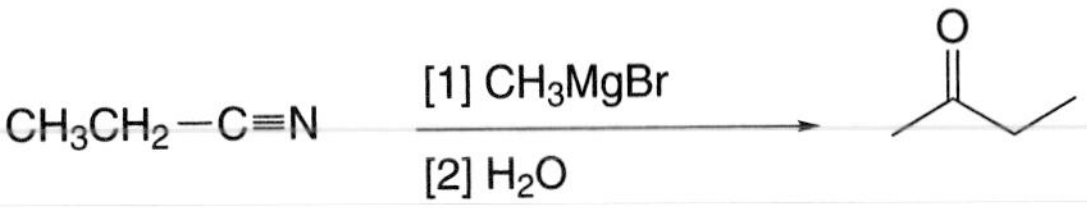

Chapter 22–16

## 22.42

a. 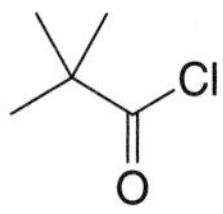 **2,2-dimethylpropanoyl chloride**

b. 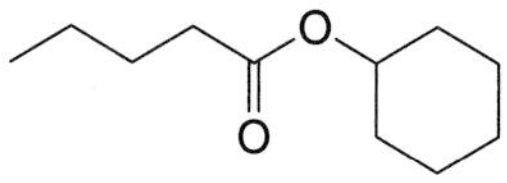 **cyclohexyl pentanoate**

c. 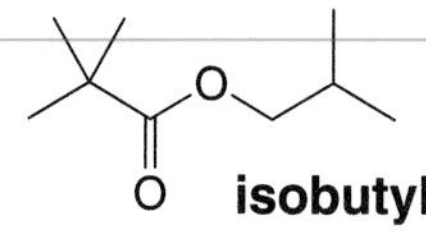 **isobutyl 2,2-dimethylpropanoate**

d. 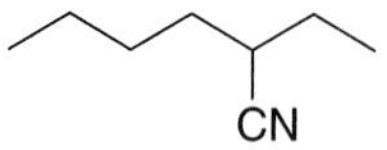 **2-ethylhexanenitrile**

e. 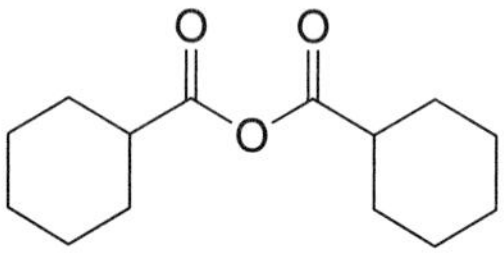 **cyclohexanecarboxylic anhydride**

f. 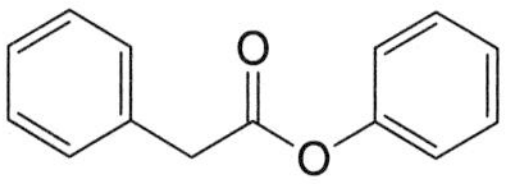 **phenyl phenylacetate**

g. 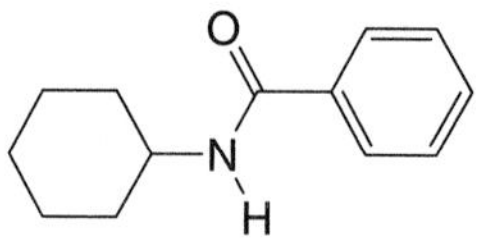 **N-cyclohexylbenzamide**

h. 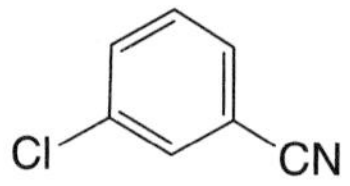 **m-chlorobenzonitrile**

i. 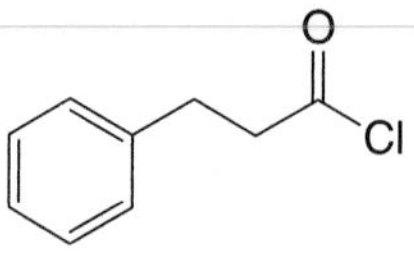 **3-phenylpropanoyl chloride**

j. 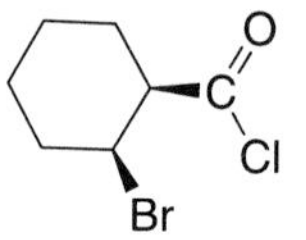 **cis-2-bromocyclohexane-carbonyl chloride**

k. 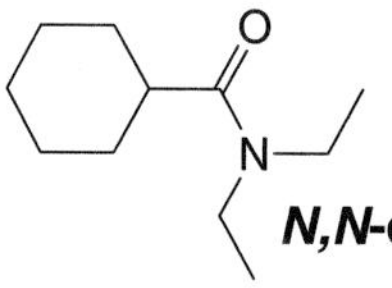 **N,N-diethylcyclohexanecarboxamide**

l. 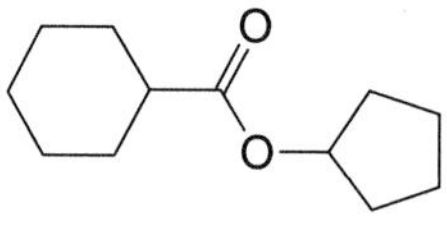 **cyclopentyl cyclohexanecarboxylate**

## 22.43

a. propanoic anhydride

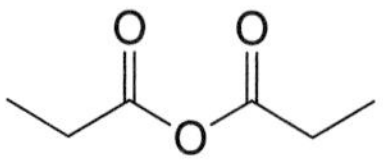

b. α-chlorobutyryl chloride

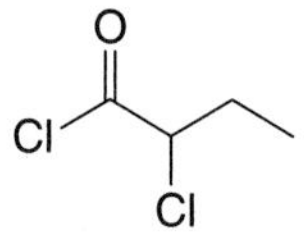

c. cyclohexyl propanoate

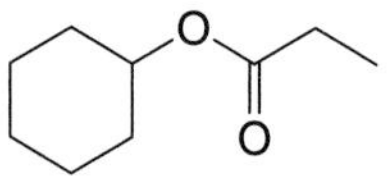

d. cyclohexanecarboxamide

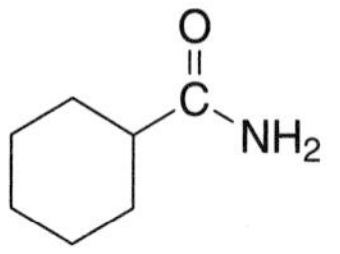

e. isopropyl formate

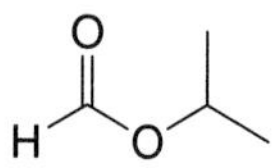

f. N-cyclopentylpentanamide

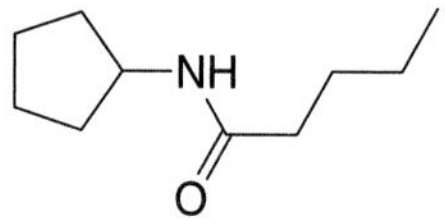

g. 4-methylheptanenitrile

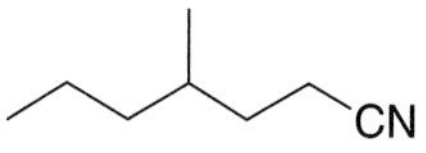

h. vinyl acetate

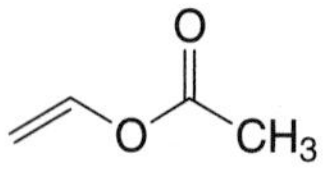

i. benzoic propanoic anhydride

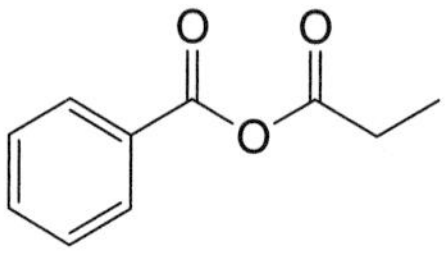

j. 3-methylhexanoyl chloride

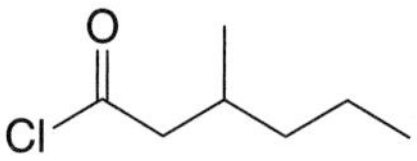

k. octyl butanoate

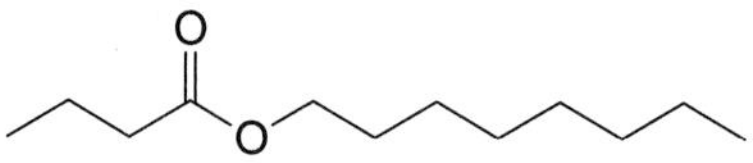

l. N,N-dibenzylformamide

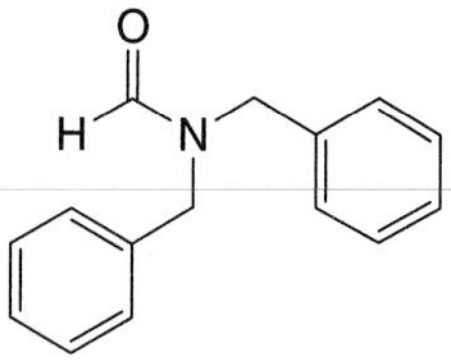

**22.44** Rank the compounds using the rules from Answer 22.12.

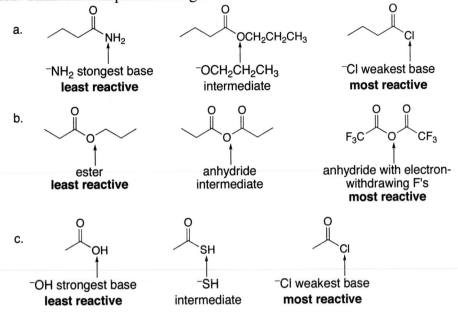

a.

$^-NH_2$ stongest base
**least reactive**

$^-OCH_2CH_2CH_3$
intermediate

$^-Cl$ weakest base
**most reactive**

b.

ester
**least reactive**

anhydride
intermediate

anhydride with electron-
withdrawing F's
**most reactive**

c.

$^-OH$ strongest base
**least reactive**

$^-SH$
intermediate

$^-Cl$ weakest base
**most reactive**

## 22.45

a. Better leaving groups make acyl compounds more reactive. **A** has an electron-withdrawing $NO_2$ group, which stabilizes the negative charge of the leaving group, whereas **B** has an electron-donating $OCH_3$ group, which destabilizes the leaving group.

**A**

**B**

an electron-withdrawing substituent    an electron-donating substituent

leaving group from **A**        one possible
resonance structure

Delocalizing the negative charge on the
$NO_2$ stabilizes the leaving group
making **A** more reactive than **B**.

leaving group from **B**        one possible
resonance structure

Adjacent negative charges destabilize
the leaving group.

resonance structures for the leaving group

b.

imidazolide

The leaving group is both resonance stabilized and aromatic
($6\pi$ electrons), making it a much better leaving group than
exists in a regular amide.

**22.46**

**Reaction as an acid:**

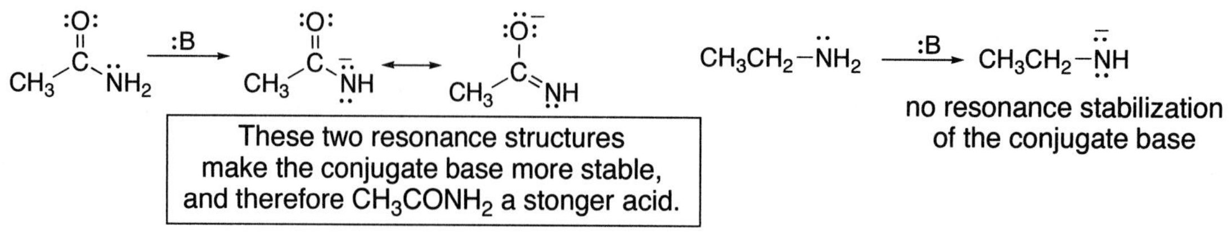

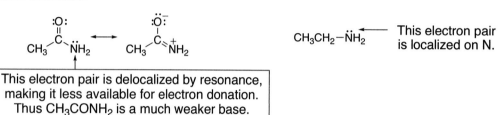

These two resonance structures
make the conjugate base more stable,
and therefore CH₃CONH₂ a stonger acid.

$CH_3CH_2-\overset{..}{N}H_2$ $\xrightarrow{:B}$ $CH_3CH_2-\overset{..}{\underset{..}{N}}H$

no resonance stabilization
of the conjugate base

**Reaction as a base:**

This electron pair is delocalized by resonance,
making it less available for electron donation.
Thus CH₃CONH₂ is a much weaker base.

$CH_3CH_2-\overset{..}{N}H_2$ ← This electron pair
is localized on N.

**22.47**

CH₃CH₂CH₂CH₂COCl

a. $\xrightarrow[\text{pyridine}]{H_2O}$ CH₃CH₂CH₂CH₂COOH + [pyridinium] Cl⁻

b. $\xrightarrow[\text{pyridine}]{CH_3CH_2OH}$ CH₃CH₂CH₂CH₂COOCH₂CH₃ + [pyridinium] Cl⁻

c. $\xrightarrow{CH_3COO^-}$ CH₃CH₂CH₂CH₂—O—...—CH₃ + Cl⁻

d. $\xrightarrow[\text{excess}]{NH_3}$ CH₃CH₂CH₂CH₂CONH₂ + $\overset{+}{N}H_4$ Cl⁻

e. $\xrightarrow[\text{excess}]{(CH_3CH_2)_2NH}$ CH₃CH₂CH₂CH₂CON(CH₂CH₃)₂
+ $(CH_3CH_2)_2\overset{+}{N}H_2$ Cl⁻

f. $\xrightarrow[\text{excess}]{C_6H_5NH_2}$ CH₃CH₂CH₂CH₂CONHC₆H₅
+ $C_6H_5\overset{+}{N}H_3$ Cl⁻

**22.48**

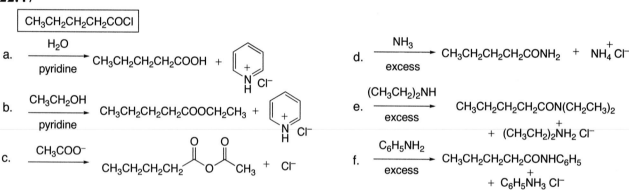

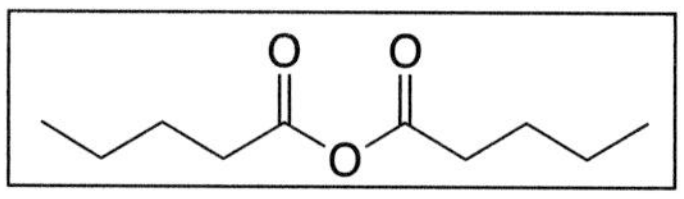

a. $\xrightarrow{SOCl_2}$ no reaction

b. $\xrightarrow{H_2O}$ 2

c. $\xrightarrow{CH_3OH}$

d. $\xrightarrow{NaCl}$ no reaction

e. $\xrightarrow[\text{excess}]{(CH_3CH_2)_2NH}$

f. $\xrightarrow[\text{excess}]{CH_3CH_2NH_2}$

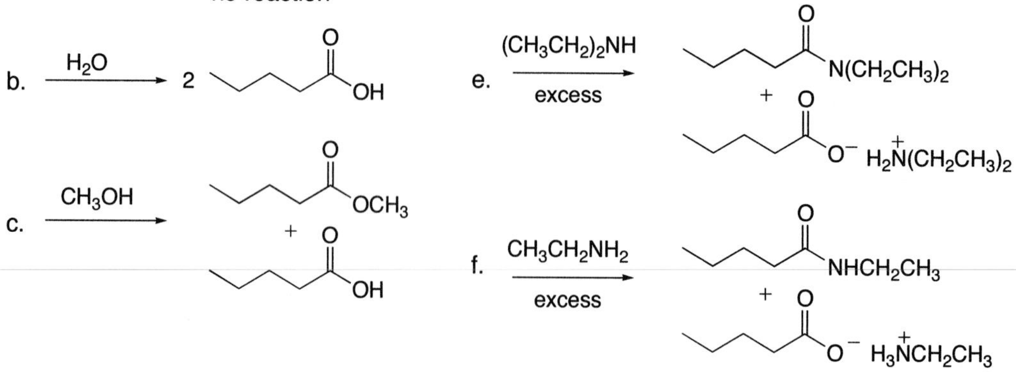

## 22.49

$C_6H_5CH_2COOH$

a. $\xrightarrow{\text{NaHCO}_3}$ [phenylacetate sodium salt] $+ H_2CO_3$

g. $\xrightarrow[\text{H}_2\text{SO}_4]{\text{CH}_3\text{OH}}$ [methyl phenylacetate, OCH$_3$]

b. $\xrightarrow{\text{NaOH}}$ [phenylacetate sodium salt] $+ H_2O$

h. $\xrightarrow[^-\text{OH}]{\text{CH}_3\text{OH}}$ [phenylacetate carboxylate, O$^-$]

c. $\xrightarrow{\text{SOCl}_2}$ [phenylacetyl chloride, Cl]

i. $\xrightarrow[\text{[2] CH}_3\text{COCl}]{\text{[1] NaOH}}$ [mixed anhydride]

d. $\xrightarrow{\text{NaCl}}$ no reaction

j. $\xrightarrow[\text{DCC}]{\text{CH}_3\text{NH}_2}$ [N-methyl phenylacetamide, NHCH$_3$]

e. $\xrightarrow[\text{(1 equiv)}]{\text{NH}_3}$ [ammonium phenylacetate, O$^-$ NH$_4^+$]

k. $\xrightarrow[\text{[2] CH}_3\text{CH}_2\text{CH}_2\text{NH}_2]{\text{[1] SOCl}_2}$ [N-propyl phenylacetamide, NHCH$_2$CH$_2$CH$_3$]

f. $\xrightarrow[\text{[2] }\Delta]{\text{[1] NH}_3}$ [phenylacetamide, NH$_2$]

l. $\xrightarrow[\text{[2] (CH}_3)_2\text{CHOH}]{\text{[1] SOCl}_2}$ [isopropyl phenylacetate, OCH(CH$_3$)$_2$]

## 22.50

$CH_3CH_2CH_2CO_2CH_2CH_3$

a. $\xrightarrow{\text{SOCl}_2}$ no reaction

b. $\xrightarrow{\text{H}_3\text{O}^+}$ [butanoic acid, OH] $+$ HO [ethanol]

c. $\xrightarrow{\text{H}_2\text{O, }^-\text{OH}}$ [butanoate, O$^-$] $+$ HO [ethanol]

d. $\xrightarrow{\text{NH}_3}$ [butanamide, NH$_2$] $+$ HO [ethanol]

e. $\xrightarrow{\text{CH}_3\text{CH}_2\text{NH}_2}$ [N-ethyl butanamide, NHCH$_2$CH$_3$] $+$ HO [ethanol]

## 22.51

[phenylacetamide, NH$_2$, O]

a. $\xrightarrow{\text{H}_3\text{O}^+}$ [CH$_2$COOH]

b. $\xrightarrow{\text{H}_2\text{O, }^-\text{OH}}$ [CH$_2$COO$^-$]

## 22.52

a. $\xrightarrow{\text{H}_3\text{O}^+}$ [CH$_2$COOH]

$C_6H_5CH_2CN$

b. $\xrightarrow{\text{H}_2\text{O, }^-\text{OH}}$ [CH$_2$COO$^-$]

c. $\xrightarrow[\text{[2] H}_2\text{O}]{\text{[1] CH}_3\text{MgBr}}$ [ketone, CH$_3$]

d. $\xrightarrow[\text{[2] H}_2\text{O}]{\text{[1] CH}_3\text{CH}_2\text{Li}}$ [ketone]

e. $\xrightarrow[\text{[2] H}_2\text{O}]{\text{[1] DIBAL-H}}$ [aldehyde, C(=O)H]

f. $\xrightarrow[\text{[2] H}_2\text{O}]{\text{[1] LiAlH}_4}$ [CH$_2$NH$_2$]

**22.53**

a. [structure: COOH, OH benzene] →SOCl₂→ [structure: COCl, OH benzene]

g.  $CH_3CH_2CH_2CH_2Br$  →[1] NaCN / [2] H₂O, ⁻OH→  $CH_3CH_2CH_2CH_2$COO⁻

b.  $C_6H_5COCl$  + [pyrrolidine N–H] (excess) → [C₆H₅CO–N pyrrolidine] + [pyrrolidine N⁺H H] Cl⁻

h. $C_6H_5CH_2COOH$ →[1] SOCl₂ / [2] CH₃CH₂CH₂CH₂NH₂ / [3] LiAlH₄ / [4] H₂O→ $C_6H_5(CH_2)_2NH(CH_2)_3CH_3$

c.  $C_6H_5CN$ →[1] CH₃CH₂CH₂MgBr / [2] H₂O→  [structure: C₆H₅–CO–CH₂CH₂CH₃]

i.  $C_6H_5CH_2CH_2CH_2CN$ →H₃O⁺→ $C_6H_5CH_2CH_2CH_2COOH$

d.  $(CH_3)_2CHCOOH$ + CH₃CH₂CHOH(CH₃) →H₂SO₄→ $(CH_3)_2CH$–CO–$OCH(CH_3)CH_2CH_3$

j.  HOOC [cis] COOH →Δ→ [maleic anhydride structure]

e. [structure: C₆H₅–NHCOCH₃] →H₂O / ⁻OH→ [CH₃COO⁻] + [C₆H₅–NH₂]

k.  $(CH_3CO)_2O$ + [cyclohexyl–NH₂] (excess) → [cyclohexyl–NHCOCH₃]

+ CH₃COO⁻ H₃N⁺–[cyclohexyl]

f. [bicyclic lactone structure] →H₃O⁺→ [cyclohexane with COOH and CH₂CH₂OH, ketone]

l.  $C_6H_5CH_2CH_2COOCH_2CH_3$ →H₂O / ⁻OH→ $C_6H_5CH_2CH_2COO^-$

+ HOCH₂CH₃

**22.54**

[cyclohexyl–CH₂Br] →NaCN→ [cyclohexyl–CH₂CN] **A** →H₃O⁺→ [cyclohexyl–CH₂COOH] **B** →SOCl₂→ [cyclohexyl–CH₂COCl] **D** →[1] (CH₃)₂CuLi / [2] H₂O→ [cyclohexyl–CH₂–CO–CH₃] **E**

**A** →[1] LiAlH₄ / [2] H₂O→ [cyclohexyl–CH₂CH₂NH₂] **C**

**B** →CH₃OH, H⁺→ [cyclohexyl–CH₂CO₂CH₃] **F** →[1] DIBAL-H / [2] H₂O→ [cyclohexyl–CH₂CHO] **G** →[1] CH₃Li / [2] H₂O→ [cyclohexyl–CH(–)–CH(OH)CH₃] **H** →PCC→ **E**

**C** →(CH₃CO)₂O→ [cyclohexyl–CH₂CH₂NHCOCH₃] **I**

**F** →[1] LiAlH₄ / [2] H₂O→ [cyclohexyl–CH₂CH₂OH] **J** →TsCl, pyridine→ [cyclohexyl–CH₂CH₂OTs] **K** →NaCN→ [cyclohexyl–CH₂CH₂CN] **L**

**L** →[1] CH₃MgBr / [2] H₂O→ [cyclohexyl–CH₂CH₂–CO–CH₃] **M**

**22.55**

a. [cyclohexane with OH and CH₃ substituents] $\xrightarrow[\text{pyridine}]{CH_3COCl}$ [cyclohexane with O-C(=O)-CH₃ and CH₃ substituents]

b. [compound with Br, H, D] $\xrightarrow{NaCN}$ [compound with CN, D, H]

c. $CH_3-\overset{\overset{H \quad D}{|}}{C}-COOH \xrightarrow[H^+]{CH_3CH_2OH} CH_3-\overset{\overset{H \quad D}{|}}{C}-COOCH_2CH_3$

d. $CH_3-\overset{O}{\overset{||}{C}}-Cl \ + \ C_6H_5-\overset{\overset{CH_3}{|}}{\underset{NH_2}{C}}{\cdots}H \longrightarrow C_6H_5-\overset{\overset{CH_3}{|}}{\underset{\underset{O}{\overset{|}{\underset{}{NH}}}}{C}}{\cdots}H \ + \ C_6H_5-\overset{\overset{CH_3}{|}}{\underset{\overset{+}{NH_3}}{C}}{\cdots}H$

(2 equiv)                  $\overset{||}{\underset{C}{}}-CH_3$            $Cl^-$

**22.56**

a. $\overset{O}{\overset{||}{Cl-C-Cl}} \xrightarrow[\text{excess}]{CH_3NH_2} CH_3NH-\overset{O}{\overset{||}{C}}-NHCH_3$

b. $\overset{O}{\overset{||}{Cl-C-Cl}} \xrightarrow{HOCH_2CH_2OH}$ [cyclic carbonate]

All H's on these C's are identical.

c. $\overset{O}{\overset{||}{Cl-C-Cl}} \xrightarrow[\text{excess}]{H_2O} HO-\overset{O}{\overset{||}{C}}-OH \ + \ HCl$

Reaction with $H_2O$ vapor in the lungs results in the formation of HCl, increasing the acidity of the lungs.

**22.57** Hydrolyze the amide and ester bonds in aspartame to draw the products.

[structure of aspartame] $\xrightarrow{H_2O}$ [aspartic acid product] + [phenylalanine product] + $CH_3OH$

aspartame                                          phenylalanine

**22.58**

a. [mechanism showing benzoyl chloride + NH₂NH₂ → tetrahedral intermediate → loss of Cl with pyridine → benzohydrazide + :Cl:⁻ + HN⁺-pyridine]

b. [mechanism showing lactone ring opening with ⁻OH → tetrahedral intermediate → carboxylic acid → proton transfer → carboxylate]

**22.59**

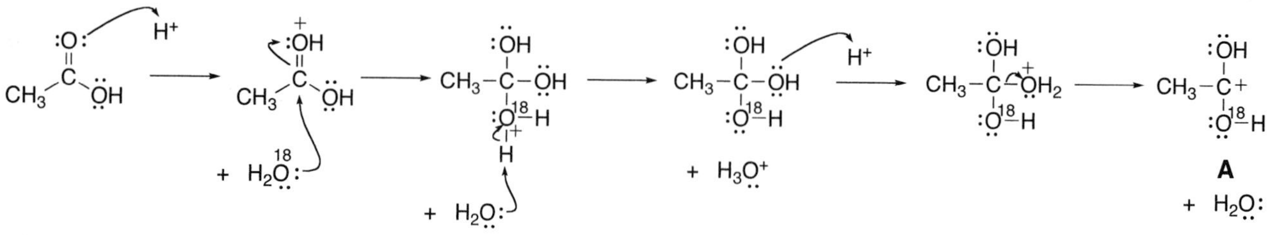

Two possibilities for **A**:

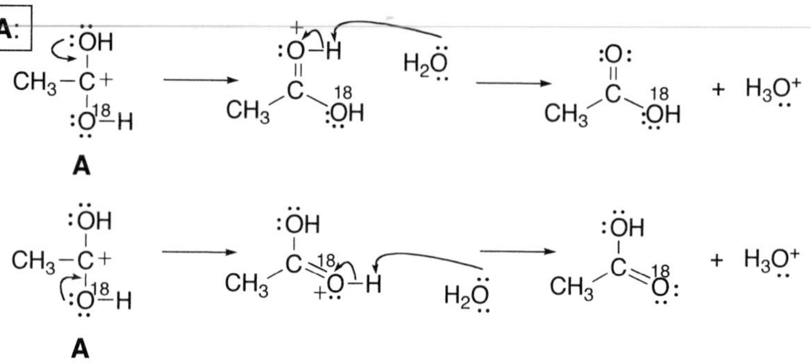

**22.60**

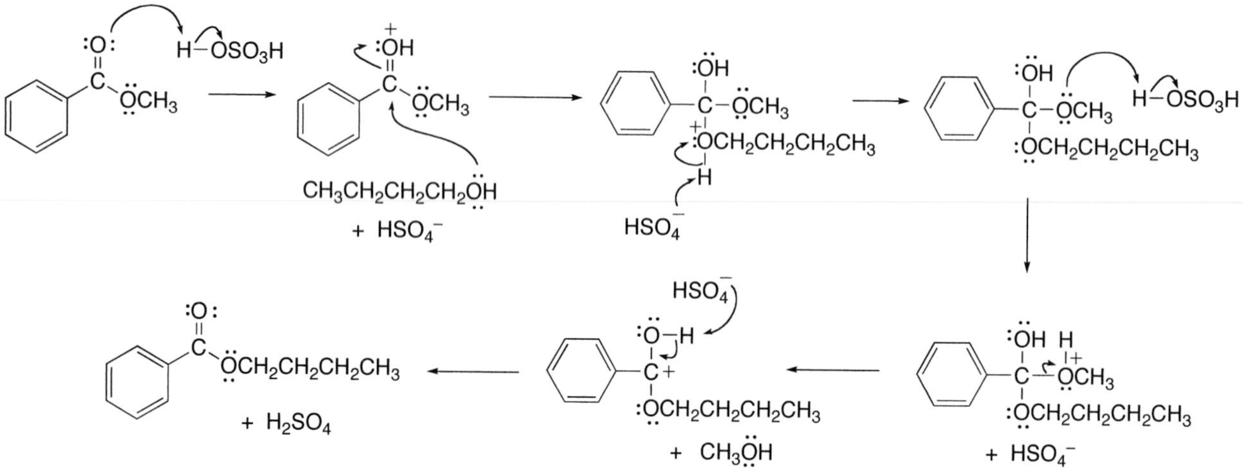

**22.61**

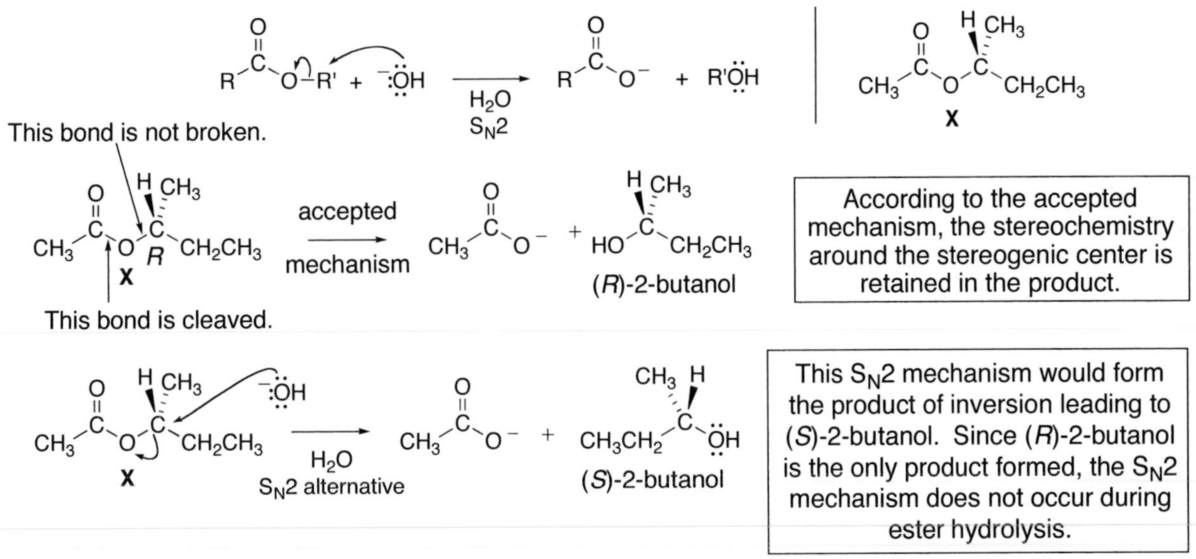

This bond is not broken.

This bond is cleaved.

accepted mechanism

(R)-2-butanol

According to the accepted mechanism, the stereochemistry around the stereogenic center is retained in the product.

SN2 alternative

(S)-2-butanol

This SN2 mechanism would form the product of inversion leading to (S)-2-butanol. Since (R)-2-butanol is the only product formed, the SN2 mechanism does not occur during ester hydrolysis.

**22.62** The mechanism is composed of two parts: hydrolysis of the acetal and intramolecular Fischer esterification of the hydroxy carboxylic acid.

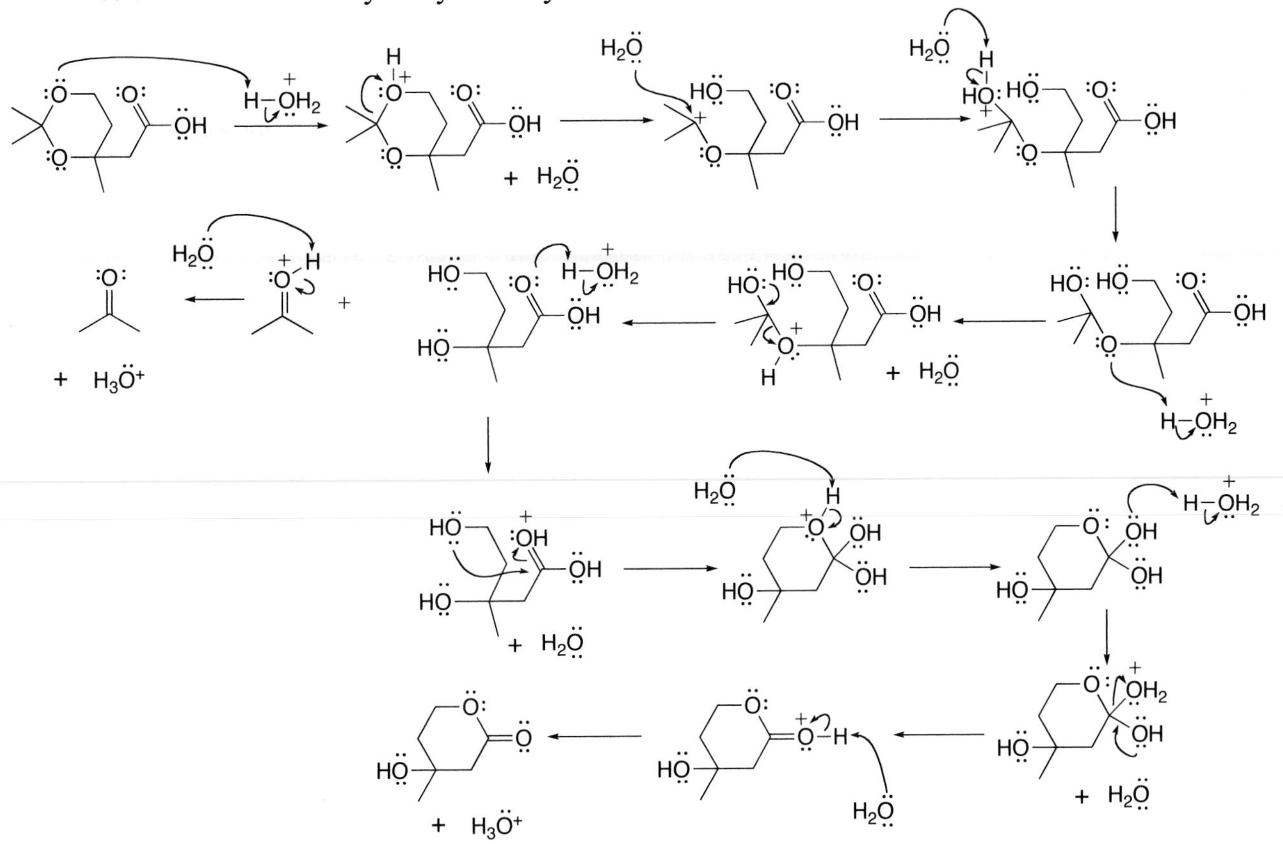

**22.63**

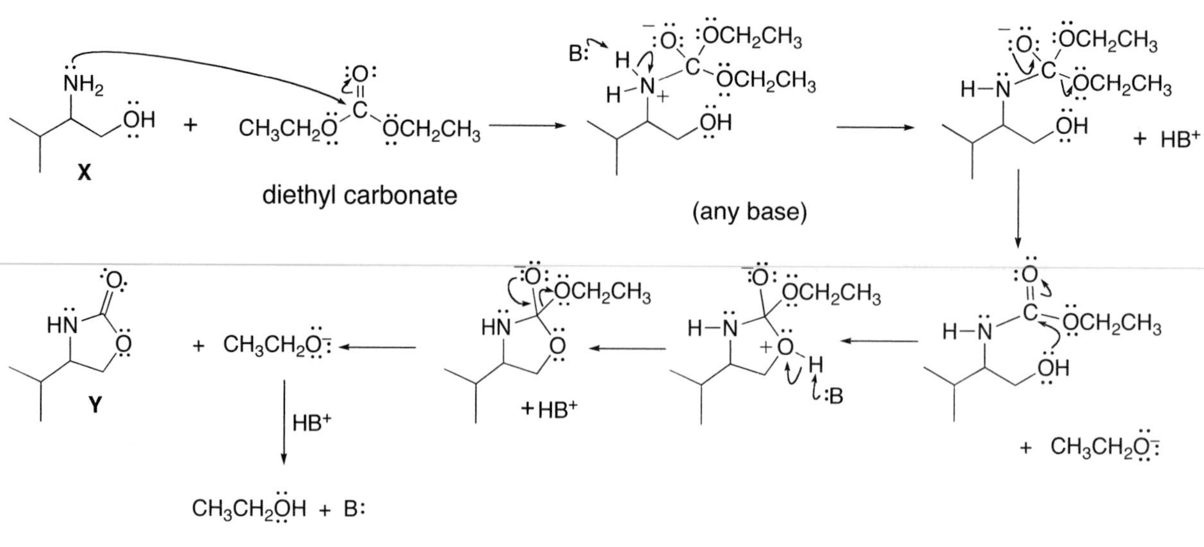

**22.64**

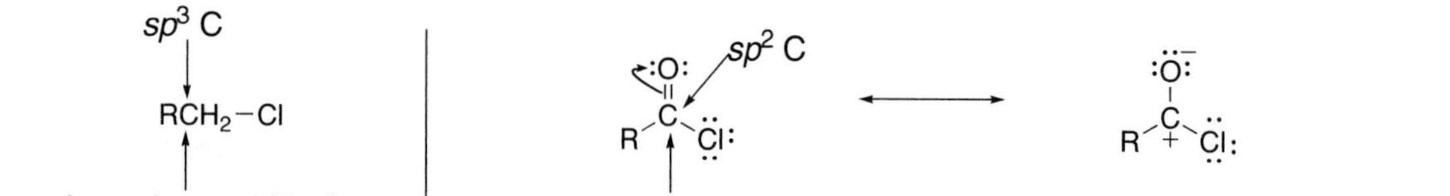

sp$^3$ C

RCH$_2$—Cl

less electrophilic C
more crowded C since it is
surrounded by four atoms

sp$^2$ C

more electrophilic C
due to electron-withdrawing O
more reactive

This resonance structure illustrates
how the electronegative O atom
withdraws more electron density
from C.

The sp$^2$ hybridized C of RCOCl is much less crowded,
and this makes nucleophilic attack easier as well.

**22.65**

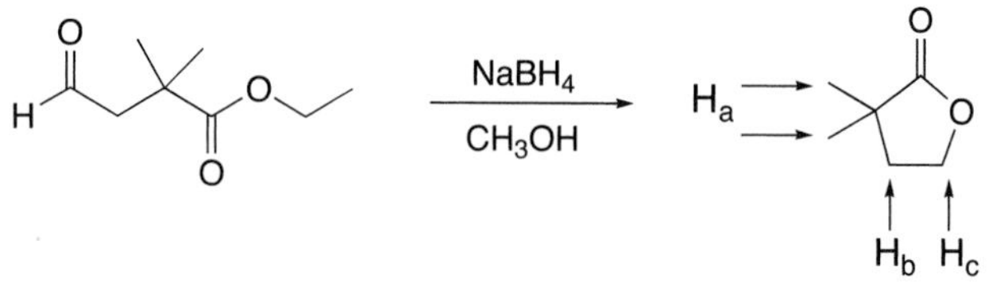

**A**
$C_6H_{10}O_2$: 2 degrees of unsaturation
IR: 1770 cm$^{-1}$ from ester C=O in a five-membered ring
$^1$H NMR: $H_a$ = 1.27 ppm (singlet, 6 H) – 2 CH$_3$ groups
$H_b$ = 2.12 ppm (triplet, 2 H) – CH$_2$ bonded to CH$_2$
$H_c$ = 4.26 ppm (triplet, 2 H) – CH$_2$ bonded to CH$_2$

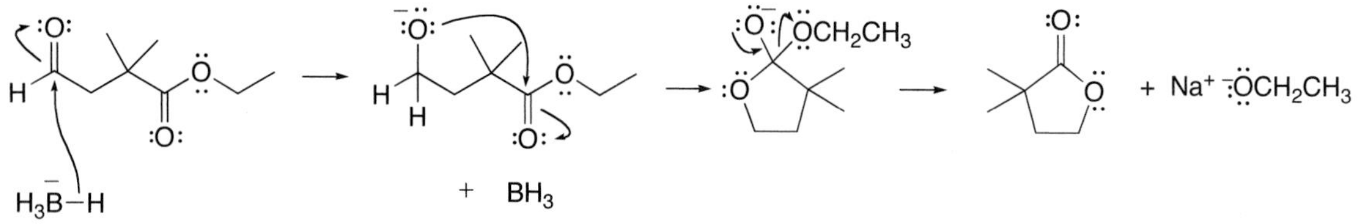

**22.66** Fischer esterification is treatment of a carboxylic acid with an alcohol in the presence of an acid catalyst to form an ester.

a. $(CH_3)_3CCO_2CH_2CH_3 \implies (CH_3)_3CCOOH$
$+ \ HOCH_2CH_3$

c.

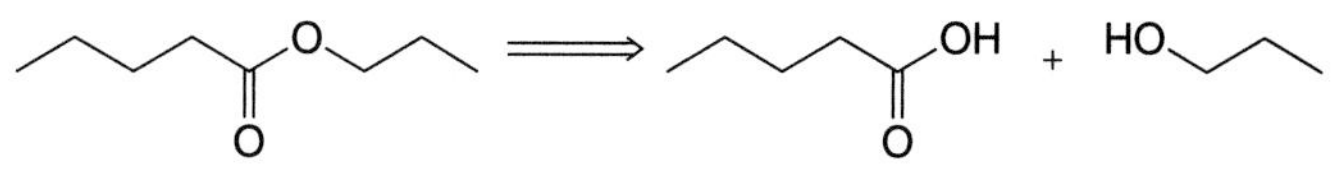

b.

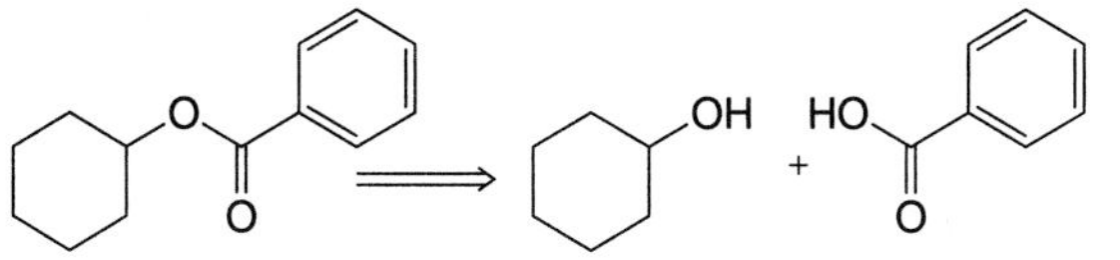

d.

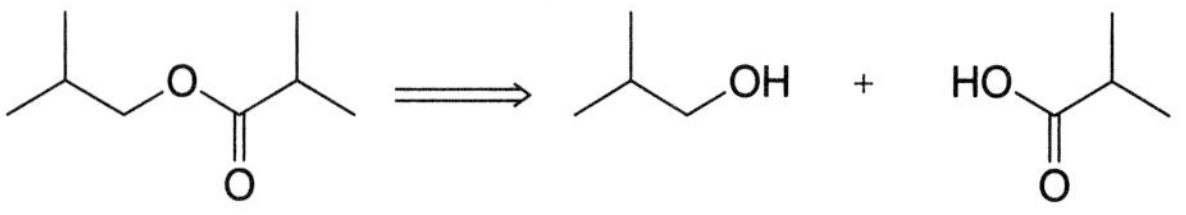

**22.67**

a.

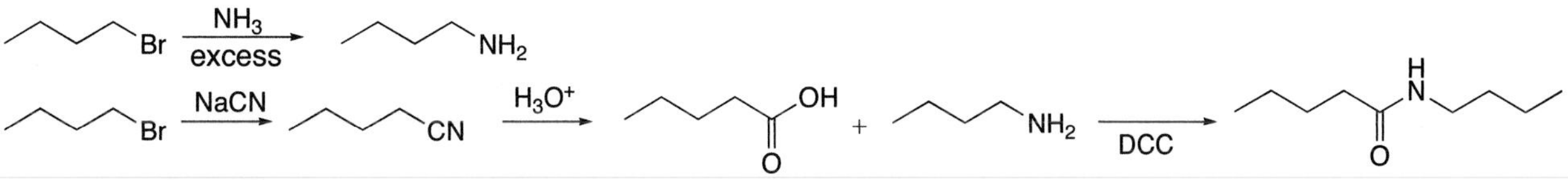

b.

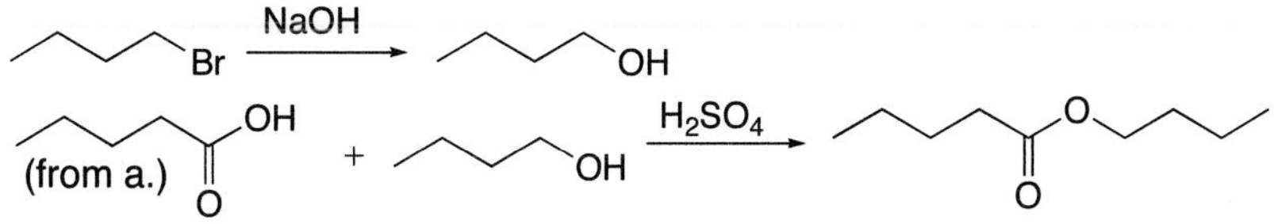

c.

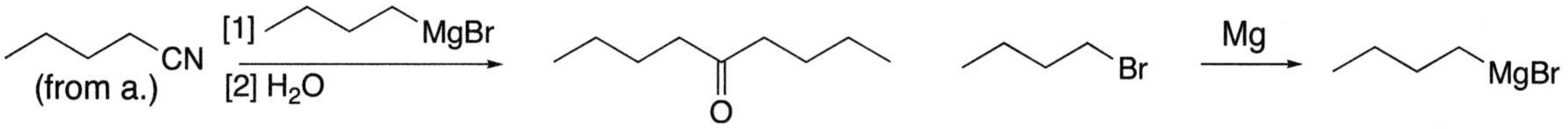

d.

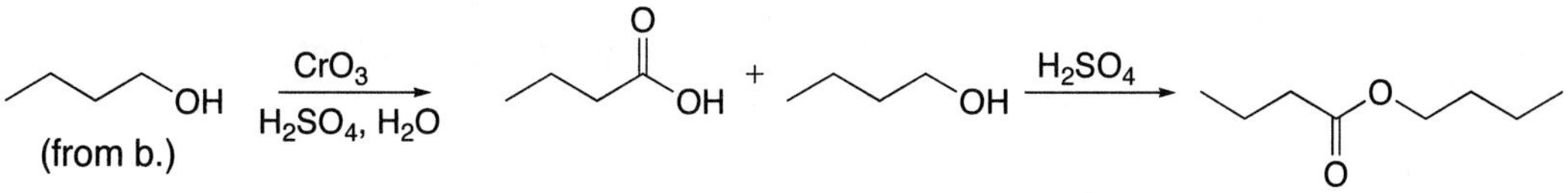

e.

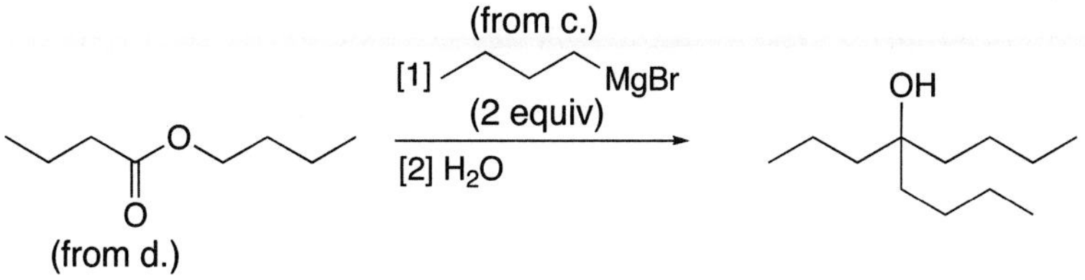

(from d.)

f.

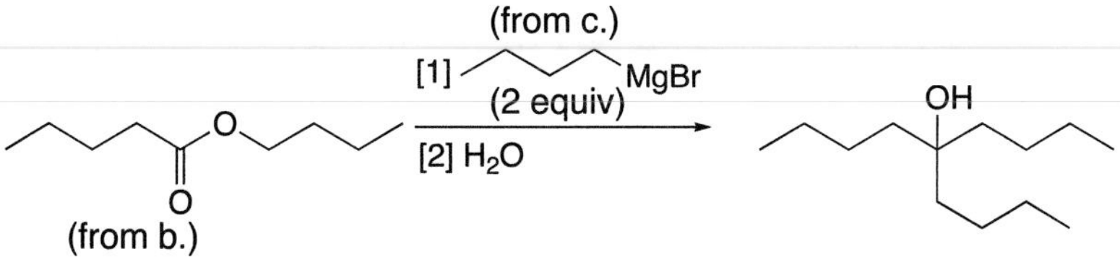

(from b.)

**22.68**

a.

b. 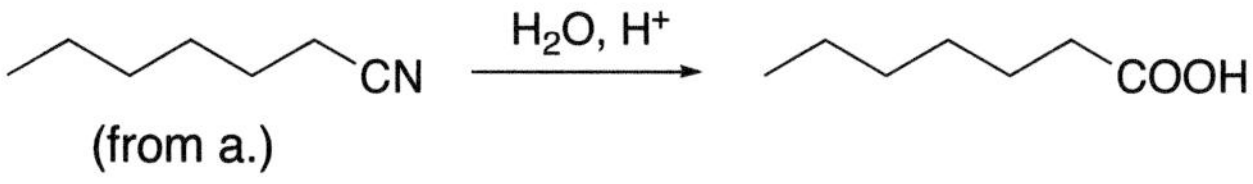
(from a.)

c. 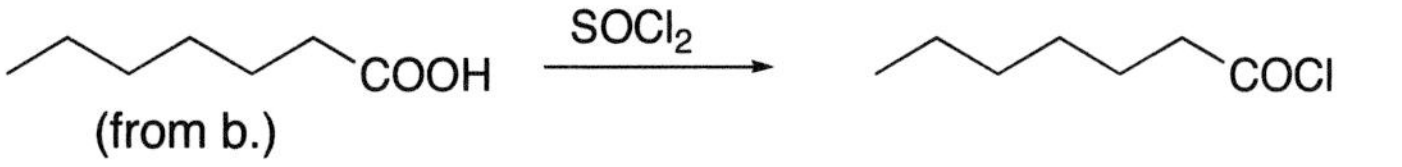
(from b.)

d. 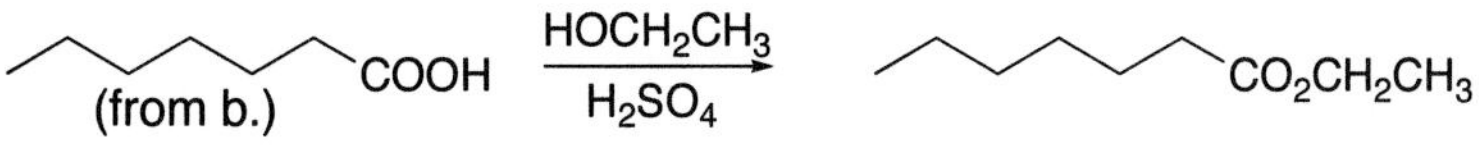
(from b.)

e.

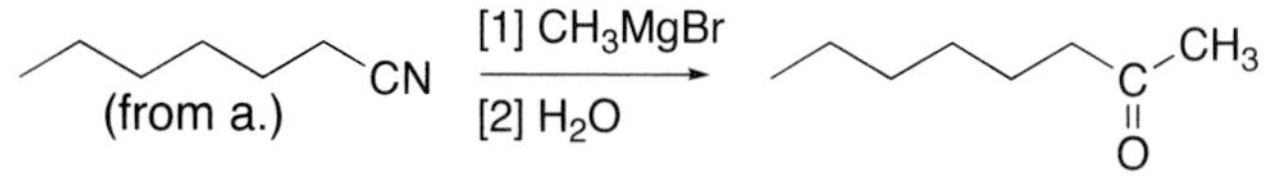

[1] CH₃MgBr / [2] H₂O

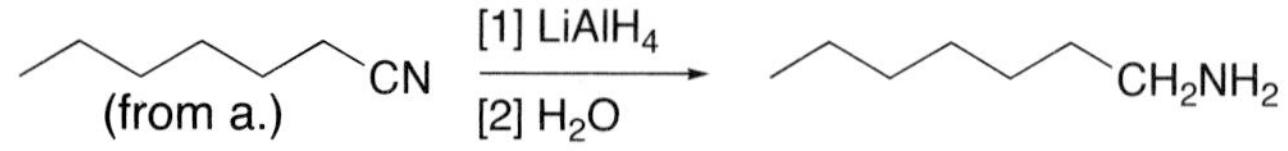

(from a.)

f.

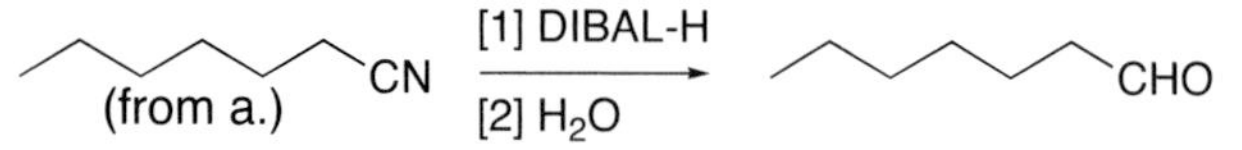

[1] DIBAL-H / [2] H₂O

(from a.)

g.

[1] LiAlH₄ / [2] H₂O

(from a.)

h.

CH₃COCl

(from g.)

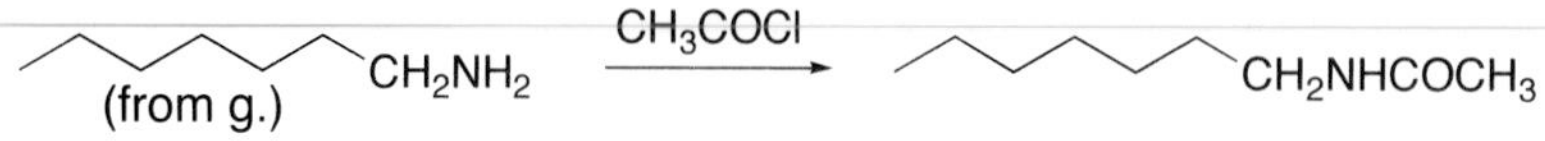

## 22.69

a. $CH_3Cl$ + NaCN ⟶ $CH_3-CN$ $\xrightarrow{H_3O^+}$ $CH_3-COOH$

$CH_3-Cl$ + Mg ⟶ $CH_3-MgCl$ $\xrightarrow[\text{[2] } H_3O^+]{\text{[1] } CO_2}$ $CH_3-COOH$

b.

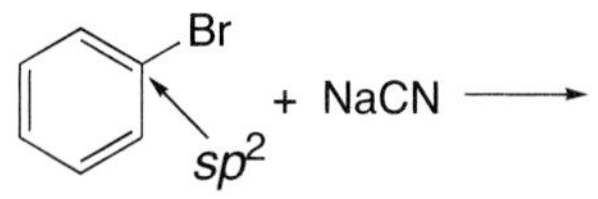

+ NaCN ⟶ This method can't be used because an $S_N2$ reaction can't be done on an $sp^2$ hybridized C.

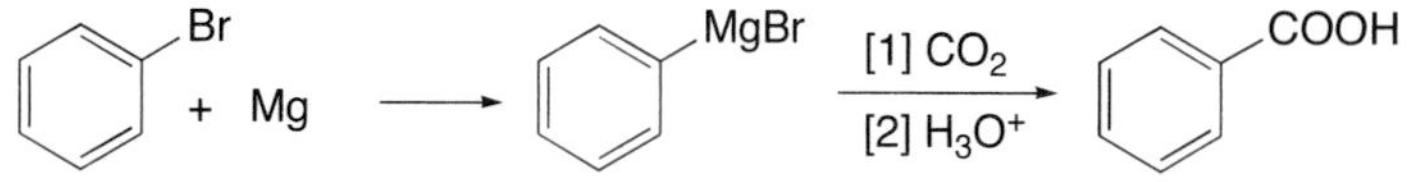

+ Mg ⟶ $\xrightarrow[\text{[2] } H_3O^+]{\text{[1] } CO_2}$

c. $(CH_3)_3CCl$ + NaCN ⟶ This method can't be used because an $S_N2$ reaction can't be done on a 3° C.

$(CH_3)_3C-Cl$ + Mg ⟶ $(CH_3)_3C-MgCl$ $\xrightarrow[\text{[2] } H_3O^+]{\text{[1] } CO_2}$ $(CH_3)_3C-COOH$

d. $HOCH_2CH_2CH_2CH_2Br$ + NaCN ⟶ $HOCH_2CH_2CH_2CH_2-CN$ $\xrightarrow{H_3O^+}$ $HOCH_2CH_2CH_2CH_2-COOH$

$HOCH_2CH_2CH_2CH_2-Br$ + Mg ⟶ This method can't be used because you can't make a Grignard reagent with an acidic OH group.

## 22.70

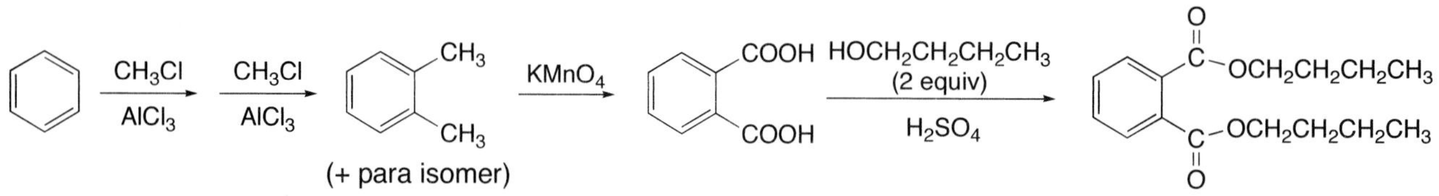

$\xrightarrow[AlCl_3]{CH_3Cl}$ $\xrightarrow[AlCl_3]{CH_3Cl}$ (+ para isomer) $\xrightarrow{KMnO_4}$ $\xrightarrow[H_2SO_4]{HOCH_2CH_2CH_2CH_3 \text{ (2 equiv)}}$ dibutyl phthalate

## 22.71

a.

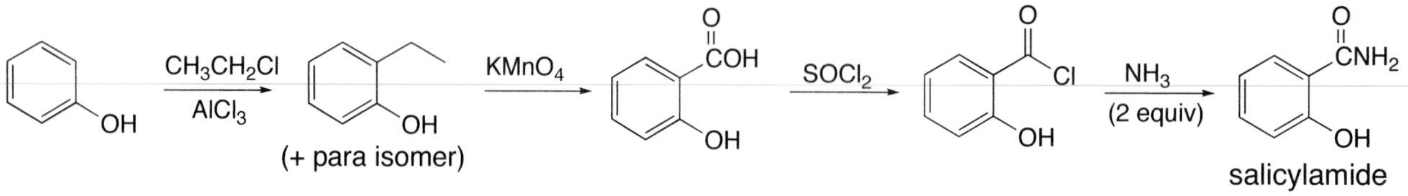

$\xrightarrow[AlCl_3]{CH_3CH_2Cl}$ (+ para isomer) $\xrightarrow{KMnO_4}$ $\xrightarrow{SOCl_2}$ $\xrightarrow[\text{(2 equiv)}]{NH_3}$ salicylamide

b.

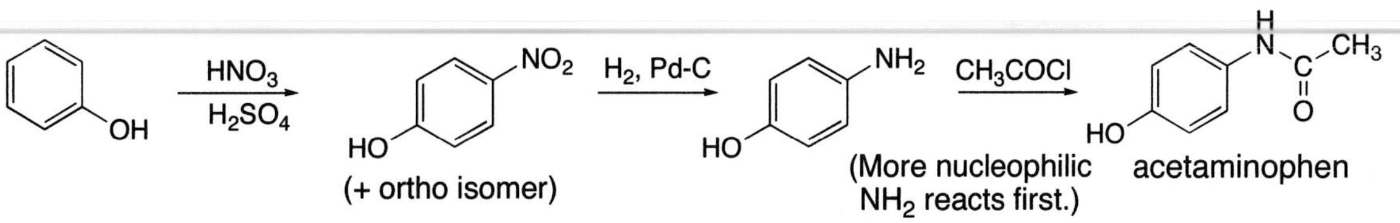

(+ ortho isomer)
(More nucleophilic
NH$_2$ reacts first.)
acetaminophen

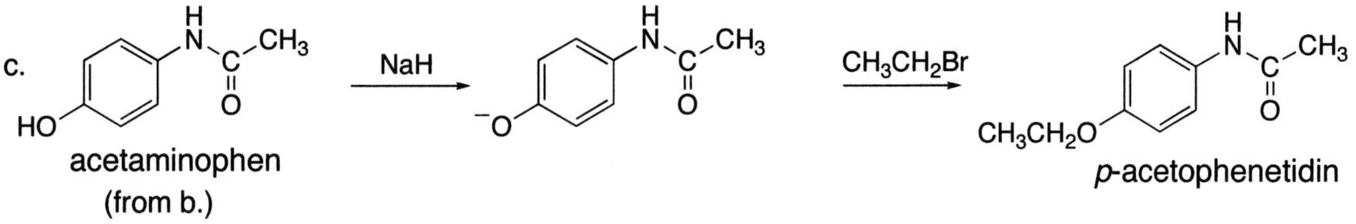

c.

acetaminophen
(from b.)

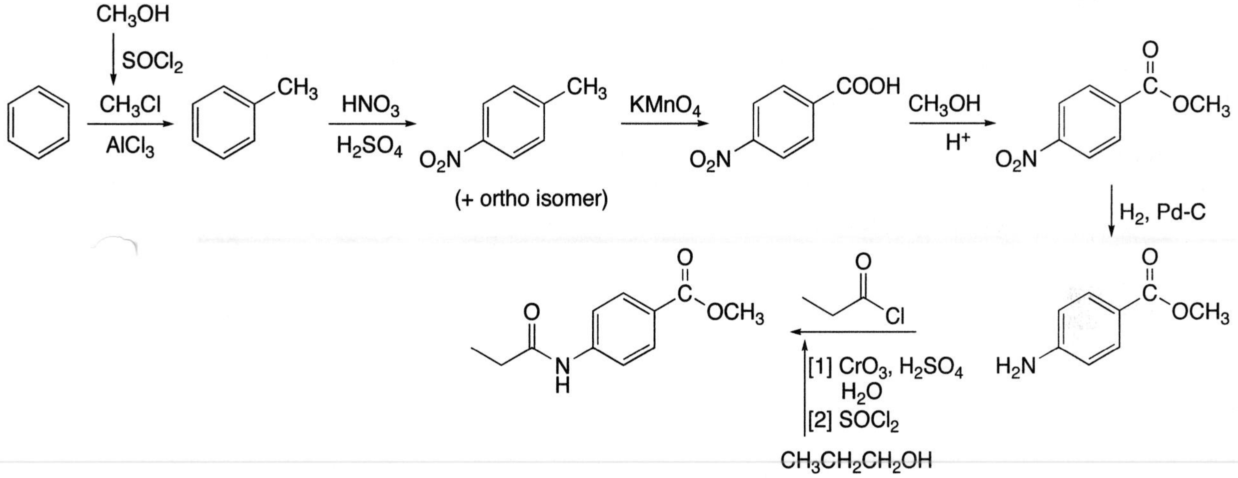

CH$_3$CH$_2$O

$p$-acetophenetidin

## 22.72

a.

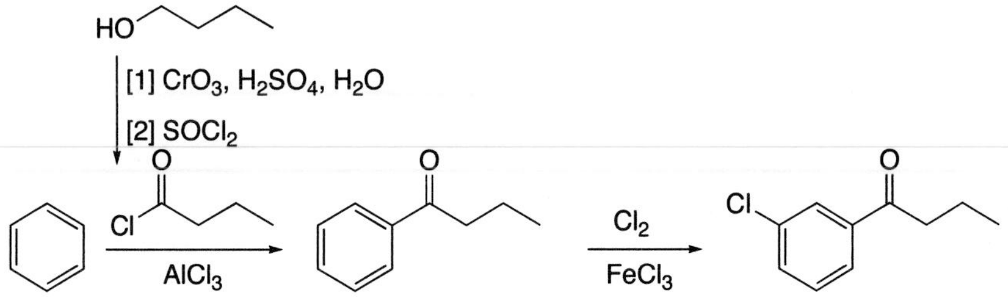

b.

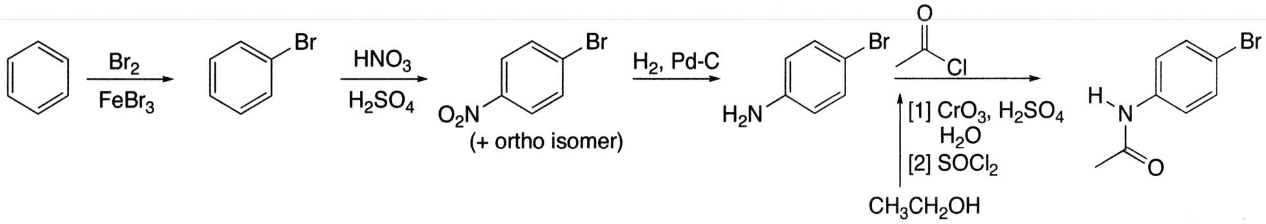

(+ ortho isomer)

c.

(+ ortho isomer)

**d.**

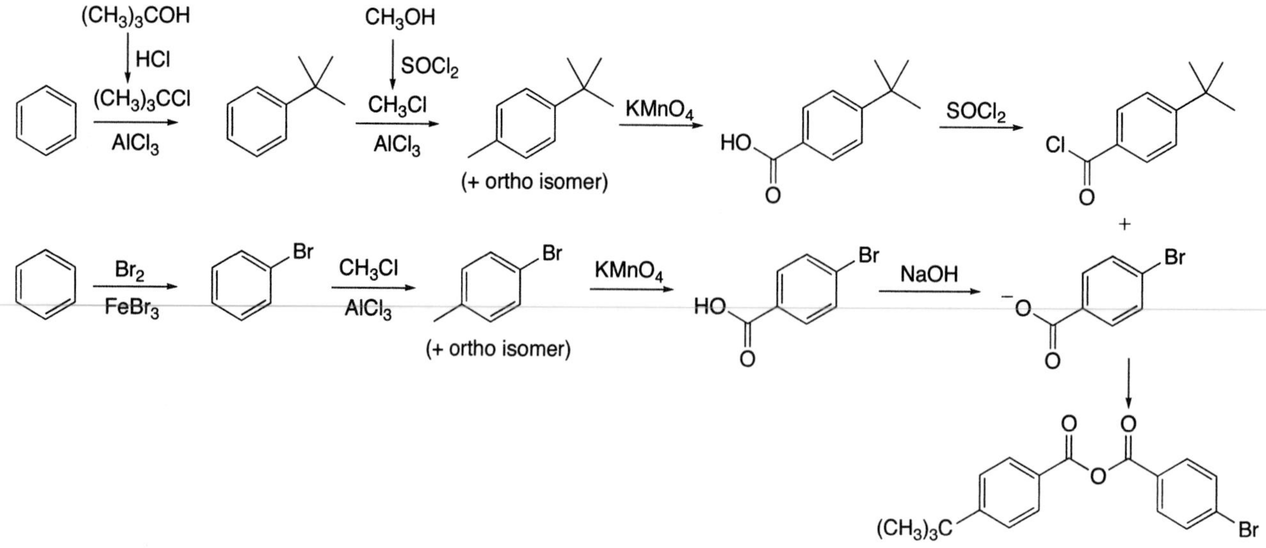

(CH₃)₃COH
HCl
(CH₃)₃CCl
AlCl₃

CH₃OH
SOCl₂
CH₃Cl
AlCl₃

(+ ortho isomer)

KMnO₄

SOCl₂

+

Br₂
FeBr₃

CH₃Cl
AlCl₃

(+ ortho isomer)

KMnO₄

NaOH

**22.73**

**a.**

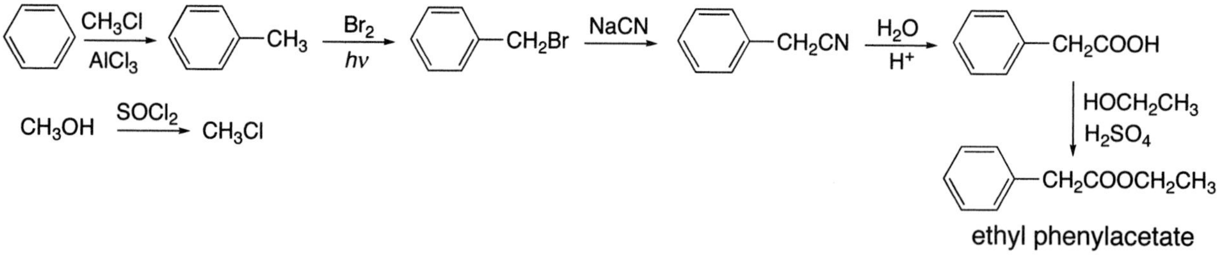

CH₃Cl
AlCl₃
—CH₃
Br₂
hv
—CH₂Br
NaCN
—CH₂CN
H₂O
H⁺
—CH₂COOH

CH₃OH
SOCl₂
CH₃Cl

HOCH₂CH₃
H₂SO₄

—CH₂COOCH₂CH₃
ethyl phenylacetate

**b.**

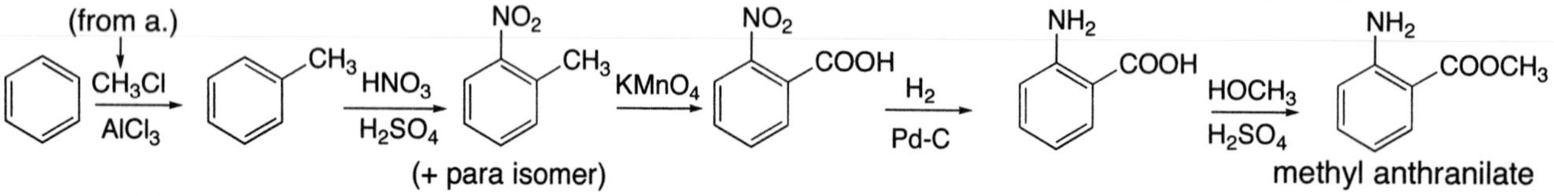

(from a.)
CH₃Cl
AlCl₃
—CH₃
HNO₃
H₂SO₄
NO₂ —CH₃
(+ para isomer)
KMnO₄
NO₂ —COOH
H₂
Pd-C
NH₂ —COOH
HOCH₃
H₂SO₄
NH₂ —COOCH₃
methyl anthranilate

**c.**

(from a.)
CH₃Cl
AlCl₃
—CH₃
KMnO₄
—COOH
[1] LiAlH₄
[2] H₂O
—OH
CH₃COOH
H₂SO₄
benzyl acetate

CH₃CH₂OH
CrO₃
H₂SO₄, H₂O
CH₃COOH

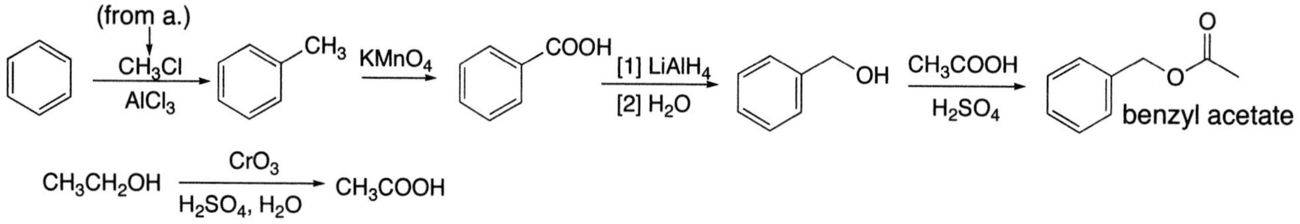

**22.74**

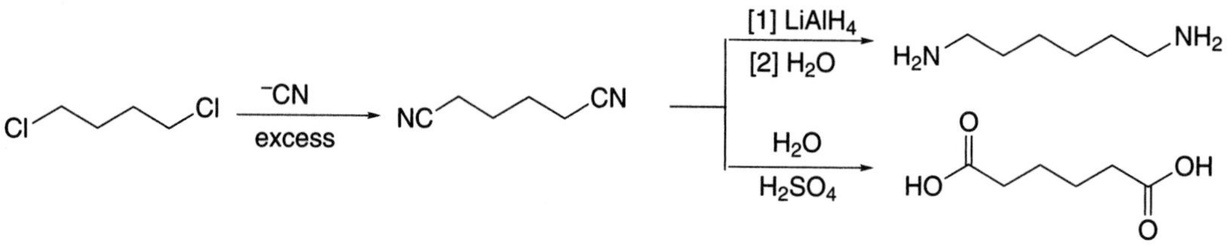

Cl〜Cl
⁻CN
excess
NC〜CN

[1] LiAlH₄
[2] H₂O
H₂N〜NH₂

H₂O
H₂SO₄
HO〜OH

**22.75**

**a.**

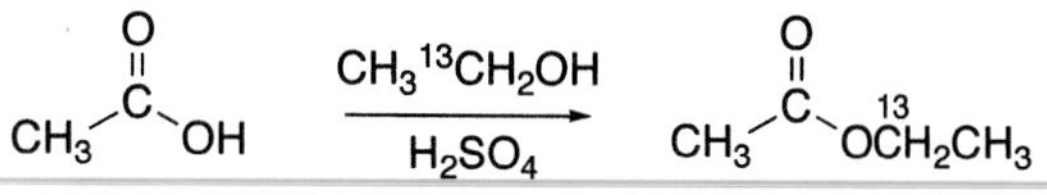

CH₃—C(=O)—OH
CH₃¹³CH₂OH
H₂SO₄
CH₃—C(=O)—¹³OCH₂CH₃

**b.** 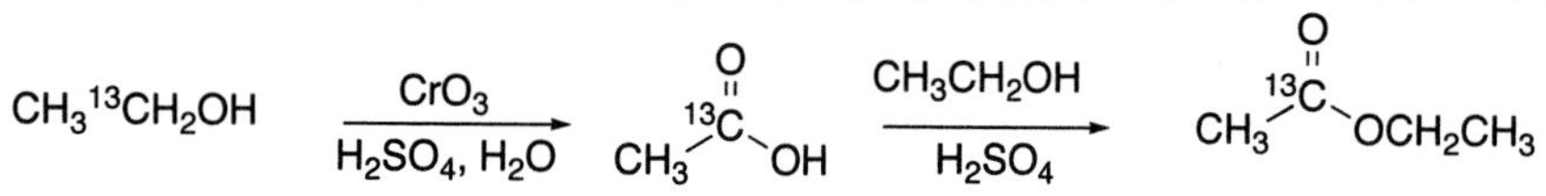 CH₃¹³CH₂OH
CrO₃
H₂SO₄, H₂O
CH₃—¹³C(=O)—OH
CH₃CH₂OH
H₂SO₄
CH₃—¹³C(=O)—OCH₂CH₃

c. 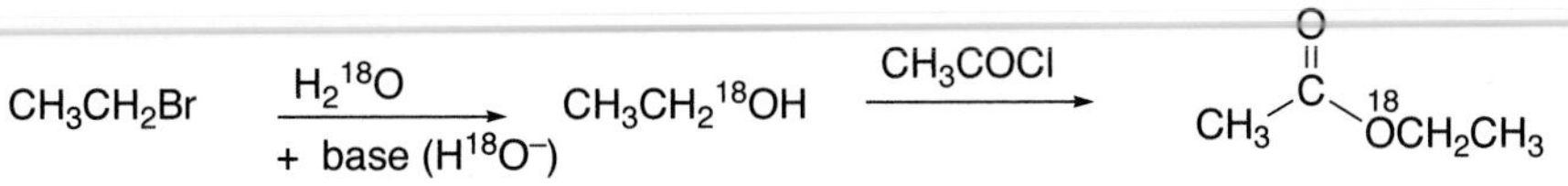 $CH_3CH_2Br$ $\xrightarrow[\text{+ base (H}^{18}\text{O}^-)]{H_2{}^{18}O}$ $CH_3CH_2{}^{18}OH$ $\xrightarrow{CH_3COCl}$ 

d. 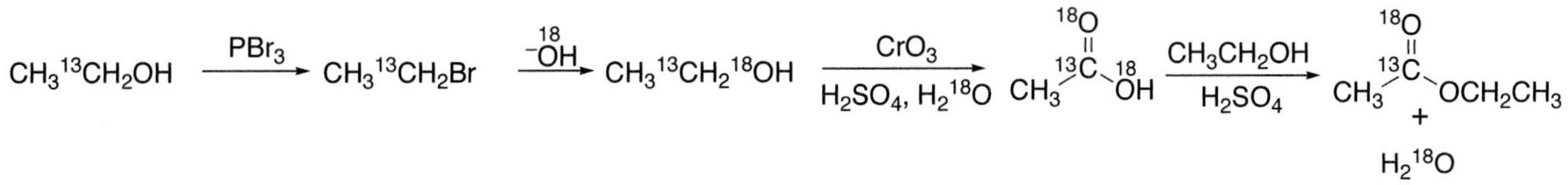 $CH_3{}^{13}CH_2OH$ $\xrightarrow{PBr_3}$ $CH_3{}^{13}CH_2Br$ $\xrightarrow{{}^{18}\text{O}^-\text{H}}$ $CH_3{}^{13}CH_2{}^{18}OH$ $\xrightarrow[H_2SO_4, H_2{}^{18}O]{CrO_3}$ 

## 22.76

a.

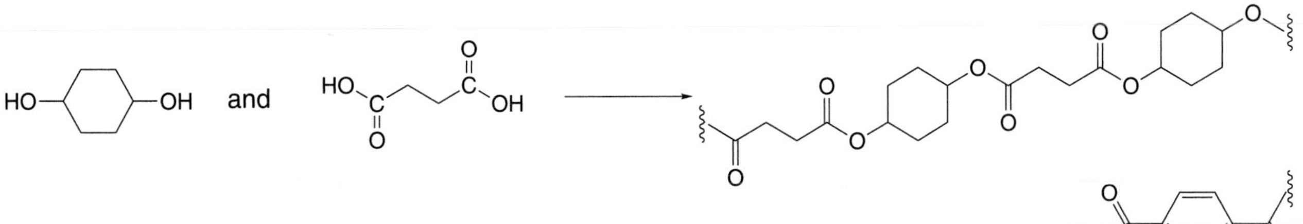

b.

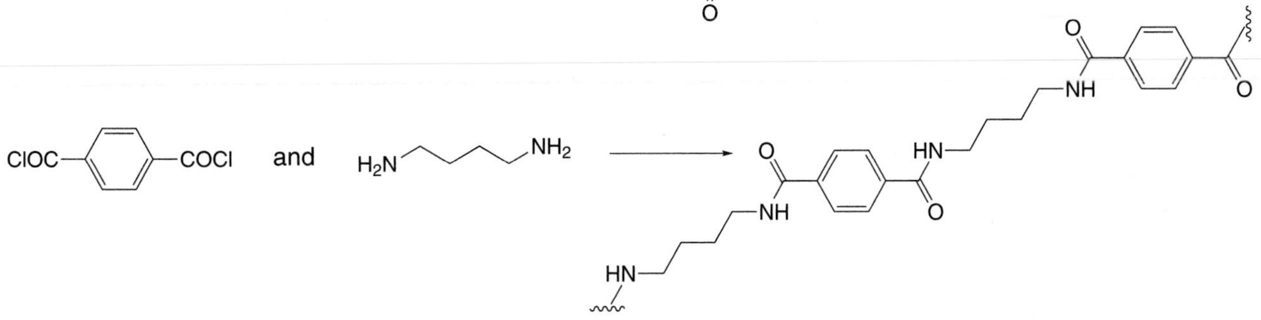

## 22.77

a.

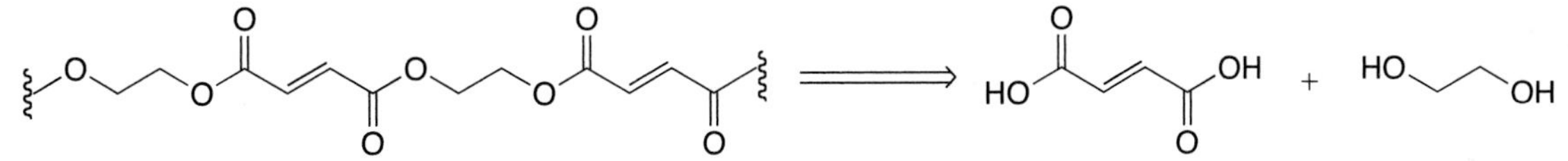

b.

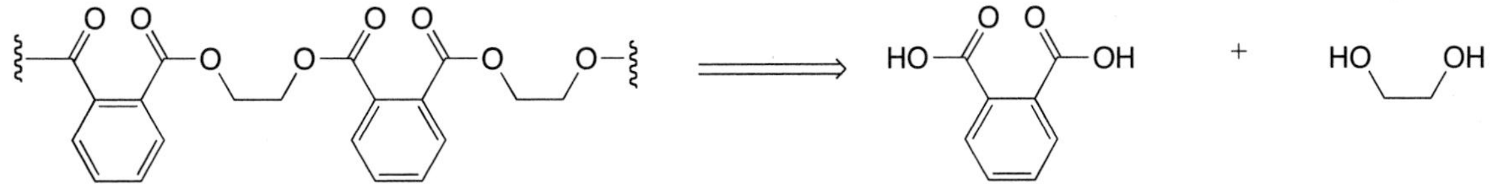

## 22.78

a.

contains a broad, strong OH absorption at 3500–2500 cm$^{-1}$

b.

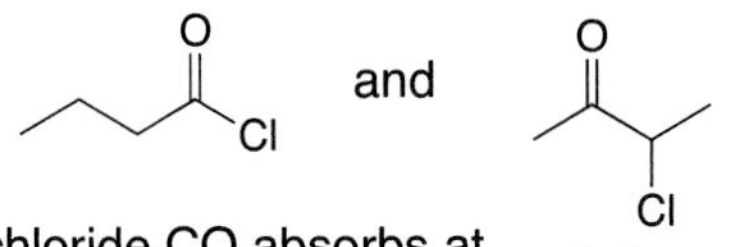

Acid chloride CO absorbs at much higher wavenumber.

ketone

c.

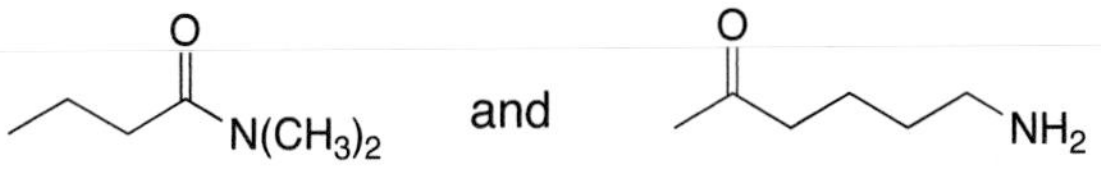

C=O at < 1700 cm$^{-1}$ due to the stabilized amide

2 NH absorptions at 3200–3400 cm$^{-1}$ C=O absorption higher frequency

d.

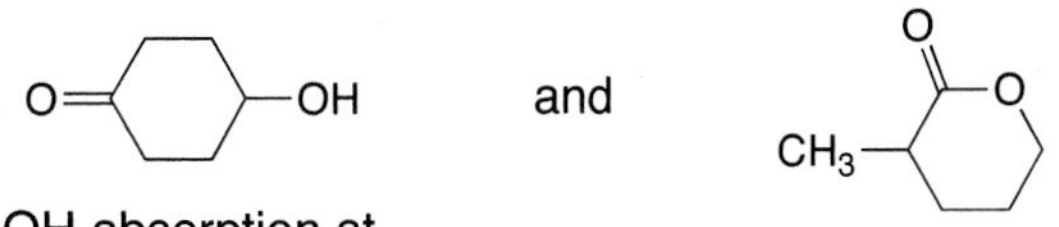

OH absorption at 3400–3200 cm$^{-1}$ + C=O

only C=O

**22.79**

a. $C_6H_5COOCH_2CH_3$   $CH_3CH_2COOCH_2CH_3$

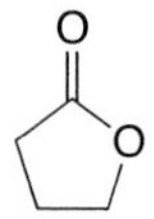

b. most resonance stabilized ———— least resonance stabilized

$CH_3CONH_2$      $CH_3COOCH_3$      $CH_3COCl$

Increasing wavenumber →        Increasing wavenumber →

---

**22.80**

a. $C_6H_{12}O_2$ → one degree of unsaturation
IR: 1738 cm$^{-1}$ → C=O
NMR: 1.12 (triplet, 3H), 1.23 (doublet, 6H),
2.28 (quartet, 2H), 5.00 (septet, 1H) ppm

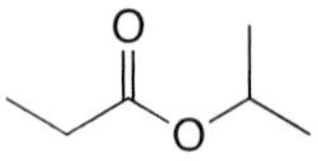

b. $C_4H_7N$
IR: 2250 cm$^{-1}$ → triple bond
NMR: 1.08 (triplet, 3H), 1.70 (multiplet, 2H),
2.34 (triplet, 2H)

$CH_3CH_2CH_2C{\equiv}N$

c. $C_8H_9NO$
IR: 3328 (NH), 1639 (conjugated amide C=O) cm$^{-1}$
NMR: 2.95 (singlet, 3H), 6.95 (singlet, 1H),
7.3–7.7 (multiplet, 5H)

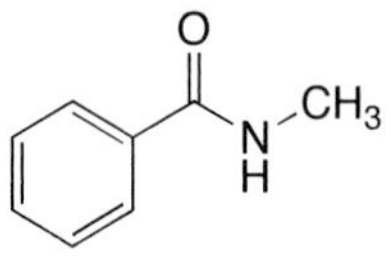

d. $C_4H_7ClO$ → one degree of unsaturation
IR: 1802 cm$^{-1}$ → C=O (high wavenumber, RCOCl)
NMR: 0.95 (triplet, 3H), 1.07 (multiplet, 2H),
2.90 (triplet, 2H)

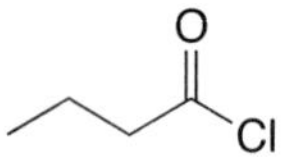

e. $C_5H_{10}O_2$ → one degree of unsaturation
IR: 1750 cm$^{-1}$ → C=O
NMR: 1.20 (doublet, 6H), 2.00 (singlet, 3H),
4.95 (septet, 1H)

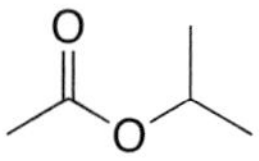

f. $C_{10}H_{12}O_2$ → five degrees of unsaturation
IR: 1740 cm$^{-1}$ → C=O
NMR: 1.2 (triplet, 3H), 2.4 (quartet, 2H),
5.1 (singlet, 2H), 7.1–7.5 (multiplet, 5H)

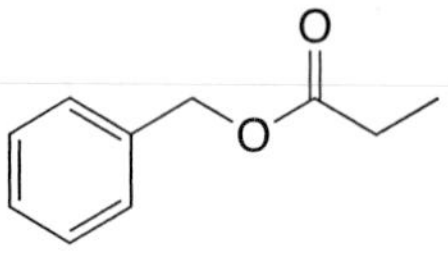

g. $C_8H_{14}O_3$ → two degrees of unsaturation
IR: 1810, 1770 cm$^{-1}$ → 2 absorptions due to
C=O (anhydride)
NMR: 1.25 (doublet, 12H), 2.65 (septet, 2H)

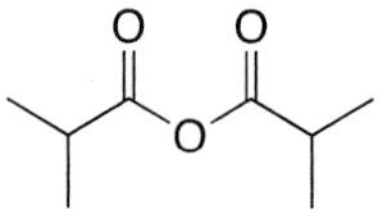

---

**22.81**

**A.** Molecular formula $C_{10}H_{12}O_2$ → five degrees of unsaturation
IR absorption at 1718 cm$^{-1}$ → C=O
NMR data (ppm):

    triplet at 1.4 (CH$_3$ adjacent to 2H's)
    singlet at 2.4 (CH$_3$)
    quartet at 4.4 (CH$_2$ adjacent to CH$_3$)
    doublet at 7.2 (2H's on benzene ring)
    doublet at 7.9 (2H's on benzene ring)

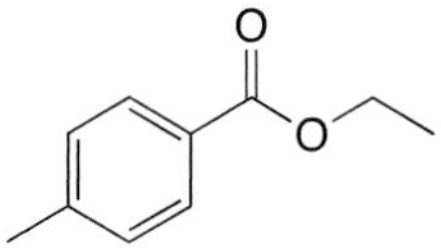

**B.** IR absorption at 1740 cm$^{-1}$ → C=O
NMR data (ppm):

singlet at 2.0 (CH$_3$)
triplet at 2.9 (CH$_2$ adjacent to CH$_2$)
triplet at 4.4 (CH$_2$ adjacent to CH$_2$)
multiplet at 7.3 (5H's, monosubstituted benzene)

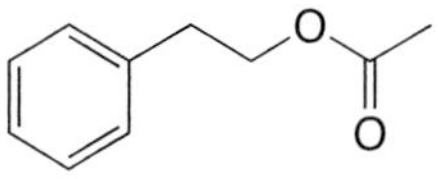

**22.82**

Molecular formula C$_{10}$H$_{13}$NO$_2$ → five degrees of unsaturation
IR absorptions at 3300 (NH) and 1680 (C=O, amide or conjugated) cm$^{-1}$
NMR data (ppm):

triplet at 1.4 (CH$_3$ adjacent to CH$_2$)
singlet at 2.2 (CH$_3$C=O)
quartet at 3.9 (CH$_2$ adjacent to CH$_3$)
doublet at 6.8 (2H's on benzene ring)
singlet at 7.2 (NH)
doublet at 7.4 (2H's on benzene ring)

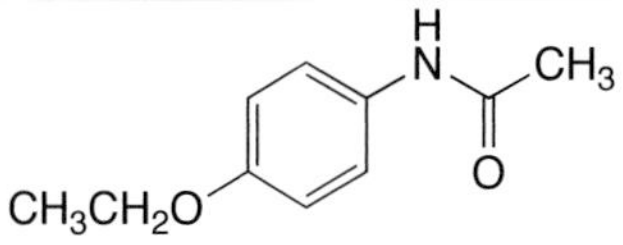

phenacetin

**22.83**

a. Molecular formula C$_6$H$_{12}$O$_2$ → one degree of unsaturation
IR absorption at 1743 cm$^{-1}$ → C=O
$^1$H NMR data (ppm):

triplet at 0.9 (3H) – CH$_3$ adjacent to CH$_2$
multiplet at 1.35 (2H) – CH$_2$
multiplet at 1.60 (2H) – CH$_2$
singlet at 2.1 (3H – from CH$_3$ bonded to C=O)
triplet at 4.1 (2H) – CH$_2$ adjacent to the electronegative O atom and another CH$_2$

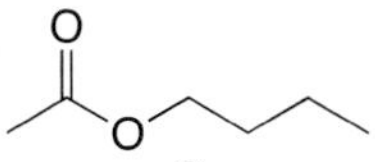

**C**

b. Molecular formula C$_6$H$_{12}$O$_2$ → one degree of unsaturation
IR absorption at 1746 cm$^{-1}$ → C=O
$^1$H NMR data (ppm):

doublet at 0.9 (6H) – 2 CH$_3$'s adjacent to CH
multiplet at 1.9 (1H)
singlet at 2.1 (3H) – CH$_3$ bonded to C=O
doublet at 3.85 (2H) – CH$_2$ bonded to electronegative O and CH

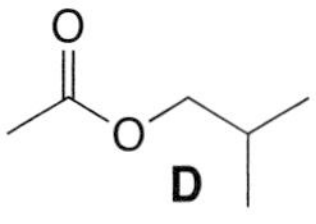

**D**

**22.84**

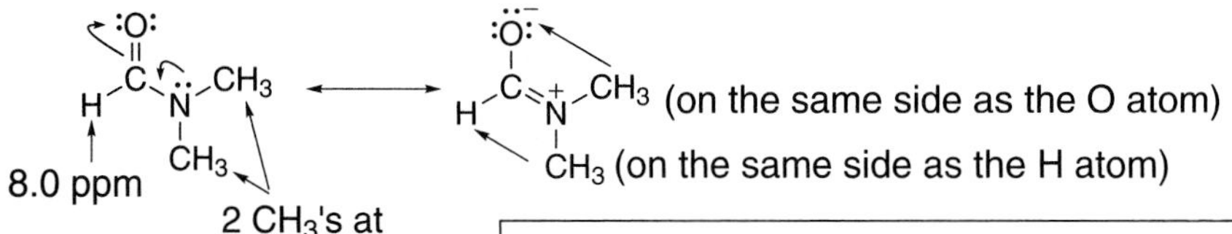

8.0 ppm

2 CH$_3$'s at
2.93 and 3.03 ppm

CH$_3$ (on the same side as the O atom)
CH$_3$ (on the same side as the H atom)

Resonance stabilization of the amide causes restricted rotation
around the C–N amide bond, so the two methyl groups on N are in
different environments. Thus they give two different NMR signals.

**22.85**

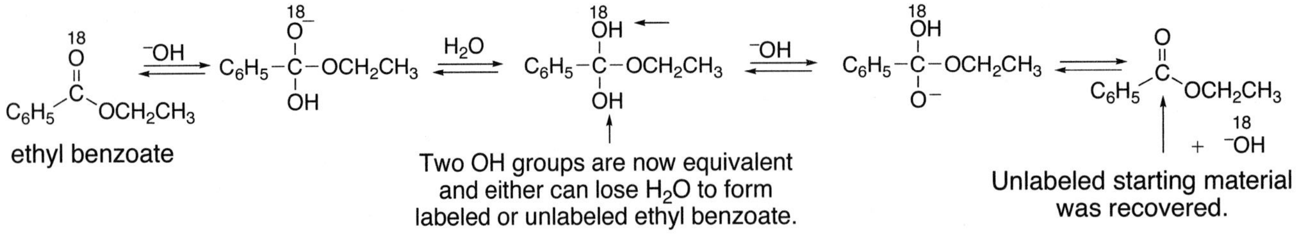

ethyl benzoate

Two OH groups are now equivalent
and either can lose $H_2O$ to form
labeled or unlabeled ethyl benzoate.

Unlabeled starting material
was recovered.

**22.86**

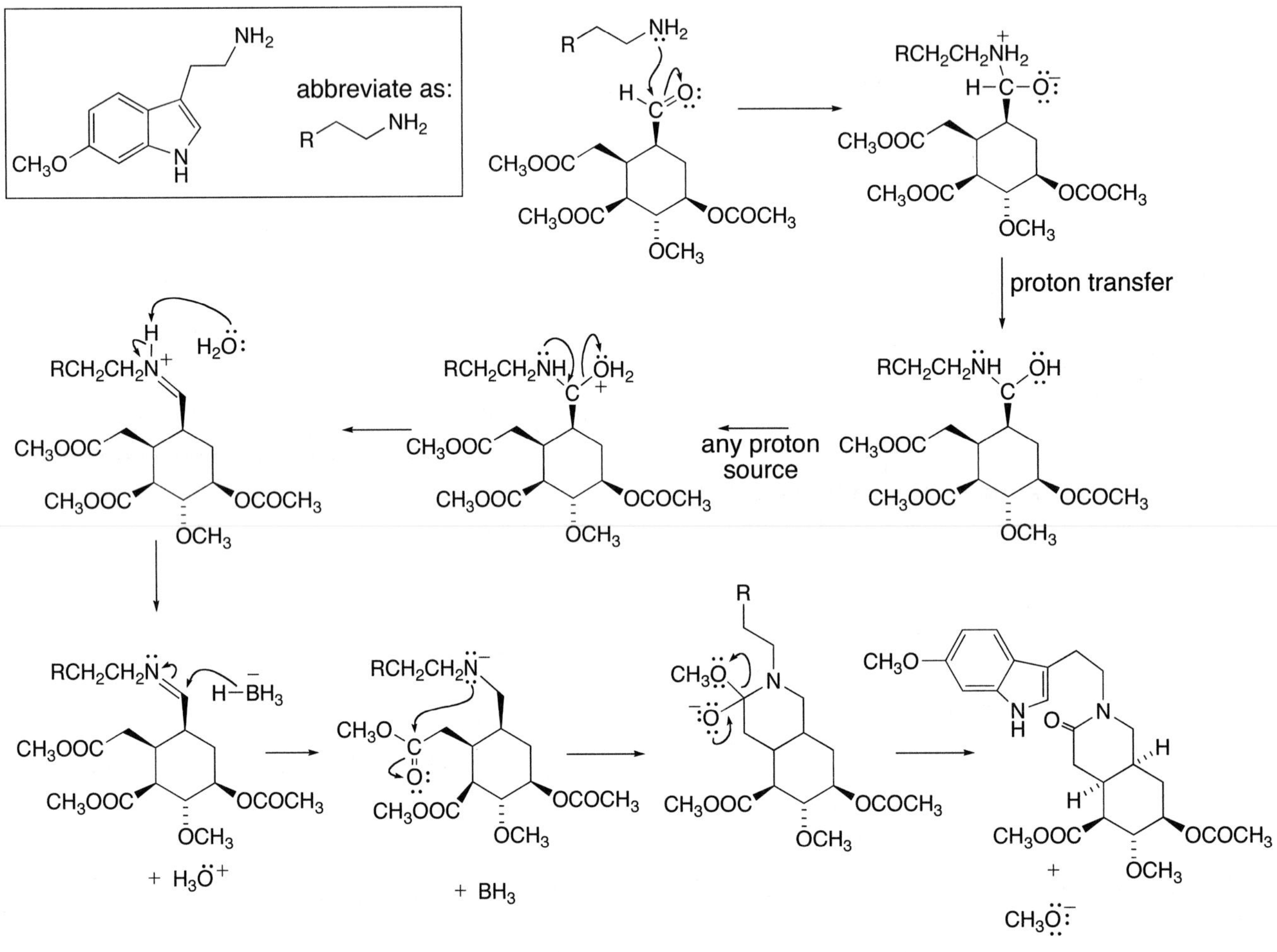

abbreviate as:

proton transfer

any proton
source

$+ H_3\ddot{O}^+$

$+ BH_3$

# Chapter 23: Substitution Reactions of Carbonyl Compounds at the α Carbon

## ♦ Kinetic versus thermodynamic enolates (23.4)

**Kinetic enolate**
- The less substituted enolate
- Favored by strong base, polar aprotic solvent, low temperature: LDA, THF, –78 °C

**Thermodynamic enolate**
- The more substituted enolate
- Favored by strong base, protic solvent, higher temperature: $NaOCH_2CH_3$, $CH_3CH_2OH$, room temperature

## ♦ Halogenation at the α carbon

### [1] Halogenation in acid (23.7A)

$$\underset{\text{CH}_3\text{COOH}}{\xrightarrow{X_2}}$$

$X_2 = Cl_2, Br_2,$ or $I_2$ — α-halo aldehyde or ketone

- The reaction occurs via enol intermediates.
- Monosubstitution of X for H occurs on the α carbon.

### [2] Halogenation in base (23.7B)

$$\underset{^-\text{OH}}{\xrightarrow{X_2 \text{ (excess)}}}$$

$X_2 = Cl_2, Br_2,$ or $I_2$

- The reaction occurs via enolate intermediates.
- Polysubstitution of X for H occurs on the α carbon.

### [3] Halogenation of *methyl* ketones in base—The haloform reaction (23.7B)

$$\underset{^-\text{OH}}{\xrightarrow{X_2 \text{ (excess)}}} \quad + \; HCX_3 \text{ haloform}$$

$X_2 = Cl_2, Br_2,$ or $I_2$

- The reaction occurs with methyl ketones, and results in cleavage of a carbon–carbon σ bond.

## ♦ Reactions of α-halo carbonyl compounds (23.7C)

### [1] Elimination to form α,β-unsaturated carbonyl compounds

$$\underset{\substack{\text{LiBr} \\ \text{DMF}}}{\xrightarrow{\text{Li}_2\text{CO}_3}}$$

- Elimination of the elements of Br and H forms a new π bond, giving an α,β-unsaturated carbonyl compound.

## [2] Nucleophilic substitution

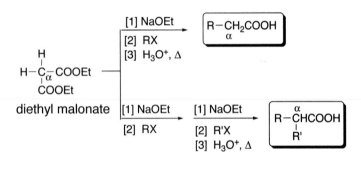

- The reaction follows an $S_N2$ mechanism, generating an α-substituted carbonyl compound.

♦ Alkylation reactions at the α carbon

### [1] Direct alkylation at the α carbon (23.8)

- The reaction forms a new C–C bond to the α carbon.
- LDA is a common base used to form an intermediate enolate.
- The alkylation in Step [2] follows an $S_N2$ mechanism.

### [2] Malonic ester synthesis (23.9)

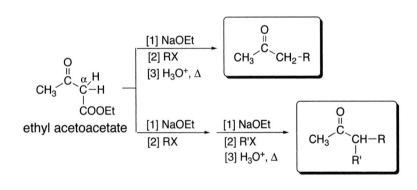

- The reaction is used to prepare carboxylic acids with one or two alkyl groups on the α carbon.
- The alkylation in Step [2] follows an $S_N2$ mechanism.

### [3] Acetoacetic ester synthesis (23.10)

- The reaction is used to prepare ketones with one or two alkyl groups on the α carbon.
- The alkylation in Step [2] follows an $S_N2$ mechanism.

## Chapter 23: Answers to Problems

**23.1** • To convert a ketone to its enol tautomer, change the C=O to C–OH, make a new double bond to an α carbon, and remove a proton at the other end of the C=C.

• To convert an enol to its keto form, find the C=C bonded to the OH. Change the C–OH to a C=O, add a proton to the other end of the C=C, and delete the double bond.

[In cases where *E* and *Z* isomers are possible, only one stereoisomer is drawn.]

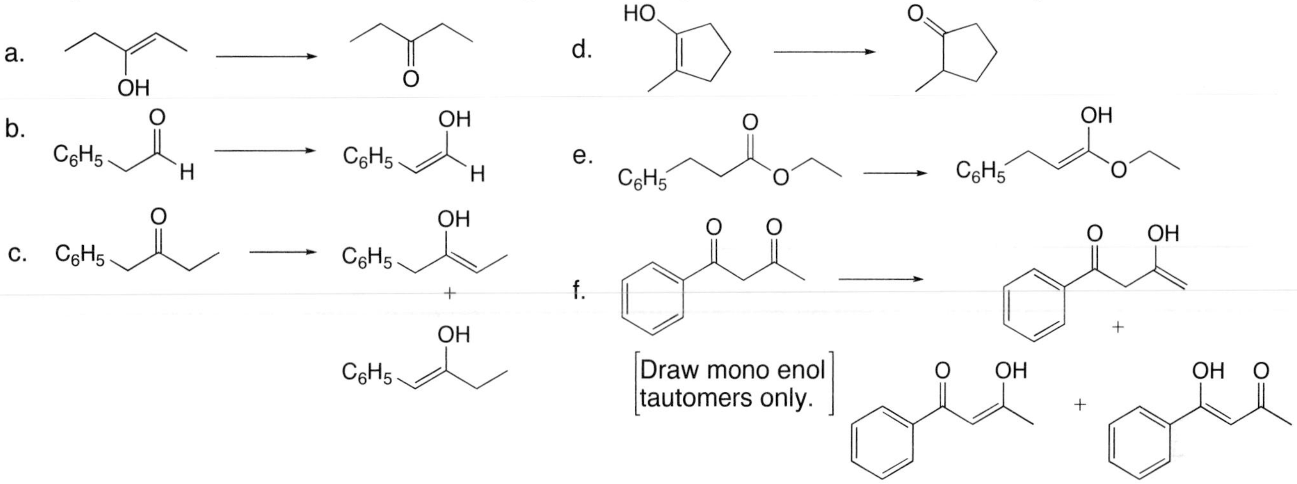
a.
b.
c.
d.
e.
f.
Draw mono enol tautomers only.

(Conjugated enols are preferred.)

**23.2**

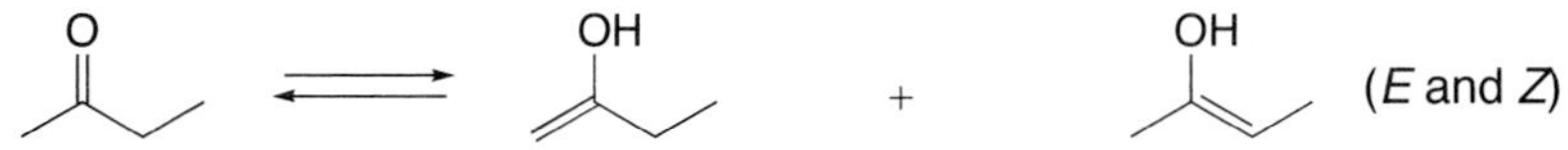
(E and Z)

2-butanone

C=C has one C bonded to it.

C=C has two C's bonded to it. The more substituted double bond is **more stable.**

**23.3**

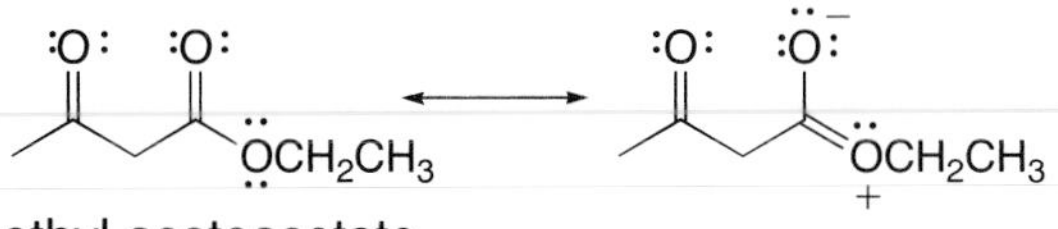

ethyl acetoacetate

The ester C=O is resonance stabilized, and is therefore less available for tautomerization. Since the carbonyl form of the ester group is stabilized by electron delocalization, less enol is present at equilibrium.

**23.4** The mechanism has two steps: protonation followed by deprotonation.

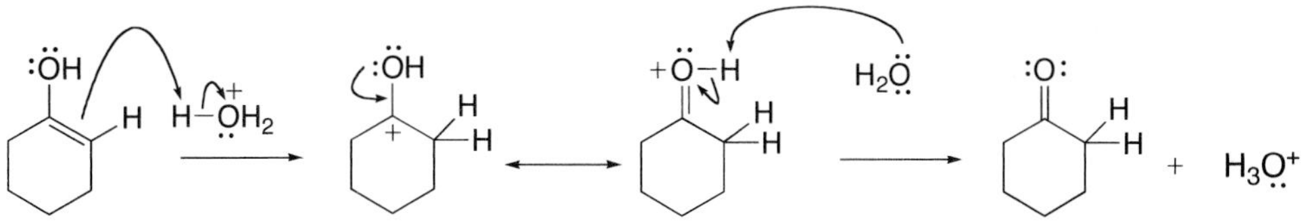

**23.5**

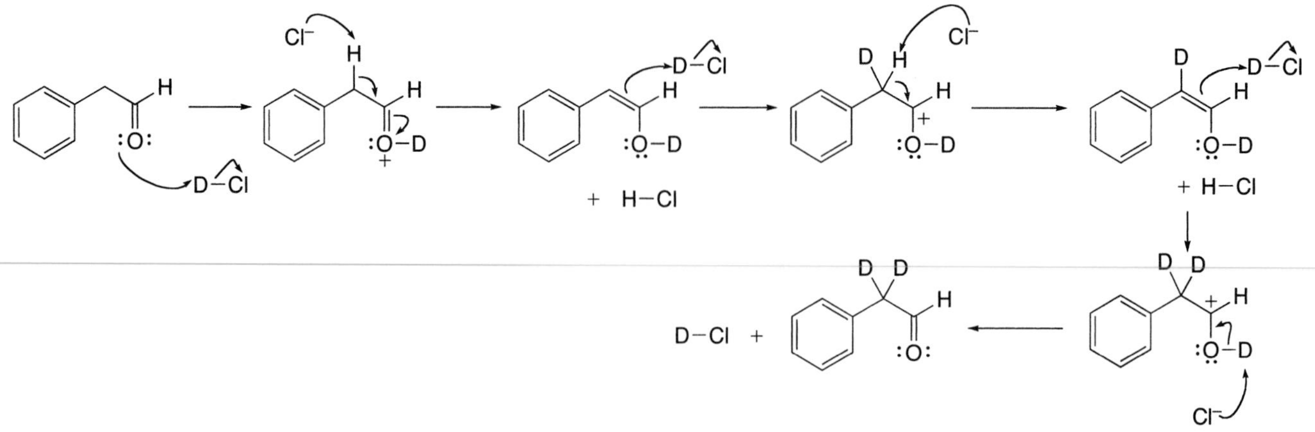

**23.6**

a.

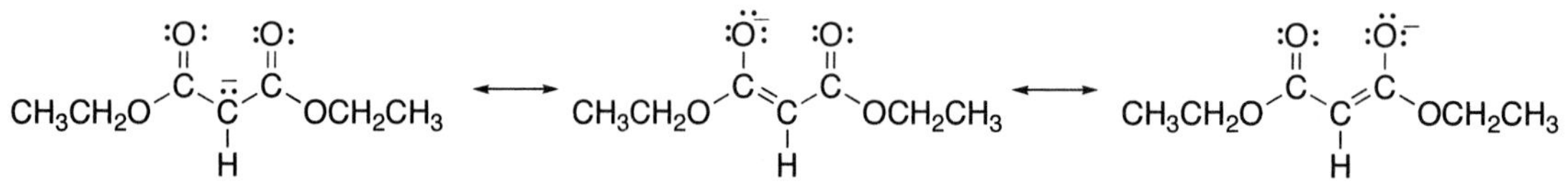

b.

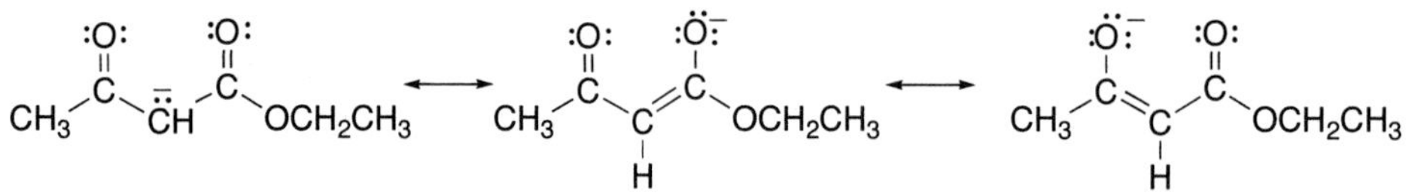

c.

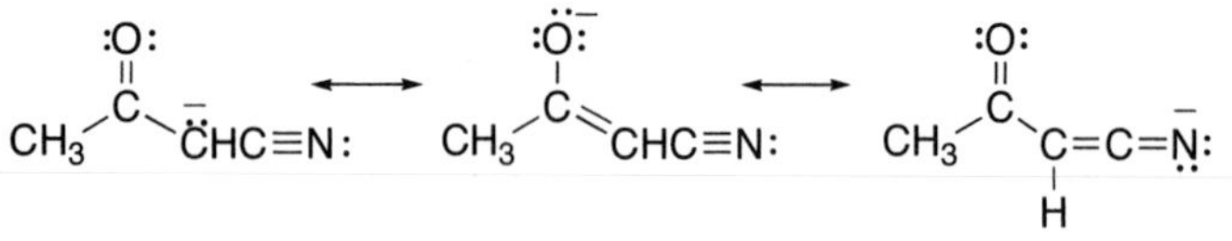

**23.7**

a.  b. $CH_3CH_2CH_2$—CN  c. 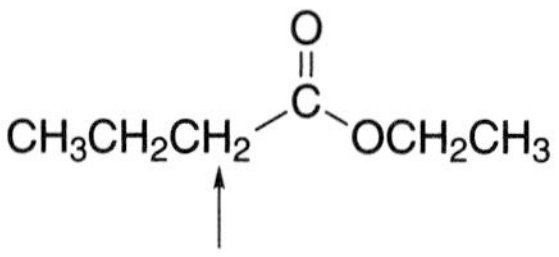  d.

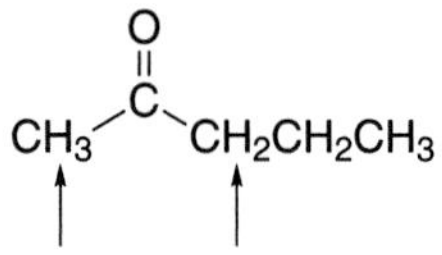

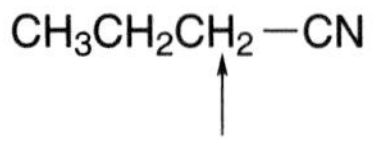

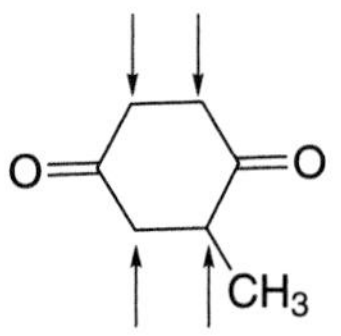

**23.8**

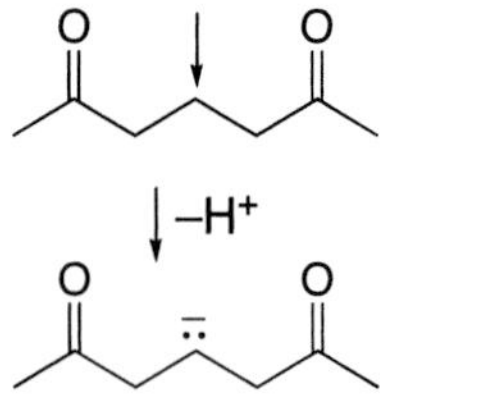

no resonance stabilization
**least acidic**

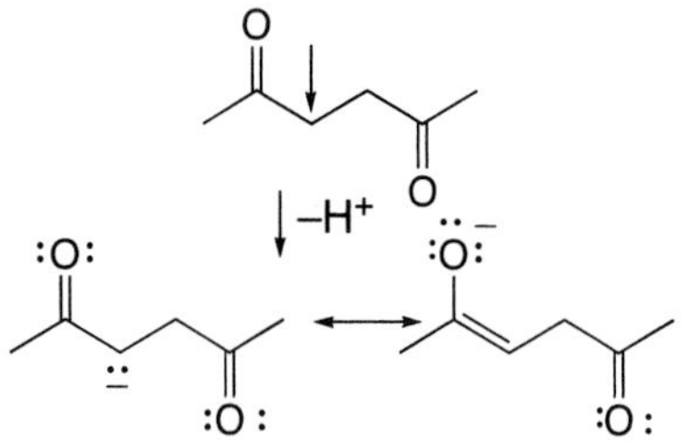

Two resonance structures
stabilize the conjugate base.
**intermediate acidity**

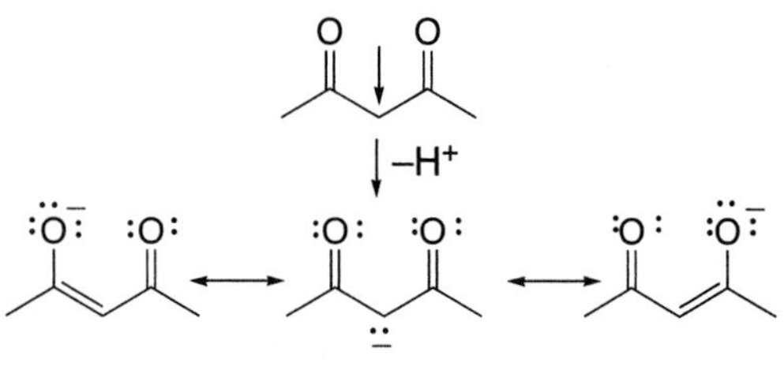

Three resonance structures
stabilize the conjugate base.
**most acidic**

**23.9** In each of the reactions, the LDA pulls off the most acidic proton.

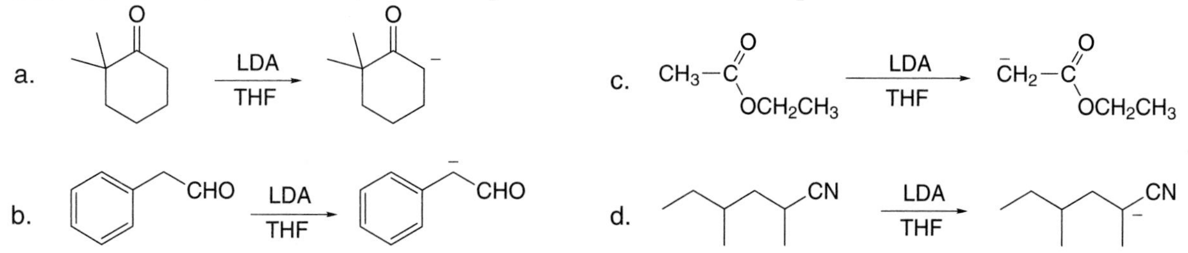

a.

c.

b.

d.

**23.10**

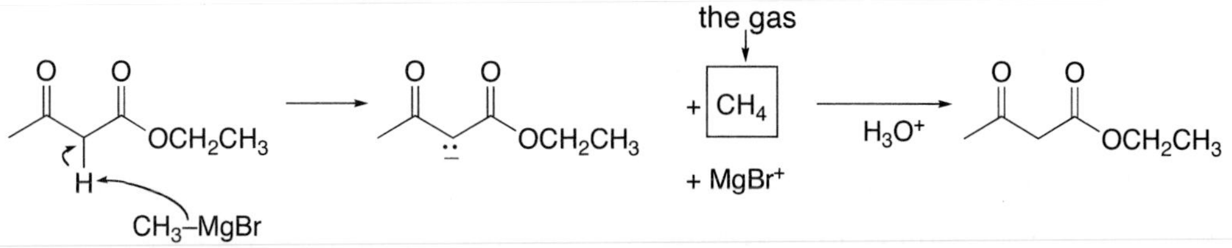

The $CH_2$ between the two C=O's contains acidic H's, so $CH_3MgBr$ reacts as a base to remove a proton. Thus, proton transfer (not nucleophilic addition) occurs.

**23.11** • LDA, THF forms the kinetic enolate by removing a proton from the less substituted C.
  • Treatment with $NaOCH_3$, $CH_3OH$ forms the thermodynamic enolate by removing a proton from the more substituted C.

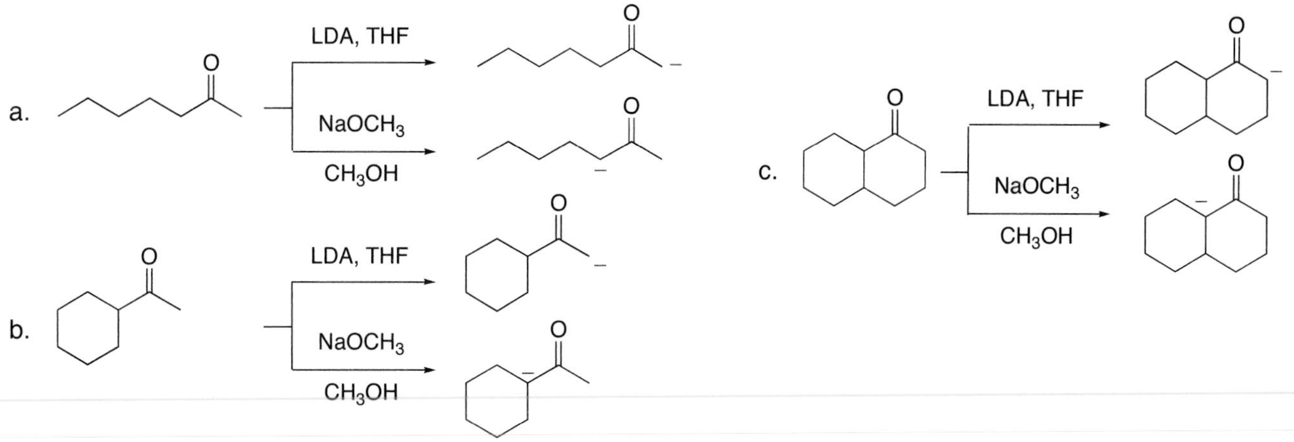

a.

b.

c.

**23.12**

a. This acidic H is removed with base to form an achiral enolate.

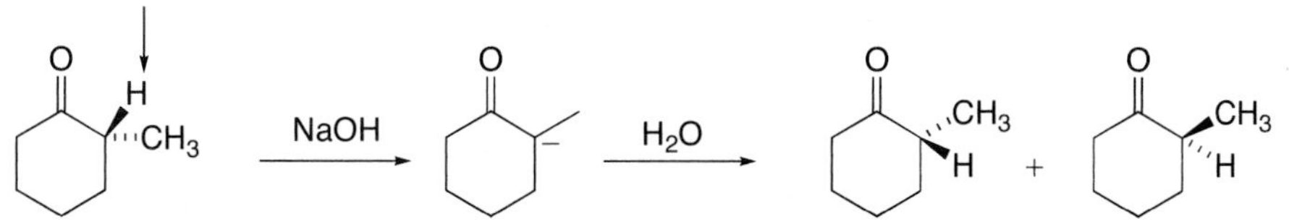

(R)-2-methylcyclohexanone     achiral

Protonation of the planar achiral enolate occurs with equal probability from two sides so a racemic mixture is formed. The racemic mixture is optically inactive.

**b.**

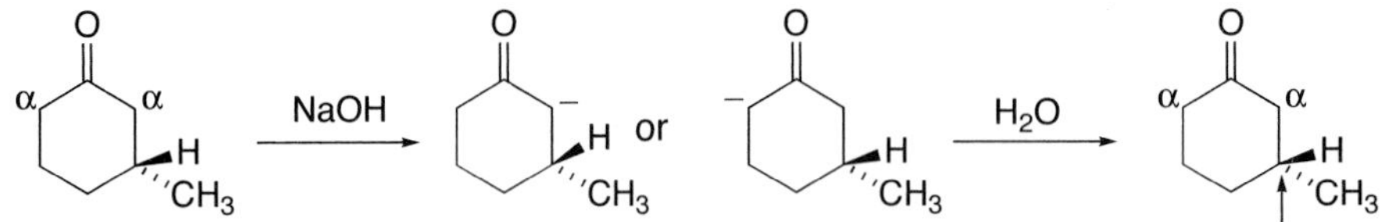

(R)-3-methylcyclohexanone

NaOH ... H or ... H₂O

This stereogenic center is not located at the α carbon, so it is not deprotonated with base. Its configuration is retained in the product, and the product remains optically active.

**23.13**

a.

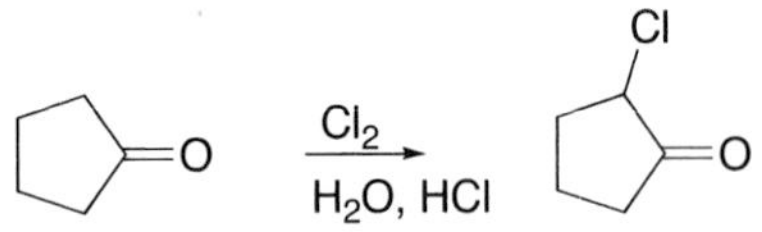

$\dfrac{Cl_2}{H_2O,\ HCl}$

c.

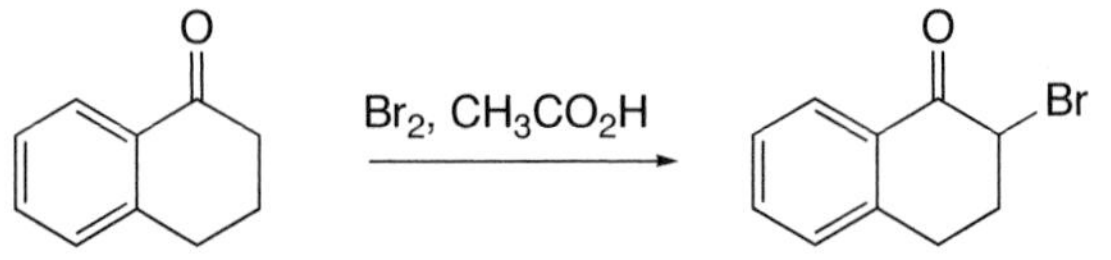

$Br_2,\ CH_3CO_2H$

b.

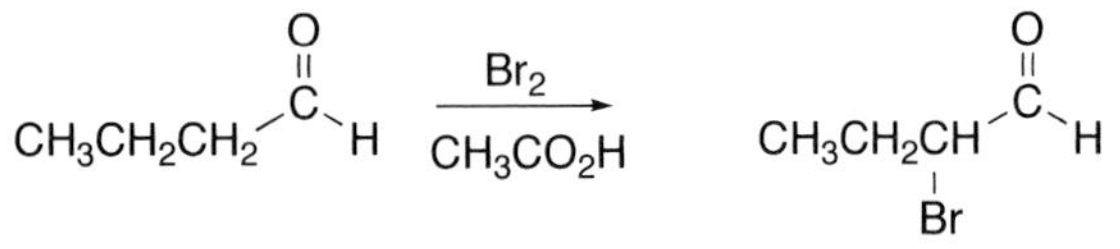

CH₃CH₂CH₂—C(=O)—H

$\dfrac{Br_2}{CH_3CO_2H}$

CH₃CH₂CH(Br)—C(=O)—H

**23.14**

a.

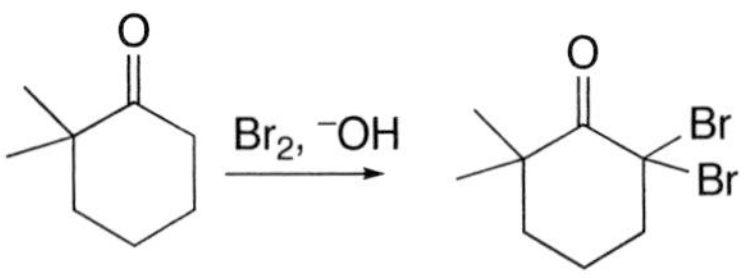

Br₂, ⁻OH

b.

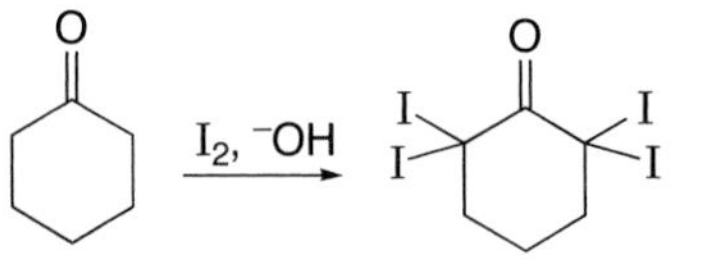

I₂, ⁻OH

c.

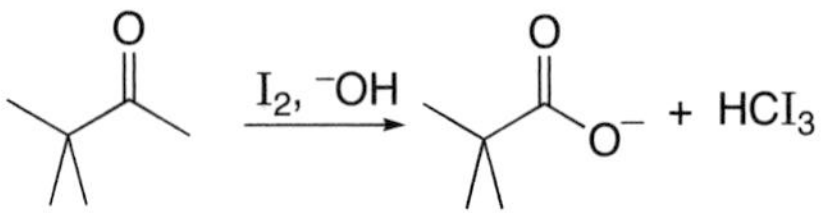

I₂, ⁻OH ... + HCI₃

**23.15**

a.

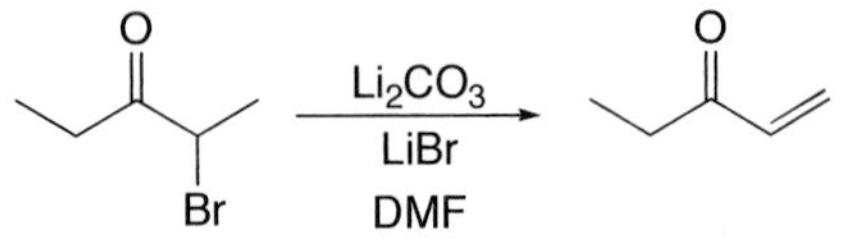

$\dfrac{Li_2CO_3,\ LiBr}{DMF}$

c.

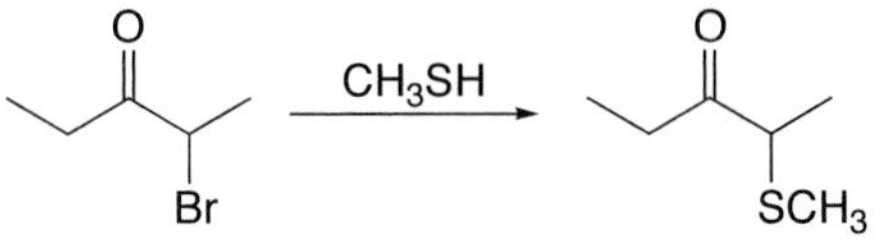

CH₃SH

b.

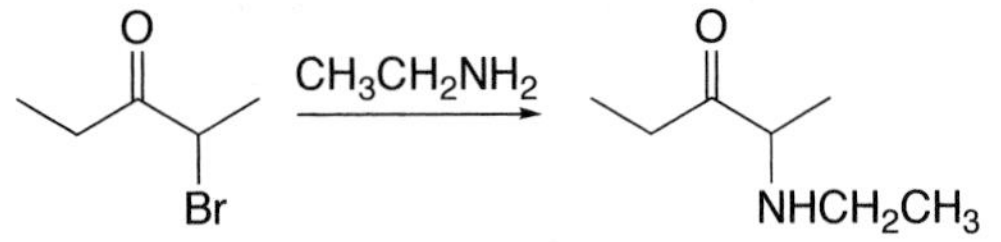

CH₃CH₂NH₂

**23.16**

a. CH₃CH₂—C(=O)—CH₂CH₃

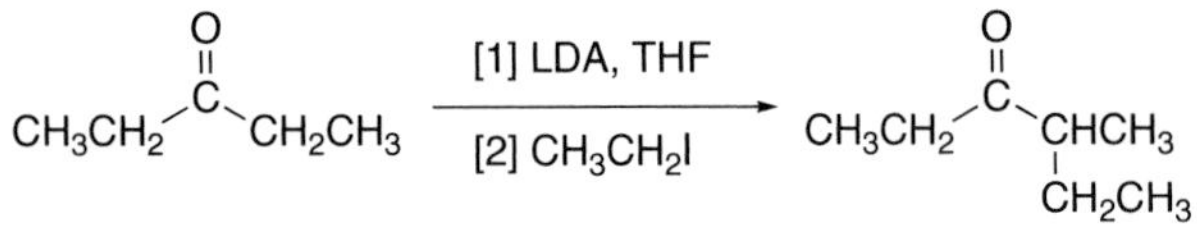

[1] LDA, THF
[2] CH₃CH₂I

CH₃CH₂—C(=O)—CHCH₃ | CH₂CH₃

c.

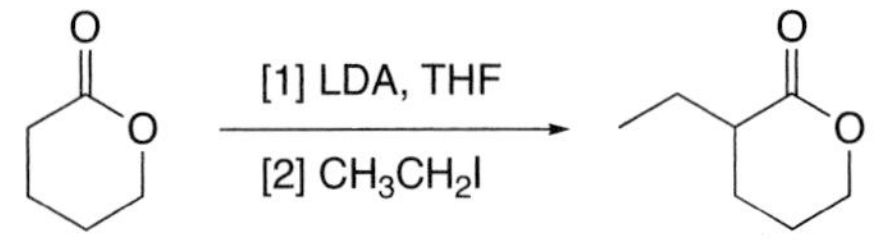

[1] LDA, THF
[2] CH₃CH₂I

b.

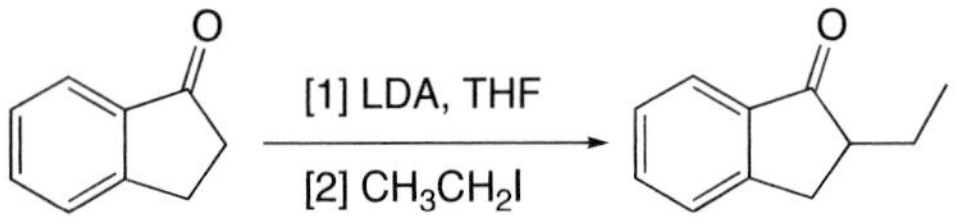

[1] LDA, THF
[2] CH₃CH₂I

d.

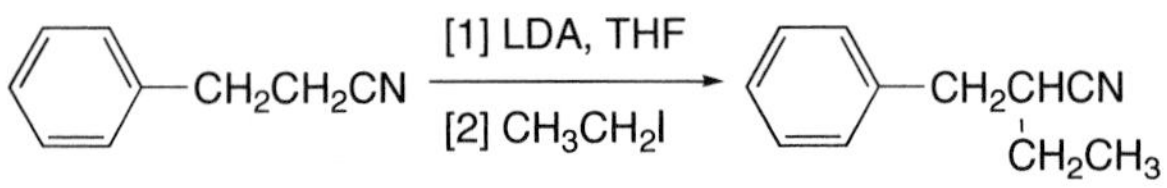

—CH₂CH₂CN

[1] LDA, THF
[2] CH₃CH₂I

—CH₂CHCN | CH₂CH₃

**23.17**

a.

[1] LDA, THF
[2] CH₃I

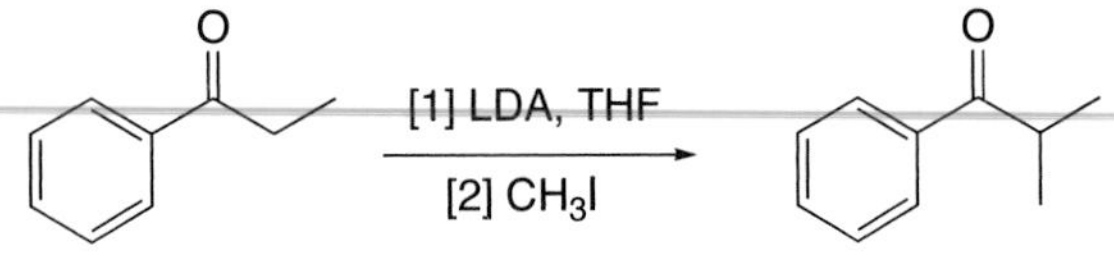

b.

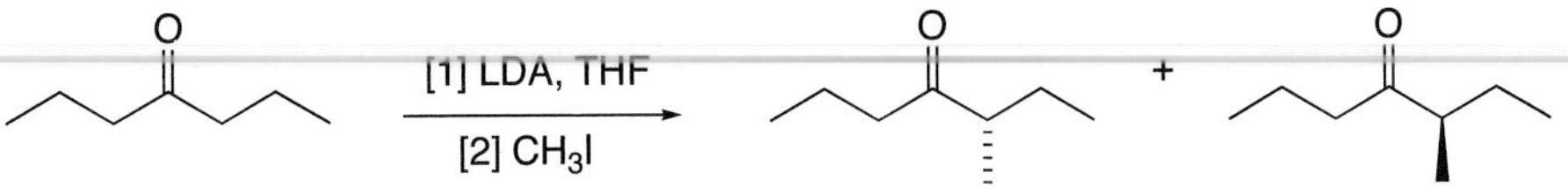

[1] LDA, THF
[2] CH₃I

$$\text{[1] LDA, THF} \quad \text{[2] CH}_3\text{I}$$

c.

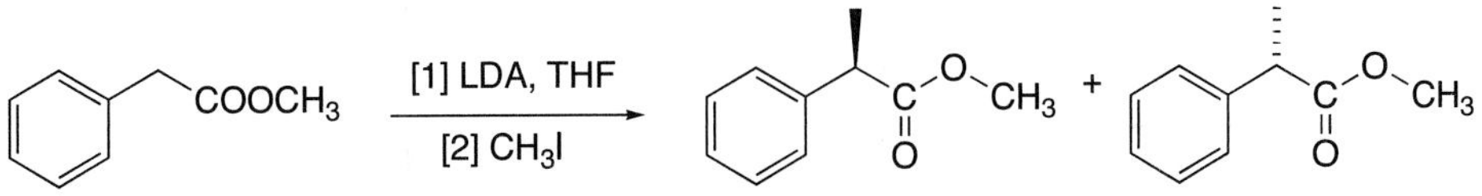

[1] LDA, THF
[2] CH₃I

**23.18**

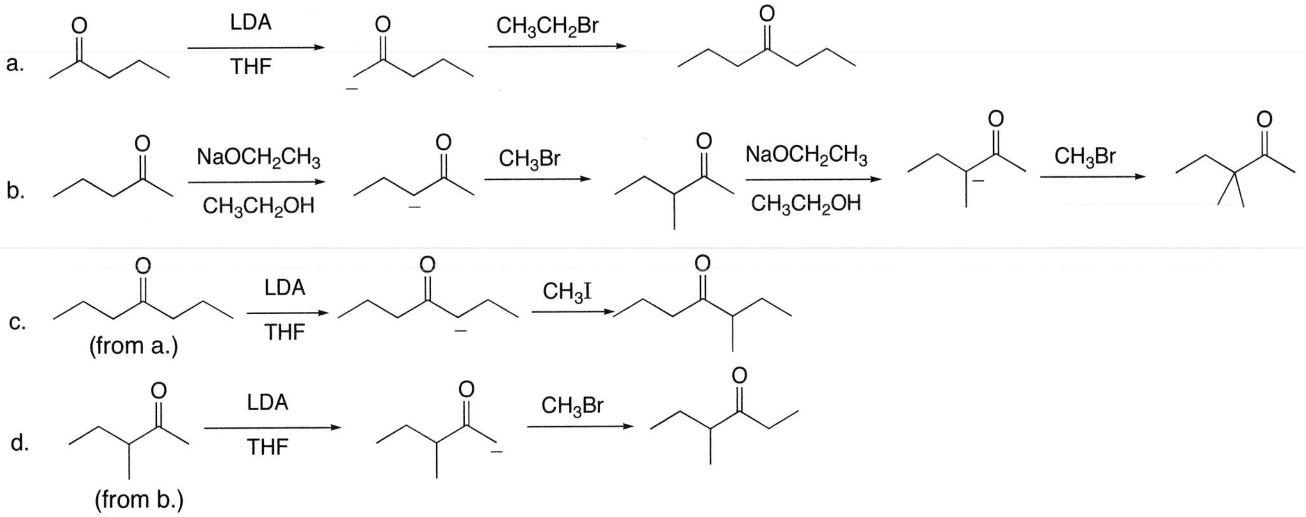

a.

LDA
THF
CH₃CH₂Br

b.

NaOCH₂CH₃
CH₃CH₂OH
CH₃Br
NaOCH₂CH₃
CH₃CH₂OH
CH₃Br

c.

(from a.)
LDA
THF
CH₃I

d.

(from b.)
LDA
THF
CH₃Br

**23.19**

LDA
THF
**A**
CH₃I
**B**
Br₂
CH₃CO₂H
**C**
Li₂CO₃
LiBr
DMF
α-methylene-
γ-butyrolactone

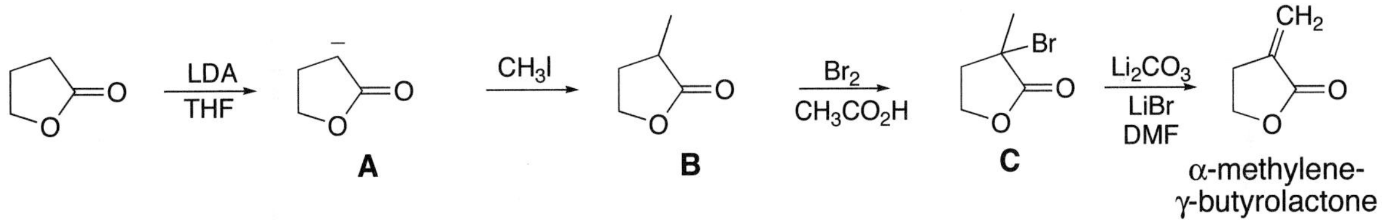

**23.20** Decarboxylation occurs only when a carboxy group is bonded to the α C of another carbonyl group.

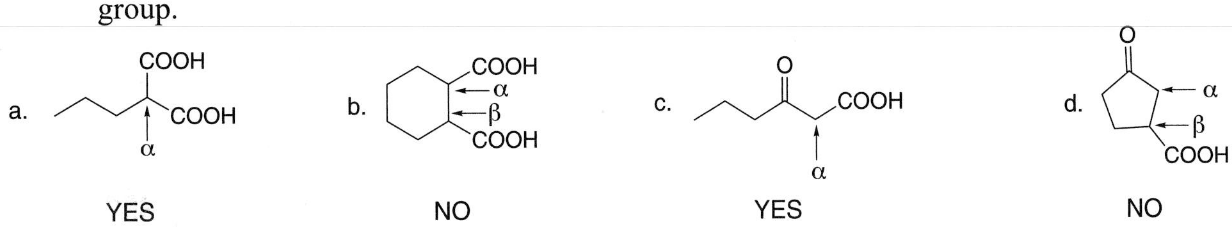

a.

YES

b.

NO

c.

YES

d.

NO

**23.21**

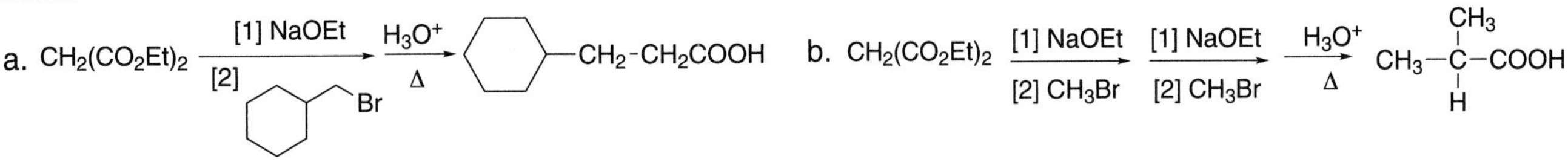

a. CH₂(CO₂Et)₂
[1] NaOEt
[2]

H₃O⁺
Δ

b. CH₂(CO₂Et)₂
[1] NaOEt
[2] CH₃Br
[1] NaOEt
[2] CH₃Br
H₃O⁺
Δ

**23.22**

a.

b.

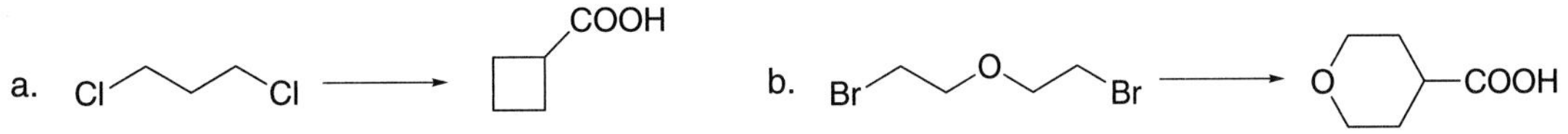

**23.23**  Locate the α C to the COOH group, and identify all of the alkyl groups bonded to it.  These
groups are from alkyl halides, and the remainder of the molecule is from diethyl malonate.

a.  $(CH_3)_2CHCH_2CH_2CH_2CH_2$ $CH_2COOH$

α

$CH_2(CO_2Et)_2$ $\xrightarrow[\text{[2] } (CH_3)_2CHCH_2CH_2CH_2CH_2Br]{\text{[1] NaOEt}}$ $\xrightarrow[\Delta]{H_3O^+}$ $(CH_3)_2CHCH_2CH_2CH_2CH_2CH_2COOH$

b. 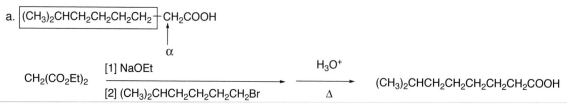 COOH

α

$CH_2(CO_2Et)_2$ $\xrightarrow[\text{[2] } CH_3CH_2CH_2Br]{\text{[1] NaOEt}}$ $\xrightarrow[\text{[2] } CH_3CH_2CH(CH_3)CH_2CH_2Br]{\text{[1] NaOEt}}$ $\xrightarrow[\Delta]{H_3O^+}$ 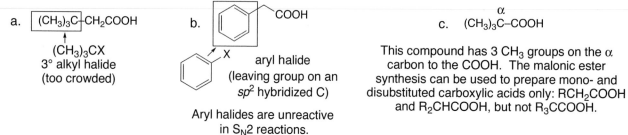 COOH

c.  $(CH_3CH_2CH_2)_2CHCOOH$

α

$CH_2(CO_2Et)_2$ $\xrightarrow[\text{[2] } CH_3CH_2CH_2Br]{\text{[1] NaOEt}}$ $\xrightarrow[\text{[2] } CH_3CH_2CH_2Br]{\text{[1] NaOEt}}$ $\xrightarrow[\Delta]{H_3O^+}$ $(CH_3CH_2CH_2)_2CHCOOH$

**23.24**  The alkyl halide must be 1° or $CH_3X$ since this is an $S_N2$ reaction.

a.  $(CH_3)_3C$ $CH_2COOH$

$(CH_3)_3CX$
3° alkyl halide
(too crowded)

b.  COOH

X  aryl halide
(leaving group on an
$sp^2$ hybridized C)

Aryl halides are unreactive
in $S_N2$ reactions.

c.  α
$(CH_3)_3C–COOH$

This compound has 3 $CH_3$ groups on the α
carbon to the COOH.  The malonic ester
synthesis can be used to prepare mono- and
disubstituted carboxylic acids only: $RCH_2COOH$
and $R_2CHCOOH$, but not $R_3CCOOH$.

**23.25**

a.  $CH_3\overset{O}{\overset{||}{C}}CH_2CO_2Et$ $\xrightarrow[\text{[2] } CH_3I]{\text{[1] NaOEt}}$ $CH_3\overset{O}{\overset{||}{C}}CH_2CH_3$
[3] $H_3O^+$, Δ

b.  $CH_3\overset{O}{\overset{||}{C}}CH_2CO_2Et$ $\xrightarrow[\text{[2] } CH_3CH_2CH_2Br]{\text{[1] NaOEt}}$ $CH_3\overset{O}{\overset{||}{C}}CH{-}CH_2{-}$
[3] NaOEt $\quad\quad\quad\quad\quad CH_2CH_2CH_3$
[4] $C_6H_5CH_2I$
[5] $H_3O^+$, Δ

**23.26** Locate the α C. All alkyl groups on the α C come from alkyl halides, and the remainder of the molecule comes from ethyl acetoacetate.

a. 

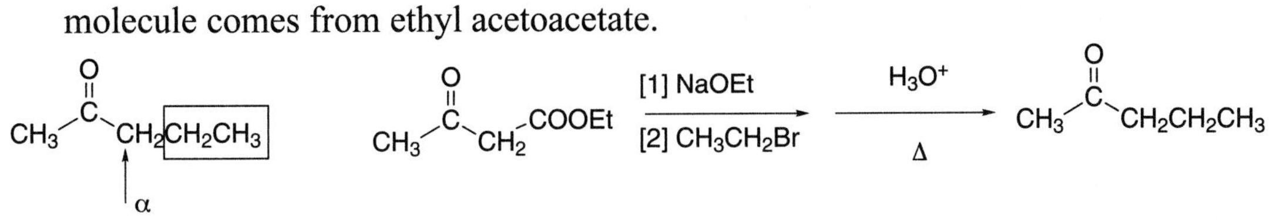

b.

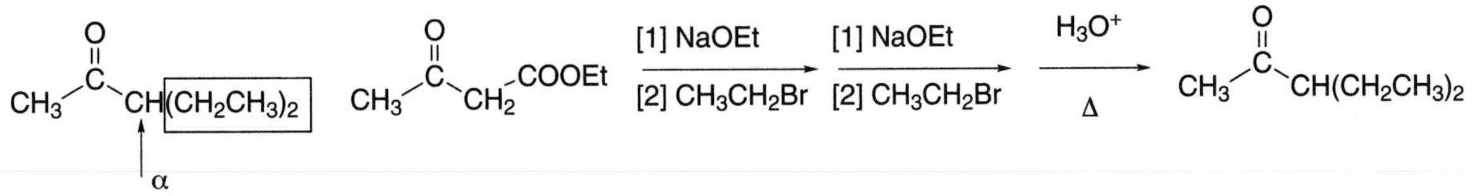

c.

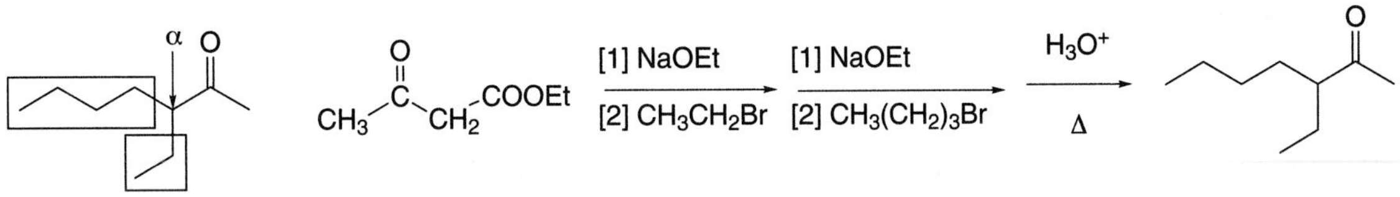

**23.27**

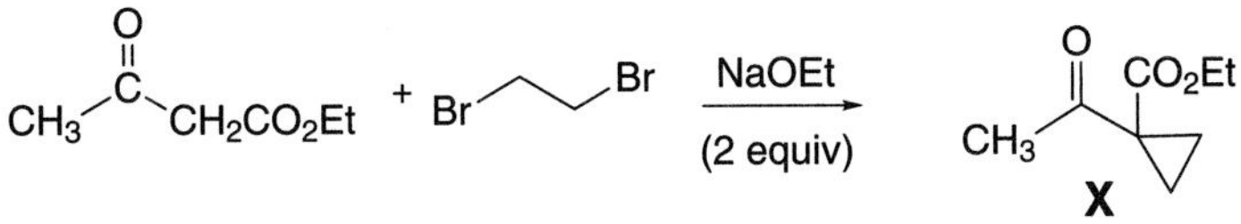

**23.28**

a.

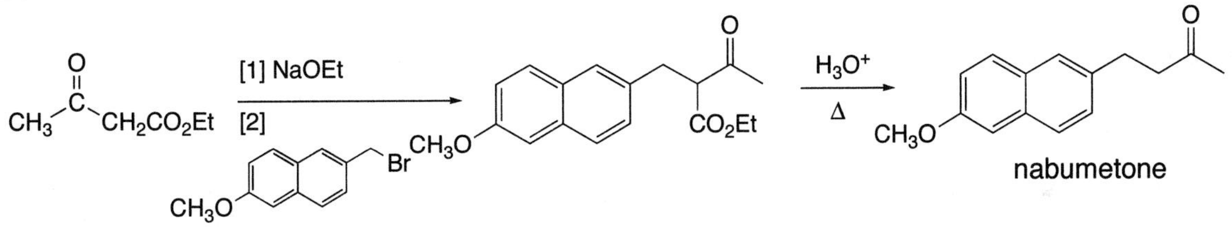

b.

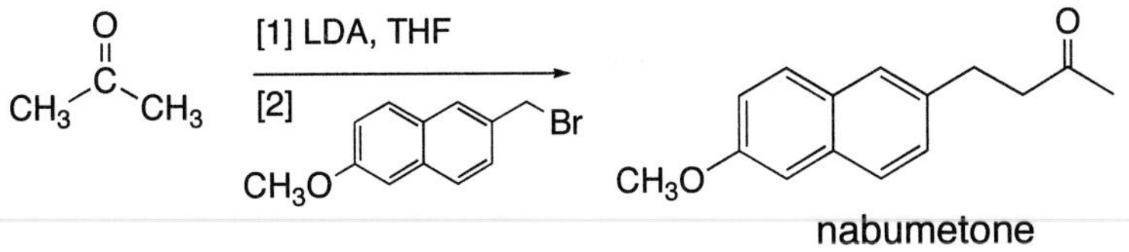

**23.29** Use the directions from Answer 23.1 to draw the enol tautomer(s). In cases where *E* and *Z* isomers can form, only one isomer is drawn.

a.

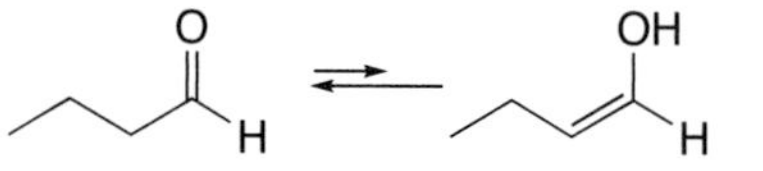

b.

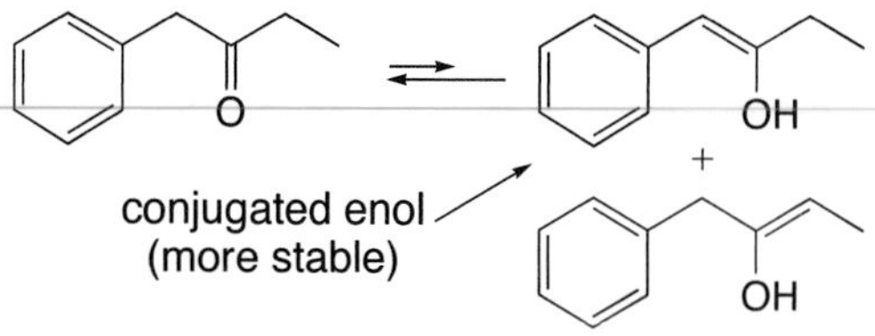

conjugated enol
(more stable)

c.

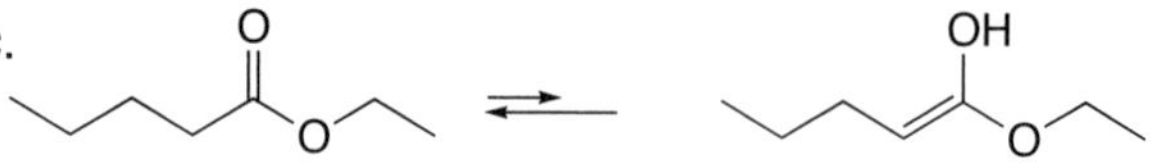

d.

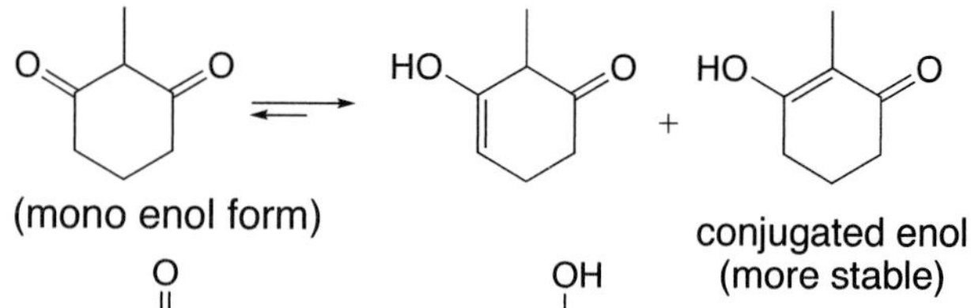

(mono enol form)

conjugated enol
(more stable)

e.

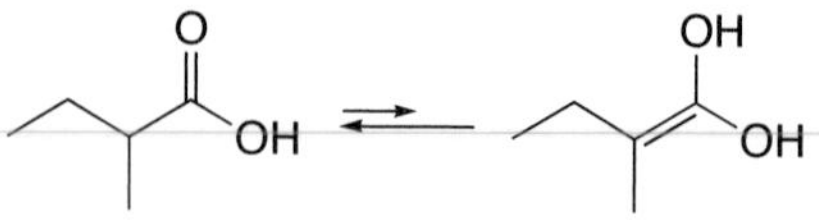

f.

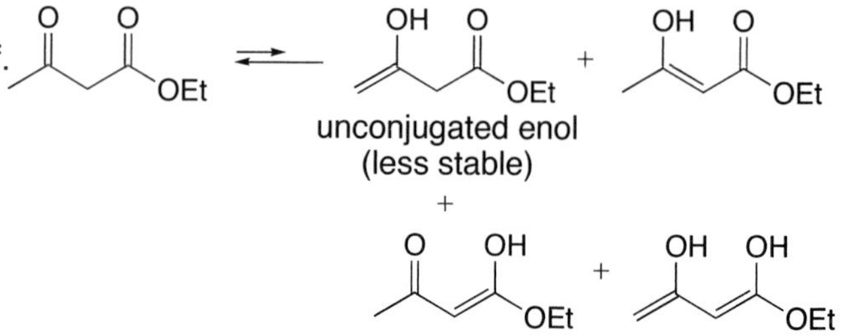

unconjugated enol
(less stable)

**23.30**

a.  CH₃CH₂CH₂CO₂CH(CH₃)₂

b.

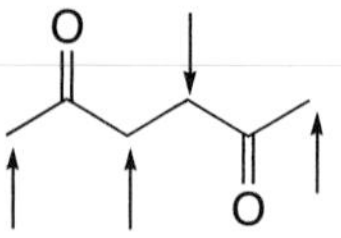

c.

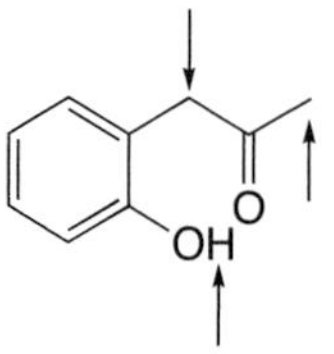

d.  CH₃O— —CH₂CN

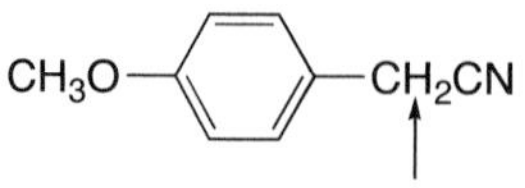

e. NC—

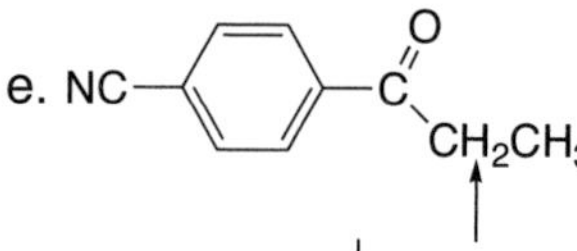

f.

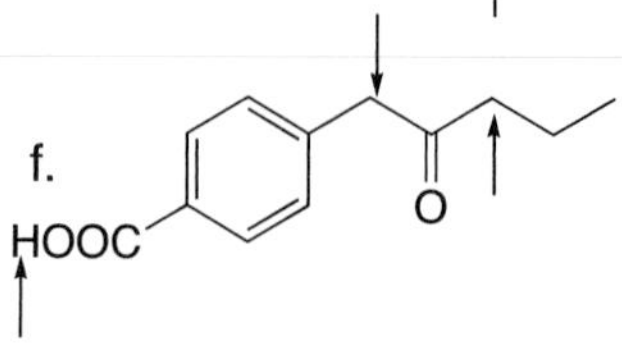

**23.31**

a.  CH₃CH₂—C
Hₐ H_b    H_c

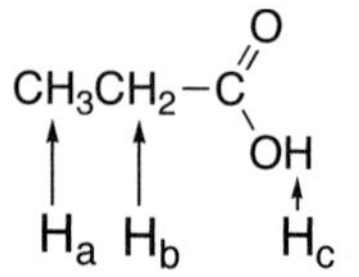

Hₐ is part of a CH₃ group =
  **least acidic**.
H_b is bonded to an α C =
  **intermediate acidity**.
H_c is bonded to O = **most acidic**.

b.

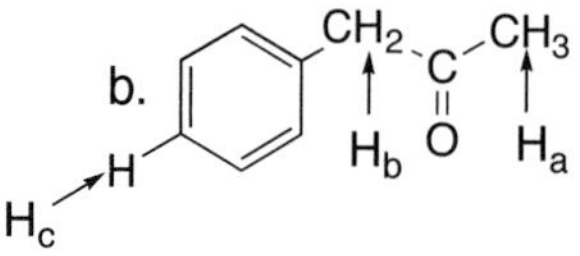

H_c is bonded to an $sp^2$ hybridized C = **least acidic**.
Hₐ is bonded to an α C = **intermediate acidity**.
H_b is bonded to an α C, and is adjacent
  to a benzene ring = **most acidic**.

**c.**

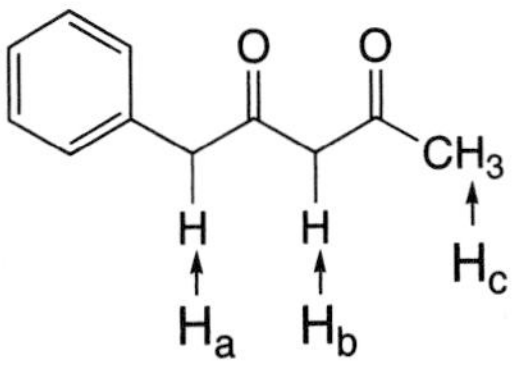

$H_c$ is bonded to an α C = **least acidic**.

$H_a$ is bonded to an α C, and is adjacent to a benzene ring = **intermediate acidity**.

$H_b$ is bonded to an α C, between two C=O groups = **most acidic**.

**d.**

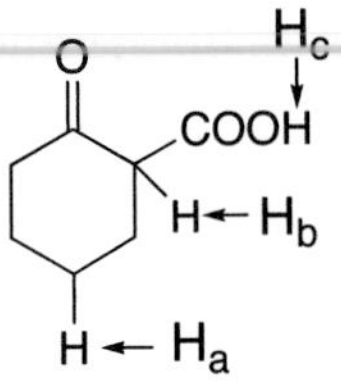

$H_a$ is bonded to an $sp^3$ hybridized C = **least acidic**.

$H_b$ is bonded to an α C = **intermediate acidity**.

$H_c$ is bonded to O = **most acidic**.

**e.**

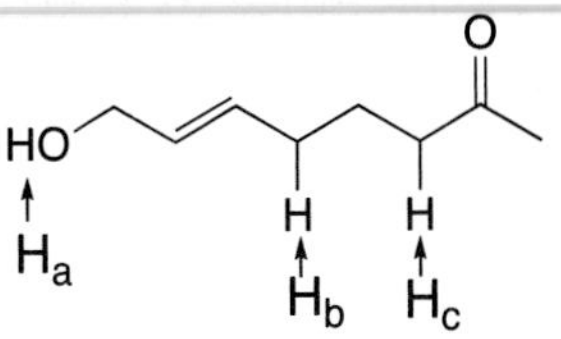

$H_b$ is bonded to an $sp^3$ hybridized C = **least acidic**.

$H_c$ is bonded to an α C = **intermediate acidity**.

$H_a$ is bonded to O = **most acidic**.

**23.32**

**a.**

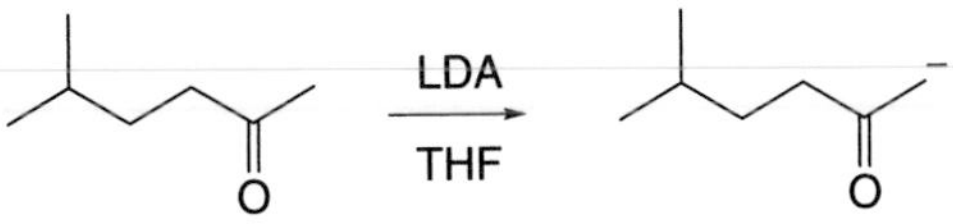

**b.**

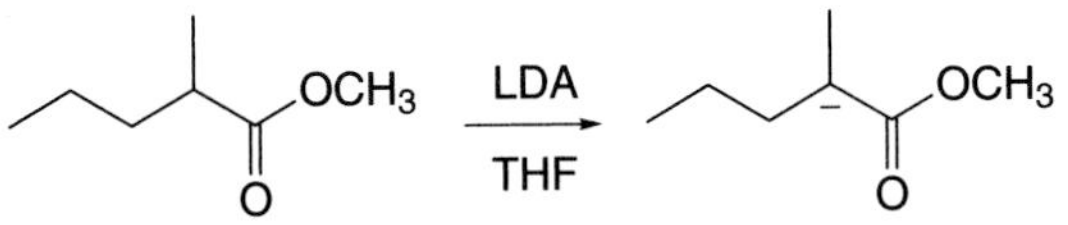

**c.**

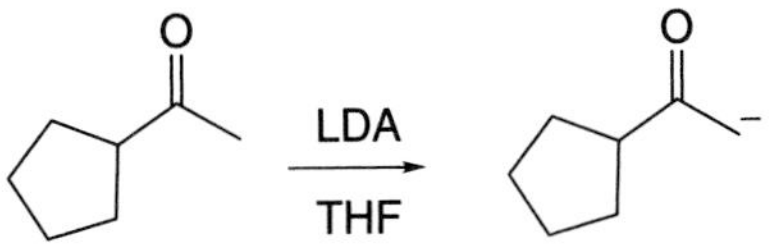

**d.**

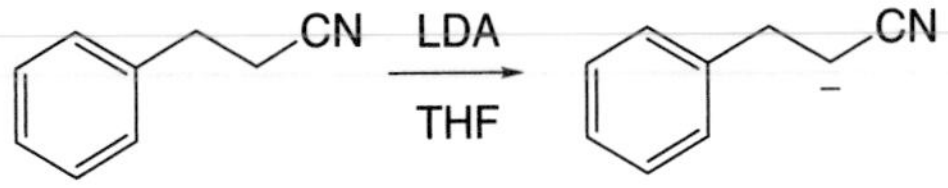

**e.**

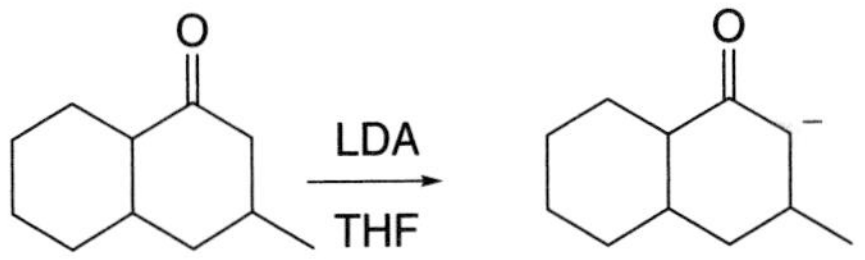

**f.**

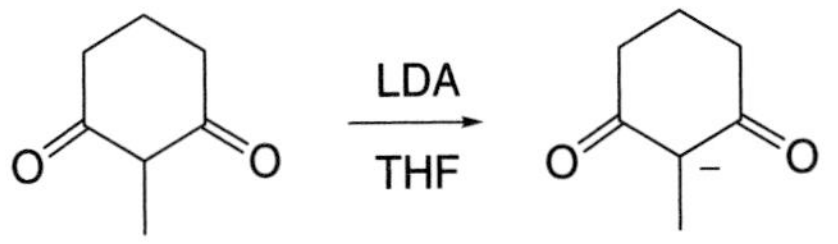

**23.33** Enol tautomers have OH groups that give a broad OH absorption at 3600–3200 cm$^{-1}$, which could be detected readily in the IR.

**23.34**

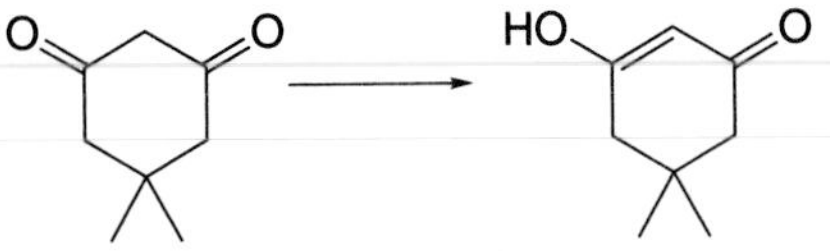

5,5-dimethyl-1,3-cyclohexanedione

5,5-Dimethyl-1,3-cyclohexanedione exists predominantly in its enol form because the C=C of the enol is conjugated with the other C=O of the dicarbonyl compound. Conjugation stabilizes this enol.

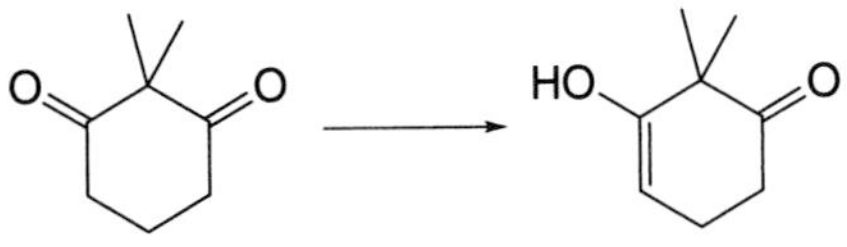

2,2-dimethyl-1,3-cyclohexanedione

The enol of 2,2-dimethyl-1,3-cyclohexanedione is not conjugated with the other carbonyl group. In this way it resembles the enol of any other carbonyl compound, and thus it is present in low concentration.

**23.35** In the presence of acid, (*R*)-α-methylbutyrophenone enolizes to form an achiral enol. Protonation of the enol from either face forms an equal mixture of two enantiomers, making the solution optically inactive.

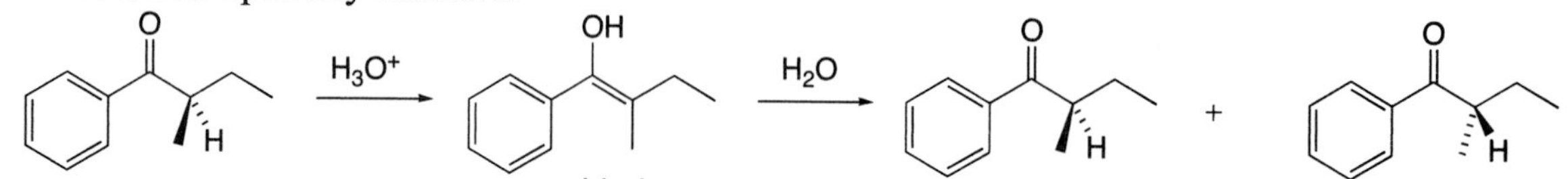

(*R*)-α-methylbutyrophenone

achiral
(*E* and *Z* isomers)

In the presence of base, (*R*)-α-methylbutyrophenone is deprotonated to form an achiral enolate, which can then be protonated from either face to form an optically inactive mixture of two enantiomers.

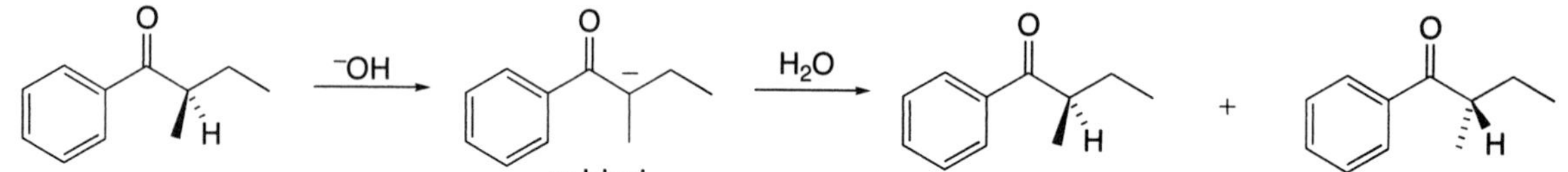

(*R*)-α-methylbutyrophenone

achiral

**23.36**

ester

The O atom of the ester OR group donates electron density by a resonance effect. The resulting resonance structure keeps a negative charge on the less electronegative C end of the enolate. This destabilizes the resonance hybrid of the conjugate base, and makes the α H's of the ester less acidic.

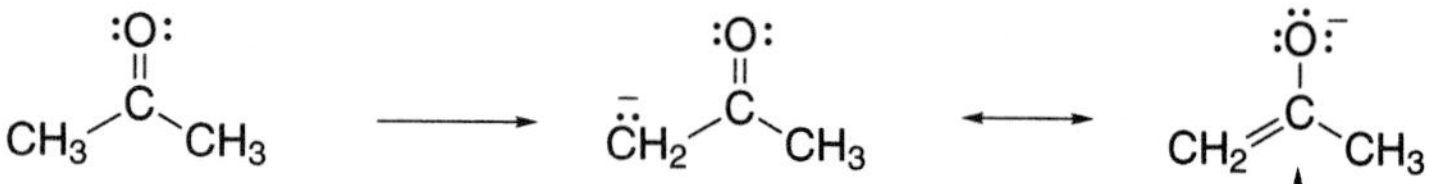

ketone

No additional resonance structures.

This structure, which places a negative charge on the O atom, is the major contributor to the hybrid, stabilizing it, and making the α H's of the ketone more acidic.

**23.37** LDA reacts with the most acidic proton. If there is any $H_2O$ present, the water would immediately react with the base:

LDA

**23.38**

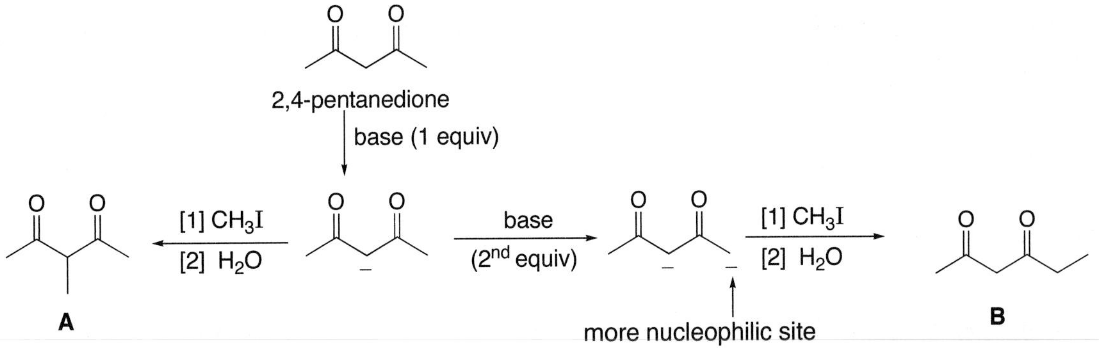
2,4-pentanedione
base (1 equiv)
[1] $CH_3I$
[2] $H_2O$
base
(2nd equiv)
[1] $CH_3I$
[2] $H_2O$
A
more nucleophilic site
B

One equivalent of base removes the most acidic proton between the two C=O's, to form **A** on alkylation with $CH_3I$.

With a second equivalent of base a dianion is formed. Since the second enolate is less resonance stabilized, it is more nucleophilic and reacts first in an alkylation with $CH_3I$, forming **B** after protonation with $H_2O$.

**23.39** The mechanism of acid-catalyzed halogenation consists of two parts: **tautomerization** of the carbonyl compound to the enol form, and **reaction of the enol with halogen.**

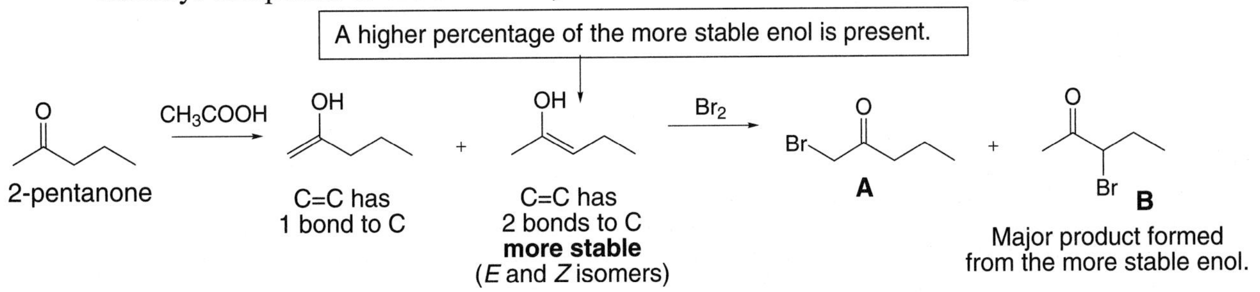
A higher percentage of the more stable enol is present.
$CH_3COOH$
$Br_2$
2-pentanone
C=C has
1 bond to C
C=C has
2 bonds to C
more stable
(E and Z isomers)
A
B
Major product formed
from the more stable enol.

**23.40** • The mechanism of acid-catalyzed halogenation (Part a) consists of two parts: **tautomerization** of the carbonyl compound to the enol form, and **reaction of the enol with halogen.**

• In the haloform reaction (Part b), the three H's of the $CH_3$ group are successively replaced by X, to form an intermediate that is oxidatively cleaved with base.

a.

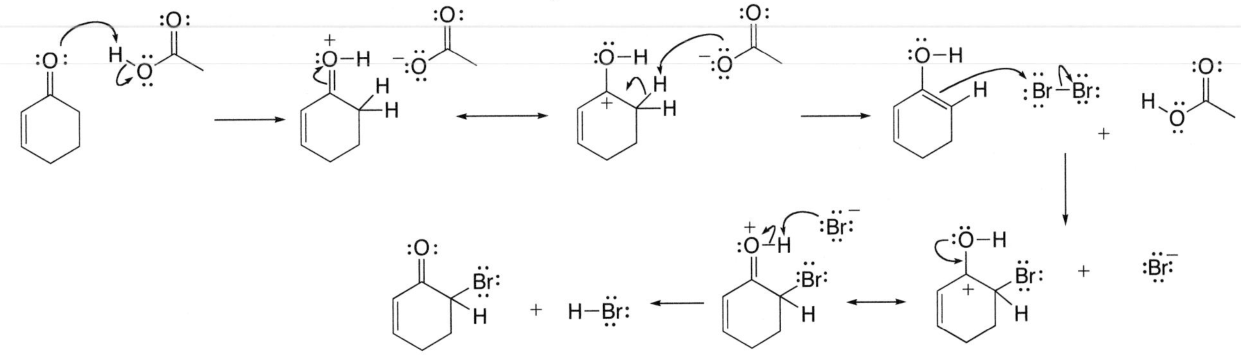

b.

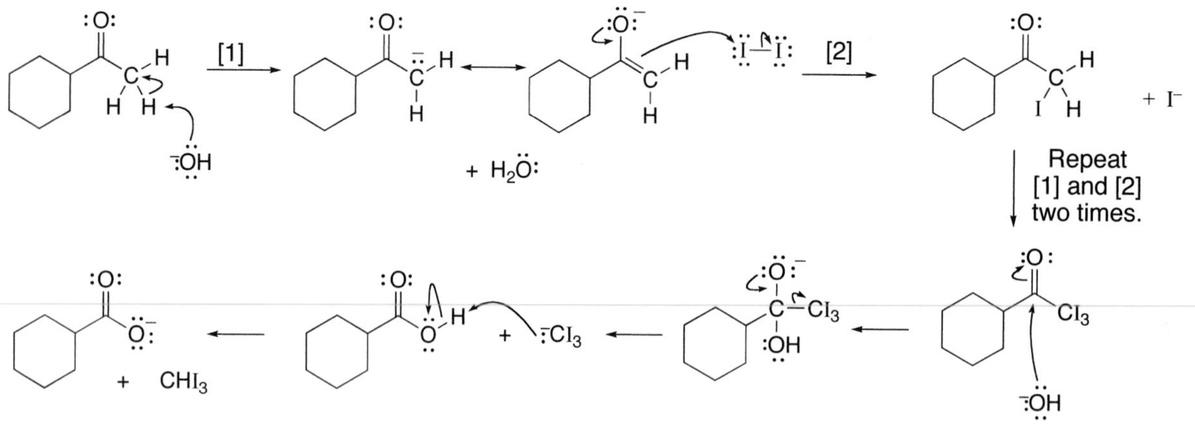

**23.41** Use the directions from Answer 23.23.

a. $\boxed{CH_3OCH_2}CH_2COOH \implies CH_3OCH_2Br$

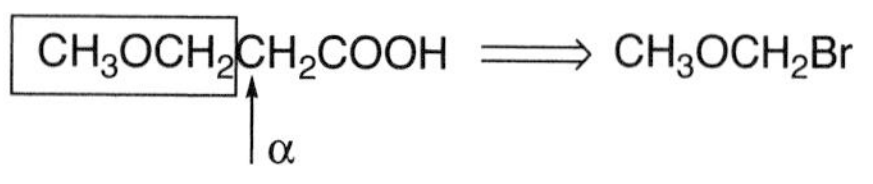

                 ↑α

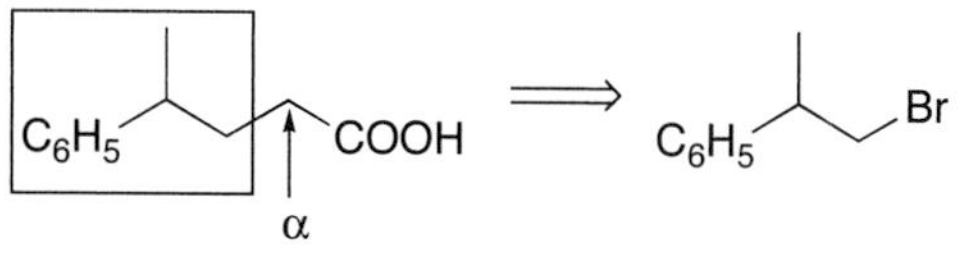

c.

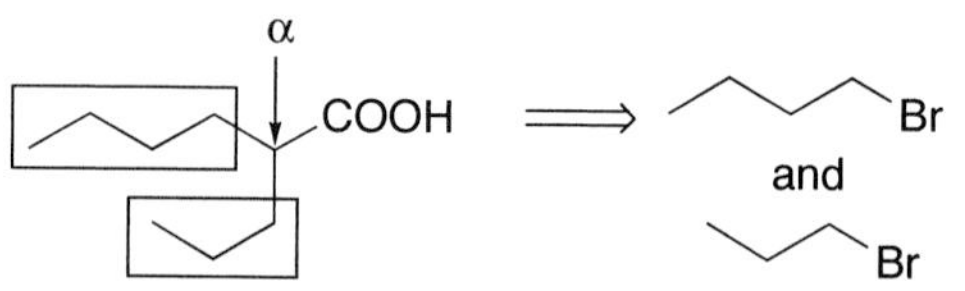

b.

**23.42**

a. $\boxed{CH_3CH_2CH_2CH_2CH_2}CH_2COOH$    $CH_2(CO_2Et)_2$    $\dfrac{\text{[1] NaOEt}}{\text{[2] } CH_3CH_2CH_2CH_2CH_2Br}$    $\dfrac{H_3O^+}{\Delta}$    $CH_3CH_2CH_2CH_2CH_2CH_2COOH$

             ↑α

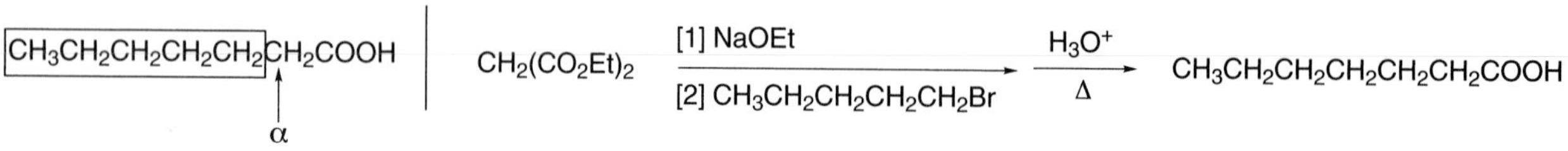

b.

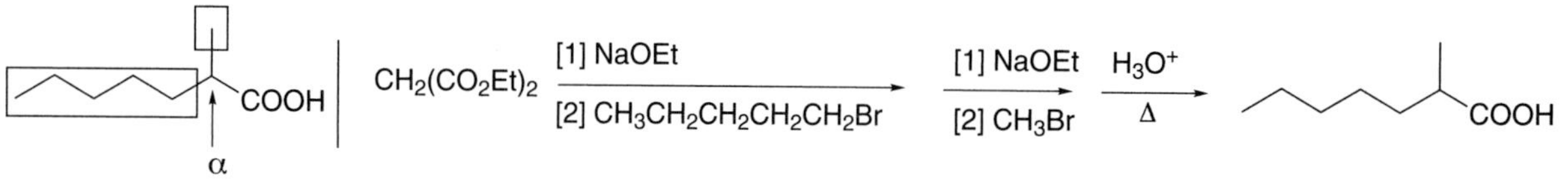

c.

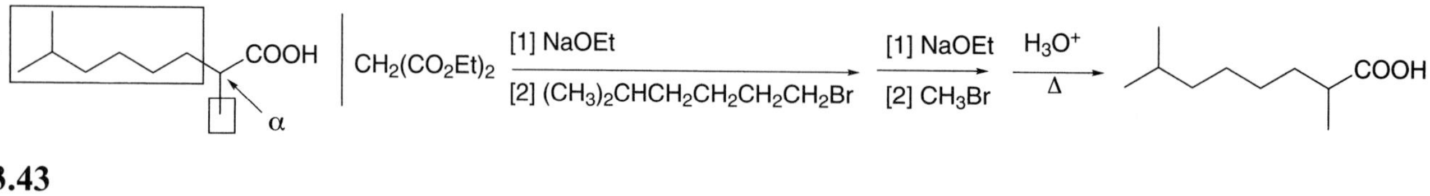

**23.43**

a. $CH_2(CO_2Et)_2$    $\dfrac{\text{[1] NaOEt}}{\substack{\text{[2] } BrCH_2CH_2CH_2CH_2CH_2Br \\ \text{[3] NaOEt}}}$    $\dfrac{H_3O^+}{\Delta}$

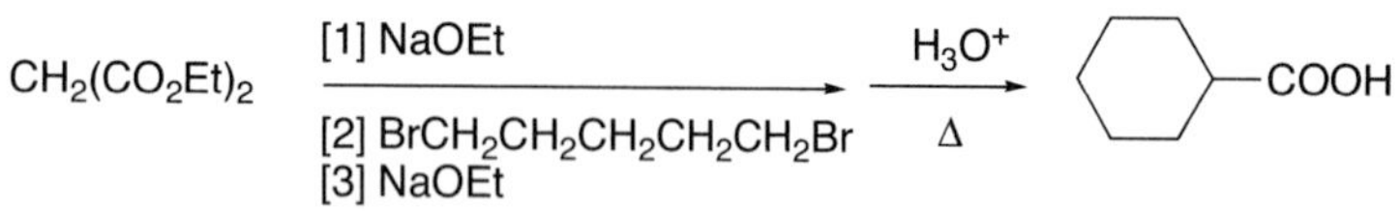

b.

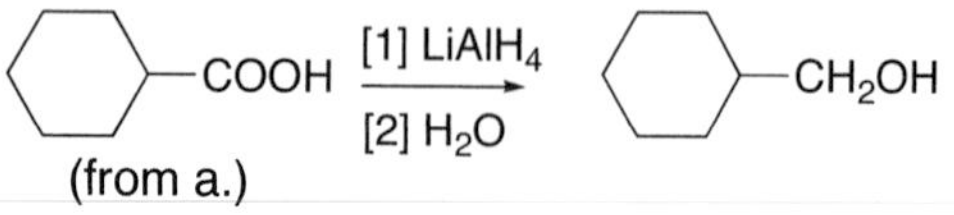

   (from a.)

c.

   (from a.)

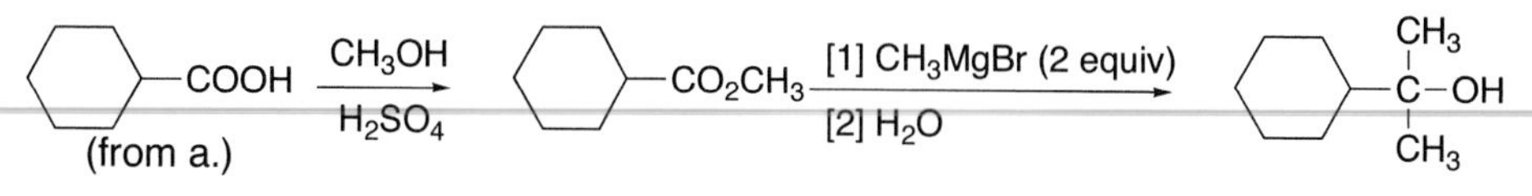

**d.** (cyclohexyl)—COOH $\xrightarrow[H_2SO_4]{CH_3CH_2OH}$ (cyclohexyl)—COOCH$_2$CH$_3$ $\xrightarrow[\text{[2] CH}_3\text{I}]{\text{[1] LDA}}$ (cyclohexyl, CH$_3$)—COOCH$_2$CH$_3$

(from a.)

## 23.44

**a.** (epoxide with CH$_3$) $\xrightarrow[\text{[2] H}_2\text{O}]{\text{[1] Na}^+ {}^-\text{CH(COOEt)}_2}$ HO–CH$_3$CHCH$_2$—CH(COOEt)$_2$

nucleophilic attack here

**b.** CH$_2$=O $\xrightarrow[\text{[2] H}_2\text{O}]{\text{[1] Na}^+ {}^-\text{CH(COOEt)}_2}$ HOCH$_2$CH(COOEt)$_2$

**c.** CH$_3$C(=O)Cl $\xrightarrow[\text{[2] H}_2\text{O}]{\text{[1] Na}^+ {}^-\text{CH(COOEt)}_2}$ CH$_3$C(=O)CH(COOEt)$_2$

**d.** CH$_3$C(=O)–O–C(=O)CH$_3$ $\xrightarrow[\text{[2] H}_2\text{O}]{\text{[1] Na}^+ {}^-\text{CH(COOEt)}_2}$ CH$_3$C(=O)CH(COOEt)$_2$

+ CH$_3$COOH

## 23.45 Use the directions from Answer 23.26.

**a.** [ketone structure, α] , CH$_3$C(=O)CH$_2$COOEt $\xrightarrow[\text{[2] Br}\text{—(isoamyl)}]{\text{[1] NaOEt}}$ $\xrightarrow[\Delta]{H_3O^+}$ [product ketone]

**b.** [ketone structure, α] , CH$_3$C(=O)CH$_2$COOEt $\xrightarrow[\text{[2] CH}_3\text{CH}_2\text{Br}]{\text{[1] NaOEt}}$ $\xrightarrow[\text{[2] Br—CH}_2\text{(cyclohexyl)}]{\text{[1] NaOEt}}$ $\xrightarrow[\Delta]{H_3O^+}$ [product ketone]

**c.** [ketone structure, α] , CH$_3$C(=O)CH$_2$COOEt $\xrightarrow[\text{[2] Br—(allyl)}]{\text{[1] NaOEt}}$ $\xrightarrow[\Delta]{H_3O^+}$ [product ketone]

**d.** [ketone structure, α] , CH$_3$C(=O)CH$_2$COOEt $\xrightarrow[\text{[2] Br—(methyl dibromide)—Br}]{\text{[1] NaOEt}}$ $\xrightarrow[\Delta]{H_3O^+}$ [product ketone]

## 23.46

**a.** CH$_3$C(=O)CH$_2$COOEt $\xrightarrow[\text{[2] CH}_3\text{CH}_2\text{Br}]{\text{[1] NaOEt}}$ $\xrightarrow[\Delta]{H_3O^+}$ CH$_3$C(=O)CH$_2$CH$_2$CH$_3$

**b.** CH$_3$C(=O)CH$_2$COOEt $\xrightarrow[\text{[2] CH}_3\text{Br}]{\text{[1] NaOEt}}$ $\xrightarrow[\text{[2] CH}_3\text{Br}]{\text{[1] NaOEt}}$ $\xrightarrow[\Delta]{H_3O^+}$ CH$_3$C(=O)CH(CH$_3$)$_2$

**c.** CH$_3$C(=O)CH(CH$_3$)$_2$ $\xrightarrow[\text{THF}]{\text{LDA}}$ {}^-CH$_2$C(=O)CH(CH$_3$)$_2$ $\xrightarrow{\text{CH}_3\text{I}}$ CH$_3$CH$_2$C(=O)CH(CH$_3$)$_2$

(from b.)

**d.** CH$_3$C(=O)CH(CH$_3$)$_2$ $\xrightarrow[\text{CH}_3\text{OH}]{\text{NaOCH}_3}$ CH$_3$C(=O){}^-C(CH$_3$)$_2$ $\xrightarrow{\text{CH}_3\text{I}}$ CH$_3$C(=O)C(CH$_3$)$_3$

(from b.)

**23.47**

a. $\xrightarrow[\text{DMF}]{\substack{\text{Li}_2\text{CO}_3 \\ \text{LiBr}}}$ (*E* and *Z*)

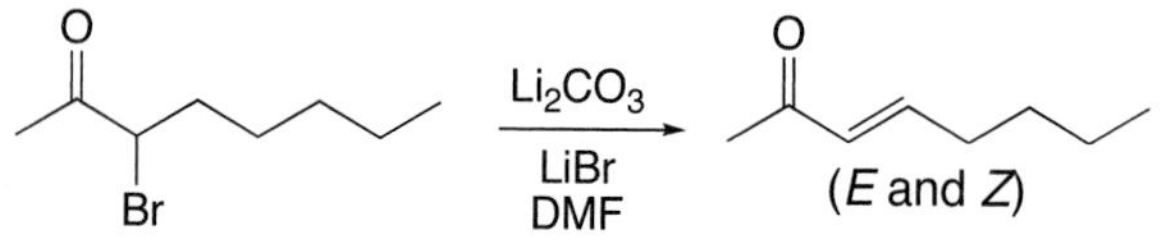

f. $\xrightarrow[\text{[2] Li}_2\text{CO}_3\text{, LiBr, DMF}]{\text{[1] Br}_2\text{, CH}_3\text{CO}_2\text{H}}$

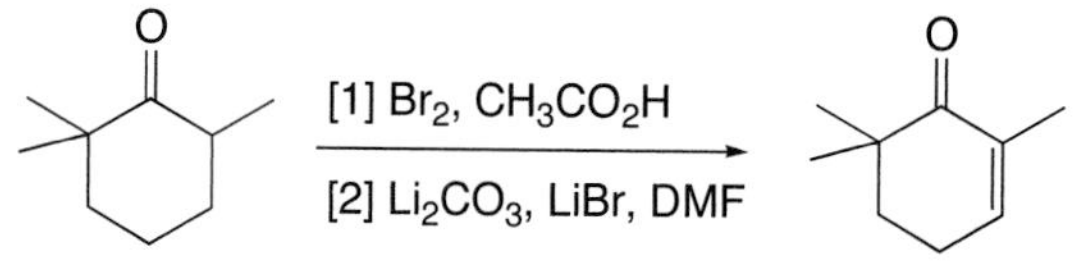

b. $\xrightarrow{\Delta}$

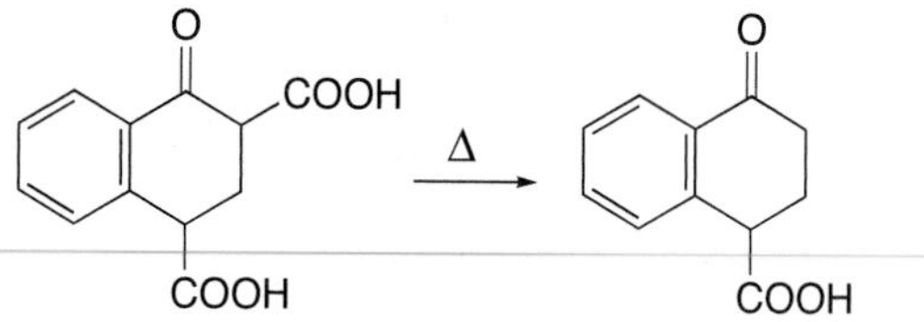

g. $\xrightarrow[^-\text{OH}]{\text{I}_2 \text{ (excess)}}$ + CHI$_3$

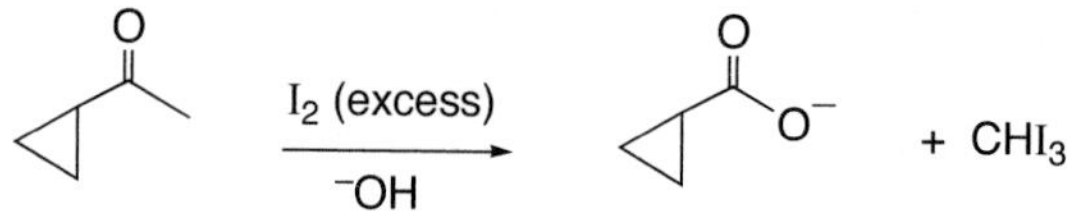

c. CH$_3$CH$_2$CH$_2$CO$_2$Et $\xrightarrow[\text{[2] CH}_3\text{CH}_2\text{I}]{\text{[1] LDA}}$ $\underset{\text{CH}_3\text{CH}_2\text{CHCO}_2\text{Et}}{\overset{\text{CH}_3\text{CH}_2}{\mid}}$

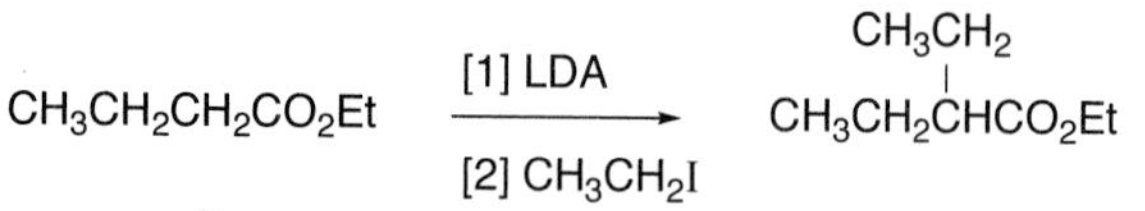

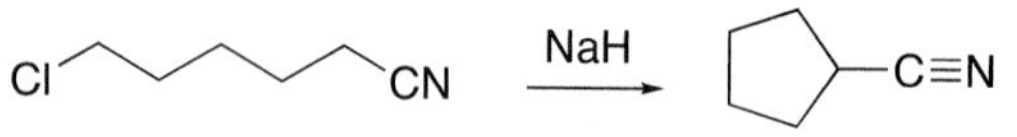

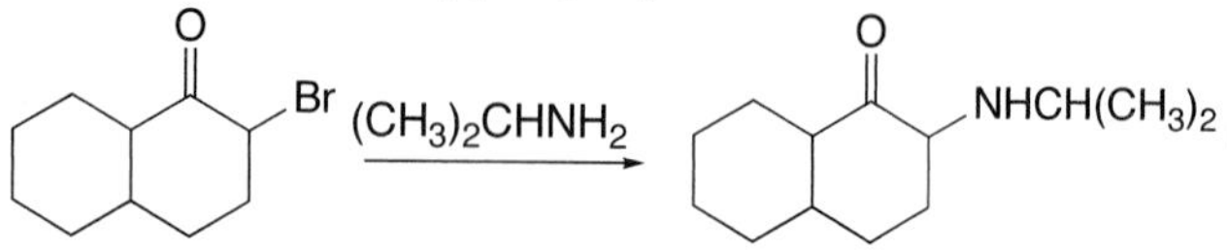

h. $\text{Cl}\diagup\diagdown\diagup\diagdown\text{CN}$ $\xrightarrow{\text{NaH}}$

d. $\xrightarrow{(\text{CH}_3)_2\text{CHNH}_2}$

i. $\xrightarrow[^-\text{OH}]{\text{Br}_2 \text{ (excess)}}$

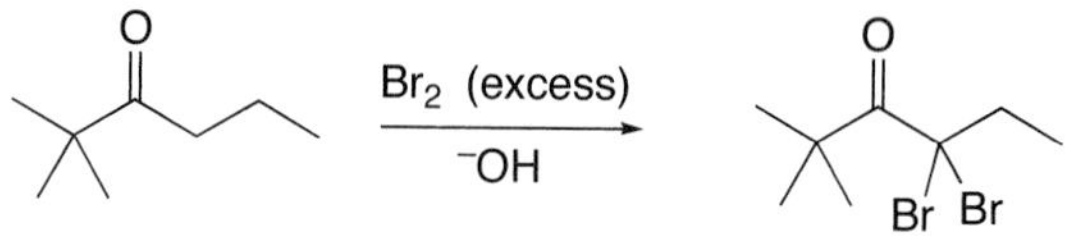

e. $\xrightarrow[\text{[2] CH}_3\text{CH}_2\text{I}]{\text{[1] LDA}}$

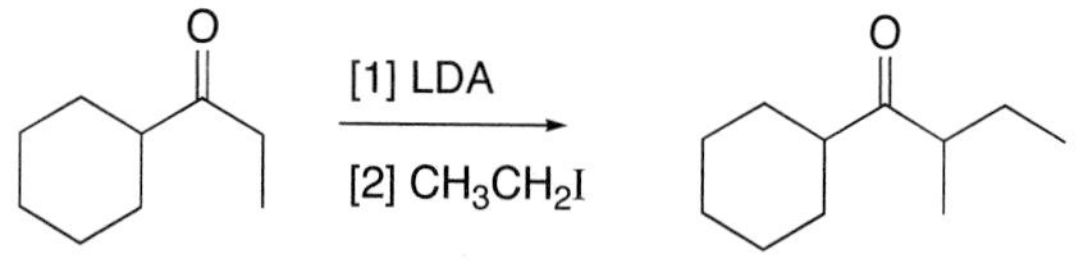

j. $\xrightarrow{\text{NaI}}$

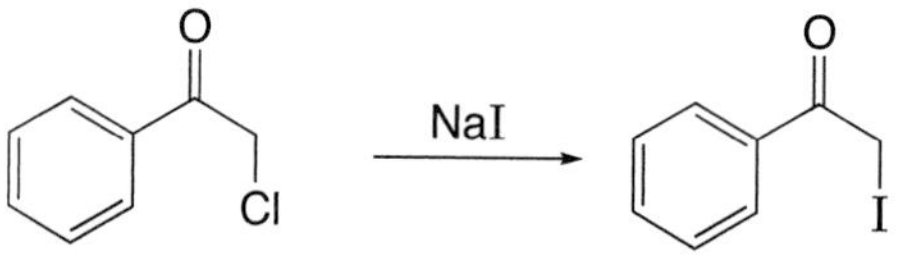

**23.48**

a. $\xrightarrow[\text{[2]}]{\text{[1] LDA}}$ +

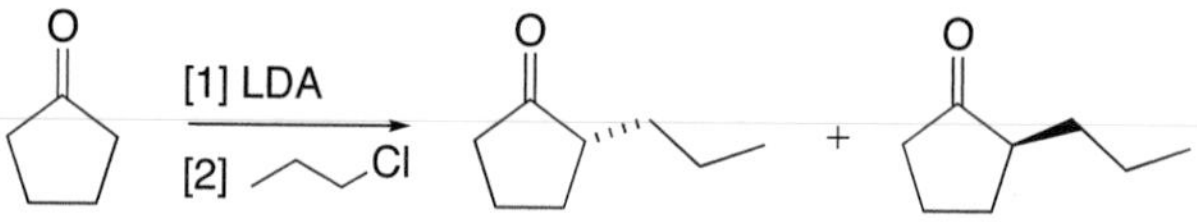

c. $\xrightarrow[\text{[2] CH}_3\text{I}]{\text{[1] LDA}}$

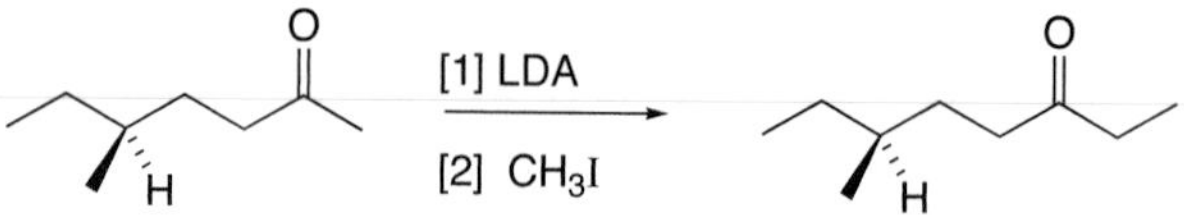

b. $\xrightarrow[\text{[2]}]{\text{[1] LDA}}$

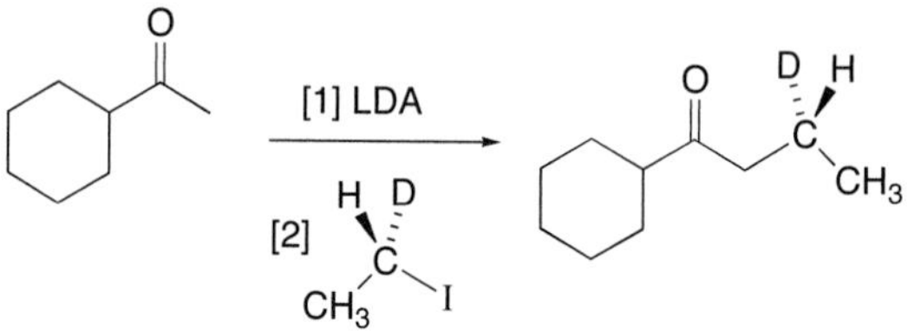

**23.49**

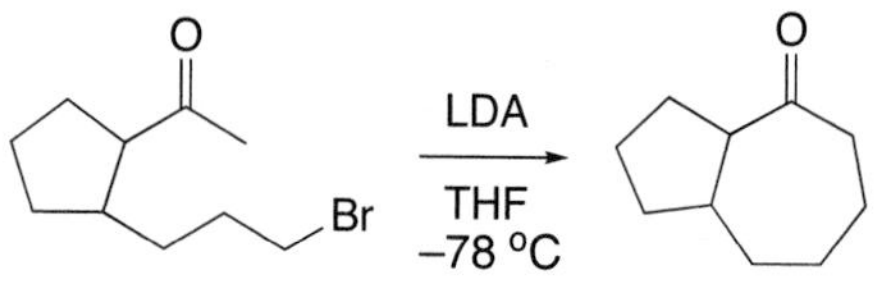

$\xrightarrow[\substack{\text{THF} \\ -78\ ^\circ\text{C}}]{\text{LDA}}$

**A**        **B**

$\xrightarrow[\substack{\text{CH}_3\text{CH}_2\text{OH} \\ \text{room temperature}}]{\text{NaOCH}_2\text{CH}_3}$

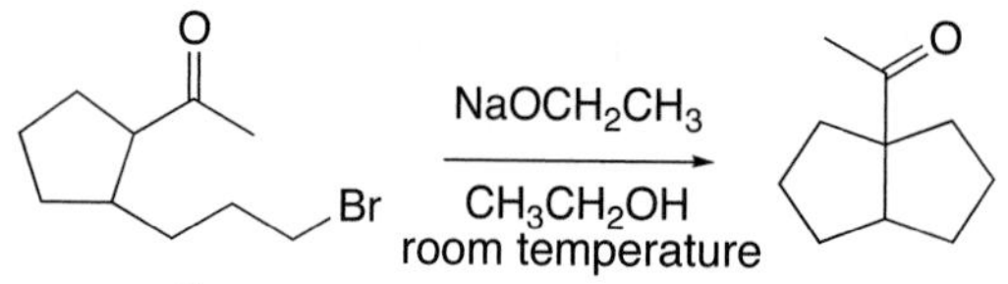

**A**        **C**

**23.50**

a.

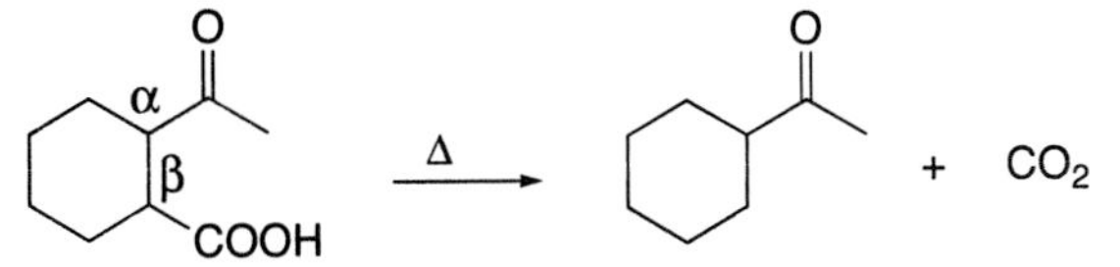

$+$ $CO_2$

In order for decarboxylation to occur readily, the COOH group must be bonded to the α C of another carbonyl group. In this case, it is bonded to the β carbon.

b. $CH_2(CO_2Et)_2$ $\xrightarrow[\text{[2] } (CH_3CH_2)_3CBr]{\text{[1] NaOEt}}$ $(CH_3CH_2)_3CCH(CO_2Et)_2$

The 3° alkyl halide is too crowded to react with the strong nucleophile by an $S_N2$ mechanism.

c.

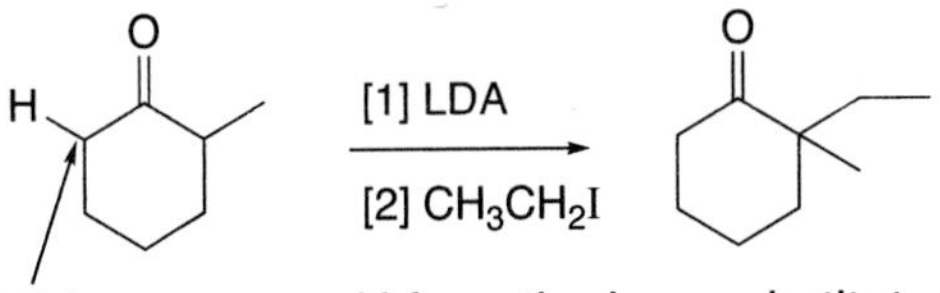

[1] LDA
[2] $CH_3CH_2I$

LDA removes a H from the less substituted C, forming the kinetic enolate. This product is from the thermodynamic enolate, which gives substitution on the more substituted α C.

**23.51**

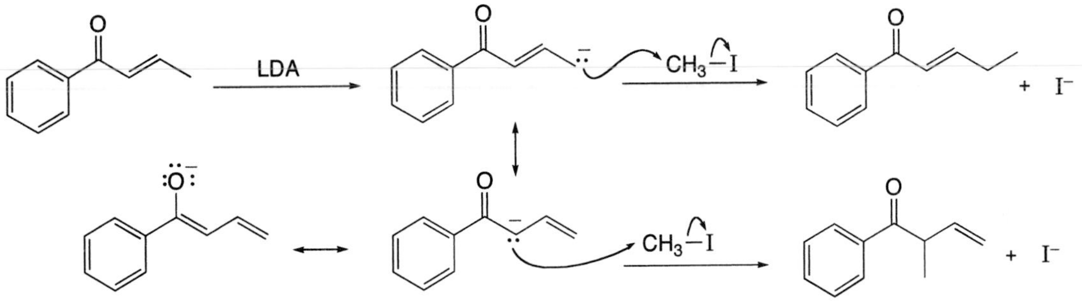

**23.52**

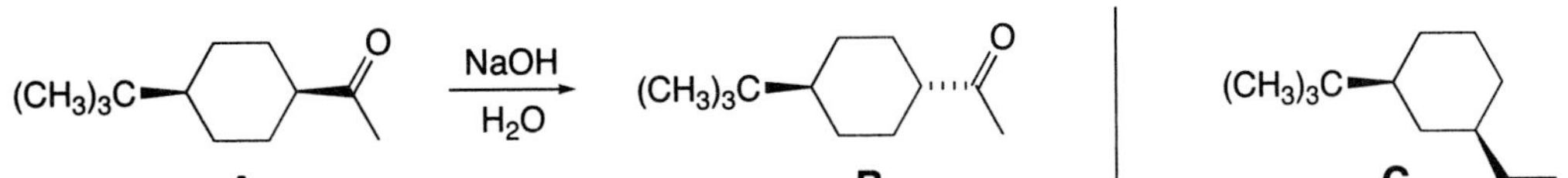

A

B

C

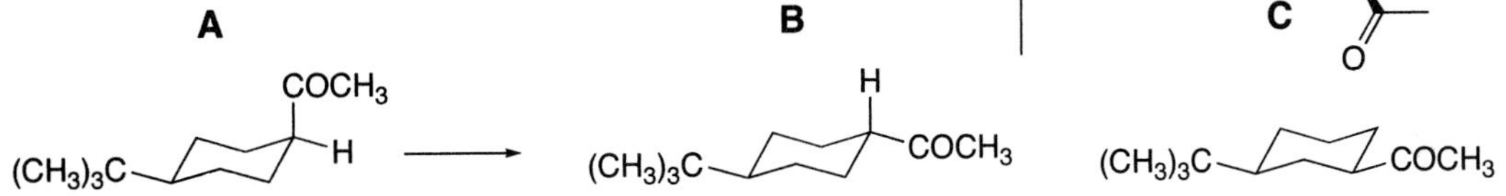

One axial and one equatorial group, **less stable**

Both groups are equatorial. **more stable**

Both groups are equatorial.

This isomerization will occur since it makes a more stable compound.

Compound **C** will not isomerize since it already has the more stable arrangement of substituents.

Isomerization occurs by way of an intermediate enolate, which can be protonated to either re-form **A**, or give **B**. Since **B** has two large groups equatorial, it is favored at equilibrium.

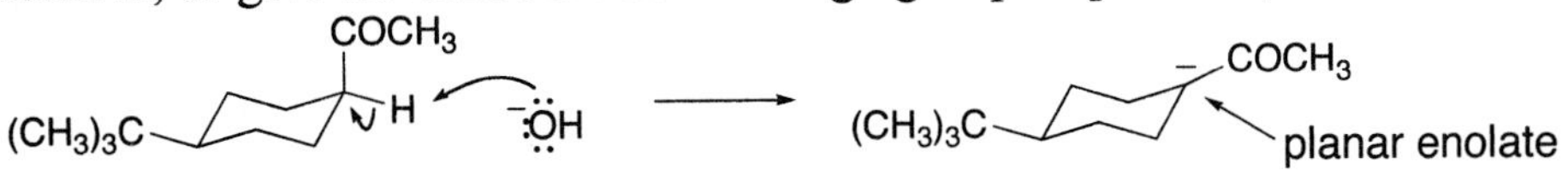

planar enolate

## 23.53

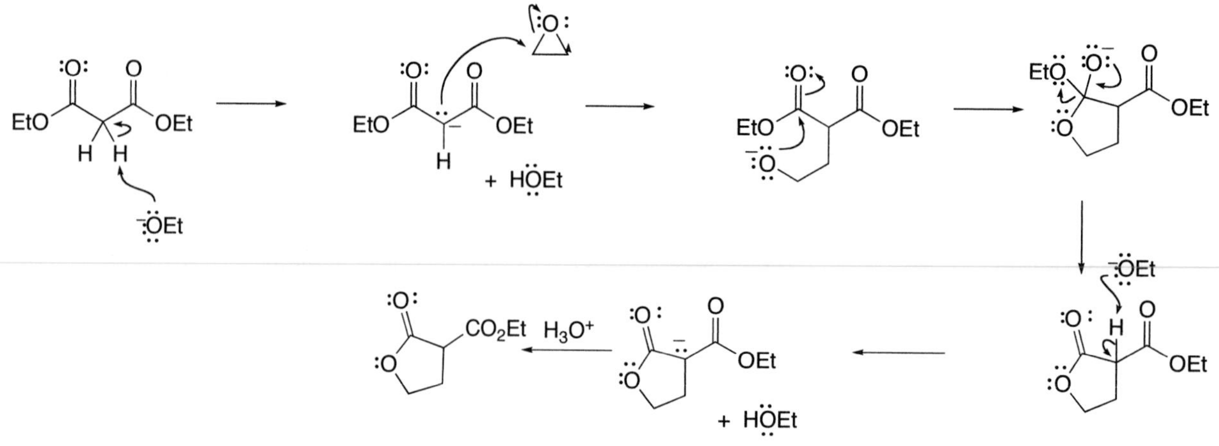

## 23.54

LDA = B:

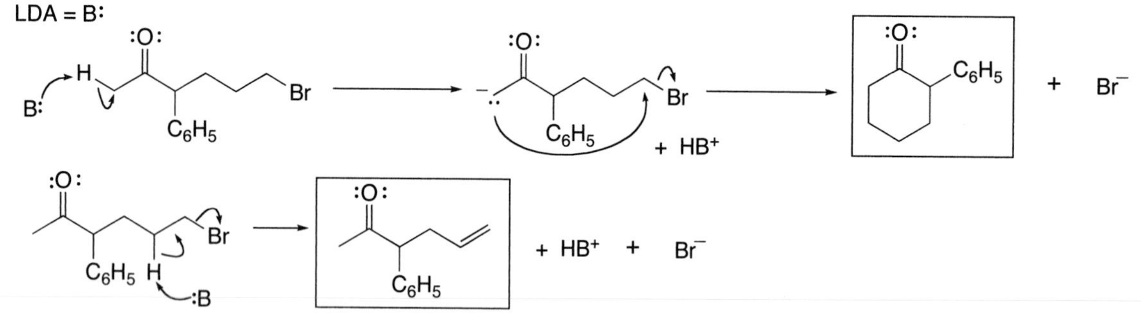

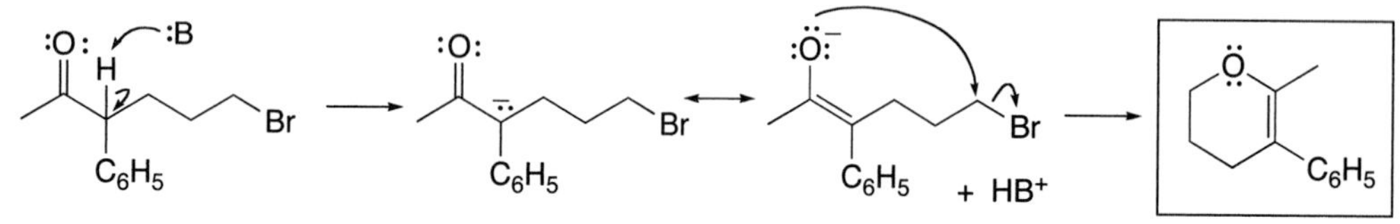

This reaction occurs with both bases [LDA and $KOC(CH_3)_3$].

+ $Br^-$

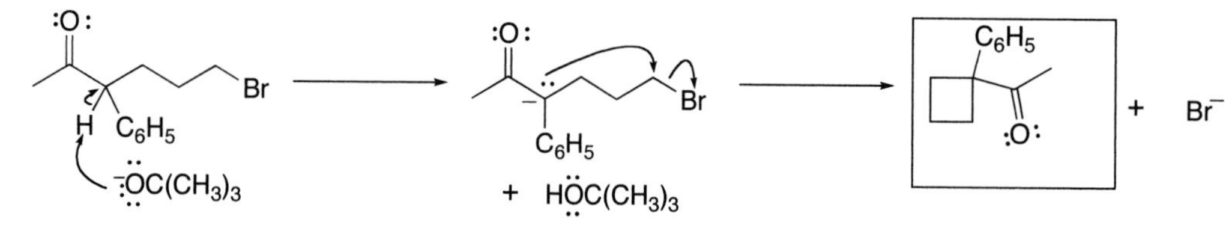

**23.55**

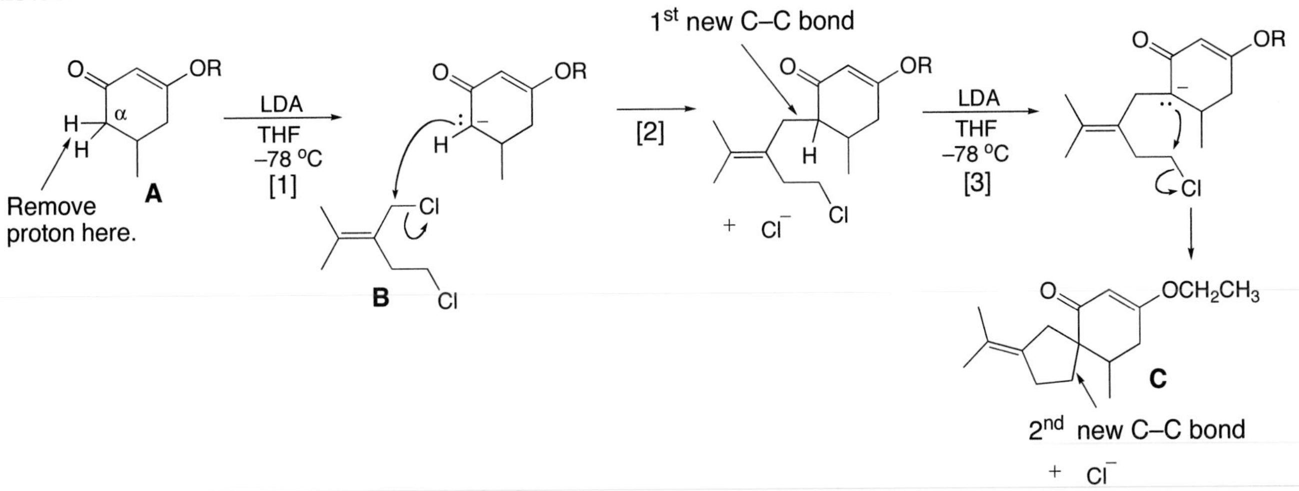

1st new C–C bond

**A** — Remove proton here.

LDA
THF
–78 °C
[1]

**B**

[2]

+ Cl⁻

LDA
THF
–78 °C
[3]

**C**

2nd new C–C bond

+ Cl⁻

**23.56**

a.

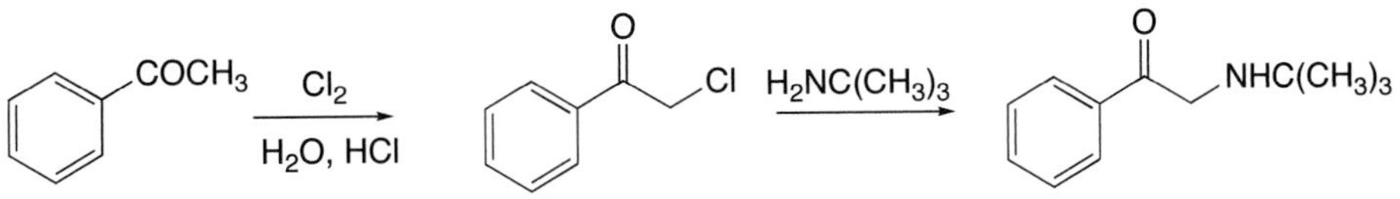

COCH₃ — $\xrightarrow[\text{H}_2\text{O, HCl}]{\text{Cl}_2}$ — CI — $\xrightarrow{\text{H}_2\text{NC(CH}_3)_3}$ — NHC(CH₃)₃

b.

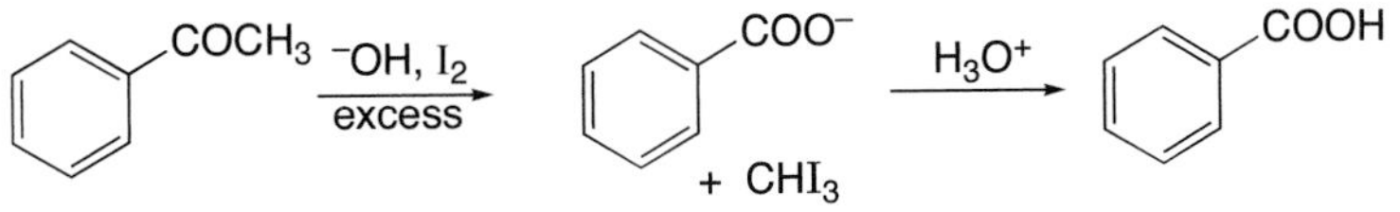

COCH₃ — $\xrightarrow[\text{excess}]{\text{$^-$OH, I}_2}$ — COO⁻ — $\xrightarrow{\text{H}_3\text{O}^+}$ — COOH

+ CHI₃

c.

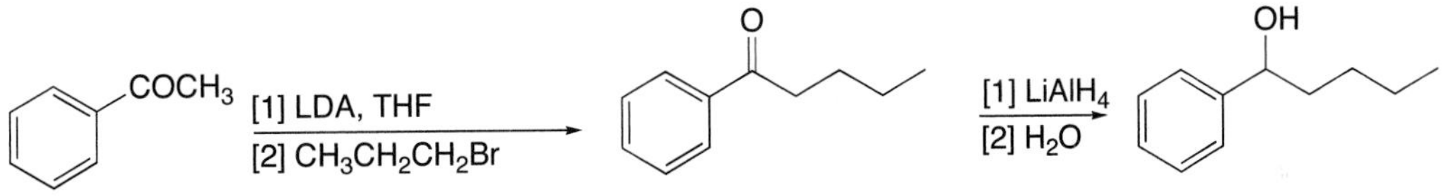

COCH₃ — $\xrightarrow[\text{[2] CH}_3\text{CH}_2\text{CH}_2\text{Br}]{\text{[1] LDA, THF}}$ — — $\xrightarrow[\text{[2] H}_2\text{O}]{\text{[1] LiAlH}_4}$ — OH

d.

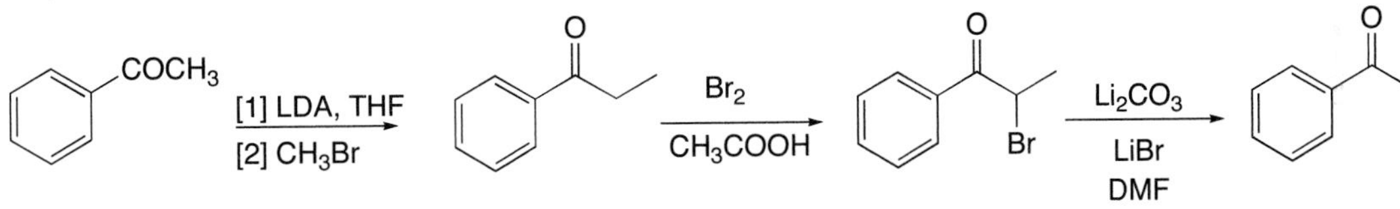

COCH₃ — $\xrightarrow[\text{[2] CH}_3\text{Br}]{\text{[1] LDA, THF}}$ — — $\xrightarrow[\text{CH}_3\text{COOH}]{\text{Br}_2}$ — Br — $\xrightarrow[\substack{\text{LiBr} \\ \text{DMF}}]{\text{Li}_2\text{CO}_3}$ —

**23.57**

a.

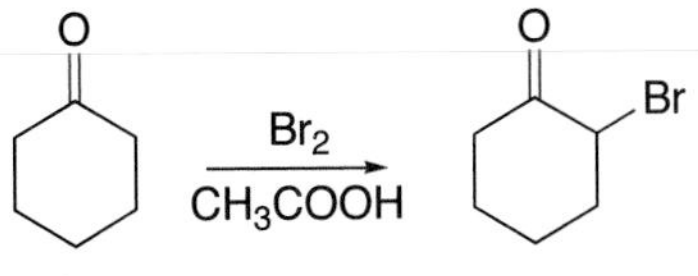

$\xrightarrow{\text{Br}_2 \atop \text{CH}_3\text{COOH}}$ — Br

b.

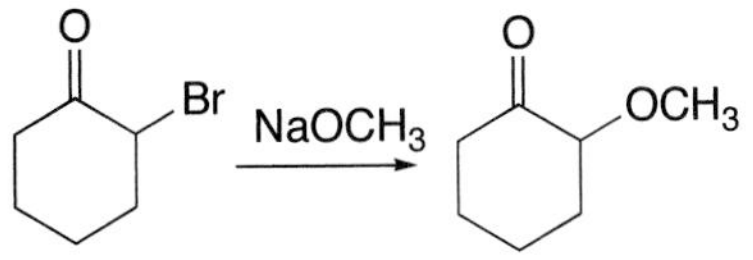

Br — $\xrightarrow{\text{NaOCH}_3}$ — OCH₃

(from a.)

c.

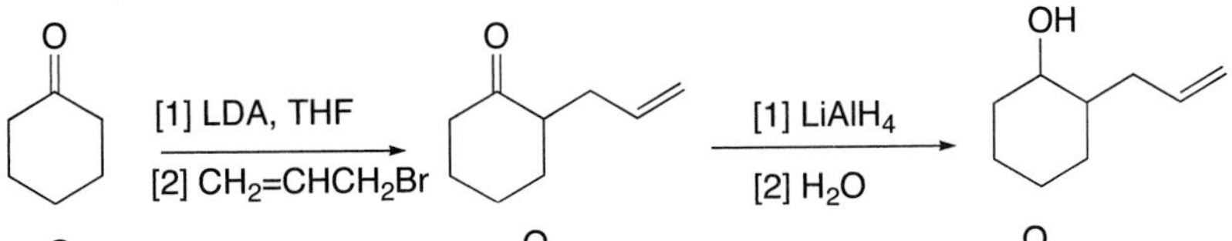

$\xrightarrow[\text{[2] CH}_2=\text{CHCH}_2\text{Br}]{\text{[1] LDA, THF}}$ — — $\xrightarrow[\text{[2] H}_2\text{O}]{\text{[1] LiAlH}_4}$ — OH

d.

$\xrightarrow[\text{[2] CH}_3\text{Br}]{\text{[1] LDA, THF}}$ — — $\xrightarrow[\text{[2] CH}_3\text{CH}_2\text{Br}]{\text{[1] LDA, THF}}$ —

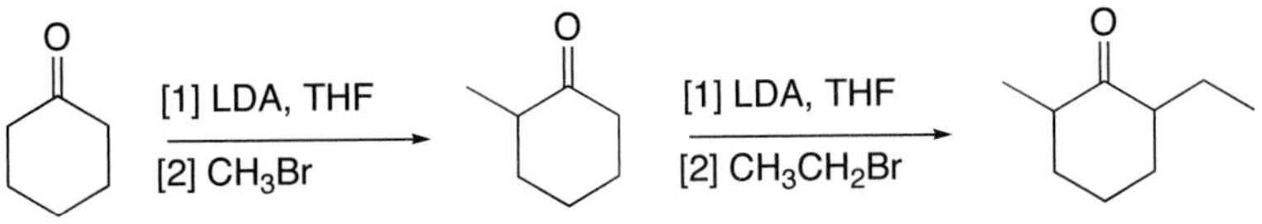

e.
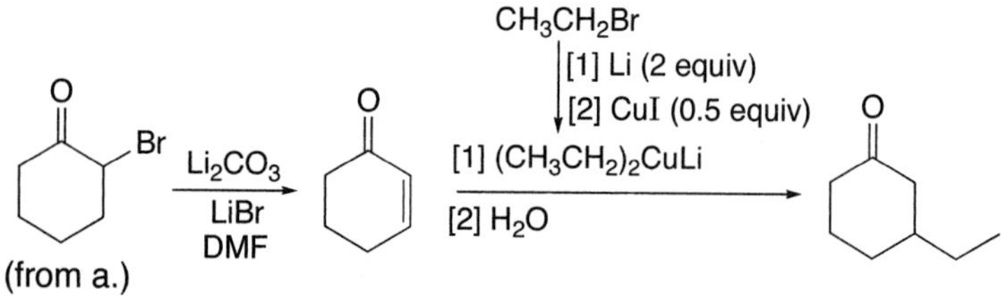
O
Br
(from a.)
Li2CO3
LiBr
DMF
CH3CH2Br
[1] Li (2 equiv)
[2] CuI (0.5 equiv)
[1] (CH3CH2)2CuLi
[2] H2O

f.
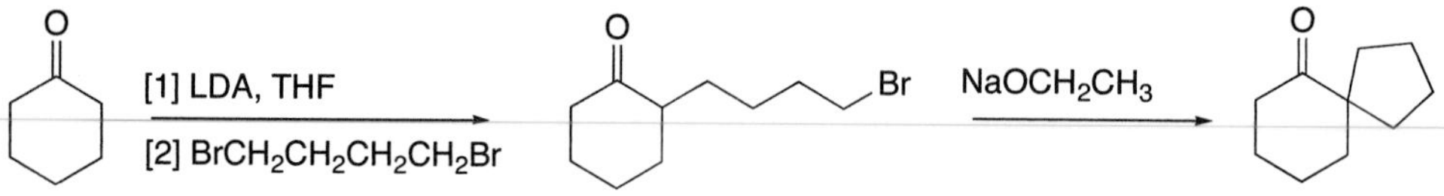
[1] LDA, THF
[2] BrCH2CH2CH2CH2Br
Br
NaOCH2CH3

g.
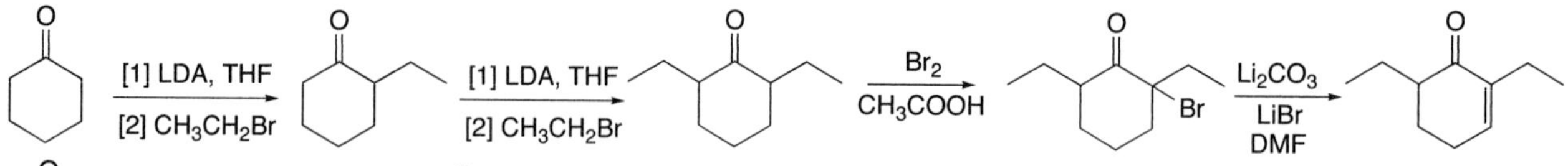
[1] LDA, THF
[2] CH3CH2Br
[1] LDA, THF
[2] CH3CH2Br
Br2
CH3COOH
Br
Li2CO3
LiBr
DMF

h.
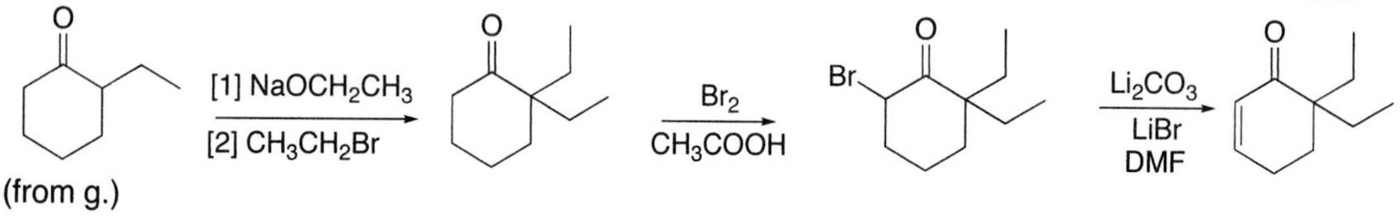
(from g.)
[1] NaOCH2CH3
[2] CH3CH2Br
Br2
CH3COOH
Br
Li2CO3
LiBr
DMF

**23.58**

a.
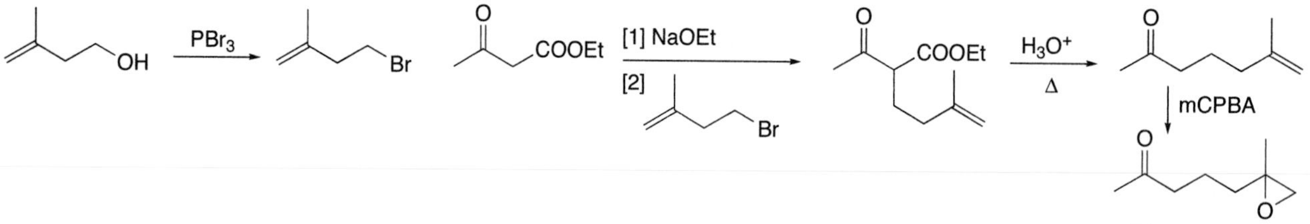
OH
PBr3
Br
COOEt
[1] NaOEt
[2]
Br
COOEt
H3O+
Δ
mCPBA

b.
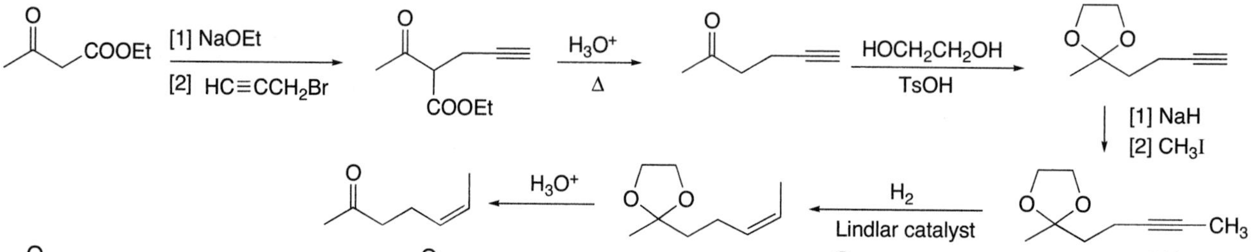
COOEt
[1] NaOEt
[2] HC≡CCH2Br
COOEt
H3O+
Δ
HOCH2CH2OH
TsOH
[1] NaH
[2] CH3I
CH3
H2
Lindlar catalyst
H3O+

c.
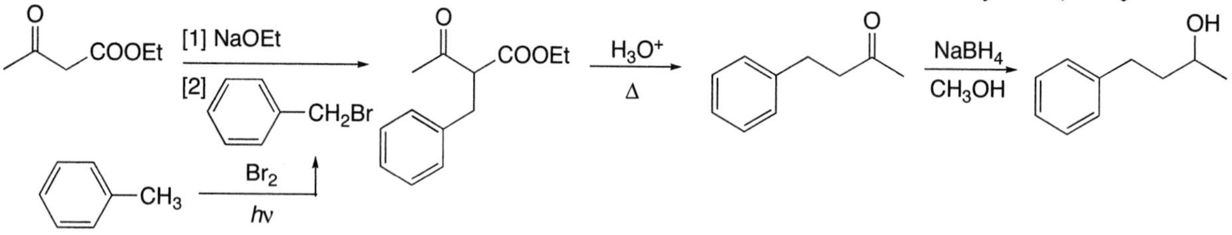
COOEt
[1] NaOEt
[2]
CH2Br
CH3
Br2
hv
COOEt
H3O+
Δ
NaBH4
CH3OH
OH

**23.59**

a.
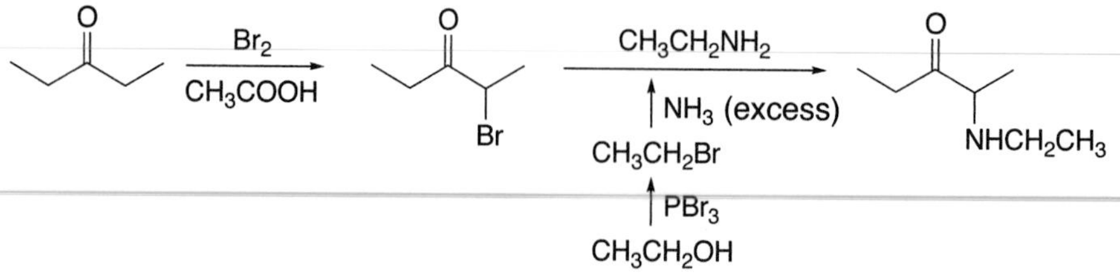
Br2
CH3COOH
Br
CH3CH2NH2
NH3 (excess)
CH3CH2Br
PBr3
CH3CH2OH
NHCH2CH3

b.

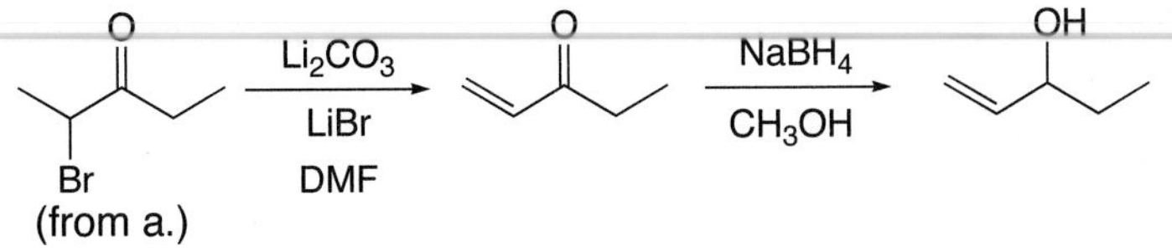

c.

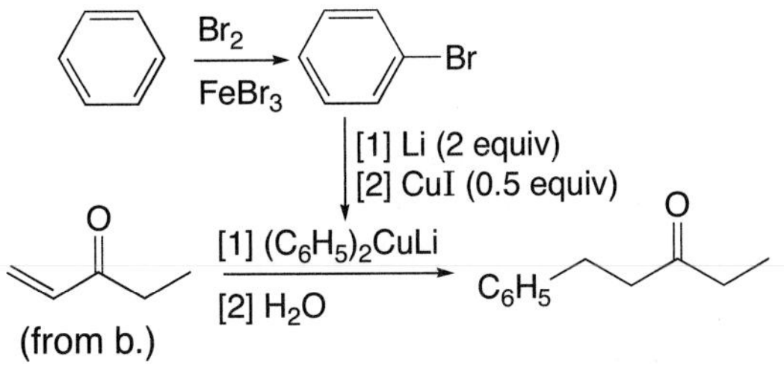

d.

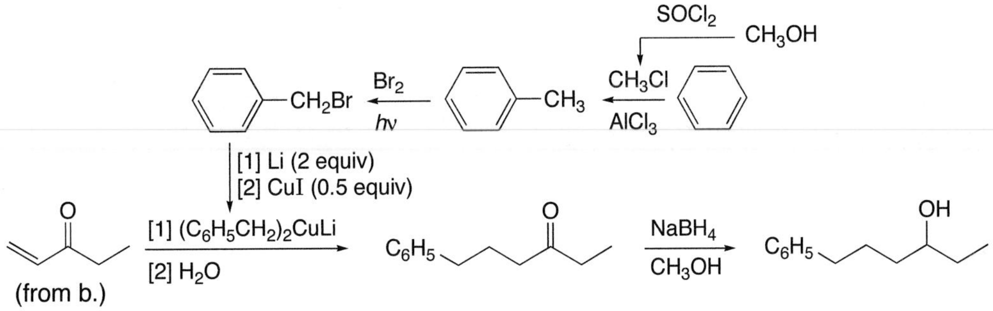

**23.60**

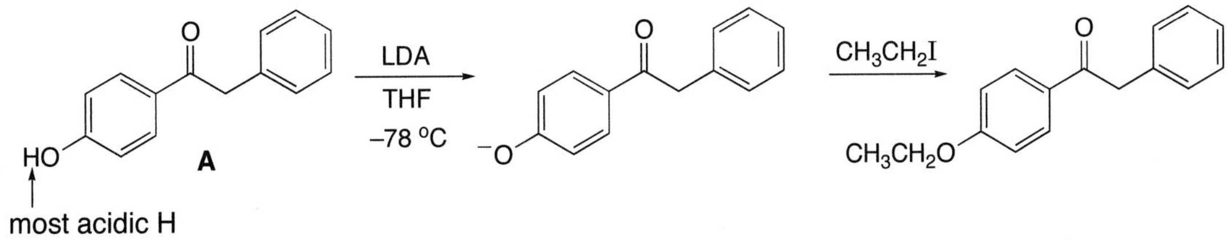

most acidic H

To synthesize the desired product, a protecting group is needed:

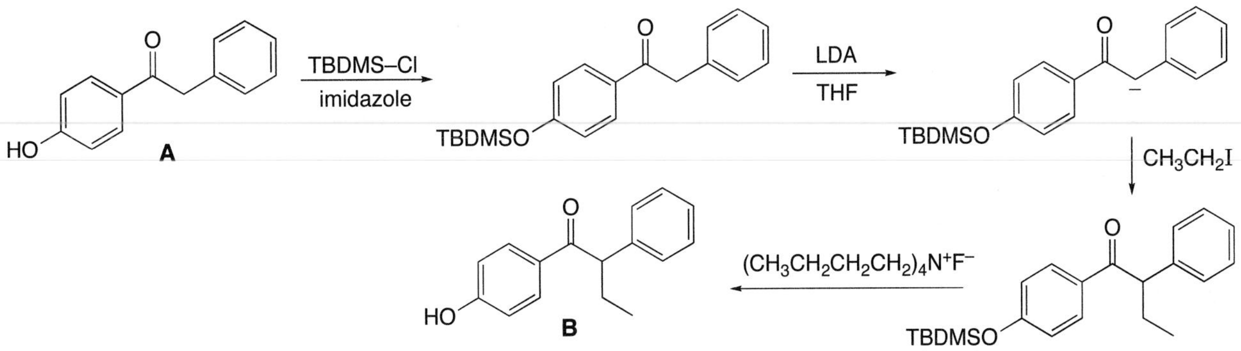

**23.61**

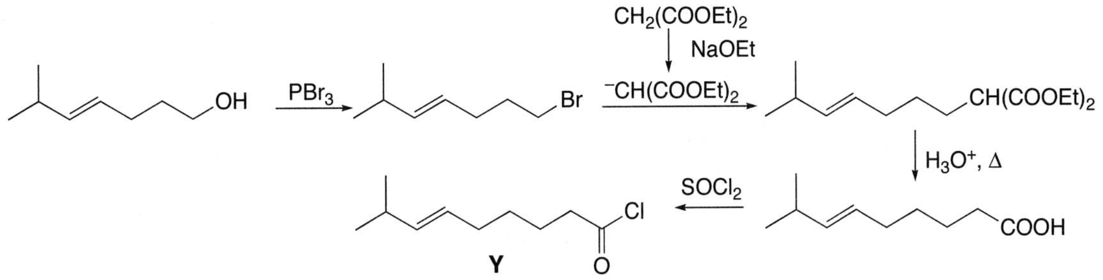

**23.62**

a.

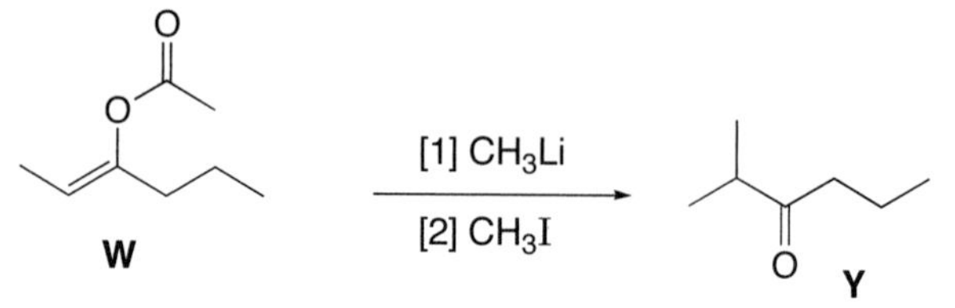

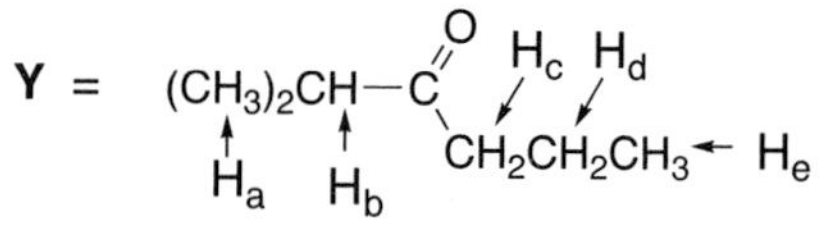

$$Y = (CH_3)_2CH-\overset{\overset{\displaystyle O}{\|}}{C}\overset{H_c\ H_d}{\underset{\underset{H_a\ \ H_b}{}}{CH_2CH_2CH_3}} \leftarrow H_e$$

$C_7H_{14}O \rightarrow$ one degree of unsaturation

IR peak at 1713 cm$^{-1}$ $\rightarrow$ C=O

$^1$H NMR signals at (ppm)
   $H_e$: triplet at 0.9 (3H)
   $H_a$: doublet at 1.1 (6H)
   $H_d$: multiplet at 1.5 (2H)
   $H_c$: triplet at 2.5 (2H)
   $H_b$: septet at 2.6 (1H)

b.

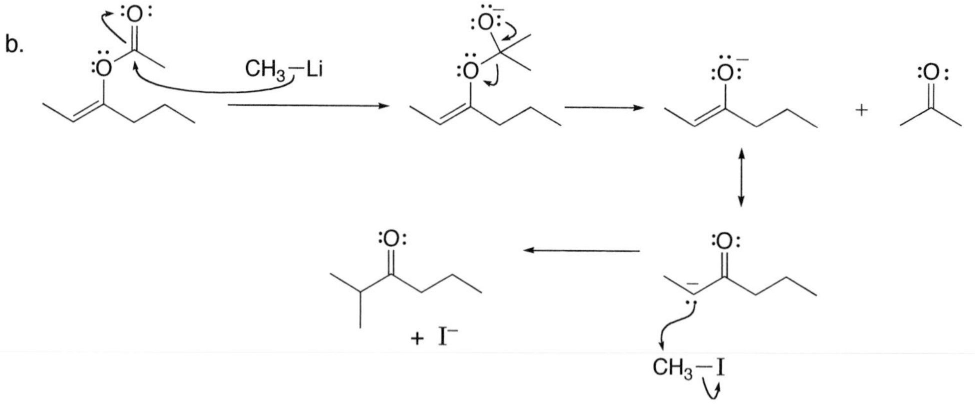

**23.63** Removal of $H_a$ with base does not generate an anion that can delocalize onto the carbonyl O atom, whereas removal of $H_b$ generates an enolate that is delocalized on O.

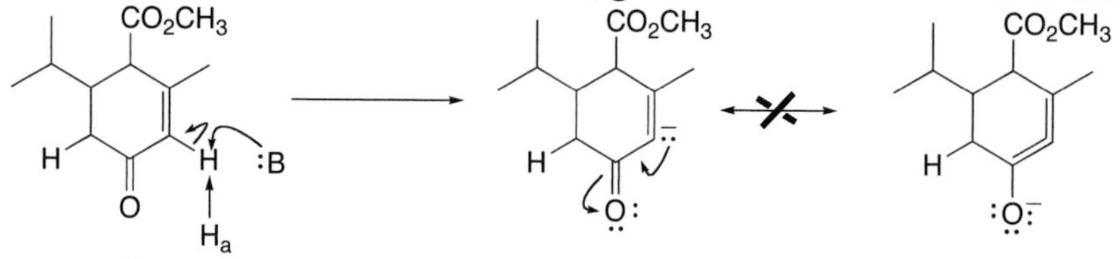

Delocalization of this sort can't occur by removal of $H_a$, making $H_a$ less acidic.

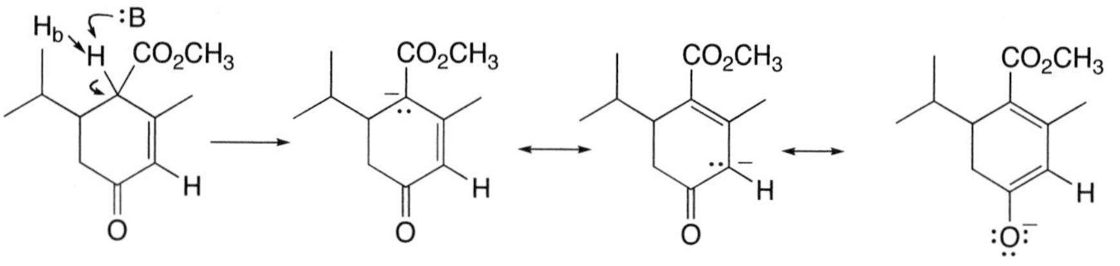

Removal of $H_b$ gives an anion that is resonance stabilized so $H_b$ is more acidic.

Mechanism:

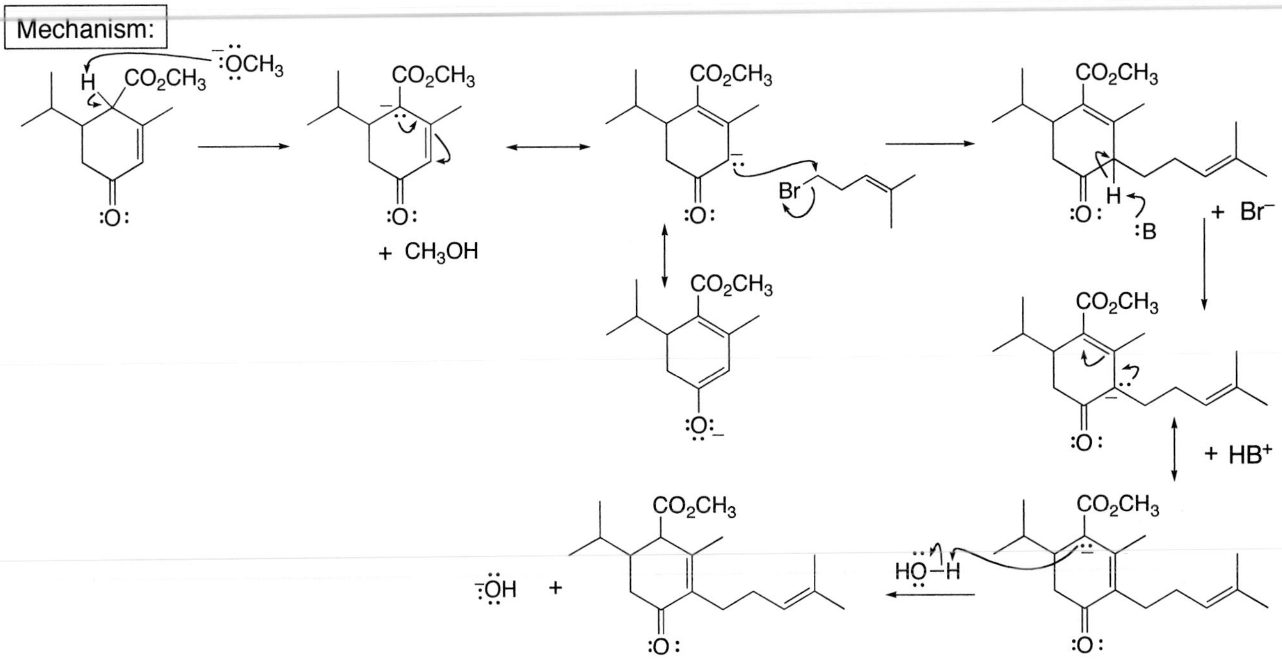

**23.64**

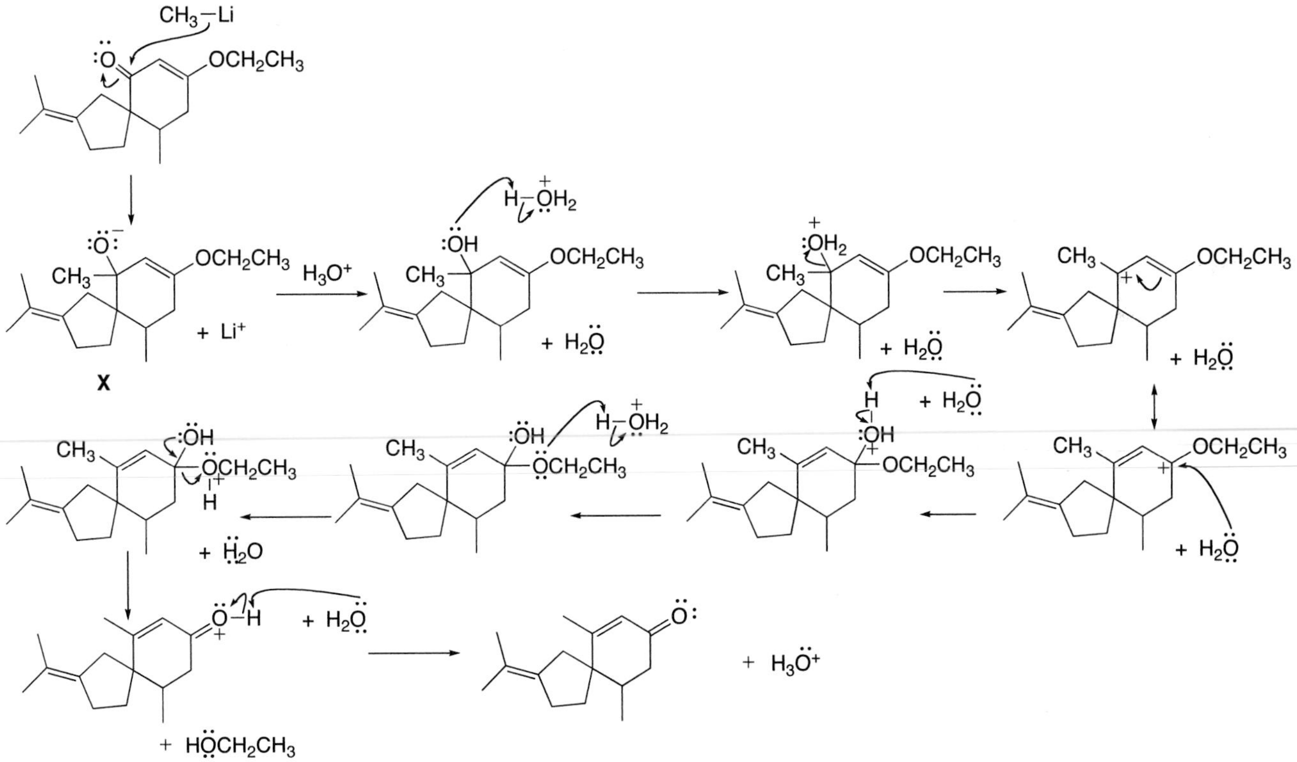

# Chapter 24: Carbonyl Condensation Reactions

♦ The four major carbonyl condensation reactions

| Reaction type | Reaction |
|---|---|
| **[1] Aldol reaction (24.1)** | 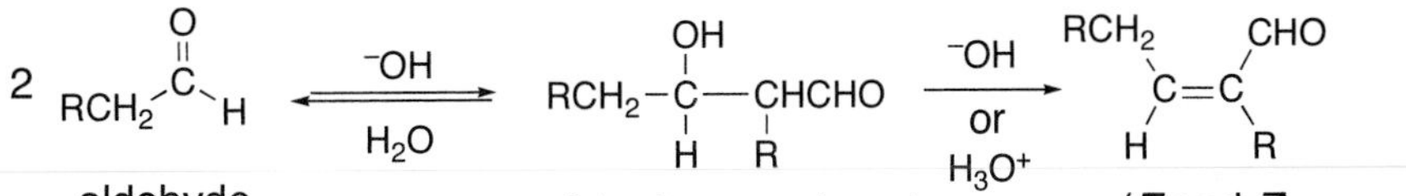 2 RCH₂CHO (aldehyde or ketone) ⇌ (⁻OH / H₂O) β-hydroxy carbonyl compound → (⁻OH or H₃O⁺) α,β-unsaturated carbonyl compound (*E* and *Z*) |
| **[2] Claisen reaction (24.5)** | 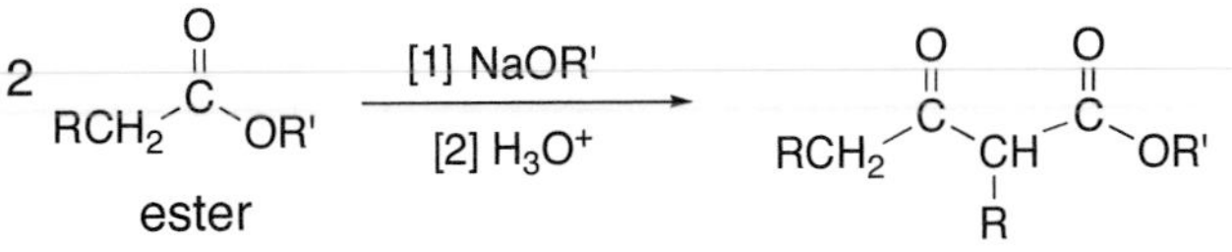 2 RCH₂CO₂OR' (ester) → ([1] NaOR', [2] H₃O⁺) β-keto ester |
| **[3] Michael reaction (24.8)** | 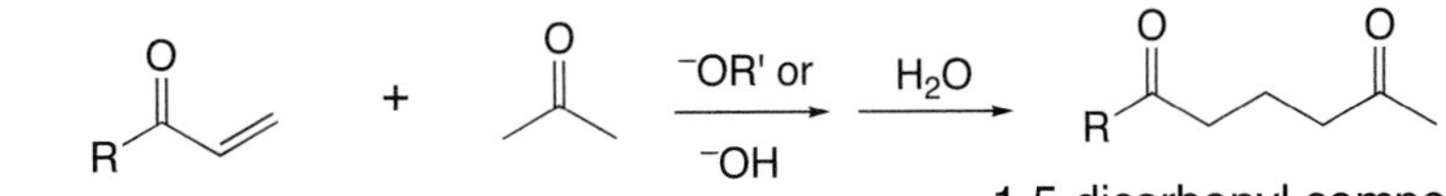 α,β-unsaturated carbonyl compound + carbonyl compound → (⁻OR' or ⁻OH, H₂O) 1,5-dicarbonyl compound |
| **[4] Robinson annulation (24.9)** |  α,β-unsaturated carbonyl compound + carbonyl compound → (⁻OH, H₂O) 2-cyclohexenone |

♦ Useful variations

## [1] Directed aldol reaction (24.3)

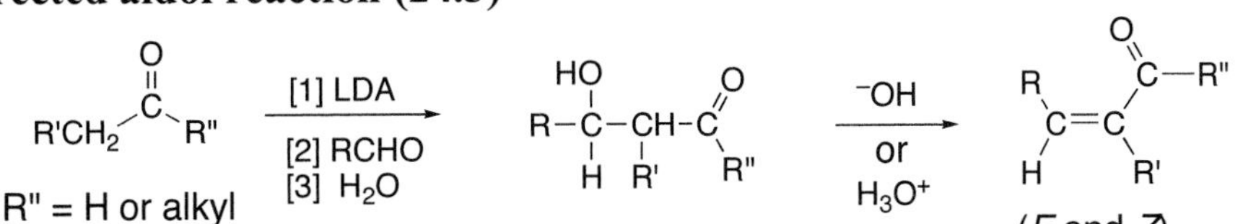

R'CH₂COR" (R" = H or alkyl) → ([1] LDA, [2] RCHO, [3] H₂O) β-hydroxy carbonyl compound → (⁻OH or H₃O⁺) α,β-unsaturated carbonyl compound (*E* and *Z*)

## [2] Intramolecular aldol reaction (24.4)

### [a] With 1,4-dicarbonyl compounds:

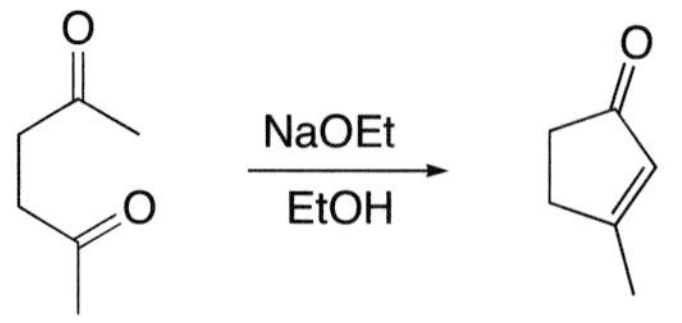

### [b] With 1,5-dicarbonyl compounds:

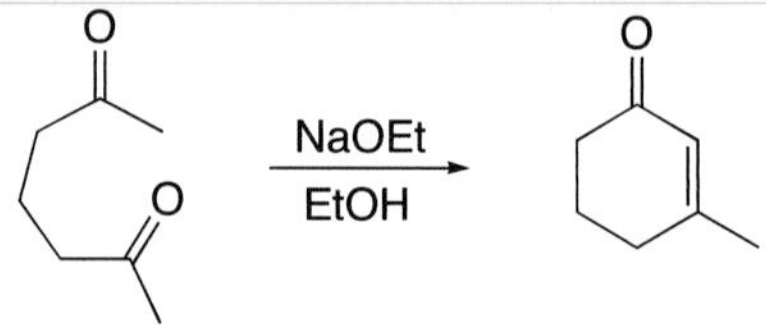

## [3] Dieckmann reaction (24.7)

### [a] With 1,6-diesters:

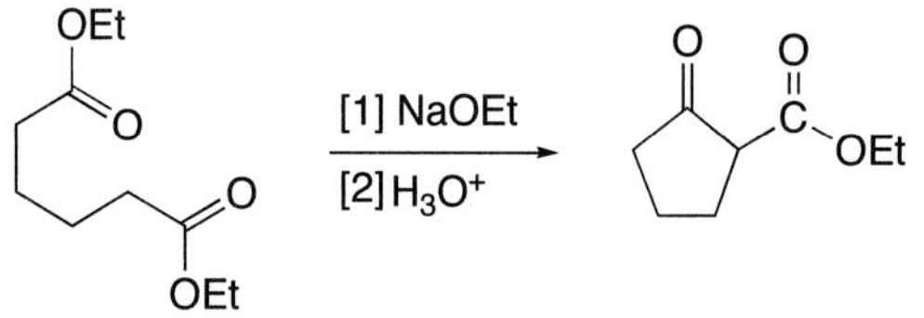

### [b] With 1,7-diesters:

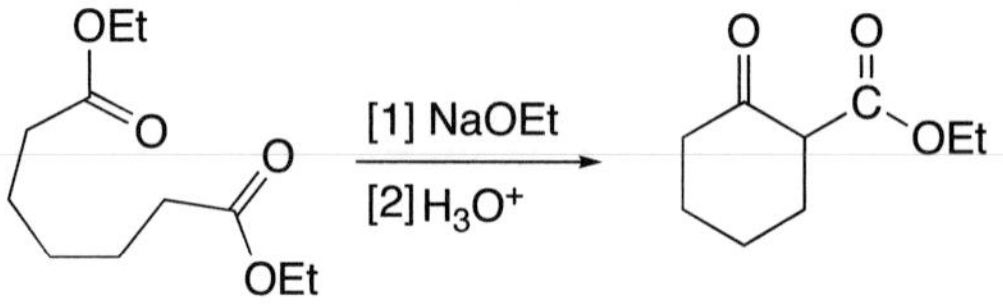

## 24.1

a.

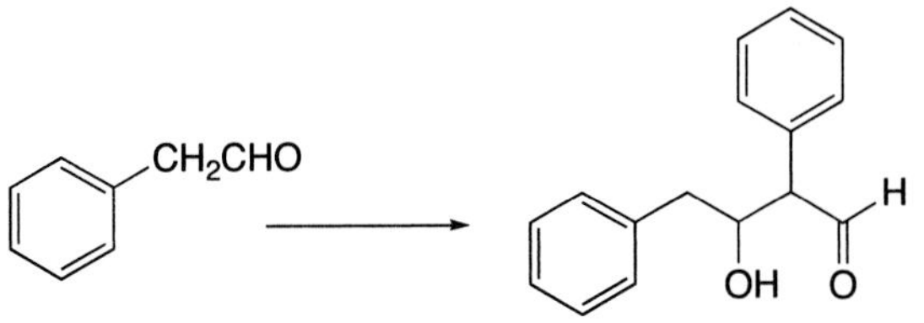

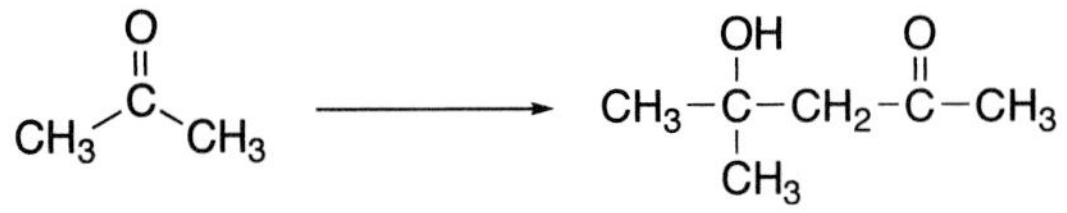

b. 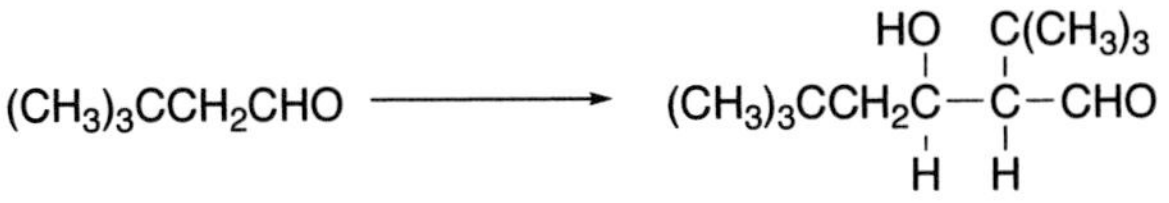  $(CH_3)_3CCH_2CHO \longrightarrow$

c.  $CH_3\overset{O}{\overset{\|}{C}}CH_3 \longrightarrow$

d.

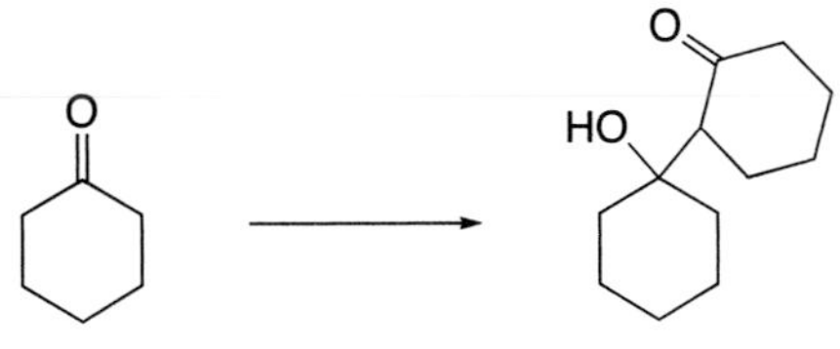

## 24.2

a.

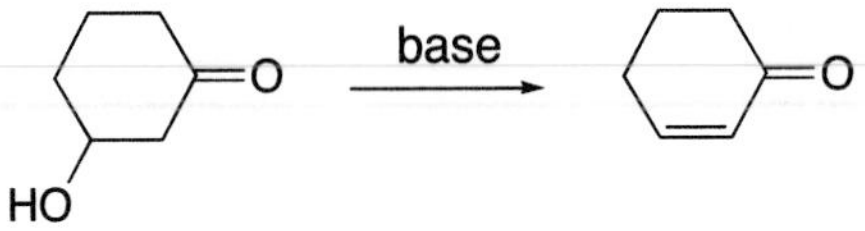

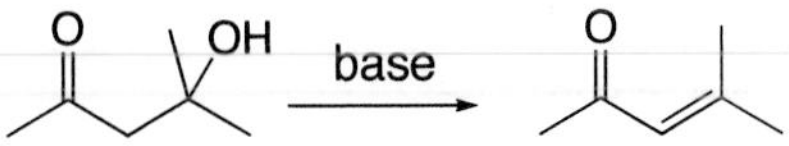

b.

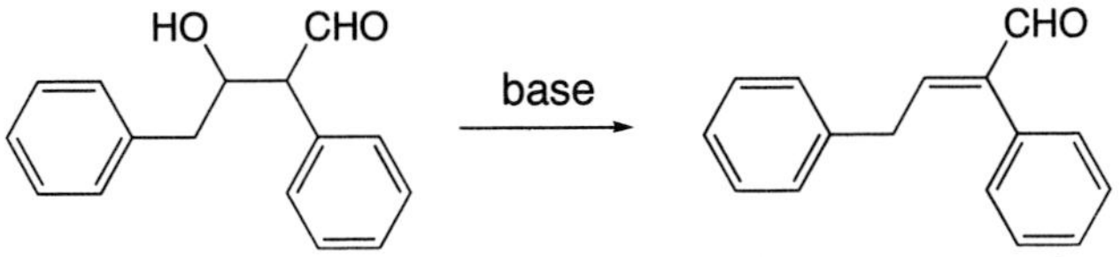

(*E* and *Z* isomers)

c.

## 24.3

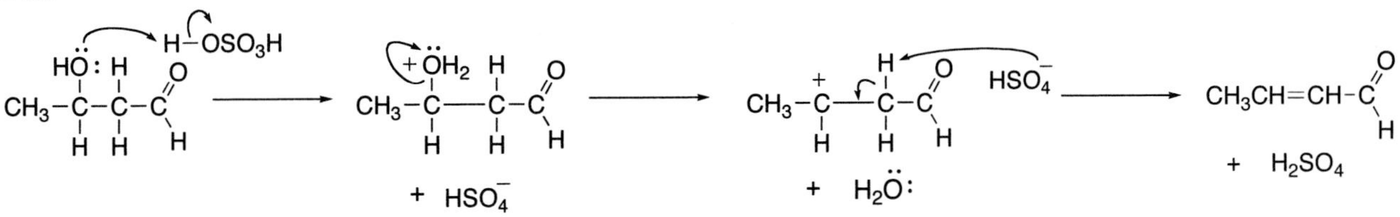

**24.4** Locate the α and β C's to the carbonyl group, and break the molecule into two halves at this bond. The α C and all of the atoms bonded to it belong to one carbonyl component. The β C and all the atoms bonded to it belong to the other carbonyl component.

a.

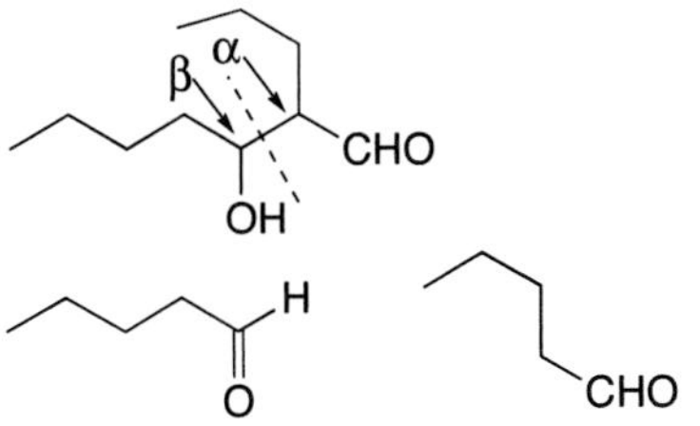

b.

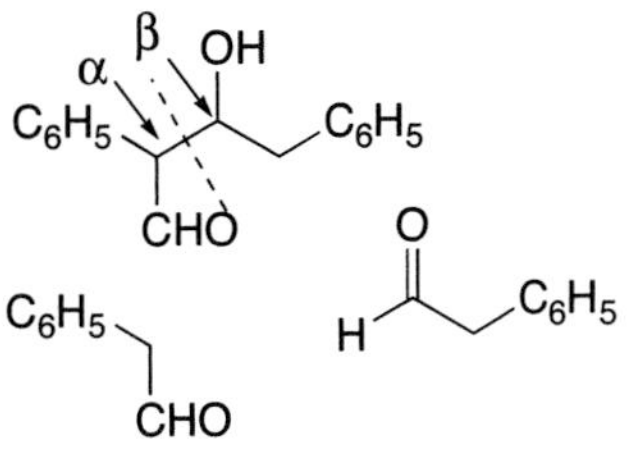

c.

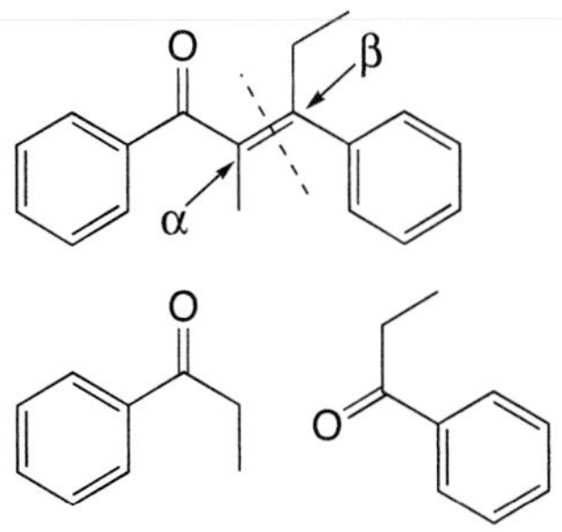

**24.5**

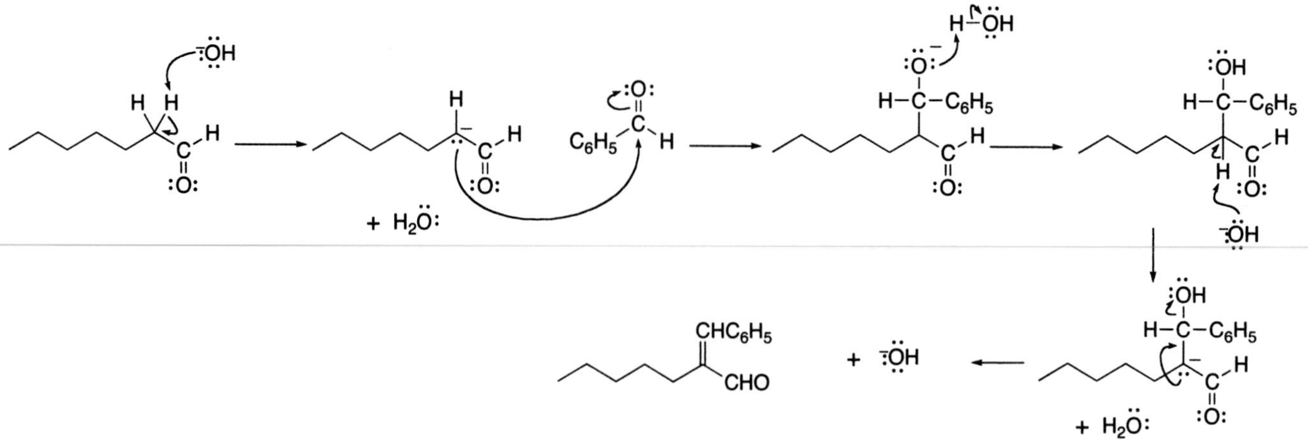

**24.6**

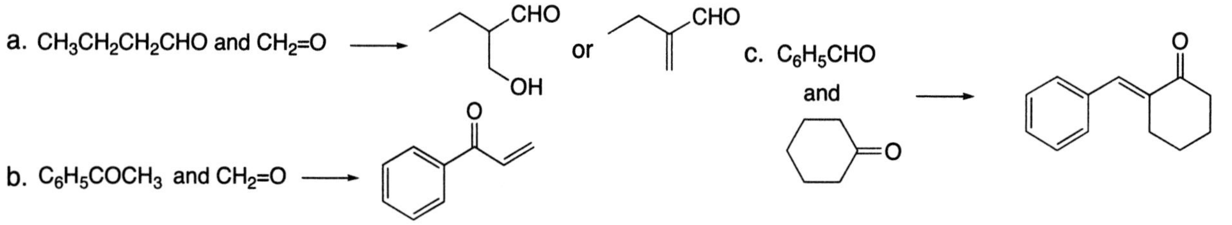

a. $CH_3CH_2CH_2CHO$ and $CH_2=O$ ⟶

or

b. $C_6H_5COCH_3$ and $CH_2=O$ ⟶

c. $C_6H_5CHO$

and ⟶

**24.7**

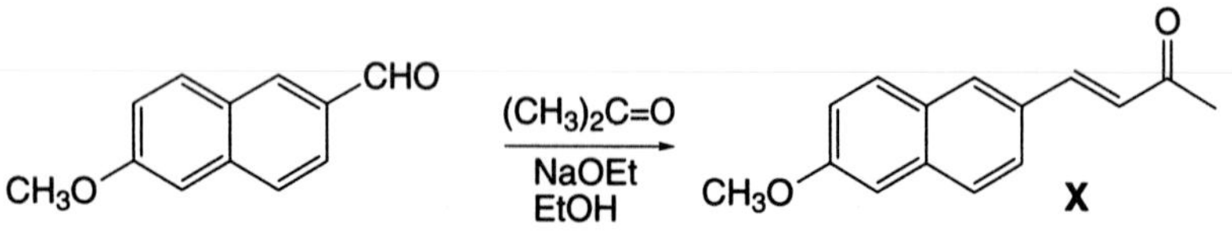

$\xrightarrow[\substack{NaOEt \\ EtOH}]{(CH_3)_2C=O}$

X

**24.8**

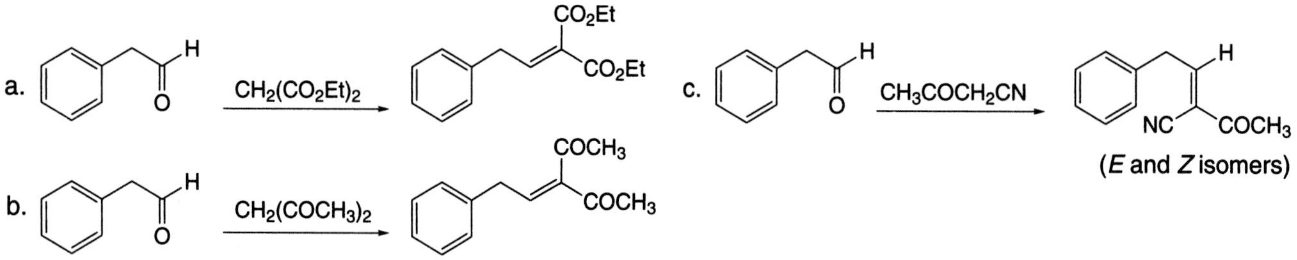

a. $\xrightarrow{CH_2(CO_2Et)_2}$

b. $\xrightarrow{CH_2(COCH_3)_2}$

c. $\xrightarrow{CH_3COCH_2CN}$

(E and Z isomers)

**24.9**

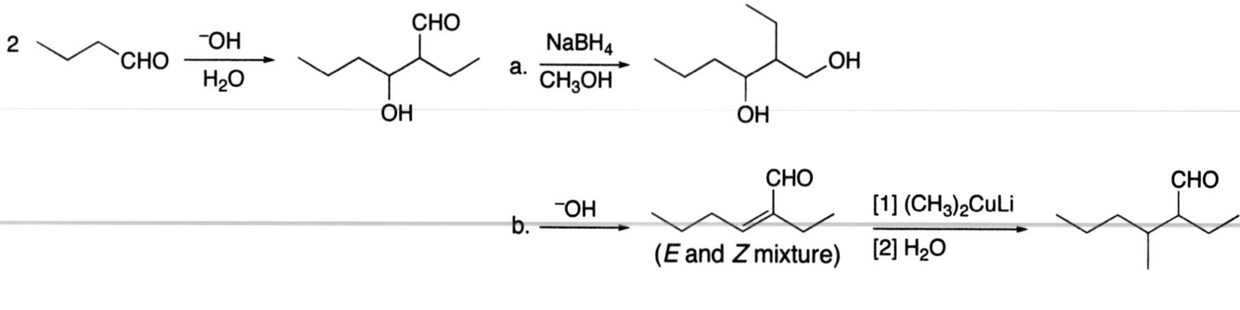

$2$ $\xrightarrow[H_2O]{^-OH}$ a. $\xrightarrow[CH_3OH]{NaBH_4}$

b. $\xrightarrow{^-OH}$

(E and Z mixture) $\xrightarrow[\text{[2] } H_2O]{\text{[1] } (CH_3)_2CuLi}}$

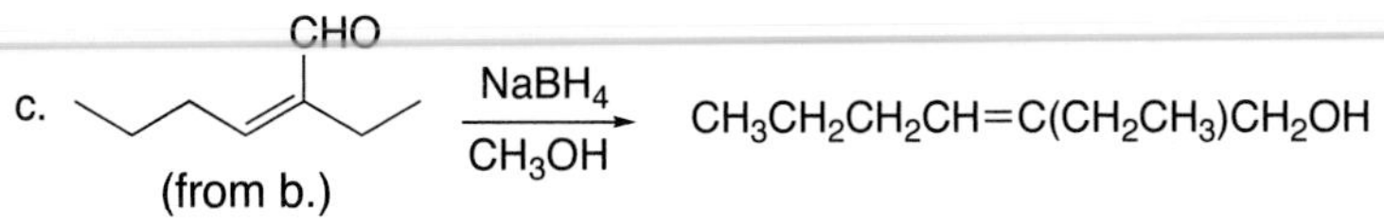

c. (CHO structure) $\xrightarrow[\text{CH}_3\text{OH}]{\text{NaBH}_4}$ $CH_3CH_2CH_2CH=C(CH_2CH_3)CH_2OH$

(from b.)

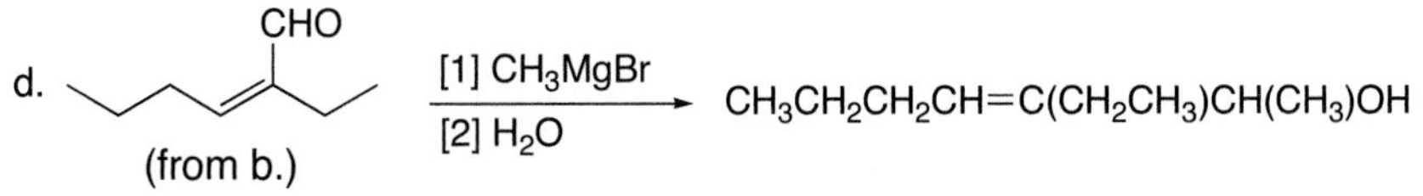

d. (CHO structure) $\xrightarrow[\text{[2] H}_2\text{O}]{\text{[1] CH}_3\text{MgBr}}$ $CH_3CH_2CH_2CH=C(CH_2CH_3)CH(CH_3)OH$

(from b.)

**24.10** Find the α and β C's to the carbonyl group and break the bond between them.

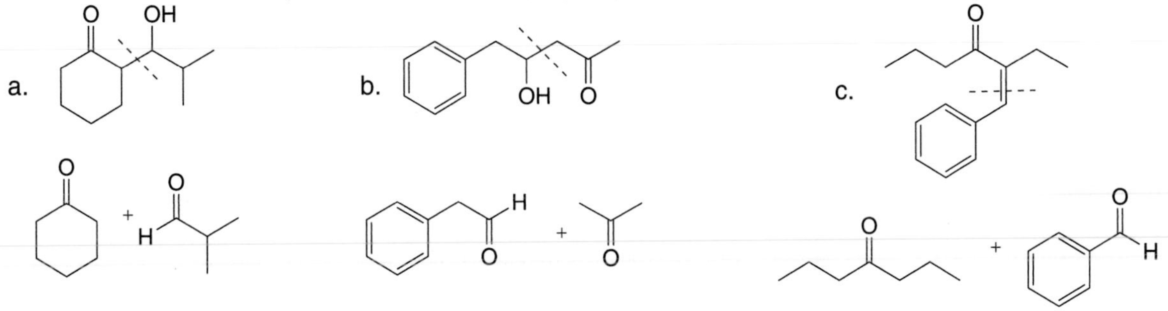

a.

b.

c.

**24.11** All enolates have a second resonance structure with a negative charge on O.

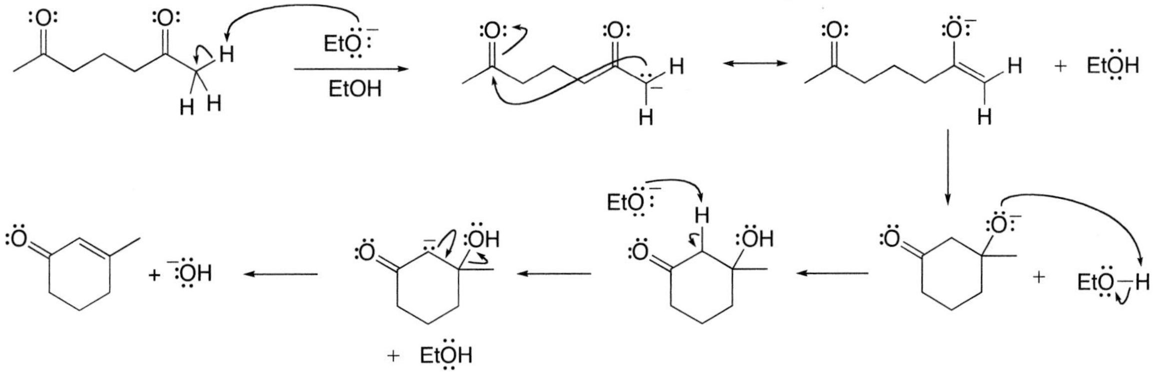

**24.12**

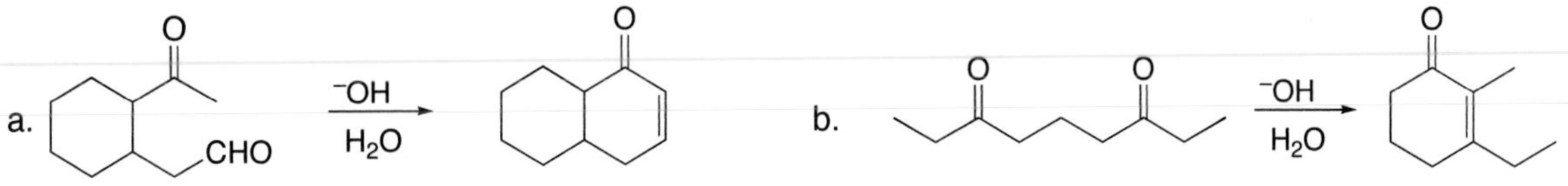

a. $\xrightarrow[\text{H}_2\text{O}]{^-\text{OH}}$

b. $\xrightarrow[\text{H}_2\text{O}]{^-\text{OH}}$

**24.13**

1-methylcyclopentene $\xrightarrow[\text{[2] (CH}_3)_2\text{S}]{\text{[1] O}_3}$ (structure) CHO $\xrightarrow[\text{H}_2\text{O}]{^-\text{OH}}$ 2-cyclohexenone

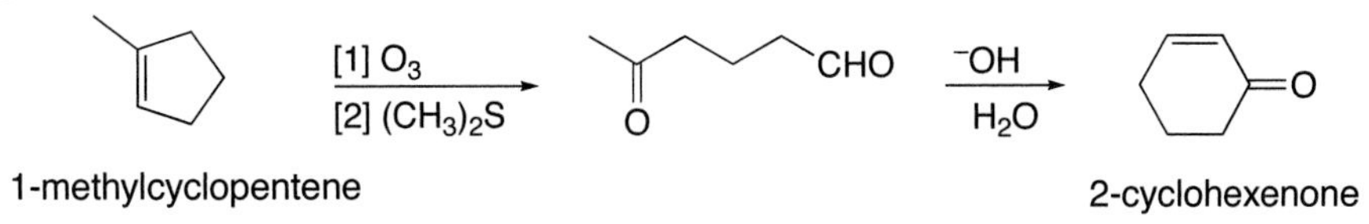

**24.14**

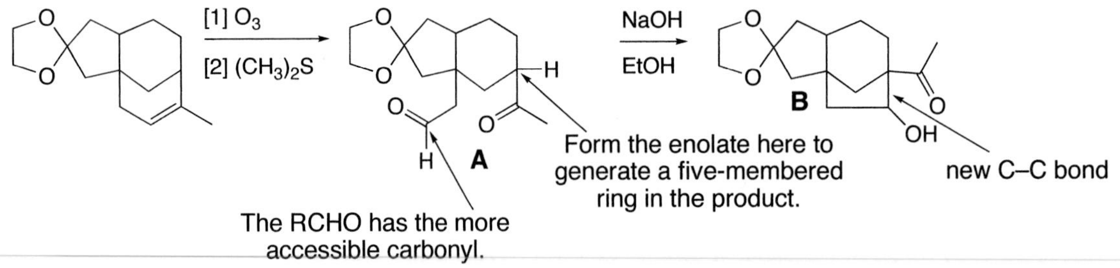

[1] O₃

[2] (CH₃)₂S

**A**

The RCHO has the more accessible carbonyl.

NaOH

EtOH

**B**

Form the enolate here to generate a five-membered ring in the product.

new C–C bond

**24.15**   Join the α C of one ester to the carbonyl C of the other ester to form the β-keto ester.

a.

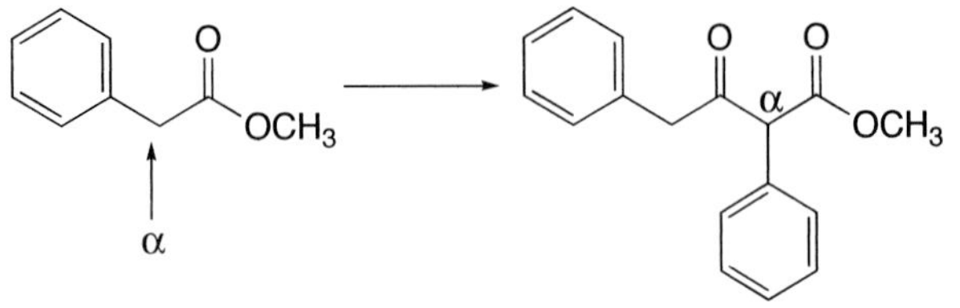

b.

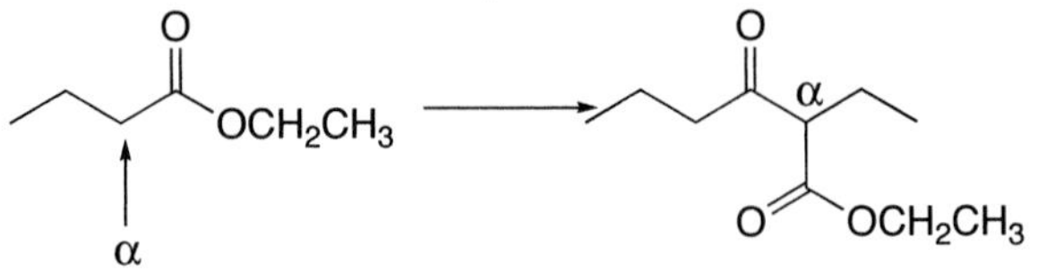

**24.16**

a.

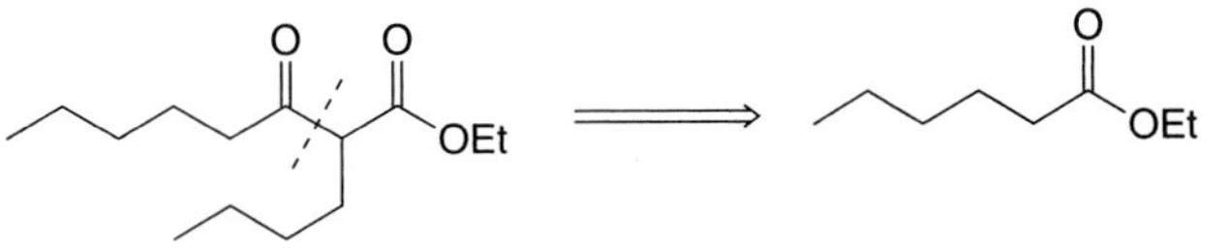

b.

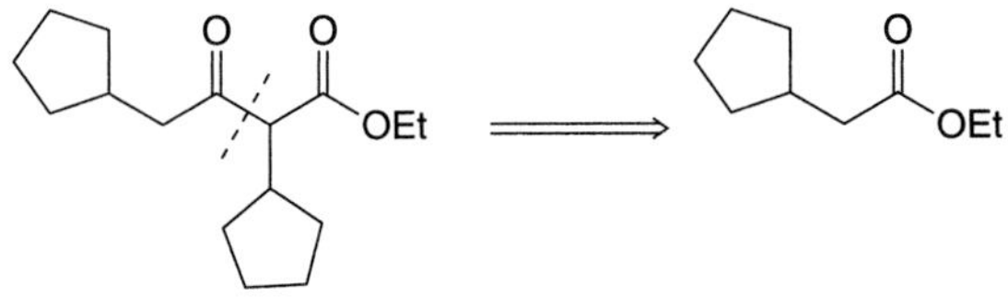

**24.17**   In a crossed Claisen reaction between an ester and a ketone, the enolate is always formed from the ketone, and the product is a β-dicarbonyl compound.

a. 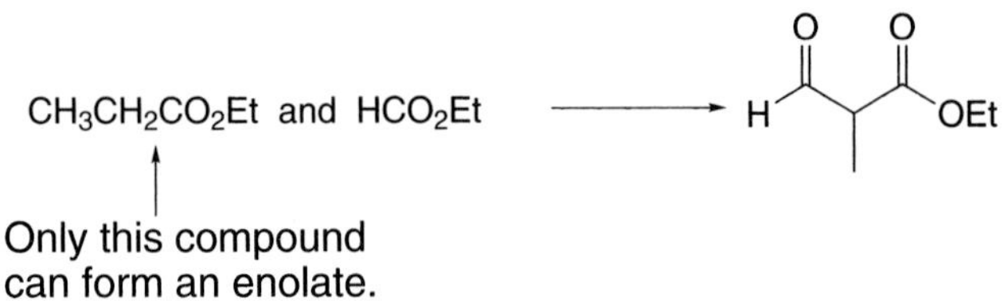   CH₃CH₂CO₂Et and HCO₂Et

Only this compound can form an enolate.

b.

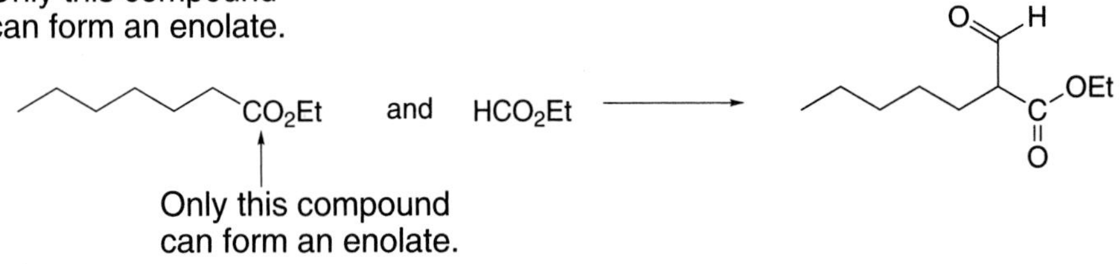

and   HCO₂Et

Only this compound can form an enolate.

c. 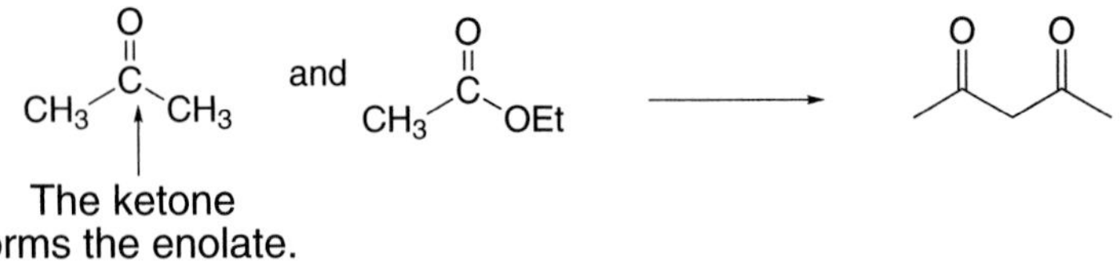   CH₃–C–CH₃   and   CH₃–C–OEt

The ketone forms the enolate.

d.

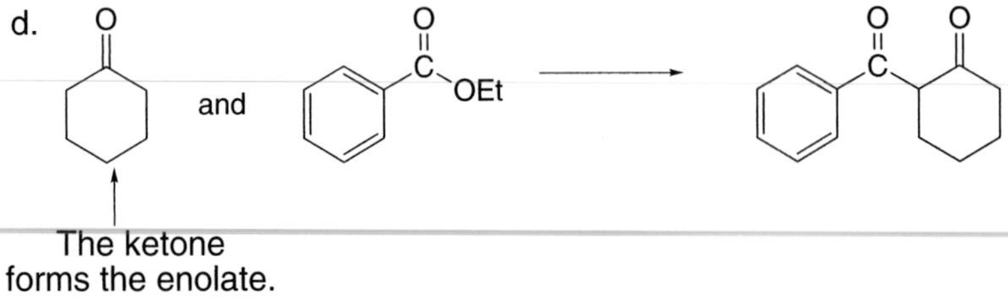

and

The ketone forms the enolate.

**24.18**

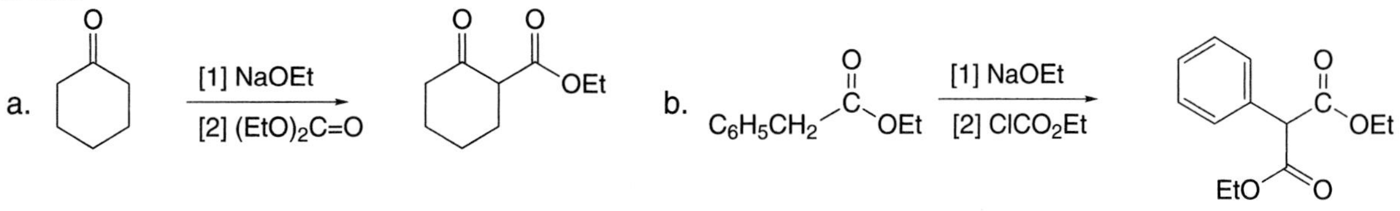

a.

b.

**24.19**

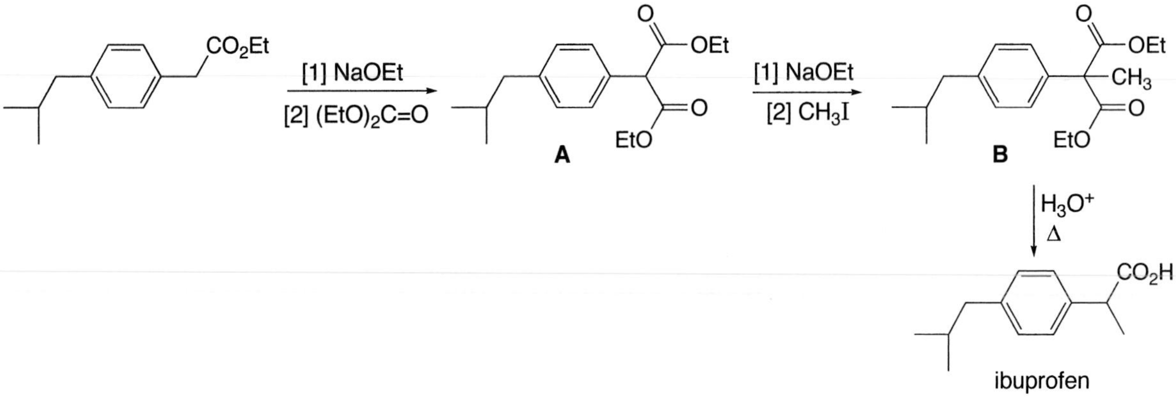

**24.20**

1,6-Diester forms a five-membered ring.

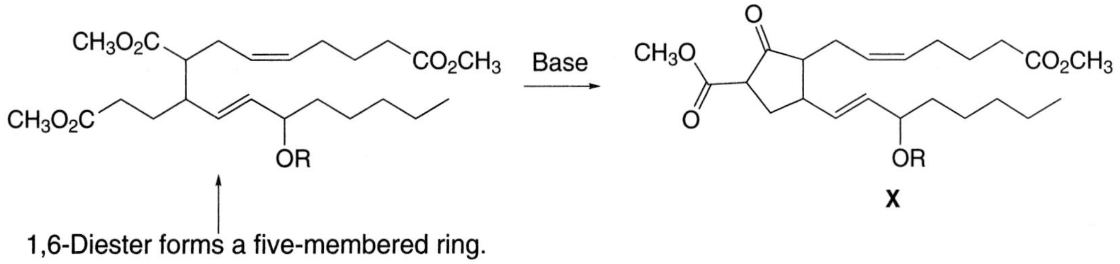

**24.21**

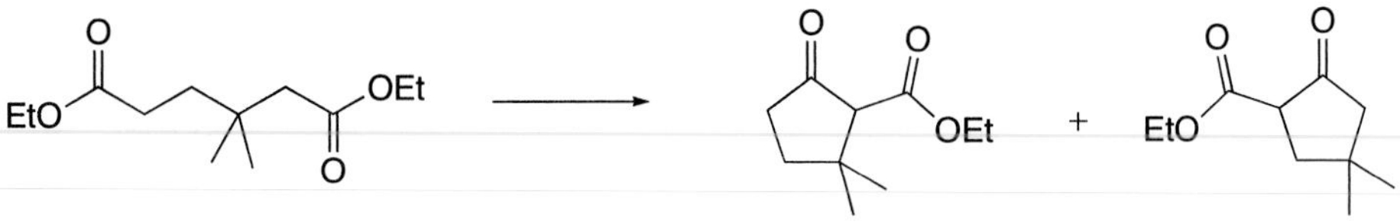

**24.22**

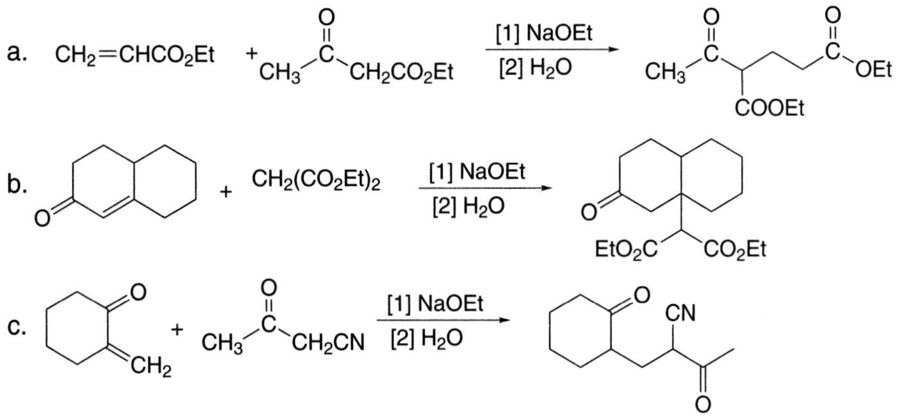

a.

b.

c.

**24.23**

a.

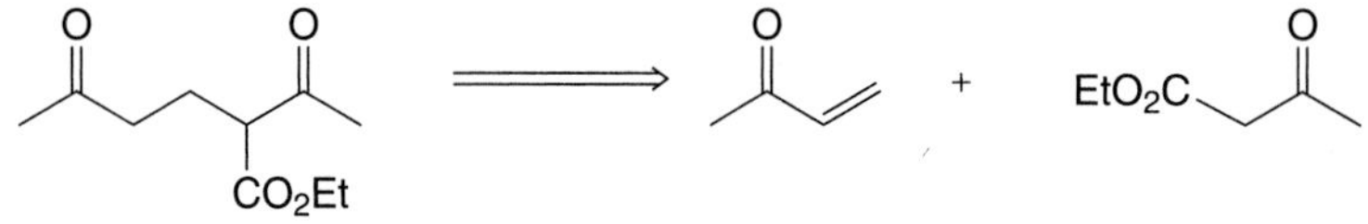

b.

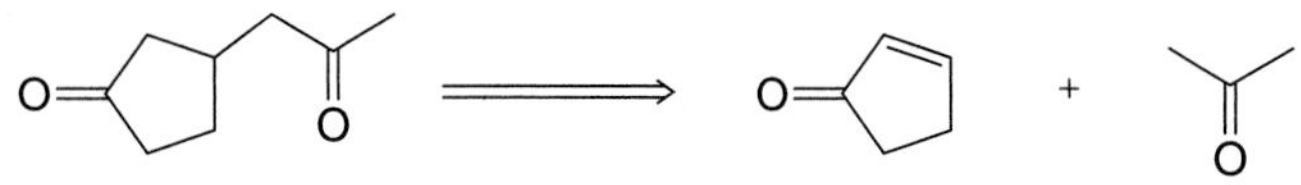

**24.24** The Robinson annulation forms a six-membered ring and three new carbon–carbon bonds: two σ bonds and one π bond.

a.

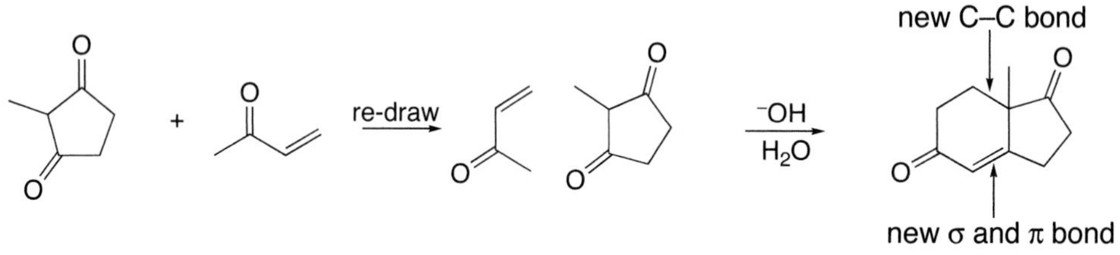

new C–C bond

re-draw

−OH
H₂O

new σ and π bond

b.

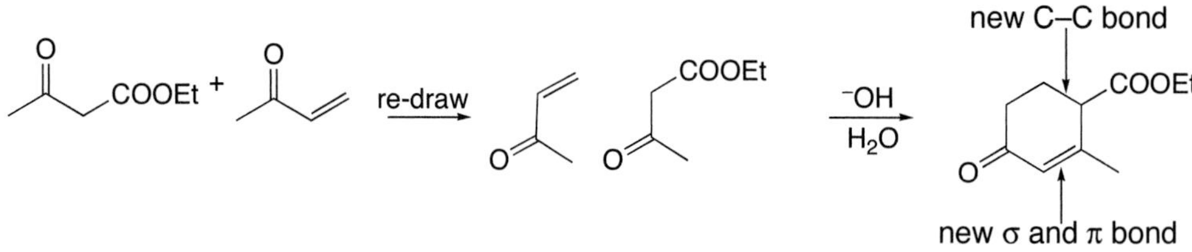

new C–C bond

COOEt +

re-draw

COOEt

−OH
H₂O

COOEt

new σ and π bond

c.

new C–C bond

re-draw

−OH
H₂O

new σ and π bond

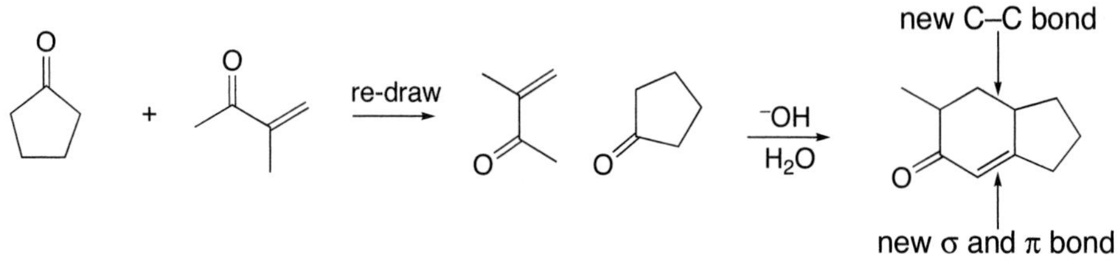

d.

EtO₂C

+

re-draw

EtOOC

−OH
H₂O

new C–C bond

COOEt

new σ and π bond

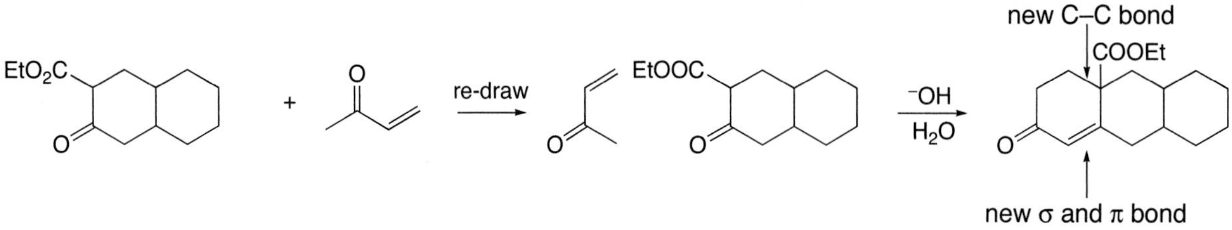

**24.25**

a.

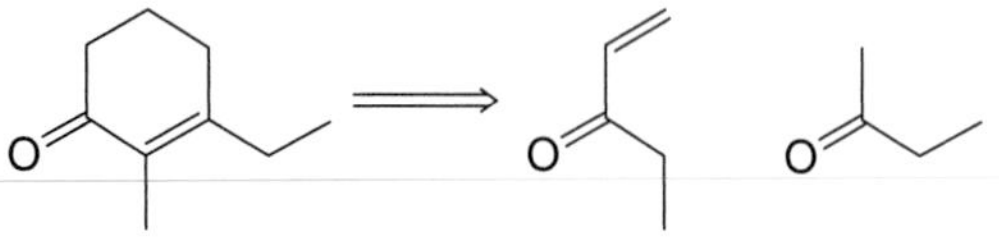

c.

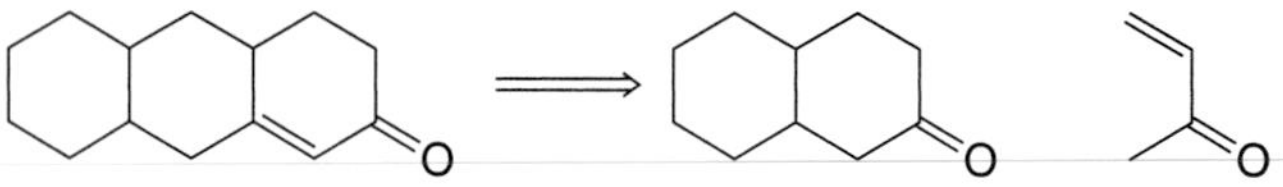

b.

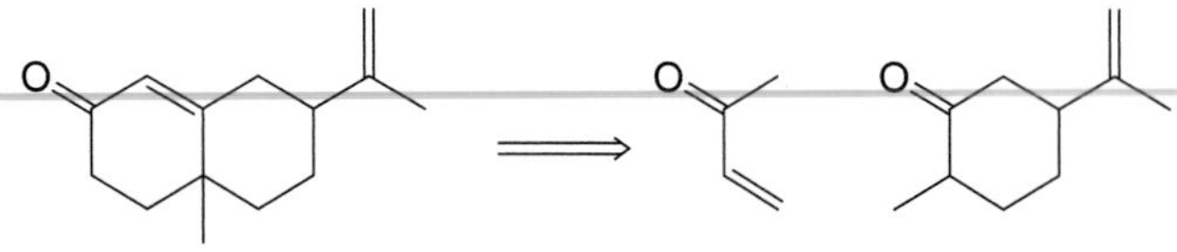

**24.26** The product of an aldol reaction is a β-hydroxy carbonyl compound or an α,β-unsaturated carbonyl compound. The latter type of compound is drawn as product unless elimination of H₂O cannot form a conjugated system.

a. 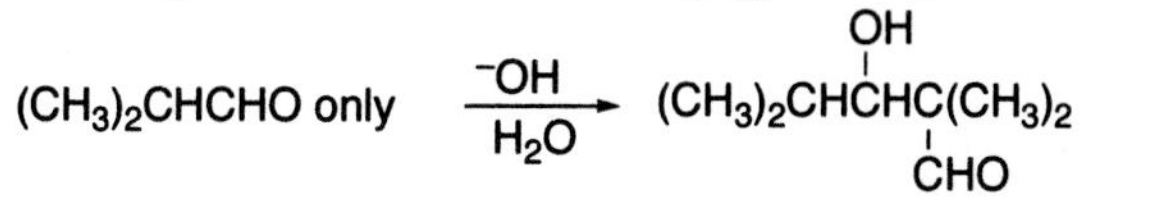 $(CH_3)_2CHCHO$ only $\xrightarrow[H_2O]{^-OH}$ $(CH_3)_2CHCHC(CH_3)_2$ with OH above and CHO below

d. 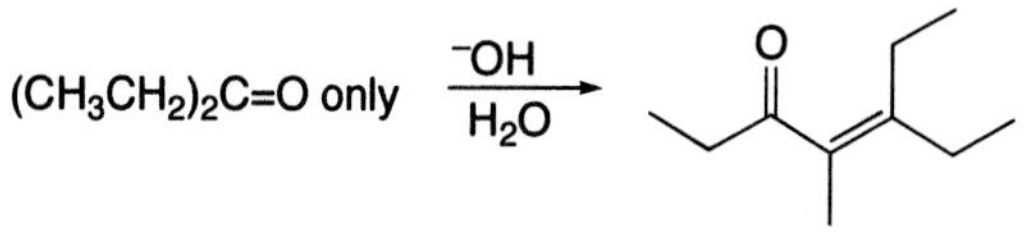 $(CH_3CH_2)_2C=O$ only $\xrightarrow[H_2O]{^-OH}$

b. 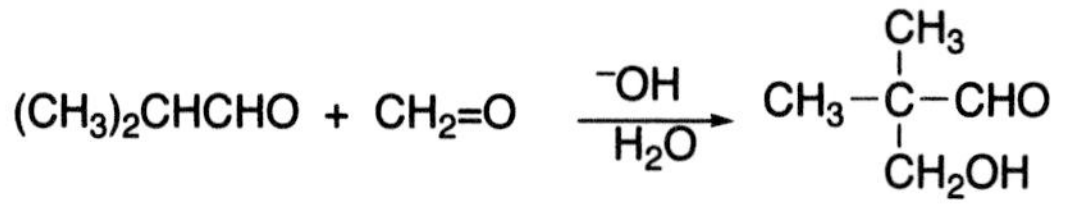 $(CH_3)_2CHCHO + CH_2=O$ $\xrightarrow[H_2O]{^-OH}$ $CH_3-\underset{CH_2OH}{\overset{CH_3}{C}}-CHO$

e. $(CH_3CH_2)_2C=O + CH_2=O$ $\xrightarrow[H_2O]{^-OH}$

c. $C_6H_5CHO + CH_3CH_2CH_2CHO$ $\xrightarrow[H_2O]{^-OH}$ → product (E and Z isomers)

f. 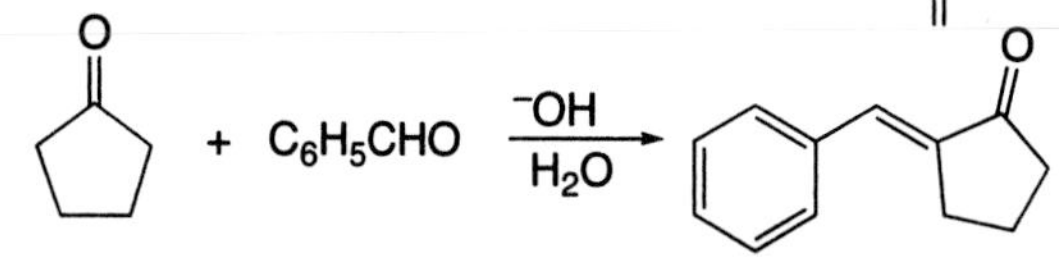 cyclopentanone $+ C_6H_5CHO$ $\xrightarrow[H_2O]{^-OH}$ → product (E and Z isomers)

**24.27**

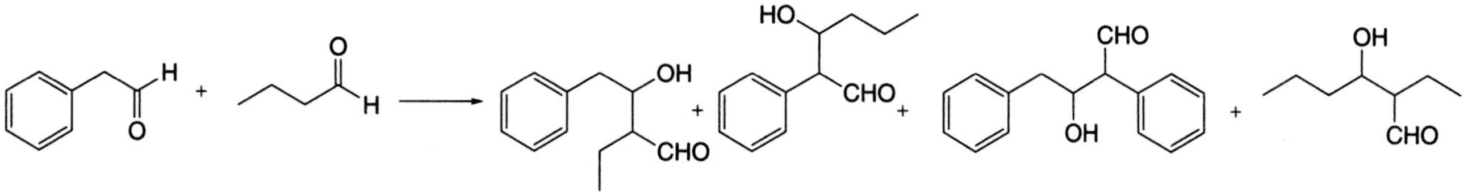

(PhCH₂CHO) + (butanal) → products with OH, CHO groups + others + OH/CHO heptane products

**24.28**

a. 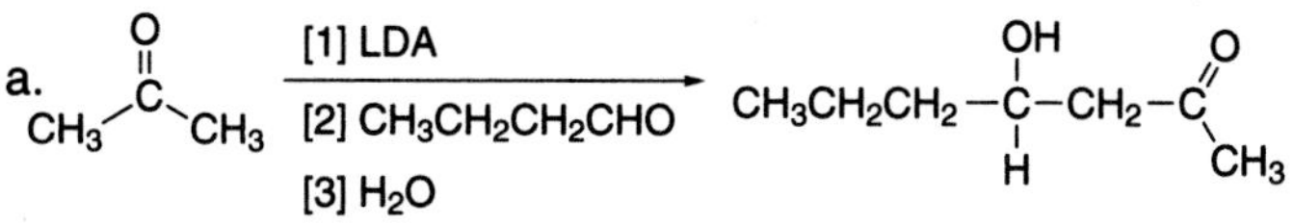 $CH_3\overset{O}{\underset{}{C}}CH_3$ $\xrightarrow[[3]\ H_2O]{[1]\ LDA,\ [2]\ CH_3CH_2CH_2CHO}$ $CH_3CH_2CH_2-\underset{H}{\overset{OH}{C}}-CH_2-\overset{O}{C}-CH_3$

b. 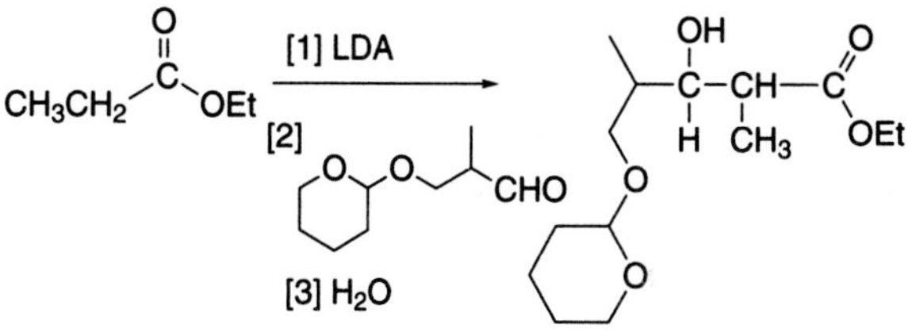 $CH_3CH_2\overset{O}{\underset{}{C}}OEt$ $\xrightarrow[[3]\ H_2O]{[1]\ LDA,\ [2]}$ product: $\underset{H}{\overset{OH}{C}}-\underset{CH_3}{\overset{}{CH}}-\overset{O}{C}OEt$

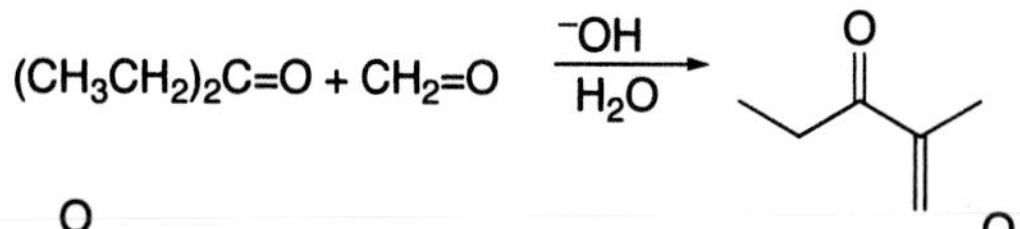

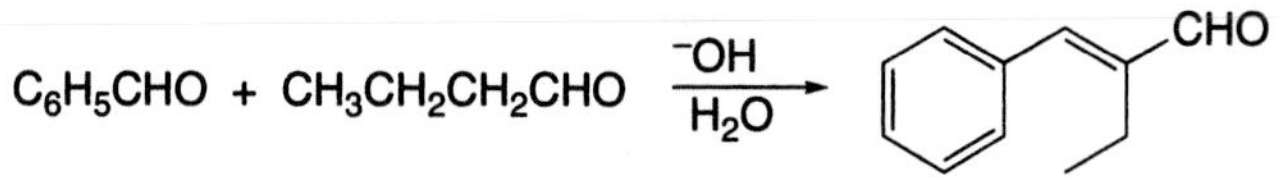

**24.29**

a. 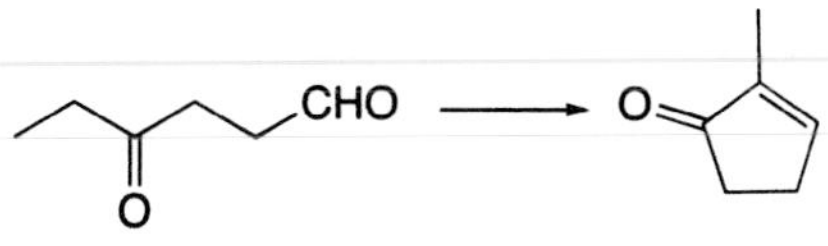 (CHO-containing chain) → methylcyclopentenone

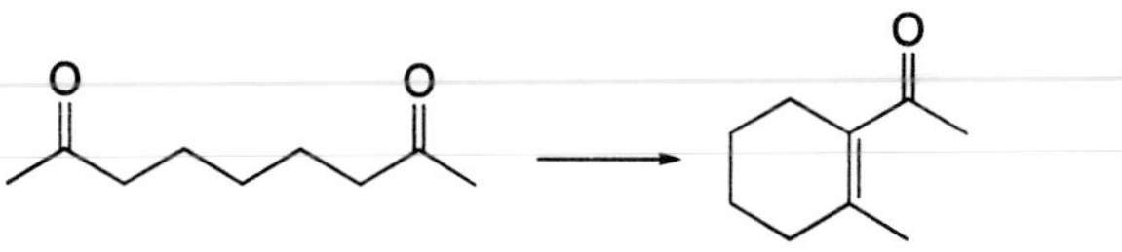

b. 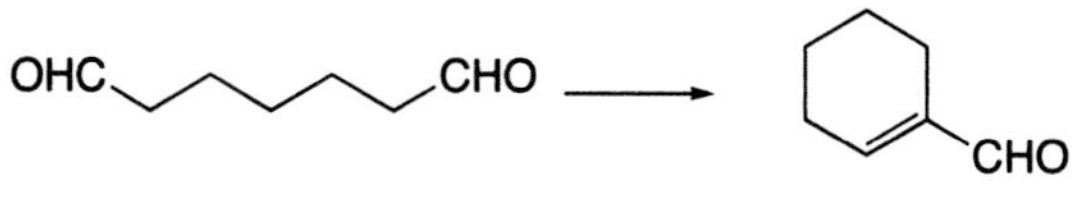 $OHC\diagdown\diagup\diagdown\diagup CHO$ → cyclohexene-CHO

c. (diketone chain) → acetylcyclohexene

**24.30** Locate the α and β C's to the carbonyl group, and break the molecule into two halves at this bond. The α C and all of the atoms bonded to it belong to one carbonyl component. The β C and all the atoms bonded to it belong to the other carbonyl component.

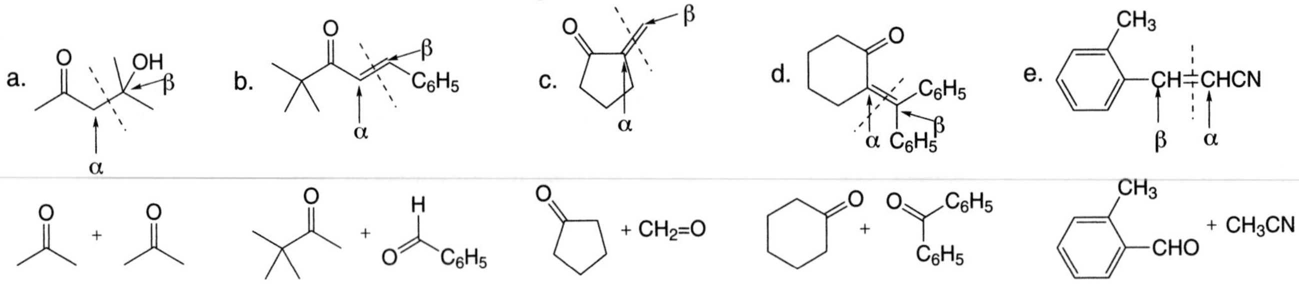
a.
b.
c.
d.
e.
α
β
$C_6H_5$
$CH_3$
CH∸CHCN
+ $CH_2$=O
+ $CH_3CN$
CHO

**24.31**

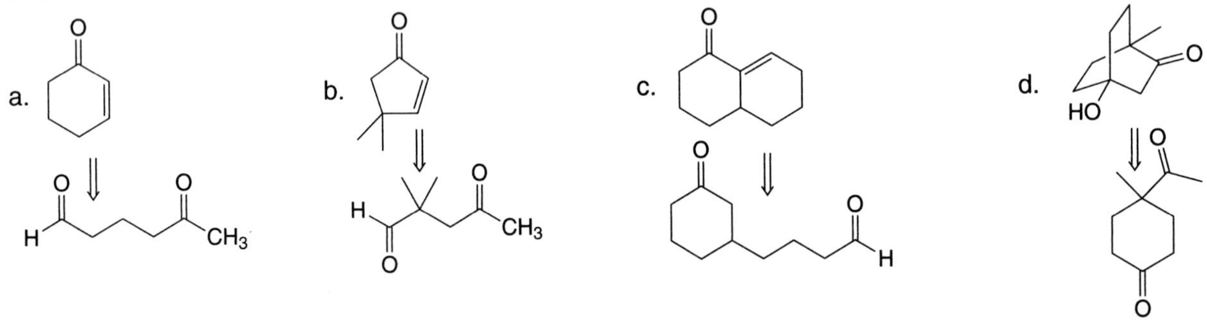
a.
b.
c.
d.
HO
$CH_3$
H

**24.32**

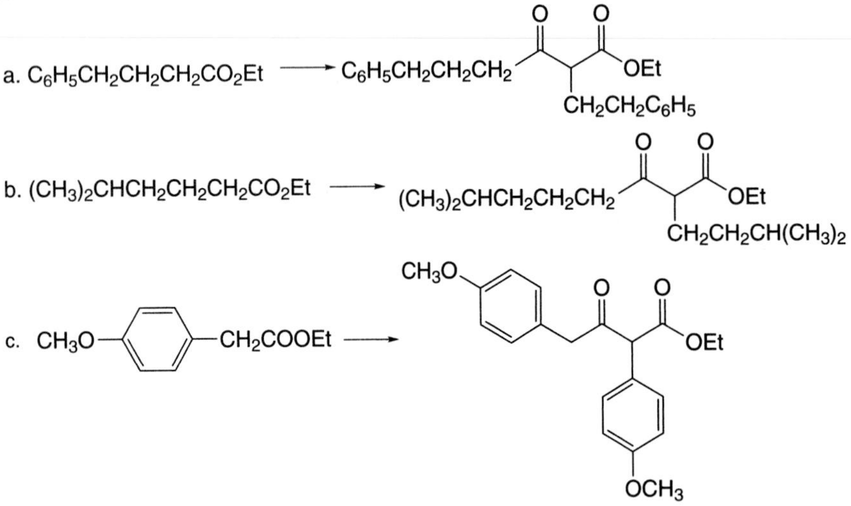
a. $C_6H_5CH_2CH_2CH_2CO_2Et$
$C_6H_5CH_2CH_2CH_2$
OEt
$CH_2CH_2C_6H_5$
b. $(CH_3)_2CHCH_2CH_2CH_2CO_2Et$
$(CH_3)_2CHCH_2CH_2CH_2$
OEt
$CH_2CH_2CH(CH_3)_2$
c. $CH_3O$
$CH_2COOEt$
$CH_3O$
OEt
$OCH_3$

**24.33**

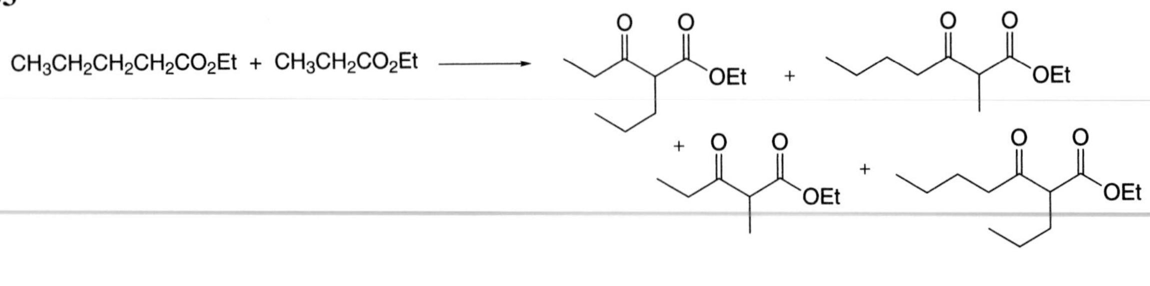
$CH_3CH_2CH_2CH_2CO_2Et$ + $CH_3CH_2CO_2Et$
OEt
+

**24.34**

a. CH$_3$CH$_2$CH$_2$CO$_2$Et only

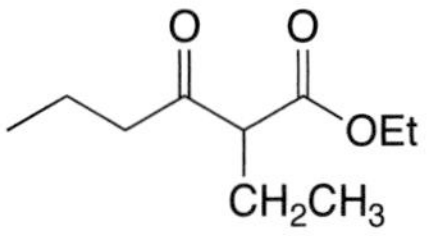

b. CH$_3$CH$_2$CH$_2$CO$_2$Et + C$_6$H$_5$CO$_2$Et

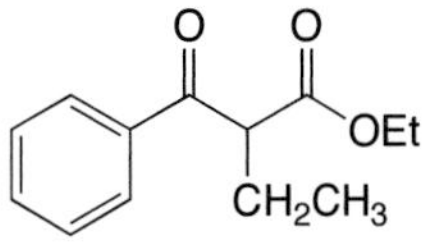

c. CH$_3$CH$_2$CH$_2$CO$_2$Et + (CH$_3$)$_2$C=O

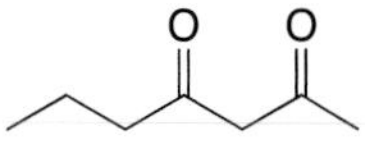

d. EtO$_2$CC(CH$_3$)$_2$CH$_2$CH$_2$CH$_2$CO$_2$Et

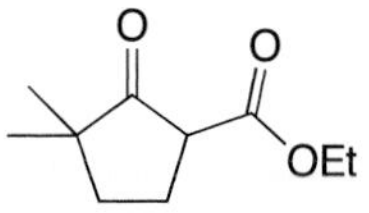

e. C$_6$H$_5$COCH$_2$CH$_3$ + C$_6$H$_5$CO$_2$Et

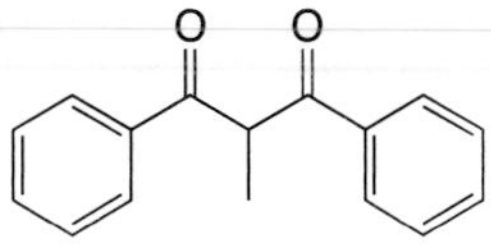

f. CH$_3$CH$_2$CO$_2$Et + (EtO)$_2$C=O

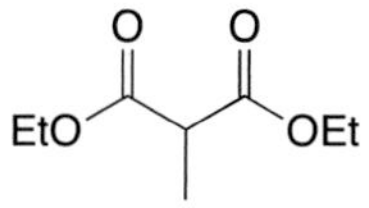

g. + HCO$_2$Et

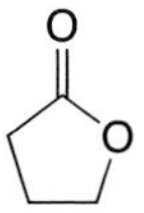

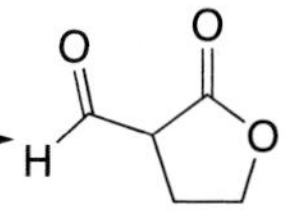

h.

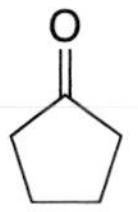

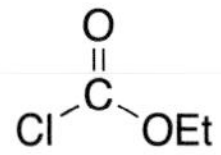

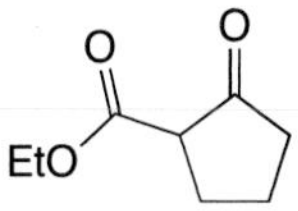

**24.35**

a.

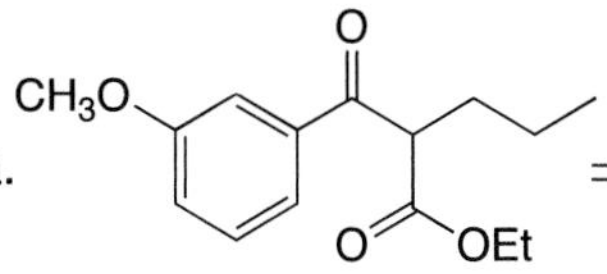

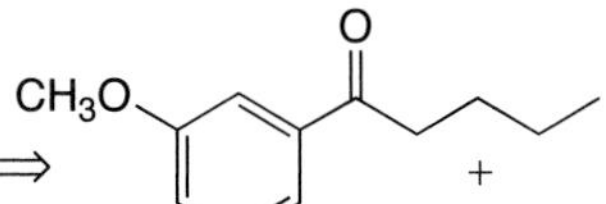

c.

b.

d. C$_6$H$_5$CH(COOEt)$_2$

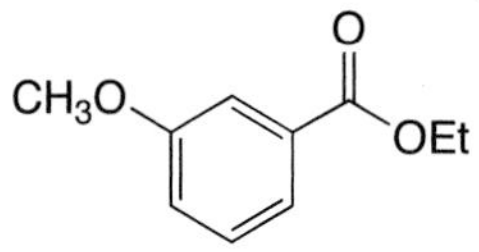

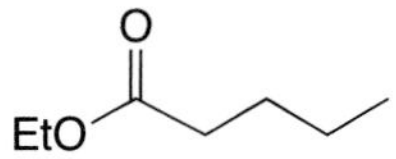

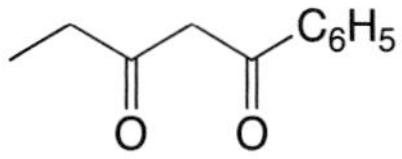

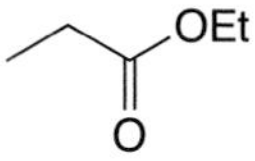

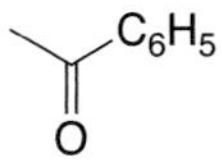

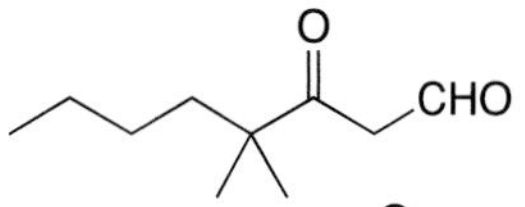

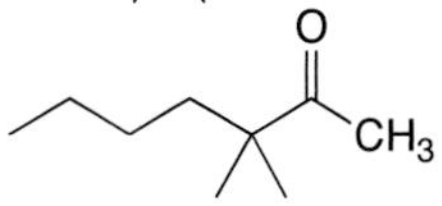

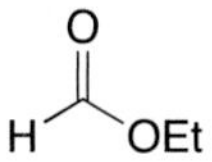

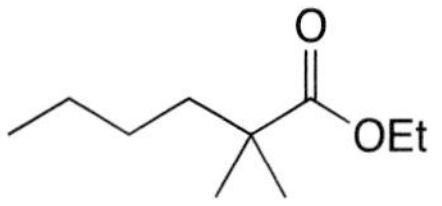

**24.36**

bond "a"    bond "b"

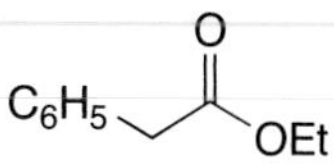

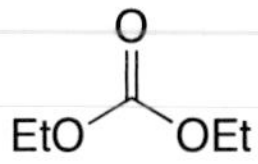

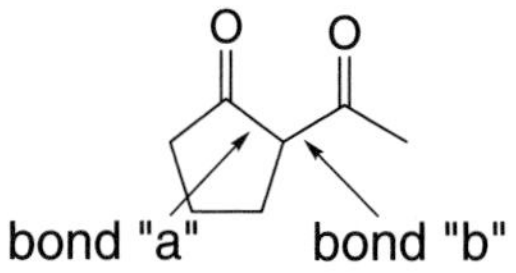

To form bond a:

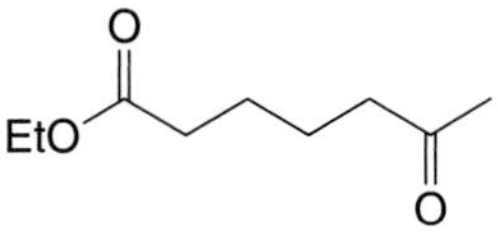

To form bond b:

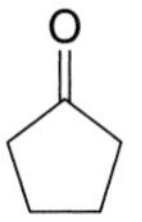

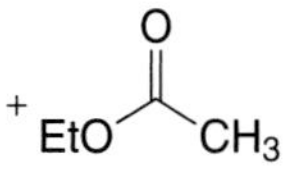

**24.37**

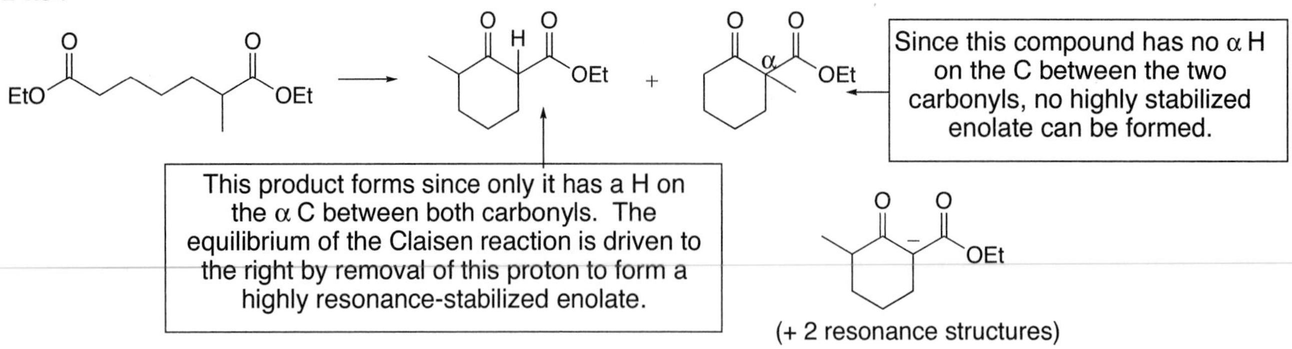
EtO
OEt
H
OEt
+
α
OEt
Since this compound has no α H on the C between the two carbonyls, no highly stabilized enolate can be formed.
This product forms since only it has a H on the α C between both carbonyls. The equilibrium of the Claisen reaction is driven to the right by removal of this proton to form a highly resonance-stabilized enolate.
OEt
(+ 2 resonance structures)

**24.38**

a. 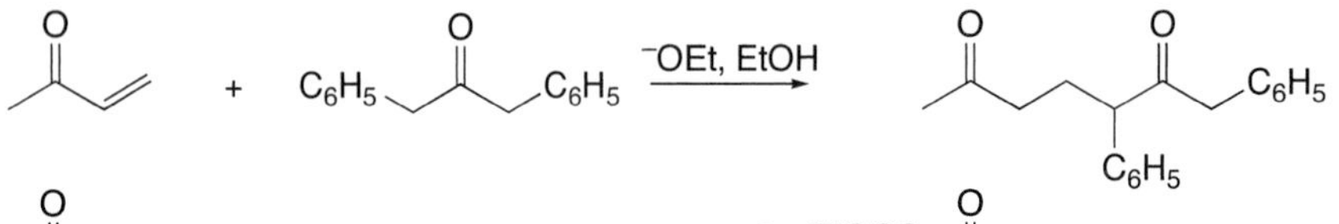
+
C6H5
C6H5
⁻OEt, EtOH
C6H5
C6H5

b. 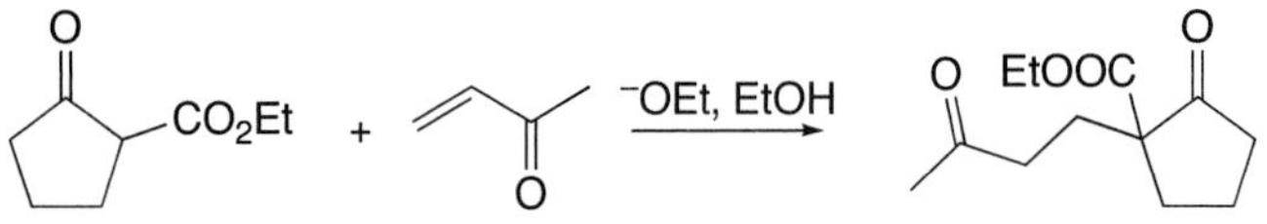
CO2Et
+
⁻OEt, EtOH
EtOOC

c. 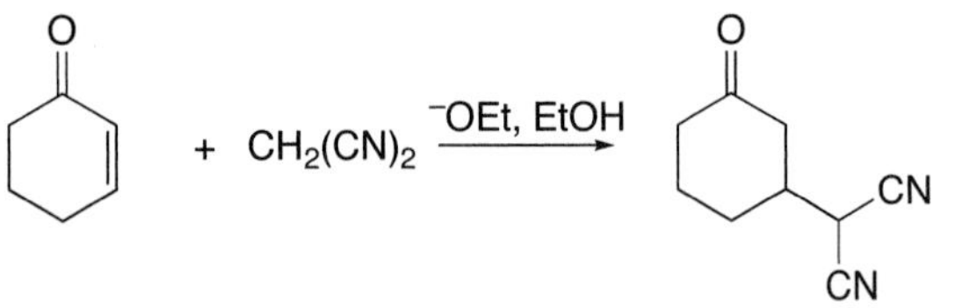
+ CH2(CN)2
⁻OEt, EtOH
CN
CN

d. 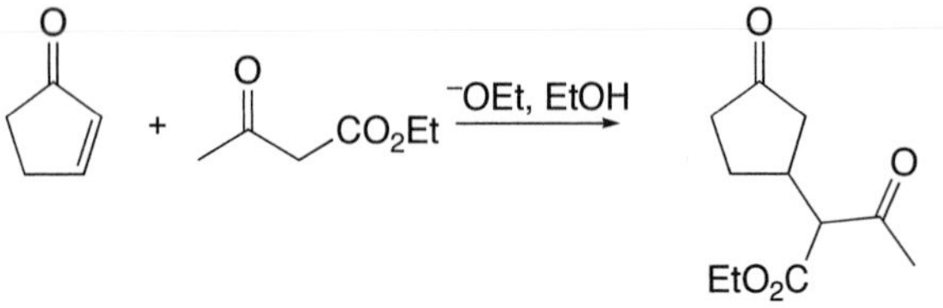
+
CO2Et
⁻OEt, EtOH
EtO2C

**24.39**

a. 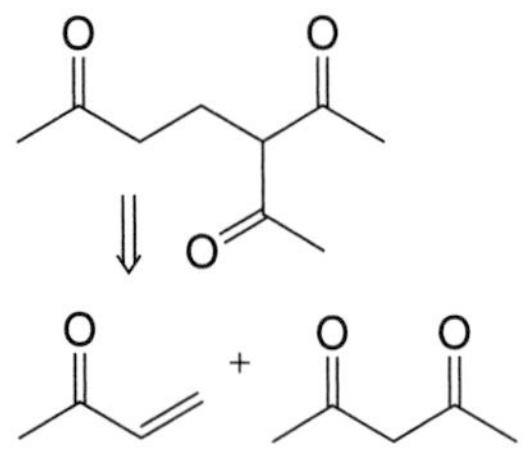

b. 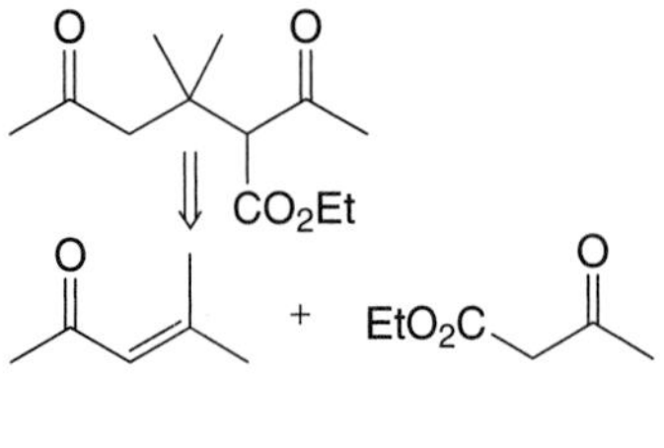
CO2Et
+
EtO2C

c. 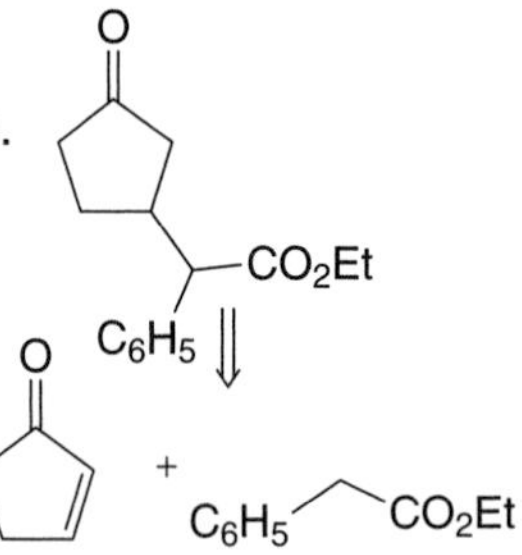
CO2Et
C6H5
+
C6H5
CO2Et

d. 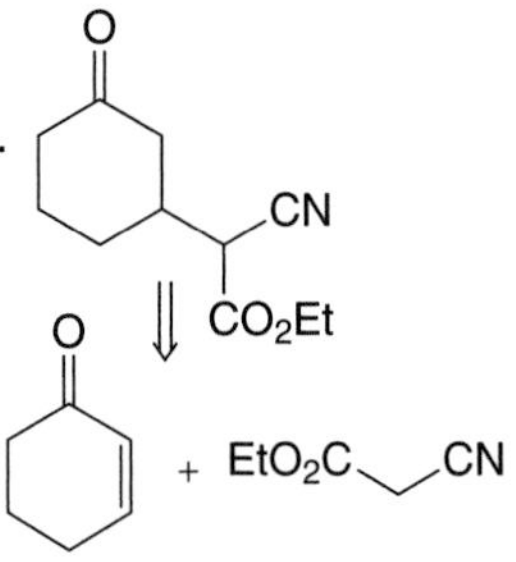
CN
CO2Et
+
EtO2C
CN

**24.40**

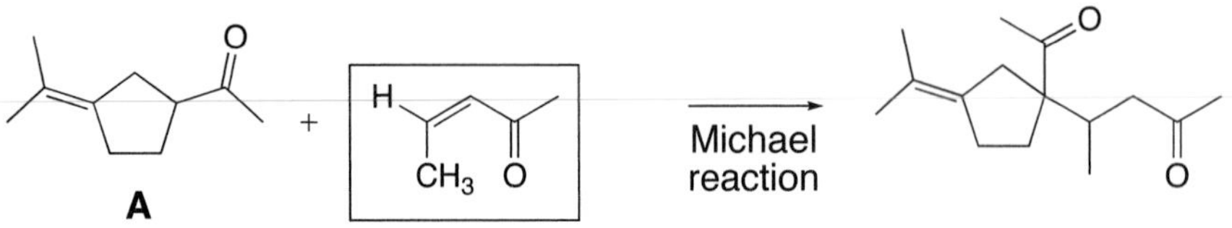
+
H
CH3
Michael
reaction

**A**

(*E* and *Z* isomers can be used.)

**24.41**

a.

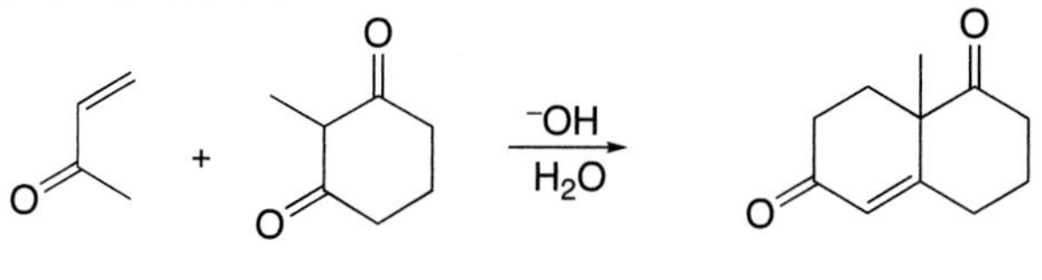

b.

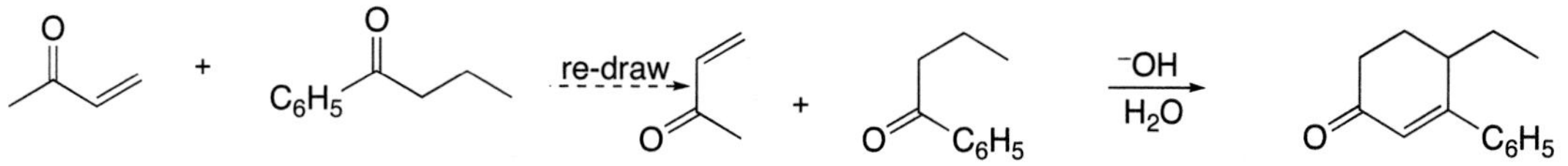

c.

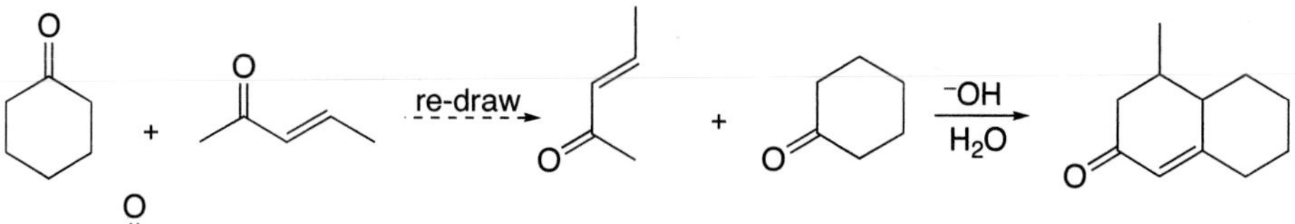

d.

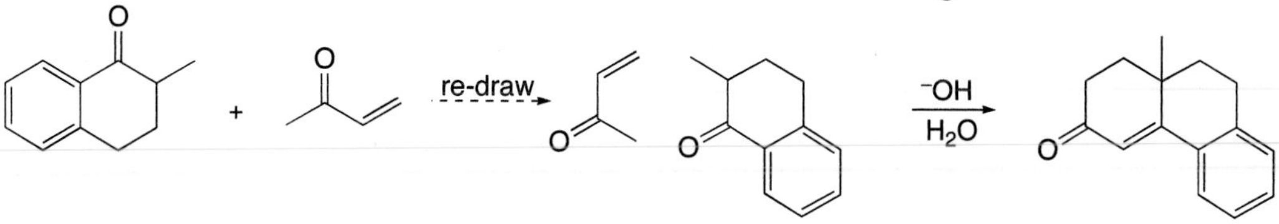

**24.42**

a.

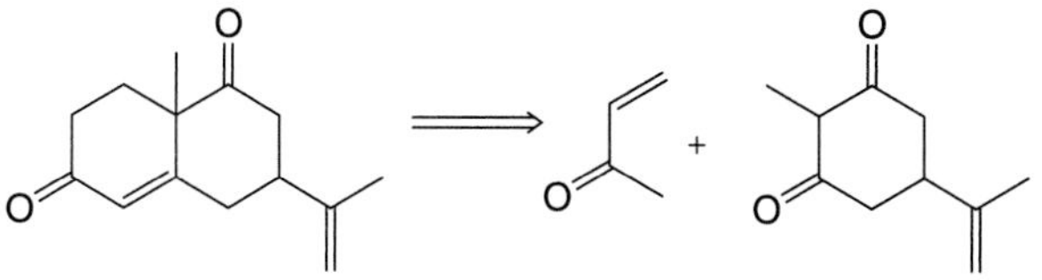

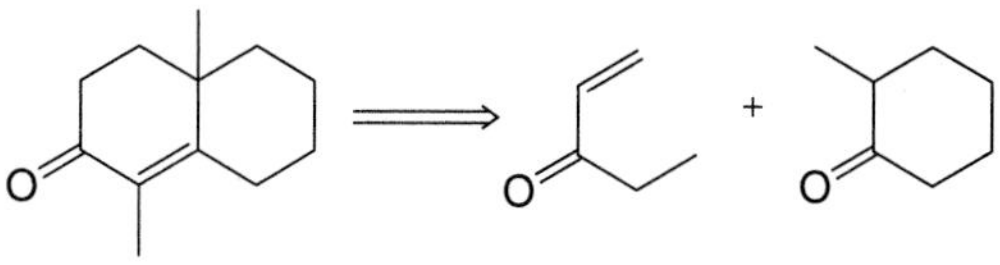

c.

b.

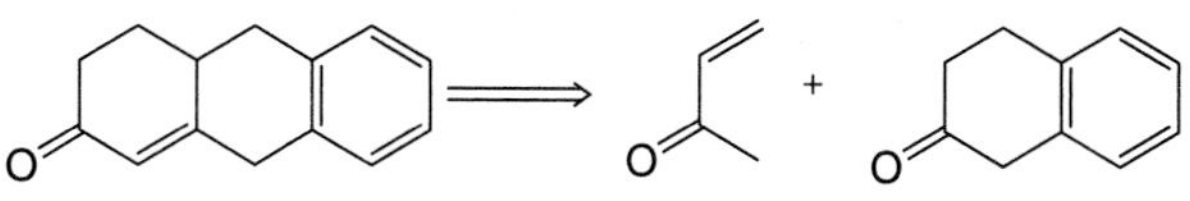

d.

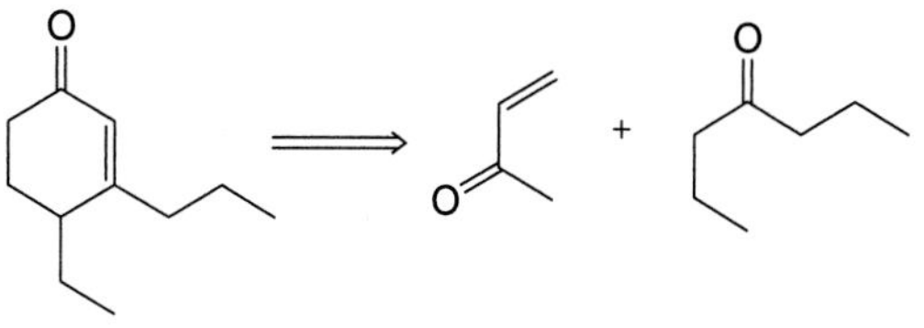

**24.43**

CH$_3$CH$_2$CH$_2$CHO

a. ⁻OH, H$_2$O → [structure: CH$_3$CH$_2$CH$_2$ / CHO with CH$_2$CH$_3$] (*E* and *Z*)

b. ⁻OH, CH$_2$=O, H$_2$O → [structure: CH$_2$=C with CHO and CH$_2$CH$_3$]

c. [1] LDA; [2] CH$_3$CHO; [3] H$_2$O → [structure with CHO, OH]

d. CH$_2$(CO$_2$Et)$_2$ / NaOEt, EtOH → [structure: CH$_3$CH$_2$CH$_2$, H, EtO$_2$C, CO$_2$Et, OH]

e. [1] CH$_3$Li; [2] H$_2$O → [structure with OH]

f. NaBH$_4$, CH$_3$OH → CH$_3$CH$_2$CH$_2$CH$_2$OH

g. H$_2$, Pd-C → CH$_3$CH$_2$CH$_2$CH$_2$OH

h. HOCH$_2$CH$_2$OH, TsOH → [dioxolane structure with H]

i. CH$_3$NH$_2$, mild acid → [structure: CH$_3$CH$_2$CH$_2$ / CH=N / CH$_3$, H]

j. (CH$_3$)$_2$NH, mild acid → CH$_3$CH$_2$CH=CHN(CH$_3$)$_2$ (*E* and *Z*)

k. CrO$_3$, H$_2$SO$_4$ → CH$_3$CH$_2$CH$_2$COOH

l. Br$_2$, CH$_3$COOH → [structure with O, H, Br]

m. Ph$_3$P=CH$_2$ → [structure: CH$_3$CH$_2$CH$_2$ / C=CH$_2$, H]

n. NaCN, HCl → CH$_3$CH$_2$CH$_2$–C–H with OH and CN

o. [1] LDA; [2] CH$_3$I → [structure with O, H]

**24.44**

a. [cyclopentanone] ⁻OH / H$_2$O → [cyclopentylidene cyclopentanone]

b. [1,3-cyclohexanedione] NaOEt, EtOH / (CH$_3$)$_2$C=O → [product with isopropylidene]

c. NCCH$_2$CO$_2$Et NaOEt, EtOH → [cyclohexylidene with CN, CO$_2$Et]

d. [furan-CHO] + CH$_3$–C(=O)–CH$_3$ ⁻OH / H$_2$O → [furan-CH=CH-C(=O)] (*E* and *Z*)

e. CH$_3$–C(=O)–CH$_3$ [1] LDA / [2] CH$_3$CH$_2$CHO / [3] H$_2$O → [product with OH, O]

f. [cyclohexanone] + C$_6$H$_5$–CH=CH–C(=O)CH$_3$ NaOCH$_3$ / CH$_3$OH → [bicyclic product with C$_6$H$_5$, O]

g. [naphthalene-CHO] + CH$_3$–C(=O)–C$_6$H$_5$ ⁻OH / H$_2$O → [naphthalene-CH=CH-C(=O)-C$_6$H$_5$] (*E* and *Z*)

h. [benzene with CH$_2$CO$_2$Et, CH$_2$CO$_2$Et] [1] NaOEt, EtOH / [2] H$_3$O⁺ → [indanone product with O, OEt]

**24.45**

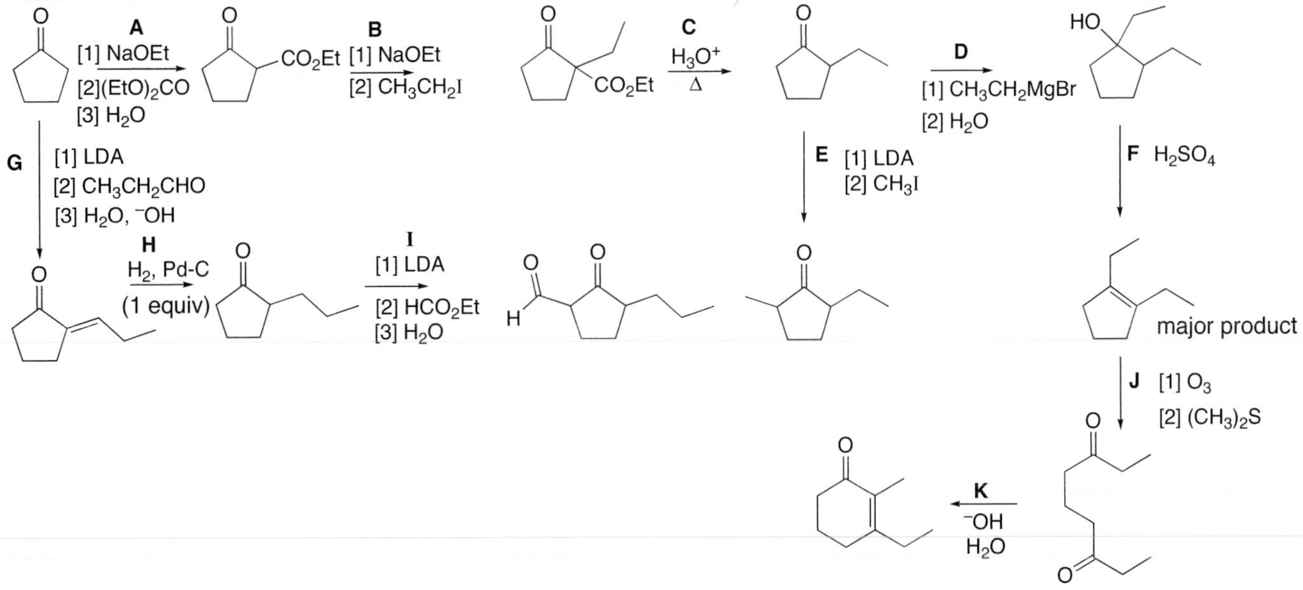

**24.46**

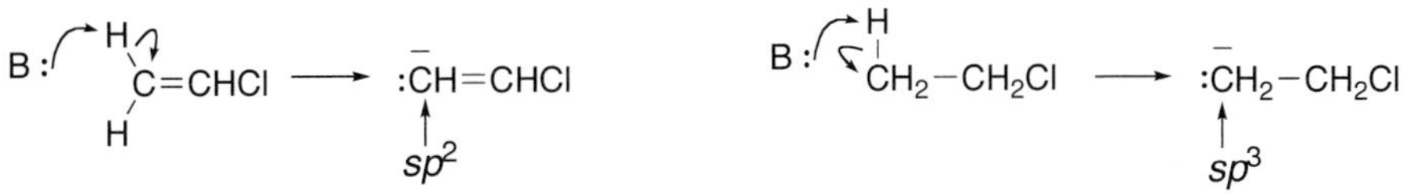

Vinyl halides undergo elimination by an E1cB mechanism more readily than alkyl halides because the carbanion intermediate formed from a vinyl halide has an $sp^2$ hybridized C, while the carbanion derived from $CH_3CH_2Cl$ is $sp^3$ hybridized. The higher percent s-character of the $sp^2$ hybridized anion makes it more stable, and therefore, it is formed more readily.

**24.47**

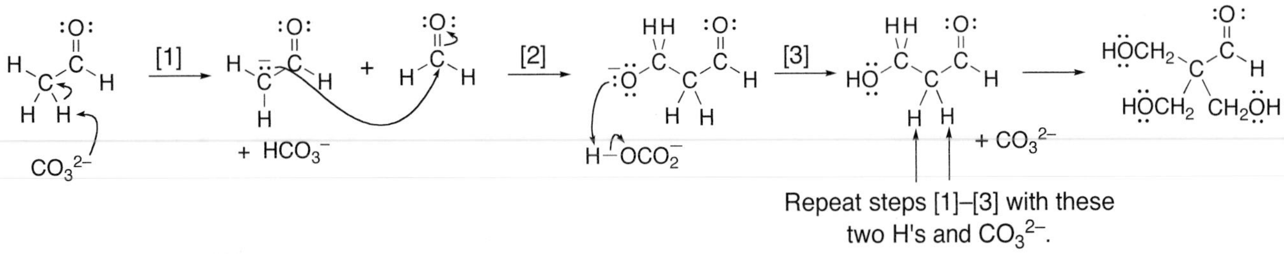

Repeat steps [1]–[3] with these two H's and $CO_3^{2-}$.

**24.48** Enolate **A** is more substituted (and more stable) than either of the other two possible enolates and attacks an aldehyde carbonyl group, which is sterically less hindered than a ketone carbonyl. The resulting ring size (five-membered) is also quite stable. That is why 1-acetylcyclopentene is the major product.

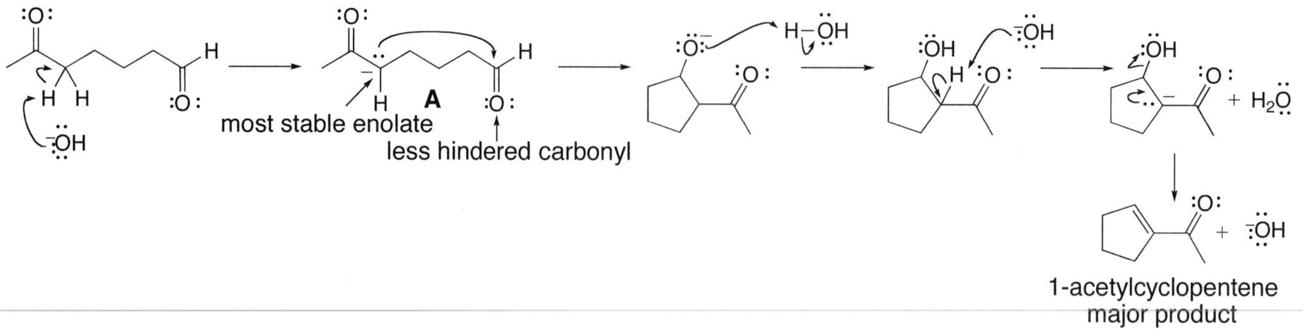
A
most stable enolate
less hindered carbonyl
1-acetylcyclopentene
major product

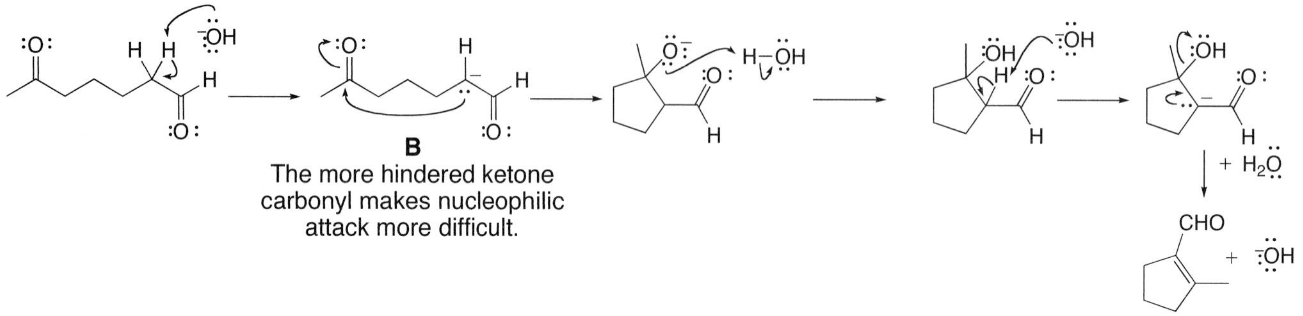
B
The more hindered ketone carbonyl makes nucleophilic attack more difficult.

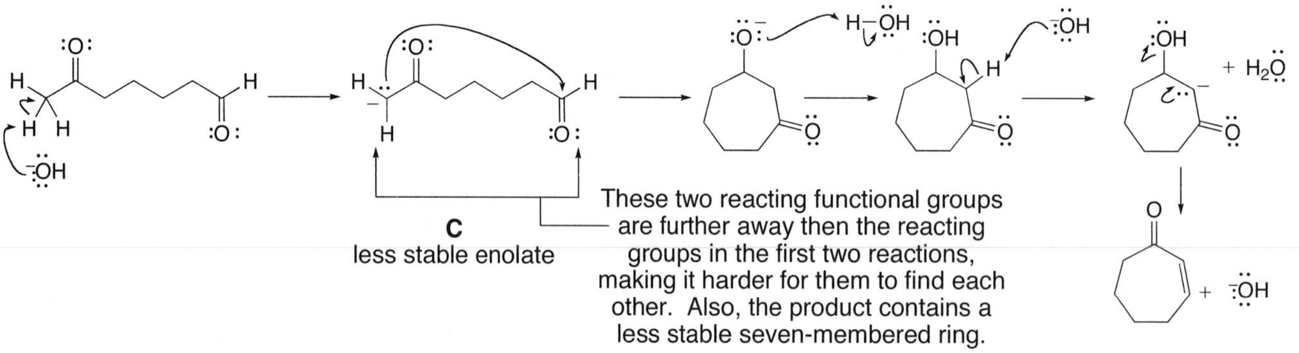
C
less stable enolate
These two reacting functional groups are further away then the reacting groups in the first two reactions, making it harder for them to find each other. Also, the product contains a less stable seven-membered ring.

**24.49** All enolates have a second resonance structure with a negative charge on O.

a.

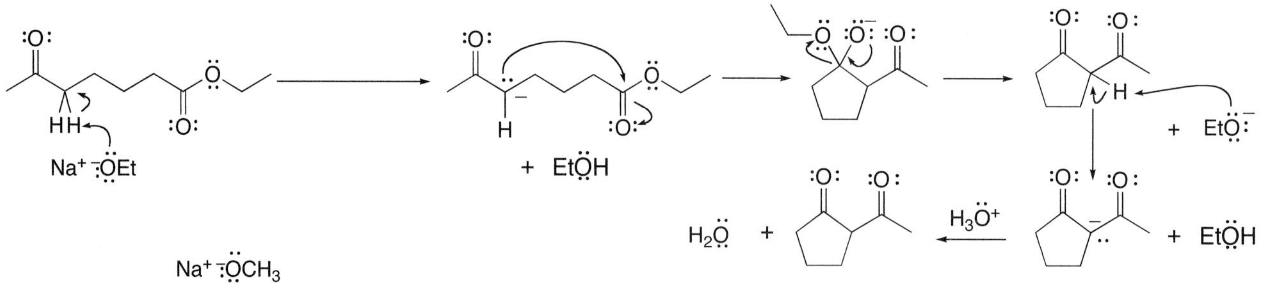

b.

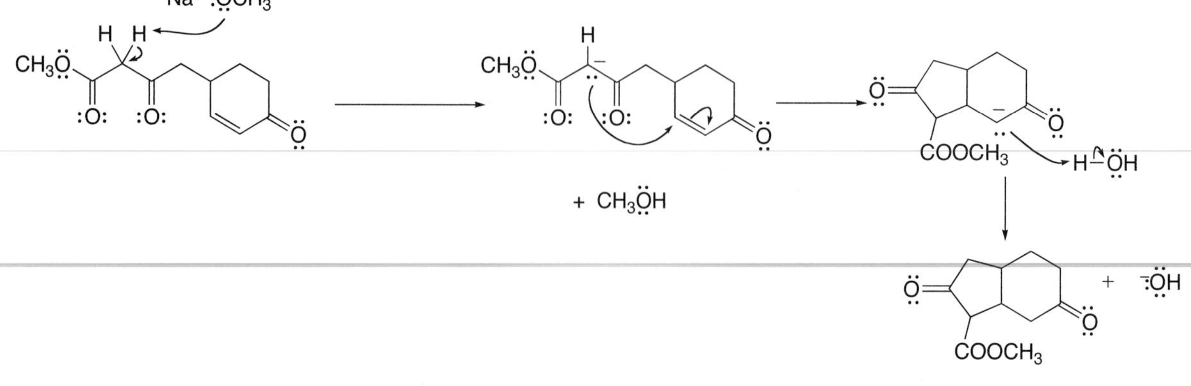

**24.50**

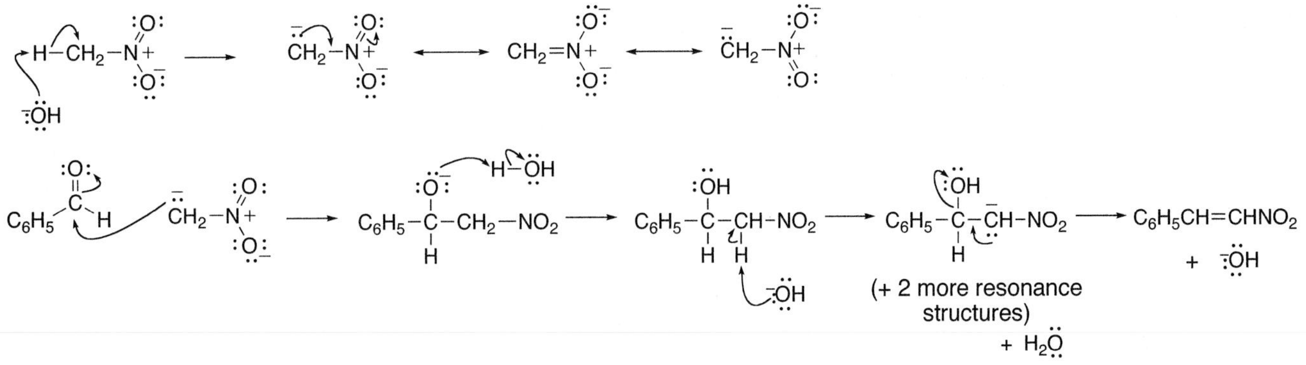

**24.51** All enolates have a second resonance structure with a negative charge on O.

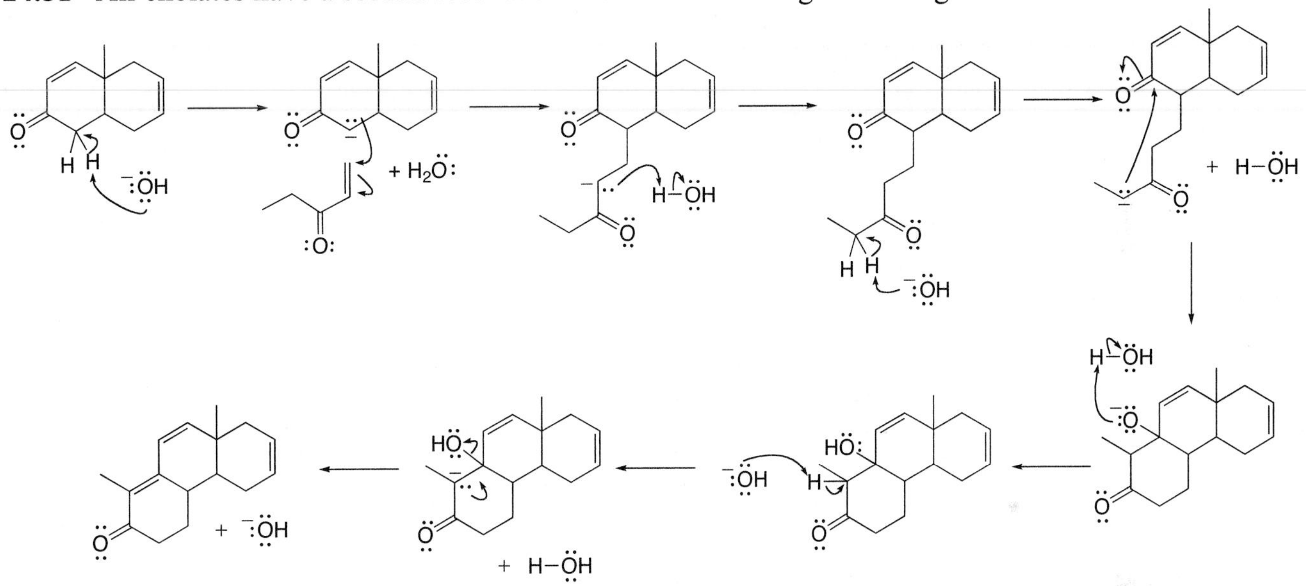

**24.52** Polymerization occurs by repeated Michael reactions.

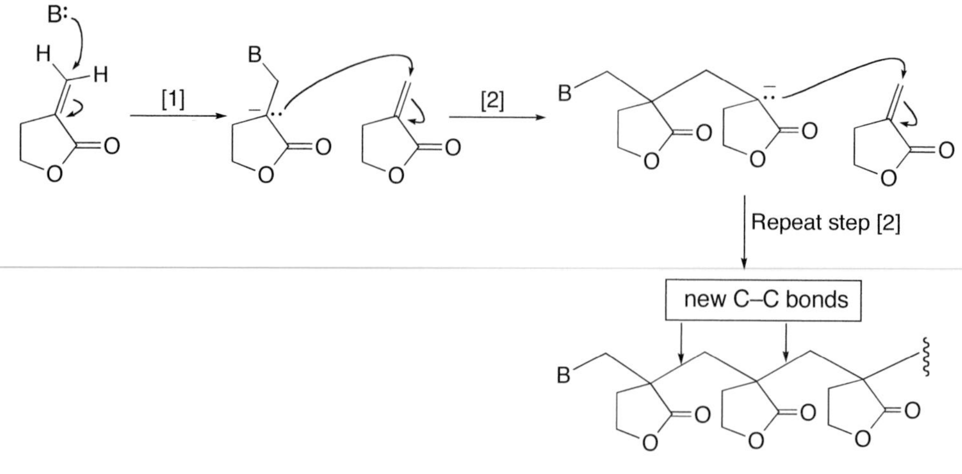

polytulipalin

**24.53**

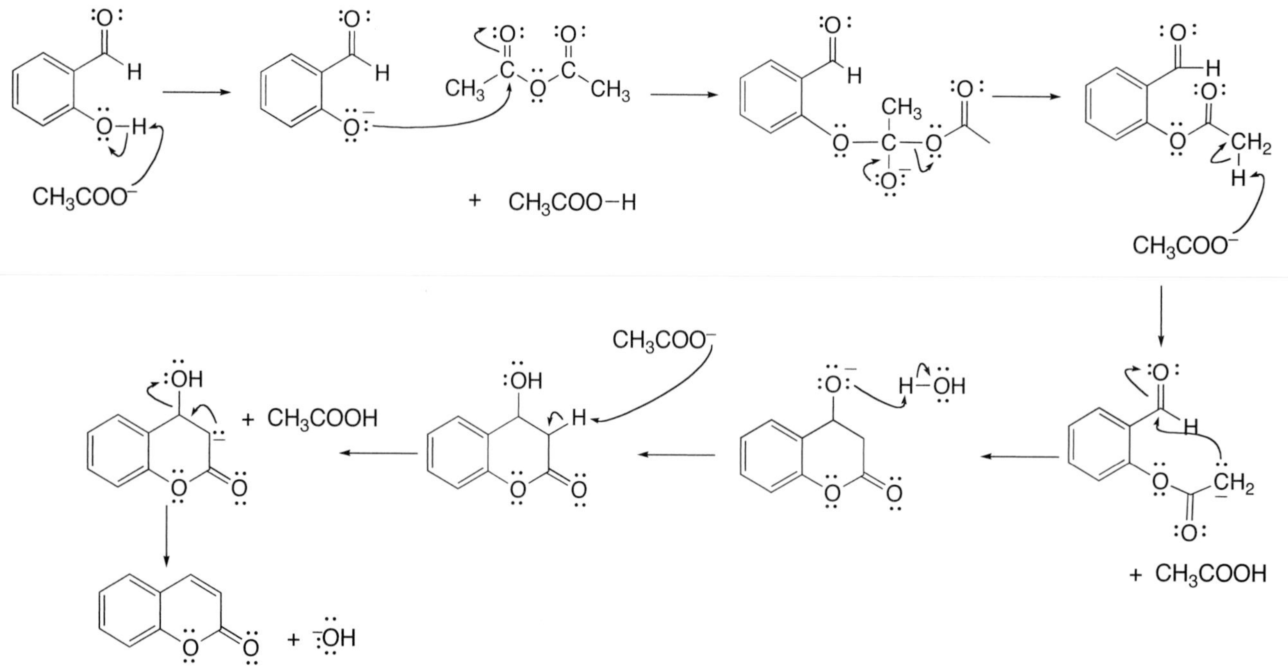

coumarin

**24.54**

a.

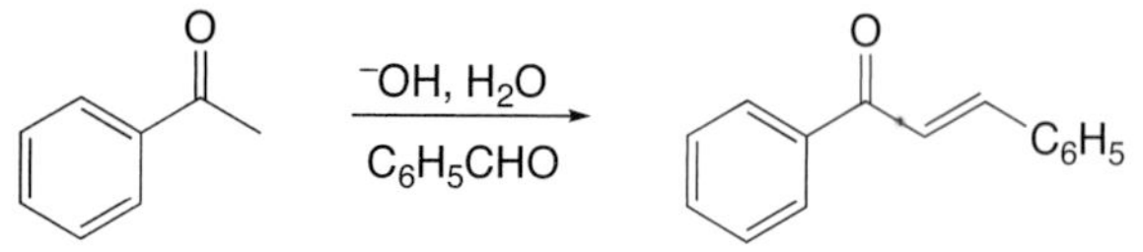

$^-$OH, $H_2O$
$C_6H_5CHO$

b.

$^-$OH, $H_2O$
$H_2C=O$

PCC

$HCO_2Et$
$^-$OEt, $CH_3CH_2OH$

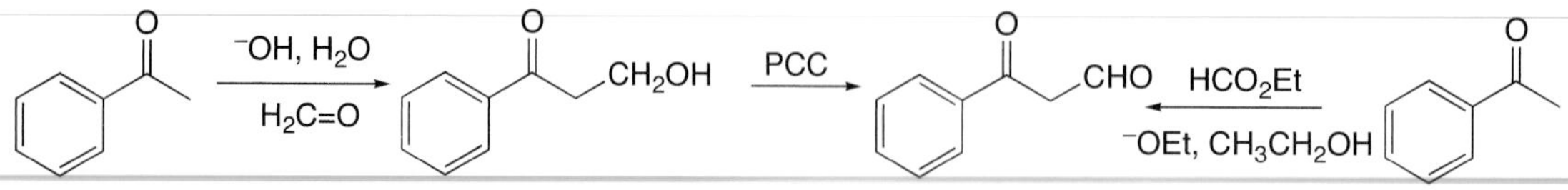

c.

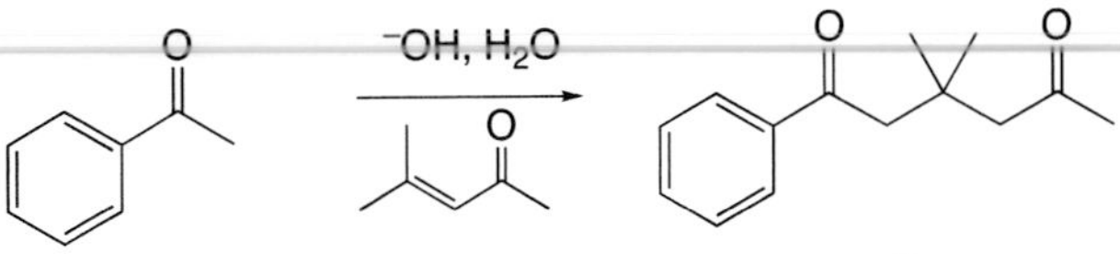

d.

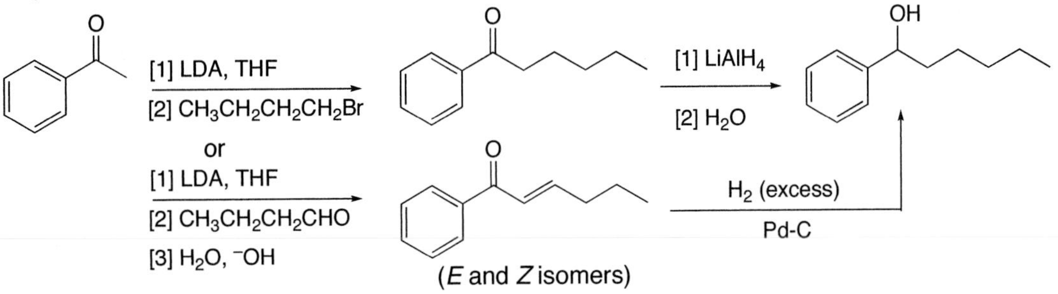

e.

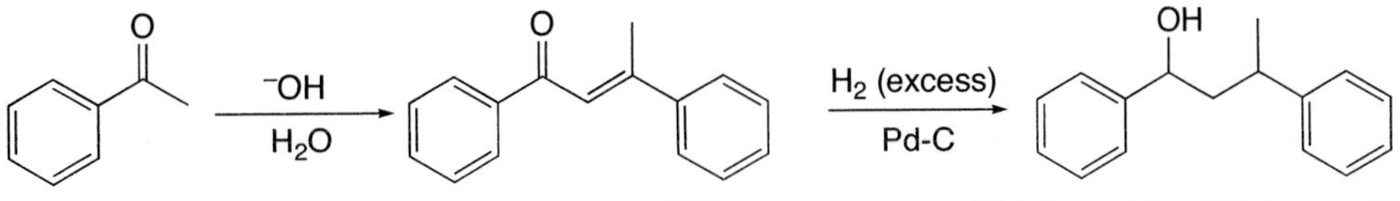

**24.55**

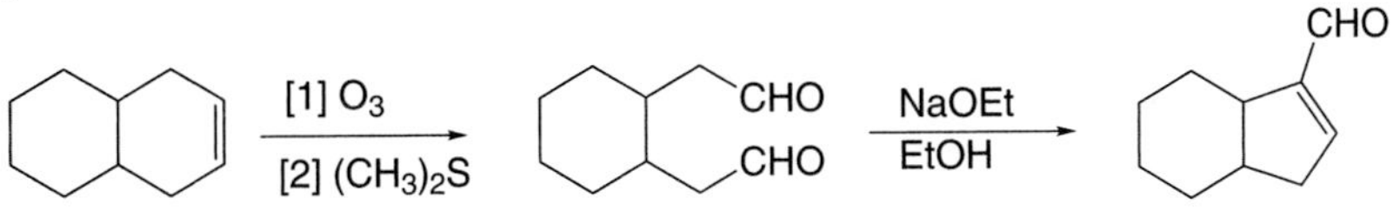

**24.56**

a.

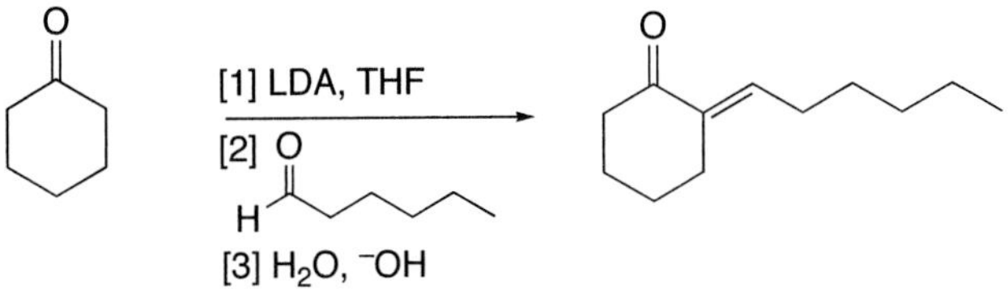

b.

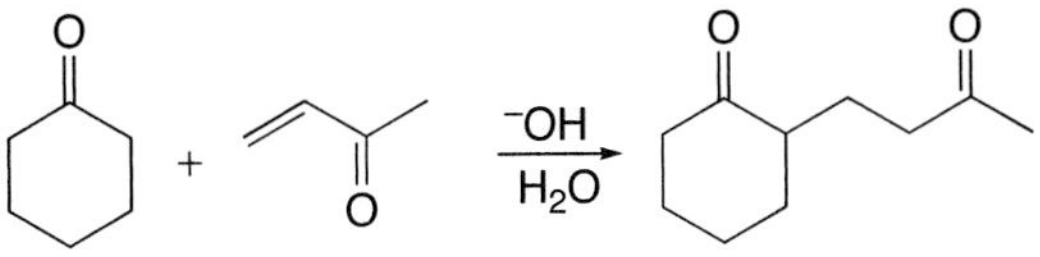

c.

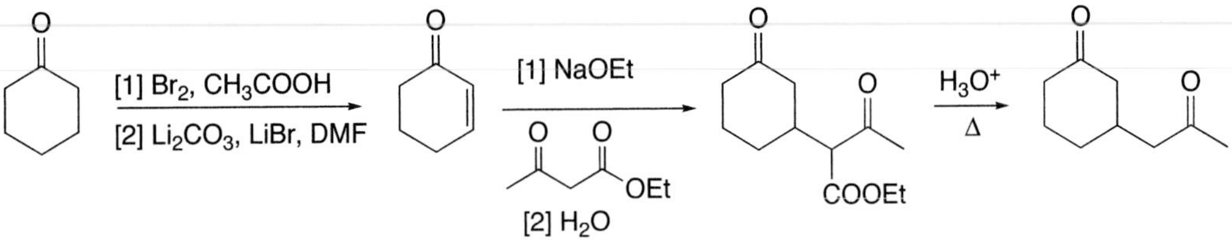

d.

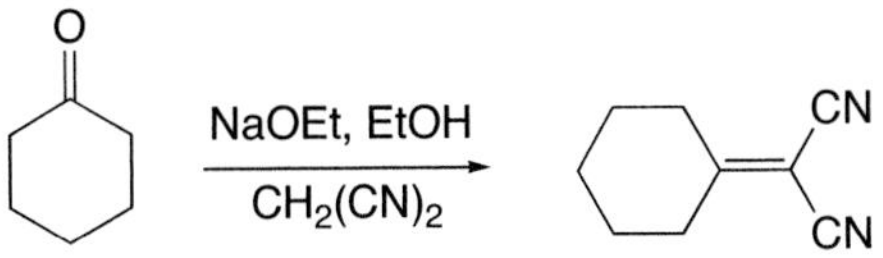

e.

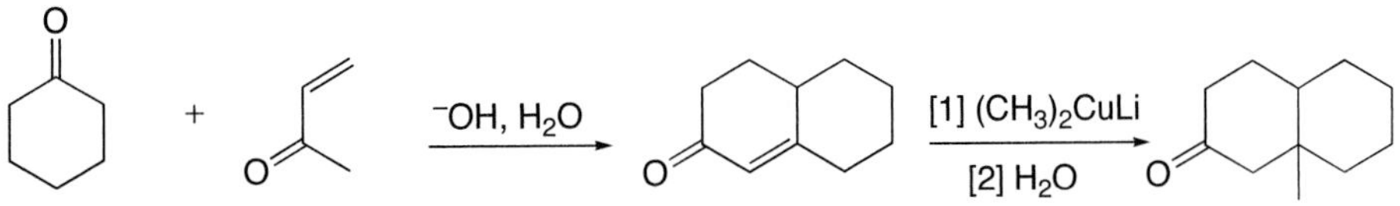

## 24.57

a.

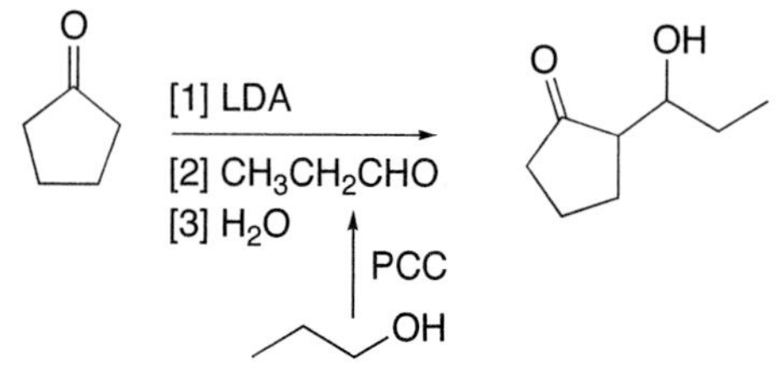
[1] LDA
[2] $CH_3CH_2CHO$
[3] $H_2O$
PCC
OH
O
OH

b.

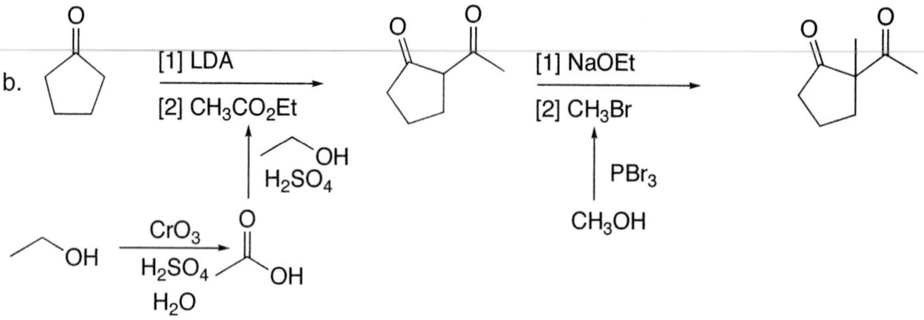
[1] LDA
[2] $CH_3CO_2Et$
OH
$H_2SO_4$
OH
$CrO_3$
$H_2SO_4$
$H_2O$
OH
[1] NaOEt
[2] $CH_3Br$
$PBr_3$
$CH_3OH$

c.

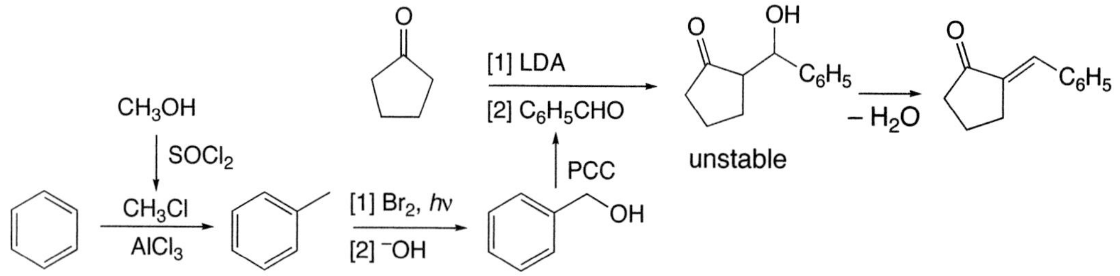
$CH_3OH$
$SOCl_2$
$CH_3Cl$
$AlCl_3$
[1] $Br_2$, hν
[2] $^-OH$
OH
PCC
[1] LDA
[2] $C_6H_5CHO$
OH
$C_6H_5$
unstable
– $H_2O$
$C_6H_5$

d.

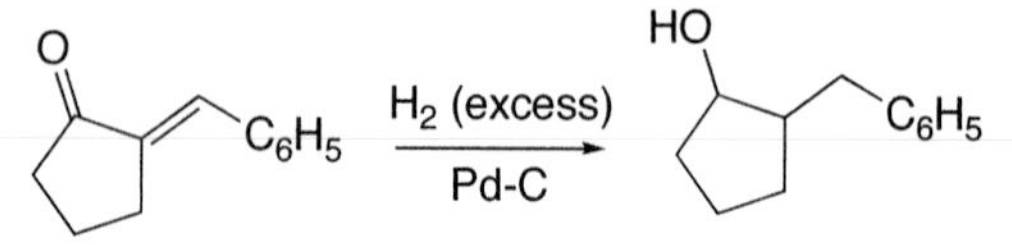
$C_6H_5$
$H_2$ (excess)
Pd-C
HO
$C_6H_5$

(from c.)

e.

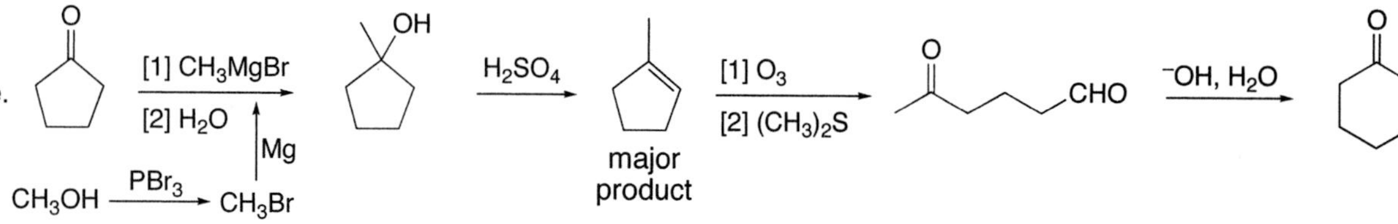
[1] $CH_3MgBr$
[2] $H_2O$
Mg
$CH_3OH$
$PBr_3$
$CH_3Br$
OH
$H_2SO_4$
major
product
[1] $O_3$
[2] $(CH_3)_2S$
CHO
$^-OH$, $H_2O$

## 24.58

a.

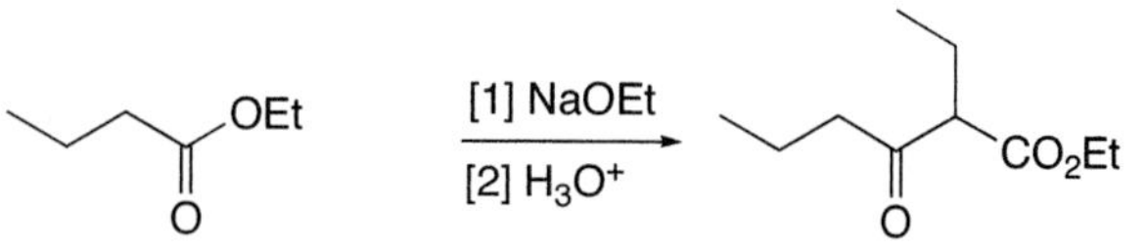
OEt
[1] NaOEt
[2] $H_3O^+$
$CO_2Et$

b.

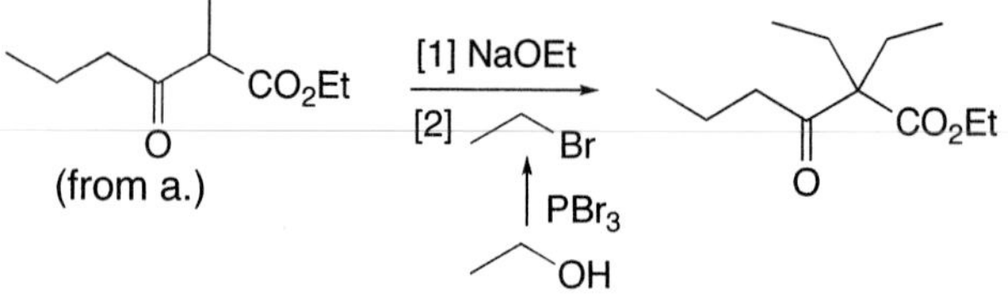
$CO_2Et$
(from a.)
[1] NaOEt
[2]
Br
$PBr_3$
OH
$CO_2Et$

c.

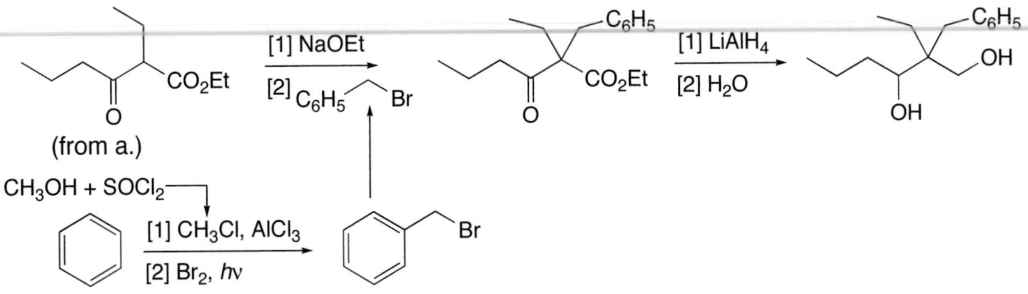

[1] NaOEt
[2] C6H5 Br

[1] LiAlH4
[2] H2O

(from a.)

CH3OH + SOCl2

[1] CH3Cl, AlCl3
[2] Br2, hv

d.

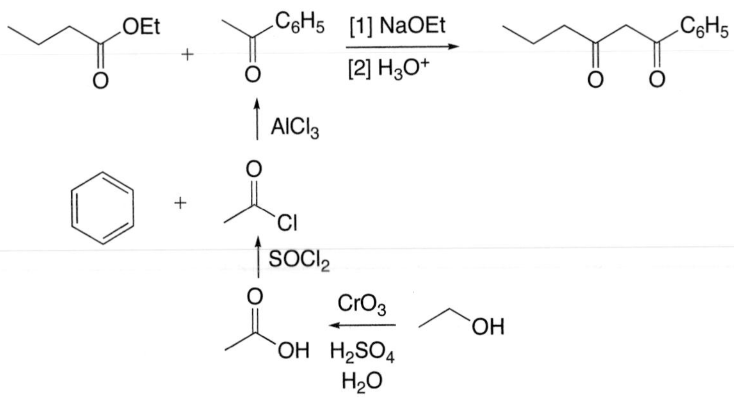

+

C6H5

[1] NaOEt
[2] H3O+

C6H5

AlCl3

+

Cl

SOCl2

CrO3
H2SO4
H2O

OH

**24.59**

a.

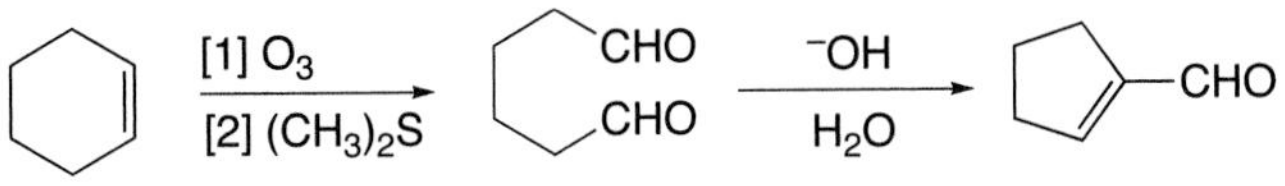

[1] O3
[2] (CH3)2S

CHO
CHO

−OH
H2O

CHO

b.

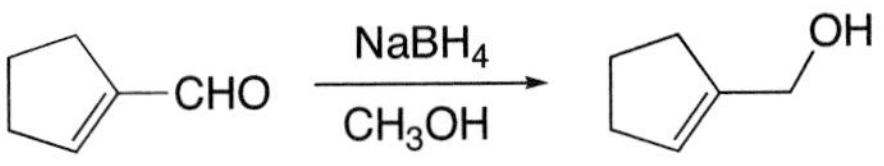

CHO

NaBH4
CH3OH

OH

(from a.)

c.

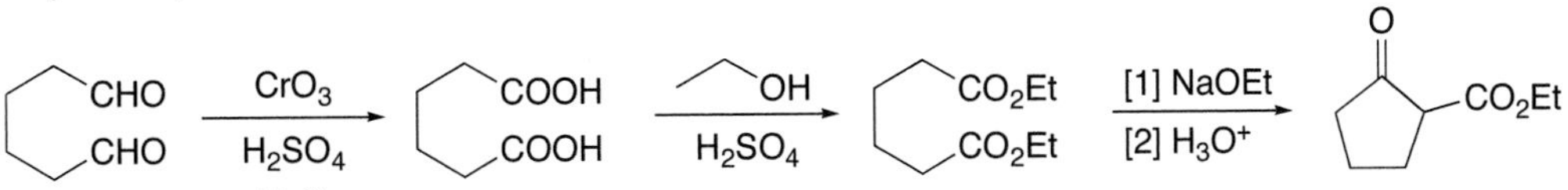

CHO
CHO

CrO3
H2SO4
H2O

COOH
COOH

OH
H2SO4

CO2Et
CO2Et

[1] NaOEt
[2] H3O+

CO2Et

(from a.)

d.

CO2Et

[1] NaOEt
[2] CH3I

CO2Et

H3O+
Δ

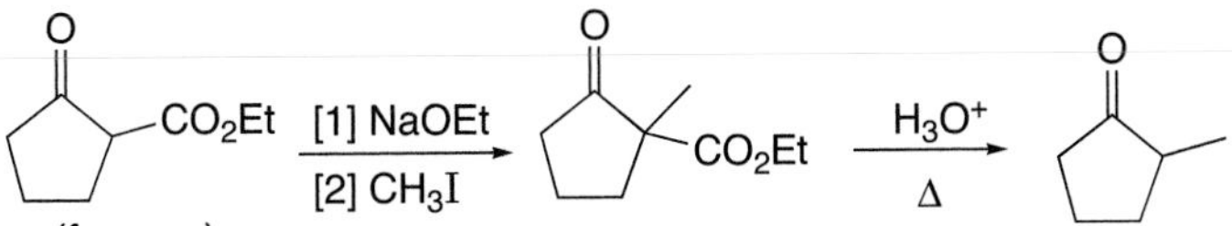

(from c.)

**24.60** All enolates have a second resonance structure with a negative charge on O.

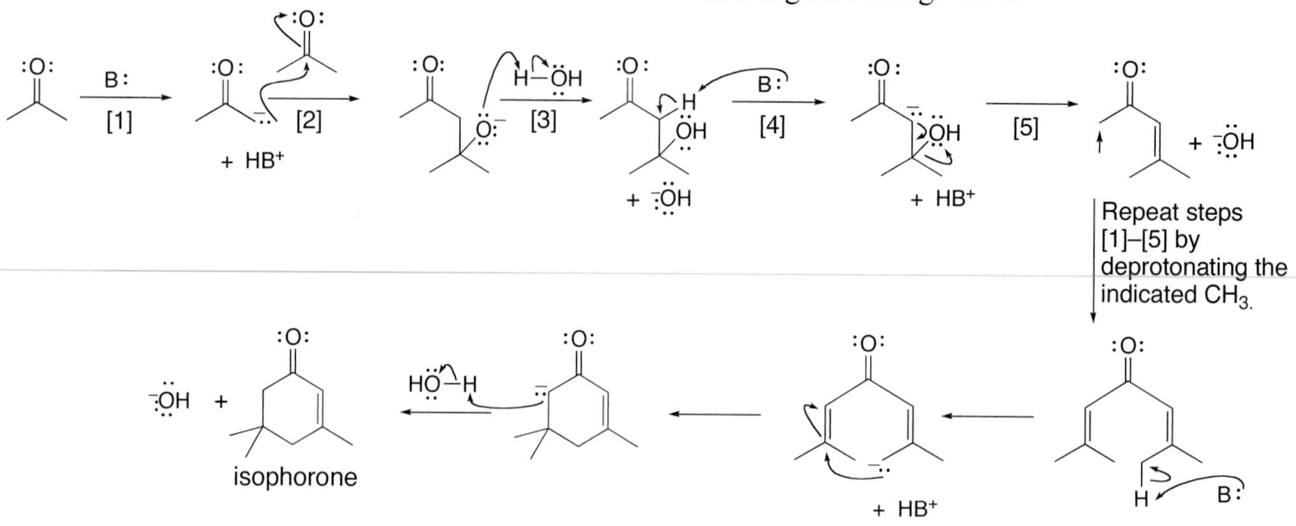
B:
[1]
+ HB+
[2]
[3]
+ :ÖH
B:
[4]
+ HB+
[5]
+ :ÖH
Repeat steps [1]–[5] by deprotonating the indicated CH3.
:ÖH +
isophorone
+ HB+
B:

**24.61** All enolates have a second resonance structure with a negative charge on O.

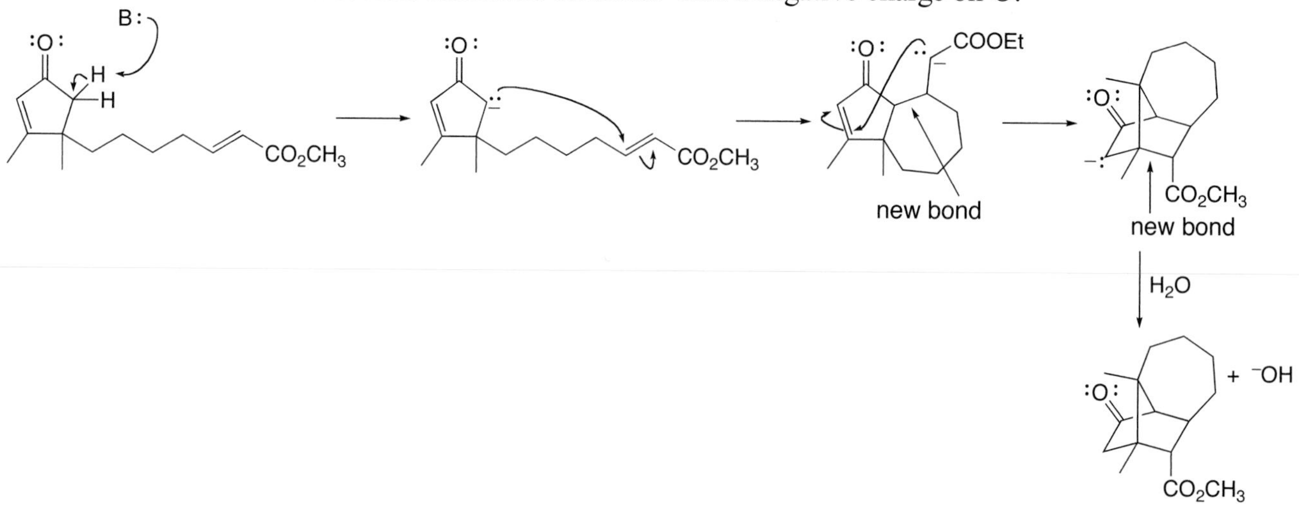
B:
CO2CH3
COOEt
new bond
new bond
H2O
+ −OH

**24.62**

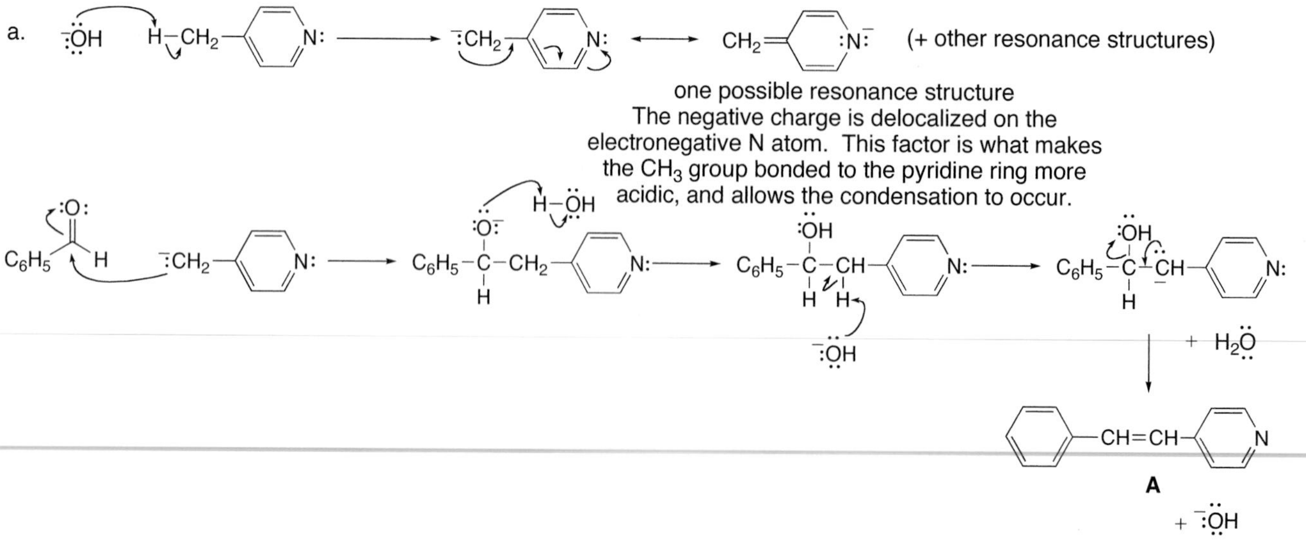
a.
(+ other resonance structures)
one possible resonance structure
The negative charge is delocalized on the electronegative N atom. This factor is what makes the CH3 group bonded to the pyridine ring more acidic, and allows the condensation to occur.
C6H5
+ H2O
A
+ :ÖH

b. The condensation reaction can occur only if the $CH_3$ group bonded to the pyridine ring has acidic hydrogens that can be removed with ⁻OH.

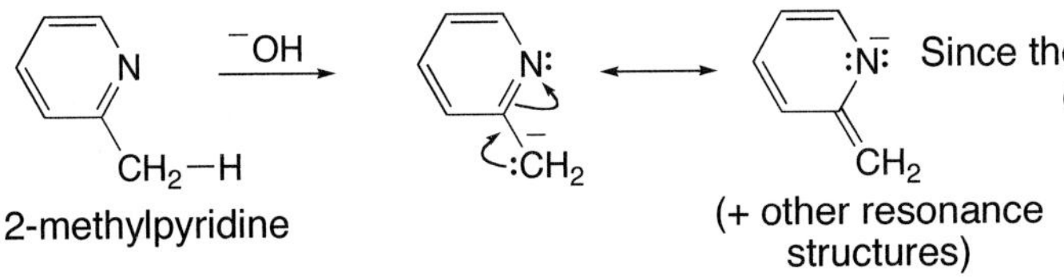

2-methylpyridine

Since the negative charge is delocalized on the N, the $CH_3$ contains acidic H's and reaction will occur.

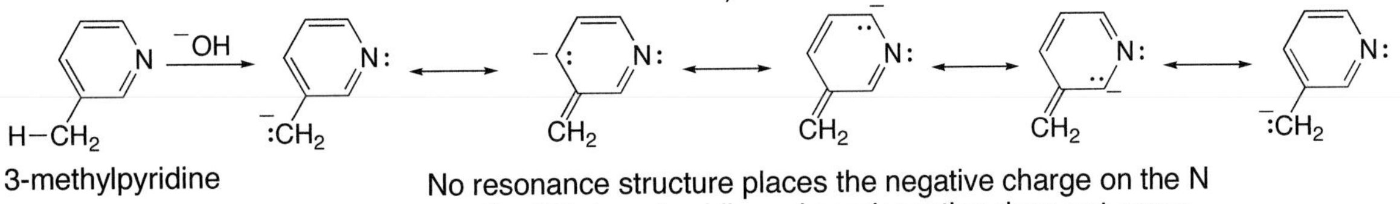

(+ other resonance structures)

3-methylpyridine

No resonance structure places the negative charge on the N so the $CH_3$ is not acidic and condensation does not occur.

## Chapter 25: Amines

### ♦ General facts

- Amines are organic nitrogen compounds having the general structure $RNH_2$, $R_2NH$, or $R_3N$, with a lone pair of electrons on N (25.1).
- Amines are named using the suffix –amine (25.3).
- All amines have polar C–N bonds. Primary (1°) and 2° amines have polar N–H bonds and are capable of intermolecular hydrogen bonding (25.4).
- The lone pair on N makes amines strong organic bases and nucleophiles (25.8).

### ♦ Summary of spectroscopic absorptions (25.5)

| Mass spectra | Molecular ion | Amines with an odd number of N atoms give an odd molecular ion. |
|---|---|---|
| **IR absorptions** | N–H | 3300–3500 cm$^{-1}$ (two peaks for $RNH_2$, one peak for $R_2NH$) |
| **$^1$H NMR absorptions** | NH | 0.5–5 ppm (no splitting with adjacent protons) |
| | CH–N | 2.3–3.0 ppm (deshielded $Csp^3$–H) |
| **$^{13}$C NMR absorption** | C–N | 30–50 ppm |

### ♦ Comparing the basicity of amines and other compounds (25.10)

- Alkylamines ($RNH_2$, $R_2NH$, and $R_3N$) are more basic than $NH_3$ because of the electron-donating R groups (25.10A).
- Alkylamines ($RNH_2$) are more basic than arylamines ($C_6H_5NH_2$), which have a delocalized lone pair from the N atom (25.10B).
- Arylamines with electron-donor groups are more basic than arylamines with electron-withdrawing groups (25.10B).
- Alkylamines ($RNH_2$) are more basic than amides ($RCONH_2$), which have a delocalized lone pair from the N atom (25.10C).
- Aromatic heterocycles with a localized electron pair on N are more basic than those with a delocalized lone pair from the N atom (25.10D).
- Alkylamines with a lone pair in an $sp^3$ hybrid orbital are more basic than those with a lone pair in an $sp^2$ hybrid orbital (25.10E).

### ♦ Preparation of amines (25.7)

### [1] Direct nucleophilic substitution with $NH_3$ and amines (25.7A)

- The mechanism is $S_N2$.
- The reaction works best for $CH_3X$ or $RCH_2X$.
- The reaction works best to prepare 1° amines and quaternary ammonium salts.

## [2] Gabriel synthesis (25.7A)

- The mechanism is $S_N2$.
- The reaction works best for $CH_3X$ or $RCH_2X$.
- Only 1° amines can be prepared.

R—X + the phthalimide anion → R—NH₂ (1° amine) + phthalate dianion

## [3] Reduction methods (25.7B)

[a] From nitro compounds

$$R-NO_2 \xrightarrow[\text{Fe, HCl or}]{\text{H}_2\text{, Pd-C or}} R-NH_2$$

Sn, HCl → R—NH₂ (1° amine)

[b] From nitriles

$$R-C\equiv N \xrightarrow[\text{[2] H}_2\text{O}]{\text{[1] LiAlH}_4} R-CH_2NH_2$$

1° amine

[c] From amides

$$R\overset{O}{\underset{}{\overset{\|}{C}}}NR'_2 \xrightarrow[\text{[2] H}_2\text{O}]{\text{[1] LiAlH}_4} RCH_2-\underset{R'}{\overset{}{N}}-R'$$

R' = H or alkyl

1°, 2°, and 3° amines

## [4] Reductive amination (25.7C)

$$\underset{R'}{\overset{R}{C}}=O + R_2''NH \xrightarrow{\text{NaBH}_3\text{CN}} R'-\overset{R}{\underset{H}{\underset{}{C}}}-\underset{R''}{N}-R''$$

R', R'' = H or alkyl

1°, 2°, and 3° amines

- Reductive amination adds one alkyl group (from an aldehyde or ketone) to a nitrogen nucleophile.
- Primary (1°), 2°, and 3° amines can be prepared.

♦ Reactions of amines

## [1] Reaction as a base (25.9)

$$R-\ddot{N}H_2 + H-A \rightleftharpoons R-\overset{+}{N}H_3 + :A^-$$

## [2] Nucleophilic addition to aldehydes and ketones (25.11)

With 1° amines:

$$\underset{R}{\overset{O}{\underset{}{\overset{\|}{C}}}}H \xrightarrow{\text{R'NH}_2} \underset{R}{\overset{NR'}{\underset{}{\overset{\|}{C}}}}H$$

R = H or alkyl

imine

With 2° amines:

$$\underset{R}{\overset{O}{\underset{}{\overset{\|}{C}}}}H \xrightarrow{\text{R'}_2\text{NH}} \underset{R}{\overset{NR'_2}{\underset{}{C}}}C$$

R = H or alkyl

enamine

## [3] Nucleophilic substitution with acid chlorides and anhydrides (25.11)

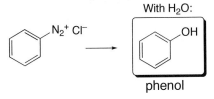

$$R-\overset{\overset{O}{\|}}{C}-Z \ + \ R'_2NH \ \longrightarrow \ R-\overset{\overset{O}{\|}}{C}-NR'_2$$

(2 equiv)

Z = Cl or OCOR
R' = H or alkyl

1°, 2°, and 3° amides

## [4] Hofmann elimination (25.12)

$$-\overset{\overset{|}{C}}{\underset{H}{|}}-\overset{\overset{|}{C}}{\underset{NH_2}{|}}- \quad \xrightarrow[\substack{[2] \ Ag_2O \\ [3] \ \Delta}]{[1] \ CH_3I \ (excess)} \quad \overset{}{C}{=}\overset{}{C}$$

alkene

- The less substituted alkene is the major product.

## [5] Reaction with nitrous acid (25.13)

With 1° amines:

$$R-NH_2 \ \xrightarrow[HCl]{NaNO_2} \ R-\overset{+}{N}{\equiv}N\text{:} \ \ Cl^-$$

alkyl diazonium salt

With 2° amines:

$$R-\overset{\overset{|}{R}}{N}-H \ \xrightarrow[HCl]{NaNO_2} \ R-\overset{\overset{|}{R}}{\ddot{N}}-\ddot{N}{=}\ddot{O}\text{:}$$

N-nitrosamine

## ◆ Reactions of diazonium salts

### [1] Substitution reactions (25.14)

With H₂O:

phenol

With CuX:

aryl chloride or
aryl bromide
X = Cl or Br

With HBF₄:

aryl fluoride

With NaI or KI:

aryl iodide

With CuCN:

benzonitrile

With H₃PO₂:

benzene

### [2] Coupling to form azo compounds (25.15)

Y =NH₂, NHR, NR₂, OH
(a strong electron-donor group)

azo compound

+ HCl

## Chapter 25: Answers to Problems

**25.1** Amines are classified as 1°, 2°, or 3° by the number of alkyl groups bonded to the *nitrogen* atom.

a. 

b. CH₃CH₂O—⟨...⟩—N—CH₃    3° amine

c. 

**25.2** The N atom of a quaternary ammonium salt is a stereogenic center when the N is surrounded by four different groups. All stereogenic centers are circled.

a. $CH_3$—$\overset{+}{N}$—$CH_2CH_2$—$\overset{+}{N}$—$CH_2CH_3$

N has 3 similar groups.

b. 

**25.3**

$CH_3$—$\ddot{N}H_2$

1.47 Å

The C–N bond is formed from two *sp³* hybridized atoms and the lone pair is localized on N.

Because the lone pair on N can be delocalized on the benzene ring, the C–N bond has partial double bond character, making it shorter. Both the C and N atoms must be *sp²* hybridized (+ have a *p* orbital) for delocalization to occur. The higher percent *s*-character in both C and N shortens the bond as well.

1.40 Å

+ 3 more resonance structures

partial double bond character

**25.4**

a. $CH_3CH_2CH(NH_2)CH_3$

2-butanamine
or
*sec*-butylamine

b. $(CH_3CH_2CH_2CH_2)_2NH$

dibutylamine

c. —N(CH₃)₂

*N,N*-dimethylcyclohexanamine

d. 

NH₂

2-methyl-5-nonanamine

e. 

NHCH₂CH₃

*N*-ethyl-3-hexanamine

f. 

CH₃

—NHCH₂CH₂CH₃

2-methyl-*N*-propylcyclopentanamine

**25.5** Aromatic amines are named as derivatives of aniline.

a. *N*-methylaniline

NHCH₃

b. *m*-ethylaniline

NH₂

CH₂CH₃

c. 3,5-diethylaniline

CH₃CH₂—⟨...⟩—NH₂

CH₂CH₃

d. *N,N*-diethylaniline

N(CH₂CH₃)₂

**25.6** An **NH₂** group named as a substituent is called an **amino group**.

a. 2,4-dimethyl-3-hexanamine    c. *N*-isopropyl-*p*-nitroaniline    e. *N,N*-dimethylethylamine    g. 1-propylcyclohexylamine

b. *N*-methylpentylamine    d. *N*-methylpiperidine    f. 2-aminocyclohexanone    h. *p*-butyl-*N*-ethylaniline

**25.7** Primary (1°) and 2° amines have higher bp's than similar compounds (like ethers) incapable of hydrogen bonding, but lower bp's than alcohols that have stronger intermolecular hydrogen bonds. Tertiary amines (3°) have lower boiling points than 1° and 2° amines of comparable molecular weight because they have no N–H bonds.

a. $(CH_3)_2CHCH_2CH_3$    $CH_3\overset{\overset{\displaystyle O}{\|}}{C}CH_2CH_3$    $(CH_3)_2CHCH_2NH_2$    b.

| alkane | ketone | amine | alkane | ether | amine |
| **lowest boiling point** | **intermediate boiling point** | N–H can hydrogen bond. **highest boiling point** | **lowest boiling point** | **intermediate boiling point** | N–H can hydrogen bond. **highest boiling point** |

**25.8** 1° Amines show *two* N–H absorptions at 3300–3500 cm⁻¹. 2° Amines show *one* N–H absorption at 3300–3500 cm⁻¹.

molecular weight = 59
one IR peak = 2° amine

$CH_3-\underset{\underset{\displaystyle H}{|}}{N}-CH_2CH_3$

**25.9** **The NH signal occurs between 0.5 and 5.0 ppm.** The protons on the carbon bonded to the amine nitrogen are deshielded and typically absorb at 2.3–3.0 ppm. The NH protons are not split.

molecular formula $C_6H_{15}N$
¹H NMR absorptions (ppm):
0.9 (singlet, 1H) ⟶ NH
1.10 (triplet, 3H) ⟶ CH₃ adjacent to CH₂
1.15 (singlet, 9H) ⟶ (CH₃)₃C
2.6 (quartet, 2H) ⟶ CH₂ adjacent to CH₃

**25.10** The atoms of 2-phenylethylamine are in bold.

a. $(CH_3CH_2)_2N$

**LSD**
lysergic acid diethyl amide

b.

**codeine**

**25.11** $S_N2$ reaction of an alkyl halide with $NH_3$ or an amine forms an amine.

a. (structure) $\xrightarrow[\text{excess}]{NH_3}$ (structure)   b. (cyclohexyl)$-NH_2$ $\xrightarrow[\text{excess}]{CH_3CH_2Br}$ (cyclohexyl)$-\overset{+}{N}(CH_2CH_3)_3$ $Br^-$

**25.12**

A          B          C

**25.13** The Gabriel synthesis converts an alkyl halide into a 1° amine by a two-step process: nucleophilic substitution followed by hydrolysis.

a. (structure) $NH_2$ $\Downarrow$ (structure) Br

b. $(CH_3)_2CHCH_2CH_2NH_2$ $\Downarrow$ $(CH_3)_2CHCH_2CH_2Br$

c. $CH_3O$-(structure)-$NH_2$ $\Rightarrow$ $CH_3O$-(structure)-$Br$

**25.14** **Nitriles are reduced to 1° amines with LiAlH₄. Nitro groups are reduced to 1° amines** using a variety of reducing agents. **Primary (1°), 2°, and 3° amides are reduced to 1°, 2°, and 3° amines** respectively, using LiAlH₄.

a.   $CH_3CHCH_2NH_2$ $\Rightarrow$ $CH_3CHCH_2NO_2$   $CH_3CHC\equiv N$   $CH_3CH\overset{O}{\overset{\|}{C}}NH_2$
    $\underset{CH_3}{}$      $\underset{CH_3}{}$      $\underset{CH_3}{}$      $\underset{CH_3}{}$

b.   (cyclohexyl)$-CH_2NH_2$ $\Rightarrow$ (cyclohexyl)$-CH_2NO_2$   (cyclohexyl)$-C\equiv N$   (cyclohexyl)$-\overset{O}{\overset{\|}{C}}-NH_2$

c.   (structure)$NH_2$ $\Rightarrow$ (structure)$NO_2$   (structure)$\overset{N}{\overset{\||}{C}}$   (structure)$\overset{O}{\overset{\|}{C}}NH_2$

**25.15** **Primary (1°), 2°, and 3° amides are reduced to 1°, 2°, and 3° amines** respectively, using LiAlH₄.

a.   (structure)$CONH_2$ $\longrightarrow$ (structure)$CH_2NH_2$

c.   (structure)$\overset{O}{\overset{\|}{C}}NHCH_3$ $\longrightarrow$ (structure)$NHCH_3$

b.   (structure)$\overset{O}{\overset{\|}{}}$N(structure) $\longrightarrow$ (structure)N(structure)

**25.16**

$(CH_3)_2CHNH_2$
$\uparrow$
**isopropylamine**

General reaction:

$R-C\equiv N \xrightarrow{[H]} R\overset{}{C}H_2NH_2$
                     $\uparrow$
The amine needs 2 H's here.

The C bonded to the N must have 2 H's to be formed by reduction of a nitrile.

**25.17** Reductive amination is a two-step method that converts aldehydes and ketones into 1°, 2°, and 3° amines. Reductive amination replaces a C=O by a C–H and C–N bond.

a. 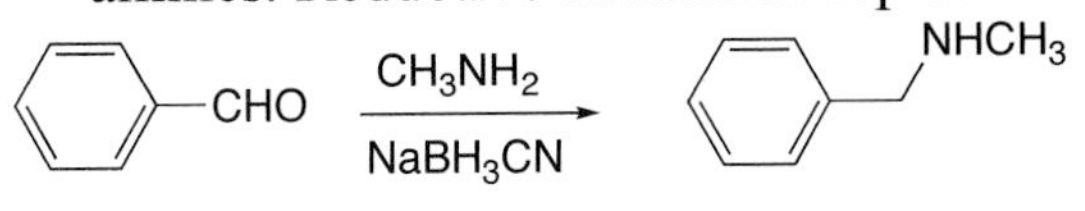  CHO $\xrightarrow[\text{NaBH}_3\text{CN}]{\text{CH}_3\text{NH}_2}$  ... NHCH$_3$

c. 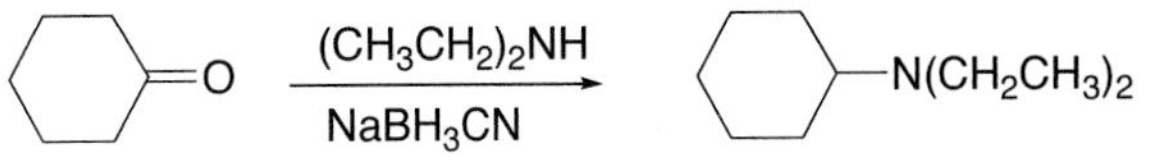 =O $\xrightarrow[\text{NaBH}_3\text{CN}]{(\text{CH}_3\text{CH}_2)_2\text{NH}}$ ... –N(CH$_2$CH$_3$)$_2$

b. 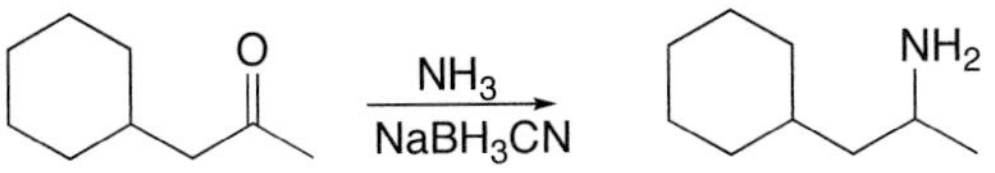 $\xrightarrow[\text{NaBH}_3\text{CN}]{\text{NH}_3}$ ... NH$_2$

d. 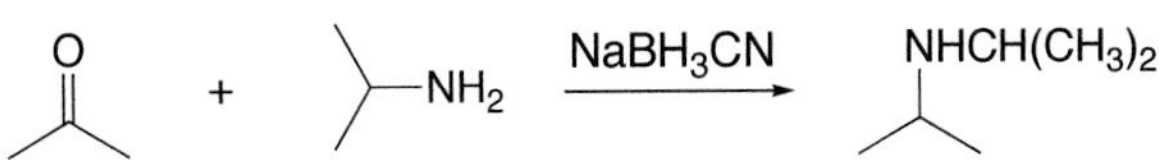 + –NH$_2$ $\xrightarrow{\text{NaBH}_3\text{CN}}$ NHCH(CH$_3$)$_2$

**25.18** In reductive amination, one alkyl group on N comes from the carbonyl compound. The remainder of the molecule comes from NH$_3$ or an amine.

a. 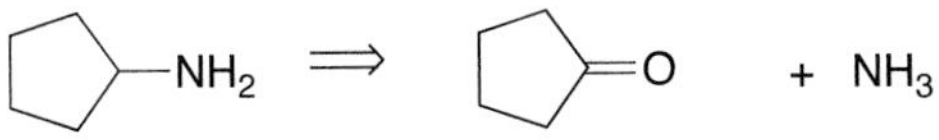 –NH$_2$ $\Longrightarrow$ =O + NH$_3$

c. 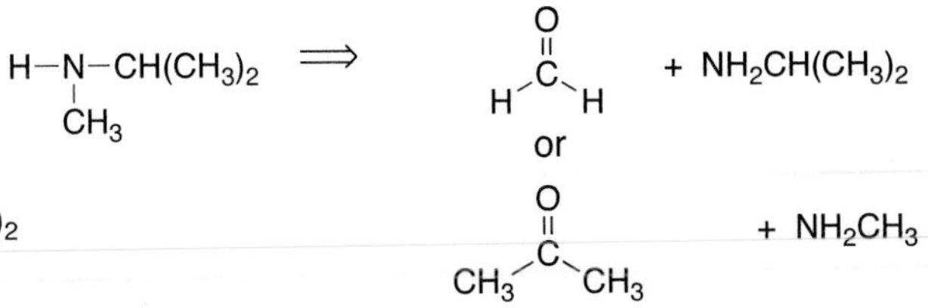 H–N–CH(CH$_3$)$_2$ (CH$_3$) $\Longrightarrow$ + NH$_2$CH(CH$_3$)$_2$
or
+ NH$_2$CH$_3$

b. 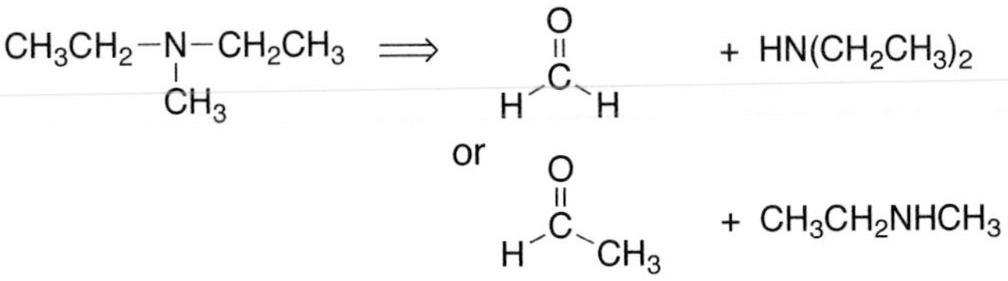 CH$_3$CH$_2$–N–CH$_2$CH$_3$ (CH$_3$) $\Longrightarrow$ + HN(CH$_2$CH$_3$)$_2$
or
+ CH$_3$CH$_2$NHCH$_3$

**25.19**

a.

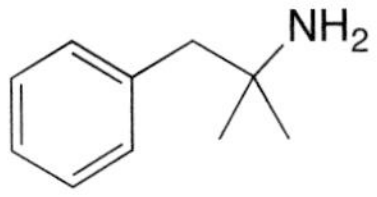

phentermine

Only amines that have a C bonded to a H and N atom can be made by reductive amination; that is, an amine must have the following structural feature:

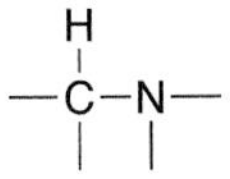

In phentermine, the C bonded to N is not bonded to a H, so it cannot be made by reductive amination.

b. 2-methyl-1-phenyl-2-propanamine

**25.20**

a. 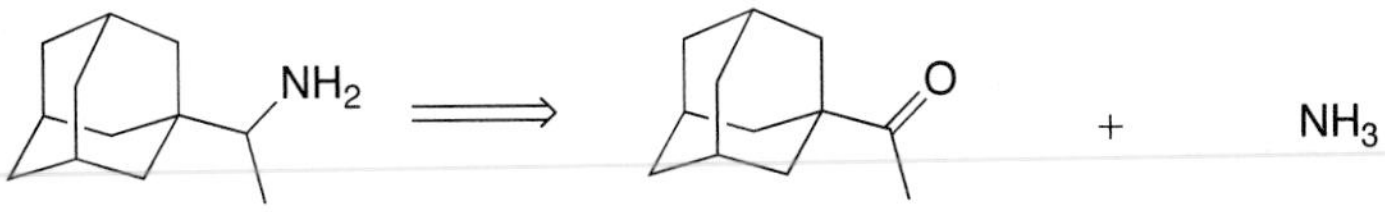 NH$_2$ $\Longrightarrow$ O + NH$_3$

b. 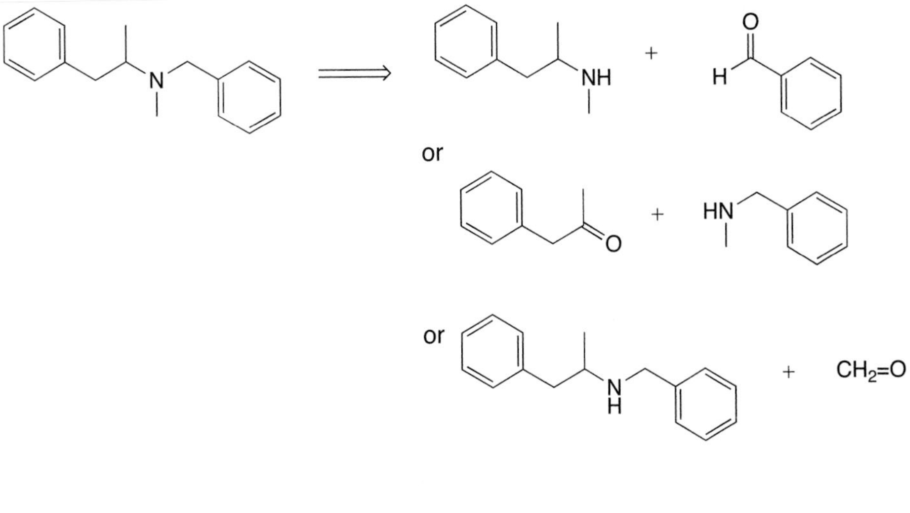 $\Longrightarrow$ NH +
or
+ HN
or
+ CH$_2$=O

**25.21** The $pK_a$ of many protonated amines is 10–11, so the $pK_a$ of the starting acid must be **less than 10** for equilibrium to favor the products. Amines are thus readily protonated by strong inorganic acids like HCl and $H_2SO_4$, and by carboxylic acids as well.

a. $CH_3CH_2CH_2CH_2-NH_2$ + HCl $\rightleftharpoons$ $CH_3CH_2CH_2CH_2-\overset{+}{N}H_3$ + Cl⁻

$pK_a = -7$        $pK_a \approx 10$
weaker acid
products favored

c. [piperidine] + $H_2O$ $\rightleftharpoons$ [protonated piperidine] + HO⁻

$pK_a = 15.7$    $pK_a \approx 10$
weaker acid
reactants favored

b. $C_6H_5COOH$ + $(CH_3)_2NH$ $\rightleftharpoons$ $(CH_3)_2\overset{+}{N}H_2$ + $C_6H_5COO^-$

$pK_a = 4.2$        $pK_a = 10.7$
weaker acid
products favored

**25.22** An amine can be separated from other organic compounds by converting it to a water-soluble ammonium salt by an acid–base reaction. In each case, the extraction procedure would employ the following steps:
- Dissolve the amine and either **X** or **Y** in $CH_2Cl_2$.
- Add a solution of 10% HCl. The amine will be protonated and dissolve in the aqueous layer, while **X** or **Y** will remain in the organic layer as a neutral compound.
- Separate the layers.

a. [C₆H₁₁–NH₂] and [C₆H₁₁–CH₃] $\xrightarrow{H-Cl}$ [C₆H₁₁–$\overset{+}{N}H_3$ Cl⁻] + [C₆H₁₁–CH₃]

                       **X**                            **X**

- **soluble in $H_2O$**       • insoluble in $H_2O$
- **insoluble in $CH_2Cl_2$**    • soluble in $CH_2Cl_2$

b. $(CH_3CH_2CH_2CH_2)_3N$ and $(CH_3CH_2CH_2CH_2)_2O$ $\xrightarrow{H-Cl}$ $(CH_3CH_2CH_2CH_2)_3\overset{+}{N}H$ Cl⁻ + $(CH_3CH_2CH_2CH_2)_2O$

                   **Y**                                 **Y**

- **soluble in $H_2O$**       • insoluble in $H_2O$
- **insoluble in $CH_2Cl_2$**    • soluble in $CH_2Cl_2$

**25.23** The weaker the conjugate acid, the higher its $pK_a$ and the stronger the base. ($pK_a$ values are for the conjugate acid of a given amine.)

a. $(CH_3CH_2)_2NH$ ($pK_a = 11.1$) and $CH_3CH_2NH_2$ ($pK_a = 10.8$)    b. $(CH_3CH_2)_3N$ ($pK_a = 11.0$) and $(CH_3)_3N$ ($pK_a = 10.6$)

   weaker conjugate acid       stronger conjugate acid       weaker conjugate acid       stronger conjugate acid
     **stronger base**           weaker base              **stronger base**           weaker base

**25.24** Primary (1°), 2°, and 3° alkylamines are more basic than $NH_3$ because of the electron-donating inductive effect of the R groups.

a. $(CH_3)_2NH$ and $NH_3$       b. $CH_3CH_2NH_2$ and $ClCH_2CH_2NH_2$

   2° alkylamine                      1° alkylamine        1° alkylamine
CH₃ groups are electron donating.     **stronger base**     Cl is electron withdrawing.
     **stronger base**                                         **weaker base**

**25.25** Arylamines are less basic than alkylamines because the electron pair on N is delocalized. Electron-donor groups add electron density to the benzene ring making the arylamine more basic than aniline. Electron-withdrawing groups remove electron density from the benzene ring, making the arylamine less basic than aniline.

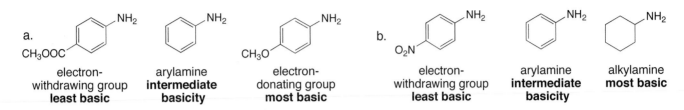

a.

| electron-withdrawing group **least basic** | arylamine **intermediate basicity** | electron-donating group **most basic** |

b.

| electron-withdrawing group **least basic** | arylamine **intermediate basicity** | alkylamine **most basic** |

**25.26** Amides are much less basic than amines because the electron pair on N is highly delocalized.

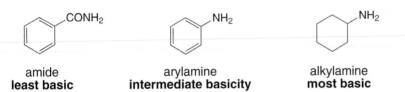

| amide **least basic** | arylamine **intermediate basicity** | alkylamine **most basic** |

**25.27**

a. Electron pair on N occupies an $sp^2$ hybrid orbital.

$sp^2$ hybridized more basic

This N is also $sp^2$ hybridized but the electron pair occupies a $p$ orpital so it can delocalize onto the aromatic ring. Delocalization makes this N less basic.

DMAP
4-(N,N-dimethylamino)pyridine

b. nicotine

$sp^3$ hybridized N stronger base

$sp^2$ hybridized N higher percent s-character weaker base

**25.28**

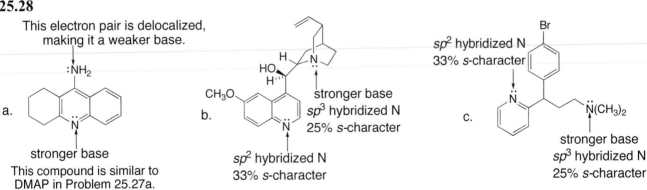

a. This electron pair is delocalized, making it a weaker base.

stronger base
This compound is similar to DMAP in Problem 25.27a.

b. stronger base
$sp^3$ hybridized N
25% s-character

$sp^2$ hybridized N
33% s-character

c. $sp^2$ hybridized N
33% s-character

stronger base
$sp^3$ hybridized N
25% s-character

**25.29**

Order of basicity: $N_b < N_a < N_c$

$N_b$ – The electron pair on this N atom is delocalized on the aromatic five-membered ring; least basic.

$N_a$ – The electron pair on this N atom is not delocalized, but is on an $sp^2$ hybridized atom.

$N_c$ – The electron pair on this N atom is on an $sp^3$ hybridized N; most basic.

**25.30** Amines attack carbonyl groups to form products of nucleophilic addition or substitution.

a. (cyclohexanone) $\xrightarrow{CH_3CH_2CH_2NH_2}$ (=NCH_2CH_2CH_3)   (cyclohexanone) $\xrightarrow{(CH_3CH_2)_2NH}$ (–N(CH_2CH_3)_2)

b. 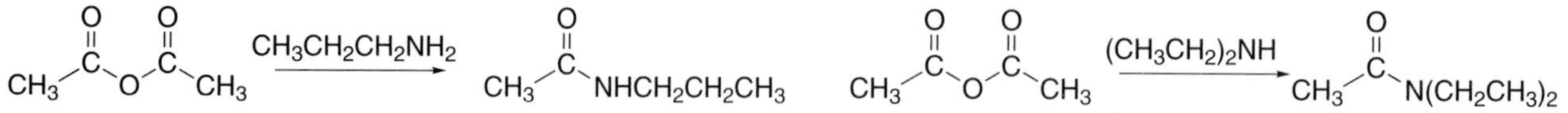 $CH_3$–C(=O)–O–C(=O)–$CH_3$ $\xrightarrow{CH_3CH_2CH_2NH_2}$ $CH_3$–C(=O)–$NHCH_2CH_2CH_3$   $CH_3$–C(=O)–O–C(=O)–$CH_3$ $\xrightarrow{(CH_3CH_2)_2NH}$ $CH_3$–C(=O)–$N(CH_2CH_3)_2$

c. (Ph)COCl $\xrightarrow{CH_3CH_2CH_2NH_2}$ (Ph)CONHCH_2CH_2CH_3   (Ph)COCl $\xrightarrow{(CH_3CH_2)_2NH}$ (Ph)CON(CH_2CH_3)_2

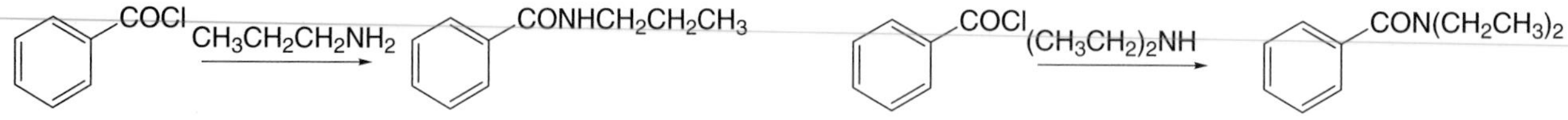

**25.31** [1] Convert the amine (aniline) into an amide (acetanilide).
  [2] **Carry out the Friedel–Crafts reaction.**
  [3] **Hydrolyze the amide** to generate the free amino group.

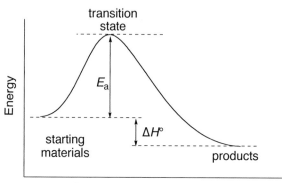

a. 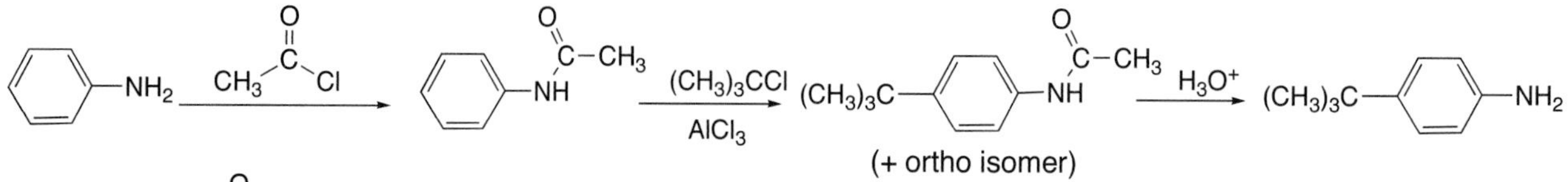 (Ph)–$NH_2$ $\xrightarrow{CH_3-C(=O)-Cl}$ (Ph)–NH–C(=O)–$CH_3$ $\xrightarrow[AlCl_3]{(CH_3)_3CCl}$ $(CH_3)_3C$–(Ph)–NH–C(=O)–$CH_3$ $\xrightarrow{H_3O^+}$ $(CH_3)_3C$–(Ph)–$NH_2$
(+ ortho isomer)

b. 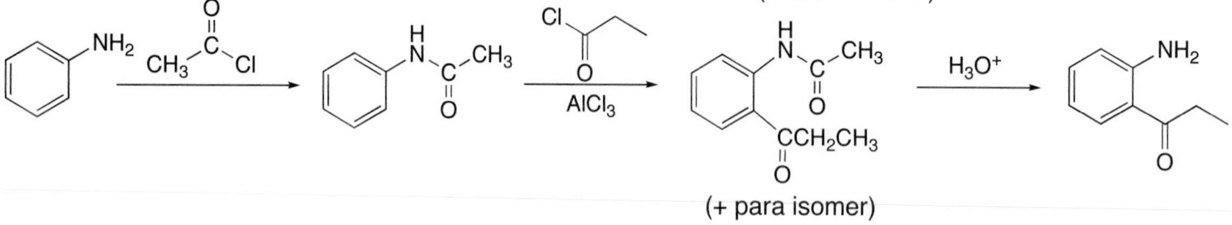 (Ph)$NH_2$ $\xrightarrow{CH_3-C(=O)-Cl}$ (Ph)N(H)–C(=O)–$CH_3$ $\xrightarrow[AlCl_3]{Cl-C(=O)CH_2CH_3}$ (Ph)(N(H)–C(=O)CH_3)(CCH_2CH_3=O) $\xrightarrow{H_3O^+}$ (Ph)($NH_2$)(C(=O)CH_2CH_3)
(+ para isomer)

**25.32**

transition state:

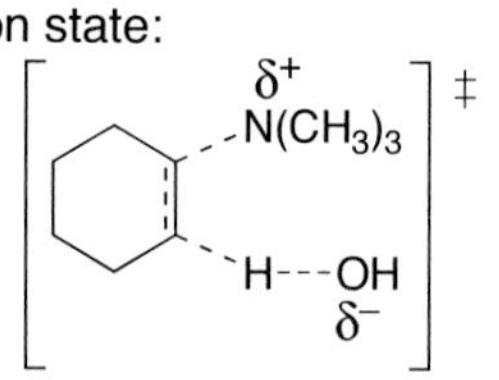

$$\left[ \begin{array}{c} \overset{\delta^+}{\text{–N(CH}_3)_3} \\ \\ \overset{}{\text{H---OH}} \\ \delta^- \end{array} \right]^{\ddagger}$$

(no 3-D geometry shown here)

*(Energy vs. Reaction coordinate diagram)*
transition state — $E_a$ — starting materials — $\Delta H°$ — products

**25.33**

a. $CH_3CH_2CH_2CH_2-NH_2$ $\xrightarrow[\substack{[2]\,Ag_2O \\ [3]\,\Delta}]{[1]\,CH_3I\,(excess)}$ $CH_3CH_2CH=CH_2$

c. 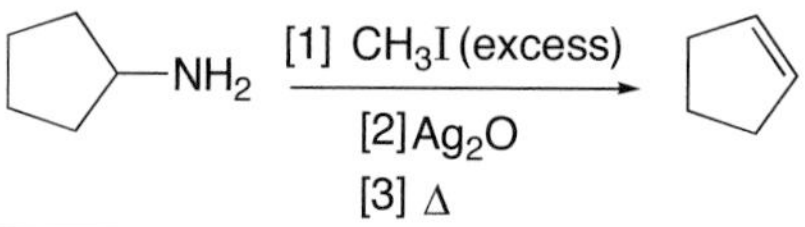 (cyclopentyl)–$NH_2$ $\xrightarrow[\substack{[2]\,Ag_2O \\ [3]\,\Delta}]{[1]\,CH_3I\,(excess)}$ (cyclopentene)

b. $(CH_3)_2CHNH_2$ $\xrightarrow[\substack{[2]\,Ag_2O \\ [3]\,\Delta}]{[1]\,CH_3I\,(excess)}$ $CH_3CH=CH_2$

**25.34** In a Hofmann elimination, the base removes a proton from the less substituted, more accessible β carbon atom, because of the bulky leaving group on the nearby α carbon.

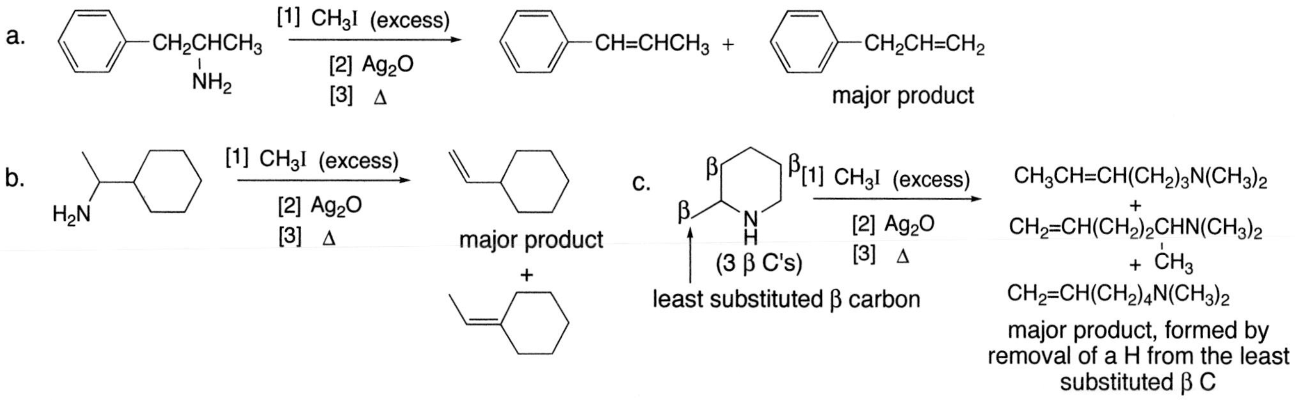

a.

CH$_2$CHCH$_3$ / NH$_2$ →[1] CH$_3$I (excess) [2] Ag$_2$O [3] Δ → CH=CHCH$_3$ + CH$_2$CH=CH$_2$ major product

b.

→[1] CH$_3$I (excess) [2] Ag$_2$O [3] Δ → major product +

c.

β β [1] CH$_3$I (excess) [2] Ag$_2$O [3] Δ

(3 β C's)

least substituted β carbon

CH$_3$CH=CH(CH$_2$)$_3$N(CH$_3$)$_2$
+
CH$_2$=CH(CH$_2$)$_2$CHN(CH$_3$)$_2$
+ CH$_3$
CH$_2$=CH(CH$_2$)$_4$N(CH$_3$)$_2$

major product, formed by removal of a H from the least substituted β C

**25.35**

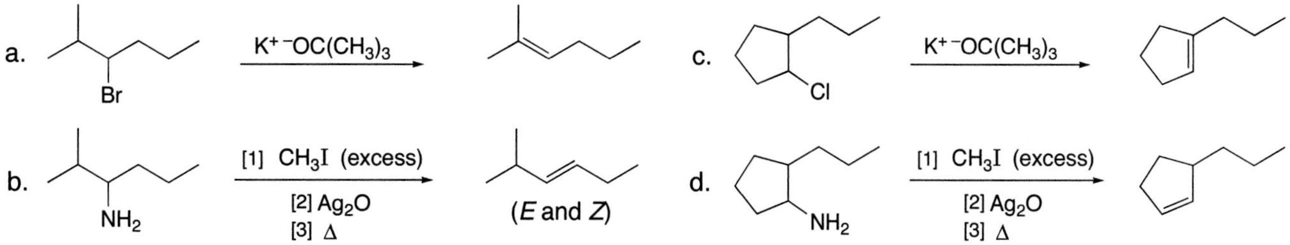

a.

→ K$^+$ $^-$OC(CH$_3$)$_3$ →

c.

→ K$^+$ $^-$OC(CH$_3$)$_3$ →

b.

NH$_2$ →[1] CH$_3$I (excess) [2] Ag$_2$O [3] Δ → (E and Z)

d.

NH$_2$ →[1] CH$_3$I (excess) [2] Ag$_2$O [3] Δ →

**25.36**

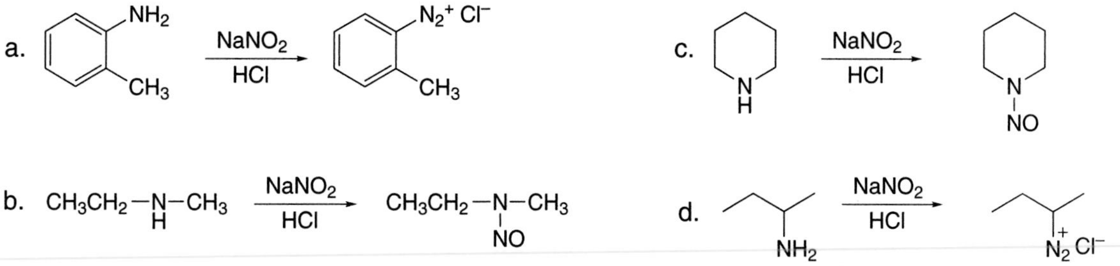

a.

NH$_2$ / CH$_3$ →NaNO$_2$ / HCl → N$_2^+$ Cl$^-$ / CH$_3$

c.

N / H →NaNO$_2$ / HCl → N / NO

b. CH$_3$CH$_2$—N—CH$_3$ / H →NaNO$_2$ / HCl → CH$_3$CH$_2$—N—CH$_3$ / NO

d.

NH$_2$ →NaNO$_2$ / HCl → N$_2^+$ Cl$^-$

**25.37**

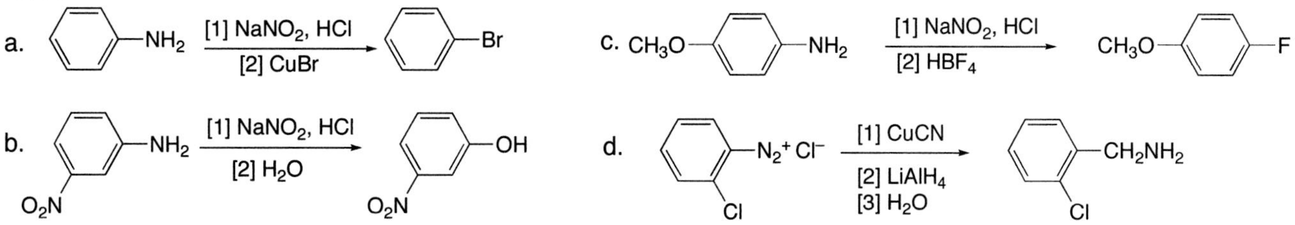

a.

—NH$_2$ →[1] NaNO$_2$, HCl [2] CuBr → —Br

c. CH$_3$O—

—NH$_2$ →[1] NaNO$_2$, HCl [2] HBF$_4$ → CH$_3$O— —F

b.

—NH$_2$ / O$_2$N →[1] NaNO$_2$, HCl [2] H$_2$O → —OH / O$_2$N

d.

—N$_2^+$ Cl$^-$ / Cl →[1] CuCN [2] LiAlH$_4$ [3] H$_2$O → —CH$_2$NH$_2$ / Cl

**25.38**

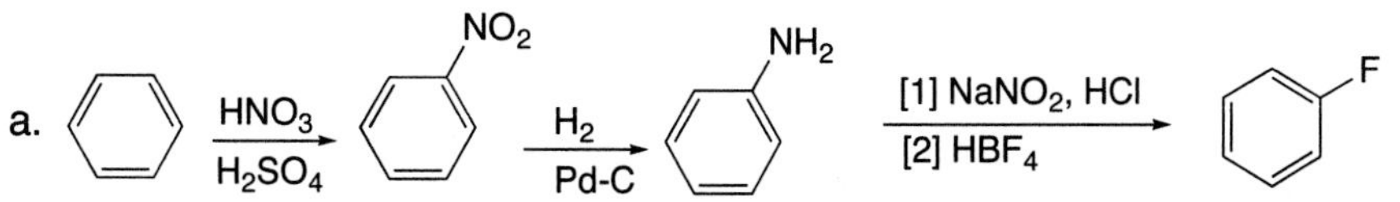

a.

→HNO$_3$ / H$_2$SO$_4$ → NO$_2$ →H$_2$ / Pd-C → NH$_2$ →[1] NaNO$_2$, HCl [2] HBF$_4$ → F

b.

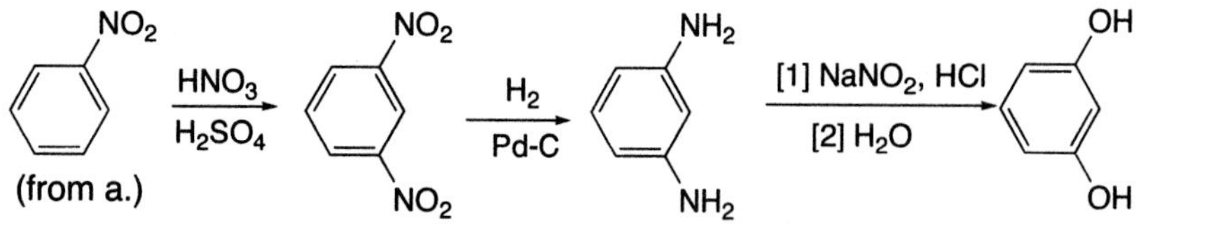

(from a.)

c.

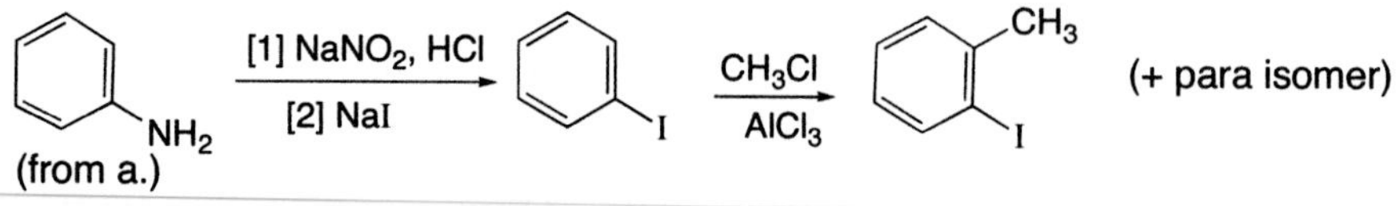

(from a.)

(+ para isomer)

d.

(from a.)

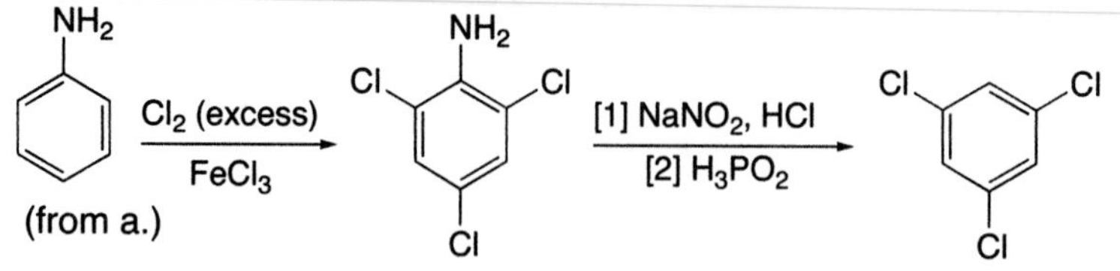

**25.39**

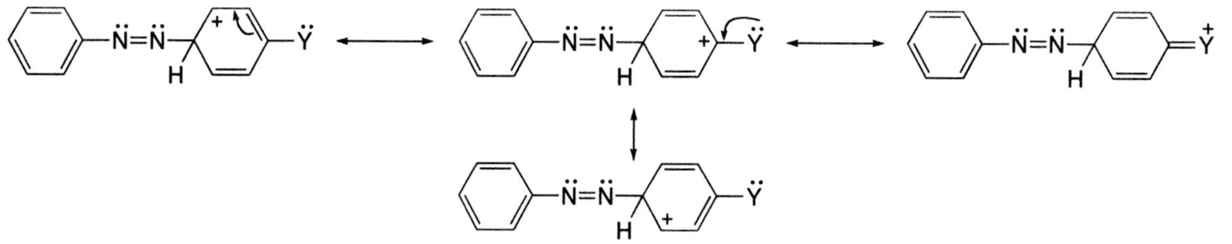

**25.40**

a.

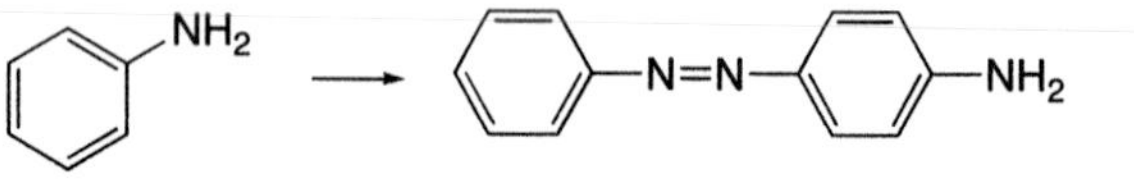

c.

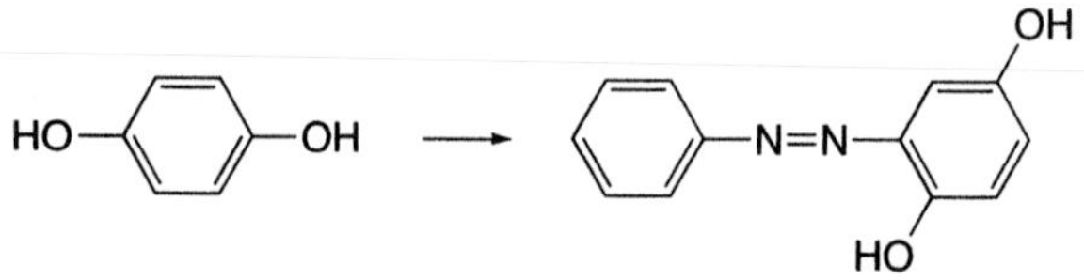

b.

**25.41** To determine what starting materials are needed to synthesize a particular azo compound, always divide the molecule into two components: **one has a benzene ring with a diazonium ion, and one has a benzene ring with a very strong electron-donor group.**

a.

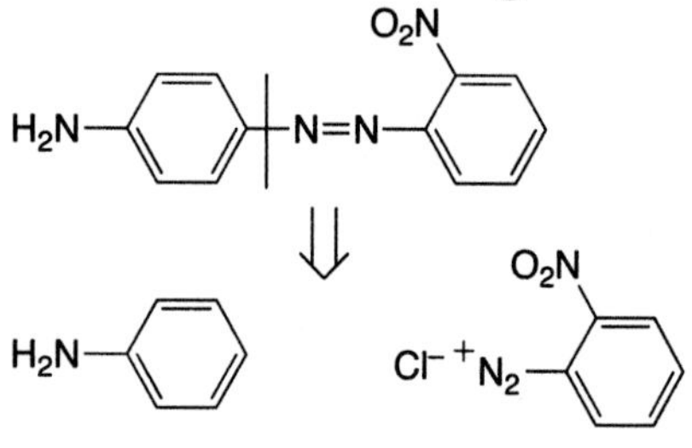

b.

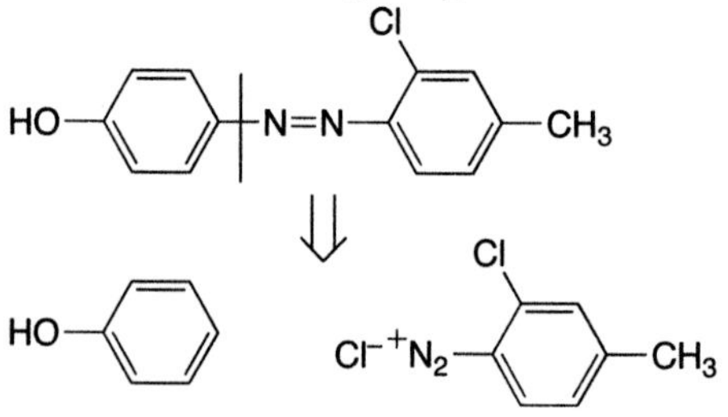

**25.42**

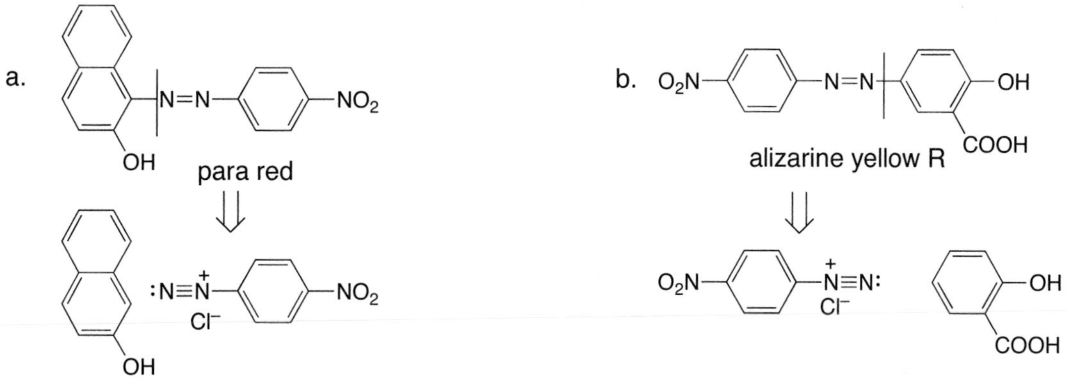

a.

para red

b.

alizarine yellow R

**25.43**

Dacron

methyl orange

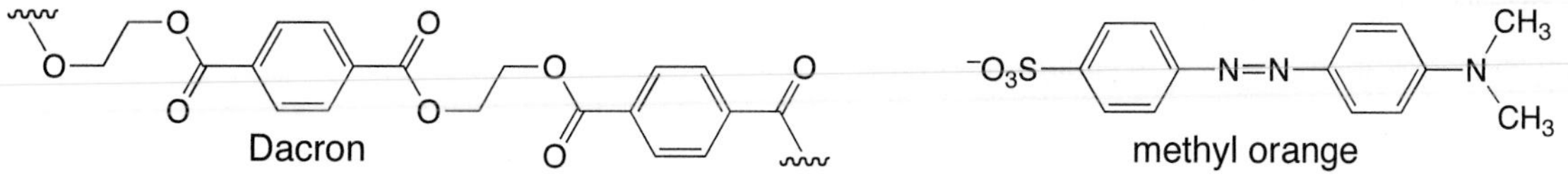

To bind to fabric, methyl orange (an anion) needs to interact with positively charged sites. Since Dacron is a neutral compound with no cationic sites on the chain, it does not bind methyl orange well.

## 25.44

a. $CH_3NHCH_2CH_2CH_2CH_3$

   N-methyl-1-butanamine
   (N-methylbutylamine)

b.

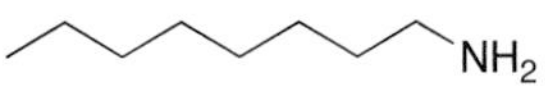

   1-octanamine
   (octylamine)

c.

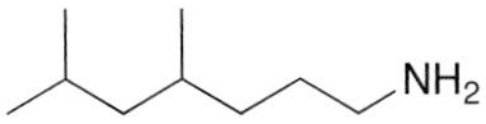

   4,6-dimethyl-1-heptanamine

d.

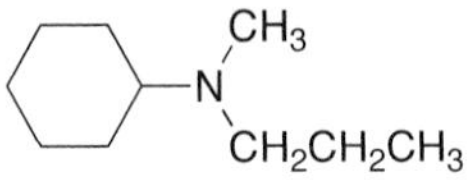

   N-methyl-N-propylcyclohexanamine

e. $(CH_3CH_2CH_2)_3N$

   tripropylamine

f. $(C_6H_5)_2NH$

   diphenylamine

g.

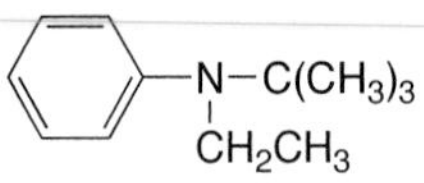

   N-tert-butyl-N-ethylaniline

h.

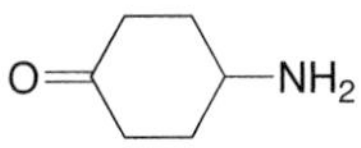

   4-aminocyclohexanone

i.

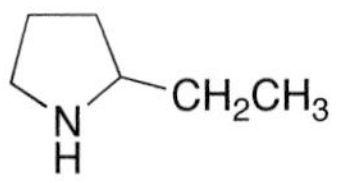

   2-ethylpyrrolidine

j. $CH_3CH_2CH_2CH(NH_2)CH(CH_3)_2$

   2-methyl-3-hexanamine

k.

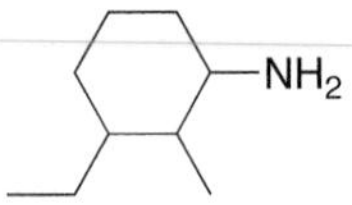

   3-ethyl-2-methylcyclohexanamine

l.

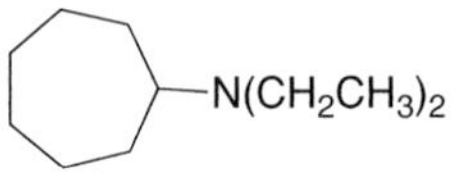

   N,N-diethylcycloheptanamine

## 25.45

a. cyclobutylamine

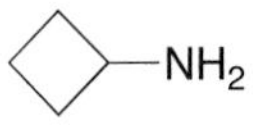

b. N-isobutylcyclopentylamine

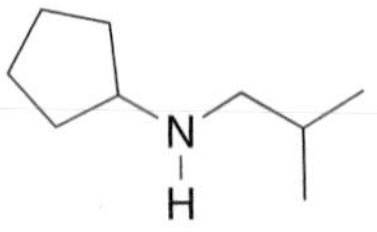

c. tri-tert-butylamine

   $N[C(CH_3)_3]_3$

d. N,N-diisopropylaniline

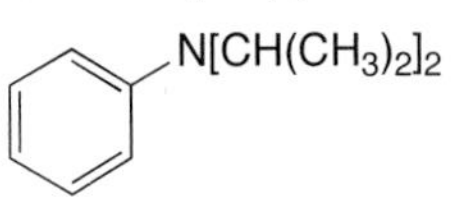

e. N-methylpyrrole

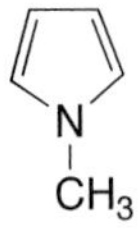

f. N-methylcyclopentylamine

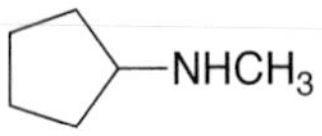

g. cis-2-aminocyclohexanol

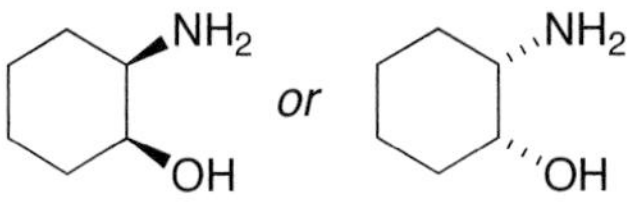

or

h. 3-methyl-2-hexanamine

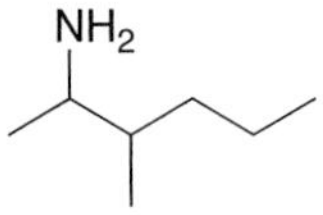

i. 2-sec-butylpiperidine

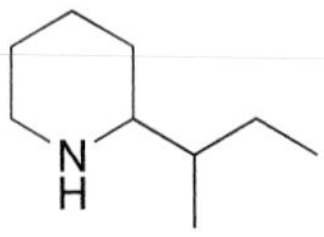

j. (S)-2-heptanamine

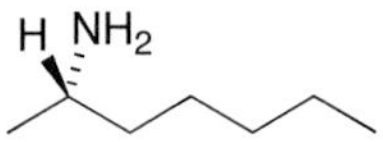

## 25.46 [* denotes a stereogenic center.]

a.

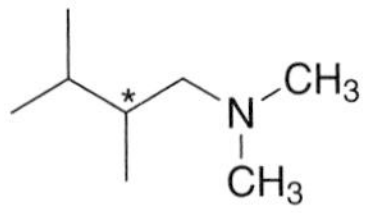

   1 stereogenic center
   2 stereoisomers

b.

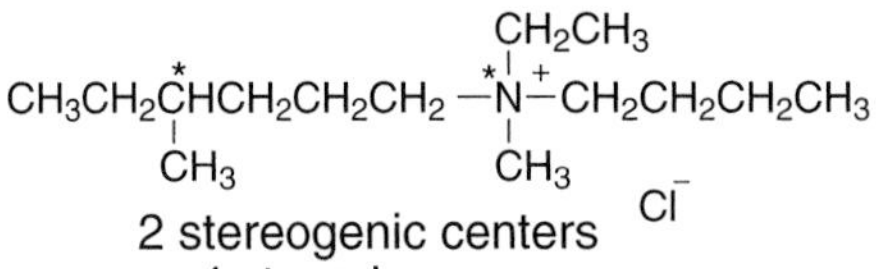

   2 stereogenic centers
   4 stereoisomers

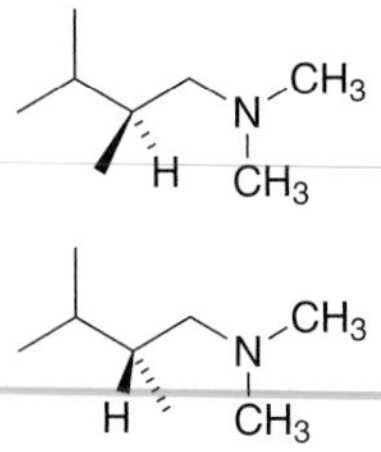

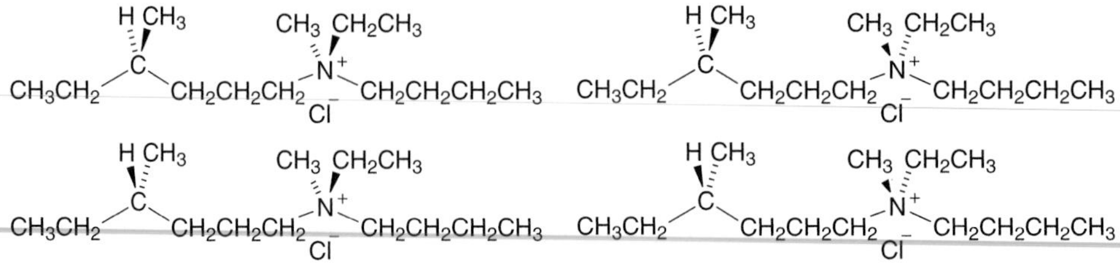

## 25.47

a. $(CH_3CH_2)_2NH$ or [pyridine structure]

$sp^3$ hybridized N    $sp^2$ hybridized N
**stronger base**    **weaker base**

b. $HCON(CH_3)_2$ or $(CH_3)_3N$

amide    alkylamine
**weaker base**   **stronger base**

c. $(CH_3CH_2)_2NH$    or    $(ClCH_2CH_2)_2NH$

2° alkylamine    2° alkylamine
**stronger base**    Cl is electron withdrawing.
           **weaker base**

d. [tetrahydroquinoline structure] or [tetrahydroisoquinoline structure]

**weaker base**      **stronger base**
(delocalized electron
pair on N)

## 25.48

a. [aniline structure] —NH₂    $NH_3$    [cyclohexylamine structure] —NH₂

arylamine   **intermediate**   alkylamine
**least basic**    **basicity**    **most basic**

b. [indole structure]   [quinoline structure]   [piperidine structure]

delocalized    $sp^2$ hybridized N   $sp^3$ hybridized N
electron pair on N   **intermediate**    **most basic**
**least basic**       **basicity**

c. [4-nitroaniline, $O_2N$]   [4-chloroaniline, Cl]   [4-methylaniline, $CH_3$]

electron-
withdrawing group    **intermediate**    electron-
**least basic**      **basicity**     donating group
                                      **most basic**

d. $(C_6H_5)_2NH$     $C_6H_5NH_2$     [cyclohexylamine]—NH₂

diarylamine    arylamine    alkylamine
**least basic**   **intermediate**   **most basic**
             **basicity**

## 25.49

a. $sp^3$ hybridized N
stronger base
$(CH_3CH_2)_2\overset{\downarrow}{N}CH_2CH_2O$—[benzoate with $NH_2$]

delocalized
electron pair

b. $sp^3$ hybridized N
stronger base
[pyridoxine structure: $\overset{\downarrow}{NH_2}$, HO, OH, $CH_3$, N]
$sp^2$ hybridized N

c. $sp^3$ hybridized N
stronger base
[dimethisoquin structure with O—CH₂—$\overset{\downarrow}{N(CH_3)_2}$, $N \leftarrow sp^2$ hybridized N]
dimethisoquin

## 25.50

[isoniazid structure with labeled nitrogens $N_c$, $NHNH_2$, $N_b$, $N_a$]
$N_b < N_a < N_c$

Order of basicity: $N_b < N_a < N_c$
$N_b$ – The electron pair on this N atom is delocalized
on the O atom; least basic.
$N_a$ – The electron pair on this N atom is not
delocalized, but is on an $sp^2$ hybridized atom.
$N_c$ – The electron pair on this N atom is on an $sp^3$
hybridized N; most basic.

**25.51** The para isomer is the weaker base because the electron pair on its $NH_2$ group can be delocalized onto the $NO_2$ group. In the meta isomer, no resonance structure places the electron pair on the $NO_2$ group, and fewer resonance structures can be drawn:

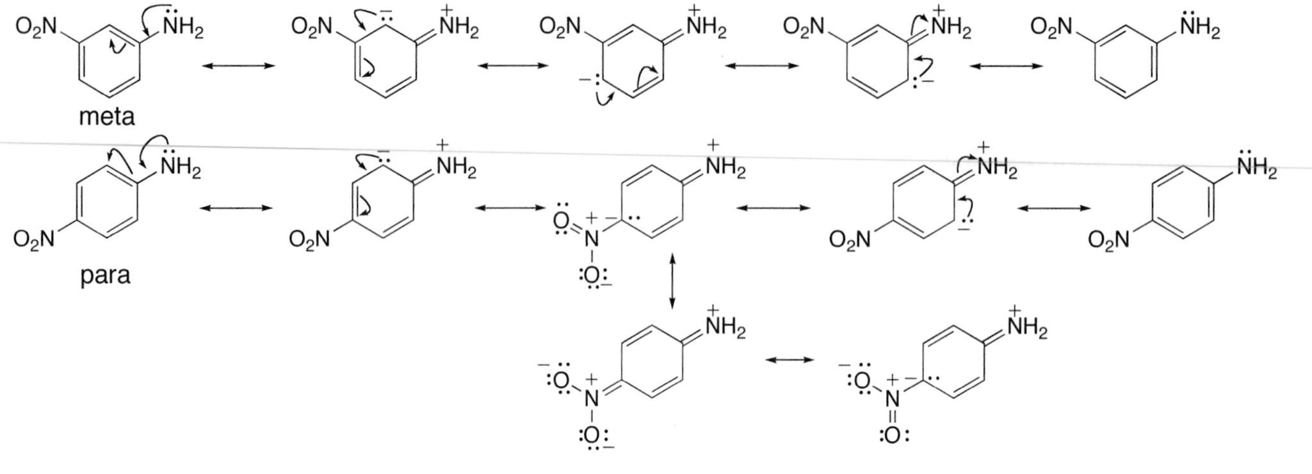

meta

para

**25.52**

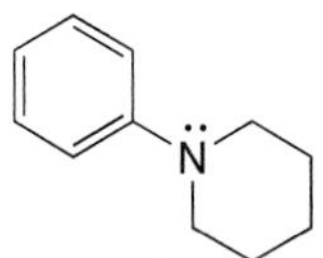

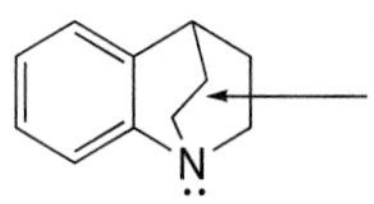

This two-carbon bridge makes it difficult for the lone pair on N to delocalize on the aromatic ring.

**A**

$pK_a$ of the conjugate acid = 5.2
stronger conjugate acid
**weaker base**
The electron pair of this arylamine is delocalized on the benzene ring, decreasing its basicity.

**B**

$pK_a$ of the conjugate acid = 7.29
weaker conjugate acid
**stronger base**

Resonance structures that place a double bond between the N atom and the benzene ring are destabilized. Since the electron pair is more localized on N, compound **B** is more basic.

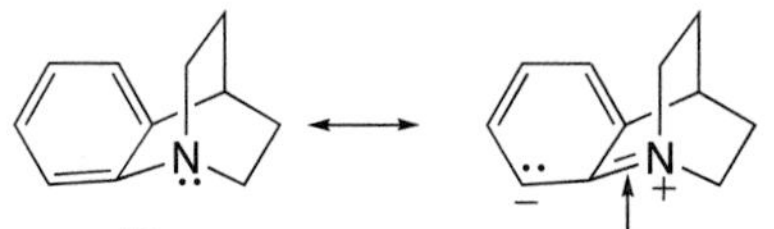

**B**

Geometry makes it difficult to have a double bond here.

**25.53**

a. $C_6H_5CH_2CH_2CH_2Br \xrightarrow[\text{excess}]{NH_3} C_6H_5CH_2CH_2CH_2NH_2$

b. $C_6H_5CH_2CH_2Br \xrightarrow{NaCN} C_6H_5CH_2CH_2CN \xrightarrow[\text{[2] } H_2O]{\text{[1] LiAlH}_4} C_6H_5CH_2CH_2CH_2NH_2$

c. $C_6H_5CH_2CH_2CH_2NO_2 \xrightarrow[\text{Pd-C}]{H_2} C_6H_5CH_2CH_2CH_2NH_2$

d. $C_6H_5CH_2CH_2CONH_2 \xrightarrow[\text{[2] } H_2O]{\text{[1] LiAlH}_4} C_6H_5CH_2CH_2CH_2NH_2$

e. $C_6H_5CH_2CH_2CHO \xrightarrow[\text{NaBH}_3CN]{NH_3} C_6H_5CH_2CH_2CH_2NH_2$

**25.54**

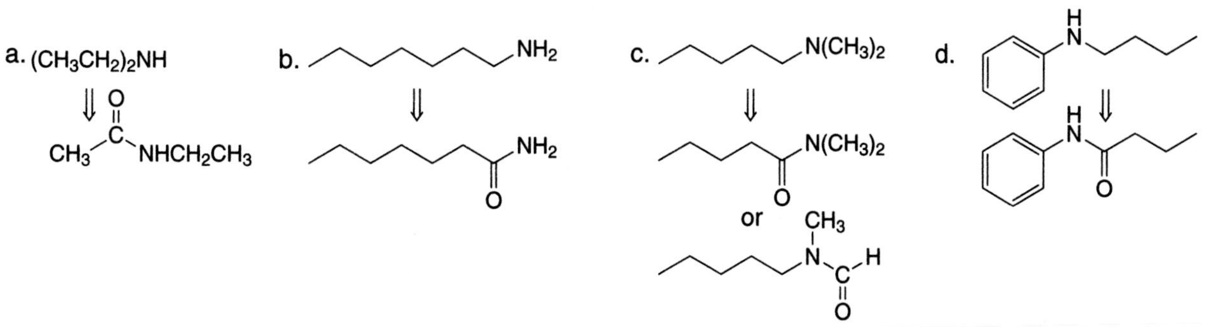

a. (CH₃CH₂)₂NH

$$\Downarrow$$

CH₃—C(=O)—NHCH₂CH₃

b. ~~~NH₂

$$\Downarrow$$

~~~C(=O)—NH₂

c. ~~~N(CH₃)₂

$$\Downarrow$$

~~~C(=O)—N(CH₃)₂

or

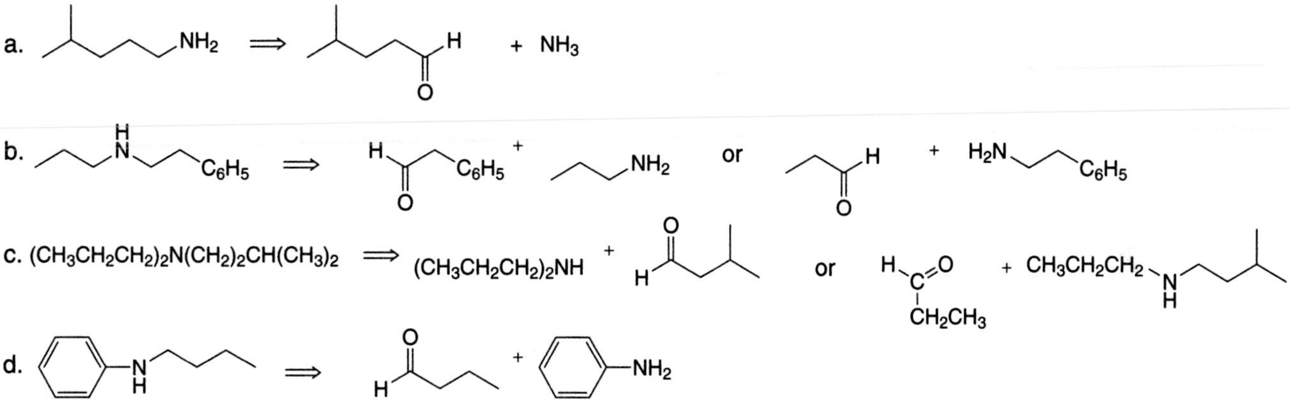

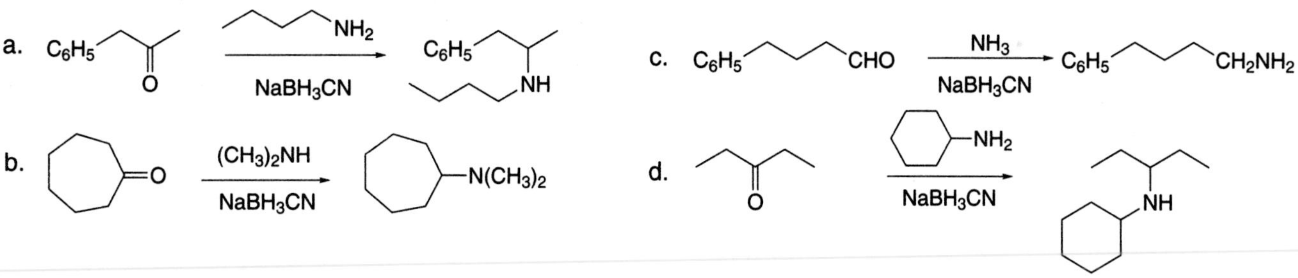

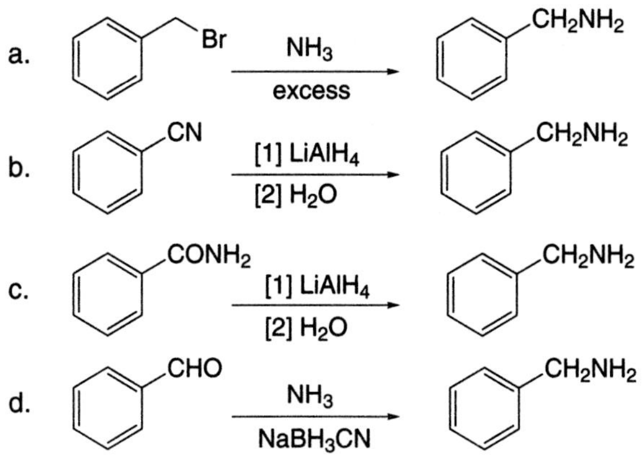

~~~N(CH₃)—CH=O

d. C₆H₅—NH—~~~ (N-butyl aniline)

$$\Downarrow$$

C₆H₅—NH—C(=O)~~~

**25.55**  Use the directions from Answer 25.18.

a. ⤳NH₂  ⟹  ⤳CHO  +  NH₃

b. ~~N(H)~~C₆H₅  ⟹  H—C(=O)—CH₂—C₆H₅  +  ~~NH₂   or   ~~CHO  +  H₂N—CH₂—C₆H₅

c. (CH₃CH₂CH₂)₂N(CH₂)₂CH(CH₃)₂  ⟹  (CH₃CH₂CH₂)₂NH  +  H—C(=O)—CH₂CH(CH₃)₂   or   H—C(=O)—CH₂CH₃  +  CH₃CH₂CH₂—N(H)—CH₂CH(CH₃)₂

d. C₆H₅—N(H)—~~~  ⟹  H—C(=O)—CH₂CH₂CH₃  +  C₆H₅—NH₂

**25.56**

a. C₆H₅—C(=O)—CH₃  + ~~~NH₂ →[NaBH₃CN]  C₆H₅—CH(CH₃)—N(H)—~~~

b. cycloheptanone + (CH₃)₂NH →[NaBH₃CN] cycloheptyl—N(CH₃)₂

c. C₆H₅~~~CHO →[NH₃ / NaBH₃CN] C₆H₅~~~CH₂NH₂

d. CH₃CH₂—C(=O)—CH₂CH₃ + cyclohexyl—NH₂ →[NaBH₃CN] CH₃CH₂CH(CH₃... )—N(H)—cyclohexyl

**25.57**

a. C₆H₅CH₂—Br →[NH₃, excess] C₆H₅—CH₂NH₂

b. C₆H₅—CN →[[1] LiAlH₄ / [2] H₂O] C₆H₅—CH₂NH₂

c. C₆H₅—CONH₂ →[[1] LiAlH₄ / [2] H₂O] C₆H₅—CH₂NH₂

d. C₆H₅—CHO →[NH₃ / NaBH₃CN] C₆H₅—CH₂NH₂

e.

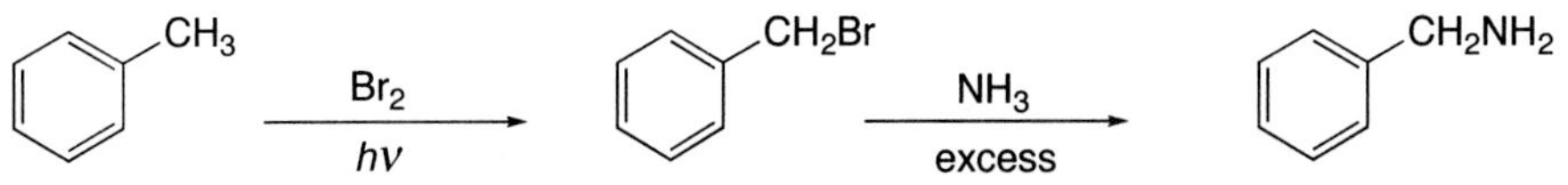

$$\text{toluene} \xrightarrow[hv]{Br_2} \text{benzyl bromide} \xrightarrow[\text{excess}]{NH_3} \text{benzylamine}$$

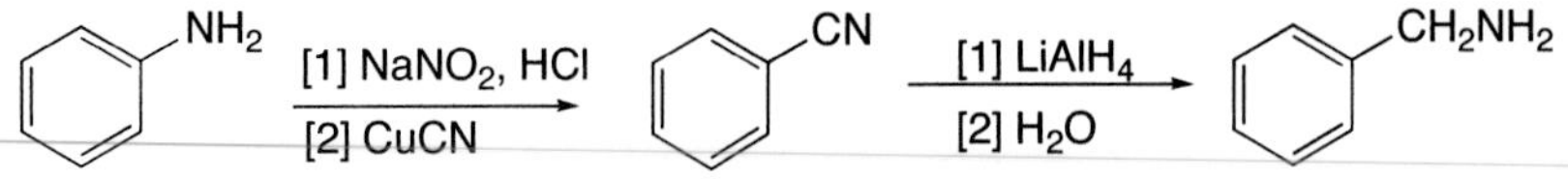

f.

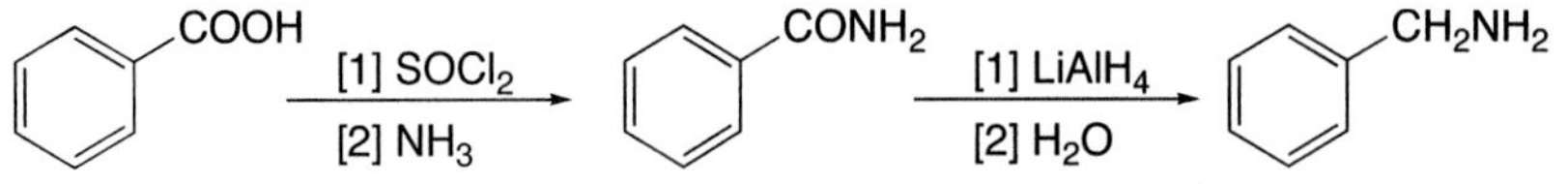

$$\text{COOH} \xrightarrow[\text{[2] NH}_3]{\text{[1] SOCl}_2} \text{CONH}_2 \xrightarrow[\text{[2] H}_2\text{O}]{\text{[1] LiAlH}_4} \text{CH}_2\text{NH}_2$$

g.

$$\text{NH}_2 \xrightarrow[\text{[2] CuCN}]{\text{[1] NaNO}_2, \text{ HCl}} \text{CN} \xrightarrow[\text{[2] H}_2\text{O}]{\text{[1] LiAlH}_4} \text{CH}_2\text{NH}_2$$

h.

$$\text{benzene} \xrightarrow[\text{H}_2\text{SO}_4]{\text{HNO}_3} \text{NO}_2 \xrightarrow[\text{Pd-C}]{\text{H}_2} \text{NH}_2 \quad \text{Then as in (g).}$$

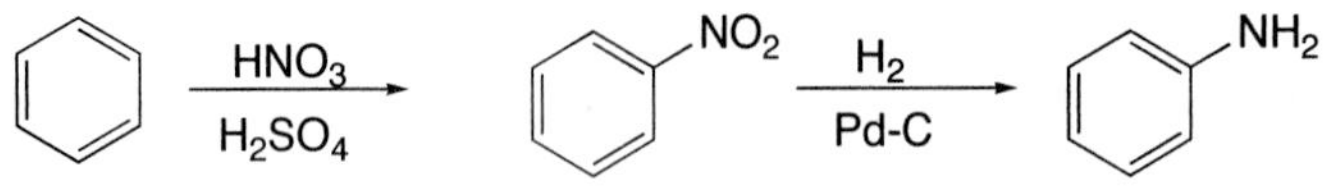

**25.58** Use the directions from Answer 25.22. Separation can be achieved because benzoic acid reacts with aqueous base and aniline reacts with aqueous acid according to the following equations:

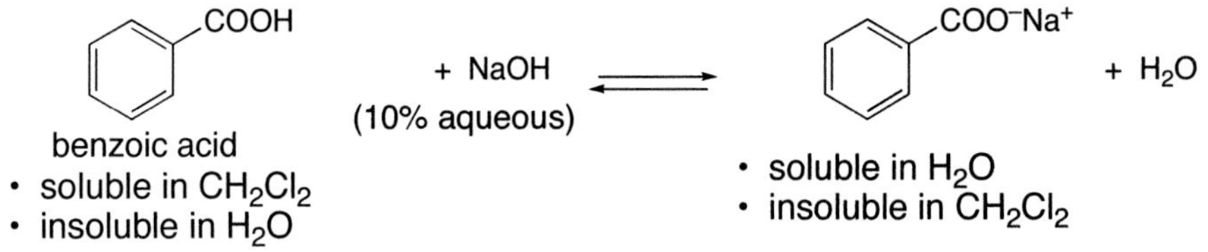

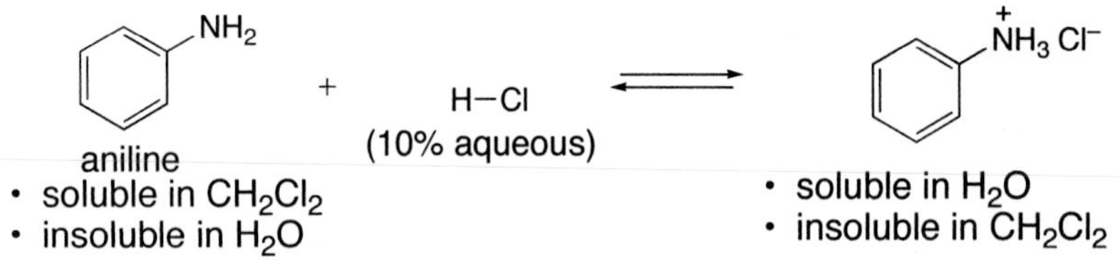

benzoic acid
• soluble in $CH_2Cl_2$
• insoluble in $H_2O$

+ NaOH
(10% aqueous)

$COO^-Na^+$ + $H_2O$
• soluble in $H_2O$
• insoluble in $CH_2Cl_2$

aniline
• soluble in $CH_2Cl_2$
• insoluble in $H_2O$

+ H–Cl
(10% aqueous)

$\overset{+}{N}H_3\ Cl^-$
• soluble in $H_2O$
• insoluble in $CH_2Cl_2$

Toluene ($C_6H_5CH_3$), on the other hand, is not protonated or deprotonated in aqueous solution so it is always soluble in $CH_2Cl_2$ and is insoluble in $H_2O$. The following flow chart illustrates the process.

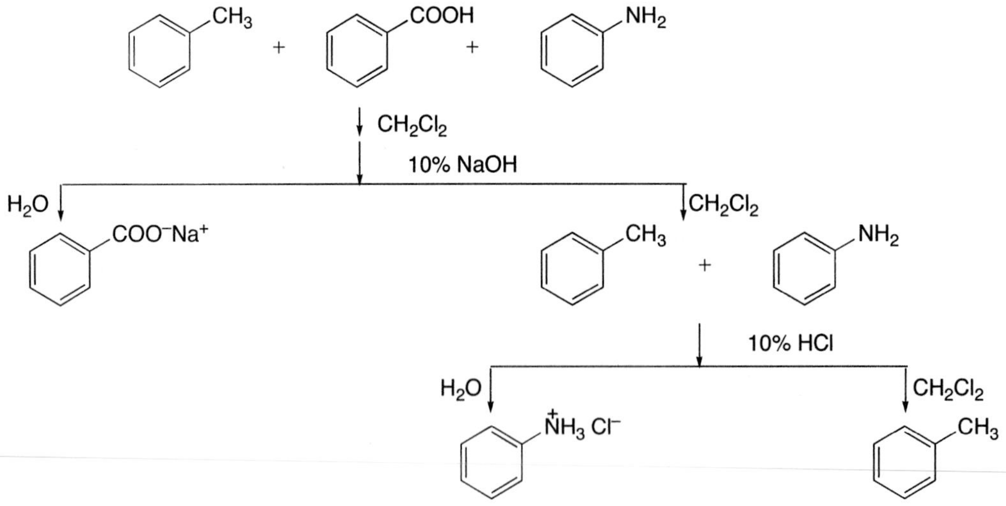

## 25.59

N-ethylaniline

a. HCl —→ [structure]

b. CH$_3$COOH —→ [structure] CH$_3$COO$^-$

c. (CH$_3$)$_2$C=O —→ [structure]

d. CH$_2$O, NaBH$_3$CN —→ [structure with CH$_3$]

e. CH$_3$I (excess) —→ [structure with CH$_3$ CH$_3$ N$^+$, I$^-$]

f. CH$_3$I (excess), followed by Ag$_2$O and Δ —→ [N(CH$_3$)$_2$ structure] + CH$_2$=CH$_2$

g. CH$_3$CH$_2$COCl —→ [structure]

h. The product in (g), then HNO$_3$, H$_2$SO$_4$ —→ [O$_2$N structure] + [NO$_2$ structure]

i. The product in (g), then [1] LiAlH$_4$; [2] H$_2$O —→ [structure]

j. The product in (h), then H$_2$, Pd-C —→ [NH$_2$ structure] + [H$_2$N structure]

## 25.60

CH$_3$—[ring]—NH$_2$  p-methylaniline

a. HCl —→ CH$_3$—[ring]—$^+$NH$_3$ Cl$^-$

b. CH$_3$COCl —→ CH$_3$—[ring]—NH—C(=O)—CH$_3$

c. (CH$_3$CO)$_2$O —→ CH$_3$—[ring]—NH—C(=O)—CH$_3$

d. excess CH$_3$I —→ CH$_3$—[ring]—$^+$N(CH$_3$)$_3$ I$^-$

e. (CH$_3$)$_2$C=O —→ CH$_3$—[ring]—N=C(CH$_3$)$_2$

f. CH$_3$COCl, AlCl$_3$ —→ CH$_3$—[ring]—$^+$NH$_2$ $\bar{A}$lCl$_3$

g. CH$_3$COOH —→ CH$_3$—[ring]—$^+$NH$_3$ CH$_3$COO$^-$

h. NaNO$_2$, HCl —→ CH$_3$—[ring]—N$_2$$^+$Cl$^-$

i. Step (b), then CH$_3$COCl, AlCl$_3$ —→ CH$_3$—[ring, NH—C(=O)—CH$_3$ and C=O CH$_3$]

j. CH$_3$CHO, NaBH$_3$CN —→ CH$_3$—[ring]—NHCH$_2$CH$_3$

## 25.61

a. CH$_3$CH$_2$CH$_2$CH$_2$NH$_2$ —[ClCOC$_6$H$_5$]→ CH$_3$CH$_2$CH$_2$CH$_2$NHCOC$_6$H$_5$

b. CH$_3$CH$_2$CH$_2$CH$_2$NH$_2$ —[O=C(CH$_2$CH$_3$)$_2$]→ CH$_3$CH$_2$CH$_2$CH$_2$N=C(CH$_2$CH$_3$)$_2$

c. CH$_3$CH$_2$CH$_2$CH$_2$NH$_2$ —[[1] CH$_3$I (excess); [2] Ag$_2$O; [3] Δ]→ CH$_3$CH$_2$CH=CH$_2$

d.  $CH_3CH_2CH_2CH_2NH_2$ $\xrightarrow[\text{NaBH}_3\text{CN}]{C_6H_5CHO}$ $CH_3CH_2CH_2CH_2NHCH_2C_6H_5$

e.  $CH_3CH_2CH_2CH_2NH_2$ $\xrightarrow[\text{NaBH}_3\text{CN}]{CH_3CHO}$ $CH_3CH_2CH_2CH_2NHCH_2CH_3$

f.  $CH_3CH_2CH_2CH_2NH_2$ $\xrightarrow[\text{excess}]{CH_3I}$ $[CH_3CH_2CH_2CH_2N(CH_3)_3]^+I^-$

## 25.62

a.  $CH_3(CH_2)_6NH_2$ $\xrightarrow[\substack{[2]Ag_2O \\ [3]\ \Delta}]{[1]\ CH_3I\ (excess)}$ $CH_3(CH_2)_4CH=CH_2$

b.  $\xrightarrow[\substack{[2]\ Ag_2O \\ [3]\ \Delta}]{[1]\ CH_3I\ (excess)}$ (E + Z) / **major product**

c.  $\xrightarrow[\substack{[2]Ag_2O \\ [3]\ \Delta}]{[1]\ CH_3I\ (excess)}$ $CH_2=CH_2$ **major product** + $(CH_3)_2CHN(CH_3)_2$  $CH_2=CHCH_3$ + $CH_3CH_2N(CH_3)_2$

d.  $\xrightarrow[\substack{[2]Ag_2O \\ [3]\ \Delta}]{[1]\ CH_3I\ (excess)}$ **major product**

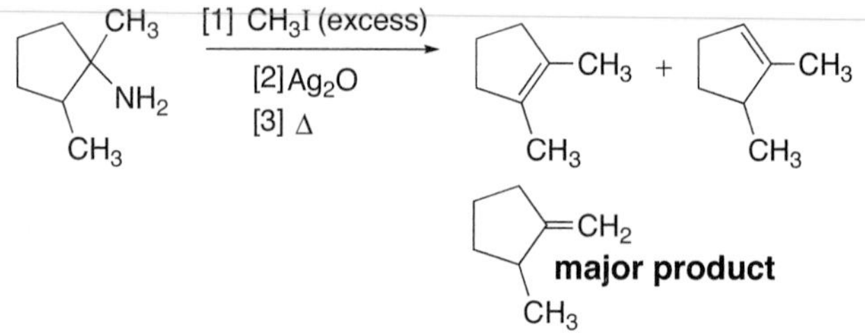

e.  $\xrightarrow[\substack{[2]Ag_2O \\ [3]\ \Delta}]{[1]\ CH_3I\ (excess)}$ **major product** +  (E + Z) +  (E + Z)

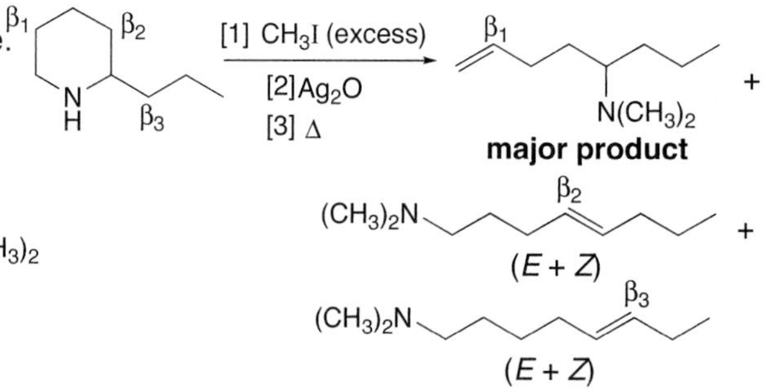

## 25.63

a.  $\xrightarrow[\text{mild acid}]{HN(CH_3)_2}$

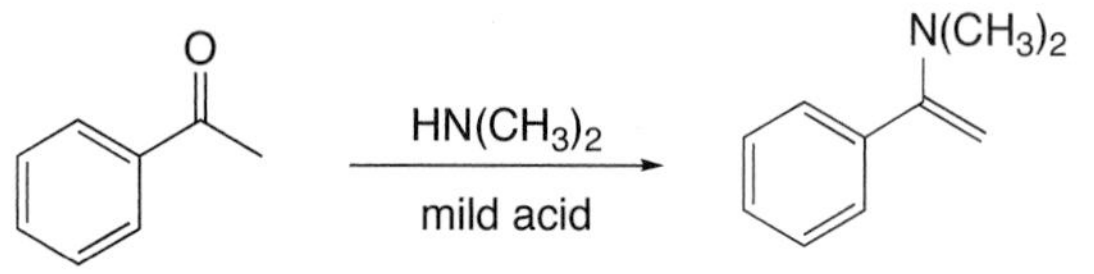

b.  $\xrightarrow[\text{mild acid}]{NH_2CH_2CH_2CH_3}$

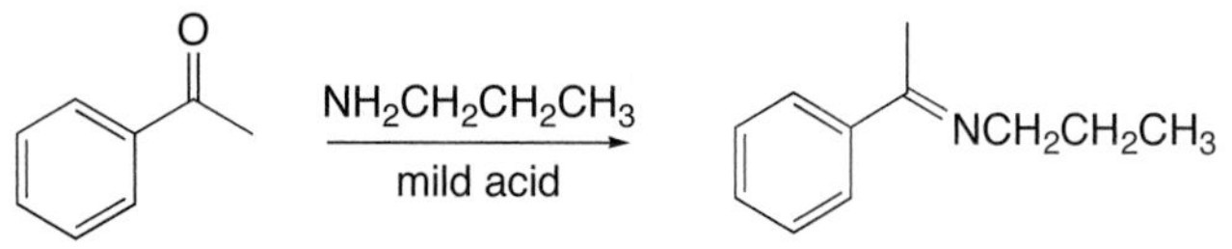

c.  $\xrightarrow[\substack{[2]\ H_2SO_4}]{[1]\ NaBH_4,\ CH_3OH}$ $\xrightarrow[\substack{[2]\ Ag_2O \\ [3]\ \Delta}]{[1]\ CH_3I\ (excess)}$ $\xleftarrow[\text{NaBH}_3\text{CN}]{NH_3}$

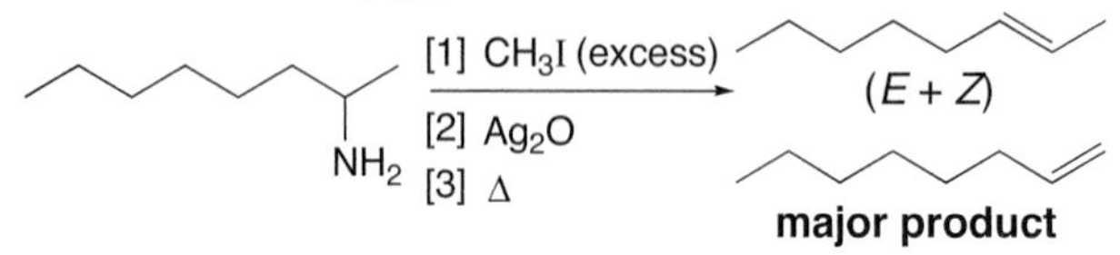

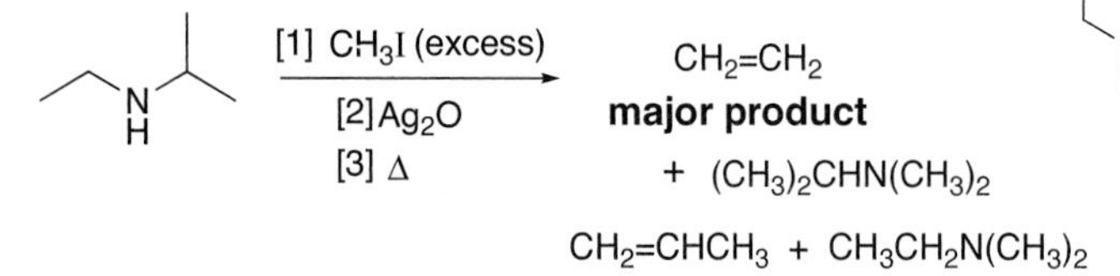

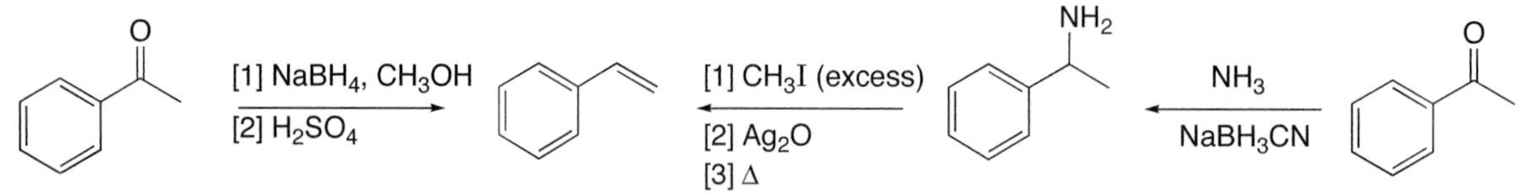

d.  $\xrightarrow{mCPBA}$ $\xrightarrow{CH_3NH_2}$
(from c.)

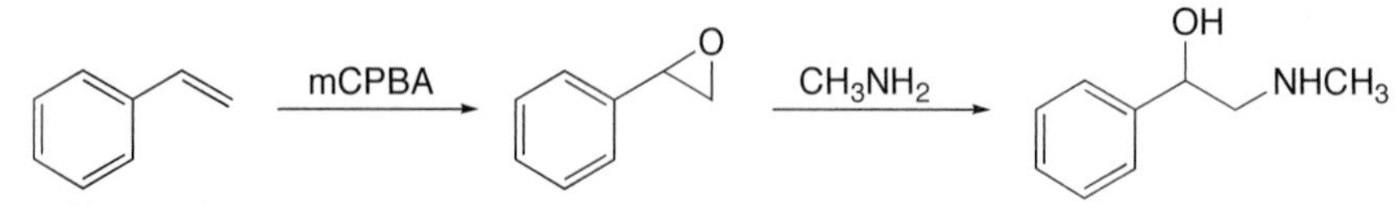

e.  $\xrightarrow[\text{NaBH}_3\text{CN}]{NH_3}$

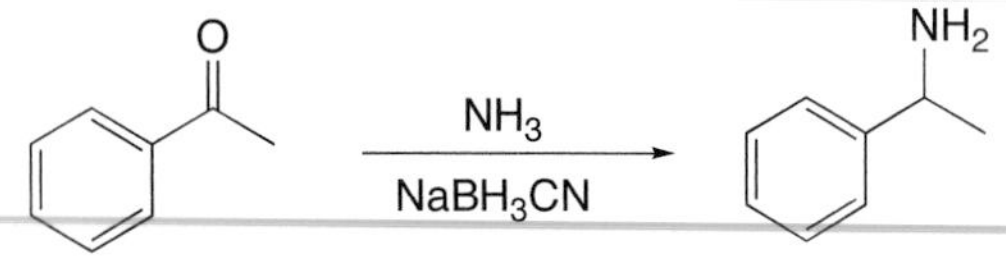

f.

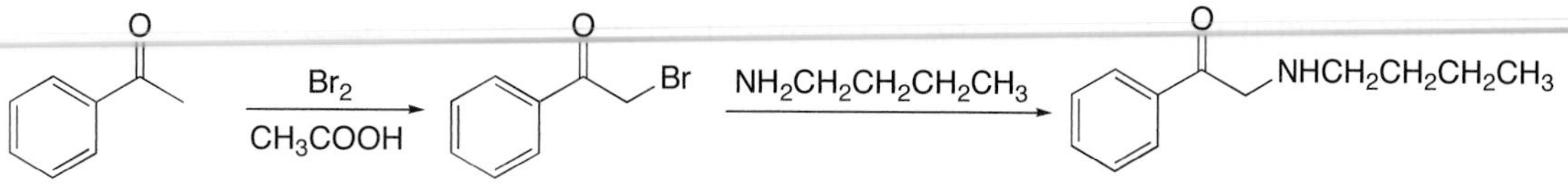

$$C_6H_5COCH_3 \xrightarrow[CH_3COOH]{Br_2} C_6H_5COCH_2Br \xrightarrow{NH_2CH_2CH_2CH_2CH_3} C_6H_5COCH_2NHCH_2CH_2CH_2CH_3$$

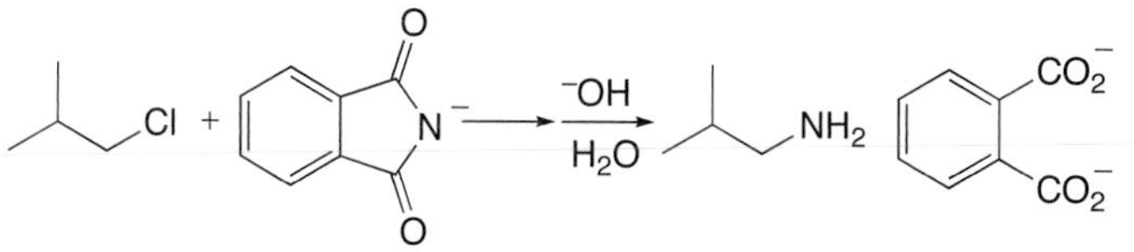

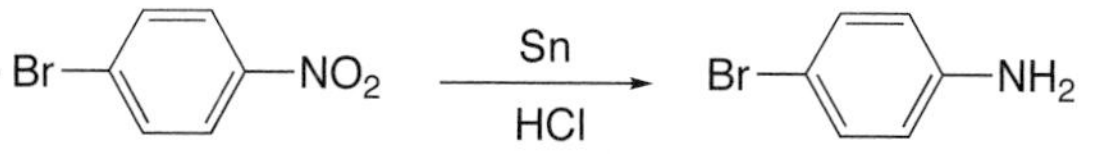

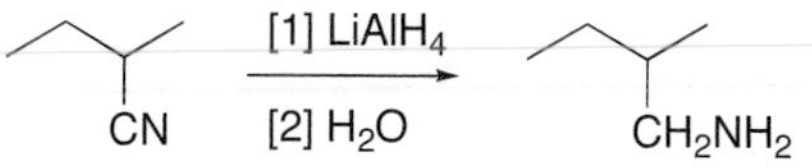

## 25.64

a.

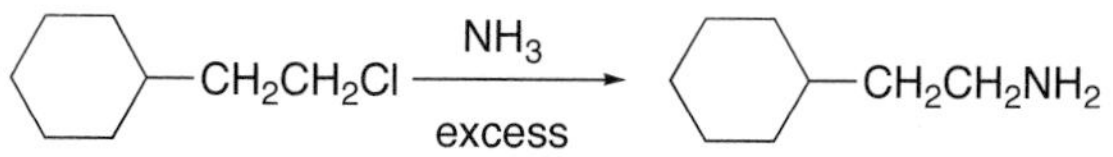

$$\text{cyclohexyl-CH}_2\text{CH}_2\text{Cl} \xrightarrow[\text{excess}]{NH_3} \text{cyclohexyl-CH}_2\text{CH}_2\text{NH}_2$$

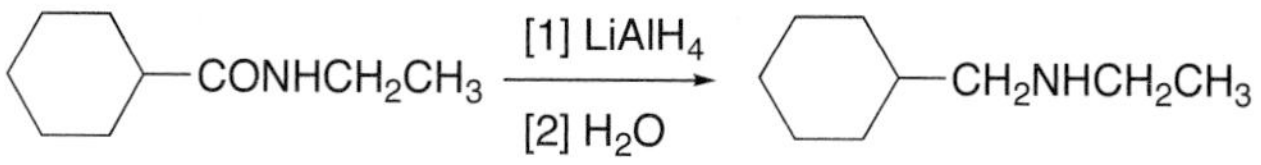

f. $C_6H_5CH_2CH_2NH_2$ + $(C_6H_5CO)_2O \longrightarrow$

$C_6H_5CH_2CH_2NHCOC_6H_5$ + $C_6H_5CH_2CH_2NH_3^+$
$C_6H_5COO^-$

b.

g.

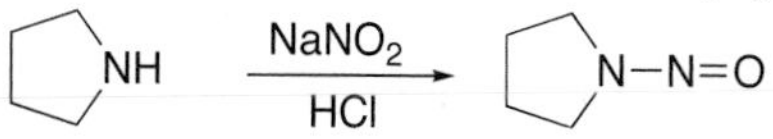

c. $Br-C_6H_4-NO_2 \xrightarrow[HCl]{Sn} Br-C_6H_4-NH_2$

h.

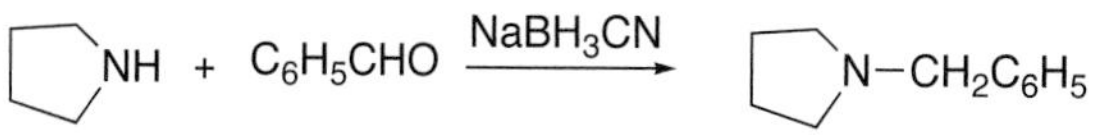

d.

i.

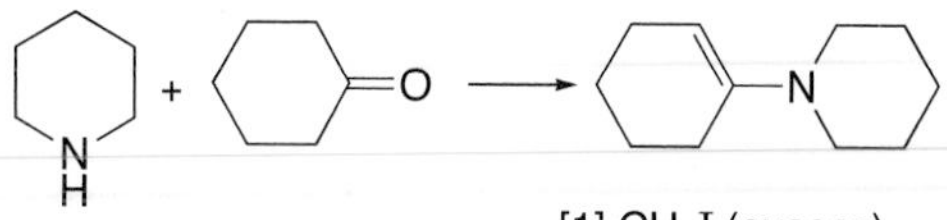

e.

j. 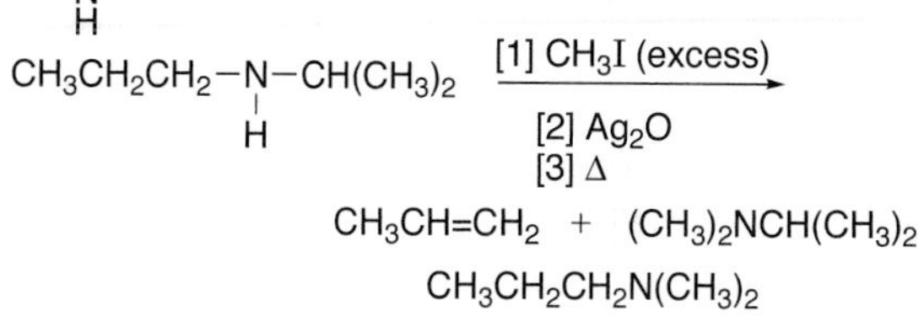 $CH_3CH_2CH_2-\underset{H}{N}-CH(CH_3)_2 \xrightarrow[\substack{[2]\ Ag_2O \\ [3]\ \Delta}]{[1]\ CH_3I\ (excess)}$

$CH_3CH=CH_2$ + $(CH_3)_2NCH(CH_3)_2$
$CH_3CH_2CH_2N(CH_3)_2$

## 25.65

a.

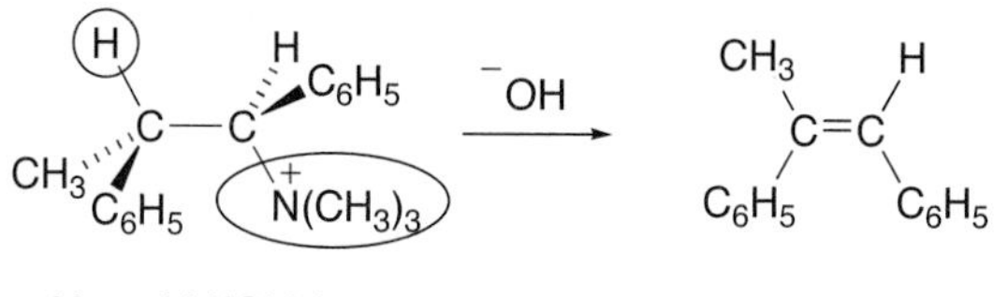

H and N(CH$_3$)$_3$
must be anti.

b.

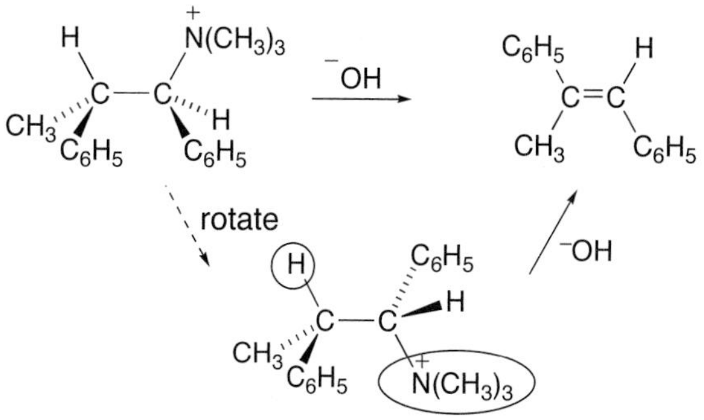

rotate

H and N(CH$_3$)$_3$
must be anti.

## 25.66

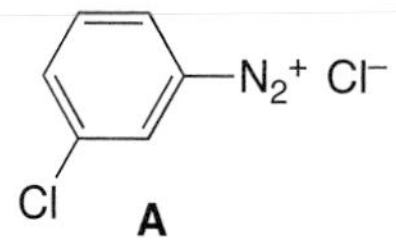

**A**

a. H$_2$O

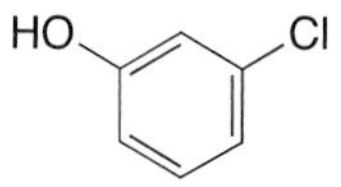

b. H$_3$PO$_2$

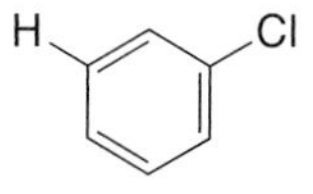

c. CuCl

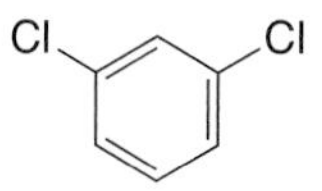

d. CuBr

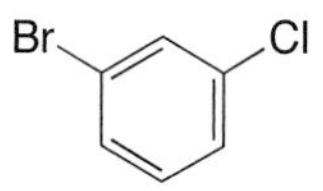

e. CuCN

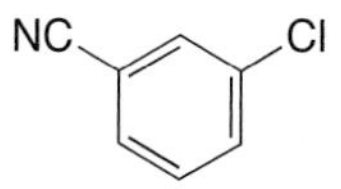

f. HBF$_4$

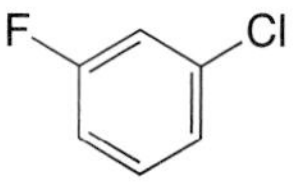

g. NaI

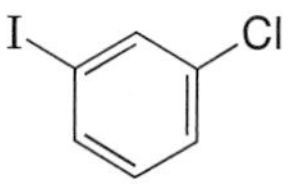

h. C$_6$H$_5$NH$_2$

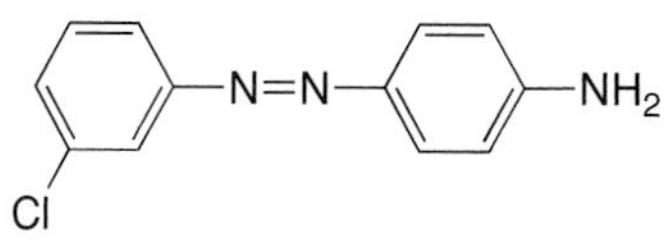

i. C$_6$H$_5$OH

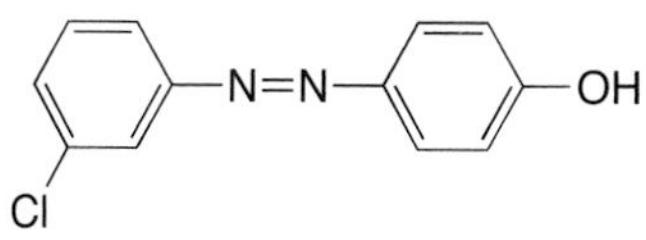

j. KI

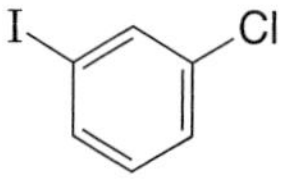

**25.67** Under the acidic conditions of the reaction, aniline is first protonated to form an ammonium salt that has a positive charge on the atom bonded to the benzene ring. The $-NH_3^+$ is now an electron-withdrawing meta director, so a significant amount of meta substitution occurs.

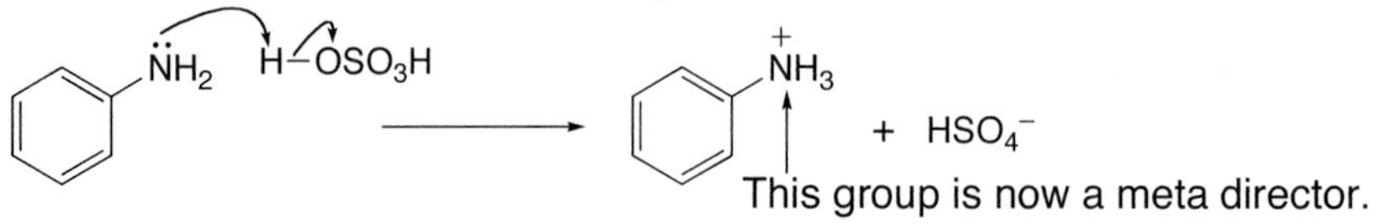

This group is now a meta director.

**25.68**

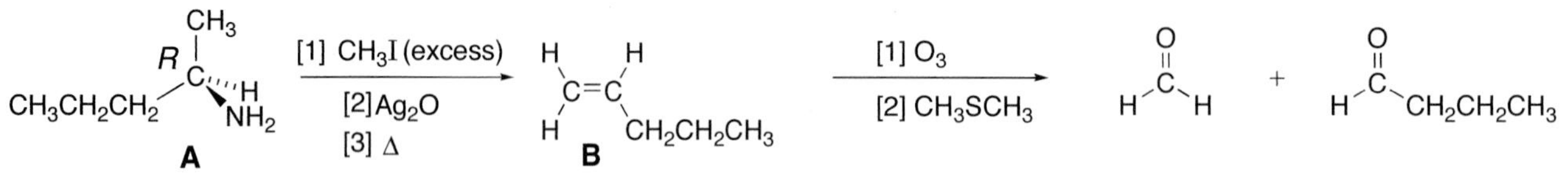

**25.69**

a.

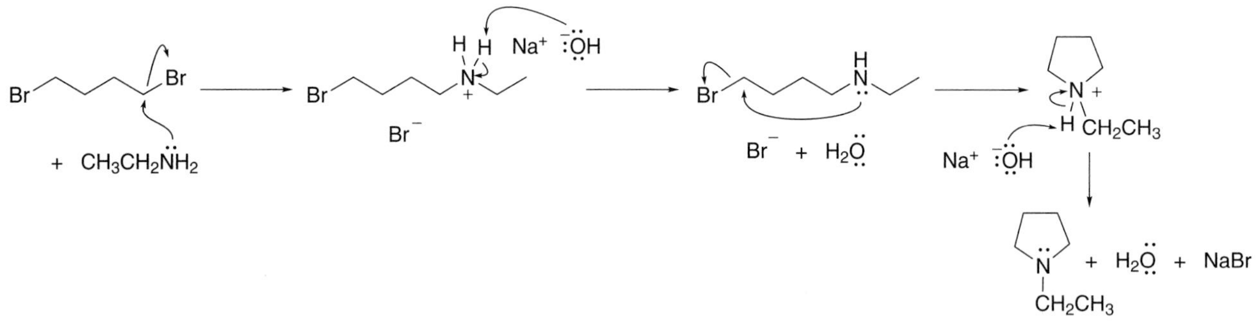

b.

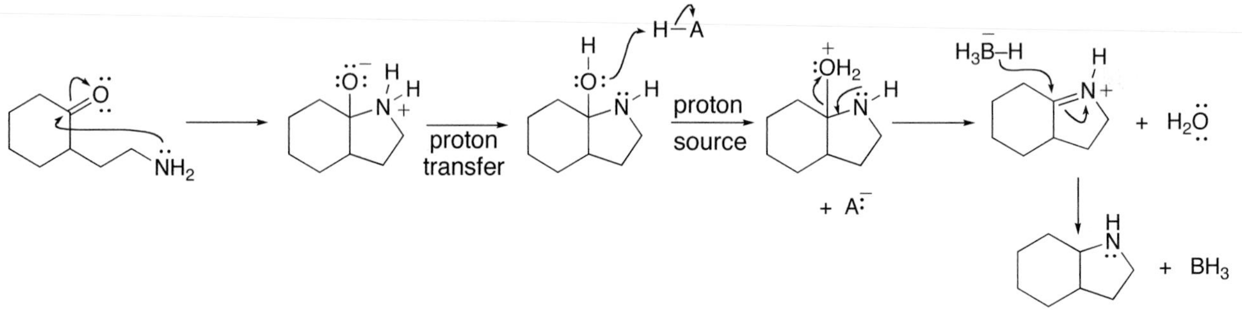

**25.70**

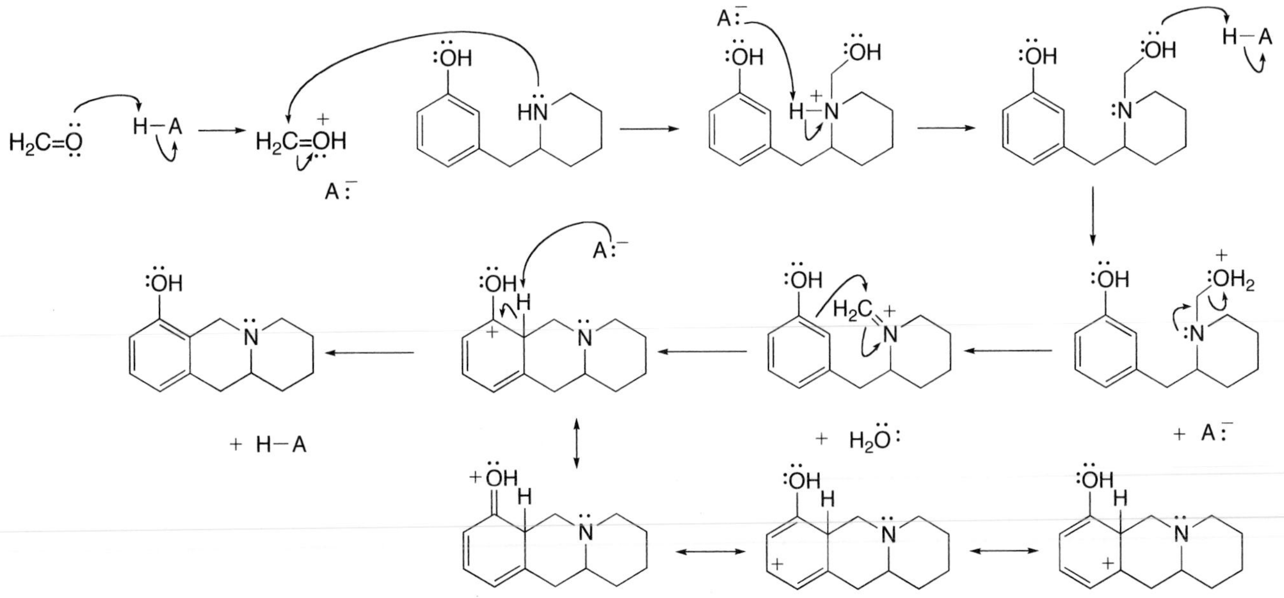

+ H–A

+ H₂Ö:

+ A:⁻

**25.71**

aryl diazonium salt

alkyl diazonium salt

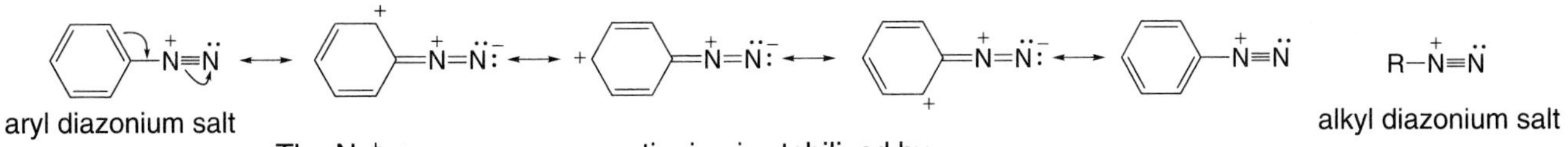

The N₂⁺ group on an aromatic ring is stabilized by
resonance, whereas the alkyl diazonium salt is not.

**25.72**

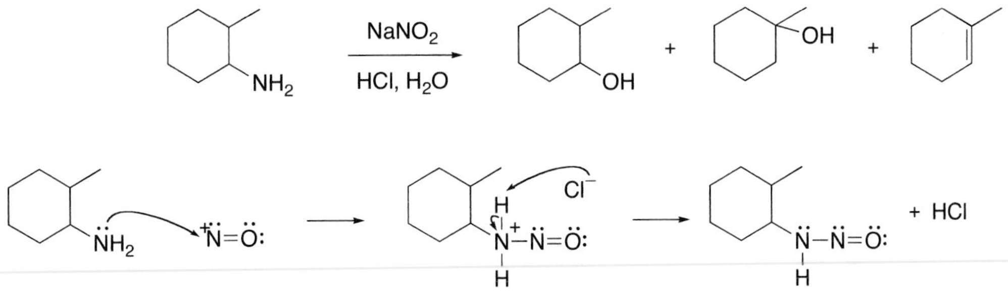

(continued on 25–24)

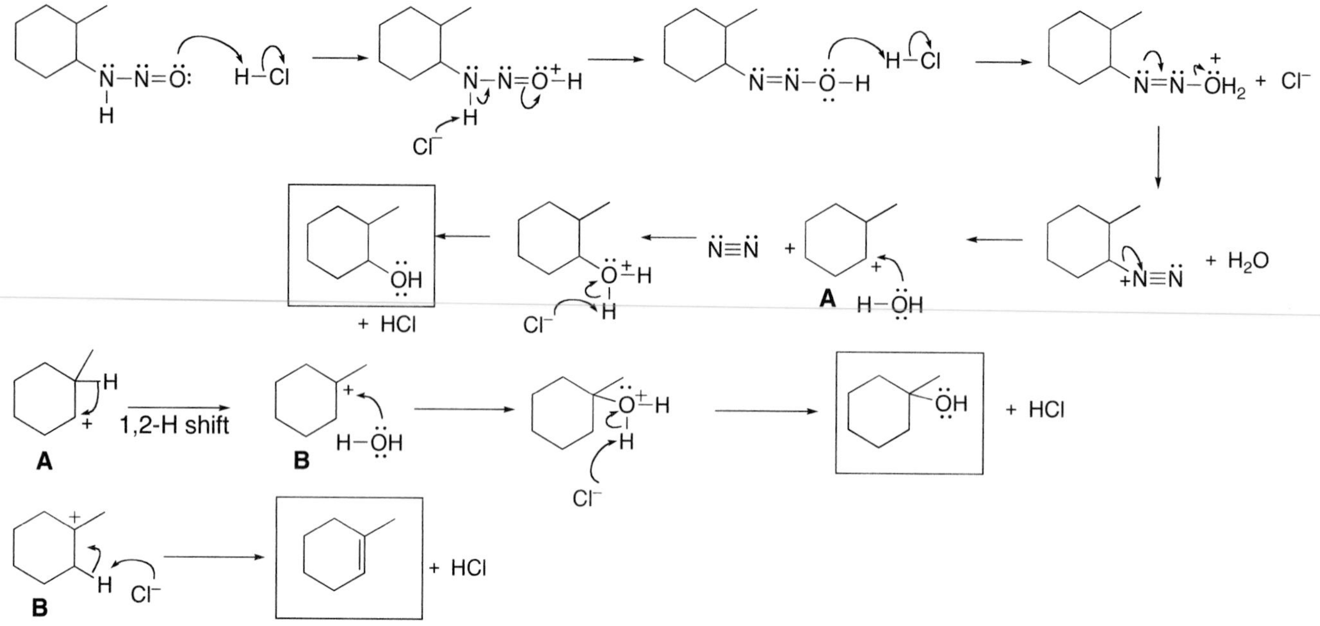

**25.73**

a.

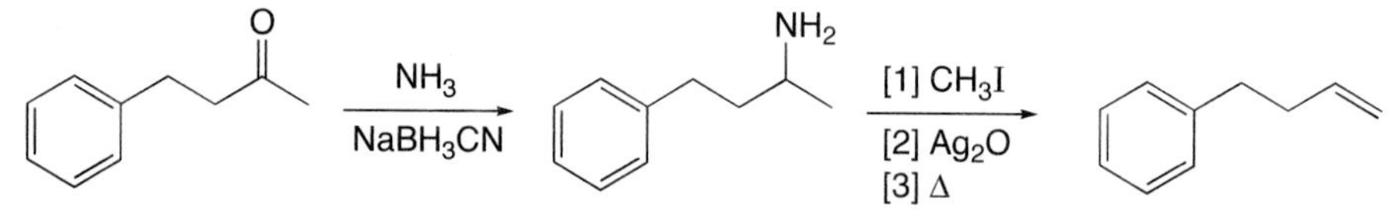

(Hofmann elimination, less substituted C=C favored)

b.

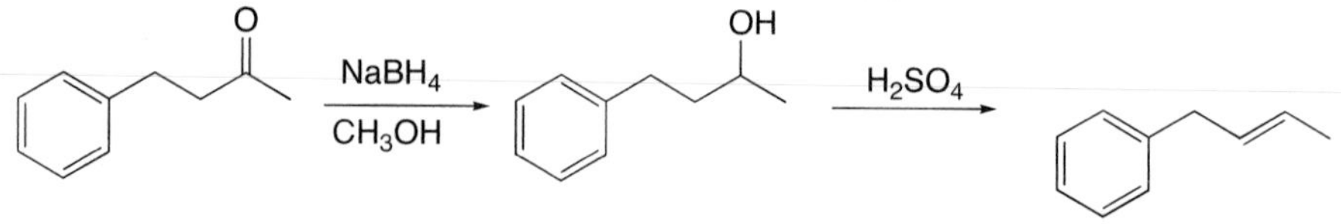

(more substituted C=C favored, more stable trans C=C favored)

**25.74**

a.

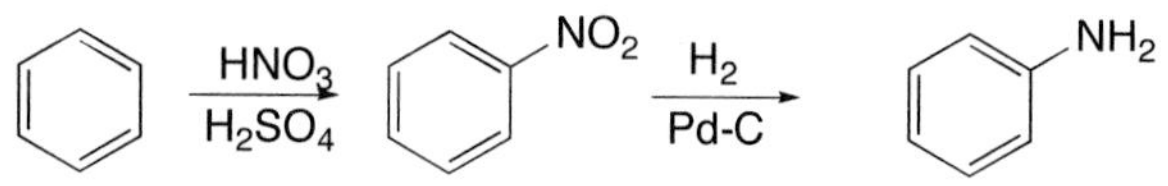

b.

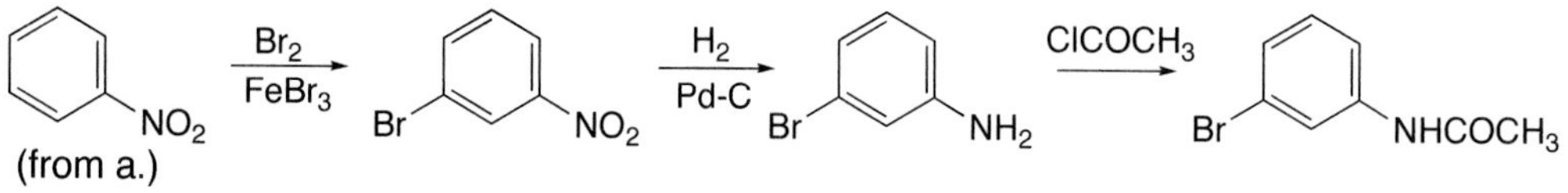

(from a.)

c.

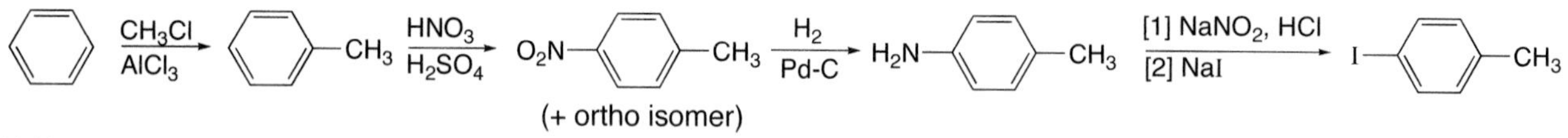

(+ ortho isomer)

d.

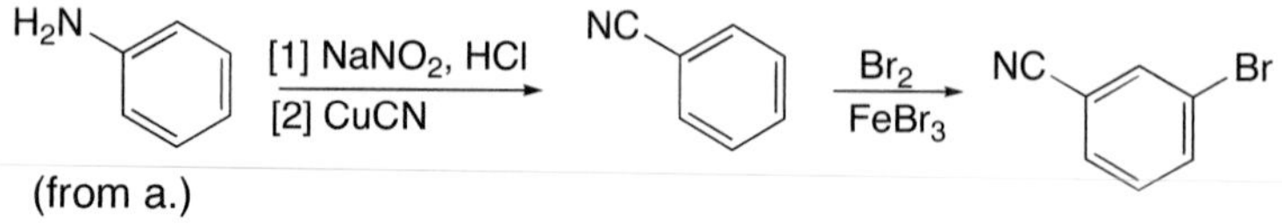

(from a.)

e.

(+ para isomer)

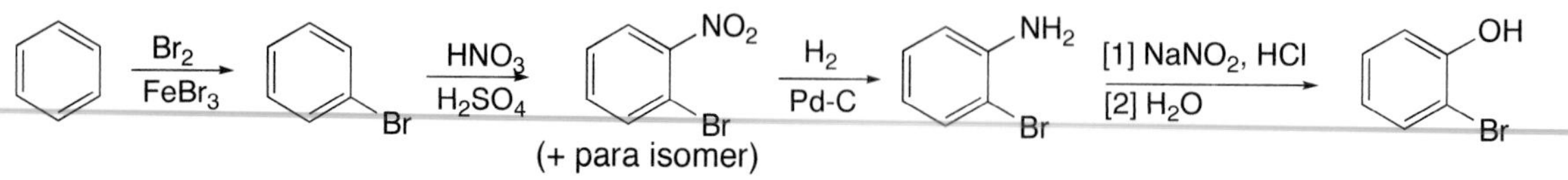

f. NH$_2$—(benzene)—CH$_3$  $\xrightarrow{\text{KMnO}_4}$  H$_2$N—(benzene)—COOH  $\xrightarrow[\text{[2] H}_2\text{O}]{\text{[1] NaNO}_2, \text{HCl}}$  HO—(benzene)—COOH

(from c.)

g. (benzene)  $\xrightarrow[\text{AlCl}_3]{\text{ClCOCH}_3}$  (acetophenone)  $\xrightarrow[\text{H}_2\text{SO}_4]{\text{HNO}_3}$  O$_2$N—(acetophenone)  $\xrightarrow[\text{Pd-C}]{\text{H}_2}$  H$_2$N—(acetophenone)  $\xrightarrow[\text{[2] NaI}]{\text{[1] NaNO}_2, \text{HCl}}$  I—(acetophenone)

h. (benzene)—NH$_2$  $\xrightarrow[\text{[2] H}_2\text{O}]{\text{[1] NaNO}_2, \text{HCl}}$  (benzene)—OH  $\xrightarrow{\text{C}_6\text{H}_5\text{N}_2{}^+\text{Cl}^-}$  (benzene)—N=N—(benzene)—OH

(from a.)   (from d, step [1])

## 25.75

a. (benzene)—NH$_2$  $\xrightarrow[\text{[2] NaCN}]{\text{[1] NaNO}_2, \text{HCl}}$  (benzene)—CN  $\xrightarrow{\text{H}_3\text{O}^+}$  (benzene)—COOH  $\xrightarrow[\text{[2] NH}_2\text{CH}_3]{\text{[1] SOCl}_2}$  (benzene)—CONHCH$_3$

b. (benzene)—NH$_2$  $\xrightarrow{\text{CH}_3\text{COCl}}$  (benzene)—NHCOCH$_3$  $\xrightarrow[\text{FeBr}_3]{\text{Br}_2}$  Br—(benzene)—NHCOCH$_3$  $\xrightarrow[\text{AlCl}_3]{\text{CH}_3\text{Cl}}$  Br—(benzene, CH$_3$)—NHCOCH$_3$  $\xrightarrow[\text{H}_2\text{O}]{{}^-\text{OH}}$  Br—(benzene, CH$_3$)—NH$_2$

(+ ortho isomer)

$\xrightarrow[\text{[2] H}_3\text{PO}_2]{\text{[1] NaNO}_2, \text{HCl}}$  CH$_3$—(benzene)—Br

c. (benzene)—COOH  $\xrightarrow[\text{H}_2\text{SO}_4]{\text{HOCH}_2\text{CH}_3}$  (benzene)—COOCH$_2$CH$_3$

(from a.)

d. (benzene)—COOH  $\xrightarrow[\text{[2] H}_2\text{O}]{\text{[1] LiAlH}_4}$  (benzene)—CH$_2$OH  $\xrightarrow[\text{FeBr}_3]{\text{Br}_2}$  Br—(benzene, Br, Br)—CH$_2$OH

(from a.)   (3x)

e. H$_2$N—(benzene)  $\xrightarrow[\text{[2] H}_2\text{O}]{\text{[1] NaNO}_2, \text{HCl}}$  HO—(benzene)  $\xrightarrow[\text{AlCl}_3]{\text{CH}_3\text{Cl}}$  HO—(benzene)—CH$_3$  $\xrightarrow{\text{C}_6\text{H}_5\text{N}_2{}^+\text{Cl}^-}$  (benzene)—N=N—(benzene, OH)—CH$_3$

(+ ortho isomer)   (from a, step [1])

## 25.76

[1] (benzene)  $\xrightarrow[\text{AlCl}_3]{\text{CH}_3\text{Cl}}$  (toluene)  $\xrightarrow[h\nu]{\text{Br}_2}$  (benzene)—CH$_2$Br  $\xrightarrow{\text{NH}_3 \text{ (excess)}}$  (benzene)—CH$_2$NH$_2$  $\xrightarrow[\text{NaBH}_3\text{CN}]{\text{H}_2\text{C=O}}$  (benzene)—CH$_2$—NH—CH$_3$

[2]  (from [1])

$\phantom{}$ Br  $\xrightarrow{\ ^-OH\ }$  OH  $\xrightarrow{PCC}$  CHO  $\xrightarrow[NaBH_3CN]{CH_3NH_2}$  N–H

[1] LiAlH$_4$
[2] H$_2$O

[3]  (from [1])

$\xrightarrow[]{KMnO_4}$  COOH  $\xrightarrow[\text{[2] } CH_3NH_2]{\text{[1] } SOCl_2}$  O NCH$_3$ H  $\xrightarrow[\text{[2] } H_2O]{\text{[1] } LiAlH_4}$  N–H

[4]  $\xrightarrow[\substack{[2]\ H_2,\ Pd\text{-}C\\ [3]\ NaNO_2,\ HCl\\ [4]\ CuCN}]{[1]\ HNO_3,\ H_2SO_4}$  CN  $\xrightarrow[\text{[2] } H_2O]{\text{[1] DIBAL-H}}$  CHO   Then route [2].

[1] LiAlH$_4$
[2] H$_2$O

NH$_2$   Then route [1].

[5]  $\xrightarrow[FeBr_3]{Br_2}$  Br  $\xrightarrow{Mg}$  MgBr  $\xrightarrow[\text{[2] } H_3O^+]{\text{[1] } CO_2}$  COOH   Then route [3].

[1] H$_2$C=O
[2] H$_2$O

OH   Then route [2].

**25.77**

a.  $\xrightarrow[H_2SO_4]{HNO_3}$  NO$_2$  $\xrightarrow[Pd\text{-}C]{H_2}$  NH$_2$  $\xrightarrow[\text{[2] } CuCN]{\text{[1] } NaNO_2,\ HCl}$  CN  $\xrightarrow{H_3O^+}$  COOH  $\xrightarrow[FeCl_3]{Cl_2}$  COOH Cl

b.  NH$_2$ (from a.)  $\xrightarrow[H_2SO_4]{CH_3COCl}$  NHCOCH$_3$  $\xrightarrow[H_2SO_4]{HNO_3}$  NHCOCH$_3$ NO$_2$ (+ ortho isomer)  $\xrightarrow[FeCl_3\ (2X)]{Cl_2}$  Cl NHCOCH$_3$ Cl NO$_2$  $\xrightarrow[H_2O]{^-OH}$  Cl NH$_2$ Cl NO$_2$  $\xrightarrow[\text{[2] } H_3PO_2]{\text{[1] } NaNO_2,\ HCl}$  Cl Cl NO$_2$

$\downarrow$ H$_2$ Pd-C

Cl Cl NH$_2$  $\xrightarrow[\text{[2] } H_2O]{\text{[1] } NaNO_2,\ HCl}$  Cl Cl OH

c.  $\xrightarrow[AlCl_3]{CH_3Cl}$  CH$_3$  $\xrightarrow[H_2SO_4]{HNO_3}$  CH$_3$–NO$_2$ (+ ortho isomer)  $\xrightarrow[Pd\text{-}C]{H_2}$  CH$_3$–NH$_2$  $\xrightarrow[\text{[2] } NaCN]{\text{[1] } NaNO_2,\ HCl}$  CH$_3$–CN  $\xrightarrow[\text{[2] } H_2O]{\text{[1] } LiAlH_4}$  CH$_3$–CH$_2$NH$_2$

d.

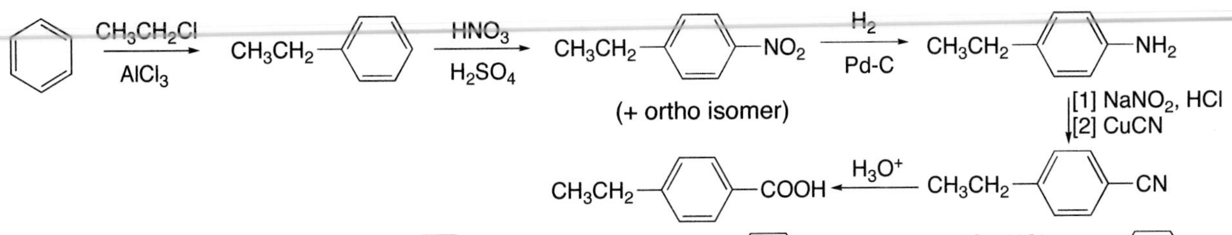

(+ ortho isomer)

e.

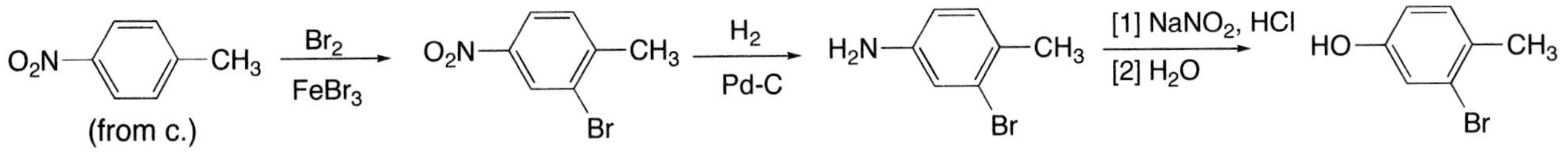

(from c.)

f.

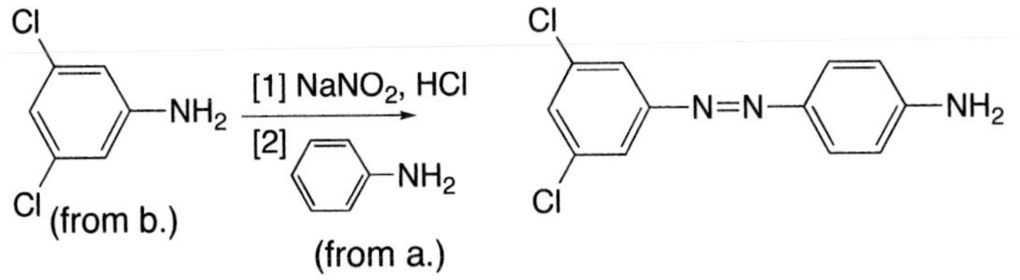

(from b.)

(from a.)

**25.78**

a.

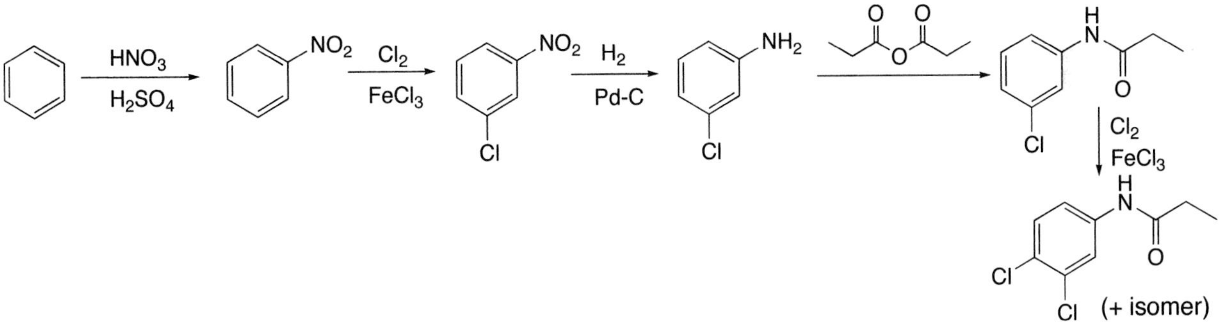

(+ isomer)

b.

(from a.)

(+ ortho isomer)

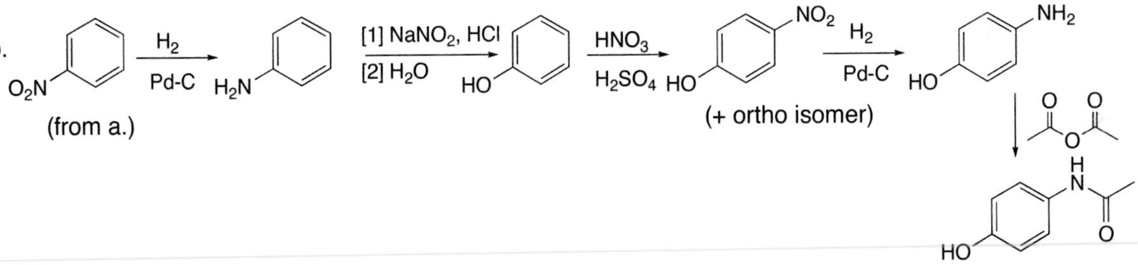

c.

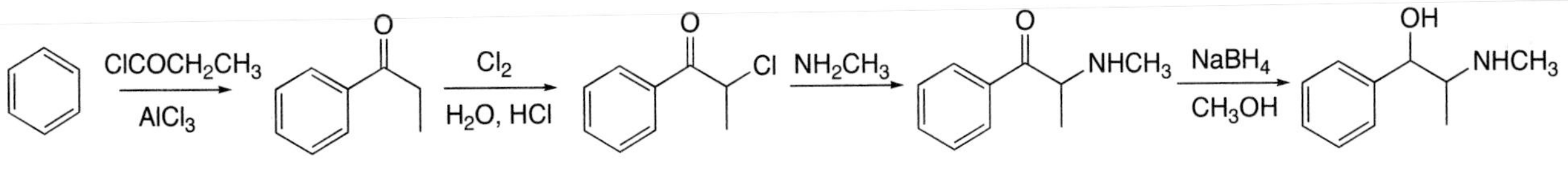

**25.79**

a.

(+ ortho isomer)

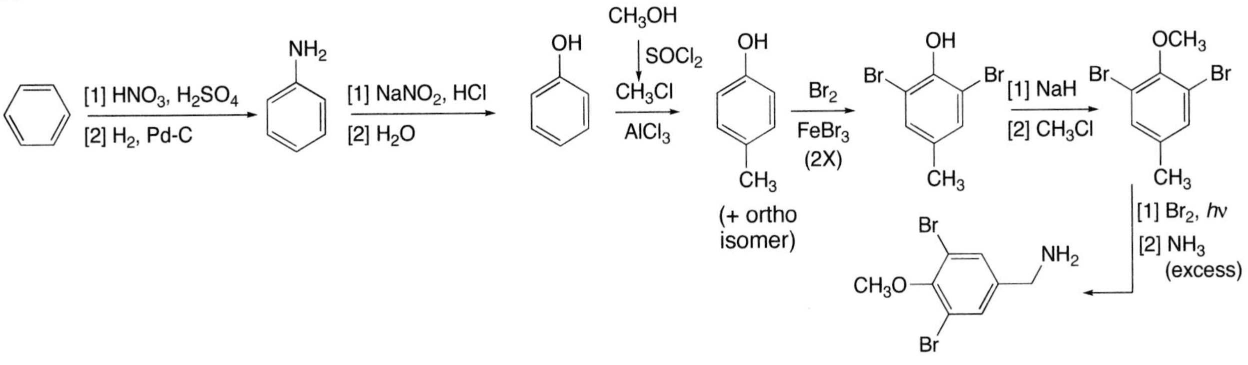

b.

CH$_3$CH$_2$OH

$\downarrow$ CrO$_3$, H$_2$SO$_4$, H$_2$O

CH$_3$COOH

$\downarrow$ SOCl$_2$

CH$_3$COCl

(from a.) $\xrightarrow{\text{SOCl}_2 / \text{CH}_3\text{COCl}}$ (acetanilide) $\xrightarrow[\text{AlCl}_3]{\text{CH}_3\text{COCl}}$ (4-acetamidoacetophenone) (+ ortho isomer) $\xrightarrow{^-\text{OH, H}_2\text{O}}$ (4-aminoacetophenone)

c.

(benzene) $\xrightarrow[\text{FeCl}_3]{\text{Cl}_2}$ (chlorobenzene)

CH$_3$OH

$\downarrow$ SOCl$_2$

$\xrightarrow[\text{[2] KMnO}_4]{\text{[1] CH}_3\text{Cl, AlCl}_3}$ HOOC—(ring)—Cl (+ortho isomer) $\xrightarrow[\text{[2] H}_2\text{O}]{\text{[1] LiAlH}_4}$ HO—(ring)—Cl $\xrightarrow{\text{PCC}}$ (4-chlorobenzaldehyde)

(from a.)
mild acid

(ring)—N=CH—(ring)—Cl

d.

(benzene) $\xrightarrow[\text{AlCl}_3]{\text{CH}_3\text{Cl}}$ (toluene) $\xrightarrow[\text{H}_2\text{SO}_4]{\text{HNO}_3}$ O$_2$N—(ring)—CH$_3$ (+ortho isomer) $\xrightarrow[\text{[2] CH}_3\text{OH, H}^+]{\text{[1] KMnO}_4}$ O$_2$N—(ring)—COOCH$_3$ $\xrightarrow[\text{Pd-C}]{\text{H}_2}$ H$_2$N—(ring)—COOCH$_3$

CH$_3$OH

$\downarrow$ SOCl$_2$

CH$_3$Cl

PCC

(CH$_3$)$_2$CHNH—(ring)—CO$_2$CH$_3$ $\xleftarrow{\text{NaBH}_3\text{CN}}$

e. O$_2$N—(ring)—CH$_3$ (from d.) $\xrightarrow[\text{[2] NaNO}_2, \text{HCl}]{\text{[1] H}_2, \text{Pd-C}}$ F—(ring)—CH$_3$ $\xrightarrow[\text{[2] SOCl}_2]{\text{[1] KMnO}_4}$ F—(ring)—C(=O)Cl (from a.) $\rightarrow$ F—(ring)—C(=O)NH—(ring)
[3] HBF$_4$

Probably a strong enough activator that the Friedel–Crafts reaction will still occur.

f. Make two parts: (4-acetamidobenzonitrile) + Cl—C(=O)—(ring)—I $\xrightarrow{\text{AlCl}_3}$ (acylated product with NHCOCH$_3$, CN, and I)

(from b.) $\xrightarrow[\text{[2] H}_2, \text{Pd-C}]{\text{[1] HNO}_3, \text{H}_2\text{SO}_4}$ (NHCOCH$_3$ / NH$_2$) (+ ortho isomer) $\xrightarrow[\text{[2] CuCN}]{\text{[1] NaNO}_2, \text{HCl}}$ (NHCOCH$_3$ / CN)

O$_2$N—(ring)—CH$_3$ (from d.) $\xrightarrow[\text{[2] NaNO}_2, \text{HCl}]{\text{[1] H}_2, \text{Pd-C}}$ I—(ring)—CH$_3$ $\xrightarrow[\text{[2] SOCl}_2]{\text{[1] KMnO}_4}$ Cl—C(=O)—(ring)—I
[3] NaI

**25.80**

molecular weight = 87
$C_5H_{13}N$
two IR peaks = 1° amine

**25.81**

[1] LiAlH$_4$
[2] H$_2$O

Compound **A**: $C_8H_7N$
IR absorption at 2230 cm$^{-1}$ → triple bond
$^1$H NMR peaks at (ppm):
  2.4 (singlet, 3H) CH$_3$
  7.2 (2H) ⎤
  7.5 (2H) ⎦ para disubstituted benzene ring

Compound **B**: $C_8H_{11}N$
IR absorption at 3370, 3290 cm$^{-1}$
two IR peaks → 1° amine
$^1$H NMR peaks at (ppm):
  1.4 (singlet, 2H) NH$_2$
  2.3 (singlet, 3H) CH$_3$
  3.8 (singlet, 2H) CH$_2$
  7.0–7.3 (4H) disubstituted benzene ring

**25.82**

Compound **A**: $C_8H_{11}N$
IR absorption at 3400 cm$^{-1}$ → 2° amine
$^1$H NMR peaks at (ppm):
  1.3 (triplet, 3H) CH$_3$ adjacent to 2 H's
  3.1 (quartet, 2H) CH$_2$ adjacent to 3 H's
  3.6 (singlet, 1H) amine H
  6.8–7.2 (multiplet, 5H) benzene ring

Compound **B**: $C_8H_{11}N$
IR absorption at 3310 cm$^{-1}$ → 2° amine
$^1$H NMR peaks at (ppm):
  1.4 (singlet, 1H) amine H
  2.4 (singlet, 3H) CH$_3$
  3.8 (singlet, 2H) CH$_2$
  7.2 (multiplet, 5H) benzene ring

Compound **C**: $C_8H_{11}N$
IR absorption at 3430 and
  3350 cm$^{-1}$ → 1° amine
$^1$H NMR peaks at (ppm):
  1.3 (triplet, 3H) CH$_3$ near CH$_2$
  2.5 (quartet, 2H) CH$_2$ near CH$_3$
  3.6 (singlet, 2H) amine H's
  6.7 (doublet, 2H) ⎤ para
  7.0 (doublet, 2H) ⎦ disubstituted benzene ring

**25.83** Guanidine is a strong base because its conjugate acid is stabilized by resonance. This resonance delocalization makes guanidine easily donate its electron pair; thus it's a strong base.

guanidine      p$K_a$ = 13.6

## 25.84

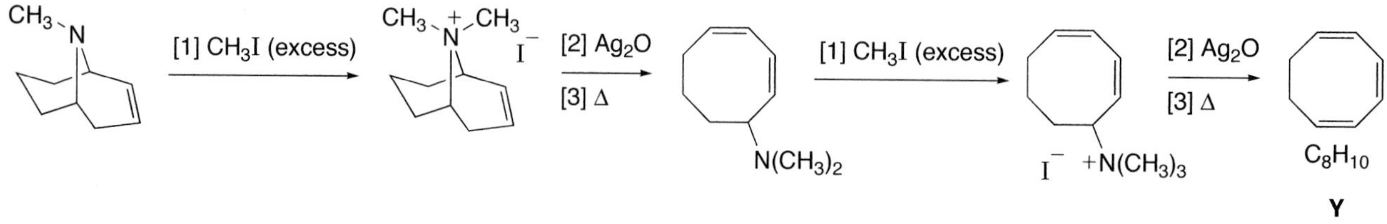

## 25.85

One possibility:

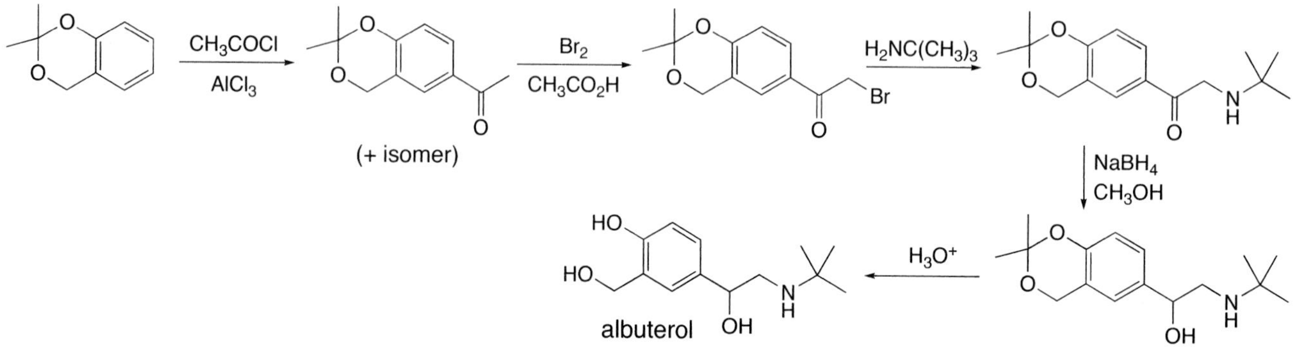

Chapter 26: Carbon–Carbon Bond-Forming Reactions in Organic Synthesis

♦ Coupling reactions

## [1] Coupling reactions of organocuprate reagents (26.1)

R'–X + R₂CuLi ⟶ │R'–R│ + RCu
+ LiX

X = Cl, Br, I

- R′X can be $CH_3X$, $RCH_2X$, $2^o$ cyclic halides, vinyl halides, and aryl halides.
- X may be Cl, Br, or I.
- With vinyl halides, coupling is stereospecific.

## [2] Suzuki reaction (26.2)

R'–X + R–B⟨Y Y ⟩ $\xrightarrow[\text{NaOH}]{\text{Pd(PPh}_3)_4}$ │R'–R│ + HO–BY₂
+ NaX

X = Br, I

- R′X is most often a vinyl halide or aryl halide.
- With vinyl halides, coupling is stereospecific.

## [3] Heck reaction (26.3)

R'–X + ⟍⟍Z $\xrightarrow[\substack{\text{P(}o\text{-tolyl)}_3 \\ (CH_3CH_2)_3N}]{\text{Pd(OAc)}_2}$ │R'⟍⟍Z│

X = Br or I

+ $(CH_3CH_2)_3\overset{+}{N}H\ X^-$

- R′X is a vinyl halide or aryl halide.
- Z = H, Ph, COOR, CN
- With vinyl halides, coupling is stereospecific.
- The reaction forms trans alkenes.

♦ Cyclopropane synthesis

## [1] Addition of dihalocarbenes to alkenes (26.4)

⟍C=C⟋ $\xrightarrow[\text{KOC(CH}_3)_3]{\text{CHX}_3}$ (cyclopropane with X X on top carbon)

- The reaction occurs with syn addition.
- The position of substituents in the alkene is retained in the cyclopropane.

## [2] Simmons–Smith reaction (26.5)

⟍C=C⟋ $\xrightarrow[\text{Zn(Cu)}]{\text{CH}_2\text{I}_2}$ (cyclopropane with H H on top carbon) + ZnI₂

- The reaction occurs with syn addition.
- The position of substituents in the alkene is retained in the cyclopropane.

♦ Metathesis (26.6)

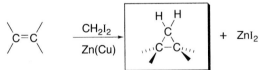

2 RCH=CH₂ $\xrightarrow[\text{catalyst}]{\text{Grubbs}}$ │RCH=CHR│ + CH₂=CH₂

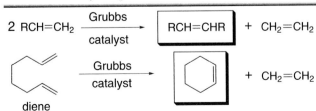

$\xrightarrow[\text{catalyst}]{\text{Grubbs}}$ (cyclohexene) + CH₂=CH₂

diene

- Metathesis works best when CH₂=CH₂, a gas that escapes from the reaction mixture, is formed as one product.

## Chapter 26: Answers to Problems

**26.1** A new C–C bond is formed in each coupling reaction.

a. 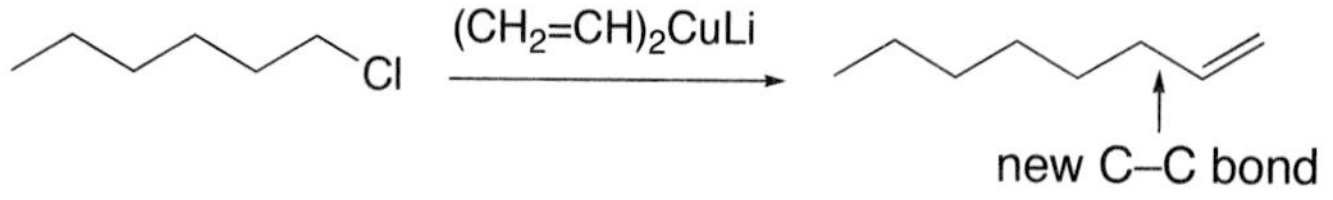
Cl
(CH2=CH)2CuLi
new C–C bond

b. 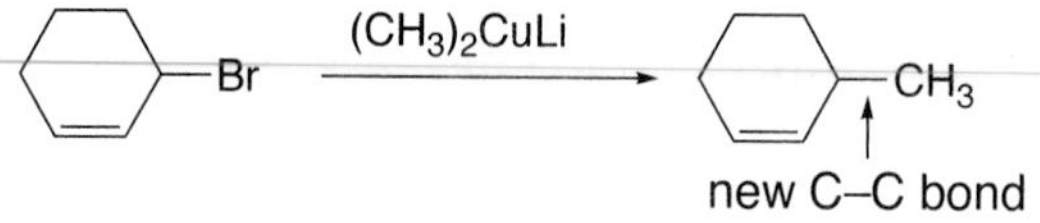
Br
(CH3)2CuLi
CH3
new C–C bond

c. 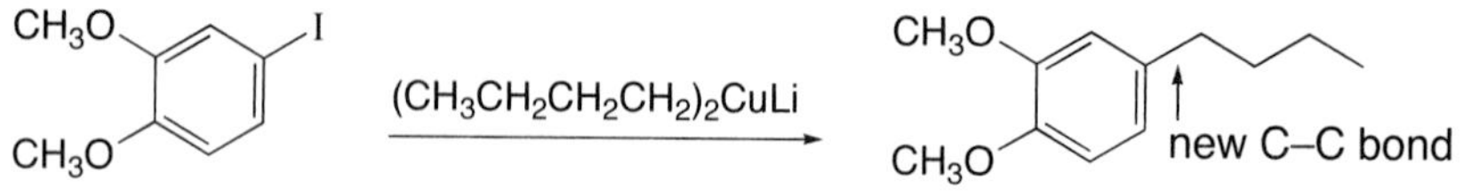
CH3O
CH3O
I
(CH3CH2CH2CH2)2CuLi
new C–C bond

d. 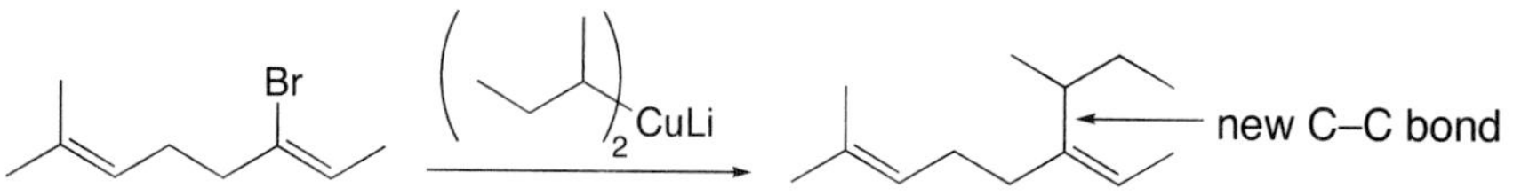
Br
CuLi
2
new C–C bond

**26.2**

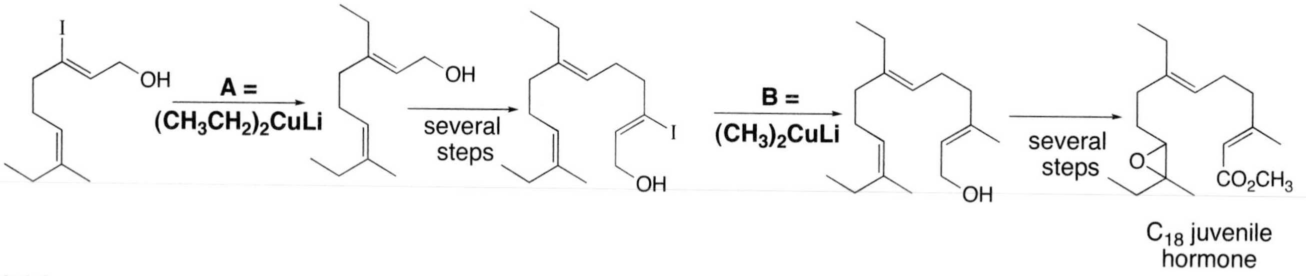
I
OH
A =
(CH3CH2)2CuLi
several steps
B =
(CH3)2CuLi
CO2CH3
C18 juvenile hormone

**26.3**

a. 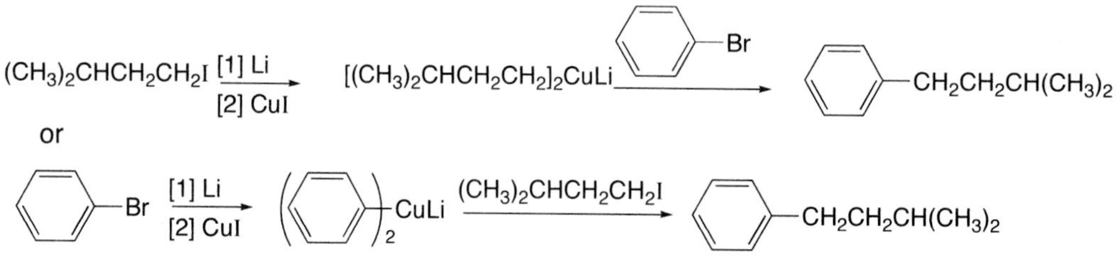
(CH3)2CHCH2CH2I
[1] Li
[2] CuI
[(CH3)2CHCH2CH2]2CuLi
Br
CH2CH2CH(CH3)2
or
CuLi
(CH3)2CHCH2CH2I

b. 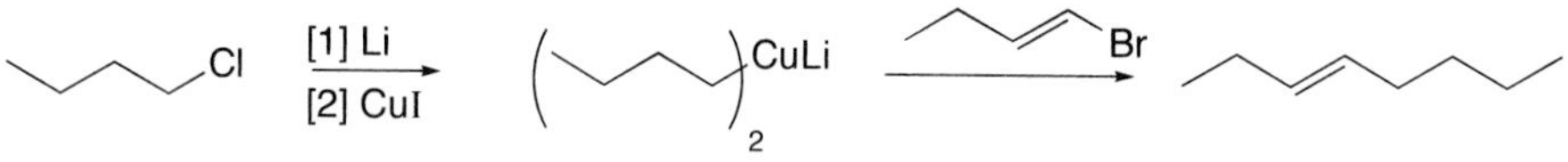
Cl
[1] Li
[2] CuI
CuLi
Br

c. 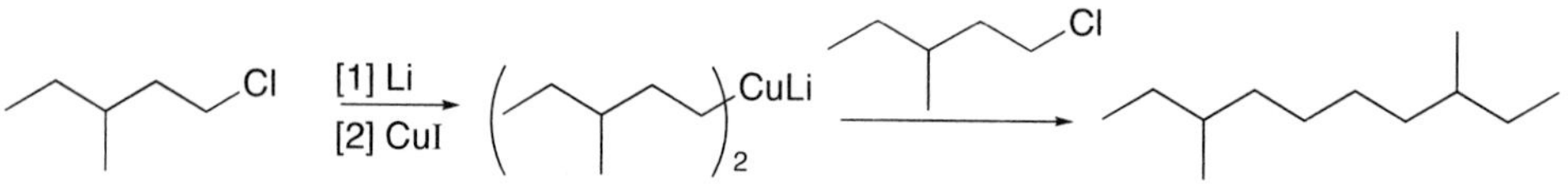
Cl
[1] Li
[2] CuI
CuLi

**26.4**

a.

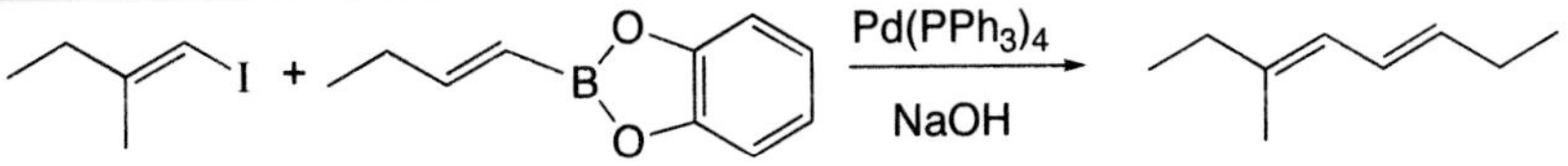

b.

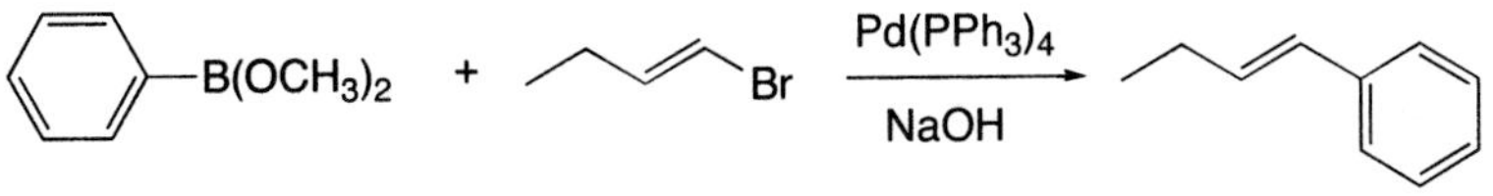

c.

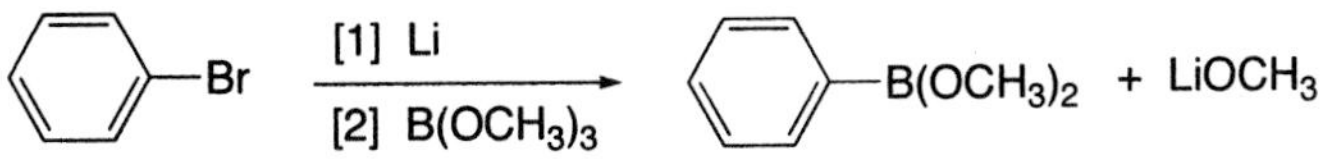

d.

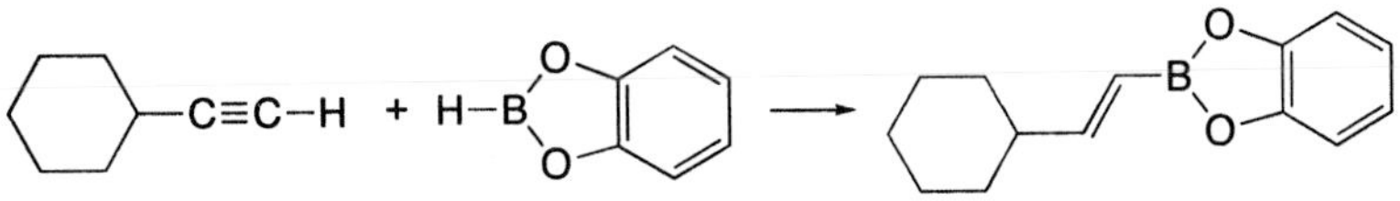

**26.5**

a. 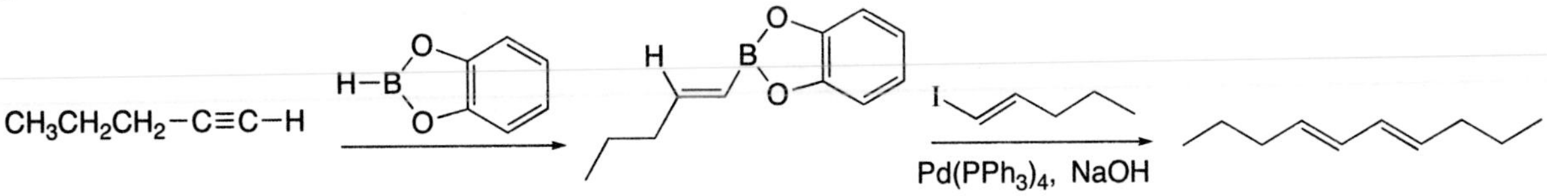 $CH_3CH_2CH_2-C\equiv C-H$

b.

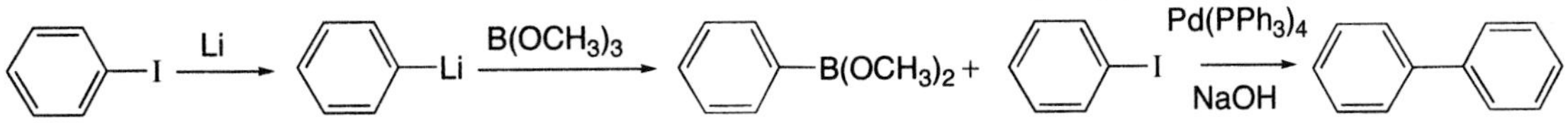

c.

(+ ortho isomer)

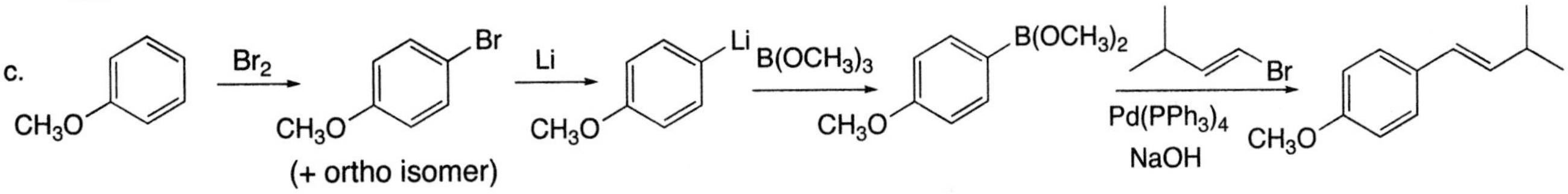

**26.6**

a.

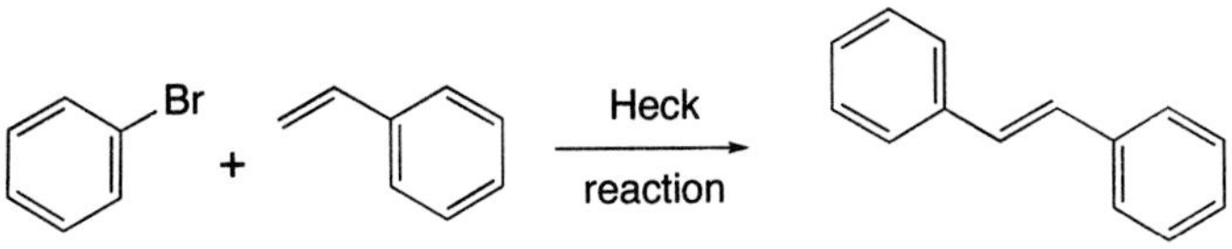

b.

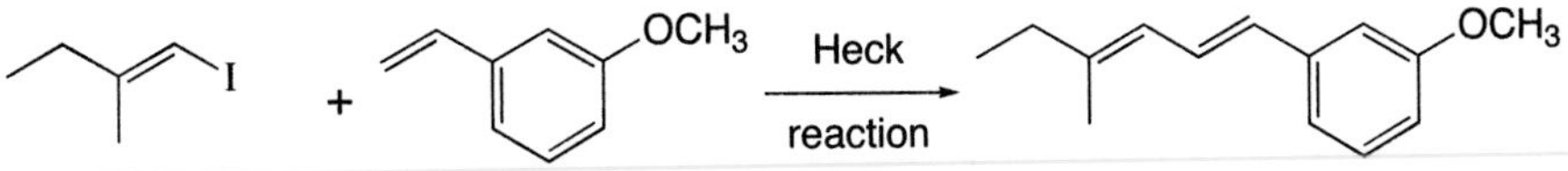

c.

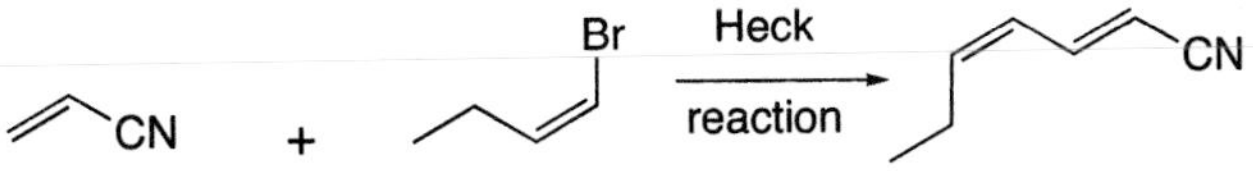

d.

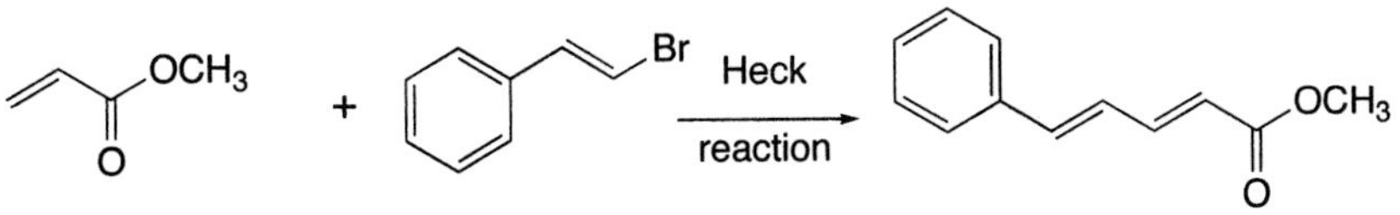

**26.7** Locate the double bond with the aryl, COOR, or CN substituent, and break the molecule into two components at the end of the C=C not bonded to one of these substituents.

a.

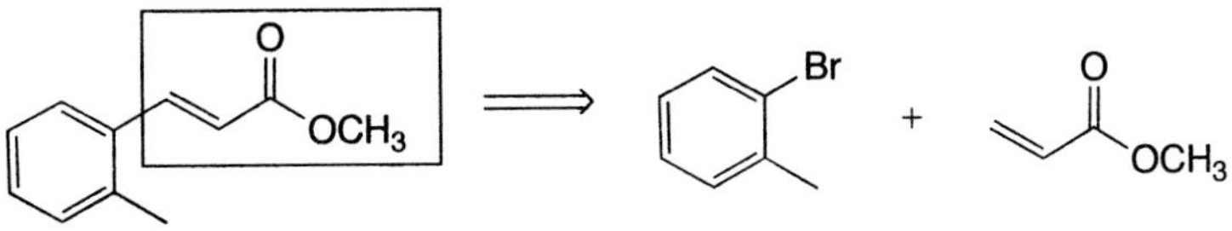

b.

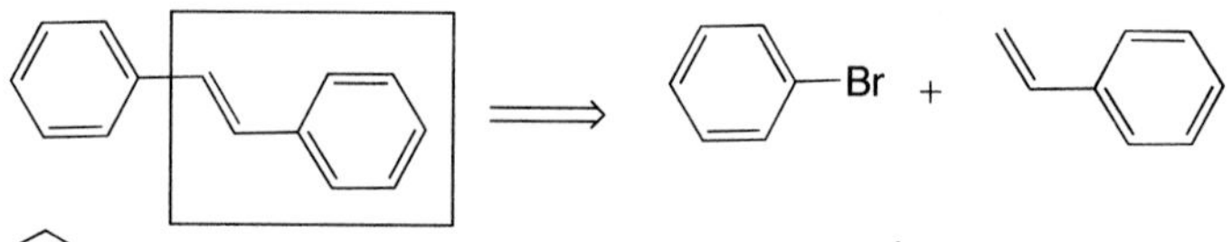

c.

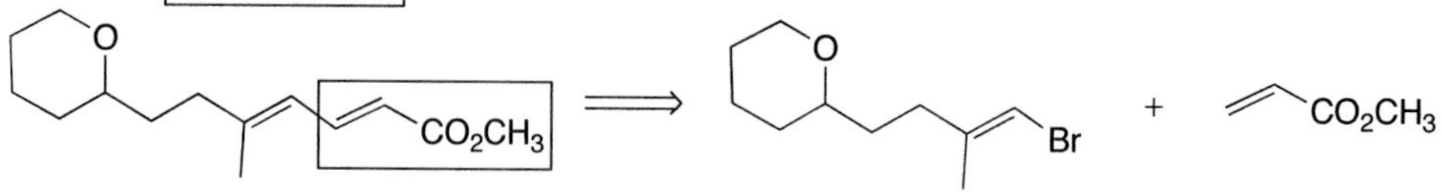

**26.8** Add the carbene carbon from either side of the alkene.

a.

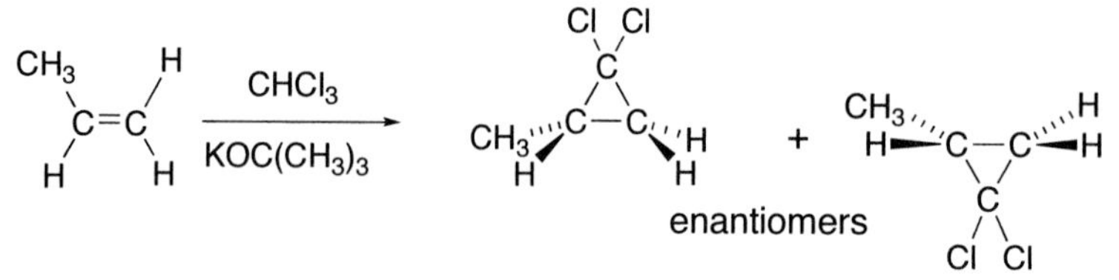

enantiomers

b.

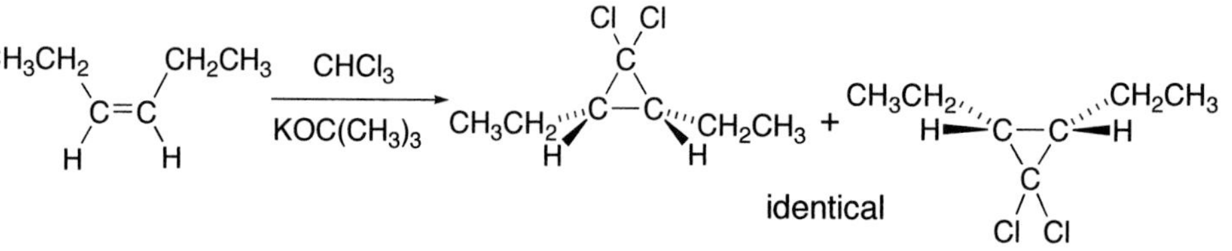

identical

c.

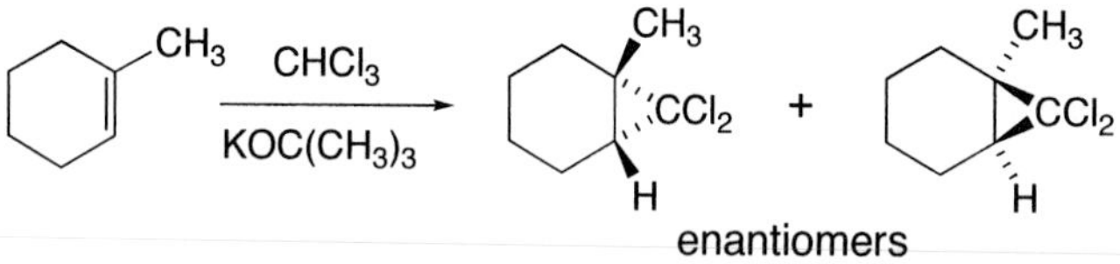

enantiomers

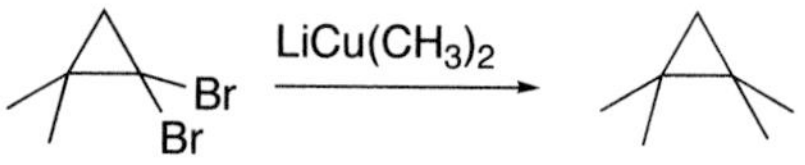

**26.9**

a.

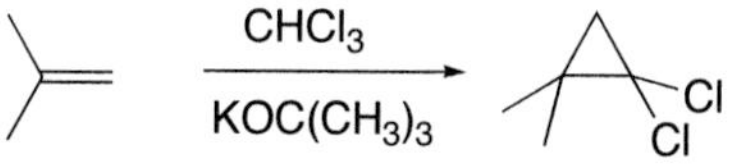

b.

c.

(from b.)

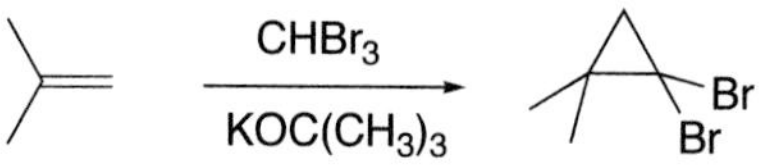

**26.10**

a.

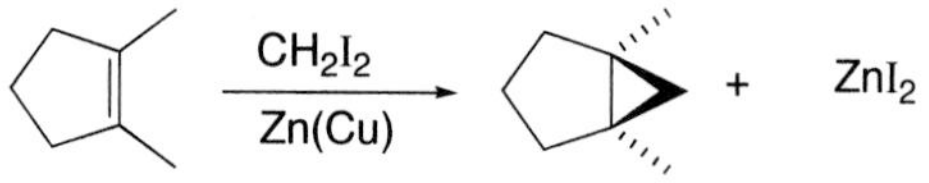

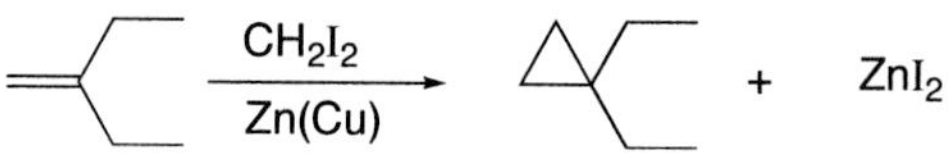

b.

c.

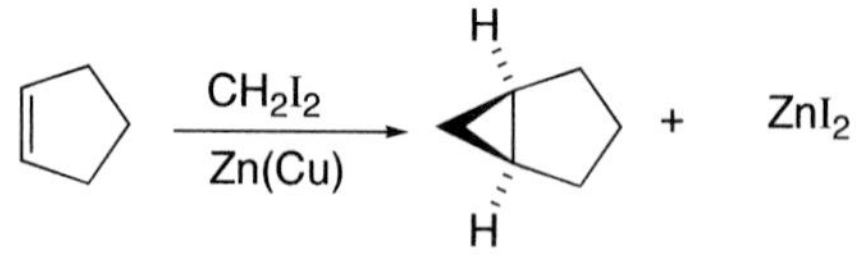

**26.11** The relative position of substituents in the reactant is retained in the product.

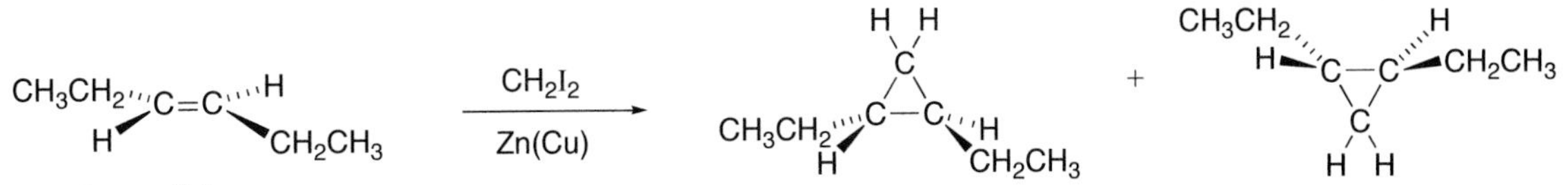

trans-3-hexene

two enantiomers of trans-1,2-diethylcyclopropane

**26.12**

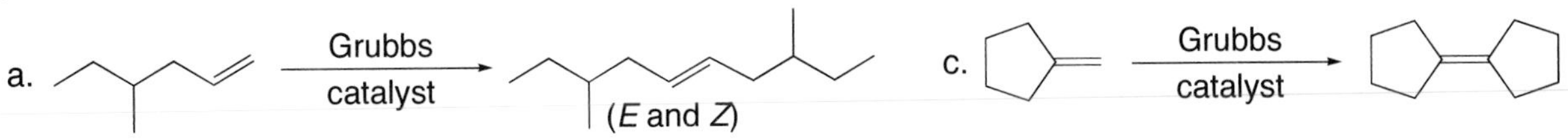

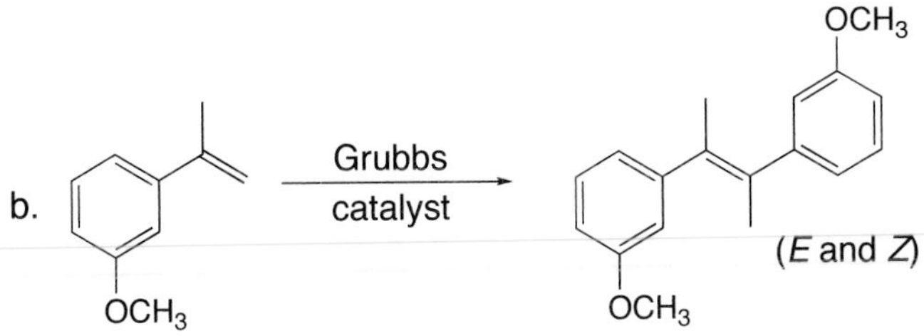

a.

(E and Z)

c.

b.

(E and Z)

(CH₂=CH₂ is also formed in each reaction.)

**26.13**

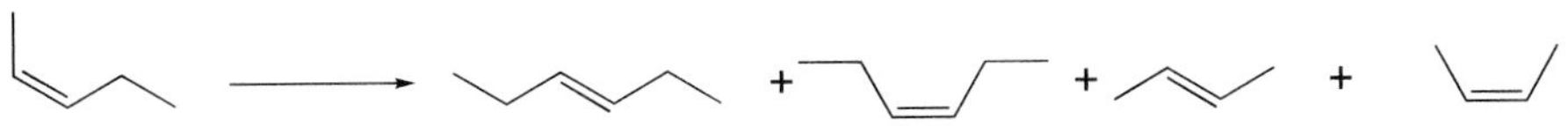

cis-2-pentene

There are four products formed in this reaction including stereoisomers, and therefore, it is not a practical method to synthesize 1,2-disubstituted alkenes.

**26.14**

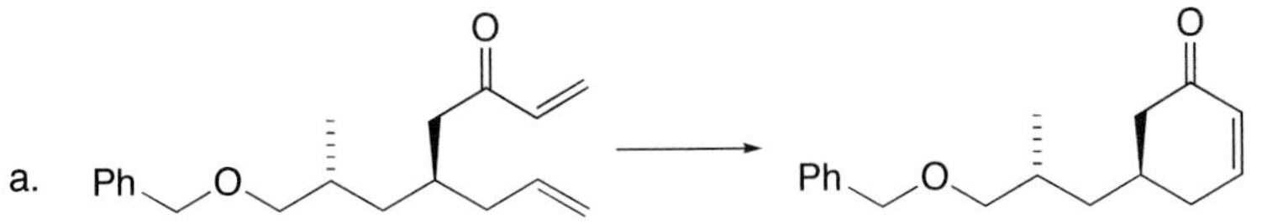

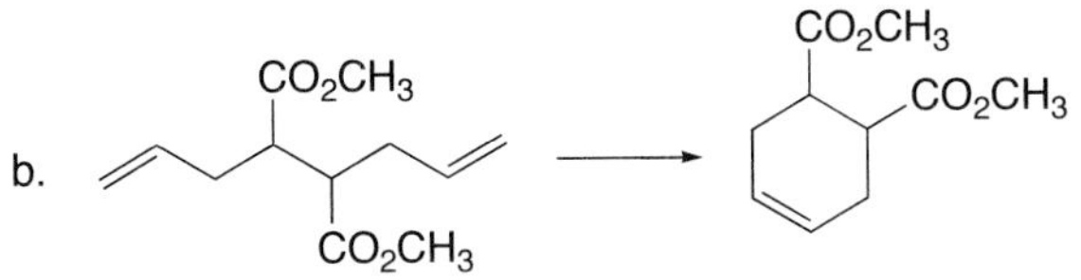

a.

b.

**26.15** High dilution conditions favor intramolecular metathesis.

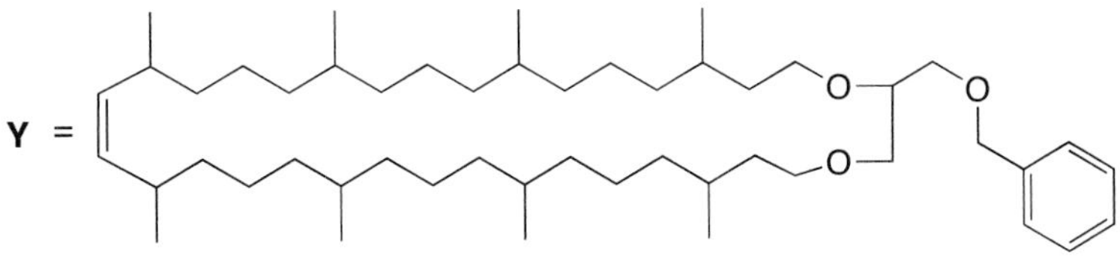

Y =

Under high dilution conditions, an intramolecular reaction is favored, resulting in **Y**. Two different molecules do not often come into contact, so two double bonds in the same molecule react.

**1**

long chain

**2**

long chain

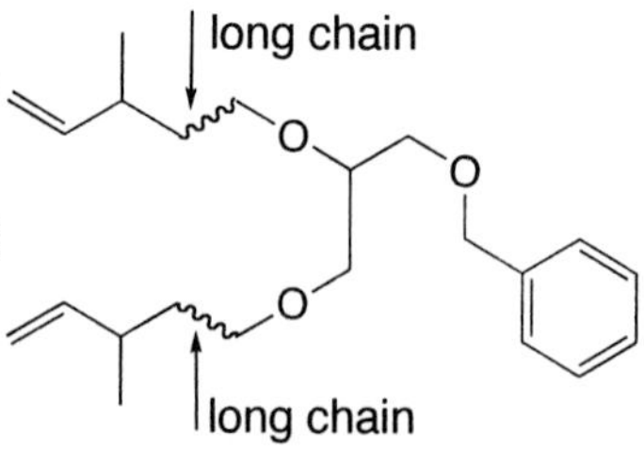

Alkenes **1** and **2** differ in proximity to the ether side chain. Three products are possible because alkene **1** can react with alkene **1** in another molecule; alkene **1** can react with alkene **2** in another molecule; and alkene **2** can react with alkene **2** in another molecule. At usual reaction concentrations, the probability of two molecules approaching each other is greater than the probability of two sites (connected by a long chain) in the same molecule, so the intermolecular reaction is favored.

**Z** =

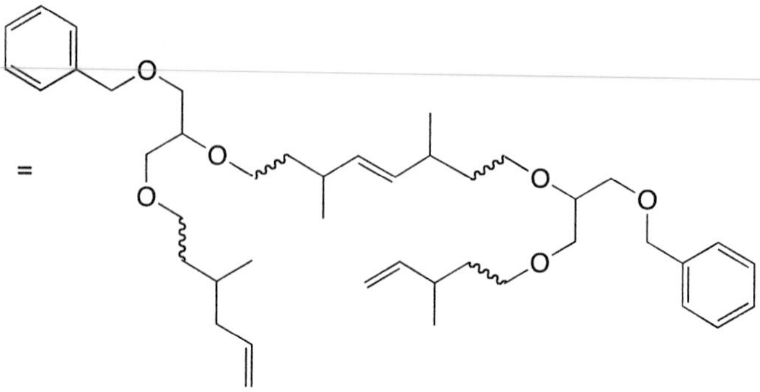

**Z'** =

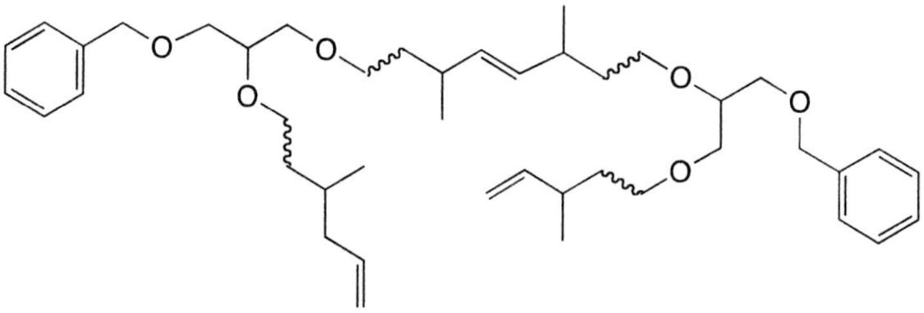

**Z"** =

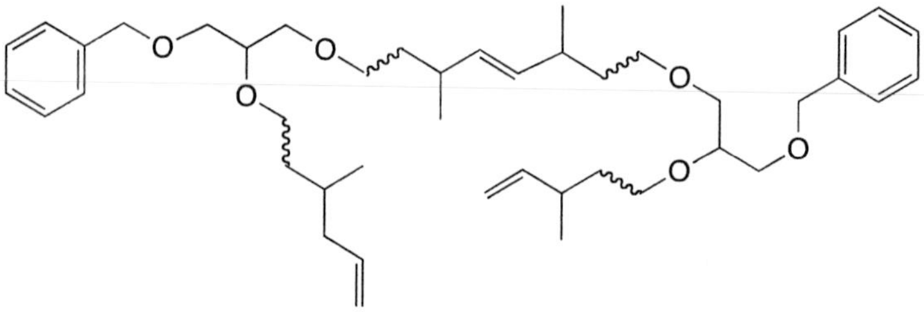

**26.16**

a.

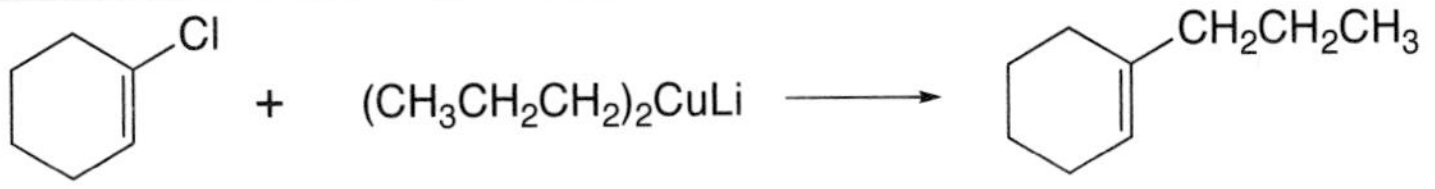

b.

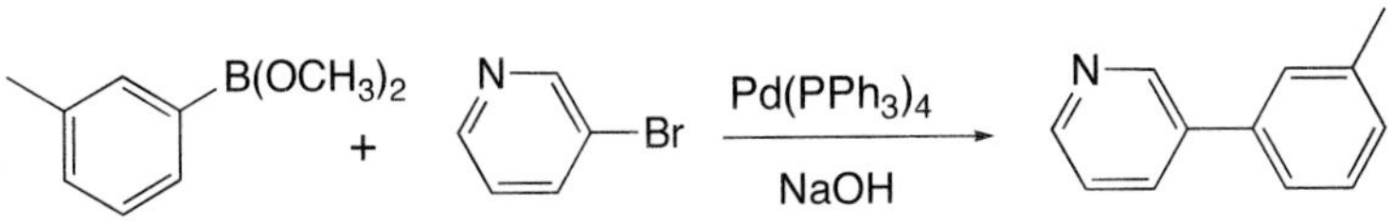

c.

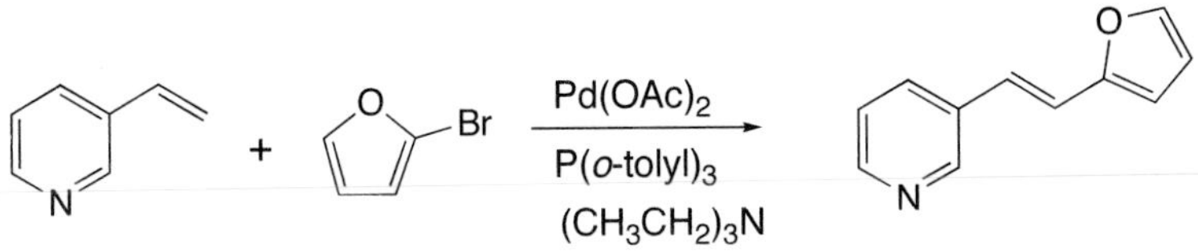

d.

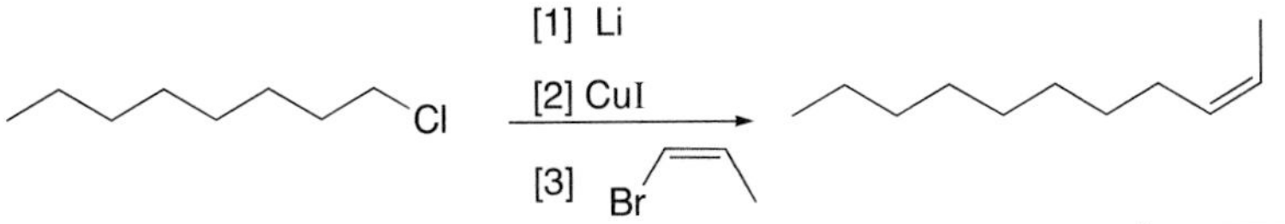

e.

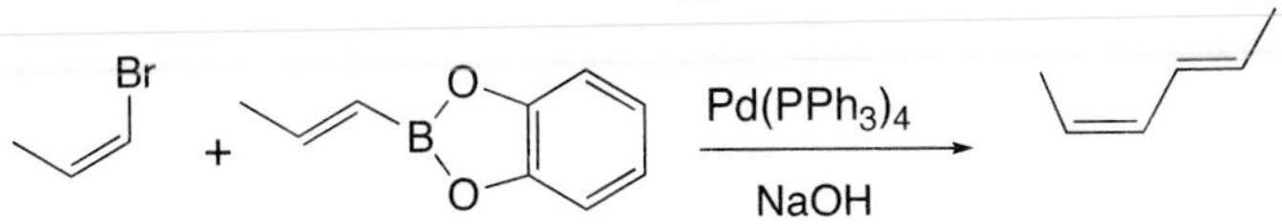

f. 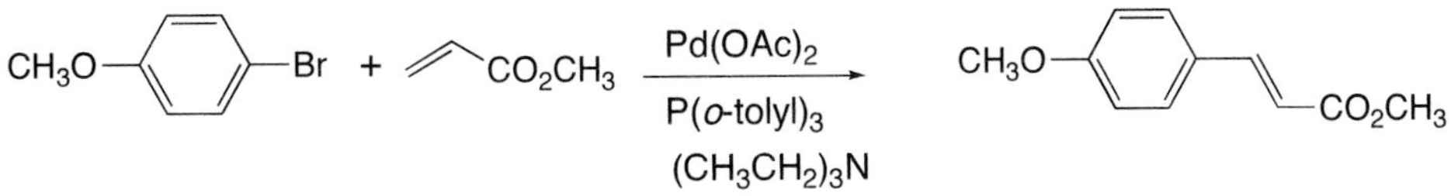 $CH_3O$—(ring)—Br + (alkene)$CO_2CH_3$ $\xrightarrow[\substack{P(o\text{-tolyl})_3 \\ (CH_3CH_2)_3N}]{Pd(OAc)_2}$ $CH_3O$—(ring)—(alkene)$CO_2CH_3$

g.

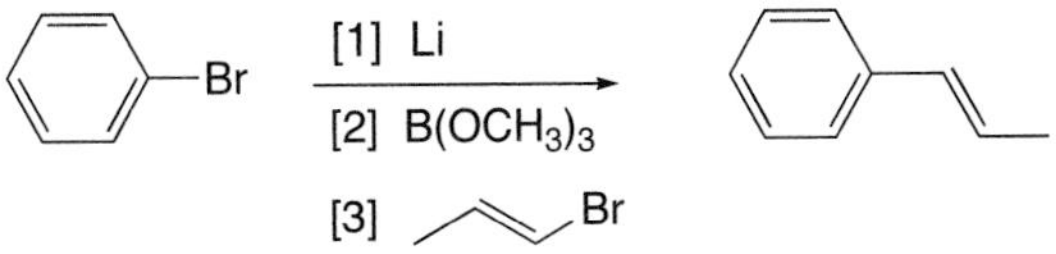

h. 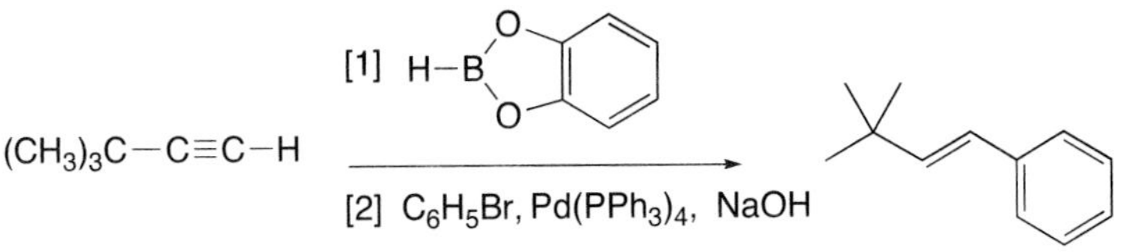 $(CH_3)_3C-C\equiv C-H$ $\xrightarrow[\text{[2] } C_6H_5Br, Pd(PPh_3)_4, \text{ NaOH}]{\text{[1] } H-B(\text{catecholborane})}$

**26.17**

a.

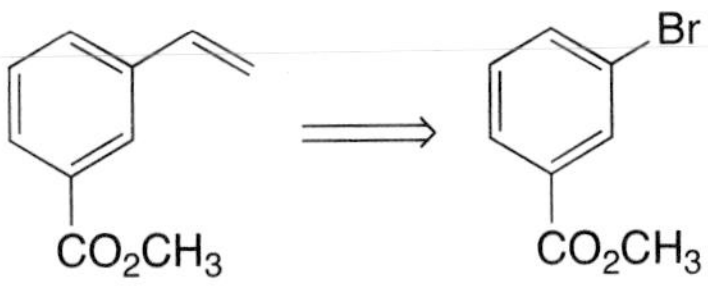

b.

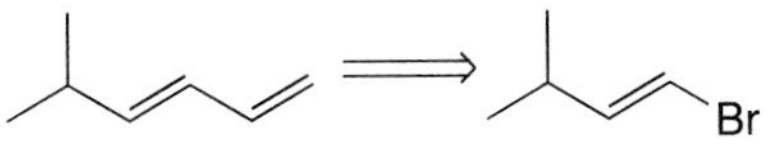

c.

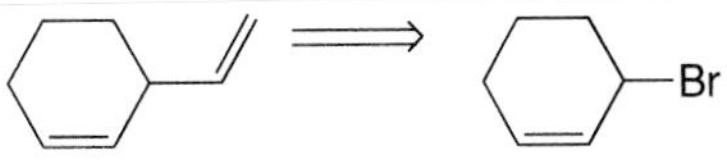

d.

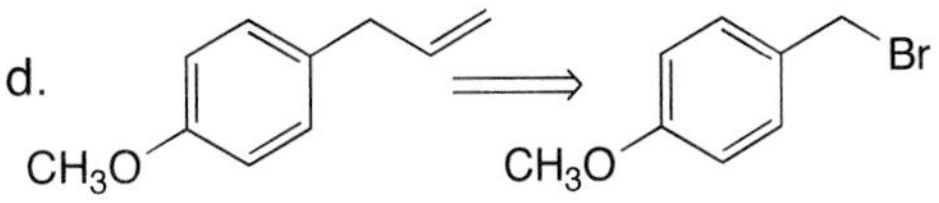

**26.18**

Each coupling reaction uses Pd(PPh₃)₄
and NaOH to form the conjugated diene.

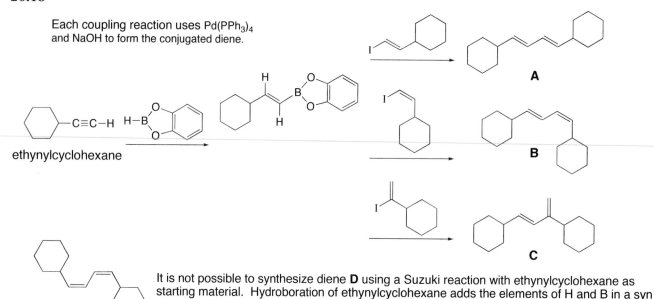

ethynylcyclohexane

**A**

**B**

**C**

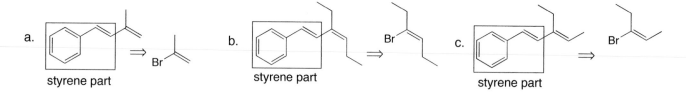

It is not possible to synthesize diene **D** using a Suzuki reaction with ethynylcyclohexane as starting material. Hydroboration of ethynylcyclohexane adds the elements of H and B in a syn fashion affording a trans vinylborane. Since the Suzuki reaction is stereospecific, one of the double bonds in the product must therefore be trans.

**D**

**26.19** Locate the styrene part of the molecule, and break the molecule into two components.

a.

styrene part

b.

styrene part

c.

styrene part

**26.20** Inversion of configuration occurs with substitution of the methyl group for the tosylate.

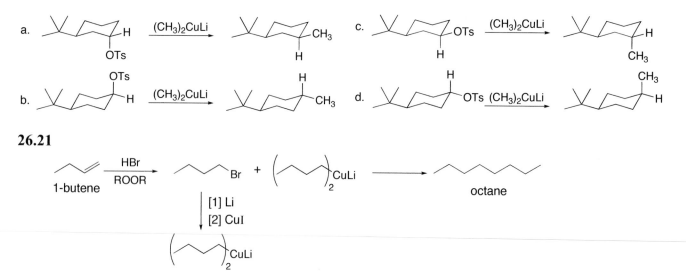

a.     (CH₃)₂CuLi

b.     (CH₃)₂CuLi

c.     (CH₃)₂CuLi

d.     (CH₃)₂CuLi

**26.21**

1-butene    HBr / ROOR

[1] Li
[2] CuI

**26.22** Add the carbene carbon from either side of the alkene.

a.

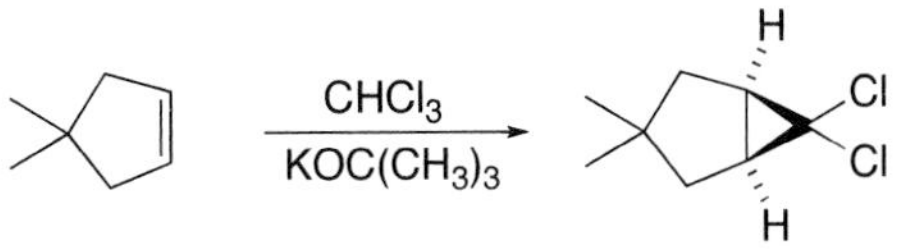

b.

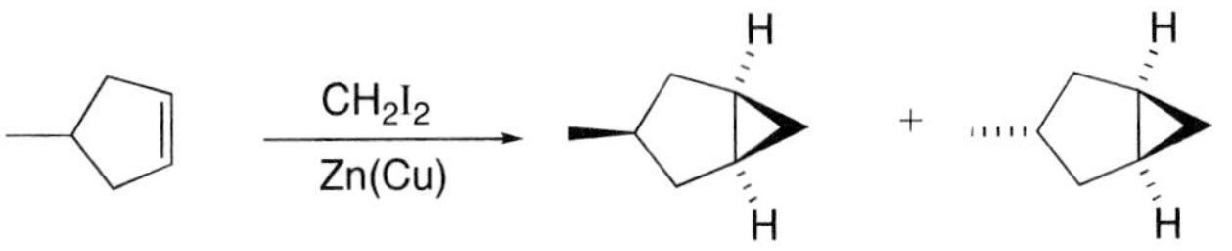

c.

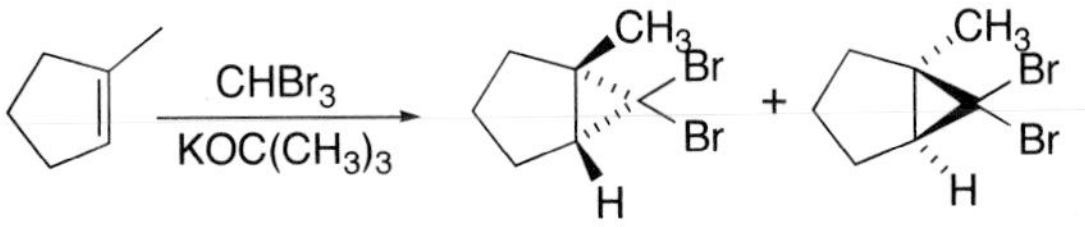

d.

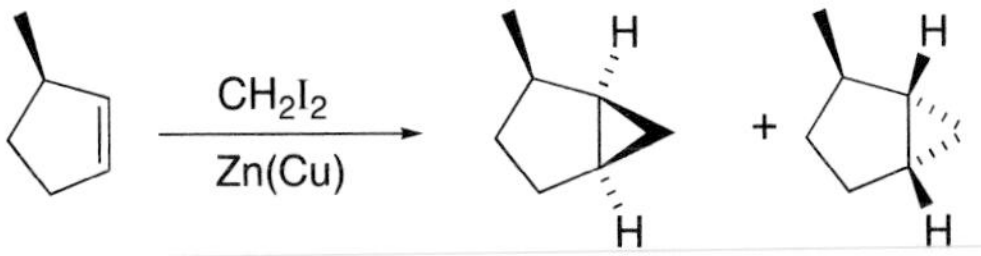

e.

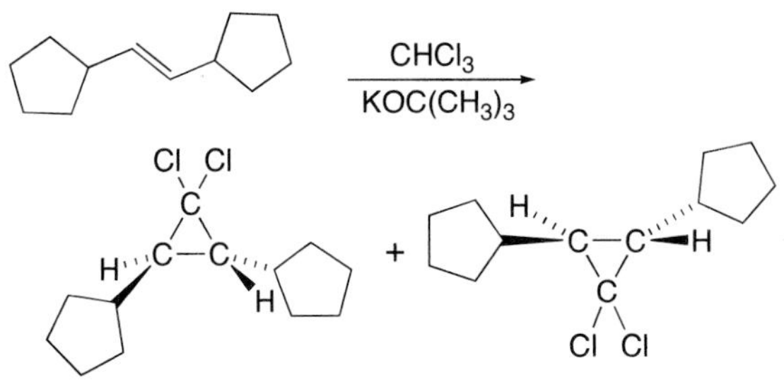

f.

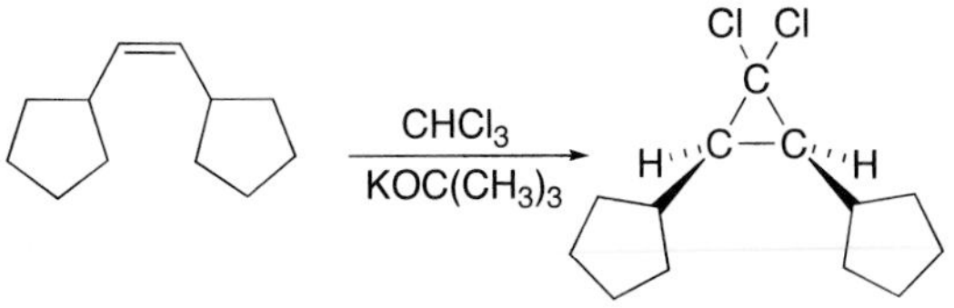

**26.23** Since the new three-membered ring has a stereogenic center on the C bonded to the phenyl group, the phenyl group can be oriented in two different ways to afford two stereoisomers. These products are diastereomers of each other.

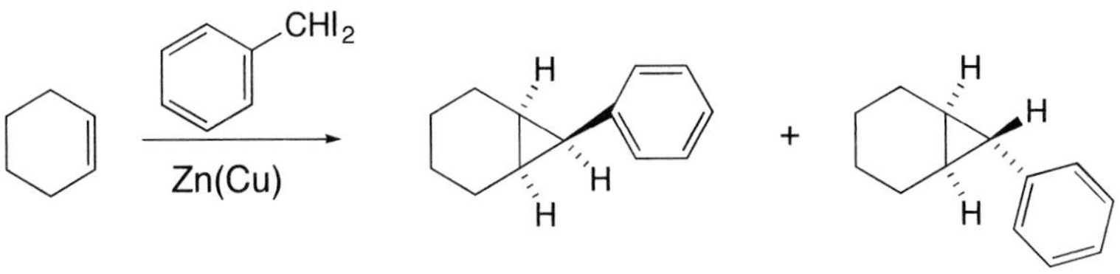

**26.24** High dilution conditions favor intramolecular metathesis.

a.

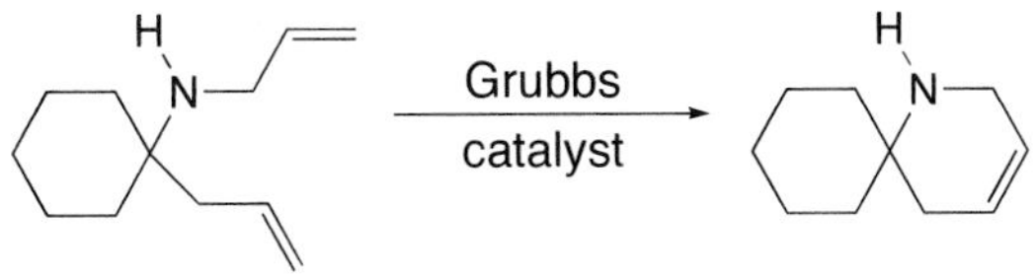

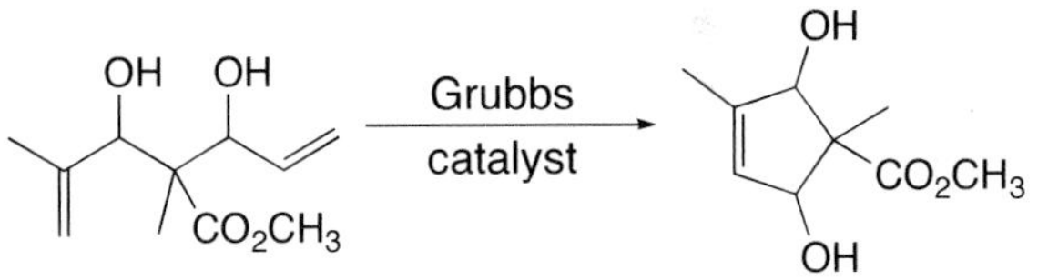

b.

c.

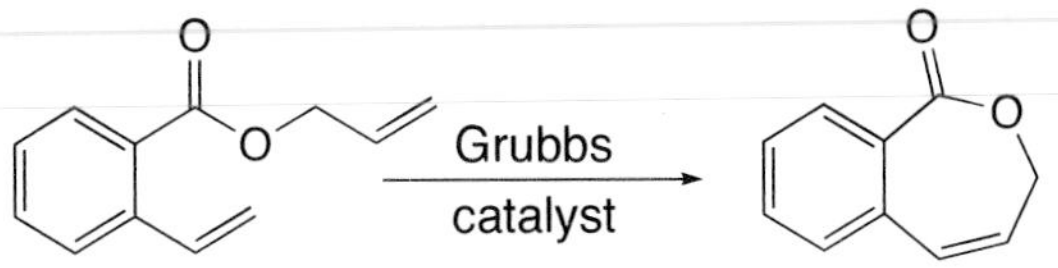

**26.25** Retrosynthetically break the double bond in the cyclic compound and add a new $=CH_2$ at each end to find the starting material.

a.

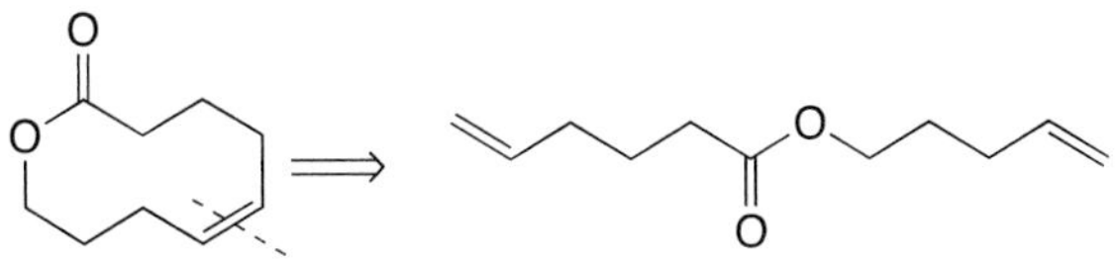

b.

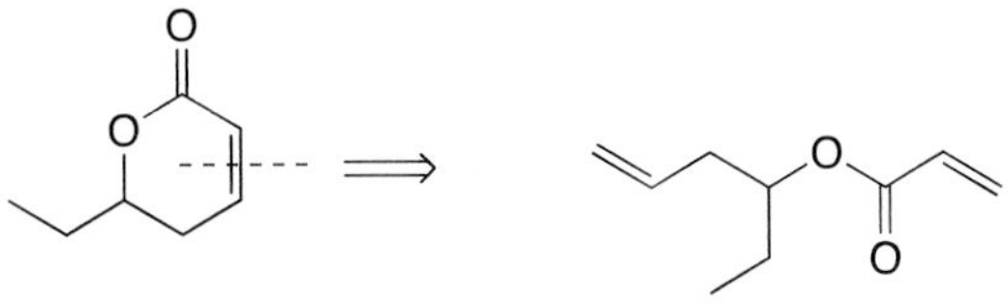

c.

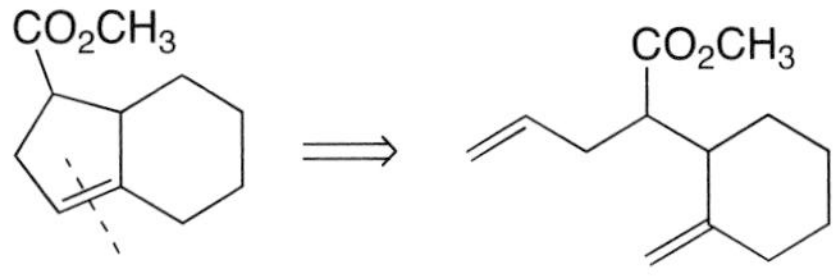

**26.26** Alkene metathesis with two different alkenes is only synthetically useful when both alkenes are symmetrically substituted; that is, the two groups on each end of the double bond are identical to the two groups on the other end of the double bond.

a.  CH$_3$CH$_2$CH=CH$_2$

   +  ⟶  CH$_3$CH$_2$CH$_2$CH=CHCH$_2$CH$_3$  +  CH$_3$CH$_2$CH=CHCH$_2$CH$_3$  +  CH$_3$CH$_2$CH$_2$CH=CHCH$_2$CH$_2$CH$_3$

   CH$_3$CH$_2$CH$_2$CH=CH$_2$     + CH$_2$=CH$_2$   (*Z* + *E*)                 (*Z* + *E*)                      (*Z* + *E*)

b.

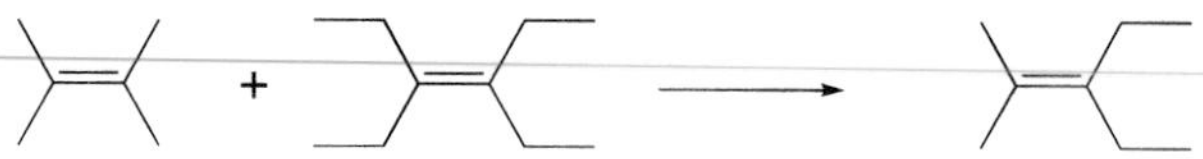

This reaction is synthetically useful since it yields only one product.

c.

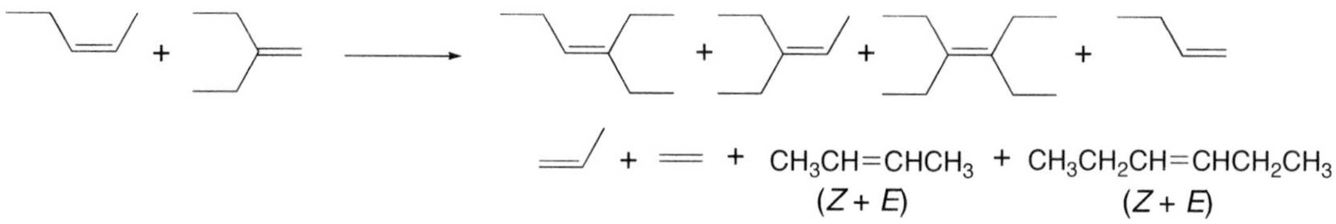

⟶  + = + CH$_3$CH=CHCH$_3$ + CH$_3$CH$_2$CH=CHCH$_2$CH$_3$

(*Z* + *E*)          (*Z* + *E*)

**26.27**

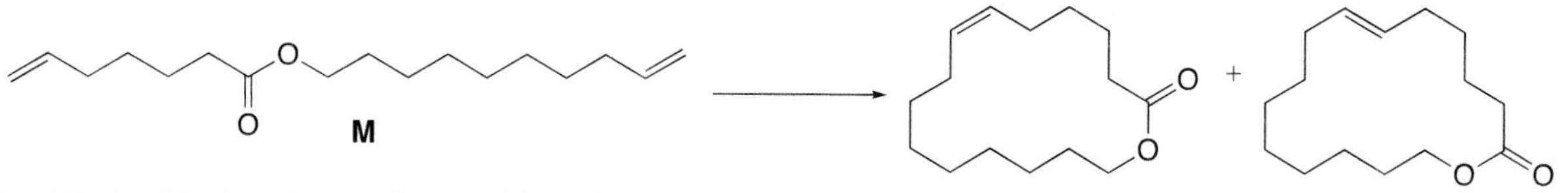

**26.28** All double bonds can have either the *E* or *Z* configuration.

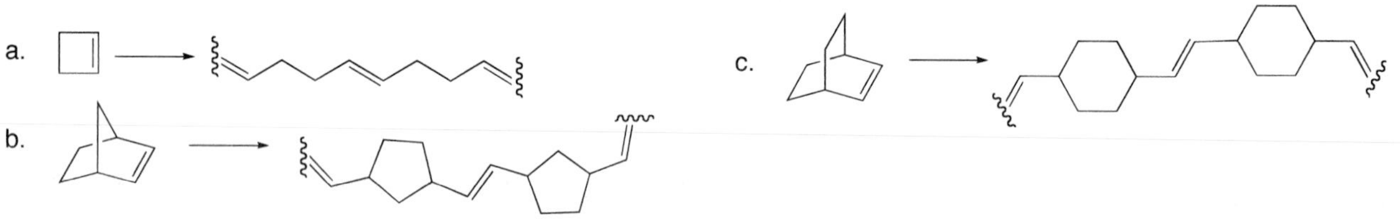

a.

c.

b.

**26.29**

a.

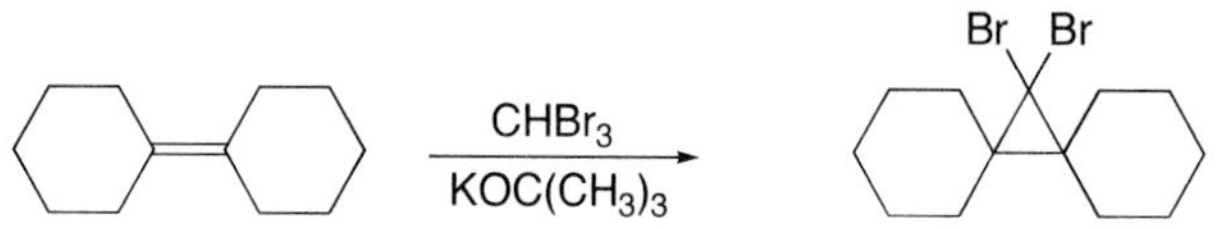

CHBr$_3$ / KOC(CH$_3$)$_3$

b.

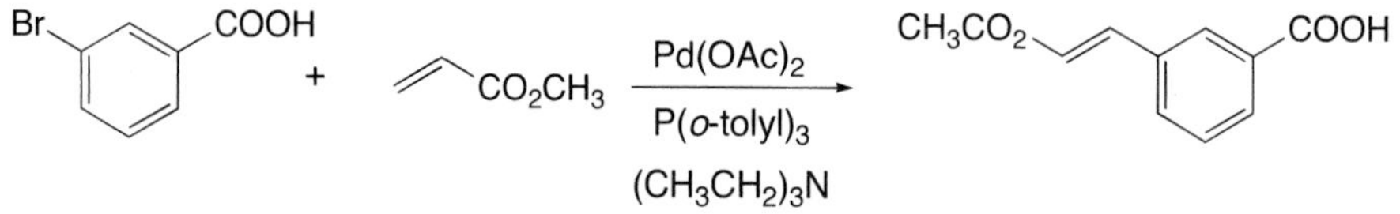

Pd(OAc)$_2$ / P(*o*-tolyl)$_3$ / (CH$_3$CH$_2$)$_3$N

c.

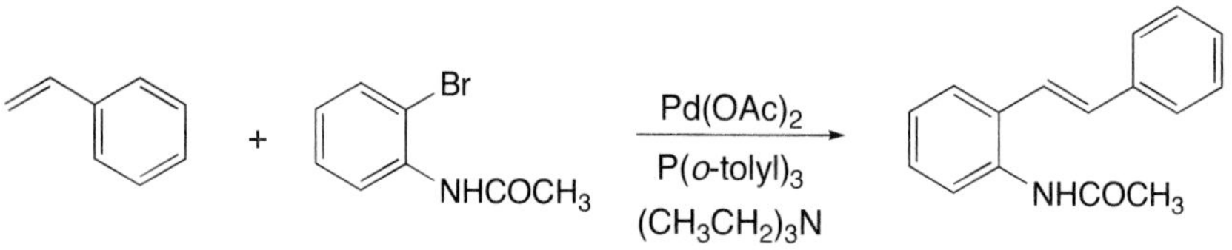

Pd(OAc)$_2$ / P(*o*-tolyl)$_3$ / (CH$_3$CH$_2$)$_3$N

d.

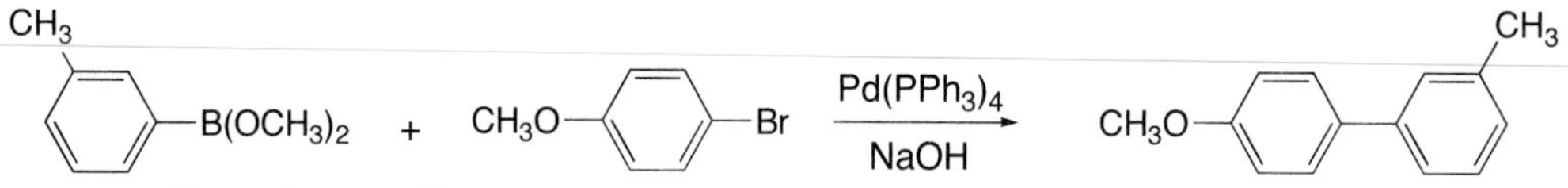

Pd(PPh$_3$)$_4$ / NaOH

e.

Pd(PPh$_3$)$_4$ / NaOH

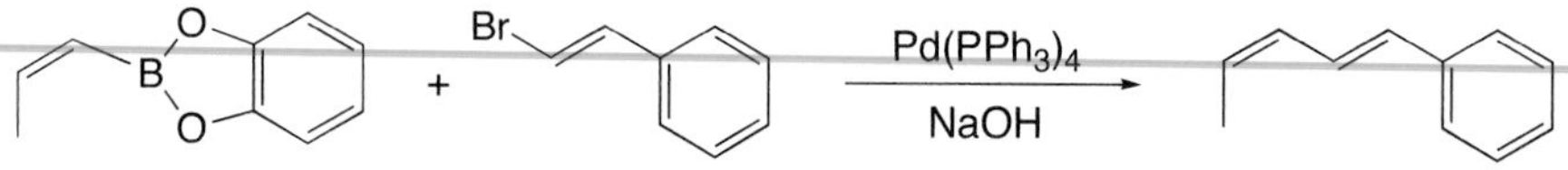

f.

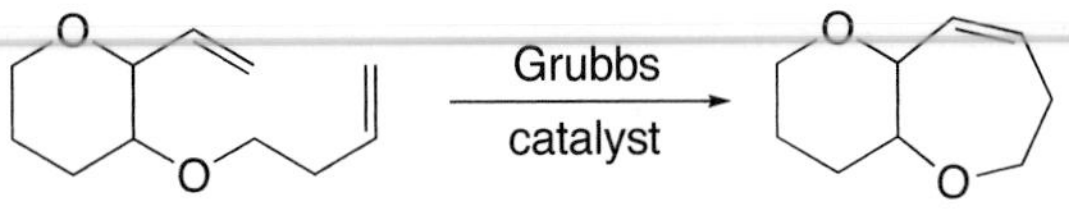

g.

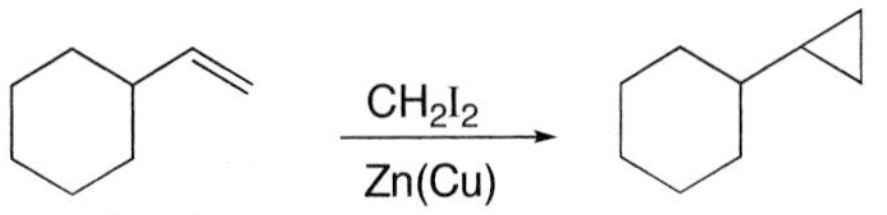

h.

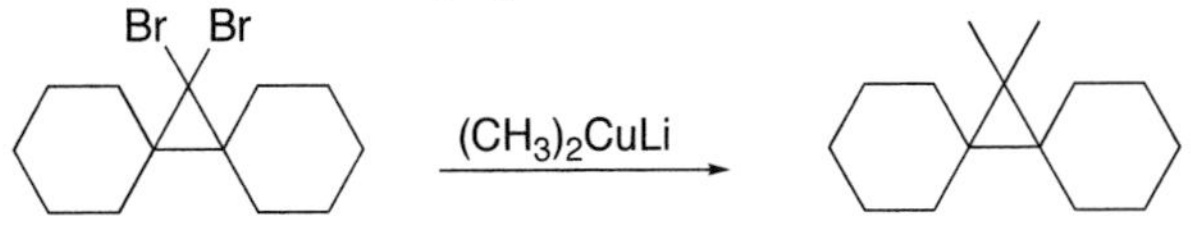

Br Br

Product in (a)

(CH₃)₂CuLi

i.

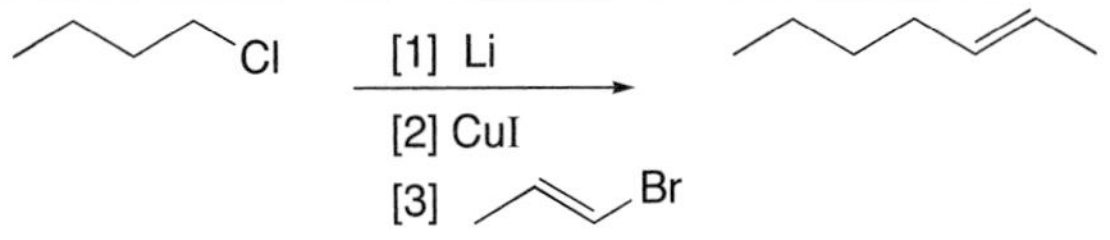

[1] Li
[2] CuI
[3] Br

j. CH₃CH₂–C≡C–H

[1] H–B (catecholborane)

[2] isopropyl–Br

Pd(PPh₃)₄
NaOH

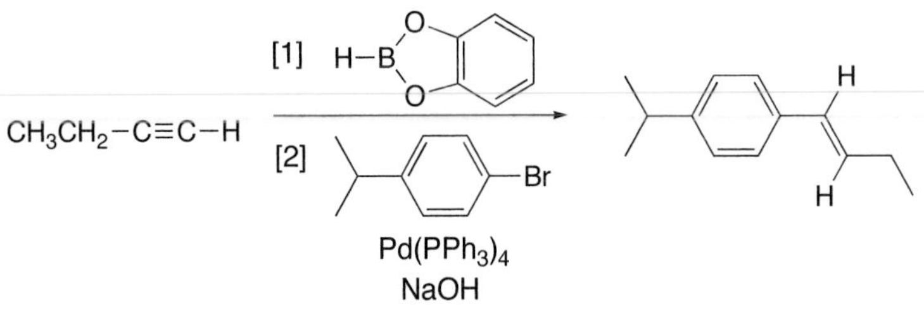

**26.30**

$$Cl_3C-C(:O:)(:O:Na^+) \xrightarrow{\Delta} Cl-C:Cl \ Na^+ + \ddot{O}=C=\ddot{O} \longrightarrow Cl-C:Cl + CO_2 + NaCl$$

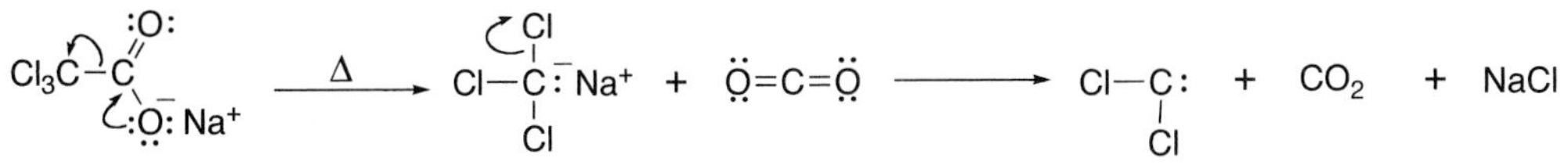

**26.31** This reaction follows the Simmons–Smith reaction mechanism illustrated in Mechanism 26.5.

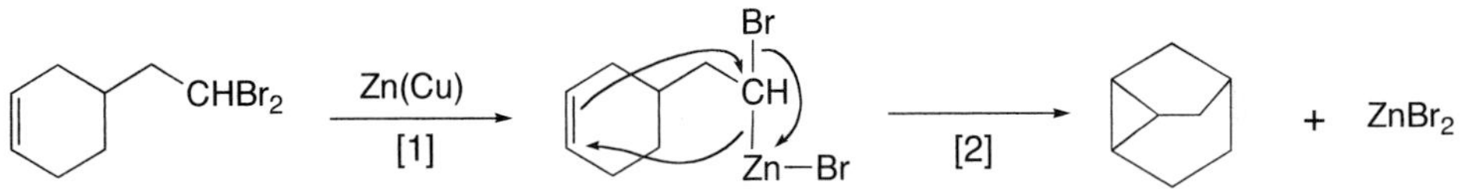

$$\text{CHBr}_2 \xrightarrow[\text{[1]}]{\text{Zn(Cu)}} \cdots \xrightarrow{\text{[2]}} \cdots + \text{ZnBr}_2$$

**26.32**

a sulfur ylide

1,4-addition of the
sulfur ylide to the
β carbon

a resonance-stabilized
enolate

methyl *trans*-chrysanthemate

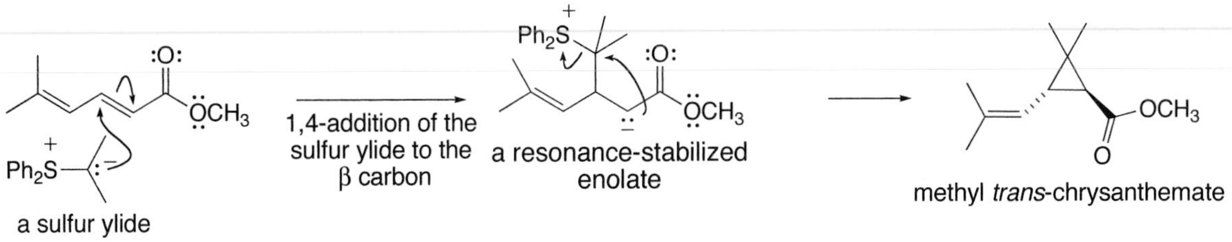

**26.33**

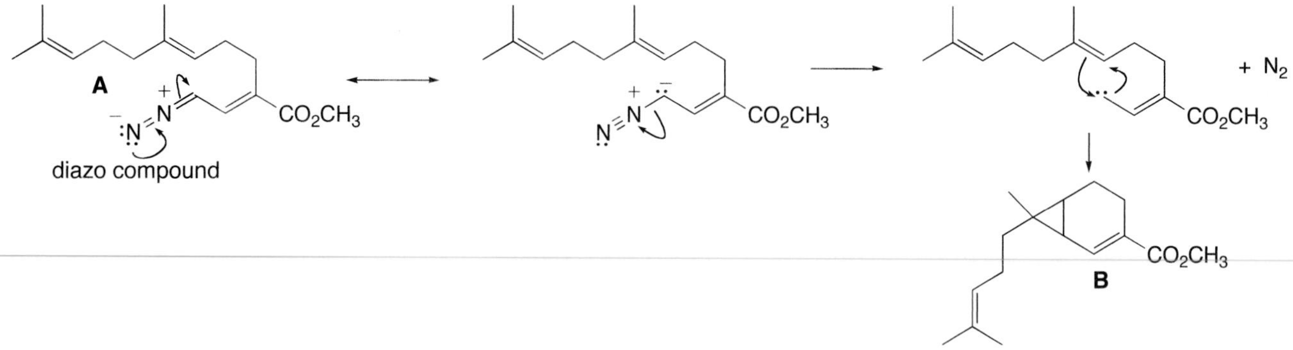

**26.34**

a.

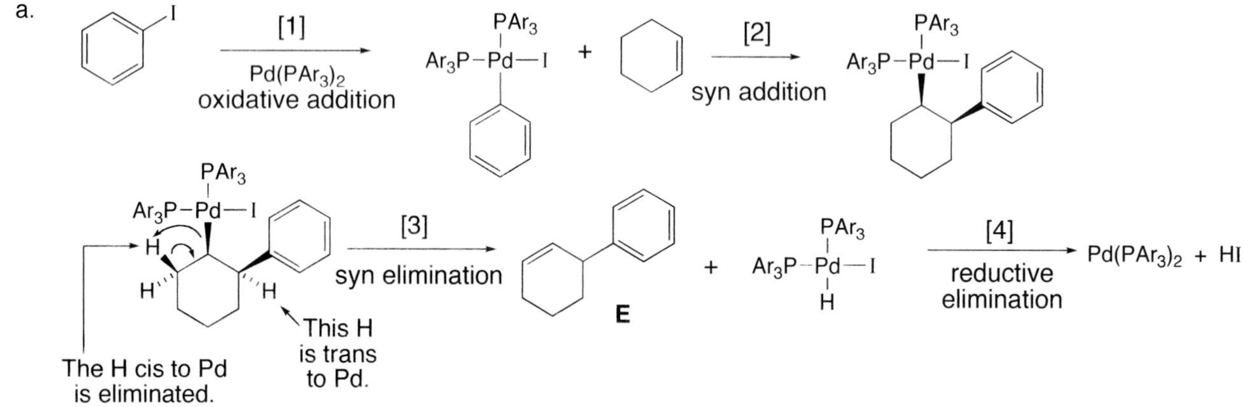

b. This suggests that the stereochemistry in Step [3] must occur with syn elimination of H and Pd to form **E**. Product **F** cannot form because the only H on the C bonded to the benzene ring is trans to the Pd species, and therefore it cannot be removed if elimination occurs in a syn fashion.

**26.35**

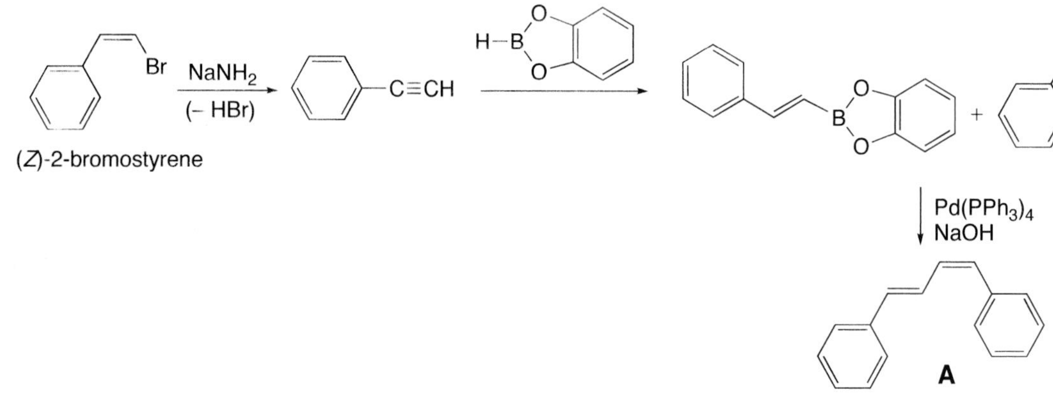

**26.36**

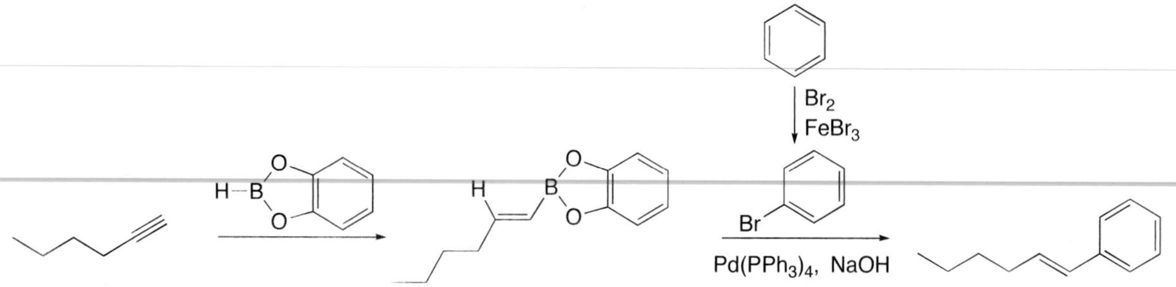

**26.37**

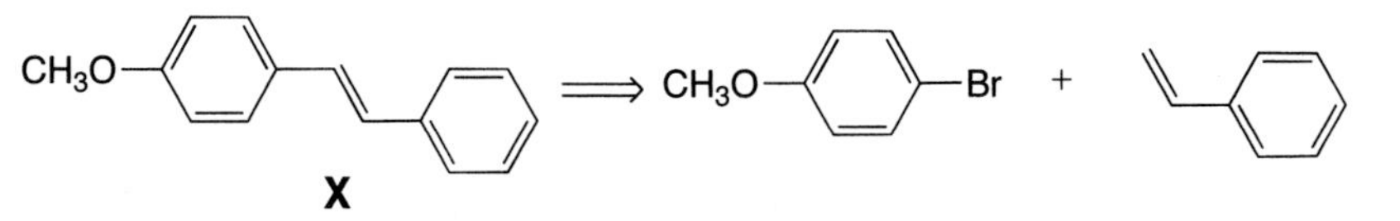

Synthesize these two components, and then use a Heck reaction to synthesize **X**.

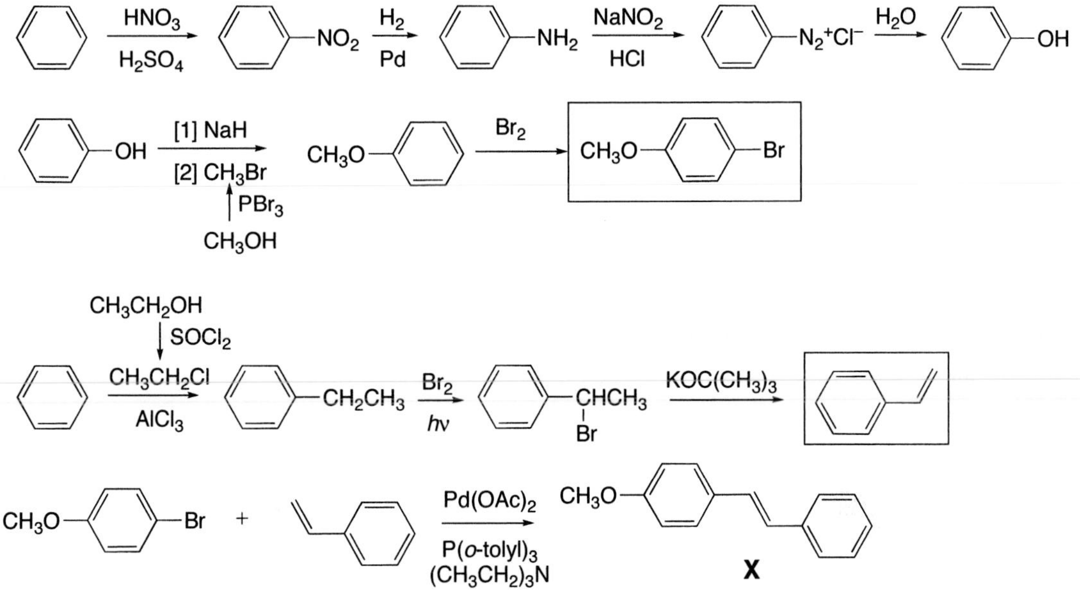

**26.38**

a.

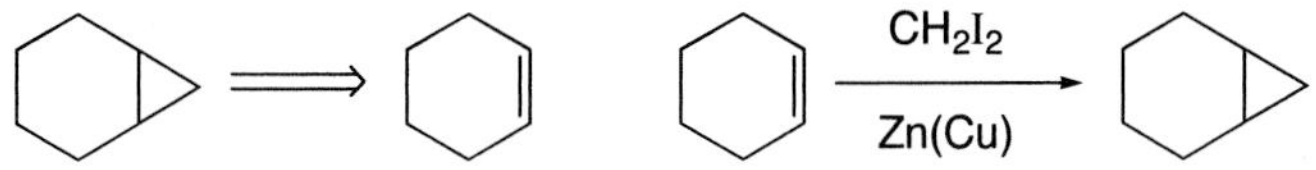

b.

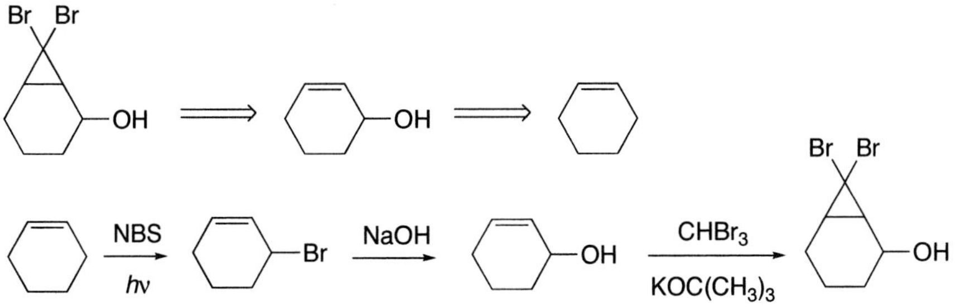

c.

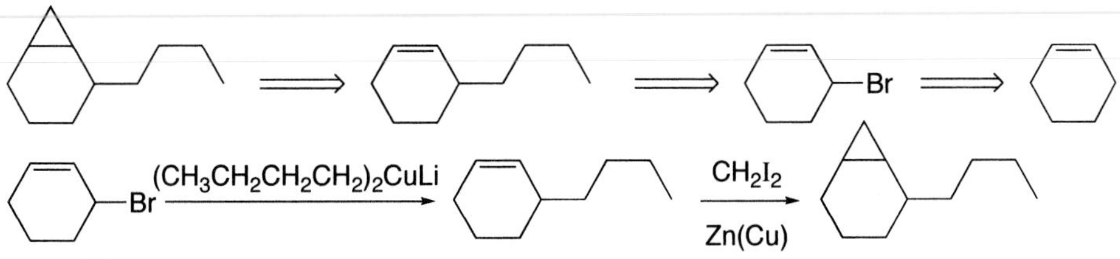

(from b.)

d.

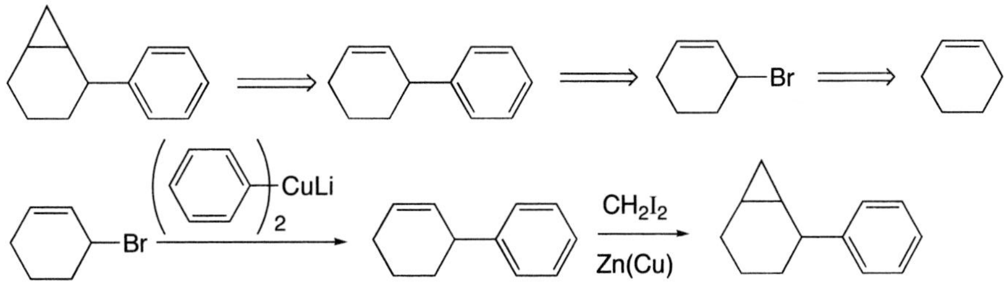

(from b.)

**26.39**

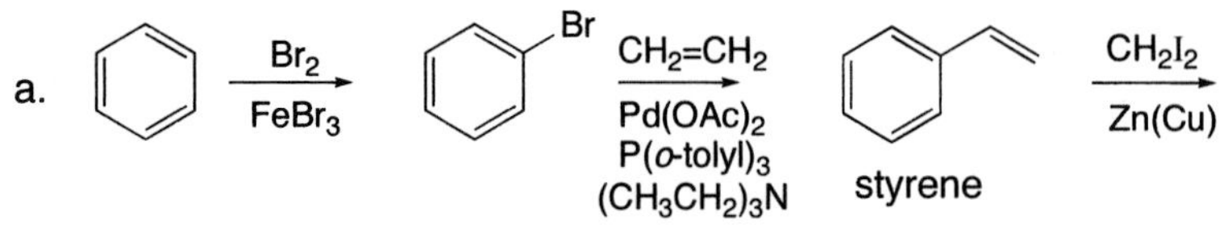

a.

For an alternate synthesis of styrene see Problem 26.37.

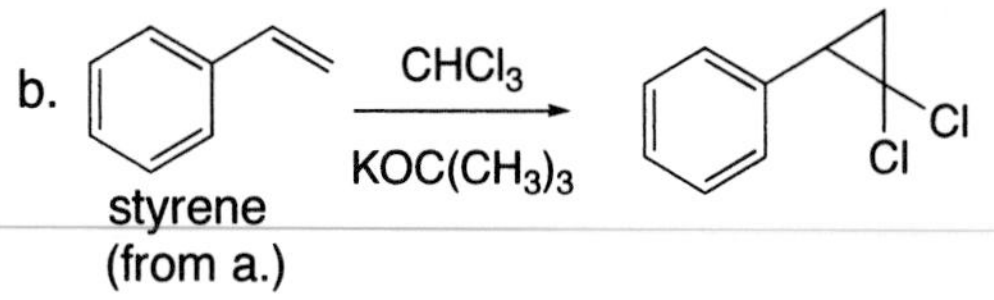

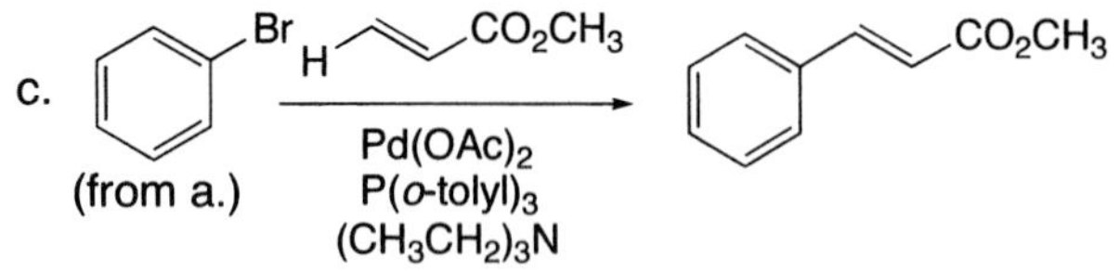

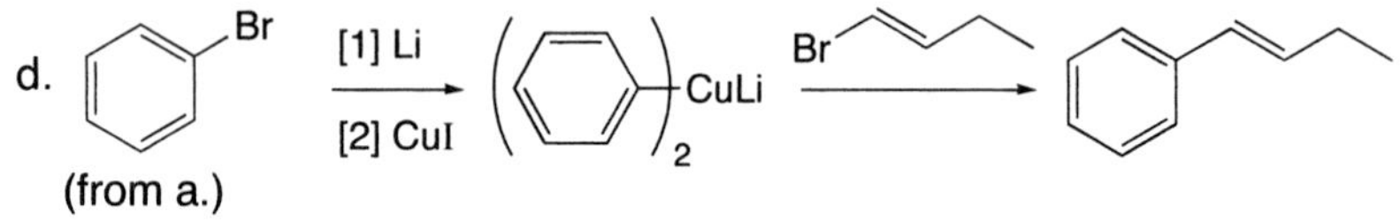

b.

styrene
(from a.)

c.

(from a.)

d.

(from a.)

**26.40**

a.

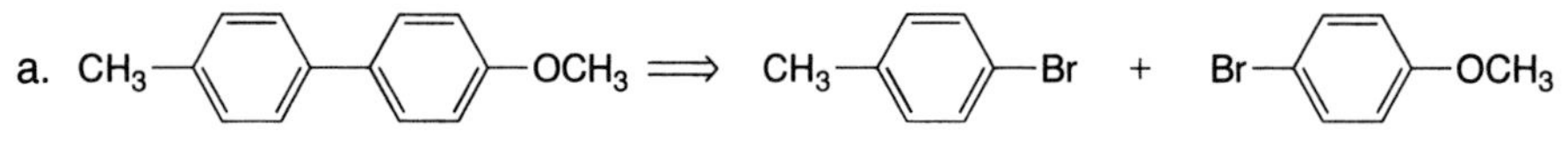

Either compound can be used to synthesize the organoborane, so two routes are possible.

Possibility [1]:

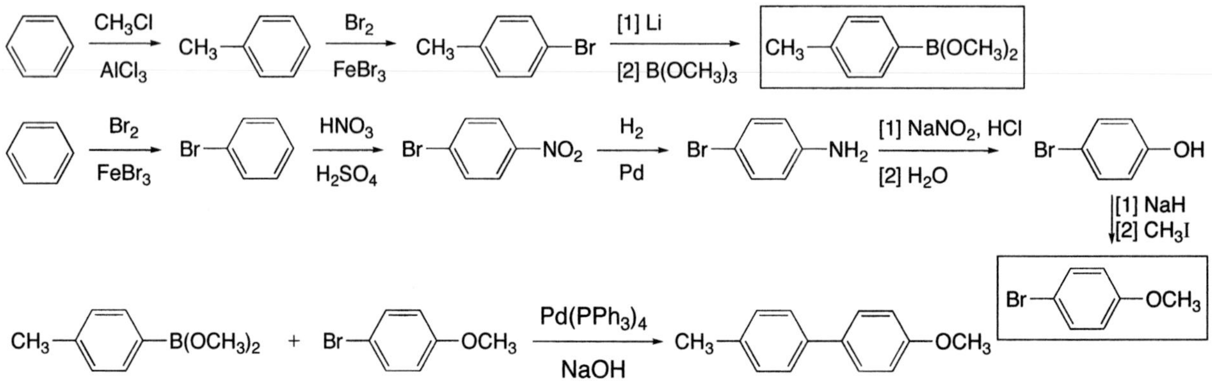

Possibility [2]:

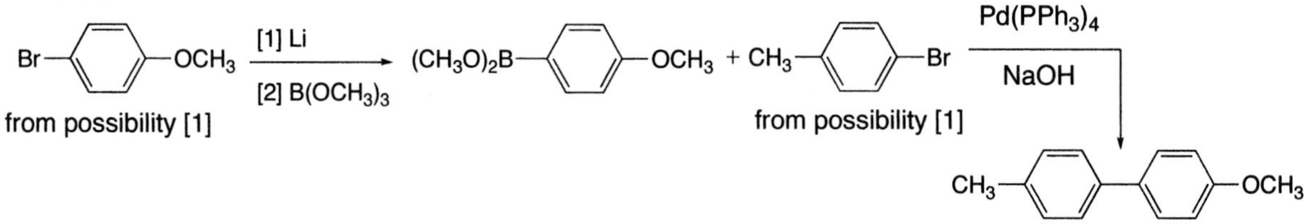

b.

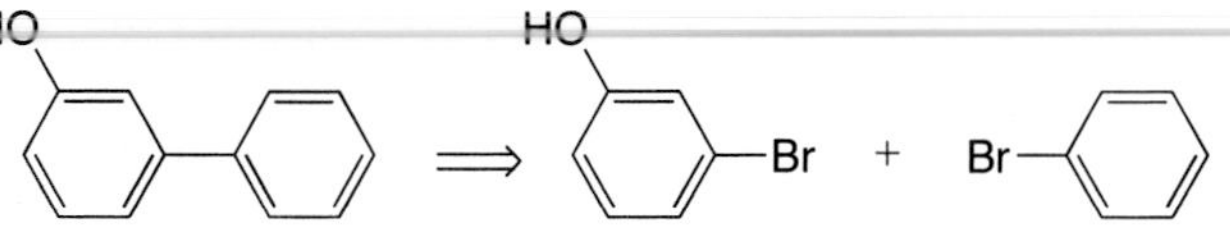

The acidic OH makes it impossible to prepare an organolithium reagent from this aryl halide, so this compound must be used as the aryl halide that couples with the organoborane from bromobenzene.

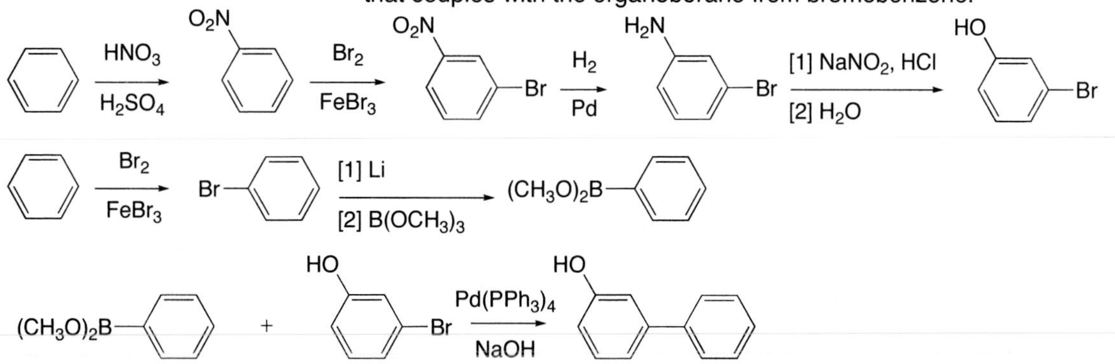

c.

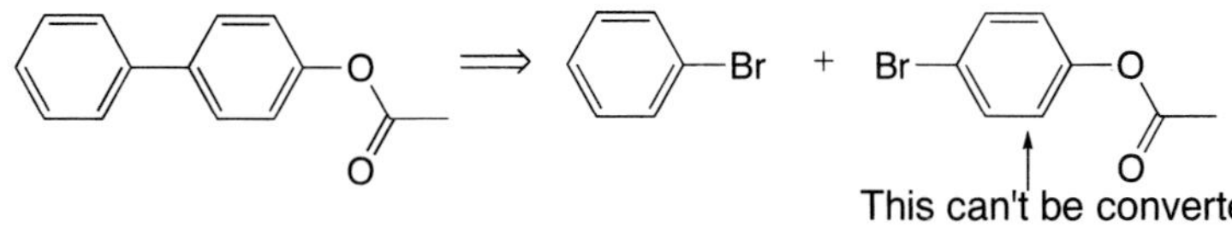

This can't be converted to an organoborane reagent via an organolithium reagent.

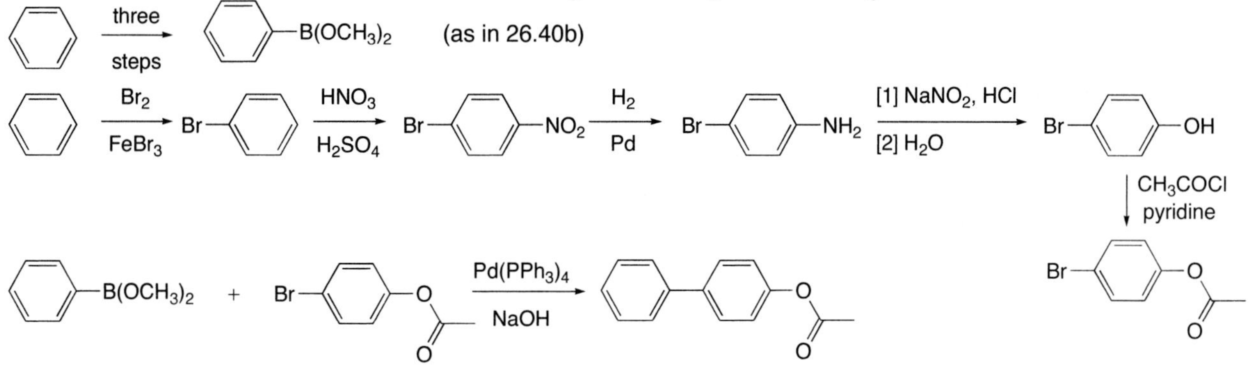

**26.41**

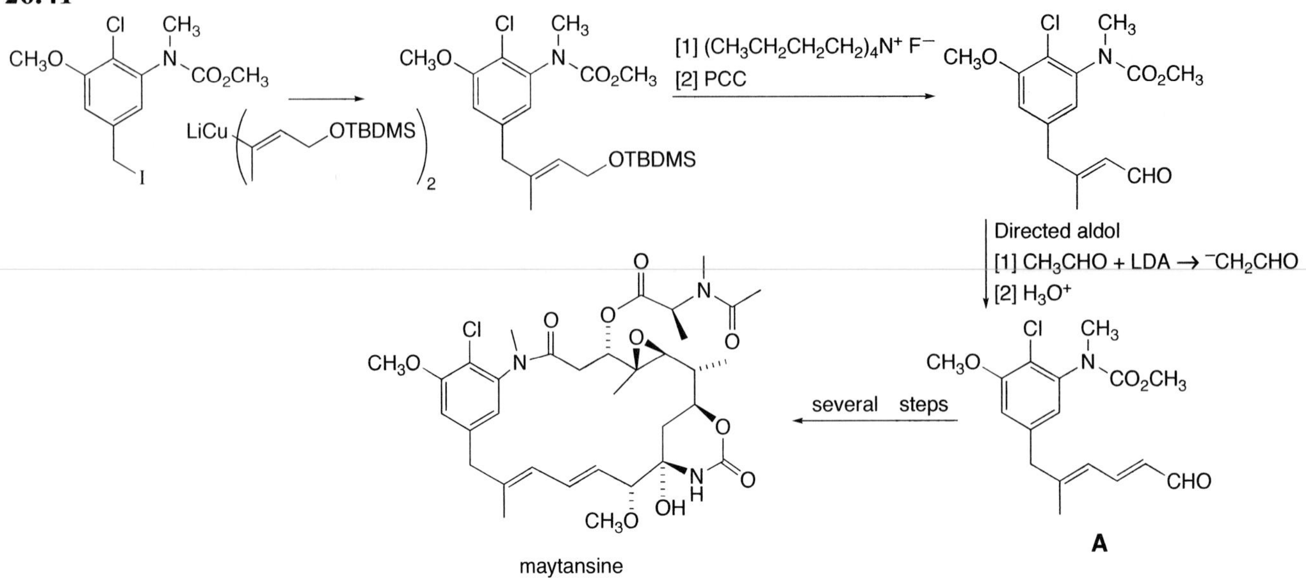

maytansine

**26.42**

a.

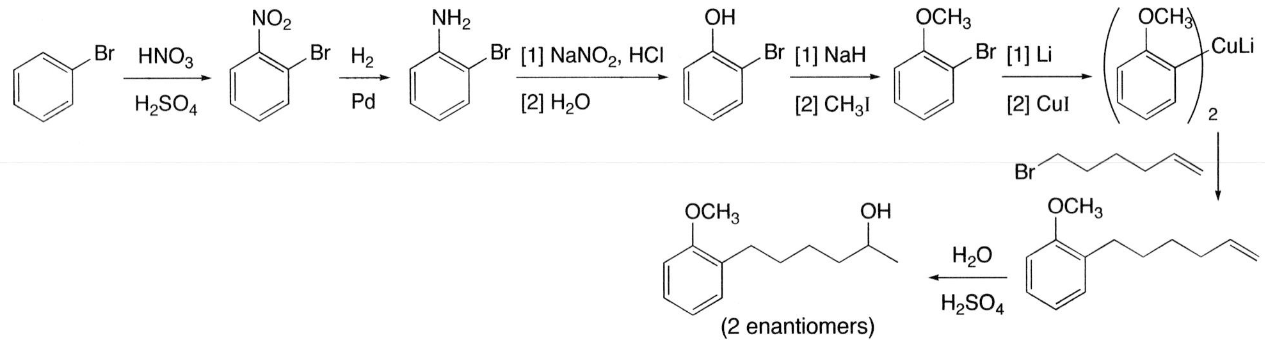

(2 enantiomers)

b.

(2 enantiomers)

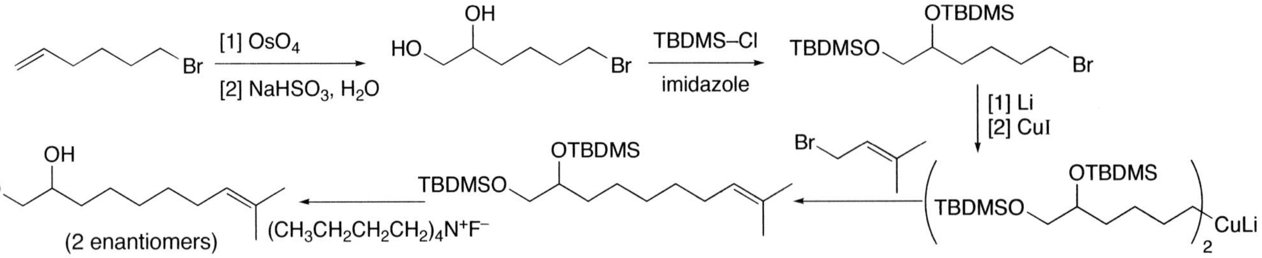

c.

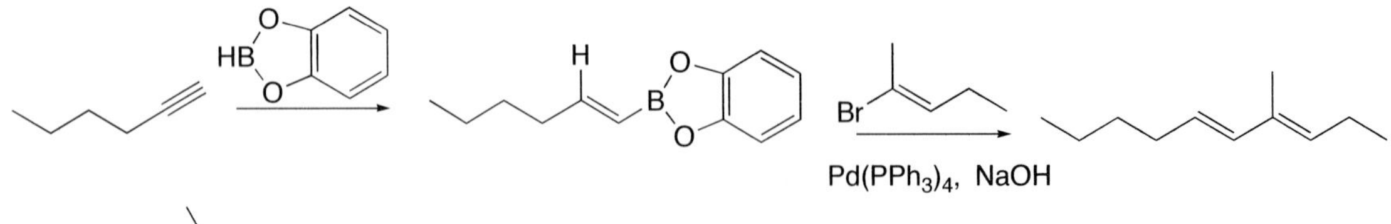

d.

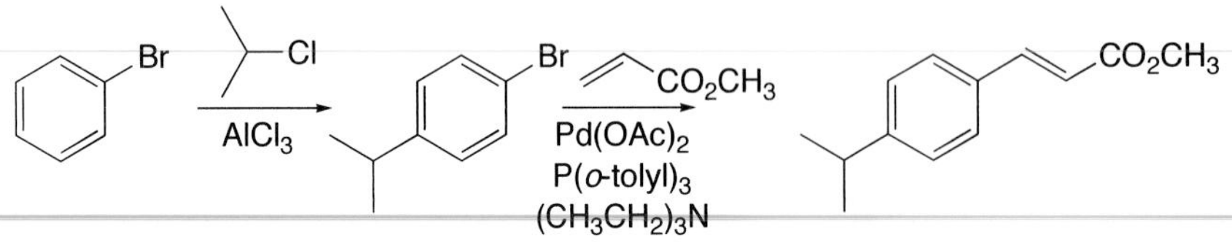

e.

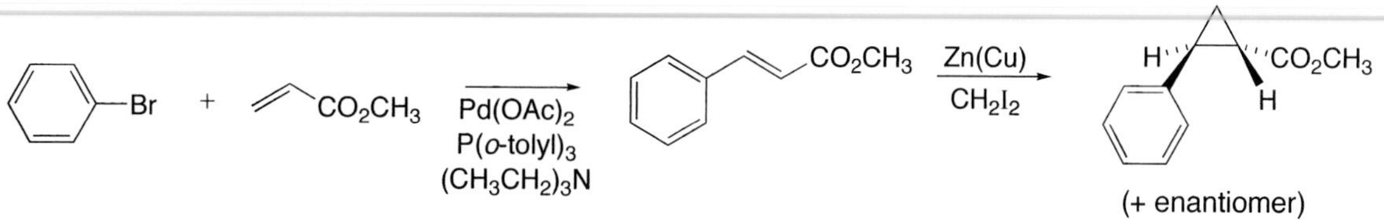
Br
CO2CH3
Pd(OAc)2
P(o-tolyl)3
(CH3CH2)3N
CO2CH3
Zn(Cu)
CH2I2
H
CO2CH3
H
(+ enantiomer)

f.

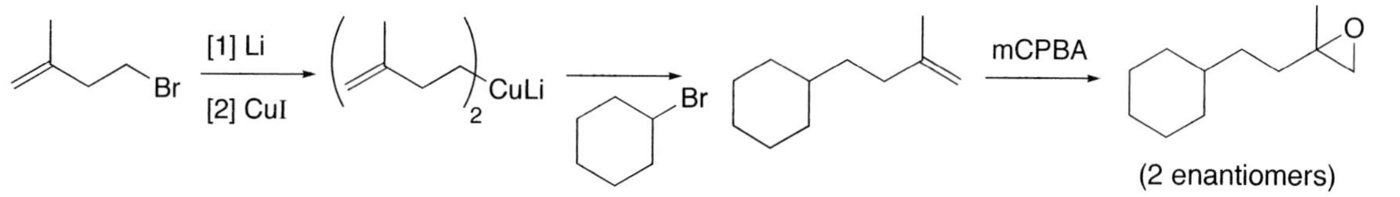
Br
[1] Li
[2] CuI
CuLi
2
Br
mCPBA
(2 enantiomers)

**26.43**

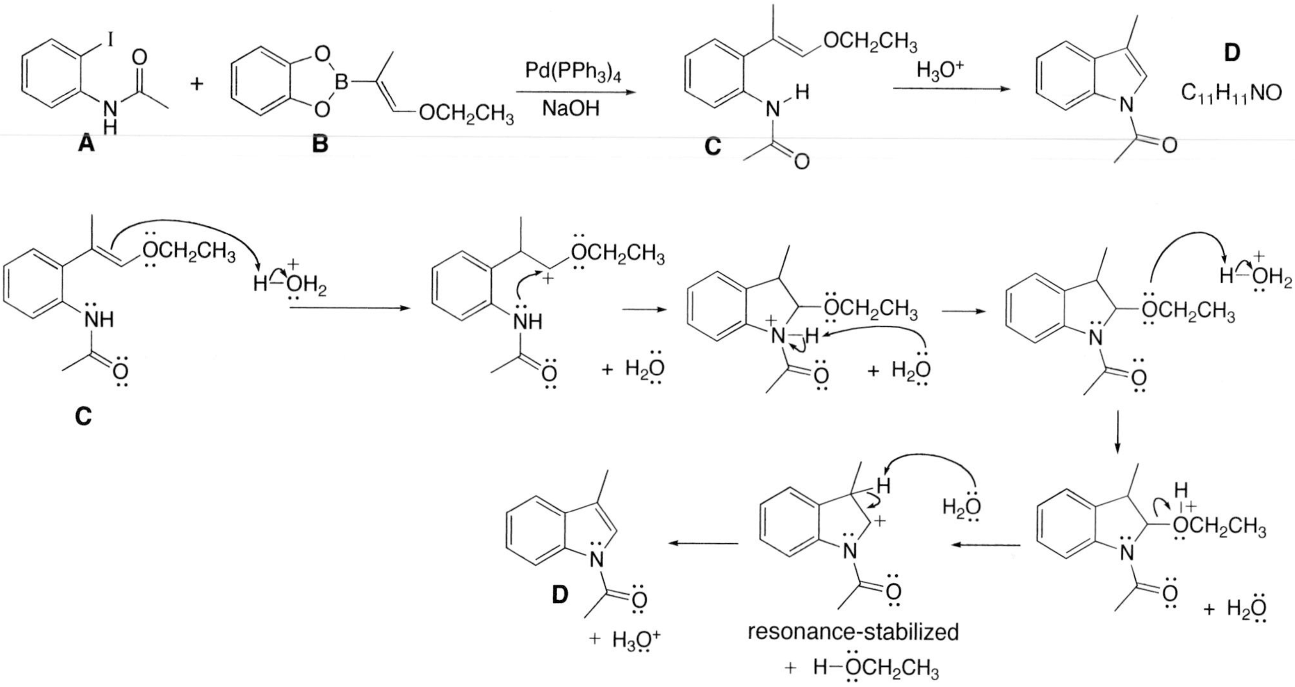
A
B
Pd(PPh3)4
NaOH
OCH2CH3
C
H3O+
D
C11H11NO
+ H2O
resonance-stabilized
+ H–OCH2CH3
+ H3O+

**26.44**

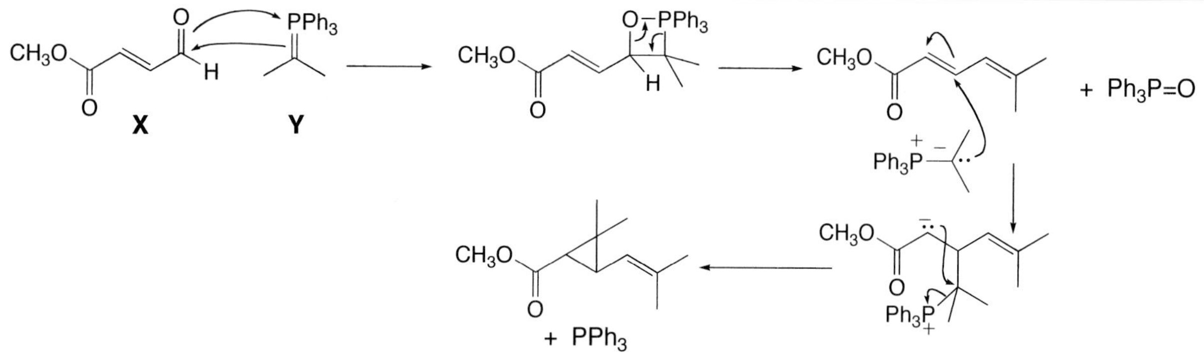
CH3O
PPh3
X
Y
O–PPh3
+ Ph3P=O
Ph3P
+ PPh3

# Chapter 27: Carbohydrates

## ♦ Important terms

- **Aldose** A monosaccharide containing an aldehyde (27.2)
- **Ketose** A monosaccharide containing a ketone (27.2)
- **D-Sugar** A monosaccharide with the O bonded to the stereogenic center furthest from the carbonyl group drawn on the right in the Fischer projection (27.2C)
- **Epimers** Two diastereomers that differ in configuration around one stereogenic center only (27.3)
- **Anomers** Monosaccharides that differ in configuration at only the hemiacetal OH group (27.6)
- **Glycoside** An acetal derived from a monosaccharide hemiacetal (27.7)

## ♦ Acyclic, Haworth, and 3-D representations for D-glucose (27.6)

**Haworth projection**

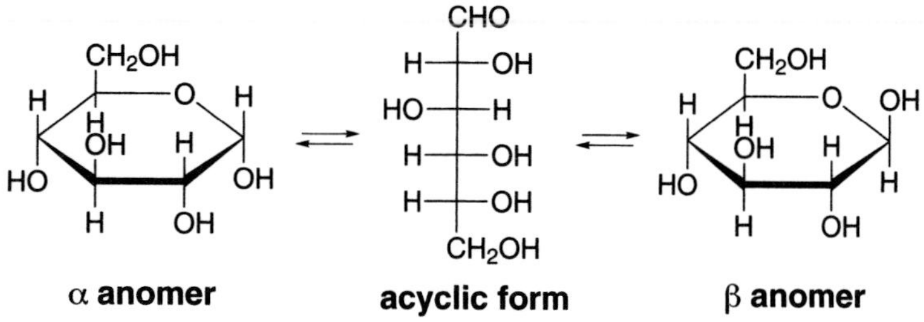


**3-D representation**

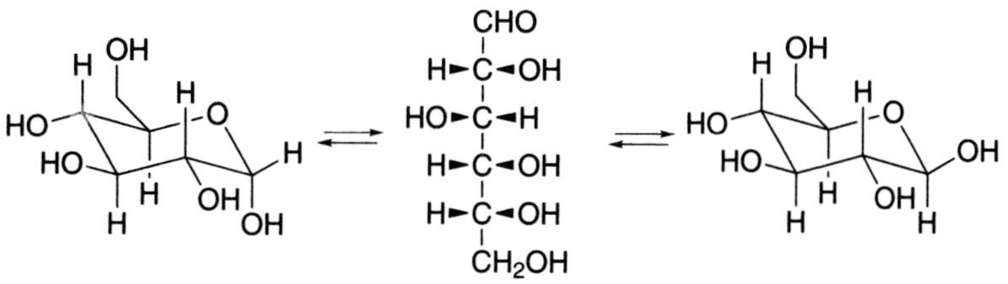

## ♦ Reactions of monosaccharides involving the hemiacetal

### [1] Glycoside formation (27.7A)

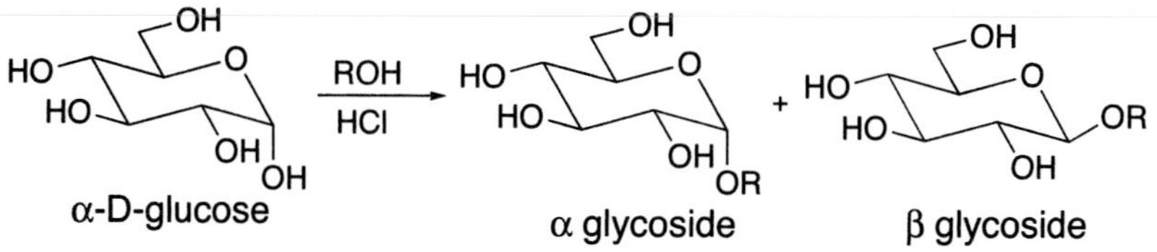


- Only the hemiacetal OH reacts.
- A mixture of α and β glycosides forms.

### [2] Glycoside hydrolysis (27.7B)

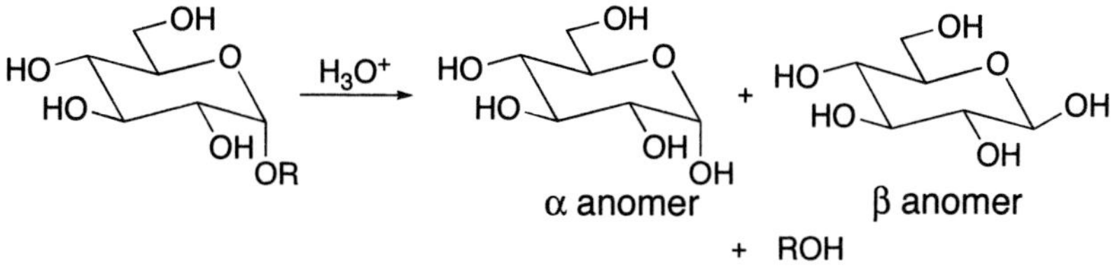


- A mixture of α and β anomers forms.

♦ **Reactions of monosaccharides at the OH groups**

## [1] Ether formation (27.8)

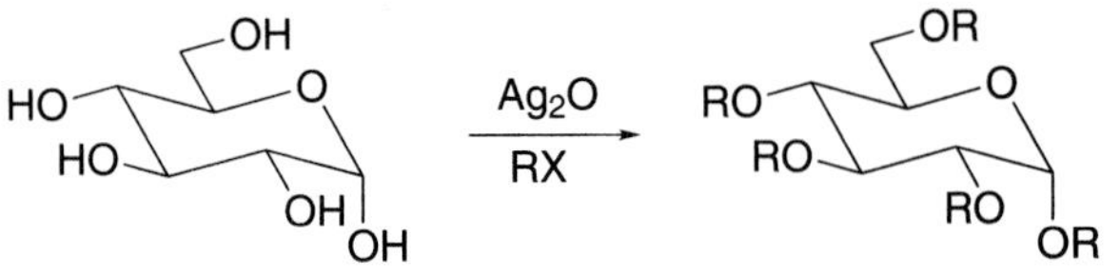

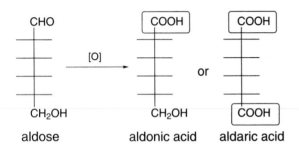

- All OH groups react.
- The stereochemistry at all stereogenic centers is retained.

## [2] Ester formation (27.8)

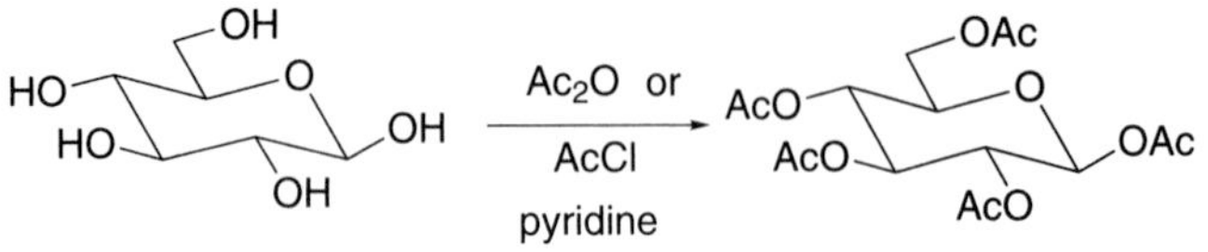

- All OH groups react.
- The stereochemistry at all stereogenic centers is retained.

♦ **Reactions of monosaccharides at the carbonyl group**

## [1] Oxidation of aldoses (27.9B)

- Aldonic acids are formed using:
  - $Ag_2O$, $NH_4OH$
  - $Cu^{2+}$
  - $Br_2$, $H_2O$
- Aldaric acids are formed with $HNO_3$, $H_2O$.

aldose      aldonic acid      aldaric acid

## [2] Reduction of aldoses to alditols (27.9A)

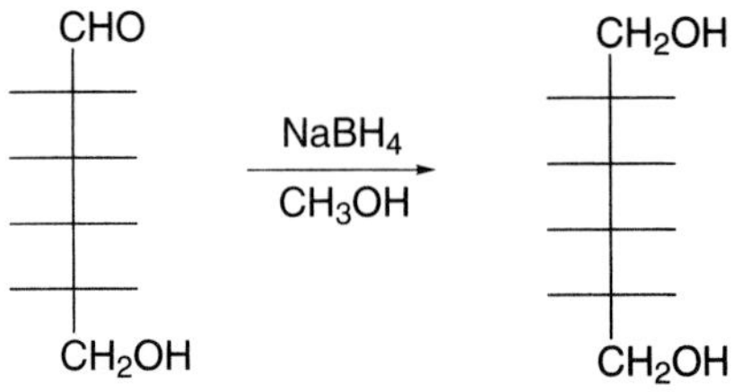

CHO     $\xrightarrow[\text{CH}_3\text{OH}]{\text{NaBH}_4}$     $CH_2OH$

aldose         alditol

## [3] Wohl degradation (27.10A)

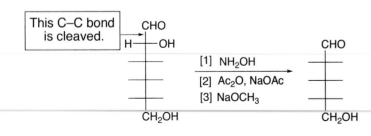

This C–C bond is cleaved.

[1] $NH_2OH$
[2] $Ac_2O$, NaOAc
[3] $NaOCH_3$

- The C1–C2 bond is cleaved to shorten an aldose chain by one carbon.
- The stereochemistry at all other stereogenic centers is retained.
- Two epimers at C2 form the same product.

## [4] Kiliani–Fischer synthesis (27.10B)

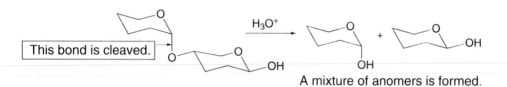

- One carbon is added to the aldehyde end of an aldose.
- Two epimers at C2 are formed.

## ◆ Other reactions

## [1] Hydrolysis of disaccharides (27.12)

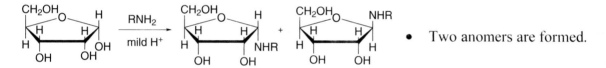

A mixture of anomers is formed.

## [2] Formation of *N*-glycosides (27.14B)

- Two anomers are formed.

## Chapter 27: Answers to Problems

**27.1** A *ketose* is a monosaccharide containing a ketone. An *aldose* is a monosaccharide containing an aldehyde. A monosaccharide is called: a *triose* if it has three C's; a *tetrose* if it has four C's; a *pentose* if it has five C's; a *hexose* if it has six C's, and so forth.

a. a ketotetrose

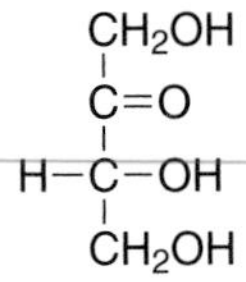

$$\begin{array}{c} CH_2OH \\ | \\ C=O \\ | \\ H-C-OH \\ | \\ CH_2OH \end{array}$$

b. an aldopentose

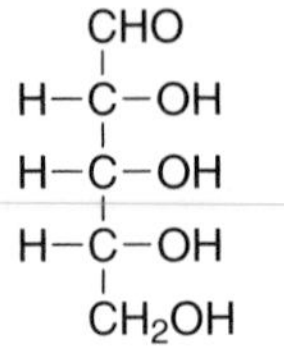

$$\begin{array}{c} CHO \\ | \\ H-C-OH \\ | \\ H-C-OH \\ | \\ H-C-OH \\ | \\ CH_2OH \end{array}$$

c. an aldotetrose

$$\begin{array}{c} CHO \\ | \\ H-C-OH \\ | \\ H-C-OH \\ | \\ CH_2OH \end{array}$$

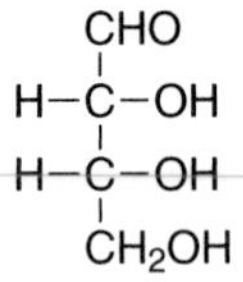

**27.2** Rotate and re-draw each molecule to place the horizontal bonds in front of the plane and the vertical bonds behind the plane. Then use a cross to represent the stereogenic center in a Fischer projection formula.

a. 
$$\begin{array}{c} COOH \\ | \\ CH_3-C-OH \\ | \\ CH_2CH_2OH \end{array} = \begin{array}{c} COOH \\ | \\ CH_3 - OH \\ | \\ CH_2CH_2OH \end{array}$$

b. 
$$\begin{array}{c} CHO \\ | \\ HO-C-H \\ | \\ CH_3 \end{array} \xrightarrow{\text{re-draw}} \begin{array}{c} CHO \\ | \\ HO-C-CH_3 \\ | \\ H \end{array} = \begin{array}{c} CHO \\ | \\ HO - CH_3 \\ | \\ H \end{array}$$

c. 
$$\begin{array}{c} OHC \quad CH_2CH_3 \\ C \\ HOCH_2 \quad H \end{array} \xrightarrow{\text{re-draw}} \begin{array}{c} CHO \\ | \\ H-C-CH_2OH \\ | \\ CH_2CH_3 \end{array} = \begin{array}{c} CHO \\ | \\ H - CH_2OH \\ | \\ CH_2CH_3 \end{array}$$

d. 
$$\begin{array}{c} OH \\ | \\ H-C-CHO \\ | \\ CH_3 \end{array} \xrightarrow{\text{re-draw}} \begin{array}{c} CHO \\ | \\ HO-C-CH_3 \\ | \\ H \end{array} = \begin{array}{c} CHO \\ | \\ HO - CH_3 \\ | \\ H \end{array}$$

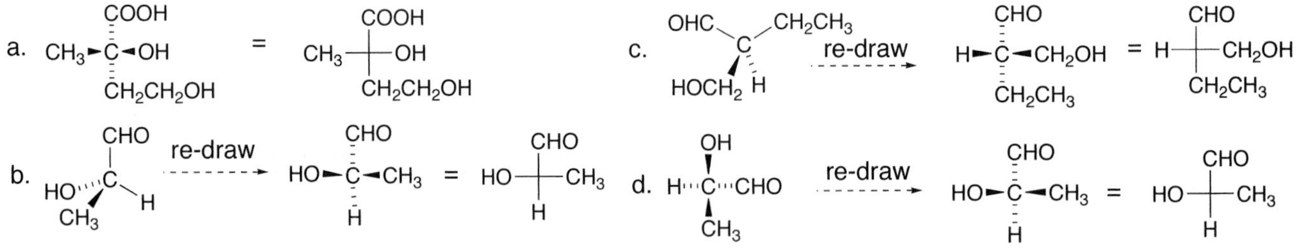

**27.3** For each molecule:

[1] Convert the Fischer projection formula to a representation with wedges and dashes.
[2] Assign priorities (Section 5.6).
[3] Determine *R* or *S* in the usual manner. Reverse the answer if priority group [4] is oriented forward (on a wedge).

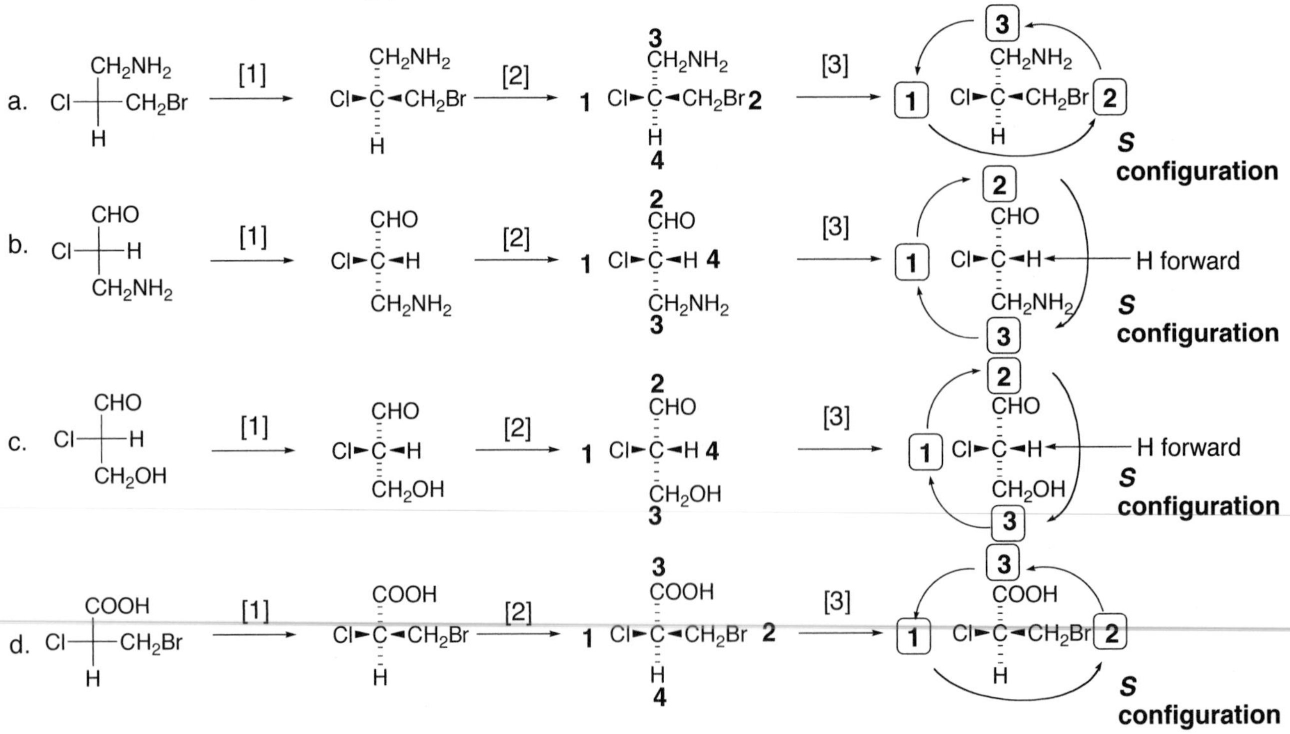

a. 
$$\begin{array}{c} CH_2NH_2 \\ | \\ Cl - CH_2Br \\ | \\ H \end{array} \xrightarrow{[1]} \begin{array}{c} CH_2NH_2 \\ | \\ Cl-C-CH_2Br \\ | \\ H \end{array} \xrightarrow{[2]} \begin{array}{c} \mathbf{3} \\ CH_2NH_2 \\ | \\ \mathbf{1}\ Cl-C-CH_2Br\ \mathbf{2} \\ | \\ H \\ \mathbf{4} \end{array} \xrightarrow{[3]} \begin{array}{c} \mathbf{3} \\ CH_2NH_2 \\ | \\ \mathbf{1}\ Cl-C-CH_2Br\ \mathbf{2} \\ | \\ H \end{array}$$
**S** configuration

b. 
$$\begin{array}{c} CHO \\ | \\ Cl - H \\ | \\ CH_2NH_2 \end{array} \xrightarrow{[1]} \begin{array}{c} CHO \\ | \\ Cl-C-H \\ | \\ CH_2NH_2 \end{array} \xrightarrow{[2]} \begin{array}{c} \mathbf{2} \\ CHO \\ | \\ \mathbf{1}\ Cl-C-H\ \mathbf{4} \\ | \\ CH_2NH_2 \\ \mathbf{3} \end{array} \xrightarrow{[3]} \begin{array}{c} \mathbf{2} \\ CHO \\ | \\ \mathbf{1}\ Cl-C-H \longleftarrow H\ forward \\ | \\ CH_2NH_2 \\ \mathbf{3} \end{array}$$
**S** configuration

c. 
$$\begin{array}{c} CHO \\ | \\ Cl - H \\ | \\ CH_2OH \end{array} \xrightarrow{[1]} \begin{array}{c} CHO \\ | \\ Cl-C-H \\ | \\ CH_2OH \end{array} \xrightarrow{[2]} \begin{array}{c} \mathbf{2} \\ CHO \\ | \\ \mathbf{1}\ Cl-C-H\ \mathbf{4} \\ | \\ CH_2OH \\ \mathbf{3} \end{array} \xrightarrow{[3]} \begin{array}{c} \mathbf{2} \\ CHO \\ | \\ \mathbf{1}\ Cl-C-H \longleftarrow H\ forward \\ | \\ CH_2OH \\ \mathbf{3} \end{array}$$
**S** configuration

d. 
$$\begin{array}{c} COOH \\ | \\ Cl - CH_2Br \\ | \\ H \end{array} \xrightarrow{[1]} \begin{array}{c} COOH \\ | \\ Cl-C-CH_2Br \\ | \\ H \end{array} \xrightarrow{[2]} \begin{array}{c} \mathbf{3} \\ COOH \\ | \\ \mathbf{1}\ Cl-C-CH_2Br\ \mathbf{2} \\ | \\ H \\ \mathbf{4} \end{array} \xrightarrow{[3]} \begin{array}{c} \mathbf{3} \\ COOH \\ | \\ \mathbf{1}\ Cl-C-CH_2Br\ \mathbf{2} \\ | \\ H \end{array}$$
**S** configuration

**27.4**

```
        CHO      R
         |      /
    H►C◄OH   S
         |
   HO►C◄H     R
         |
    H►C◄OH   R
         |
    H►C◄OH
         |
       CH₂OH
```

D-glucose

**27.5**

a. aldotetrose: 2 stereogenic centers

b. a ketohexose: 3 stereogenic centers

```
        CHO
         |
   H—C*—OH
         |
   H—C*—OH
         |
       CH₂OH
```

```
       CH₂OH
         |
       C=O
         |*
   H—C—OH
         |*
   H—C—OH
         |*
   H—C—OH
         |
       CH₂OH
```

[• = stereogenic center]

**27.6** A D sugar has the OH group on the stereogenic center furthest from the carbonyl on the right. An L sugar has the OH group on the stereogenic center furthest from the carbonyl on the left.

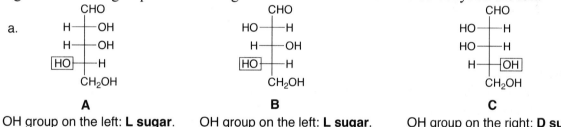

a.

**A**
OH group on the left: **L sugar**.

**B**
OH group on the left: **L sugar**.

**C**
OH group on the right: **D sugar**.

b. A and B are diastereomers.
A and C are enantiomers.
B and C are diastereomers.

**27.7** The D- notation signifies the position of the OH group on the stereogenic carbon furthest from the carbonyl group, and does not correlate with dextrorotatory or levorotatory. The latter terms describe a physical phenomenon, the direction of rotation of plane-polarized light.

**27.8** There are 32 aldoheptoses; 16 are D sugars.

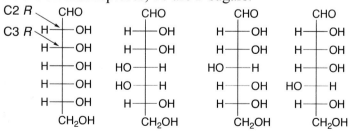

**27.9**   *Epimers* are two diastereomers that differ in the configuration around only one stereogenic center.

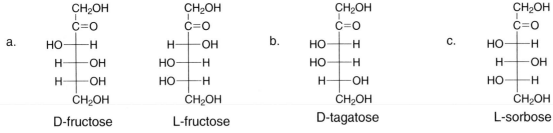

**epimers**

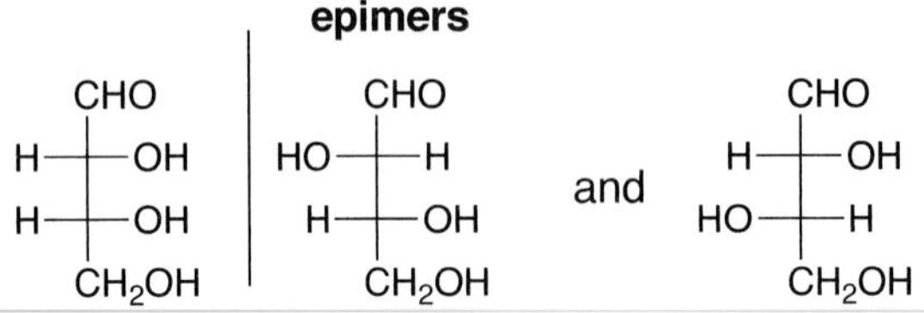

D-erythrose     D-threose          L-threose

**27.10**   a.  D-allose and L-allose: **enantiomers**
  b.  D-altrose and D-gulose: **diastereomers** but not epimers
  c.  D-galactose and D-talose: **epimers**
  d.  D-mannose and D-fructose: **constitutional isomers**
  e.  D-fructose and D-sorbose: **diastereomers** but not epimers
  f.  L-sorbose and L-tagatose: **epimers**

**27.11**

a.

b.

c.

D-fructose     L-fructose          D-tagatose          L-sorbose

**enantiomers**

**27.12**

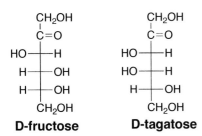

**D-fructose**     **D-tagatose**

**27.13**    Step [1]: Place the O atom in the upper right corner of a hexagon, and add the CH$_2$OH group
on the first carbon counterclockwise from the O atom.

Step [2]: Place the anomeric carbon on the first carbon clockwise from the O atom.

Step [3]: Add the substituents at the three remaining stereogenic centers, clockwise around the
ring.

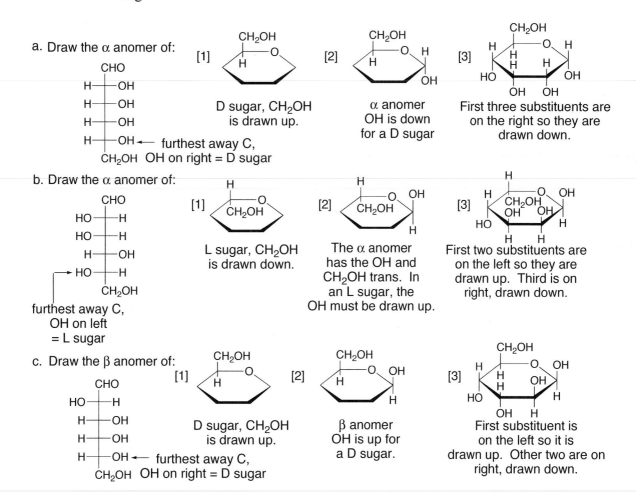

a. Draw the α anomer of:

[1] D sugar, CH$_2$OH is drawn up.

[2] α anomer OH is down for a D sugar

[3] First three substituents are on the right so they are drawn down.

b. Draw the α anomer of:

[1] L sugar, CH$_2$OH is drawn down.

[2] The α anomer has the OH and CH$_2$OH trans. In an L sugar, the OH must be drawn up.

[3] First two substituents are on the left so they are drawn up. Third is on right, drawn down.

c. Draw the β anomer of:

[1] D sugar, CH$_2$OH is drawn up.

[2] β anomer OH is up for a D sugar.

[3] First substituent is on the left so it is drawn up. Other two are on right, drawn down.

**27.14** To convert each Haworth projection into its acyclic form:

[1] Draw the C skeleton with the CHO on the top and the $CH_2OH$ on the bottom.

[2] Draw in the OH group furthest from the C=O.

A $CH_2OH$ group drawn up means a D sugar; a $CH_2OH$ group drawn down means an L sugar.

[3] Add the three other stereogenic centers, counterclockwise around the ring.

"Up" groups go on the left and "down" groups go on the right.

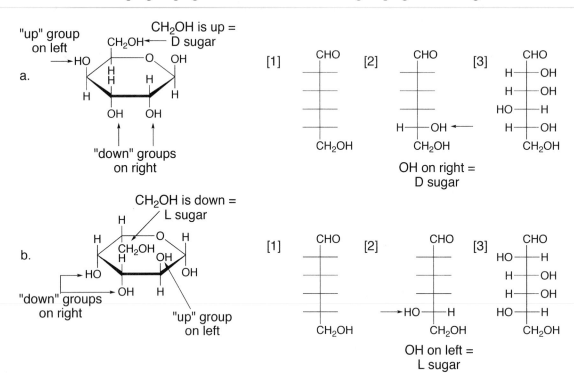

**27.15** To convert a Haworth projection into a 3-D representation with a chair cyclohexane:

[1] Draw the pyranose ring as a chair with the O as an "up" atom.

[2] Add the substituents around the ring.

**27.16** Cyclization always forms a new stereogenic center at the anomeric carbon, so two different anomers are possible.

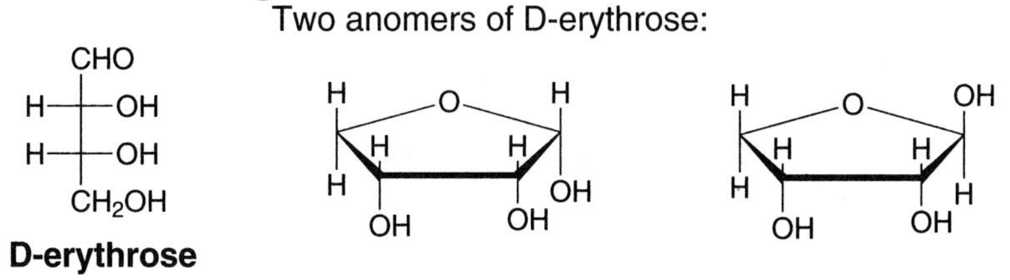

Two anomers of D-erythrose:
CHO
H
OH
H
OH
CH2OH
D-erythrose
O
H
OH


**27.17**

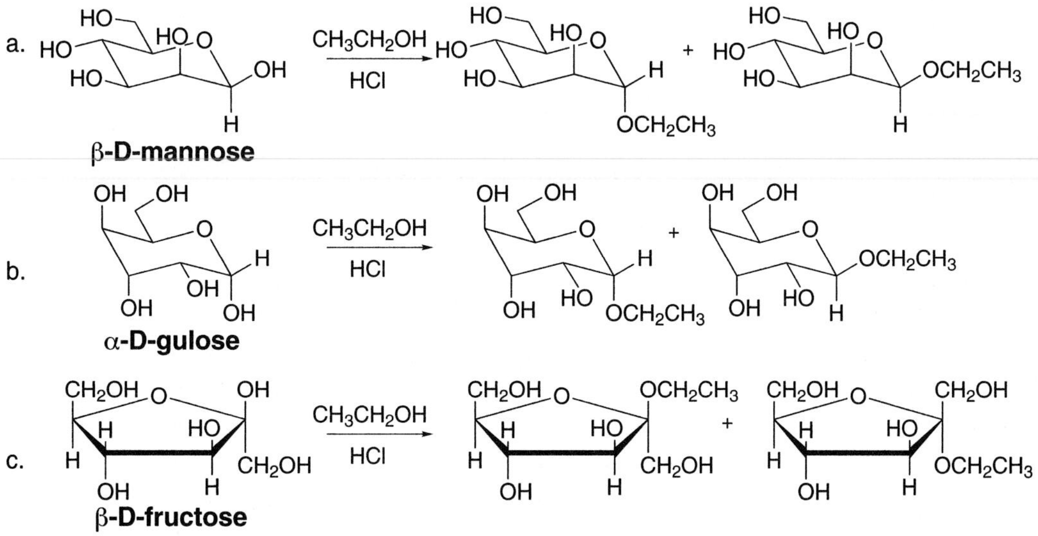

a.
β-D-mannose
CH3CH2OH
HCl
OCH2CH3
b.
α-D-gulose
c.
β-D-fructose
CH2OH


**27.18**

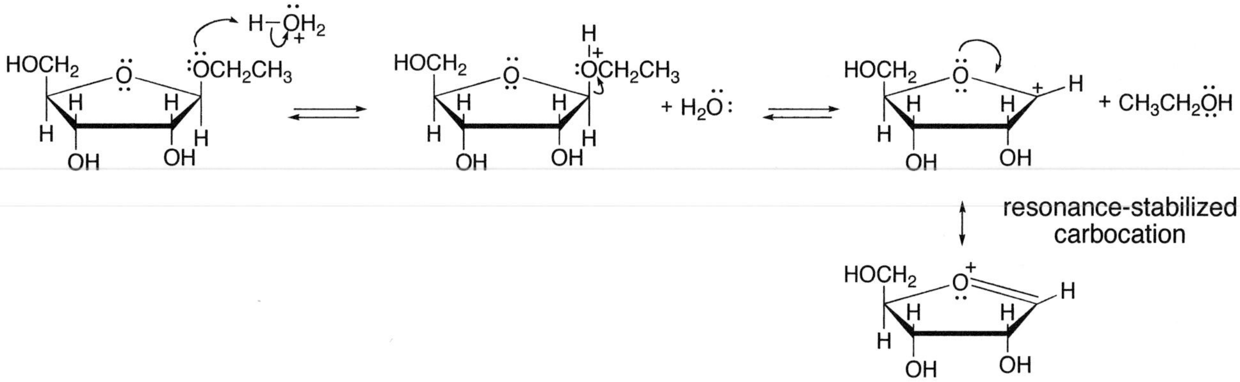

HOCH2
OCH2CH3
+ H2O:
+ CH3CH2OH
resonance-stabilized
carbocation

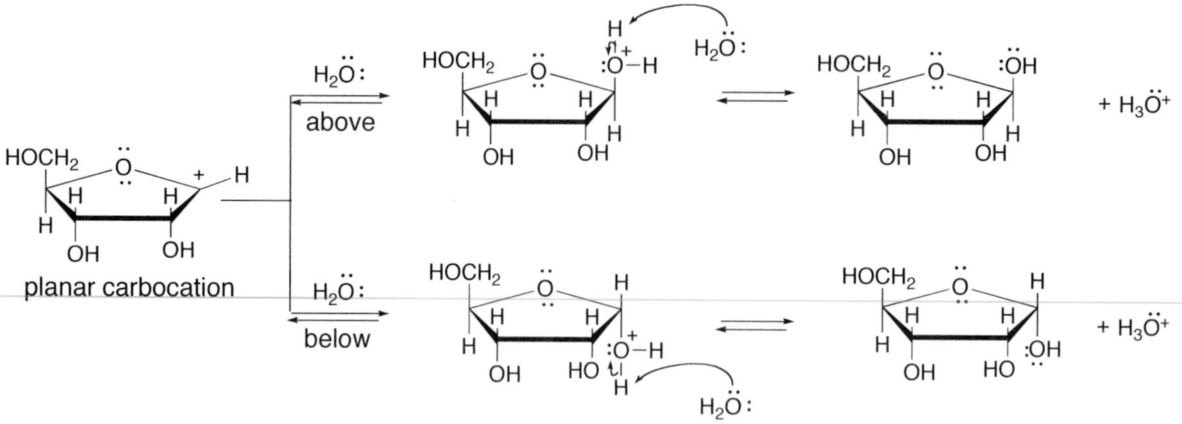


**27.19**

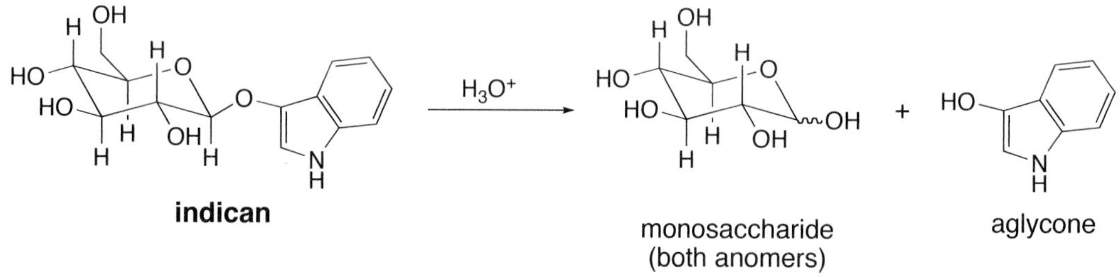


**27.20**

a.

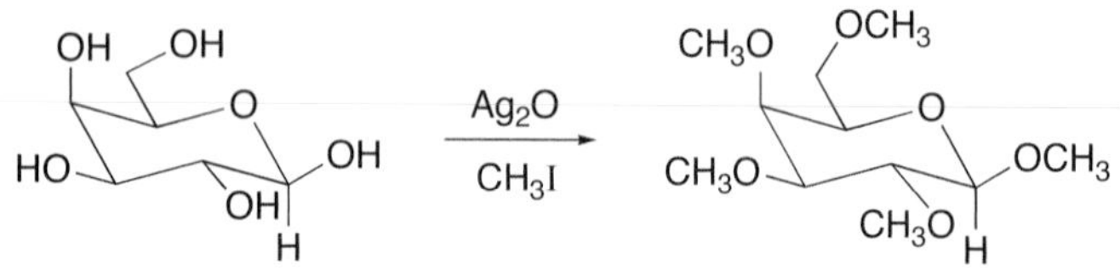

b.

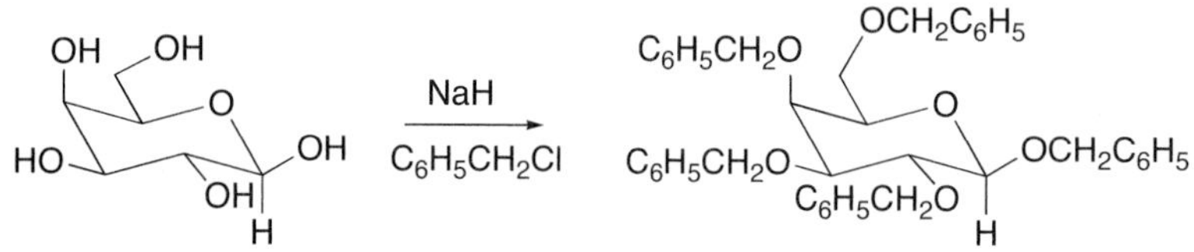

c.

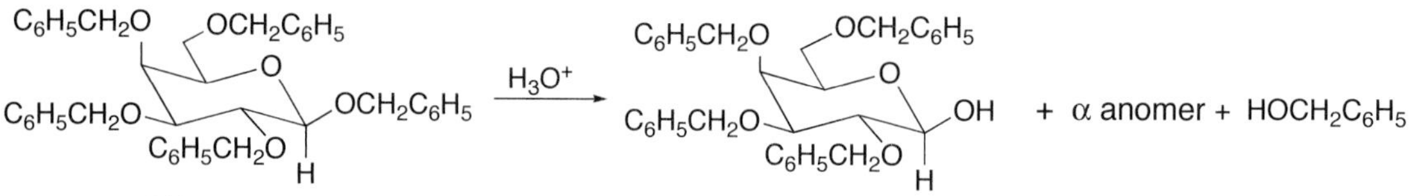


d.

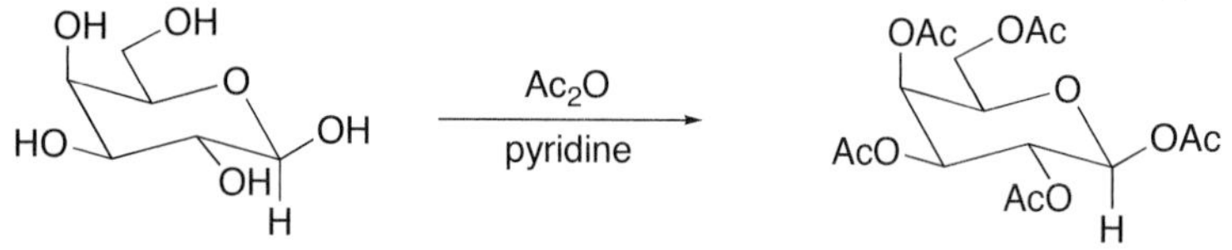

e.

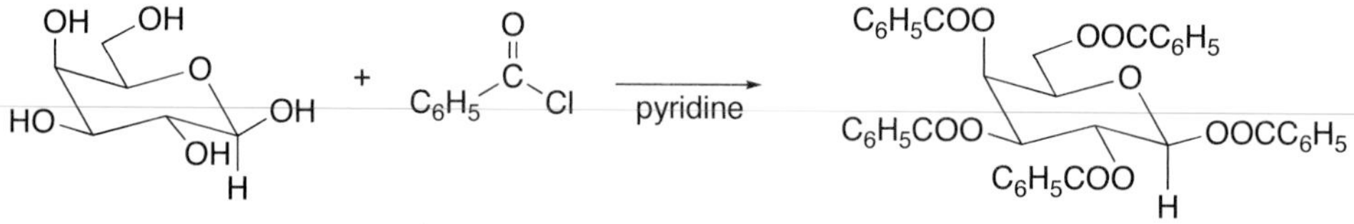

f.

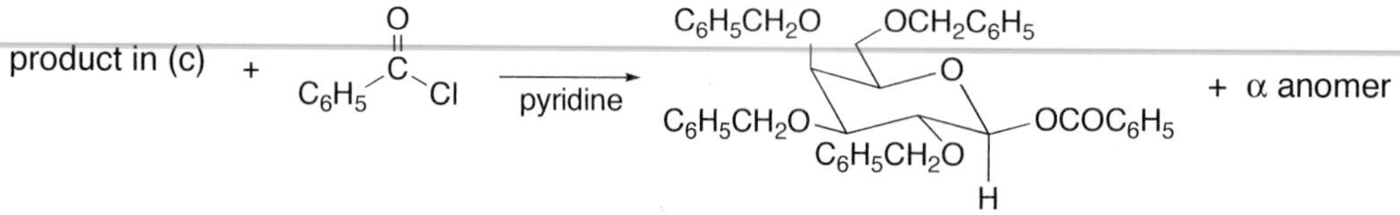

**27.21**

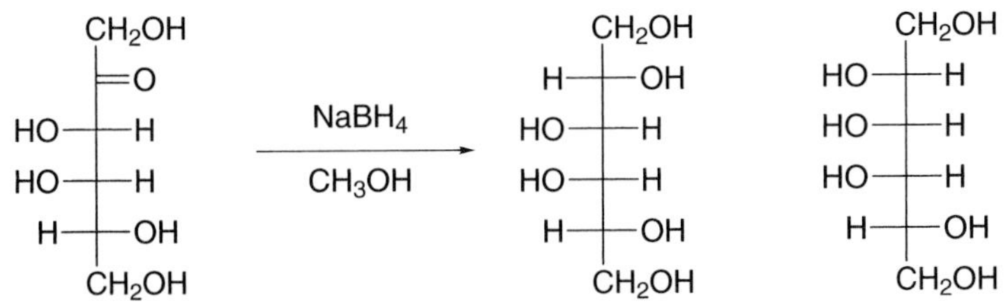

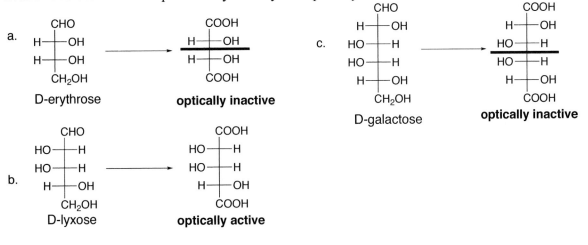

CH₂OH
=O
HO—H
HO—H
H—OH
CH₂OH

**D-tagatose**

NaBH₄
CH₃OH

CH₂OH
H—OH
HO—H
HO—H
H—OH
CH₂OH

**D-galactitol**

CH₂OH
HO—H
HO—H
HO—H
H—OH
CH₂OH

**D-talitol**

**27.22** Carbohydrates containing a hemiacetal are in equilibrium with an acyclic aldehyde, making them reducing sugars. Glycosides are acetals, so they are not in equilibrium with any acyclic aldehyde, making them nonreducing sugars.

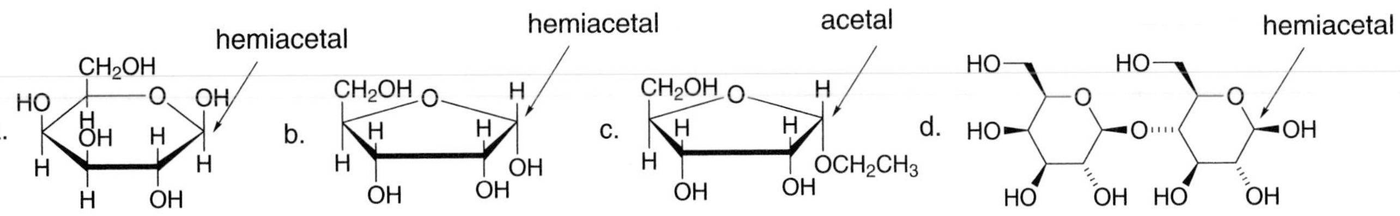

a. hemiacetal — **reducing sugar**

b. hemiacetal — **reducing sugar**

c. acetal — **nonreducing sugar**

d. hemiacetal — lactose **reducing sugar**

**27.23**

a.
CHO
HO—H
H—OH
H—OH
CH₂OH

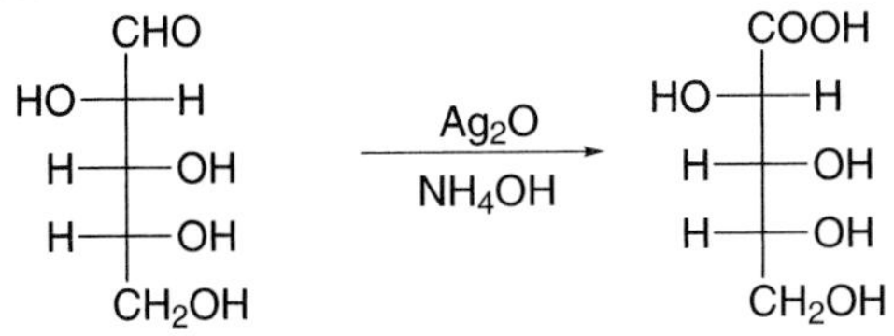

Ag₂O / NH₄OH

COOH
HO—H
H—OH
H—OH
CH₂OH

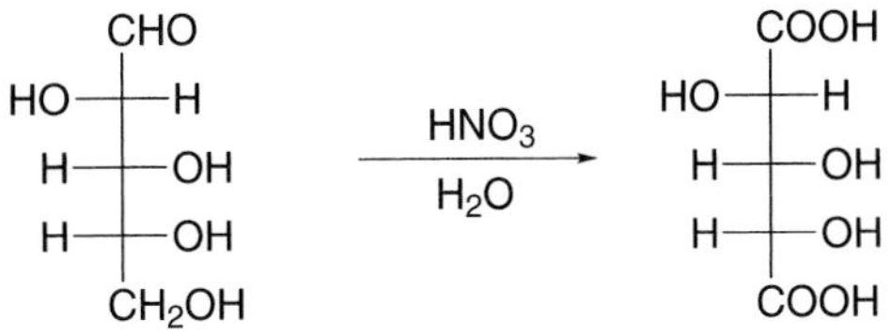

b.
CHO
HO—H
H—OH
H—OH
CH₂OH

Br₂ / H₂O

COOH
HO—H
H—OH
H—OH
CH₂OH

c.
CHO
HO—H
H—OH
H—OH
CH₂OH

HNO₃ / H₂O

COOH
HO—H
H—OH
H—OH
COOH

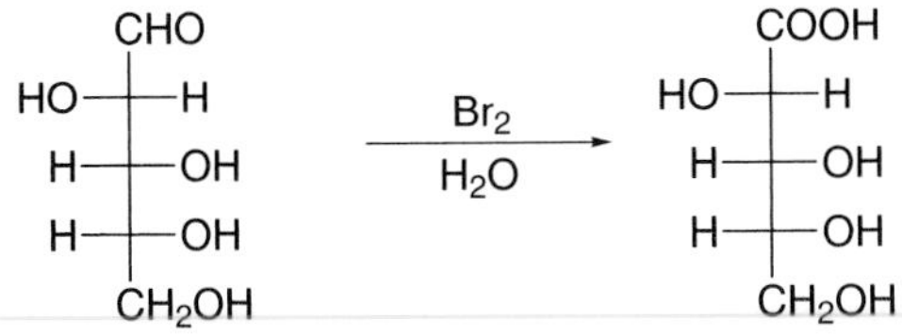

**27.24** Molecules with a plane of symmetry are optically inactive.

a.
CHO
H—OH
H—OH
CH₂OH

**D-erythrose**

→

COOH
H—OH
H—OH
COOH

**optically inactive**

b.
CHO
HO—H
HO—H
H—OH
CH₂OH

**D-lyxose**

→

COOH
HO—H
HO—H
H—OH
COOH

**optically active**

c.
CHO
H—OH
HO—H
HO—H
H—OH
CH₂OH

**D-galactose**

→

COOH
H—OH
HO—H
HO—H
H—OH
COOH

**optically inactive**

**27.25**

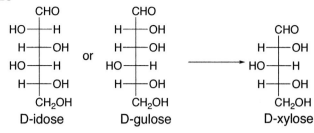

or
D-idose
D-gulose
D-xylose

**27.26**

a.
D-threose
b.
D-ribose
c.
D-galactose

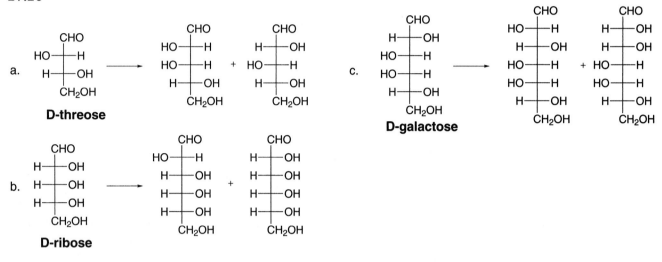

**27.27**

Possible optically inactive D-aldaric acids:
plane of symmetry
A'
A"
This OH is on the right for a D sugar.

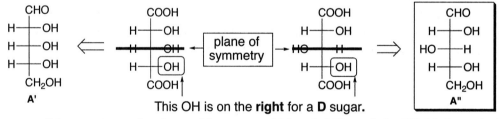

There are two possible structures for the D-aldopentose (**A'** and **A"**), and the Wohl degradation determines which structure corresponds to **A**.

Product of Wohl degradation:
This is A.
Wohl
[O]
A'
optically inactive
optically active
B
A"
Since this compound has no plane of symmetry, its precursor is B, and thus A" = A.

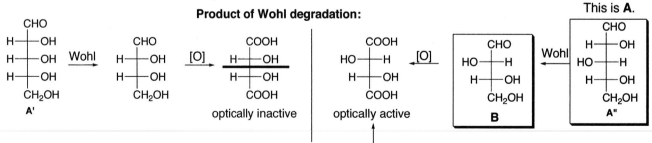

**27.28**

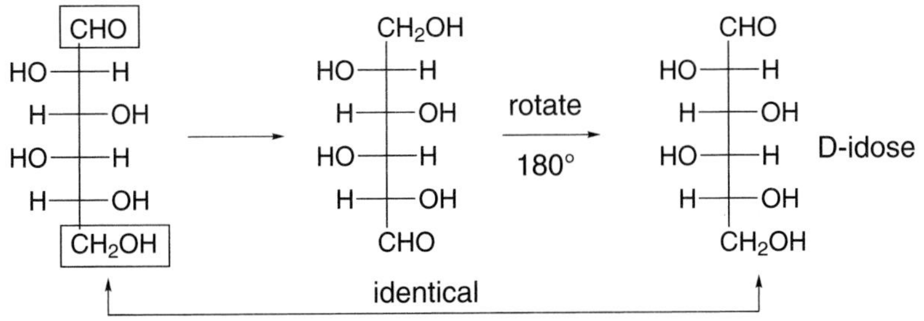

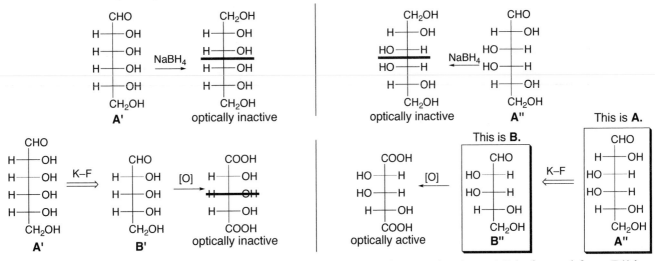

**27.29**

**Optically inactive alditols formed from NaBH₄ reduction of a D-aldohexose.**

Two D-aldohexoses (**A'** and **A''**) give optically inactive alditols on reduction.  **A''** is formed from **B''** by Kiliani–Fischer synthesis.  Since **B''** affords an optically active aldaric acid on oxidation, **B''** is **B** and **A''** is **A**.  The alternate possibility (**A'**) is formed from an aldopentose **B'** that gives an optically inactive aldaric acid on oxidation.

**27.30**

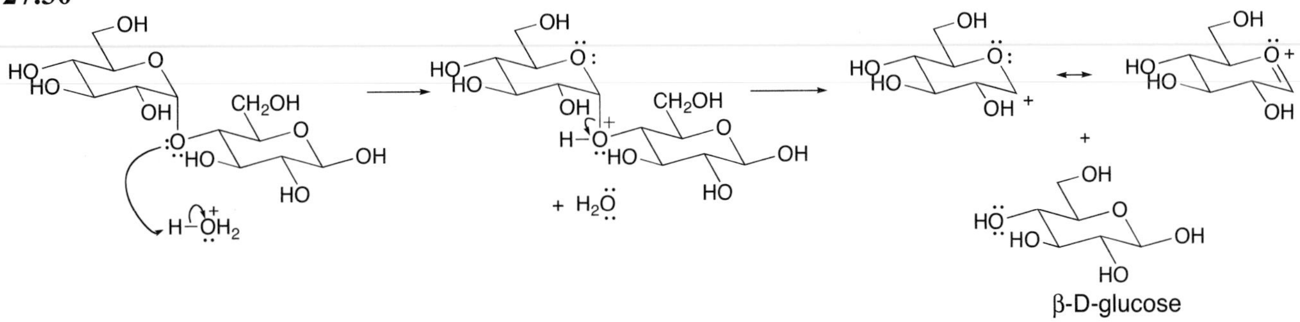

β-D-glucose

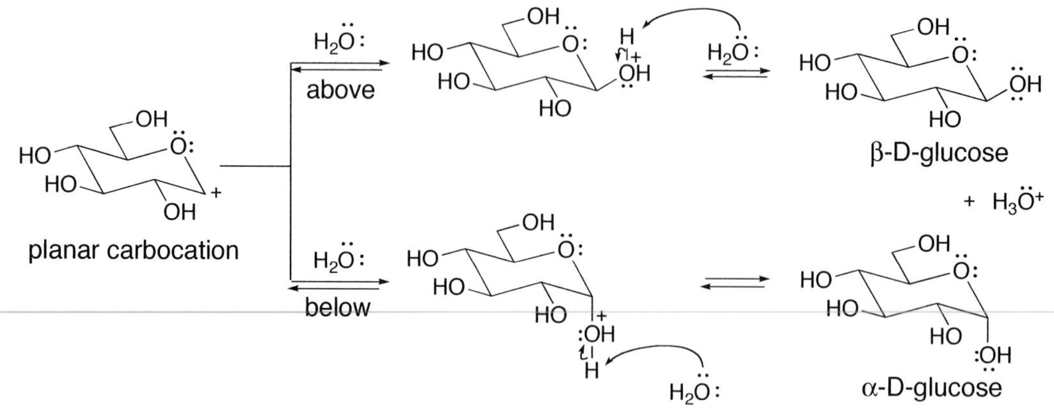

β-D-glucose

+ H₃O⁺

α-D-glucose

planar carbocation

**27.31**

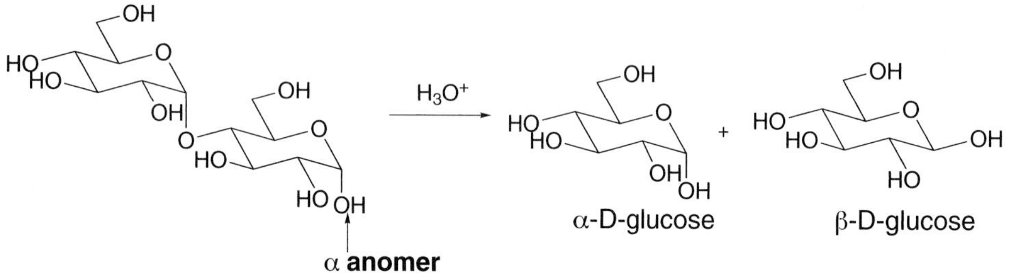

The same products are formed on hydrolysis of the α and β anomers of maltose.

α-D-glucose  β-D-glucose

α **anomer**

**27.32**

β glycoside bond

**cellobiose**

Two possible anomers here. β OH is drawn.

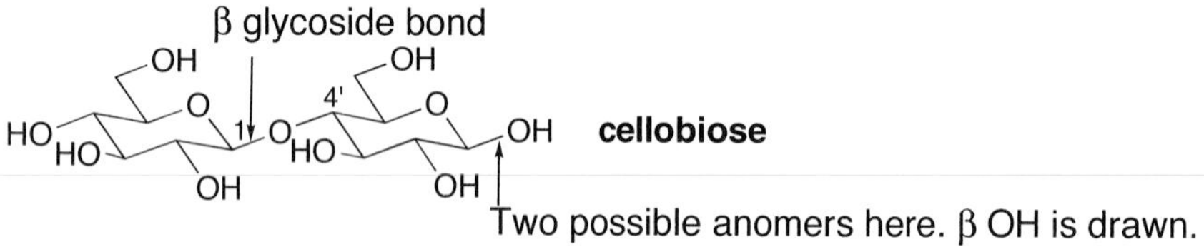

**27.33**

a.

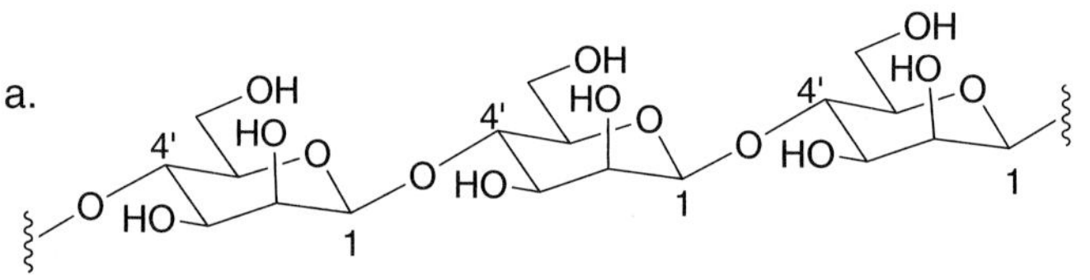

b. 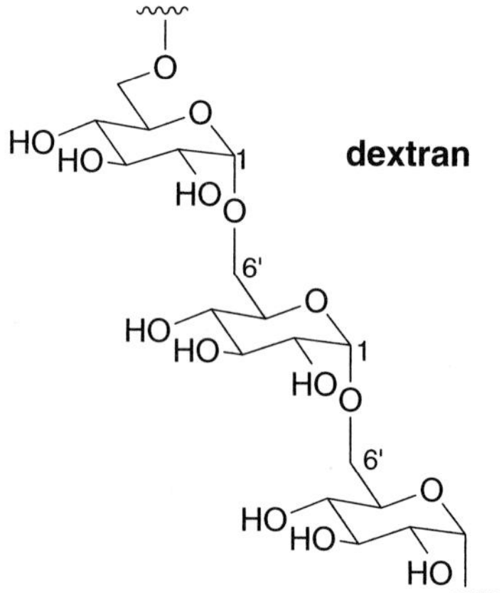 **dextran**

**27.34**

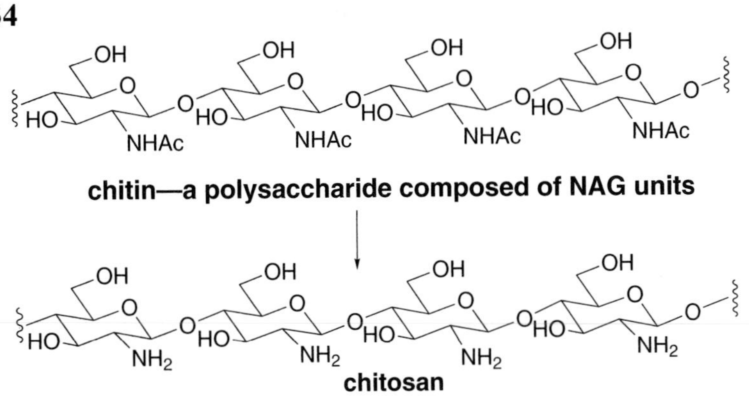

**chitin—a polysaccharide composed of NAG units**

**chitosan**

**27.35**

a.

b.

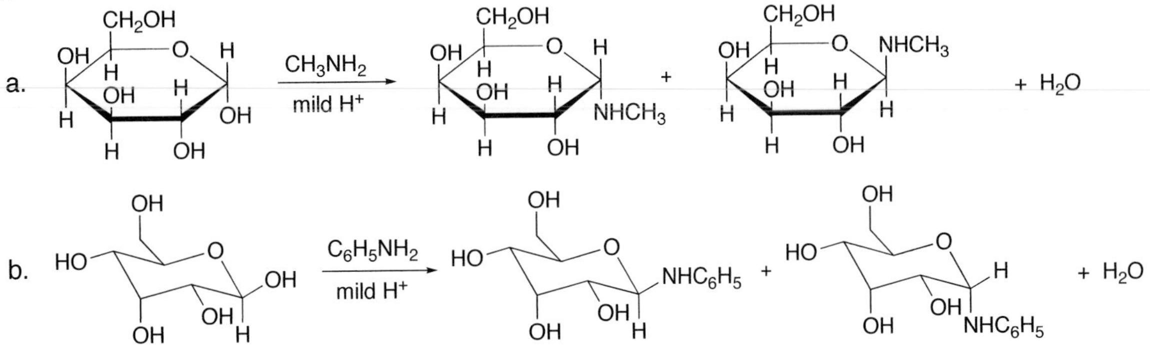

**27.36**

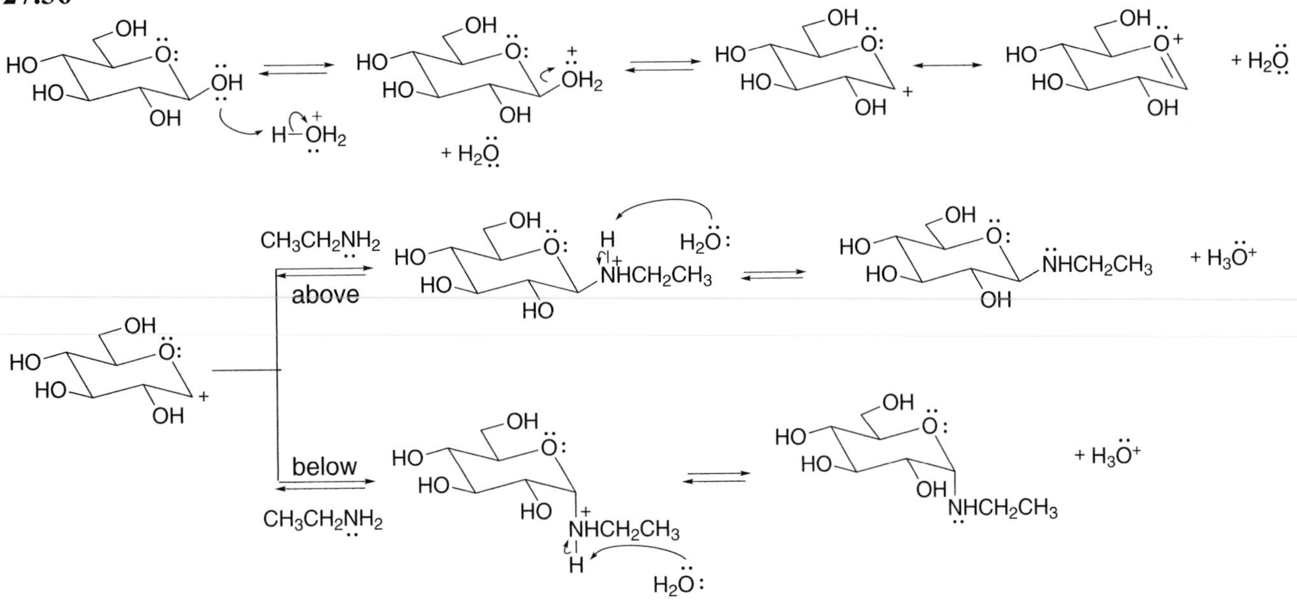

**27.37**

a.

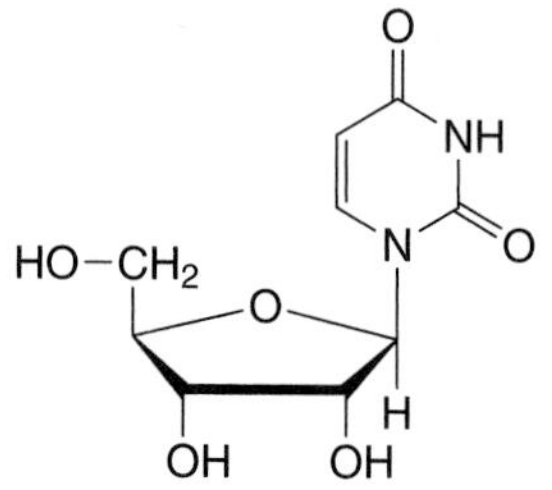

b.

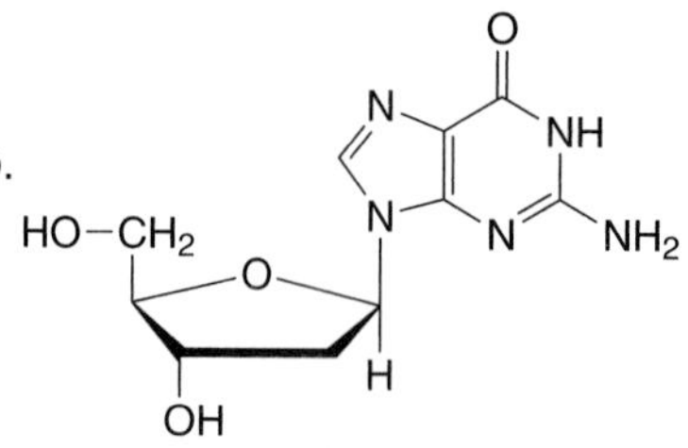

**27.38**

    a. Two purine bases (A and G) are both bicyclic bases. Therefore they are too big to hydrogen bond to each other on the inside of the DNA double helix.

    b. Hydrogen bonding between guanine and cytosine has three hydrogen bonds, whereas between guanine and thymine there are only two. This makes hydrogen bonding between guanine and cytosine more favorable.

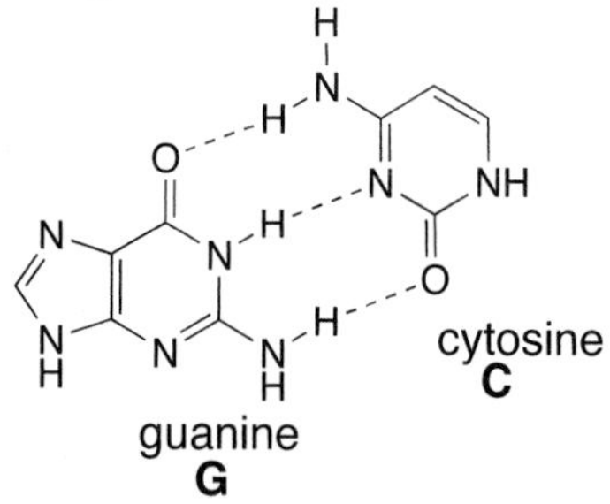

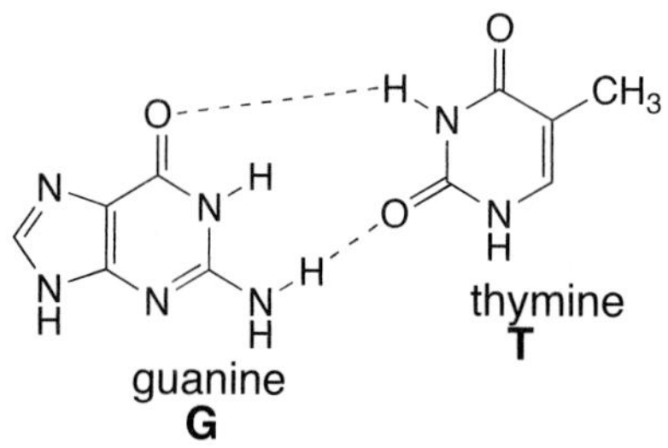

**27.39** Label the compounds with *R* or *S* and then classify.

CHO
H——OH
CH₂CH₃

**A**
**R**

a. CH₃CH₂—C—OH (C with CHO down, H up) 

**R**
**identical**

b. (structure with CHO, OH, H)

**R**
**identical**

c. (structure with H OH, CH₃CH₂, CHO)

**S**
**enantiomer**

d. CHO, HO—C—CH₂CH₃, H

**S**
**enantiomer**

**27.40** Use the directions from Answer 27.2 to draw each Fischer projection.

a. CH₃—C—Br (COOH top, H bottom) = CH₃—**S**—Br (COOH top, H bottom)

b. CH₃''C'''Cl (Br top, H bottom) --re-draw--> H—C—Br (CH₃ top, Cl bottom) = H—**S**—Br (CH₃ top, Cl bottom)

c. (C with CH₃O, OCH₂CH₃, CH₃, CH₂CH₃) --re-draw--> CH₃—C—CH₂CH₃ (OCH₃ top, OCH₂CH₃ bottom) = CH₃—**S**—CH₂CH₃ (OCH₃ top, OCH₂CH₃ bottom)

d. (C with Cl, CH₃CH₂, H, Br) --re-draw--> CH₃CH₂—C—Cl (H top, Br bottom) = CH₃CH₂—**S**—Cl (H top, Br bottom)

e. H—C—Br (CH₃ top), Cl—C—H (CH₂CH₃ bottom) = H—**S**—Br (CH₃ top), Cl—**S**—H (CH₂CH₃ bottom)

f. (C—C with H, Cl, Br, Cl, Br, H) --re-draw--> Br—C—H (Cl top, **S**), Cl—C—Br (H bottom, **R**) = Br—H (Cl top, **S**), Cl—Br (H bottom, **R**)

g. (C—C with CH₃, H, Br, Br, CH₃) --re-draw--> H—C—Br (CH₃ top), Br—C—H (CH₃ bottom) = H—**S**—Br (CH₃ top), Br—**S**—H (CH₃ bottom)

h. HO... (structure with HO H, H OH, HO H, CHO, CH₂OH) --re-draw--> HO—C—H, HO—C—H, H—C—OH (CHO top, CH₂OH bottom) = HO—**S**—H, HO—**S**—H, H—**R**—O (CHO top, CH₂O bottom)

**27.41** *Epimers* are two diastereomers that differ in the configuration around only one stereogenic center.

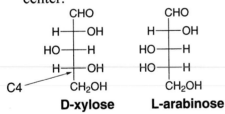

CHO
H——OH
HO——H
H——OH
C4 — CH₂OH
**D-xylose**

CHO
H——OH
HO——H
HO——H
CH₂OH
**L-arabinose**

**27.42**

a.
CHO
HO——H
H——OH
H——OH
CH₂OH
**D-arabinose**

CHO
H——OH
HO——H
HO——H
CH₂OH
**enantiomer**

b.
CHO
HO——H
HO——H
C3 — H——OH
CH₂OH
**epimer**

c.
CHO
H——OH
HO——H
H——OH
CH₂OH
**diastereomer**
**(but not epimer)**

d.
OH  O
HO——————OH
OH
**constitutional**
**isomer**

**27.43**

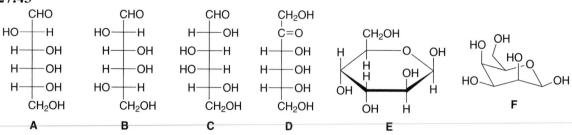

**A**

**B**

**C**

**D**

**E**

**F**

a. **A** and **B** **epimers**
b. **A** and **C** **diastereomers**
c. **B** and **C** **enantiomers**
d. **A** and **D** **constitutional isomers**
e. **E** and **F** **diastereomers**

**27.44**

**A**

**B**

a. anomers, epimers, diastereomers, reducing sugars
b.

**D-xylose**

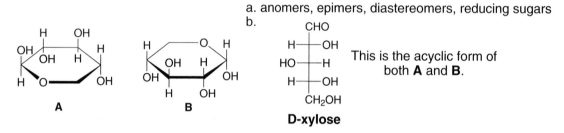

This is the acyclic form of both **A** and **B**.

**27.45** Use the directions from Answer 27.13.

a. β-D-talopyranose

← furthest away C, OH on right = D sugar
**D-talose**

[1]
D sugar, $CH_2OH$ is drawn up.

[2]
β anomer OH is up for a D sugar.

[3]

b. β-D-mannopyranose

← furthest away C, OH on right = D sugar
**D-mannose**

[1]
D sugar, $CH_2OH$ is drawn up.

[2]
β anomer OH is up for a D sugar.

[3]

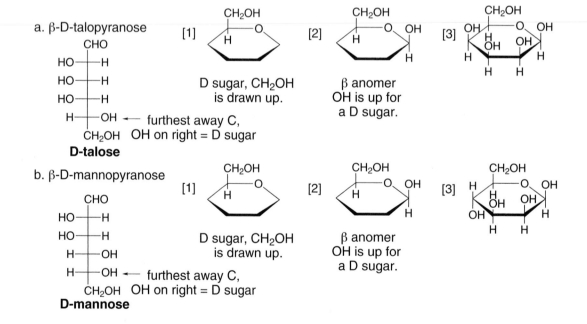

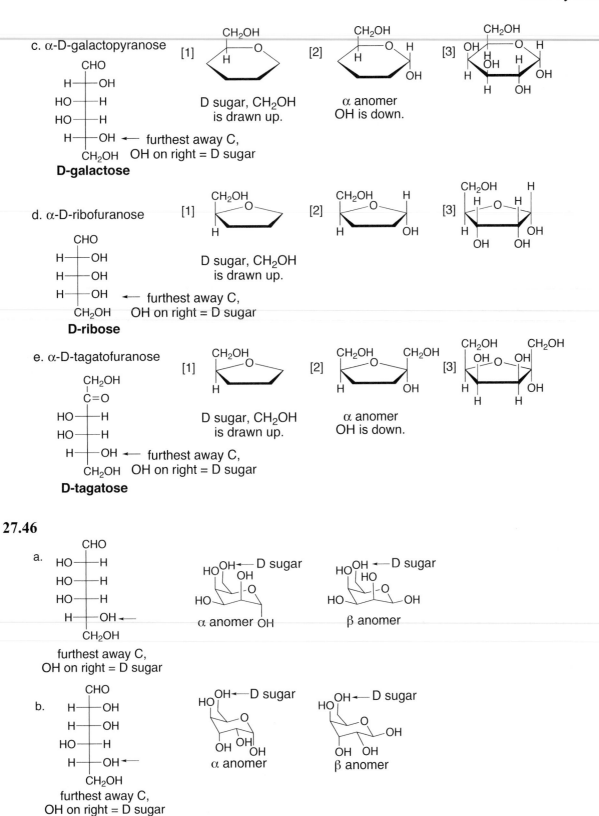

c. α-D-galactopyranose

CHO
H — OH
HO — H
HO — H
H — OH ← furthest away C,
CH₂OH   OH on right = D sugar
**D-galactose**

[1] D sugar, CH₂OH is drawn up.

[2] α anomer OH is down.

[3]

d. α-D-ribofuranose

CHO
H — OH
H — OH
H — OH ← furthest away C,
CH₂OH   OH on right = D sugar
**D-ribose**

[1] D sugar, CH₂OH is drawn up.

[2]

[3]

e. α-D-tagatofuranose

CH₂OH
C=O
HO — H
HO — H
H — OH ← furthest away C,
CH₂OH   OH on right = D sugar
**D-tagatose**

[1] D sugar, CH₂OH is drawn up.

[2] α anomer OH is down.

[3]

**27.46**

a.
CHO
HO — H
HO — H
HO — H
H — OH ←
CH₂OH
furthest away C,
OH on right = D sugar

α anomer

β anomer

b.
CHO
H — OH
H — OH
HO — H
H — OH ←
CH₂OH
furthest away C,
OH on right = D sugar

α anomer

β anomer

c.

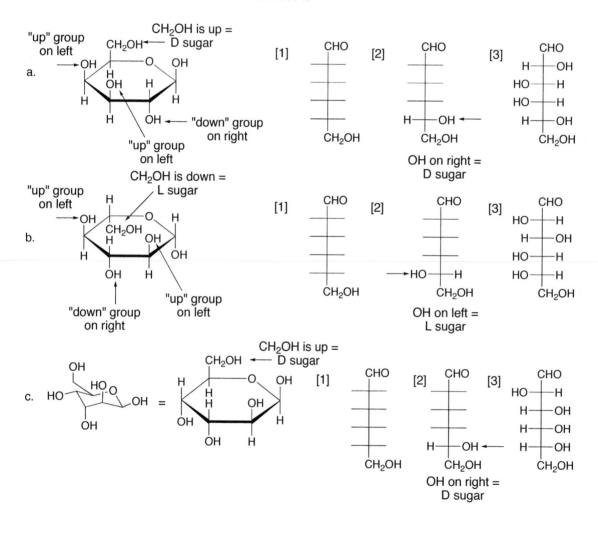

**27.47** Use the directions from Answer 27.14.

d.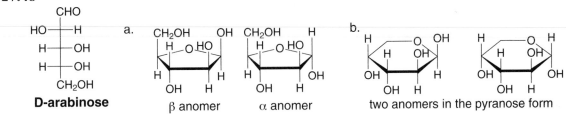

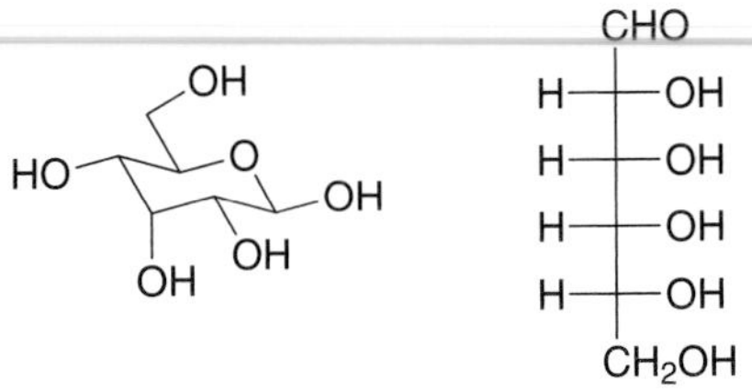

D sugar

e. HOCH₂

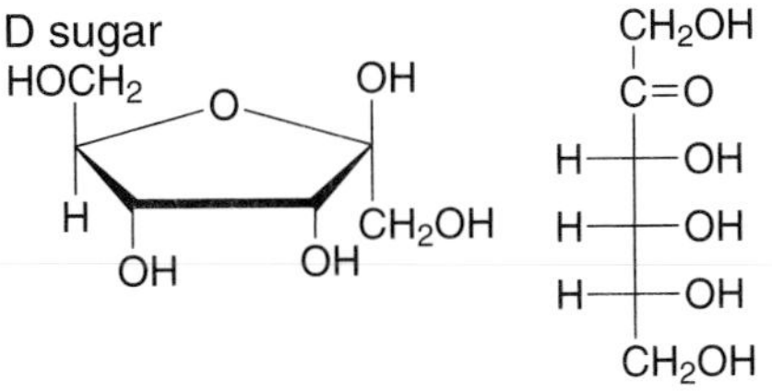

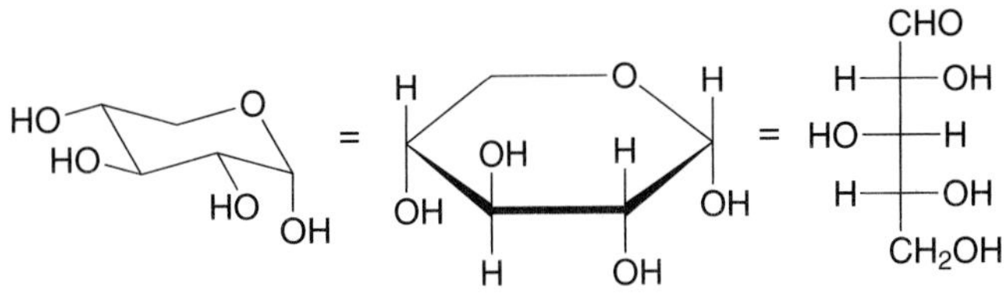

**27.48**

CHO
HO—H
H—OH
H—OH
CH₂OH
**D-arabinose**

a.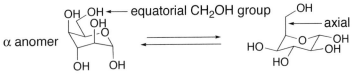

β anomer    α anomer

b.

two anomers in the pyranose form

**27.49**

Two anomers of D-idose, as well as two conformations of each anomer:

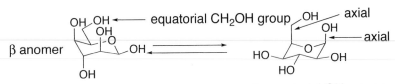

α anomer

4 axial substituents    4 equatorial OH groups

More stable conformation for the α anomer—the
CH₂OH is axial, but all other groups are equatorial.

β anomer

3 axial substituents    3 equatorial OH groups

The more stable conformation for the β anomer—
the CH₂OH is axial, as is the anomeric OH, but
three other OH groups are equatorial.

**27.50**

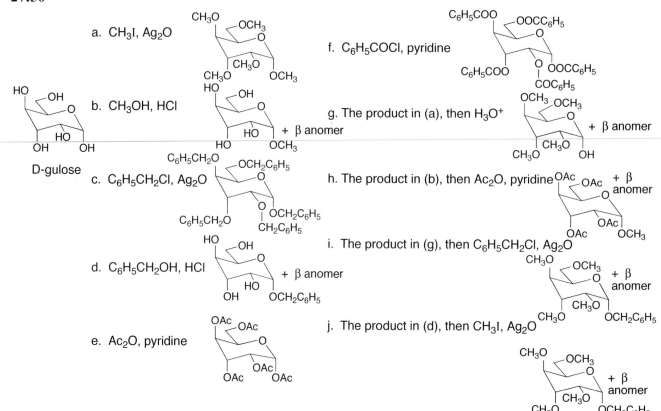

a. CH₃I, Ag₂O

D-gulose

b. CH₃OH, HCl

c. C₆H₅CH₂Cl, Ag₂O

d. C₆H₅CH₂OH, HCl

e. Ac₂O, pyridine

f. C₆H₅COCl, pyridine

g. The product in (a), then H₃O⁺    + β anomer

h. The product in (b), then Ac₂O, pyridine    + β anomer

i. The product in (g), then C₆H₅CH₂Cl, Ag₂O    + β anomer

j. The product in (d), then CH₃I, Ag₂O    + β anomer

**27.51**

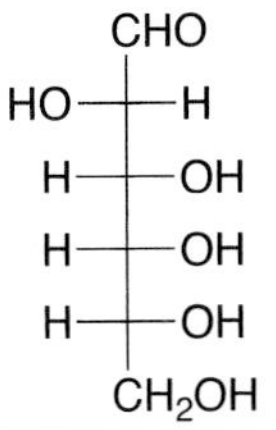

CHO
HO—H
H—OH
H—OH
H—OH
CH$_2$OH

D-altrose

a. CH$_3$OH, HCl  + β anomer

(structure with OH, OH, HO, O, OH, OCH$_3$)

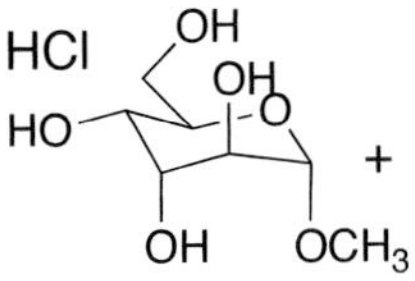

b. (CH$_3$)$_2$CHOH, HCl + β anomer

(structure with OH, OH, HO, O, OH, OCH(CH$_3$)$_2$)

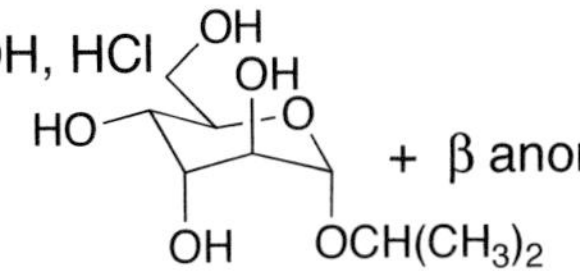

c. NaBH$_4$, CH$_3$OH

CH$_2$OH
HO—H
H—OH
H—OH
H—OH
CH$_2$OH

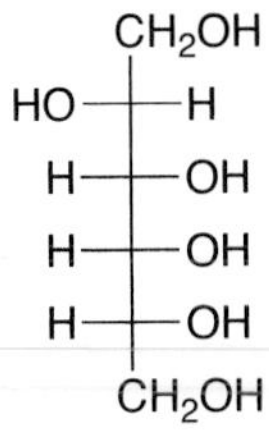

d. Br$_2$, H$_2$O

COOH
HO—H
H—OH
H—OH
H—OH
CH$_2$OH

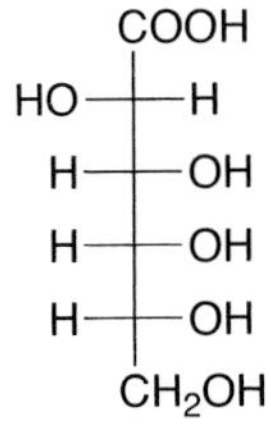

e. HNO$_3$, H$_2$O

COOH
HO—H
H—OH
H—OH
H—OH
COOH

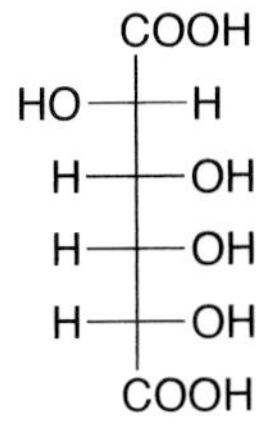

f. [1] NH$_2$OH
[2] (CH$_3$CO)$_2$O, NaOCOCH$_3$
[3] NaOCH$_3$

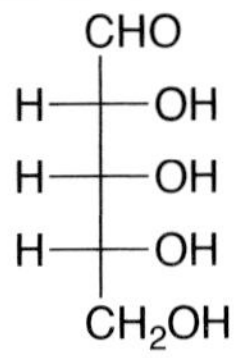

CHO
H—OH
H—OH
H—OH
CH$_2$OH

g. [1] NaCN, HCl
[2] H$_2$, Pd-BaSO$_4$
[3] H$_3$O$^+$

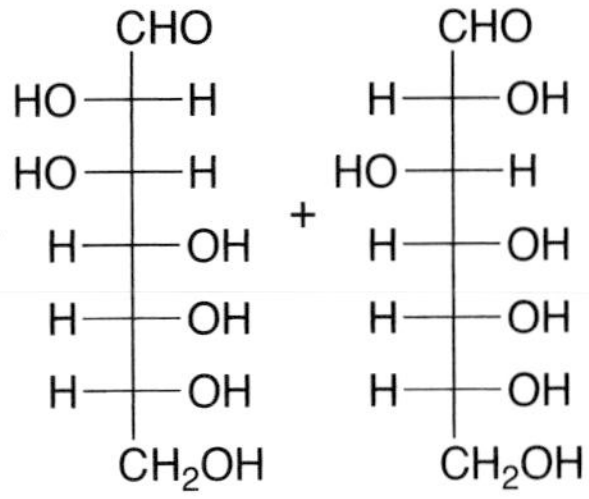

CHO
HO—H
HO—H
H—OH
H—OH
H—OH
CH$_2$OH

+

CHO
H—OH
HO—H
H—OH
H—OH
H—OH
CH$_2$OH

h. CH$_3$I, Ag$_2$O + β anomer

(structure with OCH$_3$, OCH$_3$, CH$_3$O, O, OCH$_3$, OCH$_3$)

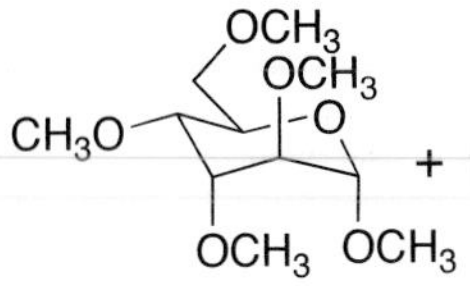

i. Ac$_2$O, pyridine + β anomer

(structure with OAc, OAc, AcO, O, OAc, OAc)

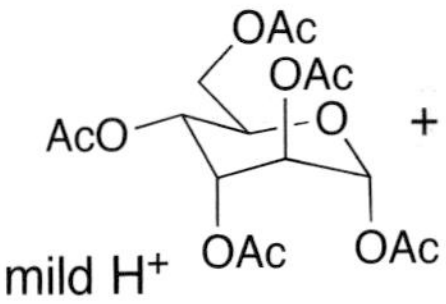

j. C$_6$H$_5$CH$_2$NH$_2$, mild H$^+$ + β anomer

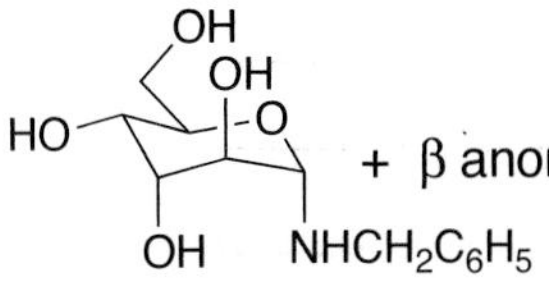

(structure with OH, OH, HO, O, OH, NHCH$_2$C$_6$H$_5$)

**27.52**

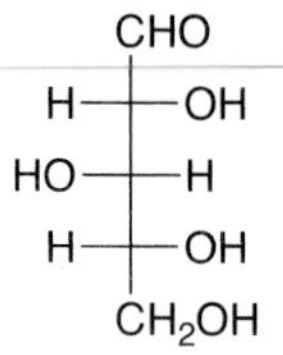

a. $CH_3OH$, HCl

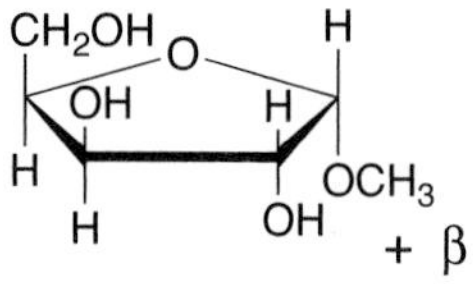

+ β anomer

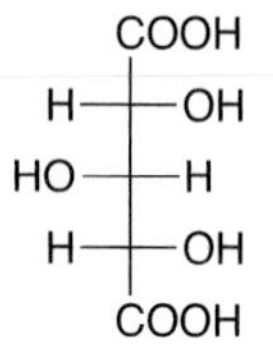

f. [1] $NH_2OH$
[2] $(CH_3CO)_2O$, NaOCOCH₃
[3] $NaOCH_3$

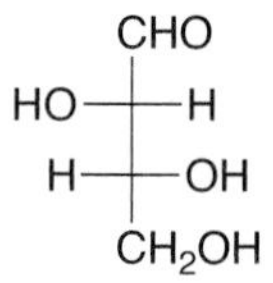

b. $(CH_3)_2CHOH$, HCl

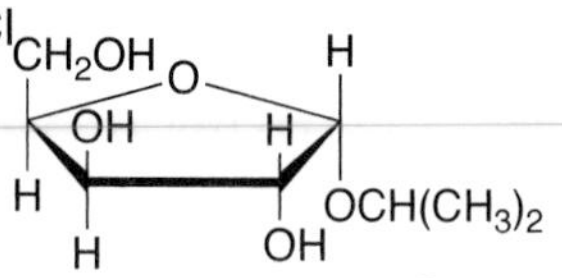

+ β anomer

g. [1] NaCN, HCl
[2] $H_2$, Pd-BaSO₄
[3] $H_3O^+$

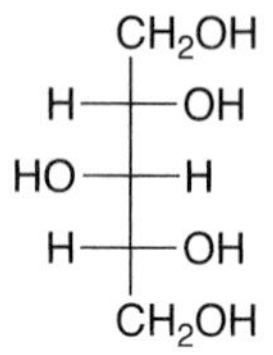

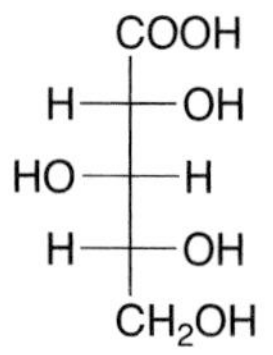

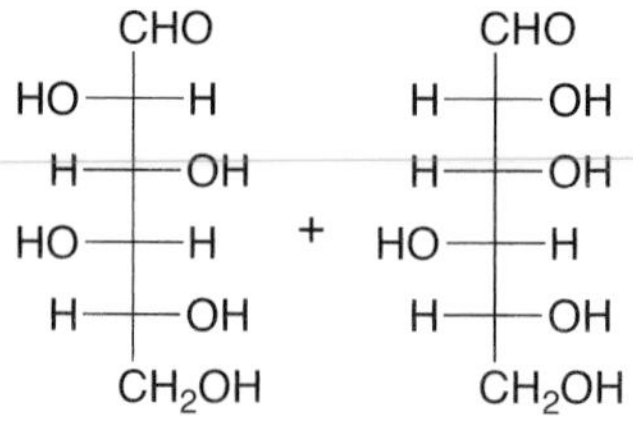

D-xylose

c. $NaBH_4$, $CH_3OH$

h. $CH_3I$, $Ag_2O$

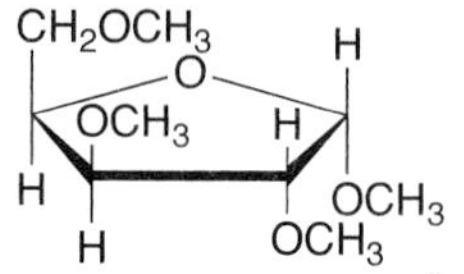

+ β anomer

d. $Br_2$, $H_2O$

i. $Ac_2O$, pyridine

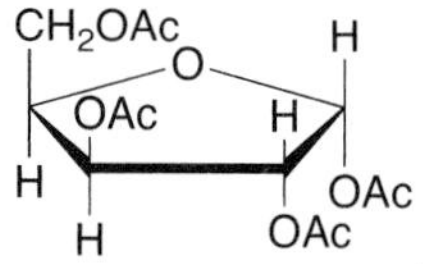

+ β anomer

j. $C_6H_5CH_2NH_2$, mild $H^+$

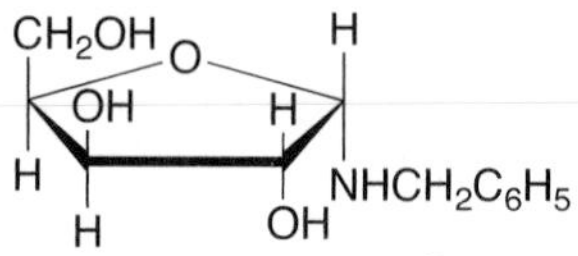

e. $HNO_3$, $H_2O$

+ β anomer

**27.53**

salicin $\xrightarrow{H_3O^+}$ monosaccharide (both anomers) + aglycon

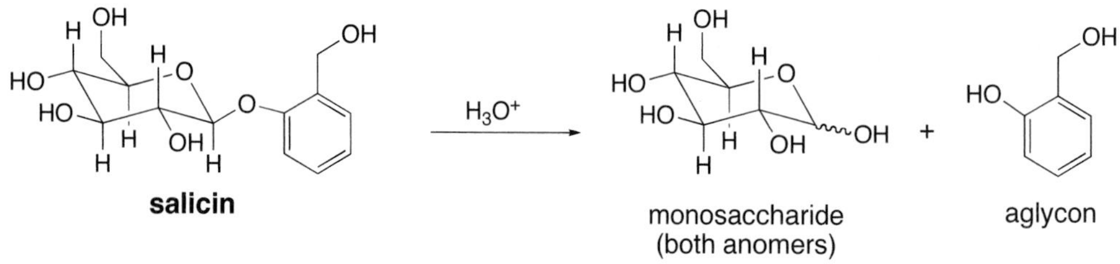

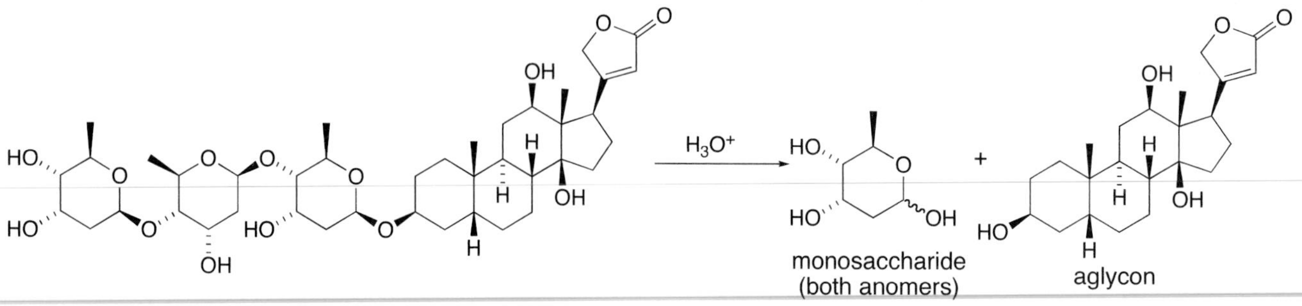

digoxin $\xrightarrow{H_3O^+}$ monosaccharide (both anomers) + aglycon

**27.54**

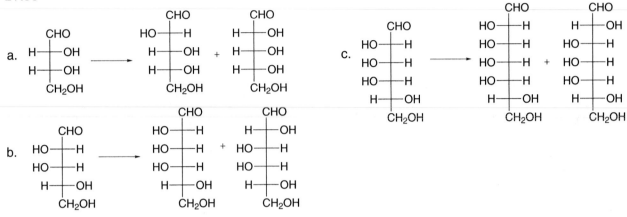

**27.55**

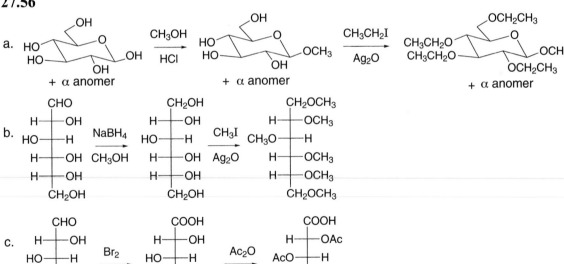

**27.56**

**27.57** Molecules with a plane of symmetry are optically inactive.

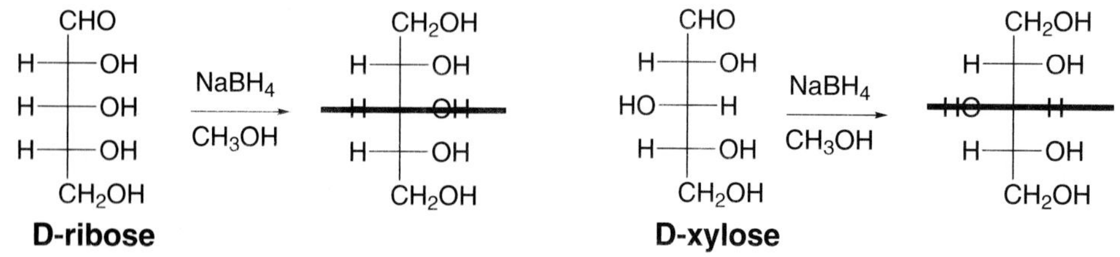

**D-ribose**        **D-xylose**

**27.58**

a.

b.

c.

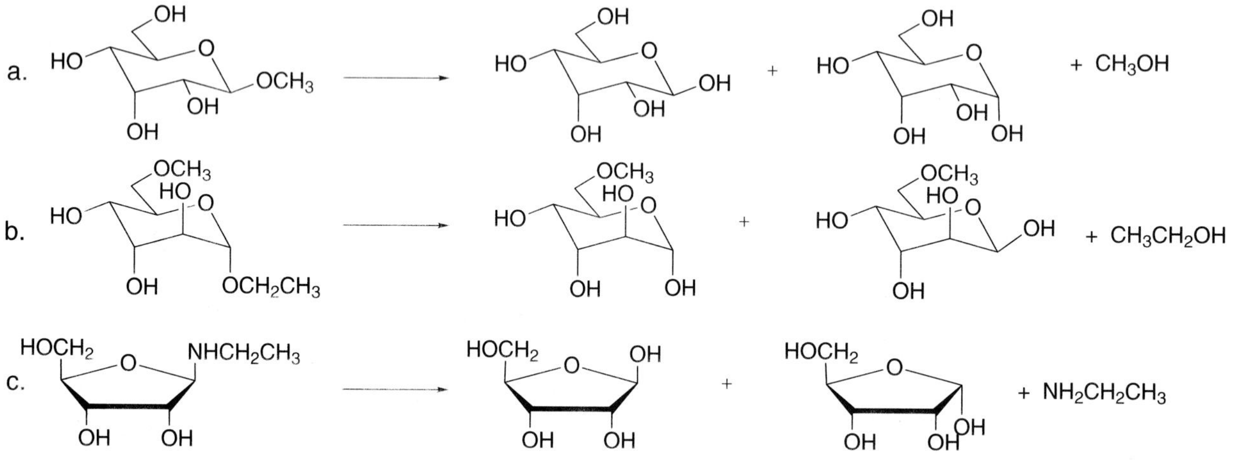

**27.59**

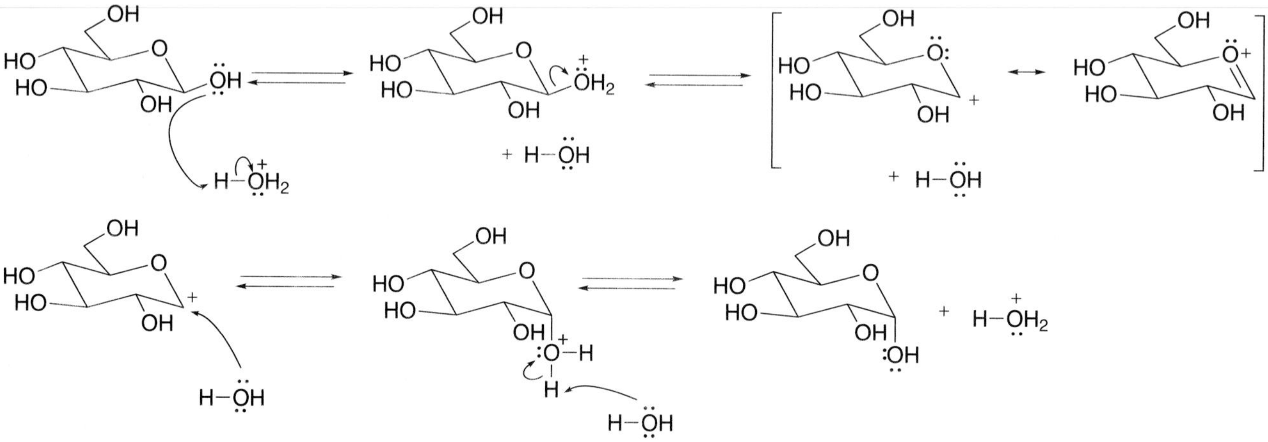

**27.60**

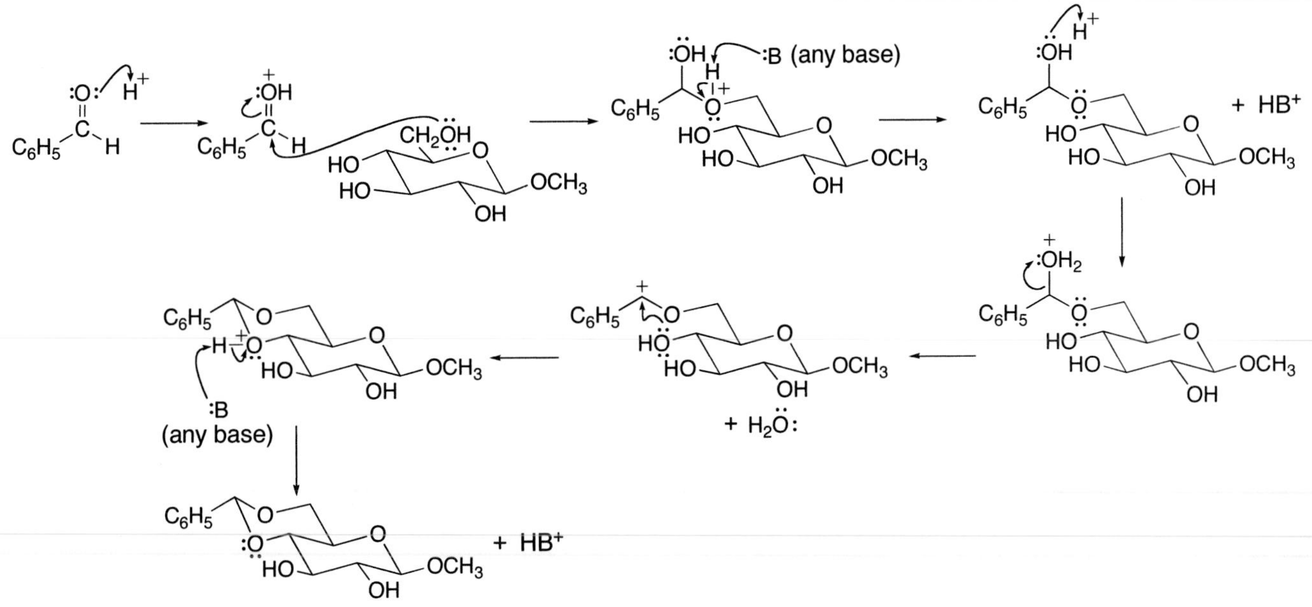

**27.61**

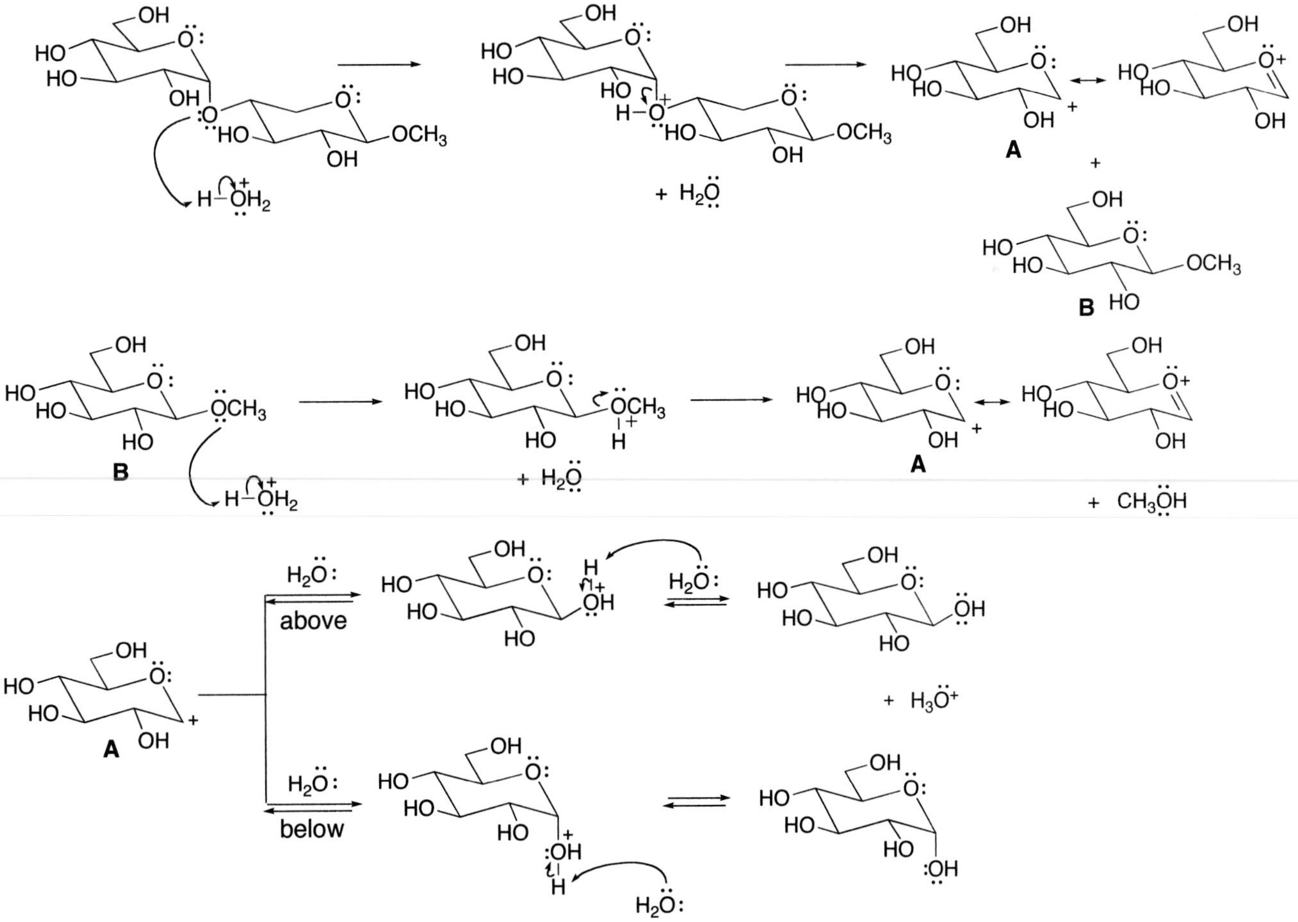

**27.62**

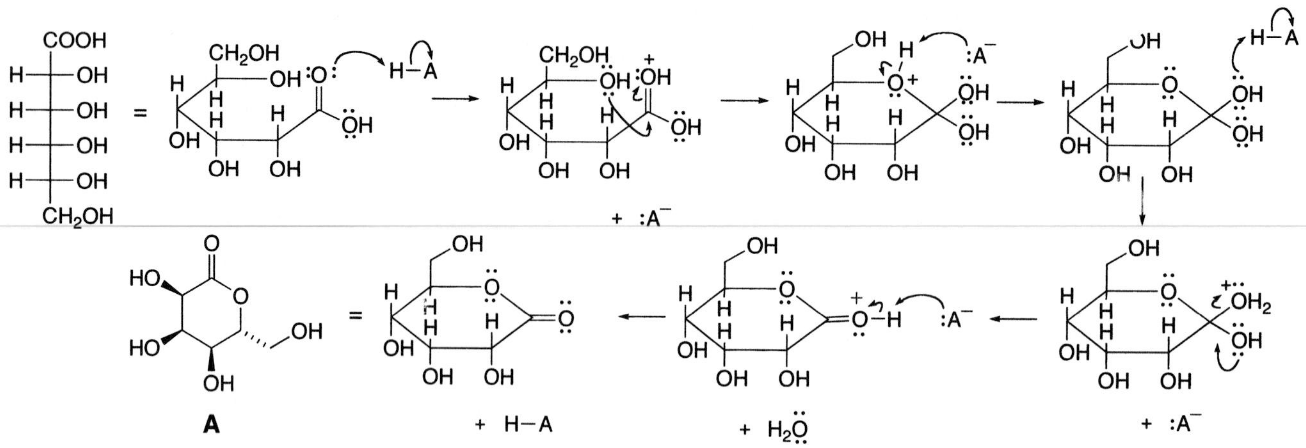

**A**

**27.63**

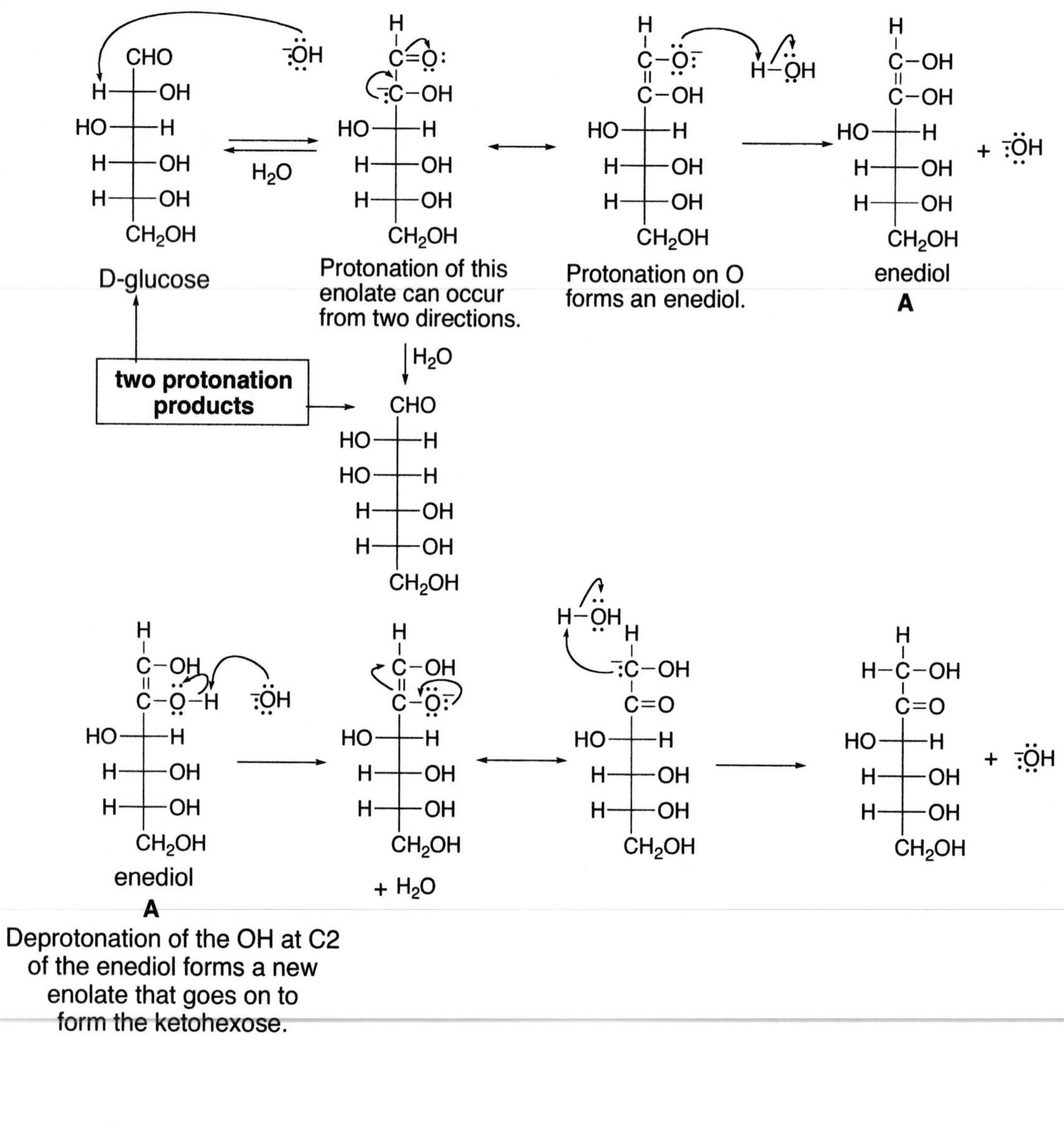

D-glucose

Protonation of this enolate can occur from two directions.

Protonation on O forms an enediol.

enediol

**A**

two protonation products

enediol

**A**

Deprotonation of the OH at C2 of the enediol forms a new enolate that goes on to form the ketohexose.

**27.64**

Two D-aldopentoses (**A'** and **A"**) yield optically active aldaric acids when oxidized.
**Optically active D-aldaric acids:**

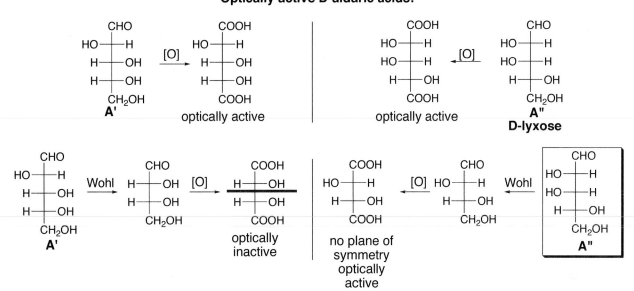

Only **A"** undergoes Wohl degradation to an aldotetrose that is oxidized to an optically active aldaric acid so **A"** is the structure of the D-aldopentose in question.

**27.65**

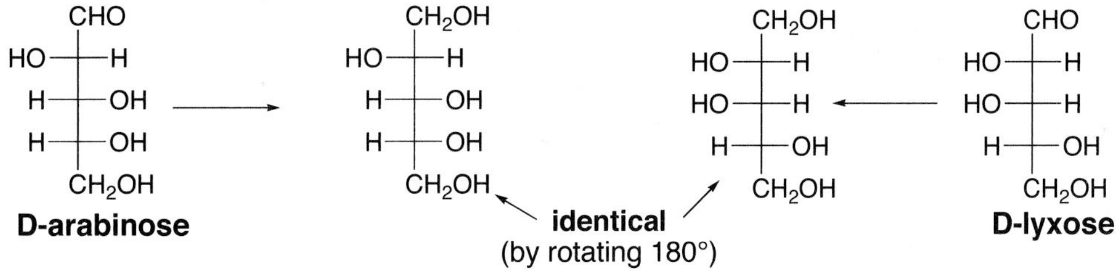

**27.66**

Only two D-aldopentoses (**A'** and **A''**) yield optically inactive aldaric acids (**B'** and **B''**).

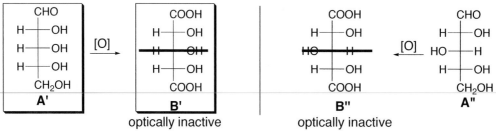

**Product of Kiliani–Fischer synthesis:**

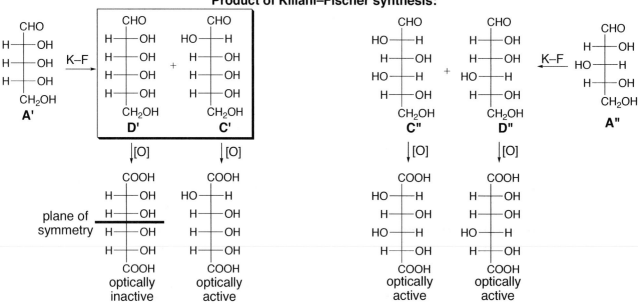

Only **A'** fits the criteria. Kiliani–Fischer synthesis of **A'** forms **C'** and **D'** which are oxidized to one optically active and one optically inactive aldaric acid. A similar procedure with **A''** forms two optically active aldaric acids. Thus, the structures of **A–D**, correspond to the structures of **A'–D'**.

**27.67**

Only two D-aldopentoses (**A'** and **A''**) are reduced to optically active alditols.

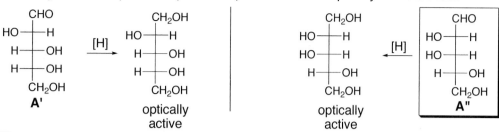

**Product of Kiliani–Fischer synthesis:**

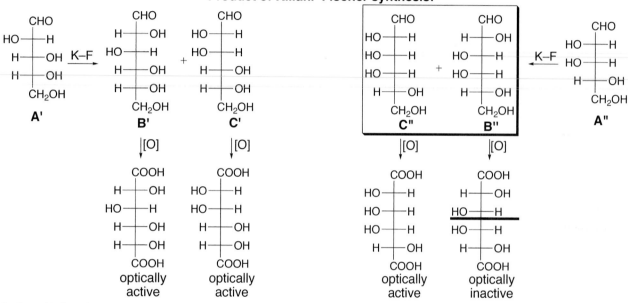

Only **A''** fits the criteria. Kiliani–Fischer synthesis of **A''** forms **B''** and **C''**, which are oxidized to one optically inactive and one optically active diacid. A similar procedure with **A'** forms two optically active diacids. Thus, the structures of **A–C** correspond to **A''–C''**.

**27.68**

**D-gulose**

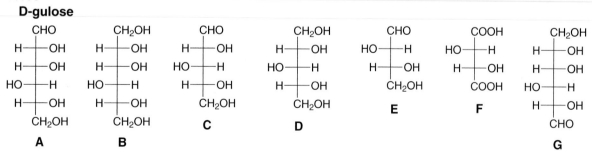

**27.69**

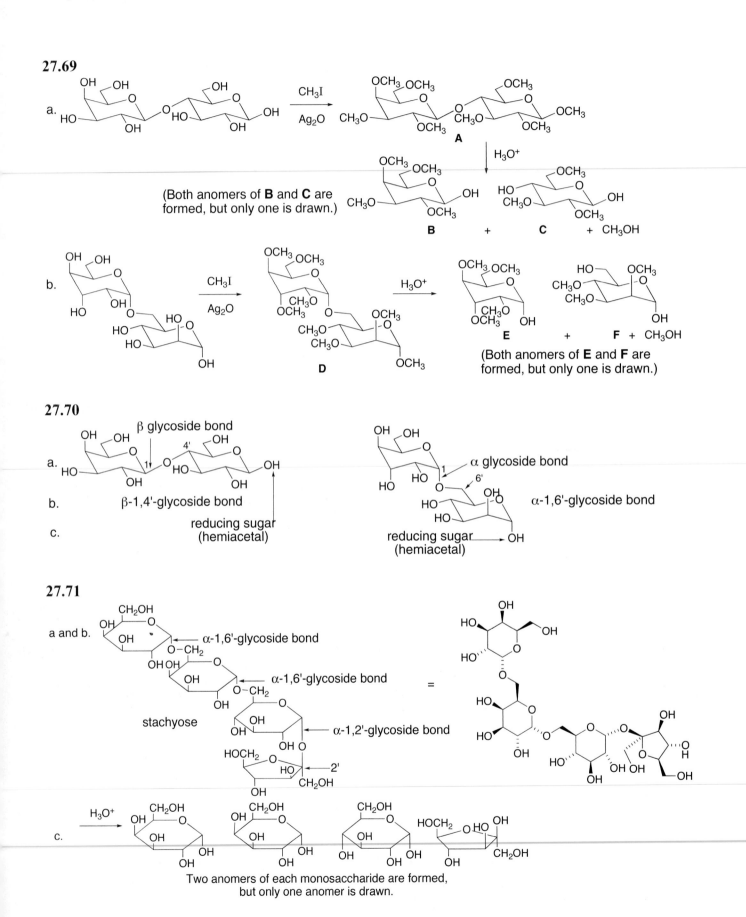

a.

(Both anomers of **B** and **C** are formed, but only one is drawn.)

**B** + **C** + CH₃OH

b.

**D**

**E** + **F** + CH₃OH

(Both anomers of **E** and **F** are formed, but only one is drawn.)

**27.70**

a. β glycoside bond

b. β-1,4'-glycoside bond

c. reducing sugar (hemiacetal)

α glycoside bond

α-1,6'-glycoside bond

reducing sugar (hemiacetal)

**27.71**

a and b. α-1,6'-glycoside bond

α-1,6'-glycoside bond

stachyose

α-1,2'-glycoside bond

=

c.

Two anomers of each monosaccharide are formed, but only one anomer is drawn.

d. Stachyose is not a reducing sugar since it contains no hemiacetal.

e.

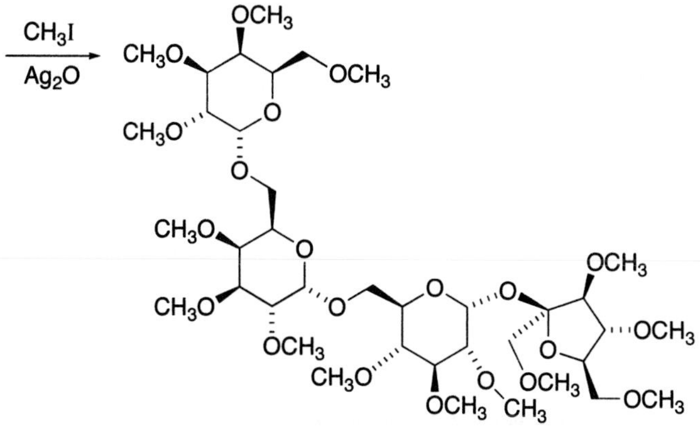

f.

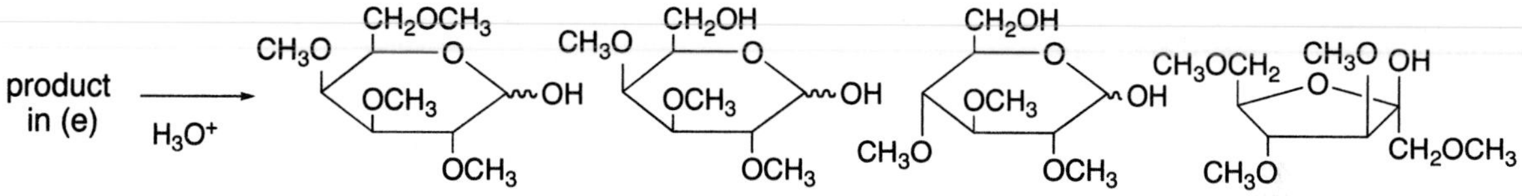

product
in (e) $\xrightarrow{H_3O^+}$

Two anomers of each
monosaccharide are formed.

**27.72**

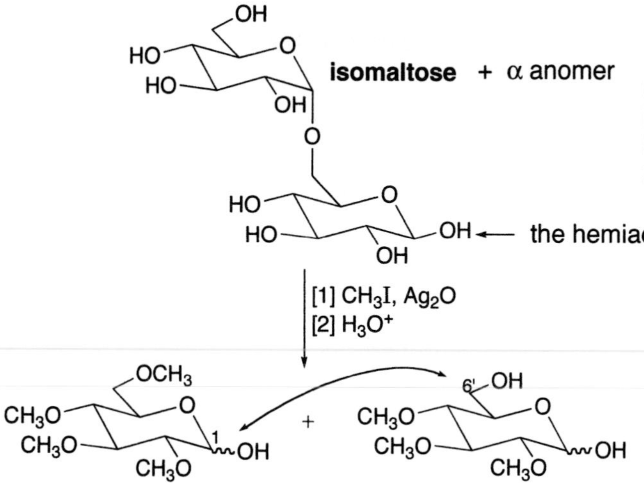

**isomaltose** + α anomer

Isomaltose must be composed of two glucose units
in an α-glycosidic linkage. Since it is a reducing sugar
it contains a hemiacetal. The free OH groups in the
hydrolysis products show where the two
monosaccharides are joined.

[1] CH$_3$I, Ag$_2$O
[2] H$_3$O$^+$

(Both anomers are present.)

**27.73**

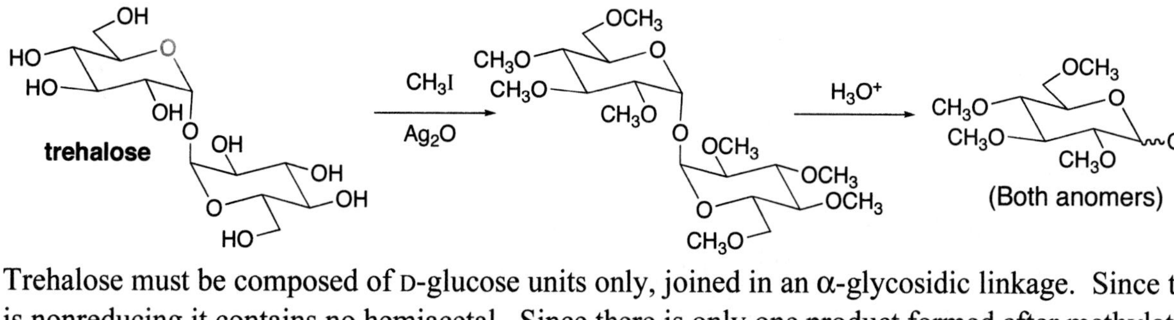

**trehalose**

Trehalose must be composed of D-glucose units only, joined in an α-glycosidic linkage. Since trehalose
is nonreducing it contains no hemiacetal. Since there is only one product formed after methylation and
hydrolysis, the two anomeric C's must be joined

**27.74**

a.

b.

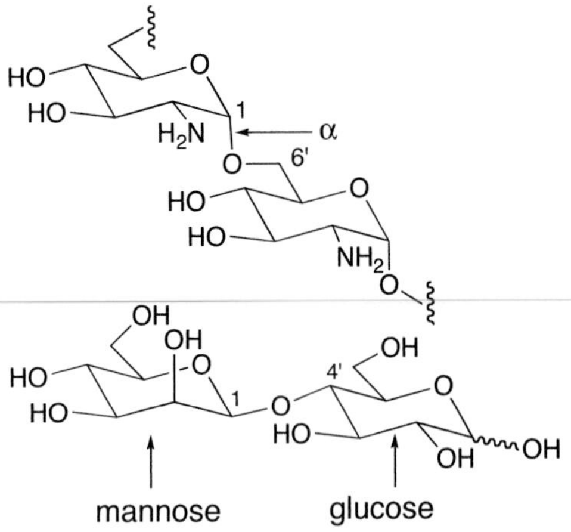

mannose    glucose

c.

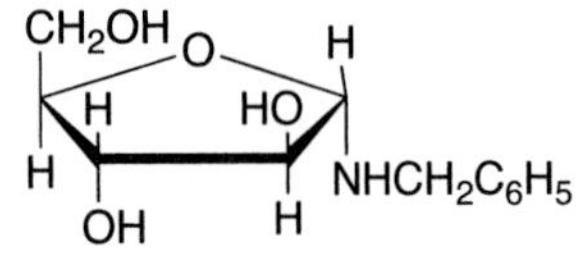

d.

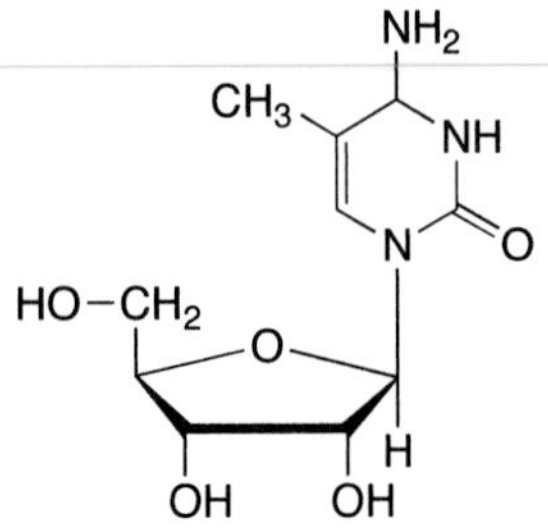

**27.75**

Ignoring stereochemistry along the way:

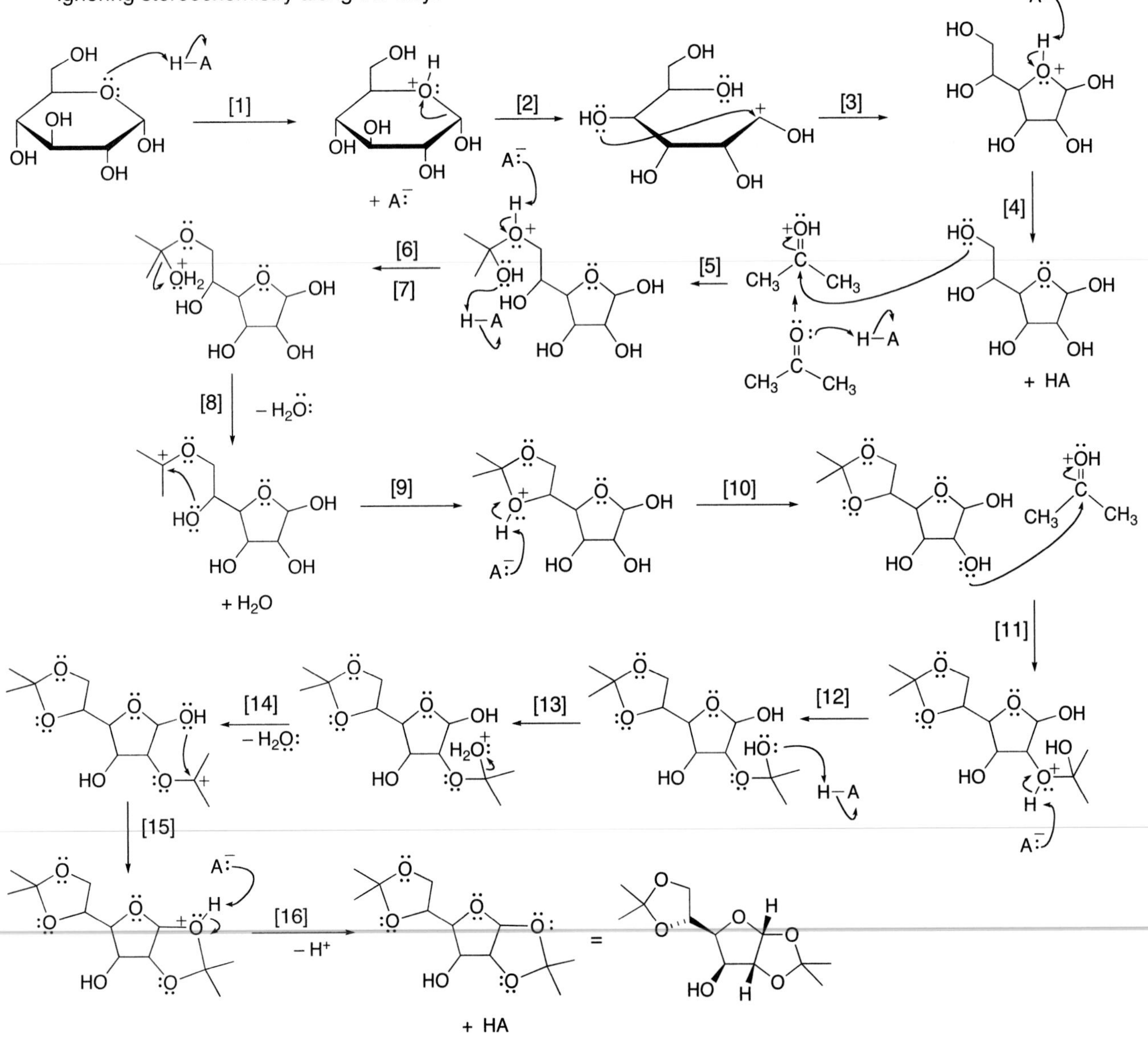

**27.76** The hydrolysis data suggest that the trisaccharide has D-galactose on one end and D-fructose on the other. D-Galactose must be joined to its adjacent sugar by a β-glycosidic linkage. D-Fructose must be joined to its adjacent sugar by an α-glycosidic linkage.

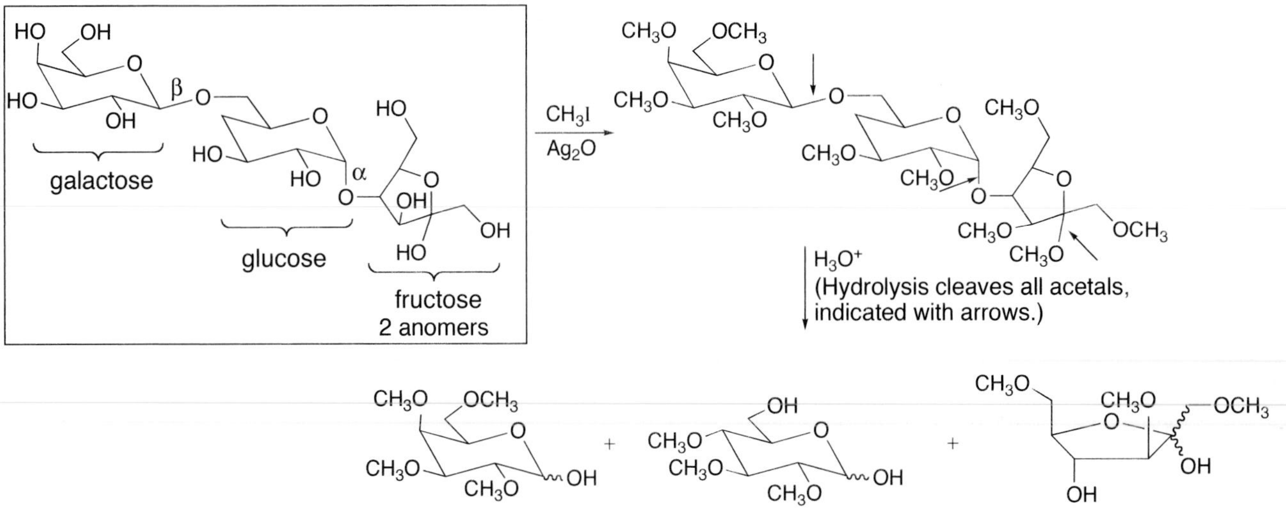

galactose

glucose

fructose
2 anomers

$CH_3I$ / $Ag_2O$

$H_3O^+$
(Hydrolysis cleaves all acetals, indicated with arrows.)

2,3,4,6-tetra-O-methyl-D-galactose    +    2,3,4-tri-O-methyl-D-glucose    +    1,3,6-tri-O-methyl-D-fructose

(Both anomers of each compound are formed.)

♦ Synthesis of amino acids (28.2)

## [1] From α-halo carboxylic acids by S_N2 reaction

$$R-\underset{\underset{Br}{|}}{CH}COOH \xrightarrow[\substack{\text{(large excess)} \\ \mathbf{S_N2}}]{NH_3} R-\underset{\underset{NH_3^+}{|}}{CH}COO^- \ + \ NH_4^+ \ Br^-$$

## [2] By alkylation of diethyl acetamidomalonate

$$\underset{CH_3}{\overset{O}{\underset{\|}{C}}}-\overset{H}{\underset{}{N}}-\overset{H}{\underset{COOEt}{\underset{|}{C}}}-COOEt \xrightarrow[\substack{\text{[2] RX} \\ \text{[3] } H_3O^+, \Delta}]{\text{[1] NaOEt}} H_2N-\overset{R}{\underset{H}{\underset{|}{C}}}-COOH$$

• Alkylation works best with unhindered alkyl halides—that is, $CH_3X$ and $RCH_2X$.

## [3] Strecker synthesis

$$\underset{R}{\overset{O}{\underset{\|}{C}}}{\overset{}{\diagdown}}_H \xrightarrow[NaCN]{NH_4Cl} R-\overset{NH_2}{\underset{H}{\underset{|}{C}}}-CN \xrightarrow{H_3O^+} R-\overset{NH_2}{\underset{H}{\underset{|}{C}}}-COOH$$

α-amino nitrile

♦ Preparation of optically active amino acids

## [1] Resolution of enantiomers by forming diastereomers (28.3A)

- Convert a racemic mixture of amino acids into a racemic mixture of N-acetyl amino acids [(S)- and (R)-CH₃CONHCH(R)COOH].
- React the enantiomers with a chiral amine to form a mixture of diastereomers.
- Separate the diastereomers.
- Regenerate the amino acids by protonation of the carboxylate salt and hydrolysis of the N-acetyl group.

## [2] Kinetic resolution using enzymes (28.3B)

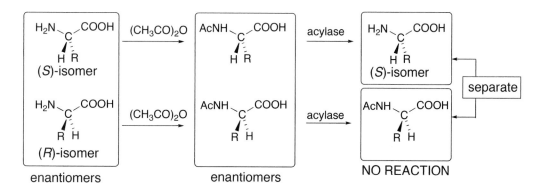

## [3] By enantioselective hydrogenation (28.4)

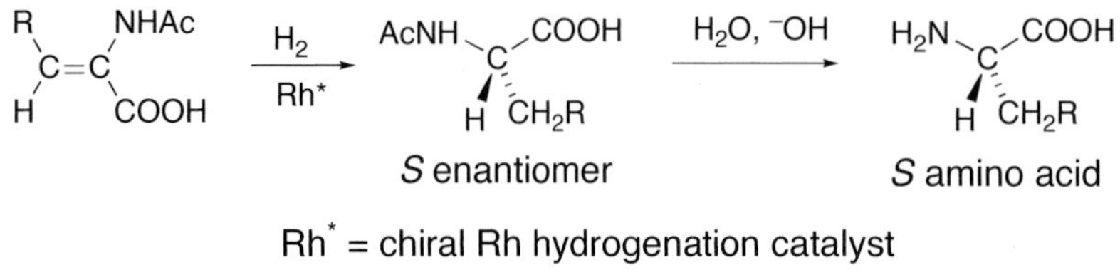

Rh* = chiral Rh hydrogenation catalyst

♦ Adding and removing protecting groups for amino acids (28.6)

### [1] Protection of an amino group as a Boc derivative

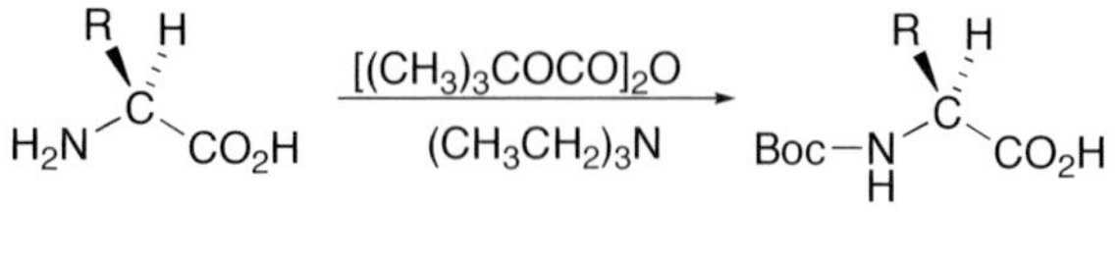

### [2] Deprotection of a Boc-protected amino acid

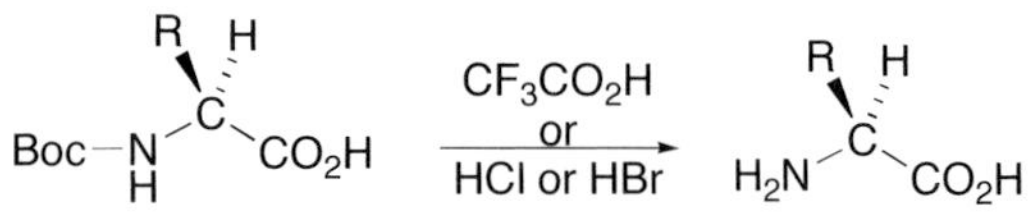

### [3] Protection of an amino group as an Fmoc derivative

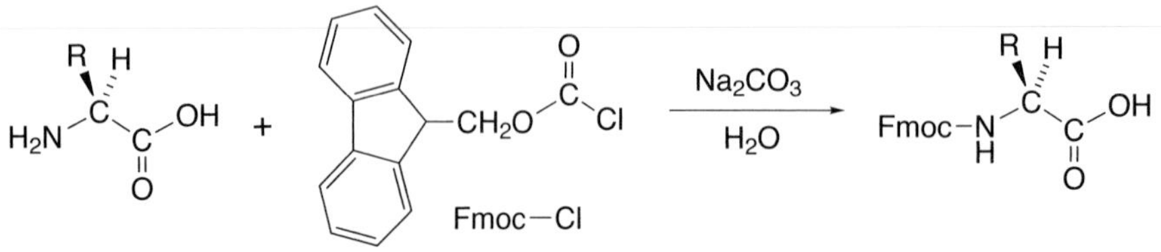

### [4] Deprotection of an Fmoc-protected amino acid

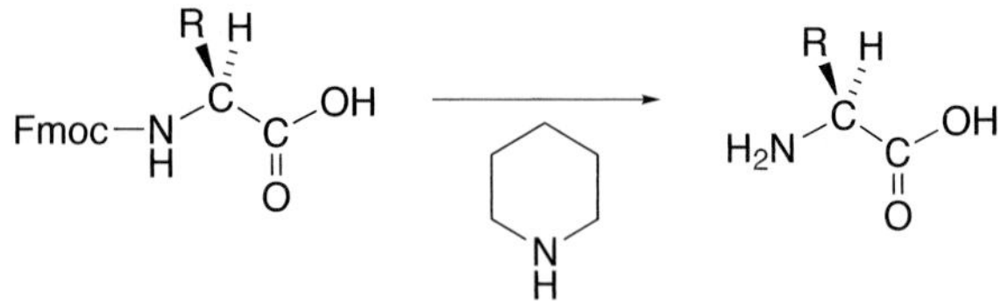

### [5] Protection of a carboxy group as an ester

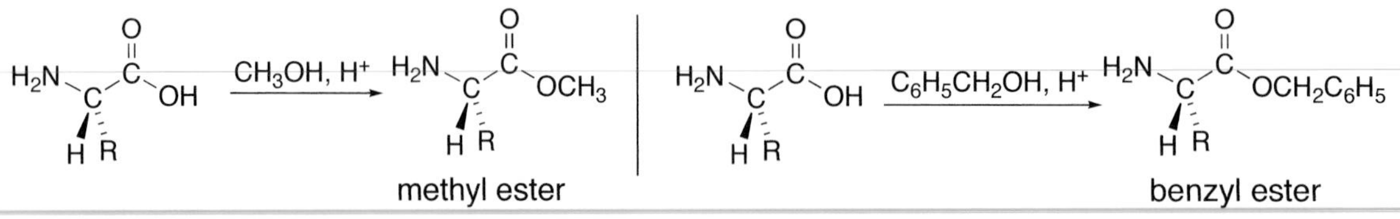

methyl ester          benzyl ester

## [6] Deprotection of an ester group

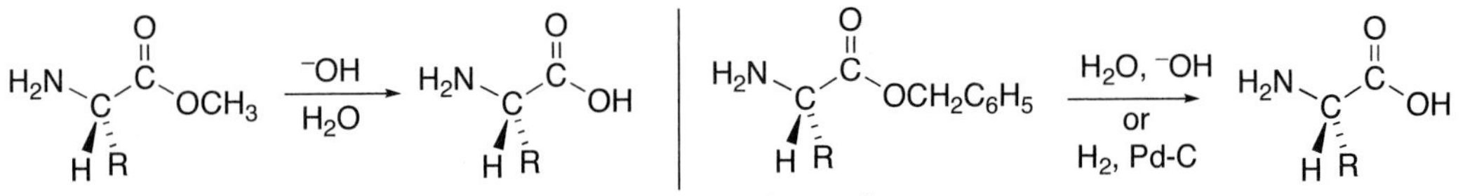

methyl ester

benzyl ester

♦ **Synthesis of dipeptides (28.6)**

## [1] Amide formation with DCC

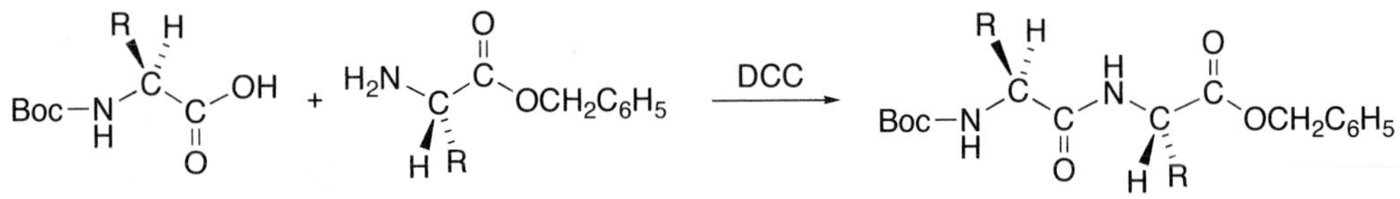

## [2] Four steps are needed to synthesize a dipeptide:

[a] **Protect** the amino group of one amino acid using a Boc or Fmoc group.
[b] **Protect** the carboxy group of the second amino acid using an ester.
[c] Form the amide bond with **DCC.**
[d] **Remove both protecting groups** in one or two reactions.

♦ **Summary of the Merrifield method of peptide synthesis (28.7)**

[1] Attach an Fmoc-protected amino acid to a polymer derived from polystyrene.
[2] Remove the Fmoc protecting group.
[3] Form the amide bond with a second Fmoc protected amino acid using DCC.
[4] Repeat steps [2] and [3].
[5] Remove the protecting group and detach the peptide from the polymer.

## Chapter 28: Answers to Problems

**28.1**

L-isoleucine

**28.2**

a. $(CH_3)_2CH-\underset{\underset{H}{|}}{\overset{\overset{NH_3^+}{|}}{C}}-COO^-$

b. $(CH_3)_2CHCH_2-\underset{\underset{H}{|}}{\overset{\overset{NH_3^+}{|}}{C}}-COO^-$

c.

d. $HOOCCH_2CH_2-\underset{\underset{H}{|}}{\overset{\overset{NH_3^+}{|}}{C}}-COO^-$

**28.3** In an amino acid, the electron-withdrawing carboxy group destabilizes the ammonium ion ($-NH_3^+$), making it more readily donate a proton; that is, it makes it a stronger acid. Also, the electron-withdrawing carboxy group removes electron density from the amino group ($-NH_2$) of the conjugate base, making it a weaker base than a $1°$ amine, which has no electron-withdrawing group.

**28.4** The most direct way to synthesize an $\alpha$-amino acid is by **$S_N2$ reaction of an $\alpha$-halo carboxylic acid with a large excess of $NH_3$.**

a. $Br-\underset{\underset{H}{|}}{CH}-COOH \xrightarrow[\text{(large excess)}]{NH_3} H_2N-\underset{\underset{H}{|}}{CH}-COO^- NH_4^+$
glycine

b. $Br-\underset{\substack{| \\ CH-CH_3 \\ | \\ CH_2 \\ | \\ CH_3}}{CH}-COOH \xrightarrow[\text{(large excess)}]{NH_3} H_2N-\underset{\substack{| \\ CH-CH_3 \\ | \\ CH_2 \\ | \\ CH_3}}{CH}-COO^- NH_4^+$
isoleucine

c. $Br-\underset{\underset{CH_2}{|}}{CH}-COOH \xrightarrow[\text{(large excess)}]{NH_3} H_2N-\underset{\underset{CH_2}{|}}{CH}-COO^- NH_4^+$

phenylalanine

**28.5**

a. $\xrightarrow{CH_3I} H_2N-\underset{\underset{H}{|}}{\overset{\overset{CH_3}{|}}{C}}-COOH$ alanine

b. $\xrightarrow{(CH_3)_2CHCH_2Cl} H_2N-\underset{\underset{H}{|}}{\overset{\overset{CH_2CH(CH_3)_2}{|}}{C}}-COOH$ leucine

c. $\xrightarrow{CH_3CH_2CH(CH_3)Br} H_2N-\underset{\underset{H}{|}}{\overset{\overset{CH(CH_3)CH_2CH_3}{|}}{C}}-COOH$ isoleucine

**28.6**

$\xrightarrow[\substack{[2]\ CH_2=O \\ [3]\ H_3O^+,\ \Delta}]{[1]\ NaOEt}$ $H_2N-\underset{\underset{H}{|}}{\overset{\overset{CH_2OH}{|}}{C}}-COOH$
serine

## 28.7

a. H₂N–CHCOOH ⟹ (CH₃)₂CH–C(=O)H
   |
   CH·CH₃
   |
   CH₃
   **valine**

b. H₂N–CHCOOH ⟹ (CH₃)₂CHCH₂–C(=O)H
   |
   CH₂
   |
   CH–CH₃
   |
   CH₃
   **leucine**

c. H₂N–CHCOOH ⟹ C₆H₅CH₂–C(=O)H
   |
   CH₂
   |
   (phenyl ring)
   **phenylalanine**

## 28.8

a. BrCH₂COOH  —(NH₃, large excess)→  NH₂CH₂COO⁻ NH₄⁺

c. CH₃CH₂CH(CH₃)CHO  —[1] NH₄Cl, NaCN  [2] H₃O⁺→  H₂N–CHCOOH
                                                      |
                                                      CH(CH₃)CH₂CH₃

b. CH₃CONH–CH(COOEt)(COOEt)  —[1] NaOEt  [2] (CH₃)₂CHCl  [3] H₃O⁺, Δ→  H₂N–C(H)(COOH)(CH(CH₃)₂)

d. CH₃CONH–CH(COOEt)(COOEt)  —[1] NaOEt  [2] BrCH₂CO₂Et  [3] H₃O⁺, Δ→  H₂N–C(H)(COOH)(CH₂CO₂H)

## 28.9 A chiral amine must be used to resolve a racemic mixture of amino acids.

a. C₆H₅CH₂CH₂NH₂
   **achiral**

b. (quinuclidine structure)
   **achiral**

c. CH₃CH₂–C(CH₃)(H)–NH₂
   **chiral**
   (can be used)

d. (strychnine-type structure)
   **chiral**
   (can be used)

## 28.10

| To begin: Convert the amino acids into *N*-acetyl amino acids (two enantiomers). |

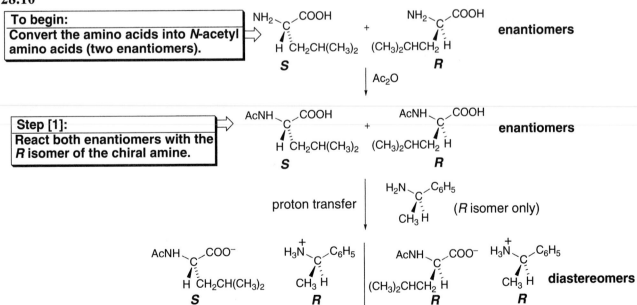

**enantiomers**

↓ Ac₂O

**Step [1]:** React both enantiomers with the *R* isomer of the chiral amine.

**enantiomers**

proton transfer    H₂N–C(C₆H₅)(CH₃)(H)   (*R* isomer only)

**diastereomers**

These salts have the *same* configuration around one stereogenic center,
but the *opposite* configuration about the other stereogenic center.

Step [2]:
Separate the diastereomers.
separate
S
R
R
R
Step [3]:
Regenerate the amino acid by hydrolysis of the amide.
(S)-leucine
(R)-leucine
The chiral amine is also regenerated.
The amino acids are now separated.

**28.11**

(mixture of enantiomers)
[1] (CH3CO)2O
[2] acylase
(S)-leucine
N-acetyl-(R)-leucine

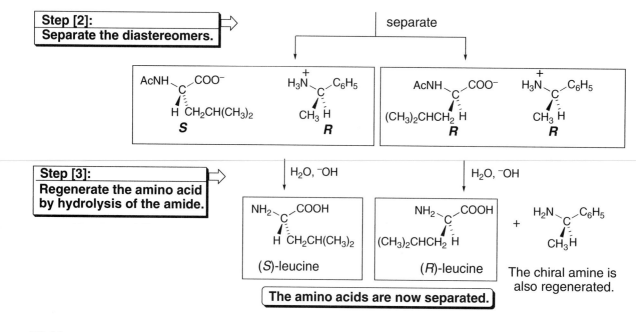

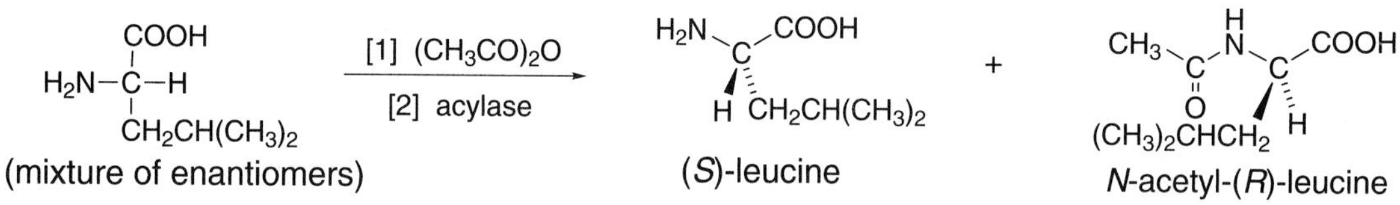

**28.12**

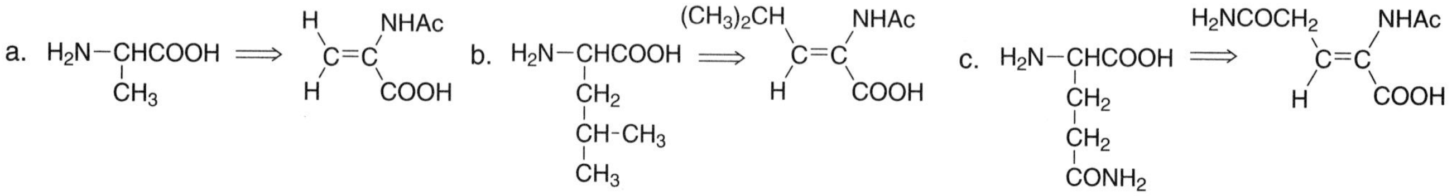

**28.13** Draw the peptide by joining adjacent COOH and $NH_2$ groups in amide bonds.

amide
C-terminal
N-terminal
Val
Glu
Val–Glu

amide
Gly
His
Leu
N-terminal
C-terminal
amide
Gly–His–Leu

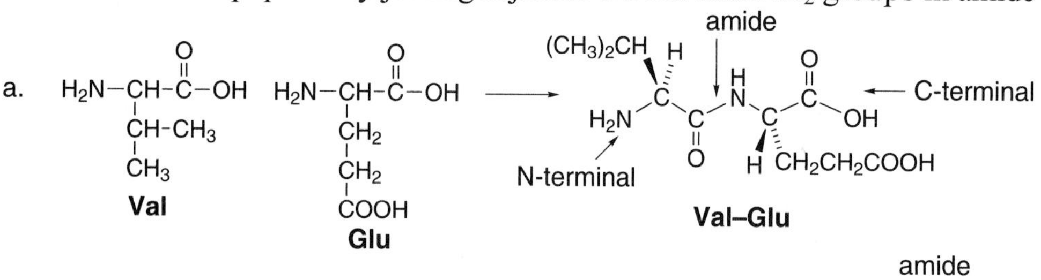

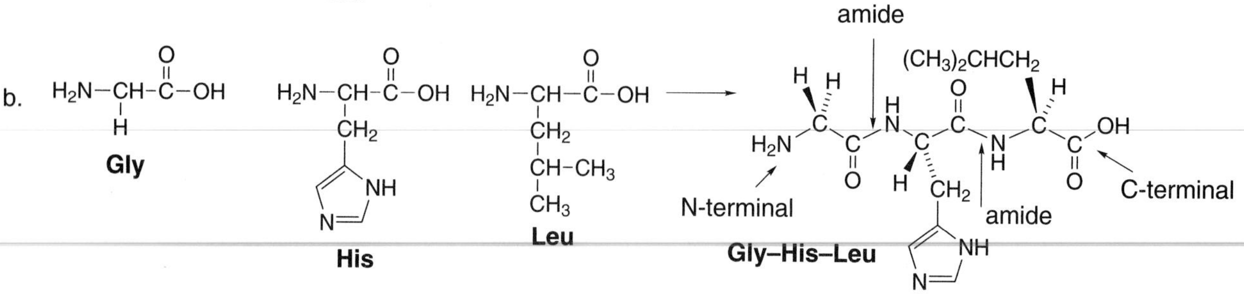

c.

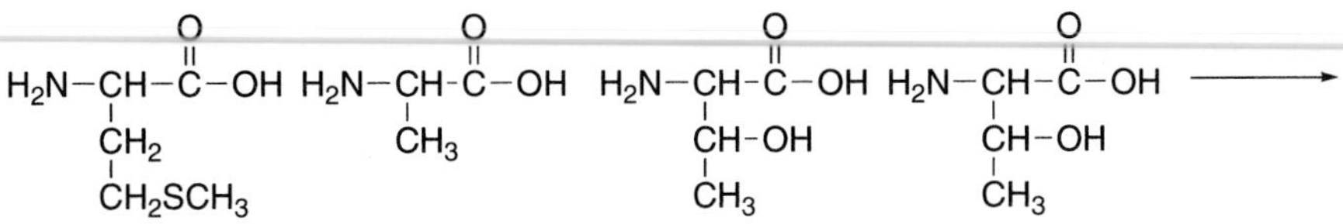

Met   Ala   Thr   Thr   ⟶

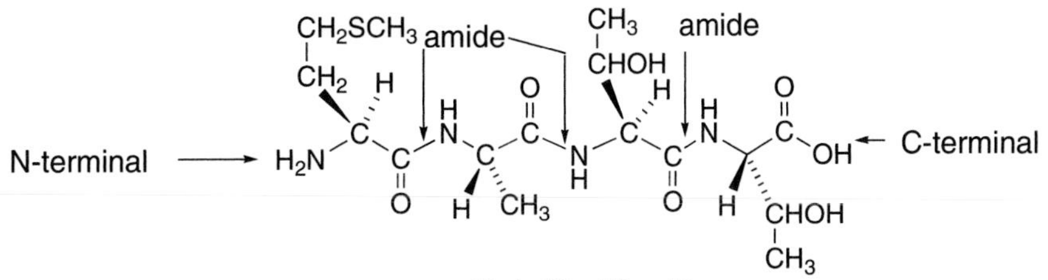

N-terminal ⟶                          ⟵ C-terminal

**Met–Ala–Thr–Thr**

**28.14**

a.

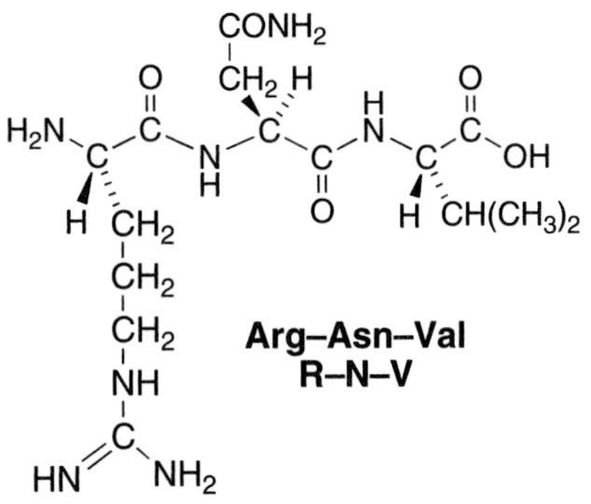

**Arg–Asn–Val**
**R–N–V**

b.

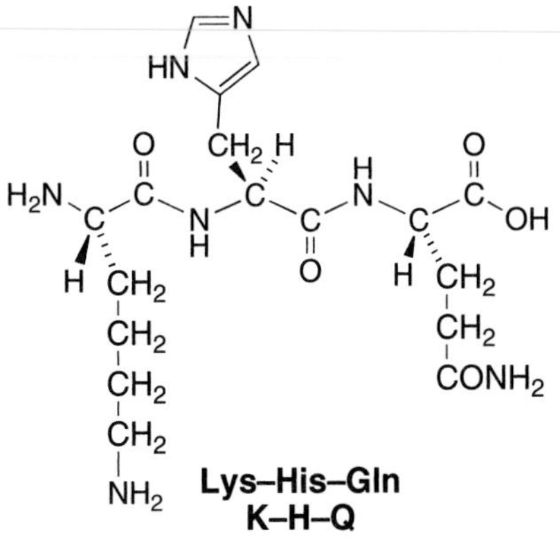

**Lys–His–Gln**
**K–H–Q**

**28.15** There are six different tripeptides that can be formed from three amino acids (A, B, C): A–B–C, A–C–B, B–A–C, B–C–A, C–A–B, C–B–A.

**28.16** The *s*-trans conformation has the two C's oriented on *opposite* sides of the C–N bond.
The *s*-cis conformation has the two C's oriented on the *same* side of the C–N bond.

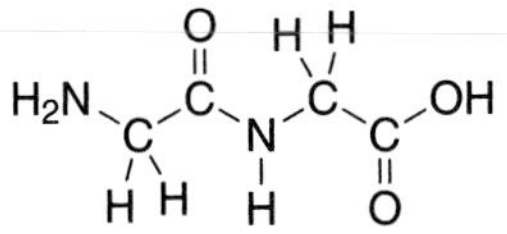

*s*-trans

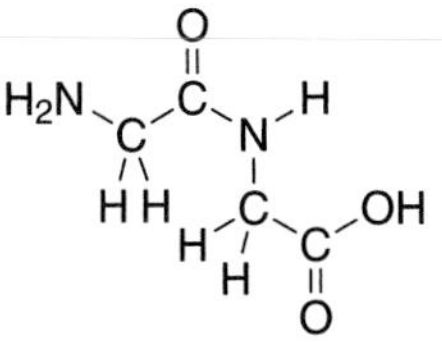

*s*-cis

**28.17**

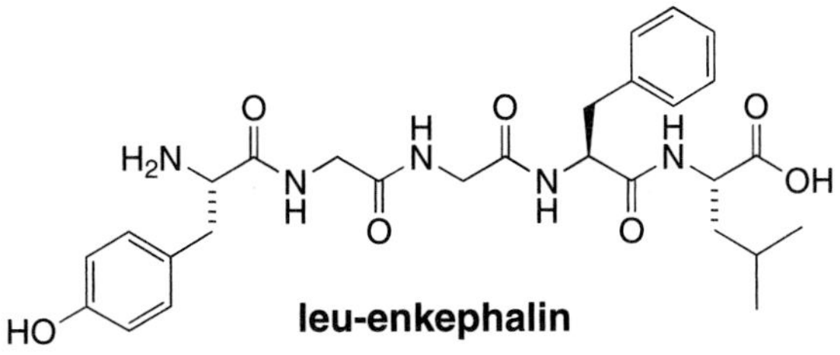

**leu-enkephalin**

**28.18**

a.

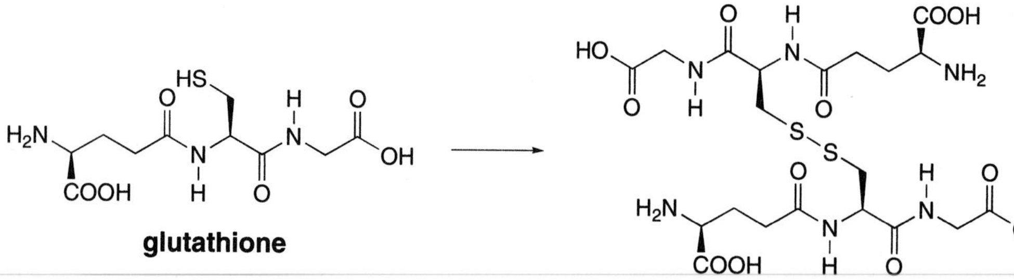

**glutathione**

b.

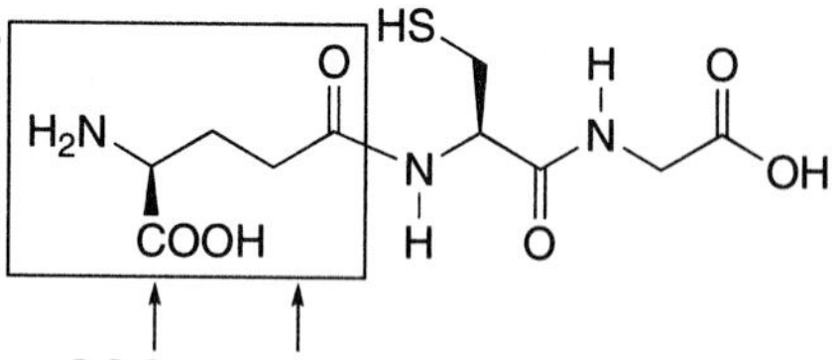

The peptide bond beween glutamic aid and its adjacent amino acid (cysteine) is formed from the COOH in the R group of glutamic acid, not the α COOH.

α COOH    This comes from the amino acid glutamic acid.

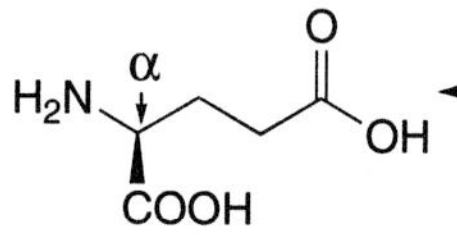

This carboxy group is used to form the amide bond in the peptide, not the α COOH, as is usual.  That's what makes glutathione's structure unusual.

glutamic acid

**28.19**

a.

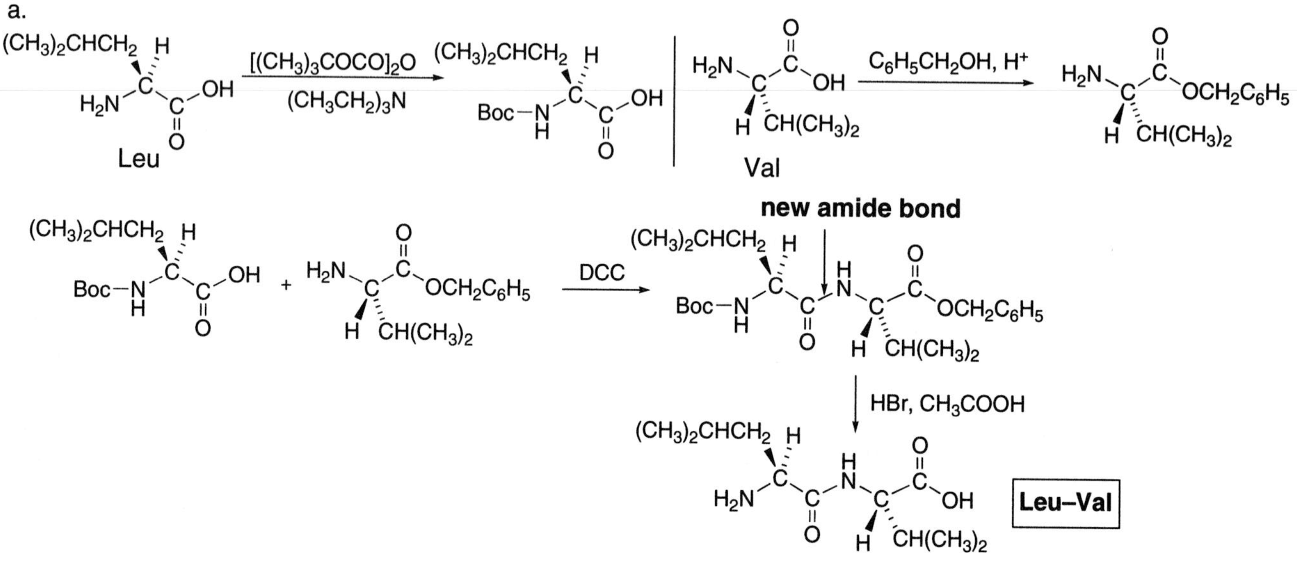

**new amide bond**

**Leu–Val**

b.

c.

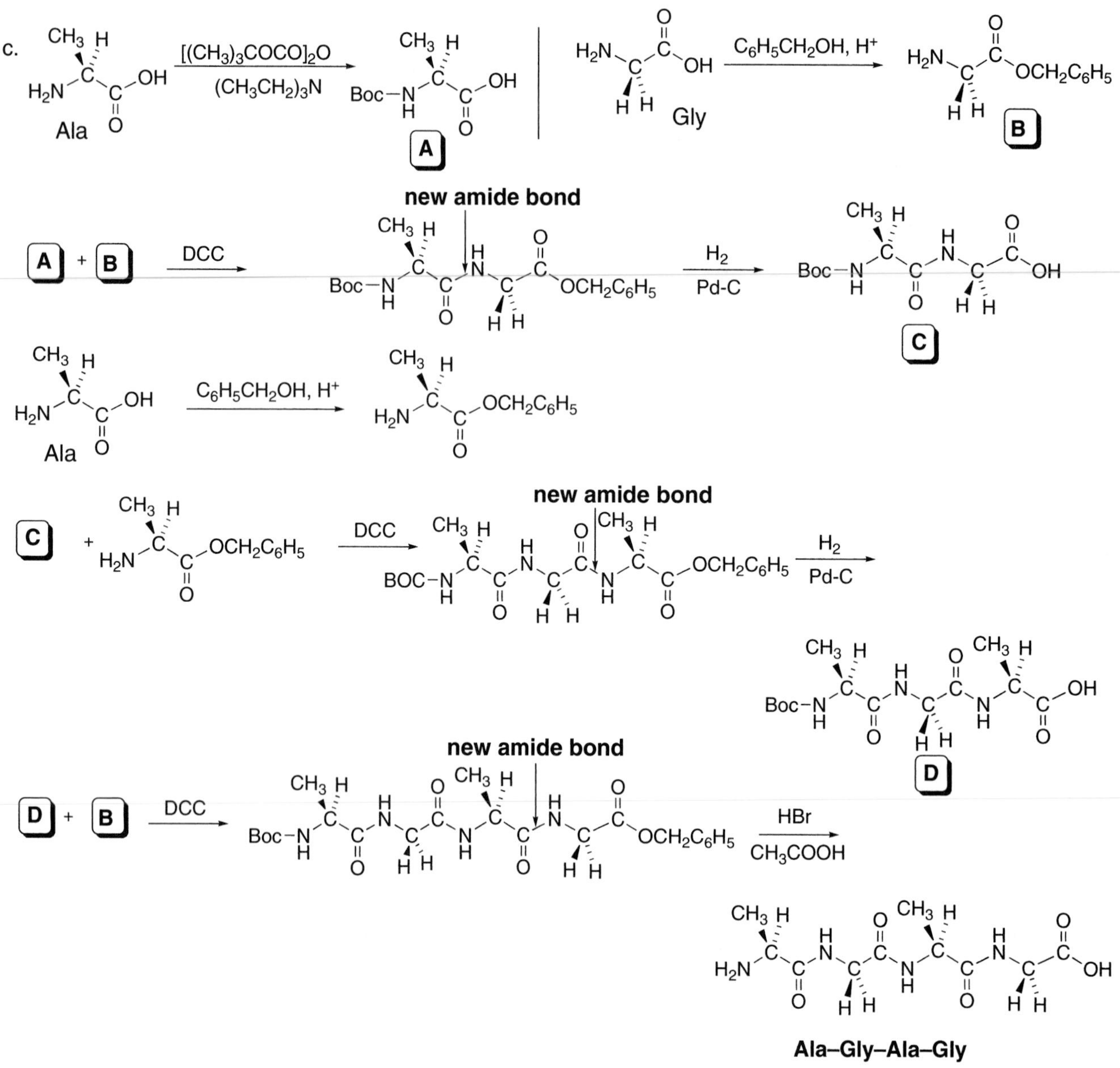
Ala
[(CH3)3COCO]2O
(CH3CH2)3N
A
Gly
C6H5CH2OH, H+
B
new amide bond
DCC
H2
Pd-C
C
Ala
C6H5CH2OH, H+
new amide bond
DCC
H2
Pd-C
D
new amide bond
DCC
HBr
CH3COOH
Ala–Gly–Ala–Gly

**28.20**

All Fmoc-protected amino acids are made by the following general reaction:

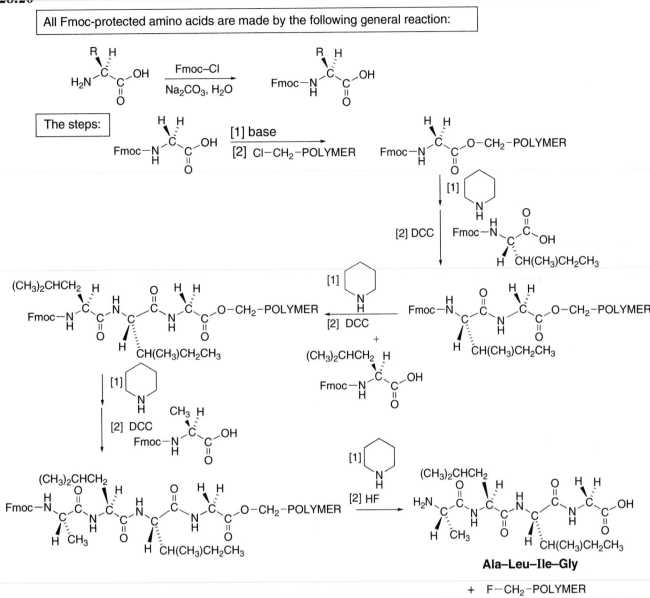

**28.21** Antiparallel β-pleated sheets are more stable then parallel β-pleated sheets because of geometry. The N–H and C=O of one chain are directly aligned with the N–H and C=O of an adjacent chain in the antiparallel β-pleated sheet, whereas they are not in the parallel β-pleated sheet. This makes the latter set of hydrogen bonds weaker.

**28.22**   In a *parallel* β-pleated sheet, the strands run in the *same* direction from the N- to C-terminal
amino acid.   In an *antiparallel* β-pleated sheet, the strands run in the *opposite* direction.

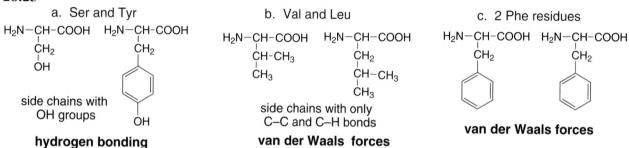

parallel                                                     antiparallel

**28.23**

a.  Ser and Tyr

H₂N–CH–COOH   H₂N–CH–COOH
    |                     |
   CH₂                   CH₂
    |
   OH

side chains with
OH groups

**hydrogen bonding**

b.  Val and Leu

H₂N–CH–COOH   H₂N–CH–COOH
    |                     |
  CH–CH₃                 CH₂
    |                     |
   CH₃                  CH–CH₃
                          |
                         CH₃

side chains with only
C–C and C–H bonds

**van der Waals forces**

c.  2 Phe residues

H₂N–CH–COOH   H₂N–CH–COOH
    |                     |
   CH₂                   CH₂

**van der Waals forces**

**28.24**   a. The R group for glycine is a hydrogen.  The R groups must be small to allow the β-pleated
sheets to stack on top of each other.  With large R groups, steric hindrance prevents stacking.

b. Silk fibers are water insoluble because most of the polar functional groups are in the interior of
the stacked sheets.  The β-pleated sheets are stacked one on top of another so few polar
functional groups are available for hydrogen bonding.

**28.25** All L-amino acids except cysteine have the **S configuration**. L-Cysteine has the *R* configuration because the R group contains a sulfur atom, which has higher priority.

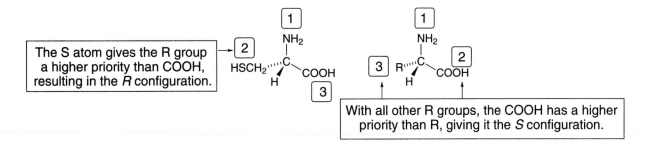

The S atom gives the R group a higher priority than COOH, resulting in the *R* configuration.

With all other R groups, the COOH has a higher priority than R, giving it the *S* configuration.

**28.26**

a.

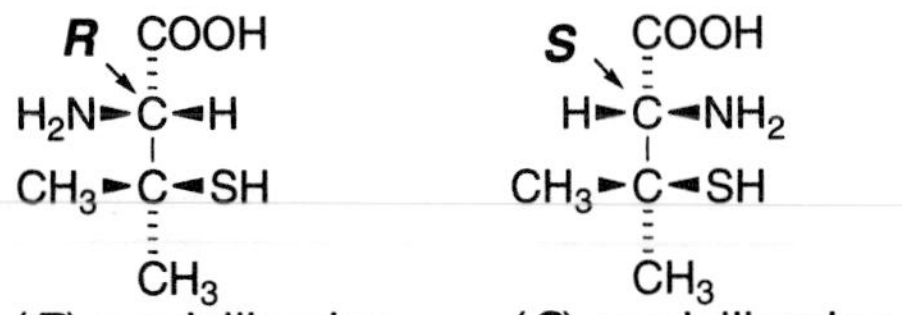

(R)-penicillamine        (S)-penicillamine

b.

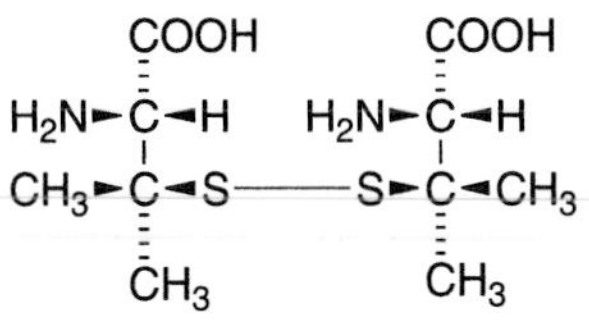

**28.27** Amino acids are insoluble in diethyl ether because amino acids are highly polar; they exist as salts in their neutral form. Diethyl ether is weakly polar, so amino acids are not soluble in it. *N*-Acetyl amino acids are soluble because they are polar but not salts.

amino acid, a salt
$H_2O$ soluble and ether insoluble

*N*-acetyl amino acid
ether soluble

**28.28** The electron pair on the N atom not part of a double bond is delocalized on the five-membered ring, making it less basic.

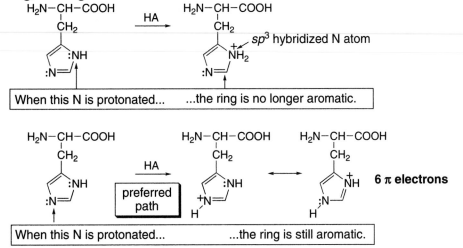

When this N is protonated...        ...the ring is no longer aromatic.

$sp^3$ hybridized N atom

preferred path

When this N is protonated...        ...the ring is still aromatic.

**6 π electrons**

**28.29**

The ring structure on tryptophan is aromatic since each atom contains a *p* orbital. Protonation of the N atom would disrupt the aromaticity, making this a less favorable reaction.

No *p* orbital on N.

This electron pair is delocalized on the bicyclic ring system (giving it 10 π electrons), making it less available for donation, and thus less basic.

**28.30**   At its isoelectric point, each amino acid is neutral.

a.
$$\underset{\text{alanine}}{\overset{\displaystyle COO^-}{\underset{\displaystyle CH_3}{H_3\overset{+}{N}-C-H}}}$$

b.
$$\underset{\text{methionine}}{\overset{\displaystyle COO^-}{\underset{\displaystyle CH_2CH_2SCH_3}{H_3\overset{+}{N}-C-H}}}$$

c.
$$\underset{\text{aspartic acid}}{\overset{\displaystyle COO^-}{\underset{\displaystyle CH_2COOH}{H_3\overset{+}{N}-C-H}}}$$

d.
$$\underset{\text{lysine}}{\overset{\displaystyle COO^-}{\underset{\displaystyle CH_2CH_2CH_2CH_2\overset{+}{N}H_3}{H_2N-C-H}}}$$

**28.31**

a.   [1] glutamic acid: use the p$K_a$'s 2.10 + 4.07
[2] lysine: use the p$K_a$'s 8.95 + 10.53
[3] arginine: use the p$K_a$'s 9.04 + 12.48
b. In general, the p$I$ of an acidic amino acid is lower than that of a neutral amino acid.
c. In general, the p$I$ of a basic amino acid is higher than that of a neutral amino acid.

**28.32**

| a. threonine | b. methionine | c. aspartic acid | d. arginine |
|---|---|---|---|
| p$I$ = 5.06 | p$I$ = 5.74 | p$I$ = 2.98 | p$I$ = 5.41 |
| (+1) charge at pH = 1 | (+1) charge at pH = 1 | (+1) charge at pH = 1 | (+2) charge at pH = 1 |

a.
$$\overset{+}{H_3}N-\overset{\displaystyle CH-COOH}{\underset{\displaystyle CH_3}{\overset{\displaystyle |}{\underset{\displaystyle CH-OH}{}}}}$$

b.
$$\overset{+}{H_3}N-CH-COOH \\ CH_2 \\ CH_2 \\ S \\ CH_3$$

c.
$$\overset{+}{H_3}N-CH-COOH \\ CH_2 \\ COOH$$

d.
$$\overset{+}{H_3}N-CH-COOH \\ CH_2 \\ CH_2 \\ CH_2 \\ NH \\ \overset{+}{C}=NH_2 \\ NH_2$$

**28.33**

| a. **valine** | b. **proline** | c. **glutamic acid** | d. **lysine** |
|---|---|---|---|
| $pI = 6.00$ | $pI = 6.30$ | $pI = 3.08$ | $pI = 9.74$ |
| (−1) charge | (−1) charge | (−2) charge | (−1) charge |
| at pH = 11 | at pH = 11 | at pH = 11 | at pH = 11 |

a. valine:

$H_2N-CH-COO^-$
$\quad\;\; CH-CH_3$
$\qquad\;\; CH_3$

b. proline:

$COO^-$ on ring carbon; $HN$ in ring

c. glutamic acid:

$H_2N-CH-COO^-$
$\qquad\; CH_2$
$\qquad\; CH_2$
$\qquad\; COO^-$

d. lysine:

$H_2N-CH-COO^-$
$\qquad\; CH_2$
$\qquad\; CH_2$
$\qquad\; CH_2$
$\qquad\; CH_2$
$\qquad\; NH_2$

**28.34**

a.

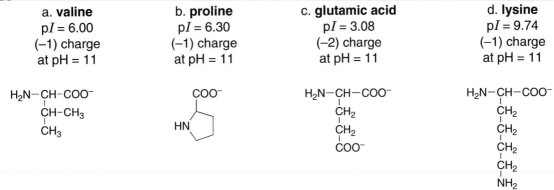

$H_2N-CH-C-OH$ with $O$ double bond and $CH_3$

**Ala** → A–A–A

b. at pH = 1

$H_3N^+$ ... $OH$ tripeptide structure

c. The $pK_a$ of the COOH of the tripeptide is higher than the $pK_a$ of the COOH group of alanine, making it less acidic. This occurs because the COOH group in the tripeptide is further away from the $-NH_3^+$ group. The positively charged $-NH_3^+$ group stabilizes the negatively charged carboxylate anion of alanine more than the carboxylate anion of the tripeptide because it is so much closer in alanine. The opposite effect is observed with the ionization of the $-NH_3^+$ group. In alanine, the $-NH_3^+$ is closer to the $COO^-$ group, so it is more difficult to lose a proton, resulting in a higher $pK_a$. In the tripeptide, the $-NH_3^+$ is further away from the $COO^-$, so it is less affected by its presence.

**28.35**

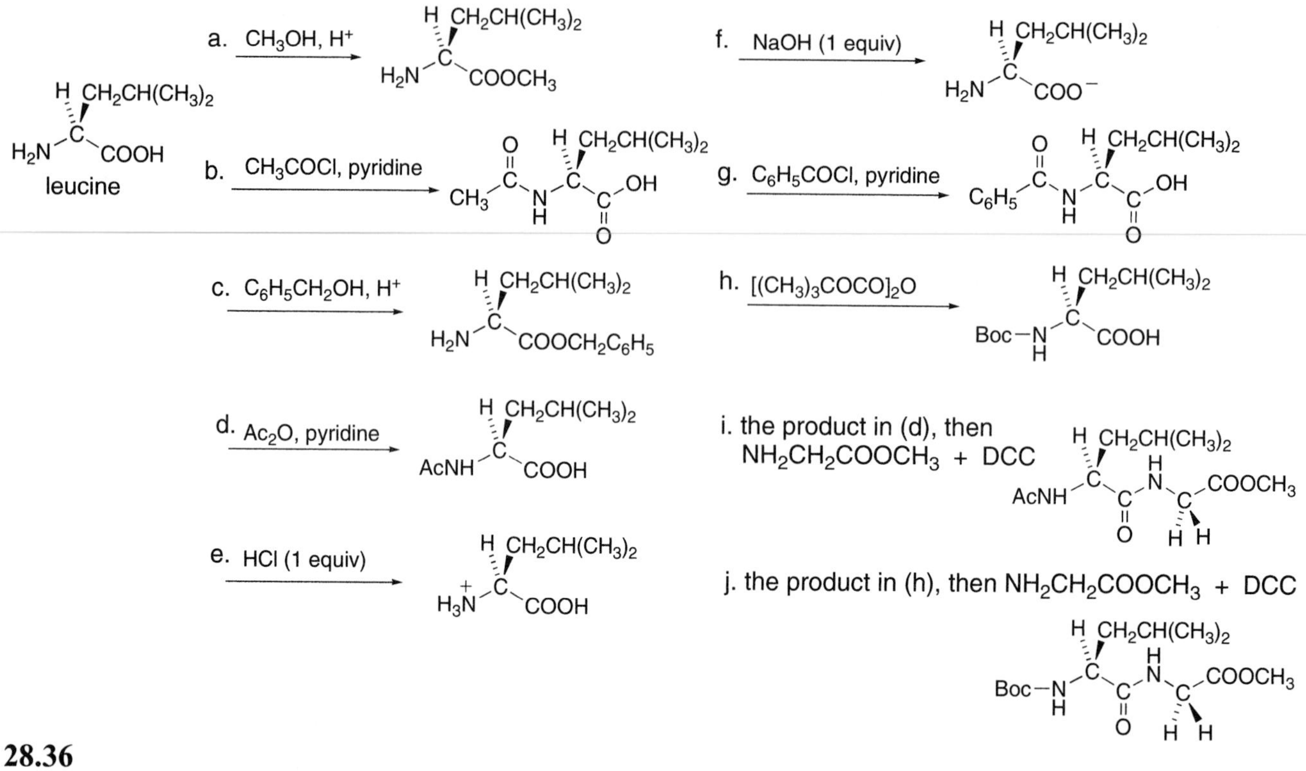

leucine

a. CH$_3$OH, H$^+$

b. CH$_3$COCl, pyridine

c. C$_6$H$_5$CH$_2$OH, H$^+$

d. Ac$_2$O, pyridine

e. HCl (1 equiv)

f. NaOH (1 equiv)

g. C$_6$H$_5$COCl, pyridine

h. [(CH$_3$)$_3$COCO]$_2$O

i. the product in (d), then NH$_2$CH$_2$COOCH$_3$ + DCC

j. the product in (h), then NH$_2$CH$_2$COOCH$_3$ + DCC

**28.36**

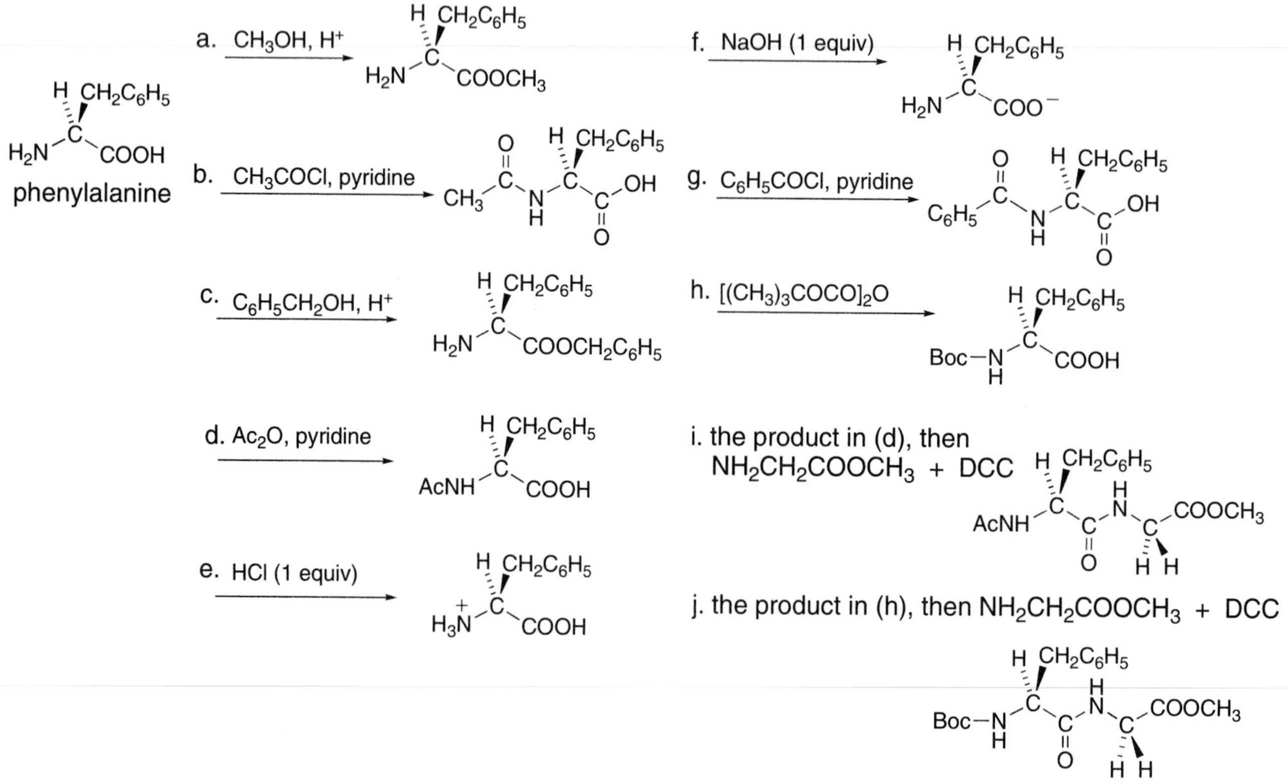

phenylalanine

a. CH$_3$OH, H$^+$

b. CH$_3$COCl, pyridine

c. C$_6$H$_5$CH$_2$OH, H$^+$

d. Ac$_2$O, pyridine

e. HCl (1 equiv)

f. NaOH (1 equiv)

g. C$_6$H$_5$COCl, pyridine

h. [(CH$_3$)$_3$COCO]$_2$O

i. the product in (d), then NH$_2$CH$_2$COOCH$_3$ + DCC

j. the product in (h), then NH$_2$CH$_2$COOCH$_3$ + DCC

**28.37**

a. $(CH_3)_2CHCH_2CHCOOH$ with Br substituent $\xrightarrow[\text{excess}]{NH_3}$ $(CH_3)_2CHCH_2CHCOO^-$ with $\overset{+}{N}H_3$ substituent $+ NH_4^+ + Br^-$

b. $CH_3CONHCH(COOEt)_2$ $\xrightarrow{\text{[1] NaOEt}}$ [2] (4-acetoxybenzyl bromide: $CH_3$-C(=O)-O-C$_6$H$_4$-CH$_2$Br) [3] $H_3O^+, \Delta$ → HO-C$_6$H$_4$-CH$_2$-CHCOOH with NH$_2$

c. [5-acetamidopentanal: OHC-CH$_2$CH$_2$CH$_2$-NHC(=O)CH$_3$] $\xrightarrow{\text{[1] NH}_4\text{Cl, NaCN}}{\text{[2] H}_3\text{O}^+}$ HOOC-CH(NH$_2$)-CH$_2$CH$_2$CH$_2$-NH$_2$

d. [methyl 4-oxobutanoate: CH$_3$O-C(=O)-CH$_2$CH$_2$-CHO] $\xrightarrow{\text{[1] NH}_4\text{Cl, NaCN}}{\text{[2] H}_3\text{O}^+}$ HO-C(=O)-CH$_2$CH$_2$-CH(NH$_2$)-COOH

e. $CH_3CONHCH(COOEt)_2$ $\xrightarrow{\text{[1] NaOEt}}$ [2] $ClCH_2CH_2CH_2CH_2NHAc$ [3] $H_3O^+, \Delta$ → $H_2N-\overset{CH_2CH_2CH_2CH_2NH_2}{\underset{H}{C}}-COOH$

**28.38**

a. Asn

$H_2N-CH(-CH_2-C(=O)-NH_2)-C(=O)-OH$ $\Longrightarrow$ $BrCH_2-C(=O)-NH_2$

b. His

$H_2N-CH(-CH_2-\text{imidazolyl})-C(=O)-OH$ $\Longrightarrow$ 2-(bromomethyl)imidazole ($CH_2Br$)

c. Trp

$H_2N-CH(-CH_2-\text{indol-3-yl})-C(=O)-OH$ $\Longrightarrow$ 3-(bromomethyl)indole ($CH_2Br$)

**28.39**

$CH_3-C(=O)-NH-C(-COOEt)(H)-COOEt$ $\xrightarrow{\text{[1] NaOEt}}$ [2] $CH_3CHO$ [3] $H_3O^+, \Delta$ → $H_2N-\overset{\overset{OH}{|}\ \overset{CHCH_3}{|}}{\underset{H}{C}}-COOH$

threonine

**28.40**

a. $CH_3CHO$ $\xrightarrow[\text{CH}_3\text{COOH}]{Br_2}$ $BrCH_2CHO$ $\xrightarrow[\text{H}_2\text{SO}_4, \text{H}_2\text{O}]{CrO_3}$ $BrCH_2COOH$ $\xrightarrow[\text{excess}]{NH_3}$ $H_3\overset{+}{N}CH_2COO^-$

glycine

b. $CH_3-C(=O)-H$ $\xrightarrow{\text{[1] NH}_4\text{Cl, NaCN}}{\text{[2] H}_3\text{O}^+}$ $CH_3-\overset{\overset{NH_2}{|}}{\underset{H}{C}}-COOH$

alanine

**28.41**

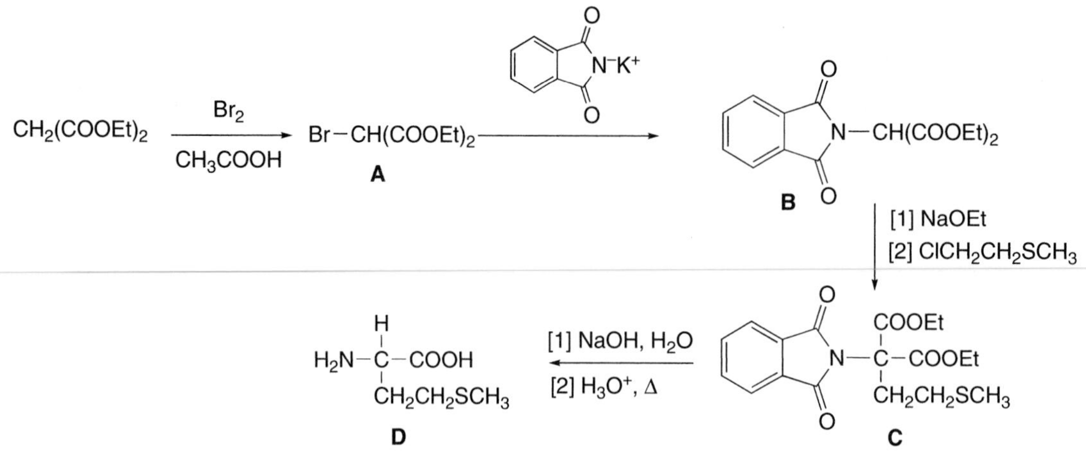
CH2(COOEt)2
Br2
CH3COOH
Br–CH(COOEt)2
A
N⁻K⁺
N–CH(COOEt)2
B
[1] NaOEt
[2] ClCH2CH2SCH3
COOEt
N–C–COOEt
CH2CH2SCH3
C
[1] NaOH, H2O
[2] H3O+, Δ
H2N–C–COOH
CH2CH2SCH3
D

**28.42**

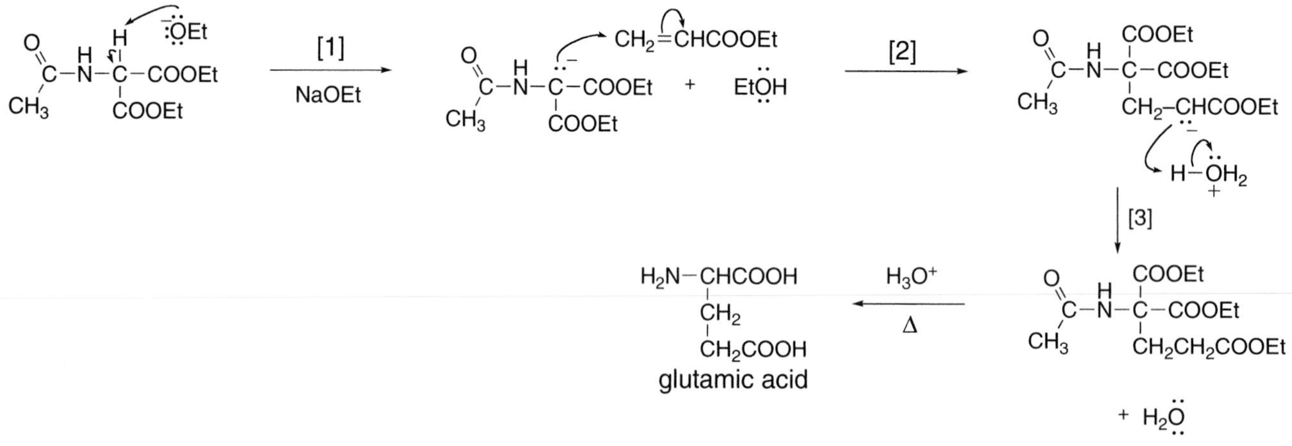
[1]
NaOEt
CH2=CHCOOEt
EtOH
[2]
[3]
H3O+
Δ
glutamic acid
+ H2O

**28.43**

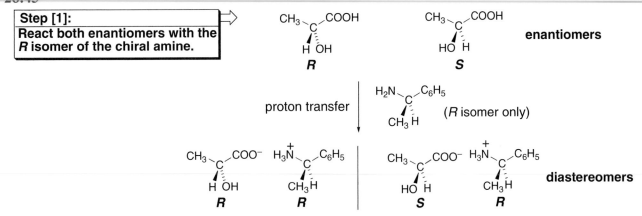

**Step [1]:**
**React both enantiomers with the**
***R* isomer of the chiral amine.**

enantiomers

proton transfer    (*R* isomer only)

diastereomers

These salts have the *same* configuration around one stereogenic center,
but the *opposite* configuration about the other stereogenic center.

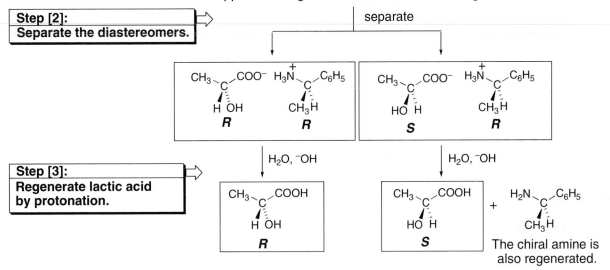

**Step [2]:**
**Separate the diastereomers.**

separate

**Step [3]:**
**Regenerate lactic acid**
**by protonation.**

The chiral amine is
also regenerated.

**28.44**

| To begin: |
| --- |
| Convert the amino acids into amino acids ester (two enantiomers). |

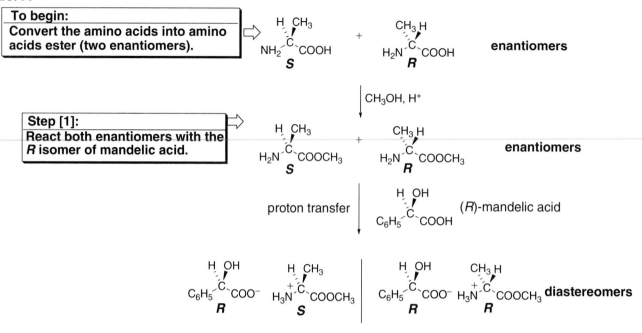

Step [1]:
React both enantiomers with the *R* isomer of mandelic acid.

These salts have the *same* configuration around one stereogenic center, but the *opposite* configuration about the other stereogenic center.

| Step [2]: |
| --- |
| Separate the diastereomers. |

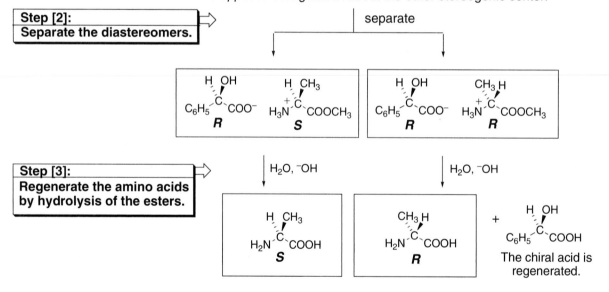

| Step [3]: |
| --- |
| Regenerate the amino acids by hydrolysis of the esters. |

**28.45**

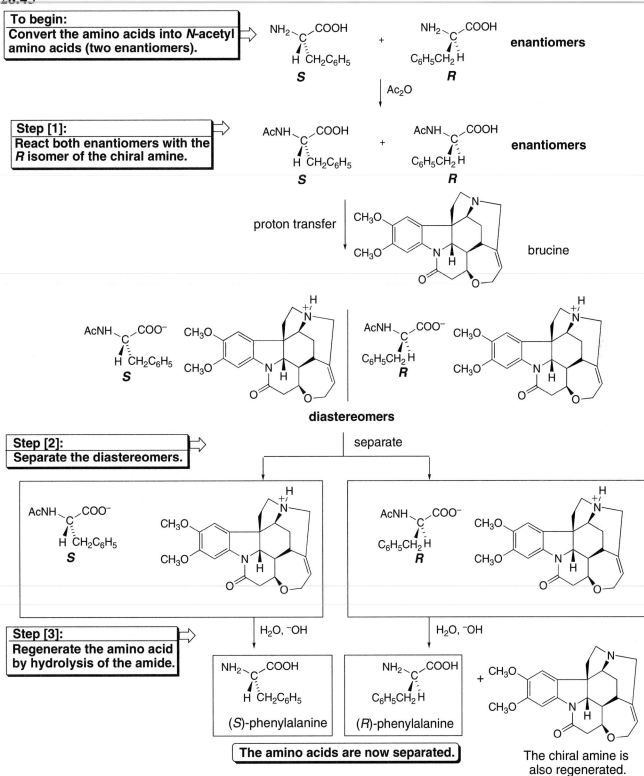

**To begin:**
Convert the amino acids into *N*-acetyl amino acids (two enantiomers).

**enantiomers**

Ac₂O

**Step [1]:**
React both enantiomers with the *R* isomer of the chiral amine.

**enantiomers**

proton transfer

brucine

**diastereomers**

**Step [2]:**
Separate the diastereomers.

separate

**Step [3]:**
Regenerate the amino acid by hydrolysis of the amide.

H₂O, ⁻OH

H₂O, ⁻OH

(*S*)-phenylalanine

(*R*)-phenylalanine

**The amino acids are now separated.**

The chiral amine is also regenerated.

**28.46**

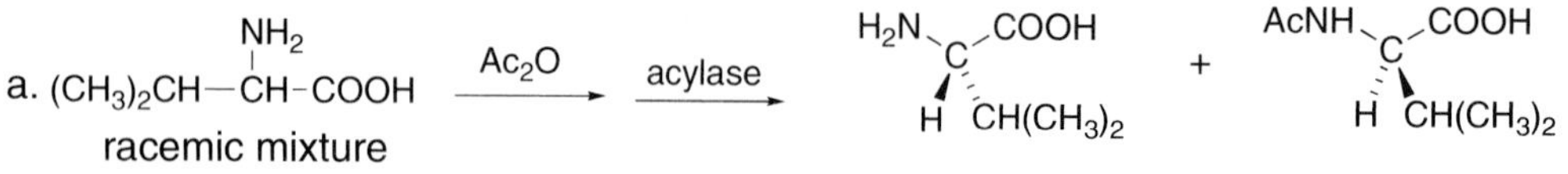

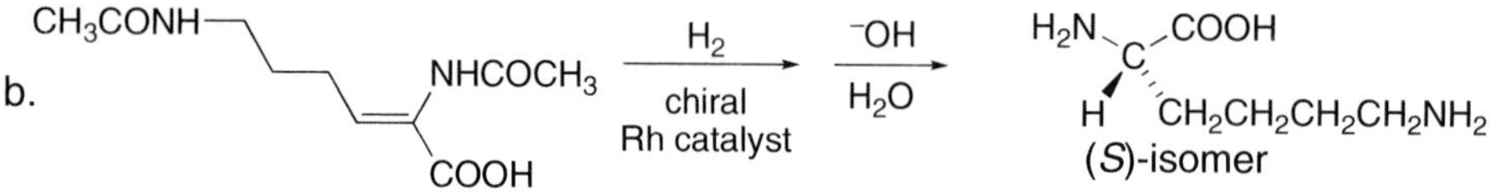

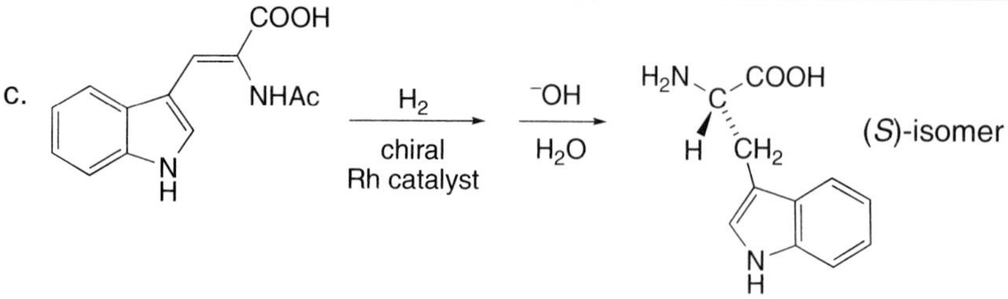

a. (CH₃)₂CH—CH—COOH  racemic mixture  →(Ac₂O)→ →(acylase)→  H₂N...COOH (H, CH(CH₃)₂)  +  AcNH...COOH (H, CH(CH₃)₂)

b. CH₃CONH— ... —NHCOCH₃, COOH  →(H₂, chiral Rh catalyst)→ →(⁻OH, H₂O)→  H₂N...COOH (H, CH₂CH₂CH₂CH₂NH₂)  (S)-isomer

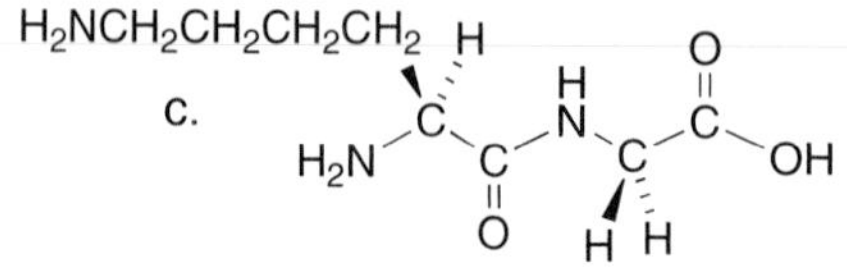

c.  COOH / NHAc (indole)  →(H₂, chiral Rh catalyst)→ →(⁻OH, H₂O)→  H₂N...COOH (H, CH₂ indole)  (S)-isomer

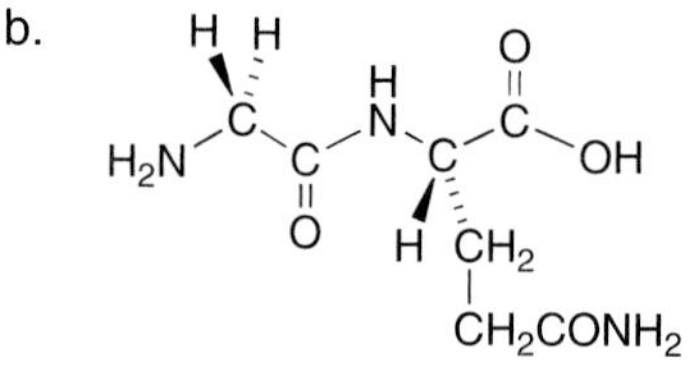

**28.47**

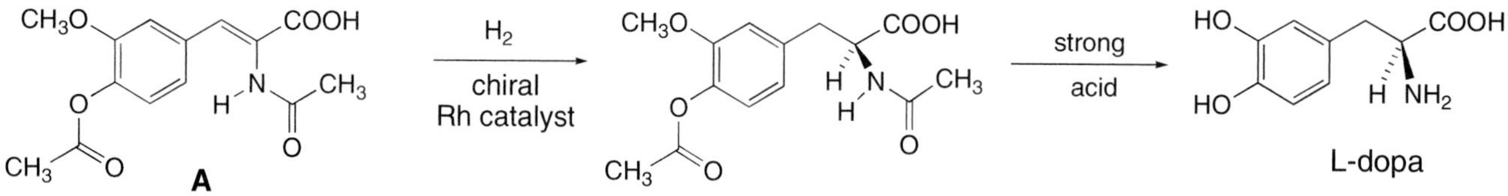

CH₃O—...—COOH, N(H)(CH₃)(C=O), OC(=O)CH₃  **A**  →(H₂, chiral Rh catalyst)→  CH₃O—...—COOH (H, N, CH₃, C=O), OC(=O)CH₃  →(strong acid)→  HO—...—COOH (H, NH₂), HO  **L-dopa**

**28.48**

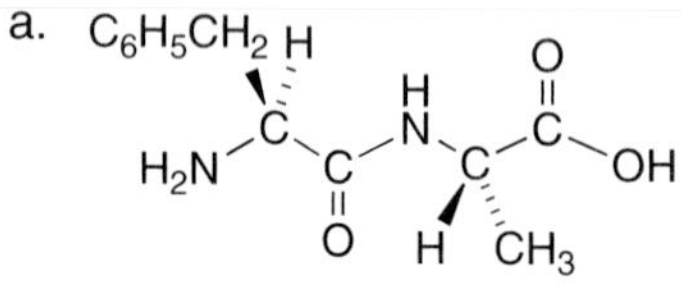

a. C₆H₅CH₂...  H₂N—C(=O)—N(H)—C(—OH)(=O) (H, CH₃)  **Phe–Ala**

b. H H / H₂N—C(=O)—N(H)—C(—OH)(=O) (H, CH₂, CH₂CONH₂)  **Gly–Gln**

c. H₂NCH₂CH₂CH₂CH₂...  H₂N—C(=O)—N(H)—C(—OH)(=O) (H H)  **Lys–Gly**

d. NH₂C(NH)NHCH₂CH₂CH₂...  H₂N—C(=O)—N(H)—C(—OH)(=O) (H, CH₂ imidazole)  **R–H**

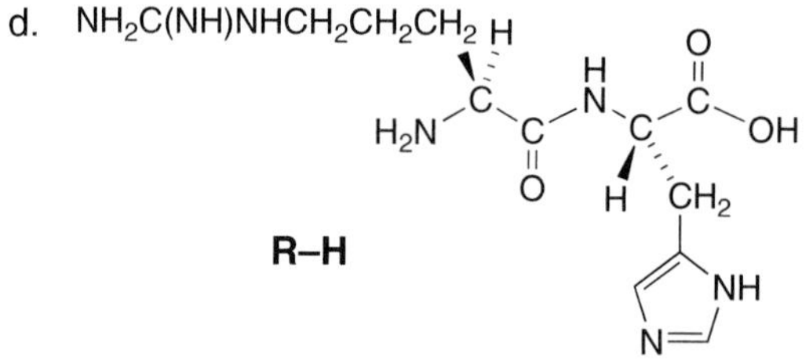

**28.49** Amide bonds are bold lines but not wedges.

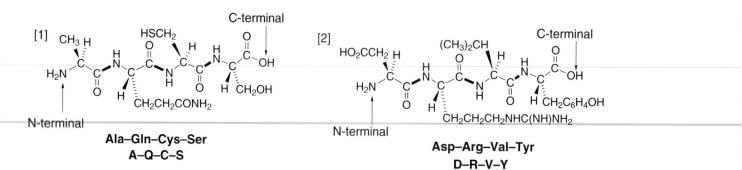

[1] CH₃ H, HSCH₂ H, H, C-terminal  H₂N—C—C(=O)—N(H)—C—C(=O)—N(H)—C—C(=O)—OH  N-terminal  CH₂CH₂CONH₂, CH₂OH  **Ala–Gln–Cys–Ser**  **A–Q–C–S**

[2] HO₂CCH₂ H, (CH₃)₂CH H, H, C-terminal  H₂N—C—C(=O)—N(H)—C—C(=O)—N(H)—C—C(=O)—N(H)—C—C(=O)—OH  N-terminal  CH₂CH₂CH₂NHC(NH)NH₂, CH₂C₆H₄OH  **Asp–Arg–Val–Tyr**  **D–R–V–Y**

**28.50** Name a peptide from the N-terminal to the C-terminal end.

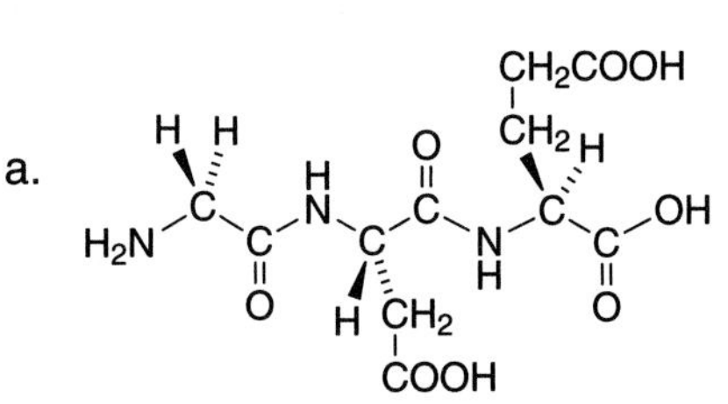

a.

**Gly–Asp–Glu**
**G–D–E**

b.

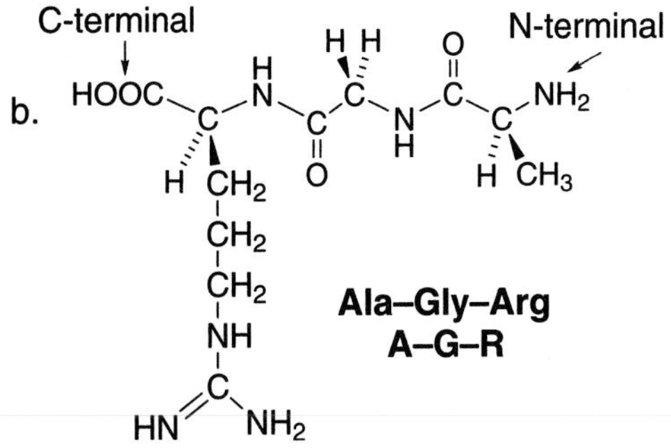

**Ala–Gly–Arg**
**A–G–R**

**28.51** A peptide C–N bond is stronger than an ester C–O bond because the C–N bond has more double bond character due to resonance. Since N is more basic than O, an amide C–N bond is more stabilized by delocalization of the lone pair on N.

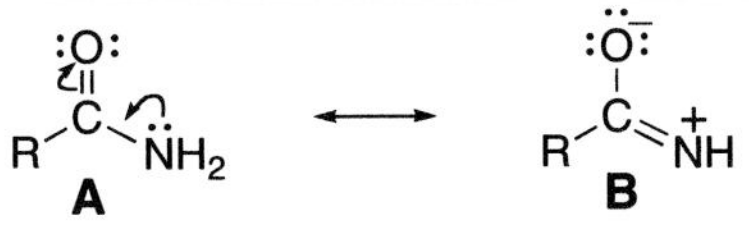

Structure **B** contributes greatly to the resonance hybrid and this shortens and strengthens the C–N bond.

**28.52** Use the principles from Answer 28.16.

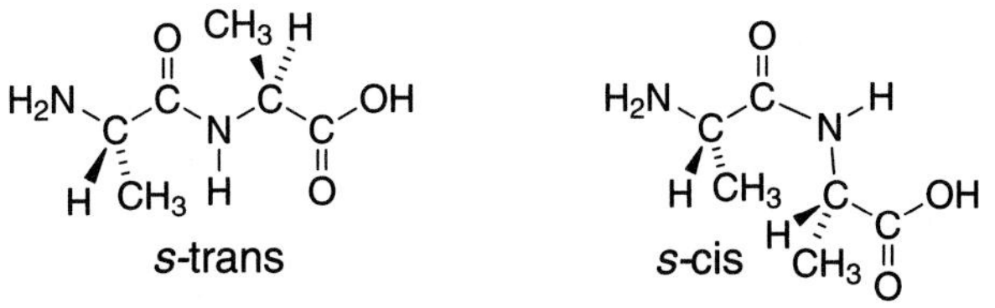

*s*-trans

*s*-cis

**28.53** **A** and **B** can react to form an amide, or two molecules of **B** can form an amide.

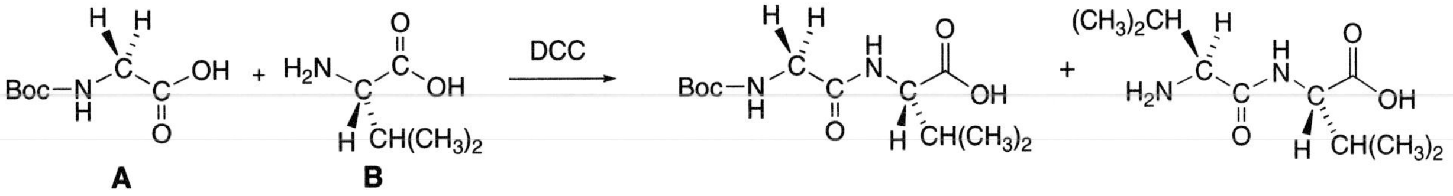

**28.54**

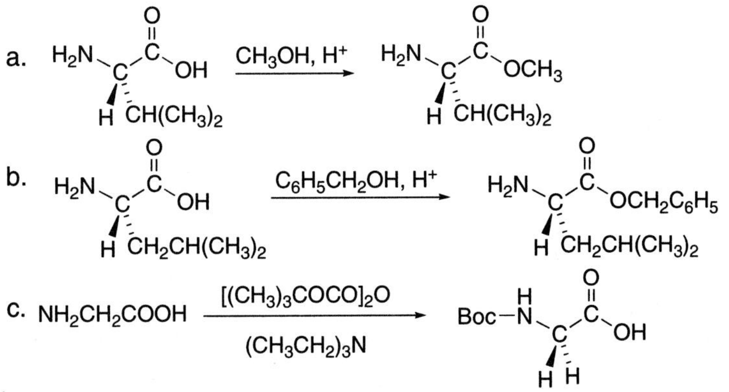

a.

b.

c.

d. 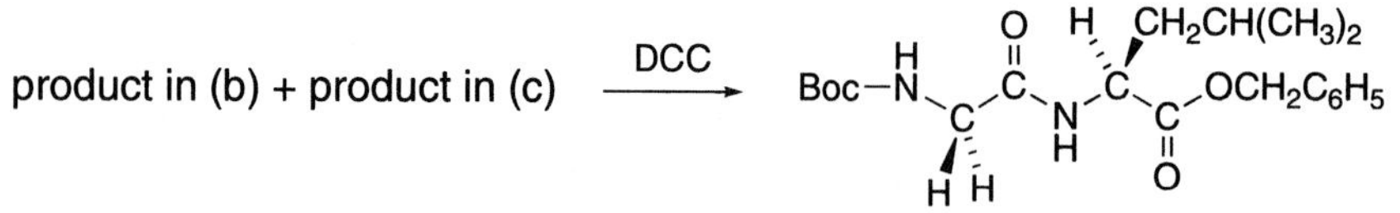 product in (b) + product in (c) $\xrightarrow{\text{DCC}}$ Boc–N–C–N–C–OCH₂C₆H₅

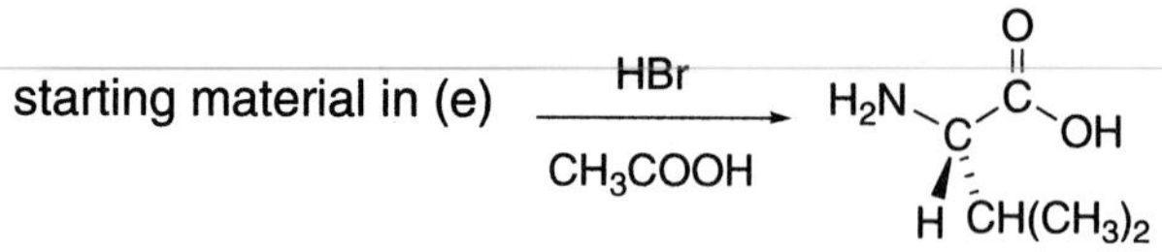

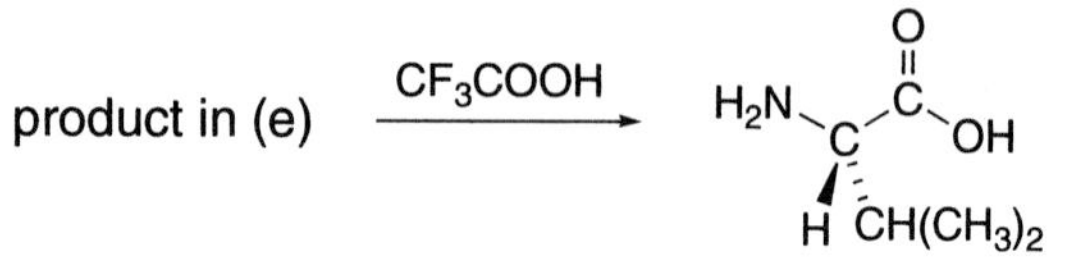

e. 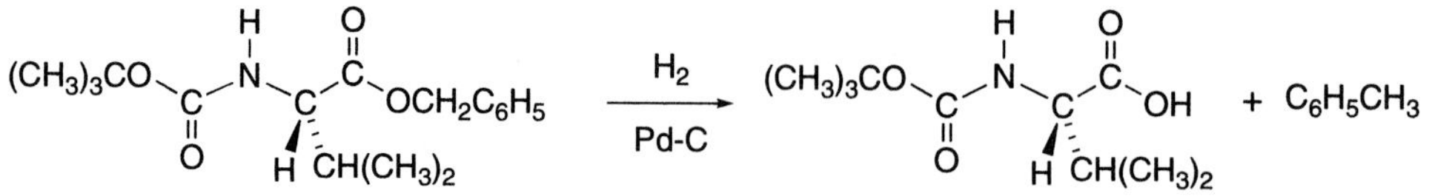 $(CH_3)_3CO$ ... $\xrightarrow[\text{Pd-C}]{H_2}$ $(CH_3)_3CO$ ... + $C_6H_5CH_3$

f. starting material in (e) $\xrightarrow[\text{CH}_3\text{COOH}]{\text{HBr}}$ $H_2N$ ...

g. product in (e) $\xrightarrow{\text{CF}_3\text{COOH}}$ $H_2N$ ...

h. + Fmoc–Cl $\xrightarrow[\text{H}_2\text{O}]{\text{Na}_2\text{CO}_3}$

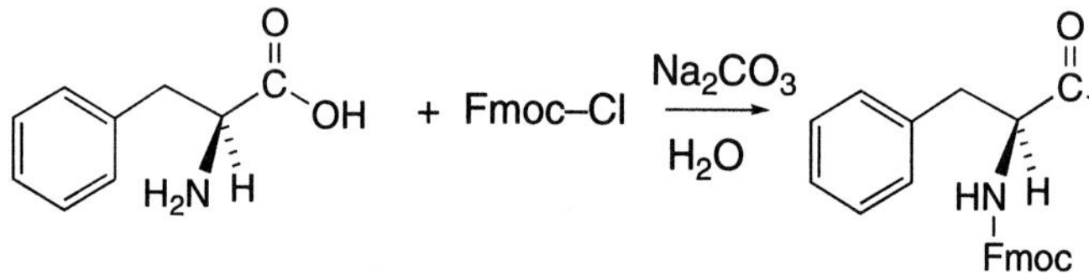

**28.55**

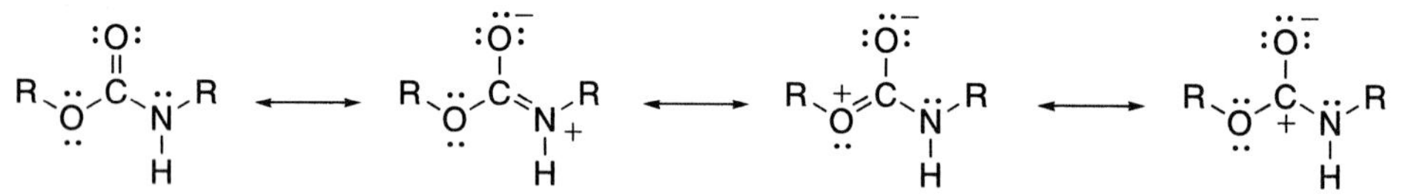

**28.56**

a. 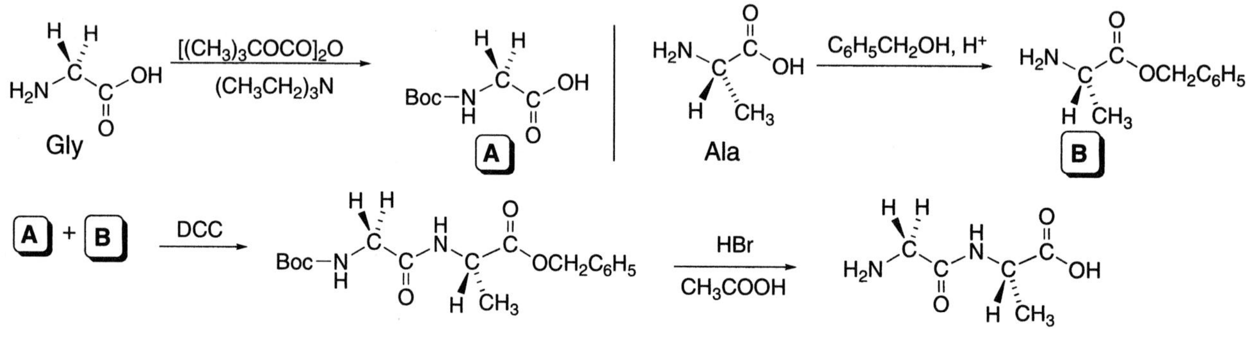 Gly $\xrightarrow[\text{(CH}_3\text{CH}_2)_3\text{N}]{[(CH_3)_3COCO]_2O}$ **A** | Ala $\xrightarrow{C_6H_5CH_2OH, H^+}$ **B**

**A** + **B** $\xrightarrow{\text{DCC}}$ $\xrightarrow[\text{CH}_3\text{COOH}]{\text{HBr}}$ **Gly–Ala**

b. 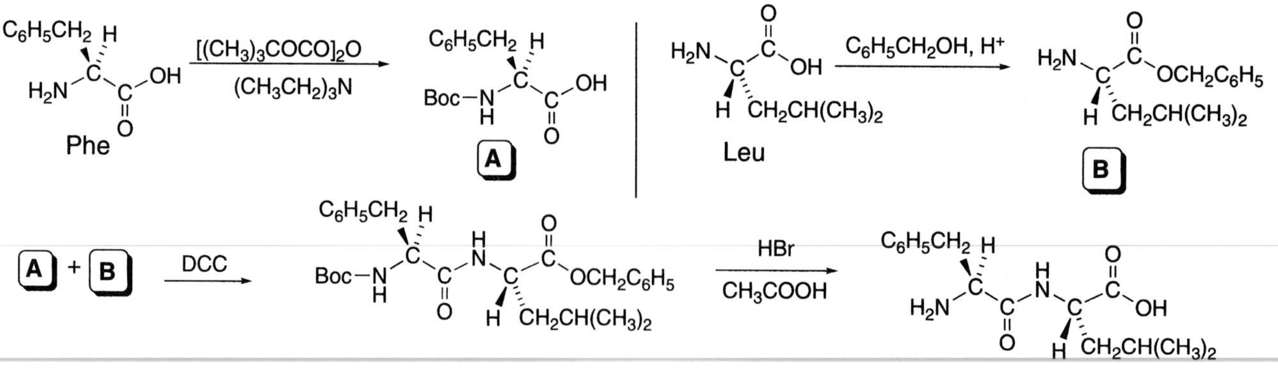 Phe $\xrightarrow[\text{(CH}_3\text{CH}_2)_3\text{N}]{[(CH_3)_3COCO]_2O}$ **A** | Leu $\xrightarrow{C_6H_5CH_2OH, H^+}$ **B**

**A** + **B** $\xrightarrow{\text{DCC}}$ $\xrightarrow[\text{CH}_3\text{COOH}]{\text{HBr}}$ **Phe–Leu**

c.

**Ile–Ala–Phe**

**28.57** Make all the Fmoc derivatives as described in Problem 28.20.

a.

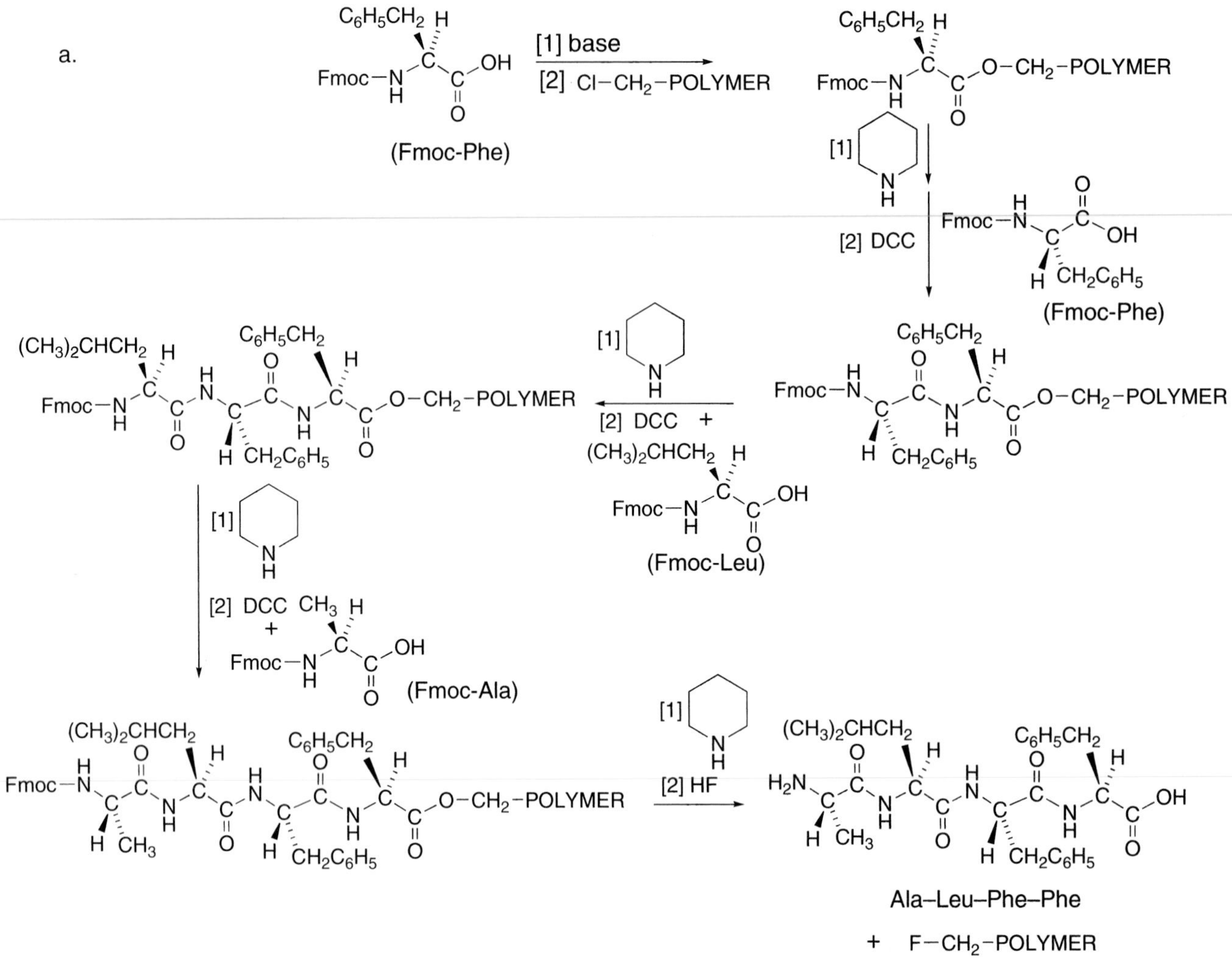

b.

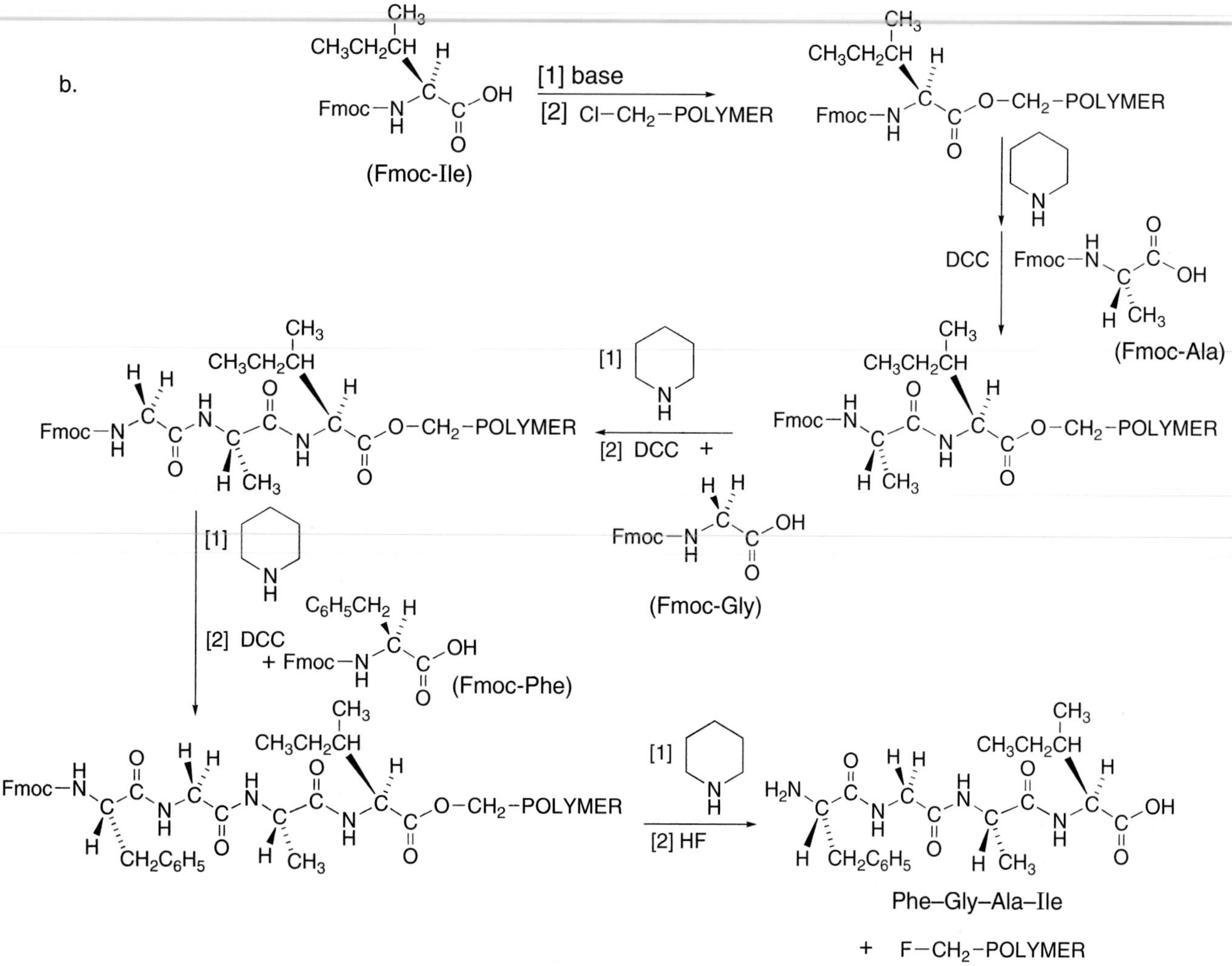

(Fmoc-Ile)

[1] base
[2] Cl–CH₂–POLYMER

(Fmoc-Ala)

(Fmoc-Gly)

(Fmoc-Phe)

Phe–Gly–Ala–Ile

+   F–CH₂–POLYMER

**28.58** An acetyl group on the NH₂ forms an amide. Although this amide does block an amino group from reaction, this amide is no different in reactivity than any of the peptide amide bonds. To remove the acetyl group after the peptide bond is formed would require harsh reaction conditions that would also cleave the amide bonds of the peptide.

*N*-acetyl amino acid

**28.59**

a. A *p*-nitrophenyl ester activates the carboxy group of the first amino acid to amide formation by converting the OH group into a good leaving group, the *p*-nitrophenoxide group, which is highly resonance stabilized. In this case the electron-withdrawing $NO_2$ group further stabilizes the leaving group.

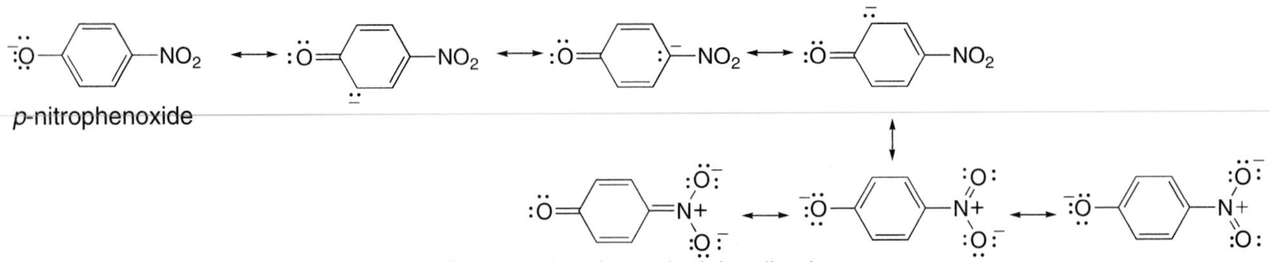

*p*-nitrophenoxide

The negative charge is delocalized on the O atom of the $NO_2$ group.

b. The *p*-methoxyphenyl ester contains an electron-donating $OCH_3$ group, making $CH_3OC_6H_4O^-$ a poorer leaving group than $NO_2C_6H_4O^-$, so this ester does not activate the amino acid to amide formation as much.

**28.60**

a.

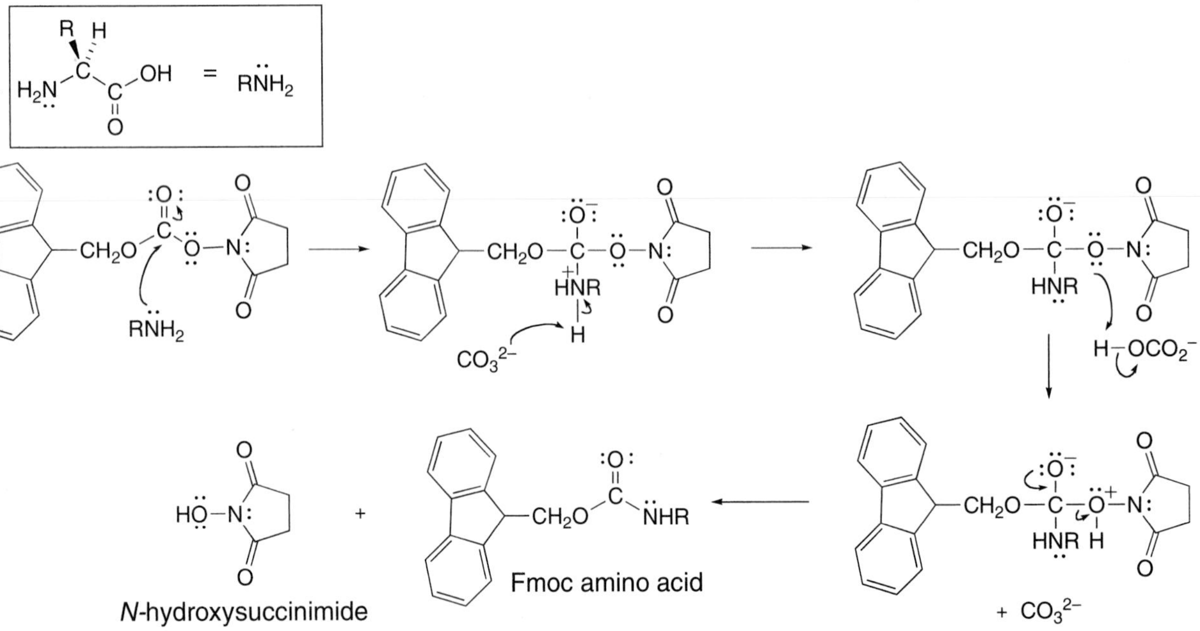

N-hydroxysuccinimide

Fmoc amino acid

b.

Fmoc amino acid

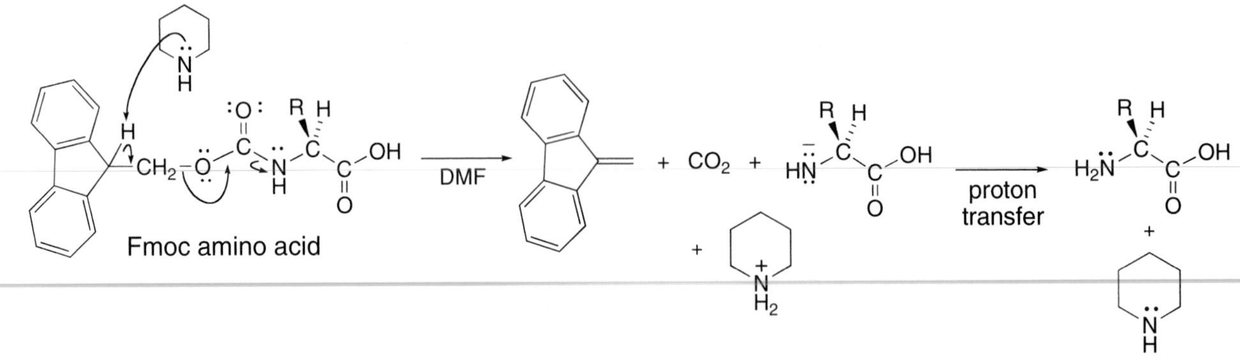

**28.61** Amino acids commonly found in the interior of a globular protein have nonpolar or weakly polar side chains: isoleucine and phenylalanine. Amino acids commonly found on the surface have COOH, $NH_2$, and other groups that can hydrogen bond to water: aspartic acid, lysine, arginine, and glutamic acid.

**28.62** The proline residues on collagen are hydroxylated to increase hydrogen bonding interactions.

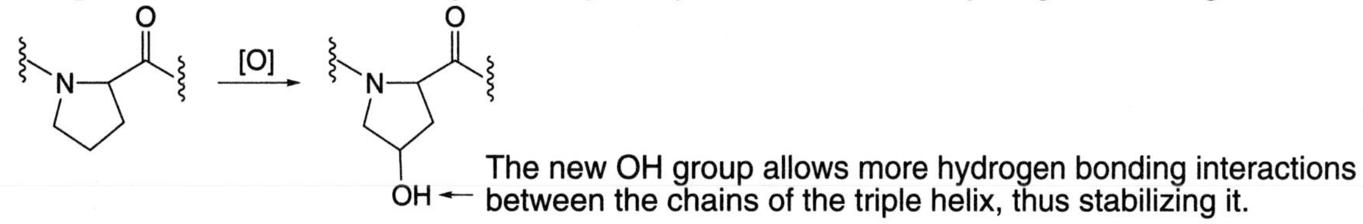
[O]
The new OH group allows more hydrogen bonding interactions between the chains of the triple helix, thus stabilizing it.

**28.63**

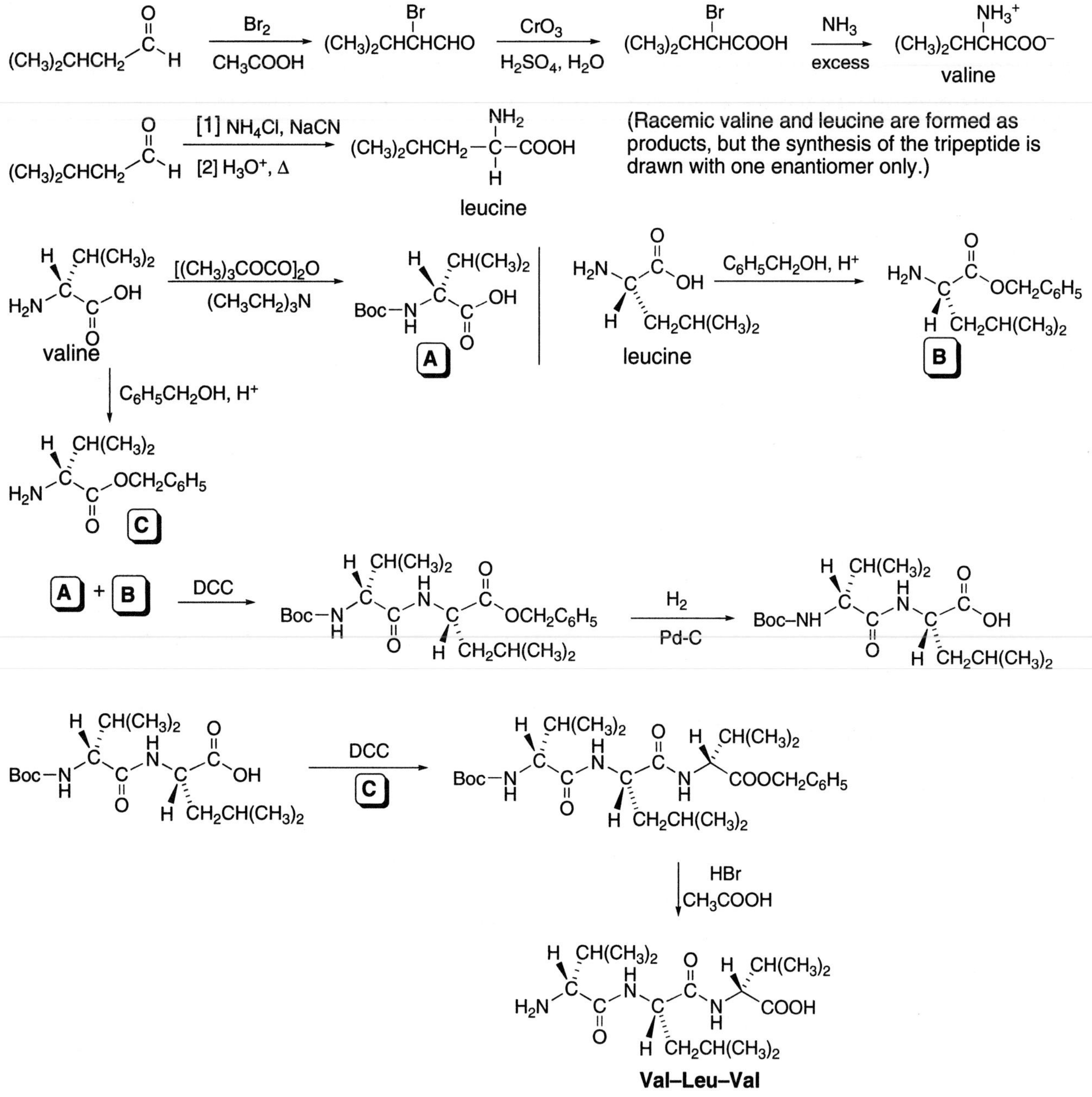
Br2
CH3COOH
CrO3
H2SO4, H2O
NH3
excess
valine
[1] NH4Cl, NaCN
[2] H3O+, Δ
leucine
(Racemic valine and leucine are formed as products, but the synthesis of the tripeptide is drawn with one enantiomer only.)
[(CH3)3COCO]2O
(CH3CH2)3N
A
C6H5CH2OH, H+
B
C
DCC
H2
Pd-C
HBr
CH3COOH
Val–Leu–Val

**28.64** Perhaps using a chiral amine R*$NH_2$ (or related chiral nitrogen-containing compound) to make a chiral imine, will now favor formation of one of the amino nitriles in the Strecker synthesis. Hydrolysis of the CN group and removal of R* would then form the amino acid.

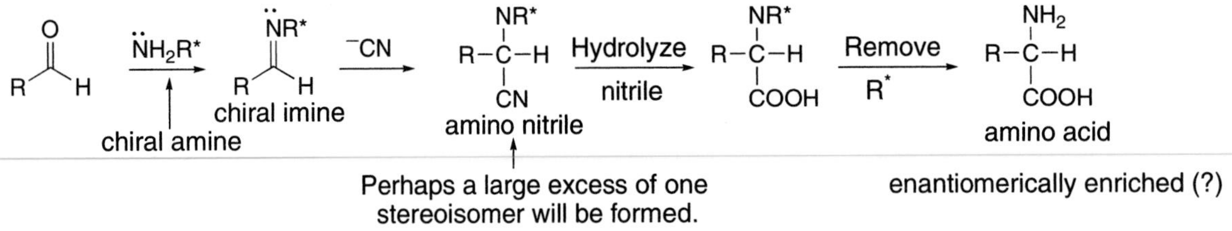

NH2R*
chiral amine
chiral imine
−CN
amino nitrile
Hydrolyze
nitrile
Remove
R*
amino acid
Perhaps a large excess of one
stereoisomer will be formed.
enantiomerically enriched (?)

Chapter 29: Lipids

♦ **Hydrolyzable lipids**

**[1] Waxes (29.2)**—Esters formed from a long-chain alcohol and a long-chain carboxylic acid

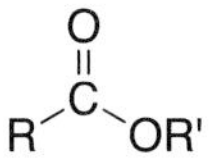

R, R' = long chains of C's

**[2] Triacylglycerols (29.3)**—Triesters of glycerol with three fatty acids

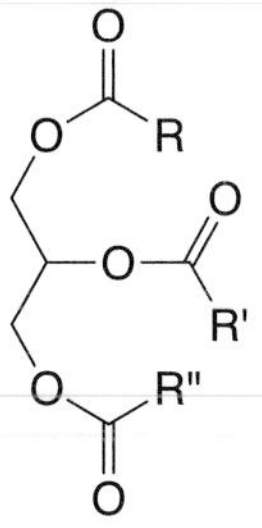

R, R', R" = alkyl groups with 11–19 C's

**[3] Phospholipids** (29.4)

[a] Phosphatidylethanolamine (cephalin)

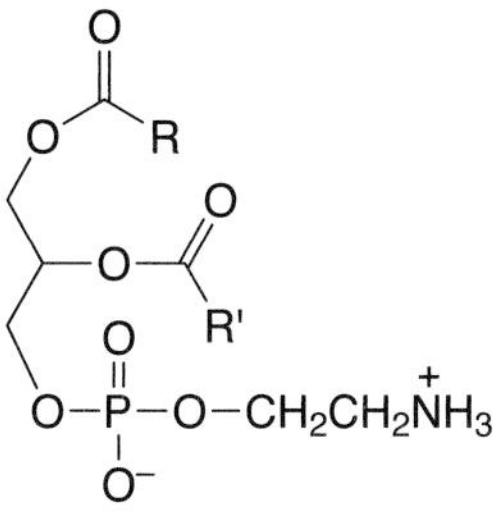

$O-P-O-CH_2CH_2\overset{+}{N}H_3$

R, R' = long carbon chain

[b] Phosphatidylcholine (lecithin)

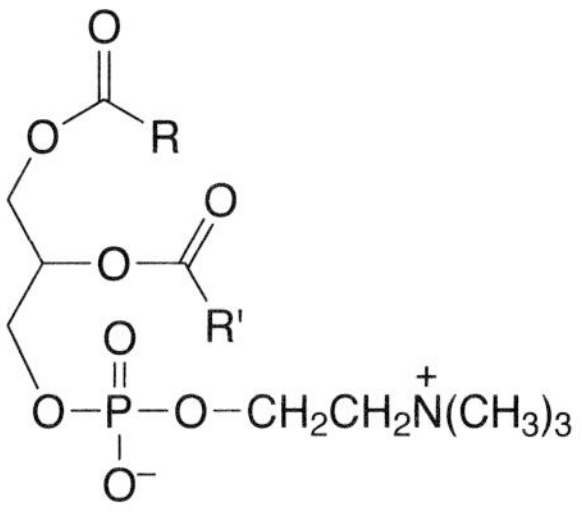

$O-P-O-CH_2CH_2\overset{+}{N}(CH_3)_3$

R, R' = long carbon chain

[c] Sphingomyelin

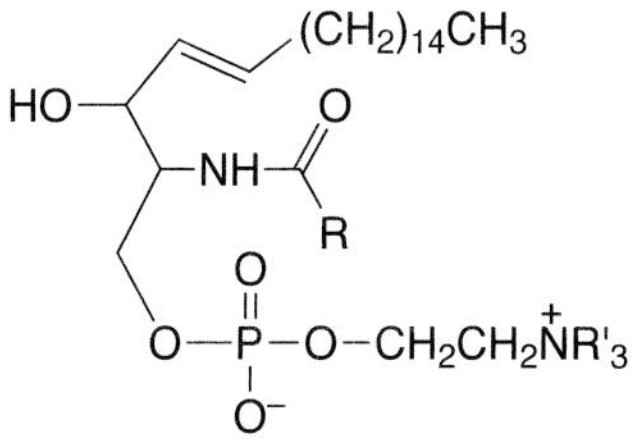

$(CH_2)_{14}CH_3$

$O-P-O-CH_2CH_2\overset{+}{N}R'_3$

R = long carbon chain
R' = H or $CH_3$

♦ **Nonhydrolyzable lipids**

**[1] Fat-soluble vitamins** (29.5)—Vitamins A, D, E, and K

**[2] Eicosanoids** (29.6)—Compounds containing 20 carbons derived from arachidonic acid.  There are four types:  prostaglandins, thromboxanes, prostacyclins, and leukotrienes.

**[3] Terpenes** (29.7)—Lipids composed of repeating five-carbon units called isoprene units

| Isoprene unit | Types of terpenes | | | |
|---|---|---|---|---|
| | [1] monoterpene | 10 C's | [4] sesterterpene | 25 C's |
| | [2] sesquiterpene | 15 C's | [5] triterpene | 30 C's |
| | [3] diterpene | 20 C's | [6] tetraterpene | 40 C's |

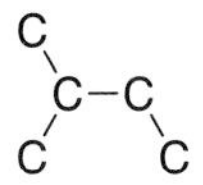

[4] **Steroids** (29.8)—Tetracyclic lipids composed of three six-membered and one five-membered ring

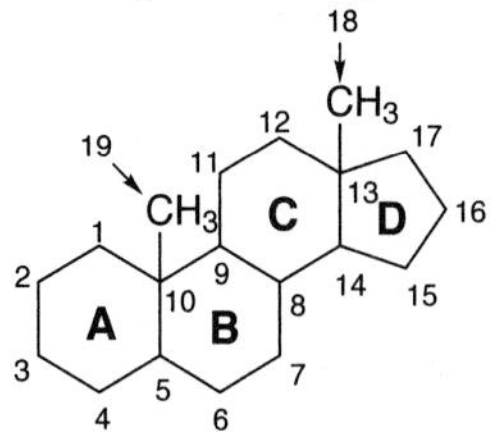

**29.1** Waxes are esters (RCOOR′) formed from a high molecular weight alcohol (R′OH) and a fatty acid (RCOOH).

a.

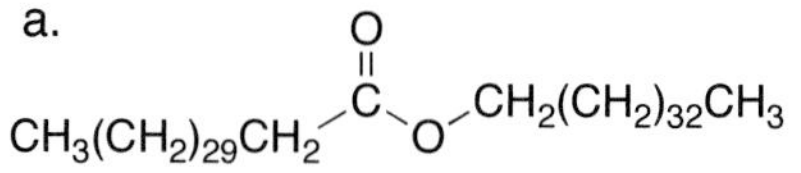

$$CH_3(CH_2)_{29}CH_2-\overset{O}{\underset{\|}{C}}-O-CH_2(CH_2)_{32}CH_3$$

b.

$$\text{—}O(CH_2)_{17}-\overset{O}{\underset{\|}{C}}\underset{O(CH_2)_{17}-C}{\overset{O}{\diagdown}}$$

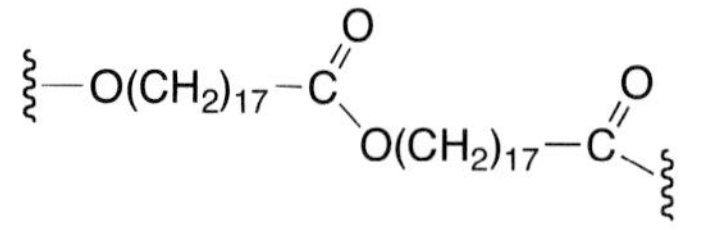

**29.2** Eicosapentaenoic acid has 20 C's and 5 C=C's. Since an increasing number of double bonds decreases the melting point, eicosapentaenoic acid should have a melting point lower than arachidonic acid; that is, < –49°C.

**29.3**

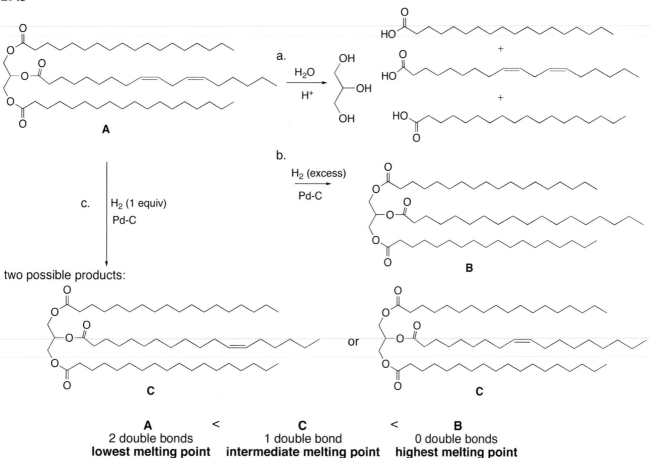

|   A   | < |   C   | < |   B   |
|-------|---|-------|---|-------|
| 2 double bonds | | 1 double bond | | 0 double bonds |
| **lowest melting point** | | **intermediate melting point** | | **highest melting point** |

**29.4**

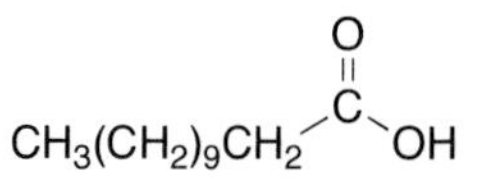

$$CH_3(CH_2)_9CH_2-\overset{O}{\underset{\|}{C}}-OH$$

lauric acid

Lauric acid is a saturated fatty acid but has only 12 C's. The carbon chain is much shorter than palmitic acid (16 C's) and stearic acid (18 C's), making coconut oil a liquid at room temperature.

**29.5**

**29.6** A lecithin is a type of phosphoacylglycerol. Two of the hydroxy groups of glycerol are esterified with fatty acids. The third OH group is part of a phosphodiester, which is also bonded to another low molecular weight alcohol.

general structure
of a lecithin

**29.7** Soaps and phosphoacylglycerols have hydrophilic and hydrophobic components. Both compounds have an ionic "head" that is attracted to polar solvents like $H_2O$. This head is small in size compared to the hydrophobic region, which consists of one or two long hydrocarbon chains. These nonpolar chains consist of only C–C and C–H bonds and exhibit only van der Waals forces.

**29.8** Phospholipids have a polar (ionic) head and two nonpolar tails. These two regions, which exhibit very different forces of attraction, allow the phospholipids to form a bilayer with a central hydrophobic region that serves as a barrier to agents crossing a cell membrane, while still possessing an ionic head to interact with the aqueous environment inside and outside the cell. Two different regions are needed in the molecule. Triacylglycerols have three polar, uncharged ester groups, but they are not nearly as polar as phospholipids. They do not have an ionic head with nonpolar tails and so they do not form bilayers. They are largely nonpolar C–C and C–H bonds so they are not attracted to an aqueous medium, making them $H_2O$ insoluble.

**29.9** Fat-soluble vitamins are hydrophobic and therefore are readily stored in the fatty tissues of the body. Water-soluble vitamins, on the other hand, are readily excreted in the urine and large concentrations cannot build up in the body.

**29.10**

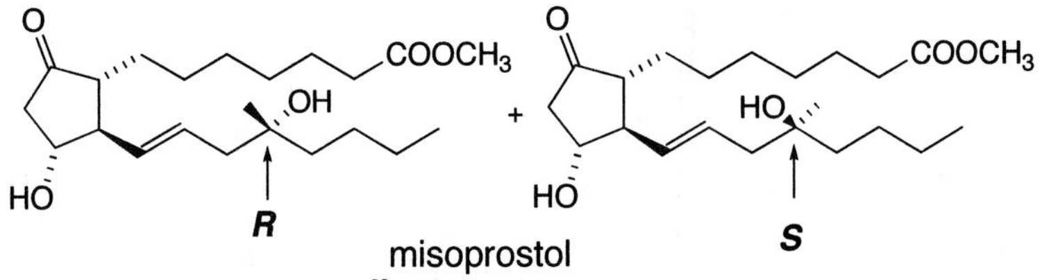

misoprostol
**diastereomers**
Only one tetrahedral stereogenic center is different in these two compounds.

**29.11**

a. 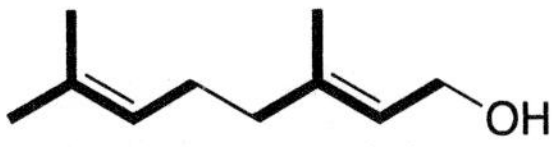 geraniol

b. 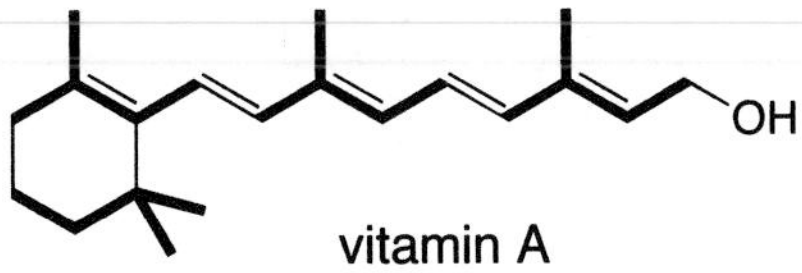 vitamin A

c. grandisol

d. 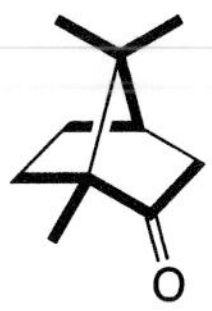 camphor

**29.12**

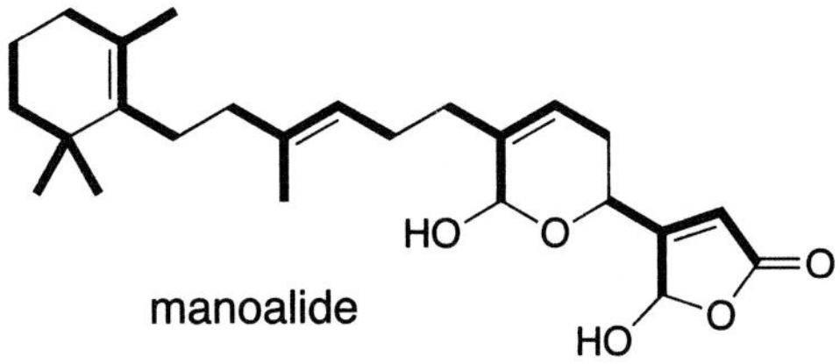

manoalide

**29.13**

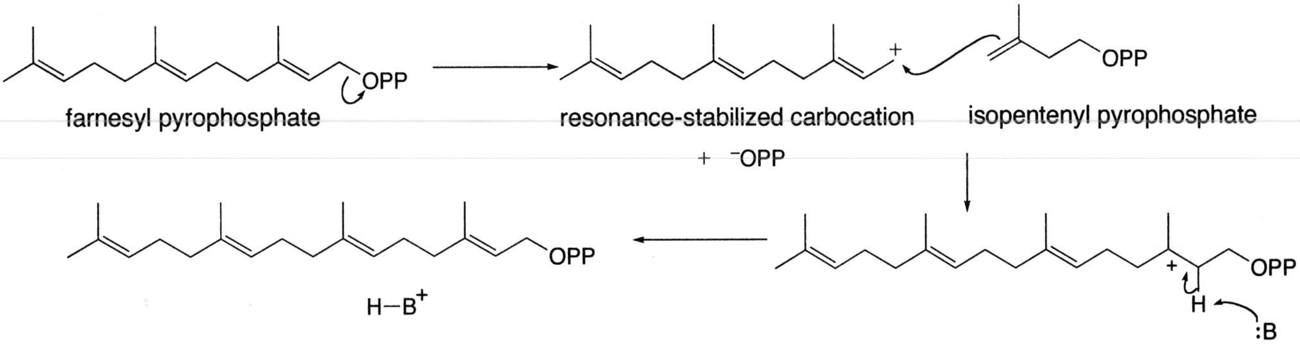

farnesyl pyrophosphate → resonance-stabilized carbocation + isopentenyl pyrophosphate

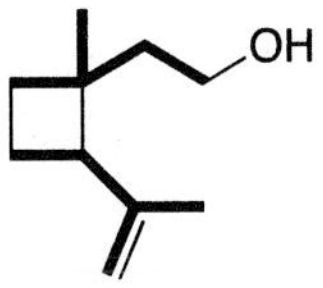

+ ⁻OPP

H—B⁺     :B

**29.14**

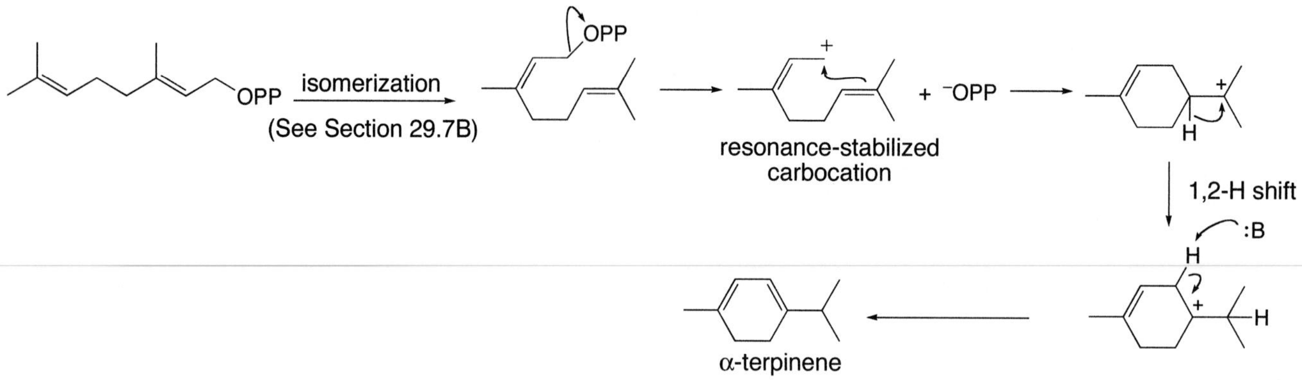

α-terpinene

**29.15**

cholesterol

equatorial OH

enantiomer

different here

diastereomer

different here

diastereomer

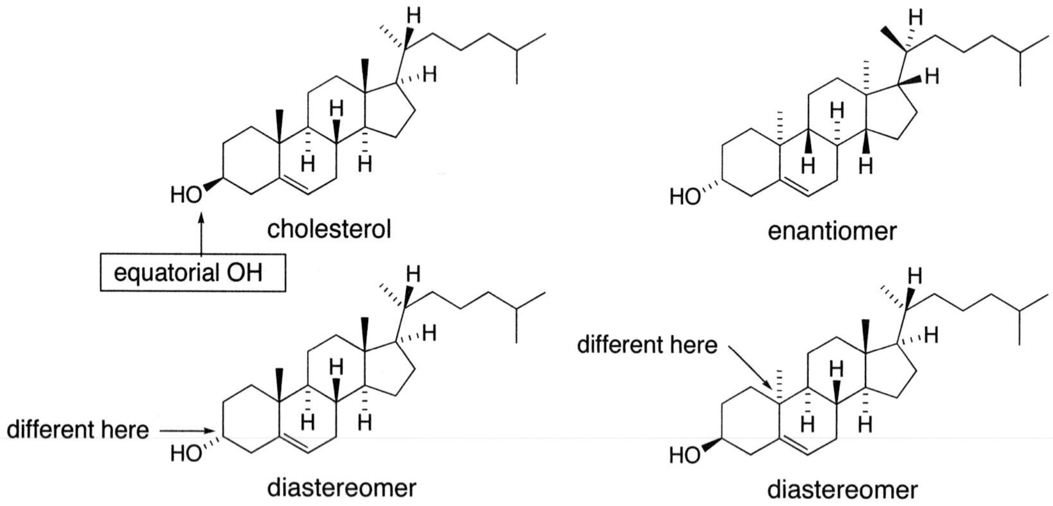

**29.16**

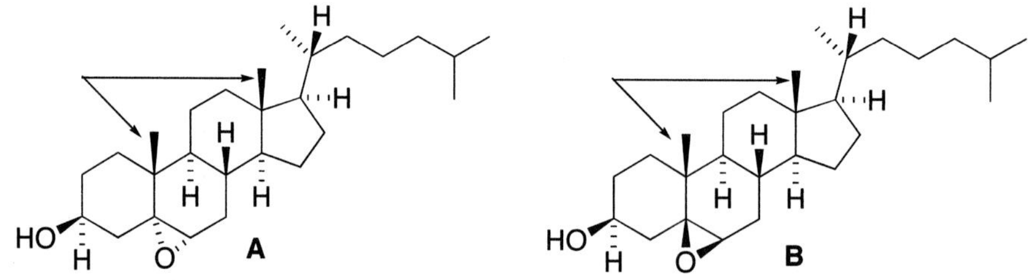

All four rings are in the same plane. The bulky $CH_3$ groups (arrows) are located above the plane. Epoxide **A** is favored, because it results from epoxidation below the plane, on the opposite side from the $CH_3$ groups that shield the top of the molecule somewhat to attack by reagents. In **B**, the epoxide ring is above the plane on the same side as the $CH_3$ groups. Formation of **B** would require epoxidation of the planar C=C from the less accessible, more sterically hindered side of the double bond. This path is thus disfavored.

**29.17**

H
H
H
H
H
O
C
O
$CH_3(CH_2)_{16}$


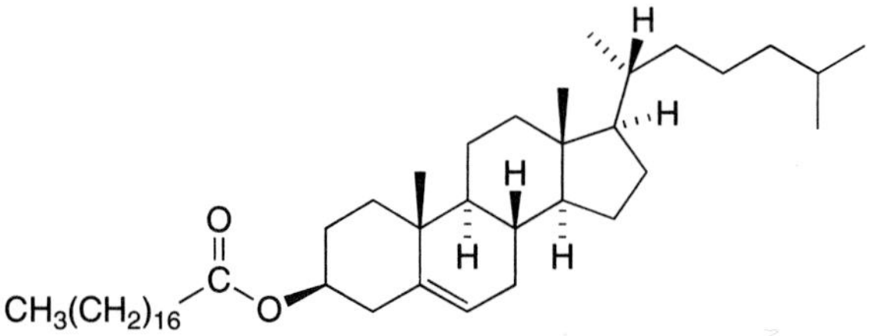

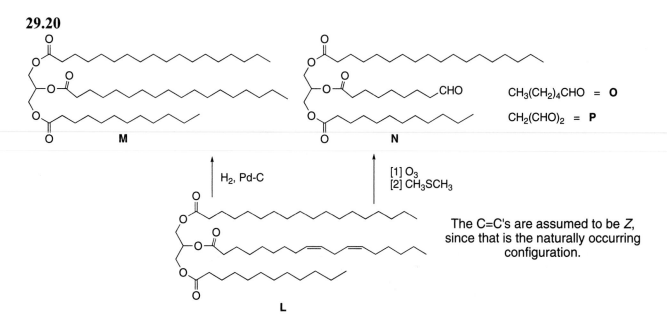

**29.18** Each compound has one tetrahedral stereogenic center (circled), so there are two stereoisomers (two enantiomers) possible. All C=C's have the *Z* configuration.

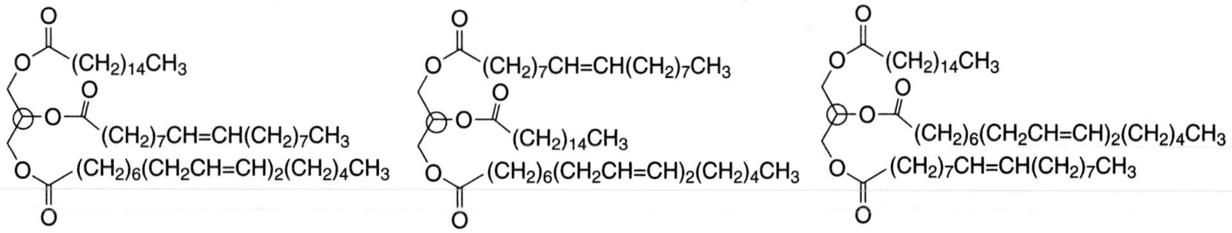

$(CH_2)_{14}CH_3$
$(CH_2)_7CH=CH(CH_2)_7CH_3$
$(CH_2)_6(CH_2CH=CH)_2(CH_2)_4CH_3$
$(CH_2)_7CH=CH(CH_2)_7CH_3$
$(CH_2)_{14}CH_3$
$(CH_2)_6(CH_2CH=CH)_2(CH_2)_4CH_3$
$(CH_2)_{14}CH_3$
$(CH_2)_6(CH_2CH=CH)_2(CH_2)_4CH_3$
$(CH_2)_7CH=CH(CH_2)_7CH_3$


**29.19**

$(CH_2)_7CH=CH(CH_2)_7CH_3$
$(CH_2)_{14}CH_3$
$(CH_2)_7CH=CH(CH_2)_7CH_3$
There is no stereogenic center in this triacylglycerol since these R groups are identical, making this triacyclglycerol optically inactive.


**29.20**

M
CHO
N
$CH_3(CH_2)_4CHO$ = **O**
$CH_2(CHO)_2$ = **P**
$H_2$, Pd-C
[1] $O_3$
[2] $CH_3SCH_3$
L
The C=C's are assumed to be *Z*, since that is the naturally occurring configuration.

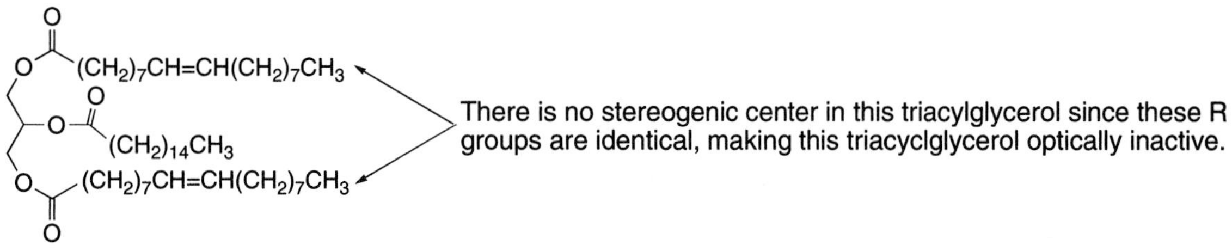

**29.21**   When R″ = CH₂CH₂NH₃⁺, the compound is called a **phosphatidylethanolamine** or **cephalin.**

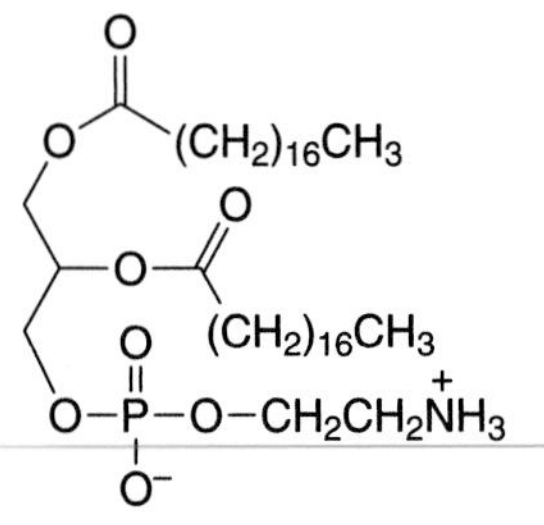

cephalin

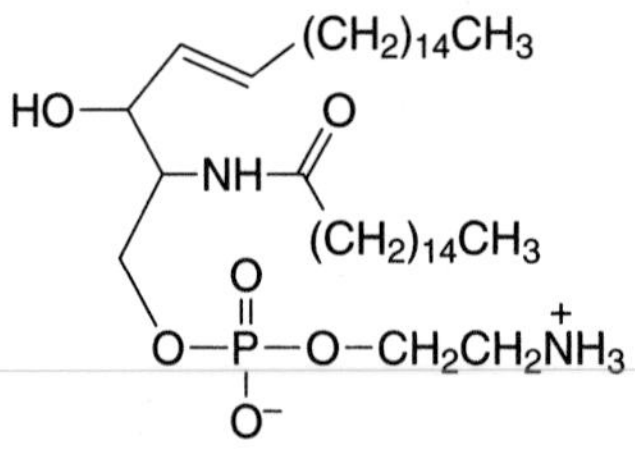

sphingomyelin

## 29.22

a. PGF₂α

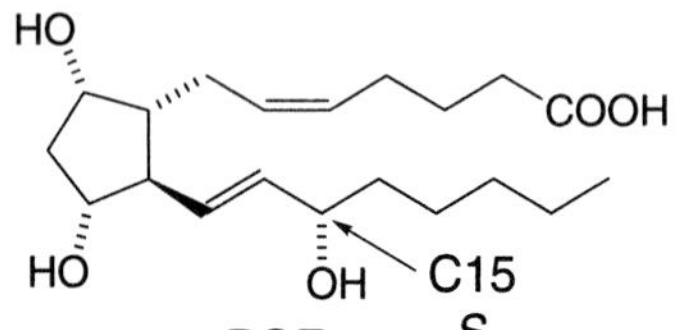

b. A

[1] (R)-CBS reagent
[2] H₂O

c. X

[1] Zn(BH₄)₂
[2] H₂O

**B** and **C**
diastereomers

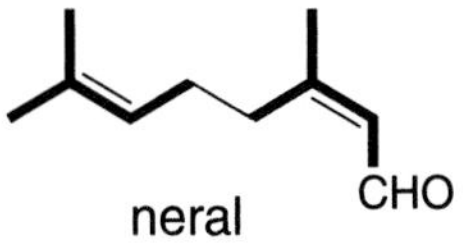

R

S

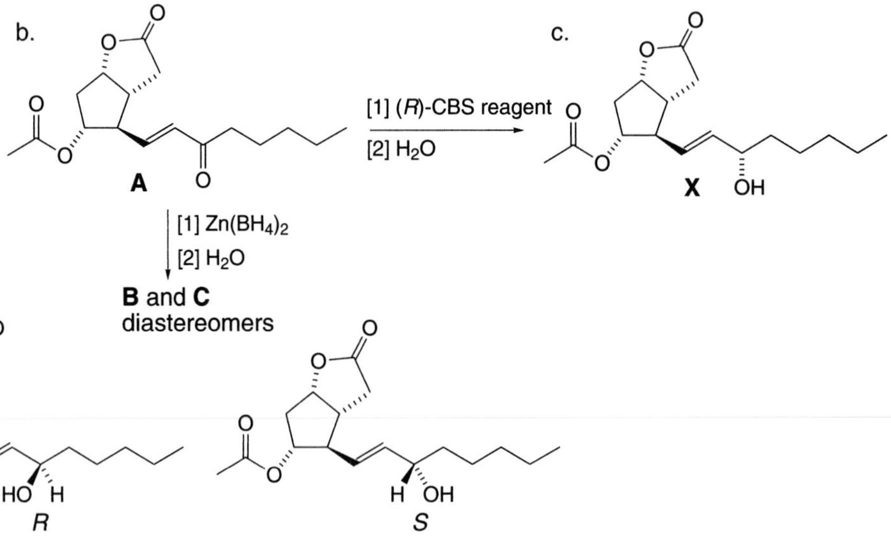

## 29.23

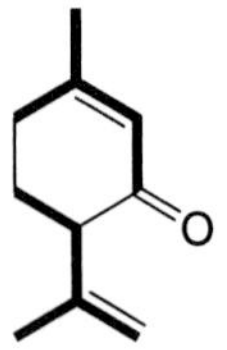

a. neral

b. carvone

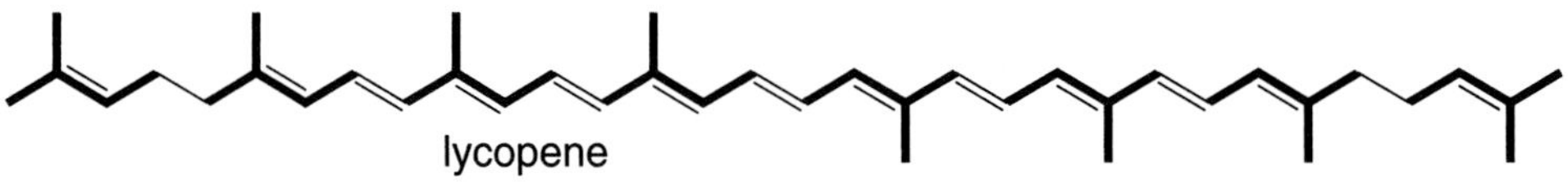

c. α-pinene

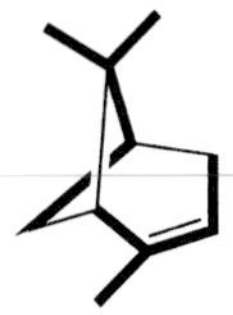

d. lycopene

e. β-carotene

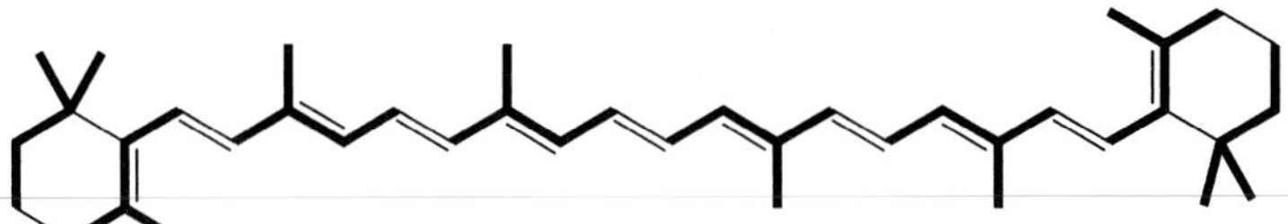

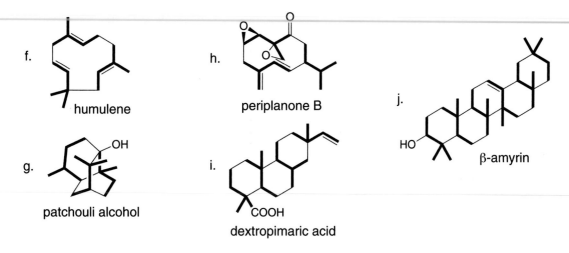

f. humulene

h. periplanone B

j. β-amyrin

g. patchouli alcohol

i. dextropimaric acid

**29.24** A *monoterpene* **contains 10 carbons** and two isoprene units; a *sesquiterpene* **contains 15 carbons** and three isoprene units, etc. See Table 29.5.

a. monoterpene CHO

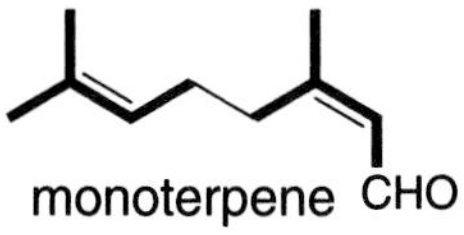

b. monoterpene

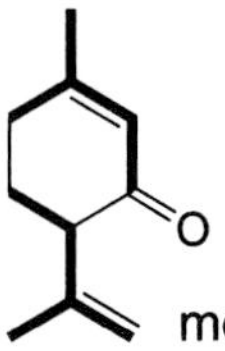

c. monoterpene

f. sesquiterpene

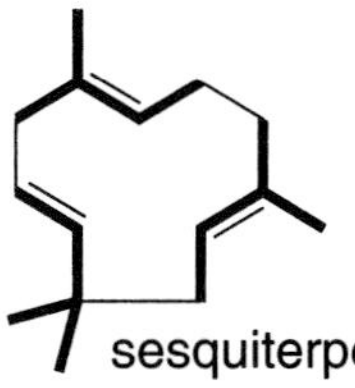

g. sesquiterpene

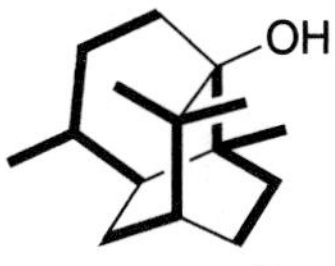

h. sesquiterpene

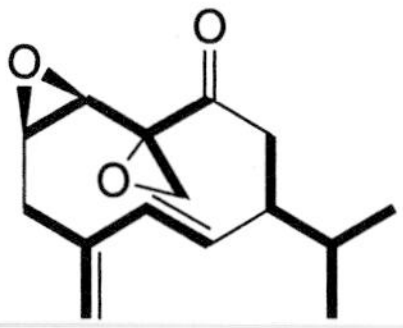

i. diterpene

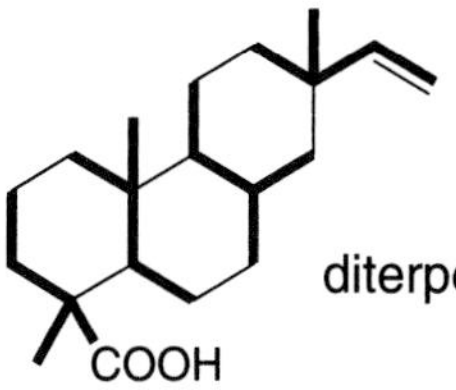

j. triterpene

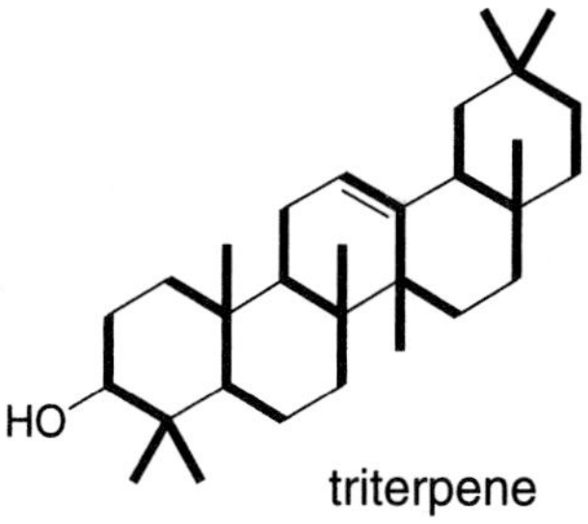

d. tetraterpene

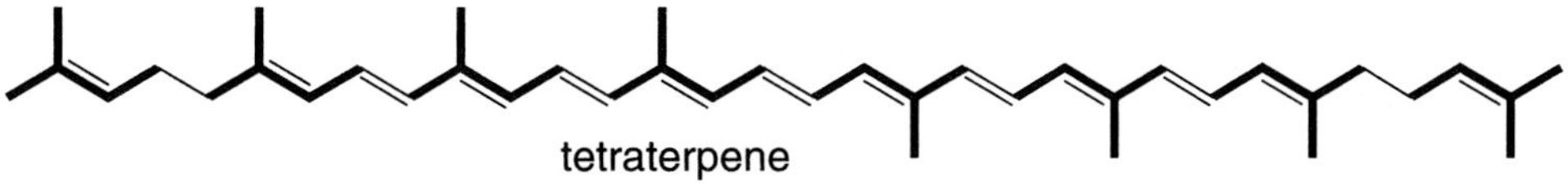

e. tetraterpene

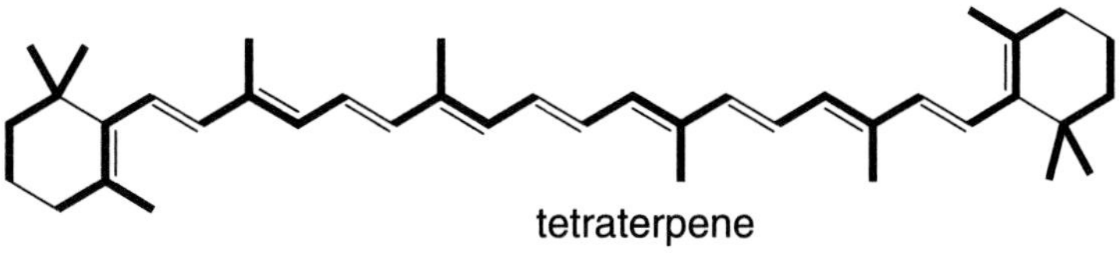

**29.25**

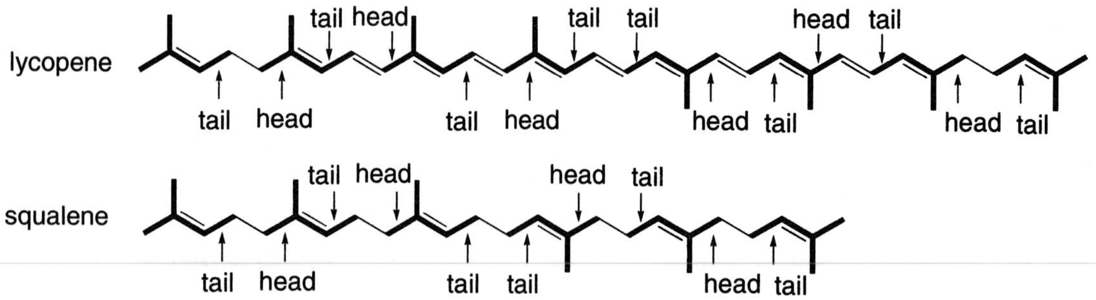

lycopene

squalene

**29.26**

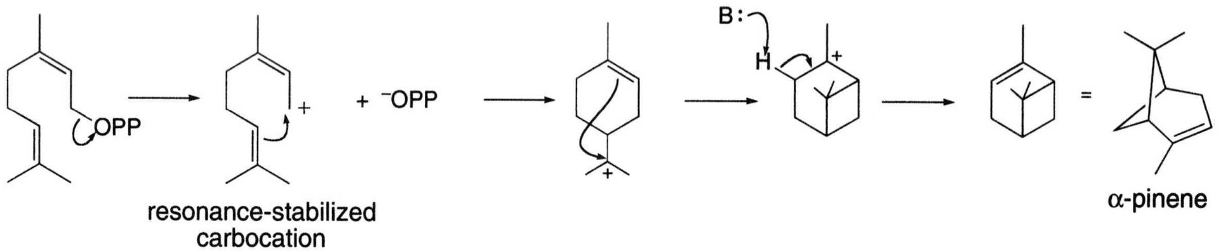

resonance-stabilized
carbocation

α-pinene

**29.27**

a.

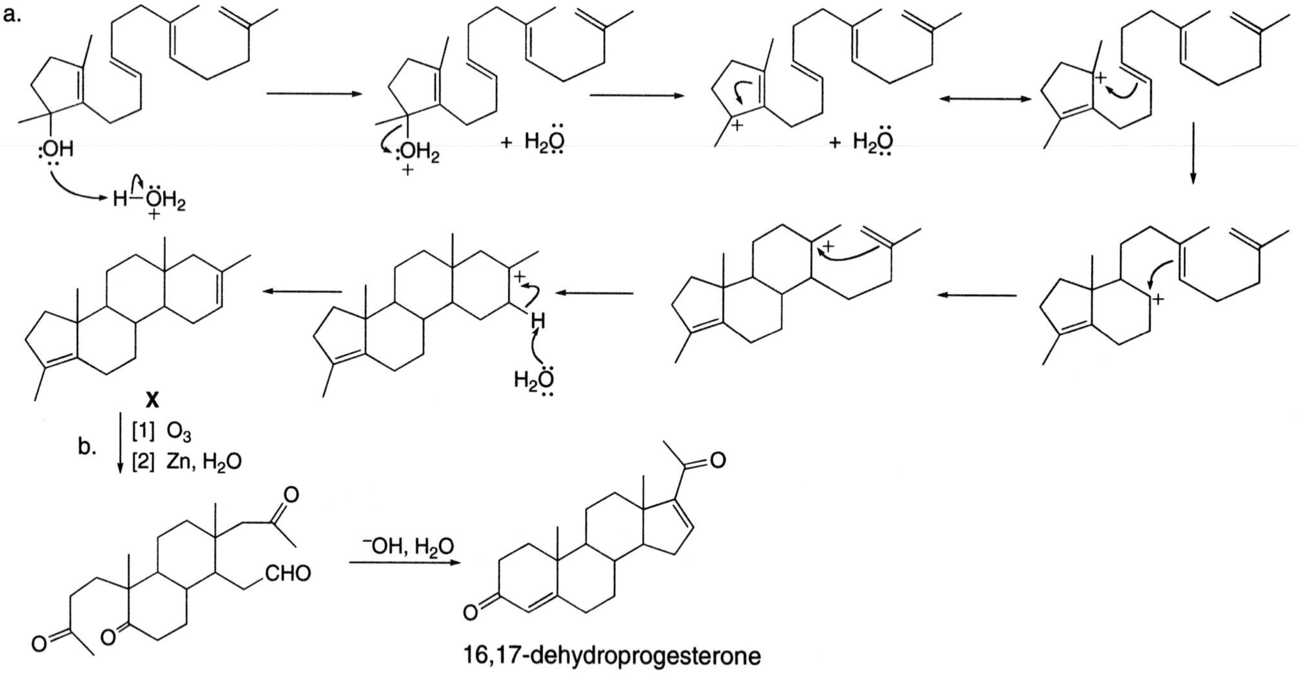

X

b. [1] O₃
[2] Zn, H₂O

16,17-dehydroprogesterone

**29.28**

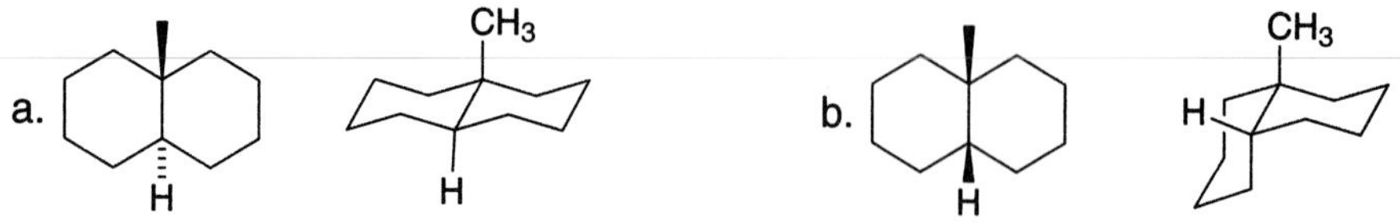

a.

b.

**29.29**

a. equatorial OH

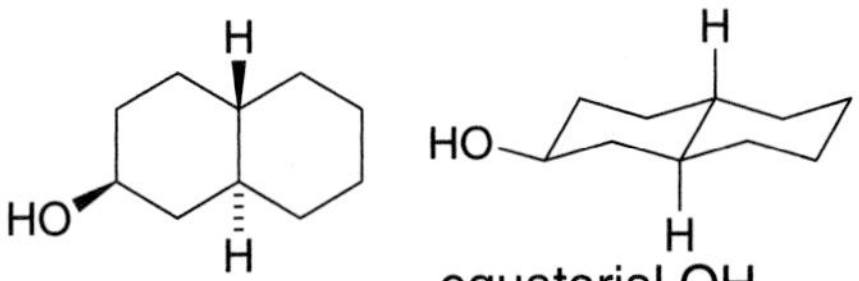

c. axial OH

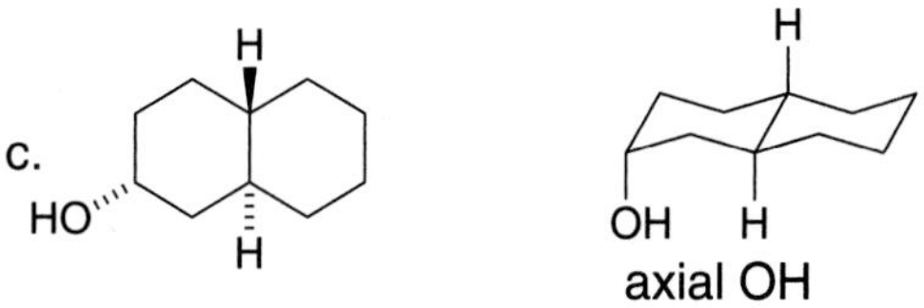

b. axial OH

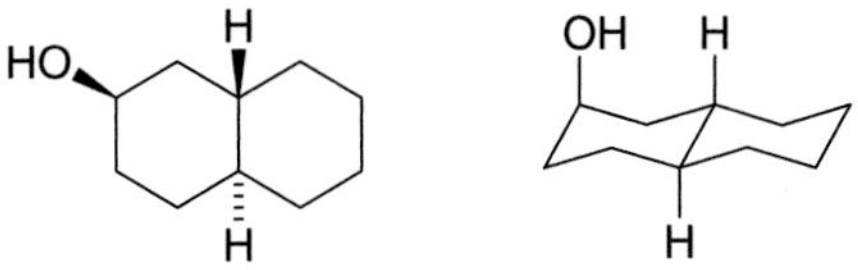

d. equatorial OH

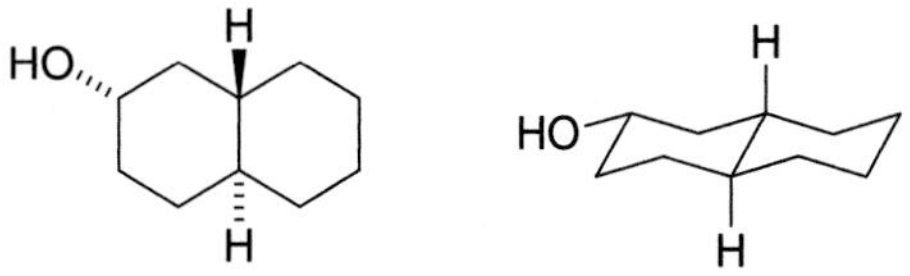

**29.30**

a. (eq)  OH (ax)  Axial reacts faster.

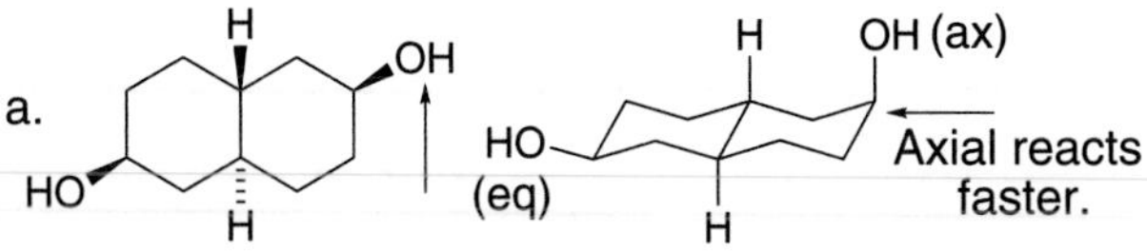

b. (eq)  OH (ax)  Axial reacts faster.

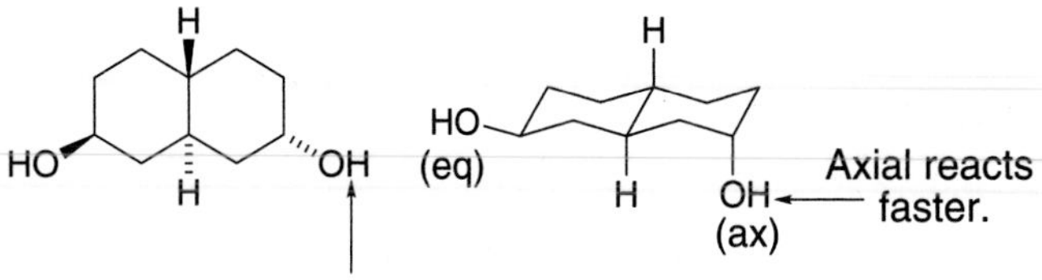

**29.31**

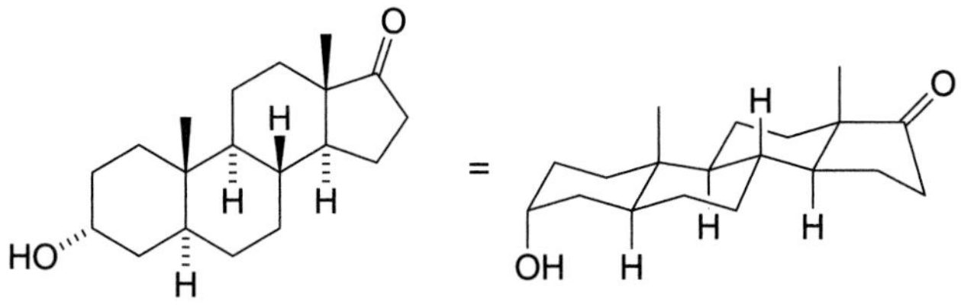

= 

**29.32**

CH₃ groups make this face more sterically hindered.

a. =  

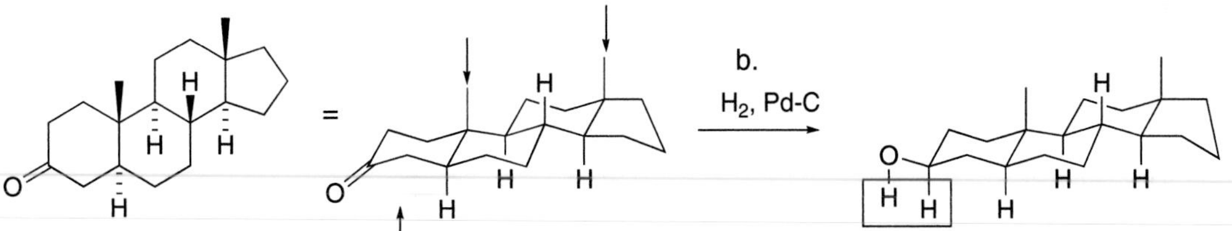

H₂, Pd-C

b.

H₂ added from below.

The bottom face is more accessible so the H₂ is added from this face to form an equatorial OH.

## 29.33

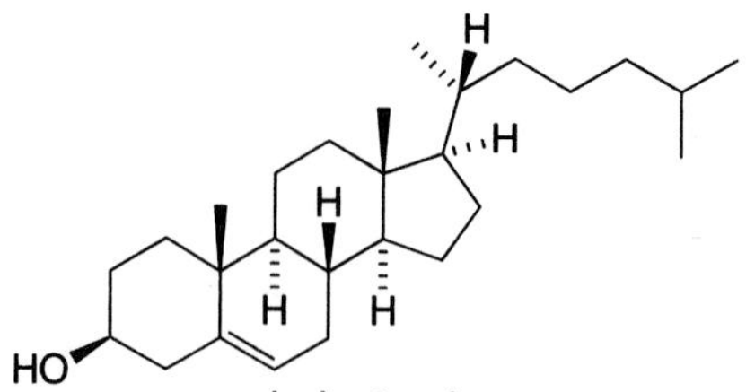

cholesterol

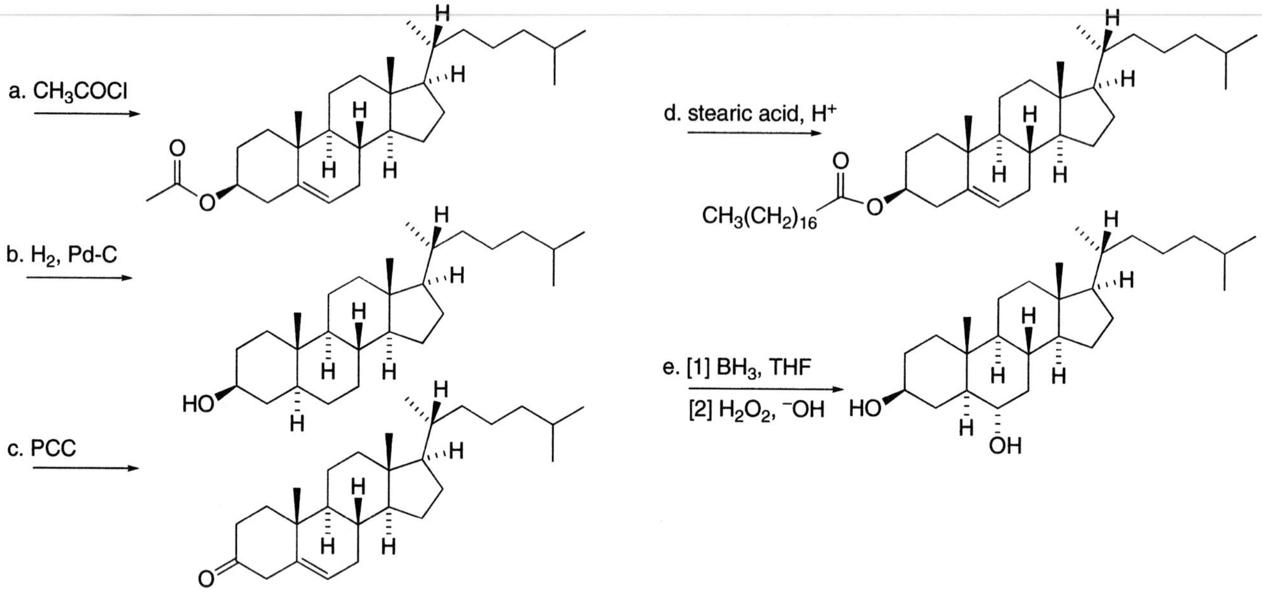
a. $CH_3COCl$
b. $H_2$, Pd-C
c. PCC
d. stearic acid, $H^+$
$CH_3(CH_2)_{16}$
e. [1] $BH_3$, THF
[2] $H_2O_2$, $^-OH$

## 29.34

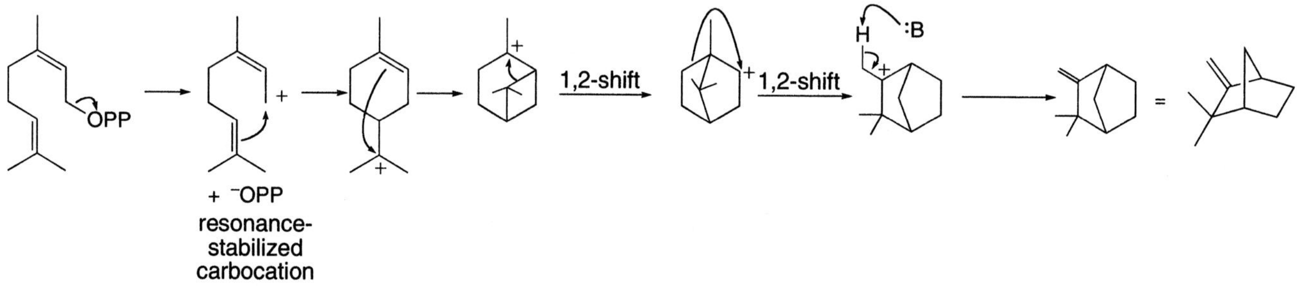
OPP
+ $^-OPP$
resonance-stabilized carbocation
1,2-shift
1,2-shift
:B

**29.35** Re-draw the starting material in a conformation that suggests the structure of the product.

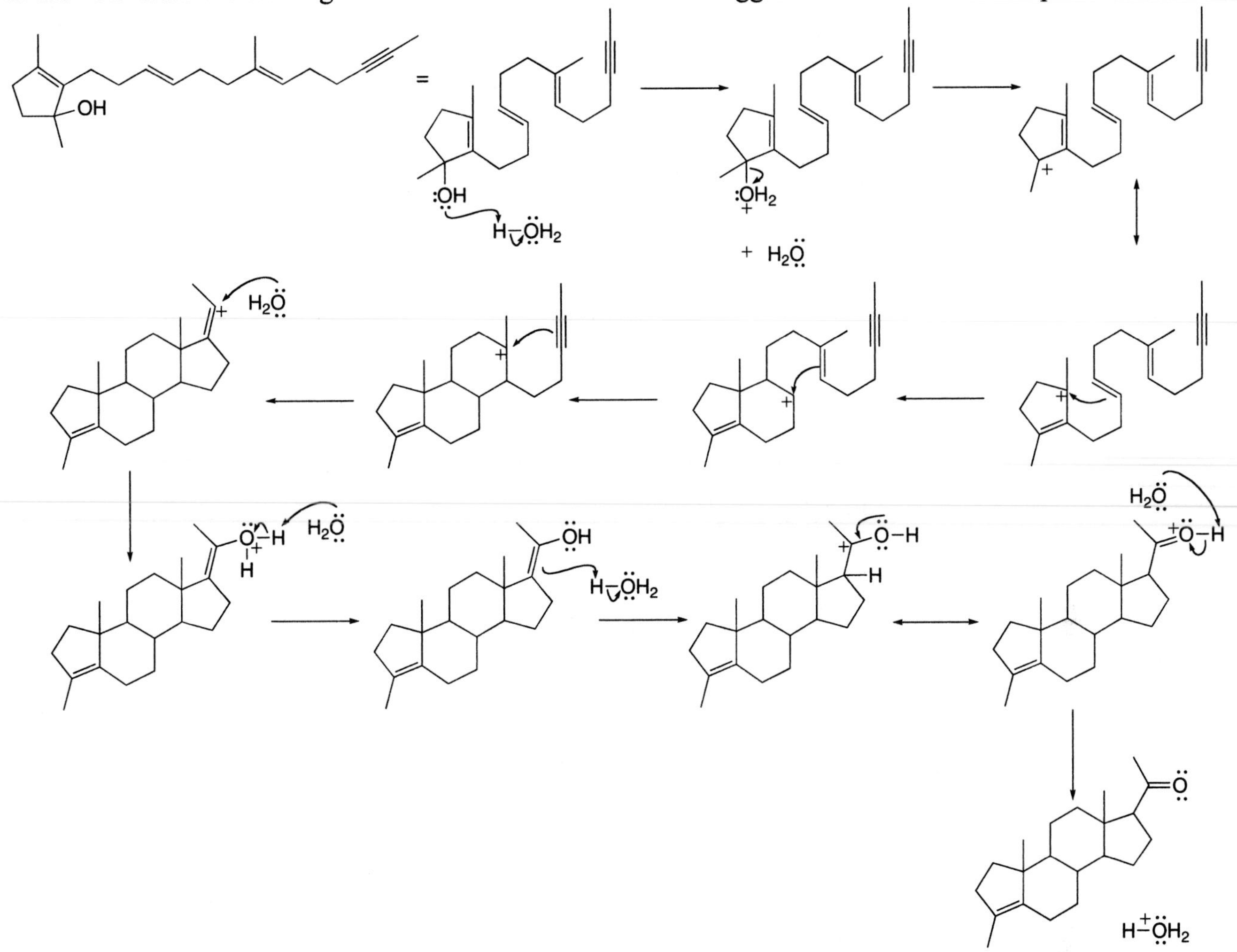

**29.36**

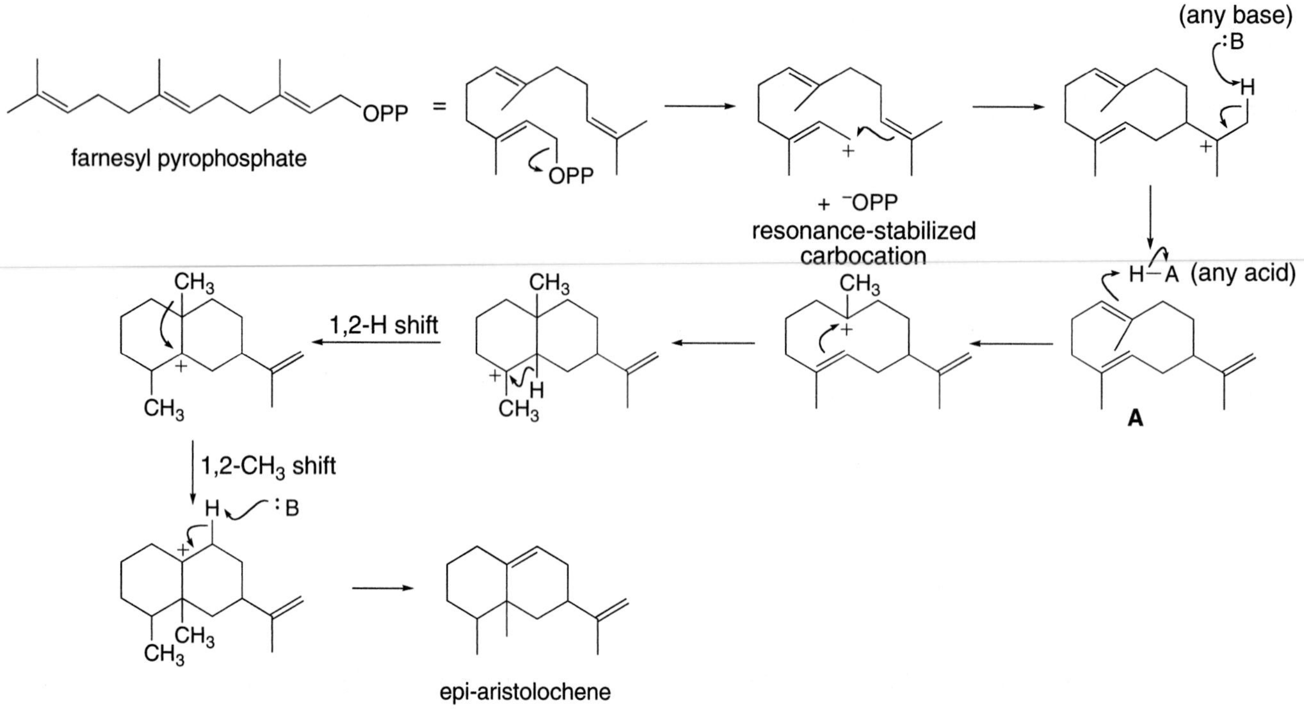

(any base)
:B
H
OPP
=
farnesyl pyrophosphate
OPP
+
+ ⁻OPP
resonance-stabilized carbocation
+
H–A (any acid)
A
CH3
+
CH3
+
H
CH3
1,2-H shift
CH3
+
CH3
1,2-CH3 shift
H
:B
+
CH3
CH3
epi-aristolochene

## Chapter 30: Synthetic Polymers

♦ **Chain-growth polymers—Addition polymers**

**[1] Chain-growth polymers with alkene starting materials (30.2)**

- General reaction:

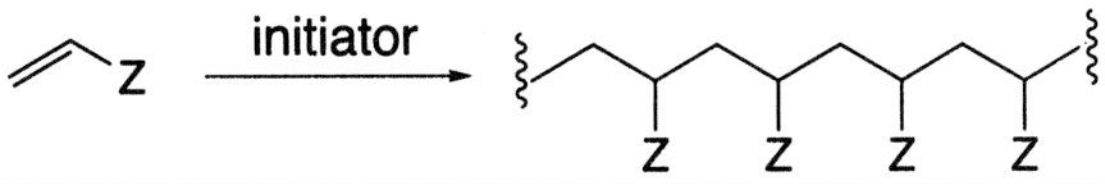

- Mechanism—three possibilities, depending on the identity of Z:

| Type | Identity of Z | Initiator | Comments |
|---|---|---|---|
| [1] radical polymerization | Z stabilizes a radical Z = R, Ph, Cl, etc. | A source of radicals (ROOR) | Termination occurs by radical coupling or disproportionation. Chain branching occurs. |
| [2] cationic polymerization | Z stabilizes a carbocation Z = R, Ph, OR, etc. | H–A or a Lewis acid (BF$_3$ + H$_2$O) | Termination occurs by loss of a proton. |
| [3] anionic polymerization | Z stabilizes a carbanion Z = Ph, COOR, COR, CN, etc. | An organolithium reagent (R–Li) | Termination occurs only when an acid or other electrophile is added. |

**[2] Chain-growth polymers with epoxide starting materials (30.3)**

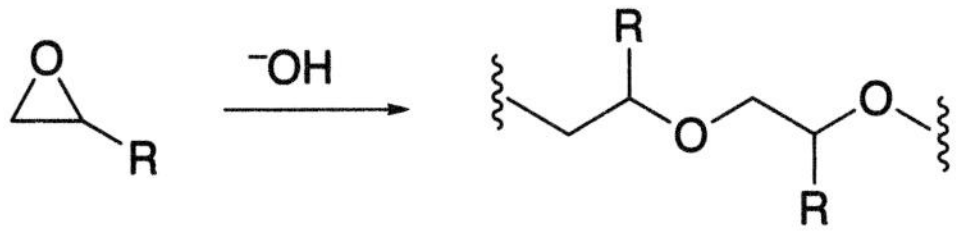

- The mechanism is S$_N$2.
- Ring opening occurs at the less substituted carbon of the epoxide.

♦ **Examples of step-growth polymers—Condensation polymers (30.6)**

| Polyamides | Polyesters |
|---|---|
| nylon 6 | polyethylene terephthalate |
| Kevlar | copolymer of glycolic and lactic acids |
| **Polyurethanes** | **Polycarbonates** |
| a polyurethane | Lexan |

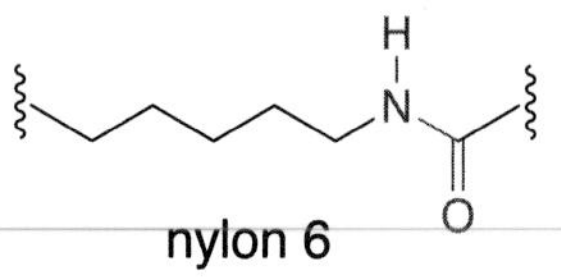

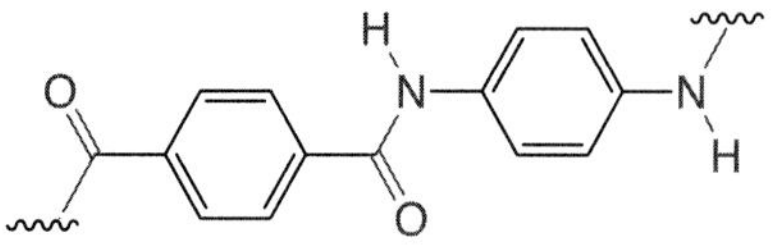

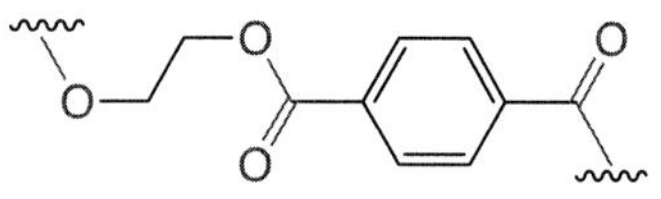

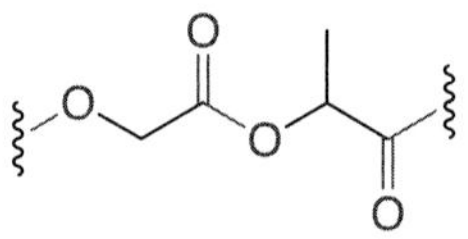

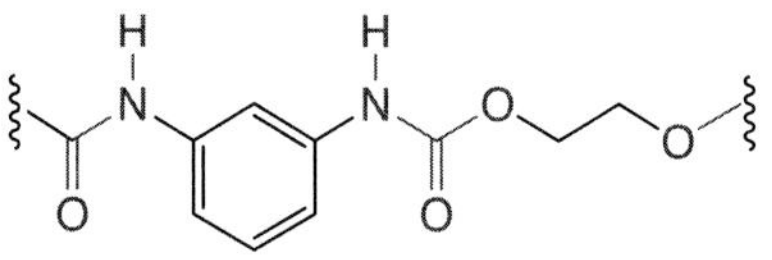

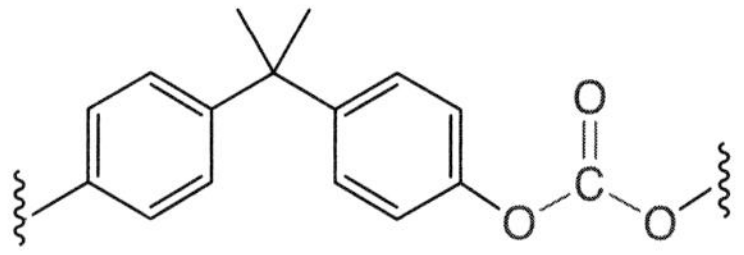

♦ **Structure and properties**

- Polymers prepared from monomers having the general structure $CH_2=CHZ$ can be **isotactic**, **syndiotactic**, or **atactic** depending on the identity of Z and the method of preparation (30.4).
- **Ziegler–Natta catalysts** form polymers without significant branching. Polymers can be isotactic, syndiotactic, or atactic depending on the catalyst. Polymers prepared from 1,3-dienes have the $E$ or $Z$ configuration depending on the monomer (30.4, 30.5).
- Most polymers contain ordered crystalline regions and less ordered amorphous regions (30.7). The greater the crystallinity, the harder the polymer.
- **Elastomers** are polymers that stretch and can return to their original shape (30.5).
- **Thermoplastics** are hard at room temperature but soften on heating so they can be molded (30.7).
- **Thermosetting polymers** are composed of complex networks of covalent bonds so they cannot be melted to form a liquid phase (30.7).

## Chapter 30: Answers to Problems

**30.1** Place brackets around the repeating unit that creates the polymer.

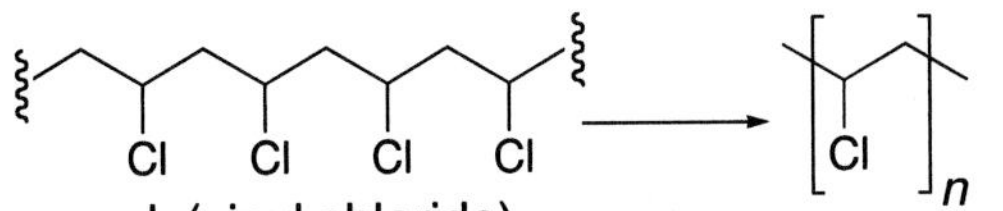

poly(vinyl chloride)

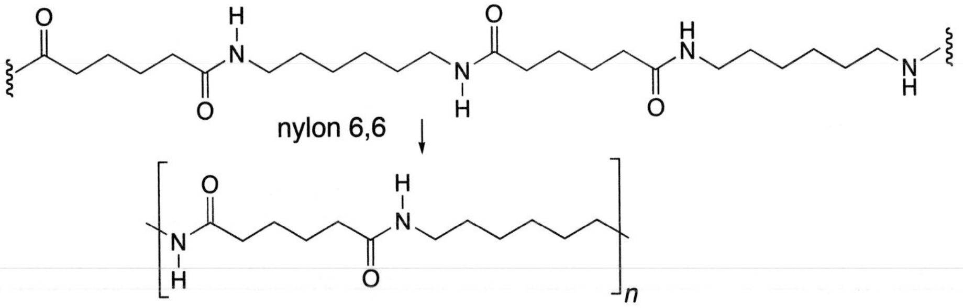

nylon 6,6

**30.2** Draw each polymer formed by chain-growth polymerization.

a.

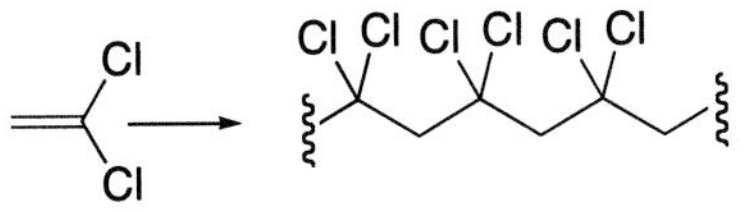

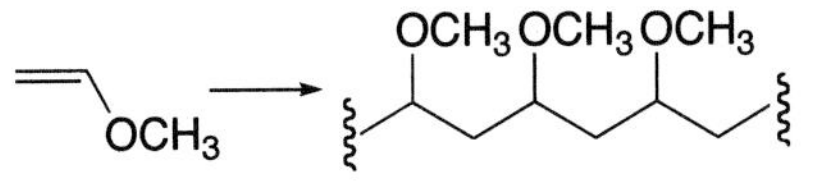

b.

c.

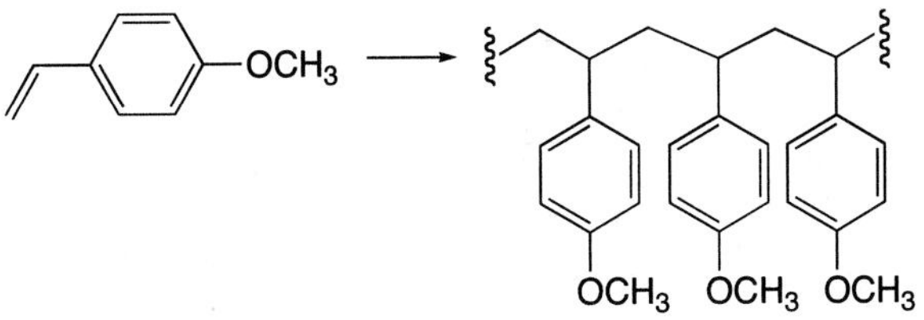

d.

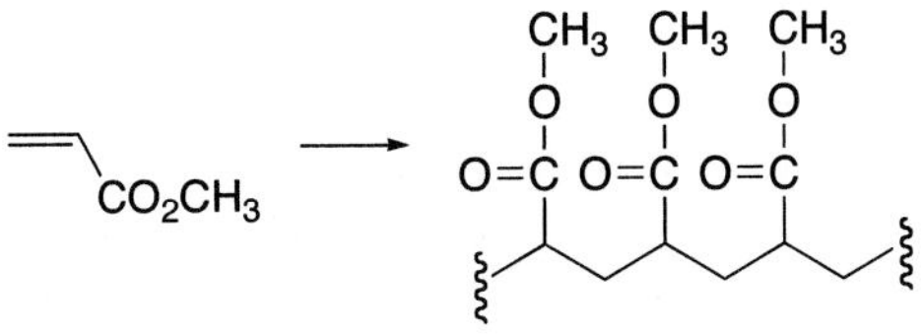

**30.3** Draw each polymer formed by radical polymerization.

a.

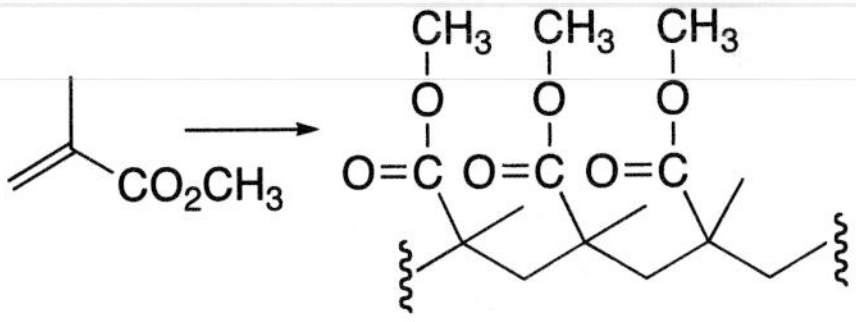

b.

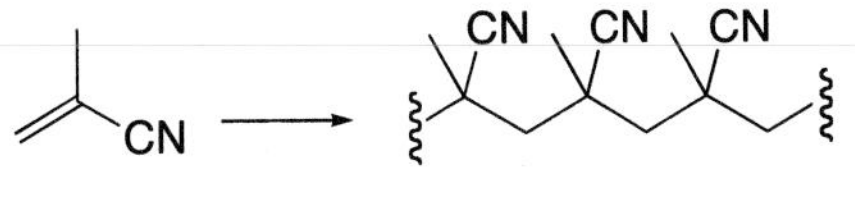

**30.4** Use Mechanism 30.1 as a model of radical polymerization.

Initiation:

Propagation:

Repeat Step [3] over and over.

Termination:

(Ac = CH₃CO–)

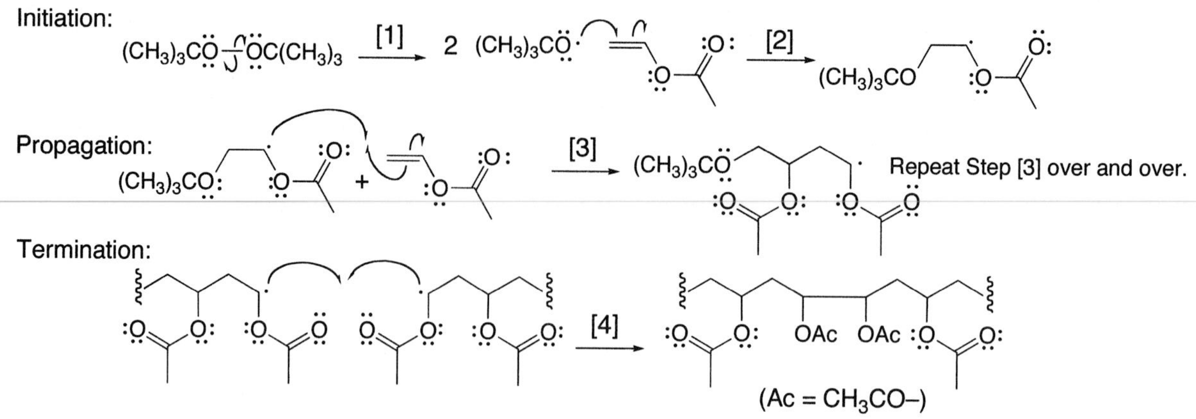

**30.5** Radical polymerization forms a long chain of polystyrene with phenyl groups bonded to every other carbon. To form branches on this polystyrene chain, a radical on a second polymer chain abstracts a H atom. Abstraction of $H_a$ forms a resonance-stabilized radical **A'**. The 2° radical **B'**, (without added resonance stabilization) is formed by abstraction of $H_b$. Abstraction of $H_a$ is favored, therefore, and this radical goes on to form products with 4° C's (**A**).

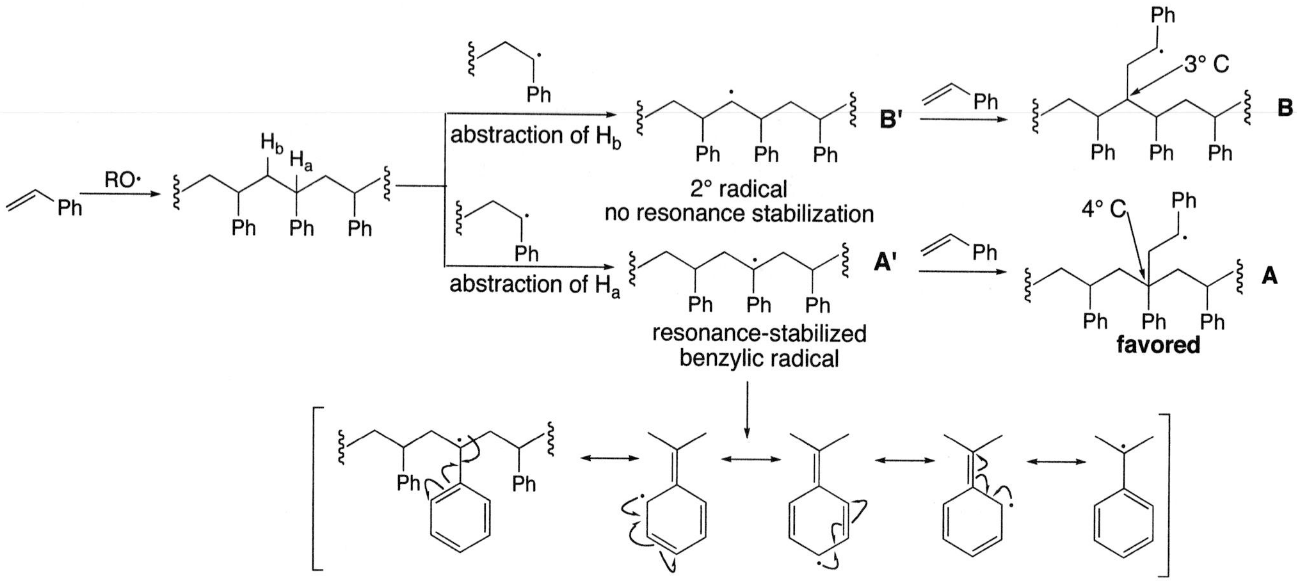

**30.6** Cationic polymerization proceeds via a carbocation intermediate. Substrates that form more stable 3° carbocations react more readily in these polymerization reactions than substrates that form less stable 1° carbocations. $CH_2=C(CH_3)_2$ will form a more substituted carbocation than $CH_2=CH_2$.

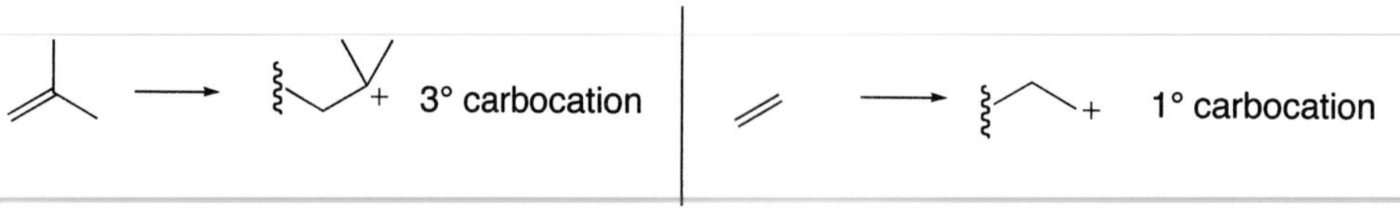

**30.7** Cationic polymerization occurs with alkene monomers having substituents that can stabilize carbocations, such as alkyl groups and other electron-donor groups. Anionic polymerization occurs with alkene monomers having substituents that can stabilize a negative charge, such as COR, COOR, or CN.

a. $CH_2=C(CH_3)COOCH_3$

electron-withdrawing group
**anion polymerization**

b. $CH_2=CHCH_3$

alkyl group
**cationic polymerization**

c. $CH_2=CHOC(CH_3)_3$

an electron-donating
resonance effect
**cationic polymerization**

d. $CH_2=CHCOCH_3$

electron-withdrawing group
**anion polymerization**

**30.8** Use Mechanism 30.4 as a model of anion polymerization.

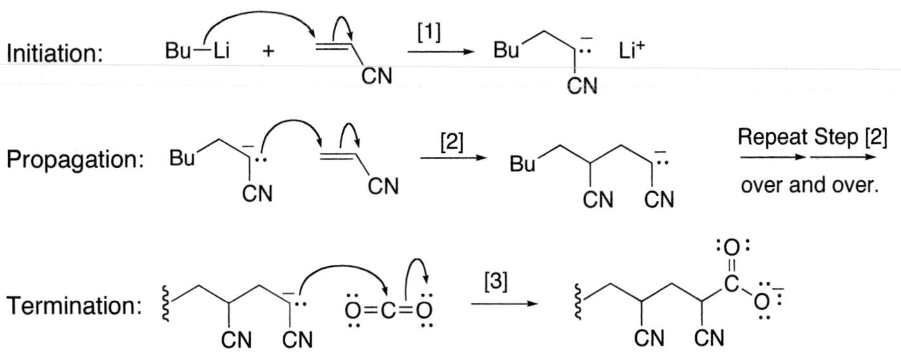

Initiation: Bu—Li + [1] Bu ⁻ Li⁺

Propagation: Bu [2] Bu — Repeat Step [2] over and over.

Termination: [3]

**30.9** Styrene ($CH_2=CHPh$) can by polymerized by all three methods of chain-growth polymerization because a benzene ring can stabilize a radical, a carbocation, and also a carbanion by resonance delocalization.

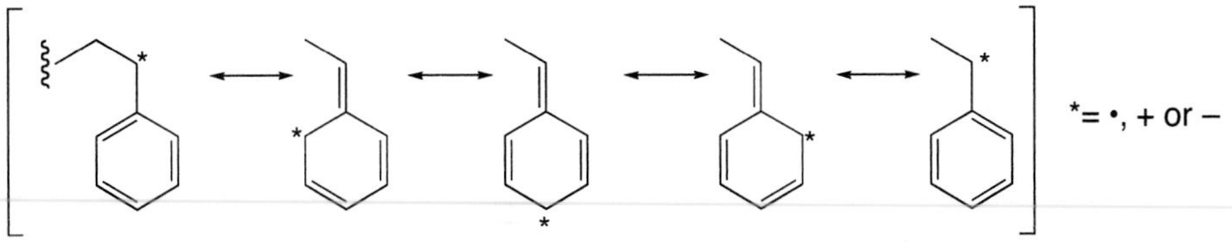

$* = •, + or −$

**30.10** Draw the copolymers formed in each reactions.

a. 

b.

**30.11**

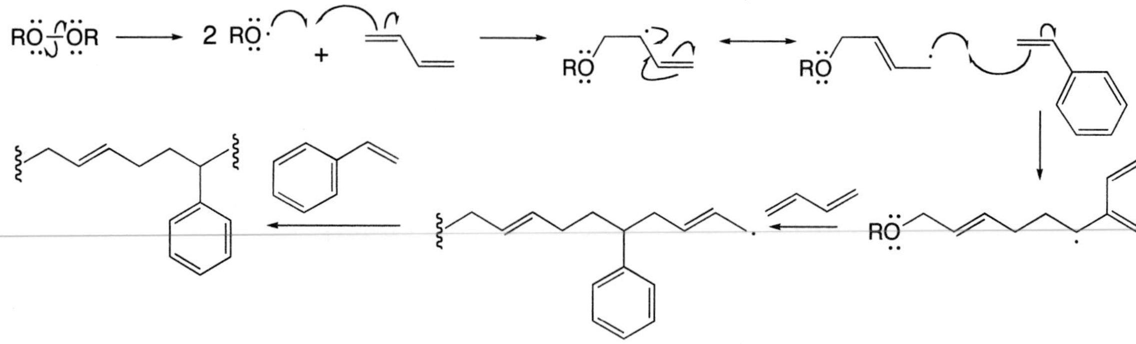

**30.12**

a.

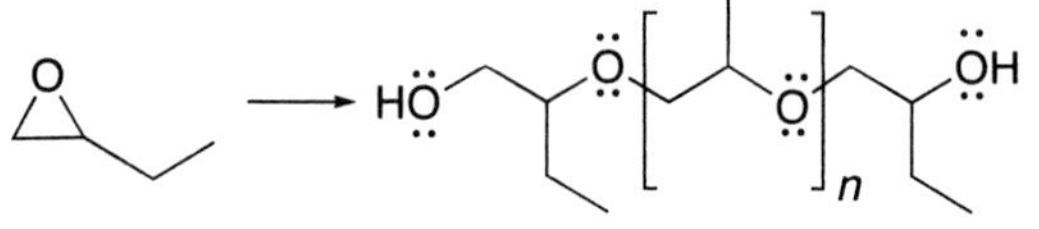

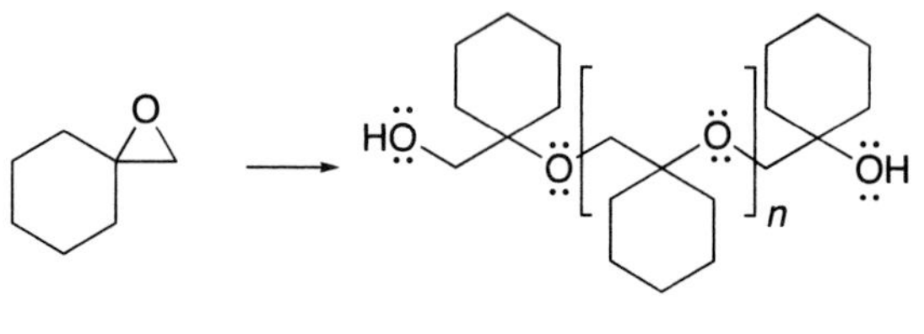

b.

c.

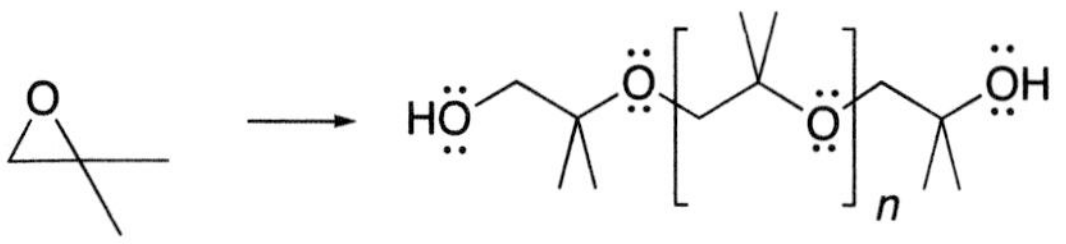

**30.13**

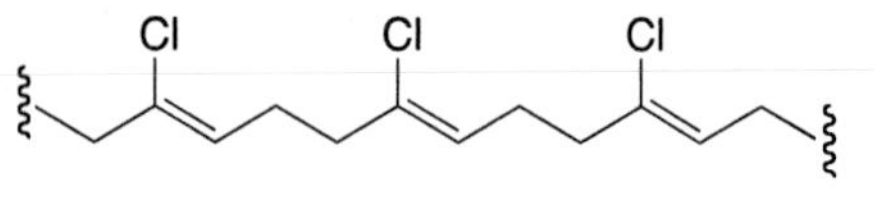

neoprene

All double bonds have the *Z* configuration.

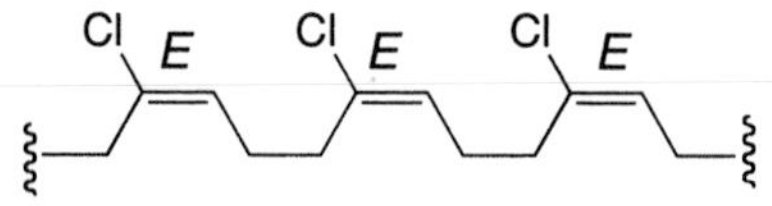

*E* configuration of each double bond.

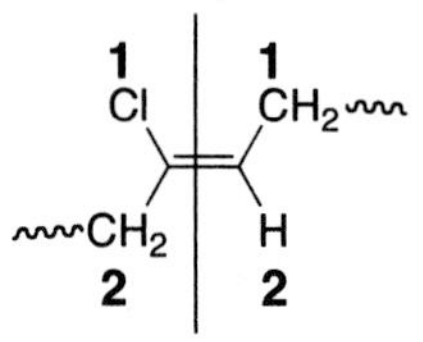

Two higher priority groups (1's) are on the same side of the double bond - - - > *Z* configuration.

**30.14**

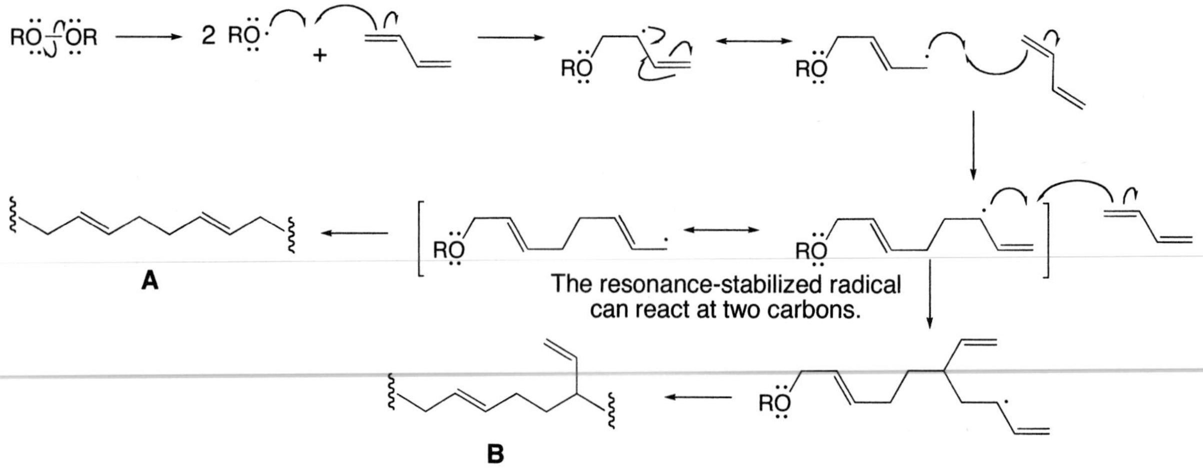

**A**

The resonance-stabilized radical can react at two carbons.

**B**

**30.15**

a.

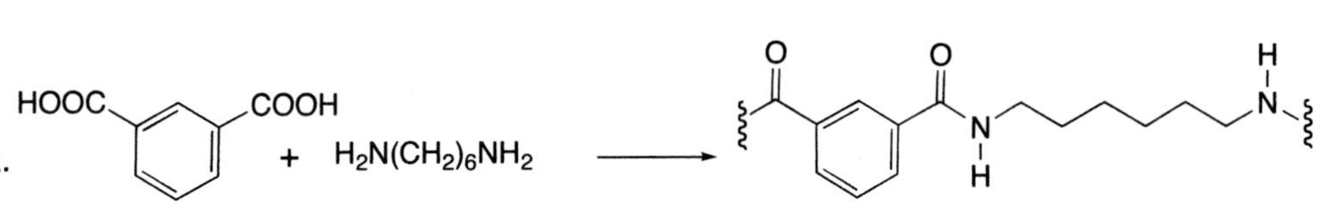

b.

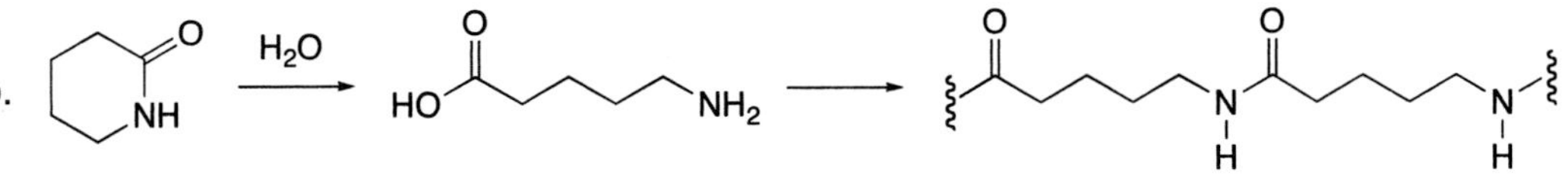

c.

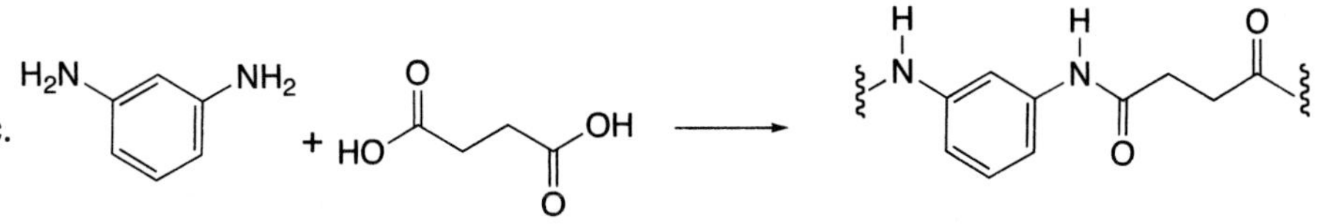

**30.16**

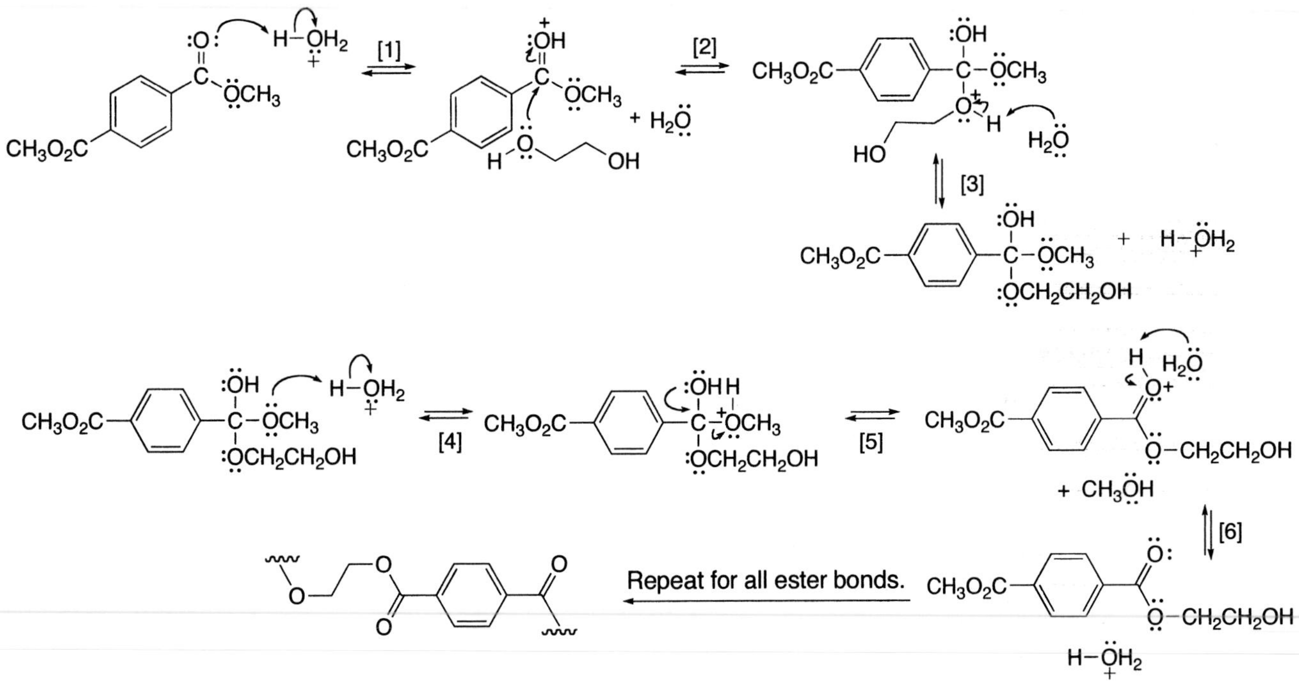

Repeat for all ester bonds.

**30.17**

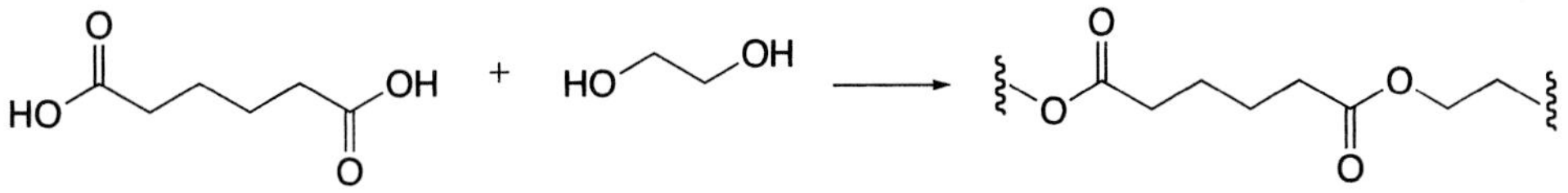

This compound is less suitable than either nylon 6,6 or PET for use in consumer products because esters are more easily hydrolyzed than amides so this polyester is less stable than the polyamide nylon. This polyester has more flexible chains than PET, and this translates into a less strong fiber.

**30.18**

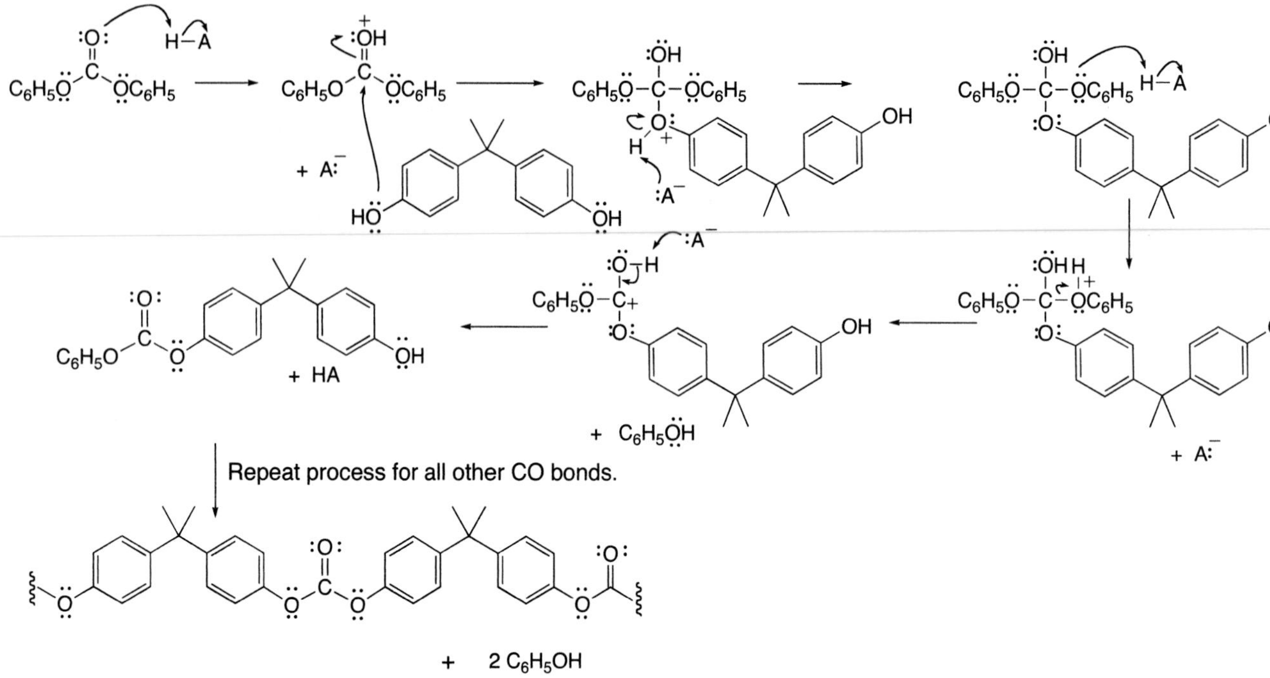

Repeat process for all other CO bonds.

+ 2 C₆H₅OH

**30.19**

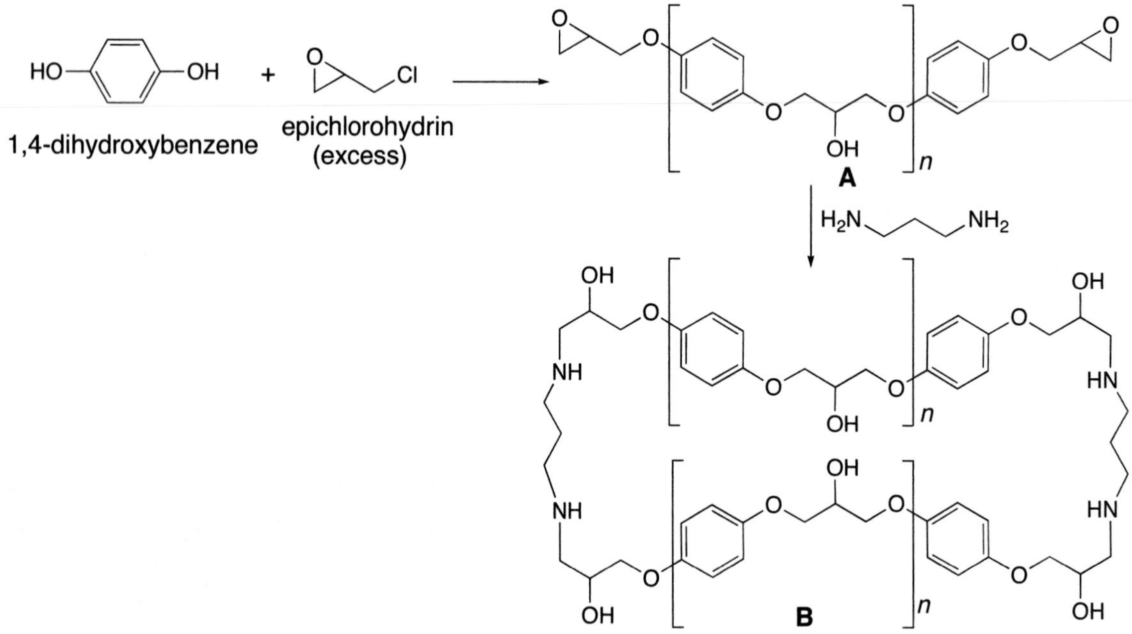

1,4-dihydroxybenzene

epichlorohydrin
(excess)

A

B

**30.20**

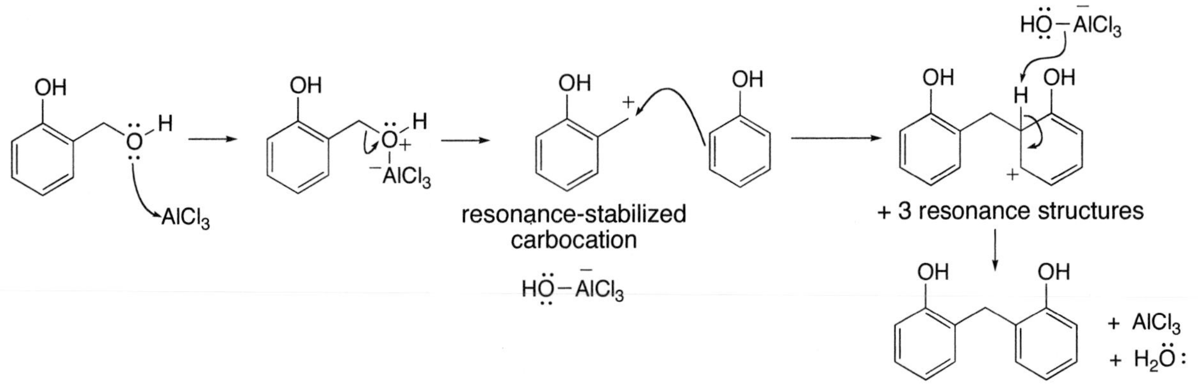

**30.21**

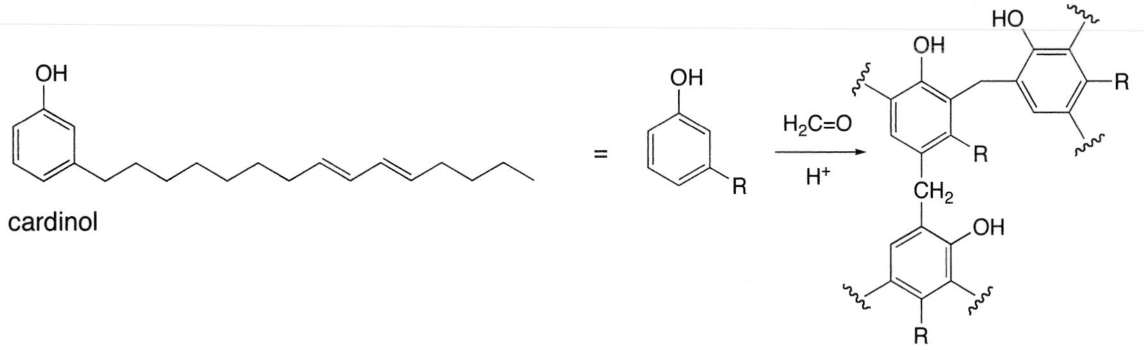

cardinol

**30.22** Chemical recycling of HDPE and LDPE is not easily done because these polymers are both long chains of $CH_2$ groups joined together in a linear fashion. Since there are only C–C bonds and no functional groups in the polymer chain, there are no easy methods to convert the polymers to their monomers. This process is readily accomplished only when the polymer backbone contains hydrolyzable functional groups.

**30.23** a. Combustion of polyethylene forms $CO_2 + H_2O$.
b. Combustion of polyethylene terephthalate forms $CO_2 + H_2O$.
c. These reactions are exothermic.
d. HDPE and PET must be separated from poly(vinyl chloride) prior to incineration because combustion of hydrocarbons (like HDPE) and oxygen-containing organics (like PET) releases only $CO_2 + H_2O$ into the atmosphere. Poly(vinyl chloride) also contains Cl atoms bonded to a hydrocarbon chain. On combustion this forms HCl, which cannot be released directly into the atmosphere, making incineration of halogen-containing polymers more laborious and more expensive.

**30.24**

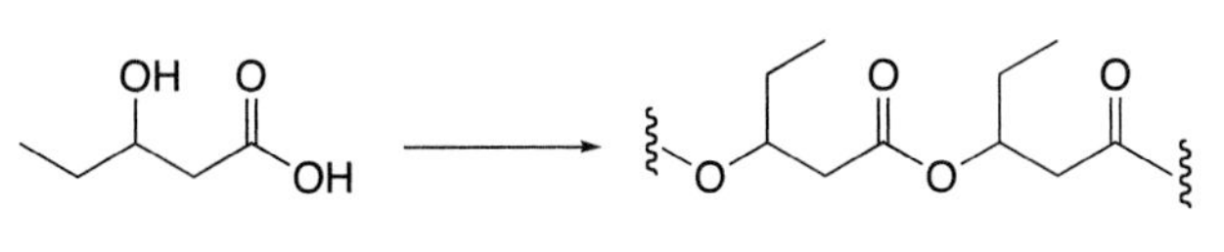

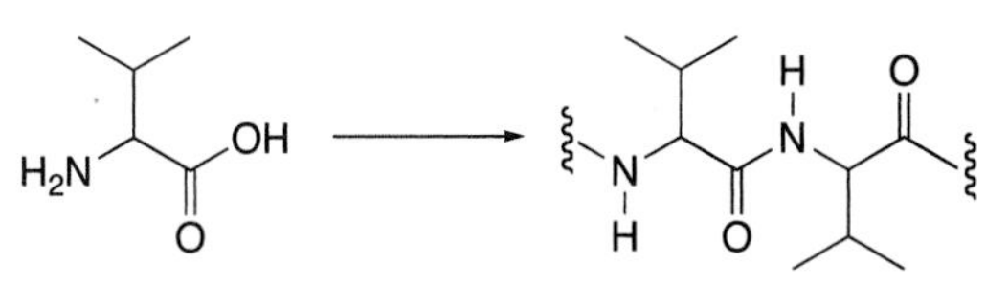

**30.25** Draw the polymer formed by chain-growth polymerization as in Answer 30.2.

a.

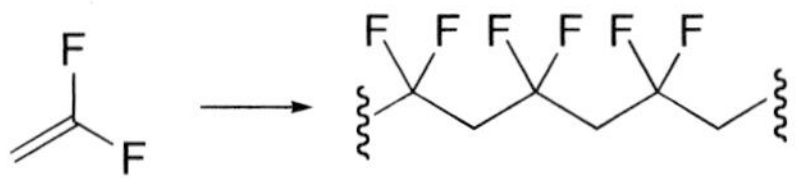

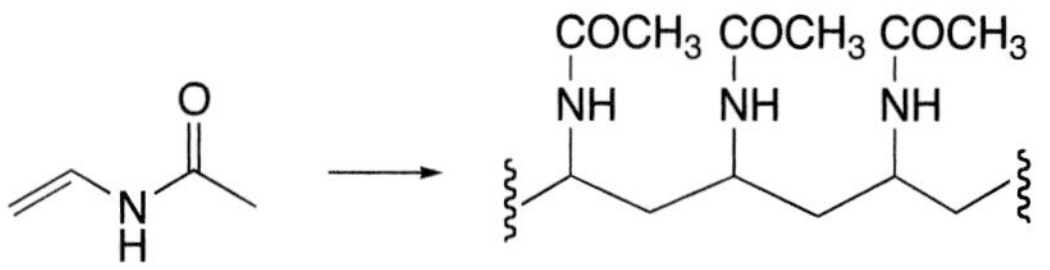

b.

c.

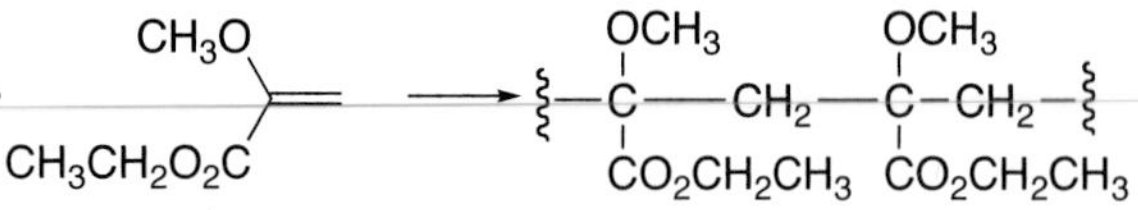

d.

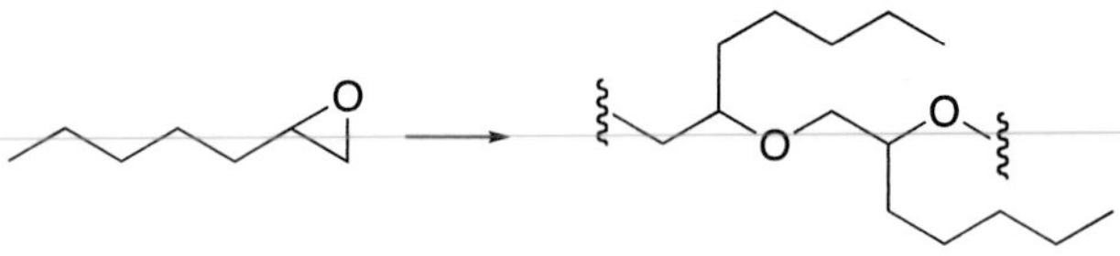

**30.26** Draw the copolymers.

a.

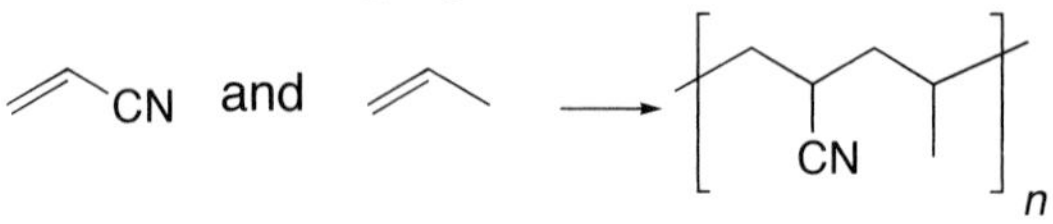

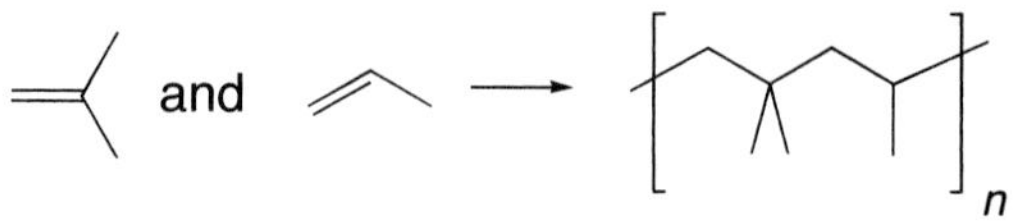

b.

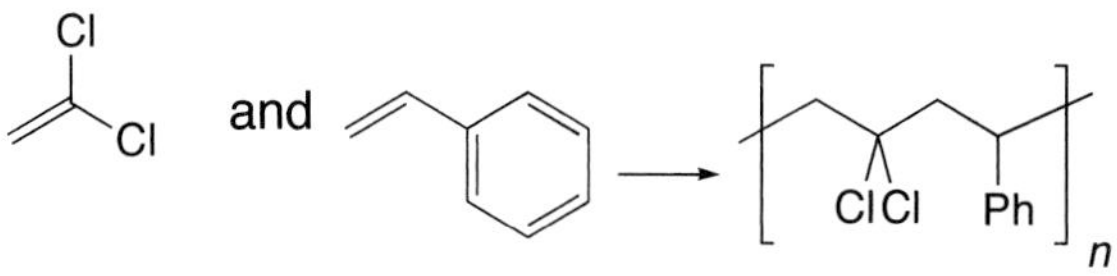

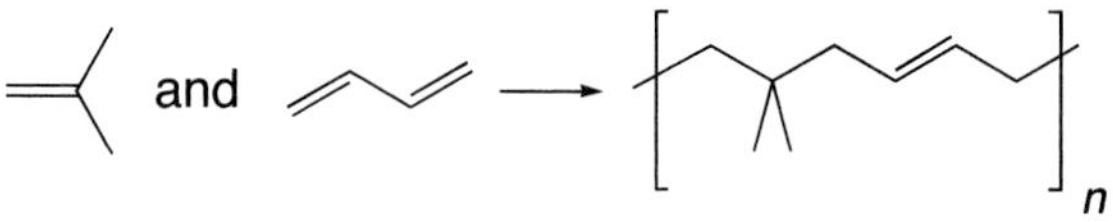

c.

d.

e.

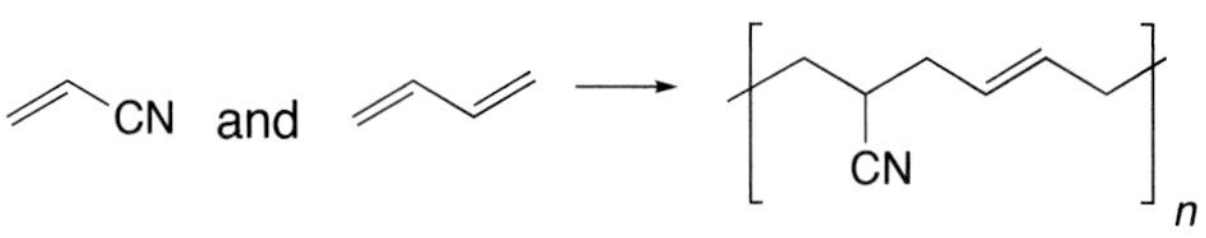

**30.27**

a.

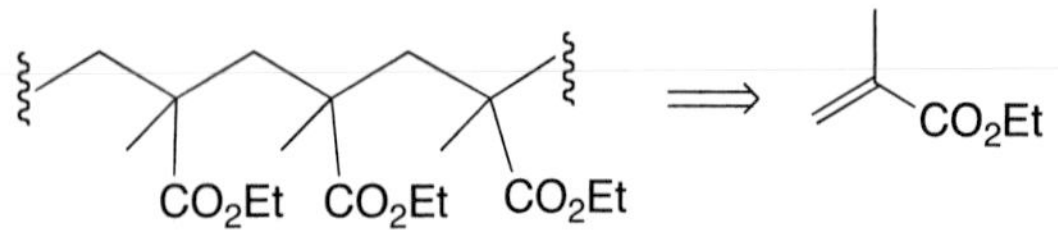

b.

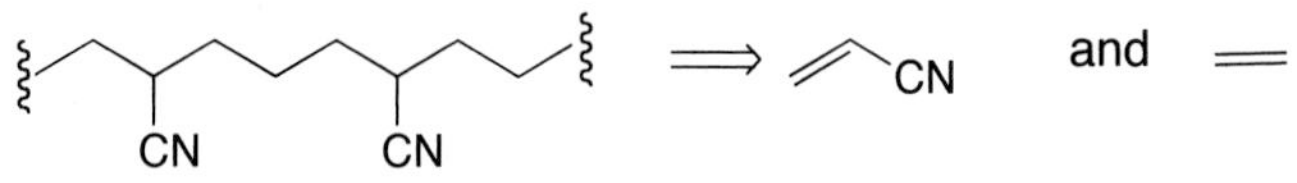

c.

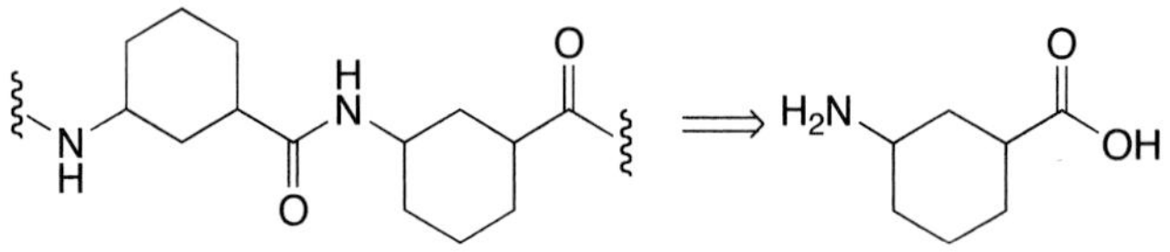

d.

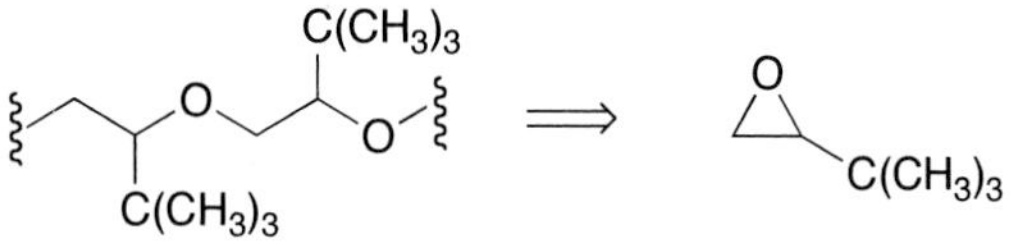

e.

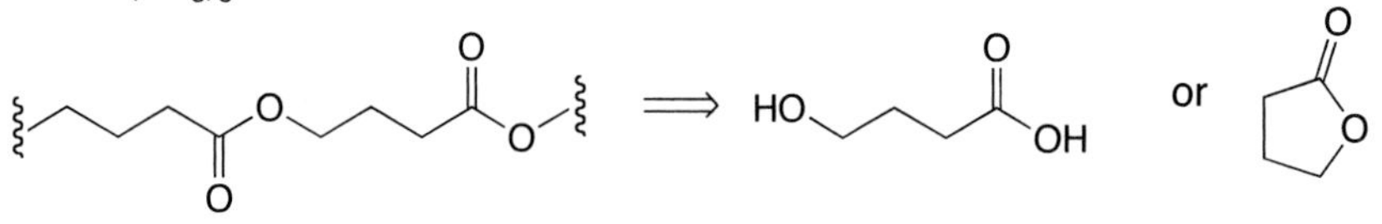

f.

**30.28**

a.

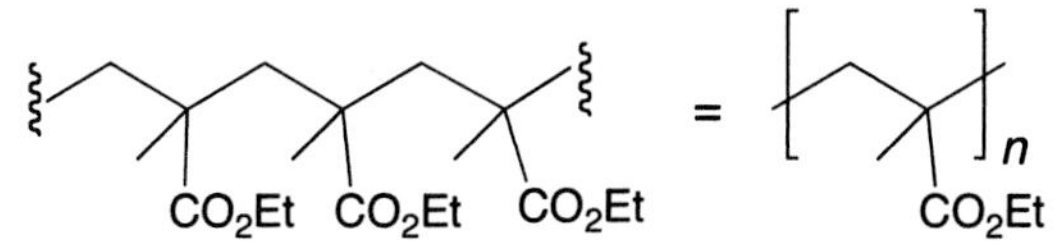

b.

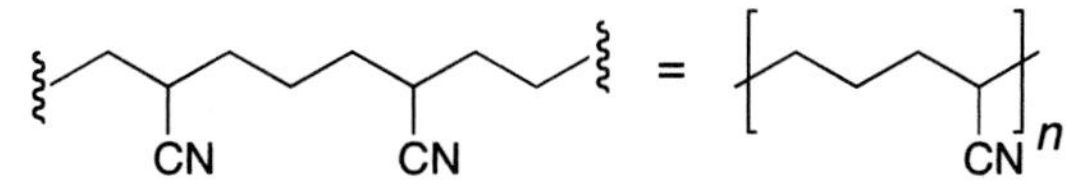

c.

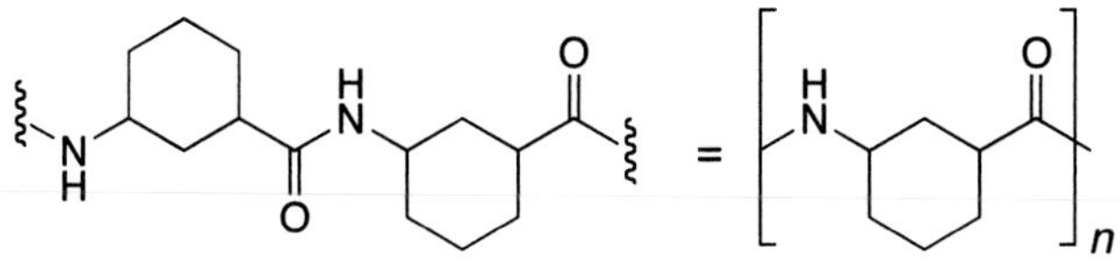

d.

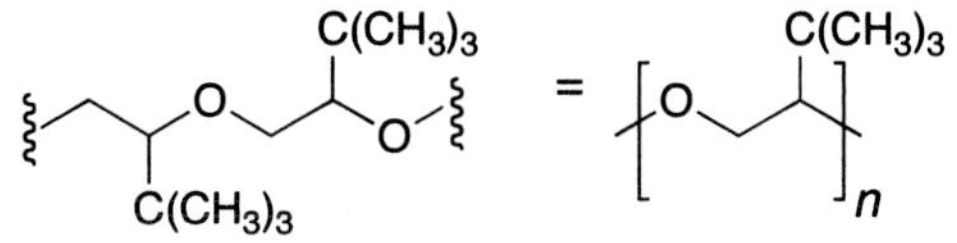

e.

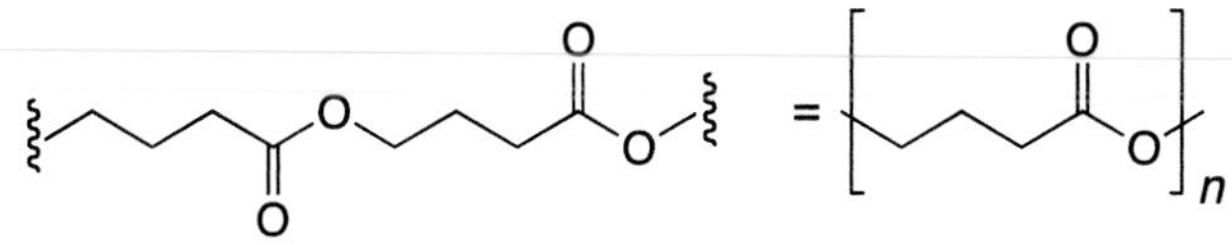

f.

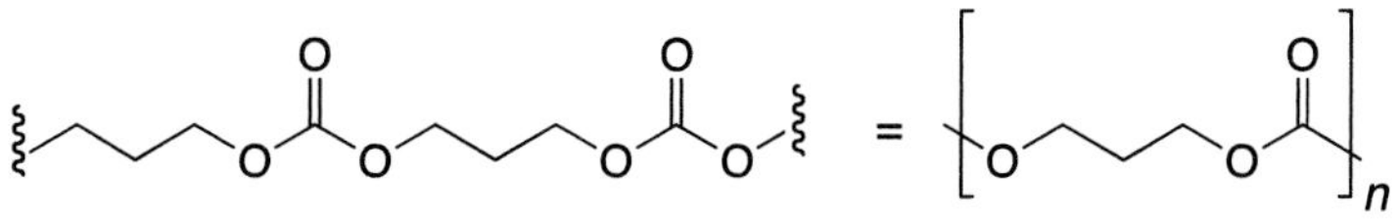

**30.29** An **isotactic polymer** has all Z groups on the same side of the carbon backbone. A **syndiotactic polymer** has the Z groups alternating from one side of the carbon chain to the other. An **atactic polymer** has the Z groups oriented randomly along the polymer chain.

a. 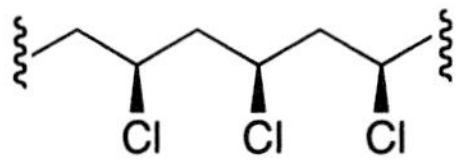   b. 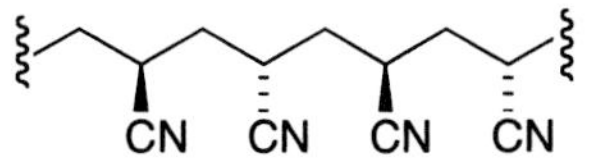   c.

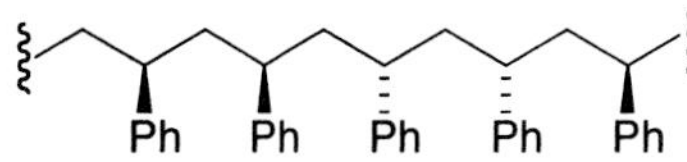

**30.30**

a.

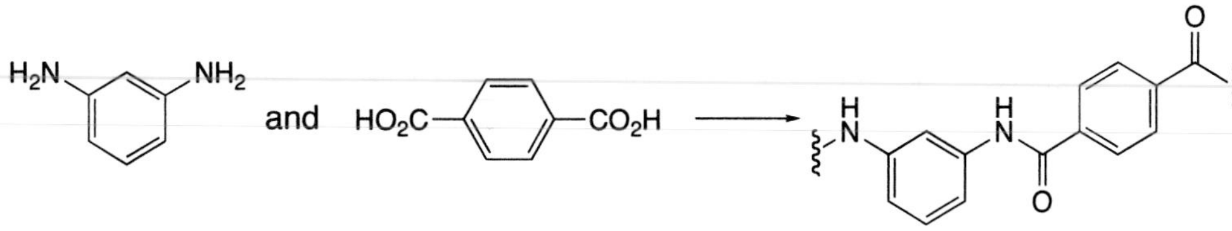

b.

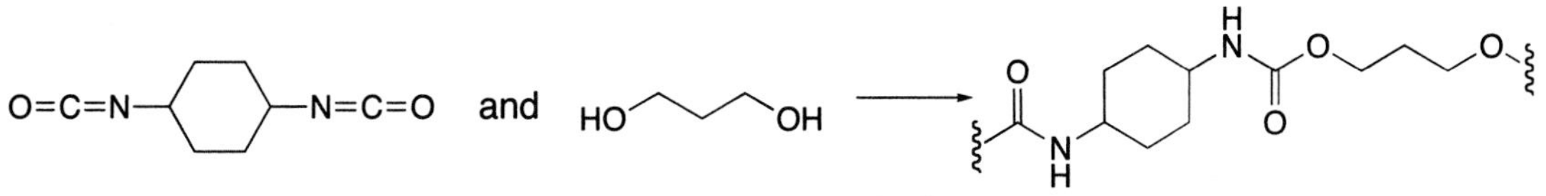

c.

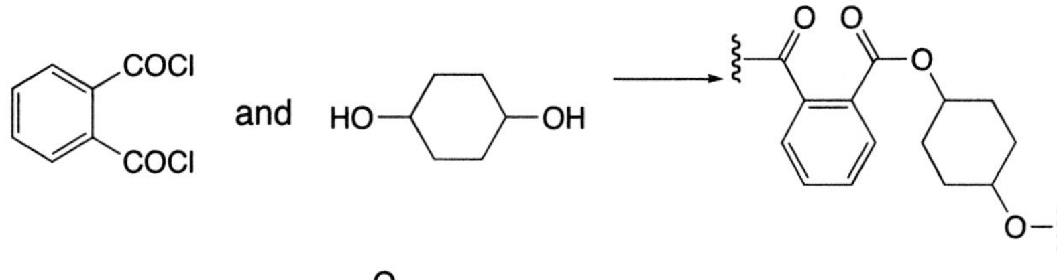

d.

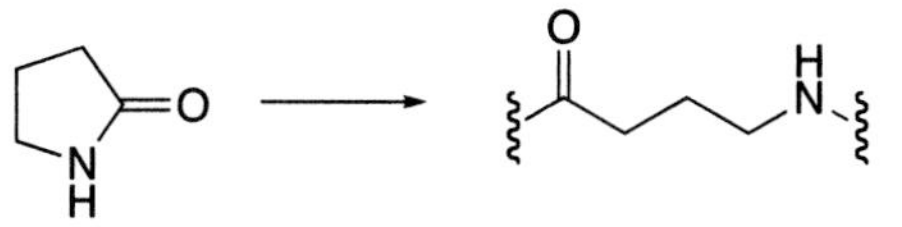

e.

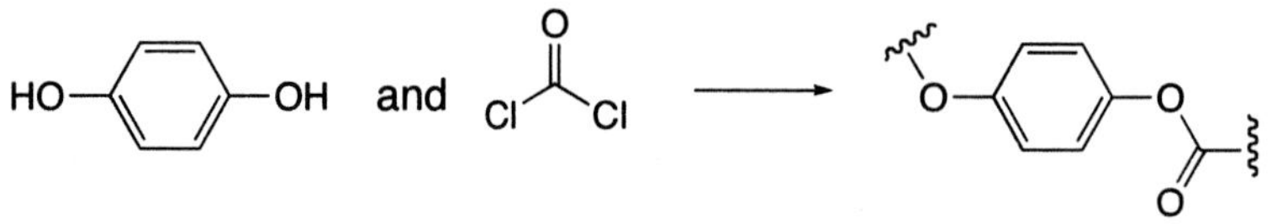

f.

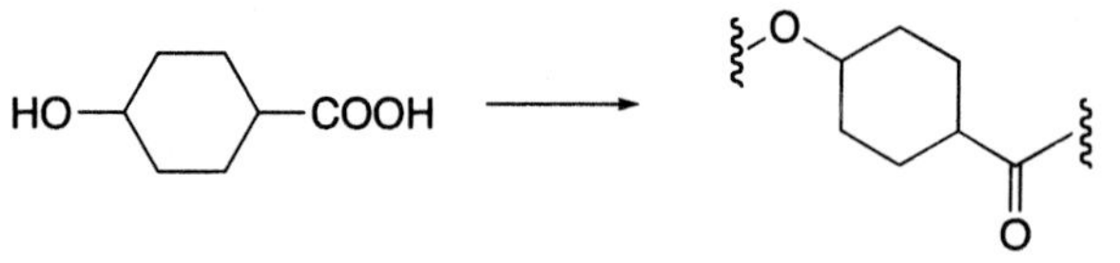

**30.31**

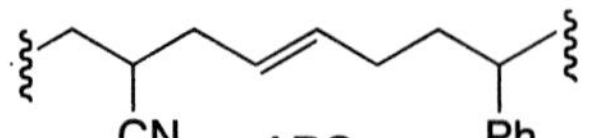

ABS

**30.32**

a.

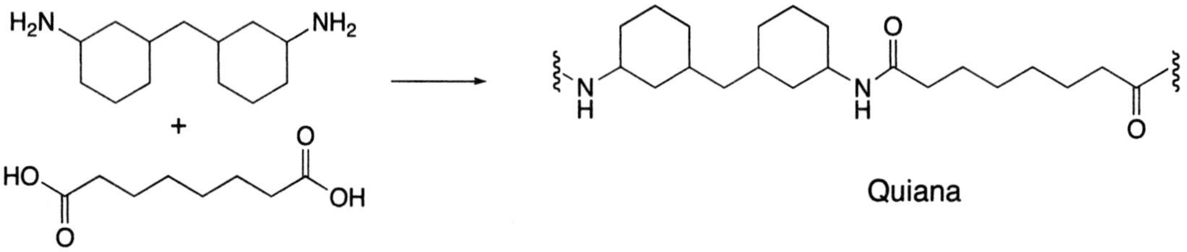

Quiana

b.

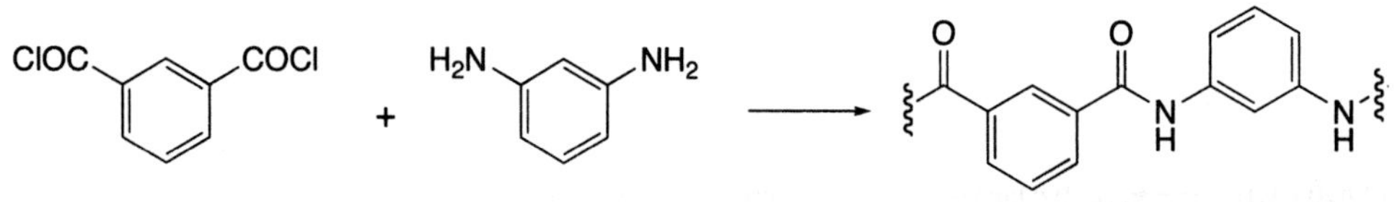

Nomex

**30.33**

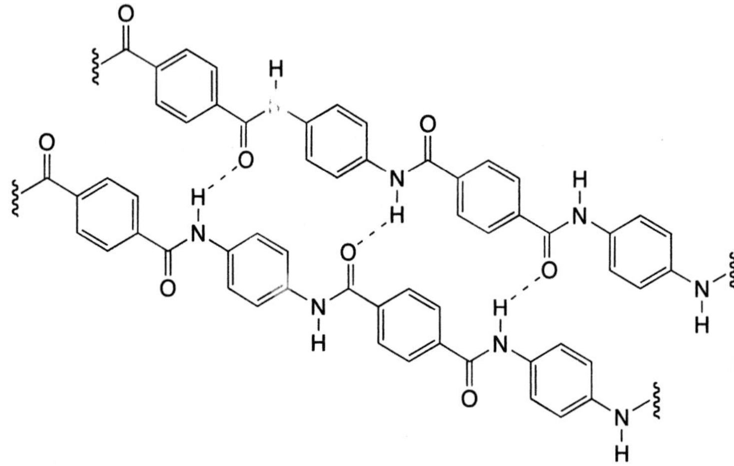

Kevlar

**30.34**

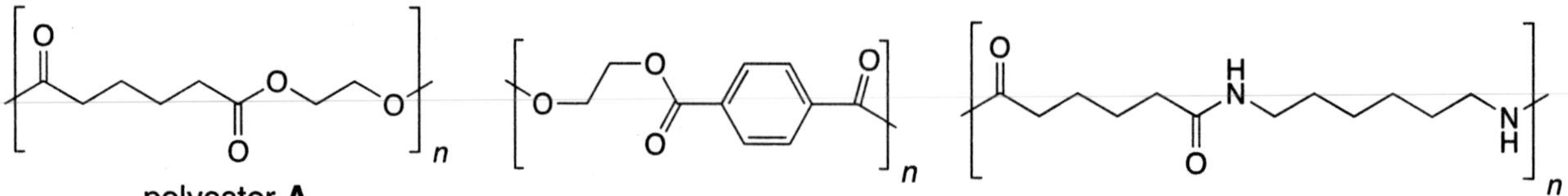

| polyester **A** | **PET** | nylon 6,6 |
|---|---|---|
| $T_g = \, < 0\ ^\circ C$ | $T_g = 70\ ^\circ C$ | $T_g = 53\ ^\circ C$ |
| $T_m = 50\ ^\circ C$ | $T_m = 265\ ^\circ C$ | $T_m = 265\ ^\circ C$ |

a. Polyester **A** has a lower $T_g$ and $T_m$ than PET because its polymer chain is more flexible. There are no rigid benzene rings so the polymer is less ordered.
b. Polyester **A** has a lower $T_g$ and $T_m$ than nylon 6,6 because the N–H bonds of nylon 6,6 allow chains to hydrogen bond to each other, which makes the polymer more ordered.
c. The $T_m$ for Kevlar would be higher than that of nylon 6,6 because in addition to extensive hydrogen bonding between chains, each chain contains rigid benzene rings. This results in a more ordered polymer.

**30.35**

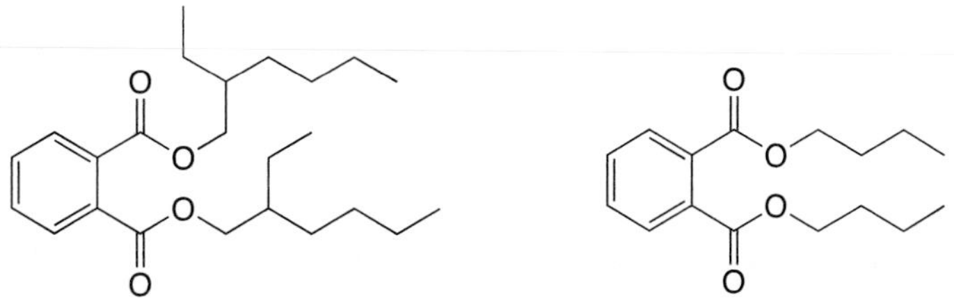

**A** dibutyl phthalate

Diester **A** is often used as a plasticizer in place of dibutyl phthalate because it has a higher molecular weight, giving it a higher boiling point. **A** should therefore be less volatile than dibutyl phthalate, so it should evaporate from a polymer less readily.

**30.36**

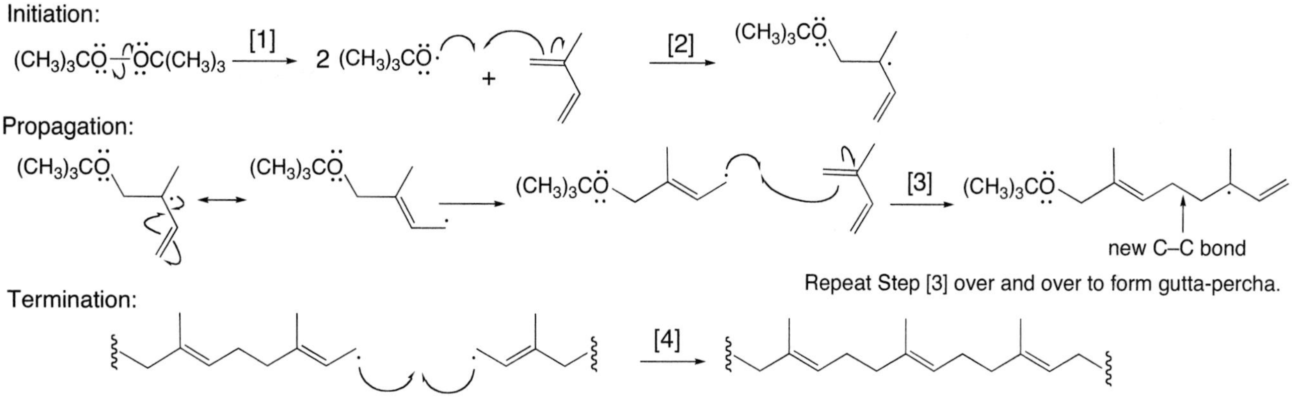
Initiation:
(CH3)3CO–OC(CH3)3
[1]
2 (CH3)3CO·
+
[2]
(CH3)3CO
Propagation:
(CH3)3CO
(CH3)3CO
(CH3)3CO
[3]
(CH3)3CO
new C–C bond
Repeat Step [3] over and over to form gutta-percha.
Termination:
[4]

**30.37**

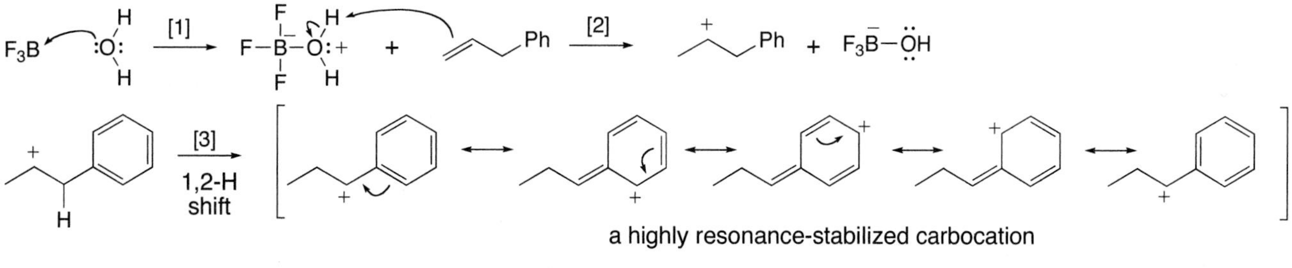
F3B
H
:O:
H
[1]
F–B–O:+
+
Ph
[2]
Ph
+ F3B–OH
[3]
1,2-H
shift
a highly resonance-stabilized carbocation

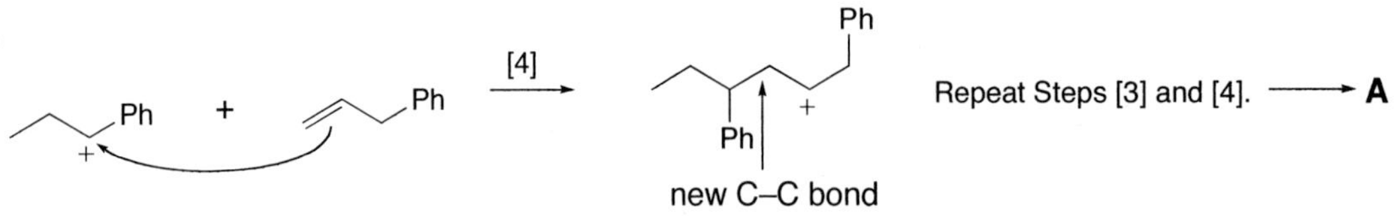

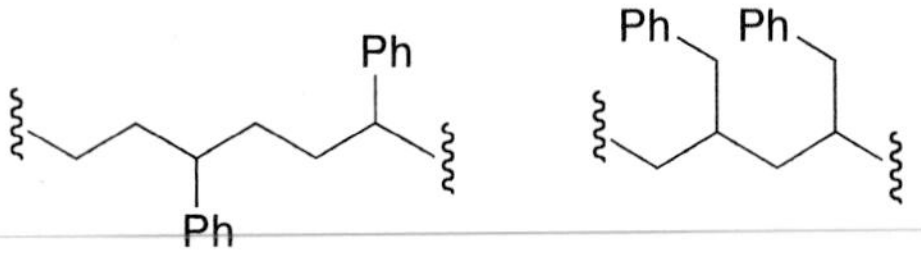

Repeat Steps [3] and [4]. ⟶ **A**

new C–C bond

**A** is the major product formed due to the 1,2-H shift (Step [3]) that occurs to form a resonance-stabilized carbocation.
**B** is the product that would form without this shift.

**A**
major product

**B**

## 30.38

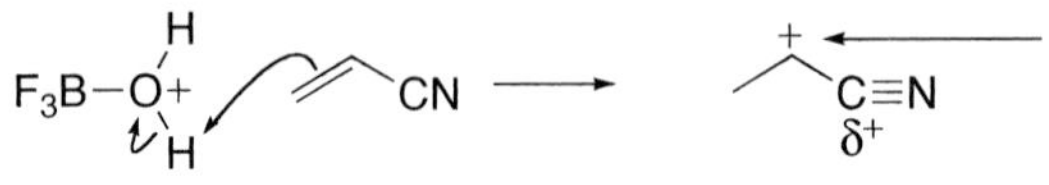

This carbocation is unstable because it is located next to an electron-withdrawing CN group that bears a $\delta^+$ on its C atom. This carbocation is difficult to form, so $CH_2=CHCN$ is only slowly polymerized under cationic conditions.

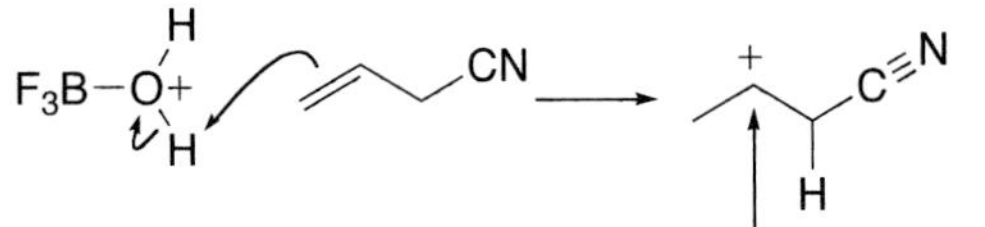

This 2° carbocation is more stable because it is not directly bonded to the electron-withdrawing CN group. As a result, it is more readily formed. Thus, cationic polymerization can occur more readily.

## 30.39

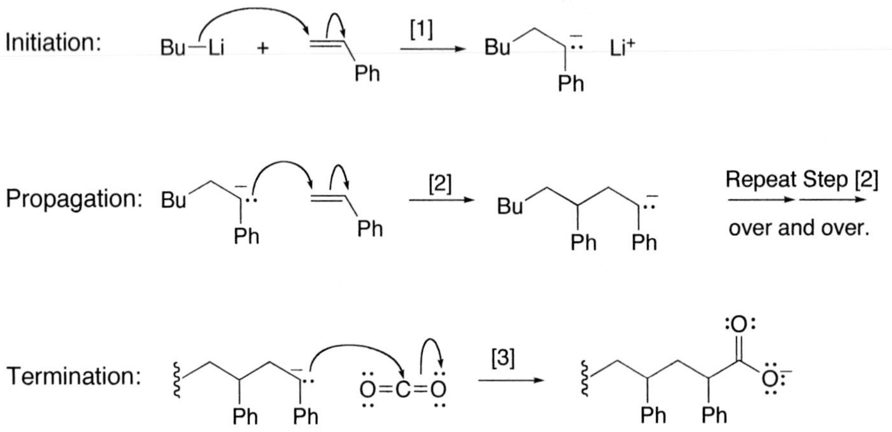

Initiation:

Propagation:

Repeat Step [2] over and over.

Termination:

**30.40** The substituent on styrene determines whether cationic or anionic polymerization is preferred. When the substituent stabilizes a carbocation, cationic polymerization will occur. When the substituent stabilizes a carbanion, anionic polymerization will occur.

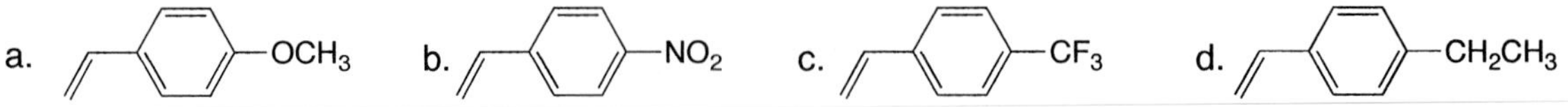

a.
—OCH$_3$
cationic
polymerization

b.
—NO$_2$
anionic
polymerization

c.
—CF$_3$
anionic
polymerization

d.
—CH$_2$CH$_3$
cationic
polymerization

**30.41** The rate of anionic polymerization depends on the ability of the substituents on the alkene to stabilize an intermediate carbanion: the better a substituent stabilizes a carbanion, the faster anionic polymerization occurs.

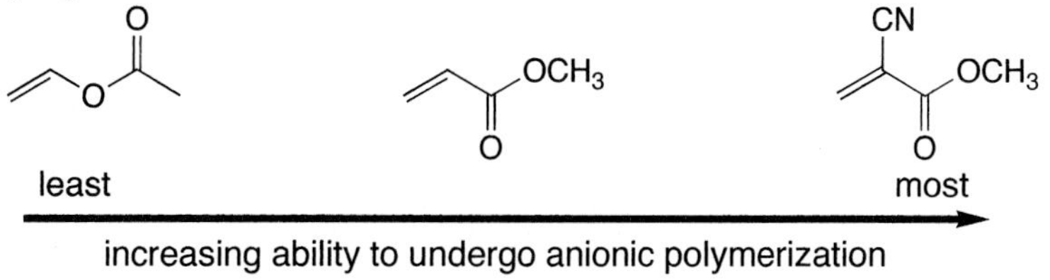

least                                                                                    most

increasing ability to undergo anionic polymerization

**30.42** The reason for this selectivity is explained in Figure 9.9. In the ring opening of an unsymmetrical epoxide under acidic conditions, nucleophilic attack occurs at the carbon atom that is more able to accept a $\delta^+$ in the transition state; that is, nucleophilic attack occurs at the more substituted carbon. The transition state having a $\delta^+$ on a C with an electron-donating $CH_3$ group is more stabilized (lower in energy), permitting a faster reaction.

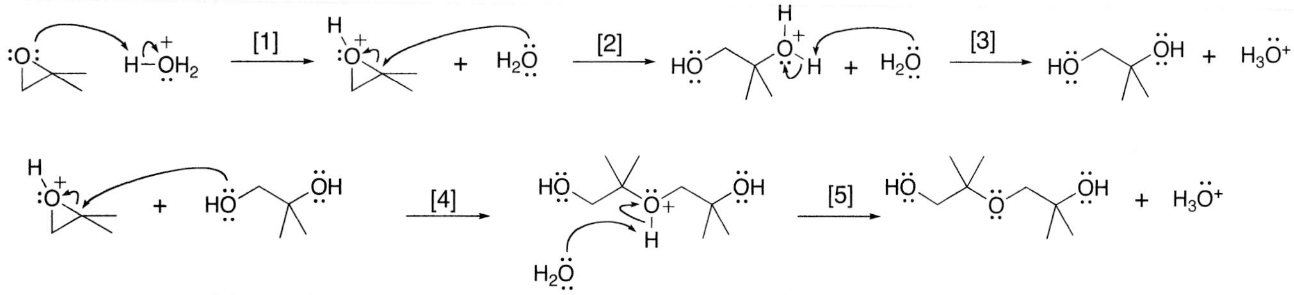

Repeat steps [4] and [5] over and over.

**30.43**

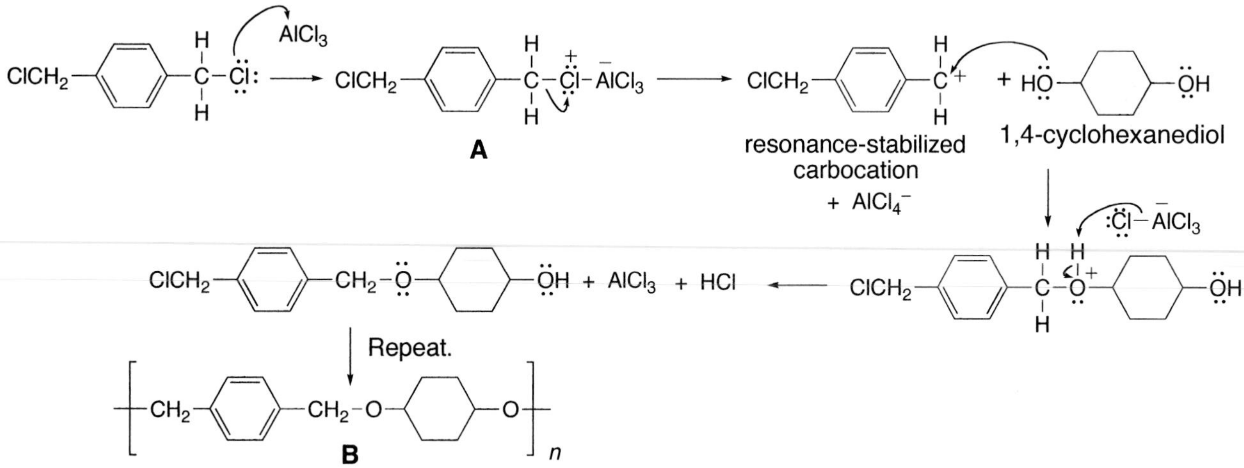

**30.44**

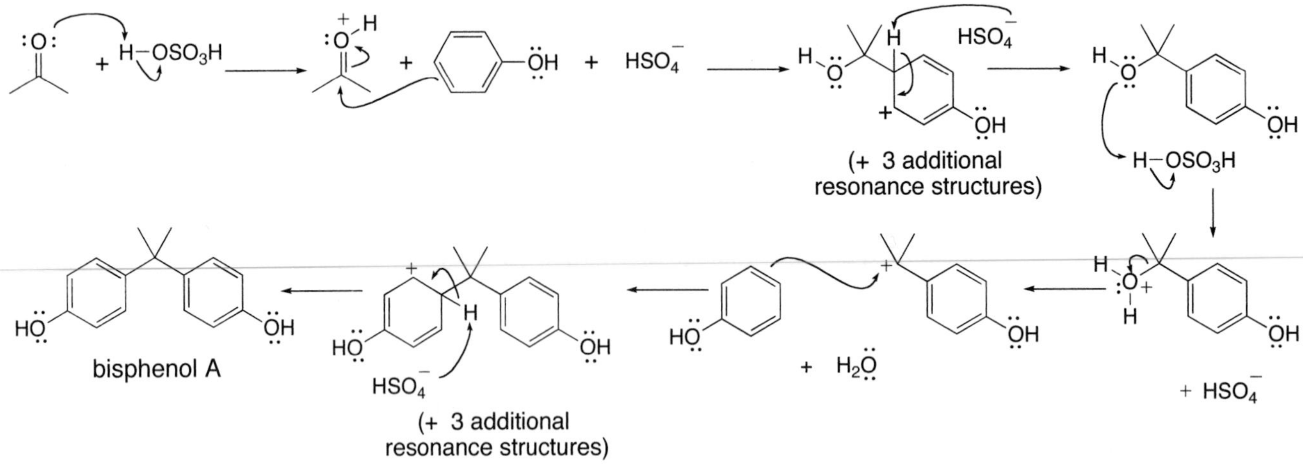

bisphenol A

**30.45**

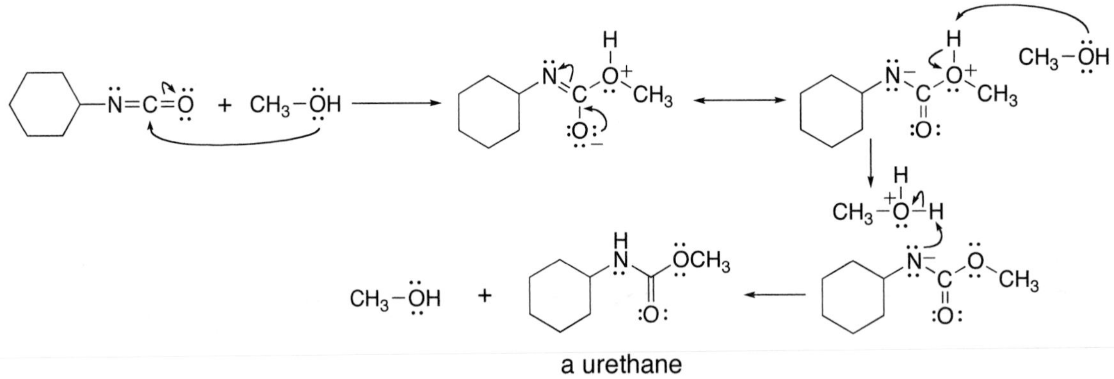

a urethane

**30.46**

melamine

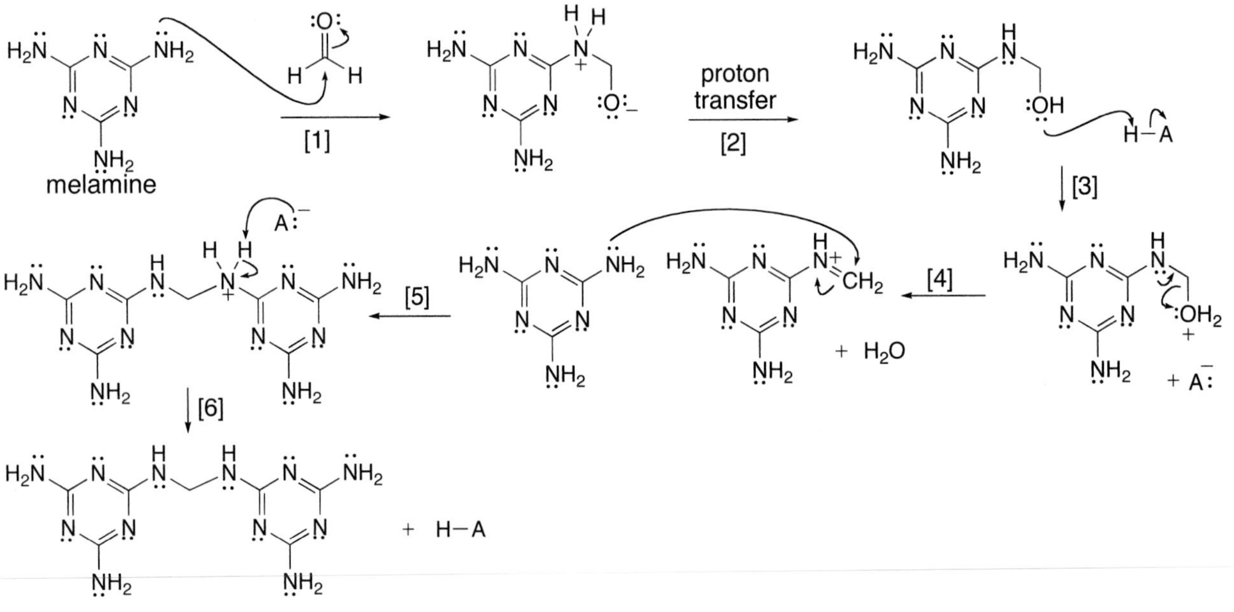

**30.47**

a.

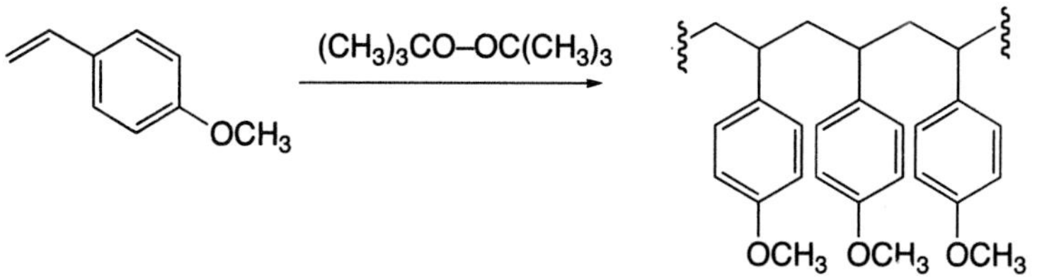

b.

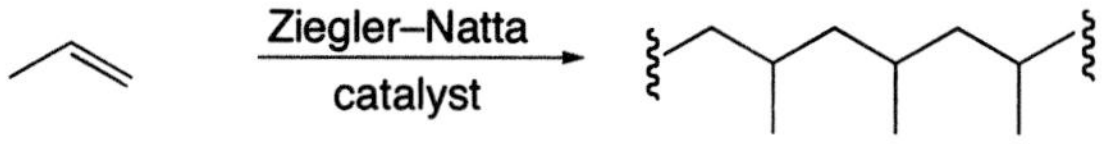

c.

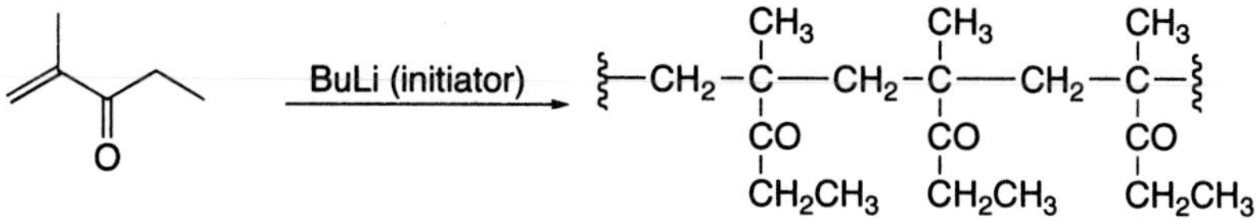

d.

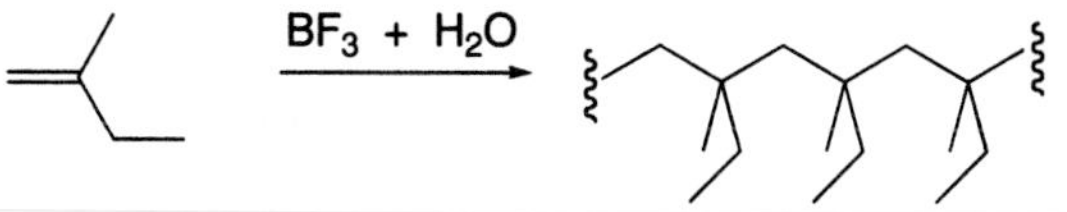

e.

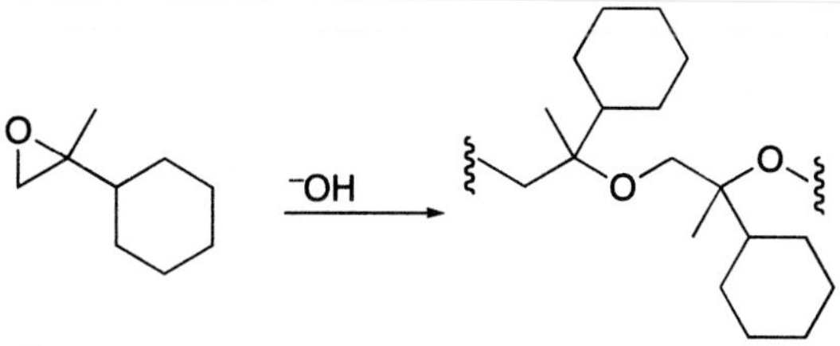

f.

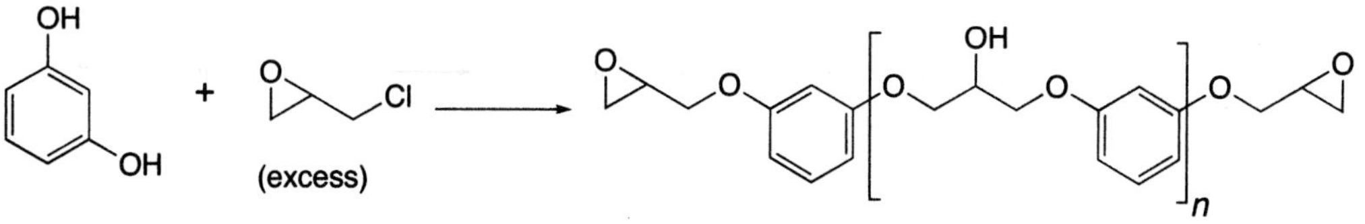

g.

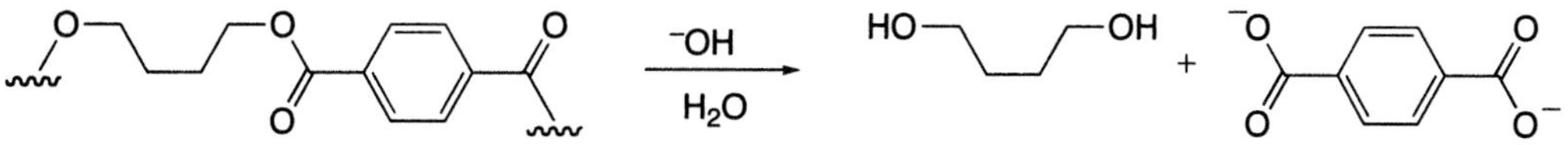

h.

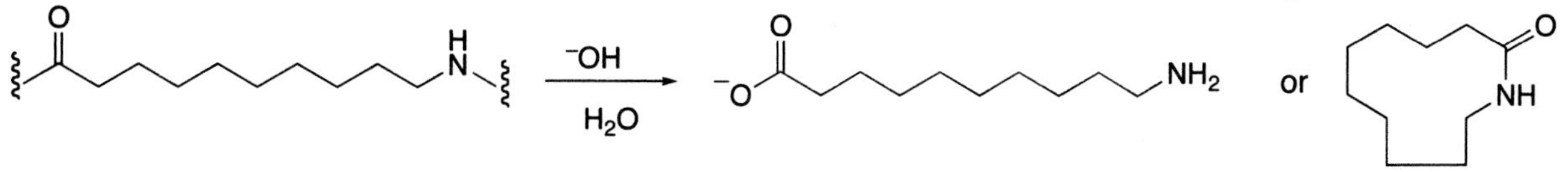

i.

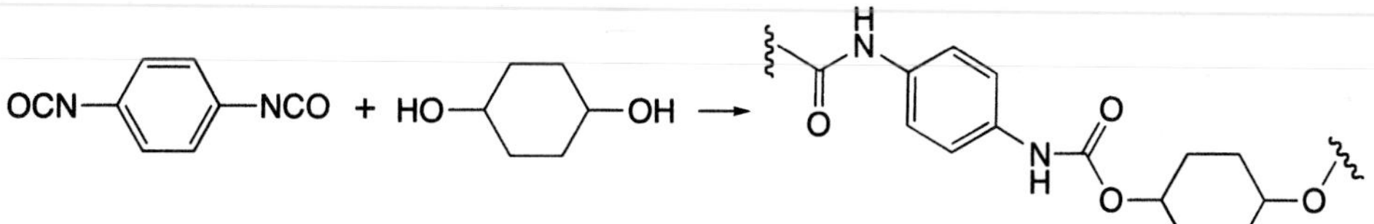

j.

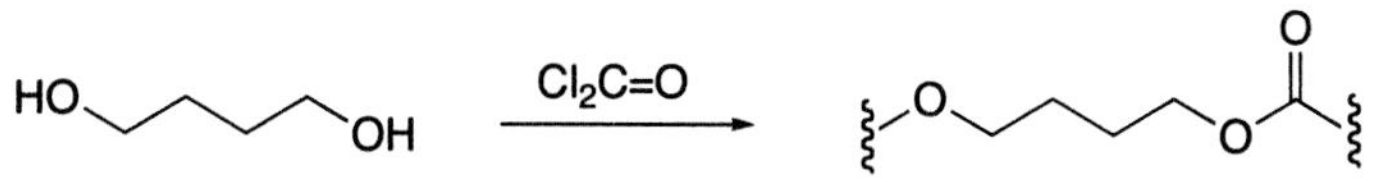

**30.48** Polyethylene bottles are resistant to NaOH because they are hydrocarbons with no reactive sites. Polyester shirts and nylon stockings both contain functional groups. Nylon contains amides and polyester contains esters, two functional groups that are susceptible to hydrolysis with aqueous NaOH. Thus, the polymers are converted to their monomer starting materials, creating a hole in the garment.

**30.49**

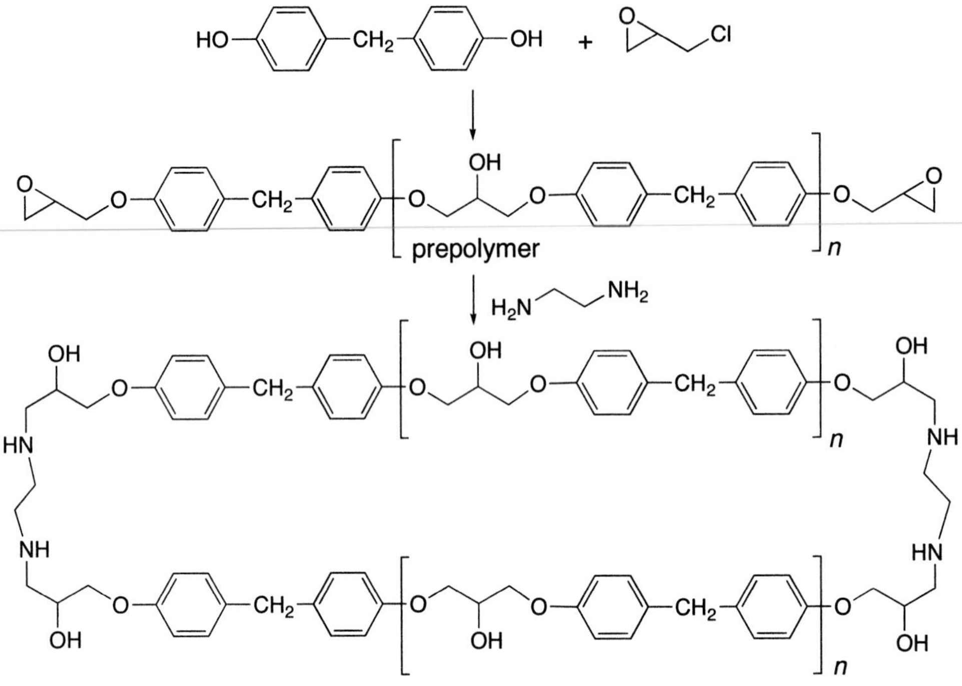

prepolymer

**30.50**

a.

vinyl alcohol    poly(vinyl alcohol)

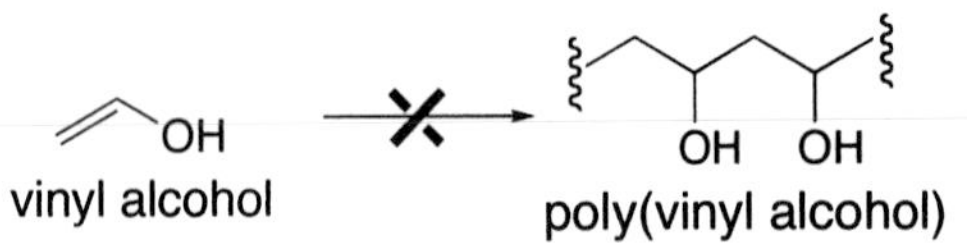

Poly(vinyl alcohol) cannot be prepared from vinyl alcohol because vinyl alcohol is not a stable monomer. It is the enol of acetaldehyde ($CH_3CHO$), and thus it can't be converted to poly(vinyl alcohol).

b.

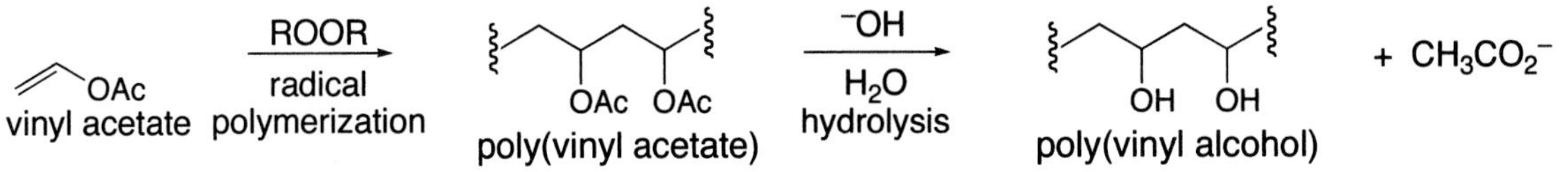

vinyl acetate    poly(vinyl acetate)    poly(vinyl alcohol)    + $CH_3CO_2^-$

c.

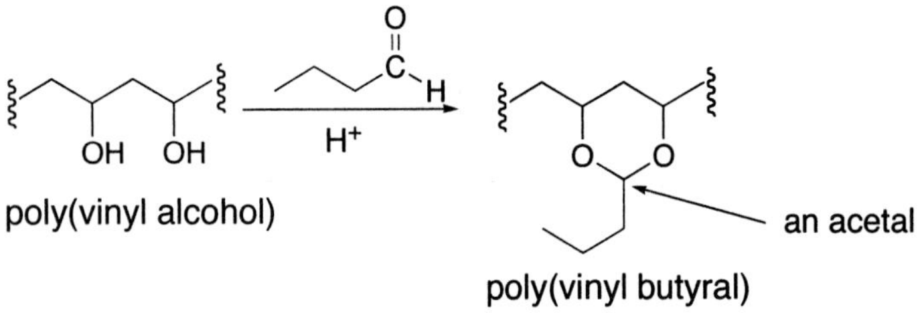

poly(vinyl alcohol)    an acetal

poly(vinyl butyral)

**30.51**

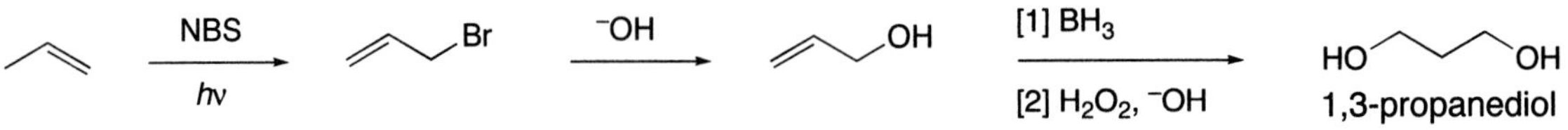

1,3-propanediol

**30.52**

terephthalic acid

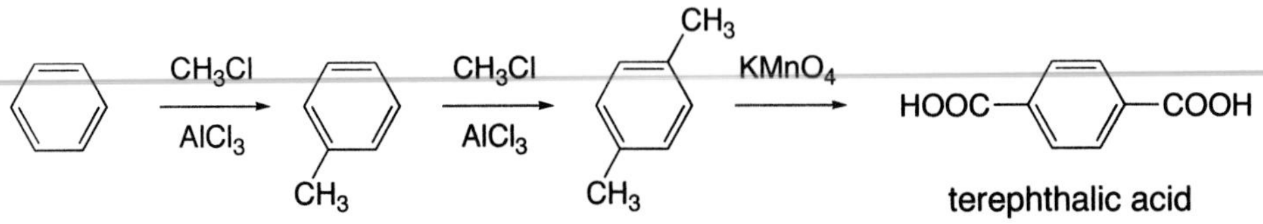

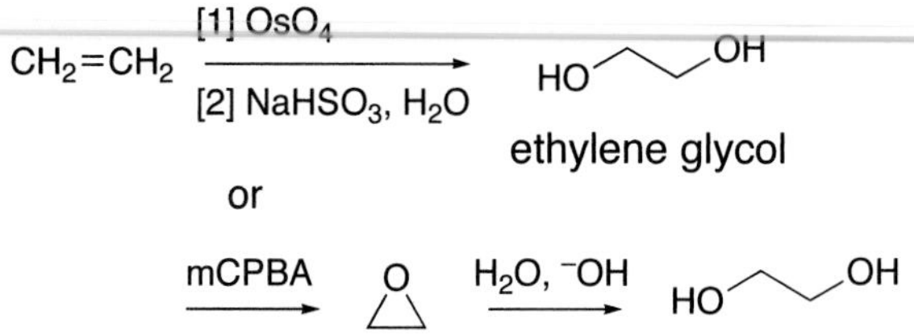

$$CH_2{=}CH_2 \xrightarrow[\text{[2] NaHSO}_3, \text{ H}_2\text{O}]{\text{[1] OsO}_4} \quad HO{\sim}OH$$

ethylene glycol

or

$$\xrightarrow{\text{mCPBA}} \triangle\!O \xrightarrow{H_2O, \ ^-OH} HO{\sim}OH$$

**30.53**

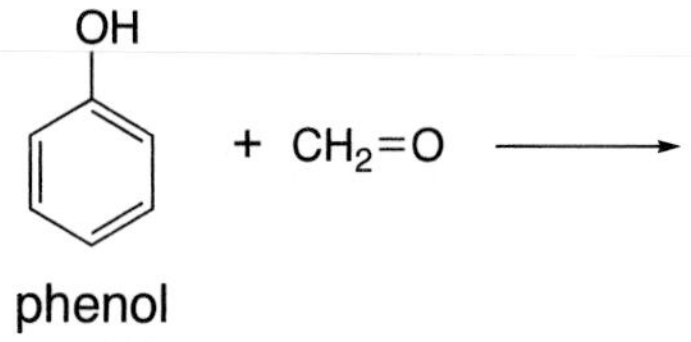

phenol

+ CH$_2$=O ⟶

**Bakelite**

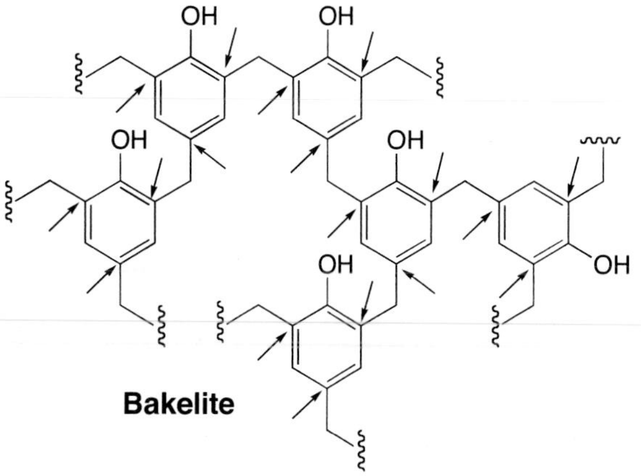

Since phenol has no substituents at any ortho or para position, an extensive network of covalent bonds can join the benzene rings together at all ortho and para positions to the OH groups.

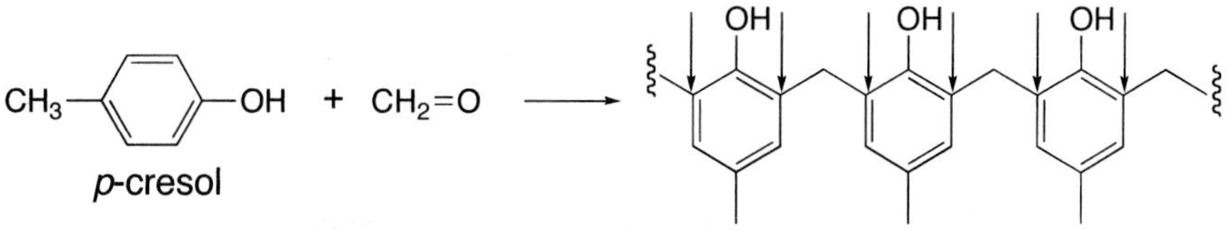

$p$-cresol + CH$_2$=O ⟶

Since $p$-cresol has a CH$_3$ group at the para position to the OH group, new bonds can only be formed at two ortho positions so that a less extensive three-dimensional network can form.

**30.54**

a.

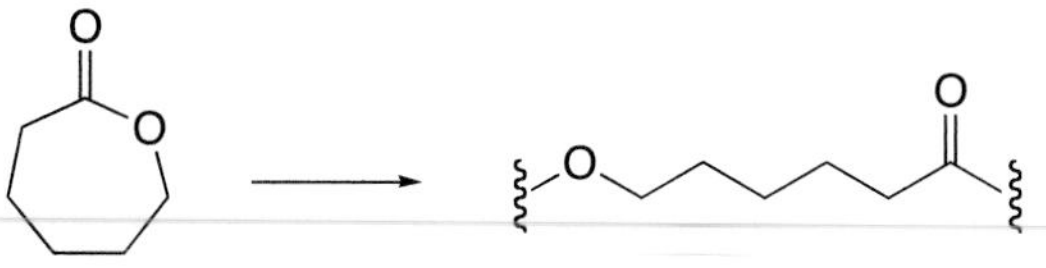

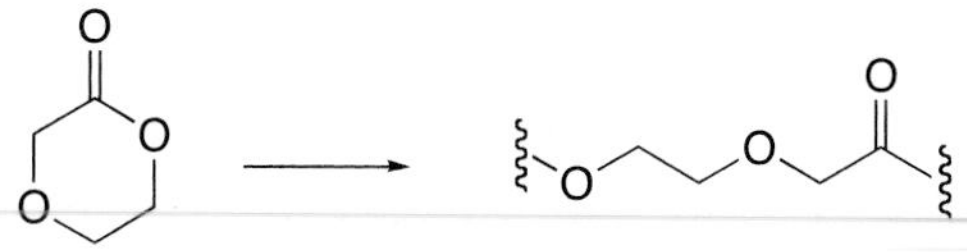

ε-caprolactone ⟶ polycaprolactone

b.

$p$-dioxanone ⟶ polydioxanone

## 30.55

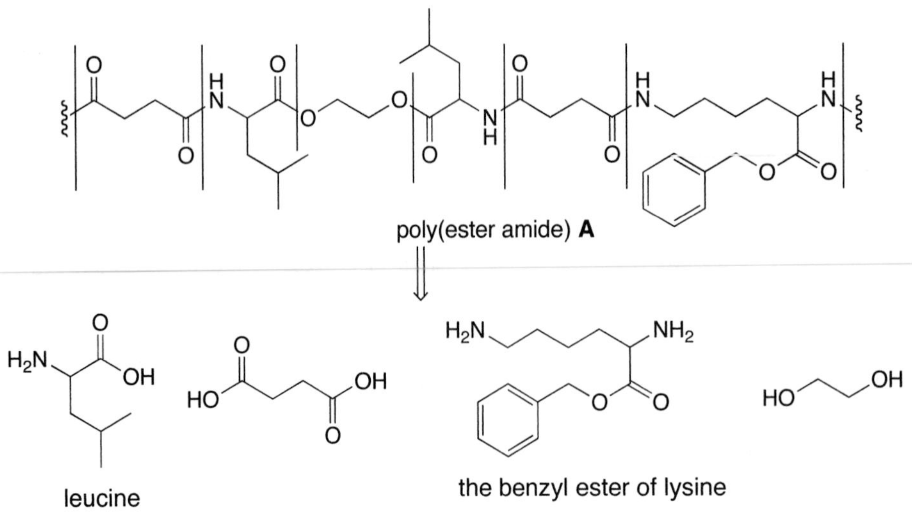
poly(ester amide) A
leucine
the benzyl ester of lysine

## 30.56

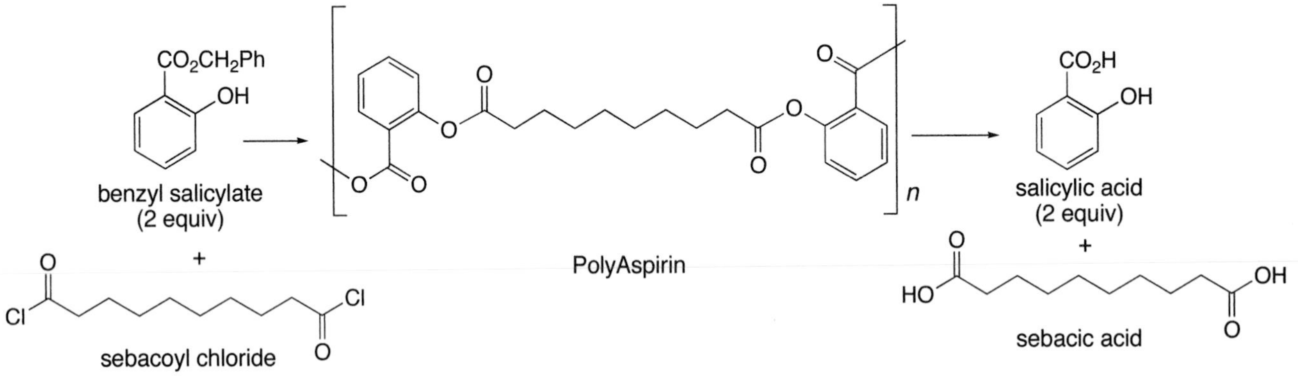
$CO_2CH_2Ph$
OH
benzyl salicylate
(2 equiv)
+
sebacoyl chloride
PolyAspirin
n
$CO_2H$
salicylic acid
(2 equiv)
+
sebacic acid

## 30.57

a.

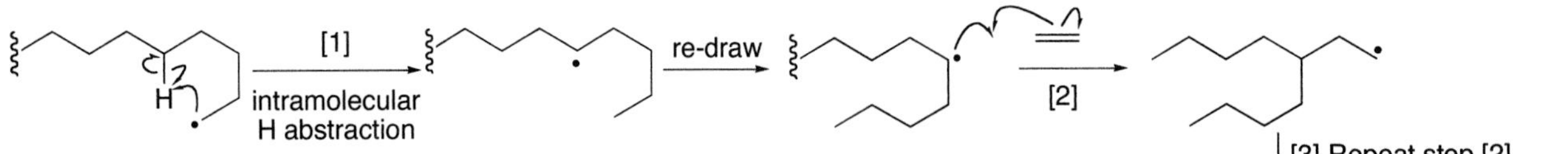
[1]
intramolecular
H abstraction
re-draw
[2]

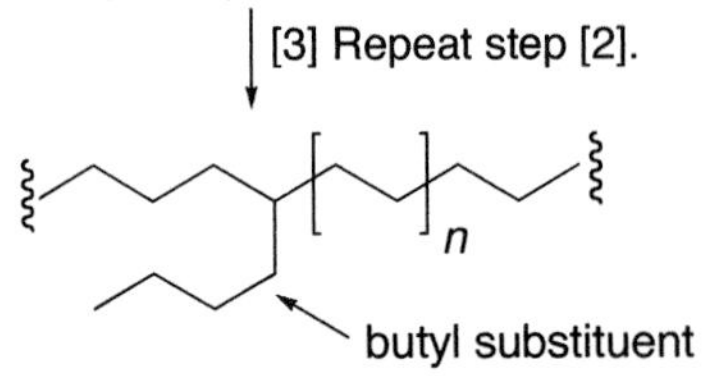
[3] Repeat step [2].
n
butyl substituent

b. Abstraction of the H is more facile than abstraction of the other H's because the H atom that is removed is six atoms from the radical. The transition state for this intramolecular reaction is cyclic, and resembles a six-membered ring, the most stable ring size. Other H's are too far away or the transition state would resemble a smaller, less stable ring.

**30.58**